Pathologies of Calcium Channels

Norbert Weiss · Alexandra Koschak
Editors

Pathologies of Calcium Channels

Springer

Editors
Norbert Weiss
Department of Physiology and Pharmacology, Hotchkiss Brain Institute
University of Calgary
Calgary
Canada

Alexandra Koschak
Department of Neurophysiology and Pharmacology
Medical University Vienna
Vienna
Austria

ISBN 978-3-642-40281-4 ISBN 978-3-642-40282-1 (eBook)
DOI 10.1007/978-3-642-40282-1
Springer Heidelberg New York Dordrecht London

Library of Congress Control Number: 2013951126

Printed on acid-free paper

Springer is part of Springer Science+Business Media (www.springer.com)

Foreword

Calcium signaling is a highly versatile system responsible for regulating a large number of diverse cell functions in both excitable and non-excitable cells. It initiates a new life at the time of fertilization and then guides many of the subsequent developmental processes that orchestrate the emergence of all the multiple cell types that constitute all the organ systems of the human body. The enormous diversity and subtlety of these different signaling systems is quite remarkable. At synaptic endings, for example, a brief highly localized pulse of calcium can trigger transmitter release within microseconds, whereas global elevations of calcium lasting for seconds to minutes activate fertilization and metabolism in liver cells. To achieve this diversity, cells have access to an extensive calcium signaling toolkit from which they can select out and assemble unique signaling systems with the spatial and temporal properties necessary to control their particular cell-specific functions.

One of the features of these cell-type-specific calcium signaling systems is their relative stability. This stability depends on the turnover of signaling components being tightly regulated, which depends on a balance between the degradation of individual components and their replacement by ongoing transcription processes. Despite this stability, it is clear that these signaling systems are remarkably flexible and this has gradually come to light from knockout experiments where experimenters often are disappointed to find no apparent change in phenotype after removing an apparent key calcium signaling component. Such negative results are usually explained by invoking compensatory responses and are not investigated further. However, when the phenotype is examined in more detail, some surprising observations have emerged to indicate that when a particular component is removed, cells can interrogate their extensive toolkit to come up with an alternative closely related component that will restore the same function with varying degrees of effectiveness.

While this flexibility of the calcium signaling system enables organisms to survive in the face of changes in individual components, these compensatory changes are usually not perfect and the modified signaling systems often results in the emergence of a multiple diseased states. There is no better way of illustrating this point than to consider the contents of this remarkably complete anthology of calcium channel pathologies. It highlights how so many human diseases are caused by defects in calcium signaling systems in both excitable and non-excitable cells.

It is extraordinary to see that mutations in voltage-gated calcium channels have been linked to so many human pathologies and new examples are continuously appearing. For example, mutations in the *CACNA1C* gene that codes for the α_{1C} subunit of the $Ca_v1.2$ L-type calcium channel, which causes Timothy syndrome as described in this volume, has just been linked to a variety of psychiatric disorders such as schizophrenia and bipolar disorder. The insights gained from understanding the genetic basis of all these debilitating diseases are providing essential new information for our quest to understand how specific cell functions are controlled in complex organ systems such as the brain.

The editors, Norbert Weiss and Alexandra Koschak, have been extraordinary successful in bringing together so many of the major researchers in this rapidly expanding field. To have all this information in a single volume is a fantastic resource for both those workers actively involved in this exciting field and for those wishing to find out more about particular diseases.

Cambridge, UK Sir Michael Berridge

Preface

Calcium ions (Ca^{2+}) represent one of the most versatile second messengers in eukaryotic cells, and changes in the intracellular Ca^{2+} concentration regulate many cellular physiological functions as diverse as gene transcription, muscle contraction, neurotransmission fertilization and more. In order to make use and regulate the amplitude, duration and subcellular localization of the Ca^{2+} signal, cells have developed a complex machinery, the Ca^{2+} signaling toolkit, initially introduced by Sir Michael Berridge, where Ca^{2+} channels play a key role in the initiation of the Ca^{2+} signal. Therefore, it is not surprising that mutations in Ca^{2+} channel genes, causing either a loss or a gain of channel function, can lead to serious (chronic) disorders. We have not only learned much within the last two decades regarding the structure/function basis of Ca^{2+} channels, but also the list of human diseases known to be associated with a defect in Ca^{2+} channels has grown considerably. This book aims to provide an informative and up-to-date account of our present understanding of human diseases associated to the various Ca^{2+} channels, from their molecular basis to the point of clinical pharmacology.

Voltage-Gated Ca $^{2+}$ Channels

Voltage-gated Ca^{2+} channels (VGCCs) are pore-forming plasma membrane proteins that convert an electrical signal into intracellular Ca^{2+} elevations. To date, 10 genes encoding the pore-forming subunits of mammalian VGCCs have been identified. 7 genes encode the high-voltage activated (HVA) channel subfamily (comprising L-type ($Ca_v1.1$ to $Ca_v1.4$), P/Q-type ($Ca_v2.1$), N-type ($Ca_v2.2$) and R-type ($Ca_v2.3$) channels) and 3 genes encode the low-voltage-activated (LVA) channel subfamily (composed exclusively of T-type ($Ca_v3.1$ to $Ca_v3.3$). In addition to the Ca_v pore-forming subunit, HVA channels contain auxiliary subunits: β (β_1 to β_4, a 55 kDa cytosolic protein of the membrane-associated guanylate kinase (MAGUK) family), $\alpha_2\delta$ ($\alpha_2\delta_1$ to $\alpha_2\delta_4$, a 170 kDa highly glycosylated extracellular protein with a single transmembrane domain), and in some cases γ (γ_1 to γ_8, a 33 kDa transmembrane protein). Calcium entry through neuronal $Ca_v2.1$ and $Ca_v2.2$ channels initiates fast neurotransmitter release at the central and peripheral synapses, respectively, by triggering the fusion of secretory vesicles with the plasma membrane through actions on the SNARE synaptic protein complex. Therefore, small changes in their biophysical properties or subcellular expression result in alteration in neuronal excitability. In Chap. 1, Daniela Pietrobon discusses the implication of $Ca_v2.1$ channels in the pathophysiology of a rare form of migraine with aura, the familial hemiplegic migraine type I. Their implication in various forms of cerebellar ataxia in then discussed in Chap. 2 by Stephanie Schorge and Kinya Ishikawa. In contrast to Ca_v2 channels which are widely expression through the nervous system, $Ca_v1.3$ and $Ca_v1.4$ channels predominate in sensory cells of the inner ear and in retinal neurons, respectively, and have been linked to sensory disorders as discussed by Alexandra Koschak, Alexandra Pinggera, Klaus Schicker and Jörg Striessnig in Chap. 3. In contrast to HVA Ca^{2+} channels, LVA channels (referred as T-type channels or Ca_v3 channels), because of their particular activation threshold around the resting membrane potential of the cells, underlie neuronal burst-firing, often associated with spike-and-wave discharges observed on electroencephalography recordings during certain forms of epileptic seizures as discussed in Chap. 4 by Stuart M. Cain, Michael E. Hildebrand and Terrance P. Snutch. Finally, Annette Dolphin discusses the implication of the voltage-gated Ca^{2+} channel $\alpha_2\delta$-subunit in various neurological disorders and most importantly its clinical relevance as therapeutic target for gabapentinoid drugs (Chap. 5). In skeletal muscle, $Ca_v1.1$ functions as both a slowly-activating, voltage-gated L-type Ca^{2+} channel and as the voltage sensor that triggers the opening of ryanodine receptor Ca^{2+} release from the sarcoplasmic reticulum during excitation–contraction (EC) coupling (voltage-induced-Ca^{2+} release (VICR) mechanism). In Chap. 6, Amaury Francois Anne Pizzoccaro, Sophie Laffray and Emmanuel Bourinet review the current knowledge on the role of the LVA T-type channels in the perception and modulation of pain. In Chap. 7, Emma Matthews and Mike Hanna specifically discuss the implication $Ca_v1.1$ channels in hypokalemic periodic paralysis. Then, Viktor Yarotskyy and Robert T. Dirksen analyse its implication in malignant hyperthermia (Chap. 8), while Julien Fauré, Joël Lunardi, Nicole Monnier and Isabelle Marty investigate this functional coupling from the side of the ryanodine receptor type I (Chap. 9). In Chap. 10, Paul W. Wirtz, Maarten J. Titulaez and Jan J. G. M. Verschuuren discuss the molecular physiopathology of the Lambert-Eaton Mysathenic Syndrome, an autoimmune disease of the neuromuscular junction and the implication of $Ca_v2.1$ channels. In contrast to the skeletal $Ca_v1.1$ channel, cardiac $Ca_v1.2$ channels trigger the opening of the ryanodine receptor via a Ca^{2+}-induced-Ca^{2+}-release mechanism (CICR), and mutations in the gene encoding $Ca_v1.2$ channels have been linked to various cardiac syndrome, including the Timothy syndrome discussed by Hua Huang, Juejin Wang and Tuck Wah Soong in Chap. 11, and the Brugada syndrome discussed by Brett Simms in Chap. 12. In Chap. 13, Matteo Mangoni specifically discusses the implication of $Ca_v1.3$ channels in sinoatrial node dysfunction, and Stephan Herzig and Jan Matthes discuss in Chap. 14 how an alteration in the expression pattern of the VGCCs β-subunit and its consequence on the L-type current leads to an alteration of the cardiac function.

As previously mentioned, Ca^{2+} influx from VGCCs controls many cellular processes including cell growth and differentiation, and not surprisingly these channels have been implication in different form of cancer. This aspect is discussed in Chap. 15 by Maria Beatrice Morelli, Sonia Liberati, Consuelo Amantini, Matteo Santoni, Massimo Nabissi, Valerio Farfariello and Giorgio Santoni in the particular case of Ca_v3 channels. Finally, Ľubica Lacinová and Lucia Lichvárová close this section by providing an up-to-date picture of the clinically relevant drugs against VGCCs and associated diseases (Chap. 16).

Transient Receptor Potential Channels

In contrast to VGCCs, the Transient Receptor Potential (TRP) superfamily of cation channels displays greater diversity in activation mechanisms and selectivity. Based on the sequence homology, Mammalian TRP channels are divided into seven subfamilies: TRP Canonical (TRPC; TRPC1 to TRPC7), TRP Vanilloid (TRPV; TRPV1 to TRPV6), TRP Melastatin (TRPM; TRPM1 to TRPM8), TRP Polycystin (TRPP; TRPP2, TRPP3 and TRPP5), TRP Mucolipin (TRPML, TRPML1 to TRPML3), and TRP Ankyrin (TRPA; TRPA1). TRP channels are membrane spanning proteins that share the common features of six putative transmembrane segments and a non-selective permeability to cations. Although the cytoplasmic amino and carboxy termini are variable between TRP subfamily members, they contain different domains (i.e., ankyrin repeats, coiled–coiled, calmodulin binding sites and more) critical for channel regulation. Functional TRP channels form either homo- or heterotetramers assembled from identical or similar TRP subunits, and play critical roles in sensory physiology like vision, taste, olfaction, hearing, touch, but also thermo- and osmosensation. Although only few TRP members have been linked to hereditary diseases, the implication of TRP channels in the pathogenesis of many disease states has emerged. In Chap. 17, Shoichi Irie and Takahisa Furukawa present the physiological role of TRPM1 channel in retinal bipolar cells, and the association of human TRPM1 mutations with congenital stationary night blindness. In Chap. 18, Meredith C. Hermosura discusses how alteration in TRPM2 and TRPM7 functions in the immune system may lead to neurodegenerative diseases such as amyotrophic lateral sclerosis and Parkinsonism dementia. Christian Grimm and Math P. Cuajungco then highlight the cellular function of TRPML channels endolysosomal system and their implication in the lysosomal storage disorder mucolipidosis type IV (Chap. 19). As previously mentioned, activation of TRP channels present a great diversity in activation mechanisms, and are thus considered as polymodal cellular sensors. This diversity is particularly well illustrated by the implication of TRP channels in the signaling pathway of various painful stimuli, as discussed in Chap. 20 by Alexandre Denadai-Souza and Nicolas Cenac.

Besides their role in neurological disorders, the contribution of TRP channels has also been described in various muscular/skeletal and cardio-vascular disorders.

In Chap. 21, Jeremy M. Sullivan, Thomas E. Lloyd and Charlotte J. Sumner present the hereditary channelopathies (Charcot-Marie-Tooth disease, muscular atrophy, dysplasia and brachyolmia) caused by mutations in the gene encoding TRPV4 channel. Two chapters then discuss the role of TRPC channels in the cardiovascular system. In Chap. 22, Nadine Kirschmer, Kristina Lorenz and Petra Eder-Negrin explain how an alteration in TRPC channel activity may cause cardiac hypertrophy, whereas in Chap. 23, Michael Poteser, Sarah Krenn, and Klaus Groschner report on the contribution of these channels in the remodeling of the vascular system.

The broad implication of TRP channels in multiple diseased states in then further illustrated with the role of TRPP2 channel in hereditary polycystic kidney disease in Chap. 24 by Andrew Streets and Albert Ong. In Chap. 25, Daniel Landau and Hanna Shalev further illustrate the role of TRP channels in kidney related diseases and discuss the contribution of TPM6 channel in the development of hypomagnesemia and hypocalcemia.

Cell proliferation and differentiation highly relies on the Ca^{2+} signal, and alteration of the Ca^{2+} homeostasis often leads to disruption of normal cell cycle and, as a result, onset and progression of cancer. This aspect is further discussed in the context of prostate cancer by George Shapovalov, Roman Skryma and Natalia Prevarskaya in Chap. 26.

Finally, Ulrich Wissenbach concludes this section by discussing the pharmacological relevance of TRP channels as therapeutic targets (Chap. 27).

Ligand-Gated Channels

The third part of this book presents the channelopathies related to ligand-gated channels. As VGCCs and TRP Channels, Ligand-Gated Ca^{2+} channels form a group of transmembrane proteins, which open in response to the binding of a chemical messenger (ligand), allowing ions to pass through the membrane. This part presents the few Ca^{2+} channelopathies caused by alteration of the activity of ligand-gated Ca^{2+} channels.

IP3 Receptors

Inositol 1,4,5-trisphosphate (IP3) receptors (IP3Rs) are localized to intracellular membranes (such as the endoplasmic reticulum, the mitochondria and the nucleus) where they mediate the mobilization of intracellular Ca^{2+} stores following binding of IP3. There are three IP3 receptors subtypes (IP3R1, IP3R2 and IP3R3) exist as homo- and heterotetramers, and play an important role in intracellular Ca^{2+} signaling in a variety of cell types including nerve cells. In Chap. 28, Masayoshi Tada, Masatoyo Nishizawa and Osamu Onodera discuss the implication of these receptors in the development of neurodegenerative disorders including spinocerebellar ataxias, and Huntington's and Alzheimer's diseases.

Ryanodine Receptors

Ryanodine Receptors (RyRs), localized in the membrane of the endoplasmic reticulum and as IP3Rs they contribute to the release of Ca^{2+}. There exist three RyR isoforms (RyR1 to RyR3). Whereas RyR1 is primarily expressed in skeletal muscle and shows a voltage-depend activation mode (see Chap. 9), RyR2 and RyR3, are predominantly expressed in the heart and the brain, respectively, and are activated by a local rise in cytosolic Ca^{2+} thereby contributing to the CICR. In the heart, Ca^{2+} release via RyR2 plays a critical role in muscle contraction, and mutations in *RyR2* have been linked to cardiac disorders. In Chap. 29, Ineke Nederend, Christian van der Werf and Arthur A. M. Wilde discuss the molecular mechanisms by which RyR2 channel contribute to the development of an inherited arrhythmia syndrome, the catecholaminergic polymorphic ventricular tachycardia.

P2X Receptors

Purinergic (P2X) receptors are plasma membrane proteins that open in response to changes in extracellular ATP. Each receptor is made up of a trimer of identical subunits (P2X1 to P2X7), or can also exist as heteromer (with the exception of P2X6 which cannot form functional homomeric receptor). They share the common structure of two transmembrane domains, a large extracellular loop and intracellular amino- and carboxy-termini. P2X receptors are widely expressed throughout the body, including the central and peripheral nervous system, smooth muscle, heart and leukocytes, and are involved in a range of physiological processes such as modulation of synaptic transmission, vascular tone, cardiac rhythm and contractility, but also nociception as described in Chap. 30 by François Rassendren and Lauriane Ulmann. In Chap. 31, Melissa L. Barron, Eryn L. Werry, Iain S. McGregor and Michael Kassiou specifically elucidate the role of P2X7 receptor in depressive and bipolar disorders.

NMDA Receptors

N-methyl-*D*-aspartate (NMDA) receptors are members of the ionotropic class of glutamate receptors. They consist of two NR1 subunits combined with two NR2 (NR2A to NR2D) or NR3 (NR3A or NR3B) subunits. Activation of NMDA receptors requires binding of glutamate or aspartate, as well as binding of co-agonist glycine for the efficient opening of the channel. NMDA receptors are non-selective cation permeable channels, including Na^+, Ca^{2+} and K^+. Calcium influx through NMDA receptors is believed to be critical in synaptic plasticity, and it is not surprising that his involvement in the pathophysiology of epilepsy has been proposed, as discussed by Mehdi Ghasemi and Ahmad Reza Dehpour in Chap. 32.

Nicotinic Acetycholine Receptors

Nicotinic acetycholine receptors (nAChRs) are plasma membrane cholinergic receptors, expressed in many neurons and on the postsynaptic side of the neuromuscular junction. Each receptor is composed of pentameric combinations of subunits (α1 to α10, β1 to β4, δ, ε and γ), and opening of the channel is triggered by the binding of the neurotransmitter acetycholine (Ach). Neuronal nicotinic receptors are found in the central nervous system and in autonomic ganglia where they regulate neurotransmitter release, cell excitability and neuronal integration. In contrast, nicotinic receptors located at the neuromuscular junctions of somatic muscles are responsible for muscular contraction. The implication of nAChRs in muscle pathophysiology is illustrated in Chap. 33 by Francesca Grassi and Sergio Fucile in the context of congenital myasthenic syndromes.

CatSper Channels

CatSper channels are sperm-specific cations channels. There are four CatSper isoforms (CatSper1 to CatSper4) surrounded by at least three auxiliary subunits (α, β and γ), and consist of six putative transmembrane segments with intracellular amino- and carboxy-termini. These channels are weakly voltage-dependent and are activated by intracellular alkalinization and several extracellular ligands, i.e., progesterone and prostaglandin E in human spermatozoa. CatSper channels are essential for hyperactivation of sperm motility required for fertility. Hence, in their chapter, Takuya Nishigaki, Ana Laura González Cota and Gerardo José Orta Salazar present the functional importance of CatSper channels in sperm physiology and discuss their implication in male infertility (Chap. 34).

Together, these expert contributions provide the current state-of-play in this exciting and medically important area of physiology.

Calgary, Canada — Norbert Weiss
Vienna, Austria — Alexandra Koschak

Acknowledgments

First, we would like to warmly thank the authors who agreed to contribute to this book, especially in these *impact factor* dominated days. We feel happy to see that the motivational factor for commitment and willingness to contribute a chapter has been the sole and unique scientific interest of bringing together the current state-of-play in this exciting area of calcium channelopathies in a single volume. Writing a book chapter is not an easy task and we apologize to the authors for stressing them by sending repeated reminders. This was the "price to pay" to publish an up-to-date volume in this rapidly expanding field.

We also thank the referees who generously agreed to review the chapters, and have helped the authors in this difficult task of writing a book chapter and, who have as well as contributed to the project giving their valuable suggestions.

At the time of writing this, we realize that more than a year has passed since we started to discuss this book project. For us it has been a rewarding scientific and human experience and we believe that all the readers, whether basic researchers, clinicians, graduate students, or hopefully also general readers interested by this exciting and medically important area of physiology will find something valuable in this book.

Norbert Weiss
Alexandra Koschak

Contents

Part I
Pathologies of Voltage-Gated Calcium Channels

Chapter 1
$Ca_v2.1$ Channels and Migraine

Daniela Pietrobon

Abstract Migraine is a common disabling brain disorder whose key manifestations are recurrent attacks of unilateral headache that may be preceded by transient neurological aura symptoms. Missense mutations in CACNA1A, the gene that encodes the pore-forming α_1 subunit of human voltage-gated $Ca_v2.1$ (P/Q-type) calcium channels, cause a rare form of migraine with aura (familial hemiplegic migraine type 1: FHM1). This chapter, first, briefly summarizes current understanding of the pathophysiological mechanisms that underlie migraine headache, migraine aura, and the onset of a migraine attack. Then, the chapter describes and discusses (i) the functional consequences of FHM1 mutations on the biophysical properties of recombinant human $Ca_v2.1$ channels and native $Ca_v2.1$ channels in neurons of knockin mouse models carrying the mild R192Q or severe S218L mutations in the orthologous gene, and (ii) the functional consequences of these mutations on neurophysiological processes in the cerebral cortex and trigeminovascular system thought to be involved in the pathophysiology of migraine (including cortical spreading depression, cortical synaptic transmission, and several trigeminal ganglion neuron functions) and the insights into migraine mechanisms obtained from the alterations of these processes in FHM1 knockin mice.

1.1 Introduction

Migraine is an episodic neurological disorder characterized by recurrent attacks of typically throbbing and unilateral, often severe, headache with certain associated features such as nausea, phonophobia, and/or photophobia; in a third of patients the headache is preceded by transient neurological symptoms, that are most

D. Pietrobon (✉)
Department of Biomedical Sciences, University of Padova and CNR Institute of Neuroscience, 35121 Padova, Italy
e-mail: daniela.pietrobon@unipd.it

N. Weiss and A. Koschak (eds.), *Pathologies of Calcium Channels*,
DOI: 10.1007/978-3-642-40282-1_1,

frequently visual, but may involve other senses (migraine with aura: MA) (Lipton et al. 2004). Migraine is remarkably common (e.g., it affects 17 % of females and 8 % of males in the European population (Stovner and Hagen 2006), very costly (EUR 1.8 billion per year in Europe) (Olesen et al. 2012), and disabling (one of the 20 most disabling diseases according to the World Health Organization: (Leonardi et al. 2005).

Migraine is a complex genetic disorder, with heritability estimates as high as 50 % and a likely polygenic multifactorial inheritance (de Vries et al. 2009). Although recent genome-wide association studies have identified a few risk factors for migraine (Anttila et al. 2010; Chasman et al. 2011; Freilinger et al. 2012), most of our current molecular understanding comes from studies of familial hemiplegic migraine (FHM), a rare monogenic autosomal dominant form of MA (Pietrobon 2007; de Vries et al. 2009; Russell and Ducros 2011).

Three FHM causative genes have been identified, all encoding ion channels or transporters (Ophoff et al. 1996; De Fusco et al. 2003; Dichgans et al. 2005). FHM type 1 (FHM1) is caused by missense mutations in CACNA1A (chromosome 19p13), that encodes the pore-forming α_1 subunit of human voltage-gated $Ca_v2.1$ (P/Q-type) calcium channels (Ophoff et al. 1996).

The FHM1 mutations produce substitutions of conserved amino acids in important functional regions of the $Ca_v2.1$ channel including the pore lining and the voltage sensors (see (Pietrobon 2010; Russell and Ducros 2011; Pietrobon 2013) for recent reviews and references). Figure 1.1 shows the location of FHM1 (and a few sporadic hemiplegic migraine, SHM1) mutations in the secondary structure of the $Ca_v2.1\alpha_1$ subunit.

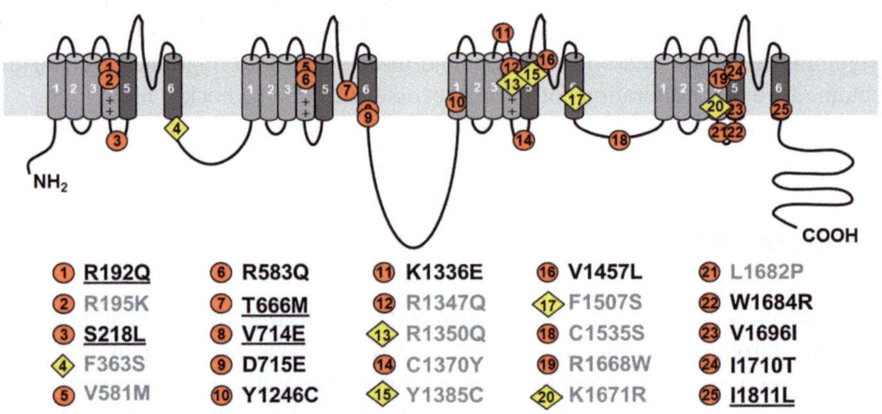

Fig. 1.1 Location of FHM1 mutations in the secondary structure of the $Ca_v2.1\alpha_1$ subunit. FHM1 mutations are indicated with *circles*; a few SHM1 mutations not found in either controls or parents and affecting conserved amino acids are indicated with rhombs. Reference sequence Genbank Ac. N. X99897. The mutations written in *black* are those whose effects on the biophysical properties of recombinant $Ca_v2.1$ channels have been studied in heterologous expression systems; the mutations underlined are those whose effects have been studied also in transfected neurons from $Ca_v2.1^{-/-}$ mice expressing human $Ca_v2.1\alpha_1$ subunits

FHM1 can be considered a model for the common forms of migraine because, apart from the motor weakness or hemiparesis during aura (and the possibly longer aura duration), typical FHM attacks resemble MA attacks (Lipton et al. 2004) and both types of attacks may alternate in patients and co-occur within families (Pietrobon 2007; de Vries et al. 2009; Russell and Ducros 2011). However, several FHM1 families show permanent cerebellar symptoms (such as slowly progressive cerebellar ataxia and/or nystagmus with cerebellar atrophy in some cases), that are usually not observed in common migraines. Pure FHM1 and FHM1 with cerebellar symptoms are usually associated with different mutations. Moreover, in addition to typical attacks, patients of some FHM1 families can have atypical severe attacks with signs of diffuse encephalopathy, impairment of consciousness (coma) or confusion, fever, prolonged hemiplegia lasting several days, and in a few cases seizures (Ducros et al. 2001; Russell and Ducros 2011).

Two different FHM1 knockin mouse models have been generated by introducing the human R192Q and S218L mutations into the orthologous *cacna1a* $Ca_v2.1$ channel gene (van den Maagdenberg et al. 2004, 2010). Whereas mutation R192Q causes in humans pure FHM, mutation S218L causes a particularly dramatic clinical syndrome, that may consists of, in addition to attacks of hemiplegic migraine, slowly progressive cerebellar ataxia and atrophy, epileptic seizures, coma or profound stupor and severe, sometimes fatal, cerebral oedema which can be triggered by only a trivial head trauma (Kors et al. 2001; van den Maagdenberg et al. 2010). Whereas homozygous R192Q and heterozygous S218L knockin mice did not exhibit an overt phenotype, homozygous S218L mice exhibited mild permanent cerebellar ataxia, spontaneous attacks of hemiparesis and/or (sometimes fatal) generalized seizures, and brain edema after only a mild head impact, thus modeling the main features of the severe S218L clinical syndrome (van den Maagdenberg et al. 2004, 2010).

$Ca_v2.1$ channels are located in presynaptic terminals and somatodendritic membranes throughout the mammalian brain and spinal cord (Westenbroek et al. 1995), and play a prominent role in initiating action potential (AP)-evoked neurotransmitter release at central nervous system synapses (Pietrobon 2005a). At many central synapses P/Q-, N-, and R-type Ca^{2+} (Ca) channels cooperate in controlling neurotransmitter release, but P/Q-type channels have a dominant role, partly because of a more efficient coupling to the exocytotic machinery (Mintz et al. 1995; Wu et al. 1999; Li et al. 2007). Moreover, at many central synapses, there is a developmental change in the Ca channel types mediating synaptic transmission, whereby the relative contribution of P/Q-type channels to release increases with postnatal age, until release becomes exclusively dependent on P/Q-type channels (Iwasaki et al. 2000). Among the presynaptic Ca channels, $Ca_v2.1$ channels are unique also in their capacity of interacting with and being modulated in a complex manner by a number of Ca-binding proteins ((Catterall and Few 2008) for review and references). As a result, $Ca_v2.1$ channels may exhibit both Ca-dependent inactivation and Ca-dependent facilitation. Ca-dependent regulation of presynaptic $Ca_v2.1$ channels may play a crucial role in short-term synaptic plasticity during trains of action potentials (Inchauspe et al. 2004; Mochida et al. 2008; Lin et al. 2012).

The somatodendritic localization of $Ca_v2.1$ channels points to additional postsynaptic roles, e.g., in neural excitability (Pineda et al. 1998; Womack et al. 2004), gene expression (Sutton et al. 1999), and cell survival (Fletcher et al. 2001).

$Ca_v2.1$ channels are expressed in all brain regions that have been implicated in the pathogenesis of migraine and/or migraine pain (cf. next Section), including the cerebral cortex (see Sect. 1.5.1), the trigeminal ganglia, and brainstem nuclei involved in the central control of nociception (see Sect. 1.5.2). Their expression is particularly high in the cerebellum, a feature that explains the cerebellar symptoms caused by several FHM1 mutations ((Pietrobon 2010) and Chapter 2 in this book).

1.2 Migraine Pathophysiology

It is generally recognized that most migraine attacks start in the brain, as suggested by the premonitory symptoms (such as difficulty with speech and reading, increased emotionality, sensory hypersensitivity), that in many patients are highly predictive of the attack although occurring up to 12 h before it, and as suggested also by the nature of some typical migraine triggers such as stress, sleep deprivation, oversleeping, hunger, prolonged sensory stimulation; moreover, psychophysical and neurophysiological studies have provided clear evidence that in the period between attacks migraineurs show hypersensitivity to sensory stimuli and abnormal processing of sensory information (see (Pietrobon and Striessnig 2003; Aurora and Wilkinson 2007; Coppola et al. 2007; Vecchia and Pietrobon 2012; Pietrobon and Moskowitz 2013) for reviews and references).

The neurophysiologial correlate of migraine aura is cortical spreading depression (CSD), a self-sustaining slowly propagating (2–5 mm/min) wave of nearly complete depolarization of a sizable population of brain cells that lasts 1–2 minutes (see (Lauritzen 1994; Pietrobon and Striessnig 2003; Ayata 2009; Charles and Brennan 2009) for reviews and references). CSD can be induced in healthy brain tissue by stimulation of the cerebral cortex with intense depolarizing stimuli. Although the mechanisms of initiation and propagation of CSD remain unclear, the initiation of the positive feedback cycle that ignites CSD and almost zeroes the neuronal membrane potential is thought to depend on the local increase of the extracellular concentration of K^+ ions above a critical value and on the activation of a net inward current at the pyramidal cell dendrites (Somjen 2001).

It is generally believed that the headache phase of migraine begins with the activation and sensitization of trigeminal sensory afferents that innervate cranial tissues, in particular the meninges and their large blood vessels; this then leads to sequential activation (and, in most patients, sensitization) of second and third order trigeminovascular neurons (in the trigeminal nucleus caudalis and specific thalamic nuclei, respectively), which in turn activate different areas of the brainstem and forebrain resulting in pain and other symptoms of migraine; the bidirectional signaling between many of these areas contributes to the complexity of the symptoms (see (Olesen et al. 2009; Levy 2010; Akerman et al. 2011; Pietrobon and Moskowitz 2013) for recent reviews and references).

Most dural trigeminovascular afferents can be activated and sensitized by an inflammatory soup and are capsaicin-sensitive, thus exhibiting properties characteristic of nociceptors in other tissues (Strassman et al. 1996; Bove and Moskowitz 1997; Strassman and Levy 2006; Levy et al. 2007; Levy 2010; Vaughn and Gold 2010; Fioretti et al. 2011; Yan et al. 2011). Activation of meningeal nociceptors in vivo leads to release of vasoactive proinflammatory peptides such as CGRP and substance P from their peripheral nerve endings, that produce vasodilation of meningeal blood vessels (mainly due to CGRP), plasma extravasation and local activation of dural mast cells, with ensuing release of cytokines and other inflammatory mediators (neurogenic inflammation) (Waeber and Moskowitz 2005; Levy 2010). Measurements of CGRP levels into the external and internal jugular venous blood have provided evidence that CGRP is released during migraine attacks, and several findings support a pivotal role of CGRP in migraine, including the efficacy of CGRP receptor antagonists in migraine treatment (reviewed in (Recober and Russo 2009; Villalon and Olesen 2009; Ho et al. 2010)). However, the mechanism of actions of CGRP during a migraine attack and the exact sites of action of CGRP receptor antagonists remain unclear and controversial. The localization of CGRP receptors in the trigeminovascular system points to multiple possible mechanisms at both peripheral and central sites (Lennerz et al. 2008).

Although a large body of indirect evidence supports the idea that a sterile meningeal inflammation is a key mechanism that may activate and sensitize perivascular meningeal afferents and lead to migraine pain (reviewed in (Waeber and Moskowitz 2005; Levy 2009, 2010; Vecchia and Pietrobon 2012; Pietrobon and Moskowitz 2013)), the endogenous processes that promote meningeal inflammation and the mechanisms underlying the sustained activation and sensitization of the trigeminovascular system resulting in the typical long-lasting throbbing headache remain unclear.

The mechanisms of the primary brain dysfunction(s) leading to the onset of a migraine attack, to CSD susceptibility and to episodic activation of the trigeminovascular pain pathway remain largely unknown and remain the major open issues in the neurobiology of migraine.

Increasing evidence from animal studies support the idea that CSD can activate trigeminal nociception and thus trigger the headache mechanisms (Vecchia and Pietrobon 2012; Pietrobon and Moskowitz 2013). A direct nociceptive effect of CSD has been demonstrated by the finding that a single CSD can lead to a long-lasting increase in ongoing activity of dural nociceptors and central second order trigeminovascular neurons (Zhang et al. 2010, 2011), and evokes alterations in the meninges and brainstem consistent with the development of pain (Bolay et al. 2002). An interesting recent study provided evidence that CSD leads to opening of neuronal pannexin channels and consequent release of proinflammatory mediators, which initiates a parenchymal inflammatory response leading to sustained release of inflammatory mediators from glia limitans; this CSD-induced inflammatory cascade may provide the sustained stimulus required for sensitization of trigeminal nociceptors and lasting pain (Karatas et al. 2013).

The idea that CSD is noxious and may trigger headache is indirectly supported by the finding that the electrical stimulation threshold for induction of CSD in the rat cortex increases after chronic treatment with five different migraine prophylactic drugs, that are effective in reducing the frequency of migraine attacks (both with and without aura) (Ayata et al. 2006). Moreover, further support to the view of CSD as a key migraine trigger has been provided by the analysis of experimental CSD in FHM knockin mouse models, that revealed a lower electrical stimulation threshold for CSD induction and a higher velocity of CSD propagation in both FHM1 and FHM2 knockin mice (van den Maagdenberg et al. 2004, 2010; Eikermann-Haerter et al. 2009b; Leo et al. 2011) (see Sect. 1.5.1). However, despite the strong support provided by animal studies, the idea that CSD may initiate the headache mechanisms in migraine is not generally accepted, mainly because it seems unable to explain some clinical observations, in particular the lack of a fixed relationship between aura and headache (reviewed in (Pietrobon and Striessnig 2003; Charles 2010; Vecchia and Pietrobon 2012; Pietrobon and Moskowitz 2013)).

The analysis of interictal cortical excitability using psychophysical, electrophysiological, and brain imaging methods has produced contradictory findings and interpretations regarding the mechanisms underlying the abnormal processing of sensory information in migraineurs (see (Pietrobon and Striessnig 2003; Aurora and Wilkinson 2007; Coppola et al. 2007; Vecchia and Pietrobon 2012) for reviews and references). Interestingly, recent transcranial magnetic stimulation studies in MA patients point to deficient regulatory mechanisms of cortical excitability and consequent reduced ability to dynamically maintain the cortical excitatory-inhibitory balance and to prevent excessive increases in cortical excitation, rather than merely hypo- or hyperexcitability, as the mechanisms underlying abnormal sensory processing (Antal et al. 2008; Conte et al. 2010; Siniatchkin et al. 2012). The molecular and cellular mechanisms underlying the abnormal regulation of cortical function and the increased susceptibility to CSD in migraine remain largely unknown. As discussed in Sect. 1.4, the functional analysis of FHM knockin mouse models supports the view of migraine as a disorder of brain excitability characterized by deficient regulation of the cortical excitatory-inhibitory balance, and gives insights into the possible underlying molecular and cellular mechanisms and their relationship to CSD susceptibility.

1.3 Effect of FHM1 Mutations on the Biophysical Properties of $Ca_v2.1$ Channels

The analysis of the P/Q-type calcium current in different neurons (including cortical and TG neurons) of R192Q and S218L FHM1 knockin mice revealed gain-of-function of the $Ca_v2.1$ current in a wide range of relatively mild depolarizations, reflecting shifted activation of mutant $Ca_v2.1$ channels to more negative voltages (van den Maagdenberg et al. 2004, 2010; Tottene et al. 2009; Inchauspe et al. 2010; Fioretti et al. 2011 Gao et al. 2012). P/Q current densities were similar in knockin and wild-type (WT) neurons at higher voltages (that elicit maximal $Ca_v2.1$ channel open probability), indicating similar densities of functional $Ca_v2.1$ channels (van den Maagdenberg et al. 2004, 2010; Tottene et al. 2009; Fioretti et al. 2011; Gao et al. 2012). The shift to lower voltages of $Ca_v2.1$ channel activation and the gain-of-function of the neuronal $Ca_v2.1$ current were about twice as large in homozygous compared to heterozygous knockin mice, revealing an allele-dosage effect consistent with dominance of the mutation in FHM1 patients (van den Maagdenberg et al. 2010). In correlation with the severity of the clinical phenotype, the gain-of-function of the P/Q Ca current at low voltages was larger in cerebellar granule cells of S218L than R192Q knockin mice (van den Maagdenberg et al. 2004, 2010). Two photon microscopy in cerebellar slices of S218L knockin mice revealed increased AP-evoked Ca transients in individual synaptic terminals of cerebellar granule cells, a finding consistent with increased open probability of mutant presynaptic $Ca_v2.1$ channels (Adams et al. 2010), that may be due to shifted activation to lower voltages, as shown for the somatic channels (van den Maagdenberg et al. 2010), and/or to basal Ca-dependent facilitation of mutant channels as suggested by the authors (Adams et al. 2010) (see sect. 1.4).

The gain-of-function effect of the FHM1 mutations on native neuronal mouse $Ca_v2.1$ channels is in agreement with the increased open probability (mainly due to shifted activation to lower voltages) of recombinant human $Ca_v2.1$ channels carrying eight different FHM1 mutations (including R192Q and S218L), that was revealed by single channel recordings (Hans et al. 1999; Tottene et al. 2002; Tottene et al. 2005; Catterall et al. 2008) (our unpublished observations). The larger gain-of-function of the P/Q Ca current in neurons of S218L compared to R192Q knockin mice at low voltages is in agreement with the lower threshold of activation of human S218L $Ca_v2.1$ compared to human R192Q $Ca_v2.1$ channels (Tottene et al. 2005).

Consistent (and significant for the majority of the 13 FHM1 mutations analyzed so far, shown in black in Fig. 1.1) shifts to lower voltages of activation of recombinant mutant channels were also revealed by measurements of whole-cell current in heterologous expression systems (Kraus et al. 1998, 2000; Hans et al. 1999; Tottene et al. 2002, 2005; Melliti et al. 2003; Mullner et al. 2004; Weiss et al. 2008; Adams et al. 2009; Serra et al. 2009; Garza-Lopez et al. 2012) and transfected neurons (where five FHM1 mutations, underlined in Fig. 1.1, were analyzed) (Tottene et al. 2002, 2005). Moreover, these studies have shown that the FHM1 mutations may also alter the inactivation properties of $Ca_v2.1$ channels; the effects were different depending on the mutation and, e.g., resulted in increased, decreased, or unaltered inactivation during train of pulses, depending on the mutation.

For certain mutations, the specific alterations of channel gating, including the magnitude (or even the presence) of the shift to lower voltages of channel activation, depended on the $Ca_v2.1\alpha_1$ splice variant and the type of coexpressed β subunit (Mullner et al. 2004; Adams et al. 2009). The use of different $Ca_v2.1$ splice variants and/or β subunits might explain some discrepancies existing in the literature regarding the functional effects of some FHM1 mutations in transfected cells: e.g., the T666 M mutation was shown to produce a negative shift of $Ca_v2.1$ activation in (Kraus et al. 1998; Tottene et al. 2002) in contrast, a lack of effect on activation gating was reported in (Barrett et al. 2005; Tao et al. 2012).

Moreover, the expression of specific $Ca_v2.1$ splice variants and/or auxiliary subunits in different neurons might underlie the recent finding of a differential effect of the R192Q mutation on activation gating of native P/Q-type Ca channels in different TG neurons of knockin mice (Fioretti et al. 2011). In small capsaicin-insensitive TG neurons characterized by expression of T-type Ca currents (CI-T neurons) there was a larger P/Q-type Ca current density following mild depolarizations and a larger AP-evoked P/Q calcium current when compared to CI-T neurons from WT mice. In striking contrast, the P/Q-type current density, voltage dependence, and kinetics were not altered by the FHM1 mutation in small capsaicin-sensitive neurons of R192Q knockin mice (Fioretti et al. 2011).

Interestingly, while in cortical pyramidal cells of R192Q KI mice the shift to lower voltages of $Ca_v2.1$ channel activation resulted in increased AP-evoked Ca current, a similar shift of mutant R192Q channels at the Calyx of Held synaptic terminals did not alter the AP-evoked Ca current (Inchauspe et al. 2010). The different durations of the AP in pyramidal cells and Calyx (1.8 vs 0.44 ms AP half width) may largely explain the differential effects of the FHM1 mutation on the AP-evoked Ca current; in fact, the AP-evoked Ca current became larger in knockin compared to WT Calyx terminals when the longer duration pyramidal cell AP was used as depolarizing stimulus (Inchauspe et al. 2010).

The findings of (Fioretti et al. 2011) and (Inchauspe et al. 2010) have important general implications for familial migraine mechanisms, in that neuron subtype-specific (and/or subcellular compartment-specific) alterations of $Ca_v2.1$ channels and/or AP-evoked Ca influx may help to explain why a mutation in a Ca channel that is widely expressed in the nervous system (Westenbroek et al. 1995) produces the specific neuronal dysfunctions leading to migraine (see Sect. 1.5).

In contrast with the unaltered densities of functional $Ca_v2.1$ channels in neurons of FHM1 knockin mice (van den Maagdenberg et al. 2004, 2010; Tottene et al. 2009; Fioretti et al. 2011), overexpression of recombinant mutant $Ca_v2.1$ channels in transfected cells leads to alterated (decreased for most mutations) density of functional channels in the membrane, that results in reduced maximal whole-cell P/Q-type Ca current in cells expressing mutant channels (Hans et al. 1999; Tottene et al. 2002; Cao et al. 2004; Barrett et al. 2005; Cao and Tsien 2005; Tottene et al. 2005; Tao et al. 2012), including neurons expressing R192Q and S218L $Ca_v2.1$ channels (Tottene et al. 2002; Cao et al. 2004; Cao and Tsien 2005; Tottene et al. 2005) (cf. Fig. 3 in (Pietrobon 2010)).

1.4 Effect of FHM1 Mutations on Modulation of $Ca_v2.1$ Channels

Studies of recombinant human $Ca_v2.1$ channels carrying five different FHM1 mutations have shown that the migraine mutations reduce G-protein-mediated inhibitory modulation of $Ca_v2.1$ channels (Melliti et al. 2003; Weiss et al. 2008; Serra et al. 2009; Garza-Lopez et al. 2012), an effect that may lead to further gain-of-function of Ca influx through mutant channels during neuromodulation. The only study that investigated voltage-dependent G-protein modulation using physiological-like stimuli revealed a reduced $G\beta\gamma$-mediated inhibition of the Ca current evoked by AP-like stimuli and a reduced facilitation of the Ca current carried by mutant channels after a train of short depolarizing pulses in cells overexpressing $G\beta\gamma$ subunits; the reduction of G-protein modulation was dependent on the duration of the AP, being larger with APs of longer duration (Serra et al. 2009).

Indirect evidence of reduced G-protein-mediated inhibition of native neuronal $Ca_v2.1$ channels in R192Q knockin mice has been recently obtained in a study investigating the effect of the FHM1 mutation on sleep regulation by adenosine receptors in vivo; in fact, the FHM1 mouse models showed reduced responsiveness to either inhibition or enhancement of adenosine receptor activation (Deboer et al. 2013).

Besides G-protein modulation, the FHM1 mutations may also reduce Ca-dependent facilitation (CDF) of $Ca_v2.1$ channels. CDF of recombinant $Ca_v2.1$ channels was largely reduced by the R192Q mutation and almost eliminated by the S218L mutation (Adams et al. 2010), suggesting that the increase in the open probability of mutant channels (larger for S218L than R192Q: (Tottene et al. 2005)) partially or completely occludes further facilitation. CDF of native $Ca_v2.1$ channels in Purkinje cells of S218L knockin mice was also reduced (Adams et al. 2010). In contrast, CDF of native $Ca_v2.1$ channels in both Purkinje cells (Adams et al. 2010) and Calyx of Held synaptic terminals (Inchauspe et al. 2010) of R192Q knockin mice was not significantly affected, suggesting that also the effect of FHM1 mutations on CDF likely depends on the specific $Ca_v2.1$ variant expressed in a given neuron or neuronal subcompartment.

1.5 Insights into Migraine Pathophysiology from FHM1 Knockin Mouse Models

1.5.1 Cortical Synaptic Transmission and Cortical Spreading Depression

In the cerebral cortex, excitatory synaptic transmission at pyramidal cell synapses in different cortical areas depends predominantly on P/Q-type Ca channels (Iwasaki et al. 2000; Koester and Sakmann 2000; Rozov et al. 2001; Ali and Nelson

2006; Zaitsev et al. 2007; Tottene et al. 2009) with a notable exception at synapses between layer five pyramidal cells and burst-firing bipolar interneurons of motor cortex (Ali and Nelson 2006). Activation of the cortex following thalamic stimulation in thalamocortical slices was completely inhibited by blocking P/Q Ca channels (Llinas et al. 2007). The Ca channel pharmacology of cortical inhibitory synapses has been investigated only at fast spiking (FS) interneurons synapses; neurotransmission was found to be exclusively dependent on P/Q-type channels in many cortical areas (Zaitsev et al. 2007; Kruglikov and Rudy 2008; Tottene et al. 2009) (cf also (Sasaki et al. 2006)), but again with the exception of layer five of the motor cortex, where it was exclusively dependent on N-type (Ali and Nelson 2006).

The analysis of cortical excitatory synaptic transmission in neuronal microcultures and in brain slices from R192Q knockin mice revealed enhanced excitatory neurotransmission, due to enhanced action potential evoked Ca influx through mutant presynaptic P/Q Ca channels and enhanced probability of glutamate release at cortical pyramidal cell synapses (Tottene et al. 2009). Short-term synaptic depression during trains of action potentials was also enhanced. Neither amplitude nor frequency of miniature excitatory postsynaptic currents were altered, indicating the absence of homeostatic compensatory mechanisms at excitatory synapses onto pyramidal cells (Tottene et al. 2009). Although indirect, evidence for gain-of-function of excitatory neurotransmission has also been obtained at parallel fibers-Purkinje cell synapses in cerebellar slices of S218L knockin mice (Adams et al. 2010) and at excitatory synapses onto dorsal suprachiasmatic nucleus neurons of R192Q knockin mice (van Oosterhout et al. 2008).

In striking contrast with the enhanced glutamatergic transmission, paired recordings of fast spiking (FS) inhibitory interneurons and layer 2/3 pyramidal cells in acute cortical slices revealed that the inhibitory GABAergic transmission at FS interneuron synapses was not altered in R192Q knockin mice, despite being initiated by P/Q Ca channels (Tottene et al. 2009). The main explanation for the unaltered GABA release at FS interneuron synapses appears to be the specific expression in FS interneurons of a $Ca_v2.1$ subtype whose gating properties are little affected by the mutation (Vecchia, Tottene and Pietrobon, unpublished observations).

The investigation of experimental CSD, elicited either by electrical stimulation of the cortex in vivo or high KCl in cortical slices, revealed a lower threshold for CSD induction and an increased velocity of CSD propagation in R192Q and S218L knockin compared to WT mice (van den Maagdenberg et al. 2004, 2010; Tottene et al. 2009). Moreover, a single CSD, elicited by brief epidural application of high KCl, produced more severe and prolonged motor deficits (including hemiplegia) in FHM1 knockin mice, and, in contrast with WT mice, CSD readily propagated into the striatum (Eikermann-Haerter et al. 2009b, 2011). The much higher propensity of CSD to propagate to the striatum in FHM1 mutants compared to WT mice may explain their motor deficits and the hemiplegia typical of FHM1 aura. In agreement with the higher incidence of migraine in females, the velocity of propagation and the frequency of CSDs, elicited by continuous epidural high KCl application, were larger in females than in males of both mutant strains; the sex difference was abrogated by ovariectomy and enhanced by orchiectomy, suggesting that female and male gonadal hormones exert reciprocal effects on CSD susceptibility (Eikermann-Haerter et al. 2009a, b).

The strength of CSD facilitation as well as the severity of the post-CSD neurological motor deficits and the propensity of CSD to propagate into subcortical structures in R192Q and S218L knockin mice were all in good correlation with the strength of the gain-of-function of the $Ca_v2.1$ channel and the severity of the clinical phenotype produced by the two FHM1 mutations (van den Maagdenberg et al. 2004, 2010; Tottene et al. 2005; Eikermann-Haerter et al. 2009b, 2011; Kors et al. 2001). Propagation of CSD to the hippocampus and thalamus and repetitive CSD events following a single CSD-inducing stimulus were observed only in S218L mutants (van den Maagdenberg et al. 2010; Eikermann-Haerter et al. 2011). These unique CSD features might account for the severe attacks with seizures, coma, and cerebral edema typical of patients with the S218L mutation. Moreover, the recent finding of increased frequency of periinfarct ischemic depolarizations during acute experimental stroke (and larger critical tissue perfusion level below which infarction ensues) in S218L knockin compared to WT mice (Eikermann-Haerter et al. 2012) suggests that the S218L mutation may lead to increased vulnerability to ischemic stroke as a consequence of increased susceptibility to ischemic depolarizations akin to CSD. The irregular firing patterns and hyperexcitability of Purkinje cells recently uncovered in S218L knockin mice likely contributes to the cerebellar ataxia produced by the S218L mutation in mice and humans (Gao et al. 2012).

Tottene et al. (2009) provided direct evidence that the gain-of-function of glutamate release at synapses onto cortical pyramidal cells may explain the facilitation of experimental CSD in FHM1 knockin mice (Tottene et al. 2009). In fact, the facilitation of CSD in acute cortical slices of R192Q knockin mice was completely eliminated (both CSD threshold and velocity became similar to those in WT slices) when glutamate release at pyramidal cell synapses was brought back to WT values by partially inhibiting P/Q channels (Tottene et al. 2009). The data are consistent with and support a model of CSD initiation in which $Ca_v2.1$-dependent release of glutamate from cortical pyramidal cell synapses and activation of NMDA receptors (and possibly postsynaptic $Ca_v2.1$ channels) play a key role in the positive feedback cycle that ignites CSD (Pietrobon 2005b; Tottene et al. 2009). This model and in general the specific requirement of $Ca_v2.1$ channels in the initiation and propagation of CSD (induced by electrical stimulation or brief pulses of high K+ in healthy tissue) are further supported by the findings that (i) after blockade of either the P/Q-type Ca channels or the NMDA receptors, CSD could not be induced in cortical slices of WT mice even with largely supra threshold depolarizing stimuli; in contrast, blockade of N- or R-type Ca channels had only a small inhibitory effect on CSD threshold and velocity of propagation (Tottene et al. 2011), and (ii) in the spontaneous mouse mutants *leaner* and *tottering*, that carry loss-of-function mutations in *cacna1a* (Pietrobon 2010), the in vivo electrical threshold for CSD initiation was greatly increased and the CSD velocity decreased compared to WT mice (Ayata et al. 2000).

In migraineurs, CSD is not induced by experimental depolarizing stimuli, but arises "spontaneously" in response to specific triggers, that somehow create in the cortex conditions for initiation of the positive feedback cycle that overwhelms the regulatory mechanisms controlling cortical $[K^+]_o$ and ignites CSD. Insights into how this might occur have been provided by the differential effect of FHM1 mutations on cortical excitatory and inhibitory synaptic transmission (Tottene et al. 2009). This finding suggests that, very likely, the neuronal circuits that dynamically adjust the balance between excitation and inhibition during cortical activity are altered in FHM1 (Tottene et al. 2009). Functional alterations in these circuits are expected to lead to dysfunctional regulation of the cortical excitatory-inhibitory balance and hence to abnormal processing of sensory information (Monier et al. 2003; Shu et al. 2003). It has been hypothesized that this dysregulation may in certain conditions (e.g., in response to migraine triggers such as intense, prolonged sensory stimulation) lead to disruption of the excitatory-inhibitory balance and hyperactivity of cortical circuits, mainly due to excessive recurrent excitation, that may create the conditions for the initiation of "spontaneous" CSDs (e.g., by increasing the extracellular $[K^+]$ above a critical value) (Tottene et al. 2009; Vecchia and Pietrobon 2012).

Impairment of the cortical circuits that dynamically adjust the excitatory-inhibitory balance during cortical activity, due to excessive recurrent glutamatergic neurotransmission, might also underlie the abnormal regulation of interictal cortical function in some common migraine subtypes, for which there is indirect evidence consistent with enhanced cortical glutamatergic neurotransmission (Prescot et al. 2009; Siniatchkin et al. 2012) and enhanced cortico–cortical or recurrent excitatory neurotransmission (Siniatchkin et al. 2007; Wilkinson et al. 2008; Conte et al. 2010; Battista et al. 2011).

The gain-of-function effects on cortical pyramidal cell $Ca_v2.1$ channels and on cortical glutamatergic synaptic transmission derived from the functional analysis of R192Q and S218L FHM1 knockin mice provide a coherent picture of molecular and cellular mechanisms that may produce increased susceptibility to CSD and thus explain the aura symptoms in FHM1 patients (Fig. 1.2). Most likely, the conclusions are valid for all FHM1 mutations, given that the aura phenotype is common to all FHM1 patients and that eight different FHM1 mutations produce a consistent gain-of-function of single channel Ca influx through recombinant human $Ca_v2.1$ channels (see Sect 1.3). The opposite conclusions derived from the functional analysis of FHM1 mutations (including the R192Q mutation) in transfected hippocampal neurons overexpressing human $Ca_v2.1\alpha1$ subunits (i.e., loss-of-function of $Ca_v2.1$ channels with consequent decreased contribution of P/Q Ca channels to glutamatergic synaptic transmission, and unaltered synaptic strength due to compensatory mechanisms (Cao et al. 2004)) provide a picture that seems unable to explain the increased susceptibility to CSD and the aura phenotype of FHM1 patients. Thus, to be able to draw meaningful conclusions regarding

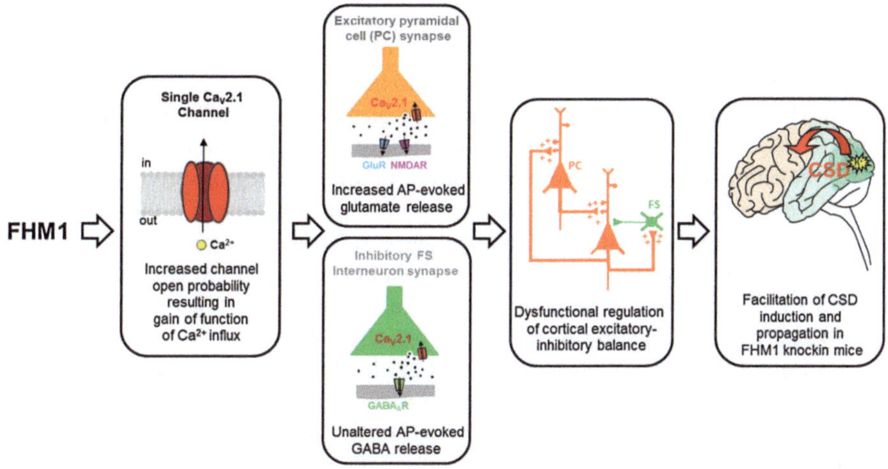

Fig. 1.2 Functional alterations in the cerebral cortex of a familial hemiplegic migraine type 1 (FHM1) knockin mouse model. Action potential (AP)-evoked glutamate release and excitatory synaptic transmission at pyramidal cell (PC) synapses are increased, due to increased AP-evoked Ca influx through presynaptic $Ca_v2.1$ channels consequent to the increased open probability and activation at lower voltages of mutant compared to WT channels. In striking contrast, AP-evoked GABA release and inhibitory synaptic transmission at fast spiking (FS) interneuron synapses are unaltered. The differential effect of the FHM1 mutation on excitatory and inhibitory synaptic transmission likely results in dysfunctional regulation of the cortical excitatory/inhibitory balance. Experimental cortical spreading depression (CSD) is facilitated, as revealed by a decreased threshold for CSD induction, an increased rate of CSD propagation and an increased propensity to propagate into subcortical structures in FHM1 mouse models

the neurophysiological processes involved in the pathogenesis of FHM1 (and probably neuronal channelopathies in general), it seems essential to study these processes in neurons expressing the channels at the native endogenous level.

1.5.2 Trigeminal Ganglion Sensory Afferents and Headache Mechanisms

P/Q-type Ca channels account for a large proportion of the Ca current of dissociated trigeminal ganglion (TG) neurons (Borgland et al. 2001; Fioretti et al. 2011; Tao et al. 2012), and are involved in the control of CGRP release from capsaicin-sensitive perivascular terminals of meningeal nociceptors (Hong et al. 1999; Akerman et al. 2003) and of glutamate release from TG neurons in culture (Xiao et al. 2008). P/Q channels are also involved in controlling tonic inhibition of trigeminal nucleus caudalis neurons with input from the dura (Ebersberger et al. 2004), and in descending inhibitory and facilitatory pathways that regulate trigeminal and spinal pain transmission (Knight et al. 2002; Urban et al. 2005).

The analysis of the pain responses in $Ca_v2.1^{-/-}$ null mice revealed a complex role of $Ca_v2.1$ channels in pain: a pronociceptive role in inflammatory and neuropathic pain but an antinociceptive role in response to acute non-injurious noxious thermal stimuli (Luvisetto et al. 2006); this complex picture likely reflects the prominent role of P/Q channels in controlling release of both excitatory and inhibitory neurotransmitters and their wide distribution in different regions involved in pain.

As mentioned in Sect. 1.3, the analysis of the P/Q-type Ca current in two defined subpopulations of small (capacitance $\leq$20 pF) TG neurons from adult R192Q knockin mice showed gain-of-function of the $Ca_v2.1$ channel in capsaicin-insensitive TG neurons expressing T-type Ca channels (CI-T neurons), but unaltered $Ca_v2.1$ channel gating properties in capsaicin-sensitive TG neurons lacking T-type Ca channels (CS neurons) (Fioretti et al. 2011). WT CI-T and CS neurons were both characterized by APs of long duration with typical shoulder and tonic or slowly adapting firing. While the rheobase (i.e., the minimal current injection that elicits APs) and the frequency of APs were not affected by the FHM1 mutation, the duration of the AP in CI-T neurons of R192Q knockin mice was prolonged, due to a delayed repolarization and more pronounced shoulder that correlated with the larger AP-evoked P/Q-type Ca current measured in these neurons (Fioretti et al. 2011). In contrast, neither the AP shape nor any other excitability property were affected by the FHM1 mutation in small capsaicin-sensitive neurons (in agreement with the unaltered P/Q channel gating and unaltered AP-evoked P/Q current in these TG neurons) (Fioretti et al. 2011).

Measurements of the Ca current in small TG neurons retrogradely labeled from the dura revealed that a major fraction of small dural afferents were CS neurons and none were CI-T neurons (Fioretti et al. 2011). Several lines of evidence indicate that most small capsaicin-sensitive dural afferents are peptidergic neurons expressing CGRP (Jansen et al. 1990; Messlinger et al. 1993; Shimizu et al. 2007). Measurements of CGRP release from dura mater in fluid-filled hemisected skulls revealed that neither basal nor K+ -evoked CGRP release were significantly different in R192Q knockin compared to WT mice (Fioretti et al. 2011); this finding is consistent with and supports lack of effect of the FHM1 mutation on presynaptic P/Q channels at the peripheral terminals of CGRP-expressing (including CS) dural afferents, as demonstrated for P/Q channels at the soma of small CS TG neurons (Fioretti et al. 2011). These data argue against the idea that the facilitation of CGRP-dependent dural vasodilation and CGRP-dependent dural mast cell degranulation contribute to the generation of migraine pain in FHM1 (Vecchia and Pietrobon 2012; Pietrobon and Moskowitz 2013).

Despite the lower percent of CGRP-expressing neurons recently reported in the trigeminal ganglion of R192Q knockin compared to WT mice (Mathew et al. 2011), the FHM1 mutation did increase evoked CGRP release from intact trigeminal ganglia (Fioretti et al. 2011) and both basal and evoked CGRP release from cultured TG neurons (Ceruti et al. 2011) of R192Q knockin mice, suggesting alternative roles. There is in vivo and in vitro evidence for non-synaptic intraganglionic release of CGRP, substance P and ATP from TG neurons cell bodies in response to depolarizing stimuli (Matsuka et al. 2001; Ulrich-Lai et al. 2001). It has been suggested that CGRP-mediated intraganglionic crosstalk between neurons and between neurons and satellite glial cells could promote and maintain a neuron-glia inflammatory cycle that might contribute to peripheral trigeminal sensitization (Villalon and Olesen 2009; Ho et al. 2010). This suggestion is mainly based on the evidence that prolonged application of CGRP to cultures of TG neurons and/or satellite glial cells leads to increased gene expression and/or membrane targeting of specific receptors (e.g., $P2X_3$) in neurons and to increased expression of inflammatory genes and release of inflammatory mediators from satellite glial cells; these inflammatory mediators can sensitize TG neurons and act back on glial cells further activating them ((Vecchia and Pietrobon 2012; Pietrobon and Moskowitz 2013) and references therein). Assuming that similar phenomena occur in the trigeminal ganglion in vivo upon prolonged elevations of CGRP, the enhanced intraganglionic CGRP release measured in FHM1 knockin mice suggests, as a possible working hypothesis, that FHM1 mutations might facilitate peripheral sensitization at the ganglion level. As a matter of fact, there is some evidence suggesting facilitation of CGRP-mediated neuron to glia crosstalk following exposure to proinflammatory stimuli in cultured TG neurons from juvenile R192Q knockin mice (Ceruti et al. 2011). Moreover, a stronger basal activation of macrophages, a larger basal release of TNFα, and an enhanced P2X3 receptor-mediated neuronal current have been recently reported in cultured TG neurons from these mice (Franceschini et al. 2013a; Nair et al. 2010). Also intact trigeminal ganglia of R192Q knockin mice appear enriched in activated macrophages both in the absence and presence of a standard inflammatory stimulus (Franceschini et al. 2013b). On the basis of these findings, it has been suggested that FHM1 mutations might lead to a basal inflammatory milieu within the trigeminal ganglion (Franceschini et al. 2013a) (but of similar levels of inflammatory cytokines in TG extracts from WT and R192Q knockin mice in (Franceschini et al. 2013b)). Interestingly, in both cultured TG neurons and intact ganglia a larger fraction of TG neurons was immunoreactive for active phosphorylated CaMKII in R192Q knockin compared to WT mice. Blockade of P/Q channels in cultured TG neurons eliminated the difference in amount of phosphoprotein between the two genotypes, suggesting facilitation of basal $Ca_v2.1$-dependent Ca signaling (Nair et al. 2010).

The functional analysis of TG neurons and dural afferents in R192Q knockin mice supports the conclusions that (i) the FHM1 mutations may lead to gain-of-function of the P/Q Ca current in certain TG neurons without affecting the P/Q current in other TG neuron subtypes, including small capsaicin-sensitive dural afferents; (ii) the FHM1 mutations lead to several gain-of-function effects at the trigeminal ganglion level, but do not alter the functional properties of small capsaicin-sensitive peptidergic dural afferents, including CGRP release from their peripheral terminals at the dura.

Again, as in the case of the cortical neurophysiological processes discussed in Sect. 1.5.1, different conclusions may be drawn from the recent functional analysis of an FHM1 mutation in transfected neonatal TG neurons expressing human mutant $Ca_v2.1\alpha1$ subunits (Tao et al. 2012). In these neurons, the P/Q current density was much smaller than in neurons expressing the WT channel, and small CGRP-expressing IB4-negative neurons showed a compensatory increase in low-voltage-activated Ca current, a lower rheobase and higher AP frequency, with unaltered AP duration, compared to neurons expressing the WT channel (Tao et al. 2012). Assuming that this subpopulation of TG neurons innervates the dura, one could predict enhanced CGRP release at the dura as a consequence of the enhanced excitability indirectly produced by the FHM1 mutation in transfected neurons. This prediction contrasts with the unaltered CGRP release measured at the dura of FHM1 knockin mice (Fioretti et al. 2011), further stressing the importance of studying neurophysiological processes potentially involved in the pathogenesis of FHM1 in neurons expressing the $Ca_v2.1$ channels at the native endogenous level.

Acknowledgements I acknowledge the support of grants from the University of Padova (Strategic Project: Physiopathology of Signaling in Neuronal Tissue), the Fondazione Cariparo (Excellence Project: Calcium Signaling in Health and Disease) and Telethon-Italy (GGP06234).

References

Adams PJ, Garcia E, David LS, Mulatz KJ, Spacey SD, Snutch TP (2009) Ca(V)2.1 P/Q-type calcium channel alternative splicing affects the functional impact of familial hemiplegic migraine mutations: implications for calcium channelopathies. Channels (Austin) 3:110–121

Adams PJ, Rungta RL, Garcia E, van den Maagdenberg AM, MacVicar BA, Snutch TP (2010) Contribution of calcium-dependent facilitation to synaptic plasticity revealed by migraine mutations in the P/Q-type calcium channel. Proc Natl Acad Sci U S A 107:18694–18699

Akerman S, Holland PR, Goadsby PJ (2011) Diencephalic and brainstem mechanisms in migraine. Nat Rev Neurosci 12:570–584

Akerman S, Williamson DJ, Goadsby PJ (2003) Voltage-dependent calcium channels are involved in neurogenic dural vasodilatation via a presynaptic transmitter release mechanism. Br J Pharmacol 140:558–566

Ali AB, Nelson C (2006) Distinct Ca2+ channels mediate transmitter release at excitatory synapses displaying different dynamic properties in rat neocortex. Cereb Cortex 16:386–393

Antal A, Lang N, Boros K, Nitsche M, Siebner HR, Paulus W (2008) Homeostatic metaplasticity of the motor cortex is altered during headache-free intervals in migraine with aura. Cereb Cortex 18:2701–2705

Anttila V, Stefansson H, Kallela M, Todt U, Terwindt GM, Calafato MS, Nyholt DR, Dimas AS, Freilinger T, Muller-Myhsok B, Artto V, Inouye M, Alakurtti K, Kaunisto MA, Hamalainen E, de Vries B, Stam AH, Weller CM, Heinze A, Heinze-Kuhn K, Goebel I, Borck G, Gobel H, Steinberg S, Wolf C, Bjornsson A, Gudmundsson G, Kirchmann M, Hauge A, Werge T, Schoenen J, Eriksson JG, Hagen K, Stovner L, Wichmann HE, Meitinger T, Alexander M, Moebus S, Schreiber S, Aulchenko YS, Breteler MM, Uitterlinden AG, Hofman A, van Duijn CM, Tikka-Kleemola P, Vepsalainen S, Lucae S, Tozzi F, Muglia P, Barrett J, Kaprio J, Farkkila M, Peltonen L, Stefansson K, Zwart JA, Ferrari MD, Olesen J, Daly M, Wessman M, van den Maagdenberg AM, Dichgans M, Kubisch C, Dermitzakis ET, Frants RR, Palotie A (2010) Genome-wide association study of migraine implicates a common susceptibility variant on 8q22.1. Nat Genet 42:869–873

Aurora SK, Wilkinson F (2007) The brain is hyperexcitable in migraine. Cephalalgia 27:1442–1453

Ayata C (2009) Spreading depression: from serendipity to targeted therapy in migraine prophylaxis. Cephalalgia 29:1095–1114

Ayata C, Jin H, Kudo C, Dalkara T, Moskowitz MA (2006) Suppression of cortical spreading depression in migraine prophylaxis. Ann Neurol 59:652–661

Ayata C, Shimizu-Sasamata M, Lo EH, Noebels JL, Moskowitz (2000) Impaired neurotransmitter release and elevated threshold for cortical spreading depression in mice with mutations in the a_{1A} subunit of P/Q type calcium channels. Neuroscience 95:639–645

Barrett CF, Cao YQ, Tsien RW (2005) Gating deficiency in a familial hemiplegic migraine type 1 mutant P/Q-type calcium channel. J Biol Chem 280:24064–24071

Battista J, Badcock DR, McKendrick AM (2011) Migraine increases centre-surround suppression for drifting visual stimuli. PLoS ONE 6:e18211

Bolay H, Reuter U, Dunn AK, Huang Z, Boas DA, Moskowitz MA (2002) Intrinsic brain activity triggers trigeminal meningeal afferents in a migraine model. Nat Med 8:136–142

Borgland SL, Connor M, Christie MJ (2001) Nociceptin inhibits calcium channel currents in a subpopulation of small nociceptive trigeminal ganglion neurons in mouse. J Physiol 536:35–47

Bove GM, Moskowitz MA (1997) Primary afferent neurons innervating guinea pig dura. J Neurophysiol 77:299–308

Cao YQ, Piedras-Renteria ES, Smith GB, Chen G, Harata NC, Tsien RW (2004) Presynaptic Ca2+ channels compete for channel type-preferring slots in altered neurotransmission arising from Ca2+ channelopathy. Neuron 43:387–400

Cao YQ, Tsien RW (2005) Effects of familial hemiplegic migraine type 1 mutations on neuronal P/Q-type Ca2+ channel activity and inhibitory synaptic transmission. Proc Natl Acad Sci U S A 102:2590–2595

Catterall WA, Dib-Hajj S, Meisler MH, Pietrobon D (2008) Inherited neuronal ion channelopathies: new windows on complex neurological diseases. J Neurosci 28:11768–11777

Catterall WA, Few AP (2008) Calcium channel regulation and presynaptic plasticity. Neuron 59:882–901

Ceruti S, Villa G, Fumagalli M, Colombo L, Magni G, Zanardelli M, Fabbretti E, Verderio C, van den Maagdenberg AM, Nistri A, Abbracchio MP (2011) Calcitonin gene-related peptide-mediated enhancement of purinergic neuron/glia communication by the algogenic factor bradykinin in mouse trigeminal ganglia from wild-type and R192Q $Ca_v2.1$ Knock-in mice: implications for basic mechanisms of migraine pain. J Neurosci 31:3638–3649

Charles A (2010) Does cortical spreading depression initiate a migraine attack? Maybe not. Headache 50:731–733

Charles A, Brennan K (2009) Cortical spreading depression-new insights and persistent questions. Cephalalgia 29:1115–1124

Chasman DI, Schurks M, Anttila V, de Vries B, Schminke U, Launer LJ, Terwindt GM, van den Maagdenberg AM, Fendrich K, Volzke H, Ernst F, Griffiths LR, Buring JE, Kallela M, Freilinger T, Kubisch C, Ridker PM, Palotie A, Ferrari MD, Hoffmann W, Zee RY, Kurth T (2011) Genome-wide association study reveals three susceptibility loci for common migraine in the general population. Nat Genet 43:695–698

Conte A, Barbanti P, Frasca V, Iacovelli E, Gabriele M, Giacomelli E, Aurilia C, Pichiorri F, Gilio F, Inghilleri M (2010) Differences in short-term primary motor cortex synaptic potentiation as assessed by repetitive transcranial magnetic stimulation in migraine patients with and without aura. Pain 148:43–48

Coppola G, Pierelli F, Schoenen J (2007) Is the cerebral cortex hyperexcitable or hyperresponsive in migraine? Cephalalgia 27:1427–1439

De Fusco M, Marconi R, Silvestri L, Atorino L, Rampoldi L, Morgante L, Ballabio A, Aridon P, Casari G (2003) Haploinsufficiency of ATP1A2 encoding the Na+/K+ pump alpha2 subunit associated with familial hemiplegic migraine type 2. Nat Genet 33:192–196

de Vries B, Frants RR, Ferrari MD, van den Maagdenberg AM (2009) Molecular genetics of migraine. Hum Genet 126:115–132

Deboer T, van Diepen HC, Ferrari MD, Van den Maagdenberg AM, Meijer JH (2013) Reduced sleep and low adenosinergic sensitivity in cacna1a R192Q mutant mice. Sleep 36:127–136

Dichgans M, Freilinger T, Eckstein G, Babini E, Lorenz-Depiereux B, Biskup S, Ferrari MD, Herzog J, van den Maagdenberg AM, Pusch M, Strom TM (2005) Mutation in the neuronal voltage-gated sodium channel SCN1A in familial hemiplegic migraine. Lancet 366:371–377

Ducros A, Denier C, Joutel A, Cecillon M, Lescoat C, Vahedi K, Darcel F, Vicaut E, Bousser MG, Tournier-Lasserve E (2001) The clinical spectrum of familial hemiplegic migraine associated with mutations in a neuronal calcium channel. N Engl J Med 345:17–24

Ebersberger A, Portz S, Meissner W, Schaible HG, Richter F (2004) Effects of N-, P/Q- and L-type calcium channel blockers on nociceptive neurones of the trigeminal nucleus with input from the dura. Cephalalgia 24:250–261

Eikermann-Haerter K, Baum MJ, Ferrari MD, van den Maagdenberg AM, Moskowitz MA, Ayata C (2009a) Androgenic suppression of spreading depression in familial hemiplegic migraine type 1 mutant mice. Ann Neurol 66:564–568

Eikermann-Haerter K, Dilekoz E, Kudo C, Savitz SI, Waeber C, Baum MJ, Ferrari MD, van den Maagdenberg AMJM, Moskowitz MA, Ayata C (2009b) Genetic and hormonal factors modulate spreading depression and transient hemiparesis in mouse models of familial hemiplegic migraine type 1. J Clin Investig 119:99–109

Eikermann-Haerter K, Hyun Lee J, Yuzawa I, Liu CH, Zhou Z, Kyoung Shin H, Zheng Y, Qin T, Kurth T, Waeber C, Ferrari MD, van den Maagdenberg AMJM, Moskowitz MA, Ayata C (2012) Migraine mutations increase stroke vulnerability by facilitating ischemic depolarizations/clinical perspective. Circulation 125:335–345

Eikermann-Haerter K, Yuzawa I, Qin T, Wang Y, Baek K, Kim YR, Hoffmann U, Dilekoz E, Waeber C, Ferrari MD, van den Maagdenberg AM, Moskowitz MA, Ayata C (2011) Enhanced subcortical spreading depression in familial hemiplegic migraine type 1 mutant mice. J Neurosci 31:5755–5763

Fioretti B, Catacuzzeno L, Sforna L, Gerke-Duncan MB, van den Maagdenberg AM, Franciolini F, Connor M, Pietrobon D (2011) Trigeminal ganglion neuron subtype-specific alterations of Ca(V)2.1 calcium current and excitability in a Cacna1a mouse model of migraine. J Physiol 589:5879–5895

Fletcher CF, Tottene A, Lennon VA, Wilson SM, Dubel SJ, Paylor R, Hosford DA, Tessarollo L, McEnery MW, Pietrobon D, Copeland NG, Jenkins NA (2001) Dystonia and cerebellar atrophy in Cacna1a null mice lacking P/Q calcium channel activity. Faseb J 15:1288–1290

Franceschini A, Hullugundi SK, van den Maagdenberg AM, Nistri A, Fabbretti E (2013a) Effects of LPS on P2X3 receptors of trigeminal sensory neurons and macrophages from mice expressing the R192Q Cacna1a gene mutation of familial hemiplegic migraine-1. Purinergic Signal 9:7–13

Franceschini A, Vilotti S, Ferrari MD, van den Maagdenberg AM, Nistri A, Fabbretti E (2013b) TNFalpha levels and macrophages expression reflect an inflammatory potential of trigeminal ganglia in a mouse model of familial hemiplegic migraine. PLoS ONE 8:e52394

Freilinger T, Anttila V, de Vries B, Malik R, Kallela M, Terwindt GM, Pozo-Rosich P, Winsvold B, Nyholt DR, van Oosterhout WP, Artto V, Todt U, Hamalainen E, Fernandez-Morales J, Louter MA, Kaunisto MA, Schoenen J, Raitakari O, Lehtimaki T, Vila-Pueyo M, Gobel H, Wichmann E, Sintas C, Uitterlinden AG, Hofman A, Rivadeneira F, Heinze A, Tronvik E, van Duijn CM, Kaprio J, Cormand B, Wessman M, Frants RR, Meitinger T, Muller-Myhsok B, Zwart JA, Farkkila M, Macaya A, Ferrari MD, Kubisch C, Palotie A, Dichgans M, van den Maagdenberg AM (2012) Genome-wide association analysis identifies susceptibility loci for migraine without aura. Nat Genet

Gao Z, Todorov B, Barrett CF, van Dorp S, Ferrari MD, van den Maagdenberg AM, De Zeeuw CI, Hoebeek FE (2012) Cerebellar ataxia by enhanced Ca(V)2.1 currents is alleviated by Ca2+ -dependent K+ -channel activators in Cacna1a(S218L) mutant mice. J Neurosci 32:15533–15546

Garza-Lopez E, Sandoval A, Gonzalez-Ramirez R, Gandini MA, Van den Maagdenberg A, De Waard M, Felix R (2012) Familial hemiplegic migraine type 1 mutations W1684R and

V1696I alter G protein-mediated regulation of Ca(V)2.1 voltage-gated calcium channels. Biochim Biophys Acta 1822:1238–1246

Hans M, Luvisetto S, Williams ME, Spagnolo M, Urrutia A, Tottene A, Brust PF, Johnson EC, Harpold MM, Stauderman KA, Pietrobon D (1999) Functional consequences of mutations in the human a_{1A} calcium channel subunit linked to familial hemiplegic migraine. J Neurosci 19:1610–1619

Ho TW, Edvinsson L, Goadsby PJ (2010) CGRP and its receptors provide new insights into migraine pathophysiology. Nat Rev Neurol 6:573–582

Hong KW, Kim CD, Rhim BY, Lee WS (1999) Effect of omega-conotoxin GVIA and omega-agatoxin IVA on the capsaicin-sensitive calcitonin gene-related peptide release and autoregulatory vasodilation in rat pial arteries. J Cereb Blood Flow Metab 19:53–60

Inchauspe CG, Martini FJ, Forsythe ID, Uchitel OD (2004) Functional compensation of P/Q by N-type channels blocks short-term plasticity at the calyx of held presynaptic terminal. J Neurosci 24:10379–10383

Inchauspe CG, Urbano FJ, Di Guilmi MN, Forsythe ID, Ferrari MD, van den Maagdenberg AM, Uchitel OD (2010) Gain of function in FHM-1 Ca(V)2.1 knock-in mice is related to the shape of the action potential. J Neurophysiol 104:291–299

Iwasaki S, Momiyama A, Uchitel OD, Takahashi T (2000) Developmental changes in calcium channel types mediating central synaptic transmission. J Neurosci 20:59–65

Jansen I, Alafaci C, Uddman R, Edvinsson L (1990) Evidence that calcitonin gene-related peptide contributes to the capsaicin-induced relaxation of guinea pig cerebral arteries. Regul Pept 31:167–178

Karatas H, Erdener SE, Gursoy-Ozdemir Y, Lule S, Eren-Kocak E, Sen ZD, Dalkara T (2013) Spreading depression triggers headache by activating neuronal Panx1 channels. Science 339:1092–1095

Knight YE, Bartsch T, Kaube H, Goadsby PJ (2002) P/Q-type calcium-channel blockade in the periaqueductal gray facilitates trigeminal nociception: a functional genetic link for migraine? J Neurosci 22:RC213

Koester HJ, Sakmann B (2000) Calcium dynamics associated with action potentials in single nerve terminals of pyramidal cells in layer 2/3 of the young rat neocortex. J Physiol 529(Pt 3):625–646

Kors EE, Terwindt GM, Vermeulen FL, Fitzsimons RB, Jardine PE, Heywood P, Love S, van den Maagdenberg AM, Haan J, Frants RR, Ferrari MD (2001) Delayed cerebral edema and fatal coma after minor head trauma: role of the CACNA1A calcium channel subunit gene and relationship with familial hemiplegic migraine. Ann Neurol 49:753–760

Kraus RL, Sinnegger M, Koschak A, Glossmann H, Stenirri S, Carrera P, Striessnig J (2000) Three new familial hemiplegic migraine mutants affect P/Q-type Ca^{2+}channel kinetics. J Biol Chem 275:9239–9243

Kraus RL, Sinnegger MJ, Glossmann H, Hering S, Striessnig J (1998) Familial hemiplegic migraine mutations change a_{1A} Ca^{2+} channel kinetics. J Biol Chem 273:5586–5590

Kruglikov I, Rudy B (2008) Perisomatic GABA release and thalamocortical integration onto neocortical excitatory cells are regulated by neuromodulators. Neuron 58:911–924

Lauritzen M (1994) Pathophysiology of the migraine aura. The spreading depression theory. Brain 117(Pt 1):199–210

Lennerz JK, Ruhle V, Ceppa EP, Neuhuber WL, Bunnett NW, Grady EF, Messlinger K (2008) Calcitonin receptor-like receptor (CLR), receptor activity-modifying protein 1 (RAMP1), and calcitonin gene-related peptide (CGRP) immunoreactivity in the rat trigeminovascular system: differences between peripheral and central CGRP receptor distribution. J Comp Neurol 507:1277–1299

Leo L, Gherardini L, Barone V, De Fusco M, Pietrobon D, Pizzorusso T, Casari G (2011) Increased susceptibility to cortical spreading depression in the mouse model of familial hemiplegic migraine type 2. PLoS Genet 7:e1002129

Leonardi M, Steiner TJ, Scher AT, Lipton RB (2005) The global burden of migraine: measuring disability in headache disorders with WHO's Classification of Functioning, Disability and Health (ICF). J Headache Pain 6:429–440

Levy D (2009) Migraine pain, meningeal inflammation, and mast cells. Curr Pain Headache Rep 13:237–240

Levy D (2010) Migraine pain and nociceptor activation–where do we stand? Headache 50:909–916

Levy D, Burstein R, Kainz V, Jakubowski M, Strassman AM (2007) Mast cell degranulation activates a pain pathway underlying migraine headache. Pain 130:166–176

Li L, Bischofberger J, Jonas P (2007) Differential gating and recruitment of P/Q-, N-, and R-type Ca2 + channels in hippocampal mossy fiber boutons. J Neurosci 27:13420–13429

Lin KH, Erazo-Fischer E, Taschenberger H (2012) Similar intracellular Ca2 + requirements for inactivation and facilitation of voltage-gated Ca2 + channels in a glutamatergic mammalian nerve terminal. J Neurosci 32:1261–1272

Lipton RB, Bigal ME, Steiner TJ, Silberstein SD, Olesen J (2004) Classification of primary headaches. Neurology 63:427–435

Llinas RR, Choi S, Urbano FJ, Shin HS (2007) Gamma-band deficiency and abnormal thalamocortical activity in P/Q-type channel mutant mice. Proc Natl Acad Sci U S A 104:17819–17824

Luvisetto S, Marinelli S, Panasiti MS, D'Amato FR, Fletcher CF, Pavone F, Pietrobon D (2006) Pain sensitivity in mice lacking the Ca(v)2.1alpha1 subunit of P/Q-type Ca2 + channels. Neuroscience 142:823–832

Mathew R, Andreou AP, Chami L, Bergerot A, van den Maagdenberg AM, Ferrari MD, Goadsby PJ (2011) Immunohistochemical characterization of calcitonin gene-related peptide in the trigeminal system of the familial hemiplegic migraine 1 knock-in mouse. Cephalalgia 31:1368–1380

Matsuka Y, Neubert JK, Maidment NT, Spigelman I (2001) Concurrent release of ATP and substance P within guinea pig trigeminal ganglia invivo. Brain Res 915:248–255

Melliti K, Grabner M, Seabrook GR (2003) The familial hemiplegic migraine mutation R192Q reduces G-protein-mediated inhibition of P/Q-type (Ca(V)2.1) calcium channels expressed in human embryonic kidney cells. J Physiol 546:337–347

Messlinger K, Hanesch U, Baumgartel M, Trost B, Schmidt RF (1993) Innervation of the dura mater encephali of cat and rat: ultrastructure and calcitonin gene-related peptide-like and substance P-like immunoreactivity. Anat Embryol (Berl) 188:219–237

Mintz IM, Sabatini BL, Regehr WG (1995) Calcium control of transmitter release at a cerebellar synapse. Neuron 15:675–688

Mochida S, Few AP, Scheuer T, Catterall WA (2008) Regulation of presynaptic Ca(V)2.1 channels by Ca2 + sensor proteins mediates short-term synaptic plasticity. Neuron 57:210–216

Monier C, Chavane F, Baudot P, Graham LJ, Fregnac Y (2003) Orientation and direction selectivity of synaptic inputs in visual cortical neurons: a diversity of combinations produces spike tuning. Neuron 37:663–680

Mullner C, Broos LA, van den Maagdenberg AM, Striessnig J (2004) Familial Hemiplegic Migraine Type 1 Mutations K1336E, W1684R, and V1696I Alter Ca_v2.1 Ca2 + Channel Gating: evidence for beta-subunit isoform-specific effects. J Biol Chem 279:51844–51850

Nair A, Simonetti M, Birsa N, Ferrari MD, van den Maagdenberg AM, Giniatullin R, Nistri A, Fabbretti E (2010) Familial hemiplegic migraine Ca(v)2.1 channel mutation R192Q enhances ATP-gated P2X3 receptor activity of mouse sensory ganglion neurons mediating trigeminal pain. Mol Pain 6:48

Olesen J, Burstein R, Ashina M, Tfelt-Hansen P (2009) Origin of pain in migraine: evidence for peripheral sensitisation. Lancet Neurol 8:679–690

Olesen J, Gustavsson A, Svensson M, Wittchen HU, Jonsson B, group Cs, European Brain C (2012) The economic cost of brain disorders in Europe. Eur J Neurol 19:155–162

Ophoff RA, Terwindt GM, Vergouwe MN, van Eijk R, Oefner PJ, Hoffman SMG, Lamerdin JE, Mohrenweiser HW, Bulman DE, Ferrari M, Haan J, Lindhout D, van Hommen G-JB, Hofker MH, Ferrari MD, Frants RR (1996) Familial hemiplegic migraine and episodic ataxia type-2 are caused by mutations in the Ca^{2+} channel gene CACNL1A4. Cell 87:543–552
Pietrobon D (2005a) Function and dysfunction of synaptic calcium channels: insights from mouse models. Curr Opin Neurobiol 15:257–265
Pietrobon D (2005b) Migraine: new molecular mechanisms. Neuroscientist 11:373–386
Pietrobon D (2007) Familial hemiplegic migraine. Neurotherapeutics 4:274–284
Pietrobon D (2010) $Ca_v2.1$ channelopathies. Pflugers Arch 460:375–393
Pietrobon D (2013) Calcium channels and migraine. Biochim Biophys Acta 1828:1655–1665
Pietrobon D, Moskowitz MA (2013) Pathophysiology of migraine. Annu Rev Physiol 75:365–391
Pietrobon D, Striessnig J (2003) Neurobiology of migraine. Nat Rev Neurosci 4:386–398
Pineda JC, Waters RS, Foehring RC (1998) Specificity in the interaction of HVA Ca2 + channel types with Ca2 + -dependent AHPs and firing behavior in neocortical pyramidal neurons. J Neurophysiol 79:2522–2534
Prescot A, Becerra L, Pendse G, Tully S, Jensen E, Hargreaves R, Renshaw P, Burstein R, Borsook D (2009) Excitatory neurotransmitters in brain regions in interictal migraine patients. Mol Pain 5:34
Recober A, Russo AF (2009) Calcitonin gene-related peptide: an update on the biology. Curr Opin Neurol 22:241–246
Rozov A, Burnashev N, Sakmann B, Neher E (2001) Transmitter release modulation by intracellular Ca2 + buffers in facilitating and depressing nerve terminals of pyramidal cells in layer 2/3 of the rat neocortex indicates a target cell-specific difference in presynaptic calcium dynamics. J Physiol 531:807–826
Russell MB, Ducros A (2011) Sporadic and familial hemiplegic migraine: pathophysiological mechanisms, clinical characteristics, diagnosis, and management. Lancet Neurol 10:457–470
Sasaki S, Huda K, Inoue T, Miyata M, Imoto K (2006) Impaired feedforward inhibition of the thalamocortical projection in epileptic Ca2 + channel mutant mice, tottering. J Neurosci 26:3056–3065
Serra SA, Fernandez-Castillo N, Macaya A, Cormand B, Valverde MA, Fernandez–Fernandez JM (2009) The hemiplegic migraine-associated Y1245C mutation in CACNA1A results in a gain of channel function due to its effect on the voltage sensor and G-protein-mediated inhibition. Pflugers Arch 458:489–502
Shimizu T, Toriumi H, Sato H, Shibata M, Nagata E, Gotoh K, Suzuki N (2007) Distribution and origin of TRPV1 receptor-containing nerve fibers in the dura mater of rat. Brain Res 1173:84–91
Shu Y, Hasenstaub A, McCormick DA (2003) Turning on and off recurrent balanced cortical activity. Nature 423:288–293
Siniatchkin M, Kroner-Herwig B, Kocabiyik E, Rothenberger A (2007) Intracortical inhibition and facilitation in migraine—a transcranial magnetic stimulation study. Headache 47:364–370
Siniatchkin M, Sendacki M, Moeller F, Wolff S, Jansen O, Siebner H, Stephani U (2012) Abnormal Changes of Synaptic Excitability in Migraine with Aura. Cereb Cortex 22:2207–2216
Somjen GG (2001) Mechanisms of spreading depression and hypoxic spreading depression-like depolarization. Physiol Rev 81:1065–1096
Stovner LJ, Hagen K (2006) Prevalence, burden, and cost of headache disorders. Curr Opin Neurol 19:281–285
Strassman AM, Levy D (2006) Response properties of dural nociceptors in relation to headache. J Neurophysiol 95:1298–1306
Strassman AM, Raymond SA, Burstein R (1996) Sensitization of meningeal sensory neurons and the origin of headaches. Nature 384:560–564

Sutton KG, McRory JE, Guthrie H, Murphy TH, Snutch TP (1999) P/Q-type calcium channels mediate the activity-dependent feedback of syntaxin-1A. Nature 401:800–804

Tao J, Liu P, Xiao Z, Zhao H, Gerber BR, Cao YQ (2012) Effects of familial hemiplegic migraine type 1 mutation T666 M on voltage-gated calcium channel activities in trigeminal ganglion neurons. J Neurophysiol 107:1666–1680

Tottene A, Conti R, Fabbro A, Vecchia D, Shapovalova M, Santello M, van den Maagdenberg AMJM, Ferrari MD, Pietrobon D (2009) Enhanced excitatory transmission at cortical synapses as the basis for facilitated spreading depression in Ca(v)2.1 knockin migraine mice. Neuron 61:762–773

Tottene A, Fellin T, Pagnutti S, Luvisetto S, Striessnig J, Fletcher C, Pietrobon D (2002) Familial hemiplegic migraine mutations increase Ca(2 +) influx through single human $Ca_v2.1$ channels and decrease maximal $Ca_v2.1$ current density in neurons. Proc Natl Acad Sci U S A 99:13284–13289

Tottene A, Pivotto F, Fellin T, Cesetti T, van den Maagdenberg AM, Pietrobon D (2005) Specific kinetic alterations of human $Ca_v2.1$ calcium channels produced by mutation S218L causing familial hemiplegic migraine and delayed cerebral edema and coma after minor head trauma. J Biol Chem 280:17678–17686

Tottene A, Urbani A, Pietrobon D (2011) Role of different voltage-gated Ca2+ channels in cortical spreading depression: specific requirement of P/Q-type Ca2+ channels. Channels (Austin) 5:110–114

Ulrich-Lai YM, Flores CM, Harding-Rose CA, Goodis HE, Hargreaves KM (2001) Capsaicin-evoked release of immunoreactive calcitonin gene-related peptide from rat trigeminal ganglion: evidence for intraganglionic neurotransmission. Pain 91:219–226

Urban MO, Ren K, Sablad M, Park KT (2005) Medullary N-type and P/Q-type calcium channels contribute to neuropathy-induced allodynia. NeuroReport 16:563–566

van den Maagdenberg AM, Pietrobon D, Pizzorusso T, Kaja S, Broos LA, Cesetti T, van de Ven RC, Tottene A, van der Kaa J, Plomp JJ, Frants RR, Ferrari MD (2004) A Cacna1a knockin migraine mouse model with increased susceptibility to cortical spreading depression. Neuron 41:701–710

van den Maagdenberg AM, Pizzorusso T, Kaja S, Terpolilli N, Shapovalova M, Hoebeek FE, Barrett CF, Gherardini L, van de Ven RC, Todorov B, Broos LA, Tottene A, Gao Z, Fodor M, De Zeeuw CI, Frants RR, Plesnila N, Plomp JJ, Pietrobon D, Ferrari MD (2010) High cortical spreading depression susceptibility and migraine-associated symptoms in Ca(v)2.1 S218L mice. Ann Neurol 67:85–98

van Oosterhout F, Michel S, Deboer T, Houben T, van de Ven RCG, Albus H, Westerhout J, Vansteensel MJ, Ferrari MD, van den Maagdenberg AMJM, Meijer JH (2008) Enhanced circadian phase resetting in R192Q $Ca_v2.1$ calcium channel migraine mice. Ann Neurol 64:315–324

Vaughn AH, Gold MS (2010) Ionic mechanisms underlying inflammatory mediator-induced sensitization of dural afferents. J Neurosci 30:7878–7888

Vecchia D, Pietrobon D (2012) Migraine: a disorder of brain excitatory-inhibitory balance? Trends Neurosci 35:507–520

Villalon CM, Olesen J (2009) The role of CGRP in the pathophysiology of migraine and efficacy of CGRP receptor antagonists as acute antimigraine drugs. Pharmacol Ther 124:309–323

Waeber C, Moskowitz MA (2005) Migraine as an inflammatory disorder. Neurology 64:S9–S15

Weiss N, Sandoval A, Felix R, Van den Maagdenberg A, De Waard M (2008) The S218L familial hemiplegic migraine mutation promotes deinhibition of Ca(v)2.1 calcium channels during direct G-protein regulation. Pflugers Arch 457:315–326

Westenbroek RE, Sakurai T, Elliott EM, Hell JW, Starr TVB, Snutch TP, Catterall WA (1995) Immunochemical identification and subcellular distribution of the a_{1A} subunits of brain calcium channels. J Neurosci 15:6403–6418

Wilkinson F, Karanovic O, Wilson HR (2008) Binocular rivalry in migraine. Cephalalgia 28:1327–1338

Womack MD, Chevez C, Khodakhah K (2004) Calcium-activated potassium channels are selectively coupled to P/Q-type calcium channels in cerebellar Purkinje neurons. J Neurosci 24:8818–8822

Wu LG, Westenbroek RE, Borst JGG, Catterall WA, Sakmann B (1999) Calcium channel types with distinct presynaptic localization couple differentially to transmitter release in single calyx-type synapses. J Neurosci 19:726–736

Xiao Y, Richter JA, Hurley JH (2008) Release of glutamate and CGRP from trigeminal ganglion neurons: role of calcium channels and 5-HT1 receptor signaling. Mol Pain 4:12

Yan J, Edelmayer RM, Wei X, De Felice M, Porreca F, Dussor G (2011) Dural afferents express acid-sensing ion channels: a role for decreased meningeal pH in migraine headache. Pain 152:106–113

Zaitsev AV, Povysheva NV, Lewis DA, Krimer LS (2007) P/Q-type, but not N-type, calcium channels mediate GABA release from fast-spiking interneurons to pyramidal cells in rat prefrontal cortex. J Neurophysiol 97:3567–3573

Zhang X, Levy D, Kainz V, Noseda R, Jakubowski M, Burstein R (2011) Activation of central trigeminovascular neurons by cortical spreading depression. Ann Neurol 69:855–865

Zhang X, Levy D, Noseda R, Kainz V, Jakubowski M, Burstein R (2010) Activation of meningeal nociceptors by cortical spreading depression: implications for migraine with aura. J Neurosci 30:8807–8814

Chapter 2
Ataxia and *CACNA1A*: Episodic or Progressive?

Stephanie Schorge and Kinya Ishikawa

Abstract Two types of cerebellar ataxia are associated with variation in the CACNA1A gene which encodes P/Q type calcium channels. As the genetic basis of these two disorders is distinct, the different clinical manifestations are likely to involve different mechanisms. Episodic ataxia can be caused by loss of a copy of the CACNA1A gene, while spinocerebellar ataxia is associated with expansions of a poly-glutamine repeat in the end of the gene (Fig. 2.1). By comparing the 'typical' clinical outcomes and the different pathological observations associated with them, it is possible to begin to determine what characteristics of CACNA1A gene activity may underlie the mechanisms in two different diseases.

2.1 Clinical Phenotypes: Similarities and Differences

Mutations in the gene CACNA1A have unusually variable clinical manifestations (recently reviewed in (Rajakulendran et al. 2012)). This gene is associated with at least four distinct clinical manifestations: Familial Hemipelagic Migraine (type 1, FHM1), Episodic Ataxia (type 2, EA2), Spinocerebellar Ataxia (type 6, SCA6) and epilepsy. FHM1 is covered elsewhere (see Chap.1). This chapter focuses on a comparison of the causes and effects of EA2 and SCA6, the two clinical disorders most strongly associated with cerebellar dysfunction.

S. Schorge (✉)
Department of Clinical and Experimental Epilepsy, UCL Institute of Neurology, Queen Square, London WC1N 3BG, UK
e-mail: s.schorge@ucl.ac.uk

K. Ishikawa
Department of Neurology, Graduate School, Tokyo Medical and Dental University, Yushima 1-Chome 5-45, Bunkyo-ku 113-8519 Tokyo, Japan
e-mail: pico.nuro@tmd.ac.jp

N. Weiss and A. Koschak (eds.), *Pathologies of Calcium Channels*,
DOI: 10.1007/978-3-642-40282-1_2,

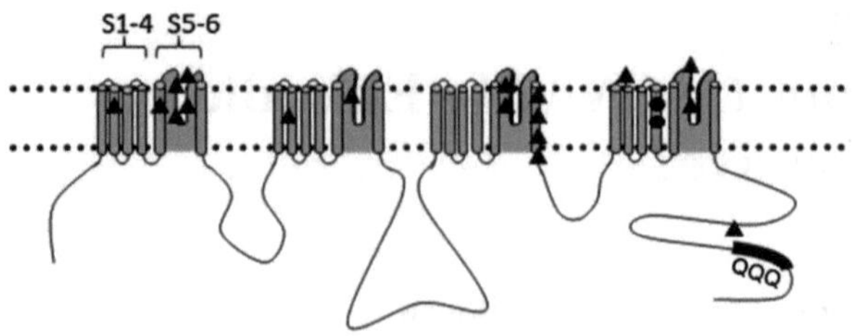

Fig. 2.1 Position of SCA6 poly glu repeat (QQQ) and missense mutations associated with EA2 (*black triangles*). Only missense mutations are shown, to indicate clustering in pore regions (S5–S6) compared to voltage sensors (S1–4). Most mutations associated with EA2 lead to truncation or loss of the protein

One key difference between these two disorders is that EA2 is episodic—that is, patients can have discrete 'episodes' or attacks of symptoms that are interspersed with periods of normal function. SCA6, in contrast, is a progressive disease, with cerebellar function slowly deteriorating over time.

By comparing the two disorders, starting with their clinical features, and linking these to cellular pathology and to genetic changes, it is possible to show how different disruptions in one gene might lead to clinically distinct outcomes. This type of analysis, which is facilitated by the ability to obtain detailed assessments of the function of P/Q channels in knock-in mouse models, is a model for how changes in DNA—even changes of individual nucleotides—can lead to altered behaviours in humans. Moreover, comparison of allelic disorders can reveal how different changes within one gene can lead to different phenotypes.

2.1.1 Clinical Overview of EA2

Episodes of EA2 are triggered by events in the environment, including alcohol, caffeine and emotional stress (some recent reviews: (Baloh et al. 1997; Jen et al. 2004, 2007; Rajakulendran et al. 2012)). Once triggered, attacks can last for hours or up to days. Attacks of ataxia associated with EA2 are characterised by incoordination, nausea, vertigo and nystagmus. While EA2 begins as an episodic disorder (usually in the second decade of life), up to half of patients go on to develop progressive cerebellar symptoms in later life. This transition to progressive cerebellar dysfunction is consistent with some overlap with SCA6, although, as discussed below, there are differences in cellular pathology. A substantial minority of EA2 patients also have migrainous features that are suggestive of some overlap with the allelic disorder, FHM1. However, in patients with EA2 the migraine is typically more vestibular, and is characterised by vertigo, rather than the classic features of migraine with aura, which are more closely associated with FHM1 (Pietrobon and Moskowitz 2012). Some EA2 patients also experience seizures, with the incidence of epilepsy in patients with EA2 being approximately

8-fold higher than in non-EA2 control populations (Rajakulendran et al. 2010a). Most EA2 is inherited in a fashion consistent with autosomal dominance in families; however, sporadic cases also occur.

A clinical hallmark of EA2 is an excellent response to treatment with acetazolamide (Griggs et al. 1978), a drug thought to work as a carbonic anhydrase inhibitor, but which—through an unknown mechanism—was serendipitously found to be effective in treating EA2. Treatment with acetazolamide is so robust that it may be used as a differential diagnosis for EA2, and has been used to validate studies of patients where the genetics are ambiguous (for example in (Rajakulendran et al. 2010a)).

2.1.2 Clinical Overview of SCA6

While a proportion of patients with EA2 develop progressive cerebellar ataxia in later decades, patients with SCA6 are characterised by a progressive deterioration of cerebellar function, but with few extracerebellar symptoms (Ishikawa et al. 1997; Takahashi et al. 2004; Zhuchenko et al. 1997). Onset of SCA6 is later than EA2, typically in the fifth decade of life. Again, in contrast to EA2, SCA6 is not notably an 'episodic' disease. Once established, cerebellar inco-ordination, including loss of balance, limb inco-ordination and difficulty with speech, is persistent, and not interrupted by periods of normal function. As with EA2, SCA6 is strongly associated with nystagmus. Patients with SCA6 do not generally suffer non-cerebellar symptoms—such as migraine, epilepsy or cognitive effects—levels higher than the control population (Ishikawa et al. 1997). This restriction to cerebellar effects implies the mechanism driving this disorder may be one confined to the cerebellum. However, it seems important to note that patients with SCA6 may also show extra-cerebellar neurological dysfunctions such as Parkinsonism, dementia. A recent study showed that SCA6 patients tend to show involvement not only in the cerebellum but also in the thalamus, putamen and globus pallidus on magnetic resonance imaging (Reetz et al. 2013).

The non-episodic nature of SCA6 is atypical of diseases associated with mutations in ion channels (channelopathies) and has drawn attention to this disorder as potentially involving a different mechanism than disorders caused by malfunctioning ion channels.

2.2 Cellular Pathologies

Both EA2 and SCA6 show signs of cerebellar degeneration, but the effects are much more consistent and pronounced in SCA6, which is characterised like many of the SCAs by pronounced cerebellar degeneration (Seidel et al. 2012).

2.2.1 Cellular Pathology in EA2

There are relatively few families with genetically confirmed EA2, and few imaging studies have been carried out. However, it has been noted that the later stages of the disorder, where progressive ataxia is seen associated with atrophy in the cerebellum, particularly of the vermis (Rajakulendran et al. 2012). One relatively large study revealed variability of cerebellar atrophy within an individual pedigree (Calandriello et al. 1997). However, this study notes the variable penetrance of atrophy in the disorder, citing other studies which have shown little or no atrophy. Data from mice with comparable genetic changes suggest that atrophy may be a late symptom, and episodes of ataxia may have an onset before cell loss is notable (Saito et al. 2009).

2.2.2 Cellular Pathology in SCA6

In contrast to EA2, degeneration of the cerebellum is a defining feature of SCA6, with patients showing pronounced loss of Purkinje cells that is similar to that seen in other spinocerebellar ataxias associated with different genes (Seidel et al. 2012). There have been several studies on the cellular pathology in tissue from patients with SCA6. In human patients, Purkinje cells are marked by the presence of protein aggregates particularly in the cytoplasm (Ishikawa et al. 1999), and these aggregates were recently shown to comprise c-terminal fragments of P/Q channels, containing the poly-glutamine repeats (Ishiguro et al. 2010) (Fig. 2.2a and b).

The P/Q type calcium channel seems to give rise to a carboxyl-terminal fragment (CTF) of 75 kilodalton in size (Kordasiewicz et al. 2006). Similar CTF also exists in L type calcium channel (Gomez-Ospina et al. 2006). The P/Q type calcium channel CTF is located in the neuronal cytoplasmic component (Ishiguro

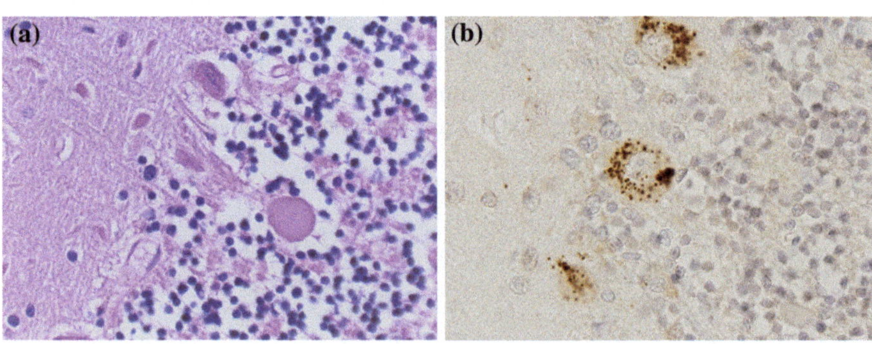

Fig. 2.2 Purkinje cell pathologies in human SCA6 subjects. **a** Hematoxylin and Eosin staining of a patient with SCA6 showing axonal swelling ("torpedo"). **b** Presence of abundant $Ca_v2.1$ aggregates demonstrated by an antibody against human-specific carboxyl-terminal portion of $Ca_v2.1$

et al. 2010), and, when it contains an expanded polyglutamine causing SCA6, can aggregate and confer toxicity against cultured cells (Takahashi et al. 2013). The CTF may also cause SCA6 by entering into the Purkinje cell nuclei (Kordasiewicz et al. 2006). Very recently, a mechanism by which the CTF is being expressed in thecerebellum and relates to SCA6 pathophysiology is uncovered (Du et al. 2013).

2.3 Molecular Genetics: The Problem with P/Q Channel Variability

The channels encoded by CACNA1A have many functions throughout the nervous system, and they are also extensively regulated by signalling cascades. The specificity and malleability of P/Q channel function represents an enormous challenge to determining the causative functional impact of individual mutations. The specificity of functions, many of which are restricted to sub-cellular compartments within neurons, means that there is growing unease with translating functional consequences assessed in non-neuronal cells (such as *Xenopus* oocytes or HEKs) to mechanisms for neurological disease.

The variability of P/Q channel function means that if one particular isoform of a channel is altered in one way (for example, by reduced inactivation), it is not clear that this effect will be replicated in different splice variants, or when one splice variant is associated with different modulating subunits. As the mRNA encoding P/Q channels is heavily modified by alternative splicing, and this splicing is enough to generate functionally and pharmacologically distinct channels (Bourinet et al. 1999; Chaudhuri et al. 2004; Krovetz et al. 2000), it is not always clear which variant is most applicable to the mechanism of disease. Moreover, the behaviour of the channels is dependent on which beta subunit is bound, and it is not clear which beta dominates in cerebellum or elsewhere in the brain (Dolphin 2009; Richards et al. 2007). Even individual channels can behave differently at different times. Channels can be modulated by G-proteins, calmodulin and phosphorylation (Dolphin 2009), meaning that the functional consequences of a mutation that disrupts one of these processes may be missed in many recording conditions. Finally, P/Q channels play distinct sub-cellular roles: they contribute to rhythmic oscillations (Llinas et al. 2007), support bi-directional plasticity in Purkinje cells of the cerebellum (Jorntell and Hansel 2006) and are probably best known for their canonical role of supporting calcium-dependent release of neurotransmitter from pre-synaptic terminals (Dolphin 2009).

For these reasons, functional characterisation of the impact of mutations is more and more reliant on ‘knock-in mice’ where the mutation is introduced to the CACNA1A gene and the endogenously produced mutant channels can be studied in the cell-type, splice variant and subunit assembly that is driven by the normal regulation of the proteins. Even so, mutations can produce conflicting data from effects in different cells challenging efforts to link changes in channel function to clinical pathology.

2.3.1 Deletions Simplify the Genetic Mechanism of EA2

EA2 was originally identified as a disorder associated with frameshift or nonsense mutations (Ophoff et al. 1996). In some cases these are associated with production of truncated proteins that had dominant negative effects in heterologous systems (Mezghrani et al. 2008; Raghib et al. 2001). However, a tremendous simplification has been introduced with the advance of methods to detect intra-genic deletions, those deletions which are too small to be seen as chromosomal aberrations, but too large to be captured in a single sequencing reaction. These methods, including MLPA, have identified many heterozygous deletions of multiple exons within CACNA1A in patients with EA2 (Labrum et al. 2009; Riant et al. 2010). In these cases, the disruption of the coding sequence, often with frameshifts introduced by missing exons, is most likely to lead to a complete loss-of-function of the affected allele. Patients with these sorts of deletions are essentially haploinsufficient for the CACNA1A gene. Because patients with deletions are clinically indistinguishable from patients with EA2 caused by missense mutations, a parsimonious explanation is that the missense mutations also lead to symptoms via a loss-of-function of the affected allele or its protein product. This explanation is supported by the frequent location of EA2 mutations in conserved elements of the P/Q channels, including in the pore-forming loops (Rajakulendran et al. 2010b) (Fig. 2.1), and is consistent with reduced currents produced by heterologously expressed channels containing missense mutations, even in non-neuronal cells (Fig. 2.3).

The realisation that EA2 may be caused by, and is often clinically indistinguishable from haploinsufficiency, vastly simplifies the comparison of functional

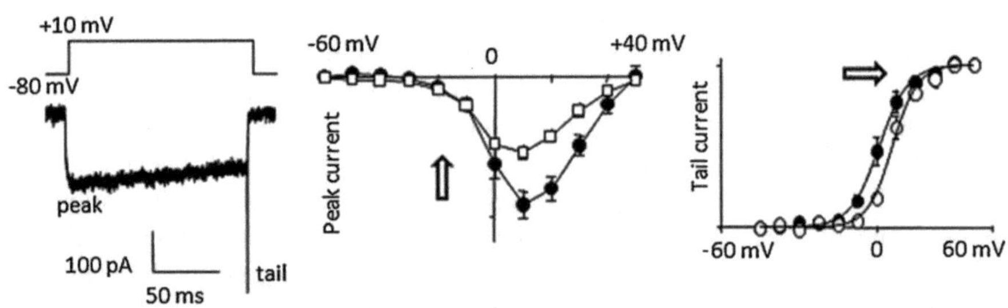

Fig. 2.3 Representative current reductions seen in human EA2 missense mutations. The left panel shows sample data from a cell expressing human CACNA1A cDNA that has been given a depolarising step from −80 to +10 mV. Current is inward, and reaches a peak that changes according to the amount of depolarisation and the driving force of calcium. After the cell is returned to −80 mV, a large amount of calcium current flows as the channels remain open and the driving force for calcium ions is strong. This produces the tail current. The middle panel shows an example of how a missense current can reduce the peak current uniformly across a range of voltage steps (reduced current density). The right panel shows how a different missense mutation changes the number of channels opened by different voltage steps, or a shift in the voltage dependence of activation, producing a shift in the tail currents measured at −80 mV. Both of these changes are consistent with a moderate loss of function of the channels. Data are adapted from (Rajakulendran et al. 2010a)

effects of mutations. In cases of halploinsufficiency, the channels produced by the remaining allele may be expected to function normally (unless a patient is unfortunate enough to inherit a second mutation), and the lost channels not to function at all. Compared to cases where missense mutations leading to a disease may have distinct effects in different cell types, channel variants or sub-cellular compartments, the functional impact of haploinsufficiency in EA2 is straightforward.

2.3.2 SCA6 Is Associated with a Poly-Glutamine Repeat in CACNA1A

The genetics of SCA6 are also more straightforward than multiple missense mutations would be: SCA6 is caused by the expansion of a polyglutamine repeat in the tail end of the CACNA1A gene.

There are a number of spinocerebellar ataxias associated with expansions of poly-glutamine repeats (recently reviewed in: (Seidel et al. 2012)). SCA6 is unusual in that the repeat is often shorter (19–33 aa) than in other SCAs (Ishikawa et al. 1997; Takahashi et al. 2004; Zhuchenko et al. 1997), and that SCA6 is more restricted to pure cerebellar manifestation. The homogeneity of the genetic cause did not initially lead to an easy interpretation of the mechanism. The position of the repeat in an ion channel has suggested that altered channel function may contribute to the pathology. Moreover, for many years the mechanism linking poly-glutamine repeats to cellular pathology was unclear. However recent work has converged on the ability of these glutamine repeat elements to disrupt calcium signalling in cells (Bezprozvanny 2011). We have argued recently that the consistent phenotype associated with these repeats, even when the gene containing them is widely expressed (such as TATA binding protein), suggests a conserved underlying mechanism (Schorge et al. 2010). The current proposed mechanism is direct binding of the poly-glutamine stretch to an intracellular calcium channel (ITPR1), and disregulation of calcium release from stores, by activating this channel (Bezprozvanny 2011).

2.3.3 ITPR1, Purkinje Cells and Spinocerebellar Ataxia

Modification of ITPR1 activity may have particularly pronounced effects on cerebellar Purkinje cells (even when the expression of a gene is widespread) because Purkinje cells use ITPR1 in a relatively unusual way. Purkinje cells are marked by a form of plasticity that depends on ITPR1 signalling, and which make them particularly sensitive to disruption of ITPR1-dependent signalling (Jorntell and Hansel 2006). The tight association between ITPR1 and Purkinje cells is supported by the recent discovery that individuals with mutations in this gene develop spinocerebellar ataxia type 15 (van de Leemput et al. 2007). Consequently, it seems disruption of ITPR1, either by poly-glutamine repeats, or by direct mutation, has severe effects on the cerebellar Purkinje cells.

The poly-glutamine repeats in CACNA1A deserve extra scrutiny. P/Q channels play an unusual role in Purkinje cells—linking glutamate receptor signalling to ITPR1 signalling—and these channels are notoriously dense in Purkinje cells. In fact, the first designation of 'P type channel' was due to their prevalence in Purkinje cells (Mintz et al. 1992). Briefly, Purkinje cells use P/Q channels, which are densely expressed in their somato-dendritic compartment, to allow calcium influx when the powerful climbing fibre synapses fire. The climbing fibre releases glutamate which activates AMPA receptors, and the AMPA receptors depolarise the cell enough to open P/Q channels (Fig. 2.4). The calcium that enters from the P type channels binds to ITPR1 receptors that are primed to release additional calcium from intracellular stores, and if coincident parallel fibres are activated while the calcium is still bound, the ITPR1-dependent calcium signal is enhanced, leading to downstream activation of a signalling cascade that leads to long-term depression (LTD). The use of P type calcium channels to link glutamate receptor activation to calcium influx appears particularly important for bi-directional plasticity in Purkinje cells (Jorntell and Hansel 2006), and the dense expression of P type calcium channels on these cells may be required for effective triggering of LTD.

Because P/Q channels are richly expressed, and play a central role in Purkinje plasticity, it is also likely that the presence of the poly-glutamine repeat in the channel may have disproportionate impact on Purkinje cells—presenting both a toxic glutamine-repeat element and a disruption of a channel that serves an

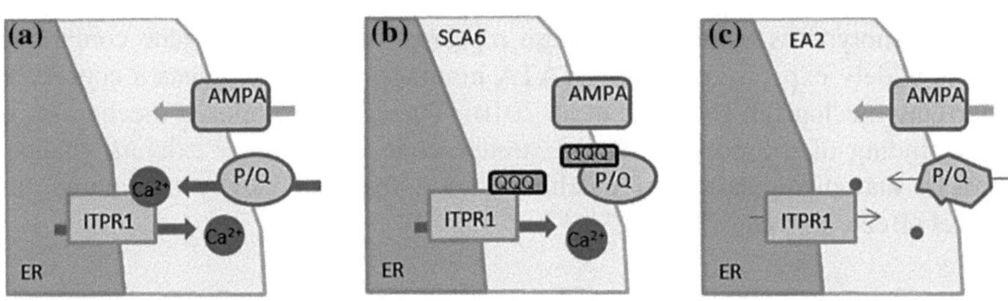

Fig. 2.4 Cartoon of how SCA6 and EA2 may be linked to altered ITPR1 function in Purkinje cells. In normal cells (**a**) depolarisation caused by activated AMPA receptors opens P/Q channels, and the calcium that enters these channels can bind to ITPR1 receptors on the endoplasmic reticulum (ER). When ITPR1 is bound to calcium, it releases more calcium from stores in response to binding of its ligand, IP3, and this serves as a coincidence detector linking AMPA-dependent depolarisation to IP3 activity. In SCA6 (**b**) the coincidence detection is disrupted as it is thought that fragments of poly-glutamine repeats (QQQ) can bind directly to ITPR1 and trigger it to release calcium from stores—even when AMPA receptors and P/Q channels are silent. In contrast, in EA2 (**c**) mutations reduce the amount of calcium entering P/Q channels even when AMPA is activated, and this means ITPR1 is less likely to be bound to calcium and will have reduced sensitivity for IP3

essential function in these cells. The poly-glutamine repeat is encoded by an alternate exon that is not included in all cells, but which is included in Purkinje cells (Ishikawa et al. 1999). One possibility is that the presence of the repeat in the mRNA alters the expression of the channels; a second possibility is that the presence of the repeat changes the behaviour of the channels in such a way as to be sufficient to lead to the atrophy of these cells. Indeed electrophysiological recordings have demonstrated that channels containing the poly-glutamine repeat may be expressed at higher levels than channels which lack the repeat in some expression systems (Piedras-Renteria et al. 2001). However, the clinical evidence suggests that the disease does not behave as a typical 'channelopathy': Unlike virtually all known channelopathies, SCA6 is not characterised by episodes of symptoms interspersed with normal behaviour. Instead it behaves similar to other poly-glutamine repeat SCAs, and the phenotype may be caused by aberrant activation of ITPR1 by the poly-glutamine repeats (Fig. 2.4b). The cerebellar restriction of clinical symptoms may be partly due to the enrichment of the channel in Purkinje cells, and partly due to the distribution of the splice variant carrying the repeat. As with other poly-glutamine-SCAs, even though CACNA1A is expressed in many cells, the phenotype is strongly associated with cerebellar degeneration and loss of Purkinje cells, and may be due to the critical role that ITPR1-dependent signalling plays in these cells.

Thus, although the genetic basis of SCA6 is more uniform than that of EA2 (which is associated with missense, nonsense or frameshift mutations), the functional impact of the genetic change in SCA6 is less clear than that of EA2, where the decrease in calcium entry (either to fewer P/Q channels or to less active, mutant channels) may lead to reduced ITPR1 signalling (Fig. 2.4c). The presence of deletions, which gives haploinsufficiency in patients with EA2, provides a clear argument for a loss-of-function mechanism leading to reduced ITPR1 activity and thence to cerebellar dysfunction. SCA6 is caused by the expansion of the poly-glutamine repeat, but how that expansion leads to cerebellar Purkinje cell death is not yet clear, but evidence is mounting for a role of toxic glutamine accumulations and possibly aberrant increases in activation of the ITPR1 signalling cascade.

2.4 Using Mouse Models to Understand Human Mechanisms

Because P/Q channels serve so many functions in neurons, it is not always possible to assess the full range of impacts of a mutation in a non-neuronal cell line. Because the channels are so variable—existing in many splice variants, and in association with many different accessory subunits—it is not always possible to transfect a single cDNA that might recapitulate the functional effects of a mature channel. For these reasons, the best insights into how mutations in channels can alter neuronal function are being obtained from mice carrying modifications in their native Cacna1a gene. Mice with mutations in Cacna1a were identified long before the gene was cloned, or even before the P/Q channel was identified. In 1962, Green and Sidman identified the 'tottering mouse' as a strain of mice that had attacks of ataxia as well as seizures. Frequently, attacks were triggered after being handled or stressed (Green and Sidman 1962). When molecular cloning identified the tottering locus as the Cacna1a gene, it was presented as an epilepsy gene, and the mouse was described as a model for studying epilepsy, in particular, absence epilepsy (Fletcher et al. 1996).

2.4.1 Insights from Mouse Models: EA2 (and Epilepsy)

Additional alleles of tottering mice, notably Rolling Nagoya, have fewer (or no) seizures, but more pronounced ataxia (Plomp et al. 2009). However, genetic evidence suggests these mice, which have dominantly inherited ataxia, may be more similar to models of FHM1 than to EA2 (Schorge and Rajakulendran 2012; Xie et al. 2007). In particular, the loss of arginines in S4 voltage sensors of the channel is more closely associated with human patients with FHM1 than with patients with EA2 (Pietrobon 2010; Xie et al. 2007). The four alleles of CACNA1A in mice that do not involve S4 arginines are characterised by episodes of ataxia and absence like seizures (Schorge and Rajakulendran 2012). Electrophysiological data has suggested that these mice all have reduced P/Q channel currents in their neurons, including in Purkinje cells (Dove et al. 1998; Fletcher et al. 1996; Kodama et al. 2006; Lorenzon et al. 1998; Miki et al. 2008; Wakamori et al. 1998). The mice also have disruptions to the normal rhythmic firing pattern of Purkinje neurons, which is linked to reduced calcium entry (Walter et al. 2006), and this disrupted firing is another possible mechanism for triggering ataxia. Moreover, a careful study showed a correlation between level of current reduction and phenotype (Saito et al. 2009). These data are supportive of evidence suggesting that the human ataxia is associated with reduced channel function—or haploinsufficiency. Because Purkinje cells have some of the highest expressions of this channel, it is not unreasonable to speculate that reduced function of the gene may affect these cells disproportionately and lead to cerebellar symptoms. However, it is unclear why these mice with reduced P/Q function almost invariably have seizures, while in humans seizures only occur in a minority ($\sim$8 %) of patients (Rajakulendran et al. 2010a). Possibly, the difference is an indication of the relative susceptibility of mutant mice to seizures (Frankel 2009). Alternatively, the distribution of CACNA1A expression in mice may be different than in humans (Schorge and Rajakulendran 2012).

2.4.2 Insights from Mouse Models: SCA6

To study the effect of the poly-glutamine repeat on CACNA1A function in vivo, Watase et al. (2008) engineered a series of mice with expanded poly-glutamine tracts (Watase et al. 2008). In humans disease onset is correlated to length of repeat, and as few as 19 repeats appear sufficient to lead to disease (Ishikawa et al. 1997, Takahashi et al. 2004; Zhuchenko et al. 1997). In mice, constructs containing up to 84 repeats were used. While mice with 30 repeats (which would be causative in humans) did not show defects in coordination, those with 84 repeats did manifest with signs of motor difficulties consistent with SCA6 (Watase et al. 2008). Similarly, in a separate study with 28 repeats included in CACNA1A, mice had only modest behavioural effects, even at 12 months of age (Saegusa et al. 2007). Expanding the CAG repeat up to 118 in new knock-in mice demonstrated exaggerated polyglutamine aggregations (Unno et al. 2012), which resembled human Purkinje cell degeneration in SCA6 (Ishikawa et al. 1999; Ishiguro et al. 2010).

Analysis of calcium currents in the Purkinje cells indicated that most parameters of the currents were similar, with a tendency for currents to be reduced, which was not correlated with symptom onset (Saegusa et al. 2007; Watase et al. 2008). These data indicate that altered P/Q currents are not likely the cause of the cerebellar defects, as the mice with shorter repeats had similar current deficits to those with longer repeats (which had more pronounced cerebellar defects). As with EA2, mice appear more resistant to SCA6 modifications in CACNA1A than humans. This may be because the cerebellar symptoms are less observable in mice, or because the channels have more compensatory backup.

The explanation, a potential gain of function of the poly-glutamine repeats, is consistent with the altered ITPR1 signalling, although this has not been directly assessed in the mice.

2.5 CACNA1A in Context

Both of the cerebellar phenotypes associated with mutations in CACNA1A overlap with other diseases in humans. In many cases the mutations that cause these overlapping phenotypes are known, thus it is possible to extrapolate from related diseases, which presumably share some mechanisms, how changes in CANCA1A lead to symptoms. For SCA6, it is possible to investigate the series of genetically defined spinocerebellar ataxias, and EA2 may be compared to the other members of the episodic ataxia family.

2.5.1 SCA6: Insights from Spinocerebellar Ataxias

There are two potential mechanisms driving the cerebellar dysfunction in SCA6, build-up of toxic glutamine repeats (which may disrupt calcium signalling in these cells) and dysfunction of the P/Q channels themselves. These two mechanisms are not mutually exclusive. It is possible that disruption of ITPR1-dependent calcium release is compounded by irregular channel function. The shortness of the repeats in SCA6 suggests that Purkinje cells are either particularly sensitive to glutamine-repeats in P/Q calcium channels, or the presence of the repeats is indeed exacerbated by some possibly subtle dysfunction of the channels. These channels are tightly regulated—in pre-synaptic terminals they are thought to be anchored into precise 'slots' (Cao et al. 2004; Cao and Tsien 2010) that provide the nanometre accuracy of their position which is required for accurate coupling of transmitter release to action potential arrival at the terminal. It is possible that the glutamine repeat in the c-terminal of the channel disrupts localisation in Purkinje cells, perhaps allowing calcium to enter in locations which are not linked tightly enough to the ITPR1 position to allow cerebellar LTD. In the end, however, given the prevalence of glutamine repeats in spinocerebellar ataxias, it is difficult to imagine that these repeats do not play a central role in the pathology of the disease, independent of whether they occur in an ion channel.

2.5.2 EA2: Insights from Other Episodic Ataxias

Of the seven reported episodic ataxia syndromes, four have now been linked to mutations in different genes. Most closely linked to EA2 is EA5, a rare form of episodic ataxia, which is associated with mutations in CACNB4 (Escayg et al. 2000), which encodes the calcium channel beta4 subunit that is thought to form a key part of P/Q channels (but see (Richards et al. 2007) for the potential importance of a different beta subunit in cerebellar 'P type' calcium channels). This ataxia also responds to acetazolamide, and has pronounced vertigo and nystagmus in the one family reported (Escayg et al. 2000). A third type of episodic ataxia, EA6, is caused by mutations in the SLC1A3 gene which encodes a glutamate uptake transporter (de Vries et al. 2009; Jen et al. 2005). This ataxia has phenotypic overlap with EA2 (it responds to acetazolamide) and with hemiplegic migraine (in the families described, it is associated with headaches and episodes of hemiparesis). For all three of these disorders, the mechanism of acetazolamide may provide a strategy for determining the conserved elements of the cascade leading to ataxia. As patients respond well to the drug, there is an implied shared underlying fault that is compensated by acetazolamide, either at the level of the cause or at the level of the symptoms.

Of the remaining episodic ataxias, EA1 is the only one linked to a single gene. However, this ataxia appears clinically more distant from EA2 than the more similar EA5 and EA6. EA1 is characterised by shorter attacks (lasting a few minutes) than EA2 and fewer vestibular symptoms, and is triggered by movements rather than by alcohol, caffeine, stress or other triggers of EA2. Unlike EA2, attacks of EA1 are not notably responsive to acetazolamide (Baloh 2012). However, EA2 can be responsive to the potassium channel blocker 4-aminopyridine (Strupp et al. 2004, 2007), and EA1 is caused by dominant negative mutations in KCNA1, which encodes shaker potassium channels Kv1.1 (Tomlinson et al. 2009). In a mouse model these channels appear to be important for regulating release of GABA onto cerebellar Purkinje cells, and preventing spontaneous action potentials (Herson et al. 2003). It appears that EA1 may be driven by the interaction between Kv1.1 and Kv1.2, with the mutant Kv1.1 subunits binding to and disrupting Kv1.2 channels (Tomlinson et al. 2009).

2.5.3 CACNA1A Context: Allelic Disorders

While several other related and unrelated genes give overlapping phenotypes with CACNA1A, there are also allelic disorders associated with this gene. That is, there are distinct diseases caused by mutations in the same gene.

The clinical genetics of CACNA1A is diverse, with migraine, seizures and muscle weakness as well as cerebellar ataxia associated with dysfunction of the channels. While there is some clinical overlap, the ability to distinguish between phenotypes is associated with mutations that have different impacts on the channels. While EA2 is most associated with loss of function, and SCA6 with toxic glutamine repeats, FHM is electrophysiologically more subtle, relying on careful assessment of the channel function in neurons (data from heterologous systems are too variable to give an unambiguous genotype–phenotype relationship for FHM1). However, it appears that FHM1 is not associated with simple loss of channel function, nor is it primarily associated with the glutamine repeats. Because, unlike EA2 and SCA6, FHM1 is not primarily cerebellar in manifestation, it is possible to speculate that cerebellar Purkinje cells are more sensitive to reduced current density (such as that in EA2 through haploinsufficiency) or to accumulation of poly-glutamine repeats (as in SCA6). The cerebellar nature of symptoms in EA2 and SCA6 suggest that other neurons throughout the human nervous system appear far less sensitive to the sorts of defects associated with these syndromes. In contrast, the changes induced by FHM1 mutations are most closely associated with headaches and may have more functional impact in cortical neurons, while sparing cerebellar Purkinje cells.

2.6 Episodic and Progressive and the Concept of 'Channelopathies'

A common dogma and a conundrum of diseases associated with ion channels is that they are episodic. Long periods of normal function are interspersed with episodes of symptoms, ataxia, migraine or seizures. Diseases associated with other channels—sodium channels, potassium channels and chloride channels—are also distinctly episodic in nature. In some cases, such as the severe seizures associated with Dravet syndrome and mutations in the sodium channel SCN1A, what begins as an episodic disease can lead to sustained deficits in cognition or developmental delay (Guerrini and Falchi 2011). In epilepsy, it is thought that severe seizures can cause lasting neurological defects, but it is not known whether attacks of ataxia lead to permanent cerebellar defects. It is possible that monitoring the outcome of EA2 patients successfully treated with acetazolamide will reveal whether blocking attacks is sufficient to protect Purkinje cells, or whether the underlying shortage of P/Q calcium channels is enough to eventually trigger atrophy.

The atrophy of Purkinje cells, a widespread feature of spinocerebellar ataxias, is usually associated with progressive rather than episodic disorders. While neural circuits appear able to compensate for defects in individual channels (until provoked into an attack by some environmental trigger) the loss of cells leads to uninterrupted disruption of the circuit, and a progressive disease.

2.7 Return to Clinical Phenotypes: Building a Genotype Mechanism for Phenotype

SCA6 is generally a pure cerebellar phenotype. Some individuals experience vertigo, suggesting there may be some involvement of brainstem, but the majority of patients have pure cerebellar ataxia. The onset is late—typically in the fifth or sixth decade of life—and progressive. Taken together these clinical details suggest the underlying mechanism is largely restricted to cerebellar neurons, and takes time to manifest. These data are consistent with a slow accumulation of toxic effects from poly-glutamine repeat elements from P/Q calcium channels having a relatively specific effect on cerebellar Purkinje cells. No treatments are currently available, but current work investigating the mechanism by which poly-glutamine repeats disrupt calcium signalling suggests an approach based on manipulating ITPR1 may be the best way forward, and their impact on Purkinje cell plasticity is already being investigated (Kimura et al. 2005). Recently it has been reported that suppression of ITPR1-dependent signalling in the related spinocerebellar ataxia, SCA2, may be effective in poly-glutamine disorders (Kasumu et al. 2012).

EA2 is a much more variable disorder, with episodes of cerebellar ataxia being accompanied by vertigo, nausea and frequently interspersed with headaches. These findings suggest that whatever deficits are caused by halploinsufficiency of

CACNA1A (or similar mutations in the gene) are more widespread in their impact than the poly-glutamine expansion that leads to SCA6. Notably, in humans, loss of P/Q channel expression appears strongly associated with disruption of the vestibular system and brainstem functions, while cognition (and other neocortical functions) is largely spared. The fact that EA2 responds to treatment with acetazolamide may mean that the consequences of reduced P/Q expression can be mitigated by regulation of pH (carbonic anhydrases are important for regulating extracellular pH); potentially, acetazolamide prevents changes in pH that are cues which trigger attacks. There is evidence that pH is elevated in EA2, which may be normalised by acetazolamide treatment, supporting a role for pH in the mechanism of the drug (Bain et al. 1992; Sappey-Marinier et al. 1999). Alternatively, acetazolamide may have an as yet unknown effect directly on the channels, although mouse data suggests this is unlikely (Kaja et al. 2008; Spacey et al. 2004). Some mutations in skeletal muscle ion channels also produce disorders that respond to acetazolamide (Matthews et al. 2011), indicating either a common pH dependence, or a similar modulation of the channels responsible by the compound. At present the mechanism of acetazolamide response is unknown, but determining how this drug effectively compensates for loss of P/Q expression is an obvious area of investigation for determining how attacks are triggered, and how future treatments may be targeted.

2.8 Summary and Future Directions

The two allelic ataxias associated with CACNA1A reveal how changes in a single gene can have distinct manifestations. The single change associated with SCA6, an expanded poly-glutamine repeat, and the pure cerebellar phenotype of this disease, are clear indications that the repeat has an effect which is important in cerebellar neurons, particularly Purkinje cells. The progressive and non-episodic nature of the disease suggests that it may not be functioning as a 'channelopathy' but as a poly-glutamine repeat disease which is particularly toxic to Purkinje cells because the poly-glutamine repeat happens to be located in a channel that is enriched in these cells.

The more diffuse nature of symptoms in EA2 is consistent with this disease having effects on some neurons outside the cerebellum. The most likely mechanistic explanation, given the prevalence of halploinsufficient patients is that neurons in the brainstem and cerebellum that are involved in balance, nausea and coordination are unable to compensate for the reduced P/Q currents as well as neurons elsewhere in the brain can compensate. The more widespread nature of the symptoms is consistent with them being associated with a function of P/Q channels that is more widespread in the brain, potentially with transmitter release. However, given the relatively small proportion of patients with cortical or thalamic symptoms, it appears that the neuronal function disrupted by haploinsufficiency of

CACNA1A may be one which neurons of the cortex and thalamus can compensate more completely than neurons of the cerebellum and vestibular system.

Future directions would involve mining ITPR1 modifying compounds for treatments of SCA6 and using the efficacy of acetazolamide to dissect the mechanism of suppression of ataxia in EA2.

References

Bain PG, O'Brien MD, Keevil SF, Porter DA (1992) Familial periodic cerebellar ataxia: a problem of cerebellar intracellular pH homeostasis. Ann Neurol 31:147–154

Baloh RW (2012) Episodic ataxias 1 and 2. Handb Clin Neurol 103:595–602

Baloh RW, Yue Q, Furman JM, Nelson SF (1997) Familial episodic ataxia: clinical heterogeneity in four families linked to chromosome 19p. Ann Neurol 41:8–16

Bezprozvanny I (2011) Role of inositol 1,4,5-trisphosphate receptors in pathogenesis of Huntington's disease and spinocerebellar ataxias. Neurochem Res 36:1186–1197

Bourinet E, Soong TW, Sutton K, Slaymaker S, Mathews E, Monteil A, Zamponi GW, Nargeot J, Snutch TP (1999) Splicing of alpha 1A subunit gene generates phenotypic variants of P- and Q-type calcium channels. Nat Neurosci 2:407–415

Calandriello L, Veneziano L, Francia A, Sabbadini G, Colonnese C, Mantuano E, Jodice C, Trettel F, Viviani P, Manfredi M et al (1997) Acetazolamide-responsive episodic ataxia in an Italian family refines gene mapping on chromosome 19p13. Brain 120(Pt 5):805–812

Cao YQ, Piedras-Renteria ES, Smith GB, Chen G, Harata NC, Tsien RW (2004) Presynaptic Ca2 + channels compete for channel type-preferring slots in altered neurotransmission arising from Ca2 + channelopathy. Neuron 43:387–400

Cao YQ, Tsien RW (2010) Different relationship of N- and P/Q-type Ca2 + channels to channel-interacting slots in controlling neurotransmission at cultured hippocampal synapses. J Neurosci 30:4536–4546

Chaudhuri D, Chang SY, DeMaria CD, Alvania RS, Soong TW, Yue DT (2004) Alternative splicing as a molecular switch for Ca2 +/calmodulin-dependent facilitation of P/Q-type Ca2 + channels. J Neurosci 24:6334–6342

de Vries B, Mamsa H, Stam AH, Wan J, Bakker SL, Vanmolkot KR, Haan J, Terwindt GM, Boon EM, Howard BD et al (2009) Episodic ataxia associated with EAAT1 mutation C186S affecting glutamate reuptake. Arch Neurol 66:97–101

Dolphin AC (2009) Calcium channel diversity: multiple roles of calcium channel subunits. Curr Opin Neurobiol 19:237–244

Dove LS, Abbott LC, Griffith WH (1998) Whole-cell and single-channel analysis of P-type calcium currents in cerebellar Purkinje cells of leaner mutant mice. J Neurosci 18:7687–7699

Du X, Wang J, Zhu H, Rinaldo L, Lamar K-M, Palmenberg AC, Hansel C, Gomez CM (2013) Second Cistron inCACNA1A gene encodes a transcription factor mediating cerebellar development and SCA6. Cell 154(1):118–133.

Escayg A, De Waard M, Lee DD, Bichet D, Wolf P, Mayer T, Johnston J, Baloh R, Sander T, Meisler MH (2000) Coding and noncoding variation of the human calcium-channel beta4-subunit gene CACNB4 in patients with idiopathic generalized epilepsy and episodic ataxia. Am J Hum Genet 66:1531–1539

Fletcher CF, Lutz CM, O'Sullivan TN, Shaughnessy JD Jr, Hawkes R, Frankel WN, Copeland NG, Jenkins NA (1996) Absence epilepsy in tottering mutant mice is associated with calcium channel defects. Cell 87:607–617

Frankel WN (2009) Genetics of complex neurological disease: challenges and opportunities for modeling epilepsy in mice and rats. Trends Genet 25:361–367

Gomez-Ospina N, Tsuruta F, Barreto-Chang O, Hu L, Dolmetsch R (2006) The C terminus of the L-typevoltage-gated calcium channel Ca(V)1.2 encodes a transcription factor. Cell 127(3):591–606

Green MC, Sidman RL (1962) Tottering–a neuromusclar mutation in the mouse. And its linkage with oligosyndacylism. J Hered 53:233–237

Griggs RC, Moxley RT 3rd, Lafrance RA, McQuillen J (1978) Hereditary paroxysmal ataxia: response to acetazolamide. Neurology 28:1259–1264

Guerrini R, Falchi M (2011) Dravet syndrome and SCN1A gene mutation related-epilepsies: cognitive impairment and its determinants. Dev Med Child Neurol 53(Suppl 2):11–15

Herson PS, Virk M, Rustay NR, Bond CT, Crabbe JC, Adelman JP, Maylie J (2003) A mouse model of episodic ataxia type-1. Nat Neurosci 6:378–383

Ishiguro T, Ishikawa K, Takahashi M, Obayashi M, Amino T, Sato N, Sakamoto M, Fujigasaki H, Tsuruta F, Dolmetsch R et al (2010) The carboxy-terminal fragment of alpha(1A) calcium channel preferentially aggregates in the cytoplasm of human spinocerebellar ataxia type 6 Purkinje cells. Acta Neuropathol 119:447–464

Ishikawa K, Fujigasaki H, Saegusa H, Ohwada K, Fujita T, Iwamoto H, Komatsuzaki Y, Toru S, Toriyama H, Watanabe M et al (1999) Abundant expression and cytoplasmic aggregations of [alpha]1A voltage-dependent calcium channel protein associated with neurodegeneration in spinocerebellar ataxia type 6. Hum Mol Genet 8:1185–1193

Ishikawa K, Tanaka H, Saito M, Ohkoshi N, Fujita T, Yoshizawa K, Ikeuchi T, Watanabe M, Hayashi A, Takiyama Y et al (1997) Japanese families with autosomal dominant pure cerebellar ataxia map to chromosome 19p13.1-p13.2 and are strongly associated with mild CAG expansions in the spinocerebellar ataxia type 6 gene in chromosome 19p13.1. Am J Hum Genet 61:336–346

Jen J, Kim GW, Baloh RW (2004) Clinical spectrum of episodic ataxia type 2. Neurology 62:17–22

Jen JC, Graves TD, Hess EJ, Hanna MG, Griggs RC, Baloh RW (2007) Primary episodic ataxias: diagnosis, pathogenesis and treatment. Brain 130:2484–2493

Jen JC, Wan J, Palos TP, Howard BD, Baloh RW (2005) Mutation in the glutamate transporter EAAT1 causes episodic ataxia, hemiplegia, and seizures. Neurology 65:529–534

Jorntell H, Hansel C (2006) Synaptic memories upside down: bidirectional plasticity at cerebellar parallel fiber-Purkinje cell synapses. Neuron 52:227–238

Kaja S, Van De Ven RC, Frants RR, Ferrari MD, Van Den Maagdenberg AM, Plomp JJ (2008) Reduced ACh release at neuromuscular synapses of heterozygous leaner Ca(v)2.1-mutant mice. Synapse 62:337–344

Kasumu AW, Liang X, Egorova P, Vorontsova D, Bezprozvanny I (2012) Chronic suppression of inositol 1,4,5-triphosphate receptor-mediated calcium signaling in cerebellar purkinje cells alleviates pathological phenotype in spinocerebellar ataxia 2 mice. J Neurosci 32:12786–12796

Kimura T, Sugimori M, Llinas RR (2005) Purkinje cell long-term depression is prevented by T-588, a neuroprotective compound that reduces cytosolic calcium release from intracellular stores. Proc Natl Acad Sci USA 102:17160–17165

Kodama T, Itsukaichi-Nishida Y, Fukazawa Y, Wakamori M, Miyata M, Molnar E, Mori Y, Shigemoto R, Imoto K (2006) A $Ca_v2.1$ calcium channel mutation rocker reduces the number of postsynaptic AMPA receptors in parallel fiber-Purkinje cell synapses. Eur J Neurosci 24:2993–3007

Kordasiewicz HB, Thompson RM, Clark HB, Gomez CM (2006) C-termini of P/Q-type Ca2+ channel alpha1Asubunits translocate to nuclei and promote polyglutamine-mediated toxicity. Hum Mol Genet 15(10):1587–99

Krovetz HS, Helton TD, Crews AL, Horne WA (2000) C-Terminal alternative splicing changes the gating properties of a human spinal cord calcium channel alpha 1A subunit. J Neurosci 20:7564–7570

Labrum RW, Rajakulendran S, Graves TD, Eunson LH, Bevan R, Sweeney MG, Hammans SR, Tubridy N, Britton T, Carr LJ et al (2009) Large scale calcium channel gene rearrangements in episodic ataxia and hemiplegic migraine: implications for diagnostic testing. J Med Genet 46:786–791

Llinas RR, Choi S, Urbano FJ, Shin HS (2007) Gamma-band deficiency and abnormal thalamocortical activity in P/Q-type channel mutant mice. Proc Natl Acad Sci USA 104:17819–17824

Lorenzon NM, Lutz CM, Frankel WN, Beam KG (1998) Altered calcium channel currents in Purkinje cells of the neurological mutant mouse leaner. J Neurosci 18:4482–4489

Matthews E, Portaro S, Ke Q, Sud R, Haworth A, Davis MB, Griggs RC, Hanna MG (2011) Acetazolamide efficacy in hypokalemic periodic paralysis and the predictive role of genotype. Neurology 77:1960–1964

Mezghrani A, Monteil A, Watschinger K, Sinnegger-Brauns MJ, Barrere C, Bourinet E, Nargeot J, Striessnig J, Lory P (2008) A destructive interaction mechanism accounts for dominant-negative effects of misfolded mutants of voltage-gated calcium channels. J Neurosci 28:4501–4511

Miki T, Zwingman TA, Wakamori M, Lutz CM, Cook SA, Hosford DA, Herrup K, Fletcher CF, Mori Y, Frankel WN et al (2008) Two novel alleles of tottering with distinct Ca(v)2.1 calcium channel neuropathologies. Neuroscience 155:31–44

Mintz IM, Venema VJ, Swiderek KM, Lee TD, Bean BP, Adams ME (1992) P-type calcium channels blocked by the spider toxin omega-Aga-IVA. Nature 355:827–829

Ophoff RA, Terwindt GM, Vergouwe MN, van Eijk R, Oefner PJ, Hoffman SM, Lamerdin JE, Mohrenweiser HW, Bulman DE, Ferrari M et al (1996) Familial hemiplegic migraine and episodic ataxia type-2 are caused by mutations in the Ca2 + channel gene CACNL1A4. Cell 87:543–552

Piedras-Renteria ES, Watase K, Harata N, Zhuchenko O, Zoghbi HY, Lee CC, Tsien RW (2001) Increased expression of alpha 1A Ca2 + channel currents arising from expanded trinucleotide repeats in spinocerebellar ataxia type 6. J Neurosci 21:9185–9193

Pietrobon D (2010) Insights into migraine mechanisms and $Ca_v2.1$ calcium channel function from mouse models of familial hemiplegic migraine. J Physiol 588:1871–1878

Pietrobon D, Moskowitz MA (2012) Pathophysiology of Migraine. Annu Rev Physiol

Plomp JJ, van den Maagdenberg AM, Kaja S (2009) The ataxic Cacna1a-mutant mouse rolling nagoya: an overview of neuromorphological and electrophysiological findings. Cerebellum 8:222–230

Raghib A, Bertaso F, Davies A, Page KM, Meir A, Bogdanov Y, Dolphin AC (2001) Dominant-negative synthesis suppression of voltage-gated calcium channel $Ca_v2.2$ induced by truncated constructs. J Neurosci 21:8495–8504

Rajakulendran S, Graves TD, Labrum RW, Kotzadimitriou D, Eunson L, Davis MB, Davies R, Wood NW, Kullmann DM, Hanna MG et al (2010a) Genetic and functional characterisation of the P/Q calcium channel in episodic ataxia with epilepsy. J Physiol 588:1905–1913

Rajakulendran S, Kaski D, Hanna MG (2012) Neuronal P/Q-type calcium channel dysfunction in inherited disorders of the CNS. Nat Rev Neurol 8:86–96

Rajakulendran S, Schorge S, Kullmann DM, Hanna MG (2010b) Dysfunction of the Ca(V)2.1 calcium channel in cerebellar ataxias. F1000 Biol Rep *2*

Reetz K, Costa AS, Mirzazade S, Lehmann A, Juzek A, Rakowicz M, Boguslawska R, Schols L, Linnemann C, Mariotti C et al (2013) Genotype-specific patterns of atrophy progression are more sensitive than clinical decline in SCA1, SCA3 and SCA6. Brain 136:905–917

Riant F, Lescoat C, Vahedi K, Kaphan E, Toutain A, Soisson T, Wiener-Vacher SR, Tournier-Lasserve E (2010) Identification of CACNA1A large deletions in four patients with episodic ataxia. Neurogenetics 11:101–106

Richards KS, Swensen AM, Lipscombe D, Bommert K (2007) Novel $Ca_v2.1$ clone replicates many properties of Purkinje cell $Ca_v2.1$ current. Eur J Neurosci 26:2950–2961

Saegusa H, Wakamori M, Matsuda Y, Wang J, Mori Y, Zong S, Tanabe T (2007) Properties of human $Ca_v2.1$ channel with a spinocerebellar ataxia type 6 mutation expressed in Purkinje cells. Mol Cell Neurosci 34:261–270

Saito H, Okada M, Miki T, Wakamori M, Futatsugi A, Mori Y, Mikoshiba K, Suzuki N (2009) Knockdown of $Ca_v2.1$ calcium channels is sufficient to induce neurological disorders observed in natural occurring Cacna1a mutants in mice. Biochem Biophys Res Commun 390:1029–1033

Sappey-Marinier D, Vighetto A, Peyron R, Broussolle E, Bonmartin A (1999) Phosphorus and proton magnetic resonance spectroscopy in episodic ataxia type 2. Ann Neurol 46:256–259

Schorge S, Rajakulendran S (2012) The P/Q channel in human disease: untangling the genetics and physiology. Wiley Interdiscip Rev Membr Transp Signal 1(3):311–320

Schorge S, van de Leemput J, Singleton A, Houlden H, Hardy J (2010) Human ataxias: a genetic dissection of inositol triphosphate receptor (ITPR1)-dependent signaling. Trends Neurosci 33:211–219

Seidel K, Siswanto S, Brunt ER, den Dunnen W, Korf HW, Rub U (2012) Brain pathology of spinocerebellar ataxias. Acta Neuropathol 124:1–21

Spacey SD, Hildebrand ME, Materek LA, Bird TD, Snutch TP (2004) Functional implications of a novel EA2 mutation in the P/Q-type calcium channel. Ann Neurol 56:213–220

Strupp M, Kalla R, Dichgans M, Freilinger T, Glasauer S, Brandt T (2004) Treatment of episodic ataxia type 2 with the potassium channel blocker 4-aminopyridine. Neurology 62:1623–1625

Strupp M, Zwergal A, Brandt T (2007) Episodic ataxia type 2. Neurotherapeutics 4:267–273

Takahashi H, Ishikawa K, Tsutsumi T, Fujigasaki H, Kawata A, Okiyama R, Fujita T, Yoshizawa K, Yamaguchi S, Tomiyasu H et al (2004) A clinical and genetic study in a large cohort of patients with spinocerebellar ataxia type 6. J Hum Genet 49:256–264

Takahashi M, Obayashi M, Ishiguro T, Sato N, Niimi Y, Ozaki K, Mogushi K, Mahmut Y, Tanaka H, Tsuruta F et al (2013) Cytoplasmic location of α1A voltage-gated calcium channel C-terminal fragment ($Ca_v2.1$-CTF)aggregate is sufficient to cause cell death. PLoS One 8(3):e50121

Tomlinson SE, Hanna MG, Kullmann DM, Tan SV, Burke D (2009) Clinical neurophysiology of the episodic ataxias: insights into ion channel dysfunction in vivo. Clin Neurophysiol 120:1768–1776

Unno T, Wakamori M, Koike M, Uchiyama Y, Ishikawa K, Kubota H, Yoshida T, Sasakawa H, Peters C, Mizusawa H, Watase K (2012) Development of Purkinje cell degeneration in a knockin mouse model revealslysosomal involvement in the pathogenesis of SCA6. Proc Natl Acad Sci U S A 109(43):17693–8

van de Leemput J, Chandran J, Knight MA, Holtzclaw LA, Scholz S, Cookson MR, Houlden H, Gwinn-Hardy K, Fung HC, Lin X et al (2007) Deletion at ITPR1 underlies ataxia in mice and spinocerebellar ataxia 15 in humans. PLoS Genet 3:e108

Wakamori M, Yamazaki K, Matsunodaira H, Teramoto T, Tanaka I, Niidome T, Sawada K, Nishizawa Y, Sekiguchi N, Mori E et al (1998) Single tottering mutations responsible for the neuropathic phenotype of the P-type calcium channel. J Biol Chem 273:34857–34867

Walter JT, Alvina K, Womack MD, Chevez C, Khodakhah K (2006) Decreases in the precision of Purkinje cell pacemaking cause cerebellar dysfunction and ataxia. Nat Neurosci 9:389–397

Watase K, Barrett CF, Miyazaki T, Ishiguro T, Ishikawa K, Hu Y, Unno T, Sun Y, Kasai S, Watanabe M et al (2008) Spinocerebellar ataxia type 6 knockin mice develop a progressive neuronal dysfunction with age-dependent accumulation of mutant $Ca_v2.1$ channels. Proc Natl Acad Sci USA 105:11987–11992

Xie G, Clapcote SJ, Nieman BJ, Tallerico T, Huang Y, Vukobradovic I, Cordes SP, Osborne LR, Rossant J, Sled JG et al (2007) Forward genetic screen of mouse reveals dominant missense mutation in the P/Q-type voltage-dependent calcium channel. CACNA1A Genes Brain Behav 6:717–727

Zhuchenko O, Bailey J, Bonnen P, Ashizawa T, Stockton DW, Amos C, Dobyns WB, Subramony SH, Zoghbi HY, Lee CC (1997) Autosomal dominant cerebellar ataxia (SCA6) associated with small polyglutamine expansions in the alpha 1A-voltage-dependent calcium channel. Nat Genet 15:62–69

Chapter 3
Role of L-Type Ca^{2+} Channels in Sensory Cells

Alexandra Koschak, Alexandra Pinggera, Klaus Schicker and Jörg Striessnig

Abstract L-type Ca^{2+} channels play an important role in sensory cells present in the inner ear and the retina. $Ca_v1.3$ predominates in sensory cells of the inner ear (inner and outer cochlear hair cells and vestibular hair cells) and $Ca_v1.4$ in retinal neurons. Their pore-forming $\alpha1$-subunits are highly homologous but functionally heterogeneous. Such variability is ensured either by differential interaction with modulatory proteins (such as Ca^{2+}-binding proteins), differences in alternative splicing, posttranslational modification (RNA-editing) or in subunit composition. We will discuss special structural features that stabilize properties required for proper function in these cells and allow fine-tuning of Ca^{2+} signals. Whereas so far only one $Ca_v1.3$ human disease mutation has been published more than 50 mutations have been reported for $Ca_v1.4$. $Ca_v1.3$ channels are currently discussed as molecular target for neuroprotection in Parkinsons Disease. Are their regulatory mechanisms also interesting for potential pharmacotherapeutic modulation? How do loss- and gain-of-function mutations on the protein level both result in impaired retinal synaptic transmission in patients carrying these mutations? We will summarize our current knowledge about the role of L-type channels for human hearing and visual disorders.

A. Koschak (✉) · K. Schicker
Center for Physiology and Pharmacology, Department of Neurophysiology and Pharmacology, Medical University Vienna, Schwarzspanierstrasse 17/I 1090 Vienna, Austria
e-mail: alexandra.koschak@meduniwien.ac.at

A. Pinggera · J. Striessnig
Institute of Pharmacy, Pharmacology and Toxicology, Center for Chemistry and Biomedicine, University of Innsbruck, Innrain 80-82/III 6020 Innsbruck, Austria

N. Weiss and A. Koschak (eds.), *Pathologies of Calcium Channels*,
DOI: 10.1007/978-3-642-40282-1_3, © Springer-Verlag Berlin Heidelberg 2014

3.1 Introduction

Voltage-gated Ca^{2+} channels (VGCCs) generate depolarization-induced Ca^{2+} signals for many important physiological functions such as muscle contraction, secretion, neurotransmission, and gene expression (for review see Catterall 2011). Like other ion channels they form part of signaling complexes together with molecules, such as receptors, kinases, phosphatases, and calmodulin (for review see, Buraei and Yang 2010; Catterall 2000). Most VGCCs exist as hetero-oligomeric assemblies of several subunits, with a so-called α1-subunit forming the Ca^{2+}-selective channel pore (Fig. 3.1). Ten different α1-subunit isoforms exist, that can associate with one of four different β-subunit isoforms and one of four different α2-δ subunits. The α1-subunits determine most of the channel's biophysical and pharmacological properties. Based on structural and pharmacological similarities of their α1-subunits VGCCs can be classified into two major families. L-type Ca^{2+} channels (LTCCs, Ca_v1 family) and non-LTCCs (Ca_v2 and Ca_v3 families). LTCCs were first described in muscle cells but are found in most electrically excitable tissues. In contrast, expression of members of the Ca_v2 family is largely restricted to neurons and endocrine cells. In neurons Ca_v2 channels form the presynaptic channels providing Ca^{2+} for fast neurotransmitter release. In contrast, in the central nervous system neuronal Ca_v1 channels are found postsynaptically whereas they are located at presynaptic release sites in (neuro-) sensory cells.

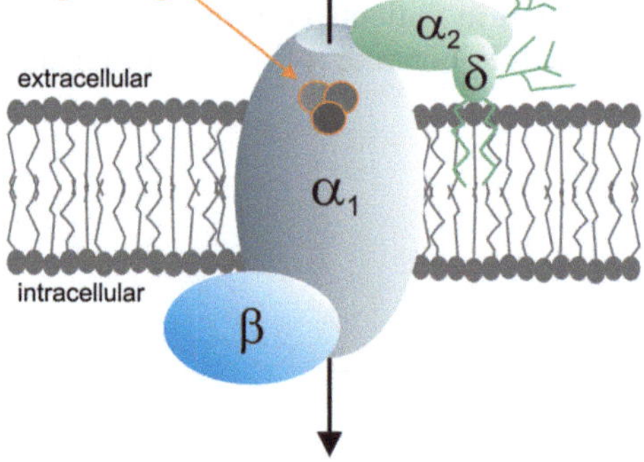

Fig. 3.1 Voltage-gated Ca^{2+} channel (VGCC) complex. VGCC consist of a pore forming, transmembrane spanning α1 subunit (*gray*), responsible for ion-conductance and –selectivity, and auxiliary β (*blue*) and $\alpha 2\delta$ (*green*) subunits. The intracellular β subunit modulates biophysical properties of the channel and is involved in membrane trafficking of the α1 subunit, as well as the extracellular $\alpha 2\delta$ subunit, which is attached through a glucosylphosphatidylinositol anchor to the extracellular side of the membrane (Bauer et al. 2010). α1-subunits contain binding domains for organic Ca^{2+} channel blockers (*orange*) such as dihydropyridines (Striessnig and Koschak 2008)

3.2 L-Type Channels in the Inner Ear

As outlined above, fast neurotransmitter release in neurons is tightly regulated by voltage-gated Ca_v2 channels. They conduct Ca^{2+} currents previously classified as P/Q-, N- and R-type, respectively (for review see Catterall 2011), and initiate the fast release of important neurotransmitters such as glutamate, GABA and acetylcholine. Ca^{2+} entering through these channels is efficiently coupled to neurotransmitter release by the third or fourth power. The channel-mediated Ca^{2+} signal is locally contained by intracellular Ca^{2+} buffers, and Ca^{2+} concentrations quickly drop within a distance of several tens of nanometers from the inner channel mouth. This process requires anchoring of downstream signaling molecules within this so-called Ca^{2+} nanodomain (Eggermann et al. 2012). Specialized presynaptic structures tether the channels to the presynaptic vesicle release machinery by direct and indirect protein interactions with the channel complex (Müller et al. 2010). The proteome of brain Ca_v2 channels has recently been reported (Müller et al. 2010).

Specialized presynaptic structures providing highly localized Ca^{2+} signals for neurotransmitter release are also present in sensory cells. However, the Ca^{2+} channels involved there are members of the Ca_v1 family which in neurons are mainly found postsynaptically in somatodendritic locations (Obermair et al. 2004). $Ca_v1.3$ predominates in sensory cells of the inner ear (inner and outer cochlear hair cells and vestibular hair cells) and, as outlined below, $Ca_v1.4$ in retinal photoreceptors. Since LTCCs are specifically inhibited by organic Ca^{2+} channel blockers, such as dihydropyridines (DHPs), neurotransmitter release in this sensory cells is DHP-sensitive. This in contrast to fast spiking neurons in which neurotransmitter release is only blocked by Ca_v2-selective blockers such as ω-conotoxin GVIA ($Ca_v2.2$) or ω-agatoxin IVA ($Ca_v2.1$). Why do sensory cells require a different set of VGCCs? One explanation is that the gating properties of LTCCs are more suitable to quickly respond to fluctuations in membrane voltage and sustain the prolonged noise- or light-induced Ca^{2+} fluxes in sensory cells. In contrast to neurons in which exocytosis is triggered by precisely timed action potentials, exocytosis in IHCs is triggered by graded changes in membrane potential induced by sound. Louder sounds elicit increasingly larger receptor potentials causing more $Ca_v1.3$ channels to be activated, increased Ca^{2+} influx and faster vesicle fusion (Goutman 2012), for review see Safieddine et al. (2012). At the same time the acoustic signal at the afferent synapses (i.e., between the IHC and the afferent fibers of the auditory nerve) is encoded with high temporal precision. Channel activity and Ca^{2+} influx therefore follow the graded changes in receptor potentials which requires that these channels must be active within the negative operating range of receptor potentials (−70 to –20 mV, Oliver et al. 2003) and only slowly inactivate so that the number of available channels does not vary too much with sound intensity. As outlined below $Ca_v1.3$ channels perfectly fulfill these criteria.

Like $Ca_v1.4$ in photoreceptors, $Ca_v1.3$ LTCCs are closely associated with specialized presynaptic active zones, so-called synaptic ribbons. These are multiprotein complexes tethering synaptic vesicles to provide a large pool of readily releasable glutamate-containing vesicles. In mature IHCs this ensures highly efficient coupling between Ca^{2+} influx through nearby $Ca_v1.3$ channels and vesicle release (Beutner and Moser 2001; Brandt et al. 2003). Ribbon-associated proteins also stabilize normal targeting and function of $Ca_v1.3$ channels (see below).

AMPA receptors are located postsynaptically on the unbranched myelinated afferent dendrite of the bipolar spiral ganglion neurons. Each spiral ganglion neuron makes only contact with one IHC synapse, whereas each IHC (through up to 20–30 ribbon synapses) can activate several ganglion neurons (for review Meyer and Moser 2010). Individual afferent fibers respond at a defined frequency, which depends on its location along the cochlear tonotopical axis (place-frequency maps, Müller et al. 2005). AMPA receptor-induced EPSCs shape the spiking of the afferent fibers which show heterogeneous spontaneous firing behavior even if they originate from the same or a neighboring IHC. High spiking rate fibers (being of larger diameter) increase spiking above their spontaneous rate already at low sound pressure but saturate quickly at higher sound intensities and are therefore tuned for high sensitivity. Low spiking rate fibers activate at higher sound intensities and show lesser tendency to saturate. Seemingly they are tuned for providing information about incremental sound intensity (Meyer and Moser 2010). Although different mechanisms may account for this heterogeneity in ganglion fibers, a presynaptic mechanism involving $Ca_v1.3$-mediated Ca^{2+} signals has recently been proposed (Meyer et al. 2009). A thorough analysis of the structure and functional properties of IHC hair cell synapses revealed that they differ in number in different cochlear regions and are most dense in IHCs of the mid-cochlear region which is most sensitive to sound. Interestingly, presynaptic Ca^{2+} signals measured by high resolution fluorescence microscopy differed substantially among synapses of an individual hair cell, regardless of location along the cochlear tonotopic axis. Variability was found for both the size as well as the voltage for half maximal activation of the Ca^{2+} signal (Meyer et al. 2009). This behavior may form a presynaptic mechanism that contributes to the heterogeneous spiking properties of individual spiral ganglion neurons driven by a given IHC (Meyer et al. 2009). It is therefore possible that the number and/or properties of presynaptic channels differ in individual ribbon synapses, and strongly suggests functional heterogeneity within $Ca_v1.3$ channels which are the only LTCCs expressed in IHCs. This heterogeneity may be ensured either by differential interaction with (modulatory) ribbon proteins, differences in structure through alternative splicing, posttranslational modification or in subunit composition. We will discuss below several molecular mechanisms that could account for functionally distinct $Ca_v1.3$ channel complexes.

3.2.1 $Ca_v1.3$ Channels are Essential for Normal Cochlear Function

Studies in the chicken cochlea provided a first hint for a dominant role of $Ca_v1.3$ for presynaptic exocytosis in hair cells because Ca^{2+} currents were DHP-sensitive and $Ca_v1.3$ $\alpha1$-subunit mRNA was abundantly expressed (Kollmar et al. 1997). Final proof for the essential role of this channel isoform in mammals was eventually obtained in $Ca_v1.3$-deficient ($Ca_v1.3^{-/-}$) mice. Patch-clamp recordings in $Ca_v1.3^{-/-}$ hair cells revealed that these channels carry 80 to >90 % (depending on the hair cell position along the longitudinal axis of the cochlea) of the total Ca^{2+} current in both, inner and outer hair cells (Brandt et al. 2003; Dou et al. 2004; Michna et al. 2003; Platzer et al. 2000). In up to 3 week old $Ca_v1.3^{-/-}$ IHCs the remaining current still triggered a tiny fast secretory component which also coupled to fast exocytosis with comparable efficacy than wild type channels (Brandt et al. 2003). However, this residual current is insufficient to support hearing and therefore $Ca_v1.3^{-/-}$ mice are completely deaf (Platzer et al. 2000). In addition to defective excitation-secretion coupling a complex developmental cochlear phenotype must also contribute to deafness in homozygous (but not heterozygous) knockouts (Dou et al. 2004; Glueckert et al. 2003; Platzer et al. 2000). Up to P3 electron microscopy revealed normal postnatal development of sensory hair cells, and their afferent as well as efferent nerve endings (Glueckert et al. 2003; Platzer et al. 2000). At P7 afferent nerve fibers started to degenerate and by P15, a loss of OHCs in apical turns was observed. In IHCs and efferent axons no ultrastructural changes occurred but intact efferent nerve fibers formed unusual direct contacts with IHCs (Brandt et al. 2003; Glueckert et al. 2003). These contacts normally disappear during the first postnatal week. At P15 spiral ganglion cells also notably degenerated and by 8 months nearly all spiral ganglion and sensory cells (including IHCs) of the organ of Corti were absent (Glueckert et al. 2003).

During development IHCs produce typical Ca^{2+} action potentials which disappear with the onset of hearing. Interestingly, these action potentials were absent in $Ca_v1.3^{-/-}$ IHCs (Brandt et al. 2003). Spontaneous firing during development therefore seems to underlie the release of synaptogenic signals (such as neurotrophic factors) required for normal cochlear development. Their absence may be one of the triggers initiating the degenerative processes described above. Moreover, the absence of $Ca_v1.3$ channels in developing IHCs was also associated with a lack of functional BK potassium—channels in the IHC membrane, presumably resulting from defective targeting (Brandt et al. 2003). Taken together these data highlight the critical role of $Ca_v1.3$ channels for normal structure and function of cochlear sensory cells. No compensatory upregulation of other VGCCs was observed. Based on the unique gating properties of $Ca_v1.3$ such upregulation would most likely not be able to compensate for $Ca_v1.3$ function.

Vestibular hair cells also have a large DHP-sensitive L-type current component. Estimates range from 35 % (Bao et al. 2003) to >80 % (Almanza et al. 2003; Boyer et al. 1998; Dou et al. 2004), and there is strong evidence that this L-type current is also mainly carried by $Ca_v1.3$. In utricular hair cells of $Ca_v1.3^{-/-}$ mice barium current (I_{Ba}) amplitude was reduced by 50 % as compared to wild type. The missing component completely accounted for the DHP-sensitive current (Dou et al. 2004). Unlike in the cochlea, the substantial non-L-type current appeared to be sufficient to maintain normal vestibular function and sensory cell viability (Dou et al. 2004).

In addition to deafness-inducing effects through compromised hair cell function and aberrant cochlear development, retrocochlear defects must also contribute to deafness in $Ca_v1.3^{-/-}$ mice. $Ca_v1.3$ deficiency is associated with a drastically reduced volume in all auditory brainstem centers (but not other brain regions) already before hearing onset (Hirtz et al. 2011). The lateral superior olive (LSO) contains fewer neurons and is abnormally shaped. The remaining LSO neurons receive functional glutamatergic input through normal dendritic trees but show an abnormal firing pattern upon depolarization which is attributed to reduced K^+-channel function. Notably, this effect was not due to inner ear dysfunction because it was also present in mice with $Ca_v1.3$ channels conditionally knocked out only in the auditory brainstem, but preserved expression in the cochlea (Satheesh et al. 2012).

A crucial role of $Ca_v1.3$ was also discovered for synaptic refinement in the auditory pathway. In general, neuronal circuits underlie a developmental process in which initially imprecisely formed synapses become refined by selective elimination of redundant immature synapses and strengthening of remaining ones. In the auditory brainstem $Ca_v1.3$ crucially controlled the refinement of inhibitory synapses in projections from the medial nucleus of the trapezoid body (MNTB) to the LSO (Hirtz et al. 2012). Normally, during the first 2 postnatal weeks the number of axons of MNTB neurons projecting to one LSO neuron declines sharply whereas the remaining ones are consolidated in mice. In $Ca_v1.3^{-/-}$ mice projections were not eliminated up to hearing onset and synaptic strengthening was strongly impaired. Moreover, the mediolateral topography was less precise and the shift from a mixed GABA/glycinergic to a purely glycinergic transmission normally seen in wild type mice before hearing onset did not occur (Hirtz et al. 2012).

3.2.2 *Functional Properties of Cochlear $Ca_v1.3$ L-Type Currents*

IHCs and OHCs are the only cells known in the mammalian organism so far that express essentially pure $Ca_v1.3$ LTCC currents. This feature allows a detailed study of $Ca_v1.3$ biophysical properties in their native environment at the synaptic ribbons. One hallmark of $Ca_v1.3$ currents discovered in IHCs is their negative activation voltage range with activation thresholds between −60 and −50 mV when physiological extracellular Ca^{2+} concentrations were used as charge carrier (Bock et al. 2011; Dou et al. 2004; Grant and Fuchs 2008; Zampini et al. 2010). This is well within the operating range of the IHC receptor potential (see above) and shows that channels operate at the foot of their current–voltage relationship.

The negative activation voltage range of $Ca_v1.3$ is an intrinsic property of their pore-forming $\alpha1$-subunits and also observed when expressed heterologously in mammalian cells (Bock et al. 2011; Huang et al. 2012; Koschak et al. 2001; Lieb et al. 2012). $Ca_v1.3$ channels show more negative activation than other LTCCs (including $Ca_v1.4$, see below) and the Ca_v2 channel family (Kiyonaka et al. 2007; Lipscombe 2002; Shcheglovitov et al. 2012). Only the typical "low voltage activated" T-type channels activate at more hyperpolarized voltages than $Ca_v1.3$ (Perez-Reyes 2003).

The low activation voltage range allows both $Ca_v1.3$ and T-type channels to carry depolarizing Ca^{2+} inward currents at threshold potentials and thereby support spontaneous pacemaking. For $Ca_v1.3$ we have already mentioned above their role for spontaneous action potentials in IHCs during development. $Ca_v1.3$ also supports diastolic depolarization in sinoatrial node cells ($Ca_v1.3^{-/-}$ mice are bradycardic and display arrhythmic sinoatrial node pacemaking, for review Striessnig and Koschak 2008), spontaneous action potentials in adrenal chromaffin cells (for review see Marcantoni et al. 2007) and contribute to the safety of pacemaking in dopaminergic neurons (for review see, Putzier et al. 2009; Striessnig and Koschak 2008). T-type currents support pacemaking in the heart (Mangoni et al. 2006) and underlie burst firing in neurons (Astori et al. 2011; Kim et al. 2001). However, in contrast to T-type currents with rapid voltage-dependent inactivation (Perez-Reyes 2003), $Ca_v1.3$ channels conduct low-voltage activated currents with very slow voltage-dependent inactivation (Koschak et al. 2001; Yang et al. 2006). As discussed above, the need for graded Ca^{2+} responses over prolonged periods of stimulation in IHCs requires slow inactivation of the underlying Ca^{2+} current. This demand favors $Ca_v1.3$ channels not only over T-type channels but also over other VGCCs, including the classical presynaptic $Ca_v2.1$ and $Ca_v2.2$ channels, which have faster voltage-dependent inactivation kinetics (Kiyonaka et al. 2007). Although slow inactivation benefits continuous Ca^{2+} supply to the synaptic vesicle pool of ribbon synapses, it may support inappropriately long depolarization and excessive Ca^{2+} influx particularly in cells where depolarizing membrane potentials exceed those in IHCs, such as in sinoatrial node cells or neurons. Indeed, $Ca_v1.3$ current components in sinoatrial node inactivate rapidly (Mangoni et al. 2003). $Ca_v1.3$ but also most other members of the Ca_v1 and Ca_v2 channel family use an additional mechanism for fine-tuning of inactivation in addition to voltage. This mechanism is Ca^{2+} dependent and therefore termed Ca^{2+}-dependent inactivation (CDI).

3.2.3 Voltage and Ca^{2+}-Dependent Inactivation of $Ca_v1.3$

CDI occurs in a calmodulin-dependent manner and was first described for Ca^{2+} currents in paramecium (Brehm and Eckert 1978). Apo-calmodulin binds to the proximal C-terminus of $\alpha1$-subunits (Fig. 3.2) and can sense Ca^{2+} entering through the pore (within the channels nanodomain) or global cellular Ca^{2+} concentrations (Tadross et al. 2008). Although the orientation of bound calmodulin and the

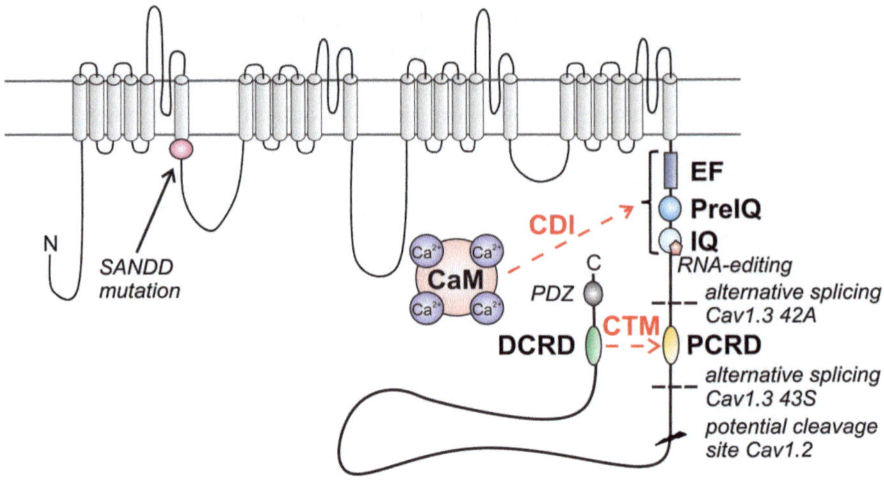

Fig. 3.2 Schematic transmembrane topology of the $Ca_v1.3$ $\alpha1$-subunit. It consists of four homologous domains, each comprising six transmembrane segments (Catterall et al. 2005); *pink circle* highlights the approximate position of the in-frame glycine insertion leading to SANDD-syndrome (Baig et al. 2011); in addition mechanisms involved in regulation of CDI are shown: apocalmodulin (CaM) interacts via EF-, preIQ-, and IQ-domain with the C-terminus, and due to a conformational change after Ca^{2+} binding initiates CDI (Erickson et al. 2003; Peterson et al. 2000), note that interaction of DCRD with PCRD forms a functional CTM, which inhibits binding of CaM to the C-terminus, thus slowing down CDI (Singh et al. 2008). RNA-editing in the IQ-motif also leads to lower affinity for CaM and therefore to a weaker CDI (Huang et al. 2012); alternative splicing in exon 42 or exon 43, leading to short transcripts, missing either both, DCRD and PCRD, or just DCRD, prevents modulatory function of CTM (Bock et al. 2011; Singh et al. 2008); long splice variants also contain a PDZ-binding domain on the very end of the C-terminus, which enables the $\alpha1$-subunit to interact with a variety of different proteins, that can act as scaffolds or modulate the biophysical properties of the channel (Calin-Jageman and Lee 2008)

relative contribution of local versus global Ca^{2+} sources vary for different channel isoforms (for review see Zamponi 2003), the basic regulatory response remains the same: calmodulin binds Ca^{2+} and the resulting conformational change promotes channel inactivation in addition to VDI. This mechanism provides channels with the possibility to adjust their inactivation rate between "fast", by utilizing maximal CDI, and "slow" by turning off CDI allowing only (slower) VDI. As described below, different molecular mechanisms were found to affect the strength of CDI in LTCCs: the regulation of intracellular Ca^{2+} levels and Ca^{2+} buffering (Grant and Fuchs 2008), intermolecular interactions with other proteins and, in the case of $Ca_v1.3$, intramolecular interactions within the C-terminus and RNA-editing. CDI is fast in $Ca_v1.3$ channels heterologously expressed in HEK293 cells (Koschak et al. 2001; Lieb et al. 2012), sinoatrial node (Mangoni et al. 2003) and adrenal chromaffin cells (Marcantoni et al. 2010) which stands in contrast to IHC currents. In adult IHCs, CDI is weak allowing primarily VDI to determine inactivation (Grant and Fuchs 2008; Platzer et al. 2000). Because VDI is also slower in

IHCs (and OHCs, Michna et al. 2003) than in other cells, overall inactivation becomes very slow and $Ca_v1.3$ currents remain permanently available at voltages within the operating range of the IHC (Grant and Fuchs 2008). This raises one important question: how do IHCs reduce CDI and VDI?

3.2.3.1 Interaction With Ca^{2+} Binding Proteins (CaBPs)

Because CDI is calmodulin-dependent, other CaBPs with insufficient Ca^{2+} sensing properties can compete with CaM binding to the channel and thereby inhibit CDI. CaBP1, CaBP2, and CaBP4 abolish CDI of $Ca_v1.3$ channels heterologously expressed in HEK293 cells (Cui et al. 2007; Schrauwen et al. 2012; Yang et al. 2006). All of them are present in IHCs (Cui et al. 2007; Schrauwen et al. 2012) but so far their overall and individual contribution to the weak CDI observed in IHCs is unclear. CaBP4-deficient mice hear normally and IHC Ca^{2+} currents were only minimally altered (Cui et al. 2007). Recently, a human mutation in CaBP2 has been identified as a novel gene associated with moderate to severe hearing loss (Schrauwen et al. 2012). Due to a frameshift, a truncated CaBP2 protein lacking the EF-hands 3 and 4 showed weaker Ca^{2+} binding properties. In functional experiments the mutation reduced the inhibitory effect of CaBP2 on CDI. CaBPs also reduced the current density of heterologously expressed channels, a property also lost for mutant CaBP2 (Schrauwen et al. 2012). These data provided indirect evidence for the important role of CaBP-dependent fine-tuning of $Ca_v1.3$ channel activity, although the effects on hearing may also be caused by CaBP2 targets other than $Ca_v1.3$.

Taken together association of CaBPs with the channel complex currently appears as the most powerful mechanism to account for the unusually weak CDI of $Ca_v1.3$ currents observed in IHCs. However, CaBPs do not explain reduced VDI. Although CaBP1 somewhat slowed VDI, other mechanisms must be responsible for this property.

3.2.3.2 Other Protein Interactions

Table 3.1 summarizes additional protein interactions with $Ca_v1.3$ channels in sensory cells. Most of them are known to affect channel function, at least upon co-expression in mammalian cells. Interesting candidates that stabilize slow VDI are Rab3-interacting molecule (RIM) and syntaxin. Coexpression of syntaxin with $Ca_v1.3$ in HEK293 cells slightly inhibited inactivation and treatment of bullfrog saccule hair cells with botulinum toxin (an inhibitor of syntaxin) increased inactivation. Unfortunately, only Ca^{2+} currents were reported (Song et al. 2003). Therefore, it remains unclear if increased inactivation was due to changes in CDI or VDI. A more detailed study was carried out with RIM. RIM2α mRNA is expressed in immature cochlear IHCs and co-localized with $Ca_v1.3$ in the same presynaptic compartment (Fig. 3.3). Co-expressed RIM proteins bound to the

Table 3.1 Protein interactions of $Ca_v1.3$ and $Ca_v1.4$ in cochlea and retina

Protein	Interaction partner	Tissue	Function	Notes/references
Bestrophin 1	$Ca_v1.3$, by auxiliary β-subunit	Retina	Alters kinetics, shifts voltage-dependent activation to more negative voltages and blocks LTCC mediated rises in I_{Ca} in retinal pigmental epithelium	*Mutations cause Best's vitelliforme macular degeneration and other retinopathies*, (Marmorstein et al. 2006; Rosenthal et al. 2006; Yu et al. 2008)
CaBPs	$Ca_v1.3$	Cochlea	CaBP1, CaBP2, and CaBP4; slow down CDI in cochlear hair cells	(Cui et al. 2007; Schrauwen et al. 2012)
	$Ca_v1.4$	Retina	CaBP4; shifts the voltage range of activation of $Ca_v1.4$ to more hyperpolarized voltages; expressed in rod and cone photoreceptors	(Haeseleer et al. 2004)
Harmonin	$Ca_v1.3$	Cochlea	Enhances ubiquitinylation of the channel, causes clustering of the channel on cell surface, suppresses $Ca_v1.3$ I_{Ba} peak current density in IHCs	(Gregory et al. 2011)
Otoferlin	$Ca_v1.3$	cochlea	May regulate exocytosis, certain mutations cause deafness, expressed in mechanosensory hair cells	(Ramakrishnan et al. 2009; Roux et al. 2006)
Retinoschisin	$Ca_v1.3$	Retina	Involved in circadian regulation of VGCCs in photoreceptors	*Likely to interact also with $Ca_v1.4$* (Shi et al. 2009a)
RIMs	Auxiliary β subunit	Cochlea	Slows VDI and CDI in IHCs	*also epressed in retina* (Gebhart et al. 2010; tom Dieck et al. 2005; Wang et al. 1997)
RIM-binding proteins	$Ca_v1.3$	Retina	Probably acting as scaffold involved in vesicle tethering and priming- fusion machinery, expressed in brain but also in IHCs and retina (in outer nuclear layer, outer plexiform layer, and ganglion cell layer)	*May similarly interact with $Ca_v1.4$ in retina, also expressed in cochlear IHCs* (Hibino et al. 2002)
Syntaxin	$Ca_v1.3$	Cochlea	Slows inactivation in cochlear hair cells	(Song et al. 2003)
Whirlin	$Ca_v1.3$	Retina	Scaffolds $Ca_v1.3$ and organizes thereby the channel in photoreceptor cells	*Mutations cause Usher-syndrome, Whirlin is also expressed in cochlea* (Kersten et al. 2010)

Only those protein interactions are listed, where co-localization in immunohistochemistry and/or co-immunoprecipitation in the corresponding tissue and/or functional evidence, has been shown. Abbreviations: LTCC, L-type Ca^{2+} channels; I_{Ca}, Ca^{2+} currents; CaBPs, Ca^{2+} binding proteins; I_{Ba}, barium currents; VGCCs, voltage gated Ca^{2+} channels; RIMs, rab3 interacting molecules; VDI, voltage-dependent inactivation; CDI, Ca^{2+} -dependent inactivation

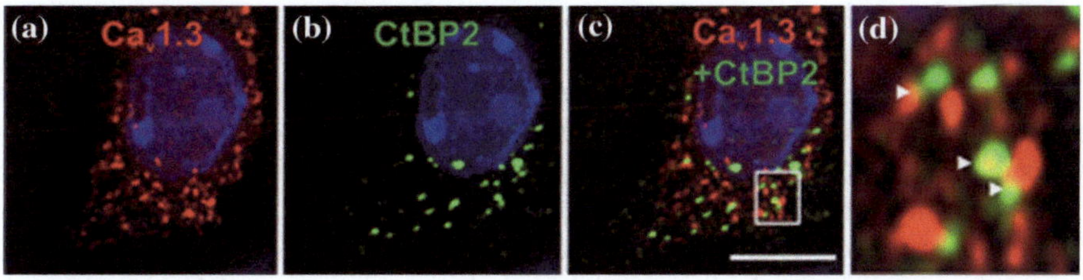

Fig. 3.3 Co-localization of $Ca_v1.3$ α1-subunit and CtBP2/Ribeye (a ribbon marker) in cochlear inner hair cells. Double staining of mouse cochlear sections (P6) with anti-Ca_v1-3 α1 and anti-CtBP2/Ribeye antibodies. **a** distribution of $Ca_v1.3$ α1-subunit (*red*), and **b** distribution of CtBP2 (*green*), on basal pole of inner hair cells, the merge in (**c**) shows partial co-localization of both proteins, **d** magnification of the box in (**c**), arrows highlighting co-localization of $Ca_v1.3$ and CtBP2 (modified from (Gebhart et al. 2010) with permission)

β-subunit of the channel complex in tsA-201 cells, slowed both CDI and VDI and thereby stabilized a larger fraction of a non-inactivating current component (Gebhart et al. 2010). Interestingly, this effect of RIM was not seen when β2a subunits were part of the channel complex (Gebhart et al. 2010). Notably, the β2a splice variant is palmitoylated at its N-terminus (Gebhart et al. 2010). In contrast to non-palmitoylated β-subunits, such as β1, β3 and β4, β2a decelerates inactivation kinetics of Ca_v1 and Ca_v2 channels which requires its palmitoylation. Although β2 is the predominant β-subunit in IHCs (Kuhn et al. 2009; Neef et al. 2009) the contribution of the palmitoylated splice variant is unfortunately unknown. Taken together RIM2 contributes to but does not fully explain the slow $Ca_v1.3$ inactivation kinetics in IHCs. The search for other protein interaction partners that slow VDI must therefore continue.

Harmonin is a scaffolding protein that is required for normal hair cell function. It has recently been found in about half of the ribbon synapses in mature IHCs. Harmonin can bind to $Ca_v1.3$ channels through the channels' C-terminal PDZ-binding motif (Fig. 3.2; Gregory et al. 2011). In deaf mice expressing a binding-deficient harmonin mutant the Ca^{2+} current in mature IHCs was slightly enhanced (33 %) as were presynaptic Ca^{2+} transients (although with large variability (Gregory et al. 2011)). Upon coexpression with $Ca_v1.3$ channels in HEK293 cells harmonin increased channel ubiquitinylation and reduced cell surface expression. Therefore, reduction of channel availability through an ubiquitin-dependent pathway by harmonin appeared as a plausible mechanism. However, it is unclear if processing of overexpressed $Ca_v1.3$ channels in HEK293 cells also reflects the situation in neurons or IHCs. Recombinant expression caused significant accumulation of channels in intracellular compartments which may be prone to artificially enhanced ubiquitinylation and proteasomal degradation. Moreover, only high molecular mass species of ubiquitinylated $Ca_v1.3$ α1 subunits were immunoprecipitated with $Ca_v1.3$ antibodies from transfected cells without evidence for non-ubiquitinylated species (Gregory et al. 2011). This is in contrast to the finding that heterologously expressed α1-subunits usually migrate very close to their predicted molecular mass (Koschak et al. 2003).

Ribeye is a self-interacting protein (Magupalli et al. 2008) that plays an essential role in ribbon synapse formation and is highly specific for ribbon synapses (Sheets et al. 2011). Its role for synaptogenesis has been studied in zebrafish by suppressing (knockdown) or enhancing (overexpression) its expression during development (Sheets et al. 2011). This approach revealed two distinct functions of Ribeye for synaptogenesis: it is crucial for inducing clustering of $Ca_v1.3$ channels at the presynapse and also for stabilizing contacts with afferent neurons.

3.2.3.3 RNA Transcript Diversity: RNA-Editing and Alternative Splicing

Another mechanism reported to contribute to inhibition of CDI is RNA-editing. Adenosine-to-inosine (A-to-I) editing is a site-selective posttranscriptional modification of double-stranded RNA by ADAR deaminases (Danecek et al. 2012). The transcriptional machinery interprets inosine as guanosine leading to A-to-G mismatches in cDNA sequences. In coding regions this can cause changes of protein sequence and increased transcript diversity (Danecek et al. 2012). RNA editing has been found to occur in the $Ca_v1.3$ C-terminal tail within the calmodulin binding domain (termed the IQ-domain based on the crucial amino acids isoleucine and glutamine (Van Petegem et al. 2005)) in an ADAR-dependent manner (Huang et al. 2012) (Fig. 3.2). Edited $Ca_v1.3$ channels exhibited reduced affinity for CaM and consequently a strong reduction of CDI (Huang et al. 2012). $Ca_v1.3$ RNA editing has so far only been demonstrated in brain and was absent in samples of rat cochlea. However, mRNA specifically isolated from IHCs or OHCs has not been analyzed so far, nor have potential age-dependent effects been studied (i.e., RNA-editing pre- and post hearing).

Alternative splicing of the long C-terminal tail has also been found to generate $Ca_v1.3$ species differing with respect to CDI. Transcripts completely lacking the IQ-domain were found in the cochlea but are believed to be expressed mainly in OHCs (Shen et al. 2006). These transcripts encode channels completely devoid of CDI (Shen et al. 2006). In contrast, other splice variants exist which are expressed in IHCs (unpublished observations) and stabilize CDI of different strengths due to the existence of an intramolecular protein–protein interaction in which a putative negatively charged α-helix in the distal C-terminus (DCRD) interacts with a putative positively charged α-helix in the proximal C-terminus (PCRD). This C-terminal modulatory domain (termed CTM) has been originally discovered in $Ca_v1.4$ channels (see below) but also exists in $Ca_v1.3$. Its structural and modulatory features have been recently reviewed (Koschak 2010; Striessnig et al. 2010).

In $Ca_v1.3$ channels full length α1-subunits ($Ca_v1.3_L$) contain both the PCRD and DCRD domain (the latter is encoded by the last exon 49) (Fig. 3.2) and allows the two domains to interact with each other as demonstrated in FRET studies (Singh et al. 2008). Interestingly, structural diversity is generated by extensive alternative splicing within the C-terminus generating short transcripts yielding either α1-subunits missing both domains (when truncated at exon 42 (Ca$_v$1.3$_{42A}$) or containing only PCRD (when truncated at exon 43 (Ca$_v$1.3$_{43S}$) (Bock et al. 2011; Singh et al. 2008)) (Fig. 3.2). In both cases the modulatory function of the CTM was lost. The functional CTM in the long variant has profound effects on channel function as evident from heterologous expression experiments: it strongly decreased the open probability of the channel, caused a further shift of its activation voltage range to more positive voltages, strongly (but not completely) inhibited CDI and slightly accelerated VDI (Bock et al. 2011; Lieb et al. 2012; Singh et al. 2008). Moreover, only the long splice variant also contains a C-terminal PDZ-motif which, in addition to harmonin, can bind a variety of PDZ-binding proteins (Table 3.1, Fig. 3.2) that modulate channel function and/or tether the channel to defined signaling pathways (such as, e.g., M1R and D2R receptors at corticostriatal synapses (Olson et al. 2005). The relative abundance of both short and long variants seems to be tissue-specific (Bock et al. 2011) indicating fine-tuning of channel function by alternative splicing. Quantitative mRNA analysis predicted expression of both the long and the short Ca$_v$1.3$_{43S}$ variant at roughly equal abundance in the brain whereas the long variant predominated in the heart (Bock et al. 2011). In the cochlea long and short variants have been detected by qualitative rt-PCR but due to the overall very low abundance of Ca$_v$1.3 α1-subunit mRNA their relative abundance is still unknown. The generation of a mutant mouse model in which the DCRD has been replaced by an HA-antibody tag thereby disrupting CTM function without truncating the C-terminus is underway (unpublished). This model should be useful to unequivocally detect the long variant in different tissues and reveal the functional significance of the CTM.

Taken together RNA-editing and C-terminal alternative splicing have profound effects on Ca$_v$1.3 channel function and may explain the slow CDI observed in IHCs. However, the slow VDI cannot be explained by these mechanisms suggesting protein interactions as a more likely mechanism as discussed above.

3.2.4 Ca$_v$1.3-Related Diseases in Humans

3.2.4.1 SANDD

All of our knowledge about the molecular and physiological properties of Ca$_v$1.3 comes from studies in rodent tissues, Ca$_v$1.3-deficient mice or heterologous expression studies with rat and human Ca$_v$1.3 channels. However, the recent discovery of a rare human disease associated with the loss of Ca$_v$1.3 function strongly indicated that Ca$_v$1.3 serves similar functions in humans and mice (Baig et al. 2011). Using positional cloning in two consanguineous families with deafness a mutation was identified in the Ca$_v$1.3 α1-subunit gene (CACNA1D). All deaf subjects also showed pronounced SAN dysfunction at rest, very similar to the SAN dysfunction found in mice (Platzer et al. 2000). This syndrome was therefore termed SANDD,

Sinoatrial Node Dysfunction and Deafness (OMIM: #614896). The mutation introduced in-frame an additional glycine residue within a highly conserved, alternatively spliced region near the channel pore (Fig. 3.2). Patch-clamp studies revealed that the mutated α1-subunit protein was still synthesized and integrated into the plasma membrane of tsA-201 cells (Baig et al. 2011) because non-linear capacitative currents (so-called ON gating currents) resulting from intramembrane movement of the channel's positively charged voltage sensors could still be recorded. However, this gating current was not accompanied by the appearance of inward ionic current indicating a complete loss of channel function. The disease also provided another interesting biochemical detail because the mutation was localized within the alternatively spliced exon 8. This allows the conclusion that the mutation-containing exon 8B is the predominant variant in human IHCs and SAN. This finding was confirmed in mouse tissues (Baig et al. 2011).

3.2.4.2 Parkinsons Disease

$Ca_V1.3$ channels are currently also discussed as a molecular target for neuroprotection in Parkinsons Disease (PD). Evidence is based on the finding that LTCC are responsible for the dendritic Ca^{2+} transients in dopaminergic substantia nigra pars compacta neurons. These transients are triggered during the spontaneous action potentials of these spontaneously firing neurons. Some indication exists that $Ca_V1.3$ channels may be mainly responsible for permanent Ca^{2+} load that could underlie the selective susceptibility of these neurons to neurodegeneration (Pfeiffer 2010). In animal models of PD, DHP Ca^{2+} channel blockers are neuroprotective (Surmeier et al. 2011). Together these finding provide a strong rationale to develop $Ca_V1.3$ selective blockers as neuroprotective agents. Drug discovery efforts are currently ongoing in pharmaceutical industry. However, one of the big questions is how save these drugs would be based on the crucial role of these channels for cochlear function and SAN pacemaking. It is reassuring that hearing impairment has so far not been reported as a symptom during intoxications with extremely high doses of brain permeable DHPs. Perhaps the unusual gating properties provide IHC Ca^{2+} channels also with pharmacological properties that would finally allow also some selectivity of blockers for brain $Ca_V1.3$ channels.

3.3 L-Type Channels in the Retina

Photoreceptors are morphologically and physiologically highly specialized light sensing cells of the retina. In darkness their membrane potential is depolarized up to −36 mV (Bader et al. 1978) due to the activity of cGMP activated cation channels (Karpen et al. 1992; Korenbrot 2012; Miller and Nicol 1979). These receptor cells signal to second order neurons via constant release of glutamate form their ribbon synapses (tom Dieck and Brandstätter 2006). This sustained release is Ca^{2+} dependent (Heidelberger and Matthews 1994; Thoreson et al. 2004) and LTCCs serve as the predominant source for Ca^{2+} entry (Bartoletti et al. 2011; Mercer and Thoreson 2011). $Ca_v1.4$ channels—which are encoded by the CACNA1F gene—are located in close vicinity to the typical horseshoe shaped ribbon synapses as has been demonstrated by their colocalization with synaptic proteins such as Bassoon, Ribeye, and Piccolo in mouse and rat retinas (Busquet et al. 2010; Mercer et al. 2011; Mercer and Thoreson 2011; Morgans 2001; Morgans et al. 2001; Specht et al. 2009). In cultured mammalian cells, heterologously expressed $Ca_v1.4$ currents activate rapidly, open at negative membrane potentials and thereby allow the channel to conduct Ca^{2+} at potentials negative to −40 mV. Only a minor fraction ($\approx$10–15 %) of channels might be available at a potential −35 mV (Hoda et al. 2006) but sufficient for triggering neurotransmitter release and tight photoreceptor tuning. $Ca_v1.4$ currents also show slow VDI accompanied by complete absence of CDI during depolarizing pulses (Hoda et al. 2006; Koschak et al. 2003). Inactivation was faster when measured at near physiological temperatures but the window current was still preserved (Peloquin et al. 2008). Data from beta2a and alpha2delta4 knock-out mice (Ball et al. 2002; Wycisk et al. 2006) inferred that $Ca_v1.4$ channels operate most likely in a complex with beta2a and alpha2delta4 as accessory subunits. Intriguingly only 2–3 channel openings are needed on average to initiate a vesicle fusion in bright light while around 1,000 LTCCs were estimated per cone terminal by Bartoletti and colleagues (Bartoletti et al. 2011). The tight coupling of Ca^{2+} channel opening and fusion events under these conditions should ensure precise timing of release which is crucial for synchronizing synaptic signaling to decrements in light intensity.

Under low light conditions the rod pathway is activated and the released glutamate binds to the metabotropic glutamate receptor mGluR6 on the postsynaptic rod bipolar cells (RBC) (Masu et al. 1995). Receptor activation in turn leads to closure of non-selective cation channels, most probably TRPM1 (Koike et al. 2010), and hyperpolarization of the RBC cell membrane. This hyperpolarization closes LTCCs in RBCs and decreases glutamate release. $Ca_v1.4$ immunostaining in mouse inner plexiform indicated that $Ca_v1.4$ LTCCs may contribute to the regulation of release at bipolar cell terminals (Berntson et al. 2003; Busquet et al. 2010; Mansergh et al. 2005; Morgans 2001). Only few RBCs seem to contact ganglion cells directly but use amacrine cells as signal relays (Strettoi et al. 1990). As shown in Fig. 3.4, glutamate release from RBCs activates AII and A17 (and probably A13, both not indicated in the figure for clarity) amacrine cells (Bloomfield 2001). A17 cells modulate the synaptic signal via a direct GABAergic feedback inhibition onto the RBC (Grimes et al. 2010). Measurements of Ca^{2+} currents in these cells revealed L-type currents with no or very slow inactivation (Grimes et al. 2009), pointing to a contribution of $Ca_v1.4$ (but maybe also $Ca_v1.3$) channels. In contrast to A17, AII cells form gap junctions to other AII cells as well as to ON cone bipolar cell axon terminals (Bloomfield and Dacheux 2001). Activation of AII cells then leads to activation of ON ganglion cells via transmitter release from ON cone bipolar cells (Demb and Singer 2012). At the same time AII amacrine cells form glycinergic synapses with OFF cone bipolar cell axon terminals and OFF ganglion cells and

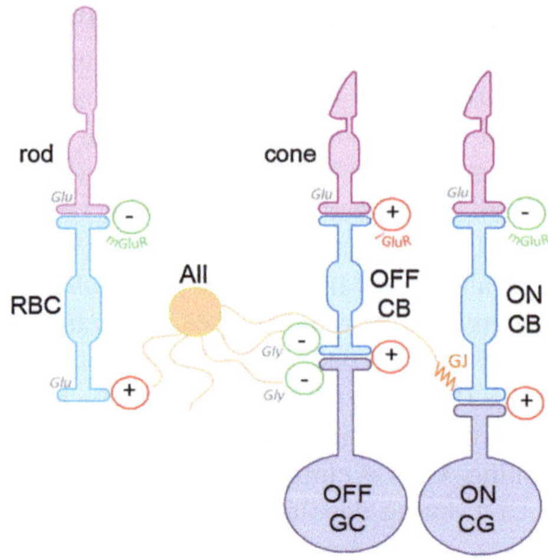

Fig. 3.4 Simplified rod photoreceptor signaling pathways. In the absence of light rod photoreceptors constantly release glutamate (Glu), which binds to metabotropic glutamate receptors (mGluR) on rod bipolar cells (RBC) resulting in a hyperpolarization and downstream a reduction of transmitter release. When rod photoreceptors are excited Glu release from the presynapse is diminished and thus RBCs are depolarized, leading to an increase in Glu release. RBCs form depolarizing contacts with AII and A17 amacrine cells (the latter mediate a feedback inhibition but are omitted in the figure for clarity). AII amacrine cells form synapses with different neurons: i. inhibiting glycinergic synapses (Gly) with OFF cone bipolar cells (OFF CB (Bloomfield and Dacheux 2001) or OFF ganglion cells (OFF GC; Arman and Sampath 2012), ii. gap junctions (GJ) with other AII cells (not shown) or ON cone bipolar cells (ON CB), thus activating these cells. + excitatory synapse (*red*), – inhibitory synapse (*green*). Also cones release glutamate and can—via ionotropic glutamate receptors (iGluR)—form excitatory synapses with OFF CB. Horizontal cells and Müller glial cells are also omitted for clarity. Adapted from Figures in Demb and Pugh (2002); Kim and Sheng (2004); Seeliger et al. (2011); Wässle (2004)

therefore actively inhibit the OFF pathway (Fig. 3.4). In addition to the RBC pathway, also direct coupling between rod and cone photoreceptors via gap junctions has been demonstrated (Seeliger et al. 2011). This pathway is used in twilight conditions (Soucy et al. 1998; Tsukamoto et al. 2001). An additional OFF pathway that has so far only been described in rodents shows direct coupling from rods to cone OFF bipolar cells (Soucy et al. 1998).

3.3.1 Ca$_v$1.4-Related Diseases in Humans

The importance of Ca$_v$1.4 channels as the most predominant LTCC in retinal neurons is well supported by the fact that mutations in the CACNA1F gene can cause several forms of human retinal diseases (OMIM: 300071, 300476, 300600), as such Åland Island Eye Disease (AIED) (Jalkanen et al. 2007), Cone-Rod

Dystrophy (CORDX3, Jalkanen et al. 2006), X-linked retinal disorder (XRD) (Hope et al. 2005), night blindness-associated Transient Tonic Downgaze (NATTD, Simonsz et al. 2009), and incomplete Congenital Stationary Night Blindness (iCSNB, CSNB2, Bech-Hansen et al. 1998; Boycott et al. 1998, 2000, 2001; Strom et al. 1998) which share a variety of clinical symptoms (for review see Stockner and Koschak 2013). Due to the X-linked condition of $Ca_v1.4$ channel dysfunction CSNB2 mainly involves males. However, also heterozygote females can be affected (Hemara-Wahanui et al. 2005; Hope et al. 2005) in case they show cellular mosaicism for healthy/mutant $Ca_v1.4$ channels due to X-inactivation of the *CACNA1F* gene. The majority of mutations were identified among patients originally diagnosed with CSNB2. Typical symptoms of CSNB2 are moderately low visual acuity, myopia, nystagmus and night blindness but one or more of these symptoms may be absent (Boycott et al. 2000). The severity of night blindness as a symptom varies and in some cases was not even reported. CSNB2 is therefore generally diagnosed on the basis of ERG abnormalities. In patients with CSNB2, dark-adapted Ganzfeld ERGs showed reduced scotopic (very little ambient light) rod b-waves and a negative bright flash ERG. This means that a large a-wave—reflecting the general health of the photoreceptors—is followed by a small b-wave—reflecting (intact) transmission from photoreceptors to bipolar cells—with a peak below ERG baseline. Oscillatory potentials can be recorded but may show abnormal implicit times (Boycott et al. 2000; Strom et al. 1998). In the light adapted ERG the cone b-wave is also much reduced (with delay of peak times, Tremblay et al. 1995). Under these photopic conditions no oscillatory potentials were elicited, and the 30 Hz flicker is reduced (Strom et al. 1998). This ERG phenotype is compatible with a defect in neurotransmission within the retina between photoreceptors and second-order neurons.

3.3.2 Human $Ca_v1.4$ Channel Mutations

Whereas only one human disease-related CACNA1D mutation has been published so far (Baig et al. 2011), more than 50 structural aberrations were identified in the *CACNA1F* gene of CSNB2 patients comprising $Ca_v1.4$ $\alpha1$-subunit missense or truncation mutations as well as insertions or deletion (Stockner and Koschak 2013). $Ca_v1.4$ mutations can be categorized on the basis of their functional effects as (i) loss-of-function (ii) impairment of the C-terminal modulator (CTM) function or (iii) gain-of-function. Here we briefly summarize $Ca_v1.4$ mutations, that have already been functionally characterized, under these criteria; the underlying structure-(dys-) function relation, however, has already been discussed elsewhere (Stockner and Koschak 2013).

The majority of mutations are predicted to form non-functional channels, often because of severe structural changes incompatible with channel function, like premature truncations. Some predicted truncated $Ca_v1.4$ channels might therefore not even be expressed on the protein level because nonsense-mediated mRNA decay eliminates mRNA containing premature stop codons in regions followed by splice sites at a distance of 50–55 nucleotides downstream (Maquat 2004). No channel activity was observed for missense mutations Ser229Pro, Gly1018Arg, Arg1060Trp, and Leu1079Pro expressed in Xenopus laevis oocytes (Hoda et al. 2005) or tsA-201 cells (Peloquin et al. 2007) although their $\alpha1$ subunit proteins were expressed at levels indistinguishable from wild type channels. Interestingly in the Leu1079Pro mutant channel current was 'rescued' in the presence of the Ca^{2+} channel activator BayK8644, although showing slightly faster inactivation kinetics compared to wild type in the presence of the activator (Hoda et al. 2005). Temperature-dependent reduced protein expression was reported in mutants Arg519Gln and Leu1375His whereas their gating properties were not (Arg519Gln) or only slightly (Leu1375His) affected (Hoda et al. 2006). Likewise no gating changes were detected for Gly674Asp and Ala928Asp (McRory et al. 2004).

The analysis of the truncation mutation Lys1602stop (named K1591X in references Hoda et al. 2006; Wahl-Schott et al. 2006) uncovered that the lack of CDI—which might be 'the' discriminating functional hallmark of $Ca_v1.4$ channel—is an intrinsic channel property that depends on the active suppression by a C-terminal inhibitory domain. How C-terminal tailoring affects LTCC function is briefly summarized in Sect. 3.2.3.3 (for review also see Striessnig et al. 2010; Striessnig 2007). In contrast to $Ca_v1.3$, the DCRD domain of $Ca_v1.4$ has even higher affinity for its upstream binding domain and therefore completely eliminates CDI (Hoda et al. 2006; Liu et al. 2010). As a consequence of this modulation, truncation mutations in the $Ca_v1.4$ C-terminus downstream of the CDI machinery (i.e., removing the modulatory domain)—either natural, by alternative splicing (Tan et al. 2012) or introduced by a disease mutation (Hoda et al. 2006; Wahl-Schott et al. 2006) are of particular interest because they undergo a complex functional change, including (i) a hyperpolarizing shift of the $Ca_v1.4$ mediated window current (compatible with a gain-of-function); and (ii) reduced Ca^{2+} influx due to occurrence of CDI (compatible with a loss-of-function, Hoda et al. 2006; Wahl-Schott et al. 2006). The net effect on the strength of Ca^{2+} signaling and retinal function by such mutations in vivo is therefore difficult to predict and requires generation of mouse mutants carrying these human mutations.

Gain-of-function mutations confer new or enhanced activity. Apparently such enhanced activity must have an unwarranted positive connotation because increased sensitivity in gain-of-function mutations does not necessarily result in improved signaling. Instead, this might result in a loss-of-control of existing Ca^{2+} signaling pathways. In the heterologous expression system, many $Ca_v1.4$ gain-of-function mutations (Gly369Arg: Hoda et al. 2005; McRory et al. 2004, Ile756Thr: Hemara-Wahanui et al. 2005 and Phe753Cys: Peloquin et al. 2007) promote enhanced Ca^{2+} entry through the channel by vastly slowed VDI which is accompanied by a strong hyperpolarizing shift in the voltage-dependence of activation. The so far most pronounced hyperpolarizing shift in $Ca_v1.4$ channel activation (around 30 mV) was found in mutation Ile756Thr, which was identified in a New Zealand family. The affected family members showed an unusual severity of the phenotype, associated with intellectual disability in males, and also heterozygote female family members had clinical and ERG abnormalities (Hemara-Wahanui et al. 2005). Currently, we do not entirely understand how a gain-of-function in the channel protein would result in a loss-of-function in retinal synaptic transmission in patients showing CSNB2 symptoms. During illumination a $Ca_v1.4$ gain-of-function could result in significant Ca^{2+} influx already at negative voltages (~ -55 mV physiologically) compared to a minor level of activation in wild type (described also above). This would mean that for a light pulse of a given intensity the dynamic range for fractional channel opening between dark and light conditions will be less pronounced compared to wild type. Obviously, this effect would also induce a change in glutamate release. Since transmitter release has been found to depend in a linear fashion on the intracellular Ca^{2+} of photoreceptors (Thoreson et al. 2004) one could envisage the following consequences for retinal network activity function in a thought experiment: in the rod pathway a reduction in the difference in transmission between darkness and light would mean less RBC depolarization at a given light intensity. AII amacrine cells might also be activated to a lesser degree and ON ganglion cell output in response to the stimulus will be decreased. At the same time the inhibition of the OFF pathway (via glycinergic synapses with OFF cone bipolar cell axon terminals and OFF ganglion cells, Fig. 3.4) will be suppressed. Overall, these effects would lead to a decrease in contrast sensitivity, a visual phenotype often observed in patients with congenital stationary night blindness (Bodis-Wollner 1980; Marmor 1986). Assuming that $Ca_v1.4$ channels are also expressed in second order neurons the matter becomes even much more complicated. In bipolar cells transmitter release has been found to depend in a highly nonlinear way on the intracellular Ca^{2+} concentration (Heidelberger et al. 1994). So a $Ca_v1.4$ gain-of-function might even offset the reduced depolarization at a given light intensity and lead to increased transmitter release and thus AII amacrine cell activity, ultimately leading to higher ON ganglion cell output. Due to the nonlinear nature of Ca^{2+}-dependent release, saturation might be reached quickly and thus also this effect could lead to a loss in contrast sensitivity.

The above scenario assumes merely functional changes but does not take into account the possibility that altered Ca^{2+} signaling by gain-of-function mutations could change retinal morphology (such as defects in early stages of synapse formation/maturation or cell death) resulting in an overall loss-of-function effect in retinal transmission. So far no reports have been published that directly test and corroborate these hypotheses in an in vivo gain-of-function mouse model.

3.3.3 $Ca_v1.4$ Rodent Models

Existing knock-out mouse models, however, indicate the important role of functional $Ca_v1.4$ channels for retinal health. In $Ca_v1.4^{-/-}$ mice, the b-wave signal is absent under scotopic conditions. At higher stimulus intensities the a-wave appeared with a threshold comparable to the wild type, but neither a b-wave nor oscillatory potentials were detected. ERG recordings obtained under photopic conditions to isolate cone response did not detect any reliable response in $Ca_v1.4^{-/-}$ mice (Mansergh et al. 2005). This finding is in accordance with the clinical status in CSNB2 patients (Boycott et al. 2000; Hope et al. 2005; Strom et al. 1998). Moreover, also visual evoked potentials which record activation in the cortex could not be detected while cortical potentials were identified in WT mice (Mansergh et al. 2005). The cacna1f^{nob2} (nob2) mouse is reported as a naturally occurring 'null' mutation for the $Ca_v1.4$ Ca^{2+} channel gene but the phenotype of this mouse is not identical to that of the targeted gene knockout model because 10 % of the mRNA is intact due to alternative splicing (Doering et al. 2008). Still, the reduction in the ERG b-wave and the reduced oscillatory potential amplitudes are pointing to the prominent role of $Ca_v1.4$ channels. Immunocytochemistry data showed unaffected photoreceptor cell bodies and inner and outer segments as well as unaffected retinal inner layers judged by normal amacrine cell labeling. However, profound loss of photoreceptor synapses, the presence of ectopic synapses deep into the outer nuclear layer and abnormal dendritic sprouting of second order neurons was detected in $Ca_v1.4^{-/-}$ and nob2 mice (Chang et al. 2006; Mansergh et al. 2005) as well as in a CSNB rat model (Gu et al. 2008; Zheng et al. 2012). Antibody staining experiments that particularly target either rod or cone synapses have not been reported so far. In histological sections outer plexiform layers (OPLs) appeared thinner than normal (Mansergh et al. 2005), a finding that also points to a distorted morphology of $Ca_v1.4$ deficient retinas. The $Ca_v1.4$ channel loss-of-function due to either (much) reduced or completely abolished Ca^{2+} influx is clearly compatible with defective retinal transmission and one could assume that a $Ca_v1.4$-mediated Ca^{2+} signal is also critical for synapse formation and/or maturation. Depending on the extent of synapse loss this could result in longer ganglion cell response latencies and therefore delayed/absent transfer of visual information to the brain.

3.3.4 $Ca_v1.4$ Channel Protein Interactions

As described above Ca^{2+} binding proteins tune $Ca_v1.3$ channels in the auditory system. They are also important regulators in the retina and mutations in Ca^{2+}-binding protein 4 (CaBP4) also cause CSNB2 (CSNB2B; OMIM 610427, Zeitz et al. 2006). Like for $Ca_v1.4$, CaBP4 is located in photoreceptor synaptic terminals, where it is directly associated with the C-terminal domain of $Ca_v1.4$ $\alpha1$ subunit. In transfected human embryonic kidney cells CaBP4 shifted the activation of $Ca_v1.4$ to hyperpolarized voltages. Mice lacking CaBP4 showed a thinner outer plexiform layer (Haeseleer et al. 2004). Both morphological and functional alterations observed were similar to $Ca_v1.4^{-/-}$ mice (Mansergh et al. 2005). Current responses of rod photoreceptor outer segments in suction electrode recordings showed faster and less sensitive responses upon light flashes in $CaBP4^{-/-}$ rods. Downstream, light flashes applied to in retinal slice preparations also showed reduced responses. These differences in rod signaling were confirmed in ERG responses (Haeseleer et al. 2004), and in addition also indicated a reduction in cone synaptic function (Maeda et al. 2005). A recent report suggested that CaBP4 increased $Ca_v1.4$ channel availability by relieving the inhibitory effects of the CTM on voltage-dependent $Ca_v1.4$ channel gating. For two CaBP4 mutants this functional effect was only partially preserved, leading to a reduction of $Ca_v1.4$ channel availability and loss-of-function (Shaltiel et al. 2012).

3.3.5 $Ca_v1.3$ and $Ca_v1.2$ Channels in the Retina

Data that emphasize the role of the LTCC subtypes different from $Ca_v1.4$ come only from studies in the non-human retina. All LTCCs have been reported to be expressed in the retina. Expression of $Ca_v1.3$ was reported in multiple retinal cells types and layers in various animal species (Busquet et al. 2010; Cristofanilli et al. 2007; Firth et al. 2001; Henderson et al. 2000; Kersten et al. 2010; Mizuno et al. 2010; Morgans 1999; Taylor and Morgans 1998; Welch et al. 2005; Xiao et al. 2007). In particular, $Ca_v1.3$ expression has been proposed in mouse cone terminals (Morgans et al. 2005). However, this finding is controversial (Xiao et al. 2007). Interestingly, $Ca_v1.3^{-/-}$ mice show a minor reduction in the b-wave (Busquet et al. 2010) so that a contribution of $Ca_v1.3$ channels on signal transmission in the outer retina cannot be completely excluded. A potential regulatory role for $Ca_v1.3$ in the retina is described in the retinal pigment epithelium (RPE) where these channels are proposed to regulate the light peak of the ERG which is caused by a depolarization of the RPE basolateral plasma membrane (Wu et al. 2007). In the cell membrane of the RPE, $Ca_v1.3$ subunits co-localize with bestrophin-1 (Uniprot O76090) which is the product of a gene mutated in Best's vitelliforme macular degeneration (OMIM 607854). Bestrophin-1 binding to β4-subunits via a proline-rich motif/SH3 interaction is suggested to accelerate time-dependent activation and decreased the activity of $Ca_v1.3$ channels and thereby might explain the changes in the light-peak amplitude observed in $Ca_v1.3^{-/-}$ mice. Kersten et al. (2010) recently demonstrated that $Ca_v1.3$ co-localizes with the Usher Syndrome associated protein Whirlin (Uniprot Q9P202) at distinct sites in the photoreceptor including the outer limiting membrane, the OPL, and the region of the connecting cilium. Already during murine development the expression patterns of $Ca_v1.3$ and Whirlin overlap significantly, not only in the eye but also in the inner ear and the central nervous system compatible with this most common form of human hereditary deaf–blindness with autosomal recessive inheritance. A PDZ-domain interaction was identified between the Whirlin N-terminus and the $Ca_v1.3$ C-terminus through which $Ca_v1.3$ is suggested to be connected to the Usher protein network. The complex, including also a calmodulin-dependent serine kinase (CASK), may participate in the regulation of neurotransmission via the organization of $Ca_v1.3$ channels in the photoreceptor cell synapses. Moreover, Mizuno and colleagues reported that $Ca_v1.3$ channels are subject to rapid glutamate-induced internalization, which may serve as a negative feedback mechanism protecting retinal neurons against glutamate-induced excitotoxicity (Mizuno et al. 2010) implying a role of $Ca_v1.3$ in other processes than retinal transmission.

Expression of $Ca_v1.2$ in the mouse retina has recently been reported to be pronounced in the inner plexiform as well as the ganglion cell layer and weak in the OPL (Busquet et al. 2010; Specht et al. 2009). In cultured chicken cone photoreceptors the expression of LTCCs is under circadian control showing nocturnal increases (Ko et al. 2004). MicroRNAs (miRNAs) influence the core oscillator in the mouse master clock, the suprachiasmatic nucleus, and also several miRNAs have been published in the mouse retina, some of which are highly expressed in the photoreceptor layer (Karali et al. 2010, 2007). In the chicken retina, Shi et al. (2009a, b) studied the regulation of $Ca_v1.2$ $\alpha 1$ subunits by microRNA-26a and found decreased protein expression and current densities after co-expression in vitro. Moreover, several signaling pathways as well as protein–Protein interactions were reported to be involved in the regulation of rhythmic expression. Ko and colleagues showed that both PI3 K–Akt and Erk signaling as well as an interaction with retinoschisin contributed to the circadian phase-dependent regulation of channel subunit trafficking (Ko et al. 2008; Shi et al. 2009a, b (interaction with $Ca_v1.3$ N-terminus)). Mutations in the retinoschisin gene RS1 cause X-linked retinoschisis (OMIM 312700), a retinal dystrophy that features disorganization of retinal cell layers, disruption of the synaptic structures and neurotransmission between photoreceptors and bipolar cells, and progressive degeneration of rod and cone photoreceptor cells.

Skeletal muscle subunit $Ca_v1.1$ $\alpha 1$ ($\alpha 1S$) staining has been reported in mouse retinas to be present postsynaptically at ON-bipolar cell dendrites where it co-localized with metabotropic glutamate receptor 6 (Specht and Sweadner 1984). This finding is somewhat unexpected and the observed immunoreactivity might still have to be judged with care due to the lack of a $Ca_v1.1$ knock-out control retina staining.

References

Almanza A, Vega R, Soto E (2003) Calcium current in type I hair cells isolated from the semicircular canal crista ampullaris of the rat. Brain Res 994:175–180

Arman AC, Sampath AP (2012) Dark-adapted response threshold of OFF ganglion cells is not set by OFF bipolar cells in the mouse retina. J Neurophysiol 107:2649–2659

Astori S, Wimmer RD, Prosser HM et al (2011) The Ca(V)3.3 calcium channel is the major sleep spindle pacemaker in thalamus. Proc Natl Acad Sci USA 108:13823–13828

Bader CR, MacLeish PR, Schwartz EA (1978) Responses to light of solitary rod photoreceptors isolated from tiger salamander retina. Proc Natl Acad Sci USA 75:3507–3511

Baig SM, Koschak A, Lieb A et al (2011) Loss of Ca(v)1.3 (CACNA1D) function in a human channelopathy with bradycardia and congenital deafness. Nat Neurosci 14:77–84

Ball SL, Powers PA, Shin H-S et al (2002) Role of the beta(2) subunit of voltage-dependent calcium channels in the retinal outer plexiform layer. Invest Ophthalmol Vis Sci 43:1595–1603

Bao H, Wong WH, Goldberg JM, Eatock RA (2003) Voltage-gated calcium channel currents in type I and type II hair cells isolated from the rat crista. J Neurophysiol 90:155–164

Bartoletti TM, Jackman SL, Babai N et al (2011) Release from the cone ribbon synapse under bright light conditions can be controlled by the opening of only a few $Ca^{(2+)}$ channels. J Neurophysiol 106:2922–2935

Bauer CS, Tran-Van-Minh A, Kadurin I, Dolphin AC (2010) A new look at calcium channel $\alpha 2\delta$ subunits. Curr Opin Neurobiol 20:563–571

Bech-Hansen NT, Naylor MJ, Maybaum TA et al (1998) Loss-of-function mutations in a calcium-channel alpha1-subunit gene in Xp11.23 cause incomplete X-linked congenital stationary night blindness. Nat Genet 19:264–267

Berntson A, Taylor WR, Morgans CW (2003) Molecular identity, synaptic localization, and physiology of calcium channels in retinal bipolar cells. J Neurosci Res 71:146–151

Beutner D, Moser T (2001) The presynaptic function of mouse cochlear inner hair cells during development of hearing. J Neurosci 21:4593–4599

Bloomfield S a, Dacheux RF (2001) Rod vision: pathways and processing in the mammalian retina. Prog Retin Eye Res 20:351–384

Bloomfield SA (2001) Plasticity of AII amacrine cell circuitry in the mammalian retina. Prog Brain Res 131:185–200

Bock G, Gebhart M, Scharinger A et al (2011) Functional properties of a newly identified C-terminal splice variant of $Ca_v1.3$ L-type Ca^{2+} channels. J Biol Chem 286:42736–42748

Bodis-Wollner I (1980) Detection of visual defects using the contrast sensitivity function. Int Ophthalmol Clin 20:135–153

Boycott KM, Maybaum TA, Naylor MJ et al (2001) A summary of 20 CACNA1F mutations identified in 36 families with incomplete X-linked congenital stationary night blindness, and characterization of splice variants. Hum Genet 108:91–97

Boycott KM, Pearce WG, Bech-Hansen NT (2000) Clinical variability among patients with incomplete X-linked congenital stationary night blindness and a founder mutation in CACNA1F. Can J Ophthalmol 35:204–213

Boycott KM, Pearce WG, Musarella MA et al (1998) Evidence for genetic heterogeneity in X-linked congenital stationary night blindness. Am J Hum Genet 62:865–875

Boyer C, Lehouelleur J, Sans A (1998) Potassium depolarization of mammalian vestibular sensory cells increases $[Ca^{2+}]$i through voltage-sensitive calcium channels. Eur J Neurosci 10:971–975

Brandt A, Striessnig J, Moser T (2003) $Ca_v1.3$ channels are essential for development and presynaptic activity of cochlear inner hair cells. J Neurosci 23:10832–10840

Brehm P, Eckert R (1978) Calcium entry leads to inactivation of calcium channel in Paramecium. Science 202:1203–1206

Buraei Z, Yang J (2010) The ß subunit of voltage-gated Ca^{2+} channels. Physiol Rev 90:1461–1506

Busquet P, Nguyen NK, Schmid E et al (2010) $Ca_v1.3$ L-type Ca^{2+} channels modulate depression-like behaviour in mice independent of deaf phenotype. Int J Neuropsychopharmacol 13:499–513

Calin-Jageman I, Lee A (2008) Ca(v)1 L-type Ca^{2+} channel signaling complexes in neurons. J Neurochem 105:573–583

Catterall WA (2000) Structure and regulation of voltage-gated Ca^{2+} channels. Annu Rev Cell Dev Biol 16:521–555

Catterall WA (2011) Voltage-gated calcium channels. Cold Spring Harb Perspect Biol 3:a003947

Catterall WA, Perez-Reyes E, Snutch TP, Striessnig J (2005) International Union of Pharmacology. XLVIII. Nomenclature and structure-function relationships of voltage-gated calcium channels. Pharmacol Rev 57:411–425

Chang B, Heckenlively JR, Bayley PR et al (2006) The nob2 mouse, a null mutation in Cacna1f: anatomical and functional abnormalities in the outer retina and their consequences on ganglion cell visual responses. Vis Neurosci 23:11–24

Cristofanilli M, Mizuno F, Akopian A (2007) Disruption of actin cytoskeleton causes internalization of Ca(v)1.3 (alpha 1D) L-type calcium channels in salamander retinal neurons. Mol Vis 13:1496–1507
Cui G, Meyer AC, Calin-Jageman I et al (2007) Ca^{2+} -binding proteins tune Ca^{2+} -feedback to $Ca_v1.3$ channels in mouse auditory hair cells. J Physiol (Lond) 585:791–803
Danecek P, Nellåker C, McIntyre RE et al (2012) High levels of RNA-editing site conservation amongst 15 laboratory mouse strains. Genome Biol 13:26
Demb JB, Pugh EN (2002) Connexin36 forms synapses essential for night vision. Neuron 36:551–553
Demb JB, Singer JH (2012) Intrinsic properties and functional circuitry of the AII amacrine cell. Vis Neurosci 29:51–60
Doering CJ, Rehak R, Bonfield S et al (2008) Modified Ca(v)1.4 expression in the Cacna1f(nob2) mouse due to alternative splicing of an ETn inserted in exon 2. PLoS ONE 3:e2538
Dou H, Vazquez AE, Namkung Y et al (2004) Null mutation of alpha1D Ca^{2+} channel gene results in deafness but no vestibular defect in mice. J Assoc Res Otolaryngol 5:215–226
Eggermann E, Bucurenciu I, Goswami SP, Jonas P (2012) Nanodomain coupling between Ca^{2+} channels and sensors of exocytosis at fast mammalian synapses. Nat Rev Neurosci 13:7–21
Erickson MG, Liang H, Mori MX, Yue DT (2003) FRET two-hybrid mapping reveals function and location of L-type Ca^{2+} channel CaM preassociation. Neuron 39:97–107
Firth SI, Morgan IG, Boelen MK, Morgans CW (2001) Localization of voltage-sensitive L-type calcium channels in the chicken retina. Clin Experiment Ophthalmol 29:183–187
Gebhart M, Juhasz-Vedres G, Zuccotti A et al (2010) Modulation of $Ca_v1.3$ Ca^{2+} channel gating by Rab3 interacting molecule. Mol Cell Neurosci 44:246–259
Glueckert R, Wietzorrek G, Kammen-Jolly K et al (2003) Role of class D L-type Ca^{2+} channels for cochlear morphology. Hear Res 178:95–105
Goutman JD (2012) Transmitter release from cochlear hair cells is phase locked to cyclic stimuli of different intensities and frequencies. J Neurosci 32:17025–17035a
Grant L, Fuchs P (2008) Calcium- and calmodulin-dependent inactivation of calcium channels in inner hair cells of the rat cochlea. J Neurophysiol 99:2183–2193
Gregory FD, Bryan KE, Pangršič T et al (2011) Harmonin inhibits presynaptic $Ca_v1.3$ Ca^{2+} channels in mouse inner hair cells. Nat Neurosci 14:1109–1111
Grimes WN, Li W, Chávez AE, Diamond JS (2009) BK channels modulate pre- and postsynaptic signaling at reciprocal synapses in retina. Nat Neurosci 12:585–592
Grimes WN, Zhang J, Graydon CW et al (2010) Retinal parallel processors: more than 100 independent microcircuits operate within a single interneuron. Neuron 65:873–885
Gu Y, Wang L, Zhou J et al (2008) A naturally-occurring mutation in Cacna1f in a rat model of congenital stationary night blindness. Mol Vis 14:20–28
Haeseleer F, Imanishi Y, Maeda T et al (2004) Essential role of Ca^{2+} -binding protein 4, a $Ca_v1.4$ channel regulator, in photoreceptor synaptic function. Nat Neurosci 7:1079–1087
Heidelberger R, Heinemann C, Neher E, Matthews G (1994) Calcium dependence of the rate of exocytosis in a synaptic terminal. Nature 371:513–515
Heidelberger R, Matthews G (1994) Dopamine enhances Ca^{2+} responses in synaptic terminals of retinal bipolar neurons. NeuroReport 5:729–732
Hemara-Wahanui A, Berjukow S, Hope CI et al (2005) A CACNA1F mutation identified in an X-linked retinal disorder shifts the voltage dependence of $Ca_v1.4$ channel activation. Proc Natl Acad Sci U S A 102:7553–7558
Henderson SR, Reuss H, Hardie RC (2000) Single photon responses in Drosophila photoreceptors and their regulation by Ca^{2+}. J Physiol (Lond) 524(Pt 1):179–194
Hibino H, Pironkova R, Onwumere O et al (2002) RIM binding proteins (RBPs) couple Rab3-interacting molecules (RIMs) to voltage-gated Ca^{2+} channels. Neuron 34:411–423
Hirtz JJ, Boesen M, Braun N et al (2011) $Ca_v1.3$ calcium channels are required for normal development of the auditory brainstem. J Neurosci 31:8280–8294

Hirtz JJ, Braun N, Griesemer D et al (2012) Synaptic refinement of an inhibitory topographic map in the auditory brainstem requires functional $Ca_v1.3$ calcium channels. J Neurosci 32:14602–14616

Hoda J-C, Zaghetto F, Koschak A, Striessnig J (2005) Congenital stationary night blindness type 2 mutations S229P, G369D, L1068P, and W1440X alter channel gating or functional expression of Ca(v)1.4 L-type Ca^{2+} channels. J Neurosci 25:252–259

Hoda J-C, Zaghetto F, Singh A et al (2006) Effects of congenital stationary night blindness type 2 mutations R508Q and L1364H on $Ca_v1.4$ L-type Ca^{2+} channel function and expression. J Neurochem 96:1648–1658

Hope CI, Sharp DM, Hemara-Wahanui A et al (2005) Clinical manifestations of a unique X-linked retinal disorder in a large New Zealand family with a novel mutation in CACNA1F, the gene responsible for CSNB2. Clinical & experimental ophthalmology 33:129–136

Huang H, Tan BZ, Shen Y et al (2012) RNA editing of the IQ domain in Ca(v)1.3 channels modulates their Ca^{2+}-dependent inactivation. Neuron 73:304–316

Jacobson LB (1969) Cluster headache: a rare cause of bradycardia. Headache 8:159–161

Jalkanen R, Bech-Hansen NT, Tobias R et al (2007) A novel CACNA1F gene mutation causes Aland Island eye disease. Invest Ophthalmol Vis Sci 48:2498–2502

Jalkanen R, Mäntyjärvi M, Tobias R et al (2006) X linked cone-rod dystrophy, CORDX3, is caused by a mutation in the CACNA1F gene. J Med Genet 43:699–704

Karali M, Peluso I, Gennarino VA et al (2010) miRNeye: a microRNA expression atlas of the mouse eye. BMC Genomics 11:715

Karali M, Peluso I, Marigo V, Banfi S (2007) Identification and characterization of microRNAs expressed in the mouse eye. Invest Ophthalmol Vis Sci 48:509–515

Karpen JW, Loney D a, Baylor D a (1992) Cyclic GMP-activated channels of salamander retinal rods: spatial distribution and variation of responsiveness. J Physiol (Lond) 448:257–274

Kersten FFJ, Van Wijk E, Van Reeuwijk J et al (2010) Association of whirlin with $Ca_v1.3$ (alpha1D) channels in photoreceptors, defining a novel member of the usher protein network. Invest Ophthalmol Vis Sci 51:2338–2346

Kim D, Song I, Keum S et al (2001) Lack of the burst firing of thalamocortical relay neurons and resistance to absence seizures in mice lacking alpha(1G) T-type Ca^{2+} channels. Neuron 31:35–45

Kim E, Sheng M (2004) PDZ domain proteins of synapses. Nat Rev Neurosci 5:771–781

Kiyonaka S, Wakamori M, Miki T et al (2007) RIM1 confers sustained activity and neurotransmitter vesicle anchoring to presynaptic Ca^{2+} channels. Nat Neurosci 10:691–701

Ko GY-P, Ko M, Dryer SE (2004) Circadian and cAMP-dependent modulation of retinal cone cGMP-gated channels does not require protein synthesis or calcium influx through L-type channels. Brain Res 1021:277–280

Ko ML, Liu Y, Shi L et al (2008) Circadian regulation of retinoschisin in the chick retina. Invest Ophthalmol Vis Sci 49:1615–1621

Koike C, Numata T, Ueda H et al (2010) TRPM1: a vertebrate TRP channel responsible for retinal ON bipolar function. Cell Calcium 48:95–101

Kollmar R, Montgomery LG, Fak J et al (1997) Predominance of the alpha1D subunit in L-type voltage-gated Ca^{2+} channels of hair cells in the chicken's cochlea. Proc Natl Acad Sci USA 94:14883–14888

Korenbrot JI (2012) Speed, sensitivity, and stability of the light response in rod and cone photoreceptors: facts and models. Prog Retin Eye Res 31:442–466

Koschak A (2010) Impact of gating modulation in $Ca_v1.3$ L-type calcium channels. Channels (Austin) 4:523–525

Koschak A, Reimer D, Huber I et al (2001) alpha 1D ($Ca_v1.3$) subunits can form l-type Ca^{2+} channels activating at negative voltages. J Biol Chem 276:22100–22106

Koschak A, Reimer D, Walter D et al (2003) $Ca_v1.4$alpha1 subunits can form slowly inactivating dihydropyridine-sensitive L-type Ca^{2+} channels lacking Ca^{2+} -dependent inactivation. J Neurosci 23:6041–6049

Kuhn S, Knirsch M, Rüttiger L et al (2009) Ba^{2+} currents in inner and outer hair cells of mice lacking the voltage-dependent Ca^{2+} channel subunits beta3 or beta4. Channels (Austin) 3:366–376

Lieb A, Scharinger A, Sartori S et al (2012) Structural determinants of $Ca_v1.3$ L-type calcium channel gating. Channels (Austin) 6:197–205

Lipscombe D (2002) L-type calcium channels: highs and new lows. Circ Res 90:933–935

Liu X, Yang PS, Yang W, Yue DT (2010) Enzyme-inhibitor-like tuning of Ca^{2+} channel connectivity with calmodulin. Nature 463:968–972

Maeda T, Lem J, Palczewski K, Haeseleer F (2005) A critical role of CaBP4 in the cone synapse. Invest Ophthalmol Vis Sci 46:4320–4327

Magupalli VG, Schwarz K, Alpadi K et al (2008) Multiple RIBEYE–RIBEYE interactions create a dynamic scaffold for the formation of synaptic ribbons. J Neurosci 28:7954–7967

Mangoni ME, Couette B, Bourinet E et al (2003) Functional role of L-type $Ca_v1.3$ Ca^{2+} channels in cardiac pacemaker activity. Proc Natl Acad Sci USA 100:5543–5548

Mangoni ME, Traboulsie A, Leoni A-L et al (2006) Bradycardia and slowing of the atrioventricular conduction in mice lacking $Ca_v3.1$/alpha1G T-type calcium channels. Circ Res 98:1422–1430

Mansergh F, Orton NC, Vessey JP et al (2005) Mutation of the calcium channel gene Cacna1f disrupts calcium signaling, synaptic transmission and cellular organization in mouse retina. Hum Mol Genet 14:3035–3046

Maquat LE (2004) Nonsense-mediated mRNA decay: splicing, translation and mRNP dynamics. Nat Rev Mol Cell Biol 5:89–99

Marcantoni A, Baldelli P, Hernandez-Guijo JM et al (2007) L-type calcium channels in adrenal chromaffin cells: role in pace-making and secretion. Cell Calcium 42:397–408

Marcantoni A, Vandael DHF, Mahapatra S et al (2010) Loss of $Ca_v1.3$ channels reveals the critical role of L-type and BK channel coupling in pacemaking mouse adrenal chromaffin cells. J Neurosci 30:491–504

Marmor MF (1986) Contrast sensitivity versus visual acuity in retinal disease. Br J Ophthalmol 70:553–559

Marmorstein LY, Wu J, McLaughlin P et al (2006) The light peak of the electroretinogram is dependent on voltage-gated calcium channels and antagonized by bestrophin (best-1). J Gen Physiol 127:577–589

Masu M, Iwakabe H, Tagawa Y et al (1995) Specific deficit of the ON response in visual transmission by targeted disruption of the mGluR6 gene. Cell 80:757–765

McRory JE, Hamid J, Doering CJ et al (2004) The CACNA1F gene encodes an L-type calcium channel with unique biophysical properties and tissue distribution. J Neurosci 24:1707–1718

Mercer AJ, Chen M, Thoreson WB (2011) Lateral mobility of presynaptic L-type calcium channels at photoreceptor ribbon synapses. J Neurosci 31:4397–4406

Mercer AJ, Thoreson WB (2011) The dynamic architecture of photoreceptor ribbon synapses: cytoskeletal, extracellular matrix, and intramembrane proteins. Vis Neurosci 28:453–471

Meyer AC, Frank T, Khimich D et al (2009) Tuning of synapse number, structure and function in the cochlea. Nat Neurosci 12:444–453

Meyer AC, Moser T (2010) Structure and function of cochlear afferent innervation. Curr Opin Otolaryngol Head Neck Surg 18:441–446

Michna M, Knirsch M, Hoda J-C et al (2003) $Ca_v1.3$ (alpha1D) Ca^{2+} currents in neonatal outer hair cells of mice. J Physiol (Lond) 553:747–758

Miller WH, Nicol GD (1979) Evidence that cyclic GMP regulates membrane potential in rod photoreceptors. Nature 280:64–66

Mizuno F, Barabas P, Krizaj D, Akopian A (2010) Glutamate-induced internalization of Ca(v)1.3 L-type Ca^{2+} channels protects retinal neurons against excitotoxicity. J Physiol (Lond) 588:953–966

Morgans CW (1999) Calcium channel heterogeneity among cone photoreceptors in the tree shrew retina. Eur J Neurosci 11:2989–2993

Morgans CW (2001) Localization of the alpha(1F) calcium channel subunit in the rat retina. Invest Ophthalmol Vis Sci 42:2414–2418

Morgans CW, Bayley PR, Oesch NW et al (2005) Photoreceptor calcium channels: insight from night blindness. Vis Neurosci 22:561–568

Morgans CW, Gaughwin P, Maleszka R (2001) Expression of the alpha1F calcium channel subunit by photoreceptors in the rat retina. Mol Vis 7:202–209

Müller CS, Haupt A, Bildl W et al (2010) Quantitative proteomics of the Ca_v2 channel nano-environments in the mammalian brain. Proc Natl Acad Sci USA 107:14950–14957

Müller M, Von Hünerbein K, Hoidis S, Smolders JWT (2005) A physiological place-frequency map of the cochlea in the CBA/J mouse. Hear Res 202:63–73

Neef J, Gehrt A, Bulankina AV et al (2009) The Ca^{2+} channel subunit beta2 regulates Ca^{2+} channel abundance and function in inner hair cells and is required for hearing. J Neurosci 29:10730–10740

Obermair GJ, Szabo Z, Bourinet E, Flucher BE (2004) Differential targeting of the L-type Ca^{2+} channel alpha 1C ($Ca_v1.2$) to synaptic and extrasynaptic compartments in hippocampal neurons. Eur J Neurosci 19:2109–2122

Oliver D, Knipper M, Derst C, Fakler B (2003) Resting potential and submembrane calcium concentration of inner hair cells in the isolated mouse cochlea are set by KCNQ-type potassium channels. J Neurosci 23:2141–2149

Olson PA, Tkatch T, Hernandez-Lopez S et al (2005) G-protein-coupled receptor modulation of striatal $Ca_v1.3$ L-type Ca^{2+} channels is dependent on a Shank-binding domain. J Neurosci 25:1050–1062

Peloquin JB, Doering CJ, Rehak R, McRory JE (2008) Temperature dependence of $Ca_v1.4$ calcium channel gating. Neuroscience 151:1066–1083

Peloquin JB, Rehak R, Doering CJ, McRory JE (2007) Functional analysis of congenital stationary night blindness type-2 CACNA1F mutations F742C, G1007R, and R1049 W. Neuroscience 150:335–345

Perez-Reyes E (2003) Molecular physiology of low-voltage-activated t-type calcium channels. Physiol Rev 83:117–161

Van Petegem F, Chatelain FC, Minor DL (2005) Insights into voltage-gated calcium channel regulation from the structure of the $Ca_v1.2$ IQ domain-Ca^{2+}/calmodulin complex. Nat Struct Mol Biol 12:1108–1115

Peterson BZ, Lee JS, Mulle JG et al (2000) Critical determinants of Ca^{2+}-dependent inactivation within an EF-hand motif of L-type Ca^{2+} channels. Biophys J 78:1906–1920

Pfeiffer RF (2010) Parkinson disease: calcium channel blockers and Parkinson disease. Nat Rev Neurol 6:188–189

Platzer J, Engel J, Schrott-Fischer A et al (2000) Congenital deafness and sinoatrial node dysfunction in mice lacking class D L-type Ca^{2+} channels. Cell 102:89–97

Putzier I, Kullmann PHM, Horn JP, Levitan ES (2009) $Ca_v1.3$ channel voltage dependence, not Ca^{2+} selectivity, drives pacemaker activity and amplifies bursts in nigral dopamine neurons. J Neurosci 29:15414–15419

Ramakrishnan NA, Drescher MJ, Drescher DG (2009) Direct interaction of otoferlin with syntaxin 1A, SNAP-25, and the L-type voltage-gated calcium channel $Ca_v1.3$. J Biol Chem 284:1364–1372

Rosenthal R, Bakall B, Kinnick T et al (2006) Expression of bestrophin-1, the product of the VMD2 gene, modulates voltage-dependent Ca^{2+} channels in retinal pigment epithelial cells. FASEB journal : official publication of the Federation of American Societies for Experimental Biology 20:178–180

Roux I, Safieddine S, Nouvian R et al (2006) Otoferlin, defective in a human deafness form, is essential for exocytosis at the auditory ribbon synapse. Cell 127:277–289

Safieddine S, El-Amraoui A, Petit C (2012) The auditory hair cell ribbon synapse: from assembly to function. Annu Rev Neurosci 35:509–528

Satheesh SV, Kunert K, Rüttiger L et al (2012) Retrocochlear function of the peripheral deafness gene Cacna1d. Hum Mol Genet 21:3896–3909

Schrauwen I, Helfmann S, Inagaki A et al (2012) A mutation in CABP2, expressed in cochlear hair cells, causes autosomal-recessive hearing impairment. Am J Hum Genet 91:636–645
Seeliger MW, Brombas A, Weiler R et al (2011) Modulation of rod photoreceptor output by HCN1 channels is essential for regular mesopic cone vision. Nat Commun 2:532
Shaltiel L, Paparizos C, Fenske S et al (2012) Complex regulation of voltage-dependent activation and inactivation properties of retinal voltage-gated $Ca_v1.4$ L-type Ca^{2+} channels by Ca^{2+} -binding protein 4 (CaBP4). J Biol Chem 287:36312–36321
Shcheglovitov A, Vitko I, Lazarenko RM et al (2012) Molecular and biophysical basis of glutamate and trace metal modulation of voltage-gated Ca(v)2.3 calcium channels. J Gen Physiol 139:219–234
Sheets L, Trapani JG, Mo W et al (2011) Ribeye is required for presynaptic Ca(V)1.3a channel localization and afferent innervation of sensory hair cells. Development 138:1309–1319
Shen Y, Yu D, Hiel H et al (2006) Alternative splicing of the Ca(v)1.3 channel IQ domain, a molecular switch for Ca^{2+} -dependent inactivation within auditory hair cells. J Neurosci 26:10690–10699
Shi L, Jian K, Ko ML et al (2009a) Retinoschisin, a new binding partner for L-type voltage-gated calcium channels in the retina. J Biol Chem 284:3966–3975
Shi L, Ko ML, Ko GY-P (2009b) Rhythmic expression of microRNA-26a regulates the L-type voltage-gated calcium channel alpha1C subunit in chicken cone photoreceptors. J Biol Chem 284:25791–25803
Simonsz HJ, Florijn RJ, Van Minderhout HM et al (2009) Nightblindness-associated transient tonic downgaze (NATTD) in infant boys with chin-up head posture. Strabismus 17:158–164
Singh A, Gebhart M, Fritsch R et al (2008) Modulation of voltage- and Ca^{2+} -dependent gating of $Ca_v1.3$ L-type calcium channels by alternative splicing of a C-terminal regulatory domain. J Biol Chem 283:20733–20744
Song H, Nie L, Rodriguez-Contreras A et al (2003) Functional interaction of auxiliary subunits and synaptic proteins with Ca(v)1.3 may impart hair cell Ca^{2+} current properties. J Neurophysiol 89:1143–1149
Soucy E, Wang Y, Nirenberg S et al (1998) A novel signaling pathway from rod photoreceptors to ganglion cells in mammalian retina. Neuron 21:481–493
Specht D, Wu S-B, Turner P et al (2009) Effects of presynaptic mutations on a postsynaptic Cacna1 s calcium channel colocalized with mGluR6 at mouse photoreceptor ribbon synapses. Invest Ophthalmol Vis Sci 50:505–515
Specht SC, Sweadner KJ (1984) Two different Na, K-ATPases in the optic nerve: cells of origin and axonal transport. Proc Natl Acad Sci USA 81:1234–1238
Stockner T, Koschak A (2013) What can naturally occurring mutations tell us about $Ca_v1.x$ channel function? Biochim Biophys Acta 1828:1598–1607
Strettoi E, Dacheux RF, Raviola E (1990) Synaptic connections of rod bipolar cells in the inner plexiform layer of the rabbit retina. J Comp Neurol 295:449–466
Striessnig J (2007) C-terminal tailoring of L-type calcium channel function. J Physiol (Lond) 585:643–644
Striessnig J, Bolz HJ, Koschak A (2010) Channelopathies in $Ca_v1.1$, $Ca_v1.3$, and $Ca_v1.4$ voltage-gated L-type Ca^{2+} channels. Pfl{ü}gers Archive: European journal of physiology 460:361–374
Striessnig J, Koschak A (2008) Exploring the function and pharmacotherapeutic potential of voltage-gated Ca^{2+} channels with gene knockout models. Channels (Austin) 2:233–251
Strom TM, Nyakatura G, Apfelstedt-Sylla E et al (1998) An L-type calcium-channel gene mutated in incomplete X-linked congenital stationary night blindness. Nat Genet 19:260–263
Surmeier DJ, Guzman JN, Sanchez-Padilla J, Goldberg JA (2011) The origins of oxidant stress in Parkinson's disease and therapeutic strategies. Antioxid Redox Sig 14:1289–1301
Tadross MR, Dick IE, Yue DT (2008) Mechanism of local and global Ca^{2+} sensing by calmodulin in complex with a Ca^{2+} channel. Cell 133:1228–1240

Tan GMY, Yu D, Wang J, Soong TW (2012) Alternative splicing at C terminus of Ca(V)1.4 calcium channel modulates calcium-dependent inactivation, activation potential, and current density. J Biol Chem 287:832–847

Taylor WR, Morgans C (1998) Localization and properties of voltage-gated calcium channels in cone photoreceptors of Tupaia belangeri. Vis Neurosci 15:541–552

Thoreson WB, Rabl K, Townes-Anderson E, Heidelberger R (2004) A highly Ca^{2+} -sensitive pool of vesicles contributes to linearity at the rod photoreceptor ribbon synapse. Neuron 42:595–605

tom Dieck S, Altrock WD, Kessels MM, et al. (2005) Molecular dissection of the photoreceptor ribbon synapse: physical interaction of Bassoon and RIBEYE is essential for the assembly of the ribbon complex. J Cell Biol 168:825–36

tom Dieck S, Brandstätter JH (2006) Ribbon synapses of the retina. Cell Tissue Res 326:339–46

Tremblay F, Laroche RG, De Becker I (1995) The electroretinographic diagnosis of the incomplete form of congenital stationary night blindness. Vision Res 35:2383–2393

Tsukamoto Y, Morigiwa K, Ueda M, Sterling P (2001) Microcircuits for night vision in mouse retina. J Neurosci 21:8616–8623

Wahl-Schott C, Baumann L, Cuny H et al (2006) Switching off calcium-dependent inactivation in L-type calcium channels by an autoinhibitory domain. Proc Natl Acad Sci U S A 103:15657–15662

Wang Y, Okamoto M, Schmitz F et al (1997) Rim is a putative Rab3 effector in regulating synaptic-vesicle fusion. Nature 388:593–598

Wässle H (2004) Parallel processing in the mammalian retina. Nat Rev Neurosci 5:747–757

Welch NC, Wood S, Jollimore C et al (2005) High-voltage-activated calcium channels in Muller cells acutely isolated from tiger salamander retina. Glia 49:259–274

Wu J, Marmorstein AD, Striessnig J, Peachey NS (2007) Voltage-dependent calcium channel $Ca_v1.3$ subunits regulate the light peak of the electroretinogram. J Neurophysiol 97:3731–3735

Wycisk KA, Budde B, Feil S et al (2006) Structural and functional abnormalities of retinal ribbon synapses due to Cacna2d4 mutation. Invest Ophthalmol Vis Sci 47:3523–3530

Xiao H, Chen X, Steele EC (2007) Abundant L-type calcium channel Ca(v)1.3 (alpha1D) subunit mRNA is detected in rod photoreceptors of the mouse retina via in situ hybridization. Mol Vis 13:764–771

Yang PS, Alseikhan BA, Hiel H et al (2006) Switching of Ca^{2+} -dependent inactivation of Ca(v)1.3 channels by calcium binding proteins of auditory hair cells. J Neurosci 26:10677–10689

Yu K, Xiao Q, Cui G et al (2008) The best disease-linked Cl- channel hBest1 regulates Ca V 1 (L-type) Ca^{2+} channels via src-homology-binding domains. J Neurosci 28:5660–5670

Zampini V, Johnson SL, Franz C et al (2010) Elementary properties of $Ca_v1.3$ Ca^{2+} channels expressed in mouse cochlear inner hair cells. J Physiol (Lond) 588:187–199

Zamponi GW (2003) Calmodulin lobotomized: novel insights into calcium regulation of voltage-gated calcium channels. Neuron 39:879–881

Zeitz C, Kloeckener-Gruissem B, Forster U et al (2006) Mutations in CABP4, the gene encoding the Ca^{2+} -binding protein 4, cause autosomal recessive night blindness. Am J Hum Genet 79:657–667

Zheng L, Yan Y, An J et al (2012) Retinal horizontal cells reduced in a rat model of congenital stationary night blindness. Neurosci Lett 521:26–30

Chapter 4
T-Type Calcium Channels and Epilepsy

Stuart M. Cain, Michael E. Hildebrand and Terrance P. Snutch

Abstract Epilepsy is characterized by spontaneous recurrent seizures in which electrical activity in particular brain regions becomes over-excitable. As different brain regions interact in cycle, one excites the next until they become locked into a self-propagating loop. Low voltage-activated T-type calcium channels underlie the burst-firing associated with spike-and-wave discharges observed on electroencephalography recordings during certain forms of epileptic seizures. It is currently estimated that around 50 million people in the world suffer from epilepsy. Of these, around 30 % of patients are resistant to current medications and up to 90 % of treatable patients experience significant side effects of the drugs. As such, there is considerable need for new approaches towards the development of novel efficacious therapeutics with fewer side effects.

4.1 T-Type Calcium Channels

Low voltage-activated T-type calcium channels are functionally distinct from other members of the voltage-gated calcium channel family. Compared to high voltage-activated (HVA) calcium channels, T-type channels have several unique biophysical properties, including small single channel conductances, fast activation and inactivation kinetics, hyperpolarized voltage-dependences of activation and inactivation and slow deactivation kinetics. The rates of T-type channel activation and inactivation are strongly voltage-dependent, creating a characteristic "crossing over" pattern in successively more depolarized traces of a

S. M. Cain · T. P. Snutch (✉)
Michael Smith Laboratories, University of British Columbia, Vancouver, BC, Canada
e-mail: snutch@msl.ubc.ca

M. E. Hildebrand
Department of Neuroscience, Carleton University, Ottawa, ON, Canada

N. Weiss and A. Koschak (eds.), *Pathologies of Calcium Channels*,
DOI: 10.1007/978-3-642-40282-1_4,

conventional square-pulse current–voltage protocol. For neuronal T-type currents, the time for half-inactivation can range from greater than 100 ms at activation thresholds (between −70 and −50 mV) to less than 20 ms at maximal activating potentials (> −30 mV) (reviewed in, Huguenard 1996). Although T-type currents exhibit strong voltage-dependent inactivation, they do not appear to inactivate in a calcium-dependent manner similar to many members of the HVA class of calcium channels.

The T-type calcium channel class is composed of three separate α_1 subunit encoding genes in the mammalian genome: *CACNA1G* ($Ca_v3.1$), *CACNA1H* ($Ca_v3.2$ and *CACNA1I* ($Ca_v3.3$) (Fig. 4.1a) (Adams and Snutch 2007). The three Ca_v subtypes have broad expression patterns, including the ovaries, placenta, heart, kidney, smooth muscle, liver, adrenal cortex and neurons throughout the peripheral and central nervous systems. Non-neuronal functions of T-type channels include smooth muscle contraction, hormone secretion and cardiac pacemaker activity (reviewed in, Catterall et al. 2005). In many cell types, multiple distinct functional variants of $Ca_v3.1$, $Ca_v3.2$ and $Ca_v3.3$ are also expressed through the process of alternative splicing (Chemin et al. 2001; David et al. 2010; Emerick et al. 2006; Mittman et al. 1999; Monteil et al. 2000; Murbartián et al. 2004; Mittman et al. 1999).

The three main recombinant T-type channel isoforms have been extensively studied and characterized (Perez-Reyes 2003). The $Ca_v3.1$ and $Ca_v3.2$ subtypes exhibit biophysical properties consistent with prototypical T-type currents, including fast activation and inactivation kinetics. In contrast, the $Ca_v3.3$ subtype exhibits significantly slower activation and inactivation kinetics, faster deactivation kinetics and a more depolarized voltage dependence of both activation and inactivation when compared with the other two T-type isoforms (Chemin et al. 2002; McRory et al. 2001). All three isoforms have a characteristically fast recovery from inactivation, with the $Ca_v3.1$ and $Ca_v3.3$ channels recovering the fastest (Klockner et al. 1999). Differences in biophysical properties, alternative splicing, and differential expression of individual Ca_v3 subtypes all allow for considerable functional specialization within the nervous system. The overlap between the voltage-dependence of T-type activation and inactivation at potentials near neuronal resting membrane potentials (McRory et al. 2001) creates the possibility for "window currents," wherein a fraction of T-type channels remain open at rest. The presence and relevance of T-type window currents (and T-type currents in general) is highly dependent on the resting membrane potential, as the channels will become completely inactivated at more depolarized potentials (Carter and Sabatini 2004). T-type window currents can also affect calcium-mediated signaling pathways as well as the membrane potential and electrical firing patterns of neurons, although inhibition of the T-type window current does not affect tonic firing frequency (Chevalier et al. 2006; Dreyfus et al. 2010; Williams et al. 1997).

The three T-type channel isoforms are differentially expressed in the brain (Fig. 4.1b) (Talley et al. 1999), likely accounting for some of the observed heterogeneity in pharmacology, modulation and biophysical properties amongst

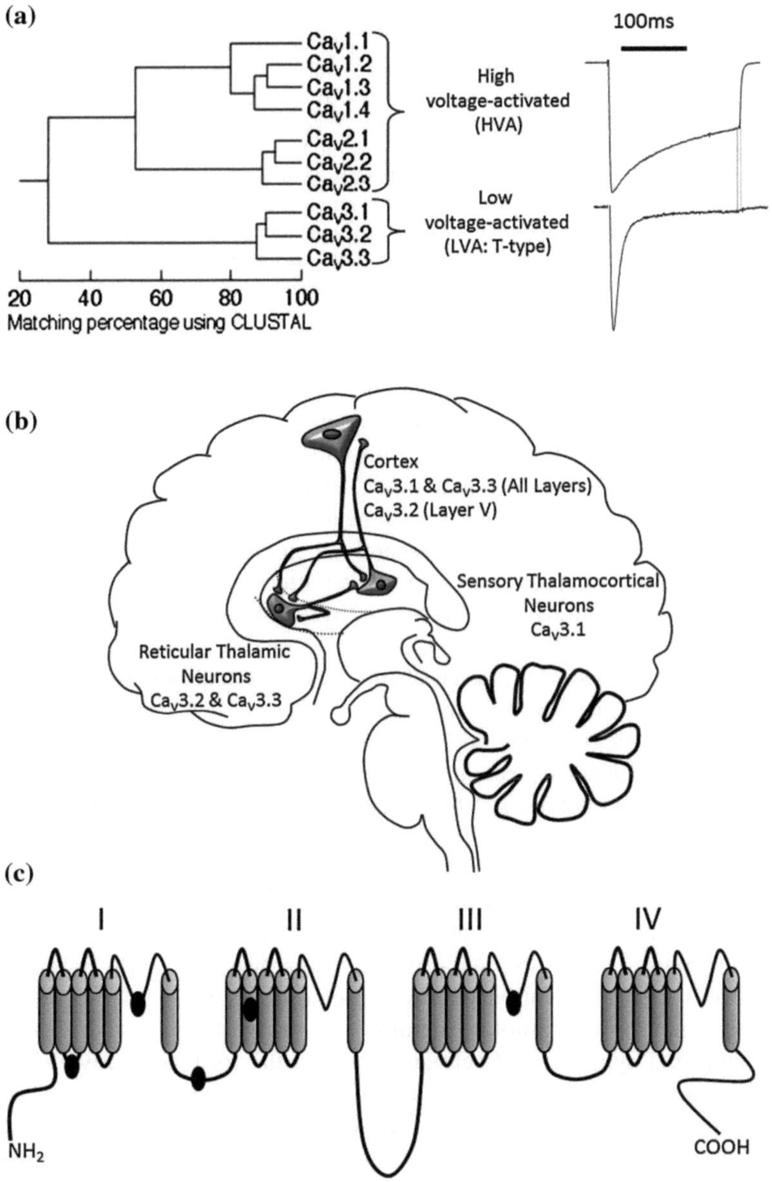
(a)
100ms
Cav1.1
Cav1.2
Cav1.3
Cav1.4
Cav2.1
Cav2.2
Cav2.3
Cav3.1
Cav3.2
Cav3.3
High voltage-activated (HVA)
Low voltage-activated (LVA: T-type)
20 40 60 80 100
Matching percentage using CLUSTAL
(b)
Cortex
Cav3.1 & Cav3.3 (All Layers)
Cav3.2 (Layer V)
Sensory Thalamocortical Neurons
Cav3.1
Reticular Thalamic Neurons
Cav3.2 & Cav3.3
(c)
I
II
III
IV
NH2
COOH

◀ **Fig. 4.1** **a** Phylogenic relationship between the 10 known calcium channel α_1 subunits (*left panel*). The Ca_v1 subunits (L-types) and Ca_v2 subunits (P/Q-type, N-type and R-type) are grouped as HVA calcium channels, and the Ca_v3 subunits form the Low Voltage-Activated (LVA) "T-type" calcium channels. Representative traces of calcium currents recorded from reticular thalamic neurons (*right panel*). The *upper trace* shows a slow inactivating HVA current and the *lower trace* shows the fast inactivating low voltage-activated (LVA) current. **b** Schematic of the primary neuronal projections in the thalamocortical network between the somatosensory cortex, the sensory thalamic relay neurons and the reticular thalamic nucleus. The distribution of the three T-type channel subtypes is indicated. **c** Schematic representation of the T-type calcium channel α_1 subunit with its four domain structure. *Black dots* denote regions of the T-type calcium channel wherein single nucleotide polymorphisms have been identified in human patients with various forms of genetic generalized epilepsy

native T-type currents (reviewed in Huguenard 1996). Studies combining immunohistochemistry (Craig et al. 1999; McKay et al. 2006) with electrophysiological recordings (Isope and Murphy 2005; Joksovic et al. 2005; Kavalali et al. 1997) reveal that the Ca_v3 channels and their functional currents are predominantly localized to the soma and dendrites of neurons, with highest expression often occurring in dendritic regions. More recently, there has been good evidence that T-type channels also contribute to neurotransmitter release at a subset of nerve terminals (Huang et al. 2011; Jacus et al. 2012; Tang et al. 2011; Todorovic and Jevtovic-Todorovic 2013).

The robust expression of T-type channels in neuronal dendrites suggests that these channels may be involved in signal integration. For example, subthreshold excitatory postsynaptic potentials (EPSPs) can cause the activation of T-type currents to boost dendritic calcium concentrations in hippocampal and cortical neurons (Magee and Johnston 1995; Markram and Sakmann 1994). This T-type activity may lead to localized feed-foward dendritic depolarization or alternatively, could activate calcium-activated outward currents to mediate dendritic hyperpolarization (Wolfart and Roeper 2002).

Besides modulating synaptic integration, T-type channel activity can also shape overall neuronal excitability. As originally demonstrated in the inferior olive, removal of T-type inactivation with hyperpolarization can initiate a spontaneous "rebound-burst" low-threshold calcium spike (Fig. 4.2b) (Llinas and Yarom 1981). Indeed, T-type channel activation is essential for regenerative low-threshold spikes and burst-firing in neurons across the CNS, including in the cerebellum, cortex, hippocampus, inferior olive, neocortex and thalamus (reviewed in Huguenard 1996). Within a subset of neurons, low threshold spikes and burst-firing can alter neuronal oscillations, causing a switch from high frequency tonic neuronal firing to a phasic firing mode characterized by regular intervals of high frequency bursts (Diana et al. 2007; Suzuki and Rogawski 1989). These T-type mediated changes in rhythmic oscillations are essential for both physiological sleep-wake gating and pathophysiological epileptic seizure activity. Of note, in the thalamus $Ca_v3.1$ is predominantly expressed in thalamocortical relay cells (TCs) at high levels, while $Ca_v3.2$ and $Ca_v3.3$ are highly expressed in thalamic reticular (nRT) neurons (Talley et al. 1999).

4.2 T-Type Calcium Channel Single Nucleotide Polymorphisms in Human Epilepsy Patients

Epilepsy is a general disorder of the nervous system that is primarily characterized by hyperexcitability and hypersynchronization of thalamic and cortical neuronal circuits. Genetic Generalized Epilepsy (GGE) represents a major group of epilepsies that have no clear etiology, are partly caused by complex non-Mendelian genetics, and includes disorders such as juvenile myoclonic epilepsy, juvenile absence epilepsy, and childhood absence epilepsy. Absence epilepsy is characterized by brief seizures that cause impairments of consciousness through 3–6 Hz spike-and-wave discharges (SWDs) mediated by oscillations within the thalamocortical circuit (reviewed in McCormick and Contreras 2001). In human studies, a number of single nucleotide polymorphisms (SNPs) have been identified in the genes encoding for the $Ca_v3.1$ and $Ca_v3.2$ T-type calcium channels in patients with GGEs (Fig. 4.1c) (Adams and Snutch 2007; Cain and Snutch 2011; Chen et al. 2003; Heron et al. 2007; Liang et al. 2007; Singh et al. 2007; Zamponi et al. 2009). Upon exogenous evaluation of T-type calcium channel biophysical properties using heterologous expression systems some of the alterations induce a gain-of-function, some a loss-of-function while others have no apparent effect on channel characteristics. While these varied effects may reflect the polygenic nature of epilepsy, some genetic variability may also represent SNPs unrelated to the epileptic condition.

4.3 Burst-Firing in Epilepsy

A key functional consequence of large-scale T-type calcium channel activation in neurons is called "burst-firing." Providing that the current density of T-type calcium channels is of sufficient magnitude in a particular neuron and that the membrane potential passes the activation threshold, calcium will surge through T-type channels with enough charge to generate a low threshold calcium spike (LTS). This depolarization lasts around 50–100 ms and peaks at around −40 mV before calcium channel inactivation combined with calcium-activated potassium channel inactivation allows the cell membrane potential to repolarize (Cain and Snutch 2010). Importantly, the depolarization is sufficient to allow activation of voltage-gated sodium and potassium channels and resulting in action potential firing on the crown of the LTS. These action potentials occur at a high frequency, presumably producing a powerful drive to interconnected neurons. The downstream consequence of burst-firing activity can include facilitated neurotransmitter release, long-term potentiation, improved signal-to-noise ratio and a reduced chance of synaptic transmission failure (Lisman 1997; Reinagel et al. 1999; Sherman 2001). It should be noted that high threshold bursts can be produced by activation of HVA calcium channels and intermittent periods of high frequency tonic firing can be observed in some neurons

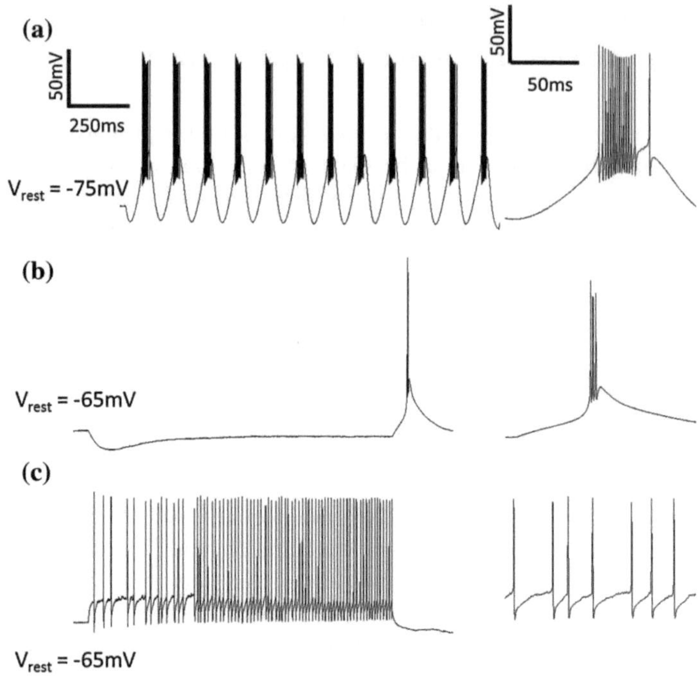

Fig. 4.2 **a** Oscillatory burst-firing recorded in a reticular thalamic neuron in response to sine-wave hyperpolarizing/depolarizing current injection. **b** Rebound burst-firing recorded in a ventroposterior medial thalamic neuron in response to a hyperpolarizing current step. **c** Tonic firing recorded in a ventroposterior medial thalamic neuron in response to a depolarizing current step. All recordings were performed in current clamp using physiological solutions at 33–35 °C. Expanded time scales of the same recordings are shown in the *right panels. Cain and Snutch, unpublished data*

(Fig. 4.2c) (Hughes and Crunelli 2005, 2007; Tennigkeit et al. 1998; Timofeev and Steriade 1998). While these non-LTS-mediated bursts may resemble those produced by T-type calcium channels with regards to action potential firing and, therefore the synaptic output they lack the LTS, which appears to have its own unique influence on neuronal excitability. When an LTS occurs it not only propagates to the neuronal cell body, but from there the calcium signal back-propagates throughout the entire dendritic tree (Errington et al. 2010). It is thought that this process prevents further out-of-phase depolarization and therefore acts as a frequency filter for the relay or propagation of only time-locked network activity that is synchronized with the LTS.

During epileptic seizures, it is thought that populations of neurons can become synchronized in their firing behavior, although de-synchronization is sometimes observed in the period immediately before the seizure starts (Jiruska et al. 2013). As seizures progress from a focal point, they propagate to interconnected brain regions which also become synchronized in their firing properties. The tendency to synchronize during seizures could be seen as normal behavior in non-epileptic brain regions triggered by the epileptic focus, or as an abnormality in an epileptogenic brain region that allows the atypical progression and propagation of the hyperexcitable activity. Burst-firing during seizures typically occurs in a repetitive pattern, appearing on an oscillating basal membrane potential as the bursts of action potentials are followed by periods of quiescence (Cain and Snutch 2010, 2013; Contreras 2006). As these synchronized cellular oscillations within specific neuronal regions recruit other regions within the same neuronal network, this leads to network oscillations as these brain regions burst in cycle. It is these large-scale depolarizations followed by periods of silence within entire brain loci that give rise to the spikes and waves that can be observed on electroencephalography (EEG) recordings during seizures, particularly absence seizures (Pinault et al. 2001; Slaght et al. 2002). It is not known how these cellular and network oscillations cause seizures, nor why some types of seizure recruit brain structures that lead to convulsions while other do not.

Burst-firing mediated by T-type calcium channels has been identified in reticular thalamic and thalamic relay neurons in a number of animal models of absence seizures. Burst-firing occurs in reticular thalamic neurons that closely correlates with spikes in spike-wave-discharges (SWDs) (Fig. 4.2a) (Pinault et al. 1998; Slaght et al. 2002) in a rat absence seizure model called Genetic Absence Epilepsy Rats from Strasbourg (GAERS) (Akman et al. 2010; Marescaux et al. 1992). Reticular thalamic neurons in this model display T-type currents (Tsakiridou et al. 1995) with increased current density due to up-regulation of the $Ca_v3.2$ channel (Talley et al. 1999, 2000), which is also mutated in this strain increasing the rate of channel recovery from inactivation (Powell et al. 2009). This hyperexcitable $Ca_v3.2$ channel is predicted to enhance calcium conductance and burst depolarizations in GAERS reticular thalamic neurons. Similarly, enhanced $Ca_v3.1$ and $Ca_v3.3$ currents have been observed in lateral geniculate and centrolateral TCR neurons (Broicher et al. 2008) in the Wistar Albino Glaxo Rats from Rijswijk (WAG/Rij) model of absence epilepsy that display spontaneously acquired seizures, (Akman et al. 2010; Coenen and Van Luijtelaar 2003). Furthermore, genetically modified mice with HVA calcium channel dysfunction display absence seizures accompanied by enhanced T-type (likely $Ca_v3.1$-mediated) calcium currents in thalamocortical relay neurons (Zhang et al. 2002) and the genetic enhancement of $Ca_v3.1$ currents in mouse thalamocortical relay neurons, induces absence seizures (Cheong et al. 2008, 2009; Ernst et al. 2009). In support, genetic ablation of $Ca_v3.1$ reduces sensitivity to some types of pharmacologically induced absence seizures (Kim et al. 2001). It appears that there are differences in the particular regions involved in different animal models and some controversies remain regarding the specific roles of those regions. This may reflect the diverse and polygenic nature of epilepsy in general with the large variety of underlying mechanisms that can lead to the genesis of seizures. For example, it may be that hyperexcitability in any thalamocortical region has the ability to cause thalamocortical oscillations and therefore, absence seizures. In addition, while burst-firing has also been identified in cortical neurons in both animal models and human brain tissue excised from epilepsy patients the evidence thus far suggests that the persistent sodium current, not a T-type calcium current, may underlie burst firing in this area (de Curtis et al. 1999; Leresche et al. 1998; Polack and Charpier 2009; Polack et al. 2009; Tryba et al. 2011), despite cortical expression of all three T-type calcium channel isoforms (Talley et al. 1999). This is of particular interest since the focus of absence seizures seems to reside in the cortex in a number of animal models (Polack et al. 2007; Zheng et al. 2012).

In addition to the thalamic role of T-type calcium channels in burst-firing and absence epilepsy, there is also a growing amount of evidence that suggests that burst-firing in the limbic areas of the brain may play a role in temporal lobe epilepsy. Systemic administration of pilocarpine is used in rats and mice as a model of complex-partial and temporal-lobe epilepsy, in the short- and long-term, respectively (Ca_valheiro 1995; Ca_valheiro et al. 1996). Several days after pilocarpine treatment animals develop chronic seizures concurrently with an up-regulation in $Ca_v3.2$ expression. Furthermore, these animals develop an increase in T-type calcium current density that correlates with an increase in the number of burst-firing neurons, predicted to originate from T-type calcium channel over-expression in the dendrites of hippocampal CA1 pyramidal neurons (Sanabria et al. 2001; Su et al. 2002; Yaari et al. 2007). Similarly, CA1 neurons also show upregulation of T-type currents following electrical kindling of the hippocampus in a different model of temporal lobe epilepsy (Faas et al. 1996). $Ca_v3.2$ knock-down mice are resistant to pilocarpine-induced seizures and do not display the seizure-induced increase in T-type currents or increased number of burst-firing neurons seen in normal pilocarpine-treated mice. Furthermore, $Ca_v3.2$ knockdown in these mice also attenuates the neuronal damage observed in the CA1 and CA3 regions following initial status epilepticus (Becker et al. 2008). Mice suffering from pilocarpine-induced seizures also display enhanced burst-firing in subiculum neurons (Wellmer et al. 2002) and in midline thalamic neurons that project to the hippocampal CA1 and entorhinal cortex with a correlating increase in $Ca_v3.2$-mediated T-type calcium currents (Graef et al. 2009).

4.4 T-Type Calcium Channel Pharmacology

The physiological functions of native T-type channels were originally difficult to characterize, as some HVA calcium channel subtypes (e.g. $Ca_v1.3$ and $Ca_v2.3$) activate in the negative membrane potential range historically used to define LVA calcium channels. Until recently, there were also no high affinity selective antagonists that could definitively distinguish between specific T-type isoforms and HVA channels. Thus, early studies which identified T-type currents based on biophysical properties and sensitivity to the inorganic antagonist, nickel (Ni^{2+}), revealed calcium currents with varying Ni^{2+}-sensitivities, biophysical properties and cell-type specific expression patterns. The subsequent identification of three T-type calcium channel isoforms demonstrated that while the $Ca_v{}^{3.2}$ T-type channel is highly sensitive to Ni^{2+} (IC50 of $\sim$12 uM), both the $Ca_v3.1$ and $Ca_v3.3$ subtypes are approximately 18–20 times less sensitive to Ni^{2+} (IC_{50}'s >200 μM for $Ca_v3.1$ and $Ca_v3.3$ (Lee et al. 1999). In fact, the HVA $Ca_v1.2$ L-type and $Ca_v2.3$ R-type channels are more sensitive to Ni^{2+} blockade than either $Ca_v3.1$ or $Ca_v3.3$ T-type channels (Lee et al. 1999; Zamponi et al. 1996) and, in retrospect, Ni^{2+} can only be effectively used in situations wherein the molecular nature of the underlying calcium channel isoforms is at least known to some extent.

Another polyvalent cation often used for studying T-type channel functions is cadmium (Cd^{2+}). Low concentrations of Cd^{2+} (20–100 μM) completely block the HVA calcium current classes, while leaving T-type currents relatively unaffected (Berrow et al. 1997; Fox et al. 1987; Tai et al. 2006). Thus, application of Cd^{2+} has been used to study LVA calcium currents largely in isolation from the HVA channels. The search for selective T-type channel antagonists present in nature initially revealed the peptide toxin, kurtoxin, purified from *Parabuthus transvaalicus* scorpion venom. Although originally described as a selective, high-affinity blocker of $Ca_v3.1$ and $Ca_v3.2$ but not of HVA calcium channels (Chuang et al. 1998), subsequent studies with native thalamic and sympathetic neurons demonstrated that kurtoxin also inhibits N-type, R-type, and L-type currents with nanomolar affinities, thereby limiting its utility as a selective T-type antagonist (Sidach and Mintz 2002).

A number of small organic agents nonspecifically inhibit T-type channels. One such compound is mibebradil, a T-type channel blocker that was used for treating hypertension and angina (Massie 1997, 1998), until it was removed from the market for a potentially fatal inhibition of essential metabolic enzymes (reviewed in, Welker et al. 1998). In some neuronal systems, nanomolar concentrations of mibefradil inhibit T-type currents but not HVA calcium currents, with increased inhibition at more depolarized potentials (McDonough and Bean 1998). However, studies in other cell types have shown that mibefradil is a nonspecific ion channel antagonist that can potently block R-type, N-type, L-type and P-type currents (Randall and Tsien 1997; Viana et al. 1997). Certain dihydropyridines (DHPs) can also block native LVA currents at low micromolar concentrations, although sensitivity is highly variable and may be partly due to the presence of dihydropyridine-sensitive $Ca_v1.3$ L-type currents that are low voltage-activated (Akaike et al. 1989) (reviewed in Yunker 2003; Yunker and McEnery 2003). Of note however, some DHPs that are less effective at blocking L-type channels ($IC_{50} \sim 40$ μM) have been shown to be reasonably potent T-type blockers ($IC_{50} \sim 1$ μM) and it remains that novel T-type specific antagonists might potentially be generated from derivatives of these compounds (Kumar et al. 2002).

4.5 Anti-Epileptic Drug Action on T-Type Calcium Channels

Calcium currents are inhibited by many of the existing anti-epileptic drugs (AEDs) available for treating seizures (Weiergraber et al. 2010). Unfortunately, the relevance of the findings can be difficult to determine, since the neuron in which the AED is tested in vitro, may not be from the region where the drug has its effect in vivo. In addition, the accurate measurement of clinical AED concentration in specific human brain areas is often not viable, resulting in the concentrations in vitro being estimated from human plasma concentrations and animal cerebral spinal fluid (CSF) to plasma concentration ratios. In effect, variable concentrations may be tested in comparison to those found in the brain of patients with epilepsy. Regardless, 100 % block of calcium channel activity is rarely seen at therapeutic concentrations of the clinically available AEDs in most in vitro studies undertaken to date. However, evidence supporting the involvement of T-type calcium channels in AED pharmacology is mounting for some of the currently used AEDs and T-type calcium channels are, therefore becoming an attractive new target for novel AEDs therapies with some small molecules showing promising results in early stage studies.

4.6 Blockade of T-Type Calcium Channels by Clinical Agents

4.6.1 Ethosuximide

Ethosuximide (ETX) is a front line anti-absence seizure treatment, with a somewhat controversial pharmacology with respect to T-type calcium channels. Studies investigating its effect in vitro have varied in the cell type investigated, isolation method and concentration of ETX and, therefore results have been somewhat inconclusive. ETX has a clinically relevant concentration of 40–100 μg/ml (280–700 μM) in CSF (Browne et al. 1975). Initial studies investigating its effect on T-type channels established that it inhibited the T-type current (now known to be $Ca_v3.1$) in acutely isolated thalamic TC neurons with block of around 22–40 % with 300 μM to 700 μM ETX, respectively, although 24 % of cells tested were unresponsive (Coulter et al. 1989a). This was later shown to be voltage-dependent with the drug being most efficacious at more hyperpolarized potentials (Coulter et al. 1989b). This was partially supported by recent work showing that ETX (750 μM) blocked the T-type current in acutely isolated TC lateral geniculate neurons, which project to the visual cortex, by 20 % in WAG/Rij rats, but only by 10 % in a nonepileptic control strain (Broicher et al. 2007). However, using computer modeling the authors concluded that the effect was not sufficient to reduce SWDs alone and predicted that it would also require inhibition of persistent sodium currents and calcium-activated potassium currents. This theory partially corroborated data from an earlier study where ETX (1 mM) was shown to have no effect on T-type or HVA currents expressed in dissociated rat TC and RTN neurons, but instead inhibited persistent sodium currents and calcium-activated potassium currents (Crunelli and Leresche 2002; Leresche et al. 1998). Similarly, little effect (<10 % inhibition) of ETX (50 0 μM) was observed on T-type or HVA currents on dissociated neocortical neurons removed from human temporal lobe epilepsy patients (Sayer et al. 1993). In a study undertaken in sensory dorsal root ganglion neurons, known to express $Ca_v3.2$ (Bourinet et al. 2005), 10 μM ETX was capable of blocking 90 % of the T-type current, although 100 μM was required for 95–99 % block of the T-type current (Kostyuk et al. 1992). However, a similar study also using DRG neurons (although in adult instead of neonatal rats) found that 100 mM ETX was required for 100 % block of the T-type current ($IC_{50} = 23$ mM), which is significantly above the clinically relevant concentration required (Todorovic and Lingle 1998). It has been suggested that as the procedure for dissociation removes most neuronal processes and T-type calcium channels are thought to be primarily dendritic in their expression. Therefore, in the studies described above, a large percentage of the T-type channels would have been removed and so attempts to assess the pharmacological profile of these channels in such preparations may not be entirely accurate with respect to in vivo effects.

In pituitary cells which express $Ca_v3.1$ and $Ca_v3.3$ (Mudado et al. 2004) ETX was found to only block the T-type current by 10 % at concentration of 2.5 mM (Herrington and Lingle 1992). However, using cloned T-type channels exogenously expressed in HEK293 cells, it was demonstrated that ETX blocked $Ca_v3.1$ by around 10 % at 100 μM, whereas 100 mM ETX was required for 100 % block ($EC_{50} = 12$ mM) (Gomora et al. 2001). Similar affinities were reported for the $Ca_v3.3$ channel.

Although the numerous in vitro studies investigating the mechanisms of ETX action on calcium channels have been inconclusive, more success has been achieved in identifying its site-of-action. Systemic injection of ethosuximide (~200–300 mg/kg) induces an almost immediate reduction in seizures (~50–80 %) within 30 min (Richards et al. 2003). However, direct microinfusion of ETX (1 mM) into the ventro-basal thalamus results in only a 25 % reduction in seizures. Microinfusion of a much lower concentration of ETX (20 nM) into the RTN reduces seizures by 70 %, however this is delayed by up to 60 min. Conversely, microinfusion of the same concentration of ETX (20 nM) into the SCX results in 90 % reduction of seizure activity within 30 min, similar to systemic administration (Manning et al. 2004). This cessation of seizures in GAERS has been shown to correlate with a reduction in hyperactivity in the somatosensory cortex (Polack and Charpier 2009) and ETX administration in this region can also reduce sensitivity to electrical kindling seizures (Gulhan Aker et al. 2009). Furthermore, intracortical infusion of ETX into the lateral somatosensory cortex WAG/Rij rats suppresses absence seizures whereas intrathalamic injection only serves to delay seizure onset (Chen et al. 2011). These studies support the theory that absence seizures initiate in the somatosensory cortex in GAERS and WAG/Rij animal models and indicate that ETX induces its primary effects in this region, although there may also be secondary action of ETX in the TC and RTN regions. Despite this, whether the anti-epileptic properties of ETX involve the blockade of T-type calcium currents remains controversial.

4.6.2 Valproic Acid

Valproic acid is a widely used AED including use as an anti-absence treatment, although fewer studies have examined the effects of this AED on calcium channels. Application of valproic acid (250 μM) to nodose ganglion neurons inhibited the T-type current by approximately 12 % (Kelly et al. 1990). Higher concentrations (500 μM) reduced the current further, however these were above the range of clinical relevance (Rogawski and Porter 1990). Similarly, valproic acid (1 mM) was shown to reduce T-type currents in dissociated TC neurons of WAG/Rij absence seizure model rats by 50 % and of an anti-epileptic control strain by 20 % (Broicher et al. 2007). However, 1 mM is significantly above the concentrations observed clinically in CSF. At the clinically relevant concentration of 200 μM valproic acid does not affect HVA or T-type currents in dissociated epileptic human temporal lobe neocortical neurons (Sayer et al. 1993).

4.6.3 Zonisamide

Zonisamide is used in the treatment of partial-onset seizures and with secondary generalization and thought to be particularly effective in progressive myoclonic epilepsies that are often resistant to AED treatment (Biton 2007; Kwan et al. 2001). In cultured cerebral cortical neurons zonisamide (500 μM) was shown to inhibit T-type currents by 60 % without affecting L-type HVA currents in the same neurons, although this is above the maximum clinically relevant concentration of 250 μM (Suzuki et al. 1992). Further studies in cultured neuroblastoma cells also demonstrated that their T-type current was inhibited by 39 % upon application of zonisamide (50 μM) and shifted the voltage-dependence of inactivation by around −20 mV (Kito et al. 1996). Using cloned T-type channels expressed exogenously in HEK293 cells, zonisamide was shown to inhibit all three T-subtypes within clinically relevant concentrations, with most efficacy at $Ca_v3.2$ (15–30 % at 50–200 μM, respectively) (Matar et al. 2009).

4.6.4 *Phenytoin*

Phenytoin is commonly used to control complex-partial and tonic–clonic seizures, but not absence seizures. It has been shown to inhibit the T-type current expressed in neuroblastoma NG-108 cultured cells by around 40 % at a clinically relevant concentration of 10 μM with no effect on HVA currents (Twombly et al. 1988). However, no effect was observed on T-type currents and variable minimal effects (1–16 %) were observed on HVA currents at clinical concentrations in acutely dissociated human neocortical neurons (Sayer et al. 1993).

4.7 Preclinical T-Type Calcium Channel Blockers in Epilepsy

4.7.1 *Z944*

Z944 was developed by Zalicus Pharmaceuticals and is currently undergoing Phase I clinical trials at the time of writing. Z944 blocks all three T-type calcium channel isoforms with submicromolar affinity when expressed exogenously and at 1 μM is capable of inducing 95 % blockade of T-type calcium currents in both reticular thalamic neurons (Tringham et al. 2012) and ventrobasal thalamic relay neurons (Cain and Snutch, unpublished data) recorded in acute brain slices. At this concentration, Z944 completely abolished burst-firing without preventing tonic firing. Systemic administration of Z944 to GAERS rats resulted in a dose-dependent reduction in the total time spent in the SWD state and also reduced the duration and spike frequency of SWDs. This was compared to ETX which demonstrated a lower percentage reduction in time spent in the SWD state at a higher dose and which had no effect on SWD duration or spike frequency.

4.7.2 *TTA Compounds*

A number of T-type blockers with submicromolar affinities for all three T-type calcium channel isoforms (TTA-P1, TTA-P2, TTA-A1, TTA-A2) have been developed by Merck Pharmaceuticals and some of these have been shown to suppress SWDs in the WAG/Rij model of absence seizures (Dreyfus et al. 2010; Kraus et al. 2010; Shipe et al. 2008; Uebele et al. 2009; Yang et al. 2008). A report indicating that TTA-A2 promotes sleep may detract from the usefulness of this and potentially the other TTA compounds as clinical anti-epileptic therapies (Kraus et al. 2010). However, structural differences between the compounds as well as clinical therapeutic ranges may vary sufficiently enough between their sleep and seizure-reducing effects for the sedative issue to be avoided.

Acknowledgments T. P. Snutch is supported by an operating grant from the Canadian Institutes of Health Research (#10677) and a Canada Research Chair in Biotechnology and Genomics-Neurobiology. S. M. Cain received support from the B.C. Epilepsy Society and the Michael Smith Foundation for Health Research.

References

Adams PJ, Snutch TP (2007) Calcium channelopathies: voltage-gated calcium channels. Subcell Biochem 45:215–251

Akaike N, Kostyuk PG, Osipchuk YV (1989) Dihydropyridine-sensitive low-threshold calcium channels in isolated rat hypothalamic neurones. J Physiol 412:181–195

Akman O, Demiralp T, Ates N, Onat FY (2010) Electroencephalographic differences between WAG/Rij and GAERS rat models of absence epilepsy. Epilepsy Res 89:185–193

Becker AJ, Pitsch J, Sochivko D, Opitz T, Staniek M, Chen CC, Campbell KP, Schoch S, Yaari Y, Beck H (2008) Transcriptional upregulation of $Ca_v3.2$ mediates epileptogenesis in the pilocarpine model of epilepsy. J Neurosci 28:13341–13353

Berrow NS, Brice NL, Tedder I, Page KM, Dolphin AC (1997) Properties of cloned rat alpha1A calcium channels transiently expressed in the COS-7 cell line. Eur J Neurosci 9:739–748

Biton V (2007) Clinical pharmacology and mechanism of action of zonisamide. Clin Neuropharmacol 30:230–240

Bourinet E, Alloui A, Monteil A, Barrère C, Couette B, Poirot O, Pages A, McRory J, Snutch TP, Eschalier A, Nargeot J (2005) Silencing of the $Ca_v3.2$ T-type calcium channel gene in sensory neurons demonstrates its major role in nociception. EMBO J 24:315–324

Broicher T, Seidenbecher T, Meuth P, Munsch T, Meuth SG, Kanyshkova T, Pape HC, Budde T (2007) T-current related effects of antiepileptic drugs and a Ca2 + channel antagonist on thalamic relay and local circuit interneurons in a rat model of absence epilepsy. Neuropharmacology 53:431–446

Broicher T, Kanyshkova T, Meuth P, Pape HC, Budde T (2008) Correlation of T-channel coding gene expression, IT, and the low threshold Ca2 + spike in the thalamus of a rat model of absence epilepsy. Mol Cell Neurosci 39:384–399

Browne TR, Dreifuss FE, Dyken PR, Goode DJ, Penry JK, Porter RJ, White BG, White PT (1975) Ethosuximide in the treatment of absence (peptit mal) seizures. Neurology 25:515–524

Cain SM, Snutch TP (2010) Contributions of T-type calcium channel isoforms to neuronal firing. Channels 4:44–51

Cain SM, Snutch TP (2011) Voltage-gated calcium channels and disease. Biofactors Oxf Engl 37:197–205

Cain SM, Snutch TP (2013) T-type calcium channels in burst-firing, network synchrony, and epilepsy. Biochim Biophys Acta 1828:1572–1578

Carter AG, Sabatini BL (2004) State-dependent calcium signaling in dendritic spines of striatal medium spiny neurons. Neuron 44:483–493

Catterall WA, Perez-Reyes E, Snutch TP, Striessnig J (2005) International Union of Pharmacology. XLVIII. Nomenclature and structure-function relationships of voltage-gated calcium channels. Pharmacol Rev 57:411–425

Cavalheiro EA (1995) The pilocarpine model of epilepsy. Ital J Neurol Sci 16:33–37

Cavalheiro EA, Santos NF, Priel MR (1996) The pilocarpine model of epilepsy in mice. Epilepsia 37:1015–1019

Chemin J, Monteil A, Bourinet E, Nargeot J, Lory P (2001) Alternatively spliced alpha(1G) (Ca(V)3.1) intracellular loops promote specific T-type Ca(2 +) channel gating properties. Biophys J 80:1238–1250

Chen Y, Lu J, Pan H, Zhang Y, Wu H, Xu K, Liu X, Jiang Y, Bao X, Yao Z, Ding K, Lo WH, Qiang B, Chan P, Shen Y, Wu X (2003) Association between genetic variation of CACNA1H and childhood absence epilepsy. Ann Neurol 54:239–243

Chen SD, Yeh KH, Huang YH, Shaw FZ (2011) Effect of intracranial administration of ethosuximide in rats with spontaneous or pentylenetetrazol-induced spike-wave discharges. Epilepsia 52:1311–1318

Cheong E, Lee S, Choi BJ, Sun M, Lee CJ, Shin HS (2008) Tuning thalamic firing modes via simultaneous modulation of T- and L-type Ca2 + channels controls pain sensory gating in the thalamus. J. Neurosci. Off. J. Soc. Neurosci. 28:13331–13340

Cheong E, Zheng Y, Lee K, Lee J, Kim S, Sanati M, Lee S, Kim YS, Shin HS (2009) Deletion of phospholipase C beta4 in thalamocortical relay nucleus leads to absence seizures. Proc Natl Acad Sci U S 106:21912–21917

Chevalier M, Lory P, Mironneau C, Macrez N, Quignard JF (2006) T-type $Ca_v3.3$ calcium channels produce spontaneous low-threshold action potentials and intracellular calcium oscillations. Eur J Neurosci 23:2321–2329

Chuang RS, Jaffe H, Cribbs L, Perez-Reyes E, Swartz KJ (1998) Inhibition of T-type voltage-gated calcium channels by a new scorpion toxin. Nat Neurosci 1:668–674

Coenen AM, Van Luijtelaar EL (2003) Genetic animal models for absence epilepsy: a review of the WAG/Rij strain of rats. Behav Genet 33:635–655

Contreras D (2006) The role of T-channels in the generation of thalamocortical rhythms. CNS Neurol Disord: Drug Targets 5:571–585

Coulter DA, Huguenard JR, Prince DA (1989a) Specific petit mal anticonvulsants reduce calcium currents in thalamic neurons. Neurosci Lett 98:74–78

Coulter DA, Huguenard JR, Prince DA (1989b) Characterization of ethosuximide reduction of low-threshold calcium current in thalamic neurons. Ann Neurol 25:582–593

Craig PJ, Beattie RE, Folly EA, Banerjee MD, Reeves MB, Priestley JV, Carney SL, Sher E, Perez-Reyes E, Volsen SG (1999) Distribution of the voltage-dependent calcium channel alpha1G subunit mRNA and protein throughout the mature rat brain. Eur J Neurosci 11:2949–2964

Crunelli V, Leresche N (2002) Block of Thalamic T-Type Ca(2 +) Channels by Ethosuximide Is Not the Whole Story. Epilepsy Curr 2:53–56

David LS, Garcia E, Cain SM, Thau E, Tyson JR, Snutch TP (2010) Splice-variant changes of the Ca(V)3.2 T-type calcium channel mediate voltage-dependent facilitation and associate with cardiac hypertrophy and development. Channels (Austin) 4:375–389

De Curtis M, Radici C, Forti M (1999) Cellular mechanisms underlying spontaneous interictal spikes in an acute model of focal cortical epileptogenesis. Neuroscience 88:107–117

Diana MA, Otsu Y, Maton G, Collin T, Chat M, Dieudonné S (2007) T-type and L-type Ca2 + conductances define and encode the bimodal firing pattern of vestibulocerebellar unipolar brush cells. J Neurosci 27:3823–3838

Dreyfus FM, Tscherter A, Errington AC, Renger JJ, Shin HS, Uebele VN, Crunelli V, Lambert RC, Leresche N (2010) Selective T-type calcium channel block in thalamic neurons reveals channel redundancy and physiological impact of I(T)window. J Neurosci 30:99–109

Emerick MC, Stein R, Kunze R, McNulty MM, Regan MR, Hanck DA, Agnew WS (2006) Profiling the array of Ca(v)3.1 variants from the human T-type calcium channel gene CACNA1G: alternative structures, developmental expression, and biophysical variations. Proteins 64:320–342

Ernst WL, Zhang Y, Yoo JW, Ernst SJ, Noebels JL (2009) Genetic enhancement of thalamocortical network activity by elevating alpha 1 g-mediated low-voltage-activated calcium current induces pure absence epilepsy. J Neurosci 29:1615–1625

Errington AC, Renger JJ, Uebele VN, Crunelli V (2010) State-dependent firing determines intrinsic dendritic Ca2 + signaling in thalamocortical neurons. J Neurosci 30:14843–14853

Faas GC, Vreugdenhil M, Wadman WJ (1996) Calcium currents in pyramidal CA1 neurons in vitro after kindling epileptogenesis in the hippocampus of the rat. Neuroscience 75:57–67

Fox AP, Nowycky MC, Tsien RW (1987) Kinetic and pharmacological properties distinguishing three types of calcium currents in chick sensory neurones. J Physiol 394:149–172

Gomora JC, Daud AN, Weiergräber M, Perez-Reyes E (2001) Block of cloned human T-type calcium channels by succinimide antiepileptic drugs. Mol Pharmacol 60:1121–1132

Graef JD, Nordskog BK, Wiggins WF, Godwin DW (2009) An acquired channelopathy involving thalamic T-type Ca2 + channels after status epilepticus. J Neurosci 29:4430–4441

Gülhan Aker R, Tezcan K, Carçak N, Sakalli E, Akin D, Onat FY (2009) Localized cortical injections of ethosuximide suppress spike-and-wave activity and reduce the resistance to kindling in genetic absence epilepsy rats (GAERS). Epilepsy Res 89:7–16

Heron SE, Khosravani H, Varela D, Bladen C, Williams TC, Newman MR, Scheffer IE, Berkovic SF, Mulley JC, Zamponi GW (2007) Extended spectrum of idiopathic generalized epilepsies associated with CACNA1H functional variants. Ann Neurol 62:560–568

Herrington J, Lingle CJ (1992) Kinetic and pharmacological properties of low voltage-activated Ca2 + current in rat clonal (GH3) pituitary cells. J Neurophysiol 68:213–232

Huang Z, Lujan R, Kadurin I, Uebele VN, Renger JJ, Dolphin AC, Shah MM (2011) Presynaptic HCN1 channels regulate $Ca_v3.2$ activity and neurotransmission at select cortical synapses. Nat Neurosci 14:478–486

Hughes SW, Crunelli V (2005) Thalamic mechanisms of EEG alpha rhythms and their pathological implications. Neuroscientist 11:357–372

Hughes SW, Crunelli V (2007) Just a phase they're going through: the complex interaction of intrinsic high-threshold bursting and gap junctions in the generation of thalamic alpha and theta rhythms. Int J Psychophysiol 64:3–17

Huguenard JR (1996) Low-threshold calcium currents in central nervous system neurons. Annu Rev Physiol 58:329–348

Isope P, Murphy TH (2005) Low threshold calcium currents in rat cerebellar Purkinje cell dendritic spines are mediated by T-type calcium channels. J Physiol 562:257–269

Jacus MO, Uebele VN, Renger JJ, Todorovic SM (2012) Presynaptic $Ca_v3.2$ channels regulate excitatory neurotransmission in nociceptive dorsal horn neurons. J Neurosci 32:9374–9382

Jiruska P, de Curtis M, Jefferys JG, Schevon CA, Schiff SJ, Schindler K (2013) Synchronization and desynchronization in epilepsy: controversies and hypotheses. J Physiol 591:787–797

Joksovic PM, Bayliss DA, Todorovic SM (2005) Different kinetic properties of two T-type Ca2 + currents of rat reticular thalamic neurones and their modulation by enflurane. J Physiol 566:125–142

Kavalali ET, Zhuo M, Bito H, Tsien RW (1997) Dendritic Ca2 + channels characterized by recordings from isolated hippocampal dendritic segments. Neuron 18:651–663

Kelly KM, Gross RA, Macdonald RL (1990) Valproic acid selectively reduces the low-threshold (T) calcium current in rat nodose neurons. Neurosci Lett 116:233–238

Kim D, Song I, Keum S, Lee T, Jeong MJ, Kim SS, McEnery MW, Shin HS (2001) Lack of the burst firing of thalamocortical relay neurons and resistance to absence seizures in mice lacking alpha(1G) T-type Ca2 + channels. Neuron 31:35–45

Kito M, Maehara M, Watanabe K (1996) Mechanisms of T-type calcium channel blockade by zonisamide. Seizure 5:115–119

Klöckner U, Lee JH, Cribbs LL, Daud A, Hescheler J, Pereverzev A, Perez-Reyes E, Schneider T (1999) Comparison of the Ca2 + currents induced by expression of three cloned alpha1 subunits, alpha1G, alpha1H and alpha1I, of low-voltage-activated T-type Ca2 + channels. Eur J Neurosci 11:4171–4178

Kostyuk PG, Molokanova EA, Pronchuk NF, Savchenko AN, Verkhratsky AN (1992) Different action of ethosuximide on low- and high-threshold calcium currents in rat sensory neurons. Neuroscience 51:755–758

Kraus RL, Li Y, Gregan Y, Gotter AL, Uebele VN, Fox SV, Doran SM, Barrow JC, Yang ZQ, Reger TS, Koblan KS, Renger JJ (2010) In vitro characterization of T-type calcium channel antagonist TTA-A2 and in vivo effects on arousal in mice. J Pharmacol Exp Ther 335:409–417

Kumar PP, Stotz SC, Paramashivappa R, Beedle AM, Zamponi GW, Rao AS (2002) Synthesis and evaluation of a new class of nifedipine analogs with T-type calcium channel blocking activity. Mol Pharmacol 61:649–658
Kwan P, Sills GJ, Brodie MJ (2001) The mechanisms of action of commonly used antiepileptic drugs. Pharmacol Ther 90:21–34
Lee JH, Gomora JC, Cribbs LL, Perez-Reyes E (1999) Nickel block of three cloned T-type calcium channels: low concentrations selectively block alpha1H. Biophys J 77:3034–3042
Leresche N, Parri HR, Erdemli G, Guyon A, Turner JP, Williams SR, Asprodini E, Crunelli V (1998) On the action of the anti-absence drug ethosuximide in the rat and cat thalamus. J Neurosci 18:4842–4853
Liang J, Zhang Y, Chen Y, Wang J, Pan H, Wu H, Xu K, Liu X, Jiang Y, Shen Y, Wu X (2007) Common polymorphisms in the CACNA1H gene associated with childhood absence epilepsy in Chinese Han population. Ann Hum Genet 71:325–335
Lisman JE (1997) Bursts as a unit of neural information: making unreliable synapses reliable. Trends Neurosci 20:38–43
Llinas R, Yarom Y (1981) Electrophysiology of mammalian inferior olivary neurones in vitro. Different types of voltage-dependent ionic conductances. J Physiol 315:549–567
Magee JC, Johnston D (1995) Synaptic activation of voltage-gated channels in the dendrites of hippocampal pyramidal neurons. Science 268:301–304
Manning JP, Richards DA, Leresche N, Crunelli V, Bowery NG (2004) Cortical-area specific block of genetically determined absence seizures by ethosuximide. Neuroscience 123:5–9
Marescaux C, Vergnes M, Depaulis A (1992) Genetic absence epilepsy in rats from Strasbourg-a review. J Neural Transm Suppl 35:37–69
Markram H, Sakmann B (1994) Calcium transients in dendrites of neocortical neurons evoked by single subthreshold excitatory postsynaptic potentials via low-voltage-activated calcium channels. Proc Natl Acad Sci U S 91:5207–5211
Massie BM (1997) Mibefradil: a selective T-type calcium antagonist. Am J Cardiol 80:23I–32I
Massie BM (1998) Mibefradil, a T-type channel-selective calcium antagonist: clinical trials in chronic stable angina pectoris. Am J Hypertens 11:95S–102S
Matar N, Jin W, Wrubel H, Hescheler J, Schneider T, Weiergräber M (2009) Zonisamide block of cloned human T-type voltage-gated calcium channels. Epilepsy Res 83:224–234
McCormick DA, Contreras D (2001) On the cellular and network bases of epileptic seizures. Annu Rev Physiol 63:815–846
McDonough SI, Bean BP (1998) Mibefradil inhibition of T-type calcium channels in cerebellar purkinje neurons. Mol Pharmacol 54:1080–1087
McKay BE, McRory JE, Molineux ML, Hamid J, Snutch TP, Zamponi GW, Turner RW (2006) Ca(V)3 T-type calcium channel isoforms differentially distribute to somatic and dendritic compartments in rat central neurons. Eur J Neurosci 24:2581–2594
McRory JE, Santi CM, Hamming KS, Mezeyova J, Sutton KG, Baillie DL, Stea A, Snutch TP (2001) Molecular and functional characterization of a family of rat brain T-type calcium channels. J Biol Chem 276:3999–4011
Mittman S, Guo J, Emerick MC, Agnew WS (1999) Structure and alternative splicing of the gene encoding alpha1I, a human brain T calcium channel alpha1 subunit. Neurosci Lett 269:121–124
Monteil A, Chemin J, Bourinet E, Mennessier G, Lory P, Nargeot J (2000) Molecular and functional properties of the human alpha(1G) subunit that forms T-type calcium channels. J Biol Chem 275:6090–6100
Mudado MA, Rodrigues AL, Prado VF, Beirão PS, Cruz JS (2004) Ca_v 3.1 and Ca_v 3.3 account for T-type Ca2 + current in GH3 cells. Braz J Med Biol Res 37:929–935
Murbartián J, Arias JM, Perez-Reyes E (2004) Functional impact of alternative splicing of human T-type Ca_v3.3 calcium channels. J Neurophysiol 92:3399–3407
Perez-Reyes E (2003) Molecular physiology of low-voltage-activated T-type calcium channels. Physiol Rev 83:117–161

Pinault D, Leresche N, Charpier S, Deniau JM, Marescaux C, Vergnes M, Crunelli V (1998) Intracellular recordings in thalamic neurones during spontaneous spike and wave discharges in rats with absence epilepsy. J Physiol 509:449–456

Pinault D, Vergnes M, Marescaux C (2001) Medium-voltage 5-9-Hz oscillations give rise to spike-and-wave discharges in a genetic model of absence epilepsy: in vivo dual extracellular recording of thalamic relay and reticular neurons. Neuroscience 105:181–201

Polack PO, Charpier S (2009) Ethosuximide converts ictogenic neurons initiating absence seizures into normal neurons in a genetic model. Epilepsia 50:1816–1820

Polack PO, Guillemain I, Hu E, Deransart C, Depaulis A, Charpier S (2007) Deep layer somatosensory cortical neurons initiate spike-and-wave discharges in a genetic model of absence seizures. J Neurosci 27:6590–6599

Polack PO, Mahon S, Chavez M, Charpier S (2009) Inactivation of the Somatosensory Cortex Prevents Paroxysmal Oscillations in Cortical and Related Thalamic Neurons in a Genetic Model of Absence Epilepsy. Cereb Cortex 19:2078–2091

Powell KL, Cain SM, Ng C, Sirdesai S, David LS, Kyi M, Garcia E, Tyson JR, Reid CA, Bahlo M, Foote SJ, Snutch TP, O'Brien TJ (2009) A $Ca_v3.2$ T-type calcium channel point mutation has splice-variant-specific effects on function and segregates with seizure expression in a polygenic rat model of absence epilepsy. J Neurosci 29:371–380

Randall AD, Tsien RW (1997) Contrasting biophysical and pharmacological properties of T-type and R-type calcium channels. Neuropharmacology 36:879–893

Reinagel P, Godwin D, Sherman SM, Koch C (1999) Encoding of visual information by LGN bursts. J Neurophysiol 81:2558–2569

Richards DA, Manning JP, Barnes D, Rombola L, Bowery NG, Caccia S, Leresche N, Crunelli V (2003) Targeting thalamic nuclei is not sufficient for the full anti-absence action of ethosuximide in a rat model of absence epilepsy. Epilepsy Res 54:97–107

Rogawski MA, Porter RJ (1990) Antiepileptic drugs: pharmacological mechanisms and clinical efficacy with consideration of promising developmental stage compounds. Pharmacol Rev 42:223–286

Sanabria ER, Su H, Yaari Y (2001) Initiation of network bursts by Ca2 + -dependent intrinsic bursting in the rat pilocarpine model of temporal lobe epilepsy. J Physiol 532:205–216

Sayer RJ, Brown AM, Schwindt PC, Crill WE (1993) Calcium currents in acutely isolated human neocortical neurons. J Neurophysiol 69:1596–1606

Sherman SM (2001) Tonic and burst firing: dual modes of thalamocortical relay. Trends Neurosci 24:122–126

Shipe WD, Barrow JC, Yang ZQ, Lindsley CW, Yang FV, Schlegel KA, Shu Y, Rittle KE, Bock MG, Hartman GD, Tang C, Ballard JE, Kuo Y, Adarayan ED, Prueksaritanont T, Zrada MM, Uebele VN, Nuss CE, Connolly TM, Doran SM, Fox SV, Kraus RL, Marino MJ, Graufelds VK, Vargas HM, Bunting PB, Hasbun-Manning M, Evans RM, Koblan KS, Renger JJ (2008) Design, synthesis, and evaluation of a novel 4-aminomethyl-4-fluoropiperidine as a T-type Ca2 + channel antagonist. J Med Chem 51:3692–3695

Sidach SS, Mintz IM (2002) Kurtoxin, a gating modifier of neuronal high- and low-threshold ca channels. J Neurosci 22:2023–2034

Singh B, Monteil A, Bidaud I, Sugimoto Y, Suzuki T, Hamano S, Oguni H, Osawa M, Alonso ME, Delgado-Escueta AV, Inoue Y, Yasui-Furukori N, Kaneko S, Lory P, Yamakawa K (2007) Mutational analysis of CACNA1G in idiopathic generalized epilepsy. Hum Mutat 28:524–525

Slaght SJ, Leresche N, Deniau JM, Crunelli V, Charpier S (2002) Activity of thalamic reticular neurons during spontaneous genetically determined spike and wave discharges. J Neurosci 22:2323–2334

Su H, Sochivko D, Becker A, Chen J, Jiang Y, Yaari Y, Beck H (2002) Upregulation of a T-type Ca2 + channel causes a long-lasting modification of neuronal firing mode after status epilepticus. J Neurosci 22:3645–3655

Suzuki S, Rogawski MA (1989) T-type calcium channels mediate the transition between tonic and phasic firing in thalamic neurons. Proc Natl Acad Sci U S A 86:7228–7232

Suzuki S, Kawakami K, Nishimura S, Watanabe Y, Yagi K, Seino M, Miyamoto K (1992) Zonisamide blocks T-type calcium channel in cultured neurons of rat cerebral cortex. Epilepsy Res 12:21–27

Tai C, Kuzmiski JB, MacVicar BA (2006) Muscarinic enhancement of R-type calcium currents in hippocampal CA1 pyramidal neurons. J Neurosci 26:6249–6258

Talley EM, Cribbs LL, Lee JH, Daud A, Perez-Reyes E, Bayliss DA (1999) Differential distribution of three members of a gene family encoding low voltage-activated (T-type) calcium channels. J Neurosci 19:1895–1911

Talley EM, Solórzano G, Depaulis A, Perez-Reyes E, Bayliss DA (2000) Low-voltage-activated calcium channel subunit expression in a genetic model of absence epilepsy in the rat. Brain Res Mol Brain Res 75:159–165

Tang AH, Karson MA, Nagode DA, McIntosh JM, Uebele VN, Renger JJ, Klugmann M, Milner TA, Alger BE (2011) Nerve terminal nicotinic acetylcholine receptors initiate quantal GABA release from perisomatic interneurons by activating axonal T-type (Ca_v3) Ca^{2+} channels and Ca^{2+} release from stores. J Neurosci 31:13546–13561

Tennigkeit F, Schwarz DW, Puil E (1998) Modulation of bursts and high-threshold calcium spikes in neurons of rat auditory thalamus. Neuroscience 83:1063–1073

Timofeev I, Steriade M (1998) Cellular mechanisms underlying intrathalamic augmenting responses of reticular and relay neurons. J Neurophysiol 79:2716–2729

Todorovic SM, Jevtovic-Todorovic V (2013) Neuropathic pain: role for presynaptic T-type channels in nociceptive signaling. Pflugers Arch. doi:10.1007/s00424-012-1211-y

Todorovic SM, Lingle CJ (1998) Pharmacological properties of T-type Ca2 + current in adult rat sensory neurons: effects of anticonvulsant and anesthetic agents. J Neurophysiol 79:240–252

Tringham E, Powell KL, Cain SM, Kuplast K, Mezeyova J, Weerapura M, Eduljee C, Jiang X, Smith P, Morrison JL, Jones NC, Braine E, Rind G, Fee-Maki M, Parker D, Pajouhesh H, Parmar M, O'Brien TJ, Snutch TP (2012) T-type calcium channel blockers that attenuate thalamic burst firing and suppress absence seizures. Sci Transl Med 4:121ra19

Tryba AK, Kaczorowski CC, Ben-Mabrouk F, Elsen FP, Lew SM, Marcuccilli CJ (2011) Rhythmic intrinsic bursting neurons in human neocortex obtained from pediatric patients with epilepsy. Eur J Neurosci 34:31–44

Tsakiridou E, Bertollini L, de Curtis M, Avanzini G, Pape HC (1995) Selective increase in T-type calcium conductance of reticular thalamic neurons in a rat model of absence epilepsy. J Neurosci 15:3110–3117

Twombly DA, Yoshii M, Narahashi T (1988) Mechanisms of calcium channel block by phenytoin. J Pharmacol Exp Ther 246:189–195

Uebele VN, Nuss CE, Fox SV, Garson SL, Cristescu R, Doran SM, Kraus RL, Santarelli VP, Li Y, Barrow JC, Yang ZQ, Schlegel KA, Rittle KE, Reger TS, Bednar RA, Lemaire W, Mullen FA, Ballard JE, Tang C, Dai G, McManus OB, Koblan KS, Renger JJ (2009) Positive allosteric interaction of structurally diverse T-type calcium channel antagonists. Cell Biochem Biophys 55:81–93

Viana F, Van den Bosch L, Missiaen L, Vandenberghe W, Droogmans G, Nilius B, Robberecht W (1997) Mibefradil (Ro 40-5967) blocks multiple types of voltage-gated calcium channels in cultured rat spinal motoneurones. Cell Calcium 22:299–311

Weiergräber M, Stephani U, Köhling R (2010) Voltage-gated calcium channels in the etiopathogenesis and treatment of absence epilepsy. Brain Res Rev 62:245–271

Welker HA, Wiltshire H, Bullingham R (1998) Clinical pharmacokinetics of mibefradil. Clin Pharmacokinet 35:405–423

Wellmer J, Su H, Beck H, Yaari Y (2002) Long-lasting modification of intrinsic discharge properties in subicular neurons following status epilepticus. Eur J Neurosci 16:259–266

Williams SR, Tóth TI, Turner JP, Hughes SW, Crunelli V (1997) The "window" component of the low threshold Ca2 + current produces input signal amplification and bistability in cat and rat thalamocortical neurones. J Physiol 505:689–705

Wolfart J, Roeper J (2002) Selective coupling of T-type calcium channels to SK potassium channels prevents intrinsic bursting in dopaminergic midbrain neurons. J Neurosci 22:3404–3413

Yaari Y, Yue C, Su H (2007) Recruitment of apical dendritic T-type Ca2 + channels by backpropagating spikes underlies de novo intrinsic bursting in hippocampal epileptogenesis. J Physiol 580:435–450

Yang ZQ, Barrow JC, Shipe WD, Schlegel KA, Shu Y, Yang FV, Lindsley CW, Rittle KE, Bock MG, Hartman GD, Uebele VN, Nuss CE, Fox SV, Kraus RL, Doran SM, Connolly TM, Tang C, Ballard JE, Kuo Y, Adarayan ED, Prueksaritanont T, Zrada MM, Marino MJ, Graufelds VK, DiLella AG, Reynolds IJ, Vargas HM, Bunting PB, Woltmann RF, Magee MM, Koblan KS, Renger JJ (2008) Discovery of 1,4-substituted piperidines as potent and selective inhibitors of T-type calcium channels. J Med Chem 51:6471–6477

Yunker AM (2003) Modulation and pharmacology of low voltage-activated ("T-Type") calcium channels. J Bioenerg Biomembr 35:577–598

Yunker AM, McEnery MW (2003) Low-voltage-activated ("T-Type") calcium channels in review. J Bioenerg Biomembr 35:533–575

Zamponi GW, Bourinet E, Snutch TP (1996) Nickel block of a family of neuronal calcium channels: subtype- and subunit-dependent action at multiple sites. J Membr Biol 151:77–90

Zamponi GW, Lory P, Perez-Reyes E (2009) Role of voltage-gated calcium channels in epilepsy. Pflugers Arch 460:395–403

Zhang Y, Mori M, Burgess DL, Noebels JL (2002) Mutations in high-voltage-activated calcium channel genes stimulate low-voltage-activated currents in mouse thalamic relay neurons. J Neurosci 22:6362–6371

Zheng TW, O'Brien TJ, Morris MJ, Reid CA, Jovanovska V, O'Brien P, van Raay L, Gandrathi AK, Pinault D (2012) Rhythmic neuronal activity in S2 somatosensory and insular cortices contribute to the initiation of absence-related spike-and-wave discharges. Epilepsia 53:1948–1958

Chapter 5
The Involvement of Calcium Channel $\alpha_2\delta$ Subunits in Diseases and as a Therapeutic Target

Annette C. Dolphin

Abstract The $\alpha_2\delta$ subunits are auxiliary subunits of voltage-gated calcium channels that may also have roles independent of calcium channels. Four mammalian $\alpha_2\delta$ subunits have been identified, and all of them have been identified to play a role in a variety of pathologies. Furthermore, two $\alpha_2\delta$ subunits ($\alpha_2\delta$-1 and $\alpha_2\delta$-2) are pharmacological targets for the gabapentinoid drugs, which are used therapeutically in the treatment of epilepsies and neuropathic pain.

5.1 Introduction

Voltage-gated calcium channels can be formed from one of ten different mammalian calcium channel $\alpha_1$subunit genes products. In the case of the high-voltage activated (HVA) Ca_v1 and Ca_v2 channels, these co-assemble with one of four $\alpha_2\delta$- and four β- subunits, all of which can exist in a number of variants owing to alternative splicing. This makes for an enormous potential diversity of properties and function in calcium channel complexes. The γ subunit γ_1 is found associated with skeletal muscle calcium channels (Jay et al. 1990), but γ subunits do not appear to form a part of cardiac (Walsh et al. 2009) or neuronal (Moss et al. 2002; Müller et al. 2010) calcium channel complexes.

Although the α_1 subunits provide the main physiological and pharmacological properties of the channels, this is modified by the auxiliary subunits. The auxiliary $\alpha_2\delta$ and β subunits also have major roles in trafficking the channels to the plasma membrane and specific domains of cells, including neurons. Here I will focus on the roles of the auxiliary $\alpha_2\delta$ subunits, since there are a surprising number of different disease processes, associated with this protein family. Furthermore, some isoforms are also involved in therapeutic intervention as a drug binding site.

A. C. Dolphin (✉)
Department of Neuroscience, Physiology and Pharmacology, University College London, Gower St, London WC1E 6BT, UK
e-mail: a.dolphin@ucl.ac.uk

N. Weiss and A. Koschak (eds.), *Pathologies of Calcium Channels*,
DOI: 10.1007/978-3-642-40282-1_5, © Springer-Verlag Berlin Heidelberg 2014

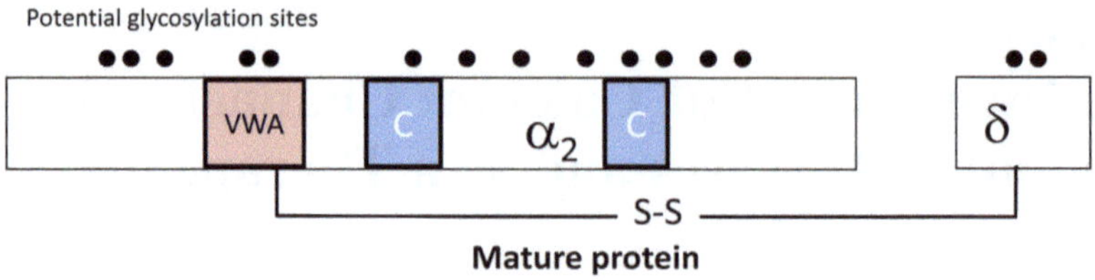

Fig. 5.1 Domain structure of $\alpha_2\delta$ subunits

5.2 The $\alpha_2\delta$ Auxiliary Subunits

5.2.1 $\alpha_2\delta$ Isoforms

There are four mammalian genes encoding $\alpha_2\delta$ subunits, termed *CACNA2D1-4*, encoding $\alpha_2\delta$-1-4 (for reviews, see Felix 1999; Davies et al. 2007; Dolphin 2012). All the $\alpha_2\delta$ proteins have a similar structure. The N-terminus has a signal sequence indicating it is extracellular, and there are a number of recognised domains in the structure, including a Von Willebrand Factor-A domain (Whittaker and Hynes 2002), and two bacterial chemosensory-like or Cache domains (Anantharaman and Aravind 2000; Dolphin 2012) (Fig. 5.1). The C-terminus possesses a hydrophobic potentially transmembrane domain (Ellis et al. 1988; Jay et al. 1991), leading to the widely accepted view that $\alpha_2\delta$ proteins are single pass type I transmembrane proteins. However, at least two of the $\alpha_2\delta$ subunits ($\alpha_2\delta$-3 and $\alpha_2\delta$-4) are strongly predicted to be glycosylphosphatidyl inositol (GPI)-anchored because of the brevity of the hydrophobic domain and its presence at the extreme C-terminus, and $\alpha_2\delta$-1 has also previously been predicted to be GPI-anchored (Pierleoni et al. 2008). We have obtained experimental evidence for this topology for the $\alpha_2\delta$ subunits $\alpha_2\delta$-1–$\alpha_2\delta$-3 (Davies et al. 2010). For all the $\alpha_2\delta$ subunits, the genes encode a pre-protein that is proteolytically cleaved (Jay et al. 1991), leaving the two subunits to remain associated by disulfide bonding. For $\alpha_2\delta$-1, the cysteines residues involved in disulfide bridge formation have been identified (Calderon-Rivera et al. 2012).

5.2.2 Association of α_1 Subunits with $\alpha_2\delta$ Subunits

All the high voltage-activated calcium channels (Ca_v1 and Ca_v2 classes) are able to associated with $\alpha_2\delta$ subunits (Tanabe et al. 1987; Witcher et al. 1993; Liu et al. 1996), although the association of the $\alpha_2\delta$ subunit with the complex is more easily disrupted than that of the β subunit (Müller et al. 2010), leading to the possibility that not all calcium channel complexes contain $\alpha_2\delta$ under all conditions. The finding that $\alpha_2\delta$ subunits may be isolated in part separately from calcium channel complexes in both muscle and brain (Jay et al. 1991; Gee et al. 1996; Müller et al. 2010) also lends support to the idea that these proteins may have other functions in tissues.

5.2.3 Mechanism of Action of $\alpha_2\delta$ Subunits with Respect to Calcium Channel Function

The $\alpha_2\delta$ subunits are all found to increase the maximum current density for the HVA Ca_v1 and Ca_v2 calcium channels (for review see Dolphin 2003). They also influence the kinetic and voltage-dependent characteristics of the calcium channel currents, increasing the inactivation rate to varying extents. In some studies, the different $\alpha_2\delta$ subunits also hyperpolarised the steady-state inactivation (Singer et al. 1991; Canti et al. 2005; Hendrich et al. 2008; Davies et al. 2010). In a study in which the cardiac calcium channel $Ca_v1.2$ was expressed in dysgenic skeletal muscle myoblasts, and $\alpha_2\delta$-1 was knocked down with siRNA, a number of consequences were recorded, including a reduction in calcium currents, a depolarizing shift in activation and slowed inactivation. An increase in action potential duration was also recorded, mainly as a result of the slowed calcium current inactivation (Tuluc et al. 2007). In $\alpha_2\delta$-1 knockout mice, a reduction in cardiac ventricular calcium channel currents and a depolarizing shift in their steady state inactivation were observed (Fuller-Bicer et al. 2009).

Both $\alpha_2\delta$ and β subunits increase plasma membrane expression of Ca_v1 and Ca_v2 channels. Evidence suggests that $\alpha_2\delta$ subunits are not effective unless β subunits are also present, and the β subunits are known to bind to the intracellular I-II linker of these channels (Pragnell et al. 1994) and protect them from endoplasmic reticulum associated proteasomal degradation (Waithe et al. 2011; Altier et al. 2011). The main mechanism for the increased current resulting from the presence of $\alpha_2\delta$ is likely to be an increase in the plasma membrane expression of Ca_v1 and Ca_v2 α_1 subunits (Canti et al. 2005), possibly due to a decrease of their turnover (Bernstein and Jones 2007), although how this occurs is still unclear. We have shown the MIDAS motif in $\alpha_2\delta$ subunits to be essential for this process (Canti et al. 2005; Hoppa et al. 2012). Mutation of this motif markedly reduced the functionality of both $\alpha_2\delta$-1 (Hoppa et al. 2012) and $\alpha_2\delta$-2 (Canti et al. 2005) subunits, with respect to increasing calcium currents in expression systems. We also found that the MIDAS mutant of $\alpha_2\delta$-2 caused intracellular retention of α_1 subunits (Canti et al. 2005). Although the location where the $\alpha_2\delta$ subunits influence the trafficking process of the calcium channel complex is not yet known, it is assumed that the $\alpha_2\delta$ proteins interact with an extracellular motif on the α_1 subunits. For example, it has been described that the α_2 subunit of $\alpha_2\delta$-1 binds to domain III of $Ca_v1.1$ (Gurnett et al. 1997). Increased trafficking may not be the only mechanism of action to increase calcium channel currents, as in an early study the reconstitution of skeletal muscle calcium channels into liposomes in vitro resulted in greater calcium flux in the presence of $\alpha_2\delta$ subunits (Gutierrez et al. 1991).

We have also studied the trafficking of $\alpha_2\delta$ subunits themselves, and our evidence indicates that $\alpha_2\delta$ subunits are likely also to interact with other cellular trafficking proteins (Davies et al. 2006; Hendrich et al. 2008; Tran-Van-Minh and Dolphin 2010). The $\alpha_2\delta$ subunits are all strongly localised to lipid raft domains (Davies et al. 2006, 2010), and substrates for phosphatidylinositol phospholipases (Davies et al. 2010), like other GPI-anchored proteins. Interestingly, a truncated $\alpha_2\delta$ subunit, lacking its C-terminal membrane anchor, despite being mainly secreted, still shows some extrinsic plasma membrane association, via an unknown binding partner (Kadurin et al. 2012).

At presynaptic terminals, overexpression of $\alpha_2\delta$-1, $\alpha_2\delta$-2 or $\alpha_2\delta$-3 markedly increased presynaptic transmitter release in response to a single action potential, while reducing calcium influx, likely as a result of reducing action potential duration (Hoppa et al. 2012). This indicates that there is a key effect of $\alpha_2\delta$ on channel localization at presynaptic active zones.

5.2.4 $\alpha_2\delta$ Subunit Splice Variants

The main $\alpha_2\delta$-1 subunit splice variant expressed in rat brain was found to be different from that in skeletal muscle (Kim et al. 1992). Three regions, termed A, B and C, were then identified from multiple sequence alignments to show alternative splicing (Angelotti and Hofmann 1996). Five different transcripts consisting of different combinations of these alternatively spliced regions were found in mouse brain, skeletal muscle, heart and smooth muscle (Angelotti and Hofmann 1996). We have found that $\alpha_2\delta$-1 ΔA + B + C is the main splice variant in dorsal root ganglion (DRG) neurons, as well as in brain (Angelotti and Hofmann 1996), but we also identified the presence of a minor splice variant ($\alpha_2\delta$-1ΔA + BΔC) in DRG neurons (Lana et al. 2012). Alternative splicing of the other $\alpha_2\delta$ subunits has also been described (Barclay and Rees 2000; Qin et al. 2002).

5.2.5 Distribution of $\alpha_2\delta$ Subunits in the Nervous System

The $\alpha_2\delta$-1, $\alpha_2\delta$-2 and $\alpha_2\delta$-3 subunits are expressed widely in both the central and peripheral nervous systems, with $\alpha_2\delta$-1 being found in many neuronal cell types (Cole et al. 2005), including DRG neurons (Newton et al. 2001; Bauer et al. 2009). The $\alpha_2\delta$-1 protein is localised in presynaptic terminal regions, and to a much lower extent in cell bodies under physiological conditions (Taylor and Garrido 2008; Bauer et al. 2009).

The expression of the $\alpha_2\delta$-1 transcript was also found to be correlated with excitatory rather than inhibitory neurons (Cole et al. 2005). In contrast, $\alpha_2\delta$-2 expression is lower in most brain regions, with restricted regions of high expression including cerebellum (Cole et al. 2005). The distribution of $\alpha_2\delta$-2 correlates partially with GABAergic neurons, including cerebellar Purkinje neurons (Barclay et al. 2001; Cole et al. 2005). The $\alpha_2\delta$-3 protein is expressed throughout the brain, particularly in the caudate-putamen (Cole et al. 2005). In contrast, $\alpha_2\delta$-4 protein is found in specific endocrine tissues, and at a low level in the brain (Qin et al. 2002). It is also present in the retina where genetic mutations result in a form of night blindness (Wycisk et al. 2006a).

5.3 Pathologies Relating to $\alpha_2\delta$ Subunits

All the $\alpha_2\delta$ subunits have been associated with pathology. This encompasses an unexpectedly large range of genetic and other pathologies, highlighting the fundamental roles of these proteins in the function of many cell types.

5.3.1 Pain

Peripheral sensory nerve damage may result in chronic pain that long outlives the injury. Nerve damage may occur due to direct physical trauma, or to nerve damage as a result of poorly regulated plasma glucose in diabetes, herpes virus infection, certain chemotherapeutic drugs and other causes. Sensory nerve injury has as one consequence a change in transcription of many genes, which may be either up or down-regulated (Newton et al. 2001; Wang et al. 2002; Xiao et al. 2002). One of the many molecular consequences of experimental peripheral nerve injury is an increase in the level of $\alpha_2\delta$-1 mRNA in damaged sensory neurons (DRGs), shown by in situ hybridization (Newton et al. 2001), microarray analysis (Wang et al. 2002; Xiao et al. 2002) and quantitative PCR (Bauer et al. 2009). There is a corresponding increase in $\alpha_2\delta$-1 protein in DRGs and spinal cord, as determined by Western blot (Luo et al. 2001) and immunohistochemistry (Bauer et al. 2009). In contrast, $Ca_v2.2$ mRNA and protein is not up-regulated following sensory nerve damage (Wang et al. 2002; Li et al. 2006), although there is a change in splicing (Altier et al. 2007). This leads to the possibility that up-regulated $\alpha_2\delta$-1 enhances $Ca_v2.2$ trafficking and presynaptic function, although it may also have other functions, beyond calcium channels.

In transgenic mice engineered to over-express $\alpha_2\delta$-1, there is a neuropathic phenotype of baseline hyperalgesia and allodynia (Li et al. 2006), indicating that the $\alpha_2\delta$-1 level in DRG neurons is a key component in determining the neuropathic phenotype. In accordance with this conclusion, we have recently found that in $\alpha_2\delta$-1 knockout mice (Fuller-Bicer et al. 2009) there are defects in mechanical and thermal sensitivity and a delay in response to nerve injury compared to wild type mice (Patel et al. 2012).

Straitjacket, the Drosophila homolog of *CACNA2D3,* was recently identified in a *Drosophila melanogaster* screen to be involved in pain sensation, since in both

Drosophila and mice mutants lacking this gene, there is impaired avoidance of noxious heat (Neely et al. 2010). The authors concluded that this was a result of impaired central processing of nociceptive signals. They also found that two intronic single nucleotide polymorphisms (SNPs) within the *CACNA2D3* gene showed association with reduced acute and chronic pain in humans, although the mechanism behind this difference is unknown, and the effect of the intronic mutations on gene expression was not determined. Indeed $\alpha_2\delta$-3 is not up-regulated in DRG neurons following peripheral nerve injury leading to neuropathic pain, rather there is a small down-regulation (Bauer et al. 2009), therefore the role of $\alpha_2\delta$-3 in this process may reflect the extensive distribution of $\alpha_2\delta$ proteins in the brain (Cole et al. 2005; Schlick et al. 2010), and widespread effect of $\alpha_2\delta$ proteins on synaptic transmission (Hoppa et al. 2012).

5.3.2 Epilepsies

Identification of the gabapentin receptor provided the first evidence that the $\alpha_2\delta$ subunits could be involved in the epilepsies, since gabapentin was already known to be efficacious in the treatment of some epilepsies, as an add-on drug to improve control of epileptic seizures (Marson et al. 2000). Gabapentin binds to both $\alpha_2\delta$-1 and $\alpha_2\delta$-2, but not to the other $\alpha_2\delta$ subunits (see Sect. 5.4 below). Subsequently, the spontaneously arising mouse mutants, *ducky*, *ducky*^{2J} and *entla* were all found to involve disruption of the *cacna2d2* gene. These mice exhibit paroxysmal dyskinesia and absence seizures in the case of the *ducky* mutations in which no full length protein is produced (Barclay et al. 2001; Brodbeck et al. 2002; Donato et al. 2006), and in the case of *entla* in which there is a duplication of exon 3 (Brill et al. 2004). Mice with a targeted *cacna2d2* gene deletion also show en epileptic/ataxic phenotype (Ivanov et al. 2004).

CACNA2D1 and *CACNA2D2* show a large number of SNPs (Klassen et al. 2011). Judging by the lack of phenotype in heterozygote mice lacking $\alpha_2\delta$-1 (Fuller-Bicer et al. 2009; Patel et al. 2012) or $\alpha_2\delta$-2 (Barclay et al. 2001), it was predicted that this would any overt epileptic phenotype will most likely involve a recessive mutation in humans. Indeed one such recessive human mutation has recently been identified (Edvardson et al. 2013).

5.3.3 Cerebellar Ataxia

The $\alpha_2\delta$-2 protein is strongly expressed in cerebellar Purkinje cells, and deletion or disruption of the *cacna2d2* gene results in cerebellar ataxia (Barclay et al. 2001; Brill et al. 2004; Ivanov et al. 2004), which is associated with altered Purkinje cell morphology (Brodbeck et al. 2002) and reduced Purkinje cell spontaneous activity

(Donato et al. 2006). Interestingly, ataxia is one of the adverse events observed with gabapentin (Beal et al. 2012) and pregabalin treatment (Zaccara et al. 2012).

5.3.4 Night Blindness

Mutations in *CACNA2D4* (encoding $\alpha 2\delta$-4 subunits) produce photoreceptor dysfunction, resulting in a type of night blindness. As well as the human mutations, a spontaneous mouse mutation in this gene has been identified, both showing autosomal recessive cone dystrophy resulting in night blindness (Wycisk et al. 2006a, b). This highlights the importance of $\alpha_2\delta$-4 in photoreceptor function.

5.3.5 Autism Spectrum Disorder

A splice site mutation in *CACNA2D3* was found to be one of a number of 'Likely Gene-Disrupting Mutations' (Iossifov et al. 2012) in autism spectrum disorders. Given the evidence that this spectrum of disorders involves synaptic dysfunction, and the involvement of mutations and copy number of variations in a number of synaptic proteins (Ting et al. 2012; Malhotra and Sebat 2012), it is perhaps not surprising that mutation in an important presynaptic protein, $\alpha_2\delta$-3, is found to be a potential cause of autism in some people.

5.3.6 Cardiac Dysfunction

The $\alpha_2\delta$-1 protein is highly expressed in skeletal, cardiac and smooth muscle (Ellis et al. 1988; Jay et al. 1991; Klugbauer et al. 1999), and is associated with the L-type calcium channels in these muscles (Jay et al. 1991; Wolf et al. 2003; Walsh et al. 2009). Mutations in *CACNA2D1* have been found to be associated with cardiac dysfunction, including Brugada (Burashnikov et al. 2010) and short QT (Templin et al. 2011) syndromes. In agreement with this, the knockout of *cacna2d1* in mice also resulted in a cardiac phenotype, the mice had a decrease in basal myocardial contractility and relaxation, and reduced cardiac calcium channel current density (Fuller-Bicer et al. 2009). These mice were also reported to have some bladder dysfunction, although whether the basis for this is of neuronal or smooth muscle origin is not yet determined (Fuller-Bicer et al. 2009).

5.3.7 *Modification of $\alpha_2\delta$ Subunit Genes in Tumour Cells*

Other roles for $\alpha_2\delta$ subunits unrelated to calcium channel function have recently been proposed. One suggestion that these proteins have multiple roles comes from the finding that the genes for both $\alpha_2\delta$-2 (Gao et al. 2000; Carboni et al. 2003; Senchenko et al. 2004) and $\alpha_2\delta$-3 (Hanke et al. 2001; Ji et al. 2005) have been localised to known tumour suppressor regions. Their suppression, by methylation-dependent silencing of transcription, has been associated with tumour susceptibility, tumour growth and poor prognosis in several different cancers (Wanajo et al. 2008; Palmieri et al. 2012; Leone et al. 2012), suggesting that they might normally play a role in intercellular communication and growth suppression (Fig. 5.2a). Furthermore, *CACNA2D2* showed increased expression in breast cancer cell lines following treatment with a demethylating agent (da Costa et al. 2011).

5.4 $\alpha_2\delta$ Subunits as a Therapeutic Target

As auxiliary subunits of voltage-gated calcium channels, the $\alpha_2\delta$ subunits would never have been investigated as *de novo* drug targets, since they were not known to have an endogenous functional ligand. They were only discovered heuristically, by identification of the protein to which gabapentin bound, as described below. In the absence of ^{3}H-gabapentin, their identification would not have been possible, because of the lack of a defined assay or known mechanism of action.

5.4.1 *Identification of $\alpha_2\delta$ Subunits as Receptors for Gabapentin*

Gabapentin (2-[1-(aminomethyl)-cyclohexyl]acetic acid) and pregabalin (*S*(+)-3-isobutyl GABA) were synthesized as analogs of the inhibitory neurotransmitter GABA, with the aim of mimicking GABA function or inhibiting GABA breakdown (Taylor et al. 2007). These drugs were found to be efficacious antiepileptic drugs, but their mechanism of action was unclear, as they were later found to have no consistent effect on GABA-A or GABA-B receptors, GABA metabolism or GABA transport (Taylor et al. 2007; Silverman 2008; Li et al. 2011).

Purification of the neuronal ^{3}H-gabapentin binding site from brain led to its subsequent identification as the $\alpha_2\delta$-1 subunit (Gee et al. 1996). ^{3}H-gabapentin was subsequently also found to bind to $\alpha_2\delta$-2 (Gong et al. 2001). A number of residues on these proteins have been shown to be involved in the binding of gabapentinoid drugs, in particular, the third arginine (R) in triple R motif, situated N-terminal to the VWA domain (Field et al. 2006; Davies et al. 2006). In a homology model of the tertiary structure of these two domains, this loop is structurally juxtaposed to

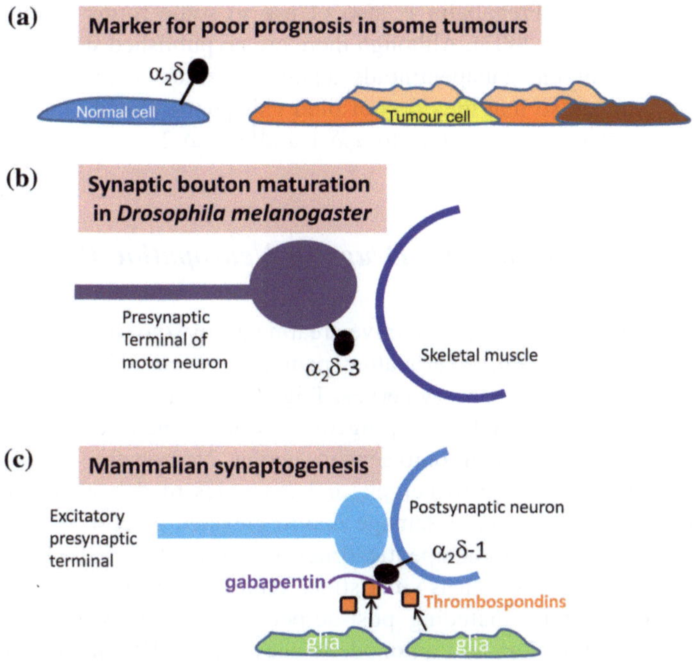

Fig. 5.2 Some examples of how $\alpha_2\delta$ subunits may affect cell function independently from its role as a calcium channel subunit. **a** Potential role for $\alpha_2\delta$ subunits in tumour progression. Reduced expression of $\alpha_2\delta$-2 and $\alpha_2\delta$-3 may be associated with increased progression (Gao et al. 2000; Carboni et al. 2003; Wanajo et al. 2008). **b** The *D.melanogaster* homolog of $\alpha_2\delta$-3 is involved in synaptic bouton formation (Kurshan et al. 2009). **c** The interaction of postsynaptic $\alpha_2\delta$-1 and thrombospondins derived from glia has been found to promote the synaptogenesis of excitatory mammalian synapses (Eroglu et al. 2009)

the first chemosensory domain (Dolphin 2012) (Fig. 5.1). It is possible that the basis of the binding of these drugs to $\alpha_2\delta$ subunits relates to the presence of these ancestral chemosensory domains. Furthermore, it is likely that an endogenous modulator might also bind to these subunits, which can be displaced by gabapentin. There is indeed evidence for the presence of an endogenous bound substance that occludes gabapentin binding, since the binding affinity for ^{3}H-gabapentin increases as the $\alpha_2\delta$ protein is purified or dialysed (Brown et al. 1998; Davies et al. 2006). It is also worth speculating that the binding of gabapentin might perturb the function(s) of the VWA domain (Dolphin 2012).

5.4.2 *Use of Gabapentinoid Drugs in Therapy of Epilepsies*

Gabapentin is effective as an adjunct therapy in several forms of epilepsy (Marson et al. 2000) and as a monotherapy for elderly adults with partial-onset seizures

(Glauser et al. 2006). Pregabalin is also of use in some forms of epilepsy (for review see Taylor et al. 2007). Although there are no published studies concerning the effectiveness of the gabapentinoids against seizures in mice in which $\alpha_2\delta$ subunits are mutated so that they do not bind gabapentin (Field et al. 2006), it is likely that this effect is via binding to $\alpha_2\delta$-1 and/or $\alpha_2\delta$-2.

5.4.3 Use of Gabapentinoid Drugs in Neuropathic Pain

Gabapentin and pregabalin are effective treatments in some patients for various forms of neuropathic pain. Neuropathic pain can be caused by direct trauma, damaging or impinging on sensory nerves. Trigeminal neuralgia which causes face and jaw pain is often caused by impingement. Cancer-induced pain can be also result from direct impingement onto sensory nerves. Neuropathic pain can also be caused by chemical damage to nerves, and examples of this would be diabetic neuropathy as a result of nerve damage due to elevated glucose levels, and neuropathy induced by chemotherapeutic drugs including taxanes, platinum-based drugs, and vinca alkaloids such as vincristine. Viral infection of sensory nerves can also result in neuralgia, including post-herpetic neuralgia (which often occurs following shingles), or human immunodeficiency virus (HIV)-induced neuralgia.

Gabapentin and pregabalin have a relatively slow onset of action when used clinically (Taylor et al. 2007; Stacey et al. 2008). Furthermore, these drugs have no effect on acute nociceptive pain perception in naive animals or humans, whereas they are effective in some patients for chronic neuropathic pain, indicating that their efficacy is state-dependent (Dickenson et al. 2005; Moore et al. 2009). It has been found that the binding of gabapentin and pregabalin to $\alpha_2\delta$-1 subunits is essential for their therapeutic effect in animal neuropathic pain models (Field et al. 2006). This result indicates that binding to $\alpha_2\delta$-2 is not important in this process, and $\alpha_2\delta$-2 was found to be down-regulated in damaged DRG neurons (Bauer et al. 2009). Pregabalin is also licensed for use in the treatment of fibromyalgia, which is defined as generalised widespread pain that may also have a neuropsychiatric component (Smith and Moore 2012).

We have recently observed changes in alternative splicing of $\alpha_2\delta$-1 in DRG neurons following spinal nerve ligation (SNL) in rats; specifically, the increased expression of a low abundance splice variant (ΔA + BΔC $\alpha_2\delta$-1) with a reduced affinity for gabapentin (Lana et al. 2012). It is possible that variable up-regulation of this splice variant in patients treated with $\alpha_2\delta$ ligand drugs has relevance to the inconsistent therapeutic efficacy of these drugs within the population.

5.4.4 Mechanism of Action of the Gabapentinoid Drugs, Following Binding to $\alpha_2\delta$ Subunits

It was initially surmised that since $\alpha_2\delta$ proteins are subunits of voltage-gated calcium channels, gabapentin would inhibit calcium currents in DRG neurons. Indeed, small inhibitory effects have been observed (Stefani et al. 1998; Martin et al. 2002; Sutton et al. 2002). However, other studies have reported no acute effect of gabapentin on expressed calcium currents (Hendrich et al. 2008; Heblich et al. 2008), and those in brain (Schumacher et al. 1998; Davies et al. 2006) and DRG neurons (Hendrich et al. 2008). In another study, it was found that there was no effect of gabapentin on calcium channel currents in wild-type mouse DRG neurons, although when DRGs from $\alpha_2\delta$-1-overexpressing mice were used, the currents were sensitive to inhibition by gabapentin (Li et al. 2006). These results suggest that gabapentin is unlikely to be a direct channel blocker but more likely to affect the plasma membrane insertion, turnover or stability of the channel complex.

5.4.5 Effects of Gabapentinoids on Calcium Channel Trafficking

We have found that chronic application of gabapentin markedly reduced calcium channel currents, both in expression systems and in DRG neurons (Hendrich et al. 2008; Tran-Van-Minh and Dolphin 2010). There was a corresponding reduction in cell surface localization of $\alpha_2\delta$ and $\alpha 1$ subunits (Hendrich et al. 2008; Tran-Van-Minh and Dolphin 2010). In a development of these findings, we further identified that gabapentin inhibited post-Golgi trafficking of $\alpha_2\delta$-2, in a process involving Rab11, which is involved in trafficking through the recycling endosome compartment (Tran-Van-Minh and Dolphin 2010). Furthermore, we observed that a chronic anti-hyperalgesic dosing regime of pregabalin markedly reduced the increase of $\alpha_2\delta$-1 in the presynaptic terminals of the injured DRGs in the dorsal horn in vivo, an effect we also interpreted as being due to impaired trafficking (Bauer et al. 2009).

5.4.6 Role of $\alpha_2\delta$ Subunits in Synaptogenesis, and Effects of Gabapentinoids

The $\alpha_2\delta$ subunits can be isolated from various tissues in the absence of $\alpha 1$ subunits (Gee et al. 1996; Müller et al. 2010), indicating that they are only loosely associated with $\alpha 1$ subunits, and suggesting the possibility that they may have functions other than as calcium channel subunits. In agreement with this, the $\alpha_2\delta$-3 proteins have been found to have a role in synaptogenesis in Drosophila, which appears to be independent of the association with calcium channels (Kurshan et al. 2009) (Fig. 5.2b). Furthermore, the $\alpha_2\delta$-1 protein has been described to be one of the binding partners of the extracellular matrix proteins of the thrombospondin family (Eroglu et al. 2009), although thrombospondins also bind to many other proteins (Kazerounian et al. 2008). Thrombospondin-induced excitatory synapse formation were found to require postsynaptic $\alpha_2\delta$-1 (Eroglu et al. 2009) (Fig. 5.2c). Gabapentin has been reported to disrupt in vitro the interaction between $\alpha_2\delta$-1 and thrombospondins, and by this means to disrupt synaptogenesis, although it did not affect pre-formed synapses (Eroglu et al. 2009) (Fig. 5.2c). This effect on synaptogenesis is unlikely to be relevant to its mechanism of action either as an antiepileptic drug, or in neuropathic pain, since any synaptic remodelling would have occurred before the onset of therapy. However, an effect of gabapentin on synaptogenesis could have major clinical implications for chronic use of this drug, if confirmed. Nevertheless, birth defects in babies following chronic gabapentin exposure in utero were found to be very uncommon from several studies (Morrow et al. 2006; Molgaard-Nielsen and Hviid 2011), indicating that synaptogenesis in humans is little affected in vivo by clinically used concentrations of these drugs.

5.5 Conclusion

The $\alpha_2\delta$ subunits, accessory subunits of voltage-gated calcium channels, play major roles in trafficking and stability of these channels, as well as influencing their biophysical properties. However, the influence of the physiologically relevant $\alpha_2\delta$ subunits on the properties of calcium currents, characteristics of action potentials and depolarization-induced calcium flux are likely to differ significantly for skeletal muscle ($Ca_v1.1$), cardiac muscle ($Ca_v1.2$) and the neuronal Ca_v2 channels. For example, the effects of $\alpha_2\delta$ subunits on numbers of calcium channels in the plasma membrane appears to be less important for $Ca_v1.2$ than for the Ca_v2 class of channels, possibly because of the lower turnover of $Ca_v1.2$ channels.

The $\alpha_2\delta$ subunits are also implicated in a number of genetic and other diseases, and form a target site for the gabapentinoid drugs. These drugs, gabapentin and pregabalin have major effects on synaptic transmission in damaged DRG neurons, in which $\alpha_2\delta$-1 is up-regulated, but they do not significantly affect skeletal muscle or cardiac function. It is therefore possible that they influence the trafficking and membrane stability of Ca_v2 channels in cells where there is a high concentration of $\alpha_2\delta$-1, and where there is little $\alpha_2\delta$-3, that does not bind gabapentin, present for compensation.

Thus, the diverse physiological and pharmacological properties of the four $\alpha_2\delta$ subunits, as well as their differential distribution in particular cell types, gives rise to a wide variety of pathologies resulting from their mutation, loss or overexpression in different disease states.

References

Altier C, Dale CS, Kisilevsky AE, Chapman K, Castiglioni AJ, Matthews EA, Evans RM, Dickenson AH, Lipscombe D, Vergnolle N, Zamponi GW (2007) Differential role of N-type calcium channel splice isoforms in pain. J Neurosci 27:6363–6373

Altier C, Garcia-Caballero A, Simms B, You H, Chen L, Walcher J, Tedford HW, Hermosilla T, Zamponi GW (2011) The Cavbeta subunit prevents RFP2-mediated ubiquitination and proteasomal degradation of L-type channels. Nat Neurosci 14:173–180

Anantharaman V, Aravind L (2000) Cache-a signalling domain common to animal Ca channel subunits and a class of prokaryotic chemotaxis receptors. Trends Biochem Sci 25:535–537

Angelotti T, Hofmann F (1996) Tissue-specific expression of splice variants of the mouse voltage-gated calcium channel a2/d subunit. FEBS Lett 397:331–337

Barclay J, Rees M (2000) Genomic organization of the mouse and human a2d2 voltage-dependent calcium channel subunit genes. Mamm Genome 11:1142–1144

Barclay J, Balaguero N, Mione M, Ackerman SL, Letts VA, Brodbeck J, Canti C, Meir A, Page KM, Kusumi K, PerezReyes E, Lander ES, Frankel WN, Gardiner RM, Dolphin AC, Rees M (2001) Ducky mouse phenotype of epilepsy and ataxia is associated with mutations in the *Cacna2d2* gene and decreased calcium channel current in cerebellar Purkinje cells. J Neurosci 21:6095–6104

Bauer CS, Nieto-Rostro M, Rahman W, Tran-Van-Minh A, Ferron L, Douglas L, Kadurin I, Sri Ranjan Y, Fernandez-Alacid L, Millar NS, Dickenson AH, Lujan R, Dolphin AC (2009) The increased trafficking of the calcium channel subunit a2d-1 to presynaptic terminals in neuropathic pain is inhibited by the a2d ligand pregabalin. J Neurosci 29:4076–4088

Beal B, Moeller-Bertram T, Schilling JM, Wallace MS (2012) Gabapentin for once-daily treatment of post-herpetic neuralgia: a review. Clin Interv Aging 7:249–255

Bernstein GM, Jones OT (2007) Kinetics of internalization and degradation of N-type voltage-gated calcium channels: Role of the alpha(2)/delta subunit. Cell Calcium 41:27–40

Brill J, Klocke R, Paul D, Boison D, Gouder N, Klugbauer N, Hofmann F, Becker CM, Becker K (2004) entla, a novel epileptic and ataxic Cacna2d2 mutant of the mouse. J Biol Chem 279:7322–7330

Brodbeck J, Davies A, Courtney J-M, Meir A, Balaguero N, Canti C, Moss FJ, Page KM, Pratt WS, Hunt SP, Barclay J, Rees M, Dolphin AC (2002) The ducky mutation in *Cacna2d2* results in altered Purkinje cell morphology and is associated with the expression of a truncated *a2d*-2 protein with abnormal function. J Biol Chem 277:7684–7693

Brown JP, Dissanayake VU, Briggs AR, Milic MR, Gee NS (1998) Isolation of the [3H]gabapentin-binding protein/alpha 2 delta Ca2 + channel subunit from porcine brain: development of a radioligand binding assay for alpha 2 delta subunits using [3H]leucine. Anal Biochem 255:236–243

Burashnikov E et al (2010) Mutations in the cardiac L-type calcium channel associated with inherited J-wave syndromes and sudden cardiac death. Heart Rhythm 7:1872–1882

Calderon-Rivera A, Andrade A, Hernandez–Hernandez O, Gonzalez-Ramirez R, Sandoval A, Rivera M, Gomora JC, Felix R (2012) Identification of a disulfide bridge essential for structure and function of the voltage-gated Ca(2 +) channel alpha(2)delta-1 auxiliary subunit. Cell Calcium 51:22–30

Canti C, Nieto-Rostro M, Foucault I, Heblich F, Wratten J, Richards MW, Hendrich J, Douglas L, Page KM, Davies A, Dolphin AC (2005) The metal-ion-dependent adhesion site in the Von Willebrand factor-A domain of alpha2delta subunits is key to trafficking voltage-gated Ca2 + channels. Proc Natl Acad Sci USA 102:11230–11235

Carboni GL, Gao B, Nishizaki M, Xu K, Minna JD, Roth JA, Ji L (2003) CACNA2D2-mediated apoptosis in NSCLC cells is associated with alterations of the intracellular calcium signaling and disruption of mitochondria membrane integrity. Oncogene 22:615–626

Cole RL, Lechner SM, Williams ME, Prodanovich P, Bleicher L, Varney MA, Gu G (2005) Differential distribution of voltage-gated calcium channel alpha-2 delta (alpha2delta) subunit

mRNA-containing cells in the rat central nervous system and the dorsal root ganglia. J Comp Neurol 491:246–269
da Costa PE, Cavalli LR, Rainho CA (2011) Evidence of epigenetic regulation of the tumor suppressor gene cluster flanking RASSF1 in breast cancer cell lines. Epigenetics 6:1413–1424
Davies A, Douglas L, Hendrich J, Wratten J, Tran-Van-Minh A, Foucault I, Koch D, Pratt WS, Saibil H, Dolphin AC (2006) The calcium channel a2d-2 subunit partitions with $Ca_v2.1$ in lipid rafts in cerebellum: implications for localization and function. J Neurosci 26:8748–8757
Davies A, Hendrich J, Van Minh AT, Wratten J, Douglas L, Dolphin AC (2007) Functional biology of the alpha(2)delta subunits of voltage-gated calcium channels. Trends Pharmacol Sci 28:220–228
Davies A, Kadurin I, Alvarez-Laviada A, Douglas L, Nieto-Rostro M, Bauer CS, Pratt WS, Dolphin AC (2010) The a2d subunits of voltage-gated calcium channels form GPI-anchored proteins, a post-translational modification essential for function. Proc Natl Acad Sci USA 107:1654–1659
Dickenson AH, Bee LA, Suzuki R (2005) Pains, gains, and midbrains. Proc Natl Acad Sci USA 102:17885–17886
Dolphin AC (2003) G protein modulation of voltage-gated calcium channels. Pharmacol Rev 55:607–627
Dolphin AC (2012) Calcium channel auxiliary alpha(2)delta and beta subunits: trafficking and one step beyond. Nat Rev Neurosci 13:542–555
Donato R, Page KM, Koch D, Nieto-Rostro M, Foucault I, Davies A, Wilkinson T, Rees M, Edwards FA, Dolphin AC (2006) The ducky2 J mutation in Cacna2d2 results in reduced spontaneous Purkinje cell activity and altered gene expression. J Neurosci 26:12576–12586
Edvardson S, Oz S, Abulhijaa FA, Taher FB, Shaag A, Zenvirt S, Dascal N, Elpeleg O (2013) Early infantile epileptic encephalopathy associated with a high voltage gated calcium channelopathy. J Med Genet 50:118–23
Ellis SB, Williams ME, Ways NR, Brenner R, Sharp AH, Leung AT, Campbell KP, McKenna E, Koch WJ, Hui A, Schwartz A, Harpold MM (1988) Sequence and expression of mRNAs encoding the a_1 and a_2 subunits of a DHP-sensitive calcium channel. Science 241:1661–1664
Eroglu C, Allen NJ, Susman MW, O'Rourke NA, Park CY, Ozkan E, Chakraborty C, Mulinyawe SB, Annis DS, Huberman AD, Green EM, LAWLER J, Dolmetsch R, Garcia KC, Smith SJ, Luo ZD, Rosenthal A, Mosher DF, Barres BA (2009) Gabapentin Receptor alpha2delta-1 Is a Neuronal Thrombospondin Receptor Responsible for Excitatory CNS Synaptogenesis. Cell 139:380–392
Felix R (1999) Voltage-dependent Ca2 + channel alpha(2)delta auxiliary subunit: Structure, function and regulation. Receptors Channels 6:351–362
Field MJ, Cox PJ, Stott E, Melrose H, Offord J, Su TZ, Bramwell S, Corradini L, England S, Winks J, Kinloch RA, Hendrich J, Dolphin AC, Webb T, Williams D (2006) Identification of the a2d-1 subunit of voltage-dependent calcium channels as a novel molecular target for pain mediating the analgesic actions of pregabalin. Proc Natl Acad Sci USA 103:17537–17542
Fuller-Bicer GA, Varadi G, Koch SE, Ishii M, Bodi I, Kadeer N, Muth JN, Mikala G, Petrashevskaya NN, Jordan MA, Zhang SP, Qin N, Flores CM, Isaacsohn I, Varadi M, Mori Y, Jones WK, Schwartz A (2009) Targeted Disruption of the Voltage-Dependent Ca2 + Channel {alpha}2/{delta}-1 Subunit. Am J Physiol Heart Circ Physiol 297:H117–H124
Gao B, Sekido Y, Maximov A, Saad M, Forgacs E, Latif F, Wei MH, Lerman M, Lee JH, Perez-Reyes E, Bezprozvanny I, Minna JD (2000) Functional properties of a new voltage-dependent calcium channel a_2d auxiliary subunit gene (CACNA2D2). J Biol Chem 275:12237–12242
Gee NS, Brown JP, Dissanayake VUK, Offord J, Thurlow R, Woodruff GN (1996) The novel anticonvulsant drug, gabapentin (Neurontin), binds to the a_2d subunit of a calcium channel. J Biol Chem 271:5768–5776
Glauser T, Ben Menachem E, Bourgeois B, Cnaan A, Chadwick D, Guerreiro C, Kalviainen R, Mattson R, Perucca E, Tomson T (2006) ILAE treatment guidelines: evidence-based analysis

of antiepileptic drug efficacy and effectiveness as initial monotherapy for epileptic seizures and syndromes. Epilepsia 47:1094–1120
Gong HC, Hang J, Kohler W, Li L, Su TZ (2001) Tissue-specific expression and gabapentin-binding properties of calcium channel alpha2delta subunit subtypes. J Membr Biol 184:35–43
Gurnett CA, Felix R, Campbell KP (1997) Extracellular interaction of the voltage-dependent Ca^{2+} channel a_2d and a_1 subunits. J Biol Chem 272:18508–18512
Gutierrez LM, Brawley RM, Hosey MM (1991) Dihydropyridine-sensitive calcium channels from skeletal muscle. I. Roles of subunits in channel activity. J Biol Chem 266:16387–16394
Hanke S, Bugert P, Chudek J, Kovacs G (2001) Cloning a calcium channel a2d-3 subunit gene from a putative tumor suppressor gene region at chromosome 3p21.1 in conventional renal cell carcinoma. Gene 264:69–75
Heblich F, Tran-Van-Minh A, Hendrich J, Watschinger K, Dolphin AC (2008) Time course and specificity of the pharmacological disruption of the trafficking of voltage-gated calcium channels by gabapentin. Channels 2:4–9
Hendrich J, Tran-Van-Minh A, Heblich F, Nieto-Rostro M, Watschinger K, Striessnig J, Wratten J, Davies A, Dolphin AC (2008) Pharmacological disruption of calcium channel trafficking by the a2d ligand gabapentin. Proc Natl Acad Sci USA 105:3628–3633
Hoppa M, Lana B, Margas W, Dolphin AC, Ryan TA (2012) a2d couples calcium channels to neurotransmitter release sites to control release probability. Nature 486:122–125
Iossifov I et al (2012) De novo gene disruptions in children on the autistic spectrum. Neuron 74:285–299
Ivanov SV, Ward JM, Tessarollo L, McAreavey D, Sachdev V, Fananapazir L, Banks MK, Morris N, Djurickovic D, Devor-Henneman DE, Wei MH, Alvord GW, Gao B, Richardson JA, Minna JD, Rogawski MA, Lerman MI (2004) Cerebellar ataxia, seizures, premature death, and cardiac abnormalities in mice with targeted disruption of the Cacna2d2 gene. Am J Pathol 165:1007–1018
Jay SD, Ellis SB, McCue AF, Williams ME, Vedvick TS, Harpold MM, Campbell KP (1990) Primary structure of the gamma subunit of the DHP-sensitive calcium channel from skeletal muscle. Science 248:490–492
Jay SD, Sharp AH, Kahl SD, Vedvick TS, Harpold MM, Campbell KP (1991) Structural characterization of the dihydropyridine-sensitive calcium channel a_2-subunit and the associated d peptides. J Biol Chem 266:3287–3293
Ji L, Minna JD, Roth JA (2005) 3p21.3 tumor suppressor cluster: prospects for translational applications. Future Oncology 1:79–92
Kadurin I, Alvarez-Laviada A, Ng SF, Walker-Gray R, D'Arco M, Fadel MG, Pratt WS, Dolphin AC (2012) Calcium currents are enhanced by alpha2delta-1 lacking its membrane anchor. J Biol Chem 1287:33554–33566
Kazerounian S, Yee KO, LAWLER J (2008) Thrombospondins in cancer. Cell Mol Life Sci 65:700–712
Kim H-L, Kim H, Lee P, King RG, Chin H (1992) Rat brain expresses an alternatively spliced form of the dihydropyridine-sensitive L-type calcium channel a2 subunit. Proc Natl Acad Sci USA 89:3251–3255
Klassen T, Davis C, Goldman A, Burgess D, Chen T, Wheeler D, McPherson J, Bourquin T, Lewis L, Villasana D, Morgan M, Muzny D, Gibbs R, Noebels J (2011) Exome sequencing of ion channel genes reveals complex profiles confounding personal risk assessment in epilepsy. Cell 145:1036–1048
Klugbauer N, Lacinova L, Marais E, Hobom M, Hofmann F (1999) Molecular diversity of the calcium channel a2-d subunit. J Neurosci 19:684–691
Kurshan PT, Oztan A, Schwarz TL (2009) Presynaptic alpha(2)delta-3 is required for synaptic morphogenesis independent of its Ca(2 +)-channel functions. Nat Neurosci 12:1415–1423
Lana B, Schlick B, Bauer CS, Rahman W, Dickenson AH, Dolphin AC (2012) Differential up-regulation of a2d-1 splice variants following nerve injury. FENS 2012 Abstracts A417
Leone PE, Gonzalez MB, Elosua C, Gomez-Moreta JA, Lumbreras E, Robledo C, Santos-Briz A, Valero JM, de la Guardia RD, Gutierrez NC, Hernandez JM, Garcia JL (2012) Integration of

global spectral karyotyping, CGH arrays, and expression arrays reveals important genes in the pathogenesis of glioblastoma multiforme. Ann Surg Oncol 19:2367–2379

Li CY, Zhang XL, Matthews EA, Li KW, Kurwa A, Boroujerdi A, Gross J, Gold MS, Dickenson AH, Feng G, Luo ZD (2006) Calcium channel alpha(2)delta(1) subunit mediates spinal hyperexcitability in pain modulation. Pain 125:20–34

Li Z, Taylor CP, Weber M, Piechan J, Prior F, Bian F, Cui M, Hoffman D, Donevan S (2011) Pregabalin is a potent and selective ligand for alpha(2)delta-1 and alpha(2)delta-2 calcium channel subunits. Eur J Pharmacol 667:80–90

Liu H, De Waard M, Scott VES, Gurnett CA, Lennon VA, Campbell KP (1996) Identification of three subunits of the high affinity w-conotoxin MVIIC-sensitive Ca^{2+} channel. J Biol Chem 271:13804–13810

Luo ZD, Chaplan SR, Higuera ES, Sorkin LS, Stauderman KA, Williams ME, Yaksh TL (2001) Upregulation of dorsal root ganglion a_2d calcium channel subunit and its correlation with allodynia in spinal nerve-injured rats. J Neurosci 21:1868–1875

Malhotra D, Sebat J (2012) CNVs: harbingers of a rare variant revolution in psychiatric genetics. Cell 148:1223–1241

Marson AG, Kadir ZA, Hutton JL, Chadwick DW (2000) Gabapentin add-on for drug-resistant partial epilepsy. Cochrane Database Syst Rev CD001415

Martin DJ, McClelland D, Herd MB, Sutton KG, Hall MD, Lee K, Pinnock RD, Scott RH (2002) Gabapentin-mediated inhibition of voltage-activated Ca2 + channel currents in cultured sensory neurones is dependent on culture conditions and channel subunit expression. Neuropharmacology 42:353–366

Molgaard-Nielsen D, Hviid A (2011) Newer-generation antiepileptic drugs and the risk of major birth defects. JAMA 305:1996–2002

Moore RA, Straube S, Wiffen PJ, Derry S, McQuay HJ (2009) Pregabalin for acute and chronic pain in adults. Cochrane Database Syst Rev CD007076

Morrow J, Russell A, Guthrie E, Parsons L, Robertson I, Waddell R, Irwin B, McGivern RC, Morrison PJ, Craig J (2006) Malformation risks of antiepileptic drugs in pregnancy: a prospective study from the UK Epilepsy and Pregnancy Register. J Neurol Neurosurg Psychiatry 77:193–198

Moss FJ, Viard P, Davies A, Bertaso F, Page KM, Graham A, Canti C, Plumpton M, Plumpton C, Clare JJ, Dolphin AC (2002) The novel product of a five-exon *stargazin*-related gene abolishes $Ca_v2.2$ calcium channel expression. EMBO J 21:1514–1523

Müller CS, Haupt A, Bildl W, Schindler J, Knaus HG, Meissner M, Rammner B, Striessnig J + , Flockerzi V, Fakler B, Schulte U (2010) Quantitative proteomics of the Ca_v2 channel nano-environments in the mammalian brain. PNAS 107:14950–14957

Neely GG et al (2010) A genome-wide Drosophila screen for heat nociception identifies alpha2delta3 as an evolutionarily conserved pain gene. Cell 143:628–638

Newton RA, Bingham S, Case PC, Sanger GJ, Lawson SN (2001) Dorsal root ganglion neurons show increased expression of the calcium channel alpha2delta-1 subunit following partial sciatic nerve injury. Brain Res Mol Brain Res 95:1–8

Palmieri C, Rudraraju B, Monteverde M, Lattanzio L, Gojis O, Brizio R, Garrone O, Merlano M, Syed N, Lo NC, Crook T (2012) Methylation of the calcium channel regulatory subunit alpha2delta-3 (CACNA2D3) predicts site-specific relapse in oestrogen receptor-positive primary breast carcinomas. Br J Cancer 107:375–381

Patel R, Bauer CS, Nieto-Rostro M, Crews K, Dickenson AH, Dolphin AC (2012) Thermal and mechanical processing in a2d-1 knockout mice. Society for Neuroscience Abstracts 2012 81.01

Pierleoni A, Martelli PL, Casadio R (2008) PredGPI: a GPI-anchor predictor. BMC Bioinformatics 9:392

Pragnell M, De Waard M, Mori Y, Tanabe T, Snutch TP, Campbell KP (1994) Calcium channel b-subunit binds to a conserved motif in the I-II cytoplasmic linker of the a_1-subunit. Nature 368:67–70

Qin N, Yagel S, Momplaisir ML, Codd EE, D'Andrea MR (2002) Molecular cloning and characterization of the human voltage-gated calcium channel a_2d-4 subunit. Mol Pharmacol 62:485–496

Schlick B, Flucher BE, Obermair GJ (2010) Voltage-activated calcium channel expression profiles in mouse brain and cultured hippocampal neurons. Neuroscience 167:786–798

Schumacher TB, Beck H, Steinhäuser C, Schramm J, Elger CE (1998) Effects of phenytoin, carbamazepine, and gabapentin on calcium channels in hippocampal granule cells from patients with temporal lobe epilepsy. Epilepsia 39:355–363

Senchenko VN, Liu J, Loginov W, Bazov I, Angeloni D, Seryogin Y, Ermilova V, Kazubskaya T, Garkavtseva R, Zabarovska VI, Kashuba VI, Kisselev LL, Minna JD, Lerman MI, Klein G, Braga EA, Zabarovsky ER (2004) Discovery of frequent homozygous deletions in chromosome 3p21.3 LUCA and AP20 regions in renal, lung and breast carcinomas. Oncogene 23:5719–5728

Silverman RB (2008) From basic science to blockbuster drug: the discovery of Lyrica. Angew Chem Int Ed Engl 47:3500–3504

Singer D, Biel M, Lotan I, Flockerzi V, Hofmann F, Dascal N (1991) The roles of the subunits in the function of the calcium channel. Science 253:1553–1557

Smith MT, Moore BJ (2012) Pregabalin for the treatment of fibromyalgia. Expert Opin Pharmacother 13:1527–1533

Stacey BR, Barrett JA, Whalen E, Phillips KF, Rowbotham MC (2008) Pregabalin for postherpetic neuralgia: placebo-controlled trial of fixed and flexible dosing regimens on allodynia and time to onset of pain relief. J Pain 9:1006–1017

Stefani A, Spadoni F, Bernardi G (1998) Gabapentin inhibits calcium currents in isolated rat brain neurons. Neuropharmacology 37:83–91

Sutton KG, Martin DJ, Pinnock RD, Lee K, Scott RH (2002) Gabapentin inhibits high-threshold calcium channel currents in cultured rat dorsal root ganglion neurones. Br J Pharmacol 135:257–265

Tanabe T, Takeshima H, Mikami A, Flockerzi V, Takahashi H, Kangawa K, Kojima M, Matsuo H, Hirose T, Numa S (1987) Primary structure of the receptor for calcium channel blockers from skeletal muscle. Nature 328:313–318

Taylor CP, Garrido R (2008) Immunostaining of rat brain, spinal cord, sensory neurons and skeletal muscle for calcium channel alpha2-delta (alpha2-delta) type 1 protein. Neuroscience 155:510–521

Taylor CP, Angelotti T, Fauman E (2007) Pharmacology and mechanism of action of pregabalin: the calcium channel alpha2-delta (alpha2-delta) subunit as a target for antiepileptic drug discovery. Epilepsy Res 73:137–150

Templin C, Ghadri JR, Rougier JS, Baumer A, Kaplan V, Albesa M, Sticht H, Rauch A, Puleo C, Hu D, Barajas-Martinez H, Antzelevitch C, Luscher TF, Abriel H, Duru F (2011) Identification of a novel loss-of-function calcium channel gene mutation in short QT syndrome (SQTS6). Eur Heart J 32:1077–1088

Ting JT, Peca J, Feng G (2012) Functional consequences of mutations in postsynaptic scaffolding proteins and relevance to psychiatric disorders. Annu Rev Neurosci 35:49–71

Tran-Van-Minh A, Dolphin AC (2010) Gabapentin inhibits the Rab11-dependent recycling of the calcium channel subunit alpha2delta-2. J Neurosci 130:12856–12867

Tuluc P, Kern G, Obermair GJ, Flucher BE (2007) Computer modeling of siRNA knockdown effects indicates an essential role of the Ca2 + channel alpha2delta-1 subunit in cardiac excitation-contraction coupling. Proc Natl Acad Sci USA 104:11091–11096

Waithe D, Ferron L, Page KM, Chaggar K, Dolphin AC (2011) b-Subunits Promote the Expression Of $Ca_v2.2$ channels by reducing their proteasomal degradation. J Biol Chem 286:9598–9611

Walsh CP, Davies A, Butcher AJ, Dolphin AC, Kitmitto A (2009) 3D structure of $Ca_v3.1$ - comparison with the cardiac L-type voltage-gated calcium channel monomer architecture. J Biol Chem 284:22310–22321

Wanajo A, Sasaki A, Nagasaki H, Shimada S, Otsubo T, Owaki S, Shimizu Y, Eishi Y, Kojima K, Nakajima Y, Kawano T, Yuasa Y, Akiyama Y (2008) Methylation of the calcium channel-related gene, CACNA2D3, is frequent and a poor prognostic factor in gastric cancer. Gastroenterology 135:580–590

Wang H, Sun H, Della PK, Benz RJ, Xu J, Gerhold DL, Holder DJ, Koblan KS (2002) Chronic neuropathic pain is accompanied by global changes in gene expression and shares pathobiology with neurodegenerative diseases. Neuroscience 114:529–546

Whittaker CA, Hynes RO (2002) Distribution and evolution of von Willebrand/integrin A domains: widely dispersed domains with roles in cell adhesion and elsewhere. Mol Biol Cell 13:3369–3387

Witcher DR, De Waard M, Sakamoto J, Franzini-Armstrong C, Pragnell M, Kahl SD, Campbell KP (1993) Subunit identification and reconstitution of the N-type Ca^{2+} channel complex purified from brain. Science 261:486–489

Wolf M, Eberhart A, Glossmann H, Striessnig J, Grigorieff N (2003) Visualization of the domain structure of an L-type Ca2 + channel using electron cryo-microscopy. J Mol Biol 332:171–182

Wycisk KA, Budde B, Feil S, Skosyrski S, Buzzi F, Neidhardt J, Glaus E, Nurnberg P, Ruether K, Berger W (2006a) Structural and functional abnormalities of retinal ribbon synapses due to Cacna2d4 mutation. Invest Ophthalmol Vis Sci 47:3523–3530

Wycisk KA, Zeitz C, Feil S, Wittmer M, Forster U, Neidhardt J, Wissinger B, Zrenner E, Wilke R, Kohl S, Berger W (2006b) Mutation in the auxiliary calcium-channel subunit CACNA2D4 causes autosomal recessive cone dystrophy. Am J Hum Genet 79:973–977

Xiao HS, Huang QH, Zhang FX, Bao L, Lu YJ, Guo C, Yang L, Huang WJ, Fu G, Xu SH, Cheng XP, Yan Q, Zhu ZD, Zhang X, Chen Z, Han ZG, Zhang X (2002) Identification of gene expression profile of dorsal root ganglion in the rat peripheral axotomy model of neuropathic pain. Proc Natl Acad Sci USA 99:8360–8365

Zaccara G, Perucca P, Gangemi PF (2012) The adverse event profile of pregabalin across different disorders: a meta-analysis. Eur J Clin Pharmacol 68:903–912

Chapter 6
T-Type Calcium Channels in Pain Neuronal Circuits

Amaury Francois, Anne Pizzoccaro, Sophie Laffray and Emmanuel Bourinet

Abstract Pain is a quite frequent complaint accompanying numerous pathologies. Among these pathological cases numerous neuropathies are retrieved with identified etiologies (chemotherapy induced peripheral neuropathies (CIPN), diabetes, surgeries) and also more diffuse syndromes such as fibromyalgia, migraine. More broadly, pain is one of the first and dramatic consequences of the majority of inherited diseases. Despite their importance for the quality of life, current therapies in symptomatic pain management are limited to drugs that are either old, or with a limited efficacy or that possess a bad benefit/risk ratio. Morphine and opioids for example have severe side effects. As no new pharmacological concept has led to new analgesics in the Past, the discovery of new medications is needed. It is necessary to identify new targets (such as ionic channels, the primary molecules of cellular excitability) in pain transmission before hoping to find specific molecules to treat different kinds of pain. Therefore studies of ion channels in pain pathways are extremely active. This is particularly true with ion channels in peripheral sensory neurons in dorsal root ganglia known now to express unique sets of these channels. Moreover, both spinal and supra spinal levels are clearly important in pain modulation with a key role of limbic areas such as thalamus filtering upcoming noxious information in their way to the cortex in the so called pain matrix representing a network of cerebral regions involved in the building of pain sensation, which comprises cortical somatic areas, insula, cingulate cortex, and

A. Francois · A. Pizzoccaro · S. Laffray · E. Bourinet (✉)
Laboratories of Excellence, Ion Channel Science and Therapeutics, Institut de Génomique Fonctionnelle, 141 rue de la Cardonille 34094 Montpellier, France
e-mail: Emmanuel.Bourinet@igf.cnrs.fr

A. Francois · A. Pizzoccaro · S. Laffray · E. Bourinet
CNRS, UMR5203, Montpellier, France

A. Francois · A. Pizzoccaro · S. Laffray · E. Bourinet
INSERM, U661, Montpellier, France

A. Francois · S. Laffray · E. Bourinet
IFR3, Universités Montpellier I&II, Montpellier, France

N. Weiss and A. Koschak (eds.), *Pathologies of Calcium Channels*,
DOI: 10.1007/978-3-642-40282-1_6,

associated with multiple other regions. Among these ion channels, we and others revealed the important role of low voltage-gated calcium channels (T-type channels) in cellular excitability in different steps of the pain pathways. These channels, by being activated nearby resting membrane potential have biophysical characteristic suited to facilitate action potential generation and rhythmicity. In this chapter we will review the current knowledge on the role of these channels in the perception and modulation of pain.

6.1 Anatomy of the Pain Pathways

Pain signaling is initiated by the detection of noxious stimuli through specialized primary nociceptors located in peripheral endings within the skin and in internal organs. The cell bodies of these neurons are contained within the dorsal root ganglia (DRG) or in the trigeminal ganglia (for cephalic sensory innervation), whereas their presynaptic nerve terminals are localized in the superficial layers (laminae I and II) of the dorsal horn of the spinal cord (for DRG) or in the brain stem (for trigeminal ganglia) (Basbaum et al. 2009). In humans, 29 pairs of DRG (at each vertebral level) and 1 pair of trigeminal ganglia contain sensory neurons. These neurons have a peculiar morphology with a pseudo unipolar axonal projection arising from the cell body and bifurcating in two branches: one very long projection targets the peripheral receptive fields, and a second projection connects to the CNS in the spinal cord or brain stem (Tandrup 1995) (Fig. 6.1). Therefore, that vast majority of the afferent neuron is comprised of axonal structures (more than 99 % of the cell membrane; Devor 1999) dedicated to detecting external stimuli in the distal receptive field. These structures serve to transduce information in the form of in fiber depolarization that gives rise to action potentials propagating along the axonal fibers up to central synapses in the CNS. The role of the sensory neuron cell body in coding sensory information is less defined (Devor 1999). More globally, peripheral sensory neurons convey a diversity of sensory modalities including pain and itch, discriminative touch, and perception of body muscle tension (proprioception). The classification of sensory fiber subtypes depends both on their function (i.e., conduction velocity, CV) and on their anatomical features (such as axonal fiber size and myelination; Erlanger 1927; Erlanger and Gasser 1930; Perl 2007; Zotterman 1939). Fast conducting Aα and Aβ fibers (CV: Aα 70–120 m/s, Aβ 70–30 m/s) have large calibers (5–20 μm) and a large cell body (>40 μm). They are heavily myelinated and correspond to proprioceptive neurons (Aα) and proprioceptive and tactile neurons (Aβ). Some subclasses of Aβ fibers also support nociceptive signals (Djouhri and Lawson 2004; Fang et al. 2005). Lightly myelinated Aδ fibers, with a slower CV (12–30 m/s) and a smaller diameter (2–5 μm) and cell body size (30–40 μm), convey tactile and nociceptive information. Finally, slow conducting C fibers (CV 0.5–2 m/s) with unmyelinated thin axons (0.4–1.2 μM) and small soma size (<25 μm) are mainly involved in

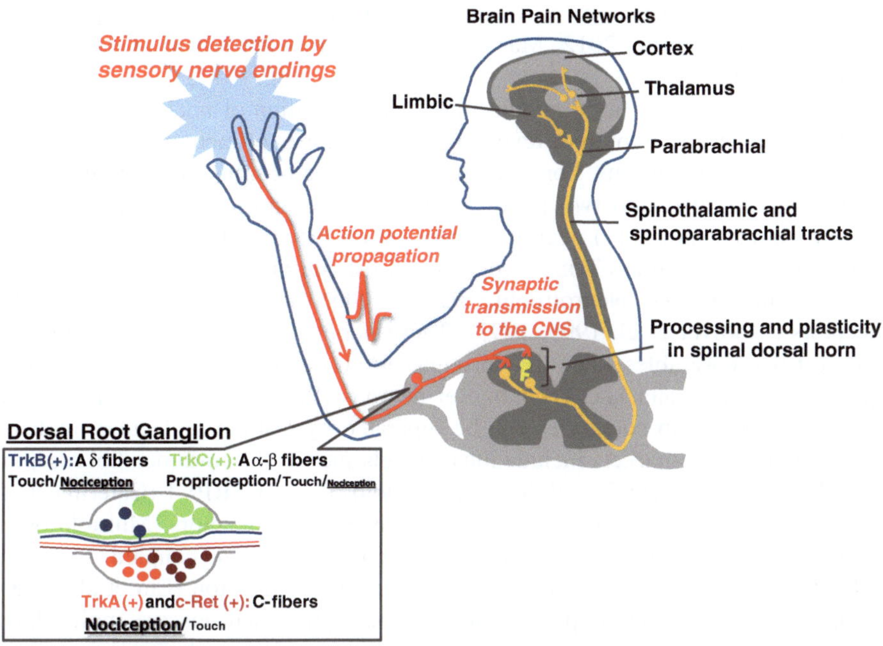

Fig. 6.1 Ascending pain neuraxis. Pain sensing neurons in the peripheral nervous system have their soma located in the dorsal root ganglia (DRG). These neurons have a peripheral axon innervating the distal territories (skin, viscera, etc.) where they detect painful stimuli leading to an action potential that travels along the fibers up to the DRG and then to the first relay in the dorsal spinal cord. Sensory neurons within the DRGs are diverse and can be separated based on the expression of neurotrophin receptors. The majority are TrkA and c-Ret positive small diameter sensory afferents that correspond to unmyelinated C-fibers mainly involved in nociception. TrkB and TrkC positive myelinated larger diameter afferents correspond to A-δ and A-α/β fibers respectively. They convey touch and proprioception signals although both of these subclasses contain nociceptive neurons. The sensory information is processed locally in neuronal circuitry within the dorsal horn of the spinal cord before being sent to the thalamus to convey nociceptive information. Following thalamic filtering the information is sent to the cortical structures of the pain matrix

detecting pain and itch signals but also participate in light touch related to tickling (Julius and Basbaum 2001; Zotterman 1939). Studies using skin-nerve preparations (Reeh 1986; Zimmermann et al. 2009) or in vivo single unit recording of peripheral nerve axons with microneurography (Reeh 1986; Serra 2010; Vallbo et al. 1999; Zimmermann et al. 2009) further highlight their diversity. To mirror these functional data gathered over the years, detailed description of the anatomy of the distal and central nerve endings required novel technical approaches. The use of genetically modified mice for specific labeling of fiber subtypes allowed researchers to address this issue, revealing that the structural organisation of distal fibers in the skin (Wu et al. 2012) and the central fibers in the dorsal horn of the spinal cord (Li et al. 2011) is extremely complex. Thus, based on these criteria, nociceptive neurons encompass a highly heterogeneous population of neurons with respect to their morphological, anatomical and electrophysiological properties (Lallemend and Ernfors 2012; Marmigere and Ernfors 2007; Reed-Geaghan and Maricich 2011).

Over the past few years the molecular characterization of nociceptive neurons has been intensively explored, revealing that a number of factors/markers define specific subsets of neurons. For example, nociceptive neurons in the adult animal have been classified into two major categories according to their expression of neurotrophin receptors: (1) neurons dependent on the neurotrophin nerve growth factor (NGF) that express TrkA receptors and (2) neurons responsive to members of the Glial Derived Neurotrophic Factor (GDNF) family that express Ret receptors (Molliver et al. 1995, 1997). These populations are even more diverse than just two ensembles since they originate from distinct lineages during development with early and late TrkA or Ret neurons (Bachy et al. 2011; Luo et al. 2007). Therefore sub categories can be separated by distinct molecular markers. Interestingly, these markers include ion channels such as the cold/menthol receptor TRPM8 (Peier et al. 2002) and the heat/vanilloid receptor TRPV1 (Caterina et al. 1997) that segregate into non overlapping classes of nociceptors. Multiple other signaling proteins such as Mrgpr/SNSR class G-protein coupled receptors (GPCRs) are largely expressed in a mutually exclusive fashion (Liu et al. 2008). Deciphering how these subpopulations of nociceptive neurons are molecularly specified and functionally diversified will greatly expand the understanding of pain biology, but this also represents a challenge in many laboratories working on molecular pain physiology.

The spinal cord dorsal horn is the essential CNS sensory processing hub connecting the periphery to the brain. In this nociceptive pathway, dorsal horn neurons integrate inputs from peripheral nociceptors, local interneurons and descending projections and transmit processed signals to the brain pain network (Basbaum and Braz 2010) (Fig. 6.1). Neurons in the superficial layers of the dorsal horn (laminae I and II) primarily receive nociceptive-specific inputs through Aδ- and C-fiber primary afferents. Laminae I and II neurons display considerable heterogeneity in molecular, functional, and morphological properties and can be divided into subpopulations based on their morphological, biochemical and electrophysiological profiles (Graham et al. 2007). Excitatory and inhibitory interneurons predominate in laminae I and II, while a subset of lamina I neurons project directly to brain pain centers which include the lateral parabrachial area, the periaquedecutal grey matter and the thalamus (spinobulbar tract). Within deeper laminae of the dorsal horn (laminae V), wide dynamic range neurons respond to both innocuous and noxious inputs and project to brain pain networks through the spinothalamic tract. These anatomical connections mediate altogether the discriminative, emotional/affective, mnemonic, and autonomic features of pain. (Casey 1999). Moreover, the integration of nociceptive inputs in the spinal cord is under the control of descending influences that have either facilitatory or inhibitory roles. These involve largely bulbo spinal pathways where descending tracts via periaqueductal grey, locus coerelus and rostromedial medulla play central roles (Millan 2002).

6.2 Calcium Channel Subtypes and Molecular Composition

While a number of ionic conductances contribute to neuronal firing, voltage-gated calcium channels are unique in being involved in both shaping the action potential and triggering downstream an array of physiological cytoplasmic processes. Calcium entry via voltage gated calcium channels triggers cytoplasmic processes including the activation of calcium dependent enzymes, gene transcription, and the release of neurotransmitter from presynaptic nerve terminals. Most neurons express multiple types of calcium channels which have been classified into high and low voltage activated channels based on their voltage-dependences of activation (Nowycky et al. 1985). High voltage-activated calcium channels include pharmacologically distinct N-, P/-, Q-, R- and L-types, all of which are heteromultimers that consist of a pore forming $\alpha1$ subunit which defines the channel type, plus one of four different types of β subunits and one of four different $\alpha2$-δ subunits as largely described in the present book. The low voltage-activated channels (also known as T-types) appear to consist only of the $\alpha1$ subunit since its sole expression in recombinant systems recapitulates the properties of native channels. Moreover, no evidence of essential auxiliary subunits has been provided to date. The present picture arising from the completion of sequencing of multiple mammalian genomes shows that ten different types of calcium channel $\alpha1$ subunits are distributed into three families (Ca_v1, Ca_v2, and Ca_v3) based on their homology of sequence. They correspond to the native calcium channel functionally identified in excitable cells by electrophysiological means. These subunits share a similar transmembrane topology, being comprised of four homologous domains, each containing six putative transmembrane helices (S1–S6) plus a re-entrant pore forming loop. The four domains are connected via large cytoplasmic linker regions which are molecular platforms for protein kinases, and form interaction sites for multiple regulatory proteins, such as G protein $\beta\gamma$ subunits or the calcium channel β subunits in the case of Ca_v1 and Ca_v2 $\alpha1$ subunits.

Different calcium channel isoforms show distinct cellular and subcellular distributions, and fulfill specific functional roles. The Ca_v1 family is more involved in excitation contraction coupling, while the Ca_v2 family members by being expressed at presynaptic terminals are more involved in the control of neurotransmitter release from nerve terminals. These two 1 families may also partake in the activation of calcium dependent enzymes and gene transcription (Wheeler et al. 2012). However, the exact roles and distributions of each channel subtypes are neuron subtype dependent, such that most types of calcium channels are expressed at various subcellular loci and do in fact support a wider range of functions. These diverse functional roles ultimately pose a challenge when designing new calcium channel therapeutics with a low risk of side effects.

At the molecular and biochemical level, T-type calcium channels are formed by a single $Ca_v\alpha1$ subunit—a $\sim$250 kDa protein that is comprised, as for the HVA calcium channel subunits of four membrane domains that are connected by cytoplasmic regions and whose N- and C-termini are also cytoplasmic (Catterall 2011). Each membrane domain contains six membrane spanning helices (S1–S6) that include a voltage sensor region plus a re-entrant p-loop motif that lines the pore of the channel and controls ion selectivity. The mammalian genome encodes three distinct T-type calcium channel $\alpha 1$ subunits, termed $Ca_v3.1$, $Ca_v3.2$ and $Ca_v3.3$ (Perez-Reyes et al. 1998; Cribbs et al. 1998; Lee et al. 1999; Perez-Reyes 2003) and which show distinct brain tissue distributions (McKay et al. 2006; Talley et al. 1999). In addition to the multiplicity of genes, all of the known $Ca_v\alpha 1$ subunits undergo alternate splicing, in some cases giving rise to channels with dramatically different functional behavior (Tang et al. 2004; Shen et al. 2006; Tan et al. 2012; Allen et al. 2010; Lipscombe et al. 2013; Bourinet et al. 1999). Each of the $Ca_v3.x$ subunits are subject to alternate splicing (Senatore and Spafford 2012; Chemin et al. 2001a; David et al. 2010) with alteration of functional properties. The consequence of splicing for $Ca_v3.x$ subcellular or tissue distribution is unknown so far.

Primary afferent neurons as well as spinal and supra spinal neurons of the pain circuitry express multiple types of voltage gated calcium channels, including the three Ca_v3 isoforms coding for T-type channels. Below, we will highlight the roles of $Ca_v3.1$, $Ca_v3.2$ and $Ca_v3.3$ in pain transmission, and their usefulness as targets for analgesics.

6.3 The Role of T-Type Calcium Channels in Pain Signaling

By virtue of their hyperpolarized voltage-activation range and window current, T-type (Ca_v3) calcium channels are ideally suited to regulate neuronal excitability, as evident from their role in the development of spike and wave discharges in the epileptic brain (for review, see Zamponi et al. 2010; Khosravani and Zamponi 2006). In addition, T-type calcium channels also support secretion from neuro-endocrine cells (Okayama et al. 2006; Gackiere et al. 2008) and are capable of associating with the synaptic vesicle release machinery (Weiss et al. 2012). In addition T-type channels have been firstly well described functionally in primary sensory neurons (Fedulova et al. 1985; Carbone and Lux 1984; Bossu et al. 1985; Fox et al. 1987). Following this initial discovery, DRG neurons subtypes have been show to deferentially express T-type calcium channels with medium sized neurons being the cells with the highest expressers followed with small putative nociceptors (Scroggs and Fox 1992). In addition kinetics differences suggested the presence of two distinct T-type channels in these neurons (Coste et al. 2007; Pinchenko et al. 2005). With the molecular identification of the $Ca_v3.x$ gene family and the subsequent in situ hybridization analysis, it was clearly shown that $Ca_v3.2$ is highly expressed in small and medium sized DRGs as well as $Ca_v3.3$ to a lower extent in small neurons (Shin et al. 2003; Talley et al. 1999) (Allen Brain Atlas:

www.brain-map.org). Along these lines, T-type calcium channels have been implicated in synaptic release in the dorsal horn of the spinal cord (Jacus et al. 2012; Todorovic and Jevtovic-Todorovic 2013). $Ca_v3.2$ calcium channels are expressed in various subpopulations of primary afferent neurons (Bourinet et al. 2005; Todorovic and Jevtovic-Todorovic 2006) altogether suggesting a role of these channels in pain processing. Consistent with this idea, systemic or intrathecal delivery of T-type calcium channel blockers such as ethosuximide and mibefradil mediates analgesia in rodents (Flatters and Bennett 2004; Dogrul et al. 2003). On the flip side, T-type calcium channel activity is increased in afferent pain fibers in a number of chronic pain conditions, such as after different types of traumatic nerve injury (Yue et al. 2013; Jagodic et al. 2008; Wen et al. 2010), metabolic nerve alteration in diabetic neuropathy (Jagodic et al. 2007; Cao et al. 2011), or toxic neuropathies induced by chemotherapies (Kawabata 2013; Okubo et al. 2011; Flatters and Bennett 2004). At least in the case of diabetic neuropathy, blocking T-type channel activity restores a normal pain phenotype (Messinger et al. 2009; Latham et al. 2009). Of particular note, $Ca_v3.2$ activity is modulated by glycosylation that is altered in diabetic situation (Orestes et al. 2013; Weiss et al. 2013) offering a molecular substrate to pharmacologically control $Ca_v3.2$ cell surface expression in a therapeutic perspective. Redox modulation is another means that effectively affect $Ca_v3.2$ function and that could be a major component of inflammatory or neuropathic pain via the action of L cysteine of the gasotransmited hydrogen sulfide (Maeda et al. 2009; Todorovic et al. 2001). As a consequence, anti-oxidants could be analgesic by targeting $Ca_v3.2$ (Lee et al. 2009; Nelson et al. 2007a, b) In vivo silencing of $Ca_v3.2$ calcium channels (but not other T-type calcium channel isoforms) via antisense oligonucleotides or siRNA reduces mechanical nociception, and tactile allodynia arising from traumatic or metabolic nerve injury (Bourinet et al. 2005; Messinger et al. 2009; Takahashi et al. 2010). This fits with observations showing that $Ca_v3.2$ channels regulate mechanosensitivity of D-hair receptors (Wang and Lewin 2011; Dubreuil et al. 2004). Furthermore, these findings on somatic nociception extend to painful situations linked to visceral origin. In a rodent model of colonic hypersensitivity mimicking the Irritable Bowel Syndrome (Bourdu et al. 2005), in vivo knockdown of $Ca_v3.2$ channels reverses pain hypersensitivity in response to colorectal distension (Marger et al. 2011). These findings are further supported by analysis of other visceral pain models (Maeda et al. 2009; Matsunami et al. 2011). Altogether, these data indicate that T-type channel membrane expression is dynamically regulated and increased under conditions of chronic pain, and that counteracting this aberrant upregulation may constitute an effective means of mediating analgesia. It is interesting to note that mice lacking $Ca_v3.2$ (Chen et al. 2003) altogether show hyposensitivity to basal nociception, to formalin induced pain in both phases, and a more limited effect on neuropathic pain (Choi et al. 2007) suggesting possible compensatory phenomenon.

Compared to the DRGs, at the spinal level little is known on the impact of T-type channels but nickel sensitive low voltage gated evoked and spontaneous neurotransmitter release has been described between primary afferent neurons and outer laminas spinal projection or interneurons (Bao et al. 1998). Recent evidences using the $Ca_v3.2$ KO mice confirmed these results (Jacus et al. 2012). Moreover the implication of $Ca_v3.2$ in the presynaptic neurotransmitter release machinery is an emerging concept (Weiss and Zamponi 2012; Weiss et al. 2012; Huang et al. 2011). Postsynaptically, spinal projection and/or interneurons also express T-type channels (Walsh et al. 2009; Ikeda et al. 2003; Ku and Schneider 2011) with a selective expression in a subclass of lamina II interneurons whose function remain to be established (Walsh et al. 2009). Considering that the vast majority of spinal neurons are interneurons, the wiring underlying the spinal networks relaying sensory signaling is clearly not fully understood. In this context, a lot of explorations remain to be performed. Are T-type channels expressed in excitatory or inhibitory inter-neurons, in projection neurons? Specific genetic inactivation within the spinal cord sparing the primary afferent neurons is still waiting for behavioral analysis.

At the supraspinal level, the thalamus is the important relay of the somatosensory information on their way to the cortex. Among the sensory modalities, noxious stimuli are mainly conveyed to the thalamus by the spinothalamic tract arising from the dorsal horn of the spinal cord and the medulla, respectively. Although the importance of the ventral posterior nucleus which projects to the primary somatosensory cortex in the localization and appreciation of the intensity of a noxious stimuli has been shown, other thalamic nuclei including several nuclei of the intralaminar thalamus are innervated by these afferent tracts (Jones 1998) and project to other cerebral areas of the pain matrix including the insula and the anterior cingulate cortex. In addition, nociceptive information reach the thalamus through indirect pathway like in the case of the lateral habenula via the lateral hypothalamus (Shelton et al. 2012) or via the amygdala (Braz et al. 2005) implicated in the emotional aspects of pain perception. The dialog between thalamus and cortex is based on reciprocal connections organizing the thalamo-cortico-thalamic loop since thalamocortical (TC) relay neurons projecting to the cortex receive a dense cortical excitatory feedback. Within this loop, the nucleus reticularis thalami (NRT) exclusively composed of GABAergic neurons, that receive excitatory input from both thalamocortical collaterals and cortical afferences and provide inhibitory input to the relay neurons, plays a key role in organizing the different modes of activity displayed by the thalamus according to the state of vigilance. The most remarkable feature characterizing the transition of the thalamocortical activity from wakefulness to sleep is the changes in the firing patterns of thalamic neurons from tonic to burst mode. This dichotomic firing mode is tightly linked to the biophysical properties of the low-threshold activated T-type calcium channels, that are highly expressed in both TC ($Ca_v3.2$) and NRT ($Ca_v3.2$ and $Ca_v3.3$) neurons (Talley et al. 1999). T channels are fully inactivated in the range of membrane potentials believed to be associated with the wake state and require the substantial and prolonged hyperpolarization associated to the sleep states to de-inactivate and promote low threshold Ca^{2+} spikes (LTS) generation and the occurrence of rhythmic high frequency bursting activities (Steriade et al. 1993). The cortical consequences of this shift in firing mode is reflected on the electroencephalogram (EEG) which is dominated during sleep by high amplitude, low frequency oscillatory activity (<15 Hz) and by contrast during activated states (waking and paradoxical sleep), by low amplitude, high frequency oscillatory activity in the gamma band (30–50 Hz). Interestingly, an increasing number of clinical studies have reported the presence of slow TC oscillations in the theta frequency band (4–8 Hz) in patients awake who present a wide variety of neurological and psychiatric conditions, including neurogenic pain (Jeanmonod et al. 1996; Llinas et al. 1999; Walton and Llinas 2010). Moreover, single unit thalamic recordings in patients suffering of neurogenic pain demonstrated the presence of high frequency action potential bursts (Lenz et al. 1989, 1994; Modesti and Waszak 1975; Rinaldi et al. 1991) associated to this theta rhythmicity (Jeanmonod et al. 1996; Sarnthein et al. 2006). The pattern of these bursts indicates that they are underlain by T type current mediated LTS. Accordingly, thalamic lesions decreasing in the EEG theta band have proven to be an effective treatment in patients with chronic pain. Experimental evidences obtained in vivo in animal studies also strengthen the hypothesis of a crucial role played by thalamic T currents in pain processing. Extracellular recordings of thalamic neurons, performed mainly in the somatosensory ventral posterior nucleus of either anesthetized or freely moving mice, have shown bursting behavior in response to acute pain stimulation (Kim et al. 2003). Moreover, in rat peripheral and central models of chronic pain, the enhanced bursting behavior was associated to spontaneous slow rhythmic oscillation (Iwata et al. 2011; Gerke et al. 2003). This feature is also present in a gain of function mutant mouse having a single amino acid change in the sensory neuron specific $Na_v1.8$ channel leading to episodic periods of spontaneous pain with a shift in EEG rhythms pattern reminiscent of the clinical situations (Blasius et al. 2011).

A more direct way to estimate the involvement of T channels in pain response is the use of genetically modified mice for the different Ca_v3 isoforms. However, experiments performed so far have involved general KO mice for the $Ca_v3.1$ or $Ca_v3.2$ channels and non-specific T channel pharmacology (Mibefradil, Ethosuximide). The $Ca_v3.3$ KO model has not been explored for pain behavior. Hyperalgesia was first reported in response to visceral stimulation in the $Ca_v3.1$ null mice (Kim et al. 2003) exhibiting a total absence of T-type currents in somatosensory TC neurons (Kim et al. 2001; Tscherter et al. 2011). In contrast and presented earlier, the global $Ca_v3.2$ KO exhibits an analgesic phenotype regarding different models of pain (Chen et al. 2010; Choi et al. 2007; Jagodic et al. 2007; Barbara et al. 2009; Francois et al. 2013). Moreover our recent work shows that the analgesic effects of *per os* administered newly developed T-type channel agonist (affecting all three Ca_v3 isoforms) are abolished in the null mice showing that the blockade of T-type channels overall mediates analgesia (Francois et al. 2013). Direct proof of central pronociceptive role of $Ca_v3.x$ channels will require either local pharmacological approaches using specific T channel antagonist (Dreyfus et al. 2009; Francois et al. 2013; Choe et al. 2011) or conditional $Ca_v3.x$ KO mice. This conclusion stands equally to unravel the impact of $Ca_v3.x$ in the other regions of the pain matrix in the ascending as well as in the descending pathways.

Although little is known on the impact of $Ca_v3.x$ channels in descending pain modulation, it should be noted that in the bulbo spinal area, namely in inhibitory periaqueductal gray gabaergic neurons, the lack of $Ca_v3.1$ KO mice perturb morphine analgesia (Park et al. 2010). These findings could indicate that PAG located $Ca_v3.1$ may limit persistent pain, but the use of TTA-A2, a $Ca_v3.x$ pan antagonist do not result in similar effects. Further work need to be done to clarify this issue.

For all the aspects presented the precise spatial and temporal resolution of $Ca_v3.x$ expression in the pain neuronal pathways both at the cellular and subcellular levels are crucial. Up to recently the state of the art on this aspect was inexistent. However, recently two independent groups reported that T-type calcium channels are concentrated in the axon initial segment (AIS), where they contribute to local subthreshold membrane depolarization and thereby influence action potential initiation. These data where based on conclusion from functional studies on isolated neurons using cell attached patch clamp recordings as well as calcium and sodium imaging. Furthermore, these studies shows that this localization is a dynamically regulated process enabling a fine tuning of intrinsic plasticity of neuronal excitability (Bender et al. 2010; Grubb and Burrone 2010; Bender and Trussell 2009). This tuning of T-type channel localization is moreover regulated by membrane receptors such as the dopamine receptors also expressed in regions enriched in $Ca_v3.2$ (Splawski et al. 2006; Bergquist and Nissbrandt 2003; Talley et al. 1999). The plastic alterations of T-type channels distribution are likely involved during chronic pain states.

6.4 T-Type Channel Pharmacology: Towards a New Class of Analgesics?

Most of the work done so far validates the $Ca_v3.2$ isoform as a target for mediating analgesia, especially by acting in the periphery. It may be possible to exploit state dependence of drug action as a means to further preferentially inhibit T-type calcium channels in highly active pain fibers. Indeed, new generation state dependent blockers such as TTA-P2 and more prominently TTA-A2 which appear to interact preferentially with inactivated T-type calcium channels, both mediate analgesia in rodent models of pain (Choe et al. 2011; Francois et al. 2013). Z123212, a mixed blocker of voltage dependent sodium channels and T-type calcium channels mediates analgesia by selectively targeting the slow inactivated state of these channels (Hildebrand et al. 2011). In this context it is interesting to note that the local anesthetic binding domain of voltage gated sodium channels is partially conserved in T-type calcium channels (Bladen and Zamponi 2012). Given that there is now a crystal structure of a bacterial voltage gated sodium channels (Payandeh et al. 2011, 2012) this knowledge may help guide further development of T-type channel blocking drugs. Finally, Z944, another state dependent T-type channel inhibitor is currently in phase I clinical trials for pain. Several pharmaceutical companies have T-type channels on their list of targets and near future will show if clinical drugs emerge.

6.5 Direct Inhibition of T-Type Channels by GPCR Ligands

T-type calcium channels are regulated though a number of different second messenger pathways in response to activation of various GPCRs (Huc et al. 2009; Iftinca and Zamponi 2009). Some GPCR agonists have been shown to directly regulate T-type channel activity, rather than acting via G protein signaling. For example, the endocannabinoid anandamide potently blocks T-type channels (Chemin et al. 2001b). Naturally occurring anandamide derivatives also inhibit T-type channels, and by doing so, mediate analgesia (Barbara et al. 2009) in normal mice, but not in $Ca_v3.2$ channel knockout mice. Along these lines, mixed T-type channel/cannabinoid receptor ligands have been shown to be efficacious in inflammatory pain (Gadotti et al. 2013; You et al. 2011). The CCR2 receptor agonist monocyte chemoattractant protein-1 (MCP-1) also directly and selectively inhibits $Ca_v3.2$ channels (You et al. 2010). T-type channel inhibition occurs at nanomolar concentrations of this ligand, and is partial with a plateau of about 50 % inhibition of current activity. MCP-1 activation of CCR2 receptors is pro-algesic whereas CCR2 receptor antagonists mediate analgesia. Some of these CCR2 receptor antagonists also block T-type channels (You et al. 2010) and it is thus possible that such a mixed $Ca_v3.2$ channel/CCR2 receptor antagonist may have synergistic effects in treating pain. The observation that agonists of two different sets of GPCRs interact directly with T-type calcium channels raises an interesting evolutionary question with regard to the conservation of GPCR binding pockets in a voltage gated ion channel.

6.6 Concluding Remarks

A number of questions concerning the role of T-type channels in pain remain unresolved. First, it is unclear precisely how T-type channels contribute to pain signaling. Possibilities include: (1) a lowering of the firing threshold for afferent pain fibers, (2) a direct contribution to neurotransmitter release at primary afferent synapses, (3) a direct function of T-type channels as mechanosensors, (4) activation of pathways such as ERK which in turn is linked to increased pain (Chen et al. 2010) and (5) perhaps via interactions with other types of ion channels such as voltage- and calcium-activated potassium channels as described for different types of CNS neurons. Second, the mechanism by which T-type calcium channel activity is enhanced in chronic pain conditions remains to be determined such as direct modulations, post-translational modifications, and altered membrane turn-over. This could potentially include the $Ca_v\alpha2\delta$ subunit which has been shown to increase T-type channel amplitude in expression systems (Dubel et al. 2004). Finally, a number of inherited mutations have been described for the $Ca_v3.2$ coding gene in rats (Powell et al. 2009) and more importantly in humans (Khosravani et al. 2004; Vitko et al. 2007; Splawski et al. 2006). Although they are linked to some forms of epilepsies of autism, it remains to be explored if these patients suffer from altered pain perception.

In summary, among the calcium channel family the T-type calcium channels and in particular the $Ca_v3.2$ isoform appear to have a critical role in the excitability of pain neuronal circuits, and as a result are vigorously pursued as therapeutic targets.

Acknowledgments Pain related work conducted in the authors' laboratory is supported by the Agence Nationale pour la Recherche, Association Française pour le Myopathies, Institut UPSA de la Douleur, and the Fondation pour la Recherche Médicale.

References

Allen SE, Darnell RB, Lipscombe D (2010) The neuronal splicing factor Nova controls alternative splicing in N-type and P-type Ca_v2 calcium channels. Channels (Austin) 4(6):483–489

Bachy I, Franck MC, Li L, Abdo H, Pattyn A, Ernfors P (2011) The transcription factor Cux2 marks development of an A-delta sublineage of TrkA sensory neurons. Dev Biol 360(1):77–86

Bao J, Li JJ, Perl ER (1998) Differences in Ca2+ channels governing generation of miniature and evoked excitatory synaptic currents in spinal laminae I and II. J Neurosci 18(21):8740–8750

Barbara G, Alloui A, Nargeot J, Lory P, Eschalier A, Bourinet E, Chemin J (2009) T-type calcium channel inhibition underlies the analgesic effects of the endogenous lipoamino acids. J Neurosci 29(42):13106–13114

Basbaum AI, Braz JM (2010) Transgenic mouse models for the tracing of "pain" pathways. In: Kruger L, Light AR (eds) Translational pain research: from mouse to man. Frontiers in neuroscience. CRC Press, Boca Raton

Basbaum AI, Bautista DM, Scherrer G, Julius D (2009) Cellular and molecular mechanisms of pain. Cell 139(2):267–284

Bender KJ, Trussell LO (2009) Axon initial segment Ca2+ channels influence action potential generation and timing. Neuron 61(2):259–271

Bender KJ, Ford CP, Trussell LO (2010) Dopaminergic modulation of axon initial segment calcium channels regulates action potential initiation. Neuron 68(3):500–511

Bergquist F, Nissbrandt H (2003) Influence of R-type ($Ca_v2.3$) and t-type ($Ca_v3.1$-3.3) antagonists on nigral somatodendritic dopamine release measured by microdialysis. Neuroscience 120(3):757–764

Bladen C, Zamponi GW (2012) Common mechanisms of drug interactions with sodium and T-type calcium channels. Mol Pharmacol 82(3):481–487

Blasius AL, Dubin AE, Petrus MJ, Lim BK, Narezkina A, Criado JR, Wills DN, Xia Y, Moresco EM, Ehlers C, Knowlton KU, Patapoutian A, Beutler B (2011) Hypermorphic mutation of the voltage-gated sodium channel encoding gene Scn10a causes a dramatic stimulus-dependent neurobehavioral phenotype. Proc Natl Acad Sci USA 108(48):19413–19418

Bossu JL, Feltz A, Thomann JM (1985) Depolarization elicits two distinct calcium currents in vertebrate sensory neurones. Pflugers Arch 403(4):360–368

Bourdu S, Dapoigny M, Chapuy E, Artigue F, Vasson MP, Dechelotte P, Bommelaer G, Eschalier A, Ardid D (2005) Rectal instillation of butyrate provides a novel clinically relevant model of noninflammatory colonic hypersensitivity in rats. Gastroenterology 128(7):1996–2008

Bourinet E, Soong TW, Sutton K, Slaymaker S, Mathews E, Monteil A, Zamponi GW, Nargeot J, Snutch TP (1999) Splicing of alpha 1A subunit gene generates phenotypic variants of P- and Q-type calcium channels. Nat Neurosci 2(5):407–415

Bourinet E, Alloui A, Monteil A, Barrere C, Couette B, Poirot O, Pages A, McRory J, Snutch TP, Eschalier A, Nargeot J (2005) Silencing of the $Ca_v3.2$ T-type calcium channel gene in sensory neurons demonstrates its major role in nociception. EMBO J 24(2):315–324

Braz JM, Nassar MA, Wood JN, Basbaum AI (2005) Parallel "pain" pathways arise from subpopulations of primary afferent nociceptor. Neuron 47(6):787–793

Cao XH, Byun HS, Chen SR, Pan HL (2011) Diabetic neuropathy enhances voltage-activated Ca2+ channel activity and its control by M4 muscarinic receptors in primary sensory neurons. J Neurochem 119(3):594–603

Carbone E, Lux HD (1984) A low voltage-activated, fully inactivating Ca channel in vertebrate sensory neurones. Nature 310(5977):501–502

Casey KL (1999) Forebrain mechanisms of nociception and pain: analysis through imaging. Proc Natl Acad Sci USA 96(14):7668–7674

Caterina MJ, Schumacher MA, Tominaga M, Rosen TA, Levine JD, Julius D (1997) The capsaicin receptor: a heat-activated ion channel in the pain pathway. Nature 389(6653):816–824

Catterall WA (2011) Voltage-gated calcium channels. Cold Spring Harbor Perspect Biol 3(8):a003947

Chemin J, Monteil A, Bourinet E, Nargeot J, Lory P (2001a) Alternatively spliced alpha(1G) (Ca(V)3.1) intracellular loops promote specific T-type Ca(2+) channel gating properties. Biophys J 80(3):1238–1250

Chemin J, Monteil A, Perez-Reyes E, Nargeot J, Lory P (2001b) Direct inhibition of T-type calcium channels by the endogenous cannabinoid anandamide. EMBO J 20(24):7033–7040

Chen CC, Lamping KG, Nuno DW, Barresi R, Prouty SJ, Lavoie JL, Cribbs LL, England SK, Sigmund CD, Weiss RM, Williamson RA, Hill JA, Campbell KP (2003) Abnormal coronary function in mice deficient in alpha1H T-type Ca2+ channels. Science 302(5649):1416–1418

Chen WK, Liu IY, Chang YT, Chen YC, Chen CC, Yen CT, Shin HS (2010) Ca(v)3.2 T-type Ca2+ channel-dependent activation of ERK in paraventricular thalamus modulates acid-induced chronic muscle pain. J Neurosci 30(31):10360–10368

Choe W, Messinger RB, Leach E, Eckle VS, Obradovic A, Salajegheh R, Jevtovic-Todorovic V, Todorovic SM (2011) TTA-P2 is a potent and selective blocker of T-type calcium channels in rat sensory neurons and a novel antinociceptive agent. Mol Pharmacol 80(5):900–910

Choi S, Na HS, Kim J, Lee J, Lee S, Kim D, Park J, Chen CC, Campbell KP, Shin HS (2007) Attenuated pain responses in mice lacking Ca(V)3.2 T-type channels. Genes Brain Behav 6(5):425–431

Coste B, Crest M, Delmas P (2007) Pharmacological dissection and distribution of NaN/Nav1.9, T-type Ca2+ currents, and mechanically activated cation currents in different populations of DRG neurons. J Gen Physiol 129(1):57–77

Cribbs LL, Lee JH, Yang J, Satin J, Zhang Y, Daud A, Barclay J, Williamson MP, Fox M, Rees M, Perez-Reyes E (1998) Cloning and characterization of alpha1H from human heart, a member of the T-type Ca2+ channel gene family. Circ Res 83(1):103–109

David LS, Garcia E, Cain SM, Thau E, Tyson JR, Snutch TP (2010) Splice-variant changes of the Ca(V)3.2 T-type calcium channel mediate voltage-dependent facilitation and associate with cardiac hypertrophy and development. Channels (Austin) 4(5):375–389

Devor M (1999) Unexplained peculiarities of the dorsal root ganglion. Pain Suppl 6:S27–S35
Djouhri L, Lawson SN (2004) Abeta-fiber nociceptive primary afferent neurons: a review of incidence and properties in relation to other afferent A-fiber neurons in mammals. Brain Res Brain Res Rev 46(2):131–145
Dogrul A, Gardell LR, Ossipov MH, Tulunay FC, Lai J, Porreca F (2003) Reversal of experimental neuropathic pain by T-type calcium channel blockers. Pain 105(1–2):159–168
Dreyfus FM, Tscherter A, Errington AC, Renger JJ, Shin HS, Uebele VN, Crunelli V, Lambert RC, Leresche N (2009) Selective T-type calcium channel block in thalamic neurons reveals channel redundancy and physiological impact of I(T)window. J Neurosci 30(1):99–109
Dubel SJ, Altier C, Chaumont S, Lory P, Bourinet E, Nargeot J (2004) Plasma membrane expression of T-type calcium channel {alpha}1 subunits is modulated by high voltage-activated auxiliary subunits. J Biol Chem 279(28):29263–29269
Dubreuil AS, Boukhaddaoui H, Desmadryl G, Martinez-Salgado C, Moshourab R, Lewin GR, Carroll P, Valmier J, Scamps F (2004) Role of T-type calcium current in identified D-hair mechanoreceptor neurons studied in vitro. J Neurosci 24(39):8480–8484
Erlanger J (1927) The interpretation of the action potential in cutuaneous and muscle nerves. Am J Physiol 1(82):11
Erlanger J, Gasser HS (1930) The action potential in fibers of slow conduction in spinal roots and somatic nerves. Am J Physiol 92(1):39
Fang X, McMullan S, Lawson SN, Djouhri L (2005) Electrophysiological differences between nociceptive and non-nociceptive dorsal root ganglion neurones in the rat in vivo. J Physiol 565(Pt 3):927–943
Fedulova SA, Kostyuk PG, Veselovsky NS (1985) Two types of calcium channels in the somatic membrane of new-born rat dorsal root ganglion neurones. J Physiol 359:431–446
Flatters SJ, Bennett GJ (2004) Ethosuximide reverses paclitaxel- and vincristine-induced painful peripheral neuropathy. Pain 109(1–2):150–161
Fox AP, Nowycky MC, Tsien RW (1987) Single-channel recordings of three types of calcium channels in chick sensory neurones. J Physiol 394:173–200
Francois A, Kerckhove N, Meleine M, Alloui A, Barrere C, Gelot A, Uebele VN, Renger JJ, Eschalier A, Ardid D, Bourinet E (2013) State dependent properties of a new T-type calcium channel blocker enhance $Ca_v3.2$ selectivity and support analgesic effects. Pain 154:283–293
Gackiere F, Bidaux G, Delcourt P, Van Coppenolle F, Katsogiannou M, Dewailly E, Bavencoffe A, Van Chuoi-Mariot MT, Mauroy B, Prevarskaya N, Mariot P (2008) $Ca_v3.2$ T-type calcium channels are involved in calcium-dependent secretion of neuroendocrine prostate cancer cells. J Biol Chem 283(15):10162–10173
Gadotti VM, You H, Petrov RR, Berger ND, Diaz P, Zamponi GW (2013) Analgesic effect of a mixed T-type channel inhibitor/CB2 receptor agonist. Mol Pain 9(1):32
Gerke MB, Duggan AW, Xu L, Siddall PJ (2003) Thalamic neuronal activity in rats with mechanical allodynia following contusive spinal cord injury. Neuroscience 117(3):715–722
Graham BA, Brichta AM, Callister RJ (2007) Moving from an averaged to specific view of spinal cord pain processing circuits. J Neurophysiol 98(3):1057–1063
Grubb MS, Burrone J (2010) Activity-dependent relocation of the axon initial segment fine-tunes neuronal excitability. Nature 465(7301):1070–1074
Hildebrand ME, Smith PL, Bladen C, Eduljee C, Xie JY, Chen L, Fee-Maki M, Doering CJ, Mezeyova J, Zhu Y, Belardetti F, Pajouhesh H, Parker D, Arneric SP, Parmar M, Porreca F, Tringham E, Zamponi GW, Snutch TP (2011) A novel slow-inactivation-specific ion channel modulator attenuates neuropathic pain. Pain 152(4):833–843
Huang Z, Lujan R, Kadurin I, Uebele VN, Renger JJ, Dolphin AC, Shah MM (2011) Presynaptic HCN1 channels regulate $Ca_v3.2$ activity and neurotransmission at select cortical synapses. Nat Neurosci 14(4):478–486
Huc S, Monteil A, Bidaud I, Barbara G, Chemin J, Lory P (2009) Regulation of T-type calcium channels: signalling pathways and functional implications. Biochim Biophys Acta 1793(6):947–952

Iftinca MC, Zamponi GW (2009) Regulation of neuronal T-type calcium channels. Trends Pharmacol Sci 30(1):32–40

Ikeda H, Heinke B, Ruscheweyh R, Sandkuhler J (2003) Synaptic plasticity in spinal lamina I projection neurons that mediate hyperalgesia. Science 299(5610):1237–1240

Iwata M, LeBlanc BW, Kadasi LM, Zerah ML, Cosgrove RG, Saab CY (2011) High-frequency stimulation in the ventral posterolateral thalamus reverses electrophysiologic changes and hyperalgesia in a rat model of peripheral neuropathic pain. Pain 152(11):2505–2513

Jacus MO, Uebele VN, Renger JJ, Todorovic SM (2012) Presynaptic $Ca_v3.2$ channels regulate excitatory neurotransmission in nociceptive dorsal horn neurons. J Neurosci 32(27):9374–9382

Jagodic MM, Pathirathna S, Nelson MT, Mancuso S, Joksovic PM, Rosenberg ER, Bayliss DA, Jevtovic-Todorovic V, Todorovic SM (2007) Cell-specific alterations of T-type calcium current in painful diabetic neuropathy enhance excitability of sensory neurons. J Neurosci 27(12):3305–3316

Jagodic MM, Pathirathna S, Joksovic PM, Lee W, Nelson MT, Naik AK, Su P, Jevtovic-Todorovic V, Todorovic SM (2008) Upregulation of the T-type calcium current in small rat sensory neurons after chronic constrictive injury of the sciatic nerve. J Neurophysiol 99(6):3151–3156

Jeanmonod D, Magnin M, Morel A (1996) Low-threshold calcium spike bursts in the human thalamus. Common physiopathology for sensory, motor and limbic positive symptoms. Brain 119(Pt 2):363–375

Jones EG (1998) Viewpoint: the core and matrix of thalamic organization. Neuroscience 85(2):331–345

Julius D, Basbaum AI (2001) Molecular mechanisms of nociception. Nature 413(6852):203–210

Kawabata A (2013) Targeting Ca(v)3.2 T-type calcium channels as a therapeutic strategy for chemotherapy-induced neuropathic pain. Nihon yakurigaku zasshi Folia pharmacologica Japonica 141(2):81–84

Khosravani H, Zamponi GW (2006) Voltage-gated calcium channels and idiopathic generalized epilepsies. Physiol Rev 86(3):941–966

Khosravani H, Altier C, Simms B, Hamming KS, Snutch TP, Mezeyova J, McRory JE, Zamponi GW (2004) Gating effects of mutations in the $Ca_v3.2$ T-type calcium channel associated with childhood absence epilepsy. J Biol Chem 279(11):9681–9684

Kim D, Song I, Keum S, Lee T, Jeong MJ, Kim SS, McEnery MW, Shin HS (2001) Lack of the burst firing of thalamocortical relay neurons and resistance to absence seizures in mice lacking alpha(1G) T-type Ca(2+) channels. Neuron 31(1):35–45

Kim D, Park D, Choi S, Lee S, Sun M, Kim C, Shin HS (2003) Thalamic control of visceral nociception mediated by T-type Ca2+ channels. Science 302(5642):117–119

Ku WH, Schneider SP (2011) Multiple T-type Ca2+ current subtypes in electrophysiologically characterized hamster dorsal horn neurons: possible role in spinal sensory integration. J Neurophysiol 106(5):2486–2498

Lallemend F, Ernfors P (2012) Molecular interactions underlying the specification of sensory neurons. Trends Neurosci 35(6):373–381

Latham JR, Pathirathna S, Jagodic MM, Choe WJ, Levin ME, Nelson MT, Lee WY, Krishnan K, Covey DF, Todorovic SM, Jevtovic-Todorovic V (2009) Selective T-type calcium channel blockade alleviates hyperalgesia in ob/ob mice. Diabetes 58(11):2656–2665

Lee JH, Daud AN, Cribbs LL, Lacerda AE, Pereverzev A, Klockner U, Schneider T, Perez-Reyes E (1999) Cloning and expression of a novel member of the low voltage-activated T-type calcium channel family. J Neurosci 19(6):1912–1921

Lee WY, Orestes P, Latham J, Naik AK, Nelson MT, Vitko I, Perez-Reyes E, Jevtovic-Todorovic V, Todorovic SM (2009) Molecular mechanisms of lipoic acid modulation of T-type calcium channels in pain pathway. J Neurosci 29(30):9500–9509

Lenz FA, Kwan HC, Dostrovsky JO, Tasker RR (1989) Characteristics of the bursting pattern of action potentials that occurs in the thalamus of patients with central pain. Brain Res 496(1–2):357–360

Lenz FA, Gracely RH, Rowland LH, Dougherty PM (1994) A population of cells in the human thalamic principal sensory nucleus respond to painful mechanical stimuli. Neurosci Lett 180(1):46–50

Li L, Rutlin M, Abraira VE, Cassidy C, Kus L, Gong S, Jankowski MP, Luo W, Heintz N, Koerber HR, Woodbury CJ, Ginty DD (2011) The functional organization of cutaneous low-threshold mechanosensory neurons. Cell 147(7):1615–1627

Lipscombe D, Andrade A, Allen SE (2013) Alternative splicing: functional diversity among voltage-gated calcium channels and behavioral consequences. Biochim Biophys Acta 1828(7): 1522–1529

Liu Y, Yang FC, Okuda T, Dong X, Zylka MJ, Chen CL, Anderson DJ, Kuner R, Ma Q (2008) Mechanisms of compartmentalized expression of Mrg class G-protein-coupled sensory receptors. J Neurosci 28(1):125–132

Llinas RR, Ribary U, Jeanmonod D, Kronberg E, Mitra PP (1999) Thalamocortical dysrhythmia: a neurological and neuropsychiatric syndrome characterized by magnetoencephalography. Proc Natl Acad Sci USA 96(26):15222–15227

Luo W, Wickramasinghe SR, Savitt JM, Griffin JW, Dawson TM, Ginty DD (2007) A hierarchical NGF signaling cascade controls Ret-dependent and Ret-independent events during development of nonpeptidergic DRG neurons. Neuron 54(5):739–754

Maeda Y, Aoki Y, Sekiguchi F, Matsunami M, Takahashi T, Nishikawa H, Kawabata A (2009) Hyperalgesia induced by spinal and peripheral hydrogen sulfide: evidence for involvement of $Ca_v3.2$ T-type calcium channels. Pain 142(1–2):127–132

Marger F, Gelot A, Alloui A, Matricon J, Ferrer JF, Barrere C, Pizzoccaro A, Muller E, Nargeot J, Snutch TP, Eschalier A, Bourinet E, Ardid D (2011) T-type calcium channels contribute to colonic hypersensitivity in a rat model of irritable bowel syndrome. Proc Natl Acad Sci USA 108(27):11268–11273

Marmigere F, Ernfors P (2007) Specification and connectivity of neuronal subtypes in the sensory lineage. Nat Rev Neurosci 8(2):114–127

Matsunami M, Kirishi S, Okui T, Kawabata A (2011) Chelating luminal zinc mimics hydrogen sulfide-evoked colonic pain in mice: possible involvement of T-type calcium channels. Neuroscience 181:257–264

McKay BE, McRory JE, Molineux ML, Hamid J, Snutch TP, Zamponi GW, Turner RW (2006) Ca(V)3 T-type calcium channel isoforms differentially distribute to somatic and dendritic compartments in rat central neurons. Eur J Neurosci 24(9):2581–2594

Messinger RB, Naik AK, Jagodic MM, Nelson MT, Lee WY, Choe WJ, Orestes P, Latham JR, Todorovic SM, Jevtovic-Todorovic V (2009) In vivo silencing of the Ca(V)3.2 T-type calcium channels in sensory neurons alleviates hyperalgesia in rats with streptozocin-induced diabetic neuropathy. Pain 145(1–2):184–195

Millan MJ (2002) Descending control of pain. Prog Neurobiol 66(6):355–474

Modesti LM, Waszak M (1975) Firing pattern of cells in human thalamus during dorsal column stimulation. Appl Neurophysiol 38(4):251–258

Molliver DC, Radeke MJ, Feinstein SC, Snider WD (1995) Presence or absence of TrkA protein distinguishes subsets of small sensory neurons with unique cytochemical characteristics and dorsal horn projections. J Comp Neurol 361(3):404–416

Molliver DC, Wright DE, Leitner ML, Parsadanian AS, Doster K, Wen D, Yan Q, Snider WD (1997) IB4-binding DRG neurons switch from NGF to GDNF dependence in early postnatal life. Neuron 19(4):849–861

Nelson MT, Joksovic PM, Su P, Kang HW, Van Deusen A, Baumgart JP, David LS, Snutch TP, Barrett PQ, Lee JH, Zorumski CF, Perez-Reyes E, Todorovic SM (2007a) Molecular mechanisms of subtype-specific inhibition of neuronal T-type calcium channels by ascorbate. J Neurosci 27(46):12577–12583

Nelson MT, Woo J, Kang HW, Vitko I, Barrett PQ, Perez-Reyes E, Lee JH, Shin HS, Todorovic SM (2007b) Reducing agents sensitize C-type nociceptors by relieving high-affinity zinc inhibition of T-type calcium channels. J Neurosci 27(31):8250–8260

Nowycky MC, Fox AP, Tsien RW (1985) Three types of neuronal calcium channel with different calcium agonist sensitivity. Nature 316(6027):440–443

Okayama S, Imagawa K, Naya N, Iwama H, Somekawa S, Kawata H, Horii M, Nakajima T, Uemura S, Saito Y (2006) Blocking T-type Ca2 + channels with efonidipine decreased plasma aldosterone concentration in healthy volunteers. Hypertension Res Off J Jpn Soc Hypertension 29(7):493–497

Okubo K, Takahashi T, Sekiguchi F, Kanaoka D, Matsunami M, Ohkubo T, Yamazaki J, Fukushima N, Yoshida S, Kawabata A (2011) Inhibition of T-type calcium channels and hydrogen sulfide-forming enzyme reverses paclitaxel-evoked neuropathic hyperalgesia in rats. Neuroscience 188:148–156

Orestes P, Osuru HP, McIntire WE, Jacus MO, Salajegheh R, Jagodic MM, Choe W, Lee J, Lee SS, Rose KE, Poiro N, Digruccio MR, Krishnan K, Covey DF, Lee JH, Barrett PQ, Jevtovic-Todorovic V, Todorovic SM (2013) Reversal of neuropathic pain in diabetes by targeting glycosylation of $Ca_v3.2$ T-type calcium channels. Diabetes (in press)

Park C, Kim JH, Yoon BE, Choi EJ, Lee CJ, Shin HS (2010) T-type channels control the opioidergic descending analgesia at the low threshold-spiking GABAergic neurons in the periaqueductal gray. Proc Natl Acad Sci USA 107(33):14857–14862

Payandeh J, Scheuer T, Zheng N, Catterall WA (2011) The crystal structure of a voltage-gated sodium channel. Nature 475(7356):353–358

Payandeh J, Gamal El-Din TM, Scheuer T, Zheng N, Catterall WA (2012) Crystal structure of a voltage-gated sodium channel in two potentially inactivated states. Nature 486(7401):135–139

Peier AM, Moqrich A, Hergarden AC, Reeve AJ, Andersson DA, Story GM, Earley TJ, Dragoni I, McIntyre P, Bevan S, Patapoutian A (2002) A TRP channel that senses cold stimuli and menthol. Cell 108(5):705–715

Perez-Reyes E (2003) Molecular physiology of low-voltage-activated t-type calcium channels. Physiol Rev 83(1):117–161

Perez-Reyes E, Cribbs LL, Daud A, Lacerda AE, Barclay J, Williamson MP, Fox M, Rees M, Lee JH (1998) Molecular characterization of a neuronal low-voltage-activated T-type calcium channel. Nature 391(6670):896–900

Perl ER (2007) Ideas about pain, a historical view. Nat Rev Neurosci 8(1):71–80

Pinchenko VO, Kostyuk PG, Kostyuk EP (2005) Influence of external pH on two types of low-voltage-activated calcium currents in primary sensory neurons of rats. Biochim Biophys Acta 1724(1–2):1–7

Powell KL, Cain SM, Ng C, Sirdesai S, David LS, Kyi M, Garcia E, Tyson JR, Reid CA, Bahlo M, Foote SJ, Snutch TP, O'Brien TJ (2009) A $Ca_v3.2$ T-type calcium channel point mutation has splice-variant-specific effects on function and segregates with seizure expression in a polygenic rat model of absence epilepsy. J Neurosci 29(2):371–380

Reed-Geaghan EG, Maricich SM (2011) Peripheral somatosensation: a touch of genetics. Curr Opin Genet Dev 21(3):240–248

Reeh PW (1986) Sensory receptors in mammalian skin in an in vitro preparation. Neurosci Lett 66(2):141–146

Rinaldi PC, Young RF, Albe-Fessard D, Chodakiewitz J (1991) Spontaneous neuronal hyperactivity in the medial and intralaminar thalamic nuclei of patients with deafferentation pain. J Neurosurg 74(3):415–421

Sarnthein J, Stern J, Aufenberg C, Rousson V, Jeanmonod D (2006) Increased EEG power and slowed dominant frequency in patients with neurogenic pain. Brain 129(Pt 1):55–64

Scroggs RS, Fox AP (1992) Calcium current variation between acutely isolated adult rat dorsal root ganglion neurons of different size. J Physiol 445:639–658

Senatore A, Spafford JD (2012) Gene transcription and splicing of T-type channels are evolutionarily-conserved strategies for regulating channel expression and gating. PLoS ONE 7(6):e37409

Serra J (2010) Microneurography: an opportunity for translational drug development in neuropathic pain. Neurosci Lett 470(3):155–157

Shelton L, Becerra L, Borsook D (2012) Unmasking the mysteries of the habenula in pain and analgesia. Prog Neurobiol 96(2):208–219

Shen Y, Yu D, Hiel H, Liao P, Yue DT, Fuchs PA, Soong TW (2006) Alternative splicing of the Ca(v)1.3 channel IQ domain, a molecular switch for Ca2+-dependent inactivation within auditory hair cells. J Neurosci 26(42):10690–10699

Shin JB, Martinez-Salgado C, Heppenstall PA, Lewin GR (2003) A T-type calcium channel required for normal function of a mammalian mechanoreceptor. Nat Neurosci 6(7):724–730

Splawski I, Yoo DS, Stotz SC, Cherry A, Clapham DE, Keating MT (2006) CACNA1H mutations in autism spectrum disorders. J Biol Chem 281(31):22085–22091

Steriade M, McCormick DA, Sejnowski TJ (1993) Thalamocortical oscillations in the sleeping and aroused brain. Science 262(5134):679–685

Takahashi T, Aoki Y, Okubo K, Maeda Y, Sekiguchi F, Mitani K, Nishikawa H, Kawabata A (2010) Upregulation of Ca(v)3.2 T-type calcium channels targeted by endogenous hydrogen sulfide contributes to maintenance of neuropathic pain. Pain 150(1):183–191

Talley EM, Cribbs LL, Lee JH, Daud A, Perez-Reyes E, Bayliss DA (1999) Differential distribution of three members of a gene family encoding low voltage-activated (T-type) calcium channels. J Neurosci 19(6):1895–1911

Tan GM, Yu D, Wang J, Soong TW (2012) Alternative splicing at C terminus of Ca(V)1.4 calcium channel modulates calcium-dependent inactivation, activation potential, and current density. J Biol Chem 287(2):832–847

Tandrup T (1995) Are the neurons in the dorsal root ganglion pseudounipolar? A comparison of the number of neurons and number of myelinated and unmyelinated fibres in the dorsal root. J Comp Neurol 357(3):341–347

Tang ZZ, Liang MC, Lu S, Yu D, Yu CY, Yue DT, Soong TW (2004) Transcript scanning reveals novel and extensive splice variations in human l-type voltage-gated calcium channel, $Ca_v1.2$ alpha1 subunit. J Biol Chem 279(43):44335–44343

Todorovic SM, Jevtovic-Todorovic V (2006) The role of T-type calcium channels in peripheral and central pain processing. CNS Neurol Disord: Drug Targets 5(6):639–653

Todorovic SM, Jevtovic-Todorovic V (2013) Neuropathic pain: role for presynaptic T-type channels in nociceptive signaling. Pflugers Arch 465(7):921–927

Todorovic SM, Jevtovic-Todorovic V, Meyenburg A, Mennerick S, Perez-Reyes E, Romano C, Olney JW, Zorumski CF (2001) Redox modulation of T-type calcium channels in rat peripheral nociceptors. Neuron 31(1):75–85

Tscherter A, David F, Ivanova T, Deleuze C, Renger JJ, Uebele VN, Shin HS, Bal T, Leresche N, Lambert RC (2011) Minimal alterations in T-type calcium channel gating markedly modify physiological firing dynamics. J Physiol 589(Pt 7):1707–1724

Vallbo AB, Olausson H, Wessberg J (1999) Unmyelinated afferents constitute a second system coding tactile stimuli of the human hairy skin. J Neurophysiol 81(6):2753–2763

Vitko I, Bidaud I, Arias JM, Mezghrani A, Lory P, Perez-Reyes E (2007) The I-II loop controls plasma membrane expression and gating of Ca(v)3.2 T-type Ca2+ channels: a paradigm for childhood absence epilepsy mutations. J Neurosci 27(2):322–330

Walsh MA, Graham BA, Brichta AM, Callister RJ (2009) Evidence for a critical period in the development of excitability and potassium currents in mouse lumbar superficial dorsal horn neurons. J Neurophysiol 101(4):1800–1812

Walton KD, Llinas RR (2010) Central pain as a thalamocortical dysrhythmia: a thalamic efference disconnection? In: Kruger L, Light AR (eds) Translational pain research: from mouse to man. Frontiers in neuroscience. CRC Press, Boca Raton

Wang R, Lewin GR (2011) The $Ca_v3.2$ T-type calcium channel regulates temporal coding in mouse mechanoreceptors. J Physiol 589(Pt 9):2229–2243

Weiss N, Zamponi GW (2012) Control of low-threshold exocytosis by T-type calcium channels. Biochim Biophys Acta 1828 (7):1579–1586

Weiss N, Hameed S, Fernandez-Fernandez JM, Fablet K, Karmazinova M, Poillot C, Proft J, Chen L, Bidaud I, Monteil A, Huc-Brandt S, Lacinova L, Lory P, Zamponi GW, De Waard M

(2012) A Ca(v)3.2/syntaxin-1A signaling complex controls T-type channel activity and low-threshold exocytosis. J Biol Chem 287(4):2810–2818
Weiss N, Black SA, Bladen C, Chen L, Zamponi GW (2013) Surface expression and function of Ca3.2 T-type calcium channels are controlled by asparagine-linked glycosylation. Pflugers Arch. Epub 2013/03/19
Wen XJ, Xu SY, Chen ZX, Yang CX, Liang H, Li H (2010) The roles of T-type calcium channel in the development of neuropathic pain following chronic compression of rat dorsal root ganglia. Pharmacology 85(5):295–300
Wheeler DG, Groth RD, Ma H, Barrett CF, Owen SF, Safa P, Tsien RW (2012) Ca(V)1 and Ca(V)2 channels engage distinct modes of Ca(2+) signaling to control CREB-dependent gene expression. Cell 149(5):1112–1124
Wu H, Williams J, Nathans J (2012) Morphologic diversity of cutaneous sensory afferents revealed by genetically directed sparse labeling. eLife 1:e00181
You H, Altier C, Zamponi GW (2010) CCR2 receptor ligands inhibit $Ca_v3.2$ T-type calcium channels. Mol Pharmacol 77(2):211–217
You H, Gadotti VM, Petrov RR, Zamponi GW, Diaz P (2011) Functional characterization and analgesic effects of mixed cannabinoid receptor/T-type channel ligands. Mol Pain 7:89
Yue J, Liu L, Liu Z, Shu B, Zhang Y (2013) Upregulation of T-type Ca2+ channels in primary sensory neurons in spinal nerve injury. Spine 38(6):463–470
Zamponi GW, Lory P, Perez-Reyes E (2010) Role of voltage-gated calcium channels in epilepsy. Pflugers Arch 460 (2):395–403
Zimmermann K, Hein A, Hager U, Kaczmarek JS, Turnquist BP, Clapham DE, Reeh PW (2009) Phenotyping sensory nerve endings in vitro in the mouse. Nat Protoc 4(2):174–196
Zotterman Y (1939) Touch, pain and tickling: an electro-physiological investigation on cutaneous sensory nerves. J Physiol 95(1):1–28

Chapter 7
$Ca_v1.1$ Channel and Hypokalemic Periodic Paralysis

Emma Matthews and Michael G. Hanna

Abstract Hypokalaemic periodic paralysis is a rare inherited autosomal dominant neuromuscular disorder due predominantly to dysfunction of the alpha subunit of the $Ca_v1.1$ ion channel, although a significant minority of cases are due to dysfunction of another sarcolemmal ion channel, $Na_v1.4$. Hypokalaemic periodic paralysis has been phenotypically described for several centuries but it was not until 1994 that the first causative gene CACNA1S was identified, followed later by a second gene, SCN4A. Electrophysiologic studies attempted to understand how mutations in these genes affected channel function to account for the described phenotype, but early studies were frustratingly inconclusive. Not least because, a satisfactory explanation eluded researchers as to how two ion channels with very different roles could cause the same disease. In 2007, however, an aberrant gating pore current was identified in several $Na_v1.4$ mutations that revolutionised the hypothesis of the pathogenesis of hypokalemic periodic paralysis. In this chapter we review the evolution of our current understanding of this important skeletal muscle channelopathy.

7.1 Clinical Features

Periodic attacks of transient muscle paralysis followed by complete recovery have been recorded for hundreds of years. Many were considered to represent hysteria. In 1727, Musgrave used the term "periodic palsy" to describe the case of a young woman (Musgrave 1727). This has been attributed as one of the earliest descriptions of a case of periodic paralysis although some have criticised atypical features

E. Matthews · M. G. Hanna (✉)
Department of Molecular Neuroscience UCL Medical Research Council Centre for Neuromuscular Disease, UCL, Institute of Neurology and National Hospital for Neurology and Neurosurgery Queen Square London, London, UK
e-mail: m.hanna@ucl.ac.uk

N. Weiss and A. Koschak (eds.), *Pathologies of Calcium Channels*,
DOI: 10.1007/978-3-642-40282-1_7,

of this case such as a disturbance of speech (Viets 1951). In the late nineteenth century more definitive reports appeared (Goldflam 1890; Westphal 1885) and in 1941 Talbot reviewed 400 sporadic and familial cases of "periodic paralysis" (Talbot 1941). The precipitating cause of the attacks of paralysis was initially unclear, however, and several hypotheses were postulated including an unidentified metabolic toxin or muscle ischaemia (Biemond and Daniels 1934). It was not until the 1930s that it was realised that serum potassium levels were low during an attack of paralysis (Allott and McArdle 1937) and conversely muscle strength could be restored by the administration of potassium heralding the first treatment.

The phenotype of hypokalemic periodic paralysis as described in these early reports has little changed over time. Typically, attacks of flaccid muscle paralysis with areflexia begin in the second decade of life. They characteristically occur during the night or early hours of the morning and last from a few hours to several days (Miller et al. 2004). Predominantly, the limb muscles are affected. Rarely are respiratory, facial or bulbar muscles involved (Kil et al. 2010). Attacks often follow rest after strenuous exercise or can be precipitated by factors that lower serum potassium, e.g. large carbohydrate meals. Gentle exercise may help to abort an attack. In the early stages of the disease recovery of muscle strength between attacks appears to be complete. As a result muscle strength and all other parameters are often normal if a patient is examined interictally. Even today many patients are still misdiagnosed as "hysterical" as a consequence. It is reported that attacks of paralysis become less frequent with increasing age. Permanent proximal myopathy does develop with time however, and proximal weakness can then be found on examination. The extent of this and its age of onset vary greatly from one individual to another (Fouad et al. 1997; Sternberg et al. 2001). It is unclear whether the number of paralytic attacks actually directly influences this myopathy. There is some anecdotal data to suggest myopathy is a process independent of the attacks (Biemond and Daniels 1934; Buruma and Bots 1978; Links et al. 1990) and studies on recently engineered mouse models lend some support to this (Hayward et al. 2008; Wu et al. 2012).

Accordingly, muscle biopsies taken from patients with periodic paralysis often display non-specific myopathic features. Additionally, vacuoles are a common finding. The exact mechanism underlying production of these vacuoles or their composition is unknown but they are not an exclusive finding to hypoPP, occurring in other types of primary periodic paralysis (Pearson 1964; Miller et al. 2004; Venance et al. 2006), and myopathy associated with secondary causes of hypokalemia (Finsterer et al. 1998).

HypoPP is a skeletal muscle disorder with no systemic manifestations. The heart is primarily normal (Schipperheyn et al. 1978). However, the low serum potassium levels experienced during an attack can provoke the same changes in cardiac conductance as any other cause of hypokalemia. These include flattened ST segments, u waves and prolonged QT interval which if untreated can cause significant cardiac arrhythmia (Hecht et al. 1997; Kim et al. 2005).

7.2 Diagnosis

A clear history, especially if combined with a positive family history, is often enough to suggest a diagnosis of hypoPP but advancements in neurophysiologic studies and ultimately genetics have helped to achieve a more definitive diagnosis.

7.3 Provocative Tests

The realisation that serum potassium levels were low during an attack of paralysis was used to devise the first diagnostic tests. Recording hypokalemia during a spontaneous attack can be difficult as attacks of paralysis are often unpredictable and the frequency varies greatly. It may be many weeks or months between attacks in any individual. As such, physicians sought to provoke attacks of weakness by infusing insulin and dextrose which drives serum potassium intracellularly. If successful, an attack of paralysis would ensue and could then be demonstrated to recover with administration of potassium (Hyland 1949). Such provocative tests although carefully supervised were not without some risk however, of potential cardiac arrhythmias secondary to the induced hypokalemia. A final caution in relying on serum potassium levels for diagnosis is that the levels may fall during a spontaneous attack but not be necessarily out with the normal range; so ideally a pre-attack baseline is required. These tests are largely historical and not performed routinely today as they have generally been replaced by electrophysiological studies.

7.4 Electrophysiologic Studies

Electrophysiology is a technique used to record the electrical activity of muscle. Early studies demonstrated that the weak muscles of an afflicted individual during an attack were electrically inexcitable (Hyland 1949) supporting the fact the weakness was an involuntary manifestation of disease. A motor unit action potential is reflective of the stimulus received from a motor nerve to all the muscle fibres innervated by that nerve and their subsequent contraction. More detailed electrophysilogical studies described a reduction in motor unit action potential size and duration with a progressive reduction in the number of motor units which could be voluntarily activated in a weak muscle precipitated by administration of glucose (Engel et al. 1965). This is the electrical reflection of symptoms described by the patient and the observed clinical weakness. Patients also describe weakness typically occurring following rest after physical exertion, and these studies further illustrated that electrical changes could be provoked by exercise. Following repeated abduction of the little finger for 5–7.5 min the amplitude of the

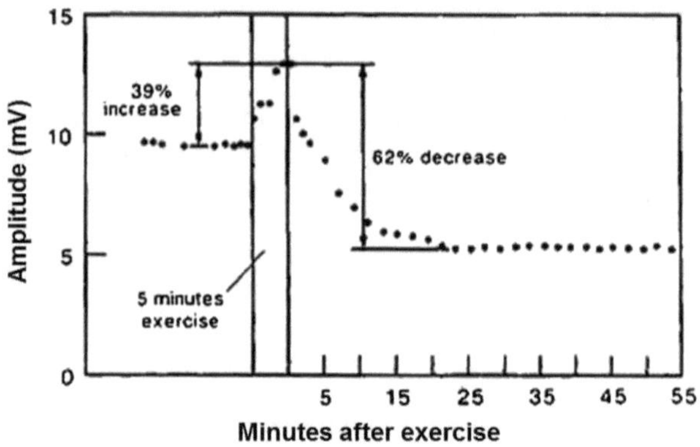

Fig. 7.1 Effect of exercise on amplitude of compound muscle action potentials in a patient with primary periodic paralysis. *Dots* indicate amplitude in response to a single stimulus. Reproduced with permission from McManis et al. 1986

compound motor unit action potential (CMAP) was seen to initially increase in size over the first few minutes but subsequently decrease over 35–40 min to between 8 and 41 % below the baseline level (Engel et al. 1965).

Later Mcmanis standardised this exercise test in a cohort of patients with both primary (including hypokalemic and hypercalcemic disorders) and secondary periodic paralysis (McManis et al. 1986). He demonstrated a typical pattern of initial increase in CMAP with subsequent slow progressive decrement in all groups of periodic paralysis but not in controls (see Fig. 7.1). This important diagnostic test has become known as the long exercise or Mcmanis test and is one of the principle investigations used today in the diagnosis of periodic paralysis (Fournier et al. 2004; Tan et al. 2011). Spontaneous attacks of paralysis almost never occur when a patient is in the examination room, so the ability to provoke a focal attack of muscle weakness without disabling widespread weakness or the potential disadvantages of inducing systemic hypokalemia was significant progress in diagnostic ability. The long exercise test does not however, distinguish between the type of periodic paralysis and false negatives can occur (McManis et al. 1986).

7.5 Genetics

In 1994, the first genetic mutations underlying hypokalemic periodic paralysis were identified. Initial studies on affected families mapped the locus to chromosome 1q31-32 and ultimately the CACNA1S gene (Fontaine et al. 1994; Ptacek et al. 1994; Jurkat-Rott et al. 1994). This gene codes for the alpha subunit of the voltage gated calcium channel $Ca_V1.1$ (or dihydropyridine receptor) that is expressed in skeletal muscle. It comprises four transmembrane domains (DI–IV), each domain having six segments (S1-6) with the fourth segment (S4) of all domains acting as the channel voltage sensors. The first mutations described were all point mutations that resulted in substitutions of arginine residues in two of the channel's voltage sensing segments in DII and DIV—R528H, R1239H and

R1239G. However, one French family did not map to the identified locus and genetic heterogeneity was implicated (Plassart et al. 1994).

Five years later a family was reported who did not have any identifiable mutations in CACNA1S but did carry a point mutation in the SCN4A gene that codes for the alpha subunit of the sarcolemmal voltage gated sodium channel $Na_v1.4$ (Bulman et al. 1999) 4. This channel has a similar structure to $Ca_v1.1$ in that it also consists of four transmembrane domains each having six subunits. It was also of note that the mutation described, R669H, again neutralised an arginine residue in one of the channel voltage sensors (DII, S4). Numerous other mutations in both genes have subsequently been reported (see Table 7.1). All but two in CACNA1S, V876E in DIII-S3 (Ke et al. 2009) and H916Q in the DIII S4-5 interlinking loop (Li et al. 2012), are voltage sensor mutations.

Table 7.1 Mutations in CACNA1S and SCN4A associated with hypokalemic periodic paralysis

Mutation	Gene	Protein position	Effect on channel gating	Reference
R528G/H	CACNA1S	DII/S4	Slower kinetics of activation Reduced current density	(Lapie et al. 1996; Morrill et al. 1999)
V876E	CACNA1S	DIII/S3	Unknown	(Ke et al. 2009)
R897S	CACNA1S	DIII/S4	Unknown	(Chabrier et al. 2008)
R900S	CACNA1S	DIII/S4	Unknown	(Matthews et al. 2009)
H916Q	CACNA1S	DIII/S4-5 loop	Unknown	(Li et al. 2012)
R1239G/H	CACNA1S	DIV/S4	Slower kinetics of activation Reduced current density	(Morrill et al. 1999)
R225 W	SCN4A	DI/S4	Unknown	(Matthews et al. 2009)
R669H	SCN4A	DII/S4	Enhanced slow inactivation Enhanced fast inactivation Reduced current density	(Struyk et al. 2000; Kuzmenkin et al. 2002)
R672C/G/H/S	SCN4A	DII/S4	Enhanced inactivation Reduced current density	(Kim et al. 2004; Jurkat-Rott et al. 2000; Bendahhou et al. 2001)
R1129Q	SCN4A	DIII/S4	Unknown	(Hong et al. 2010)
R1132Q	SCN4A	DIII/S4	Enhanced fast and slow inactivation	(Carle et al. 2006)
R1135H	SCN4A	DIII/S4	Unknown	(Matthews et al. 2009)

The first two mutated residues that were described, R528 and R1239, in CACNA1S account for the majority of cases of hypoPP with CACNA1S mutations overall being implicated in approximately 70–80 %. SCN4A mutations account for another 10 % leaving a significant minority without genetic characterization (Sternberg et al. 2001) suggesting the possibility of further genetic heterogeneity. Overwhelmingly, the majority of mutations disrupt the voltage sensors, accounting for 90 % of all hypoPP cases in one series (Matthews et al. 2009). This finding is integral to the current hypothesis of the disease pathomechanism.

7.6 Pathomechanisms

Both $Ca_v1.1$ and $Na_v1.4$ channels are expressed exclusively in skeletal muscle although isoforms are prevalent in other tissues. Both consist of a core α subunit and one or more accessory subunits. In each case, the α-subunits form the ion conducting pore of the channel. While ion selectivity, activation and inactivation are thought to be processes which are intrinsic to α subunits the full physiological function of these channels, as is common to most ion channels, is dependent on the assembly of several proteins, including the accessory subunits (Catterall 1995). To date only mutations in the α-subunits have been associated with muscle disease.

The α subunits of each channel have similar structures (four domains each with six transmembrane subunits) consistent with a shared evolutionary history linked to a single ancestral channel (Cannon 2007). The four domains fold together with the S5 and S6 segments from each domain assembling to create a central channel pore (alpha pore) through which ions enter the cell (Varadi et al. 1999; Catterall 1995). Although structurally related the channels have very different permeabilities with $Ca_v1.1$ most permeable to calcium ions and $Na_v1.4$ being highly selective for sodium. A small number of amino acid residues in the P loops between segments S5 and S6 determine the ion selectivity for each respective channel. For calcium channels four glutamic acid residues (EEEE) have been identified as the key amino acids that determine selectivity for calcium (Tang et al. 1993; Mikala et al. 1993).

The amino acids occupying the comparable positions in the voltage-gated sodium channels (aspartate, glutamate, lysine, alanine, DEKA) have similarly been implicated in setting the ion selectivity of the channel (Heinemann et al. 1992; Favre et al. 1996).

Both $Na_v1.4$ and $Ca_v1.1$ are activated by depolarisation of the muscle membrane. In the case of $Na_v1.4$ the activation is rapid, and the main role of the channel is to allow sodium ions to enter the cell, further depolarizing it, and allowing more voltage-gated channels to open. Upon depolarisation by $Na_v1.4$, $Ca_v1.1$ channels are activated. These channels, although designated voltage-gated calcium channels, are physically coupled to ryanodine receptors (RyR1) in the sarcoplasmic reticulum, and the activation of $Ca_v1.1$ leads to opening of RyR1 and release of calcium from intracellular stores. It is thought that the activation of

RyR1 via physical coupling, not via calcium influx, is the main function of $Ca_v1.1$ in healthy muscle cells (Catterall 1995; Brini 2004). However, $Ca_v1.1$ proteins still contain a channel pore that is opened in response to depolarization, and which allows calcium to enter the muscle cell.

In common with many voltage gated channels, both $Na_v1.4$ and $Ca_v1.1$ use positively charged segments within the pore-forming α subunits to detect changes in membrane voltage (Catterall 2010). The S4 segments of both channels contain positively charged residues (arginines or lysines) at every third position surrounded by hydrophobic residues. These segments with their abundance of electrical charge behave as the voltage sensors (Yang et al. 1996) by moving towards the cytoplasmic side of the membrane in response to membrane depolarisation. It is this outward movement that produces a conformational change at the intracellular surface of the channels (Catterall 1995), and opens the central pore of the channel to allow influx of ions.

Initial electrophysiological experiments focused on the consequences of the ion channel mutations on the gating properties and conductance of this ion selective pore. Each mutation had an overall "loss of function" effect, with relatively minimal alteration of channel gating and variable reduction in current density (see Table 7.1).

However, studies concentrating on native muscle taken from affected patients had indentified several other abnormalities. Hypokalemia is known to occur during an attack of paralysis. Healthy muscle fibres will hyperpolarise when placed in a low potassium solution but fibres from hypoPP patients depolarised and became inexcitable (Rudel et al. 1984; Ruff 1999). Rudel hypothesised the abnormal depolarisation was contributed to by an increase in membrane sodium conductance. However, the sodium channel blocker TTX, known to inhibit $Na_v1.4$ channels, had no effect suggesting if increased sodium conductance was present it was not via the $Na_v1.4$ channel alpha pore (Rudel et al. 1984). Ruff also identified an aberrant inward depolarising current in diseased muscle fibres that was activated when the fibres were placed in a low potassium extracellular solution. He demonstrated this current was further insensitive to the calcium channel blocker nitrendipine suggesting a pathway or trigger for the pathological depolarisation that was independent of ion conductance via either the $Ca_v1.1$ or $Na_v1.4$ channels ion selective pores (Ruff 1999).

Furthermore, he showed the muscle fibres had a reduced membrane conductance to potassium (Ruff 1999). This finding was substantiated in another study using native muscle fibres from hypoPP patients in which reduced KATP current was demonstrated (Tricarico et al. 1999). Further parallels were drawn from barium toxicity. Barium blocks inward rectifying potassium channels (Standen and Stanfield 1978) and reduces the twitch force of the skeletal muscles of mammals when these fibres are placed in a low potassium solution (Gallant 1983).

Overall, it was suggested that a reduced membrane conductance to potassium, specifically a reduced outward current of the inward rectifying potassium channels may cause the extracellular hypokalemia seen during paralytic attacks.

What was unclear was how a relatively moderate reduction in current density of the $Ca_v1.1$ and $Na_v1.4$ channels could produce such a detrimental effect on potassium channel function. Additionally, what was the source of the depolarising inward current if it was not via the alpha pore of either $Ca_v1.1$ or $Na_v1.4$, the only two channels to be genetically implicated in the synthesis of hypoPP.

7.7 The Anomalous Gating Pore Current

Each domain of the $Ca_v1.1$ and $Na_v1.4$ channels mirrors the structure of a voltage gated Shaker potassium channel. Exploration of the shaker potassium channel function illustrated key S4 arginine residues moved through the membrane in response to changes in voltage acting as a "voltage sensor" (Starace and Bezanilla 2001). Movement of these amino acids was through a pathway termed the omega pore. The omega pore in health is intended only to allow passage or movement of the VSD arginines and not as a route of entry or exit for other ions. It was proposed the omega pore was not uniform in diameter and consequently could be occluded by residence of different arginines depending on whether the membrane was depolarised or hyperpolarised and consequently whether the VSD had moved extracellularly to open the alpha pore or was resting intracellularly as the alpha pore was closed (Bezanilla 2002). However, reliance on the arginine residue to block the passage of other ions suggested removal of the arginine could open a "gate" that was subsequently demonstrated to allow cations to enter the cell via an anomalous route (Tombola et al. 2005) described as a gating pore current.

In 2007 S4 arginine mutations of $Na_v1.4$ associated with hypokalemic periodic paralysis were also shown to introduce an anomalous gating pore current (Sokolov et al. 2007; and Cannon 2007). The substituted amino acid directly influenced this. For histidine substitutions, a proton selective current was detected, but for glycine substitutions other small monovalent cations were also conducted into the cell via this aberrant pathway. Gating pore currents were further demonstrated in other hypoPP $Na_v1.4$ mutations (Struyk et al. 2008b; Francis et al. 2011).

This anomalous pathway for entry of protons and other cations into the cell was proposed as a unifying mechanism by which two separate ion channels may disrupt the excitability of the sarcolemma with the same result. The convergence of mutations on the voltage sensors of $Na_v1.4$ and $Ca_v1.1$ genetically supported this. $Ca_v1.1$ mutations account for the majority of cases of hypoPP however, and there remained some speculation as the demonstration of gating pore currents had not been attempted in this channel due to technical restraints. Recently, experiments examining the muscle fibres of a genetically engineered mouse model carrying the R528H CACNA1S mutation have illustrated findings consistent with a gating pore current (Wu et al. 2012) strengthening the argument that this is the principle pathomechanism of hypoPP.

In simple terms, it is proposed that the abnormal influx of protons and other ions such as sodium may influence the pH and sodium homeostasis of the cell both

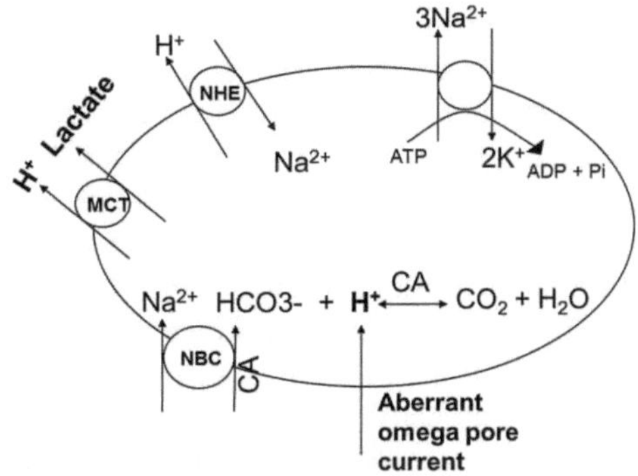

Fig. 7.2 Potential influence of an aberrant gating pore current on other cell transporters. *MCT* monocarboxylate transporter *NHE* Sodium hydrogen antiport exchanger *NBC* sodium-dependant bicarbonate transporter *CA*carbonic anhydrase. Reproduced with permission from Matthews and Hanna (2010)

directly and via secondary activation of other ion transporters or exchangers (Jurkat-Rott and Lehmann-Horn 2007) (see Fig. 7.2). Muscle fibre resting membrane potential (V_{REST}) is determined by the membrane permeability to K^+ ions. The evidence of reduced potassium permeability in native muscle fibres from hypoPP patients (Ruff 1999), impaired KATP channel conductance (Tricarico et al. 1999) and the effects of barium toxicity (Standen and Stanfield 1978; Gallant 1983; and Cannon 2008a) all argue for disturbed potassium conductance as the trigger for paradoxical membrane depolarisation in low extracellular potassium environment and the gating pore current has been shown to enhance the probability of membrane depolarisation in reduced extracellular potassium (Struyk et al. 2008b; Jurkat-Rott et al. 2009; Cannon 2010). The exact mechanistic link however, between an anomalous influx of protons and other cations, their influence on the intracellular ionic and acidic environments and ultimately the dysfunction of IRK channels requires clarification. An initial hypothesis centres on the knowledge that IRK channels are known to be inhibited by an acidic pH (Struyk and Cannon 2008a) which may be produced by a proton selective gating pore current.

7.8 Treatment of Hypokalemic Periodic Paralysis

Once the association was made between attacks of skeletal muscle weakness and low serum potassium levels in the 1930s, administration of potassium salts became the first significant treatment. However, although there was some benefit in terms of symptom control it was not clear whether this actually reduced the overall frequency of paralytic attacks or whether there was any influence on the development of proximal myopathy.

In the late 1950s, Conn performed extensive experiments on two patients with hypokalemic periodic paralysis and reported that the attacks of paralysis were not only associated with hypokalemia but also hypernatremia. He proposed that it was even possible the hypernatremia was the primary stimulus for the paralytic attacks with hypokalemia being secondary (Conn et al. 1957). While he did not suggest hyperaldosteronism was the cause per se he did identify raised aldosterone levels during attacks in his patients although others did not corroborate these findings (Poskanzer and Kerr 1961). Conn suggested a low sodium diet was beneficial in reducing both paralytic attacks and myopathy. He also proposed carbonic anhydrase inhibitors may be effective due to their ability to increase sodium excretion in the urine. In practical terms the restrictive sodium diet proposed by Conn was difficult for patients to tolerate and others did not find sodium restriction to be as beneficial (Poskanzer and Kerr 1961).

Based on his research however, spironolactone, a known inhibitor of aldosterone, was tried as therapy for hypokalemic periodic paralysis. In the initial patient studied, subjective and objective observations not only reduced paralytic attacks, but also improved inter-attack muscle strength was noted. This treatment with spironolactone did produce a slight increase in sodium excretion with normal potassium serum levels being maintained but the authors argued that the effects on sodium and potassium balance were too minimal to adequately explain the full mechanism of action of spironolactone in producing such significant symptom reduction in hypokalemic periodic paralysis (Poskanzer and Kerr 1961). Despite this the main mechanism of benefit from spironolactone and other potassium sparing diuretics is believed to be their ability to maintain serum potassium levels within the normal or high normal range. There is anecdotal evidence to support their use and they remain common current day therapies although there is no randomised trial data to substantiate this.

Although Conn proposed carbonic anhydrase inhibitors may be useful therapies due to their ability to increase urinary sodium excretion, they also increase potassium and bicarbonate excretion (Matthews and Hanna 2010). It was for this reason acetazolamide was initially given to patients with hyperkalaemic periodic paralysis in which attacks of paralysis occur in conjunction with raised serum potassium levels (McArdle 1962). Despite seeming to be an unlikely candidate for treatment of hypokalaemic periodic paralysis, a prophylactic benefit was subsequently also reported in this disorder (Resnick et al. 1968; Griggs et al. 1970). For more than half a century, carbonic anhydrase inhibitors such as acetazolamide and dichlorphenamide have been popular treatments with numerous case reports and small series relaying positive outcomes (Matthews et al. 2011). A placebo controlled crossover trial has been undertaken using dichlorphenamide (Tawil et al. 2000) and a larger multi-centre, multinational trial is currently underway although not yet reported on.

Carbonic anhydrase inhibitors primarily act by inhibiting the enzyme carbonic anhydrase from catalysing the reversible conversion of water and carbon dioxide to protons and bicarbonate with corresponding effects on pH balance. In addition to anecdotal clinical reports, experimental work also supported a rationale for their benefit in terms of effect on pH (Kuzmenkin et al. 2002), improved sarcolemmal conductance of potassium via calcium activated potassium channels (Tricarico et al. 2000, 2004) and observation of a reduced number of vacuoles seen in the muscle biopsies of potassium depleted rats administered acetazolamide (Tricarico et al. 2008).

Not all patients were noted to respond to acetazolamide however and some in fact reported worsening of their symptoms. Initially, it was proposed that this lack of benefit or detrimental effect was related to disease-causing gene, with SCN4A mutation carriers more likely to have a poor outcome. This was debated with conflicting reports of response to treatment with various SCN4A mutations (Venance et al. 2004). A recent retrospective analysis of acetazolamide treatment in two large cohorts of hypoPP patients indicated two main outcomes. First, acetazolamide appeared to be beneficial for only 50–60 % of patients overall. Second, rather than a gene-specific effect, there tentatively appears to be a mutation-specific effect with arginine to histidine substitutions more likely to respond to acetazolamide than those with other amino acid substitutes (Matthews et al. 2011). This observation is potentially supported by the differences in gating pore ionic conductance between histidine and other substitutions, i.e., histidine allows a proton selective current, whereas others allow additional small cations to be conducted. The exact mechanisms by which acetazolamide may act and how these relate to the gating pore current are still speculative to some degree and require further exploration (Matthews and Hanna 2010).

7.9 Future Considerations

The identification of an anomalous gating pore current in hypoPP has revolutionized our understanding of this skeletal muscle channelopathy. Although many questions remain unanswered, the implication that an anomalous current may have secondary consequences on other membrane channels, transporters and exchangers points to a host of novel potential future therapeutic targets. Already, bumetanide has been proposed due to its effects on the Na–K–2Cl transporter (Wu et al. 2013). In experiments exposing muscle fibres from the $Na_v1.4$ mutant hypoPP mouse model to low potassium solution bumetanide was shown to significantly reverse the decline in tetanic force that was observed and even prevent a decline in force if infused at the same time as lowering the potassium concentration in the solution. It was also shown to be vastly superior to acetazolamide. It remains to be seen whether the effect can be replicated in humans or whether bumetanide treatment would be tolerable, but it is a promising step towards improving treatment options for those with hypoPP.

Isoforms of both $Ca_v1.1$ and $Na_v1.4$ are expressed in numerous other tissues and mutations of these associated with other human diseases, e.g. hemiplegic migraine, long QT cardiac syndromes and epilepsy syndromes. Whilst none of these display such an exclusive relationship between phenotype and voltage sensor mutations, voltage sensor mutations do occur. Continued advances in our understanding and treatment of hypoPP may ultimately be extrapolated and applied to a variety of disabling human diseases.

References

Allott EN, McArdle B (1937) Clin Sci 3:229

Bendahhou S, Cummins TR, Griggs RC, Fu YH, Ptácek LJ (2001) Sodium channel inactivation defects are associated with acetazolamide-exacerbated hypokalemic periodic paralysis. Ann Neurol 50(3):417–20

Bezanilla F (2002) Voltage sensor movements. J Gen Physiol 120:465–473

Biemond A, Daniels AP (1934) Familial periodic paralysis and its transition into spinal muscular atrophy. Brain 57:91–108

Brini M (2004) Ryanodine receptor defects in muscle genetic diseases. Biochem Biophys Res Commun 322:1245–1255

Bulman DE, Scoggan KA, Van O, Nicolle MW, Hahn AF, Tollar LL et al. (1999) A novel sodium channel mutation in a family with hypokalemic periodic paralysis. Neurology 53:1932–1936

Buruma OJ, Bots GT (1978) Myopathy in familial hypokalaemic periodic paralysis independent of paralytic attacks. Acta Neurol Scand 57:171–179

Cannon SC (2010) Voltage-sensor mutations in channelopathies of skeletal muscle. J Physiol 588:1887–1895

Cannon SC (2007) Physiologic principles underlying ion channelopathies. Neurotherapeutics 4:174–183

Carle T, Lhuillier L, Luce S, Sternberg D, Devuyst O, Fontaine B et al (2006) Gating defects of a novel Na+ channel mutant causing hypokalemic periodic paralysis. Biochem Biophys Res Commun 348(2):653–61

Catterall WA (1995) Structure and function of voltage-gated ion channels. Annu Rev Biochem 64:493–531

Catterall WA (2010) Ion channel voltage sensors: structure, function, and pathophysiology. Neuron 67:915–928

Chabrier S, Monnier N, Lunardi J (2008) Early onset of hypokalaemic periodic paralysis caused by a novel mutation of the CACNA1S gene. J Med Genet 45(10):686–8

Conn JW, Fajans SS, Louis LH, Streeten DH, Johnson RD (1957) Intermittent aldosteronism in periodic paralysis; dependence of attacks on retention of sodium, and failure to induce attacks by restriction of dietary sodium. Lancet 272:802–805

Engel AG, Lambert EH, Rosevear JW, Tauxe WN (1965) Clinical and electromyographic studies in a patient with primary hypokalemic periodic paralysis. Am J Med 38:626–640

Favre I, Moczydlowski E, Schild L (1996) On the structural basis for ionic selectivity among Na+, K+, and Ca_2+ in the voltage-gated sodium channel. Biophys J 71:3110–3125

Finsterer J, Hess B, Jarius C, Stollberger C, Budka H, Mamoli B (1998) Malnutrition-induced hypokalemic myopathy in chronic alcoholism. J Toxicol Clin Toxicol 36:369–373

Fontaine B, Vale-Santos J, Jurkat-Rott K, Reboul J, Plassart E, Rime CS et al (1994) Mapping of the hypokalaemic periodic paralysis (HypoPP) locus to chromosome 1q31-32 in three European families. Nat Genet 6:267–272

Fouad G, Dalakas M, Servidei S, Mendell JR, Van den BP, Angelini C et al (1997) Genotype-phenotype correlations of DHP receptor alpha 1-subunit gene mutations causing hypokalemic periodic paralysis. Neuromuscul Disord 7:33–38

Fournier E, Arzel M, Sternberg D, Vicart S, Laforet P, Eymard B et al (2004) Electromyography guides toward subgroups of mutations in muscle channelopathies. Ann Neurol 56:650–661

Francis DG, Rybalchenko V, A, Cannon SC (2011) Leaky sodium channels from voltage sensor mutations in periodic paralysis, but not paramyotonia. Neurology 76:1635–1641

Gallant EM (1983) Barium-treated mammalian skeletal muscle: similarities to hypokalaemic periodic paralysis. J Physiol 335:577–590

Goldflam S (1890) Ueber eine eigentümliche Form von periodischer familiärer, wahrscheinlich auto-intoxicatorischer paralyse. Wien Med Presse 31:1418

Griggs RC, Engel WK, Resnick JS (1970) Acetazolamide treatment of hypokalemic periodic paralysis: prevention of attacks and improvement of persistent weakness. Ann Intern Med 73:39–48

Hayward LJ, Kim JS, Lee MY, Zhou H, Kim JW, Misra K et al (2008) Targeted mutation of mouse skeletal muscle sodium channel produces myotonia and potassium-sensitive weakness. J Clin Invest 118:1437–1449

Hecht ML, Valtysson B, Hogan K (1997) Spinal anesthesia for a patient with a calcium channel mutation causing hypokalemic periodic paralysis. Anesth Analg 84:461–464

Heinemann SH, Terlau H, Stuhmer W, Imoto K, Numa S (1992) Calcium channel characteristics conferred on the sodium channel by single mutations. Nature 356:441–443

Hong D, Luan X, Chen B, Zheng R, Zhang W, Wang Z et al (2010) Both hypokalaemic and normokalaemic periodic paralysis in different members of a single family with novel R1129Q mutation in SCN4A gene. J Neurol Neurosurg Psychiatry 81(6):703–4

Hyland HH (1949) Periodic paralysis. Can Med Assoc J 61:586–588

Jurkat-Rott K, Lehmann-Horn F (2007) Do hyperpolarization-induced proton currents contribute to the pathogenesis of hypokalemic periodic paralysis, a voltage sensor channelopathy? J Gen Physiol 130:1–5

Jurkat-Rott K, Lehmann-Horn F, Elbaz A, Heine R, Gregg RG, Hogan K et al (1994) A calcium channel mutation causing hypokalemic periodic paralysis. Hum Mol Genet 3:1415–1419

Jurkat-Rott K, Mitrovic N, Hang C, Kouzmekine A, Iaizzo P, Herzog J et al (2000) Voltage-sensor sodium channel mutations cause hypokalemic periodic paralysis type 2 by enhanced inactivation and reduced current. Proc Natl Acad Sci USA 97(17):9549–54

Jurkat-Rott K, Weber MA, Fauler M, Guo XH, Holzherr BD, Paczulla A et al (2009) K+-dependent paradoxical membrane depolarization and Na+ overload, major and reversible contributors to weakness by ion channel leaks. Proc Natl Acad Sci U S A 106:4036–4041

Ke T, Gomez CR, Mateus HE, Castano JA, Wang QK (2009) Novel CACNA1S mutation causes autosomal dominant hypokalemic periodic paralysis in a South American family. J Hum Genet 54:660–664

Kil TH, Kim JB (2010) Severe respiratory phenotype caused by a de novo Arg528Gly mutation in the CACNA1S gene in a patient with hypokalemic periodic paralysis. Eur J Paediatr Neurol 14:278–281

Kim MK, Lee SH, Park MS, Kim BC, Cho KH, Lee MC et al (2004) Mutation screening in Korean hypokalemic periodic paralysis patients: a novel SCN4A Arg672Cys mutation. Neuromuscul Disord 14(11):727–31

Kim JB, Lee KY, Hur JK (2005) A Korean family of hypokalemic periodic paralysis with mutation in a voltage-gated calcium channel (R1239G). J Korean Med Sci 20:162–165

Kuzmenkin A, Muncan V, Jurkat-Rott K, Hang C, Lerche H, Lehmann-Horn F et al (2002) Enhanced inactivation and pH sensitivity of Na(+) channel mutations causing hypokalaemic periodic paralysis type II. Brain 125:835–843

Lapie P, Goudet C, Nargeot J, Fontaine B, Lory P (1996) Electrophysiological properties of the hypokalaemic periodic paralysis mutation (R528H) of the skeletal muscle alpha 1s subunit as expressed in mouse L cells. FEBS Lett 382(3):244–8

Li FF, Li QQ, Tan ZX, Zhang SY, Liu J, Zhao EY et al (2012) A novel mutation in CACNA1S gene associated with hypokalemic periodic paralysis which has a gender difference in the penetrance. J Mol Neurosci 46:378–383

Links TP, Zwarts MJ, Wilmink JT, Molenaar WM, Oosterhuis HJ (1990) Permanent muscle weakness in familial hypokalaemic periodic paralysis. Clinical, radiological and pathological aspects. Brain 113(Pt 6):1873–1889

Matthews E, Hanna MG (2010) Muscle channelopathies: does the predicted channel gating pore offer new treatment insights for hypokalaemic periodic paralysis? J Physiol 588:1879–1886

Matthews E, Labrum R, Sweeney MG, Sud R, Haworth A, Chinnery PF et al (2009) Voltage sensor charge loss accounts for most cases of hypokalemic periodic paralysis. Neurology 72:1544–1547

Matthews E, Portaro S, Ke Q, Sud R, Haworth A, Davis MB et al (2011) Acetazolamide efficacy in hypokalemic periodic paralysis and the predictive role of genotype. Neurology 77(22):1960–1964

McArdle B (1962) Adynamia episodica heredltaria and its treatment. Brain 85:121–148

McManis PG, Lambert EH, Daube JR (1986) The exercise test in periodic paralysis. Muscle Nerve 9:704–710

Mikala G, Bahinski A, Yatani A, Tang S, Schwartz A (1993) Differential contribution by conserved glutamate residues to an ion-selectivity site in the L-type Ca_2+ channel pore. FEBS Lett 335:265–269

Miller TM, da Silva MR, Miller HA, Kwiecinski H, Mendell JR, Tawil R et al (2004) Correlating phenotype and genotype in the periodic paralyses. Neurology 63:1647–1655

Morrill JA et al (1999) Effects of mutations causing hypokalaemic periodic paralysis on the skeletal muscle L-type Ca_2+ channel expressed in Xenopus laevis oocytes. J Physiol

Musgrave W (1727) A periodic palsy. Philos Trans R Soc Lond B Biol Sci 3:33–34

Pearson CM (1964) The periodic paralyses: differential features and pathological observations in permanent myopathic weakness. Brain 87:341–354

Plassart E, Elbaz A, Santos JV, Reboul J, Lapie P, Chauveau D et al (1994) Genetic heterogeneity in hypokalemic periodic paralysis (hypoPP). Hum Genet 94:551–556

Poskanzer DC, Kerr DN (1961) Periodic paralysis with response to spironolactone. Lancet 2:511–513

Ptacek LJ, Tawil R, Griggs RC, Engel AG, Layzer RB, Kwiecinski H et al (1994) Dihydropyridine receptor mutations cause hypokalemic periodic paralysis. Cell 77:863–868

Resnick JS, Engel WK, Griggs RC, Stam AC (1968) Acetazolamide prophylaxis in hypokalemic periodic paralysis. N Engl J Med 278:582–586

Rudel R, Lehmann-Horn F, Ricker K, Kuther G (1984) Hypokalemic periodic paralysis: in vitro investigation of muscle fiber membrane parameters. Muscle Nerve 7:110–120

Ruff RL (1999) Insulin acts in hypokalemic periodic paralysis by reducing inward rectifier K+ current. Neurology 53:1556–1563

Schipperheyn JJ, Buruma OJ, Voogd PJ (1978) Hypokalaemic periodic paralysis and cardiomyopathy. Acta Neurol Scand 58:374–378

Sokolov S, Scheuer T, Catterall WA (2007) Gating pore current in an inherited ion channelopathy. Nature 446:76–78

Standen NB, Stanfield PR (1978) A potential and time-dependent blockade of inward rectification in frog skeletal muscle fibres by barium and strontium ions. J Physiol 280:169–191

Starace DM, Bezanilla F (2001) Histidine scanning mutagenesis of basic residues of the S4 segment of the shaker k+ channel. J Gen Physiol 117:469–490

Sternberg D, Maisonobe T, Jurkat-Rott K, Nicole S, Launay E, Chauveau D et al (2001) Hypokalaemic periodic paralysis type 2 caused by mutations at codon 672 in the muscle sodium channel gene SCN4A. Brain 124:1091–1099

Struyk AF, Cannon SC (2007) A Na+channel mutation linked to hypokalemic periodic paralysis exposes a proton-selective gating pore. J Gen Physiol 130:11–20

Struyk AF, Cannon SC (2008) Paradoxical depolarization of BA2 + - treated muscle exposed to low extracellular K + : insights into resting potential abnormalities in hypokalemic paralysis. Muscle Nerve 37:326–337

Struyk AF, Markin VS, Francis D, Cannon SC (2008) Gating pore currents in DIIS4 mutations of $Na_v1.4$ associated with periodic paralysis: saturation of ion flux and implications for disease pathogenesis. J Gen Physiol 132:447–464

Talbot JH (1941) Periodic paralysis, a clinical syndrome. Medicine (Baltimore) 20:85–143

Tan SV, Matthews E, Barber M, Burge JA, Rajakulendran S, Fialho D et al (2011) Refined exercise testing can aid DNA-based diagnosis in muscle channelopathies. Ann Neurol 69:328–340

Tang S, Mikala G, Bahinski A, Yatani A, Varadi G, Schwartz A (1993) Molecular localization of ion selectivity sites within the pore of a human L-type cardiac calcium channel. J Biol Chem 268:13026–13029

Tawil R, McDermott MP, Brown R, Jr., Shapiro BC, Ptacek LJ, McManis PG et al (2000) Randomized trials of dichlorphenamide in the periodic paralyses: working Group on periodic paralysis. Ann Neurol 47:46–53

Tombola F, Pathak MM, Isacoff EY (2005) Voltage-sensing arginines in a potassium channel permeate and occlude cation-selective pores. Neuron 45:379–388

Tricarico D, Barbieri M, Camerino DC (2000) Acetazolamide opens the muscular KCa_2+ channel: a novel mechanism of action that may explain the therapeutic effect of the drug in hypokalemic periodic paralysis. Ann Neurol 48:304–312

Tricarico D, Barbieri M, Mele A, Carbonara G, Camerino DC (2004) Carbonic anhydrase inhibitors are specific openers of skeletal muscle BK channel of K+-deficient rats. FASEB J 18:760–761

Tricarico D, Lovaglio S, Mele A, Rotondo G, Mancinelli E, Meola G et al (2008) Acetazolamide prevents vacuolar myopathy in skeletal muscle of K+-depleted rats. Br J Pharmacol 154:183–190

Tricarico D, Servidei S, Tonali P, Jurkat-Rott K, Camerino DC (1999) Impairment of skeletal muscle adenosine triphosphate-sensitive K+ channels in patients with hypokalemic periodic paralysis. J Clin Invest 103:675–682

Varadi G, Strobeck M, Koch S, Caglioti L, Zucchi C, Palyi G (1999) Molecular elements of ion permeation and selectivity within calcium channels. Crit Rev Biochem Mol Biol 34:181–214

Venance SL, Cannon SC, Fialho D, Fontaine B, Hanna MG, Ptacek LJ et al (2006) The primary periodic paralyses: diagnosis, pathogenesis and treatment. Brain 129:8–17

Venance SL, Jurkat-Rott K, Lehmann-Horn F, Tawil R (2004) SCN4A-associated hypokalemic periodic paralysis merits a trial of acetazolamide. Neurology 63:1977

Viets HR (1951) Periodic paralysis. JAMA 147:595

Westphal C (1885) Ü ber einen merkwürdigen Fall von periodischer Lähmung aller vier Extremitäten mit gleichzeitigem Erlöschen der elektrischen Erregbarkeit während der Lähmung. Berl klinWchenschr 22:489–511

Wu F, Mi W, Cannon SC (2013) Bumetanide prevents transient decreases in muscle force in murine hypokalemic periodic paralysis. Neurology 80(12):1110–1116

Wu F, Mi W, Hernandez-Ochoa EO, Burns DK, Fu Y, Gray HF et al (2012) A calcium channel mutant mouse model of hypokalemic periodic paralysis. J Clin Invest 122:4580–4591

Yang N, George AL Jr, Horn R (1996) Molecular basis of charge movement in voltage-gated sodium channels. Neuron 16:113–122

Chapter 8
$Ca_v1.1$ in Malignant Hyperthermia

Viktor Yarotskyy and Robert T. Dirksen

Abstract $Ca_v1.1$ is the pore-forming subunit of the dihydropyridine-sensitive L-type Ca^{2+} channel in skeletal muscle. $Ca_v1.1$ functions as both a slowly activating, voltage-gated L-type Ca^{2+} channel and as the voltage sensor that triggers the opening of ryanodine receptor Ca^{2+} release from the sarcoplasmic reticulum during excitation–contraction (EC) coupling. To date, five different point mutations in $Ca_v1.1$ (R174W, R1086C, R1086H, R1086S, and T1354S) have been linked to malignant hyperthermia (MH) in humans, a potentially lethal autosomal dominantly inherited pharmacogenetic disorder of skeletal muscle Ca^{2+} regulation associated with an uncontrolled increase in muscle activity and heat production. Interestingly, functional studies for three of these identified MH mutations in $Ca_v1.1$ (R174W, R1086H, and T1354S) indicate that each mutation enhances susceptibility to MH via fundamentally different mechanisms. This chapter describes the dual functionality of $Ca_v1.1$ in skeletal muscle as both as Ca^{2+} channel and voltage sensor for EC coupling, the genetics, and pathophysiology of MH, and the mechanisms by which MH mutations in $Ca_v1.1$ alter the Ca^{2+} channel and voltage sensor functions in a manner that results in an increased susceptibility to MH.

8.1 Introduction

It is difficult to underestimate the importance of $Ca_v1.1$ channel in muscle physiology. The primary role of $Ca_v1.1$ is to transform an electrical signal in the transverse tubule (t-tubule) membrane (an action potential) into an intracellular

V. Yarotskyy (✉) · R. T. Dirksen (✉)
Department of Pharmacology and Physiology, University of Rochester Medical Center, Rochester, USA
e-mail: viktor_yarotskyy@urmc.rochester.edu

R. T. Dirksen
e-mail: robert_dirksen@urmc.rochester.edu

N. Weiss and A. Koschak (eds.), *Pathologies of Calcium Channels*,
DOI: 10.1007/978-3-642-40282-1_8,

chemical signal (a global myoplasmic Ca^{2+} transient) used to drive skeletal muscle contraction. This function of the $Ca_v1.1$ channel, referred to as being a voltage sensor for excitation–contraction (EC) coupling, is orchestrated through a complex bidirectional mechanical interaction with the type I ryanodine receptor (RYR1) Ca^{2+} release channel located in the adjacent terminal cistern of the sarcoplasmic reticulum (SR) (Dirksen 2002). During an MH crisis, the $Ca_v1.1$-RYR1 interaction is unusually susceptible to the ability of triggering agents (e.g., isoflurane, succinylcholine) to cause uncontrolled muscle contractures and hypermetabolism that result in severe hyperthermia, acidosis, hypercapnia/hypoxia, rhabdomyolysis, and hyperkalemia (Rosenberg et al. 2013). Cardiac arrhythmias, severe tissue damage, multiorgan failure, and eventually death will result if the MH crisis is not quickly recognized and appropriately treated. Treatment of MH includes discontinuation of triggering agent, whole body cooling to control hyperthermia, ventilation with 100 % O_2, and IV administration of the MH antidote dantrolene. MH susceptibility in humans is linked to mutations in both RYR1 and $Ca_v1.1$, demonstrating that these crises reflect a dysfunction in EC coupling and the proper control of SR Ca^{2+} storage and release. In this chapter, we consider the functional role of $Ca_v1.1$ as both a Ca^{2+} permeable L-type Ca^{2+} channel and voltage sensor for EC coupling through a prism of the $Ca_v1.1$- RYR1 bidirectional signaling interaction. Specifically, we focus on the mechanisms by which mutations in $Ca_v1.1$ promote MH susceptibility by altering $Ca_v1.1$- RYR1 communication.

8.2 Role of $Ca_v1.1$ in Excitation–Contraction (EC) Coupling

The $Ca_v1.x$ family of dihydropyridine-sensitive (or L-type) voltage-gated Ca^{2+} channels consists of four channel subtypes: $Ca_v1.1$—exclusively expressed in skeletal muscle, $Ca_v1.2$—cardiac type, but also expressed in other tissues including neurons, $Ca_v1.3$—found in neuronal and secretory glands cells, and $Ca_v1.4$—expressed in retinal photoreceptors and bipolar cells. $Ca_v1.x$ channels are sensitive to three clinically relevant drug classes: dihydropyridines, phenylalkylamines, and benzothiazepines, which together with the respective biophysical properties of each channel, were used to categorize the different $Ca_v1.x$ channel subtypes prior to the molecular cloning era. In general, $Ca_v1.x$ channels are relatively slower to activate and inactivate compared to those within the $Ca_v2.x$ or $Ca_v3.x$ families. However, the kinetic properties of channels even just within the $Ca_v1.x$ family are quite variable. For example, the activation kinetics of $Ca_v1.1$ channels in skeletal muscle are >10x slower compared to that of cardiac $Ca_v1.2$ channels. In addition, as opposed to $Ca_v1.2$ and $Ca_v1.3$ channels, $Ca_v1.1$ and $Ca_v1.4$ channels essentially lack Ca-dependent inactivation and exhibit extremely slow voltage-dependent inactivation.

$Ca_v1.1$, the first cloned pore-forming subunit of a voltage-dependent Ca^{2+} channel (Tanabe et al. 1987), is expressed within the junctional membranes of the t-tubule system in skeletal muscle. The $Ca_v1.1$ protein consists of intracellular N- and C-termini and four homologous transmembrane repeat domains connected by three intracellular linkers; I–II loop, II–III loop, and III–IV loop (Fig. 8.1). Each of the four repeats exhibit six transmembrane segments (S1 through S6). The fourth transmembrane segment of each domain (S4) contains positively charged amino acids, every third amino acid that together form the core voltage sensing module of the channel. In addition, the loops between the fifth and sixth segment of each domain (S5–S6 loops) arrange to form the channel pore and Ca^{2+} selectivity filter.

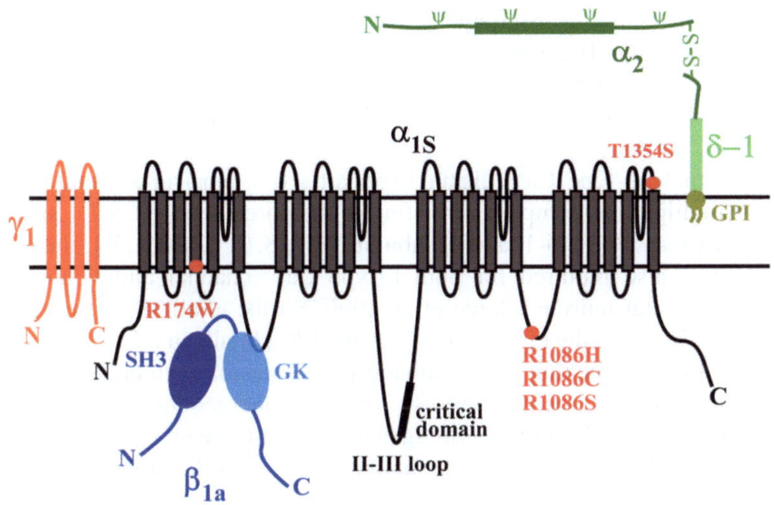

Fig. 8.1 Subunit structure of the $Ca_v1.1$ channel. The $Ca_v1.1$ channel consists of one major pore-forming α_{1S} subunit and three auxiliary β_{1a}, γ_1, and $\alpha_2\delta$-1 subunits. The pore-forming α_{1S} subunit is a polypeptide folded in a four homologous repeat domains, with each repeat domain consisting of six transmembrane segments (S1 through S6). The S1–S4 parts of each domain comprise the voltage sensing module, where S4 segments contain positively charged residues every third amino acid that serve as the primary voltage sensor while S1–S3 provides a supportive substructure for S4 movement across the transmembrane voltage drop. The linker between the S5–S6 segments of each repeat form the pore-lining and selectivity filter region of the channel. The N- and C-termini and loops linking each of the four repeats are located on the intracellular aspect of the channel. The guanylate cyclase (GK) domain of the auxiliary β_{1a} subunit (*blue*) interacts with a highly conserved region of the intracellular loop between domains I–II of $Ca_v1.1$. The linker between repeats II and III (i.e., "II–III loop") is critical for mechanical coupling between the $Ca_v1.1$ channel and the type I ryanodine receptor. The extracellular $\alpha_2\delta$-1 subunit (*green*) forms a complex whereby the δ-1 subunit is covalently associated with the heavily glycosylated α_2 subunit via disulfide linkage. While the δ-1 subunit was originally thought to be anchored to the membrane via a single transmembrane segment, recent evidence indicates that $\alpha_2\delta$-1 membrane association is mediated through a glycophosphatidylinositol (GPI) anchor attached to the δ-1 subunit (Davies et al. 2010). The γ_1 subunit (*orange*) contains four transmembrane domains. MH mutations in the pore-forming α_{1S} subunit are shown as red balls with red text indicating the identified amino acid residue substitutions

Elegant freeze fracture studies have shown that $Ca_v1.1$ is grouped into units of four intramembrane particles (or "tetrads") that are specifically aligned and directly opposed to the four subunits of every other RYR1 homotetrameric release channel in the membrane of the adjacent SR terminal cisternae (Block et al. 1988). Skeletal myotubes derived from $Ca_v1.1$-null mice lack L-type Ca^{2+} currents, intramembrane charge movements or "gating currents," voltage-gated SR Ca^{2+} release, and tetrads, all of which are restored upon re-expression of $Ca_v1.1$ (Tanabe et al. 1988; Adams et al. 1990; Takekura et al. 1994). Thus, $Ca_v1.1$ is essential for L-type Ca^{2+} channel activity, voltage sensor function, EC coupling, and tetrad formation in skeletal muscle. Indeed, the disposition of $Ca_v1.1$ tetrads specifically aligned with each of the four corners of alternate RYR1 homotetrameric SR Ca^{2+} release channels provides strong structural evidence for direct $Ca_v1.1$-RYR1 conformational coupling. Subsequent $Ca_v1.1$–$Ca_v1.2$ chimeric studies revealed that the intracellular II–III loop, and specifically a critical region that includes amino acids 720–765 in the central part of this loop (Fig. 8.1), is required for $Ca_v1.1$-RYR1 conformational coupling (Tanabe et al. 1990; Nakai et al. 1998; Kugler et al. 2004).

The complete L-type Ca^{2+} channel (or voltage sensor) in skeletal muscle is actually a multiprotein complex consisting of the pore-forming $Ca_v1.1$ subunit (or α_{1S}) plus auxiliary β_{1a}, $\alpha_2\delta$-1, and γ_1 subunits (Fig. 8.1). Like $Ca_v1.1$, the auxiliary β_{1a} subunit is also required for both L-type Ca^{2+} channel and voltage sensor function in skeletal muscle (Gregg et al. 1996; Strube et al. 1996; Schredelseker et al. 2005). Similar to the effect of Ca^{2+} channel β subunits in other tissues, the β_{1a} subunit enhances trafficking of α_{1S} subunit to the plasma membrane. In addition, like $Ca_v1.1$ and RYR1, the auxiliary β_{1a} subunit is also required for voltage-gated Ca^{2+} release during EC coupling (Gregg et al. 1996; Strube et al. 1996; Beurg et al. 1999), at least in part because the β_{1a} subunit coordinates the proper formation and positioning of tetrads (Schredelseker et al. 2005). Effects of the $\alpha_2\delta$-1 and γ_1 subunits on $Ca_v1.1$ function are more modest, with channel activation kinetics being slowed by the $\alpha_2\delta$-1 subunit (Obermair et al. 2005, 2008; Gach et al. 2008) and the γ_1 subunit modulating the voltage dependence of channel inactivation (Freise et al. 2000). Finally, the ability of RYR1 to enhance $Ca_v1.1$ Ca^{2+} current magnitude, expression, activation kinetics, modulation by DHP agonists, and divalent conductance (Nakai et al. 1996; Avila and Dirksen 2000) via retrograde coupling demonstrates that RYR1 is an important allosteric modulator of the $Ca_v1.1$ channel, analogous to that of a more conventional Ca^{2+} channel auxiliary subunit.

In summary, depolarization of the transverse tubule membrane drives rearrangements of the $Ca_v1.1$ S4 segments that result in a conformational change in II–III loop critical domain and β_{1a} subunit that promote RYR1 channel opening and SR Ca^{2+} release during a process referred to as orthograde $Ca_v1.1$-RYR1 coupling (for an excellent review see (Bannister and Beam 2012)). In addition, the junctional expression of RYR1, in turn, enhances L-type Ca^{2+} channel function (e.g., conductance and gating kinetics) via a process referred to as retrograde

$Ca_v1.1$-RYR1 coupling (Nakai et al. 1996; Avila and Dirksen 2000). Thus, the signaling between the $Ca_v1.1$ channel complex and RYR1 is bidirectional, such that the channel activity associated with each protein is strongly dependent upon this unique interaction. As a result, $Ca_v1.1$ functions both as an L-type Ca^{2+} channel and a voltage sensor that drives RYR1-mediated SR Ca^{2+} release during EC coupling. While the essential role of the $Ca_v1.1$ complex as a voltage sensor for EC coupling is widely appreciated, the importance of $Ca_v1.1$ Ca^{2+} channel function, though still unclear, has recently attracted renewed interest (Bannister and Beam 2012).

8.3 Malignant Hyperthermia

Malignant hyperthermia (MH) is a potentially fatal pharmacogenetic disorder in which susceptible individuals experience a life-threatening hypermetabolic reaction of skeletal muscle following exposure to certain volatile anesthetics (e.g., desflurane, enflurane, ether, halothane, isoflurane, methoxyflurane, and sevoflurane) and depolarizing skeletal muscle relaxants (e.g., succinylcholine) (Rosenberg et al. 2013). Following exposure to one of these triggering agents, an MH susceptible individual can undergo a hypermetabolic crisis characterized by skeletal muscle rigidity, a dramatic rise in core body temperature, respiratory acidosis, hypermetabolism, rhabdomyolysis, hyperkalemia, and cardiac arrhythmias. MH crises are lethal if not reversed immediately by removal of triggering agents, cooling the patient, and administration of dantrolene, the only FDA-approved antidote for an MH reaction (Larach et al. 2008). Recent reports indicate that a subset of MH susceptible individuals, particularly young children, have experienced similar nonanesthetic or "awake" MH-like reactions during strenuous exercise, emotional stress, febrile illness, exposure to environmental heat stress, or a combination of these triggers (Hopkins et al. 1991; Tobin et al. 2001; Wappler et al. 2001; Davis et al. 2002; Capacchione and Muldoon 2009; Nishio et al. 2009; Groom et al. 2011; Lavezzi et al. 2013).

Standardized *in vitro* caffeine and halothane contracture tests were developed in North America and Europe, and together with the clinical presentation of an MH event, are used as the "gold-standard" for diagnosis of MH susceptibility (Rosenberg et al. 2013). The contracture test determines the sensitivity of a muscle biopsy to contractures induced by graded concentrations of caffeine and halothane. If certain contracture thresholds are reached in the presence of low concentrations of caffeine and halothane, then a diagnosis of MH susceptibility is rendered. Enhanced sensitivity to either caffeine or halothane identifies patients as being MH susceptible according to the North American protocol, while sensitivity to both caffeine and halothane are required for such a designation according to the European protocol. Sensitivity to only caffeine or halothane is identified as "MH equivocal" in the European protocol.

8.4 Genetics of MH Susceptibility

MH follows an autosomal dominant mode of inheritance (Rosenberg et al. 2013). To date, MH susceptibility has been definitively linked to mutations in the genes that encode RYR1 (MHS1: *RYR1*) and $Ca_V1.1$ (MHS5: *CACNA1S*). Additional MH gene loci have been linked to chromosomal loci 17q11.2–q24 (MHS2), 7q21–q22 (MHS3), 3q13 (MHS4), and 5p (MHS6), though specific genes and mutations at these loci have not been identified. Up to 70 % of all MH cases are caused by mutations in RYR1, while only ~1 % of cases result from $Ca_V1.1$ mutations (Rosenberg et al. 2013).

Malignant hyperthermia (MH) crises are initiated by exposure of susceptible individuals to triggering agents including volatile anesthetics, depolarizing muscle relaxants, strenuous exercise, or exposure to excessive environmental heat stress. Each of these seemingly distinct triggers converge to promote uncontrolled SR Ca^{2+} release in skeletal muscle to an extent sufficient to result in the hypermetabolism, contractures, and damage to skeletal muscle that are characteristic of MH (MacLennan and Phillips 1992; Rosenberg et al. 2013). Since the $Ca_V1.1$–RYR1 bidirectional interaction plays such a central role in controlling SR Ca^{2+} release under normal conditions (see Sect. 8.2), it is not unexpected that MH mutations have been identified in both RYR1 and $Ca_V1.1$.

Over 200 point mutations and small in-frame deletions in RYR1 have been linked to MH in humans (Rosenberg et al. 2013). While only a small fraction of these MH mutations in RYR1 have been functionally characterized, the majority appear to destabilize the RYR1 resting closed state, and thus hypersensitize the RYR1 release channel to activation by triggering agents. Indeed, many RYR1 mutations are located at interdomain/subunit interfaces that are important for maintaining allosteric domain–domain stability (Ikemoto and Yamamoto 2000). MH mutations located at these interfaces are proposed to disrupt these important domain–domain interactions, ultimately leading to a destabilization of the channel closed state. As a result, MH mutations at these domain interfaces make it energetically more favorable for any RYR1 activator (e.g., voltage, caffeine, halothane, 4-chloro-m-cresol) to drive the channel into the open state (Ikemoto and Yamamoto 2000; Kobayashi et al. 2005, 2004). Accordingly, some MH mutations in RYR1 are thought to destabilize or "unzip" important domain–domain interactions, and thus render the release channel hypersensitive to activation (Kobayashi et al. 2005). This mechanism is supported by X-ray crystallography studies confirming that many identified MH mutations in RYR1 are indeed located at interdomain and intersubunit interfaces (Van Petegem 2012; Tung et al. 2010).

Interestingly, mice that lack type 1 calsequestrin (CASQ1-null), a high-capacity, moderate affinity Ca^{2+} binding protein in the terminal SR, were recently shown to exhibit lethal halothane- and heat-induced MH-like responses (Dainese et al. 2009; Protasi et al. 2009). CASQ1 increases the Ca^{2+} capacity of the SR, modulates RYR1 luminal Ca^{2+} sensitivity (Beard et al. 2002) and also influences terminal SR lumen size, geometry, RYR1 content and SR/transverse tubule interactions (Protasi et al. 2011). The increased MH susceptibility of CASQ1-null mice is proposed to result in part from reduced CASQ1-mediated inhibition of RYR1 Ca^{2+} release (Dainese et al. 2009) resulting in increased sensitization of RYR1 to activation and greater susceptibility to deep SR store depletion sufficient to activate store-operated Ca^{2+} entry (Canato et al. 2010). Indeed, CASQ1 deficiency enhances store-operated Ca^{2+} entry in skeletal muscle (Zhao et al. 2010). Although loss-of-function mutations in the *CASQ1* gene have yet to be linked to MH in humans, genes involved in controlling SR Ca^{2+} content and coordinating store-operated Ca^{2+} entry in skeletal muscle represent intriguing potential novel candidate MH loci.

8.5 MH Mutations in $Ca_v1.1$

Five mutations of three different $Ca_v1.1$ residues have been linked to MH in humans. The three mutated residues involve very different regions of the channel including the intracellular III–IV loop (R1086C, R1086H, and R1086S), the extracellular pore region of repeat IV (T1354S), and a positively charged residue in the S4 voltage sensing region of repeat I (R174W) (Fig. 8.1). The identification and functional analyses of these mutations in $Ca_v1.1$ have provided important new insights into both $Ca_v1.1$ structure/function and the pathophysiological mechanisms of MH (Fig. 8.2). The results of these studies are discussed in detail in the following subsections.

8.5.1 Mutations to R1086 in the Intracellular III–IV Linker

A point mutation of a highly conserved arginine to a histidine residue (R1086H) in the intracellular III–IV loop of $Ca_v1.1$ was found to be responsible for MH susceptibility in a large French kindred, where the proband died from a fulminant MH crisis during anesthesia (Monnier et al. 1997). Results from contracture tests and genetic analyses of family members tightly linked MH susceptibility in this family to a G3333A substitution in the *CACNA1S* gene, which results in a R1086H mutation in the intracellular III–IV loop of $Ca_v1.1$ (Monnier et al. 1997). The identical mutation was subsequently identified in a second independent family (Stewart et al. 2001). The relevance of mutations in R1086 to MH susceptibility is underscored by the fact that subsequent studies linked MH in other families to additional amino acid substitutions of the same residue (R1086S and R1086C) (Toppin et al. 2010; Jurkat-Rott et al. 2000).

The R1086H mutation in $Ca_v1.1$ was the first MH mutation identified in a protein other than RYR1. While the identification of an MH-linked mutation in a second protein essential for skeletal muscle EC coupling was not unexpected, the location of this mutation in the $Ca_v1.1$ III–IV loop was surprising since prior structure/function studies did not identify this region as being uniquely important

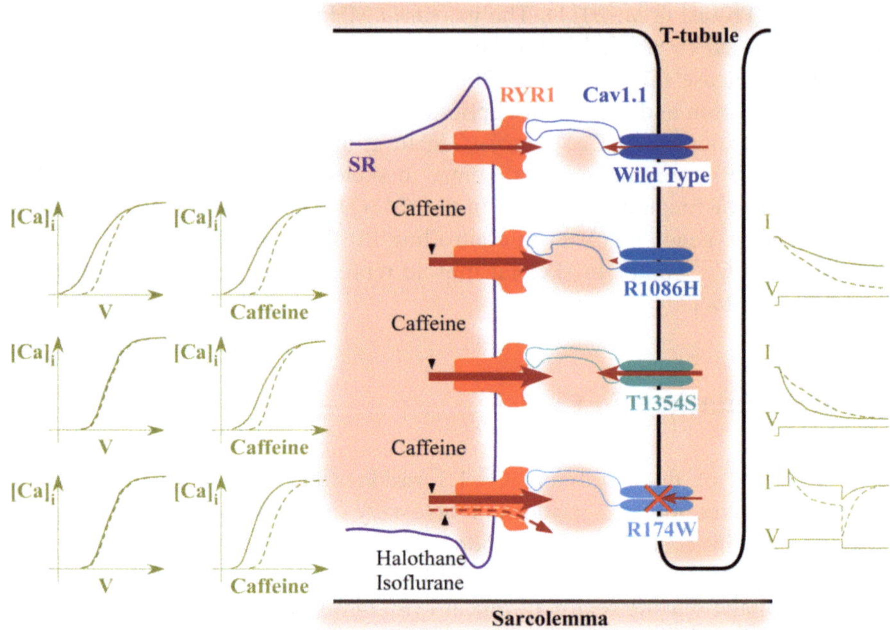

Fig. 8.2 Proposed mechanisms by which MH mutations alter $Ca_v1.1$ and RYR1 function. This schematic depicts how specific MH mutations in $Ca_v1.1$ affect both Ca^{2+} release through RYR1 and $Ca_v1.1$ gating. Wild-type and mutant $Ca_v1.1$ channels (*blue shades*) are expressed within the t-tubule membrane. $Ca_v1.1$ is mechanically coupled via the II–III loop with RYR1 Ca^{2+} release channels (*orange*) expressed in the terminal cisternae of the sarcoplasmic reticulum. Red arrows show the direction of net Ca^{2+} flux and arrow thickness reflects the relative magnitude of Ca^{2+} ion flux. Effects of MH mutations in $Ca_v1.1$ on the sensitivity of voltage- and caffeine-induced Ca^{2+} release are shown in the left panels. Dashed curves represent data obtained from myotubes expressing WT $Ca_v1.1$ and solid curves depict data obtained from myotubes expressing MH mutant $Ca_v1.1$ channels. The R1086H mutation produces a hyperpolarizing shift in the voltage dependence of RYR1 Ca^{2+} release. The T1354S and R174W mutations either accelerate the activation or ablate $Ca_v1.1$ Ca^{2+} currents, respectively. "Caffeine" and "Halothane Isoflurane" labels with black arrows represent enhanced sensitivity RYR1 to activation by these agents. Expression of each of the MH mutants in $Ca_v1.1$ in myotubes results in increased sensitivity of RYR1 to activation by caffeine. Expression of the R174W mutation in $Ca_v1.1$ in myotubes is proposed to increase SR Ca^{2+} leak through a ryanodine-insensitive pathway. Effects of MH mutations in $Ca_v1.1$ on L-type Ca^{2+} channel gating are shown in the right panels. The R1086H mutation reduces $Ca_v1.1$ current density. For the T1354S mutation, the rate of channel activation is faster compared that of the wild-type $Ca_v1.1$ channel. The R174W mutation results in the ablation of ionic Ca^{2+} currents, but not gating currents

for EC coupling, voltage gating, or ion selection/permeation. For example, transferring the III–IV loop of $Ca_v1.2$ into $Ca_v1.1$ did not significantly alter either the Ca^{2+} channel or voltage sensor functions of $Ca_v1.1$ (Tanabe et al. 1990). However, chimeric studies fail to reveal important functional domains that are highly conserved between the two homologous proteins. Thus, the functional studies of the R1086H MH mutation provided important new insights into a previously unappreciated role of the III-IV loop in $Ca_v1.1$-RYR1 signaling (Weiss et al. 2004).

Functional studies of the R1086H mutant expressed in $Ca_v1.1$-null myotubes revealed several alterations consistent with the mutation being linked to increased MH susceptibility (Weiss et al. 2004) (see Fig. 8.2, row 2). First of all, the sensitivity of caffeine-induced Ca^{2+} release was ~5-fold higher in R1086H-expressing myotubes compared to that of wild-type RYR1-expressing myotubes. In addition, the voltage dependence of RYR1 Ca^{2+} release was also shifted to more negative voltages in R1086H-expressing myotubes. These results indicate that the R1086H mutation in $Ca_v1.1$ increases the sensitivity of the SR Ca^{2+} release machinery to activation by RYR1 triggers (e.g., caffeine and voltage). The R1086H mutation also resulted in ~2-fold reduction in Ca^{2+} current density through $Ca_v1.1$ that was attributable to reduced channel open probability rather than an effect on either channel expression or retrograde coupling with RYR1. Indeed, a similar reduction in channel conductance was observed for the analogous mutation introduced into $Ca_v1.2$. Together, these results suggest that the III–IV linker acts as a negative regulatory module for RYR1 activation by caffeine/voltage and that the R1086H mutation promotes MH susceptibility by disrupting this critical negative allosteric regulatory mechanism (Weiss et al. 2004). There are two potential mechanisms by which the III–IV loop could influence RYR1 channel sensitivity to activation: (i) the III–IV loop could directly bind RYR1 as suggested previously from biochemical experiments (Leong and MacLennan 1998) or (ii) the III–IV loop could influence the structural conformation of the critical domain within the adjacent II–III loop that is required for proper $Ca_v1.1$-RYR1 coupling.

Overall, results from the functional analysis of the R1086H MH mutation in $Ca_v1.1$ are remarkably similar to those reported for MH-linked mutations in RYR1 where the mutations increased the sensitivity of the SR Ca^{2+} release mechanism to activation by both pharmacological (caffeine) and endogenous (voltage sensor) activators (Tong et al. 1997; Dirksen and Avila 2004; Yang et al. 2003; Avila and Dirksen 2001). MH mutations in RYR1 are thought to enhance channel sensitivity to activation by triggers by disrupting important interdomain interactions that destabilize the channel closed state (see Sect. 8.4). Apparently, the R1086H mutation in $Ca_v1.1$ accomplishes a similar destabilization of the RYR1 closed state by disrupting a negative regulatory module contained within the III–IV loop of $Ca_v1.1$.

8.5.2 T1354S Mutation in the Extracellular Pore Region of Repeat IV

A novel conservative point mutation (T1354S) in $Ca_v1.1$ located in the extracellular linker between the pore loop and S6 transmembrane domain of repeat IV was identified in a MH susceptible family of Italian origin (Pirone et al. 2010). The location of this particular MH mutation was unexpected since it was unclear how a residue on the extracellular face of $Ca_v1.1$ would influence the RYR1 Ca^{2+} release channel sensitivity to activation. However, functional analyses of the T1354S mutant following expression in $Ca_v1.1$-null myotubes revealed a novel mechanism by which a mutation in $Ca_v1.1$ can increase MH susceptibility (Pirone et al. 2010).

The T1354S mutation in $Ca_v1.1$ produced two major changes in L-type Ca^{2+} channel function and RYR1-dependent Ca^{2+} release (Fig. 8.2, row 3). The first effect was a significant increase in the rate of depolarization-induced Ca^{2+} current activation through $Ca_v1.1$ (Pirone et al. 2010). Although the rate of Ca^{2+} current activation was $\sim$33 % faster for the T1354S mutant channel compared to that of wild-type $Ca_v1.1$, no difference was found between the magnitudes or voltage dependences of either the L-type Ca^{2+} current or voltage-gated Ca^{2+} release. Second, T1354S-expressing myotubes also exhibited increased sensitivity of caffeine-induced Ca^{2+} release (similar to that observed previously for R1086H (Weiss et al. 2004)) and an increase in action potential evoked Ca^{2+} release in the presence of subthreshold caffeine concentrations. Consistent with a Ca^{2+} influx-dependent amplification mechanism in T1354S-expressing myotubes, this increase in action potential evoked release was abolished upon removal of extracellular Ca^{2+}. These findings suggest that the T1354S mutation in $Ca_v1.1$ promotes MH susceptibility through an amplification of SR Ca^{2+} release that results from the combined effects of an increased Ca^{2+} influx via more rapidly activating T1354S channels and an enhanced caffeine sensitivity of RYR1 Ca^{2+} release.

It is worth noting that a similar dramatic increase in the rate of $Ca_v1.1$ channel activation is also observed for the embryonic $Ca_v1.1$ splice isoform that involves the excision of exon 29 located within the adjacent extracellular S3–S4 loop of repeat IV (Tuluc and Flucher 2011). However, the embryonic $Ca_v1.1$ splice isoform also exhibits a significant hyperpolarizing shift in the voltage dependence of channel activation and an increase in maximal channel conductance (Tuluc and Flucher 2011). Interestingly, expression of the embryonic $Ca_v1.1$ splice isoform is increased in adult patients with myotonic dystrophy and the expression level of this splice isoform correlates strongly with the severity of muscle weakness in these patients (Tang et al. 2012). Thus, structure/function studies of $Ca_v1.1$-related disease mutants/splice variants have revealed an unexpected importance of the extracellular regions of repeat IV in $Ca_v1.1$ channel properties and have also led to significant advancements in our understanding of the pathomechanisms that underlie MH and myotonic dystrophy.

8.5.3 R174W Mutation in the S4 Region of Repeat I

Sequencing the entire *CACNA1S* coding region from 50 confirmed MH patients in whom RYR1 mutations were excluded resulted in the identification of a novel R174W mutation in $Ca_v1.1$ (Carpenter et al. 2009). Consistent with the R174W variant being pathogenic, the mutation was identified in the proband, its presence was concordant with disease in other family members, and the variant was not detected in control samples. The R174W mutation results in the neutralization of a highly conserved and positively charged residue in the innermost portion of the S4 voltage sensing region of repeat I. As detailed below, functional studies revealed that the impact of the R174W mutation on $Ca_v1.1$ channel function is considerably more severe than that observed previously for either the R1086H or T1354S MH mutants. Surprisingly, unlike the other $Ca_v1.1$ MH mutants, the R174W mutation did not significantly alter orthograde $Ca_v1.1$-RYR1 signaling during EC coupling. Rather, the R174W mutation in $Ca_v1.1$ is proposed to promote MH susceptibility by causing an increase in SR Ca^{2+} leak (Fig. 8.2, row 4) (Eltit et al. 2012).

Expression of the R174W mutant in $Ca_v1.1$-null myotubes revealed a complete ablation of ionic Ca^{2+} currents that occurred in the absence of a change in the magnitude or voltage dependence of either channel gating (i.e., gating currents) or depolarization-induced Ca^{2+} release (Eltit et al. 2012). Thus, while the R174W mutation ablates the Ca^{2+} channel function of $Ca_v1.1$, it does not impact its function as a voltage sensor for EC coupling. Nevertheless, R174W-expressing myotubes exhibited ~2-fold increase in sensitivity to caffeine-induced Ca^{2+} release and enhanced responses to halothane/isoflurane exposure, consistent with the mutation being responsible for enhanced susceptibility to MH.

How does the R174W mutation enhance RYR1 release channel sensitivity to activation by caffeine if it does not alter $Ca_v1.1$-RYR1 orthograde coupling (like R1086H) or increase Ca^{2+} influx through $Ca_v1.1$ (like T1354S)? Additional experiments found that resting myoplasmic Ca^{2+} levels were significantly elevated and SR Ca^{2+} content reduced in R174W-expressing myotubes, suggesting that the R174W mutant enhances the basal rate of SR Ca^{2+} leak. Consistent with this idea, elevated resting Ca^{2+} levels in R174W-expressing myotubes were normalized by addition of dantrolene (Eltit et al. 2012). Since low concentrations of Ca^{2+} are known to promote RYR1 activity, the R174W-mediated elevation in resting Ca^{2+} may act as a sensitizing co-agonist that potentiates RYR1 activation by other triggers such as caffeine and volatile anesthetics.

While the R1086H and T1354S mutations in $Ca_v1.1$ act by either destabilizing the RYR1 closed state or enhancing $Ca_v1.1$-mediated Ca^{2+} entry, the R174W mutation appears to disrupt the ability of $Ca_v1.1$ to suppress basal RYR1 Ca^{2+} leak. In support of this unique hypothesis for MH pathogenesis, prior studies reported that $Ca_v1.1$-null myotubes exhibit a ryanodine-insensitive SR Ca^{2+} leak pathway that also results in increased levels of resting myoplasmic Ca^{2+} and reduced SR Ca^{2+} content (Eltit et al. 2011, 2010). However, the validity of this hypothesis must await future structural and/or single channel evidence for a separate ryanodine-insensitive Ca^{2+} permeation pathway in RYR1 channels as proposed and schematically depicted in Fig. 8.2.

8.6 Conclusion

Admittedly, mutations in $Ca_v1.1$ account for only a small fraction ($\sim 1\ \%$) of all MH cases in humans. However, detailed functional characterization of MH-linked mutations in $Ca_v1.1$ has resulted in tremendous advances in our understanding of both $Ca_v1.1$ structure/function and MH pathogenesis. For example, such studies have revealed regions of $Ca_v1.1$ important of orthograde coupling (III–IV loop) and the rate of $Ca_v1.1$ channel activation (repeat IV pore region). In addition, these studies have also identified three distinct pathogenic mechanisms by which mutations in $Ca_v1.1$ increase MH susceptibility (RYR1 closed state destabilization, increased $Ca_v1.1$ Ca^{2+} entry, and enhanced SR Ca^{2+} leak). Future genetic studies of additional MH families are almost certain to not only unearth new MH-linked mutations $Ca_v1.1$, but ultimately, also lead to new fundamental insights into $Ca_v1.1$ function and MH pathogenesis.

Acknowledgements This work was supported by a research grant (AR059646 to RTD) and a mentored careerdevelopment award (AR060831 to VY) from the National Institutes of Health.

References

Adams BA, Tanabe T, Mikami A, Numa S, Beam KG (1990) Intramembrane charge movement restored in dysgenic skeletal muscle by injection of dihydropyridine receptor cDNAs. Nature 346:569–572

Avila G, Dirksen RT (2000) Functional impact of the ryanodine receptor on the skeletal muscle L-type Ca(2 +) channel. J Gen Physiol 115:467–480

Avila G, Dirksen RT (2001) Functional effects of central core disease mutations in the cytoplasmic region of the skeletal muscle ryanodine receptor. J Gen Physiol 118:277–290

Bannister RA, Beam KG (2012) Ca(V)1.1: the atypical prototypical voltage-gated Ca(2 +) channel. Biochim Biophys Acta 1828:1587–1597

Beard NA, Sakowska MM, Dulhunty AF, Laver DR (2002) Calsequestrin is an inhibitor of skeletal muscle ryanodine receptor calcium release channels. Biophys J 82:310–320

Beurg M, Ahern CA, Vallejo P, Conklin MW, Powers PA, Gregg RG, Coronado R (1999) Involvement of the carboxy-terminus region of the dihydropyridine receptor beta1a subunit in excitation-contraction coupling of skeletal muscle. Biophys J 77:2953–2967

Block BA, Imagawa T, Campbell KP, Franzini-Armstrong C (1988) Structural evidence for direct interaction between the molecular components of the transverse tubule/sarcoplasmic reticulum junction in skeletal muscle. J Cell Biol 107:2587–2600

Canato M, Scorzeto M, Giacomello M, Protasi F, Reggiani C, Stienen GJ (2010) Massive alterations of sarcoplasmic reticulum free calcium in skeletal muscle fibers lacking calsequestrin revealed by a genetically encoded probe. Proc Natl Acad Sci U S A 107:22326–22331

Capacchione JF, Muldoon SM (2009) The relationship between exertional heat illness, exertional rhabdomyolysis, and malignant hyperthermia. Anesth Analg 109:1065–1069

Carpenter D, Ringrose C, Leo V, Morris A, Robinson RL, Halsall PJ, Hopkins PM, Shaw MA (2009) The role of CACNA1S in predisposition to malignant hyperthermia. BMC Med Genet 10:104

Dainese M, Quarta M, Lyfenko AD, Paolini C, Canato M, Reggiani C, Dirksen RT, Protasi F (2009) Anesthetic- and heat-induced sudden death in calsequestrin-1-knockout mice. FASEB J 23:1710–1720

Davies A, Kadurin I, Alvarez-Laviada A, Douglas L, Nieto-Rostro M, Bauer CS, Pratt WS, Dolphin AC (2010) The alpha2delta subunits of voltage-gated calcium channels form GPI-anchored proteins, a posttranslational modification essential for function. Proc Natl Acad Sci U S A 107:1654–1659

Davis M, Brown R, Dickson A, Horton H, James D, Laing N, Marston R, Norgate M, Perlman D, Pollock N, Stowell K (2002) Malignant hyperthermia associated with exercise-induced rhabdomyolysis or congenital abnormalities and a novel RYR1 mutation in New Zealand and Australian pedigrees. Br J Anaesth 88:508–515

Dirksen RT (2002) Bi-directional coupling between dihydropyridine receptors and ryanodine receptors. Front Biosci 7:d659–d670

Dirksen RT, Avila G (2004) Distinct effects on Ca2 + handling caused by malignant hyperthermia and central core disease mutations in RyR1. Biophys J 87:3193–3204

Eltit JM, Bannister RA, Moua O, Altamirano F, Hopkins PM, Pessah IN, Molinski TF, Lopez JR, Beam KG, Allen PD (2012) Malignant hyperthermia susceptibility arising from altered resting coupling between the skeletal muscle L-type Ca2 + channel and the type 1 ryanodine receptor. Proc Natl Acad Sci U S A 109:7923–7928

Eltit JM, Li H, Ward CW, Molinski T, Pessah IN, Allen PD, Lopez JR (2011) Orthograde dihydropyridine receptor signal regulates ryanodine receptor passive leak. Proc Natl Acad Sci U S A 108:7046–7051

Eltit JM, Yang T, Li H, Molinski TF, Pessah IN, Allen PD, Lopez JR (2010) RyR1-mediated Ca2 + leak and Ca2 + entry determine resting intracellular Ca2 + in skeletal myotubes. J Biol Chem 285:13781–13787

Freise D, Held B, Wissenbach U, Pfeifer A, Trost C, Himmerkus N, Schweig U, Freichel M, Biel M, Hofmann F, Hoth M, Flockerzi V (2000) Absence of the gamma subunit of the skeletal muscle dihydropyridine receptor increases L-type Ca2 + currents and alters channel inactivation properties. J Biol Chem 275:14476–14481

Gach MP, Cherednichenko G, Haarmann C, Lopez JR, Beam KG, Pessah IN, Franzini-Armstrong C, Allen PD (2008) Alpha2delta1 dihydropyridine receptor subunit is a critical element for excitation-coupled calcium entry but not for formation of tetrads in skeletal myotubes. Biophys J 94:3023–3034

Gregg RG, Messing A, Strube C, Beurg M, Moss R, Behan M, Sukhareva M, Haynes S, Powell JA, Coronado R, Powers PA (1996) Absence of the beta subunit (cchb1) of the skeletal muscle dihydropyridine receptor alters expression of the alpha 1 subunit and eliminates excitation-contraction coupling. Proc Natl Acad Sci U S A 93:13961–13966

Groom L, Muldoon SM, Tang ZZ, Brandon BW, Bayarsaikhan M, Bina S, Lee H-S, Sambuughin N, Dirksen RT (2011) Identical de novo mutation in the type I ryanodine receptor gene associated with fatal, stress-induced malignant hyperthermia in two unrelated families. Anesthesiology 115:938–945

Hopkins PM, Ellis FR, Halsall PJ (1991) Evidence for related myopathies in exertional heat stroke and malignant hyperthermia. Lancet 338:1491–1492

Ikemoto N, Yamamoto T (2000) Postulated role of inter-domain interaction within the ryanodine receptor in Ca(2 +) channel regulation. Trends Cardiovasc Med 10:310–316

Jurkat-Rott K, McCarthy T, Lehmann-Horn F (2000) Genetics and pathogenesis of malignant hyperthermia. Muscle Nerve 23:4–17

Kobayashi S, Bannister ML, Gangopadhyay JP, Hamada T, Parness J, Ikemoto N (2005) Dantrolene stabilizes domain interactions within the ryanodine receptor. J Biol Chem 280:6580–6587

Kobayashi S, Yamamoto T, Parness J, Ikemoto N (2004) Antibody probe study of Ca2 + channel regulation by interdomain interaction within the ryanodine receptor. Biochem J 380:561–569

Kugler G, Weiss RG, Flucher BE, Grabner M (2004) Structural requirements of the dihydropyridine receptor alpha1S II–III loop for skeletal-type excitation-contraction coupling. J Biol Chem 279:4721–4728

Larach MG, Brandom BW, Allen GC, Gronert GA, Lehman EB (2008) Cardiac arrests and deaths associated with malignant hyperthermia in north america from 1987 to 2006: a report from the north american malignant hyperthermia registry of the malignant hyperthermia association of the United States. Anesthesiology 108:603–611

Lavezzi WA, Capacchione JF, Muldoon SM, Sambuughin N, Bina S, Steele D, Brandom BW (2013) Case report: death in the emergency department: an unrecognized awake malignant hyperthermia-like reaction in a six-year-old. Anesth Analg 116:420–423

Leong P, MacLennan DH (1998) The cytoplasmic loops between domains II and III and domains III and IV in the skeletal muscle dihydropyridine receptor bind to a contiguous site in the skeletal muscle ryanodine receptor. J Biol Chem 273:29958–29964

MacLennan DH, Phillips MS (1992) Malignant hyperthermia. Science 256:789–794

Monnier N, Procaccio V, Stieglitz P, Lunardi J (1997) Malignant-hyperthermia susceptibility is associated with a mutation of the alpha 1-subunit of the human dihydropyridine-sensitive L-type voltage-dependent calcium-channel receptor in skeletal muscle. Am J Hum Genet 60:1316–1325

Nakai J, Dirksen RT, Nguyen HT, Pessah IN, Beam KG, Allen PD (1996) Enhanced dihydropyridine receptor channel activity in the presence of ryanodine receptor. Nature 380:72–75

Nakai J, Tanabe T, Konno T, Adams B, Beam KG (1998) Localization in the II-III loop of the dihydropyridine receptor of a sequence critical for excitation-contraction coupling. J Biol Chem 273:24983–24986

Nishio H, Sato T, Fukunishi S, Tamura A, Iwata M, Tsuboi K, Suzuki K (2009) Identification of malignant hyperthermia-susceptible ryanodine receptor type 1 gene (RYR1) mutations in a child who died in a car after exposure to a high environmental temperature. Leg Med (Tokyo) 11:142–143

Obermair GJ, Kugler G, Baumgartner S, Tuluc P, Grabner M, Flucher BE (2005) The Ca2 + channel alpha2delta-1 subunit determines Ca2 + current kinetics in skeletal muscle but not targeting of alpha1S or excitation-contraction coupling. J Biol Chem 280:2229–2237

Obermair GJ, Tuluc P, Flucher BE (2008) Auxiliary Ca(2 +) channel subunits: lessons learned from muscle. Curr Opin Pharmacol 8:311–318

Pirone A, Schredelseker J, Tuluc P, Gravino E, Fortunato G, Flucher BE, Carsana A, Salvatore F, Grabner M (2010) Identification and functional characterization of malignant hyperthermia mutation T1354S in the outer pore of the Cavalpha1S-subunit. Am J Physiol Cell Physiol 299:C1345–C1354

Protasi F, Paolini C, Canato M, Reggiani C, Quarta M (2011) Lessons from calsequestrin-1 ablation in vivo: much more than a Ca(2 +) buffer after all. J Muscle Res Cell Motil 32:257–270

Protasi F, Paolini C, Dainese M (2009) Calsequestrin-1: a new candidate gene for malignant hyperthermia and exertional/environmental heat stroke. J Physiol 587:3095–3100

Rosenberg H, Sambuughin N, Dirksen R (2013) Malignant hyperthermia susceptibility. In: GeneReviews at GeneTests: medical genetics information resource [database online]. Copyright, University of Washington, Seattle, 1997–2013

Schredelseker J, Di Biase V, Obermair GJ, Felder ET, Flucher BE, Franzini-Armstrong C, Grabner M (2005) The beta 1a subunit is essential for the assembly of dihydropyridine-receptor arrays in skeletal muscle. Proc Natl Acad Sci U S A 102:17219–17224

Stewart SL, Hogan K, Rosenberg H, Fletcher JE (2001) Identification of the Arg1086His mutation in the alpha subunit of the voltage-dependent calcium channel (CACNA1S) in a North American family with malignant hyperthermia. Clin Genet 59:178–184

Strube C, Beurg M, Powers PA, Gregg RG, Coronado R (1996) Reduced Ca2 + current, charge movement, and absence of Ca2 + transients in skeletal muscle deficient in dihydropyridine receptor beta 1 subunit. Biophys J 71:2531–2543

Takekura H, Bennett L, Tanabe T, Beam KG, Franzini-Armstrong C (1994) Restoration of junctional tetrads in dysgenic myotubes by dihydropyridine receptor cDNA. Biophys J 67:793–803

Tanabe T, Beam KG, Adams BA, Niidome T, Numa S (1990) Regions of the skeletal muscle dihydropyridine receptor critical for excitation-contraction coupling. Nature 346:567–569

Tanabe T, Beam KG, Powell JA, Numa S (1988) Restoration of excitation-contraction coupling and slow calcium current in dysgenic muscle by dihydropyridine receptor complementary DNA. Nature 336:134–139

Tanabe T, Takeshima H, Mikami A, Flockerzi V, Takahashi H, Kangawa K, Kojima M, Matsuo H, Hirose T, Numa S (1987) Primary structure of the receptor for calcium channel blockers from skeletal muscle. Nature 328:313–318

Tang ZZ, Yarotskyy V, Wei L, Sobczak K, Nakamori M, Eichinger K, Moxley RT, Dirksen RT, Thornton CA (2012) Muscle weakness in myotonic dystrophy associated with misregulated splicing and altered gating of Ca(V)1.1 calcium channel. Hum Mol Genet 21:1312–1324

Tobin JR, Jason DR, Challa VR, Nelson TE, Sambuughin N (2001) Malignant hyperthermia and apparent heat stroke. JAMA 286:168–169

Tong J, Oyamada H, Demaurex N, Grinstein S, McCarthy TV, MacLennan DH (1997) Caffeine and halothane sensitivity of intracellular Ca2 + release is altered by 15 calcium release channel (ryanodine receptor) mutations associated with malignant hyperthermia and/or central core disease. J Biol Chem 272:26332–26339

Toppin PJ, Chandy TT, Ghanekar A, Kraeva N, Beattie WS, Riazi S (2010) A report of fulminant malignant hyperthermia in a patient with a novel mutation of the CACNA1S gene. Can J Anaesth 57:689–693

Tuluc P, Flucher BE (2011) Divergent biophysical properties, gating mechanisms, and possible functions of the two skeletal muscle Ca(V)1.1 calcium channel splice variants. J Muscle Res Cell Motil 32:249–256

Tung CC, Lobo PA, Kimlicka L, Van Petegem F (2010) The amino-terminal disease hotspot of ryanodine receptors forms a cytoplasmic vestibule. Nature 468:585–588

Van Petegem F (2012) Ryanodine receptors: structure and function. J Biol Chem 287:31624–31632

Wappler F, Fiege M, Steinfath M, Agarwal K, Scholz J, Singh S, Matschke J, Schulte Am Esch (2001) Evidence for susceptibility to malignant hyperthermia in patients with exercise-induced rhabdomyolysis. Anesthesiology 94:95–100

Weiss RG, O'Connell KM, Flucher BE, Allen PD, Grabner M, Dirksen RT (2004) Functional analysis of the R1086H malignant hyperthermia mutation in the DHPR reveals an unexpected influence of the III-IV loop on skeletal muscle EC coupling. Am J Physiol Cell Physiol 287:C1094–C1102

Yang T, Ta TA, Pessah IN, Allen PD (2003) Functional defects in six ryanodine receptor isoform-1 (RyR1) mutations associated with malignant hyperthermia and their impact on skeletal excitation-contraction coupling. J Biol Chem 278:25722–25730

Zhao X, Min CK, Ko JK, Parness J, Kim do H, Weisleder N, Ma J (2010) Increased store-operated Ca2 + entry in skeletal muscle with reduced calsequestrin-1 expression. Biophys J 99:1556–1564

Chapter 9
Ryanodine Receptor 1 and Associated Pathologies

Julien Fauré, Joël Lunardi, Nicole Monnier and Isabelle Marty

Abstract In skeletal muscle a rise in the cytosolic calcium concentration is the first trigger able to initiate the contraction of the sarcomere. Intracellular calcium levels are tightly controlled by channels and pumps, and it is not surprising that many inherited skeletal muscle disorders arise from mutations altering the players regulating calcium ions concentration (Betzenhauser and Marks 2010). In this chapter, we will focus on the pathologies linked to the sarcoplasmic reticulum calcium channel-RyR1 mutations.

9.1 The Muscle Excitation-Contraction Coupling: Calcium Release Complex and RyR1

Skeletal muscle contraction is initiated by a burst of calcium out of its major store, the sarcoplasmic reticulum. This rise in intracellular calcium concentration is triggered by a post synaptic action potential which propagates along the muscle fiber and reaches plasma membrane invaginations called transverse tubules. Transverse tubules (T-tubules) form regular invaginations along the muscle axis and make close and periodic contacts with the sarcoplasmic reticulum (SR) that is wrapped around the myofibril apparatus. These peculiar membrane structures are called triads (Fig. 9.1). Triads are formed by the tight apposition of two terminal cisternae of SR on both sides of a T-tubule, and are specifically localized at the A-I

J. Fauré · J. Lunardi · N. Monnier · I. Marty (✉)
Inserm U836—Grenoble Institut des Neurosciences—Equipe Muscle et Pathologies, Bat EJ Safra, Chemin Fortuné Ferrini, 38700 La Tronche, France
and Université Joseph Fourier, 38000 Grenoble, France
e-mail: isabelle.marty@ujf-grenoble.fr

J. Fauré · J. Lunardi · N. Monnier
CHRU de Grenoble, Biochimie Génétique et Moléculaire, 38000 Grenoble, France

N. Weiss and A. Koschak (eds.), *Pathologies of Calcium Channels*,
DOI: 10.1007/978-3-642-40282-1_9, © Springer-Verlag Berlin Heidelberg 2014

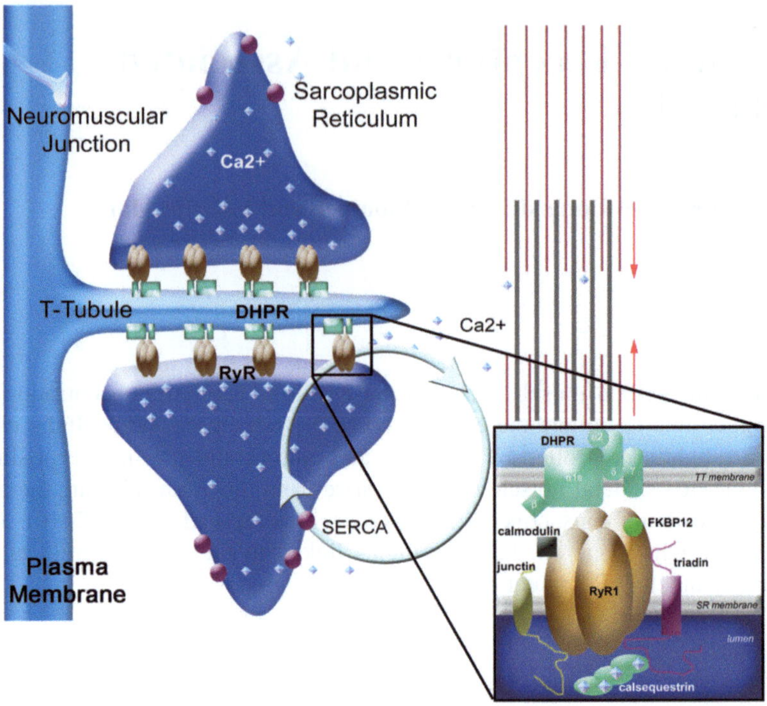

Fig. 9.1 The calcium release complex in the triad of skeletal muscle

bands transition in mammalian skeletal muscle. The transformation of an action potential into a massive calcium release, a process called the Excitation-Contraction (EC) coupling, is performed by a macromolecular complex, the Calcium Release Complex (CRC). This complex is anchored both in the T-tubules membrane and in the SR membrane, and has specific triad localization. The main players of CRC are the dihydropyridine receptor (DHPR), an L-type calcium channel of the T-tubules membrane, and the ryanodine receptor (RyR1), the sarcoplasmic reticulum calcium channel (Fig. 9.1). When an incoming action potential reaches the T-tubules, it activates the voltage gated DHPR, which functions both as a calcium channel and a voltage sensor. This first step triggers a conformational modification in the alpha 1 subunit of DHPR ($Ca_V1.1$) which is in direct contact with RyR1. The physical coupling allows a structural modification in the RyR1 channel upon DHPR activation that results in its opening and in a massive calcium efflux from the terminal cisternae of SR. The released calcium interacts with the myofilaments and induces the muscle contraction. Muscle relaxation occurs when RyR1 closes as a result of high cytosolic calcium concentration, and calcium is reuptake into the sarcoplasmic reticulum by the SERCA (Sarcoplasmic Endoplasmic Reticulum ATPase) calcium pump.

When analyzed by electron microscopy (EM), triad membranes appear as regular structures, with a 15 nm space between T-tubule and SR membrane. In this space, the cytosolic part of the 2.3 MDa RyR1 tetramer is so large that it forms an electron dense structure, the characteristic "feet" early observed between T-tubule and SR terminal cisternae (Franzini-Armstrong et al. 1983). The protein was identified thanks to its binding to ryanodine, an alkaloid extracted from the tropical plant *Ryana speciosa*, and shown to be the intracellular calcium channel of muscle cells (Inui et al. 1987). In mammals, three isoforms of ryanodine receptor have been cloned from three different genes arising from a common ancestor. RyR1 is the major skeletal muscle isoform, but is also expressed in lymphocytes and dendritic cells (Bracci et al. 2007; Uemura et al. 2007) or in specific areas of the central nervous system (Giannini et al. 1995). RyR2 is predominantly expressed in the heart, but it is also in the main brain isoform. RyR3 is weakly expressed in a large panel of tissues, among which skeletal muscle and brain. The repertoire of ryanodine receptors may vary in gene composition and expression among species. In teleost fishes *RYR2* and *RYR3* are conserved but the *RYR1* gene has been duplicated in α and β isoforms, respectively expressed in fast- or slow-twitch muscles (Franck et al. 1998; Hirata et al. 2007), whereas in the genome of *Caenorhabditis elegans* and *Drosophila melanogaster*, only one *RYR* homologous gene is found (Hasan et al. 1992; Maryon et al. 1996).

The human *RYR1* gene is located on chromosome 19q13.2. It spans 153 865 base pairs, and contains 106 exons. *RYR1* is alternatively spliced for the exons 70 and 83, but the functional role of these splicing events is unclear. The 560 kDa protein RyR1 consists of 5038 amino acid that share 60–70 % identity with RyR2 and RyR3. The functional RyR1 channel is a homotetramer whose structure is so far only partially solved. Cryo-electron microcopy images have allowed 9.6 Å resolution reconstitutions of the 3D structure of the channel (Serysheva et al. 2008) in different activation states (Samsó et al. 2009). More recently, portions of the molecule have been crystalized and their structure identified at atomic resolution (Tung et al. 2010; Yuchi et al. 2012).

Nevertheless, in the absence of precise localization, most of the functional domains that have been identified for RyR1 are mapped on the linear sequence of the protein (Fig. 9.2): E–C domains allowing interaction with the DHPR or regulatory proteins of the CRC such as calmodulin and FKBP12, modulatory regions allowing the interaction with kinases (PKC and PKA), ATP, and calcium ions. The major part (4,000 first amino acids) of the RyR1 sequence is cytosolic, and forms the so-called "feet" of the molecule observed in EM. A number of modulators of RyR1 have their binding domain in this huge cytosolic region: Ca^{2+}, ATP, Mg^{2+} as well as pharmacological agents (ryanodine, caffeine, halothane, dantrolene, or ruthenium red). It is also in the cytosolic part that RyR1 undergoes phosphorylation and nitrosylation, both modifications being involved in the regulation of the channel activity (Lanner et al. 2010). The last 1,000 amino acids, the C-terminal part of the protein, contain 6–8 transmembrane helices responsible for SR membrane anchoring and form the pore domain of the channel (Du et al. 2002). In its isolated form, the RyR1 channel is activated by Ca^{2+} or ryanodine at low

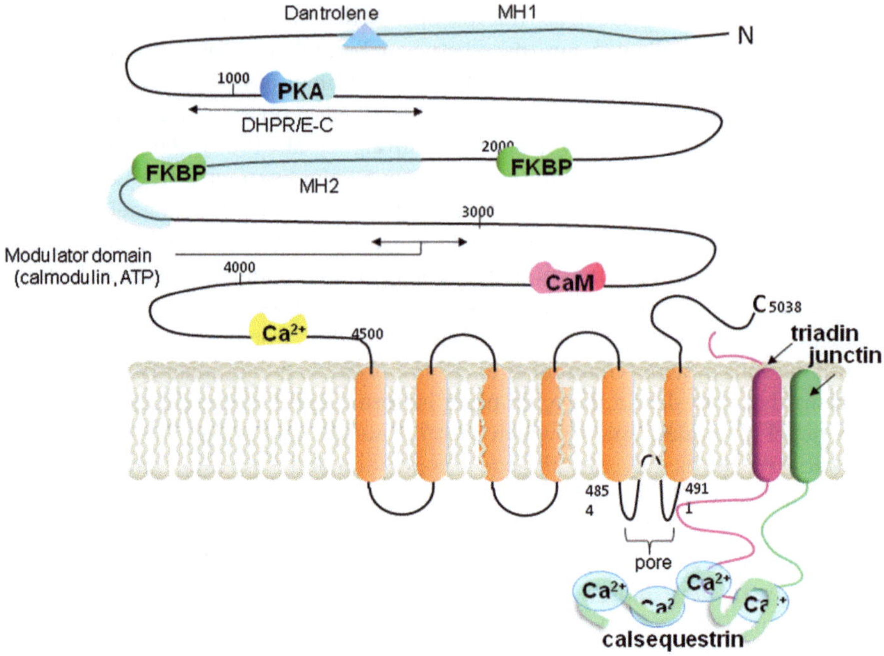

Fig. 9.2 Schematic representation of the structure of RyR1 showing the different functional domains on the *linear* structure

concentration (μM), by ATP, caffeine, or halothane. On the other hand, elevated concentrations (>100 μM) of Ca^{2+}, Mg^{2+}, and ryanodine inhibit its activity (Meissner 1994). Hence the sensibility of RyR1 to different ranges of calcium concentration allows the channel to be in a closed state at resting calcium levels (50–150 nM) and to be inactivated when cytosolic calcium concentration rise above the mM range, preventing a deleterious prolonged activation (Fill and Copello 2002).

9.2 Implication of RYR1 in Muscle Pathologies

Two classes of pathologies that differ in their presentation have been so far associated to the RYR1 gene: The first one is triggered by external stimuli and the second one appears at birth (Fig. 9.3). The first pathology ever associated to RyR1 was Malignant Hyperthermia (MH) in 1991 (Gillard et al. 1991). Mutations in RyR1 were further identified in association with Central Core Disease (CCD), a congenital myopathy of moderate severity, characterized by the presence of "cores", abnormalities in the muscle fiber revealed either by histology or electron microscopy studies (Magee et al. 1956). The description of severe forms of

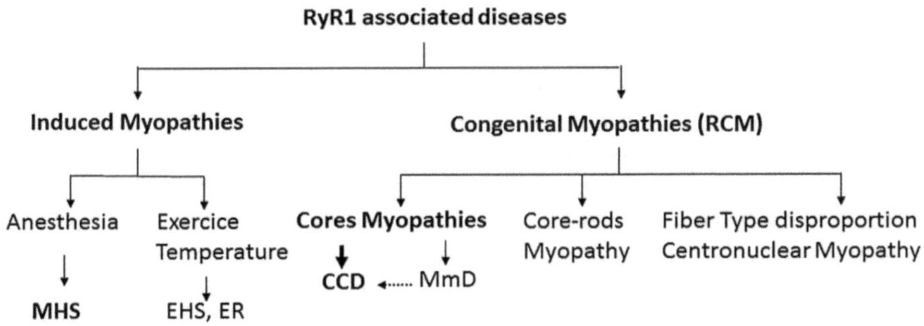

Fig. 9.3 Pathologies associated to RyR1 dysfunction

myopathies with cores and of other myopathies with histological abnormalities has then led to the notion that a whole class of myopathies, now called RyR1-Related Congenital Myopathies or RyR1-RCM, can be related to RyR1 dysfunction.

9.2.1 Induced Myopathies

9.2.1.1 Malignant Hyperthermia

Malignant Hyperthermia (OMIM:145600) is a pharmacogenetic disease of the skeletal muscle with a dominant mode of inheritance. The MH crisis is triggered by exposition to inhalational anesthetics, like halothane or isoflurane, and results in a hypermetabolic state of skeletal muscles. It is actually the clinical expression of an otherwise infra-clinic metabolic myopathy (Rosenberg et al. 2007). The exposition of susceptible individuals to the triggering agent leads to an increase in the myoplasmic calcium concentration followed by a generalized muscle contracture and by hyperthermia. A MH crisis will rapidly result in acidosis of both respiratory and metabolic origin, in a progressive degradation of muscle with a major rhabdomyolysis, together with hyperkalemia responsible for cardiac arrhythmia, hypercapnia and central hyperthermia. Hence, a MH crisis can be lethal in absence of the prompt administration of sodium dantrolene, a calcium channel blocker.

The exact prevalence of MH susceptibility is difficult to evaluate because patients do not present with symptoms apart from exposition to the triggering agent. Empirically, it has been estimated that each year 1 MH crisis arise among a population of 2 million people in Europe or North-America. This estimation probably underestimates the genetic MH susceptibility in the general population. Indeed, it has been shown that some patients undergo a MH crisis only at the second or third exposition to the triggering agent. Based on the genetic studies, it was calculated that the frequency of *RYR1* mutations responsible for MH could reach 1/2,000 in the general population (Monnier et al. 2002).

In vitro studies demonstrated that contraction of skeletal muscles fibers from MH susceptible patients in response to caffeine or halothane was enhanced compared to control. This observation was pivotal in the development of a biologic testing allowing the identification of individuals at risk for MH, the in vitro contracture test (IVCT). Two protocols are prevailing today, one used in North America and the other in Europe, and although they are based on the very same principle they differ in their experimental details and interpretation. In Europe, the contraction level of muscle fibers bathed in buffers containing increasing doses of caffeine or halothane allows to define phenotypes as normal (MHN), equivocal (MHE, when fibers contract in response to low doses of only one of the triggering agent), or susceptible (MHS, when muscle contracts in response to both agents). Noticeably, patients tested as MHE are regarded as clinically susceptible to MH. In the North American protocol, individuals are considered as susceptible to MH whenever the fibers contract with caffeine or halothane. Positive testing in IVCT is still the gold standard for MH diagnosis (EMHG 1984), but despite a 99 % sensitivity and a 94 % specificity, this test can still lead to false positive results. In Japan, a muscle biopsy is also required to test MH susceptibility, but the protocol uses the measure of calcium induced calcium release on skinned fibers (Ibarra et al. 2006). Independently of the protocol used to test MH susceptibility, genetic analysis is mandatory to identify the mutation thus allowing less invasive testing of relatives.

The *RYR1* gene was linked to MH in 1990 (MacLennan et al. 1990; McCarthy et al. 1990), and the first mutation described 1 year later (Gillard et al. 1991). To date, more than 160 mutations have been associated to MH (Monnier et al. 2005; Robinson et al. 2006; Sambuughin et al. 2005). The mutations are clustered in three regions of the gene: MH1, corresponding to the N-terminal part of the protein between amino acids 35 and 614; MH2, a central zone between amino acids 2,163 and 2,458; and MH3, localized in the C-terminal transmembrane domain (Fig. 9.2). In the European population, exhaustive sequencing showed that 60–75 % of the MH susceptible families have a *RYR1* mutation (Robinson et al. 2006), suggesting the involvement of other genes. MH genetic heterogeneity was confirmed when the gene encoding the α1-subunit of the DHPR, was also linked to the pathology, with different mutations being described in *CACNA1S* (Monnier et al. 1997; Pirone et al. 2011; Stewart et al. 2001; Toppin et al. 2010). However, genetic detection of a *RYR1* mutation has been accepted as an established cause of MH only for 30 well-characterized *RYR1* variants (http://www.emhg.org), and so far only the identification in a patient of one of these 30 mutations can replace a positive IVCT test as a formal MH diagnosis (Urwyler et al. 2001).

RYR1 mutations associated to MH are inherited on a dominant mode and are for most of them missense mutations or small DNA deletion leading to in-frame amino acids deletions. Some mutations are highly recurrent and patients with homozygous RyR1 MH mutation have been characterized. Their phenotype was similar to heterozygous patients, except for elevated plasma creatine kinase levels (Monnier et al. 2002). Although MH is considered as a non-symptomatic syndrome without exposition to the triggering agent, clinical manifestations such as myalgias, cramps, and rhabdomyolysis have been reported among few MH susceptible patients, most of them after physical exercise (Wappler et al. 2001).

9.2.1.2 Exertional Heat Stroke

Exertional Heat Stroke (EHS) is characterized by hyperthermia, central neurologic disorders, and variable levels of rhabdomyolysis during, or after, an intense or prolonged physical exercise. EHS is considered as exercise-induced hyperthermia.

EHS should be distinguished from Heat Stroke (HS) arising from exposure to overheated environment and Exercise-linked Rhabdomyolysis (ER), which usually happens after an intense exercise inducing skeletal muscles damages resulting to mechanical or metabolic stress. Notably, ER can precede EHS.

MHS, EHS, and ER are characterized by abnormal metabolic states with enhanced oxidative stress and increased cytosolic calcium levels (Capacchione et al. 2009). A common genetic defect has been hypothesized to explain this hypermetabolic state. Accordingly, a few studies have shown that mutations in the *RYR1* gene are detected in ER or EHS patients (Tobin et al. 2001; Wappler et al. 2001), without a clear evidence of clustering (Fig. 9.4). Our laboratory has recently shown in collaboration with the medical health service of the French Army that 15 % of the patients presenting with an EHS phenotype had a positive IVCT (unpublished data). Moreover, 25 % of these EHS/MHS patients had a *RYR1* variation. Without further analysis of these *RYR1* gene variations, it is however difficult to demonstrate their role in the pathophysiology of EHS. It can still be concluded form this study that individuals with *RYR1* MH-related variations could be at risk of EHS during intensive physical exercise, which is in accordance with the usual recommendation made to MHS patients not to practice prolonged and intensive physical activities (Hopkins et al. 2007).

9.2.2 Congenital Myopathies

9.2.2.1 A Heterogeneous Family of Pathologies

Congenital myopathies (CM) are a group of muscle diseases with heterogeneity in clinical, genetic, and histological presentations whose discovery has been tightly correlated with the development of histology and electron microscopy in the 1960s. The estimated incidence of CM is 6/100,000, which would rank CM among the most frequent myopathies. CM are often associated with a neonatal hypotonia of variable severity and a non-progressive muscle weakness. A primary diaphragm dysfunction can be the major sign for severe dominant phenotypes. Molecular genetic studies have revealed several genes responsible for the various forms of CM: *ACTA1*, *NEB*, *RYR1*, *SEPN1*, *MTM1*, *DNM2* … A first attempt of CM classification has been performed based exclusively on the histological

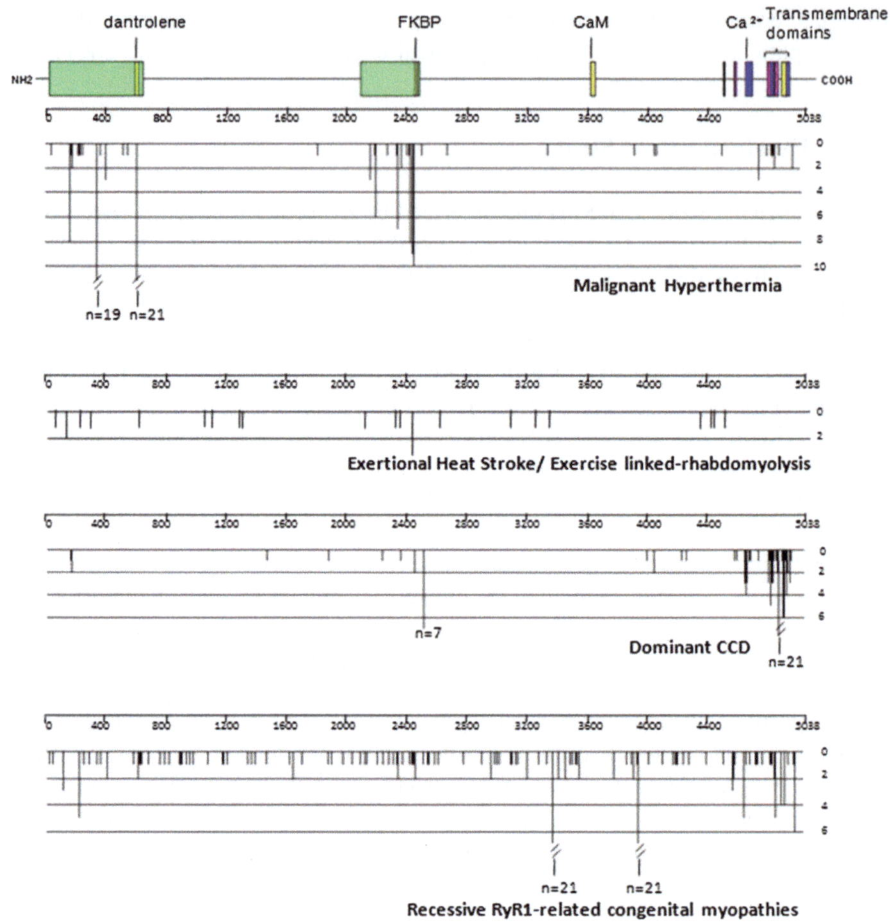

Fig. 9.4 Occurrence of RyR1 mutations in association with different pathologies. Each mutation is represented by a *vertical bar* on the sequence of RyR1, the length of which correspond to the number of independent probands identified with the same mutation, and is classified by disease

observations. However, the recent molecular studies have shown the existence of various clinical presentations with overlapping histological and genetic bases (Dubowitz and Swery 2007). The *RYR1* gene was initially associated only with dominant forms of Central Core Disease (CCD), but is now also linked to Multi-minicore Disease (MmD), Core Myopathies with rods, CentroNuclear Myopathy (CNM), and Congenital Fiber Type disproportion (CFTD) (Wilmshurst et al. 2010; Clarke et al. 2010). Altogether, the pathologies involving the *RYR1* gene, although variable in their clinical signs, should be classified as RyR1-Related Congenital Myopathies (RyR1-RCM).

9.2.2.2 Dominant Myopathies Linked to *RYR1*

Central Core Disease and Other Moderate Forms

In 1956, Shy and Magee described a new non-progressive myopathy, inherited on a dominant mode and affecting 5 patients from 2–65 years of age in three generations of the same family. The initial clinical description showed a congenital hypotonia, a delayed motor acquisition during childhood, and a slow evolution toward legs proximal muscles weakness (Shy and Magee 1956). Histological analysis revealed the presence in central portion of the fiber of numerous zones devoid of oxidative and phosphorylative activities (Fig. 9.5a), spanning along the entire longitudinal axis of the muscle. In 1958 the name "Central Core Disease"

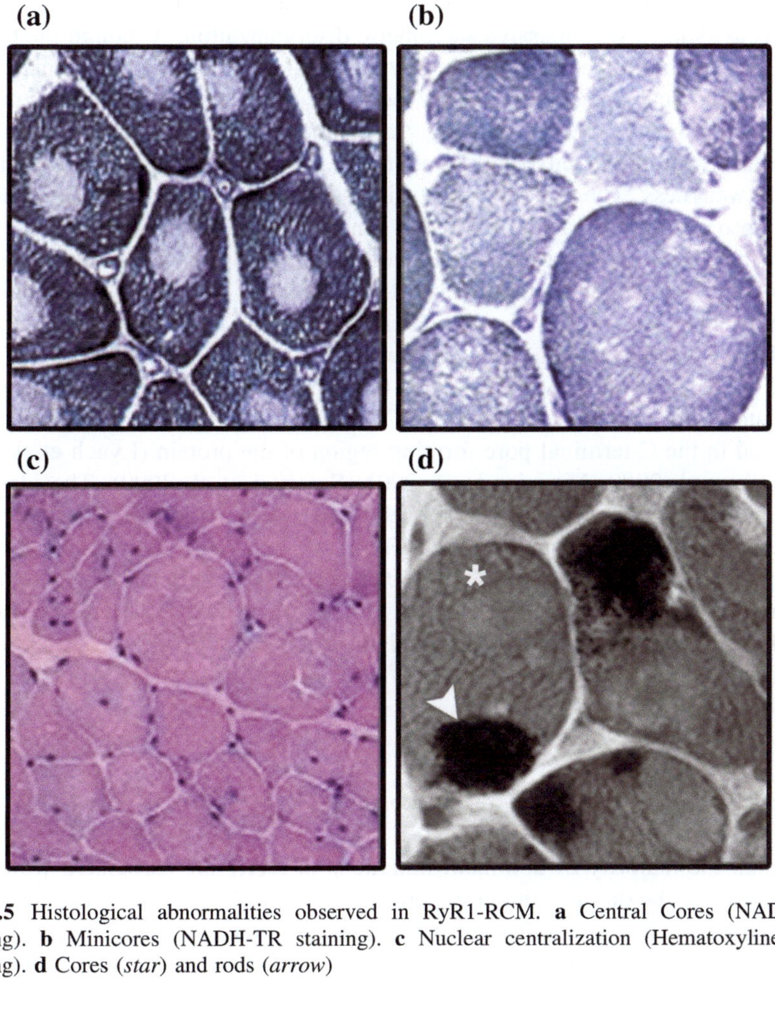

Fig. 9.5 Histological abnormalities observed in RyR1-RCM. **a** Central Cores (NADH-TR staining). **b** Minicores (NADH-TR staining). **c** Nuclear centralization (Hematoxyline-eosin staining). **d** Cores (*star*) and rods (*arrow*)

Table 9.1 Major histological presentations in RyR1–RCM

	Dominant RyR1–RCM ($n = 136$) (%)	Recessive RyR1–RCM ($n = 106$) (%)
Central cores	74,3	17
Cores + rods	5,9	9,4
Multiple cores	11,8	43,4
Centralized nucleus	0	12,3
CFTD	0	2,8
Atypical	8	15,1

was proposed for such a clinical presentation (Greenfield et al. 1958). Cores morphology observed in electron microscopy is in agreement with histological data, showing absence of mitochondria and of sarcoplasmic reticulum in these regions associated with massive sarcomeric disorganization. Although the Central Core Disease initially described by Shy and Magee is the main dominant form, other histological presentations have been described (cores and rods, multiple cores, minicores) (Table 9.1 and Fig. 9.5) (Romero et al. 2005), associated with moderate hypotonia in early childhood, delay in motor milestone acquisition, and muscle weakness. The walking ability is usually preserved and the evolution is slowly progressive, although orthopedic complications often give rise to scoliosis. In less than 10 % of the cases, a severe neonatal hypotonia is associated to delay in motor function acquisition, however the evolution is usually favorable. The involvement of the 19q13.1 locus, which contains the *RYR1* gene, has been proposed since 1990 by linkage analyses performed in families with dominant forms of CCD (Haan et al. 1990). To date, tenths of mutations in *RYR1* have been associated with dominants forms of CCD, with more than 85 % of them being localized in the C-terminal pore forming region of the protein (Lynch et al. 1999; Monnier et al. 2000; Monnier et al. 2001; Scacheri et al. 2000). The identified CCD mutations are until now exclusively missense mutations or in-frame micro deletions of one or few amino acid. From our experience on a cohort of 116 patients with moderate forms of CCD, 40 % of the 71 mutations we have found are recurrent or highly recurrent (Fig. 9.4). Mutation p.Arg4861His was for instance identified in 13 % of the patients, and mutation p.Arg2508Cys in 6 % of the patients. *De novo* mutations are very frequent (25 %) in these dominant CCD forms.

Severe Dominant Neonatal Cores Myopathy

Although the majority of dominant mutations of *RYR1* are associated with moderate forms of the disease, in our experience 15 % of them were associated with a severe neonatal hypotonia that led to respiratory failure and could be lethal in the first weeks. Half of these patients required respiratory assistance, 2/3 did not acquire walking and 1/3 showed signs of multiple arthrogyposis from birth. These severe forms of CCD were associated with de novo missense mutations localized in the C-terminal domain (Fig. 9.2), half of them being clustered in the last luminal loop forming the pore of the RyR1 channel.

Inter- and Intra familial Phenotypic Heterogeneity

Recurrent mutations have allowed the study of interfamilial phenotypic heterogeneity for CCD. Among the 83 dominant mutations our laboratory has identified, 5 are responsible for both severe and moderate clinical signs. For instance the p.Arg4861Cys mutation detected in 30 % of de novo severe forms has also been identified in 3 % of the moderate forms of CCD. Along the same line, the p.Ile4898Thr mutation associated to moderate phenotypes has been associated to a lethal de novo myopathy in twins (Hernandez-Lain et al. 2011). Such variability in phenotypic expression can be observed among members of the same family and in 8 % of the families we have explored, clinical signs worsened from one generation to the next. Only once was the aggravation associated to another variation in the *RYR1* gene (Monnier et al. 2008) showing that intrafamilial variability could arise from genetic background or variations in modifying genes.

9.2.2.3 Recessive Mutations Associated to RyR1

The description of recessive forms of congenital myopathies linked to RyR1 is more recent (Ferreiro et al. 2002a; Monnier et al. 2003; Romero et al. 2003). Congenital Multi-minicore Disease (MmD) of recessive mode of inheritance was defined on histological bases by the observation of multiple small regions devoid of oxidative activity, of only few sarcomers length, in both type I and II fibers (Ferreiro et al. 2000). The selenoprotein N gene (*SEPN1*) was first associated with 50 % of the severe so-called "classical" forms of MmD (Ferreiro et al. 2002b). However, *RYR1* mutations were further identified in families with MmD forms associated with external ophtamloplegia, as an indication of the genetic heterogeneity of MmD (Monnier et al. 2003).

Further molecular studies showed that the number of MmD patients harboring RyR1 mutations was more important than initially suspected (Klein et al. 2012; Monnier et al. 2008; Zhou et al. 2006). Recessive RyR1 mutations have been identified in 30 % of the severe MmD cases we have analyzed. Clinical signs of MmD comprise neonatal hypotonia, severe respiratory failure, sucking and swallowing difficulties, eye muscle involvement (ophtalmoplegia, ophtalmoparesis, ptosis), arthrogryposis, and skeletal deformations in 1/3 of the cases (scoliosis, cyphoscoliosis, hyperlordosis). The generalized muscle weakness leads to lethality in 30 % of the severe cases. The clinical signs associated to moderate recessive forms of MmD linked to RyR1 mutations are similar to those of moderate dominant forms of CCD, except for eye muscle involvement, usually not associated to CCD.

Histologically, the analysis of MmD can be confusing because of the large spectrum of signs, which can even be associated to other forms of congenital myopathies. First the cores can vary in size, number or morphology, depending on the muscle explored or the age of the patient. Cores presentation can vary for a same patient on biopsies performed at different ages. Cores can also be associated with rods, or nuclei internalization, that are the main sign for the Centronuclear Myopathy (CNM). Nevertheless, opposite to CNM, internalizations of nuclei are usually associated in MmD with myofibrilar disorganization. At last, nonspecific histological presentations have also been reported, with a few centralized nuclei, local myofibrilar disorganization, predominance or hypotrophy of type I fibers. In rare cases a fiber type disproportion has also been described (Fig. 9.5).

Whereas mutations in the *RYR1* gene leading to CCD are clustered in hot spots, mutations linked to MmD are spread all along the coding sequence (Fig. 9.4). Our laboratory has entirely sequenced the *RYR1* gene in 106 families with recessive RyR1-RCM. Seventy five percent of the mutations we have detected are missense or small duplication/deletion mutations, among which 8 % are also known to be responsible for MH when expressed at the heterozygous state. The remaining 25 % of the mutations are nonsense mutation or are mutations affecting mRNA or protein stability, leading to a quantitative defect of RyR1 in the affected muscle (Monnier et al. 2008). A compound heterozygous status with 2 associated missense mutations was found for 58 % of the patient with a recessive form of RyR1-RCM. In 37 % of these cases a missense mutation was associated with a mutation affecting the quantity of RyR1 protein (nonsense or splicing mutation). Two mutations affecting RyR1 quantities were associated with only 5 % of the cases, (Klein et al. 2012; Monnier et al. 2003; Monnier et al. 2008; Zhou et al. 2006). However, a RyR1-RCM with complete loss of RyR1 protein has never been described so far, in good agreement with studies performed on mice showing that deletion of the *RYR1* gene is lethal (Takeshima et al. 1994).

9.2.2.4 Complexity of RyR1-RCM Diagnosis

RyR1-RCM are among the most frequent congenital myopathies. Historically, RyR1-RCM were classified into clearly distinct entities: MH, which manifestation follows exposition to inhalational anesthetics, and CCD, characterized by muscle weakness and histologically diagnosed after analysis of muscle biopsies. The identification of core structures on muscle sections has remained a strong hint of a pathology linked to the *RYR1* gene, especially with the increasing number of recessive mutations involved in MmD. The presence of cores as a sign of Core Myopathy should nonetheless be taken with care. A study realized in our laboratory on a panel of 242 Malignant Hyperthermia patients without any muscle weakness has shown the presence of cores in 20 % of the biopsies (Monnier et al. 2005). These patients harbored a *RYR1* mutation, cores had been observed on muscle sections, but they did not show clinical signs of a myopathy. They should therefore be classified as "MHS patients with cores" rather than MH/CCD patients as it is sometimes stated in the literature. The "CCD/MH" phenotype should exclusively be attributed to patients that present with clear signs of both pathologies, on the histological and clinical point of view, which mean they have both a muscle weakness together with a positive susceptibility to MH.

With the increasing heterogeneity of clinical presentations that can be linked to *RYR1* mutations, histopathology analysis has to be backed by other techniques to refine the diagnosis. Interestingly, the skeletal muscles MRI profile from RyR1-RCM patients has been described as typical and could be used as a strong indication for investigations of the *RYR1* gene. The *gluteus maximus* from limb girdle, the *vastus lateralis, adductor magnus,* and *sartorius* from tight as well as the *soleus* from the leg are often altered, whereas the *rectus femoris, biceps femoris, gracilus, and gastrocnemius* are preserved (Jungbluth et al. 2004; Quijano-Roy et al. 2011). Genetic studies are nevertheless mandatory to show the involvement of *RYR1*. Although RyR1 mutations are frequent and spread all over the coding sequence, about 40 % of patients with well-characterized forms of CCD analyzed in our laboratory have no mutation in their entire RyR1 coding sequence, showing also a genetic heterogeneity of this congenital myopathy.

The ongoing introduction of massive parallel sequencing (or Next Generation Sequencing, NGS) in clinical diagnosis will most probably modify the strategies used to investigate heterogeneous muscle diseases. With such powerful sequencing capacity, it will be tempting to systematically sequence *RYR1* gene for any clinical presentation with minimal signs of RyR1-RCM. This approach will probably be useful but must be balanced with the complexity of the pathophysiological mechanisms so far revealed. With an expected increase in the number of detected RyR1 variants, it will soon be mandatory to have clear information regarding the functional effect for each of these variations in order to unambiguously confirm their role in the pathology: disease causing mutation, modifying factor or polymorphisms. The detailed investigation of pathophysiological mechanisms for RyR1-RCM is thus a key step in the diagnosis process of these myopathies.

9.3 Pathophysiological Mechanisms Associated to RyR1 Mutations

Several mechanisms have been proposed to explain the impact of missense RyR1 mutations (Betzenhauser and Marks 2010; Lanner et al. 2010; MacLennan and Zvaritch 2011), as a result of functional studies performed using different tools. A number of studies used cells produced from the patients (primary muscle cultures, myotubes produced from patient's fibroblasts transformed into muscle progenitors by MyoD expression, or immortalized lymphocytes). Other studies used expression of mutant RyR1 in non-muscle cell lines (HEK cells) or in mouse muscle cells devoid of endogenous RyR1 (primary culture from RyR1 KO mouse muscles). Some mutations have been studied with the different expression systems, and the proposed mechanisms are different, indicating that each expression system has its limitation and drawback. Therefore the actual pathophysiological mechanisms resulting from a precise mutation are probably more complex than the simplified and schematic mechanisms described thereafter.

9.3.1 RyR1 Gain of Function

Some of the studies led to the conclusion that mutant forms of RyR1 have a gain of function, i.e., the channel is releasing an increased amount of calcium in some conditions. This gain of function could have two expressions. First it can result in a hypersensitivity of the channel to pharmacologic agents (halothane, caffeine,...), which is characteristic of the so-called "MH mutation." This is typically what is assayed when performing IVCT, and correspond to the shift in the dose response curve representing the amplitude of calcium released by increasing concentrations of caffeine or chlorocresol. It has been proposed that these MH mutations lower the threshold of RyR1 activation by luminal calcium in the presence of triggering agents, thus inducing massive and physiologically mis-regulated calcium efflux (Tong et al. 1997). This calcium overload would then be responsible for the generalized muscle contraction and the hypermetabolic state typical for the MH crisis. The second expression of a RyR1 gain of function is a calcium leak in physiological situations where the channel should be closed, at very low (<μM) or very high (>mM) calcium concentration. If this calcium leak is not compensated by an increased calcium flux toward the SR, it will result in a reduction in the amount of calcium stored in the SR. As a consequence the amount of calcium released upon stimulation will be reduced, and so will be the muscle contraction (Tong et al. 1999). Depending on the expression system used to assay such mutations (muscle cells equipped with all the calcium channels and pumps, or non-muscle cells expressing only few of them) and on the type of expression performed (transient expression or long term, chronic expression) a compensation of the calcium leak can be in place or not. This compensation can be performed for example by overexpression of SERCA pumps or of the membrane calcium channel involved in SOCE (Store Operated Calcium Entry), both mechanisms leading to an increase in the calcium flux toward the SR. But the compensatory overexpression of proteins will only occur upon chronic alteration. As a consequence, discordant results have been reported for the same mutation. For example, an increase in the basal cytosolic calcium concentration has been observed in some expression systems but not in other for the p.R2435L mutation (Dirksen et al. 2004; Ghassemi et al. 2009). The "physiological" calcium leak does not always lead to a detectable increase in the cytosolic calcium concentration at rest, but it is always translated in a reduction in the amount of calcium stored, explaining why mutations resulting in a gain of function in RyR1 at rest could be also associated with congenital myopathies (Ghassemi et al. 2009).

The molecular mechanisms leading to RyR1 gain of function and to calcium leak has been related to redox state of the muscle fiber. For many years, RyR1 has been known as sensitive to oxidizing agents (Eu et al. 2000), and has been proposed as a redox sensor in the skeletal muscle cell (Hidalgo et al. 2005). The amino acids involved in this sensibility have been identified (Aracena-Parks et al. 2006), and it has been shown that oxidative modification like S-nitrosylation of RyR1 results in a reduction in its association to FKBP12 (Aracena et al. 2005; Zissimopoulos et al. 2007). FKBP12 stabilizes RyR1 closed state (Brillantes et al. 1994), and dissociation of FKBP12 from RyR1 leads to a leaky calcium channel. This calcium leak increases the production of oxygen or nitrogen reactive species, which in turn increase the oxidative modifications of RyR1 and the calcium leak, in a deleterious circle. This mechanism has been observed in association with other pathologies such as Duchenne Muscular dystrophy (Bellinger et al. 2009) and aging (Anderson et al. 2011), where it amplified the muscle weakness but it has also been observed as a direct consequence of a RyR1 mutation, for example the p.Y522S mutation (Durham et al. 2008).

9.3.2 RyR1 Loss of Function

A second mechanism has been described to explain the effect of missense RyR1 mutation associated with congenital myopathies. An alteration of the E–C coupling, i.e., the coupling between the DHPR and RyR1, has been proposed to be at the origin of the pathology, and has been named "E–C uncoupling" (Dirksen et al. 2002). The observed effect of different mutations showed that membrane depolarization induced a reduction in calcium release although resting calcium levels were normal, excluding the presence of a leaky RyR1. The p.I4898T mutant form of RyR1 is for instance not able to mediate calcium efflux when stimulated by the DHPR (the coupling mechanism called "orthograde"). However the presence of the p.I4898T mutation does not alter the regulation of DHPR by RyR1, also called "retrograde coupling," a mechanism that requires the sole presence of RyR1 channel in front of the DHPR, showing that the RyR1 tetramer is correctly targeted to the triad (Avila et al. 2001). More than an "E–C uncoupling," this mechanism is in fact an alteration of the permeability of RyR to calcium, which has now been clearly demonstrated in a mouse model reproducing this p.I4898T mutation (Loy et al. 2011). The consequence of this kind of mutation, lethal at the homozygous state in mouse, is a muscle weakness, which most probably correlates to inefficient coupling between the excitation stimulus mediated by DHPR and RyR1 channel opening. A physical "E–C uncoupling," namely the abolition of the interaction between DHPR and RyR1, has never been observed up to now, either because it too damaging for the cells, or because it relies on multiple contact points, involving different regions of RyR1, and could not be altered by a single point mutation.

9.3.3 Reduction in RyR1 Amount

The description of patients with mutations affecting RyR1 expression level has led to the most recent pathophysiological mechanism hypothesized (Monnier et al. 2008; Zhou et al. 2006). This mechanism is mostly associated to recessive forms, i.e., with two simultaneous RyR1 mutations. In this case, the presence of a "null" mutation (resulting in the complete absence of protein production from one allele) leads to the expression of the second RyR1 variation in a hemizygous context. The severity of the pathology in such a case depends on the nature of the expressed second mutation as well as on the residual expression level of the hypomorphic allele. In these patients, all the RyR1 expressed at the protein level could be mutated, as there is only one allele able to produce a protein. Therefore there is not only a reduction in the amount of channel expressed in the SR membrane, but also the expression of mutant RyR1 monomer in high proportion. This elevated percentage of mutant monomer can be very damaging, although its impact would have been milder if expressed together with a normal RyR1 allele. The overall reduction in RyR1 expression level could also have an impact on the stoichiometry of proteins of the CRC and its function. If the expression level of regulatory proteins (triadin, junctin, calmodulin…) is unchanged, reduced amounts of RyR1 could lead to an excess of these free regulatory proteins, which could have a negative effect on CRC function.

The tetrameric organization of RyR1 and its physical interaction with DHPR add complexity to the interpretation of the effect of mutations. The balance of combinations between normal or mutant monomers of RyR1 arising from dominant forms of myopathies, or between mutant monomers in recessive forms, is a factor that may influence the apparent discrepancy of clinical outcome and histological presentation for identical genotypes (MacLennan and Zvaritch 2011).

9.4 Conclusion

In the past decade, the class of pathologies involving RyR1 has expanded beyond its "classical" manifestation, MH, and CCD. While the complete analysis of RyR1 sequence became more accurate despite the complexity of the gene, it appeared that variations present on almost every exon of *RYR1* could be associated to some form of myopathy. The heterogeneity of RyR1-related myopathies described so far may increase when routine massive parallel sequencing will be available in diagnosis, pointing to the necessity of understanding the intimate mechanisms leading to mis-regulation of calcium release in these pathologies. So far, the mouse models expressing some of the RyR1 mutations associated to human pathologies have confirmed important observations: morphologic alterations due to RyR1 mutation can have various forms, and structural abnormalities linked to different congenital myopathies (CCD, MmD, rods myopathies) could share the same etiology (MacLennan and Zvaritch 2011). These models, as well as future mice lines, will be pivotal in deciphering RyR1 pathophysiological mechanisms, together with cellular models reproducing the function of RyR1 in the calcium release complex. Although general mechanisms leading to RyR1 dysfunction seem to be understood, a detailed analysis of cellular and physiological models are still required to decipher the precise pathophysiology of RyR1-RCM. Several questions are in particular still unsolved such as: why is the characteristic feature of RyR1-RCM a depletion of mitochondria from focal zones of the muscle cells? Why are the myofibrillar apparatus of these cells also disorganized? How do the cellular default progress and why is the pathology usually not progressive?

Finally, it is tempting to speculate that if the *RYR1* gene is associated to so many forms of muscle dysfunction due to calcium efflux dysregulation, it could also be associated to pathologies based on neuronal dysfunction. RyR1 is expressed in some neurons and recent data showed that a mutant form of RyR1 has indeed an impact on neuronal function (De Crescenzo et al. 2012). The future exploration of neuronal disorders with next generation sequencing technologies will probably provide new directions to explore in the pathophysiology of RyR1.

References

Andersson DC, Betzenhauser MJ, Reiken S, Meli AC, Umanskaya A, Xie W, Shiomi T, Zalk R, Lacampagne A, Marks AR (2011) Ryanodine receptor oxidation causes intracellular calcium leak and muscle weakness in aging. Cell Metab 14:196–207

Aracena P, Tang W, Hamilton SL, Hidalgo C (2005) Effects of S-glutathionylation and S-nitrosylation on calmodulin binding to triads and FKBP12 binding to type 1 calcium release channels. Antioxid Redox Signal 7:870–881

Aracena-Parks P, Goonasekera SA, Gilman CP, Dirksen RT, Hidalgo C, Hamilton SL (2006) Identification of cysteines involved in S-nitrosylation, S-glutathionylation, and oxidation to disulfides in ryanodine receptor type 1. J Biol Chem 281:40354–40368

Avila G, O'Brien JJ, Dirksen RT (2001) Excitation-contraction uncoupling by a human central core disease mutation in the ryanodine receptor. Proc Natl Acad Sci USA 98:4215–4220

Bellinger AM, Reiken S, Carlson C, Mongillo M, Liu X, Rothman L, Matecki S, Lacampagne A, Marks AR (2009) Hypernitrosylated ryanodine receptor calcium release channels are leaky in dystrophic muscle. Nat Med 15:325–330

Betzenhauser MJ, Marks AR (2010) Ryanodine receptor channelopathies. Pflugers Arch 460:467–480

Bracci L, Vukcevic M, Spagnoli G, Ducreux S, Zorzato F, Treves S (2007) Ca2+ signaling through ryanodine receptor 1 enhances maturation and activation of human dendritic cells. J Cell Sci 120:2232–2240

Brillantes AB, Ondrias K, Scott A, Kobrinsky E, Ondriasová E, Moschella MC, Jayaraman T, Landers M, Ehrlich BE, Marks AR (1994) Stabilization of calcium release channel (ryanodine receptor) function by FK506-bindingprotein. Cell 77:513–523

Capacchione JF, Muldoon SM (2009) The relationship between exertional heat illness, exertional rhabdomyolysis, and malignant hyperthermia. Anesth Analg 109:1065–1069

Clarke NF, Waddell LB, Cooper ST, Perry M, Smith RL, Kornberg AJ, Muntoni F, Lillis S, Straub V, Bushby K, Guglieri M, King MD, Farrell MA, Marty I, Lunardi J, Monnier N, North KN (2010) Recessive mutations in *RYR1* are a common cause of congenital fiber type disproportion. Hum Mutat 31:E1544–E1550

De Crescenzo V, Fogarty KE, Lefkowitz JJ, Bellve KD, Zvaritch E, MacLennan DH, Walsh JV Jr (2012) Type 1 ryanodine receptor knock-in mutation causing central core disease of skeletal muscle also displays a neuronal phenotype. Proc Natl Acad Sci USA 109:610–615

Dirksen RT, Avila G (2002) Altered ryanodine receptor function in central core disease: leaky or uncoupled Ca(2+) release channels? Trends Cardiovasc Med 12:189–197

Dirksen RT, Avila G (2004) Distinct effects on Ca2+ handling caused by malignant hyperthermia and central core disease mutations in RyR1. Biophys J 87:3193–3204

Du GG, Sandhu B, Khanna VK, Guo XH, MacLennan DH (2002) Topology of the Ca2+ release channel of skeletal muscle sarcoplasmic reticulum (RyR1). Proc Natl Acad Sci USA 9926:16725–16730

Dubowitz V, Sewry C (2007) Mucle biopsy, a pratical approach. Saunders Elsevier, St.Louis, pp 407–443

Durham WJ, Aracena-Parks P, Long C, Rossi AE, Goonasekera SA, Boncompagni S, Galvan DL, Gilman CP, Baker MR, Shirokova N, Protasi F, Dirksen R, Hamilton SL (2008) RyR1 S-nitrosylation underlies environmental heat stroke and sudden death in Y522S RyR1 knockin mice. Cell 133:53–65

Eu JP, Sun J, Xu L, Stamler JS, Meissner G (2000) The skeletal muscle calcium release channel: coupled O2 sensor and NO signaling functions. Cell 102:499–509

European Malignant Hyperpyrexia Group (1984) Br J Anaesth 56:1267–1269

Ferreiro A, Estournet B, Chateau D, Romero NB, Laroche C, Odent S, Toutain A, Cabello A, Fontan D, dos Santos HG, Haenggeli CA, Bertini E, Urtizberea JA, Guicheney P, Fardeau M (2000) Multi-minicore disease–searching for boundaries: phenotype analysis of 38 cases. Ann Neurol 48:745–757

Ferreiro A, Monnier N, Romero NB, Leroy JP, Bönnemann C, Haenggeli CA, Straub V, Voss WD, Nivoche Y, Jungbluth H, Lemainque A, Voit T, Lunardi J, Fardeau M, Guicheney P (2002a) A recessive form of central core disease, transiently presenting as multi-minicore disease, is associated with a homozygous mutation in the ryanodine receptor type 1 gene. Ann Neurol 51:750–759

Ferreiro A, Quijano-Roy S, Pichereau C, Moghadaszadeh B, Goemans N, Bönnemann C, Jungbluth H, Straub V, Villanova M, Leroy JP, Romero NB, Martin JJ, Muntoni F, Voit T, Estournet B, Richard P, Fardeau M, Guicheney P (2002b) Mutations of the selenoprotein N gene, which is implicated in rigid spine muscular dystrophy, cause the classical phenotype of multiminicore disease: reassessing the nosology of early-onset myopathies. Am J Hum Genet 71:739–749

Fill M, Copello JA (2002) Ryanodine receptor calcium release channels. Physiol Rev 82:893–922

Franck JP, Morrissette J, Keen JE, Londraville RL, Beamsley M, Block BA (1998) Cloning and characterization of fiber type-specific ryanodine receptor isoforms in skeletal muscles of fish. Am J Physiol 275:C401–C415

Franzini-Armstrong C, Nunzi G (1983) Junctional feet and particles in the triads of a fast-twitch muscle fibre. J Muscle Res Cell Motil 4:233–252

Ghassemi F, Vukcevic M, Xu L, Zhou H, Meissner G, Muntoni F, Jungbluth H, Zorzato F, Treves S (2009) A recessive ryanodine receptor 1 mutation in a CCD patient increases channel activity. Cell Calcium 45:192–197

Giannini G, Conti A, Mammarella S, Scrobogna M, Sorrentino V (1995) The ryanodine receptor/calcium channel genes are widely and differentially expressed in murine brain and peripheral tissues. J Cell Biol 128:893–904

Gillard EF, Otsu K, Fujii J, Khanna VK, de Leon S, Derdemezi J, Britt BA, Duff CL, Worton RG, MacLennan DH (1991) A substitution of cysteine for arginine 614 in the ryanodine receptor is potentially causative of human malignant hyperthermia. Genomics 11:751–755

Greenfield JG, Cornmant T, Shy GM (1958) The prognostic value of the muscle biopsy in the floppy infant. Brain 81:461–484

Haan EA, Freemantle CJ, McCure JA, Friend KL, Mulley JC (1990) Assignment of the gene for central core disease to chromosome 19. Hum Genet 86:187–190

Hasan G, Rosbash M (1992) Drosophila homologs of two mammalian intracellular Ca(2+)-release channels: identification and expression patterns of the inositol 1,4,5-triphosphate and the ryanodine receptor genes. Development 116:967–975

Hernandez-Lain A, Husson I, Monnier N, Farnoux C, Brochier G, Lacène E, Beuvin M, Viou M, Manéré L, Claeys KG, Fardeau M, Lunardi J, Voit T, Romero NB (2011) De novo *RYR1* heterozygous mutation (I4898T) causing lethal core-rod myopathy in twins. Eur J Med Genet 54:29–33

Hidalgo C, Donoso P, Carrasco MA (2005) The ryanodine receptors Ca2+ release channels: cellular redox sensors? IUBMB Life 57:315–322

Hirata H, Watanabe T, Hatakeyama J, Sprague SM, Saint-Amant L, Nagashima A, Cui WW, Zhou W, Kuwada JY (2007) Zebrafish relatively relaxed mutants have a ryanodine receptor defect, show slow swimming and provide a model of multi-minicore disease. Development 134:2771–2781

Hopkins PM (2007) Is there a link between malignant hyperthermia and exertional heat illness? Br J Sports Med 41:283–284, discussion 284

Ibarra MCA, Wu S, Murayama K, Minami N, Ichihara Y, Kikuchi H, Noguchi S, Hayashi YK, Ochiai R, Nishino I (2006) Malignant hyperthermia in Japan: mutation screening of the entire ryanodine receptor type 1 gene coding region by direct sequencing. Anesthesiology 104:1146–1154

Inui M, Saito A, Fleischer S (1987) Purification of the ryanodine receptor and identity with feet structures of junctional terminal cisternae of sarcoplasmic reticulum from fast skeletal muscle. J Biol Chem 262:1740–1747

Jungbluth H, Davis MR, Müller C, Counsell S, Allsop J, Chattopadhyay A, Messina S, Mercuri E, Laing NG, Sewry CA, Bydder G, Muntoni F (2004) Magnetic resonance imaging of muscle in congenital myopathies associated with *RYR1* mutations. Neuromuscul Disord 14:785–790

Klein A, Lillis S, Munteanu I, Scoto M, Zhou H, Quinlivan R, Straub V, Manzur AY, Roper H, Jeannet PY, Rakowicz W, Jones DH, Jensen UB, Wraige E, Trump N, Schara U, Lochmuller H, Sarkozy A, Kingston H, Norwood F, Damian M, Kirschner J, Longman C, Roberts M, Auer-Grumbach M, Hughes I, Bushby K, Sewry C, Robb S, Abbs S, Jungbluth H, Muntoni F (2012) Clinical and genetic findings in a large cohort of patients with ryanodine receptor 1 gene-associated myopathies. Hum Mutat 33:981–988

Lanner JT, Georgiou DK, Joshi AD, Hamilton SL (2010) Ryanodine receptors: structure, expression, molecular details, and function in calcium release. Cold Spring Harb Perspect Biol 2:a003996

Loy RE, Orynbayev M, Xu L, Andronache Z, Apostol S, Zvaritch E, MacLennan DH, Meissner G, Melzer W, Dirksen RT (2011) Muscle weakness in Ryr1I4895T/WT knock-in mice as a result of reduced ryanodine receptor Ca2+ ion permeation and release from the sarcoplasmic reticulum. J Gen Physiol 137:43–57

Lynch PJ, Tong J, Lehane M, Mallet A, Giblin L, Heffron JJ, Vaughan P, Zafra G, MacLennan DH, McCarthy TV (1999) A mutation in the transmembrane/luminal domain of the ryanodine receptor is associated with abnormal Ca2+ release channel function and severe central core disease. Proc Natl Acad Sci USA 96:4164–4169

MacLennan DH, Duff C, Zorzato F, Fujii J, Phillips M, Korneluk RG, Frodis W, Britt BA, Worton RG (1990) Ryanodine receptor gene is a candidate for predisposition to malignant hyperthermia. Nature 343:559–561

MacLennan DH, Zvaritch E (2011) Mechanistic models for muscle diseases and disorders originating in the sarcoplasmic reticulum. Biochim Biophys Acta 1813:948–964

Magee KR, Shy GM (1956) A new congenital non-progressive myopathy. Brain 79:610-621

Maryon EB, Coronado R, Anderson P (1996) Unc-68 encodes a ryanodine receptor involved in regulating C. elegans body-wall muscle contraction. J Cell Biol 134:885–893

McCarthy TV, Healy JM, Heffron JJ, Lehane M, Deufel T, Lehmann-Horn F, Farrall M, Johnson K (1990) Localization of the malignant hyperthermia susceptibility locus to human chromosome 19q12-13.2. Nature 343:562–564

Meissner G (1994) Ryanodine receptor/Ca2+ release channels and their regulation by endogenous effectors. Annu Rev Physiol 56:485–508

Monnier N, Procaccio V, Stieglitz P, Lunardi J (1997) Malignant-hyperthermia susceptibility is associated with a mutation of the alpha 1-subunit of the human dihydropyridine-sensitive L-type voltage-dependent calcium-channel receptor in skeletal muscle. Am J Hum Genet 60:1316–1325

Monnier N, Romero NB, Lerale J, Nivoche Y, Qi D, MacLennan DH, Fardeau M, Lunardi J (2000) An autosomal dominant congenital myopathy with cores and rods is associated with a neomutation in the *RYR1* gene encoding the skeletal muscle ryanodine receptor. Hum Mol Genet 9:2599–2608

Monnier N, Romero NB, Lerale J, Landrieu P, Nivoche Y, Fardeau M, Lunardi J (2001) Familial and sporadic forms of central core disease are associated with mutations in the C-terminal domain of the skeletal muscle ryanodine receptor. Hum Mol Genet 10:2581–2592

Monnier N, Krivosic-Horber R, Payen JF, Kozak-Ribbens G, Nivoche Y, Adnet P, Reyford H, Lunardi J (2002) Presence of two different genetic traits in malignant hyperthermia families: implication for genetic analysis, diagnosis, and incidence of malignant hyperthermia susceptibility. Anesthesiology 97:1067–1074

Monnier N, Ferreiro A, Marty I, Labarre-Vila A, Mezin P, Lunardi J (2003) A homozygous splicing mutation causing a depletion of skeletal muscle *RYR1* is associated with multi-minicore disease congenital myopathy with ophthalmoplegia. Hum Mol Genet 12:1171–1178

Monnier N, Kozak-Ribbens G, Krivosic-Horber R, Nivoche Y, Qi D, Kraev N, Loke J, Sharma P, Tegazzin V, Figarella-Branger D, Roméro N, Mezin P, Bendahan D, Payen JF, Depret T, Maclennan DH, Lunardi J (2005) Correlations between genotype and pharmacological, histological, functional, and clinical phenotypes in malignant hyperthermia susceptibility. Hum Mutat 26:413–425

Monnier N, Marty I, Faure J, Castiglioni C, Desnuelle C, Sacconi S, Estournet B, Ferreiro A, Romero N, Laquerriere A, Lazaro L, Martin JJ, Morava E, Rossi A, Van der Kooi A, de Visser M, Verschuuren C, Lunardi J (2008) Null mutations causing depletion of the type 1 ryanodine receptor (*RYR1*) are commonly associated with recessive structural congenital myopathies with cores. Hum Mutat 29:670–678

Pirone A, Schredelseker J, Tuluc P, Gravino E, Fortunato G, Flucher BE, Carsana A, Salvatore F, Grabner M (2010) Identification and functional characterization of malignant hyperthermia mutation T1354S in the outer pore of the Cavalpha1S-subunit. Am J Physiol Cell Physiol 299:C1345–C1354

Quijano-Roy S, Carlier RY, Fischer D (2011) Muscle imaging in congenital myopathies. Semin Pediatr Neurol 18:221–229

Robinson R, Carpenter D, Shaw MA, Halsall J, Hopkins P (2006) Mutations in *RYR1* in malignant hyperthermia and central core disease. Hum Mutat 27:977–989

Romero NB, Monnier N, Viollet L, Cortey A, Chevallay M, Leroy JP, Lunardi J, Fardeau M (2003) Dominant and recessive central core disease associated with *RYR1* mutations and fetal akinesia. Brain 126:2341–2349

Romero NB, Herasse M, Monnier N, Leroy JP, Fischer D, Ferreiro A, Viollet L, Eymard B, Laforêt P, Monges S, Lubieniecki F, Taratuto AL, Guicheney P, Lunardi J, Fardeau M (2005) Clinical and histopathological aspects of central core disease associated and non-associated with *RYR1* locus. Acta Myol 24:70–73

Rosenberg H, Davis M, James D, Pollock N, Stowell K (2007) Malignant hyperthermia. Orphanet J Rare Dis 2:21

Sambuughin N, Holley H, Muldoon S, Brandom BW, de Bantel AM, Tobin JR, Nelson TE, Goldfarb LG (2005) Screening of the entire ryanodine receptor type 1 coding region for sequence variants associated with malignant hyperthermia susceptibility in the north american population. Anesthesiology 102:515–521

Samsó M, Feng W, Pessah IN, Allen PD (2009) Coordinated movement of cytoplasmic and transmembrane domains of RyR1 upon gating. PLoS Biol 7:e85

Scacheri PC, Hoffman EP, Fratkin JD, Semino-Mora C, Senchak A, Davis MR, Laing NG, Vedanarayanan V, Subramony SH (2000) A novel ryanodine receptor gene mutation causing both cores and rods in congenital myopathy. Neurology 55:1689–1696

Serysheva II, Ludtke SJ, Baker ML, Cong Y, Topf M, Eramian D, Sali A, Hamilton SL, Chiu W (2008) Subnanometer-resolution electron cryomicroscopy-based domain models for the cytoplasmic region of skeletal muscle RyR channel. Proc Natl Acad Sci U S A 105:9610–9615

Shy GM, Magee KR (1956) A new congenital non-progressive myopathy. Brain 79:610–621

Stewart SL, Hogan K, Rosenberg H, Fletcher JE (2001) Identification of the Arg1086His mutation in the alpha subunit of the voltage-dependent calcium channel (CACNA1S) in a North American family with malignant hyperthermia. Clin Genet 59:178–184

Takeshima H, Iino M, Takekura H, Nishi M, Kuno J, Minowa O, Takano H, Noda T (1994) Excitation-contraction uncoupling and muscular degeneration in mice lacking functional skeletal muscle ryanodine-receptor gene. Nature 369:556–559

Tobin JR, Jason DR, Challa VR, Nelson TE, Sambuughin N (2001) Malignant hyperthermia and apparent heat stroke. JAMA 286:168–169

Tong J, Oyamada H, Demaurex N, Grinstein S, McCarthy TV, MacLennan DH (1997) Caffeine and halothane sensitivity of intracellular Ca2+ release is altered by 15 calcium release channel (ryanodine receptor) mutations associated with malignant hyperthermia and/or central core disease. J Biol Chem 272:26332–26339

Tong J, McCarthy TV, MacLennan DH (1999) Measurement of resting cytosolic Ca2+ concentrations and Ca2+ store size in HEK-293 cells transfected with malignant hyperthermia or central core disease mutant Ca2+ release channels. J Biol Chem 274:693–702

Toppin PJ, Chandy TT, Ghanekar A, Kraeva N, Beattie WS, Riazi S (2010) A report of fulminant malignant hyperthermia in a patient with a novel mutation of the CACNA1S gene. Can J Anaesth 57:689-693

Tung CC, Lobo PA, Kimlicka L, Van Petegem F (2010) The amino-terminal disease hotspot of ryanodine receptors forms a cytoplasmic vestibule. Nature 468:585-588

Uemura Y, Liu TY, Narita Y, Suzuki M, Ohshima S, Mizukami S, Ichihara Y, Kikuchi H, Matsushita S (2007) Identification of functional type 1 ryanodine receptors in human dendritic cells. Biochem Biophys Res Commun 362:510-515

Urwyler A, Deufel T, McCarthy T, West S (2001) European Malignant Hyperthermia Group. Guidelines for molecular genetic detection of susceptibility to malignant hyperthermia. Br J Anaesth 86:283-287

Wappler F, Fiege M, Steinfath M, Agarwal K, Scholz J, Singh S, Matschke J, Schulte Am Esch J (2001) Evidence for susceptibility to malignant hyperthermia in patients with exercise-induced rhabdomyolysis. Anesthesiology 94:95-100

Wilmshurst JM, Lillis S, Zhou H, Pillay K, Henderson H, Kress W, Müller CR, Ndondo A, Cloke V, Cullup T, Bertini E, Boennemann C, Straub V, Quinlivan R, Dowling JJ, Al-Sarraj S, Treves S, Abbs S, Manzur AY, Sewry CA, Muntoni F, Jungbluth H (2010) *RYR1* mutations are a common cause of congenital myopathies with central nuclei. Ann Neurol 68:717-726

Yuchi Z, Lau K, Van Petegem F (2012) Disease mutations in the ryanodine receptor central region: crystal structures of a phosphorylation hot spot domain. Structure 20:1201-1211

Zhou H, Yamaguchi N, Xu L, Wang Y, Sewry C, Jungbluth H, Zorzato F, Bertini E, Muntoni F, Meissner G, Treves S (2006) Characterization of recessive *RYR1* mutations in core myopathies. Hum Mol Genet 15:2791–2803

Zissimopoulos S, Docrat N, Lai FA (2007) Redox sensitivity of theryanodine receptor interaction with FK506-binding protein. J Biol Chem 282:6976–6983

Chapter 10
The Lambert-Eaton Myasthenic Syndrome

Paul W. Wirtz, Maarten J. Titulaer and Jan J. G. M. Verschuuren

Abstract The Lambert-Eaton myasthenic syndrome (LEMS) is an autoimmune disease of the neuromuscular junction. The main symptom in LEMS is proximal muscle weakness in the legs, which is the presenting symptom in most patients. Autonomic dysfunction is also part of the clinical spectrum. Neurological examination generally reveals low or absent tendon reflexes. The muscle weakness and autonomic dysfunction are caused by pathogenic autoantibodies directed against voltage-gated calcium channels (VGCC) present on the presynaptic nerve terminal, causing a decreased release of acetylcholine from nerve terminals. Half of the patients have an associated tumour, usually a small cell lung carcinoma, which expresses VGCC on its surface. If the diagnosis of LEMS is suspected from the distinctive clinical features, confirmation of the diagnosis depends on detection of VGCC autoantibodies or characteristic electrodiagnostic findings. Treatment of LEMS is based on drugs which act at the neuromuscular synapse by increasing either the release or the concentration of acetylcholine, drugs which suppress the immune response or treatment of the underlying tumour.

P. W. Wirtz (✉)
Department of Neurology, Haga Teaching Hospital, The Hague, The Netherlands
e-mail: p.wirtz@hagaziekenhuis.nl

M. J. Titulaer
Department of Neurology, Erasmus Medical Center, Rotterdam, The Netherlands

M. J. Titulaer
Department of Neurology, Hospital Clinic, The University of Barcelona, Institut d'Investigacions Biomèdiques August Pi i Sunyer [IDIBAPS], Barcelona, Spain

J. J. G. M. Verschuuren
Department of Neurology, Leiden University Medical Center, Leiden, The Netherlands

N. Weiss and A. Koschak (eds.), *Pathologies of Calcium Channels*,
DOI: 10.1007/978-3-642-40282-1_10,

10.1 Introduction

In 1953, Anderson and colleagues described a 47-year-old man with progressive muscle weakness of the legs, subsequently of the arms and neck, with transient diplopia and swallowing difficulties (Anderson et al. 1953). Neurological examination revealed proximal muscle weakness and diminished tendon reflexes. Prolonged apnea followed succinylcholine administration during general anaesthesia, but the symptoms improved on edrophonium and neostigmine. The patient improved dramatically after removal of a small cell lung carcinoma (SCLC). The authors concluded to "a possible relation between the carcinoma of the lung and myasthenia". This was the first description of a patient with the Lambert-Eaton myasthenic syndrome (LEMS).

Lambert, Eaton and Rooke noticed that the small group of patients with the clinical features of myasthenia and a lung carcinoma had electromyographical (EMG) evidence of a defect in neuromuscular transmission that was different from that of myasthenia gravis (MG) (Lambert et al. 1956). These findings were of low amplitude of the compound muscle action potential (CMAP), a further decline during repetitive nerve stimulation (RNS) at a rate of 1–10 Hz, but a marked facilitation, up to 10 times the initial amplitude, during stimulation at higher frequencies or during voluntary contraction. As the authors stated, these EMG findings resembled that of MG, "however the depression of the initial response and the phenomenon of facilitation were more marked than is usually observed in the latter disorder". Until today, these electrophysiological abnormalities are used in establishing the diagnosis of LEMS.

However, insights in other parts of the disease have changed considerably. Research in the 1980–1990s into the pathophysiology made LEMS to be the second disease, with MG, to meet all criteria for an antibody-mediated autoimmune disease. These criteria by Witebsky, Rose et al. from 1957 were derived from the Koch's postulates (Witebsky et al. 1957), and adapted in 1990 and 1993 by Drachman and Rose, respectively (Drachman 1990; Rose and Bona 1993). In the last decennium, new information became available about the HLA association, and important developments have been published about the clinical picture, the relation with respect to tumours (frequency, timing and risk assessment) and about treatment.

10.2 Pathophysiology

Elmqvist and Lambert showed a decreased release of acetylcholine from motor nerve terminals in a microphysiological study on intercostal muscle of a patient (Elmqvist and Lambert 1968). The miniature endplate potentials, as produced by spontaneously released quanta of acetylcholine, were normal in frequency and amplitude, but the quantal content after stimulation was reduced, as shown by a reduced endplate potential amplitude. This reduction of number of acetylcholine quanta released from the nerve terminal in response to a stimulus proved that the defect in neuromuscular transmission was presynaptic.

10.2.1 Antibody-Mediated Autoimmune Disease

Lambert and Rooke already noted that not all patients with this myasthenic syndrome had a carcinoma (Lambert and Rooke 1965). Gutmann et al. described the concurrence with several autoimmune disorders (Gutmann et al. 1972) and organ-specific antibodies found in other autoimmune diseases were shown in 24 % of LEMS patients with SCLC and 44 % of LEMS patients with no detectable tumour (O'Neill et al. 1988). Autoimmune diseases were more frequently found in family members of LEMS patients without a tumour as well (Wirtz et al. 2004a). Furthermore, LEMS without SCLC was associated with HLA-B8 (Willcox et al. 1985; Wirtz et al. 2001). The HLA genotype is considered the most important genetic marker of susceptibility to many autoimmune diseases. HLA-B8 was known to be associated with other autoimmune diseases like MG and coeliac disease as well. In LEMS with SCLC, no HLA association was found (Wirtz et al. 2005b). Clinical improvement of LEMS after immunosuppressive therapy was another observation supportive of an autoimmune aetiology (Streib and Rothner 1981). More recently, transient neonatal LEMS caused by passive placental transfer of maternal antibodies was described (Lecky 2006; Reuner 2008).

An important step in identifying Voltage-gated calcium channels (VGCC) as the target antigen in LEMS were freeze fracture electron microscopy studies of motor nerve terminals in LEMS patients (Fukunaga et al. 1983). These showed loss of active zone particles, believed to be the morphological representation of VGCC. Injection of LEMS serum IgG into mice produced the same clinical and electrophysiological signs (Lang et al. 1987; Fukunaga et al. 1983). Moreover, IgG could be demonstrated on the motor nerve terminal at the position where VGCC are thought to be concentrated (Fukuoka et al. 1987).

Several studies which followed showed evidence that the neuromuscular as well as the autonomic symptoms are caused by the antibodies against P/Q-type VGCC, which are found in most patients with LEMS. Antibodies can be detected by radioimmunoprecipitation of VGCC extracted from human or mammalian cerebellum that are labelled with ^{125}I-ω-Conotoxin (w-CmTx) MVIIC, a snail toxin that binds to the P/Q-type VGCC and inhibits neuromuscular transmission (Motomura et al. 1995; Lennon et al. 1995). SCLC cells are derived from the neuroectoderm and secrete a variety of neuropeptides including serotonin and ACTH. SCLC express different types of VGCC on their surface, and LEMS IgG reduces the numbers of VGCC in these cells and in other cell lines expressing the calcium channels (Pinto et al. 1998). Active immunisation with VGCC-peptides results in mild myasthenic symptoms in rats (Komai et al. 1999). Autonomic dysfunction is likely to be caused by antibodies against, particularly, the P/Q-type VGCC as well. P/Q-type and N-type VGCC are present at the presynaptic nerve terminals of the autonomic nervous system, both parasympathetic and sympathic. Mice injected with LEMS IgG showed a reduction in transmitter release subserved by P/Q-type VGCC, and to a lesser extent by N-type VGCC (Waterman et al. 1997).

Altogether, these data prove LEMS to be an antibody-mediated autoimmune disorder.

10.2.2 VGCC and Other Antigens

The α_{1A}-subunit, which forms the central ion pore in P/Q-type VGCC, presumably is a functional important immunogen in LEMS. This was shown in a study in which the effect of LEMS serum on calcium influx in human embryonic kidney cells transfected with different cloned α_1-subunits was recorded, the cells expressing the ω_{1A}-subunit being the most effectively blocked (Pinto et al. 1998). In part of LEMS patients antibodies to the S5-6 extracellular region of the four domains that form the α_{1A}-subunit were found (Takamori et al. 1998; Parsons et al. 2002). Antibody response to domain I and III was similar in LEMS without tumour and paraneoplastic LEMS, but response to domain IV was more common in LEMS without tumour (Pellkofer et al. 2008). Even the intracellular portion of the β-subunit of the VGCC is found to be an antigenic target in more than half of the LEMS patients (Verschuuren et al. 1998; Raymond et al. 1999).

VGCC link to the muscle-derived synapse organiser laminin β to organise presynaptic active zones in motor nerve terminals, mice with mutations hindering this link showed a loss of aggregation of active zones, but no clinical or electrophysiological features (Nishimune et al. 2004; Chen et al. 2011).

Mutations in the CACNA1A gene, which codes for the α-subunit of P/Q-type VGCC, cause hemiplegic migraine or episodic ataxia type 2. Ataxia is seen in both disorders, as P/Q-type VGCC is also expressed in Purkinje cells in the cerebellum. A small part of the LEMS patients has cerebellar ataxia, often being paraneoplastic of origin as well (Titulaer et al. 2008a). A post-mortem study showed a marked reduction in P/Q-type VGCC in the cerebellum of a LEMS patient with SCLC and cerebellar degeneration (Fukuda et al. 2003). Mice with P/Q-type VGCC CACNA1A mutations showed, similar to humans, ataxia and muscle weakness and the characteristic electrophysiological abnormalities of LEMS (Kaja et al. 2007). These observations suggests that in a small proportion of LEMS patients the immune response extends to the cerebellum, although it is unknown if the epitopes are similar in the neuromuscular synapse and the cerebellum.<Query ID="Q5" Please note that part labels a, b and c are mentioned in Fig caption, but not given in the art work for Fig. 10.1" –>

The presence of VGCC antibodies against different parts of the calcium channel is presumably the result of presentation of these antigens after the initial immune-mediated attack on the VGCC. In line with this, antibodies against other synapse-associated proteins are found in LEMS patients, like synaptotagmin, a synaptic vesicle protein (Leveque et al. 1992). Especially in LEMS patients with SCLC, antibodies against Hu and SOX are found (Mason et al. 1997; Sabater et al. 2008; Titulaer et al. 2009). Hu proteins are expressed in neurons and play an important role in the development and maintenance of the nervous system. SOX proteins are transcription factors that have critical roles in the regulation of developmental processes (Bylund et al. 2003; Sandberg et al. 2005). Both proteins are expressed by SCLC, like VGCC. This suggests that the immune response elicited by the tumour is directed against several SCLC-related proteins. SOX antibodies had a sensitivity of 67 % and a specificity of 95 % to discriminate between LEMS with SCLC and non-tumour LEMS. Neither clinical picture nor difference in survival was observed between SOX positive and SOX negative SCLC patients (Titulaer et al. 2009).

10.3 Clinical Characteristics

10.3.1 Symptoms

The main symptom in LEMS is proximal muscle weakness in the legs, which is the presenting symptom in most patients (O'Neill et al. 1988; Titulaer et al. 2008a).Weakness is often accompanied by muscle aching or stiffness (O'Neill et al. 1988). Limb weakness can be worse in hot weather (O'Neill et al. 1988). Weakness in the arms is less prominent, but does occur in most patients. During the course of the disease, weakness of extraocular muscle can occur, presenting as a ptosis or diplopia, being often transiently present. Bulbar symptoms are also often mildly present. Generally, weakness in LEMS tends to develop in a caudocranial direction, while in the opposite direction in MG (Wirtz et al. 2002b). In contrast to MG, it is very rare to find isolated extraocular muscle weakness in LEMS, although it has been reported (Wirtz et al. 2002b; Rudnicki 2007; Titulaer et al. 2008c). Cerebellar ataxia is seen in some LEMS patients (Titulaer et al. 2008a).

Most patients have signs of autonomic dysfunction, most frequently a dry mouth (O'Neill et al. 1988; Titulaer et al. 2008a; Pellkofer et al. 2009a). Other frequent manifestations of autonomic dysfunction are male impotence and constipation. Less frequent autonomic signs are orthostatic hypotension, micturition difficulties and dry eyes (Titulaer et al. 2011a).

Neurological examination generally reveals low or absent tendon reflexes. Immediately after contraction of the muscle being tested the reflexes may increase. This post-exercise facilitation is the characteristic of LEMS. This facilitation may also be seen when testing muscle strength, the strength increasing after initial voluntary contraction. However, being present in only 40 % of patients it is not a very sensitive sign (Odabasi et al. 2002; Oh et al. 2005). This facilitation can mask the reduced tendon reflexes. Therefore, tendon reflexes should be tested after a period of rest if a diagnosis of LEMS is suspected.

10.3.2 Small Cell Lung Cancer

Small cell lung cancer (SCLC) is a neuroendocrine tumour strongly related to smoking. SCLC is found in about half of the patients with LEMS (O'Neill et al.

1988; Nakao et al. 2002; Wirtz et al. 2004b). LEMS has been described in patients with several other types of tumours, but is unclear if there is a causal relation (Wirtz et al. 2002a). SCLC is found after the onset of LEMS in most patients, generally within 2 years. Using chest X-ray only is inadequate for the detection of SCLC (Titulaer et al. 2008b). Screening with CT-thorax or PET scans detected SCLC in 91 % of patients within 3 months after the diagnosis of LEMS (Titulaer et al. 2008b). Specific symptoms do not distinguish between patients with and without underling SCLC, but LEMS associated with SCLC has a more progressive course (Wirtz et al. 2005c).

10.3.3 Differential Diagnosis

LEMS can be difficult to recognise initially. In most patients with LEMS an alternative diagnosis is made at first (Titulaer et al. 2011a). Diagnostic delay is often long, particularly in patients without a tumour (Wirtz et al. 2004b).

MG is the most common alternative diagnosis made in patients with LEMS (O'Neill et al. 1988). Like LEMS, MG is an acquired autoimmune disorder with a defective neuromuscular transmission, characterised by variable weakness. However, 90 % of MG patients start with oculobulbar symptoms, as opposed to only 5 % in LEMS (Wirtz et al. 2002b). Generally, muscle weakness in MG evolves in a craniocaudal direction, as opposed to the reverse in LEMS (Wirtz et al. 2002b). Decreased tendon reflexes and autonomic dysfunction are features of LEMS, and not of MG (O'Neill et al. 1988). MG can be distinguished from LEMS by different findings after RNS and by autoantibody testing.

Myopathies often present with proximal, symmetric muscle weakness, like LEMS. Especially, inclusion body myositis is common in older patients and typically presents with weakness of quadriceps muscles. Creatine kinase (CK) is substantially elevated in many myopathies, but not in LEMS. Autonomic symptoms point towards LEMS. In some patients, a psychiatric disorder is considered, because the patient's symptoms often sound disproportionate to the rather limited signs on neurological examination.

If only standard nerve conduction without RNS in a patient with LEMS is performed, the low CMAP amplitude can be mistaken for an axonal polyneuropathy or motor neuron disease. Some patients with a rather acute development of symptoms may wrongly be diagnosed with Guillain-Barré syndrome, although sensory symptoms or radicular pain are not found in patients with LEMS. In motor neuron disease, fasciculations, atrophy and asymmetry contrast the signs and symptoms in LEMS.

Especially, in patients with a SCLC the symptoms of LEMS could be misinterpreted as a consequence of cachexia or chemotherapy.

10.4 Diagnosis

If the diagnosis of LEMS is suspected from the distinctive clinical features, confirmation of the diagnosis depends on detection of VGCC autoantibodies or characteristic electrodiagnostic findings.

10.4.1 Anti-Voltage-Gated Calcium Channel Antibodies

Using a standard radioimmunoassay, antibodies against ^{125}I-ω-CmTx MVIIC labelled (P/Q-type) VGCCs are detected in significant titers in 85–90 % of LEMS patients (Motomura et al. 1997). About 33 % of these sera are also positive for ^{125}I-ω-CgTx GVIA labelled (N-type) VGCC antibodies (Motomura et al. 1997). Some patients also have antibodies to L-type VGCC (Johnston et al. 1994). Anti-P/Q-type VGCC antibodies were detectable in 1–4 % of patients with SCLC and no neurological symptoms (Titulaer et al. 2009; Lang et al. 1998) Also, 24 % of patients with SCLC and paraneoplastic cerebellar degeneration had anti-P/Q-type VGCC antibodies, half of whom had no manifestations of LEMS (Mason et al. 1997). In clinical practice, only P/Q-type VGCC antibodies are important in the diagnostic phase, and not the more rare N-type of L-type VGCC antibodies.

10.4.2 Electrodiagnostic Features

Eaton and Lambert first defined the classical electrophysiological characteristics of LEMS (Eaton and Lambert 1957). They demonstrated reduced amplitude of the resting CMAP obtained by supramaximal nerve stimulation in patients with LEMS. Further decrement was seen with slow RNS at rates less than 10 per second. However, following high rate RNS (>10 Hz) or 10 s of maximal voluntary contraction, there was an increase in CMAP amplitude, typically over 100 %. These results suggested that the neuromuscular transmission disorder was presynaptic. The combination of a low first CMAP, decrement on low rate RNS, and increment after high rate RNS in a patient with acquired proximal muscle weakness is pathognomonic for LEMS (Fig. 10.1). RNS may be abnormal in other neuromuscular disorders. If other neuromuscular disorders are considered as well, nerve conduction studies and needle EMG must be performed. Needle EMG demonstrates increased low amplitude, short duration motor unit action potentials. Amplitudes may enlarge with sustained contraction.

Although weakness in LEMS predominates in proximal muscles, RNS by stimulating the ulnar nerve at the wrist with recording from the abductor digiti minimi is very sensitive in detecting typical electrophysiological abnormalities in LEMS (Maddison et al. 1998). Almost all LEMS patients had incremental CMAP

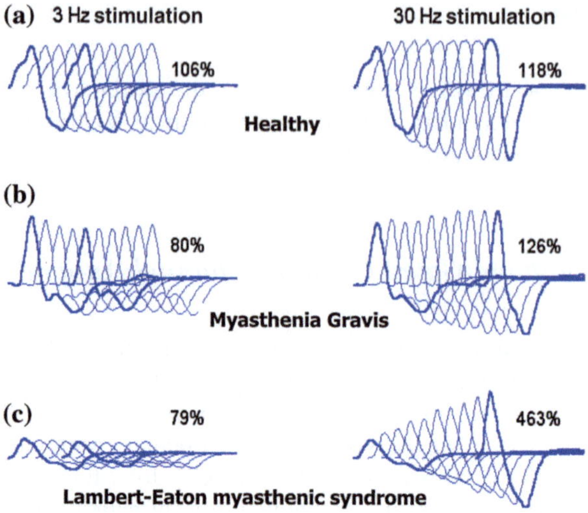

Fig. 10.1 Repetitive nerve stimulation studies. Repetitive nerve stimulation studies of the adductor digiti minimi muscle in a healthy person (**a**), in a patient with myasthenia gravis (MG) with acetylcholine receptor antibodies (**b**), or in a patient with Lambert-Eaton myasthenic syndrome (LEMS) (**c**). The left-hand column shows the results after low frequency 3 Hz nerve stimulation. The right-hand side shows the recordings obtained directly after high frequency 30 Hz nerve stimulation. Both patients with MG and LEMS show decrement at low frequency stimulation (**b**, **c**). Only the patient with LEMS shows increment more than 60 % after maximal voluntary contraction (**c**). Courtesy to Prof. Gert van Dijk for providing these figures

responses over 100 % in abductor digiti minimi, abductor pollicis brevis, or anconeus after 10 s of voluntary contraction (Maddison et al. 1998). In another study, an increment of 60 % rather than 100 % was sufficient to diagnose LEMS with the same specificity over healthy control patients (Oh et al. 2005). The increment of the CMAP amplitude is in fact a normalisation of the previous low CMAP amplitude. Thus, the higher the first, original CMAP is the less the increment that can be obtained.

In clinical practice, the differential diagnosis is often whether a patient has LEMS or MG. The post-synaptic disorder of MG characteristically shows a decremental response in CMAP amplitude, which is found in LEMS during low-frequency RNS as well. In MG, however, no significant incremental pattern is found, because the original CMAP amplitude is not or only marginally decreased. Oh observed that in MG, a 60 % increment was observed in only 4 of 538 patients by high frequency RNS and in none by post-exercise testing (Oh et al. 2005). The use of a 60 % increment showed a sensitivity of 97 % for the diagnosis of LEMS, while the specificity in excluding MG was 99 %.

Single-fibre EMG detects defects of neuromuscular transmission with high sensitivity. Abnormal jitter and blocking, typical of defective neuromuscular transmission, is seen in all LEMS patients (O'Neill et al. 1988). However, it lacks specificity and cannot reliably be used in differentiating from MG, although the degree of jitter and blocking seen in LEMS is usually less when high frequency nerve stimulation is used, a finding more specific to LEMS (Trontelj and Stålberg 1991; Sanders 1992).

10.4.3 DELTA-P Score

Diagnosis of LEMS alerts the physician to look for an SCLC. A clinical score was developed that predicts the presence of SCLC by using only clinical criteria at the time of diagnosis of LEMS (Table 10.1) (Titulaer et al. 2011b). This simple Dutch-English LEMS Tumour Association Prediction Score (DELTA-P score) proved to be easy, sensitive, specific and reproducible. The probability for SCLC in a LEMS patient can be calculated as early as 3 months from onset of LEMS using only clinically defined signs and symptoms. A DELTA-P score of 0–1 gives a risk of 0–2.6 %, rising up steeply to 83.9 % and up to 100 % respectively, with a DELTA-P score of 3–6. The score can be used to determine the frequency of screening for SCLC (Fig. 10.2).

10.5 Treatment and Prognosis

Treatment of LEMS is based on drugs which act at the neuromuscular synapse by increasing either the release or the concentration of acetylcholine, drugs which suppress the immune response or treatment of the underlying tumour.

Table 10.1 DELTA-P score. This score helps to estimate the chance for the presence of a small cell lung carcinoma in a patients presenting with clinical signs and symptoms of Lambert-Eaton myasthenic syndrome

	Categories	<3 months of onset	Score
D	**D**ysarthria, dysphagia, chewing, neck weakness: bulbar weakness	absent	0
		present	1
E	**E**rectile dysfunction	female	0
		male: absent	0
		male: present	1
L	**L**oss of Weight	absent or <5 %	0
		≥5 %	1
T	**T**obacco use at onset	absent	0
		present	1
A	**A**ge of onset	<50 years	0
		≥50 years	1
P	Karnofsky **P**erformance Score	70–100	0
		0–60	1
	DELTA-P score		0–6

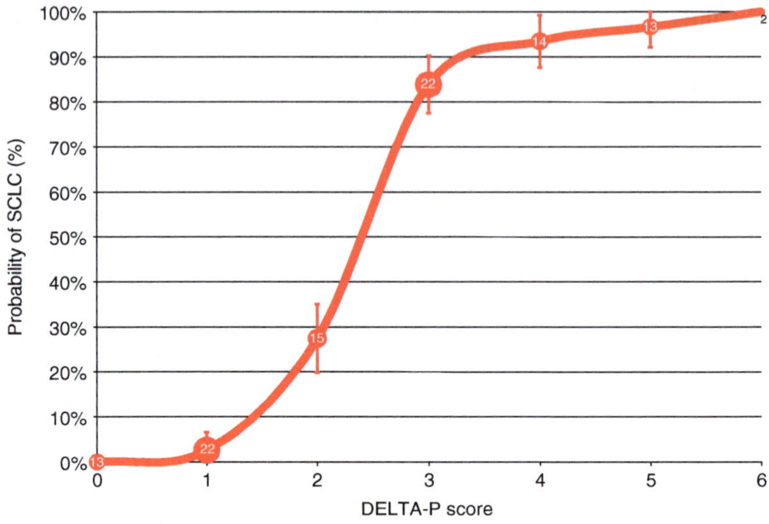

Fig. 10.2 Predicted percentage of small cell lung carcinoma in patients with Lambert-Eaton myasthenic syndrome, based on the Dutch-English LEMS Tumour Association Prediction (DELTA-P) score (Titulaer et al. 2011b). DELTA-P score is calculated as a sum score according to the different categories listed below. The DELTA-P score can vary from 0 to 6. Point sizes proportionate to the number of patients with a specific score, also represented by the percentage inside the circle. Vertical bars indicate SEM

10.5.1 3,4-Diaminopyridine and Pyridostigmine

Aminopyridines increase the release of acetylcholine by blocking voltage-gated K+ channels, which prolongs the action potential. The beneficial effect of 3,4-diaminopyridine (3,4-DAP), both orally and intravenously, was shown in four randomised trials (McEvoy et al. 1989; Sanders et al. 2000; Wirtz et al. 2009; Oh et al. 2009). This effect was significant on both muscle strength and resting CMAP amplitude. Although two trials reported an additional effect of pyridostigmine as well (McEvoy et al. 1989; Sanders et al. 2000), this effect was not quantified or studied during the randomised phase of the studies. The only trial comparing the effects of 3,4-DAP with pyridostigmine and studying the effects of the combination of the two drugs found no significant effects of pyridostigmine alone or added to 3,4-DAP (Wirtz et al. 2009). Most patients commence 3,4-DAP treatment at 10 mg 3–4 times daily, increasing up to a 100 mg maximum per day. Symptomatic improvement usually occurs within an hour of ingestion.

3,4-DAP was well tolerated and serious adverse events were uncommon in clinical trials (McEvoy et al. 1989; Sanders et al. 2000; Wirtz et al. 2009; Oh et al. 2009). In general, side effects are mild and occur especially at higher doses. Most frequently reported were perioral or acral paresthesias. Fatigue, insomnia and epigastric distress were reported by some patients. Epileptic seizures have been described in patients using 100 mg a day (McEvoy et al. 1989; Sanders 1998) Based on these observations, some authors report to limit the dose to 80 mg a day (Sanders et al. 2000). However, in all patients described having seizures during treatment with 3,4-DAP, it was not made clear how the daily dose was divided during the day, which is an important aspect since seizures are most likely related to peak-dose concentrations. During intravenous treatment with 3,4-DAP, patients reported transient pain at the site of intravenous administration of 3,4-DAP (Wirtz et al. 2009; Lundh et al. 1984).

10.5.2 Immunosuppressive Therapy

When response to above mentioned therapy is insufficient, immunosuppressive therapy can be used. Plasmapheresis and intravenous immunoglobulins have a swift, but rather short lasting effect, and are used for patients with severe symptoms requiring prompt treatment (Bain et al. 1996; Newsom-Davis and Murray 1984). The efficacy of immunoglobulins was significantly shown in a cross-over study (Bain et al. 1996). The improvement in muscle strength after administration of immunoglobulins was maximal after 2 to 4 weeks and then subsided (Bain et al. 1996). To obtain a long lasting clinical response immunosuppressive drugs like prednisone or azathioprine are used, most often in combination. The choice and dosing of the drugs are similar to MG. In LEMS patients without tumour, this was used in 80 of 114 (70 %), whilst 46 of 104 (44 %) of patients with SCLC needed these drugs (Titulaer et al. 2011a). Effectiveness of this combination therapy has been demonstrated only in a retrospective study (Newsom-Davis and Murray 1984), but is supported by the positive results of the combined treatment in a closely related disorder, MG (Palace et al. 1998). Other immunosuppressive drugs like ciclosporin or methotrexate have been used effectively in LEMS patients as well (Maddison et al. 2001). Recently, rituximab was described to be effective in some LEMS patients with severe weakness, who were resistant to treatment with several different immunosuppressive drugs (Maddison et al. 2011; Pellkofer et al. 2009b).

10.5.3 Tumour Therapy

In LEMS patients with SCLC, the third treatment option is tumour therapy. To achieve tumour control and improve the condition and survival of the patient, patients with SCLC are usually treated with chemotherapy, like cisplatinum and etoposide (Jansen-Heijen et al. 2011). LEMS tends to improve as well with chemotherapy. Chemotherapy has a powerful immunosuppressive effect, which will contribute to the remission of LEMS. Furthermore, it removes an important antigenic stimulus for the immune system, thereby improving the symptoms of LEMS. In a series of 11 LEMS patients with SCLC, LEMS in 10 patients improved after treatment for their SCLC (Chalk et al. 1990). If remission of symptoms is incomplete, additional immunosuppressive treatment may induce improvement. There are indications that the immune response against VGCC helps to control tumour growth (Maddison et al. 1999). This implies that immunosuppressive treatment of LEMS patients with SCLC could potentially allow the tumour to escape immune-mediated suppression and enhance its growth. Clinical follow-up studies of these patients, however, did not indicate that immunosuppressive treatment is contraindicated (Chalk et al. 1990). If myasthenic weakness severely impairs daily function, the evidence favours trying to achieve short-term symptomatic benefits and not withhold immunosuppressive treatment because of a theoretically disadvantageous influence on tumour growth.

10.5.4 Prognosis

In a long term, outcome study of 47 LEMS patients without SCLC, sustained clinical remission was achieved by 20 (43 %) of whom only four remained in remission without the need for immunosuppression (Maddison et al. 2001). The only independent predictor of sustained clinical remission was initial pretreatment clinical score, and not anti-VGCC antibody titers or EMG results. Muscle strength scores were improved in 88 % of patients.

In SCLC patients with LEMS, survival seems favourable in comparison to SCLC patients without LEMS, although still poor (Maddison et al. 1999; Wirtz et al. 2005a). These studies suggest that the immune response in LEMS against the SCLC may retard tumour growth, but only rarely will cure the patient from the tumour. Tumour eradication by an autoimmune response is difficult to prove, but has been described in some case reports of SCLC-associated paraneoplastic autoimmunity (Gill et al. 2003).

10.6 Conclusion

LEMS is a neuromuscular autoimmune disease, in which the characteristic proximal muscle weakness is caused by pathogenic autoantibodies directed against P/Q-type VGCC present on the presynaptic nerve terminal. Half of patients with LEMS have an associated SCLC, which also expresses functional VGCC. Insights in the pathophysiology have made LEMS a model for autoimmunity and tumour immunology.

References

Anderson HJ, Churchill-Davidson HC, Richardson AT (1953) Bronchial neoplasm with myasthenia; prolonged apnoea after administration of succinylcholine. Lancet 265:1291–1293

Bain PG, Motomura M, Newsom-Davis J, Misbah SA, Chapel HM, Lee ML, Vincent A, Lang B (1996) Effects of intravenous immunoglobulin on muscle weakness and calcium-channel autoantibodies in the Lambert-Eaton myasthenic syndrome. Neurology 47:678–683

Bylund M, Andersson E, Novitch BG, Muhr J (2003) Vertebrate neurogenesis is counteracted by Sox1–3 activity. Nat Neurosci 6:1162–1168

Chalk CH, Murray NM, Newsom-Davis J, O'Neill JH, Spiro SG (1990) Response of the Lambert-Eaton myasthenic syndrome to treatment of associated small-cell lung carcinoma. Neurology 40:1552–1556

Chen J, Billings SE, Nishimune H (2011) Calcium channels link the muscle-derived synapse organizer laminin $\beta 2$ to bassoon and CAST/Erc2 to organize presynaptic active zones. J Neurosci 31:512–525

Drachman DB (1990) How to recognize an antibody-mediated autoimmune disease: criteria. Res Publ Assoc Res Nerv Ment Dis 68:183–186

Eaton LM, Lambert EH (1957) Electromyography and electric stimulation of nerves in diseases of motor unit; observations on myasthenic syndrome associated with malignant tumours. J Am Med Assoc 30:1117–1124

Elmqvist D, Lambert EH (1968) Detailed analysis of neuromuscular transmission in a patient with myasthenic syndrome sometimes associated with bronchogenic carcinoma. Mayo Clin Proc 43:689–713

Fukuda T, Motomura M, Nakao Y, Shiraishi H, Yoshimura T, Iwanaga K, Tsujihata M, Eguchi K (2003) Reduction of P/Q-type calcium channels in the postmortem cerebellum of paraneoplastic cerebellar degeneration with Lambert-Eaton myasthenic syndrome. Ann Neurol 53:21–28

Fukunaga H, Engel AG, Lang B, Newsom-Davis J, Vincent A (1983) Passive transfer of Lambert-Eaton myasthenic syndrome with IgG from man to mouse depletes the presynaptic membrane active zones. Proc Natl Acad Sci U S A 80:7636–7640

Fukuoka T, Engel AG, Lang B, Newsom-Davis J, Vincent A (1987) Lambert-Eaton myasthenic syndrome: II, Immunoelectron microscopy localization of IgG at the mouse motor end-plate. Ann Neurol 22:200–211

Gill S, Murray N, Dalmau J, Thiessen B (2003) Paraneoplastic sensory neuronopathy and spontaneous regression of small cell lung cancer. Can J Neurol Sci 30:269–271

Gutmann L, Crosby TW, Takamori M, Martin JD (1972) The Eaton-Lambert syndrome and autoimmune disorders. Am J Med 53:354–356

Janssen-Heijnen ML, Maas HA, Siesling S, Koning CC, Coebergh JW, Groen HJ (2011) Treatment and survival of patients with small-cell lung cancer: small steps forward, but not for patients > 80. Ann Oncol 23:954–960

Johnston I, Lang B, Leys K, Newsom-Davis J (1994) Heterogeneity of calcium channel autoantibodies detected using a small-cell lung cancer line derived from a Lambert-Eaton myasthenic syndrome patient. Neurology 44:334–338

Kaja S, van de Ven RC, van Dijk JG, Verschuuren JJ, Arahata K, Frants RR, Ferrari MD, van den Maagdenberg AM, Plomp JJ (2007) Severely impaired neuromuscular synaptic transmission causes muscle weakness in the cacna1a-mutant mouse rolling Nagoya. Eur J Neurosci 25:2009–2020

Komai K, Iwasa K, Takamori M (1999) Calcium channel peptide can cause an autoimmune-mediated model of Lambert-Eaton myasthenic syndrome in rats. J Neurol Sci 166:126–130

Lambert EH, Eaton LM, Rooke ED (1956) Defect of neuromuscular conduction associated with malignant neoplasms. Am J Physiol 187:612–613

Lambert EH, Rooke ED (1965) Myasthenic state and lung cancer. In: Brain L and Nooris E. The remote effects of cancer on the nervous system. Grune & Stratton :67–80

Lang B, Newsom-Davis J, Peers C, Wray DW (1987) Selective Action of Lambert-Eaton Myasthenic Syndrome Antibodies on Ca-2+ Channels in the Neuroblastoma X Glioma Hybrid Cell-Line Ng108-15. J Physiol 394:43, London

Lang B, Waterman S, Pinto A, Jones D, Moss F, Boot J, Brust P, Williams M, Stauderman K, Harpold M, Motomura M, Moll JW, Vincent A, Newsom-Davis J (1998) The role of autoantibodies in Lambert-Eaton myasthenic syndrome. Ann N Y Acad Sci 841:596–605

Lecky BR (2006) Transient neonatal Lambert-Eaton syndrome. J Neurol Neurosurg Psychiatry 77:1094

Lennon VA, Kryzer TJ, Griesmann GE, O'Suilleabhain PE, Windebank AJ, Woppmann A, Miljanich GP, Lambert EH (1995) Calcium-channel antibodies in the Lambert-Eaton syndrome and other paraneoplastic syndromes. N Engl J Med 332:1467–1674

Leveque C, Hoshino T, David P, Shoji-Kasai Y, Leys K, Omori A, Lang B, el Far O, Sato K, Martin-Moutot N, Newsom-Davis J, Takahashi M, Seagar MJ (1992) The synaptic vesicle protein synaptotagmin associates with calcium channels and is a putative Lambert-Eaton myasthenic syndrome antigen. Proc Natl Acad Sci U S A 89:3625–3629

Lundh H, Nilsson O, Rosén I (1984) Treatment of Lambert-Eaton myasthenic syndrome: 3,4-diaminopyridine and pyridostigmine. Neurol 34:1324–1330

Maddison P, Lang B, Mills K, Newsom-Davis J (2001) Long term outcome in Lambert-Eaton myasthenic syndrome without lung cancer. J Neurol Neurosurg Psychiatry 70:212–217

Maddison P, McConville J, Farrugia ME, Davies N, Rose M, Norwood F, Jungbluth H, Robb S, Hilton-Jones D (2011) The use of rituximab in myasthenia gravis and Lambert-Eaton myasthenic syndrome. J Neurol Neurosurg Psychiatry 82:671–673

Maddison P, Newsom-Davis J, Mills KR (1998) Distribution of electrophysiological abnormality in Lambert-Eaton myasthenic syndrome. J Neurol Neurosurg Psychiatry 65:213–217

Maddison P, Newsom-Davis J, Mills KR, Souhami RL (1999) Favourable prognosis in Lambert-Eaton myasthenic syndrome and small-cell lung carcinoma. Lancet 353:117–118

Mason WP, Graus F, Lang B, Honnorat J, Delattre JY, Valldeoriola F, Antoine JC, Rosenblum MK, Rosenfeld MR, Newsom-Davis J, Posner JB, Dalmau J (1997) Small-cell lung cancer, paraneoplastic cerebellar degeneration and the Lambert-Eaton myasthenic syndrome. Brain 120:1279–1300

McEvoy KM, Windebank AJ, Daube JR, Low PA (1989) 3,4-Diaminopyridine in the treatment of Lambert-Eaton myasthenic syndrome. N Engl J Med 321:1567–1571

Motomura M, Johnston I, Lang B, Vincent A, Newsom-Davis J (1995) An improved diagnostic assay for Lambert-Eaton myasthenic syndrome. J Neurol Neurosurg Psychiatry 58:85–87

Motomura M, Lang B, Johnston I, Palace J, Vincent A, Newsom-Davis J (1997) Incidence of serum anti-P/Q-type and anti-N-type calcium channel autoantibodies in the Lambert-Eaton myasthenic syndrome. J Neurol Sci 147:35–42

Nakao YK, Motomura M, Fukudome T, Fukuda T, Shiraishi H, Yoshimura T, Tsujihata M, Eguchi K (2002) Seronegative Lambert-Eaton myasthenic syndrome: study of 110 Japanese patients. Neurol 10:1175–1773

Newsom-Davis J, Murray NM (1984) Plasma exchange and immunosuppressive drug treatment in the Lambert-Eaton myasthenic syndrome. Neurol 34:480–485

Nishimune H, Sanes JR, Carlson SS (2004) A synaptic laminin-calcium channel interaction organizes active zones in motor nerve terminals. Nature 432:580–587

Odabasi Z, Demirci M, Kim DS, Lee DK, Ryan HF, Claussen GC, Tseng A, Oh SJ (2002) Postexercise facilitation of reflexes is not common in Lambert-Eaton myasthenic syndrome. Neurol 59:1085–1087

Oh SJ, Claussen GG, Hatanaka Y, Morgan MB (2009) 3,4-Diaminopyridine is more effective than placebo in a randomized, double-blind, cross-over drug study in LEMS. Muscle Nerve 40:795–800

Oh SJ, Kurokawa K, Claussen GC, Ryan HF Jr (2005) Electrophysiological diagnostic criteria of Lambert-Eaton myasthenic syndrome. Muscle Nerve 32:515–520

O'Neill JH, Murray NMF, Newsom-Davis J (1988) The Lambert-Eaton myasthenic syndrome - A review of 50 cases. Brain 111:577–596

Palace J, Newsom-Davis J, Lecky B (1998) A randomized double-blind trial of prednisolone alone or with azathioprine in myasthenia gravis. Myasthenia Gravis Study Group Neurol 50:1778–1783

Parsons KT, Kwok WW (2002) Linear B-cell epitopes in Lambert-Eaton myasthenic syndrome defined by cell-free synthetic peptide binding. J Neuroimmunol 126:190–195

Pellkofer HL, Armbruster L, Krumbholz M, Titulaer MJ, Verschuuren JJ, Schumm F, Voltz R (2008) Lambert-eaton myasthenic syndrome differential reactivity of tumour versus non-tumour patients to subunits of the voltage-gated calcium channel. J Neuroimmunol 204:136–139

Pellkofer HL, Armbruster L, Linke R, Schumm F, Voltz R (2009a) Managing non-paraneoplastic Lambert-Eaton myasthenic syndrome: clinical characteristics in 25 German patients. J Neuroimmunol 217:90–94

Pellkofer HL, Voltz R, Kuempfel T (2009b) Favorable response to rituximab in a patient with anti-VGCC-positive Lambert-Eaton myasthenic syndrome and cerebellar dysfunction. Muscle Nerve 40:305–308

Pinto A, Gillard S, Moss F, Whyte K, Brust P, Williams M, Stauderman K, Harpold M, Lang B, Newsom-Davis J, Bleakman D, Lodge D, Boot J (1998) Human autoantibodies specific for the alpha1A calcium channel subunit reduce both P-type and Q-type calcium currents in cerebellar neurons. Proc Natl Acad Sci U S A 95:8328–8333

Raymond C, Walker D, Bichet D, Iborra C, Martin-Moutot N, Seagar M, De Waard M (1999) Antibodies against the beta subunit of voltage-dependent calcium channels in Lambert-Eaton myasthenic syndrome. Neurosci 90:269–277

Reuner U (2008) Transient neonatal Lambert-Eaton syndrome. J Neurol 255:1827

Rose NR, Bona C (1993) Defining criteria for autoimmune diseases (Witebsky's postulates revisited). Immunol Today 14:426–430

Rudnicki SA (2007) Lambert-Eaton myasthenic syndrome with pure ocular weakness. Neurol 68:1863–186

Sabater L, Titulaer M, Saiz A, Verschuuren J, Güre AO, Graus F (2008) SOX1 antibodies are markers of paraneoplastic Lambert-Eaton myasthenic syndrome. Neurol 70:924–928

Sandberg M, Kallstrom M, Muhr J (2005) Sox21 promotes the progression of vertebrate neurogenesis. Nat Neurosci 8:995–1001

Sanders DB (1992) The effect of firing rate on neuromuscular jitter in Lambert-Eaton myasthenic syndrome. Muscle Nerve 15:256–258

Sanders DB (1998) 3,4-Diaminopyridine (DAP) in the treatment of Lambert-Eaton myasthenic syndrome (LEMS). Ann N Y Acad Sci 841:811–816

Sanders DB, Massey JM, Sanders LL, Edwards LJ (2000) A randomized trial of 3,4-diaminopyridine in Lambert-Eaton myasthenic syndrome. Neurol 54:603–607

Streib EW, Rothner AD (1981) Eaton-Lambert myasthenic syndrome: long-term treatment of three patients with prednisone. Ann Neurol 10:448–453

Takamori M, Iwasa K, Komai K (1998) Antigenic sites of the voltage-gated calcium channel in Lambert-Eaton myasthenic syndrome. Ann N Y Acad Sci 841:625–635

Titulaer MJ, Klooster R, Potman M, Sabater L, Graus F, Hegeman IM, Thijssen PE, Wirtz PW, Twijnstra A, Smitt PA, van der Maarel SM, Verschuuren JJ (2009) SOX antibodies in small-cell lung cancer and Lambert-Eaton myasthenic syndrome: frequency and relation with survival. J Clin Oncol 27:4260–4267

Titulaer MJ, Lang B, Verschuuren JJ (2011a) Lambert-Eaton myasthenic syndrome: from clinical characteristics to therapeutic strategies. Lancet Neurol 10:1098–1107

Titulaer MJ, Maddison P, Sont JK, Wirtz PW, Hilton-Jones D, Klooster R, Willcox N, Potman M, Sillevis Smitt PA, Kuks JB, Roep BO, Vincent A, van der Maarel SM, van Dijk JG, Lang B, Verschuuren JJ (2011b) Clinical dutch-english Lambert-Eaton myasthenic syndrome (LEMS) tumour association prediction score accurately predicts small-cell lung cancer in the LEMS. J Clin Oncol 29:902–908

Titulaer MJ, Wirtz PW, Kuks JB, Schelhaas HJ, van der Kooi AJ, Faber CG, van der Pol WL, de Visser M, Sillevis Smitt PA, Verschuuren JJ (2008a) The Lambert-Eaton myasthenic syndrome 1988-2008: A clinical picture in 97 patients. J Neuroimmunol 201–202:153–158

Titulaer MJ, Wirtz PW, Willems LN, van Kralingen KW, Smitt PA, Verschuuren JJ (2008b) Screening for small-cell lung cancer: a follow-up study of patients with Lambert-Eaton myasthenic syndrome. J Clin Oncol 26:4276–4281

Titulaer MJ, Wirtz PW, Wintzen AR, Verschuuren JJ (2008c) Re: Lambert-Eaton myasthenic syndrome with pure ocular weakness. Neurol 70:86

Trontelj JV, Stålberg E (1991) Single motor end-plates in myasthenia gravis and LEMS at different firing rates. Muscle Nerve 14:226–232

Verschuuren JJ, Dalmau J, Tunkel R, Lang B, Graus F, Schramm L, Posner JB, Newsom-Davis J, Rosenfeld MR (1998) Antibodies against the calcium channel beta-subunit in Lambert-Eaton myasthenic syndrome. Neurol 50:475–479

Waterman SA, Lang B, Newsom-Davis J (1997) Effect of Lambert-Eaton myasthenic syndrome antibodies on autonomic neurons in the mouse. Ann Neurol 42:147–156

Willcox N, Demaine AG, Newsom-Davis J, Welsh KI, Robb SA, Spiro SG (1985) Increased frequency of IgG heavy chain marker Glm(2) and of HLA-B8 in Lambert-Eaton myasthenic syndrome with and without associated lung carcinoma. Hum Immunol 14:29–36

Wirtz PW, Bradshaw J, Wintzen AR, Verschuuren JJ (2004a) Associated autoimmune diseases in patients with the Lambert-Eaton myasthenic syndrome and their families. J Neurol 251:1255–1259

Wirtz PW, Lang B, Graus F, van den Maagdenberg AM, Saiz A, de Koning Gans PA, Twijnstra A, Verschuuren JJ (2005a) P/Q-type calcium channel antibodies, Lambert-Eaton myasthenic syndrome and survival in small cell lung cancer. J Neuroimmunol 164:161–165

Wirtz PW, Roep BO, Schreuder GM, van Doorn PA, van Engelen BG, Kuks JB, Twijnstra A, de Visser M, Visser LH, Wokke JH, Wintzen AR, Verschuuren JJ (2001) HLA class I and II in Lambert-Eaton myasthenic syndrome without associated tumour. Hum Immunol 62:809–813

Wirtz PW, Smallegange TM, Wintzen AR, Verschuuren JJ (2002a) Differences in clinical features between the Lambert-Eaton myasthenic syndrome with and without cancer: an analysis of 227 published cases. Clin Neurol Neurosurg 104:902–908

Wirtz PW, Sotodeh M, Nijnuis M, Van Doorn PA, Van Engelen BG, Hintzen RQ, De Kort PL, Kuks JB, Twijnstra A, De Visser M, Visser LH, Wokke JH, Wintzen AR, Verschuuren JJ (2002b) Difference in distribution of muscle weakness between myasthenia gravis and the Lambert-Eaton myasthenic syndrome. J Neurol Neurosurg Psychiatry 73:766–768

Wirtz PW, van Dijk JG, van Doorn PA, van Engelen BG, van der Kooi AJ, Kuks JB, Twijnstra A, de Visser M, Visser LH, Wokke JH, Wintzen AR, Verschuuren JJ (2004b) The epidemiology of the Lambert-Eaton myasthenic syndrome in the Netherlands. Neurol 63:397–398

Wirtz PW, Verschuuren JJ, van Dijk JG, de Kam ML, Schoemaker RC, van Hasselt JG, Titulaer MJ, Tjaden UR, den Hartigh J, van Gerven JM (2009) Efficacy of 3,4-diaminopyridine and pyridostigmine in the treatment of Lambert-Eaton myasthenic syndrome: a randomized, double-blind, placebo-controlled, crossover study. Clin Pharmacol Ther 86:44–48

Wirtz PW, Willcox N, van der Slik AR, Lang B, Maddison P, Koeleman BP, Giphart MJ, Wintzen AR, Roep BO, Verschuuren JJ (2005b) HLA and smoking in prediction and prognosis of small cell lung cancer in autoimmune Lambert-Eaton myasthenic syndrome. J Neuroimmunol 159:230–237

Wirtz PW, Wintzen AR, Verschuuren JJ (2005c) Lambert-Eaton myasthenic syndrome has a more progressive course in patients with lung cancer. Muscle Nerve 32:226–229

Witebsky E, Rose NR, Terplan K, Paine JR, Egan RW (1957) Chronic thyroiditis and autoimmunization. J Am Med Assoc 164:1439–1447

Chapter 11
Alternative Exon Effect on Phenotype of $Ca_v1.2$ Channelopathy: Implications in Timothy Syndrome

Hua Huang, Juejin Wang and Tuck Wah Soong

Abstract Timothy Syndrome is a multiorgan disorder that results from de novo, gain-of-function mutations within the voltage-gated $Ca_v1.2$ calcium channel. While G406R mutation occurs within exon 8, G402S and G406R are located in the mutually exclusive exon 8a. All three mutations similarly produced a drastic slowdown of channel inactivation. Excessive Ca^{2+} currents flowing through the $Ca_v1.2$ channels in different tissues underlie pathologies such as arrhythmia, dysregulation of functions in the endocrine and immune systems, autism spectral disorder, and other developmental defects such as syndactyly. Here, we discuss the diverse patterns and regulation of alternative splicing of the $Ca_v1.2$ channels, the functional impacts of the TS mutations on channel properties to relate to clinical features, and the possible therapeutic approaches in the future.

11.1 Introduction

The voltage-gated calcium channel $Ca_v1.2$ are known to be expressed diversely in different organs and perform critical physiological functions ranging from maintaining long-term potential (LTP), contraction of cardiac and vascular myocytes to

H. Huang · T. W. Soong (✉)
Department of Physiology, Yong Loo Lin School of Medicine, Singapore, Singapore
e-mail: phsstw@nus.edu.sg

J. Wang
Department of Physiology, Nanjing Medical University, Nanjing, China

T. W. Soong
NUS Graduate School for Integrative Sciences and Engineering, singapore, singapore

T. W. Soong
Neurobiology/Ageing Programme, National University of Singapore, Singapore, Singapore

T. W. Soong
National Neuroscience Institute, Singapore, Singapore

N. Weiss and A. Koschak (eds.), *Pathologies of Calcium Channels*,
DOI: 10.1007/978-3-642-40282-1_11,

hormonal secretion. It is therefore not surprising that drastic alteration of channel properties leads to multiorgan failure as observed in the Timothy Syndrome. TS mutations were found to be located within a pair of mutually exclusive exon 8a and 8. Conventionally, between exons 8a and 8, the exon upstream in the genomic DNA sequence was named as exon 8a and the downstream, exon 8 (Soldatov 1994; Tang et al. 2007; Zuhlke et al. 1998). However, some other reports named exon 8 as 8A and exon 8a, 8 (Splawski et al. 2004, 2005). For the purpose of consistency, we will adopt the nomenclature of exon 8a and 8 throughout the whole discussion. Mutually exclusive exon 8a and 8 represent only a type of alternative splicing in the context of complex splicing pattern of $Ca_v1.2$ channels transcripts and in human; 20 out of 56 exons were identified to undergo alternative splicing. To offer a more holistic picture, this review starts with a discussion of functional roles and regulation of alternative splice pattern within $Ca_v1.2$ channels. Subsequently, we will discuss the recent progress in understanding the mechanisms that underlie the drastic functional effects of the TS mutations on channel properties. The discussion is further supplemented with a review of clinical features of TS and lastly, the application of cutting-edge induced pluripotent stem cells (iPSC) as a potential therapeutic approach in future.

11.2 Functional Impacts of Alternative Splicing on $Ca_v1.2$ Calcium Channels

11.2.1 Introduction

The transcripts of $Ca_v1.2$ channels undergo complex alternative splicing patterns (Fig. 11.1) that are often developmentally and tissue-specifically regulated, generating a variety of channel variants with unique or subtle changes in channel properties that could be customized for various physiological or pathophysiological conditions. Identifying and characterizing different splice isoforms not only aid in the understanding of basic calcium channel physiology but also provide important answers on how native $Ca_v1.2$ currents support normal physiological processes, and on how dysregulation of alternative splicing in channels leads to diseases. The splice variants implicated in disease could be potential targets for future drug design and therapy.

11.2.2 N-Terminus

The expressions of exon 1a and 1b of human $Ca_v1.2$ channels are under the control of different promoters upstream of the respective exons (Dai et al. 2002; Pang et al. 2003; Saada et al. 2003). Exon 1a lies upstream of exon 1b in the genomic

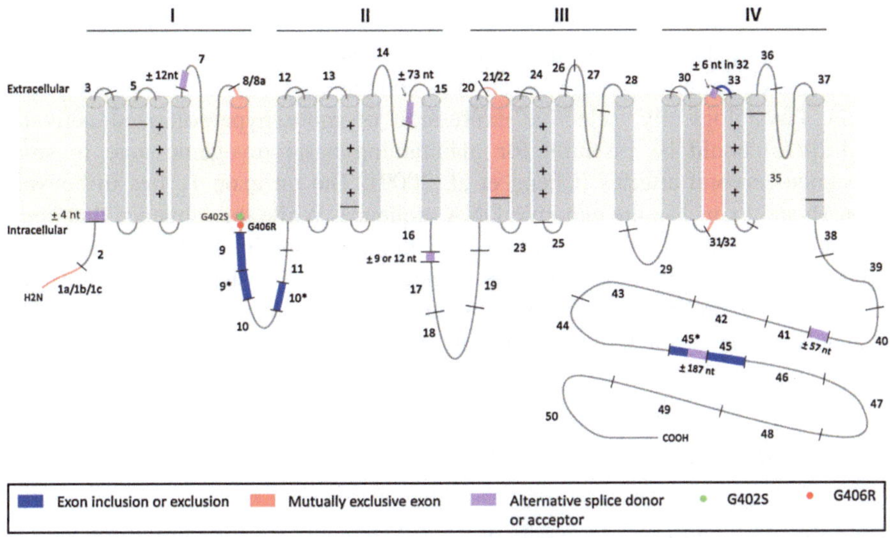

Fig. 11.1 Schematic representation of the pore-forming Ca$_v$1.2-subunit adapted from Liao et al. (2005), highlighting identified sites of alternative splicing. The site of G402S and G406R mutations are indicated by green and red circles, respectively. The channel is composed of four domain repeats, each containing six transmembrane segments. The S5 and S6 segments and the associated re-entrant loop serve as the selective filter, while S1 through to S4 segments function collaboratively as voltage sensor. The four domains are linked by three intracellular loop domains including I–II loop, II–III loop, and III–IV loops. The exons are indicated numerically and the exon boundaries between exons are marked by lines

mapping of the *CACNA1C* gene and activation of the respective promoter induces the expression of respective exon 1a or 1b in a tissue-specific manner. While exon 1a containing Ca$_v$1.2 transcripts (Ca$_v$1.2[e1a]) are selectively expressed in the adult and neonatal cardiac muscles (Cheng et al. 2007; Dai et al. 2002; Tang et al. 2004), alternative use of the other promoter upstream of exon 1b drives the expression of Ca$_v$1.2[e1b] mainly in smooth muscle (Saada et al. 2003) and more ubiquitously in other tissues including heart, brain, uterus, small intestine, kidney, and adrenal gland (Pang et al. 2003; Tang et al. 2004). Functionally, the inclusion of exon 1a allows the cardiac Ca$_v$1.2 current to be regulated by PKC-mediated mechanism (McHugh et al. 2000; Shistik et al. 1998). However, whether exon 1a is subjected to phosphorylation by PKC remains controversial (McHugh et al. 2000; Shistik et al. 1998; Yang et al. 2009). On the other hand, the presence of exon 1a is also required for β-subunit-mediated enhancement of maximal open probability of Ca$_v$1.2 channels (Shistik et al. 1999).

A novel exon 1c was more recently identified in rat arterial myocytes and to a much lower level in cardiac myocytes (Cheng et al. 2007). In comparison, exon 1a was not detected and 1b showed very low expression in arterial smooth muscles (Cheng et al. 2007). Consistently, shRNA targeting exon 1c reduced the Ca$_v$1.2 current recorded from arterial myocytes and induced more efficient vasodilatation as compared to the knockdown of exon 1b (Bannister et al. 2011). Biophysically, the rat $Ca_v1.2$[e1c] has more negative window current and lower current density due to less efficient membrane insertion as compared to $Ca_v1.2$[e1b] (Cheng et al. 2007). Physiologically, selective expression of more hyperpolarized activated $Ca_v1.2$[e1c] could be essential for maintaining basal myogenic tone in small resistance cerebral arteries (Cheng et al. 2009). The rat exon 1c lies in between exon 1b and exon 2 in the genomic DNA sequence. While the sequence that codes for rat exon 1c was not found in human, a human genomic DNA sequence at the same position shares high sequence homology to the upstream 5' untranslated region of rat exon 1c, indicating possibly another alternate exon 1 in human $Ca_v1.2$ channels besides exon 1a and 1b (Cheng et al. 2009).

11.2.3 IS6, IIIS2, and IVS3 Segments

The IS6, IIIS2, and IVS3 segments of $Ca_v1.2$ channels are coded by three pairs of mutually exclusive exons: 8a/8, 21/22, and 31/32. Electrophysiologically, alternate inclusion of exon 8a results in more hyperpolarized shift in the steady-state-inactivation, slower voltage-dependent inactivation, and twice less sensitivity to DHP inhibition as compared to exon 8 (Welling et al. 1997; Zuhlke et al. 1998). While $Ca_v1.2$[e8a] is strongly expressed in cardiac muscles and brain, $Ca_v1.2$[e8] prevails in vascular smooth muscles (Splawski et al. 2004; Tang et al. 2007; Welling et al. 1997). De novo mutation G406R in exon 8 was associated with Timothy syndrome type 1 (TS-1), a multi-organ, developmental dysfunction characterized by symptoms including lethal arrhythmia, congenital heart disease, cognitive abnormality, autism, immune deficiency, intermittent hypoglycemia, and webbing of fingers and toes (Splawski et al. 2004). Biophysically, G406R mutation almost completely abolishes voltage-dependent inactivation (VDI) and significantly slows down calcium-dependent inactivation (CDI). Long-lasting channel opening therefore explains certain pathological features such as prolonged QT interval, leading to arrhythmia while chronic calcium overload in neurons could result in cognitive impairment and mental retardation. Subsequently, two other de novo mutations, G402S and G406R in exon 8a were identified in a more severe variant of TS, known as TS-2. Similarly, G402S and G406R mutations in exon 8a also slowed down channel inactivation. The extreme long prolongation of the QT interval and more severe mental retardation observed in TS-2 patients as compared to TS-1 could be explained by the dominant expression of $Ca_v1.2$[e8a] in both brain and heart.

On the other hand, inclusion of exon 21 or 22 did not alter the gating properties of the channel but the sensitivity of the $Ca_v1.2$ channel to DHP was significantly reduced with inclusion of exon 21 (Zuhlke et al. 1998) and the expression of $Ca_v1.2$[e21] is specifically upregulated in the aorta (Tang et al. 2007). Lastly, $Ca_v1.2$ channels with either exon 31 or 32 display similar electrophysiological and pharmacological properties (Tang et al. 2007; Zuhlke et al. 1998). Interestingly, a developmental switch from exon 31 to exon 32 was observed from fetal to adult heart (Diebold et al. 1992). Moreover, the expression level of $Ca_v1.2$[e31] increased significantly in cardiac hypertrophy induced by myocardial infarction in the adult rat (Gidh-Jain et al. 1995). The reversion of adult channel isoform to a fetal form could therefore be a useful genetic marker for cardiac failure.

11.2.4 I–II Loop

The cassette exon 9* codes for 25 amino acids immediately downstream of exon 9 in the I–II loop domain of the channel. The expression pattern of exon 9* is spatially regulated with higher abundance in vascular myocytes but limited expression in brain and heart (Cheng et al. 2009; Liao et al. 2004; Nystoriak et al. 2009). Inclusion of exon 9* induces hyperpolarizing shift in the window current closer to the resting membrane potential of vascular myocytes, thereby allowing the channels to be activated with slight depolarizing stimuli. The tonic calcium influx could thus be critical for maintaining the basal vascular tone (Liao et al. 2004). Downregulation of $Ca_v1.2$[e9*] with anti-sense RNA alone resulted in drastically reduced constriction of cerebral arteries in response to K^+-induced depolarization (Nystoriak et al. 2009). A recent discovery showed splice-variant selective modulation of $Ca_v1.2$ channels by galectin-1, a lectin that binds beta-galactoside that is expressed in arterial smooth muscle, but not in cardiac muscle. The mechanism by which $Ca_v1.2_{\Delta e9*}$ channels, that lacked exon 9*, are down-regulated is via the masking of the export signal upon galectin-1 binding and the overall consequence leads to the dominant expression of $Ca_v1.2_{9*}$ variants on the cell surface (Wang et al. 2011).

11.2.5 IVS3–IVS4 Linker

The extracellular IVS3–IVS4 linker represents another hotspot for alternative splicing. The peptide sequence of this loop is encoded by distal sequence of mutually exclusive exon 31/32, alternate exon 33, and proximal part of constitutive exon 34 (Tang et al. 2004). In addition, alternate use of splice donor site within exon 32 results in deletion of 6 nucleotides and consequently, shortening of two amino acids within IVS3–IVS4 linker. The combinatorial assortment of different alternative splice exons including 31/32, 32 ±6nt, and ±32 therefore could generate 8 combinations just within this 24 amino acid loop region. While the $Ca_v1.2$[e32, e33] is the dominant isoform in human brain and heart, exon 33 is developmentally downregulated in human heart, resulting in significant expression of $Ca_v1.2$[e32, −e33] in the adult human heart (Tang et al. 2004). Biophysically, the length of IVS3–IVS4 linker serves to regulate the activation potential of $Ca_v1.2$ channel; the longer amino acid sequence the more depolarized is the activation potential (Tang et al. 2004). In addition, deletion of exon 33 alone resulted in functional variations including hyperpolarized window current, faster voltage-dependent inactivation and higher sensitivity to DHP blockers (Liao et al. 2007). Thus, similar to exon 9*, expression of $Ca_v1.2[-e33]$ in vascular myocytes serves to customize the $Ca_v1.2$ isoforms with lower activation threshold to the maintaining basal vasotone and more sensitive contractility in arteries.

11.2.6 C-Terminus

Two cassette exon 45a and 45 are located at the C-terminus of the channel. Alternative inclusion or exclusion of either or both exons did not result in observable changes in the electrophysiological properties of the channel (Klockner et al. 1997). However, interestingly, the proximal section of exon 45a was shown to be important for the hypoxia and carbon monoxide (CO)-induced reversible inhibition of $Ca_v1.2$ current (Fearon et al. 1997, 2000; Scragg et al. 2008). The mechanism appeared to involve redox modification of the three cysteine residues including C1789, C1790, and C1810 located at the proximal section of exon 45a as a result of reactive oxygen species (ROS) production specifically from complex III of mitochondria (Scragg et al. 2008). Carbon monoxide is produced under conditions of stress, such as myocardial infarction. The upregulation of heme oxygenase (HO-1) in the arterial and ventricular myocytes results in increase CO production, a product of heme catabolism (Lakkisto et al. 2002). The decrease of the calcium influx through $Ca_v1.2_{45a}$ variant could be an inherent adaptive mechanism that prevents further cell death during ischemic heart attack and myocardial infarction. Indeed, ischemia-induced necrotic cell death could be similarly prevented by L-type channel blockers or reducing external Ca^{2+} (Uemura et al. 2005).

11.2.7 Other Aberrant Alternative Splicing Patterns

Beyond the above splice variants that have been relatively better characterized, additional alternative splicing patterns have also been described (Liao et al. 2005). More interestingly, aberrant splicing patterns that result in production of nonfunctional channels have been reported. Exclusion of both exon 8 and 8a was detected on the transcript level that could potentially result in the deletion of the critical IS6 segment (Tang et al. 2004). Exclusion of exon 19 and exclusion of both exon 17 and 18 have been identified in the rabbit $Ca_v1.2$ channels (Wielowieyski et al. 2001). In silico prediction suggested that both splice forms lead to frameshifting of downstream sequence and early termination in exon 20 which codes for part of the II–III loop region, resulting in the production of hemichannels with only transmembrane domain I and domain II. The expression of the hemichannels could be confirmed on both the transcript and protein level in rabbit heart (Wielowieyski et al. 2001). Although the exact function of such membrane protein is currently unknown, similar truncated two-domain channels have also been reported for $Ca_v2.1$ and $Ca_v2.2$ channels. Functionally, co-transfection with the truncated $Ca_v2.1$ and $Ca_v2.2$ suppressed the expression of full-length channels via either competing for the binding to the β-subunit (Arikkath et al. 2002) or through activation of the unfolded protein response system (Page et al. 2004; Raghib et al. 2001), respectively.

11.2.8 Full-Length $Ca_v1.2$ Clones

More importantly, studies that reported on the full-length sequence of $Ca_v1.2$ transcripts allow a more comprehensive understanding of tissue-specific expression of dominant splice isoform. While $Ca_v1.2$[e1a, 8a, −9*, 21/22, 32, 33] has been identified to be the major isoform expressed in the rat heart, $Ca_v1.2$[e1b, 8, ± 9*, 21, 32, ±33] is dominantly expressed in the aorta (Tang et al. 2007). Interestingly, additional information such as exon linkage could also be revealed via examining the full-length clones. For example, the expression of exons 1a and 8a are ~100 % linked in rat heart, meaning the expression of exon 1a coincides with the expression of exon 8a (Tang et al. 2007). On the other hand, in the small resistant arterial myocytes, expression of exon 9* occurs in 72 % of clones containing exon 1c but only 33 % in clones that contained exon 1b (Cheng et al. 2009).

11.2.9 Altered Alternative Splicing Patterns in Pathology

$Ca_v1.2$ transcripts have been shown to undergo differential splicing patterns in various pathological conditions, leading to the generation of various channel isoforms with altered channel properties that possibly underlie certain aspects of disease development or adaptation. Analysis of vascular smooth muscle cells samples extracted from human arterial region of atherosclerotic plaques revealed downregulation of exon 9* and 41A and upregulation of exon 22, 32, and 33 in the diseased tissue as compared to the nearby normal tissue (Tiwari et al. 2006). The altered splicing pattern was associated with increased calcium influx that could possibly contribute toward proliferation and migration of VSMC during the progression of atherosclerosis (Tiwari et al. 2006).

On the other hand, altered alternative splicing pattern was also observed in spontaneously hypertensive rats (SHRs) as compared to the normotensive Wistar Kyoto Rats (WKYs). Particularly, the amount of aberrant splice isoforms whereby the mutually exclusive exons 21 and 22 were both included increased drastically in the hypertensive rats as compared to the normal rats (Tang et al. 2008).

Interestingly, although $Ca_v1.2_{21+22}$ channels upon transiently transfected in HEK 293 cells did not yield any measurable current, co-transfection of such splice variants with normal functional channels resulted in substantial reduction of observed current density (Tang et al. 2008). Although native Ca^{2+} current recorded from both SHR and WKY ventricular myocytes did not differ substantially in the gating properties and pharmacological sensitivity toward DHP blockers (Tang et al. 2008), it is yet to be determined if in the native system, upregulation of aberrantly spliced channel isoforms serve to modulate the expression or activity of calcium channel as a compensatory mechanism to counteract hypertension.

11.2.10 Regulatory Mechanism of Alternative Splicing

Although $Ca_v1.2$ transcripts are subject to extensive alternative splicing, often in a developmental and tissue-specific regulated manner, only several recent studies have started to elucidate how alternative splicing of the $Ca_v1.2$ transcripts could be regulated by splicing factors such as the Fox protein family. Expression of Fox 1, 2, and 3 proteins in the mouse cortex increased from embryonic day 12 through to embryonic day 18, correlating with increasing exclusion of exon 9* and progressive inclusion of exon 33 (Tang et al. 2009). As demonstrated using heterologously transfected cell lines, downregulation of Fox 2 protein promoted inclusion of exon 9* while repressed inclusion of exon 33. The reverse experiment over-expressing either Fox 1 or Fox 2 decreased inclusion of exon 9* and increased inclusion of exon 33. Terminologically, when splicing occurs in a particular exon, it means the exon is included in the final mature transcript and when the splicing is repressed for a particular exon, it means exon is skipped and therefore excluded. While Fox 1 and 2 binds to the Fox binding element downstream of exon 33 to promote its splicing, the regulation of splicing of exon 9* is more complicated and requires binding of two Fox binding elements; one upstream and one within exon 9* (Tang et al. 2009). In addition, in silico prediction suggested that exon 32 and exon 37 were also subjected to Fox 2 regulation, while exon 32 is a mutually exclusive to exon 31, exon 37 is constitutively expressed (Zhang et al. 2008).

On the other hand, the splicing pattern of mutually exclusive exon 8 and 8a was found to be controlled by another regulatory factor known as polypyrimidine tract-binding protein (PTB). PTB represses the splicing of exon 8a by binding to the regulatory element upstream of exon 8a, resulting in the dominant expression of exon 8 during embryonic mouse brain. Downregulation of PTB and also its neuronal homologue nPTB during neuronal differentiation therefore mediate a switch from exon 8 to 8a splicing (Tang et al. 2011). More importantly, understanding the spatial and temporal regulation of exon 8/8a splicing in human via correlation with the activity of PTB would offer better understanding of the spectrum of defects in Timothy syndrome due to either exon 8 or 8a mutations.

11.3 Functional Impacts of Timothy Syndrome Mutations

11.3.1 Introduction

The gain-of-function G406R in exon 8 and G402S and G406R in exon 8a significantly diminish the voltage-dependent inactivation (VDI) kinetic of $Ca_v1.2$ channels, leading to the multiorgan defects observed in Timothy syndrome. Numerous studies following the initial reports by Splawski et al. (2004, 2005) have been performed in order to understand the drastic change of channel properties due to such mutations (Barrett and Tsien 2008; Cheng et al. 2011; Depil et al. 2011; Raybaud et al. 2006; Thiel et al. 2008; Yarotskyy et al. 2009). More detailed analysis of elementary currents revealed that G406R enhanced the 'mode 2' gating of the channels, leading to more frequent and prolonged channel opening (Erxleben et al. 2006). Further electrophysiological recording showed that G406R display more hyperpolarized activation (Raybaud et al. 2006) and slower deactivation (Yarotskyy et al. 2009). Subsequent in silico modeling demonstrated that slower inactivation kinetic and deactivation were sufficient to cause prolonged cardiac action potential duration, leading to cardiac arrhythmia (Yarotskyy et al. 2009). Despite the well-established biophysical impacts of G402S and G406R mutations, the mechanism underlying such effects of the mutations on channel properties remains unclear. Several hypotheses have been proposed.

11.3.2 Structure–Function Implications of Timothy Syndrome Mutations

Structural prediction indicates that G402 is situated close to the cytoplasmic end of the IS6 helical structure while G406 is located within a short stretch of nonhelical motif at the start of the cytoplasmic I–II loop (Fig. 11.2). The process of VDI is initiated upon voltage-dependent conformational rearrangement of the voltage-sensing domain, which includes the S1–S4 segments (Swartz 2008), leading to the opening of the S6 gate which comprises of the distal part of four S6 segments (Liu et al. 1997; Xie et al. 2005). Inactivation, which is coupled to activation, is a process that involves the occlusion of the gate by the I–II loop via a 'hinge lid' mechanism.

It has been suggested that the small glycine residues offer flexibility to allow for the I–II loop to interact with the intracellular pore of the channel (Bannister et al. 2011; Depil et al. 2011; Raybaud et al. 2006; Splawski et al. 2004; Stotz et al. 2004). Mutations of respective glycines to more bulky residues therefore impede the movement of the I–II loop, leading to slower channel inactivation (Fig. 11.3). In addition, the glycine residue at 402 position of IS6 contributes to the formation of G/A/G/A motif with the alanine or glycine residues found at the same positions in the IIS6, IIIS6, and IVS6 segments. Each of the residues of the G/A/G/A motif interacts with the corresponding bulky hydrophobic residues in the adjacent S6

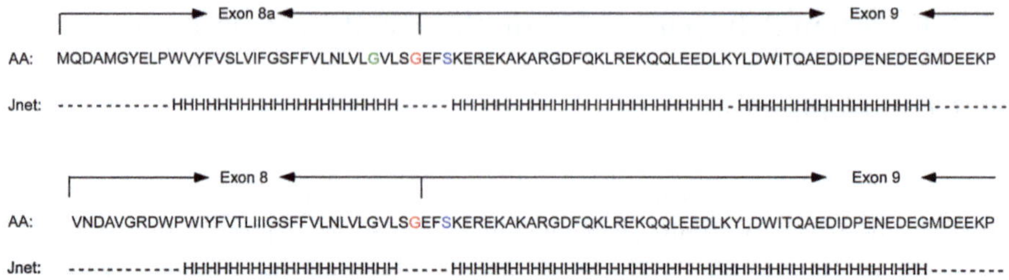

Fig. 11.2 The secondary structure prediction of $Ca_v1.2$ IS6 and immediate downstream loop domain. The secondary structure was predicted by Jpred 3 Secondary Structure Prediction Server (Cole et al. 2008). H, helix structure. Top panel, modeling with exon 8a and part of exon 9 sequences, the exon boundaries were highlighted above. G402 (*green*) lie in the helical structure with G406 (*red*) located in the short random structure. The putative phosphorylated serine 409, as a result of G406R mutation, is colored blue. Bottom panel, modeling with exon 8 and exon 9 sequences

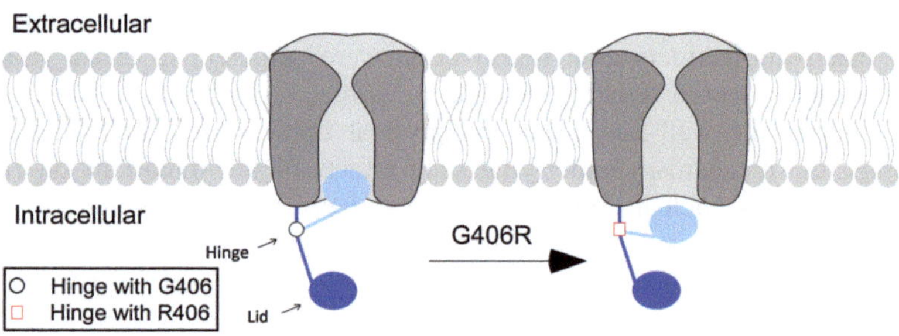

Fig. 11.3 Proposed model showing G406R mutation disrupts the flexible hinge domain via introduction of bulky arginine side chain. The additional steric hindrance impedes the movement of the lid domain in the process of pore occlusion, therefore delaying the channel inactivation

helices to stabilize the inactivation gate. Mutating G402 to a more bulky serine residue could therefore destabilize the tight packing of the S6 helices in the close state, resulting in slower deactivation (Depil et al. 2011) (Fig. 11.4).

11.3.3 Creation of New Phosphorylation Site Due to G406R Mutation

It has also been shown that increase of frequency and duration of channel opening associated with G406R mutation was due to the creation of a novel target for calmodulin-dependent protein kinase II (CaMKII) at position S409 in the I–II loop region (Erxleben et al. 2006). The phosphorylation of S409 could be demonstrated in vitro using $Ca_v1.2$ peptides and purified CaMKII. Pharmacological inhibition of CaMKII or S409A mutation prevented the increase of mode 2 gating in the

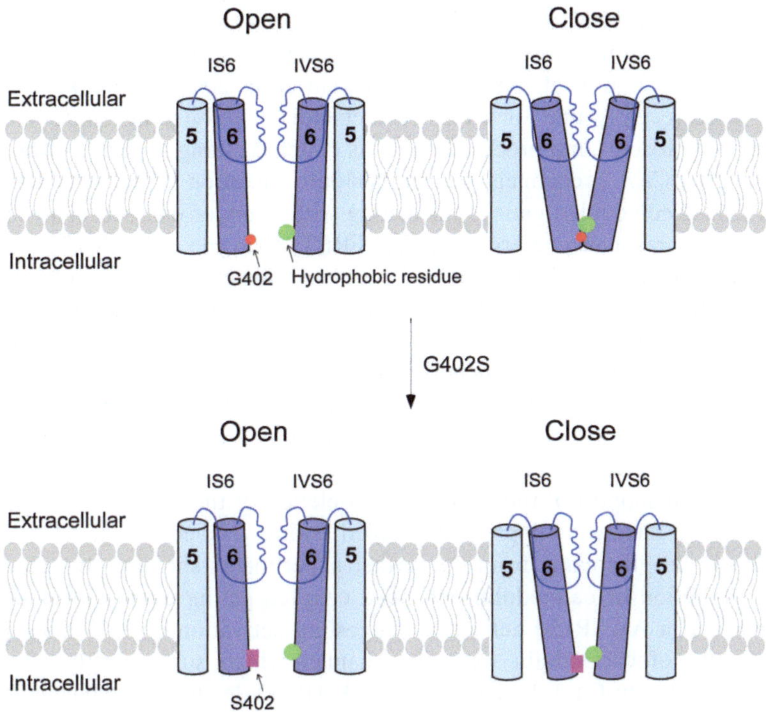

Fig. 11.4 Illustration of the disruption of tight packing of S6 inner gate via G402S mutation (adapted from Depil et al. 2011). Top panel, G402 in IS6 and hydrophobic residue located at the IVS6 residue form adhesion point to allow for tight closure of the inner gates. Bottom panel, G402S mutation introduce additional steric hindrance that could possibly disrupt the tight packing of the inner gate during channel closure

presence of G406R (Erxleben et al. 2006). Moreover, mutating S409 to the negatively charged glutamate residue, mimicking phosphorylation, similarly slowed down VDI (Yarotskyy et al. 2009). Additional support for the phosphorylation hypothesis was derived from results of experiments using rat ventricular myocyte transfected with the G406R $Ca_v1.2$ construct. Such myocyte model of TS displayed proarrhythmic features such as slower inactivating Ca^{2+} current, prolongation of action potential duration, and after-depolarizations which were rescued by inhibiting the CaMKII (Thiel et al. 2008).

11.3.4 Enhanced Coupled Gating as a Result of G406R Mutation

Adding to the complexity, $Ca_v1.2$ channels were found to dimerize through the C-terminal preIQ-IQ motifs in complex with multiple Ca^{2+}-bound calmodulin molecules (Fallon et al. 2009). The use of extended peptide of preIQ-IQ region improved the previous derived structure of IQ peptide and calmodulin in a simple 1:1 complex (Fallon et al. 2005; Van Petegem et al. 2005). Remarkably, it was shown more recently that such clustered channels exhibit transient, coordinated opening and closing, in a process known as coupled gating (Navedo et al. 2010). Interestingly, $Ca_v1.2$ channels bearing G406R mutation ($Ca_v1.2$_G406R) displayed enhanced coupled gating, correlating with stronger interaction as observed by fluorescent resonance energy transfer (FRET) (Navedo et al. 2010). Although the mechanism is currently unclear, several factors appeared to be involved. While coupled gating was enhanced by activation of PKC and displacing calmodulin from the C-terminus of the channel, it was inhibited by the absence of scaffolding protein known as A-Kinase Anchoring Proteins 150 (AKAP150) (Navedo et al. 2010). AKAP150 has been shown to traffic PKC to the cell membrane (Klauck et al. 1996) and to also interact with the C-terminus of the channel via leucine zipper (LZ) motifs to facilitate interactions between $Ca_v1.2$ carboxyl tails (Oliveria et al. 2007). In support of this mechanism, deletion of the AKAP150 binding site within $Ca_v1.2$ channels suppressed coupled gating (Navedo et al. 2010). More mechanistic insights came from the further experiments showing that pharmacological inhibition of calmodulin enhanced coupled gating even with deletion of PKC but not in AKAP150 null cells, suggesting activation of PKC triggered the displacement of calmodulin from the channel which subsequently allows for interaction between $Ca_v1.2$ channels and AKAP150, leading to enhanced coordination of channel activity (Navedo et al. 2010).

The coupled gating mechanism may be critical for excitation–contraction (EC) coupling in ventricular myocytes as only the coordinated opening of several calcium channels existing in clusters could generate sufficient Ca^{2+} influx to induce the opening of apposed ryanodine receptors (RyRs) that are located in the sarcoplasmic reticulum (SR) (Inoue and Bridge 2003). The Ca^{2+}-induced Ca^{2+} release (CICR) from SR then increase the global Ca^{2+} concentration to trigger muscle contraction. However, excessive Ca^{2+} influx in TS could prolong ventricular contraction and lead to arrhythmia. Indeed, in a more physiological setting, analysis of ventricular myocytes of mice model of TS which carry the $Ca_v1.2$_G406R mutation revealed enhanced coupled gating, correlating with slower inactivation of macroscopic whole cell current and also arrhythmia, typical of Timothy syndrome. Further deletion of AKAP150 in $Ca_v1.2$_G406R rescued all the phenotypic abnormalities to the wildtype therefore suggesting that $Ca_v1.2$ bearing G406R preferentially coupled to AKAP150 and such interaction stabilized the open conformation of the channel and increased the propensity of coupled gating (Cheng et al. 2011). A more recent study further suggested that optically induced coupling of wildtype (WT) $Ca_v1.2$ and $Ca_v1.2$_G406R channels induced WT channels to behave like a TS channel with prolonged sustained channel opening, offering a possible explanation to the autosomal dominant effect of TS mutations (Dixon et al. 2012).

11.4 Clinical Features

11.4.1 Introduction

In 1992, patients with syndactyly (webbing of fingers and toes) and suffering from a novel form of arrhythmia syndrome were reported (Reichenbach et al. 1992). It was then postulated that this syndrome might be a new form of long QT disorder associated with sudden death and syndactyly (Marks et al. 1995a, b). In 2004, Splawski et al. formally named this disorder as Timothy Syndrome (TS) (Splawski et al. 2004), and now it is also known as long QT syndrome 8 (LQT8).

$Ca_v1.2$ calcium channels are widely distributed in multiple organs, especially in the cardiovascular and nervous systems and they play essential roles during development and in physiological functions. Dysfunctions of the $Ca_v1.2$ channels will lead to multiorgan disorders, recognized as $Ca_v1.2$ channelopathies, including TS (Liao and Soong 2010). TS mutant channels with gain-of-function mutation of the α_1-subunit of the $Ca_v1.2$ channels are characterized by a slowing of their voltage-dependent inactivation (VDI) property (Splawski et al. 2004, 2005). The net result was predicted to be an increase in cellular Ca^{2+} influx that in turn contributes to failure in multisystems, but especially in cardiac and neural tissues. The two forms of TS are type 1 (classic) and type 2 (atypical), which were caused by mutations in an alternative transcript variant of the exon 8 or exon 8a in *CACNA1C* gene, respectively (Splawski et al. 2004, 2005). Moreover, TS-1 patients with somatic mosaicism were presented with less severe clinical features than the typical TS-1 patients (Etheridge et al. 2011).

11.4.2 Timothy Syndrome Type 1

Phenotypic features of classic TS-1 (type 1) include cardiac arrhythmia, cutaneous syndactyly, craniofacial and neuropsychiatric disorders (Marks et al. 1995a, b; Reichenbach et al. 1992; Splawski et al. 2004, 2005).

11.4.2.1 Cardiac Manifestations

It has been demonstrated experimentally that the G406R mutant $Ca_v1.2$ channels have an impaired VDI, which led to a sustained increase of Ca^{2+} influx that in cardiomyocytes resulted in the prolongation of the action potential duration (APD). TS patients' electrocardiographs (ECG) display a longer QT interval and they could develop cardiac arrhythmia and may die from it (Splawski et al. 2004). A rate-corrected QT interval (QTc) in TS patients is between 480 ms and 700 ms (Marks et al. 1995a, b; Reichenbach et al. 1992; Splawski et al. 2004). In contrast to other long QT syndromes, TS patients also commonly display atrioventricular

(AV) block and T-wave alternans (TWA). About 80 % of classic TS patients have 2:1 AV block (Marks et al. 1995a, b; Gillis et al. 2011; Splawski et al. 2004; Krause et al. 2011; Etheridge et al. 2011), which is not a result of malfunction of the AV node, but is more likely caused by the severe prolongation of ventricular repolarization and refractory periods; while TWA can be found in ~70 % of TS patients (Gillis et al. 2011; Splawski et al. 2004) and these TWAs displayed on a beat-to-beat basis, both positive and negative T waves. Ventricular tachyarrhythmia (ventricular tachycardia and ventricular fibrillation) is the leading cause of TS mortality, and accounts for 80 % of death of TS patients. The average age at death with classic TS is 2.5 years in children (Splawski et al. 2011).

Furthermore, congenital heart defects are present in approximately 61 % of individuals and include patent ductus arteriosus (PDA), patent foramen ovale (PFO), ventricular septal defect (VSD), tetralogy of Fallot (TOF), and hypertrophic cardiomyopathy (HCM) (Marks et al. 1995a; Splawski et al. 2004).

11.4.2.2 Extracardiac Manifestations

Other phenotypic features of TS-1 include bilateral cutaneous syndactyly of toes two and three or cutaneous syndactyly that involve the index, middle, ring, and little fingers (Splawski et al. 2011). The craniofacial findings include baldness followed by thin scalps, ears that are low-set, flat nasal bridge, and small upper jaw (Splawski et al. 2004).

Moreover, in approximately 80 % of individuals there are neuropsychiatric involvement. Developmental delays in the central nervous system were observed in TS patients, and that may lead to a series of symptoms such as generalized impairments in cognition, language, or motor abilities (Splawski et al. 2004, 2005). Children with TS were impaired in all areas of adaptive functions affecting daily living and including communication skills and socialization. Some children babbled and were unable to produce speech sounds during infancy, while others had difficulties in articulating expressive language. Splawski et al. evaluated five children for autism and among them, three were diagnosed with autism spectrum disorders (ASD), and one with severe delays in the development of language (Splawski et al. 2004).

In addition, episodic hypoglycemia and hypocalcemia have also be described in the individuals with TS (Splawski et al. 2004), which may be due to the irregular function of endocrinal organ because of TS-mutant $Ca_v1.2$ channels, correlating with the essential role of $Ca_v1.2$ channels in regulating insulin release (Nitert et al. 2008; Schulla et al. 2003). Frequent infections, such as in the sinus, ear, or respiratory system, secondary to altered immune responses and intermittent hypoglycemia, were also found in TS patients, suggesting an important role $Ca_v1.2$ channels may play in the immune system is yet to be fully elucidated.

11.4.3 Timothy Syndrome Type 2

Due to the dominant expression of *CACNA1C* exon 8a (about 80 %) in cardiac and brain tissue, mutation of exon 8a in TS-2 were associated with more severe cardiac and neurological defects (Splawski et al. 2005). Two cases of TS-2 were reported and patients suffered from multiple arrhythmias and sudden death owing to extreme prolongation of the QT interval (QTc ranging from 620 to 730 ms), but syndactyly was not manifested (Splawski et al. 2005). One severe intellectually disabled child had nemaline rods detected in the muscle biopsy and this condition was considered to arise secondarily to prolonged immobility rather than to being a primary muscle disorder (Splawski et al. 2005). However, in these two TS-2 individuals, their deaths were not related to tachyarrhythmia. Instead, one patient died from severe infection while the other died from complications arising from intractable hypoglycemia. More recently, the mouse model with TS-2 (G406R in exon 8) has displayed some traits of autism, indicating TS-2 mutation can influence multiple and distinct behavioral defects (Bader et al. 2011).

11.5 Genetic Diagnosis of Timothy Syndrome

Mutations found in *CACNA1C*, the gene encoding the $Ca_v1.2$ calcium channel, are known to be associated with TS as revealed by genetic screenings (Splawski et al. 2004, 2005). While all patients with the classical TS-1 have been identified with the G406R mutation in exon 8 of the *CACNA1C* gene (Splawski et al. 2004), patients with atypical TS-2 phenotypes carried two mutations, G406R or G402S, in exon 8a (Splawski et al. 2005). However, the milder phenotype resulting from somatic cell mosaicism could be hard to detect (Etheridge et al. 2011). Mosaicism implies that in an organism from a single zygote there is the presence of genetically distinct cell lines (Youssoufian and Pyeritz 2002). For instance, as described by Splawski et al., in 2004, G406R mutation in exon 8 could be detected in the oral muscosa sample but not in the blood sample while the germline mosaicism in the mother could have gone unnoticed (Splawski et al. 2004). Careful screening of parental tissue in family with incidence of TS, prenatal DNA screening, and monitoring of the fetal echocardiogram are therefore extremely important. Furthermore, in addition to the two widely known de novo G406R or G402S mutations, recently, another mutation in exon 38 was reported in a patient with TS but who did not have the G406R or G402S mutations (Gillis et al. 2011), thus expanding the molecular basis of TS.

11.6 Potential Therapeutic Approaches in Future

The advent of induced pluripotent stem cells (iPSC), a type of pluripotent stem cells artificially reprogrammed from somatic cells has created much excitement because of the possibility of producing unique patient- and disease- specific human iPSC (hiPSC) lines (Takahashi et al. 2007; Takahashi and Yamanaka 2006; Yu et al. 2007). The Dolmetsch's group has generated iPSCs, which were reprogrammed from human skin cells from TS patients, and differentiated them into cardiomyocytes and cortical neuronal precursor cells. Analysis of such cells uncovered cellular phenotypes associated with TS which could be reversed with roscovitine which is both an atypical L-type—channel blocker and a cyclin-dependent kinase inhibitor (Yazawa et al. 2011; Pasca et al. 2011). The iPSC approach could potentially be a powerful tool for diagnosis and prognosis of TS. It could also provide a cellular model of TS to enable detailed analysis and discovery of the pathomechanisms of the disease and to serve as substrates for testing of potential drugs for therapy (Huttner and Rakic 2011).

11.7 Conclusion

Timothy Syndrome is a life-threatening multiorgan disorder correlating with diverse expression patterns and physiological roles of $Ca_V1.2$ channels. Given the autosomal dominant effect of TS mutations, careful paternal screening especially in the germline and prenatal monitoring of the fetus are therefore critical. The de novo nature and somatic mosaicism of TS mutations also imply the possibility of milder tissue-specific symptoms in the population that could have evaded the typical clinical diagnosis. Recent progress in research has provided mechanistic insights toward the better understanding of the impacts of TS mutations on channel properties and the ensuing physiological consequences. Deriving iPSCs from affected patients and subsequent differentiation of such cells into multiple cell lines would hopefully allow better understanding of the multiorgan manifestations of such disorder and opportunity of possible discovery of drugs for therapeutic treatment in future.

References

Arikkath J, Felix R, Ahern C, Chen CC, Mori Y, Song I, Shin HS, Coronado R, Campbell KP (2002) Molecular characterization of a two-domain form of the neuronal voltage-gated P/Q-type calcium channel alpha(1)2.1 subunit. FEBS Lett 532(3):300–308

Bader PL, Faizi M, Kim LH, Owen SF, Tadross MR, Alfa RW, Bett GC, Tsien RW, Rasmusson RL, Shamloo M (2011) Mouse model of Timothy syndrome recapitulates triad of autistic traits. Proc Natl Acad Sci U S A 108(37):15432–15437

Bannister JP, Thomas-Gatewood CM, Neeb ZP, Adebiyi A, Cheng X, Jaggar JH (2011) Ca(V)1.2 channel N-terminal splice variants modulate functional surface expression in resistance size artery smooth muscle cells. J Biol Chem 286(17):15058–15066

Barrett CF, Tsien RW (2008) The Timothy syndrome mutation differentially affects voltage— and calcium-dependent inactivation of $Ca_v1.2$ L-type calcium channels. Proc Natl Acad Sci U S A 105(6):2157–2162

Cheng EP, Yuan C, Navedo MF, Dixon RE, Nieves-Cintron M, Scott JD, Santana LF (2011) Restoration of normal L-type Ca2 + channel function during Timothy syndrome by ablation of an anchoring protein. Circ Res 109(3):255–261

Cheng X, Liu J, Asuncion-Chin M, Blaskova E, Bannister JP, Dopico AM, Jaggar JH (2007) A novel Ca(V)1.2 N terminus expressed in smooth muscle cells of resistance size arteries modifies channel regulation by auxiliary subunits. J Biol Chem 282(40):29211–29221

Cheng X, Pachuau J, Blaskova E, Asuncion-Chin M, Liu J, Dopico AM, Jaggar JH (2009) Alternative splicing of $Ca_v1.2$ channel exons in smooth muscle cells of resistance-size arteries generates currents with unique electrophysiological properties. Am J Physiol Heart Circ Physiol 297(2):H680–H688

Cole C, Barber JD, Barton GJ (2008) The Jpred 3 secondary structure prediction server. Nucleic Acids Res 36(Web Server issue):W197–201

Dai B, Saada N, Echetebu C, Dettbarn C, Palade P (2002) A new promoter for alpha1C subunit of human L-type cardiac calcium channel Ca(V)1.2. Biochem Biophys Res Commun 296(2):429–433

Depil K, Beyl S, Stary-Weinzinger A, Hohaus A, Timin E, Hering S (2011) Timothy mutation disrupts the link between activation and inactivation in Ca(V)1.2 protein. J Biol Chem 286(36):31557–31564

Diebold RJ, Koch WJ, Ellinor PT, Wang JJ, Muthuchamy M, Wieczorek DF, Schwartz A (1992) Mutually exclusive exon splicing of the cardiac calcium channel alpha 1 subunit gene generates developmentally regulated isoforms in the rat heart. Proc Natl Acad Sci U S A 89(4):1497–1501

Dixon RE, Yuan C, Cheng EP, Navedo MF, Santana LF (2012) Ca2 + signaling amplification by oligomerization of L-type $Ca_v1.2$ channels. Proc Natl Acad Sci U S A 109(5):1749–1754

Erxleben C, Liao Y, Gentile S, Chin D, Gomez-Alegria C, Mori Y, Birnbaumer L, Armstrong DL (2006) Cyclosporin and Timothy syndrome increase mode 2 gating of $Ca_v1.2$ calcium channels through aberrant phosphorylation of S6 helices. Proc Natl Acad Sci U S A 103(10):3932–3937

Etheridge SP, Bowles NE, Arrington CB, Pilcher T, Rope A, Wilde AA, Alders M, Saarel EV, Tavernier R, Timothy KW, Tristani-Firouzi M (2011) Somatic mosaicism contributes to phenotypic variation in Timothy syndrome. Am J Med Genet A 155A(10):2578–2583

Fallon JL, Baker MR, Xiong L, Loy RE, Yang G, Dirksen RT, Hamilton SL, Quiocho FA (2009) Crystal structure of dimeric cardiac L-type calcium channel regulatory domains bridged by Ca2 + * calmodulins. Proc Natl Acad Sci U S A 106(13):5135–5140

Fallon JL, Halling DB, Hamilton SL, Quiocho FA (2005) Structure of calmodulin bound to the hydrophobic IQ domain of the cardiac Ca(v)1.2 calcium channel. Structure 13(12):1881–1886

Fearon IM, Palmer AC, Balmforth AJ, Ball SG, Mikala G, Schwartz A, Peers C (1997) Hypoxia inhibits the recombinant alpha 1C subunit of the human cardiac L-type Ca2 + channel. J Physiol 500(Pt 3):551–556

Fearon IM, Varadi G, Koch S, Isaacsohn I, Ball SG, Peers C (2000) Splice variants reveal the region involved in oxygen sensing by recombinant human L-type Ca(2 +) channels. Circ Res 87(7):537–539

Gidh-Jain M, Huang B, Jain P, Battula V, el-Sherif N (1995) Reemergence of the fetal pattern of L-type calcium channel gene expression in non infarcted myocardium during left ventricular remodeling. Biochem Biophys Res Commun 216(3):892–897

Gillis J, Burashnikov E, Antzelevitch C, Blaser S, Gross G, Turner L, Babul-Hirji R, Chitayat D (2011) Long QT, syndactyly, joint contractures, stroke and novel CACNA1C mutation:

expanding the spectrum of Timothy syndrome. Am J Med Genet A. doi:10.1002/ajmg.a.34355
Huttner A, Rakic P (2011) Diagnosis in a dish: your skin can help your brain. Nat Med 17(12):1558–1559
Inoue M, Bridge JH (2003) Ca2 + sparks in rabbit ventricular myocytes evoked by action potentials: involvement of clusters of L-type Ca2 + channels. Circ Res 92(5):532–538
Klauck TM, Faux MC, Labudda K, Langeberg LK, Jaken S, Scott JD (1996) Coordination of three signaling enzymes by AKAP79, a mammalian scaffold protein. Science 271(5255):1589–1592
Klockner U, Mikala G, Eisfeld J, Iles DE, Strobeck M, Mershon JL, Schwartz A, Varadi G (1997) Properties of three COOH-terminal splice variants of a human cardiac L-type Ca2 + -channel alpha1-subunit. Am J Physiol 272(3 Pt 2):H1372–H1381
Krause U, Gravenhorst V, Kriebel T, Ruschewski W, Paul T (2011) A rare association of long QT syndrome and syndactyly: timothy syndrome (LQT 8). Clin Res Cardiol 100(12):1123–1127
Lakkisto P, Palojoki E, Backlund T, Saraste A, Tikkanen I, Voipio-Pulkki LM, Pulkki K (2002) Expression of heme oxygenase-1 in response to myocardial infarction in rats. J Mol Cell Cardiol 34(10):1357–1365
Liao P, Soong TW (2010) $Ca_v1.2$ channelopathies: from arrhythmias to autism, bipolar disorder, and immunodeficiency. Pflugers Arch 460(2):353–359
Liao P, Yong TF, Liang MC, Yue DT, Soong TW (2005) Splicing for alternative structures of $Ca_v1.2$ Ca2 + channels in cardiac and smooth muscles. Cardiovasc Res 68(2):197–203
Liao P, Yu D, Li G, Yong TF, Soon JL, Chua YL, Soong TW (2007) A smooth muscle $Ca_v1.2$ calcium channel splice variant underlies hyperpolarized window current and enhanced state-dependent inhibition by nifedipine. J Biol Chem 282(48):35133–35142
Liao P, Yu D, Lu S, Tang Z, Liang MC, Zeng S, Lin W, Soong TW (2004) Smooth muscle-selective alternatively spliced exon generates functional variation in $Ca_v1.2$ calcium channels. J Biol Chem 279(48):50329–50335
Liu Y, Holmgren M, Jurman ME, Yellen G (1997) Gated access to the pore of a voltage-dependent K + channel. Neuron 19(1):175–184
Marks ML, Trippel DL, Keating MT (1995a) Long QT syndrome associated with syndactyly identified in females. Am J Cardiol 76(10):744–745
Marks ML, Whisler SL, Clericuzio C, Keating M (1995b) A new form of long QT syndrome associated with syndactyly. J Am Coll Cardiol 25(1):59–64
McHugh D, Sharp EM, Scheuer T, Catterall WA (2000) Inhibition of cardiac L-type calcium channels by protein kinase C phosphorylation of two sites in the N-terminal domain. Proc Natl Acad Sci U S A 97(22):12334–12338
Navedo MF, Cheng EP, Yuan C, Votaw S, Molkentin JD, Scott JD, Santana LF (2010) Increased coupled gating of L-type Ca2 + channels during hypertension and Timothy syndrome. Circ Res 106(4):748–756
Nitert MD, Nagorny CL, Wendt A, Eliasson L, Mulder H (2008) $Ca_v1.2$ rather than $Ca_v1.3$ is coupled to glucose-stimulated insulin secretion in INS-1 832/13 cells. J Mol Endocrinol 41(1):1–11
Nystoriak MA, Murakami K, Penar PL, Wellman GC (2009) Ca(v)1.2 splice variant with exon 9* is critical for regulation of cerebral artery diameter. Am J Physiol Heart Circ Physiol 297(5):H1820–H1828
Oliveria SF, Dell'Acqua ML, Sather WA (2007) AKAP79/150 anchoring of calcineurin controls neuronal L-type Ca2 + channel activity and nuclear signaling. Neuron 55(2):261–275
Page KM, Heblich F, Davies A, Butcher AJ, Leroy J, Bertaso F, Pratt WS, Dolphin AC (2004) Dominant-negative calcium channel suppression by truncated constructs involves a kinase implicated in the unfolded protein response. J Neurosci 24(23):5400–5409
Pang L, Koren G, Wang Z, Nattel S (2003) Tissue-specific expression of two human Ca(v)1.2 isoforms under the control of distinct 5' flanking regulatory elements. FEBS Lett 546(2–3):349–354

Pasca SP, Portmann T, Voineagu I, Yazawa M, Shcheglovitov A, Pasca AM, Cord B, Palmer TD, Chikahisa S, Nishino S, Bernstein JA, Hallmayer J, Geschwind DH, Dolmetsch RE (2011) Using iPSC-derived neurons to uncover cellular phenotypes associated with Timothy syndrome. Nat Med 17(12):1657–1662

Raghib A, Bertaso F, Davies A, Page KM, Meir A, Bogdanov Y, Dolphin AC (2001) Dominant-negative synthesis suppression of voltage-gated calcium channel $Ca_v2.2$ induced by truncated constructs. J Neurosci 21(21):8495–8504

Raybaud A, Dodier Y, Bissonnette P, Simoes M, Bichet DG, Sauve R, Parent L (2006) The role of the GX9GX3G motif in the gating of high voltage-activated Ca2 + channels. J Biol Chem 281(51):39424–39436

Reichenbach H, Meister EM, Theile H (1992) The heart-hand syndrome. A new variant of disorders of heart conduction and syndactylia including osseous changes in hands and feet. Kinderarztl Prax 60(2):54–56

Saada N, Dai B, Echetebu C, Sarna SK, Palade P (2003) Smooth muscle uses another promoter to express primarily a form of human $Ca_v1.2$ L-type calcium channel different from the principal heart form. Biochem Biophys Res Commun 302(1):23–28

Schulla V, Renstrom E, Feil R, Feil S, Franklin I, Gjinovci A, Jing XJ, Laux D, Lundquist I, Magnuson MA, Obermuller S, Olofsson CS, Salehi A, Wendt A, Klugbauer N, Wollheim CB, Rorsman P, Hofmann F (2003) Impaired insulin secretion and glucose tolerance in beta cell-selective Ca(v)1.2 Ca2 + channel null mice. EMBO J 22(15):3844–3854

Scragg JL, Dallas ML, Wilkinson JA, Varadi G, Peers C (2008) Carbon monoxide inhibits L-type Ca2 + channels via redox modulation of key cysteine residues by mitochondrial reactive oxygen species. J Biol Chem 283(36):24412–24419

Shistik E, Ivanina T, Blumenstein Y, Dascal N (1998) Crucial role of N terminus in function of cardiac L-type Ca2 + channel and its modulation by protein kinase C. J Biol Chem 273(28):17901–17909

Shistik E, Keren-Raifman T, Idelson GH, Blumenstein Y, Dascal N, Ivanina T (1999) The N terminus of the cardiac L-type Ca(2 +) channel alpha(1C) subunit. The initial segment is ubiquitous and crucial for protein kinase C modulation, but is not directly phosphorylated. J Biol Chem 274(44):31145–31149

Soldatov NM (1994) Genomic structure of human L-type Ca2 + channel. Genomics 22(1):77–87

Splawski I, Timothy KW, Decher N, Kumar P, Sachse FB, Beggs AH, Sanguinetti MC, Keating MT (2005) Severe arrhythmia disorder caused by cardiac L-type calcium channel mutations. Proc Natl Acad Sci U S A 102 (23):8089–8096; discussion 8086–8088

Splawski I, Timothy KW, Priori SG, Napolitano C, Bloise R (2011) Timothy Syndrome. Gene Reviews. doi: NBK1403[bookaccession]

Splawski I, Timothy KW, Sharpe LM, Decher N, Kumar P, Bloise R, Napolitano C, Schwartz PJ, Joseph RM, Condouris K, Tager-Flusberg H, Priori SG, Sanguinetti MC, Keating MT (2004) Ca(V)1.2 calcium channel dysfunction causes a multisystem disorder including arrhythmia and autism. Cell 119(1):19–31

Stotz SC, Jarvis SE, Zamponi GW (2004) Functional roles of cytoplasmic loops and pore lining transmembrane helices in the voltage-dependent inactivation of HVA calcium channels. J Physiol 554(Pt 2):263–273

Swartz KJ (2008) Sensing voltage across lipid membranes. Nature 456(7224):891–897

Takahashi K, Tanabe K, Ohnuki M, Narita M, Ichisaka T, Tomoda K, Yamanaka S (2007) Induction of pluripotent stem cells from adult human fibroblasts by defined factors. Cell 131(5):861–872

Takahashi K, Yamanaka S (2006) Induction of pluripotent stem cells from mouse embryonic and adult fibroblast cultures by defined factors. Cell 126(4):663–676

Tang ZZ, Hong X, Wang J, Soong TW (2007) Signature combinatorial splicing profiles of rat cardiac- and smooth-muscle $Ca_v1.2$ channels with distinct biophysical properties. Cell Calcium 41(5):417–428

Tang ZZ, Liang MC, Lu S, Yu D, Yu CY, Yue DT, Soong TW (2004) Transcript scanning reveals novel and extensive splice variations in human l-type voltage-gated calcium channel, $Ca_v1.2$ alpha1 subunit. J Biol Chem 279(43):44335–44343

Tang ZZ, Liao P, Li G, Jiang FL, Yu D, Hong X, Yong TF, Tan G, Lu S, Wang J, Soong TW (2008) Differential splicing patterns of L-type calcium channel $Ca_v1.2$ subunit in hearts of Spontaneously Hypertensive Rats and Wistar Kyoto Rats. Biochim Biophys Acta 1783(1):118–130

Tang ZZ, Sharma S, Zheng S, Chawla G, Nikolic J, Black DL (2011) Regulation of the mutually exclusive exons 8a and 8 in the $Ca_v1.2$ calcium channel transcript by polypyrimidine tract-binding protein. J Biol Chem 286(12):10007–10016

Tang ZZ, Zheng S, Nikolic J, Black DL (2009) Developmental control of $Ca_v1.2$ L-type calcium channel splicing by Fox proteins. Mol Cell Biol 29(17):4757–4765

Thiel WH, Chen B, Hund TJ, Koval OM, Purohit A, Song LS, Mohler PJ, Anderson ME (2008) Proarrhythmic defects in Timothy syndrome require calmodulin kinase II. Circulation 118(22):2225–2234

Tiwari S, Zhang Y, Heller J, Abernethy DR, Soldatov NM (2006) Atherosclerosis-related molecular alteration of the human $Ca_v1.2$ calcium channel alpha1C subunit. Proc Natl Acad Sci U S A 103(45):17024–17029

Uemura K, Adachi-Akahane S, Shintani-Ishida K, Yoshida K (2005) Carbon monoxide protects cardiomyogenic cells against ischemic death through L-type Ca2 + channel inhibition. Biochem Biophys Res Commun 334(2):661–668

Van Petegem F, Chatelain FC, Minor DL Jr (2005) Insights into voltage-gated calcium channel regulation from the structure of the $Ca_v1.2$ IQ domain-Ca2 +/calmodulin complex. Nat Struct Mol Biol 12(12):1108–1115

Wang J, Thio SS, Yang SS, Yu D, Yu CY, Wong YP, Liao P, Li S, Soong TW (2011) Splice variant specific modulation of $Ca_v1.2$ calcium channel by galectin-1 regulates arterial constriction. Circ Res 109(11):1250–1258

Welling A, Ludwig A, Zimmer S, Klugbauer N, Flockerzi V, Hofmann F (1997) Alternatively spliced IS6 segments of the alpha 1C gene determine the tissue-specific dihydropyridine sensitivity of cardiac and vascular smooth muscle L-type Ca2 + channels. Circ Res 81(4):526–532

Wielowieyski PA, Wigle JT, Salih M, Hum P, Tuana BS (2001) Alternative splicing in intracellular loop connecting domains II and III of the alpha 1 subunit of $Ca_v1.2$ Ca2 + channels predicts two-domain polypeptides with unique C-terminal tails. J Biol Chem 276(2):1398–1406

Xie C, Zhen XG, Yang J (2005) Localization of the activation gate of a voltage-gated Ca2 + channel. J Gen Physiol 126(3):205–212

Yang L, Doshi D, Morrow J, Katchman A, Chen X, Marx SO (2009) Protein kinase C isoforms differentially phosphorylate Ca(v)1.2 alpha(1c). Biochemistry 48(28):6674–6683

Yarotskyy V, Gao G, Peterson BZ, Elmslie KS (2009) The Timothy syndrome mutation of cardiac $Ca_v1.2$ (L-type) channels: multiple altered gating mechanisms and pharmacological restoration of inactivation. J Physiol 587 (Pt 3):551–565

Yazawa M, Hsueh B, Jia X, Pasca AM, Bernstein JA, Hallmayer J, Dolmetsch RE (2011) Using induced pluripotent stem cells to investigate cardiac phenotypes in Timothy syndrome. Nature 471(7337):230–234

Youssoufian H, Pyeritz RE (2002) Mechanisms and consequences of somatic mosaicism in humans. Nat Rev Genet 3(10):748–758

Yu J, Vodyanik MA, Smuga-Otto K, Antosiewicz-Bourget J, Frane JL, Tian S, Nie J, Jonsdottir GA, Ruotti V, Stewart R, Slukvin II, Thomson JA (2007) Induced pluripotent stem cell lines derived from human somatic cells. Science 318(5858):1917–1920

Zhang C, Zhang Z, Castle J, Sun S, Johnson J, Krainer AR, Zhang MQ (2008) Defining the regulatory network of the tissue-specific splicing factors Fox-1 and Fox-2. Genes Dev 22(18):2550–2563

Zuhlke RD, Bouron A, Soldatov NM, Reuter H (1998) Ca2 + channel sensitivity towards the blocker isradipine is affected by alternative splicing of the human alpha1C subunit gene. FEBS Lett 427(2):220–224

Chapter 12
Brugada Syndrome and Voltage-Gated Calcium Channels

Brett Simms

Abstract Brugada Syndrome is an inherited disease affecting ion channels of the heart leading to arrhythmia and sudden cardiac death. Brugada Syndrome typically manifests in middle age men after or during periods of rest, with the vast majority of sudden death occurring in individuals without structural heart defects. Cardiac ion channel dysfunction in Brugada Syndrome is well recognized; however, how voltage-gated calcium channel genes, specifically *CACNA1C*, *CACNB2b*, and *CACNA2D1* are implicated, is a recent development. This chapter addresses why VGCCs are important for the ventricular action potential and how Brugada Syndrome is visualized on ECG, in particular the two types of elevated ST segments commonly used for clinical diagnosis. We also examine voltage-gated calcium channel mutations linked to Brugada Syndrome and how specific mutations cause a loss-of-function of the channel complex.

12.1 Brugada Syndrome is an Idiopathic, Cardiac Ion Channel Disease

Brugada Syndrome(BS) is an inherited cardiac illness characterized by a defect in repolarization of the right ventricular outflow tract, leading to phase 2 (discussed below) reentry and polymorphic ventricular tachycardia (Napolitano and Antzelevitch 2011). BS has an estimated prevalence of 1/10,000–5/10,000 in Europe, and 12/10,000 in Southeast Asia (Hermida et al. 2000; Sinner et al. 2009; Miyasaka et al. 2001). Men account for 80 % of BS patients, with most arrhythmic

B. Simms (✉)
Department of Physiology and Pharmacology, Hotchkiss Brain Institute, university of Calgary, Calgary, Canada
e-mail: basimms@ucalgary.ca

N. Weiss and A. Koschak (eds.), *Pathologies of Calcium Channels*,
DOI: 10.1007/978-3-642-40282-1_12, © Springer-Verlag Berlin Heidelberg 2014

events occurring between the ages of 40 and 45 years and typically after large meals, in periods of rest, or during sleep (Sacher et al. 2008; Antzelevitch et al. 2005; Probst et al. 2010). BS is responsible for 4 % of all sudden deaths and 20 % of sudden deaths in people without structural heart problems (Antzelevitch et al. 2005). Emerging clinical evidence suggests that sudden unexpected nocturnal death syndrome (SUNDS) and sudden infant death syndrome (SIDS) may in fact be an infant forms of BS (Kanter et al. 2011). The most common gene mutated in BS patients is *SCN5A* ($Na_v1.5$), which is also aberrant in several SUNDS and SIDS patients (Skinner et al. 2005; Vatta et al. 2002b). Further genetic screening is needed, but it is likely that SUNDS, SIDS, and BS require common ion channel backgrounds, which when disrupted can lead to sudden cardiac death.

BS was first described by Pedro and Josep Brugada who recognized a previously undiagnosed patient population with structurally normal hearts, and who exhibited idiopathic ventricular fibrillation (Brugada and Brugada 1992). Excessive use of tricyclic antidepressants, lithium, diphenhydramine, and cocaine have all been shown to promote BS-like ECG patterns in individuals without a prior history of cardiac disease (Brahmi et al. 2007; Darbar et al. 2005; Lopez-Barbeito et al. 2005; Littmann et al. 2000). Several instances of fever induced BS have also been reported (Suzuki et al. 2012; Lamelas et al. 2012). The broad range of environmental factors capable of promoting BS-like symptoms is explained by the fact that any condition which causes a gain of repolarization, or loss of depolarization during ventricular contraction, can manifest as BS-like on ECG (Shimizu 2005). We will address what a BS pattern looks like on ECG in the next section. Despite environmental aggravates causing BS-like phenomena, BS is considered as inherited disorder and is linked to mutations in several ion channel genes including: voltage-gated sodium (*SCN5A/SCN1B/SCN3B*), potassium (*KCND3/KCNE3/MiRP2/KCNJ8*), hyperpolarization activated cation (*HCN4*), and voltage-gated calcium channel subunits (*CACNA1C/CACNB2b/CACNA2D1*), the latter of which is the focus of this chapter (Antzelevitch et al. 2007; Burashnikov et al. 2010; Delpon et al. 2008; Hu et al. 2009; Watanabe et al. 2008; Giudicessi et al. 2011; Rook et al. 1999; Barajas-Martinez et al. 2012). It should be noted that a gain-of-function of $K_v4.3$ (I_{TO}) or conversely a loss-of-function of $Na_v1.5$, or $Ca_v1.2$ (I_{NA}/I_{CA}) underlie the BS phenotype. Pharmacological treatment therefore targets some of these channels (i.e., Quinidine for block of I_{TO}) but direct block of voltage-dependent calcium channels (VGCCs) is contraindicative for BS (Yatani et al. 1993). Beta adrenergic stimulation (isoproterenol) or phosphodiesterase inhibition (cilostazol) on the other hand is helpful for managing I_{Ca} in BS patients (Riera et al. 2007; Tsuchiya et al. 2002). Only the implantable cardioverter defibrillator (ICD)—an electrical impulse generator designed to detect and correct arrhythmias—is considered an effective long-term treatment for BS (McGregor and Chen 2004).

12.2 Understanding the Ventricular Action Potential, ECG, and BS Diagnosis

BS is an electrical disease of the ventricles and therefore it is worth revisiting the ventricular action potential (VAP) to understand how ion channels contribute to the VAP, and how a VAP is displayed on an electrocardiograph (ECG). The VAP which was first modeled by G.W. Beeler in 1977 proposed that two depolarizing and two repolarizing currents govern ventricular contraction (Beeler and Reuter 1977). Beeler's model although simplistic was correct in the fact that it identified time and voltage sensitive currents whose coordinated activity shaped the VAP. Figure 12.1 illustrates the five phases of the VAP. An inward rush of sodium through $Na_v1.5$ gives the initial peak of VAP depolarization (Phase 0), which is recessed by the exit of potassium through $K_v4.3$ channels and voltage-dependent inactivation (VDI) of $Na_v1.5$ (Phase 1) (Bennett et al. 1995; Levi and DeFelice 1986). The plateau (Phase 2) of the VAP is maintained by the opening and prolonged influx of calcium through $Ca_v1.2$ channels, and efflux of potassium by the delayed rectifier hERG (I_{Kr}) (Levi and DeFelice 1986; Cohen and Lederer 1987; Sanguinetti et al. 1995). $Ca_v1.2$ conductance during the VAP provides the initial influx of calcium necessary to trigger calcium release from the sarcorplasmic reticulum causing muscle fibers in the heart to physically contract, in other words excitation–contraction coupling (Romey et al. 1988) (for reviews see (Benitah et al. 2010; Dulhunty 2006)). During repolarization of the VAP (Phase 3) $Ca_v1.2$ channels close limiting depolarization, while the slow outward rectifier KCNQ1 (I_{Ks}) opens, driving the membrane potential down toward resting levels (Barhanin

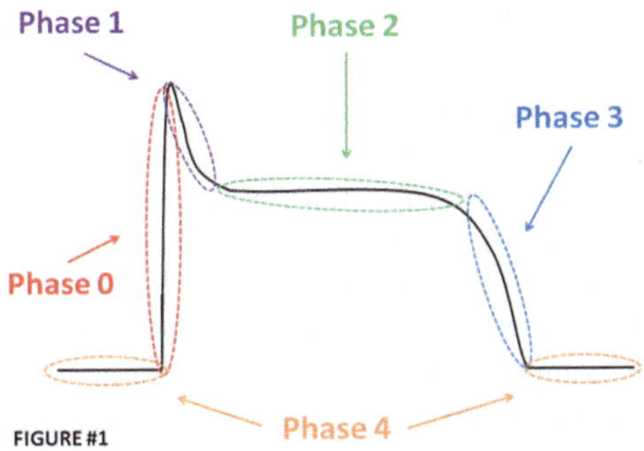

Fig. 12.1 Ion channel conductances which contribute to the five phases of the VAP. The VAP begins with Phase 0 (red) which is the rapid inward current carried by $Na_v1.5$ (I_{Na}) leading to immediate depolarization of the ventricles. In Phase 1(purple) partial repolarization of the ventricles occurs as $Na_v1.5$ channels close and Kv4.3 (I_{TO}) open, permitting potassium to exit. During Phase 2 (green) sustained inward calcium via $Ca_v1.2$ (I_{Ca}) and outward delayed potassium currents through hERG maintain the VAP plateau. Phase 3 (blue) leads to complete repolarization of the VAP as $Ca_v1.2$ channels close, and slow outward potassium currents (KCNQ1) activate. Phase 4 (orange) is maintenance of the resting membrane potential by inward rectifying Kir2.1 channels

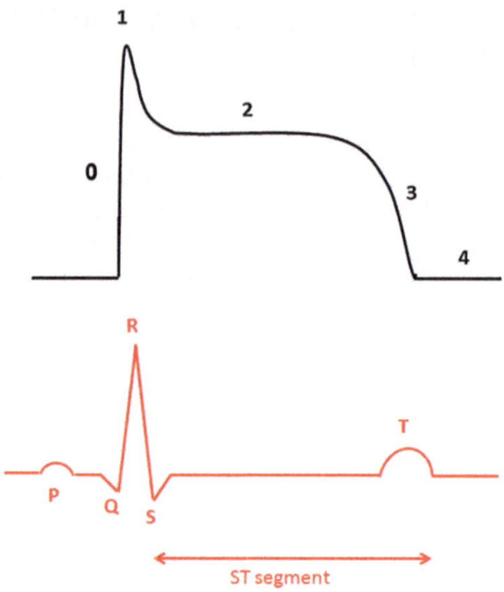

Fig. 12.2 The VAP can be visualized on ECG. The VAP encompasses the QRST portion of the ECG. The preceding P wave represents depolarization of the atria. Phase 0 and 1 of the VAP are depicted by the QRS complex which represents depolarization of the ventricles and repolarization of the atria, but because ventricular depolarization is so large, atrial repolarization is masked on ECG. The plateau phase of the VAP (Phase 2) is represented by the ST segment on ECG. Phase 3 of the VAP is represented by the T-wave on an ECG and signifies ventricular repolarization

et al. 1996). The fourth phase of the VAP is the maintenance of resting membrane potential by Kir2.1 (I_{K1}) an inward rectifying potassium channel (Nakamura et al. 1998).

Figure 12.2 equates the electrical activity of the VAP to that of the whole heart beat as seen on ECG. The first ECG recorded was by William Einthoven in 1903, but not until 1942 was the ECG given its modern 12-lead (with precordial leads) configuration by Emanual Goldberger (Rivera-Ruiz et al. 2008; Goldberger 1946). Precordial leads, also denoted V_1, V_2, V_3, are placed laterally across the fourth intercostal space and are vital for BS diagnosis. Note that in Fig. 12.2 the VAP encompasses the QRST portion of the ECG. Phases 0 and 1 of the VAP are depicted by the QRS complex which represents depolarization of the ventricles. Phase 2 or maintenance of ventricular contraction is represented by the ST segment of the ECG. It is this segment of the ECG which is most critical for BS diagnosis. BS patients present with abnormally elevated ST segments, commonly called J-waves, in one or all of the ECG precordial leads (Berne and Brugada 2012). A J-wave signifies early repolarization or premature ventricular contraction and if left untreated can lead to arrhythmias and sudden cardiac death. Finally, Phase 3 of the VAP is represented by a T-wave on an ECG and represents proper ventricular repolarization in healthy individuals.

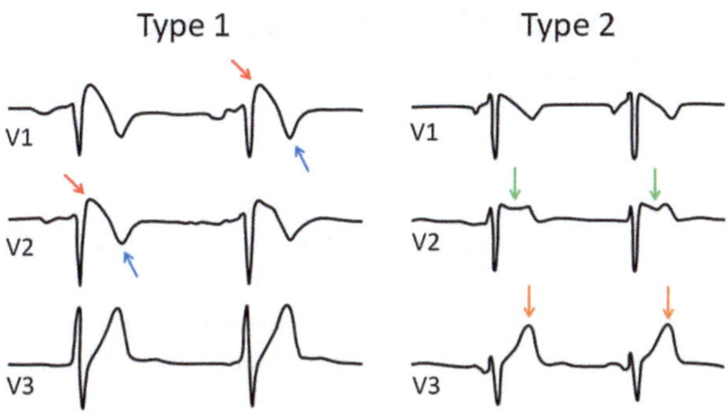

Fig. 12.3 There are two ECG patterns used to diagnosis Brugada Syndrome. The Type I ECG pattern displays an elevated, coved ST segment (red arrow) followed by a negative T-wave (blue arrow). In this trace, the diagnostic pattern appears on both the V1 and V2 precordial leads. The Type II pattern displays an elevated ST segment with a saddleback appearance (green arrow) and in this example, a positive T-wave (orange arrow). The saddleback ST segment is most clear on the V2 lead, and positive T-wave on the V3 lead. These example traces are modeled after (Berne and Brugada 2012) Fig. 12.1

With VAP and ECG background in place we can now visit the extended definition of BS which as the field, is ever evolving. The current criteria for diagnosis of BS includes an elevated ST segment of the right precordial leads in the presence, or absence of a sodium channel blocker with one of the following: Ventricular fibrillation (VF), inducibility of VT with programmed electrical stimulation, polymorphic ventricular tachycardia (PVT), a family history of sudden cardiac death under the age of 45, or syncope (Antzelevitch et al. 2005). This lengthy definition for BS is necessary clinically because of related J-wave syndromes with similar electrical characteristics (Ogawa et al. 2005). BS patients can present with one of two types of elevated ST segment differentially characterized by shape and degree of elevation. The first type displays a coved ST elevation (>2 mm) followed by a negative T-wave, the second an elevated ST segment (>1 mm) with a saddleback appearance and a positive, or bi-phasic T-wave (Berne and Brugada 2012; Bayes de Luna et al. 2012). Figure 12.3 highlights the two types of ECG patterns commonly observed. Also common in clinical diagnosis is the "unmasking" of BS patterns on ECG by sodium channel block (Fowler and Priori 2009). Challenging cardiac sodium channels with a blocker differentially reveals elevated ST segments in BS patients, but not healthy individuals.

Clearly voltage-gated calcium channels play a role in the VAP, but how are these channels implicated in BS The next section will discuss the studies which proved pharmacologically and genetically that *CACNA1C*, *CACNB2b*, and *CACNA2D1* are disease alleles of (BS).

12.3 Pharmacology and Genetics Link Voltage-Gated Calcium Channels to BS

It is accepted that voltage-gated sodium channel block unmasks ST elevation in BS patients and implicates $Na_v1.5$ dysfunction in this disease. There is an abundance of literature demonstrating that mutations in *SCN5A*, or pharmacological block of $Na_v1.5$ promote ST elevation in BS patients (Rook et al. 1999; Yan and Antzelevitch 1999). Several well written reviews summarize the molecular (Rook et al. 2012; Wilde and Brugada 2011) and pharmacological (Antzelevitch and Fish 2006) association between $Na_v1.5$ and BS. Unraveling how, or if voltage-gated calcium channels were implicated in BS has only recently begun.

The first clue that VGCCs were involved in BS came from work done by Miyazaki et al. who showed that ST elevation in the precordial leads of three BS patients could be exacerbated by β adrenergic blockers, α adrenergic stimulants, and the muscarinic receptor stimulant, acetylcholine (Miyazaki et al. 1996). This data agreed with previously published literature which showed that L-type calcium channels could be stimulated and suppressed by β and α adrenergic stimulation, respectively (Petit-Jacques et al. 1993; Hess et al. 1986; Keung and Karliner 1990). Around this time it was also demonstrated that β adrenergic stimulation could potentiate excitation–contraction coupling, again suggesting the importance of L-type calcium channels in the VAP (Hussain and Orchard 1997; Viatchenko-Karpinski and Gyorke 2001). The first direct evidence that L-type calcium channels could contribute to the BS phenotype came from Fish et al. who showed that L-type channel block with terfenadine (also verapamil) was sufficient to dramatically increase ST segment elevation in dissociated ventricular myocytes (Fish and Antzelevitch 2004). Verapamil has since been disputed in its ability to promote J-waves in all BS patients, but this likely reflects the variety of compensatory ion channel mutations possible in individual patients (Chinushi et al. 2006; Chinushi et al. 2009).

The landmark study which first showed genetic evidence that VGCC mutations were linked to BS came from Antzelevitch et al. who showed that two point mutations (A39 V/G490R) in the pore forming $Ca_v1.2$ channel and one in the accessory $Ca_v\beta_{2b}$ (S481L) subunit, reduced L-type channel function in CHO-K1 cells (Antzelevitch et al. 2007). Burashnikov et al. have recently expanded the number of known BS mutations found in VGCC subunits to nine for $Ca_v1.2$, ten for $Ca_v\beta_{2b}$, and four for $Ca_v\alpha_2\delta_1$ (Burashnikov et al. 2010; Cordeiro et al. 2009). Figure 12.4 summarizes the location of currently known BS mutations in VGCC subunits. We will now examine the types of mutations found at the DNA level in BS patients and the mechanisms of action for some of these VGCC mutations.

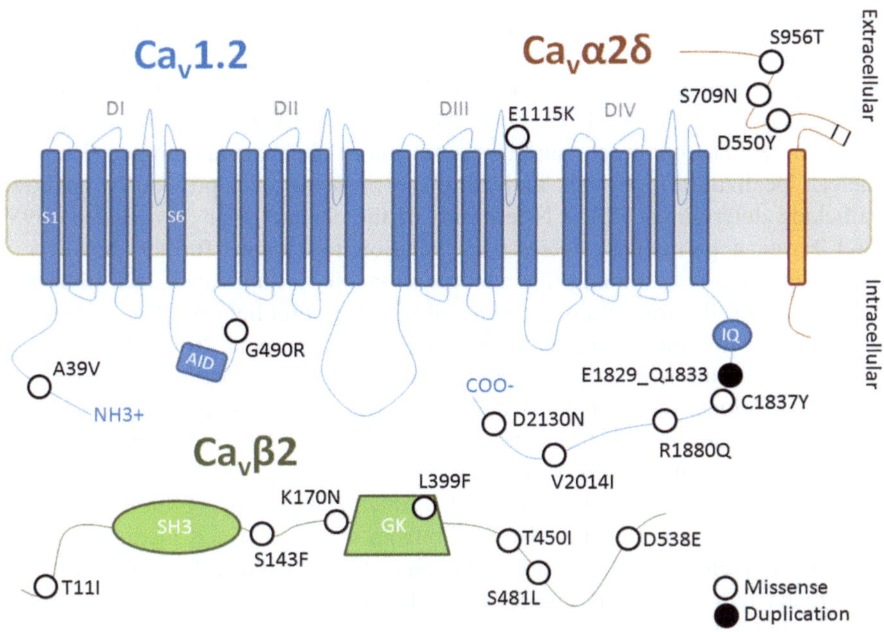

Fig. 12.4 The nature and location of BS mutations in VGCC subunits. The Ca$_v$1.2 pore forming subunit (blue) has four domains (DI-IV), each with six transmembrane segments (S1-S6). The alpha interaction domain (AID) and IQ domain of Ca$_v$1.2 are highlighted for reference in the I-II and C-terminal linkers, respectively. The Ca$_v\beta2$ subunit is displayed in green with its SH3 and GK-like domains highlighted. The Ca$_v\alpha2\delta$ subunit (orange) is cleaved after translation into $\alpha2$ and δ components (see Chap. 1 for review), which are then disulfide linked (vertical black lines) in the intact protein. All currently known BS mutations are shown in this diagram with missense mutations (open circles) and a duplication event (dark circle) noted. This figure is modeled after Burashnikov et al. (2010) Fig. 12.3

12.4 Brugada Syndrome Mutations in VGCCs: There Location and Functional Consequence

BS mutations in VGCC subunits promote a loss of channel function by: 1) decreasing conductance via gating malfunctions or pore aberrations, or 2) by decreasing surface expression of the channel complex. BS mutations in VGCC subunits thus reduce I_{Ca} in Phase 2 of the VAP, which is seen as a J-wave on ECG. At the molecular level most BS mutations are missense mutations, but a duplication event has been reported (see Fig. 12.4). We will now discuss some of the characterized BS mutations found in VGCC subunits.

One of the first BS mutations described for Ca$_v$1.2 was A39V, an N-terminal mutation which causes the channel to distribute in a peri-nuclear manner (Antzelevitch et al. 2007). Peri-nuclear localization is equivalent to reduced surface expression because channel complexes remain intracellular and do not travel to the cell membrane. Co-expression of $Ca_v\beta$ with wild-type channel typically and robustly increases the surface expression of the entire channel complex by $Ca_v\beta$ binding the intracellular I-II linker of $Ca_v1.2$ (Bichet et al. 2000; Fang and Colecraft 2011; Altier et al. 2011). In the case of A39V-$Ca_v1.2$ however, $Ca_v\beta_{2b}$ could not reconstitute proper surface expression of the channel complex (i.e., peri-nuclear localization), perhaps indicating the presence of a previously unknown trafficking determinant in the N-terminus of the channel. For this reason A39V-$Ca_v1.2$ was a particularly interesting mutation to explore further; however, a follow up paper reexamining this effect was unable to duplicate a lack of surface expression in the rat brain isoform of $Ca_v1.2$, highlighting the phenotypic restriction of this particular mutation (Simms and Zamponi 2012). Interestingly, N-terminal mutations recently discovered in $Na_v1.5$ have been shown to exhibit trafficking defects reminiscent of A39 V-$Ca_v1.2$, perhaps suggesting that the N-terminus contains motifs more generally associated with trafficking of cardiac voltage-gated channels (Clatot et al. 2012).

Another VGCC mutation identified by Antzelevitch et al. (G490R) located in the I-II linker of $Ca_v1.2$ may also suffer from a lack of $Ca_v\beta$ mediated surface expression given its prime location to influence $Ca_v\beta$ mediated forward trafficking, and greatly reduced current density (Antzelevitch et al. 2007). However, because $Ca_v\beta$ and the I-II linker are also involved in voltage-dependent gating of the channel, G490R-$Ca_v1.2$ may have a functional abnormality yet to be revealed at the single channel level, and so improper trafficking cannot be confirmed (Dafi et al. 2004).

One further BS mutation with greatly reduced current density is a result of a duplication of five amino acids in the distal C-terminus of $Ca_v1.2$ (Burashnikov et al. 2010). Much like G490R-$Ca_v1.2$ the exact mechanism for why E1829_Q1833-dup-$Ca_v1.2$ has severely reduced current is unknown, but could be because the duplication event created a sequence which reduces membrane trafficking of the channel. The C-terminus of $Ca_v1.2$ contains critical trafficking and ER retention motifs in various regions including the residues in and around E1829-Q1833 (Altier et al. 2011; Fang and Colecraft 2011). Much like the I-II linker, however, the C-terminus of $Ca_v1.2$ is intimately tied to channel function and participates in voltage and calcium-dependent gating, so whether E1829_Q1833-$Ca_v1.2$ is defective in trafficking remains to be seen. (Kobrinsky et al. 2005; Pitt et al. 2001; Erickson et al. 2001).

BS mutations do not always effect channel trafficking, but in many cases reduce channel function once at the cell membrane. VGCCs preferentially pass calcium because of sequence specific pore loops or calcium selectivity filters which are located between S5 and S6 of each domain (Catterall 2000; Yang et al. 1993). A single point mutation in a "p loop" residue for instance can transform a calcium selective channel into one which prefers sodium (Heinemann et al. 1992). The BS mutation E1115K-$Ca_v1.2$ has reduced calcium flux through its pore because of such a mutation (Burashnikov et al. 2010). The E1115K mutation does not alter voltage or calcium-dependent gating of the channel, but strictly reduces single channel conductance of calcium. Extrapolating this cellular phenotype to whole heart, it is easy to understand why a calcium channel which no longer prefers calcium could promote premature repolarization of the ventricles. In other words, less calcium influx through $Ca_v1.2$ would mean excitation–contraction coupling would be weakened in the ventricles and so the ST segment would be shortened. E1115 K-$Ca_v1.2$ is a unique mechanism for loss-of-function of VGCCs in BS, however, an analogs mutation in the pore of $Na_v1.5$ has been identified suggesting commonality between channel families (Vatta et al. 2002a).

So far we have considered BS mutations exclusive to the pore forming $Ca_v1.2$. Although the pore is arguably the most important component of any channel, BS mutations have been characterized for the accessory $Ca_v\beta_{2b}$ subunit as well. Among the many functional attributes $Ca_v\beta$ subunits impart on high voltage activated calcium channels, regulation of voltage-dependent inactivation (VDI) is among the most prominent (Buraei and Yang 2010). $Ca_v\beta$ subunits modulate VDI of channels in a manner which depends on both the composition, and length of the $Ca_v\beta$ N-terminus (Herzig et al. 2007). Therefore, that the BS mutation T11I-$Ca_v\beta_{2b}$ mildly accelerates VDI and increases steady-state inactivation of $Ca_v1.2$ is not surprising, yet offers an explanation for why this mutation is associated with BS (Cordeiro et al. 2009). A channel complex which inactivates quicker and is more sensitive to steady-state inactivation could undoubtedly shorten ventricular contraction. The T11I-$Ca_v\beta_{2b}$ mutation brings up another interesting point to consider, which is, that even subtle functional deficits can promote a disease phenotype in the right context. The dramatic loss of current density and trafficking described for some of the aforementioned mutations are impressive experimentally, but dramatic losses in $Ca_v1.2$ function are lethal early in life, at least in animal models (Weissgerber et al. 2006). Recall that most BS patients first show disease symptoms in middle age, and 20 % of them have structurally normal hearts. In essence it could be argued that BS mutations should be subtle and that dramatic losses of cell surface expression seen in expression systems, for example, might not occur in heart.

With the subtlety of T11I-$Ca_v\beta_{2b}$ in mind, we will close out this section by considering an interesting duo of offsetting polymorphisms found in $Ca_v1.2$ and $Ca_v\beta_{2b}$. The BS mutation V2014I-$Ca_v1.2$, and the rare polymorphism D601E-$Ca_v\beta_{2b}$, were isolated from the same patient and together emphasize the importance of multiple genetic variations in disease progression. Burashnikov et al. showed that while V2014I-$Ca_v1.2$ had reduced current density in CHO cells—likely attributed to reduced open probability—D601E-$Ca_v\beta_{2b}$ actually increased channel activity by slowing the kinetics of inactivation (Burashnikov et al. 2010; Hu et al. 2010). In other words, D601E-$Ca_v\beta_{2b}$ is capable of prolonging calcium influx through $Ca_v1.2$ channels, which is especially important if V2014I-$Ca_v1.2$ has compromised conductance. Furthermore, without D601E-$Ca_v\beta_{2b}$ prolonging conductance of V2014I-$Ca_v1.2$ it is possible the functional deficit of the mutant channel may have been too severe for this patient to live in middle age. This rare combination of polymorphisms may be the first example of allelic compensation between cardiac VGCC subunits in heart, although examples do exist for other proteins involved in cardiac calcium signaling (Rizzuto and Pozzan 2003).

BS mutations in VGCC subunits decrease either $Ca_v1.2$ surface expression, or conductance once at the cell surface. As a result I_{Ca} is decreased during Phase 2 of the VAP and in turn causes premature ventricular repolarization, or J-waves on ECG. Although most characterized BS abnormalities are missense mutations in $Ca_v1.2$ and $Ca_v\beta2$, a duplication event in the C-terminus of $Ca_v1.2$ has been identified at the DNA level.

12.5 Conclusion

BS is an inherited disorder which effects 1/10,000-5/10,000 people and shows a preference for men. Electrically BS is characterized by premature ventricular contraction which if left untreated by ICD, can lead to arrthymia and sudden cardiac death. BS is verified on ECG by the presence of an elevated ST segment, or J-wave, and can be unveiled by sodium channel blockade. SUNDS and SIDS are likely variants of BS and future genetic evidence could link mutations in VGCC subunits to all three diseases.

$Ca_v1.2$ channel activity is important for prolonged depolarization of the VAP, a fact supported by recent pharmacological and genetic evidence. Missense and duplication mutations have been found in *CACNA1C*, *CACNB2b*, and *CACNA2D1* which directly link these genes to BS. While BS mutations can reduce $Ca_v1.2$ activity in a variety of ways, a loss of surface expression, or a reduction in conductance once at the cell membrane, explain the lack of I_{Ca} in ventricles of BS patients.

References

Altier C, Garcia-Caballero A, Simms B, You H, Chen L, Walcher J, Tedford HW, Hermosilla T, Zamponi GW (2011) The Ca_vbeta subunit prevents RFP2-mediated ubiquitination and proteasomal degradation of L-type channels. Nat Neurosci 14 (2):173-180

Antzelevitch C, Brugada P, Borggrefe M, Brugada J, Brugada R, Corrado D, Gussak I, LeMarec H, Nademanee K, Perez Riera AR, Shimizu W, Schulze-Bahr E, Tan H, Wilde A (2005) Brugada syndrome: report of the second consensus conference. Heart Rhythm 2 (4):429-440

Antzelevitch C, Fish JM (2006) Therapy for the Brugada syndrome. Handb Exp Pharmacol (171):305-330

Antzelevitch C, Pollevick GD, Cordeiro JM, Casis O, Sanguinetti MC, Aizawa Y, Guerchicoff A, Pfeiffer R, Oliva A, Wollnik B, Gelber P, Bonaros EP, Jr., Burashnikov E, Wu Y, Sargent JD, Schickel S, Oberheiden R, Bhatia A, Hsu LF, Haissaguerre M, Schimpf R, Borggrefe M, Wolpert C (2007) Loss-of-function mutations in the cardiac calcium channel underlie a new clinical entity characterized by ST-segment elevation, short QT intervals, and sudden cardiac death. Circulation 115 (4):442-449

Barajas-Martinez H, Hu D, Ferrer T, Onetti CG, Wu Y, Burashnikov E, Boyle M, Surman T, Urrutia J, Veltmann C, Schimpf R, Borggrefe M, Wolpert C, Ibrahim BB, Sanchez-Chapula JA, Winters S, Haissaguerre M, Antzelevitch C (2012) Molecular genetic and functional association of Brugada and early repolarization syndromes with S422L missense mutation in KCNJ8. Heart Rhythm 9 (4):548-555

Barhanin J, Lesage F, Guillemare E, Fink M, Lazdunski M, Romey G (1996) K(V)LQT1 and lsK (minK) proteins associate to form the I(Ks) cardiac potassium current. Nature 384 (6604):78-80

Bayes de Luna A, Brugada J, Baranchuk A, Borggrefe M, Breithardt G, Goldwasser D, Lambiase P, Riera AP, Garcia-Niebla J, Pastore C, Oreto G, McKenna W, Zareba W, Brugada R, Brugada P (2012) Current electrocardiographic criteria for diagnosis of Brugada pattern: a consensus report. J Electrocardiol 45 (5):433-442

Beeler GW, Reuter H (1977) Reconstruction of the action potential of ventricular myocardial fibres. J Physiol 268 (1):177-210

Benitah JP, Alvarez JL, Gomez AM (2010) L-type Ca(2 +) current in ventricular cardiomyocytes. J Mol Cell Cardiol 48 (1):26-36

Bennett PB, Yazawa K, Makita N, George AL, Jr. (1995) Molecular mechanism for an inherited cardiac arrhythmia. Nature 376 (6542):683-685

Berne P, Brugada J (2012) Brugada syndrome 2012. Circ J 76 (7):1563-1571

Bichet D, Cornet V, Geib S, Carlier E, Volsen S, Hoshi T, Mori Y, De Waard M (2000) The I-II loop of the Ca2 + channel alpha1 subunit contains an endoplasmic reticulum retention signal antagonized by the beta subunit. Neuron 25 (1):177-190

Brahmi N, Thabet H, Kouraichi N, Driss I, Amamou M (2007) [Brugada syndrome and other cardiovascular abnormalities related to tricyclic antidepressants ans related drugs intoxication]. Arch Mal Coeur Vaiss 100 (1):28-33

Brugada P, Brugada J (1992) Right bundle branch block, persistent ST segment elevation and sudden cardiac death: a distinct clinical and electrocardiographic syndrome. A multicenter report. J Am Coll Cardiol 20 (6):1391-1396

Buraei Z, Yang J (2010) The Beta subunit of voltage-gated Ca2 + channels. Physiol Rev 90 (4):1461-1506

Burashnikov E, Pfeiffer R, Barajas-Martinez H, Delpon E, Hu D, Desai M, Borggrefe M, Haissaguerre M, Kanter R, Pollevick GD, Guerchicoff A, Laino R, Marieb M, Nademanee K, Nam GB, Robles R, Schimpf R, Stapleton DD, Viskin S, Winters S, Wolpert C, Zimmern S, Veltmann C, Antzelevitch C (2010) Mutations in the cardiac L-type calcium channel associated with inherited J-wave syndromes and sudden cardiac death. Heart Rhythm 7 (12):1872-1882

Catterall WA (2000) Structure and regulation of voltage-gated Ca2 + channels. Annu Rev Cell Dev Biol 16:521-555

Chinushi M, Iijima K, Tagawa M, Komura S, Furushima H, Aizawa Y (2009) Effects of verapamil on anterior ST segment and ventricular fibrillation cycle length in patients with Brugada syndrome. J Electrocardiol 42 (4):367-373

Chinushi M, Tagawa M, Nakamura Y, Aizawa Y (2006) Shortening of the ventricular fibrillatory intervals after administration of verapamil in a patient with Brugada syndrome and vasospastic angina. J Electrocardiol 39 (3):331-335

Clatot J, Ziyadeh-Isleem A, Maugenre S, Denjoy I, Liu H, Dilanian G, Hatem SN, Deschenes I, Coulombe A, Guicheney P, Neyroud N (2012) Dominant-negative effect of SCN5A N-terminal mutations through the interaction of $Na_v1.5$ alpha-subunits. Cardiovasc Res

Cohen NM, Lederer WJ (1987) Calcium current in isolated neonatal rat ventricular myocytes. J Physiol 391:169-191

Cordeiro JM, Marieb M, Pfeiffer R, Calloe K, Burashnikov E, Antzelevitch C (2009) Accelerated inactivation of the L-type calcium current due to a mutation in CACNB2b underlies Brugada syndrome. J Mol Cell Cardiol 46 (5):695-703

Dafi O, Berrou L, Dodier Y, Raybaud A, Sauve R, Parent L (2004) Negatively charged residues in the N-terminal of the AID helix confer slow voltage dependent inactivation gating to $Ca_v1.2$. Biophys J 87 (5):3181-3192

Darbar D, Yang T, Churchwell K, Wilde AA, Roden DM (2005) Unmasking of brugada syndrome by lithium. Circulation 112 (11):1527-1531

Delpon E, Cordeiro JM, Nunez L, Thomsen PE, Guerchicoff A, Pollevick GD, Wu Y, Kanters JK, Larsen CT, Hofman-Bang J, Burashnikov E, Christiansen M, Antzelevitch C (2008) Functional effects of KCNE3 mutation and its role in the development of Brugada syndrome. Circ Arrhythm Electrophysiol 1 (3):209-218

Dulhunty AF (2006) Excitation-contraction coupling from the 1950s into the new millennium. Clin Exp Pharmacol Physiol 33 (9):763-772

Erickson MG, Alseikhan BA, Peterson BZ, Yue DT (2001) Preassociation of calmodulin with voltage-gated Ca(2 +) channels revealed by FRET in single living cells. Neuron 31 (6):973-985

Fang K, Colecraft HM (2011) Mechanism of auxiliary β-subunit-mediated membrane targeting of L-type ($Ca_v1.2$) channels. J Physiol

Fish JM, Antzelevitch C (2004) Role of sodium and calcium channel block in unmasking the Brugada syndrome. Heart Rhythm 1 (2):210-217

Fowler SJ, Priori SG (2009) Clinical spectrum of patients with a Brugada ECG. Curr Opin Cardiol 24 (1):74-81

Giudicessi JR, Ye D, Tester DJ, Crotti L, Mugione A, Nesterenko VV, Albertson RM, Antzelevitch C, Schwartz PJ, Ackerman MJ (2011) Transient outward current (I(to)) gain-of-function mutations in the KCND3-encoded Kv4.3 potassium channel and Brugada syndrome. Heart Rhythm 8 (7):1024-1032

Goldberger E (1946) Significance of downward T waves in precordial leads of normal children. Am J Dis Child 71:618-621

Heinemann SH, Terlau H, Stuhmer W, Imoto K, Numa S (1992) Calcium channel characteristics conferred on the sodium channel by single mutations. Nature 356 (6368):441-443

Hermida JS, Lemoine JL, Aoun FB, Jarry G, Rey JL, Quiret JC (2000) Prevalence of the brugada syndrome in an apparently healthy population. Am J Cardiol 86 (1):91-94

Herzig S, Khan IF, Grundemann D, Matthes J, Ludwig A, Michels G, Hoppe UC, Chaudhuri D, Schwartz A, Yue DT, Hullin R (2007) Mechanism of Ca(v)1.2 channel modulation by the amino terminus of cardiac beta2-subunits. Faseb J 21 (7):1527-1538

Hess P, Lansman JB, Nilius B, Tsien RW (1986) Calcium channel types in cardiac myocytes: modulation by dihydropyridines and beta-adrenergic stimulation. J Cardiovasc Pharmacol 8 Suppl 9:S11-21

Hu D, Barajas-Martinez H, Burashnikov E, Springer M, Wu Y, Varro A, Pfeiffer R, Koopmann TT, Cordeiro JM, Guerchicoff A, Pollevick GD, Antzelevitch C (2009) A mutation in the beta 3 subunit of the cardiac sodium channel associated with Brugada ECG phenotype. Circ Cardiovasc Genet 2 (3):270-278

Hu D, Barajas-Martinez H, Nesterenko VV, Pfeiffer R, Guerchicoff A, Cordeiro JM, Curtis AB, Pollevick GD, Wu Y, Burashnikov E, Antzelevitch C (2010) Dual variation in SCN5A and CACNB2b underlies the development of cardiac conduction disease without Brugada syndrome. Pacing Clin Electrophysiol 33 (3):274-285

Hussain M, Orchard CH (1997) Sarcoplasmic reticulum Ca2 + content, L-type Ca2 + current and the Ca2 + transient in rat myocytes during beta-adrenergic stimulation. J Physiol 505 (Pt 2):385-402

Kanter RJ, Pfeiffer R, Hu D, Barajas-Martinez H, Carboni MP, Antzelevitch C (2011) Brugada-like syndrome in infancy presenting with rapid ventricular tachycardia and intraventricular conduction delay. Circulation 125 (1):14-22

Keung EC, Karliner JS (1990) Complex regulation of calcium current in cardiac cells. Dependence on a pertussis toxin-sensitive substrate, adenosine triphosphate, and an alpha 1-adrenoceptor. J Clin Invest 85 (3):950-954

Kobrinsky E, Tiwari S, Maltsev VA, Harry JB, Lakatta E, Abernethy DR, Soldatov NM (2005) Differential role of the alpha1C subunit tails in regulation of the $Ca_v1.2$ channel by membrane potential, beta subunits, and Ca2 + ions. J Biol Chem 280 (13):12474-12485

Lamelas P, Labadet C, Spernanzoni F, Saubidet CL, Alvarez PA (2012) Brugada electrocardiographic pattern induced by fever. World J Cardiol 4 (3):84-86

Levi R, DeFelice LJ (1986) Sodium-conducting channels in cardiac membranes in low calcium. Biophys J 50 (1):5-9

Littmann L, Monroe MH, Svenson RH (2000) Brugada-type electrocardiographic pattern induced by cocaine. Mayo Clin Proc 75 (8):845-849

Lopez-Barbeito B, Lluis M, Delgado V, Jimenez S, Diaz-Infante E, Nogue-Xarau S, Brugada J (2005) Diphenhydramine overdose and Brugada sign. Pacing Clin Electrophysiol 28 (7):730-732

McGregor M, Chen J (2004) Should the implantable cardiac defibrillator be used for primary prevention of sudden death? A review of the issues relevant to hospital decision making. Can J Cardiol 20 (12):1199-1204

Miyasaka Y, Tsuji H, Yamada K, Tokunaga S, Saito D, Imuro Y, Matsumoto N, Iwasaka T (2001) Prevalence and mortality of the Brugada-type electrocardiogram in one city in Japan. J Am Coll Cardiol 38 (3):771-774

Miyazaki T, Mitamura H, Miyoshi S, Soejima K, Aizawa Y, Ogawa S (1996) Autonomic and antiarrhythmic drug modulation of ST segment elevation in patients with Brugada syndrome. J Am Coll Cardiol 27 (5):1061-1070

Nakamura TY, Artman M, Rudy B, Coetzee WA (1998) Inhibition of rat ventricular IK1 with antisense oligonucleotides targeted to Kir2.1 mRNA. Am J Physiol 274 (3 Pt 2):H892-900

Napolitano C, Antzelevitch C (2011) Phenotypical manifestations of mutations in the genes encoding subunits of the cardiac voltage-dependent L-type calcium channel. Circ Res 108 (5):607-618

Ogawa M, Kumagai K, Yamanouchi Y, Saku K (2005) Spontaneous onset of ventricular fibrillation in Brugada syndrome with J wave and ST-segment elevation in the inferior leads. Heart Rhythm 2 (1):97-99

Petit-Jacques J, Bois P, Bescond J, Lenfant J (1993) Mechanism of muscarinic control of the high-threshold calcium current in rabbit sino-atrial node myocytes. Pflugers Arch 423 (1-2):21-27

Pitt GS, Zuhlke RD, Hudmon A, Schulman H, Reuter H, Tsien RW (2001) Molecular basis of calmodulin tethering and Ca2 + -dependent inactivation of L-type Ca2 + channels. J Biol Chem 276 (33):30794-30802

Probst V, Veltmann C, Eckardt L, Meregalli PG, Gaita F, Tan HL, Babuty D, Sacher F, Giustetto C, Schulze-Bahr E, Borggrefe M, Haissaguerre M, Mabo P, Le Marec H, Wolpert C, Wilde AA (2010) Long-term prognosis of patients diagnosed with Brugada syndrome: Results from the FINGER Brugada Syndrome Registry. Circulation 121 (5):635-643

Riera AR, Zhang L, Uchida AH, Schapachnik E, Dubner S, Ferreira C (2007) The management of Brugada syndrome patients. Cardiol J 14 (1):97-106

Rivera-Ruiz M, Cajavilca C, Varon J (2008) Einthoven's string galvanometer: the first electrocardiograph. Tex Heart Inst J 35 (2):174-178

Rizzuto R, Pozzan T (2003) When calcium goes wrong: genetic alterations of a ubiquitous signaling route. Nat Genet 34 (2):135-141

Romey G, Garcia L, Rieger F, Lazdunski M (1988) Targets for calcium channel blockers in mammalian skeletal muscle and their respective functions in excitation-contraction coupling. Biochem Biophys Res Commun 156 (3):1324-1332

Rook MB, Bezzina Alshinawi C, Groenewegen WA, van Gelder IC, van Ginneken AC, Jongsma HJ, Mannens MM, Wilde AA (1999) Human SCN5A gene mutations alter cardiac sodium channel kinetics and are associated with the Brugada syndrome. Cardiovasc Res 44 (3):507-517

Rook MB, Evers MM, Vos MA, Bierhuizen MF (2012) Biology of cardiac sodium channel $Na_v1.5$ expression. Cardiovasc Res 93 (1):12-23

Sacher F, Meregalli P, Veltmann C, Field ME, Solnon A, Bru P, Abbey S, Jais P, Tan HL, Wolpert C, Lande G, Bertault V, Derval N, Babuty D, Lacroix D, Boveda S, Maury P, Hocini M, Clementy J, Mabo P, Lemarec H, Mansourati J, Borggrefe M, Wilde A, Haissaguerre M, Probst V (2008) Are women with severely symptomatic brugada syndrome different from men? J Cardiovasc Electrophysiol 19 (11):1181-1185

Sanguinetti MC, Jiang C, Curran ME, Keating MT (1995) A mechanistic link between an inherited and an acquired cardiac arrhythmia: HERG encodes the IKr potassium channel. Cell 81 (2):299-307

Shimizu W (2005) Acquired forms of the Brugada syndrome. J Electrocardiol 38 (4 Suppl):22-25

Simms BA, Zamponi GW (2012) The Brugada syndrome mutation A39 V does not affect surface expression of neuronal rat $Ca_v1.2$ channels. Mol Brain 5:9

Sinner MF, Pfeufer A, Perz S, Schulze-Bahr E, Monnig G, Eckardt L, Beckmann BM, Wichmann HE, Breithardt G, Steinbeck G, Fabritz L, Kaab S, Kirchhof P (2009) Spontaneous Brugada electrocardiogram patterns are rare in the German general population: results from the KORA study. Europace 11 (10):1338-1344

Skinner JR, Chung SK, Montgomery D, McCulley CH, Crawford J, French J, Rees MI (2005) Near-miss SIDS due to Brugada syndrome. Arch Dis Child 90 (5):528-529

Suzuki M, Shiroshita-Takeshita A, Kuruihara T, Takatsuki S, Satoh T, Sato Y, Fukuda K, Hori S (2012) Fever-induced ST-segment elevation in a syncopal patient with Brugada syndrome. Am J Emerg Med 30 (1):263 e261-265

Tsuchiya T, Ashikaga K, Honda T, Arita M (2002) Prevention of ventricular fibrillation by cilostazol, an oral phosphodiesterase inhibitor, in a patient with Brugada syndrome. J Cardiovasc Electrophysiol 13 (7):698-701

Vatta M, Dumaine R, Antzelevitch C, Brugada R, Li H, Bowles NE, Nademanee K, Brugada J, Brugada P, Towbin JA (2002a) Novel mutations in domain I of SCN5A cause Brugada syndrome. Mol Genet Metab 75 (4):317-324

Vatta M, Dumaine R, Varghese G, Richard TA, Shimizu W, Aihara N, Nademanee K, Brugada R, Brugada J, Veerakul G, Li H, Bowles NE, Brugada P, Antzelevitch C, Towbin JA (2002b) Genetic and biophysical basis of sudden unexplained nocturnal death syndrome (SUNDS), a disease allelic to Brugada syndrome. Hum Mol Genet 11 (3):337-345

Viatchenko-Karpinski S, Gyorke S (2001) Modulation of the Ca(2 +)-induced Ca(2 +) release cascade by beta-adrenergic stimulation in rat ventricular myocytes. J Physiol 533 (Pt 3):837-848

Watanabe H, Koopmann TT, Le Scouarnec S, Yang T, Ingram CR, Schott JJ, Demolombe S, Probst V, Anselme F, Escande D, Wiesfeld AC, Pfeufer A, Kaab S, Wichmann HE, Hasdemir C, Aizawa Y, Wilde AA, Roden DM, Bezzina CR (2008) Sodium channel beta1 subunit mutations associated with Brugada syndrome and cardiac conduction disease in humans. J Clin Invest 118 (6):2260-2268

Weissgerber P, Held B, Bloch W, Kaestner L, Chien KR, Fleischmann BK, Lipp P, Flockerzi V, Freichel M (2006) Reduced cardiac L-type Ca2 + current in Ca(V)beta2-/- embryos impairs cardiac development and contraction with secondary defects in vascular maturation. Circ Res 99 (7):749-757

Wilde AA, Brugada R (2011) Phenotypical manifestations of mutations in the genes encoding subunits of the cardiac sodium channel. Circ Res 108 (7):884-897

Yan GX, Antzelevitch C (1999) Cellular basis for the Brugada syndrome and other mechanisms of arrhythmogenesis associated with ST-segment elevation. Circulation 100 (15):1660-1666

Yang J, Ellinor PT, Sather WA, Zhang JF, Tsien RW (1993) Molecular determinants of Ca2 + selectivity and ion permeation in L-type Ca2 + channels. Nature 366 (6451):158-161

Yatani A, Wakamori M, Mikala G, Bahinski A (1993) Block of transient outward-type cloned cardiac K + channel currents by quinidine. Circ Res 73 (2):351-359

Chapter 13
$Ca_v1.3$ Channels and Sino-Atrial Node Dysfunction

Pietro Mesirca and Matteo E. Mangoni

Abstract Cardiac pacemaker activity controls the heartbeat in everyday life. The heart impulse originates in the sino-atrial node, which is formed by a small population of myocytes generating automatic action potentials. Automaticity is due to the presence of the slow diastolic depolarization, which leads the membrane voltage at the end of an action potential toward the threshold of the following action potential. The functional role of voltage-dependent Ca^{2+} channels (VDCCs) to the generation of the slow diastolic depolarization and its regulation by the autonomic nervous system has been matter of debate for almost 30 years. During the last 10 years, however, increasing evidence obtained from genetically modified mice, genomic analysis, and human genetics clearly demonstrated that L-type $Ca_v1.3$ channels play a major role in the genesis of cardiac automaticity. First, studies on mice lacking $Ca_v1.3$ channels have shown that $Ca_v1.3$ loss-of-function leads to bradycardia and atrioventricular conduction block in vivo. Second, patch clamp recordings of pacemaker cells demonstrated that $Ca_v1.3$ channels bring critical inward current in a voltage range spanning the diastolic depolarization. Third, genomic analysis demonstrated that $Ca_v1.3$ channels are widely expressed in pacemaker tissue of mice, rabbits, and humans. Fourth, two congenital diseases of pacemaking have been attributed to $Ca_v1.3$ loss-of-function. In this chapter, we will discuss recent advances in the understanding of the physiological relevance of $Ca_v1.3$ channels in cardiac pacemaker activity and review the current knowledge in the involvement of these channels in the determination of heart automaticity and atrioventricular conduction in humans.

P. Mesirca · M. E. Mangoni (✉)
Département de Physiologie, Universités de Montpellier 1 & 2, CNRS, UMR-5203, Institut de Génomique Fonctionnelle, LabEx ICST F-34094 Montpellier, France
e-mail: matteo.mangoni@igf.cnrs.fr

P. Mesirca · M. E. Mangoni
INSERM, U661 F-34094 Montpellier, France

N. Weiss and A. Koschak (eds.), *Pathologies of Calcium Channels*,
DOI: 10.1007/978-3-642-40282-1_13, © Springer-Verlag Berlin Heidelberg 2014

13.1 Introduction: Ion Channels and Dysfunction in the Generation of the Cardiac Impulse

The heart automaticity is a fundamental physiologic and naturally intriguing phenomenon for the physiologist. The readout of cardiac automaticity is heart rate, which is a primary determinant of the capability of the organism to fulfill the physiological demand of the environment. In the adult heart of higher vertebrates, automaticity is generated in the sino-atrial node (SAN) by specialized “pacemaker” cells having low contractility and generating a periodic electrical oscillation. The two major rhythmogenic centers of the heart, the atrioventricular node (AVN), and the Purkinje fibers network are normally devoted to impulse conduction but can generate viable pacemaker activity in case of SAN failure.

Nevertheless, failure of generating the cardiac impulse underlies SAN bradycardia and disease. In addition to bradycardia, clinical studies have established that elevated heart rates also constitute an important independent risk factor in cardiovascular disease (Gillman et al. 1993; Bohm et al. 2010; Diaz et al. 2005). SAN disease accounts for more than 450,000 electronic pacemaker implantations each year in Europe and North America. SAND is characterized by a combination of symptoms including fatigue, syncope, and atrial fibrillation. Bradycardia, SAN arrest or exit blocks are all typical features of SAND (Mangrum and DiMarco 2000). In some cases, alternating periods of bradycardia and tachyarrhythmias can be observed in SND patients (Bertram et al. 1996). In a substantial number of cases, SND is associated with acquired cardiac conditions, such as heart failure, ischemia, cardiomyopathy, or administration of antiarrhythmic drugs. However, in a significant percentage of patients, SND is unrelated to structural abnormalities of the heart, but shows familial legacy (Sarachek and Leonard 1972; Lehmann and Klein 1978; Mackintosh and Chamberlain 1979). Mutations in genes regulating ion channels involved in sino-atrial automaticity such as L-type $Ca_v1.3$ (Mangoni et al. 2003; Baig et al. 2011), hyperpolarization-activated (f-) channels HCN4 (Milanesi et al. 2006), as well as ryanodine receptors (RYRs) (Neco et al. 2012), ankyrin-B, and T-box transcription factors (Postma et al. 2008) are associated with various forms of previously unexplained tachy-brady syndromes and conduction defects (Mangoni and Nargeot 2008; Pfeufer et al. 2010).

13.2 Current Models of Generation of Pacemaker Activity: Which Role for L-Type VDCCs?

It is generally accepted that pacemaker activity requires a complex functional interplay between ion channels, membrane transporters, and intracellular Ca^{2+} release (Mangoni and Nargeot 2008). However, the relative importance of these mechanisms is still hotly debated. Among ion channels families, both T-type and L-type voltage-gated Ca^{2+} channels (VDCCs) are highly expressed in spontaneously active cells of the SAN and AVN. To date, two models for the generation of heart automaticity have been highlighted. In the first, "pacemaker" f-(HCN) channels mediating the hyperpolarization-activated non-selective cation current (I_f) are considered the key player of heart rate control: the so-called I_f–based model of pacemaking (Lakatta and DiFrancesco 2009). The I_f–based pacemaker model postulates that heart rate is regulated on a beat-to-beat basis by intracellular 3'5' cyclic adenosine monophosphate (cAMP) concentration. cAMP controls the open probability of f-channels in such a way that the degree of activation of f-channels permanently tunes the steepness of the diastolic depolarization. Recently, an alternative model of heart rate control has been proposed. According to this view, pacemaker cells spontaneously generate rhythmic ryanodine receptor (RYR)-dependent local Ca^{2+}-induced Ca^{2+} release (LCICR) with a mechanism which is independent from membrane voltage (Lakatta and DiFrancesco 2009). Spontaneous LCICR stimulates the activity of the Na^+/Ca^{2+} exchanger that drives the membrane voltage to the action potential threshold. In the spontaneous LCICR–based model of pacemaking, cAMP controls heart rate by regulating the phase and frequency of LCICR. Both the I_f–and LCR–based model of pacemaking do not adequately consider the role of L-type VDCCs in pacemaking. The I_f–based model of pacemaking postulates that $I_{Ca,L}$ controls the upstroke phase of the action potential. The LCICR–based model of pacemaking assumes that L-type VDCCs are not involved in the pacemaker mechanism per se, but constitute an important mechanism for controlling the Ca^{2+} load of the sarcoplasmic reticulum (DiFrancesco 2010; Lakatta et al. 2010).

Both these views, however, cannot explain a substantial amount of experimental and genetic evidence on $Ca_v1.3$ channels. We will discuss this in-depth.

13.3 How do L-type VDCCs Contribute to Pacemaking? Historical Perspective

H. Reuter was the first to show the existence of Ca^{2+} currents in the heart automatic tissue by voltage-clamp recording of spontaneously active sheep Purkinje (Reuter 1967). At first, Ca^{2+} currents were not considered as part of the heart pacemaker mechanism but were indicated as connected to the control of action potential plateau phase and myocyte contraction. The cardiac Ca^{2+} current was named I_{si}, for "slow" or "secondary inward" current, to distinguish "slow" Ca^{2+} currents from the "fast" Na^+ current (Noble 1984). I_{si} was found in all cardiac cells including the multicellular rabbit SAN preparation (Brown et al. 1979). During the first half of the 1980s it became evident that I_{si} was a mixed current composed by an inward Ca^{2+} current and a secondary Ca^{2+}–dependent processes identified as the Na^+/Ca^{2+} exchanger (I_{NCX}) (Noble 1984). In multicellular rabbit SAN preparations and in amphibian sinus venosus (Giles and Shibata 1985; Shibata and Giles 1985) I_{si} was found to be highly sensitive to adrenaline, a property that fitted with the high sensitivity of SAN pacemaking to β-adrenergic receptor agonist (Brown et al. 1979). As a consequence, the Japanese School leaded by A. Noma and H. Irisawa proposed that I_{si} played a major role in the generation of the diastolic depolarization and in the positive chronotropic effect of the sympathetic nervous system (Noma et al. 1980, 1983). Expression of I_{si} explained also the Ca^{2+}–dependence of the upstroke phase of the SAN action potential (Noble 1984). The biophysical and pharmacologic interpretation of I_{si} has been changed substantially by the use of the patch-clamp technique (Hamill et al. 1981). L-type Ca^{2+} channels are now known as the ionic basis of the SAN I_{si}. Hagiwara, Irisawa, and Kameyama were the first to report the expression of $I_{CaL,}$ in isolated SAN pacemaker cells (Hagiwara et al. 1988) and to describe its kinetic and pharmacologic properties. In particular, Hagiwara and co-workers defined $I_{Ca,L}$ as a "high"-threshold Ca^{2+} current-activated from about –30 mV and distinguished from $I_{Ca,T}$, a "low" threshold Ca^{2+} current activated at –50 mV. Based on the observation that, contrary to $I_{Ca,T}$, $I_{Ca,L}$ was sensitive to isoproterenol, Hagiwara and coworkers proposed that $I_{Ca,L}$ importantly contributed to the positive chronotropic effect of the sympathetic nervous system.

During the 1990s several groups have tried to assess the relevance of $I_{Ca,L}$ in the generation of the diastolic depolarization (see Irisawa et al. for review (Irisawa et al. 1993)). In this respect, different authors have disagreed as to the relative importance of $I_{Ca,L}$ with respect to I_f (Petit-Jacques et al. 1993; Zaza et al. 1996). In the rabbit SAN $I_{Ca,L}$ density is expressed heterogeneously, which is reflected at the SAN tissue level with a differential contribution of $I_{Ca,L}$ to the pacemaker action potential. Indeed, nifedipine blocks automaticity in the center of the SAN (Kodama et al. 1997). In contrast, tissue samples from the SAN periphery show moderate negative chronotropism following perfusion of nifedipine (Kodama et al. 1997). These results are consistent with the view that the pacemaker action potential in the center of the rabbit SAN depends from $I_{Ca,L,}$ as opposed to the periphery, where I_{Na} is robustly expressed, thereby compensating for the blockade of $I_{Ca,L}$. The differential role of $I_{Ca,L}$ in the center and the periphery of the SAN underlines the problem of separating the possible contribution of $I_{Ca,L}$ to the diastolic depolarization from that of the upstroke phase of the action potential.

A major problem in evaluating the contribution of $I_{Ca,L}$ in pacemaking is the difficulty in transposing the steady-state properties of the current into the effective current transported by the channels during the pacemaker action potential cycle. Doerr and co-workers tried to evaluate the contribution of $I_{Ca,L}$ in the pacemaker cycle in rabbit SAN cells, by using the action potential clamp technique (Doerr et al. 1989) and nifedipine to block $I_{Ca,L}$. They reported a net nifedipine-sensitive current measurable during the early diastolic depolarization as well a long lasting component during the plateau phase. Verhejick and co-workers have recorded the net DHP-sensitive $I_{Ca,L}$ at different times during the diastolic depolarization and the action potential upstroke. Contrary to expectations, they demonstrated that $I_{Ca,L}$ activates during the early diastolic depolarization, and then progressively increases up to the threshold of the action potential upstroke. They also demonstrated that activation of $I_{Ca,L}$ in the diastolic depolarization is due to a "low voltage"-activated $I_{Ca,L}$ These observations have been the first to challenge the view that only $I_{Ca,T}$ and I_f can be activated at negative voltages during the diastolic depolarization and paved the way to the description of the functional role of $Ca_v1.3$ channels in pacemaking.

13.4 Expression and Distribution of $Ca_v1.3$ in the Rodent and Human Heart

In the heart, expression of $Ca_v1.3$ is restricted to the supraventricular regions, the SAN, the AVN, and the atria (Mangoni et al. 2003; Marionneau et al. 2005; Tellez et al. 2006; Chandler et al. 2009). Quantitatively, rhythmogenic centers express higher amounts of $Ca_v1.3$. Expression of $Ca_v1.3$ in the atria does have functional consequences. First, it has been shown that loss-of-function of $Ca_v1.3$ induces susceptibility to atrial fibrillation in $Ca_v1.3^{-/-}$ mice (Zhang et al. 2005). Second, $Ca_v1.3$ channels regulate the atrial action potential duration via functional cross-talk with Ca^{2+}-activated K^+ channels (Xu et al. 2005). It is possible, but not yet demonstrated that atrial fibrillation in $Ca_v1.3^{-/-}$ mice are caused by dysregulation of Ca^{2+}-activated K^+ channels.

Besides atrial tissue, expression of $Ca_v1.3$ channels is now considered as a hallmark of the pacemaker function. In restricted genomic studies of ion channels expression in the mouse (Marionneau et al. 2005) and human hearts (Chandler et al. 2009), high expression of mRNA coding for $Ca_v1.3$ groups with other ion channels and proteins that are highly expressed in SAN and AVN. These include the HCN4 and the $Ca_v3.1$ channel subunits, the Ca^{2+} channel accessory subunit α_2–σ_2, as well as the $Na_v1.1$ and $Na_v1.3$ Na^+ channel isoforms that are typically expressed in neurons (Fig. 13.1). Importantly, in the rabbit SAN, $Ca_v1.3$ shows an increasing gradient of expression from the SAN periphery to the center suggesting that $Ca_v1.3$ expression is higher in the SAN center, which is the dominant pacemaker region (Tellez et al. 2006). The existence of "low" voltage-activated $I_{Ca,L}$ in rabbit Purkinje fibers suggests that $Ca_v1.3$ is also functionally expressed in this tissue (Mangoni and Nargeot 2008). The specific role of $Ca_v1.3$ channels in Purkinje fibers automaticity is still unknown. On the other hand, $Ca_v1.3$ is poorly or not expressed in ventricular tissue. This is in agreement with the preserved ventricular function in mice with loss-of-function of $Ca_v1.3$

13.5 Contribution of $Ca_v1.3$ Channels to in Vivo Heart Rate in Genetically modified Model Mice

Substantial genetic evidence data support the view that $Ca_v1.3$-mediated $I_{Ca,L}$ contribute to the diastolic depolarization (Mangoni et al. 2003, 2006a). Involvement of $Ca_v1.3$ channels in pacemaking in vivo is also strongly suggested by

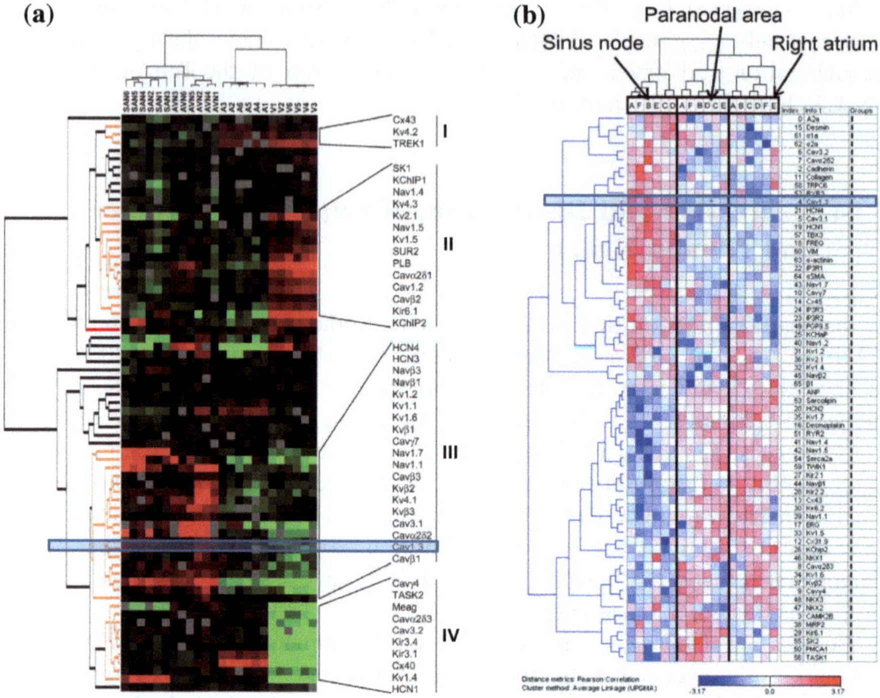

Fig. 13.1 Expression of $Ca_v1.3$ in heart chambers and rhythmogenic centers. **a** Two-way hierarchical agglomerative clustering applied to 71 selected genes (*vertically*) and to 6 pools of SAN (SAN1–SAN6), AVN (AVN1–AVN6), atria (A1–A6), and ventricles (V1–V6) from the mouse heart. The input consisted of the ratio for each pool and gene versus HPRT. Each gene is represented by a single row of colored boxes and each pool by a single column. The entire gene clustering is shown on the *left*. Four selected gene clusters are shown on the *right* (I, II, III, IV) containing relevant genes to the nodal and working tissue discrimination (I, III) and to the atria and ventricle discrimination (II, IV). Each color patch in the resulting visual map represents the gene expression level, with a continuum of expression levels from *dark green (lowest)* to *bright red (highest)*. Missing values are coded as *silver* [Adapted from (Marionneau et al. 2005), with permission]. **b** 2-way hierarchical cluster analysis (Pearson distances) of 66 "indicator" mRNAs between sinus node, paranodal area, and right atrium in human heart. *Red squares* indicate values above average for a given transcript, *white squares* indicate average values, and *blue squares* indicate values below average. These calculations were performed on normalized log10 expression values [Adapted from (Chandler et al. 2009), with permission]

in vivo ECG recordings showing that DHPs induce bradycardia in anesthetized mice (Lande et al. 2001; Sinnegger-Brauns et al. 2004). Probably the best demonstration of the importance of $Ca_v1.3$ channels in SAN physiology is the observation that mice lacking $Ca_v1.3$ channels ($Ca_v1.3^{-/-}$) are bradycardic and present major dysfunction of SAN pacemaking at the cellular level (Platzer et al. 2000; Mangoni et al. 2003). Bradycardia persists after block of the autonomic nervous system and intact atria and SANs from $Ca_v1.3^{-/-}$ mice have slower pacing rate than wild-type counterparts (Platzer et al. 2000; Zhang et al. 2002).

Two studies demonstrated that $Ca_v1.3$ channels play a major role in automaticity of isolated SAN (Zhang et al. 2002) and in pacemaker cells (Mangoni et al. 2003). Indeed, pacemaker cells from $Ca_v1.3^{-/-}$ mice show irregular pacemaking. Furthermore, some $Ca_v1.3^{-/-}$ SAN cells do not show pacemaker activity (Mangoni et al. 2003).

In conclusion, insights gained from $Ca_v1.3^{-/-}$ mice show a distinction in the functional role of $Ca_v1.3$ and $Ca_v1.2$ channels, the first contributing to automaticity, and $Ca_v1.2$ channels triggering myocardial contraction. Indeed, inactivation of $Ca_v1.3$-mediated $I_{Ca,L}$ impairs pacemaking and atrioventricular conduction, but has no effect on myocardial contractility (Matthes et al. 2004). The differential roles of $Ca_v1.3$ and $Ca_v1.2$ channels has been shown pharmacologically by employing a knockin mouse strain in which $Ca_v1.2$ channels are insensitive to DHPs ($Ca_v1.2DHP^{-/-}$ mouse) (Sinnegger-Brauns et al. 2004). Experiments in these mice show that the in vivo bradycardic effect induced by DHPs is not changed in $Ca_v1.2DHP^{-/-}$ mice indicating that the dominant L-type VDCC isoform participating to automaticity is $Ca_v1.3$ (Sinnegger-Brauns et al. 2004).

13.6 Mechanistic Links Between $Ca_v1.3$ Channels and the Generation of the Diastolic Depolarization in SAN Cells

The kinetic and activation properties of $Ca_v1.3$–mediated $I_{Ca,L}$ are distinct from those of $Ca_v1.2$–mediated $I_{Ca,L}$. Native $Ca_v1.3$ channels activate at negative potentials from about −50 mV (Fig. 13.2). In SAN cells, loss of $Ca_v1.3$ channels shifts the activation of total $I_{Ca,L}$ to more positive potentials (Mangoni et al. 2003). Such a shift eliminates all $I_{Ca,L}$ in the voltage range corresponding to that of the diastolic depolarization (Mangoni et al. 2003). Knockout of $Ca_v1.3$ channels reduces $I_{Ca,L}$ density by about 70 % in SAN (Mangoni et al. 2003) and by almost 100 % in AVN (Marger et al. 2011) cells. Residual $I_{Ca,L}$ in SAN mediated by $Ca_v1.2$ (Mangoni et al. 2003; Christel et al. 2012). Consistent with this hypothesis, $I_{Ca,L}$ in pacemaker cells from $Ca_v1.3$ knockout mice is more sensitive to DHPs (Mangoni et al. 2003) and has faster inactivation kinetics (Zhang et al. 2002). β-adrenergic receptors activation negatively shifts the activation of $Ca_v1.3$–mediated $I_{Ca,L}$, to −55 mV (Mangoni et al. 2003). Interestingly, this threshold is comparable with that observed for the nifedipine-sensitive $I_{Ca,L}$ measured in spontaneously beating rabbit SAN cells (Verheijck et al. 1999). This indicates the functional existence and importance of $Ca_v1.3$ channels also in rabbit SAN cells. The maximal pacing rate in $Ca_v1.3^{-/-}$ in the presence of isoproterenol is slightly slower than that of wild-type hearts (Matthes et al. 2004). This observation can be explained by taking into consideration that stimulation of $I_{Ca,L}$ by saturating concentrations of noradrenaline cannot compensate for the lack of $I_{Ca,L}$ in the diastolic depolarization range (Mangoni et al. 2003). Beside the fact that $Ca_v1.3$

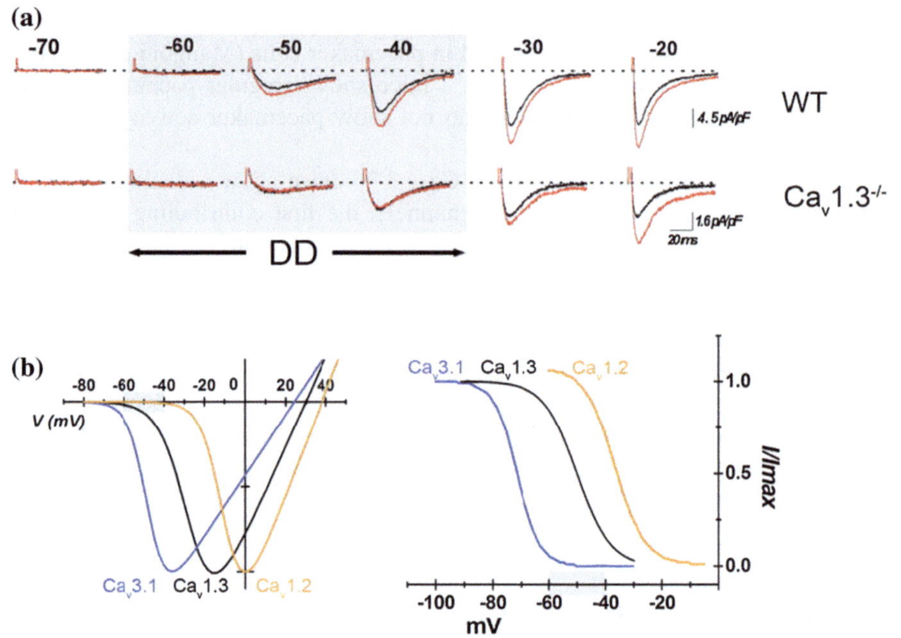

Fig. 13.2 $Ca_v1.3$ channels activate in the diastolic depolarization range. **a** voltage-dependent Ca^{2+} currents in pacemaker SAN cells from wild type (WT) and $Ca_v1.3^{-/-}$ mice. Currents are evoked from a holding potential of −80 mV at the indicated test potentials. Current stimulation by BAK K 8644 is used to detect activation of $I_{Ca,L}$ at a given test voltage. In $Ca_v1.3^{-/-}$ SAN cells, BAY K 8644 (*red traces*) has no effect at test potentials negative to −30 mV. This observation indicates that inactivation of $Ca_v1.3$ channels abolishes a $I_{Ca,L}$ component which has intermediate activation threshold between $I_{Ca,T}$ and $Ca_v1.2$-mediated $I_{Ca,L}$. $Ca_v1.3$-mediated $I_{Ca,L}$ is activated in the diastolic depolarization range (DD, *shaded gray box*). [From (Mangoni et al. 2003), with permission] **b** Current-to-voltage relationships (*left*) and steady-state inactivation (*right*) of native SAN $Ca_v3.1$, $Ca_v1.3$, and $Ca_v1.2$ channels. [Adapted from Mangoni et al. (Mangoni et al. 2006b), with permission]

and $Ca_v1.2$ channels have different activation properties, recent evidence also shows that these channels are differentially regulated by membrane voltage. Indeed, contrary to $Ca_v1.2$ channels, $Ca_v1.3$ channels undergo voltage-dependent facilitation (Mangoni et al. 2000; Christel et al. 2012). Voltage-dependent facilitation increases $Ca_v1.3$ current at slow heart rate, when the channel recovers from inactivation and is maximal at voltages spanning the diastolic depolarization (Mangoni et al. 2000). In comparison to $Ca_v1.3$, $Ca_v1.2$ channels undergo little voltage-dependent facilitation. We have proposed that voltage-dependent facilitation constitutes a safeguard mechanism which increases the availability of $I_{Ca,L}$ during excessive heart rate slowing (Mangoni et al. 2000; Christel et al. 2012). Conversely, reduction in voltage-dependent facilitation at high heart rates may prevent excessive tachycardia (Mangoni et al. 2000). The lack of voltage-dependent facilitation in $Ca_v1.2$ renders these channels less capable of stimulating normal pacemaking despite a significant contribution (30–40 %) to the total whole cell $I_{Ca,L}$ in SAN cells.

Another characteristic of $Ca_v1.3$ channels is their colocalization with RYRs of the sarcoplasmic reticulum in primary SAN pacemaker cells (Christel et al. 2012). In comparison, relatively poor colocalization between $Ca_v1.2$ channels and sarcomeric RYRs is found in SAN cells (Christel et al. 2012). Characterization of interactions between $Ca_v1.3$ and RYRs will reveal additional insights as to how sarcoplasmic reticulum Ca^{2+} release stimulates pacemaking.

The strong colocalization of $Ca_v1.3$ with RYR2 may be relevant for the functional role of RYR-mediated Ca^{2+} release in pacemaking (Vinogradova et al. 2002). During the late phase of the diastolic depolarization, RYR-mediated Ca^{2+} release promotes NCX activation, which accelerates reaching the threshold of the SAN action potential upstroke. Close apposition of $Ca_v1.3$ with RYRs may facilitate SR Ca^{2+} release since $I_{Ca,L}$ stimulates RYR open probability (Christel et al. 2012).

We recently developed a numerical model of SAN automaticity that implements both voltage-dependent facilitation of $Ca_v1.3$ and preferential coupling of these channels to RYR-dependent Ca^{2+} release (Christel et al. 2012). This model suggests that facilitation stimulates SAN pacemaker activity. In particular, numerical simulations predict that the slope of rise of diastolic RYR-dependent Ca^{2+} release increases as a function of $Ca_v1.3$-mediated $I_{Ca,L}$ half-activation voltage.

In conclusion, differences in activation/inactivation, voltage-dependent facilitation and co-localization with RYRs contribute to the differential roles of $Ca_v1.2$ and $Ca_v1.3$ in controlling SAN automaticity.

13.7 Functional Role of $Ca_v1.3$ Channels in AVN

$Ca_v1.3^{-/-}$ mice show strong slowing of the atrioventricular conduction (Platzer et al. 2000; Zhang et al. 2002; Matthes et al. 2004; Marger et al. 2011). Freely moving $Ca_v1.3^{-/-}$ mice show 1st and 2nd degree atrio-ventricular blocks (Platzer et al. 2000; Marger et al. 2011). Some mice also display complete atrioventricular block with dissociated atrial and ventricular rhythms (Mesirca et al., unpublished observations). Second degree atrioventricular block is observed also in anesthetized animals (Zhang et al. 2002) and in isolated $Ca_v1.3^{-/-}$ hearts (Matthes et al. 2004). The prominent impact of $Ca_v1.3$ inactivation in atrioventricular conduction of $Ca_v1.3^{-/-}$ mice suggests that these channels play an important role in AVN physiology. Others and we have shown that $Ca_v1.3$ channels play a critical role in pacemaking of AVN cells (Marger et al. 2011; Zhang et al. 2011). Indeed, under normal conditions $Ca_v1.3^{-/-}$ AVN cells do not show automaticity even in the presence of saturating doses of isoproterenol (Marger et al. 2011). In $Ca_v1.3^{-/-}$ AVN cells, pacemaking can be initiated by injection of negative current. However, this "initiated" automaticity is relatively slow and irregular. Furthermore, no diastolic depolarization could be recorded even in current-injected cells and the membrane voltage during the diastolic phase showed only sub-threshold oscillations of the membrane potential. This observation indicates that $Ca_v1.3$ channels are critical for generating the diastolic depolarization of AVN pacemaker cells (Marger et al. 2011). Third, while initiated pacemaking of current-injected $Ca_v1.3^{-/-}$ AVNC cells is sensitive to isoproterenol, maximal rates are still lower than those of wild-type AVN cells. An interesting observation is that quiescent $Ca_v1.3^{-/-}$ AVN cells show positive resting membrane potential. It is likely that the loss of crosstalk between $Ca_v1.3$ channels and SK2 K^+ channels (Zhang et al. 2008) in $Ca_v1.3^{-/-}$ AVN cells induces depolarization of the membrane resting potential. The lack of spontaneous automaticity in $Ca_v1.3$−/− AVNC cells in vitro may not imply unexcitability in vivo . Indeed, $Ca_v1.3^{-/-}$ AVNCs are still able to fire I_{Na}-dependent action potentials (Marger et al. 2011). It is possible that in the intact AVN, $Ca_v1.3^{-/-}$ cells are also sufficiently hyperpolarized to enable I_{Na}-dependent action potentials to be triggered by SAN impulse. In vivo, both the SAN and AVN is subject of the hyperpolarizing load imposed by the electrical coupling with the right atrium. This phenomenon explains why I_{Kr} blockers completely stop pacemaker activity of isolated SANCs, but not in rabbit intact atrio-sinus preparations (Verheijck et al. 2002) or pacemaking of intact mouse hearts (Clark et al. 2004). Hyperpolarizing load by the right atrium could also explain why in another $Ca_v1.3^{-/-}$ mouse line, Zhang and co-workers (Zhang et al. 2011) were able to record residual AVN automaticity in intact atrial preparations containing the AVN. Finally, it is worth of note that the AVN is a highly heterogeneous structure in which different cell types, possibly serving different conduction pathways, can co-exist. Cell–cell interactions between automatic and non-automatic AVNCs, together with the alternative use of different conduction pathways could also explain why $Ca_v1.3^{-/-}$ mice can also explain the variability in the degree of atrioventricular dysfunction observed in $Ca_v1.3^{-/-}$ mice.

13.8 $Ca_v1.3$ Channels and Human Disease of Heart Automaticity

The importance of $Ca_v1.3$ channels in the generation of pacemaker activity in model mice has been demonstrated in the early 2000s. However, some authors questioned the relevance of these channels in pacemaking of large mammals and humans. Reason for this stood for the high heart rate in mice in comparison to that of human heart. In other words, it has been suggested that $Ca_v1.3$ channels might have been a form of adaptation in mice due to the necessity of keeping high heart rates. However, the discovery of two congenital pathologies of heart automaticity and atrioventricular conduction that could be attributed to $Ca_v1.3$ downregulation or loss-of-function rendered this argument outdated. These pathologies underscore the importance of $Ca_v1.3$ channels in the generation of heart automaticity, but also highlight their physiological relevance in the determination of heart rate and atrioventricular conduction in humans.

13.8.1 Involvement of $Ca_v1.3$ in Congenital Heart Block (CHB)

Beside diseases due to pure SAN dysfunction, bradycardia is also observed in association with atrioventricular block. An example of this type of disease is congenital heart block (CHB) in infants (Hu et al. 2004). CHB is characterized by progressive partial or complete atrioventricular block and affects foetuses and newborns (see Ref. (Boutjdir 2000) for review). CHB is generally detected just before or immediately after birth and can carry severe consequences (Waltuck and Buyon 1994). In the clinical practice, treatment of CHB can necessitate the implantation of an electronic pacemaker. CHB is due to production by pregnant mothers of auto-antibodies against intracellular soluble ribonucleoproteins 48-kDa SSB/La, 52-kDa SSA/Ro, and 60-kDa SSA/Ro (Waltuck and Buyon 1994). While the aetiology of this disease has remained obscure for long time, there is now strong evidence that loss-of-function of $Ca_v1.3$ and probably $Ca_v3.1$ channel underlie this pathology. Indeed, Hu et al. (Hu et al. 2004) have reported inhibition of $I_{Ca,L}$ and $I_{Ca,T}$ by IgGs isolated from mothers having CHB-affected children. SAN bradycardia and CHB can be explained at least in part by downregulation of $Ca_v1.3$ and $Ca_v3.1$ channels by maternal antibodies (Hu et al. 2004). This hypothesis has been validated in vitro by a study showing that the $Ca_v1.3$ channel protein is expressed in the human fetal heart and that anti-Ro/La antibodies can effectively inhibit $Ca_v1.3$-mediated $I_{Ca,L}$ expressed in tsA201 cells (Qu et al. 2005).

13.8.2 $Ca_v1.3$ in Congenital SAN Dysfunction

As discussed above $Ca_v1.3^{-/-}$ mice are phenotypically characterized by deafness and bradycardia. The discovery of two consanguineous families from Pakistan with congenital deafness and bradycardia constituted a major insight into the functional role of $Ca_v1.3$ channels in the determination of heart rate in humans. It also demonstrated that $Ca_v1.3$ channels are a conserved pacemaker mechanism between mice and humans (Baig et al. 2011). Because of the association between deafness and bradycardia, this newly described disease was named Sino-atrial Node Dysfunction and Deafness syndrome (SANDD). After proper scaling of the average heart rate for a human subject in respect to that of mice, the similarity between SAN dysrhythmia observed in $Ca_v1.3^{-/-}$ mice and affected SANDD individuals is striking (Fig. 13.3). Indeed, patients with SANDD present with SAN bradycardia at rest. Bradycardia can be deep, with mean heart rates of about 35 beat per minutes, especially during nighttime. SANDD patients also exhibit variable degree of atrioventricular block and dissociated rhythms (Fig. 13.3). This last observation can be explained by our recent result showing that $Ca_v1.3$ is important for automaticity of AVN cells (Marger et al. 2011). $Ca_v1.3$ loss-of-function in SANDD patients is due to a homozygous insertion of 3 base pairs in the

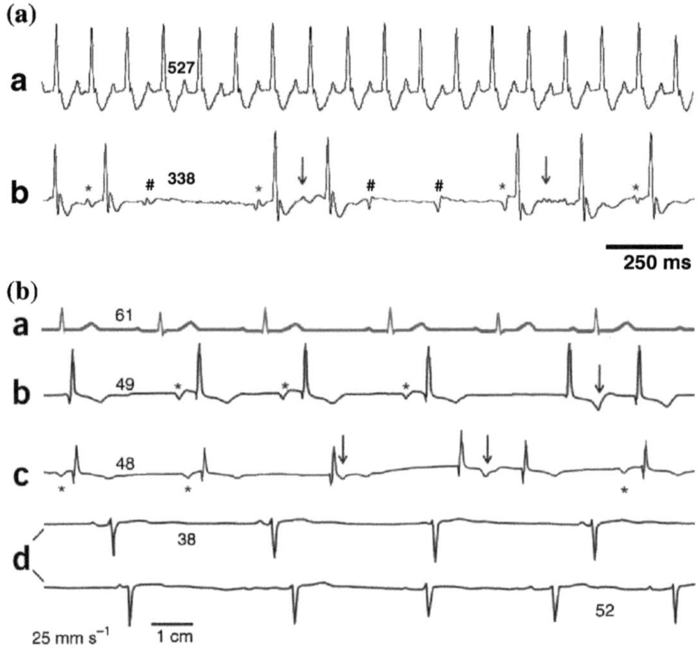

Fig. 13.3 **a** ECG sample recordings from wild type (a) and $Ca_v1.3^{-/-}$ mice (b). **b** ECG recordings from an unaffected person (**a**) and three individuals who are homozygous for the *CACNA1D* SANDD mutation (**b–d**). Asterisks mark P waves that precede QRS complexes; arrows indicate waveforms that suggest P waves coinciding with T waves; hashes indicate not conducted P waves. In $Ca_v1.3^{-/-}$ mice as in affected individuals, both P–P and R–R intervals are variable (SAN arrhythmia). Numbers indicate heart rate (bpm) calculated from the corresponding beat-to-beat R–R interval. [Adapted from (Baig et al. 2011), with permission]

alternatively spliced exon 8B, which is preferentially used in the SAN and the inner hear (Baig et al. 2011). Selective usage of exon 8B of the *CACNA1D* may explain why the SANDD mutation does not cause a systemic disease. The SANDD insertion mutation renders $Ca_v1.3$ channels non-conductive. In a heterologous expression system, mutated $Ca_v1.3$ channels are correctly inserted in the plasma membrane. Gating currents can also be recorded from mutated SANDD channels but the gating movement does not lead to current activation.

Beside SANDD, human mutations in the gene encoding the multifunctional scaffolding protein, ankyrin B, cause reductions in $Ca_v1.3$-mediated $I_{Ca,L}$ in SAN cells and atrial myocytes, which is associated with bradycardia and atrial fibrillation, respectively (Cunha et al. 2011; Le Scouarnec et al. 2008). With respect to defects in $Ca_v1.3$ causing SAN dysfunction, loss of the rapid activation kinetics and negative activation thresholds of $Ca_v1.3$ weakens the diastolic depolarization, thus slowing pacemaking (Mangoni et al. 2003).

13.9 Concluding Remarks

The relevance of $Ca_v1.3$ channels in the generation and regulation of cardiac pacemaking, atrioventricular conduction, and heart rate determination is now well established. $Ca_v1.3$ channels activate in the diastolic depolarization range facilitating reaching of the action potential threshold. Activation, inactivation, and voltage-dependent facilitation of $Ca_v1.3$ channels are designed to promote pacemaking. Importantly, the functional role of $Ca_v1.3$ channels is conserved between rodents and humans. $Ca_v1.3$ channels also play a major role in atrioventricular conduction as underscored by the presence of dissociated rhythms in $Ca_v1.3^{-/-}$ mice and SANDD patients (Baig et al. 2011). Future studies will further address the role of $Ca_v1.3$ channels in pacemaker activity and in particular their importance with respect to other ion channels involved in automaticity such as HCN4 and RyRs.

References

Baig SM, Koschak A, Lieb A, Gebhart M, Dafinger C, Nurnberg G, Ali A, Ahmad I, Sinnegger-Brauns MJ, Brandt N, Engel J, Mangoni ME, Farooq M, Khan HU, Nurnberg P, Striessnig J, Bolz HJ (2011) Loss of Ca(v)1.3 (CACNA1D) function in a human channelopathy with bradycardia and congenital deafness. Nat Neurosci 14 (1):77–;84. doi:nn.2694 [pii]10.1038/nn.2694

Bertram H, Paul T, Beyer F, Kallfelz HC (1996) Familial idiopathic atrial fibrillation with bradyarrhythmia. Eur J Pediatr 155(1):7–10

Bohm M, Swedberg K, Komajda M, Borer JS, Ford I, Dubost-Brama A, Lerebours G, Tavazzi L (2010) Heart rate as a risk factor in chronic heart failure (SHIFT): the association between heart rate and outcomes in a randomised placebo-controlled trial. Lancet 376(9744):886–894. doi:S0140-6736(10)61259-7 [pii]10.1016/S0140-6736(10)61259-7

Boutjdir M (2000) Molecular and ionic basis of congenital complete heart block. Trends Cardiovasc Med 10(3):114–122

Brown HF, DiFrancesco D, Noble SJ (1979) How does adrenaline accelerate the heart? Nature 280(5719):235–236

Chandler NJ, Greener ID, Tellez JO, Inada S, Musa H, Molenaar P, Difrancesco D, Baruscotti M, Longhi R, Anderson RH, Billeter R, Sharma V, Sigg DC, Boyett MR, Dobrzynski H (2009) Molecular architecture of the human sinus node: insights into the function of the cardiac pacemaker. Circulation 119(12):1562–1575

Christel CJ, Cardona N, Mesirca P, Herrmann S, Hofmann F, Striessnig J, Ludwig A, Mangoni ME, Lee A (2012) Distinct localization and modulation of $Ca_v1.2$ and $Ca_v1.3$ L-type Ca2+ channels in mouse sinoatrial node. J Physiol 590(Pt 24):6327–6342. doi:jphysiol.2012.239954 [pii]10.1113/jphysiol.2012.239954

Clark RB, Mangoni ME, Lueger A, Couette B, Nargeot J, Giles WR (2004) A rapidly activating delayed rectifier K+ current regulates pacemaker activity in adult mouse sinoatrial node cells. Am J Physiol Heart Circ Physiol 286(5):H1757–H1766

Cunha SR, Hund TJ, Hashemi S, Voigt N, Li N, Wright P, Koval O, Li J, Gudmundsson H, Gumina RJ, Karck M, Schott JJ, Probst V, Le Marec H, Anderson ME, Dobrev D, Wehrens XH, Mohler PJ (2011) Defects in ankyrin-based membrane protein targeting pathways underlie atrial fibrillation. Circulation 12411):1212–1222. doi:CIRCULATIONAHA.111.023986 [pii]10.1161/CIRCULATIONAHA.111.023986

Diaz A, Bourassa MG, Guertin MC, Tardif JC (2005) Long-term prognostic value of resting heart rate in patients with suspected or proven coronary artery disease. Eur Heart J 26(10):967–974. doi:ehi190 [pii]10.1093/eurheartj/ehi190

DiFrancesco D (2010) The role of the funny current in pacemaker activity. Circ Res 106(3):434–446. doi:106/3/434 [pii]10.1161/CIRCRESAHA.109.208041

Doerr T, Denger R, Trautwein W (1989) Calcium currents in single SA nodal cells of the rabbit heart studied with action potential clamp. Pflugers Arch 413(6):599–603

Giles WR, Shibata EF (1985) Voltage clamp of bull-frog cardiac pace-maker cells: a quantitative analysis of potassium currents. J Physiol 368:265–292

Gillman MW, Kannel WB, Belanger A, D'Agostino RB (1993) Influence of heart rate on mortality among persons with hypertension: the Framingham Study. Am Heart J 125(4):1148–1154

Hagiwara N, Irisawa H, Kameyama M (1988) Contribution of two types of calcium currents to the pacemaker potentials of rabbit sino-atrial node cells. J Physiol 395:233–253

Hamill OP, Marty A, Neher E, Sakmann B, Sigworth FJ (1981) Improved patch-clamp techniques for high-resolution current recording from cells and cell-free membrane patches. Pflugers Archiv-European J Physiol 391(2):85–100

Hu K, Qu Y, Yue Y, Boutjdir M (2004) Functional basis of sinus bradycardia in congenital heart block. Circ Res 94(4):e32–e38

Irisawa H, Brown HF, Giles W (1993) Cardiac pacemaking in the sinoatrial node. Physiol Rev 73(1):197–227

Kodama I, Nikmaram MR, Boyett MR, Suzuki R, Honjo H, Owen JM (1997) Regional differences in the role of the Ca2+ and Na+ currents in pacemaker activity in the sinoatrial node. Am J Physiol 272(6 Pt 2):H2793–H2806

Lakatta EG, DiFrancesco D (2009) What keeps us ticking: a funny current, a calcium clock, or both? J Mol Cell Cardiol 47(2):157–170

Lakatta EG, Maltsev VA, Vinogradova TM (2010) A coupled SYSTEM of intracellular Ca2+ clocks and surface membrane voltage clocks controls the timekeeping mechanism of the heart's pacemaker. Circ Res 106(4):659–673. doi:106/4/659 [pii]10.1161/CIRCRESAHA.109.206078

Lande G, Kyndt F, Baro I, Chabannes D, Boisseau P, Pony JC, Escande D, Le Marec H (2001) Dynamic analysis of the QT interval in long QT1 syndrome patients with a normal phenotype. Eur Heart J 22(5):410–422

Le Scouarnec S, Bhasin N, Vieyres C, Hund TJ, Cunha SR, Koval O, Marionneau C, Chen B, Wu Y, Demolombe S, Song LS, Le Marec H, Probst V, Schott JJ, Anderson ME, Mohler PJ (2008) Dysfunction in ankyrin-B-dependent ion channel and transporter targeting causes human sinus node disease. Proc Natl Acad Sci USA 105(40):15617–15622. doi:0805500105 [pii]10.1073/pnas.0805500105

Lehmann H, Klein UE (1978) Familial sinus node dysfunction with autosomal dominant inheritance. Br Heart J 40(11):1314–1316

Mackintosh AF, Chamberlain DA (1979) Sinus node disease affecting both parents and both children. Eur J Cardiol 10(2):117–122

Mangoni ME, Couette B, Bourinet E, Platzer J, Reimer D, Striessnig J, Nargeot J (2003) Functional role of L-type $Ca_v1.3$ Ca2+ channels in cardiac pacemaker activity. Proc Natl Acad Sci USA 100(9):5543–5548

Mangoni ME, Couette B, Marger L, Bourinet E, Striessnig J, Nargeot J (2006a) Voltage-dependent calcium channels and cardiac pacemaker activity: From ionic currents to genes. Prog Biophys Mol Biol 90(1–3):38–63

Mangoni ME, Fontanaud P, Noble PJ, Noble D, Benkemoun H, Nargeot J, Richard S (2000) Facilitation of the L-type calcium current in rabbit sino-atrial cells: effect on cardiac automaticity. Cardiovasc Res 48(3):375–392

Mangoni ME, Nargeot J (2008) Genesis and regulation of the heart automaticity. Physiol Rev 88(3):919–982

Mangoni ME, Traboulsie A, Leoni AL, Couette B, Marger L, Le Quang K, Kupfer E, Cohen-Solal A, Vilar J, Shin HS, Escande D, Charpentier F, Nargeot J, Lory P (2006b) Bradycardia and slowing of the atrioventricular conduction in mice lacking $Ca_v3.1$/alpha1G T-type calcium channels. Circ Res 98(11):1422–1430

Mangrum JM, DiMarco JP (2000) The evaluation and management of bradycardia. N Engl J Med 342(10):703–709

Marger L, Mesirca P, Alig J, Torrente A, Dubel S, Engeland B, Kanani S, Fontanaud P, Striessnig J, Shin HS, Isbrandt D, Ehmke H, Nargeot J, Mangoni ME (2011) Functional roles of Ca(v)1.3, Ca(v)3.1 and HCN channels in automaticity of mouse atrioventricular cells: insights into the atrioventricular pacemaker mechanism. Channels (Austin) 5(3):251–261. doi:15266 [pii]

Marionneau C, Couette B, Liu J, Li H, Mangoni ME, Nargeot J, Lei M, Escande D, Demolombe S (2005) Specific pattern of ionic channel gene expression associated with pacemaker activity in the mouse heart. J Physiol 562(Pt 1):223–234

Matthes J, Yildirim L, Wietzorrek G, Reimer D, Striessnig J, Herzig S (2004) Disturbed atrio-ventricular conduction and normal contractile function in isolated hearts from $Ca_v1.3$-knockout mice. Naunyn Schmiedebergs Arch Pharmacol 369(6):554–562

Milanesi R, Baruscotti M, Gnecchi-Ruscone T, DiFrancesco D (2006) Familial sinus bradycardia associated with a mutation in the cardiac pacemaker channel. N Engl J Med 354(2):151–157

Neco P, Torrente AG, Mesirca P, Zorio E, Liu N, Priori SG, Napolitano C, Richard S, Benitah JP, Mangoni ME, Gomez AM (2012) Paradoxical effect of increased diastolic Ca2+ release and decreased sinoatrial node activity in a mouse model of catecholaminergic polymorphic ventricular tachycardia. Circulation 126(4):392–401. doi:CIRCULATIONAHA.111.075382 [pii]10.1161/CIRCULATIONAHA.111.075382

Noble D (1984) The surprising heart: a review of recent progress in cardiac electrophysiology. J Physiol 353:1–50

Noma A, Irisawa H, Kokobun S, Kotake H, Nishimura M, Watanabe Y (1980) Slow current systems in the A-V node of the rabbit heart. Nature 285(5762):228–229

Noma A, Morad M, Irisawa H (1983) Does the "pacemaker current" generate the diastolic depolarization in the rabbit SA node cells? Pflugers Arch 397(3):190–194

Petit-Jacques J, Bois P, Bescond J, Lenfant J (1993) Mechanism of muscarinic control of the high-threshold calcium current in rabbit sino-atrial node myocytes. Pflugers Arch 423(1–2):21–27

Pfeufer A, van Noord C, Marciante KD, Arking DE, Larson MG, Smith AV, Tarasov KV, Muller M, Sotoodehnia N, Sinner MF, Verwoert GC, Li M, Kao WH, Kottgen A, Coresh J, Bis JC, Psaty BM, Rice K, Rotter JI, Rivadeneira F, Hofman A, Kors JA, Stricker BH, Uitterlinden AG, van Duijn CM, Beckmann BM, Sauter W, Gieger C, Lubitz SA, Newton-Cheh C, Wang TJ, Magnani JW, Schnabel RB, Chung MK, Barnard J, Smith JD, Van Wagoner DR, Vasan RS, Aspelund T, Eiriksdottir G, Harris TB, Launer LJ, Najjar SS, Lakatta E, Schlessinger D, Uda M, Abecasis GR, Muller-Myhsok B, Ehret GB, Boerwinkle E, Chakravarti A, Soliman EZ, Lunetta KL, Perz S, Wichmann HE, Meitinger T, Levy D, Gudnason V, Ellinor PT, Sanna S, Kaab S, Witteman JC, Alonso A, Benjamin EJ, Heckbert SR (2010) Genome-wide association study of PR interval. Nat Genet 42(2):153–159. doi:ng.517 [pii]10.1038/ng.517

Platzer J, Engel J, Schrott-Fischer A, Stephan K, Bova S, Chen H, Zheng H, Striessnig J (2000) Congenital deafness and sinoatrial node dysfunction in mice lacking class D L-type Ca2+ channels. Cell 102(1):89–97

Postma AV, van de Meerakker JB, Mathijssen IB, Barnett P, Christoffels VM, Ilgun A, Lam J, Wilde AA, Lekanne Deprez RH, Moorman AF (2008) A gain-of-function TBX5 mutation is associated with atypical Holt-Oram syndrome and paroxysmal atrial fibrillation. Circ Res 102(11):1433–1442. doi:CIRCRESAHA.107.168294 [pii]10.1161/CIRCRESAHA.107.168294

Qu Y, Baroudi G, Yue Y, Boutjdir M (2005) Novel molecular mechanism involving alpha1D ($Ca_v1.3$) L-type calcium channel in autoimmune-associated sinus bradycardia. Circulation 111(23):3034–3041

Reuter H (1967) The dependence of slow inward current in Purkinje fibres on the extracellular calcium-concentration. J Physiol 192(2):479–492

Sarachek NS, Leonard JL (1972) Familial heart block and sinus bradycardia. Classification and natural history. Am J Cardiol 29(4):451–458

Shibata EF, Giles WR (1985) Ionic currents that generate the spontaneous diastolic depolarization in individual cardiac pacemaker cells. Proc Natl Acad Sci U S A 82(22):7796–7800

Sinnegger-Brauns MJ, Hetzenauer A, Huber IG, Renstrom E, Wietzorrek G, Berjukov S, Cavalli M, Walter D, Koschak A, Waldschutz R, Hering S, Bova S, Rorsman P, Pongs O, Singewald N, Striessnig JJ (2004) Isoform-specific regulation of mood behavior and pancreatic beta cell and cardiovascular function by L-type Ca 2 + channels. J Clin Invest 113(10):1430–1439

Tellez JO, Dobrzynski H, Greener ID, Graham GM, Laing E, Honjo H, Hubbard SJ, Boyett MR, Billeter R (2006) Differential expression of ion channel transcripts in atrial muscle and sinoatrial node in rabbit. Circ Res 99(12):1384–1393

Verheijck EE, van Ginneken AC, Wilders R, Bouman LN (1999) Contribution of L-type Ca2+ current to electrical activity in sinoatrial nodal myocytes of rabbits. Am J Physiol 276(3 Pt 2):H1064–H1077

Verheijck EE, Wilders R, Bouman LN (2002) Atrio-sinus interaction demonstrated by blockade of the rapid delayed rectifier current. Circulation 105(7):880–885

Vinogradova TM, Bogdanov KY, Lakatta EG (2002) Beta-Adrenergic stimulation modulates ryanodine receptor Ca(2+) release during diastolic depolarization to accelerate pacemaker activity in rabbit sinoatrial nodal cells. Circ Res 90(1):73–79

Waltuck J, Buyon JP (1994) Autoantibody-associated congenital heart block: outcome in mothers and children. Ann Intern Med 120(7):544–551

Xu Y, Zhang Z, Timofeyev V, Sharma D, Xu D, Tuteja D, Dong PH, Ahmmed GU, Ji Y, Shull GE, Periasamy M, Chiamvimonvat N (2005) The effects of intracellular Ca2+ on cardiac K+ channel expression and activity: novel insights from genetically altered mice. J Physiol 562(Pt 3):745–758

Zaza A, Robinson RB, DiFrancesco D (1996) Basal responses of the L-type Ca2+ and hyperpolarization-activated currents to autonomic agonists in the rabbit sino-atrial node. J Physiol 491(Pt 2):347–355

Zhang Q, Timofeyev V, Lu L, Li N, Singapuri A, Long MK, Bond CT, Adelman JP, Chiamvimonvat N (2008) Functional roles of a Ca2+ -activated K+ channel in atrioventricular nodes. Circ Res 102(4):465–471. doi:CIRCRESAHA.107.161778 [pii]10.1161/CIRCRESAHA.107.161778

Zhang Q, Timofeyev V, Qiu H, Lu L, Li N, Singapuri A, Torado CL, Shin HS, Chiamvimonvat N (2011) Expression and roles of $Ca_v1.3$ (alpha1D) L-type Ca(2)+ channel in atrioventricular node automaticity. J Mol Cell Cardiol 50(1):194–202. doi:S0022-2828(10)00380-9 [pii]10.1016/j.yjmcc.2010.10.002

Zhang Z, He Y, Tuteja D, Xu D, Timofeyev V, Zhang Q, Glatter KA, Xu Y, Shin HS, Low R, Chiamvimonvat N (2005) Functional roles of $Ca_v1.3$(alpha1D) calcium channels in atria: insights gained from gene-targeted null mutant mice. Circulation 112(13):1936–1944

Zhang Z, Xu Y, Song H, Rodriguez J, Tuteja D, Namkung Y, Shin HS, Chiamvimonvat N (2002) Functional Roles of Ca(v)1.3 (alpha(1D)) calcium channel in sinoatrial nodes: insight gained using gene-targeted null mutant mice. Circ Res 90(9):981–987

Chapter 14
Auxiliary β-Subunits of L-Type Ca^{2+} Channels in Heart Failure

Jan Matthes and Stefan Herzig

Abstract The cardiac L-type Ca^{2+} channel consists of an ion-conducting channel pore $Ca_v1.2$ and auxiliary subunits, namely $\alpha_2\delta$- and β-subunits ($Ca_v\beta$). Four $Ca_v\beta$ isoforms and several splice variants are known to date. $Ca_v\beta$ subunits modulate L-type Ca^{2+} channels by significantly increasing its activity with $Ca_v\beta_2$ isoforms having the strongest effect. Furthermore, $Ca_v\beta$ subunits are involved in modulation of membrane expression of the channel pore $Ca_v1.2$. In human heart failure, density of Ca^{2+} currents mediated by L-type Ca^{2+} channels is unchanged and the activity of single L-type Ca^{2+} channels is significantly increased suggesting a reduced channel expression. The "heart failure phenotype" of single-channel gating might be explained by an altered expression pattern of $Ca_v\beta$ subunits. Indeed cardiac $Ca_v\beta_2$ isoforms are upregulated in human heart failure. Targeted overexpression of $Ca_v\beta_2$ subunits in murine hearts mimicked L-type Ca^{2+} channel features typical for human heart failure and furthermore induced cardiac hypertrophy and contractile dysfunction. Lowering cardiac $Ca_v\beta_2$ expression attenuated pressure-induced ventricular hypertrophy in rats and appeared to be well tolerated in mice. These findings strongly support the idea that $Ca_v\beta_2$ subunits are of pathophysiological relevance for development of cardiac hypertrophy and heart failure and a promising target for future treatment options.

14.1 Cardiac L-Type Ca^{2+} Channels

L-type Ca^{2+} channels belong to the family of voltage-gated calcium channels and are involved in intracellular processes such as contraction, secretion, neurotransmission, or gene expression by mediating calcium influx in response to membrane

J. Matthes · S. Herzig (✉)
Department of Pharmacology, University of Cologne, Cologne, Germany
e-mail: stefan.herzig@uni-koeln.de

N. Weiss and A. Koschak (eds.), *Pathologies of Calcium Channels*,
DOI: 10.1007/978-3-642-40282-1_14, © Springer-Verlag Berlin Heidelberg 2014

depolarisation (Catterall et al. 2005). The L-type subfamily includes channels containing $Ca_v1.1$, $Ca_v1.2$, $Ca_v1.3$, or $Ca_v1.4$ pore subunits (formerly α1S, α1C, α1D, and α1F, respectively; Ertel et al. 2000) and among others its members differ regarding gating kinetics, tissue distribution, and physiological functions (Catterall et al. 2005). The cardiac L-type Ca^{2+} channel consists of an ion-conducting channel pore $Ca_v1.2$ and auxiliary subunits, namely $\alpha_2\delta$- and β-subunits (Bodi et al. 2005). The $\alpha_2\delta$-subunit is composed of two parts both coded by the same gene: the extracellular α_2-part and the transmembrane δ-part, post-translationally cleaved but remaining associated by disulfide bonds. To date, four different isoforms ($\alpha_2\delta_{1\text{-}4}$) encoded by separate genes have been identified with at least $\alpha_2\delta_1$ and $\alpha_2\delta_2$ being expressed in the heart (Klugbauer et al. 2003; Bodi et al. 2005). In heterologous expression systems coexpression of the $\alpha_2\delta$-subunit affects $Ca_v1.2$ expression as shown by increased channel density, charge movement, and maximum of drug binding (e.g., of the dihydropyridine calcium-channel blocker (S)-isradipine). Four β-subunit isoforms ($Ca_v\beta_{1\text{-}4}$) have also been described, all of which are expressed in human myocardium (Dolphin 2003; Hullin et al. 2003; Foell et al. 2004). β-subunits significantly modulate L-type Ca^{2+} channels as described in detail in the following sections.

14.2 Expression and Function of Cardiac L-Type Ca^{2+} Channel $Ca_v\beta$ Subunits

14.2.1 Expression of $Ca_v\beta$ Subunits in the Heart

In general, all four isoforms $Ca_v\beta_{1\text{-}4}$ have been found at mRNA and protein level to be expressed in cardiac preparations of different species (Chu et al. 2004; Foell et al. 2004; Buraei and Yang 2010). In murine hearts, expression of $Ca_v\beta$ isoforms other than $Ca_v\beta_2$ normally seems to be rather negligible (Link et al. 2009; Meissner et al. 2011). This is supported by findings from murine knockout models of $Ca_v\beta$ genes (Cacnb1-4) with only $Ca_v\beta_2$ knockouts showing morphological and functional impairment of the heart (Gregg et al. 1996; Burgess et al. 1997; Namkung et al. 1998; Murakami et al. 2003 Weissgerber et al. 2006). However, cardiac $Ca_v\beta_2$ gene deletion induced in adult mice caused only a moderate decrease of I_{CaL} and was well tolerated (Meissner et al. 2011). In rabbit heart, $Ca_v\beta_2$ protein has also been shown to predominate with only low levels of $Ca_v\beta_3$ and $Ca_v\beta_1$ yet below detection limits (Reimer et al. 2000). In contrast, in the human heart, $Ca_v\beta_1$ and $Ca_v\beta_3$ subunits seem to be expressed at relevant levels although still less abundant than $Ca_v\beta_2$ isoforms (Hullin et al. 2003, 2007; Foell et al. 2004). In addition to the differences between several species, the subcellular distribution of $Ca_v\beta$ subunits has to be considered. In murine hearts, Serikov et al. found $Ca_v\beta_2$ proteins to be predominantly located in sarcolemmal fractions whereas $Ca_v\beta_1$ and $Ca_v\beta_3$ were mostly cytosolic (Serikov et al. 2002). In canine heart, $Ca_v\beta_2$ protein expression was higher in t-tubular sarcolemma compared to non t-tubular sarcolemmal membrane (Foell et al. 2004). The same was true for $Ca_v\beta_{1b}$ while $Ca_v\beta_{1a}$ and $Ca_v\beta_4$ dominated in non t-tubular sarcolemma. At least for the pore-forming cardiac L-type Ca^{2+} channel subunit $Ca_v1.2$, it has been indicated that subpopulations are specifically located to caveolae, i.e., sarcolemmal microdomains characterized by caveolin-3 (Balijepalli et al. 2006). Recent work suggests that this subcellular distribution is related to differential regulation and distinct functions of L-type Ca^{2+} channels (Makarewich et al. 2012).

14.2.2 $Ca_v\beta$ Subunits Differentially Modulate L-Type Ca^{2+} Channel Activity

The whole-cell Ca^{2+} current I_{CaL} can be defined as $I_{CaL} = N \cdot i \cdot P_o \cdot f_{active}$ (with N being the number of channels in the investigated membrane, i the unitary current through a single channel, P_o the probability of an available channel to be in the open state, and f_{active} the fraction of channels being available, i.e., those that contribute to I_{CaL}). Thus, both the activity (Sects. 14.2.2 and 14.2.3) and the number of channels in the plasmalemma (Sect. 14.2.4) are critical for regulation of Ca^{2+} influx via L-type Ca^{2+} channels.

As well as enhancing whole cell Ca^{2+} current (I_{CaL}) density, $Ca_v\beta$ isoforms affect I_{CaL} kinetics (reviewed by Buraei and Yang 2010) suggesting an influence on single-channel gating. Indeed $Ca_v\beta$ overexpression enhances open probability (P_o) of single L-type Ca^{2+} channels compared to no $Ca_v\beta$ co-expressed and $Ca_v\beta_{2a}$ appears to have the strongest stimulatory effect on P_o (Colecraft et al. 2002; Hullin et al. 2003, 2007; Herzig et al. 2007). Using an adenoviral approach, Colecraft et al. expressed different $Ca_v\beta$ isoforms in primary cultures of adult rat ventricle cells (Colecraft et al. 2002). Comparison of open probability revealed a rank order of $Ca_v\beta_2 > Ca_v\beta_3 > Ca_v\beta_1$. In addition to the common caveats regarding overexpression approaches, two possible confounders must be kept in mind: first the use of native cardiomyocytes and thus the expression of endogenous $Ca_v\beta$ subunits at relevant levels, and second the experimental expression of non-rat $Ca_v\beta$ subunits in rat cardiomyocytes possibly giving rise to artificial interactions. Using exclusively human L-type Ca^{2+} channel subunits expressed in HEK293 cells, we obtained the same ranking as Colecraft et al. but found all investigated $Ca_v\beta$ isoforms to increase open probability compared to coexpression of no $Ca_v\beta$ (Hullin et al. 2007). There was a differential effect on L-type Ca^{2+} channel activity even within the group of $Ca_v\beta_2$ isoforms: $Ca_v\beta_{2a}$ and $Ca_v\beta_{2b}$ showed stronger enhancement of channel activity than $Ca_v\beta_{2c}$ and $Ca_v\beta_{2d}$. Similar results obtained in another study furthermore revealed that the effect on single-channel gating depends on the N-terminal length of the particular $Ca_v\beta_2$ (subsequently confirmed with a set of $Ca_v\beta_1$ isoforms) (Herzig et al. 2007; Jangsangthong et al. 2010). Of note, we recently found that within a time range of minutes several $Ca_v\beta$ isoforms can switch at the channel as distinguishable by a corresponding switch of the respective typical gating behaviors (Jangsangthong et al. 2011). Proteins interacting with a $Ca_v\beta$ subunit might hamper its stimulation of L-type Ca^{2+} channel activity as suggested for Rem, a member of the RGK protein family (Sect. 14.2.4) (Yang et al. 2010). Taken together, single L-type Ca^{2+} channel behavior is differentially modulated by the particular $Ca_v\beta$ isoforms associated with the channel pore. Thus, composition of the heterologous channel complex appears to be one critical factor of Ca^{2+} current regulation.

14.2.3 Role of $Ca_v\beta$ Phosphorylation for L-Type Ca^{2+} Channel Activity

β-adrenergic stimulation leads to an increased I_{CaL} via enhanced cAMP levels and PKA activity (reviewed by McDonald et al. 1994; Kamp and Hell 2000; van der Heyden et al. 2005). At the single-channel level, a hallmark feature of β-adrenergic stimulation is an increased mode 2 gating, i.e., an increased number of long-lasting channel openings and an increased open probability (Tsien et al. 1986; Yue et al. 1990; Kamp and Hell 2000). To date, it remains unclear which phosphorylation sites are necessary. Though there is a bulk of potential and at least some proven phosphorylation sites (van der Heyden et al. 2005) several studies indicated that phosphorylation of serine 1928 of the (rabbit or rat) $Ca_v1.2$ pore subunit is the mediator of β-adrenergic I_{CaL} stimulation (De Jongh et al. 1996; Mitterdorfer et al. 1996; Gao et al. 1997; Hulme et al. 2006), but more recent studies question this (Ganesan et al. 2006; Lemke et al. 2008). Using mass spectrometry Emrick et al. showed that other sites on the skeletal muscle $Ca_v1.1$ pore (S1575, T1579) are phosphorylated in vivo (Emrick et al. 2010). These sites are conserved in cardiac $Ca_v1.2$ of several species (S1700, T1704; van der Heyden et al. 2005) and seem to be another target of interest regarding β-adrenergic regulation of cardiac L-type Ca^{2+} channels (Fuller et al. 2010). In vivo phosphorylation of $Ca_v\beta_2$ subunits has also been shown (Haase et al. 1993, 1996). Bünemann et al. found that impeding phosphorylation of S478 or S479 of a rat $Ca_v\beta_{2a}$ by mutagenesis (S $\rightarrow$ A) completely abolished PKA-mediated I_{CaL} increase in tsA-201 cells while exchanging S459 to alanine did not (Bünemann et al. 1999). In contrast, recent studies with (guinea pig or murine) cardiomyocytes showed that deletion of the corresponding phosphorylation sites by an adenoviral approach in vitro (Miriyala et al. 2008) or by mutagenesis in vivo (Brandmayr et al. 2012) did not abolish PKA-dependent I_{CaL} modulation. In addition, $Ca_v\beta$ phosphorylation does not exclusively involve PKA-specific sites. For example, Grueter et al. showed that T498 of a rat $Ca_v\beta_{2a}$ is phosphorylated by Ca^{2+}/calmodulin-dependent protein kinase II (CaMKII) and that mutation of T498 to alanine prevents CaMKII-mediated facilitation of recombinant L-type Ca^{2+} channels (Grueter et al. 2006). This CaMKII-resistant mutation furthermore reverses $Ca_v\beta_2$-induced increase of single-channel activity in rabbit ventricular myocytes (Koval et al. 2010). Besides modulation of L-type Ca^{2+} channel activity $Ca_v\beta$ phosphorylation seems to be involved in plasmalemmal expression of L-type Ca^{2+} channels (Sect. 14.2.4). It has to be kept in mind that β-adrenergic regulation of L-type Ca^{2+} channels is even more complex: regulatory mechanisms include scaffolding (e.g., by A kinase anchoring protein AKAP15; Catterall 2010), dephosphorylation and cAMP degradation (e.g., by protein phosphatases PP1 or PP2A, and phospho-diesterases PDE3 and PDE4; Herzig and Neumann 2000; Dai et al. 2009), or compartmentalisation (e.g., by caveolin-3; Best and Kamp 2012; Makarewich et al. 2012).

14.2.4 *Effect of $Ca_v\beta$ on Plasmalemmal Localization of L-Type Ca^{2+} Channels*

$Ca_v\beta$ is involved in regulation of plasmalemmal expression of the L-type Ca^{2+} channel pore $Ca_v1.2$ (reviewed in Dolphin 2003; Correll et al. 2008; Buraei and Yang 2010). Two main mechanisms have been described. First, $Ca_v\beta$ seems to act as a chaperone that allows trafficking of the pore subunit of voltage-gated Ca^{2+} channels from the endoplasmic reticulum to the plasmalemma (Dolphin 2003). While this has mainly described for $Ca_v2.x$ (i.e., P/Q-, N- and R-type Ca^{2+} channels), there are experimental findings suggesting this for $Ca_v1.2$, too (Chien et al. 1995). Members of the RGK protein family (Ras-related GTP-binding proteins Rem-Gem-Kir but also Rad) reduce I_{CaL} density obviously by interfering with $Ca_v\beta$-mediated $Ca_v1.2$ trafficking (reviewed by Correll et al. 2008). More recent data suggest that—as the second main mechanism—RGK proteins also influence the dwell-time of L-type Ca^{2+} channel pore subunits in the plasmalemma (Yang et al. 2010). Indeed, $Ca_v\beta$ has been shown to prevent ubiquitination and proteasomal degradation of L-type Ca^{2+} channels, i.e., to affect the time the channel pore spends in the plasmalemma (Altier et al. 2011). Other studies indicated that homodimerisation of the $Ca_v\beta$ SH3 domain leads to dynamin-dependent endocytosis of $Ca_v1.2$ (Gonzalez-Gutierrez et al. 2007; Miranda-Laferte et al. 2011). Akt-dependent phosphorylation of a $Ca_v\beta_2$ at S625 antagonizes $Ca_v1.2$ protein degradation thus indicating $Ca_v\beta$ phosphorylation to be involved in both channel activity (Sect. 14.2.3) and expression, though, obviously, involving distinct phosphorylation sites (Catalucci et al. 2009) (Fig. 14.1).

14.3 L-Type Ca^{2+} Channels and I_{CaL} in Heart Failure

Most studies found that ventricular I_{CaL} density in heart failure is similar compared to non-failing myocytes (Table 14.1; Mukherjee and Spinale 1998; Richard et al. 1998; Pitt et al. 2006; Bénitah et al. 2010). This is also true for human cardiac

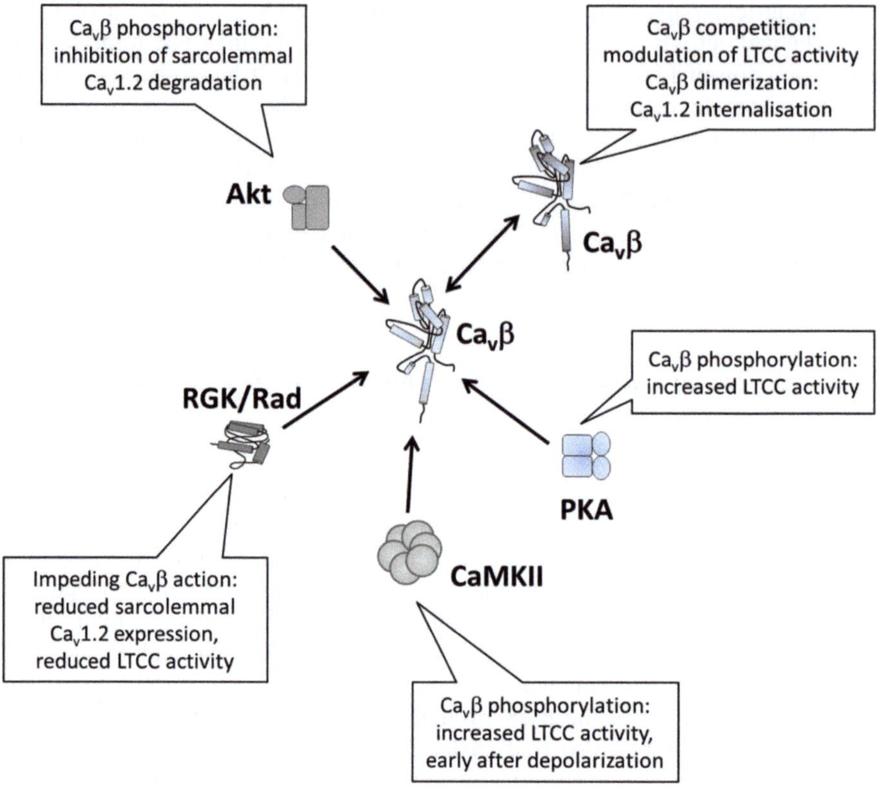

Fig. 14.1 Interaction partners of Ca$_v\beta$. The scheme displays interaction partners that have been shown to directly interfere with Ca$_v\beta$ effects on L-type Ca^{2+} channel activity and/or expression (cp. Sects. 14.2.2, 14.2.3, 14.2.4). Akt: protein kinase B; CaMKII: Ca^{2+}/calmodulin-dependent kinase II; LTCC: L-type Ca^{2+} channel; PKA: protein kinase A; RGK/Rad: members of the Ras Gem/Kir protein family including Rad

preparations (Beuckelmann and Erdmann 1992; Beuckelmann et al. 1992; Mewes and Ravens 1994; Beuckelmann 1997; Schröder et al. 1998; Chen et al. 2002, 2008; but: Ouadid et al. 1995). In apparent contradiction activity of single L-type Ca^{2+} channels in myocytes from human failing ventricles is increased (Fig. 14.2; Schröder et al. 1998). Since I_{CaL} density is based on both expression and activity (Sect. 14.2), this discrepancy might be resolved by a decreased number of channels in the membrane. Indeed, binding studies have suggested a reduced number of L-type Ca^{2+} channels in human heart failure (Tomaselli and Marban 1999). This is further supported by functional data: though in ventricular myocytes from non-failing and failing human hearts I_{CaL} density was similar under basal conditions the L-type Ca^{2+} channel agonist BayK 8644 caused a higher increase of I_{CaL} density in non-failing myocytes suggesting a higher number of available channels (Chen et al. 2002). Studies on tachycardia-induced heart failure in dogs as well showed a decreased sarcolemmal L-type Ca^{2+} channel expression (He et al. 2001;

Table 14.1 $Ca_v\beta$ expression, L-type Ca^{2+} currents and further findings from studies on human heart failure

Study	Disease	$Ca_v\beta$ RNA expression	$Ca_v\beta$ protein expression	Whole cell I_{CaL}	Single-channel I_{CaL}	Further findings
Beuckelmann et al. (1992)	Human heart failure, DCM and ICM	–	–	$\leftrightarrow$	–	Impaired Ca^{2+} transients
Beuckelmann and Erdmann (1992)	Human heart failure	–	–	$\leftrightarrow$	–	Impaired Ca^{2+} transients
Mewes and Ravens (1994)	Human heart failure, DCM and ICM	–	–	$\leftrightarrow$	–	Unaffected I_{CaL} response to adenylyl cyclase activation, unaffected activation and inactivation kinetics
Ouadid et al. (1995)	Human heart failure	–	–	$\downarrow$	–	Blunted I_{CaL} response to β-adrenergic stimulation
Haase et al. (1996)	Human HOCM	–	$\uparrow Ca_v\beta$	–	–	–
Beuckelmann (1997)	Human heart failure, DCM and ICM	–	–	$\leftrightarrow$	–	Impaired Ca^{2+} induced Ca^{2+} release
Schröder et al. (1998)	Human heart failure, DCM and ICM	$\leftrightarrow Ca_v\beta$	–	$\leftrightarrow$	$\uparrow$	Blunted I_{CaL} response to cAMP and phosphatase inhibition
Hullin et al. (1999)	Human failing allografts	$\downarrow Ca_v\beta$	–	–	–	–
Chen et al. (2002)	Human heart failure, mainly ICM	–	–	$\leftrightarrow$	–	Blunted I_{CaL} response to cAMP, phosphatase inhibition and BayK 8644; leftward shift of I_{CaL} activation
Hullin et al. (2007)	Human heart failure, DCM	$\uparrow Ca_v\beta_2$ $\leftrightarrow Ca_v\beta_1$, $Ca_v\beta_3$	$\uparrow Ca_v\beta_2$ $\leftrightarrow Ca_v\beta_1$ $\downarrow Ca_v\beta_3$	–	–	–
Chen et al. (2008)	Human heart failure, mainly ICM	–	–	$\leftrightarrow$	–	Blunted I_{CaL} response to cAMP, phosphatase inhibition and BayK 8644

DCM dilative cardiomyopathy; *ICM* ischemic cardiomyopathy; *HOCM* hypertrophic obstructive cardiomyopathy

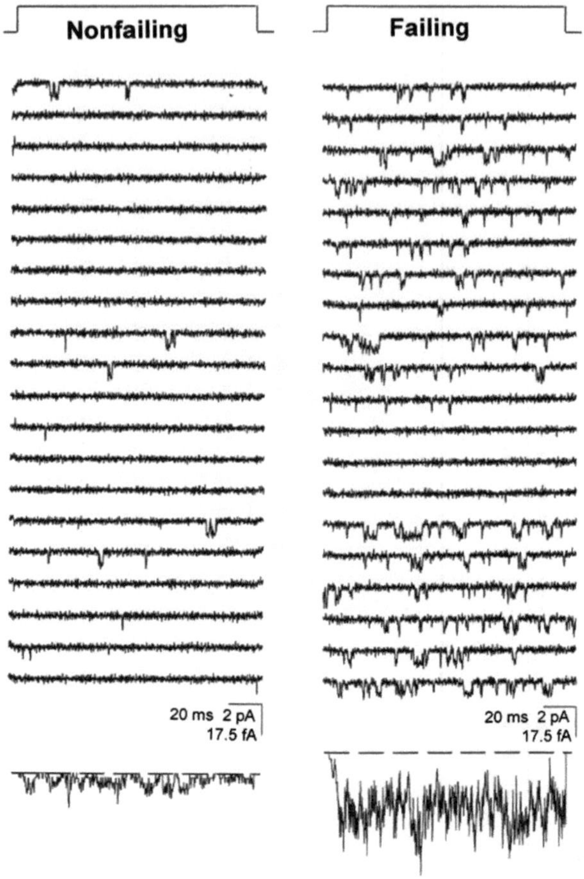

Fig. 14.2 "Heart failure phenotype" of L-type Ca^{2+} gating. Representative consecutive traces showing single L-type Ca^{2+} channel activity obtained in human ventricular myocytes from an explanted terminally failing heart and a non-failing donor heart (not transplanted for technical reasons). Channel openings were evoked by depolarizing steps (150 ms, 120 mV amplitude). *Bottom rows* ensemble average current of all 240 (*left*) or 300 (*right*) traces. *Scale bars* 20 ms and 2 pA (unitary current traces) or 17.5 fA (ensemble averages). Figure was taken from the original publication (Schröder et al. *Circulation* 1998)

Balijepalli et al. 2003). Of interest, in this canine model a loss of t-tubular sarcolemma was observed. Since a loss of ventricular t-tubules has been shown to be accompanied by reduced synchrony of SR Ca^{2+} release (Louch et al. 2004), this might in part explain why despite mostly unchanged I_{CaL} density the excitation–contraction coupling gain (i.e., ratio of Ca^{2+} released from sarcoplasmic reticulum per Ca^{2+} influx via the sarcolemma) was markedly impaired in heart failure (Gomez et al. 1997; Cannell et al. 2006; Bito et al. 2008; Heinzel et al. 2008). Though contribution of slowed I_{CaL} decay has repeatedly been discussed as a reason for the prolonged action potential observed in human failing ventricles data on I_{CaL} kinetics in heart failure are rather heterogeneous and do not confirm this interpretation so far (Bénitah et al. 2002; Beuckelmann et al. 1992; Chen et al. 2002; Gwathmey et al. 1987, 1990; Mewes and Ravens 1994; Pitt et al. 2006; Tomaselli and Marban 1999). The finding by Chen et al. of I_{CaL} activation to be shifted toward more hyperpolarised potentials might reflect an increased basal stimulation of L-type Ca^{2+} channels in human heart failure (Chen et al. 2002, 2008; Schröder et al. 1998).

The molecular mechanisms underlying the increased activity of single L-type Ca^{2+} channels in human heart failure are not yet fully elucidated. Blunted response of I_{CaL} and single L-type Ca^{2+} channel activity to adrenergic stimulation (Table 14.1) indicates an increased basal stimulation, e.g., by enhanced phosphorylation levels (Schröder et al. 1998; Chen et al. 2002, 2008). Particular $Ca_v\beta$ subunits (namely $Ca_v\beta_2$) can cause single-channel gating similar to the "heart-failure phenotype" observed with human ventricular myocytes (Sect. 14.2.2) and $Ca_v\beta_2$ is a putative target of (hyper-) phosphorylation (Sect. 14.2.3). Thus, the role of $Ca_v\beta$ subunits in heart failure is subject to intensive research (Sect. 14.4).

14.4 Role of $Ca_v\beta$ Subunits in Heart Failure

14.4.1 Expression of $Ca_v\beta$ Subunits in Human Heart Failure

Studies on expression of $Ca_v\beta$ subunits in human heart failure have been heterogeneous and rather inconclusive so far (Table 14.1; Hersel et al. 2002; Hullin et al. 2007). In Northern blot analyses Schröder et al. could not find a change in $Ca_v\beta$ mRNA expression in hearts from patients suffering from ischemic or dilative cardiomyopathy possibly due to large scatter, especially for the values from non-failing control hearts (Schröder et al. 1998). Hullin et al. examined $Ca_v\beta$ expression in biopsies of cardiac allografts that developed diastolic heart failure (Hullin et al. 1999). Compared to non-failing allografts, 75 % reduction of $Ca_v\beta$ mRNA was detected in hearts with ventricular dysfunction while mRNA levels of the channel pore were unchanged. Using specific antibodies Haase et al. tested heart samples from humans suffering from hypertrophic obstructive cardiomyopathy (Haase et al. 1996). Their immunoprecipitation experiments suggested an upregulation of $Ca_v\beta_2$ protein—along with the pore-forming L-type Ca^{2+} channel subunit. We performed a more detailed analysis of human cardiac specimens (Hullin et al. 2007). Comparison of mRNA copy numbers obtained by real time PCR revealed that the overall mRNA expression of neither $Ca_v\beta_1$ nor $Ca_v\beta_3$ subunits was altered in failing myocardium. Of note, $Ca_v\beta_2$ expression (in particular of the $Ca_v\beta_{2a}$ splice variant but also the quantitatively predominant $Ca_v\beta_{2b}$) was increased. At the protein level, $Ca_v\beta_2$ upregulation was confirmed while $Ca_v\beta_3$ was significantly reduced. Expression of the pore-forming subunit was unchanged. Since among $Ca_v\beta$ subunits $Ca_v\beta_2$ isoforms have turned out to be the strongest stimulators of L-type Ca^{2+} channel activity an (overall and/or relatively) increased $Ca_v\beta_2$ expression might underlie the "heart-failure phenotype" of single-channel gating, i.e., enhancement of open probability and fraction of active sweeps (Schröder et al. 1998; Colecraft et al. 2002; Hullin et al. 2003, 2007). Animal studies addressing the role of $Ca_v\beta$ in heart-failure development are described in the Sect. 14.4.2, (Table 14.2).

14.4.2 Animal Studies on the Role of $Ca_v\beta$ in Cardiac Hypertrophy and Heart Failure

Ca^{2+} influx via L-type Ca^{2+} channels affects both contraction and gene transcription, and impaired ventricular Ca^{2+} cycling including diastolic Ca^{2+} overload have early been suggested to be involved in development of heart failure (Beuckelmann et al. 1992; Bers 2008). Several animal studies have addressed the role of L-type Ca^{2+} channels in general and $Ca_v\beta$ in particular in heart-failure development (Table 14.2). Muth et al. created a mouse model of increased sarcolemmal Ca^{2+} influx by cardiac overexpression of the human L-type Ca^{2+} channel pore $Ca_v1.2$ (formerly α1C; Muth et al. 1999, 2001). Transgenic mice displayed no cardiac hypertrophy at an age of 2 months, a slightly increased heart to body weight ratio without contractile dysfunction at an age of 4 months but severe cardiac hypertrophy and impaired contraction at an age $\geq$8 months (Muth et al. 1999; Bodi et al. 2003). Despite a 3-fold increased expression of L-type Ca^{2+} channel pore subunits I_{CaL} density was enhanced by only 50 % (Muth et al. 1999, 2001; Song et al. 2002; Groner et al. 2004). We were able to resolve this discrepancy by showing a reduction of the single-channel activity by 60 % in 4 months old $Ca_v1.2$ transgenics (Groner et al. 2004). As a putative underlying mechanism we detected a significantly decreased protein expression of the strong stimulatory $Ca_v\beta_2$ subunits. Of note, at an age of $\geq$8 months (i.e., when already showing overt heart failure) single-channel activity was significantly increased compared to non-failing hearts from 4 months old transgenics although I_{CaL} density still was only moderately increased (Muth et al. 2001; Hullin et al. 2007). Again $Ca_v\beta$ expression pattern can explain single-channel activity; compared to age-matched wildtype littermates $Ca_v\beta_2$ protein expression was significantly increased by ca. 4-fold in $Ca_v1.2$ transgenics (Hullin et al. 2007). In summary, this mouse model showed increased activity of single L-type Ca^{2+} channels and enhanced $Ca_v\beta_2$ expression in failing hearts thus resembling findings of human heart failure (Tables 14.1 and 14.2). To further elucidate the pathophysiological relevance of cardiac $Ca_v\beta_2$ we aimed for a mouse model with cardiac-specific overexpression of a (rat) $Ca_v\beta_{2a}$ subunit. Since mice constitutively overexpressing $Ca_v\beta_2$ subunits are not viable (David T. Yue, personal communication) we developed genetically engineered mice that allowed for induction of cardiac transgene expression at an adult age (Hullin et al. 2007). Within 48 h after treatment with the inducing drug ventricular $Ca_v\beta_2$ protein expression was clearly increased in transgenic mice compared to either untreated transgenic or treated wildtype littermates. Continuous application of the inducing drug by miniosmotic pumps and thus $Ca_v\beta_2$ overexpression over a period of 4 weeks also caused a significant increase of single L-type Ca^{2+} channel activity (Fig. 14.3). This was mainly based upon a 3-fold increase of open probability. Of note, maximum whole cell I_{CaL} density was only moderately increased by 23 % or even unchanged at potentials positive to 0 mV (Beetz et al. 2009). In accordance with previous in vitro studies on effects of $Ca_v\beta_2$ overexpression inactivation and—as also shown

Table 14.2 $Ca_v\beta$ expression, L-type Ca^{2+} currents and further findings from selected animal studies

Study	Disease/model/intervention	$Ca_v\beta$ RNA expression	$Ca_v\beta$ protein expression	Whole cell I_{CaL}	Single-channel I_{CaL}	Further findings
Hullin et al. (2007)	Constitutive $Ca_v1.2$ overexpression, mouse	–	↑ $Ca_v\beta_2$, $Ca_v\beta_3$	–	↑	–
Hullin et al. (2007); Beetz et al. (2009)	Induced $Ca_v\beta_{2a}$ overexpression, mouse	↑ rat $Ca_v\beta_{2a}$ ↔ murine $Ca_v\beta_{1\text{-}3}$	↑ $Ca_v\beta_2$	(↑)	↑	Blunted I_{CaL} response to cAMP and phosphatase inhibition
Beetz et al. (2009)	Constitutive $Ca_v1.2$ and induced $Ca_v\beta_{2a}$ overexpression, mouse	↑ rat $Ca_v\beta_{2a}$ ↔ murine $Ca_v\beta_{1\text{-}3}$	–	(↑)	↑	Cardiac fibrosis, increased Ctgf expression, arrhythmia
Nakayama et al. (2007)	Dox-regulated $Ca_v\beta_{2a}$ overexpression, mouse	–	↑ $Ca_v\beta_2$	↑	–	Initially increased contractility, contractile dysfunction and premature death depending on $Ca_v\beta_2$ expression level
Wang et al. (2010)	Dox-regulated $Ca_v\beta_{2a}$ overexpression, mouse	–	–	↑	–	Blunted I_{CaL} response to β-adrenergic stimulation, leftward shift of I_{CaL} activation
Tang et al. (2010)	Dox-regulated $Ca_v\beta_{2a}$ overexpression, mouse	–	–	↑	–	Blunted I_{CaL} response to β-adrenergic stimulation, reduced ecc eficiency
Zhang et al. (2010)	Dox-regulated $Ca_v\beta_{2a}$ overexpression, MI by coronary artery occlusion, mouse	–	–	↑	–	Decreased survival post MI, regained I_{CaL} increase by β-adrenergic stimulation post MI
Chen et al. (2011)	Dox-regulated $Ca_v\beta_{2a}$ overexpression, TAC, mouse	–	–	↑	–	Exaggerated hypertrophy after TAC
Cingolani et al. (2007)	$Ca_v\beta_2$ silencing by shRNA, TAC, rat	–	–	↓	–	Attenuated hypertrophy after TAC
Meissner et al. (2011)	Induced $Ca_v\beta_2$ knockout, mouse	–	↓ $Ca_v\beta_2$	(↓)	–	No effect on heart size during up to 12 weeks

Ctgf connective tissue growth factor; *Dox* doxycycline; *ecc* excitation–contraction coupling; *MI* myocardial infarction; *TAC* transverse aortic constriction; *shRNA* short hairpin RNA

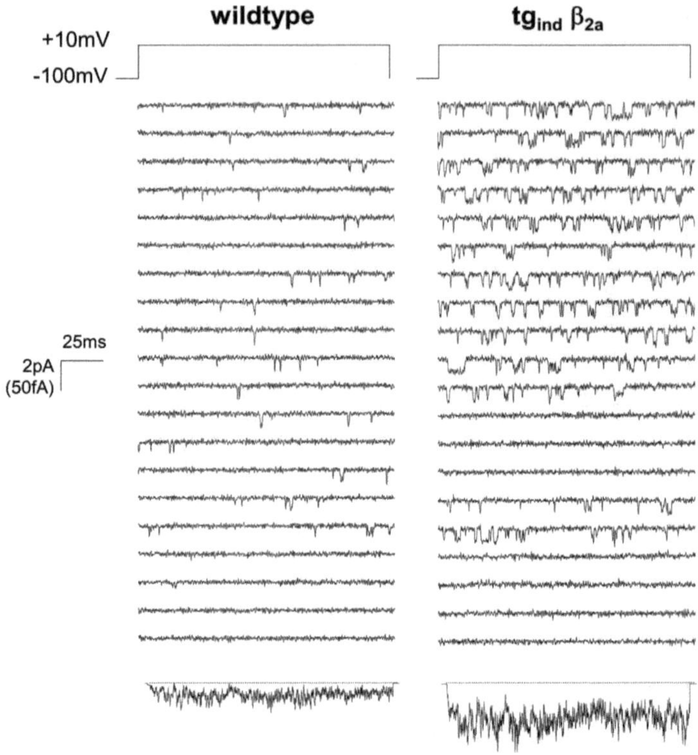

Fig. 14.3 $Ca_v\beta$ overexpression in murine hearts mimics "heart failure phenotype" of L-type Ca^{2+} gating. Representative consecutive traces showing single L-type Ca^{2+} channel activity obtained in murine ventricular myocytes from a wildtype and a $Ca_v\beta$ overexpressing heart (tg_{ind} β_{2a}). Cells were isolated after applying the transgene inducing drug tebufenozide over a period of 4 weeks. Channel openings were evoked by depolarizing steps (150 ms, from −100 mV to +10 mV). *Bottom rows* ensemble average current of all traces from the particular recording (≥180). *Scale bars* 25 ms and 2 pA (unitary current traces) or 50 fA (ensemble averages). Figure was taken from the original publication (Beetz et al. 2009)

for human heart failure—activation kinetics revealed a shift towards more hyperpolarized potentials (Perez-Reyes et al. 1992; Chen et al. 2002; Colecraft et al. 2002). As in human heart failure the response of I_{CaL} to β-adrenergic stimulation (here: 8-Br-cAMP in the presence of a phosphatase inhibitor) was blunted in mice overexpressing $Ca_v\beta_2$. Taken together our mouse model of inducible $Ca_v\beta_2$ overexpression mimicked many L-type Ca^{2+} channel features observed with human failing cardiomyocytes (Tables 14.1 and 14.2). We then examined the effect of inducing additional $Ca_v\beta_2$ overexpression in mice transgenic for the L-type Ca^{2+} channel pore. $Ca_v1.2$ transgenics at an age of 4 months were used, i.e., when normally showing $Ca_v\beta_2$ downregulation and still no hypertrophy or contractile dysfunction (Muth et al. 2001; Bodi et al. 2003; Hullin et al. 2007). The induced overexpression of $Ca_v\beta_2$ caused a 33 % increase of ventricular I_{CaL}

density (61 % compared to wildtype) and a significant increase of single-channel activity, too. $Ca_v\beta_2$ overexpression on either genetic background (wildtype or $Ca_v1.2$-transgenic) did not alter hemodynamic function in vivo. In $Ca_v1.2$-transgenic mice the additional overexpression of $Ca_v\beta_2$ aggravated cardiac fibrosis and furthermore caused arrhythmia in the majority of analyzed animals. In this context an in vitro study is noteworthy. In adult rabbit ventricular myocytes $Ca_v\beta_{2a}$ overexpression by viral infection significantly increased the risk of early after depolarisations (EADs) and reduced cell survival (Koval et al. 2010). This was obviously mediated by $Ca_v\beta_{2a}$ phosphorylation or binding by CaMKII since expression of CaMKII-resistant $Ca_v\beta_2$ mutants (T498A and L493A) significantly reduced the number of cells displaying EADs and reconstituted normal survival rate of cultured cardiomyocytes. Furthermore, single L-type Ca^{2+} channel activity that was increased by wildtype $Ca_v\beta_{2a}$ overexpression was significantly reduced when expressing the $Ca_v\beta_{2a}$ mutations resistant to modulation by CaMKII.

Another independently developed mouse model of $Ca_v\beta_2$ overexpression has been extensively investigated in the last years (Nakayama et al. 2007; Table 14.2). Using doxycycline-regulated cardiac overexpression of a rat $Ca_v\beta_2$, the authors were able to examine $Ca_v\beta_2$ effects over a period of several months. Even at the lowest level of $Ca_v\beta_2$ overexpression (i.e. 3-fold at the protein level) maximum ventricular I_{CaL} density was nearly doubled. At the same age (i.e., 14 weeks) contractile properties like cellular fractional shortening or contractility of working heart preparations were improved. At an age of 6–12 months fibrosis, ventricular dysfunction, and premature lethality emerged. In animals with high $Ca_v\beta_2$ expression levels (i.e., 7.4-fold) signs of cardiomyopathy already occurred at an age of 12–14 weeks and fractional shortening in vivo was significantly reduced at 4 months of age. Treating mice overexpressing $Ca_v\beta_2$ at high levels with the Ca^{2+} channel blocker verapamil for 14 weeks prevented hypertrophy and cardiac failure. β-adrenergic stimulation by isoproterenol infusion (and thus putatively a further increase of ventricular Ca^{2+} influx) aggravated cardiac dysfunction and mortality. Even in the absence of isoproterenol treatment the β-adrenoceptor blocker metoprolol normalized cardiac hypertrophy and cellular fractional shortening and attenuated ventricular fibrosis. Nakayama et al. found Ca^{2+}- and mitochondrial-dependent necrosis to underlie development of heart failure in their mouse model. A subsequent study using isolated ventricular myocytes from the same mouse line showed that cardiomyocyte death induced by β_1-adrenoceptor activation was mediated by both L-type Ca^{2+} channel dependent and independent pathways (Wang et al. 2010). Excitation–contraction coupling efficiency and excitation–contraction coupling gain were reduced in isolated ventricular myocytes from these $Ca_v\beta_2$ transgenic mice (Tang et al. 2010). Furthermore, cardiac contractility in vivo was significantly less affected by acute isoproterenol treatment. Accordingly, in isolated ventricular myocytes isoproterenol effects on Ca^{2+} transients, I_{CaL} density, fractional shortening, and SR Ca^{2+} load were abolished (cp. Wang et al. 2010). Zhang et al. subjected these $Ca_v\beta_2$ overexpressing mice to myocardial infarction (MI) by permanent coronary artery occlusion at an age of 4–6 months (Zhang et al. 2010). In myocytes isolated from $Ca_v\beta_2$ transgenic mice

I_{CaL} was decreased after myocardial infarction but the remaining I_{CaL} was still even higher than in wildtype myocytes without MI. As expected from previous studies I_{CaL} response to isoproterenol application was blunted in myocytes isolated from $Ca_v\beta_2$ overexpressing mice before MI. Surprisingly, I_{CaL} increase due to β-adrenergic stimulation was reconstituted after MI in transgenic myocytes, while it did not change in wildtype cells. Peak Ca^{2+} transient amplitude was still higher in myocytes obtained from transgenic mice following MI and its stimulation by isoproterenol was preserved. One week after MI contraction was similarly impaired in both wildtype and $Ca_v\beta_2$ transgenic mice but further declined during the next 5 weeks in the transgenic group only. Furthermore, at 6 weeks after MI cardiomyocyte hypertrophy of $Ca_v\beta_2$ overexpressing mice exceeded that of wildtype animals and survival rate was significantly worse. In summary, the authors concluded that—taking wildtype mice without MI as reference—$Ca_v\beta_2$ overexpression maintains Ca^{2+} handling following MI while this depresses rather than improves pump function and outcome. Chen et al. subjected the $Ca_v\beta_2$ transgenic mice introduced by Nakayama et al. to transverse aortic constriction (TAC; Chen et al. 2011). Four weeks after TAC cardiac hypertrophy and fibrosis were significantly more pronounced in transgenic compared to wildtype mice and only the former developed contractile dysfunction. The authors furthermore overexpressed the rat $Ca_v\beta_{2a}$ in cultured adult feline (AFVM) and neonatal rat ventricular myocytes (NRVM) by adenoviral infection to investigate pathological signaling involved in the observed in vivo and ex vivo effects. In both cell types $Ca_v\beta_{2a}$ overexpression-induced cardiomyocyte hypertrophy within 4 days. In case of AFVM enhancement of Ca^{2+} handling (e.g., increase of I_{CaL}, Ca^{2+} transient amplitude and fractional Ca^{2+} release) and fractional shortening depended on multiplicity of infection. Coexpression of GFP-labeled NFAT3 (Nuclear factor of activated T-cells 3, a well-known Ca^{2+}/calcineurin-dependent mediator of cardiac hypertrophy) revealed an increased translocation of NFAT3 to the nucleus in AFVM overexpressing $Ca_v\beta_2$. In another set of experiments, HDAC5 (histone deacetylase 5, a Ca^{2+}/CaMKII-dependent repressor of hypertrophic signaling) was dislocated from the nucleus in $Ca_v\beta_2$ overexpressing AFVM. Thus, the authors identified two pathways that likely underlie the hypertrophic response to enhanced Ca^{2+} entry via L-type Ca^{2+} channels. Another study has shown that enhancing Ca^{2+} entry into murine cardiomyocytes by in vivo overexpression of the T-type Ca^{2+} channel pore $Ca_v3.1$ did not cause cardiac hypertrophy (Jaleel et al. 2008). Though compared to $Ca_v3.1$ transgenic animals net Ca^{2+} influx during an action potential was slightly (but not statistically significant) higher in cardiomyocytes from $Ca_v\beta_2$ overexpressing mice, this suggests that the particular mode of Ca^{2+} entry is of importance for induction of cardiac disease.

Studies on the RGK protein Rad gave indirect hints toward a role for $Ca_v\beta$ subunits in the development of heart failure. Rad was reduced at the mRNA and protein level in human failing hearts (Chang et al. 2007). The same was true for cardiomyocytes from mice subjected to pressure-induced cardiac hypertrophy by transverse aortic constriction (TAC). Of interest, mice deficient for Rad were more susceptible for hypertrophy following TAC and showed increased heart to body weight ratio, cross-sectional cardiomyocyte area and expression of the hypertrophy markers atrial natriuretic factor and brain natriuretic peptide. Similar to our findings with induced cardiac $Ca_v\beta_2$ overexpression Rad deficient mice developed severe cardiac fibrosis and mice expressing a dominant negative Rad isoform showed ventricular arrhythmia (Yada et al. 2007; Beetz et al. 2009; Zhang et al. 2011). Given the $Ca_v\beta_2$ impeding effects of RGK proteins (Sects. 14.2.2 and 14.2.4), it is tempting to speculate that the effects observed in Rad deficient mice are due to an increased action of $Ca_v\beta_2$ subunits.

In 2007, Cingolani et al. published an exciting study using short hairpin RNA to induce knockdown of $Ca_v\beta_2$ (Cingolani et al. 2007). Five days after in vivo injection of lentiviral plasmids that code for the $Ca_v\beta_2$-silencing shRNA the I_{CaL} density was reduced by 60 % in isolated rat ventricular myocytes. $Ca_v\beta_2$ knockdown in cultured neonatal rat ventricular myocytes reduced Ca^{2+} transient amplitude by 34 % and attenuated phenylephrine-induced hypertrophy by 54 %. Subjecting rats treated with a non-silencing lentiviral vector to transverse aortic constriction (TAC) caused a 2-fold increase of $Ca_v\beta_2$ mRNA within 4 weeks while in rats treated with $Ca_v\beta_2$ shRNA a 23 % decrease of $Ca_v\beta_2$ mRNA was observed at the same time after TAC. The expression of the pore-forming L-type Ca^{2+} channel pore was unchanged. Of interest, in $Ca_v\beta_2$ shRNA treated rats ventricular hypertrophy and expression of hypertrophy markers like atrial natriuretic peptide (ANP) were significantly attenuated compared to rats treated with non-silencing lentiviral vectors. Neither rats with cardiac $Ca_v\beta_2$ knockdown nor rats treated with non-silencing vectors developed contractile dysfunction compared to sham-operated animals. Taken together, this study indicated that lowering ventricular Ca^{2+} influx via L-type Ca^{2+} channels by reducing expression of $Ca_v\beta_2$ subunits is antihypertrophic in vitro and in vivo. Goonasekera et al. showed that reducing expression of the L-type Ca^{2+} channel pore in murine hearts by either constitutive heterogeneous or loxP mediated induced knockout led to a ventricular I_{CaL} decrease by only 25 % but caused impaired cellular Ca^{2+} handling, cardiac hypertrophy, and pump failure (Goonasekera et al. 2012). A $>$70 % cardiac knockdown of $Ca_v\beta_2$ induced in mice at an age of 10–12 weeks resulted in a similar reduction of ventricular I_{CaL} by $<$29 % but did not cause any obvious impairment (e.g., of heart rate or heart to body weight ratio) for up to 12 weeks (Meissner et al. 2011).

14.5 Conclusion

Taken together it has been shown that increased cardiac expression of $Ca_v\beta_2$ subunits in mice mimicked features typical for human heart failure and furthermore induced cardiac hypertrophy and contractile dysfunction. Lowering cardiac $Ca_v\beta_2$ expression attenuated pressure-induced ventricular hypertrophy in rats and appeared to be well tolerated in mice. These findings strongly support the idea that

$Ca_v\beta_2$ subunits are of pathophysiological relevance for development of cardiac hypertrophy and heart failure and a promising target for future treatment options.

Acknowledgments We thank Lori Borgal for language editing of the manuscript. The permissions of AHA Journals to show Fig. 14.2 taken from Schröder et al. (Circulation 98(10):969–976) and by Oxford Journals to show Fig. 14.3 taken from Beetz et al. (Cardiovasc Res 84(3):396–406; doi: 10.1093/cvr/cvp251) are appreciated.

References

Altier C, Garcia-Caballero A, Simms B, You H, Chen L, Walcher J, Tedford HW, Hermosilla T, Zamponi GW (2011) The $Ca_v\beta$ subunit prevents RFP2-mediated ubiquitination and proteasomal degradation of L-type channels. Nat Neurosci 14(2):173–180

Balijepalli RC, Foell JD, Hall DD, Hell JW, Kamp TJ (2006) Localization of cardiac L-type Ca(2 +) channels to a caveolar macromolecular signaling complex is required for beta(2)-adrenergic regulation. Proc Natl Acad Sci U S A 103(19):7500–7505

Balijepalli RC, Lokuta AJ, Maertz NA, Buck JM, Haworth RA, Valdivia HH, Kamp TJ (2003) Depletion of T-tubules and specific subcellular changes in sarcolemmal proteins in tachycardia-induced heart failure. Cardiovasc Res 59(1):67–77

Beetz N, Hein L, Meszaros J, Gilsbach R, Barreto F, Meissner M, Hoppe UC, Schwartz A, Herzig S, Matthes J (2009) Transgenic simulation of human heart failure-like L-type Ca2 + -channels: implications for fibrosis and heart rate in mice. Cardiovasc Res 84(3):396–406

Bénitah JP, Alvarez JL, Gómez AM (2010) L-type Ca(2 +) current in ventricular cardiomyocytes. J Mol Cell Cardiol 48(1):26–36

Bénitah JP, Gómez AM, Fauconnier J, Kerfant BG, Perrier E, Vassort G, Richard S (2002) Voltage-gated Ca^{2+} currents in the human pathophysiologic heart: a review. Basic Res Cardiol 97(Suppl 1):I11–I18

Bers DM (2008) Calcium cycling and signaling in cardiac myocytes. Annu Rev Physiol 70:23–49

Best JM, Kamp TJ (2012) Different subcellular populations of L-type Ca2 + channels exhibit unique regulation and functional roles in cardiomyocytes. J Mol Cell Cardiol 52(2):376–387

Beuckelmann DJ (1997) Contributions of Ca(2 +)-influx via the L-type Ca(2 +)-current and Ca(2 +)-release from the sarcoplasmic reticulum to [Ca2 +]i-transients in human myocytes. Basic Res Cardiol 92(Suppl 1):105–110

Beuckelmann DJ, Erdmann E (1992) Ca(2 +)-currents and intracellular [Ca2 +]i-transients in single ventricular myocytes isolated from terminally failing human myocardium. Basic Res Cardiol 87(Suppl 1):235–243

Beuckelmann DJ, Näbauer M, Erdmann E (1992) Intracellular calcium handling in isolated ventricular myocytes from patients with terminal heart failure. Circulation 85(3):1046–1055

Bito V, Heinzel FR, Biesmans L, Antoons G, Sipido KR (2008) Crosstalk between L-type Ca2 + channels and the sarcoplasmic reticulum: alterations during cardiac remodelling. Cardiovasc Res 77(2):315–324

Bodi I, Mikala G, Koch SE, Akhter SA, Schwartz A (2005) The L-type calcium channel in the heart: the beat goes on. J Clin Invest 115(12):3306–3317

Bodi I, Muth JN, Hahn HS, Petrashevskaya NN, Rubio M, Koch SE, Varadi G, Schwartz A (2003) Electrical remodeling in hearts from a calcium-dependent mouse model of hypertrophy and failure: complex nature of K + current changes and actionpotential duration. J Am Coll Cardiol 41(9):1611–1622

Brandmayr J, Poomvanicha M, Domes K, Ding J, Blaich A, Wegener JW, Moosmang S, Hofmann F (2012) Deletion of the C-terminal phosphorylation sites in the cardiac β-subunit does not affect the basic β-adrenergic response of the heart and the Ca(v)1.2 channel. J Biol Chem 287(27):22584–22592

Bünemann M, Gerhardstein BL, Gao T, Hosey MM (1999) Functional regulation of L-type calcium channels via protein kinase A-mediated phosphorylation of the beta(2) subunit. J Biol Chem 274(48):33851–33854

Buraei Z, Yang J (2010) The ß subunit of voltage-gated Ca2 + channels. Physiol Rev 90(4):1461–1506

Burgess DL, Jones JM, Meisler MH, Noebels JL (1997) Mutation of the Ca2 + channel beta subunit gene Cchb4 is associated with ataxia and seizures in the lethargic (lh) mouse. Cell 88(3):385–392

Cannell MB, Crossman DJ, Soeller C (2006) Effect of changes in action potential spike configuration, junctional sarcoplasmic reticulum micro-architecture and altered t-tubule structure in human heart failure. J Muscle Res Cell Motil 27(5–7):297–306

Catalucci D, Zhang DH, DeSantiago J, Aimond F, Barbara G, Chemin J, Bonci D, Picht E, Rusconi F, Dalton ND, Peterson KL, Richard S, Bers DM, Brown JH, Condorelli G (2009) Akt regulates L-type Ca2 + channel activity by modulating Cavalpha1 protein stability. J Cell Biol 184(6):923–933

Catterall WA (2010) Signaling complexes of voltage-gated sodium and calcium channels. Neurosci Lett 486(2):107–116

Catterall WA, Perez-Reyes E, Snutch TP, Striessnig J (2005) International union of pharmacology. XLVIII. Nomenclature and structure-function relationships of voltage-gated calcium channels. Pharmacol Rev 57(4):411–425

Chang L, Zhang J, Tseng YH, Xie CQ, Ilany J, Brüning JC, Sun Z, Zhu X, Cui T, Youker KA, Yang Q, Day SM, Kahn CR, Chen YE (2007) Rad GTPase deficiency leads to cardiac hypertrophy. Circulation 116(25):2976–2983

Chen X, Nakayama H, Zhang X, Ai X, Harris DM, Tang M, Zhang H, Szeto C, Stockbower K, Berretta RM, Eckhart AD, Koch WJ, Molkentin JD, Houser SR (2011) Calcium influx through $Ca_v1.2$ is a proximal signal for pathological cardiomyocyte hypertrophy. J Mol Cell Cardiol 50(3):460–470

Chen X, Piacentino V 3rd, Furukawa S, Goldman B, Margulies KB, Houser SR (2002) L-type Ca2 + channel density and regulation are altered in failing human ventricular myocytes and recover after support with mechanical assist devices. Circ Res 91(6):517–524

Chen X, Zhang X, Harris DM, Piacentino V 3rd, Berretta RM, Margulies KB, Houser SR (2008) Reduced effects of BAY K 8644 on L-type Ca2 + current in failing human cardiac myocytes are related to abnormal adrenergic regulation. Am J Physiol Heart Circ Physiol 294(5):H2257–H2267

Chien AJ, Zhao X, Shirokov RE, Puri TS, Chang CF, Sun D, Rios E, Hosey MM (1995) Roles of a membrane-localized beta subunit in the formation and targeting of functional L-type Ca2 + channels. J Biol Chem 270(50):30036–30044

Chu PJ, Larsen JK, Chen CC, Best PM (2004) Distribution and relative expression levels of calcium channel beta subunits within the chambers of the rat heart. J Mol Cell Cardiol 36(3):423–434

Cingolani E, Ramirez Correa GA, Kizana E, Murata M, Cho HC, Marbán E (2007) Gene therapy to inhibit the calcium channel beta subunit: physiological consequences and pathophysiological effects in models of cardiac hypertrophy. Circ Res 101(2):166–175

Colecraft HM, Alseikhan B, Takahashi SX, Chaudhuri D, Mittman S, Yegnasubramanian V, Alvania RS, Johns DC, Marbán E, Yue DT (2002) Novel functional properties of Ca(2 +) channel beta subunits revealed by their expression in adult rat heart cells. J Physiol 541(Pt 2):435–452

Correll RN, Pang C, Niedowicz DM, Finlin BS, Andres DA (2008) The RGK family of GTP-binding proteins: regulators of voltage-dependent calcium channels and cytoskeleton remodeling. Cell Signal 20(2):292–300

Dai S, Hall DD, Hell JW (2009) Supramolecular assemblies and localized regulation of voltage-gated ion channels. Physiol Rev 89(2):411–452

De Jongh KS, Murphy BJ, Colvin AA, Hell JW, Takahashi M, Catterall WA (1996) Specific phosphorylation of a site in the full-length form of the alpha 1 subunit of the cardiac L-type calcium channel by adenosine 3′,5′-cyclic monophosphate-dependent protein kinase. Biochemistry 35(32):10392–10402

Dolphin AC (2003) Beta subunits of voltage-gated calcium channels. J Bioenerg Biomembr 35(6):599–620

Emrick MA, Sadilek M, Konoki K, Catterall WA (2010) Beta-adrenergic-regulated phosphorylation of the skeletal muscle Ca(V)1.1 channel in the fight-or-flight response. Proc Natl Acad Sci U S A 107(43):18712–18717

Ertel EA, Campbell KP, Harpold MM, Hofmann F, Mori Y, Perez-Reyes E, Schwartz A, Snutch TP, Tanabe T, Birnbaumer L, Tsien RW, Catterall WA (2000) Nomenclature of voltage-gated calcium channels. Neuron 25(3):533–535

Foell JD, Balijepalli RC, Delisle BP, Yunker AM, Robia SL, Walker JW, McEnery MW, January CT, Kamp TJ (2004) Molecular heterogeneity of calcium channel beta-subunits in canine and human heart: evidence for differential subcellular localization. Physiol Genomics 17(2):183–200

Fuller MD, Emrick MA, Sadilek M, Scheuer T, Catterall WA (2010) Molecular mechanism of calcium channel regulation in the fight-or-flight response. Sci Signal 3(141):ra70

Ganesan AN, Maack C, Johns DC, Sidor A, O'Rourke B (2006) Beta-adrenergic stimulation of L-type Ca2 + channels in cardiac myocytes requires the distal carboxyl terminus of alpha1C but not serine 1928. Circ Res 98(2):e11–18

Gao T, Yatani A, Dell'Acqua ML, Sako H, Green SA, Dascal N, Scott JD, Hosey MM (1997) cAMP-dependent regulation of cardiac L-type Ca2 + channels requires membrane targeting of PKA and phosphorylation of channel subunits. Neuron 19(1):185–196

Gómez AM, Valdivia HH, Cheng H, Lederer MR, Santana LF, Cannell MB, McCune SA, Altschuld RA, Lederer WJ (1997) Defective excitation-contraction coupling in experimental cardiac hypertrophy and heart failure. Science 276(5313):800–806

Gonzalez-Gutierrez G, Miranda-Laferte E, Neely A, Hidalgo P (2007) The Src homology 3 domain of the beta-subunit of voltage-gated calcium channels promotes endocytosis via dynamin interaction. J Biol Chem 282(4):2156–21562

Goonasekera SA, Hammer K, Auger-Messier M, Bodi I, Chen X, Zhang H, Reiken S, Elrod JW, Correll RN, York AJ, Sargent MA, Hofmann F, Moosmang S, Marks AR, Houser SR, Bers DM, Molkentin JD (2012) Decreased cardiac L-type Ca^{2+} channel activity induces hypertrophy and heart failure in mice. J Clin Invest 122(1):280–290

Gregg RG, Messing A, Strube C, Beurg M, Moss R, Behan M, Sukhareva M, Haynes S, Powell JA, Coronado R, Powers PA (1996) Absence of the beta subunit (cchb1) of the skeletal muscle dihydropyridine receptor alters expression of the alpha 1 subunit and eliminates excitation-contraction coupling. Proc Natl Acad Sci U S A 93(24):13961–13966

Groner F, Rubio M, Schulte-Euler P, Matthes J, Khan IF, Bodi I, Koch SE, Schwartz A, Herzig S (2004) Single-channel gating and regulation of human L-type calcium channels in cardiomyocytes of transgenic mice. Biochem Biophys Res Commun 314(3):878–884

Grueter CE, Abiria SA, Dzhura I, Wu Y, Ham AJ, Mohler PJ, Anderson ME, Colbran RJ (2006) L-type Ca2 + channel facilitation mediated by phosphorylation of the beta subunit by CaMKII. Mol Cell 23(5):641–650

Gwathmey JK, Copelas L, MacKinnon R, Schoen FJ, Feldman MD, Grossman W, Morgan JP (1987) Abnormal intracellular calcium handling in myocardium from patients with end-stage heart failure. Circ Res 61(1):70–76

Gwathmey JK, Slawsky MT, Hajjar RJ, Briggs GM, Morgan JP (1990) Role of intracellular calcium handling in force-interval relationships of human ventricular myocardium. J Clin Invest 85(5):1599–1613

Haase H, Bartel S, Karczewski P, Morano I, Krause EG (1996) In-vivo phosphorylation of the cardiac L-type calcium channel beta-subunit in response to catecholamines. Mol Cell Biochem 163–164:99–106

Haase H, Karczewski P, Beckert R, Krause EG (1993) Phosphorylation of the L-type calcium channel beta subunit is involved in beta-adrenergic signal transduction in canine myocardium. FEBS Lett 335(2):217–222

He J, Conklin MW, Foell JD, Wolff MR, Haworth RA, Coronado R, Kamp TJ (2001) Reduction in density of transverse tubules and L-type Ca(2 +) channels in canine tachycardia-induced heart failure. Cardiovasc Res 49(2):298–307

Heinzel FR, Bito V, Biesmans L, Wu M, Detre E, von Wegner F, Claus P, Dymarkowski S, Maes F, Bogaert J, Rademakers F, D'hooge J, Sipido K (2008) Remodeling of T-tubules and reduced synchrony of Ca2 + release in myocytes from chronically ischemic myocardium. Circ Res 102(3):338–346

Hersel J, Jung S, Mohacsi P, Hullin R (2002) Expression of the L-type calcium channel in human heart failure. Basic Res Cardiol 97(Suppl 1):I4–I10

Herzig S, Khan IF, Gründemann D, Matthes J, Ludwig A, Michels G, Hoppe UC, Chaudhuri D, Schwartz A, Yue DT, Hullin R (2007) Mechanism of Ca(v)1.2 channel modulation by the amino terminus of cardiac beta2-subunits. FASEB J 21(7):1527–1538

Herzig S, Neumann J (2000) Effects of serine/threonine protein phosphatases on ion channels in excitable membranes. Physiol Rev 80(1):173–210

Hullin R, Asmus F, Ludwig A, Hersel J, Boekstegers P (1999) Subunit expression of the cardiac L-type calcium channel is differentially regulated in diastolic heart failure of the cardiac allograft. Circulation 100(2):155–163

Hullin R, Khan IF, Wirtz S, Mohacsi P, Varadi G, Schwartz A, Herzig S (2003) Cardiac L-type calcium channel beta-subunits expressed in human heart have differential effects on single channel characteristics. J Biol Chem 278(24):21623–21630

Hullin R, Matthes J, von Vietinghoff S, Bodi I, Rubio M, D'Souza K, Friedrich Khan I, Rottländer D, Hoppe UC, Mohacsi P, Schmitteckert E, Gilsbach R, Bünemann M, Hein L, Schwartz A, Herzig S (2007) Increased expression of the auxiliary beta(2)-subunit of ventricular L-type Ca(2) + channels leads to single-channel activity characteristic of heart failure. PLoS ONE 2(3):e292

Hulme JT, Westenbroek RE, Scheuer T, Catterall WA (2006) Phosphorylation of serine 1928 in the distal C-terminal domain of cardiac $Ca_v1.2$ channels during beta1-adrenergic regulation. Proc Natl Acad Sci U S A 103(44):16574–16579

Jaleel N, Nakayama H, Chen X, Kubo H, MacDonnell S, Zhang H, Berretta R, Robbins J, Cribbs L, Molkentin JD, Houser SR (2008) Ca2 + influx through T- and L-type Ca2 + channels have different effects on myocyte contractility and induce unique cardiac phenotypes. Circ Res 103(10):1109–1119

Jangsangthong W, Kuzmenkina E, Böhnke AK, Herzig S (2011) Single-channel monitoring of reversible L-type Ca(2 +) channel Ca(V)α(1)-Ca(V)β subunit interaction. Biophys J 101(11):2661–2670

Jangsangthong W, Kuzmenkina E, Khan IF, Matthes J, Hullin R, Herzig S (2010) Inactivation of L-type calcium channels is determined by the length of the N terminus of mutant beta(1) subunits. Pflugers Arch 459(3):399–411

Kamp TJ, Hell JW (2000) Regulation of cardiac L-type calcium channels by protein kinase A and protein kinase C. Circ Res 87(12):1095–1102

Klugbauer N, Marais E, Hofmann F (2003) Calcium channel alpha2delta subunits: differential expression, function, and drug binding. J Bioenerg Biomembr 35(6):639–647

Koval OM, Guan X, Wu Y, Joiner ML, Gao Z, Chen B, Grumbach IM, Luczak ED, Colbran RJ, Song LS, Hund TJ, Mohler PJ, Anderson ME (2010) $Ca_v1.2$ beta-subunit coordinates CaMKII-triggered cardiomyocyte death and afterdepolarizations. Proc Natl Acad Sci U S A 107(11):4996–5000

Lemke T, Welling A, Christel CJ, Blaich A, Bernhard D, Lenhardt P, Hofmann F, Moosmang S (2008) Unchanged beta-adrenergic stimulation of cardiac L-type calcium channels in Ca v 1.2 phosphorylation site S1928A mutant mice. J Biol Chem 283(50):34738–34744

Link S, Meissner M, Held B, Beck A, Weissgerber P, Freichel M, Flockerzi V (2009) Diversity and developmental expression of L-type calcium channel beta2 proteins and their influence on calcium current in murine heart. J Biol Chem 284(44):30129–30137

Louch WE, Bito V, Heinzel FR, Macianskiene R, Vanhaecke J, Flameng W, Mubagwa K, Sipido KR (2004) Reduced synchrony of Ca2 + release with loss of T-tubules-a comparison to Ca2 + release in human failing cardiomyocytes. Cardiovasc Res 62(1):63–73

Makarewich CA, Correll RN, Gao H, Zhang H, Yang B, Berretta RM, Rizzo V, Molkentin JD, Houser SR (2012) A caveolae-targeted L-type Ca^2 + channel antagonist inhibits hypertrophic signaling without reducing cardiac contractility. Circ Res 110(5):669–674

McDonald TF, Pelzer S, Trautwein W, Pelzer DJ (1994) Regulation and modulation of calcium channels in cardiac, skeletal, and smooth muscle cells. Physiol Rev 74(2):365–507

Meissner M, Weissgerber P, Londoño JE, Prenen J, Link S, Ruppenthal S, Molkentin JD, Lipp P, Nilius B, Freichel M, Flockerzi V (2011) Moderate calcium channel dysfunction in adult mice with inducible cardiomyocyte-specific excision of the cacnb2 gene. J Biol Chem 286(18):15875–15882

Mewes T, Ravens U (1994) L-type calcium currents of human myocytes from ventricle of non-failing and failing hearts and from atrium. J Mol Cell Cardiol 26(10):1307–1320

Miranda-Laferte E, Gonzalez-Gutierrez G, Schmidt S, Zeug A, Ponimaskin EG, Neely A, Hidalgo P (2011) Homodimerization of the Src homology 3 domain of the calcium channel β-subunit drives dynamin-dependent endocytosis. J Biol Chem 286(25):22203–22210

Miriyala J, Nguyen T, Yue DT, Colecraft HM (2008) Role of Ca_vbeta subunits, and lack of functional reserve, in protein kinase A modulation of cardiac Ca_v1.2 channels. Circ Res 102(7):e54–e64

Mitterdorfer J, Froschmayr M, Grabner M, Moebius FF, Glossmann H, Striessnig J (1996) Identification of PK-A phosphorylation sites in the carboxyl terminus of L-type calcium channel alpha 1 subunits. Biochemistry 35(29):9400–9406

Mukherjee R, Spinale FG (1998) L-type calcium channel abundance and function with cardiac hypertrophy and failure: a review. J Mol Cell Cardiol 30(10):1899–1916

Murakami M, Yamamura H, Suzuki T, Kang MG, Ohya S, Murakami A, Miyoshi I, Sasano H, Muraki K, Hano T, Kasai N, Nakayama S, Campbell KP, Flockerzi V, Imaizumi Y, Yanagisawa T, Iijima T (2003) Modified cardiovascular L-type channels in mice lacking the voltage-dependent Ca2 + channel beta3 subunit. J Biol Chem 278(44):43261–43267

Muth JN, Bodi I, Lewis W, Varadi G, Schwartz A (2001) A Ca(2 +)-dependent transgenic model of cardiac hypertrophy: a role for protein kinase Calpha. Circulation 103(1):140–147

Muth JN, Yamaguchi H, Mikala G, Grupp IL, Lewis W, Cheng H, Song LS, Lakatta EG, Varadi G, Schwartz A (1999) Cardiac-specific overexpression of the alpha(1) subunit of the L-type voltage-dependent Ca(2 +) channel in transgenic mice. Loss of isoproterenol-induced contraction. J Biol Chem 274(31):21503–21506

Nakayama H, Chen X, Baines CP, Klevitsky R, Zhang X, Zhang H, Jaleel N, Chua BH, Hewett TE, Robbins J, Houser SR, Molkentin JD (2007) Ca2 + - and mitochondrial-dependent cardiomyocyte necrosis as a primary mediator of heart failure. J Clin Invest 117(9):2431–2444

Namkung Y, Smith SM, Lee SB, Skrypnyk NV, Kim HL, Chin H, Scheller RH, Tsien RW, Shin HS (1998) Targeted disruption of the Ca2 + channel beta3 subunit reduces N- and L-type Ca2 + channel activity and alters the voltage-dependent activation of P/Q-type Ca2 + channels in neurons. Proc Natl Acad Sci U S A 95(20):12010–12015

Ouadid H, Albat B, Nargeot J (1995) Calcium currents in diseased human cardiac cells. J Cardiovasc Pharmacol 25(2):282–291

Perez-Reyes E, Castellano A, Kim HS, Bertrand P, Baggstrom E, Lacerda AE, Wie XY, Birnbaumer L (1992) Cloning and expression of a cardiac/brain beta subunit of the L-type calcium channel. J Biol Chem 267(3):1792–1797

Pitt GS, Dun W, Boyden PA (2006) Remodeled cardiac calcium channels. J Mol Cell Cardiol 41(3):373–388

Reimer D, Huber IG, Garcia ML, Haase H, Striessnig J (2000) beta subunit heterogeneity of L-type Ca(2 +) channels in smooth muscle tissues. FEBS Lett 467(1):65–69

Richard S, Leclercq F, Lemaire S, Piot C, Nargeot J (1998) Ca2 + currents in compensated hypertrophy and heart failure. Cardiovasc Res 37(2):300–311

Schröder F, Handrock R, Beuckelmann DJ, Hirt S, Hullin R, Priebe L, Schwinger RH, Weil J, Herzig S (1998) Increased availability and open probability of single L-type calcium channels from failing compared with nonfailing human ventricle. Circulation 98(10):969–976

Serikov V, Bodi I, Koch SE, Muth JN, Mikala G, Martinov SG, Haase H, Schwartz A (2002) Mice with cardiac-specific sequestration of the beta-subunit of the L-type calcium channel. Biochem Biophys Res Commun 293(5):1405–1411

Song LS, Guia A, Muth JN, Rubio M, Wang SQ, Xiao RP, Josephson IR, Lakatta EG, Schwartz A, Cheng H (2002) Ca(2 +) signaling in cardiac myocytes overexpressing the alpha(1) subunit of L-type Ca(2 +) channel. Circ Res 90(2):174–181

Tang M, Zhang X, Li Y, Guan Y, Ai X, Szeto C, Nakayama H, Zhang H, Ge S, Molkentin JD, Houser SR, Chen X (2010) Enhanced basal contractility but reduced excitation-contraction coupling efficiency and beta-adrenergic reserve of hearts with increased $Ca_v1.2$ activity. Am J Physiol Heart Circ Physiol 299(2):H519–H528

Tomaselli GF, Marbán E (1999) Electrophysiological remodeling in hypertrophy and heart failure. Cardiovasc Res 42(2):270–283

Tsien RW, Bean BP, Hess P, Lansman JB, Nilius B, Nowycky MC (1986) Mechanisms of calcium channel modulation by beta-adrenergic agents and dihydropyridine calcium agonists. J Mol Cell Cardiol 18(7):691–710

Wijnhoven TJ, van der Heyden MA, Opthof T (2005) Molecular aspects of adrenergic modulation of cardiac L-type Ca2 + channels. Cardiovasc Res 65(1):28–39

Wang W, Zhang H, Gao H, Kubo H, Berretta RM, Chen X, Houser SR (2010) {beta}1-Adrenergic receptor activation induces mouse cardiac myocyte death through both L-type calcium channel-dependent and -independent pathways. Am J Physiol Heart Circ Physiol 299(2):H322–H331

Weissgerber P, Held B, Bloch W, Kaestner L, Chien KR, Fleischmann BK, Lipp P, Flockerzi V, Freichel M (2006) Reduced cardiac L-type Ca2 + current in Ca(V)beta2-/- embryos impairs cardiac development and contraction with secondary defects in vascular maturation. Circ Res 99(7):749–757

Yada H, Murata M, Shimoda K, Yuasa S, Kawaguchi H, Ieda M, Adachi T, Murata M, Ogawa S, Fukuda K (2007) Dominant negative suppression of Rad leads to QT prolongation and causes ventricular arrhythmias via modulation of L-type Ca2 + channels in the heart. Circ Res 101(1):69–77

Yang T, Xu X, Kernan T, Wu V, Colecraft HM (2010) Rem, a member of the RGK GTPases, inhibits recombinant $Ca_v1.2$ channels using multiple mechanisms that require distinct conformations of the GTPase. J Physiol 588(Pt 10):1665–1681

Yue DT, Herzig S, Marban E (1990) Beta-adrenergic stimulation of calcium channels occurs by potentiation of high-activity gating modes. Proc Natl Acad Sci U S A 87(2):753–757

Zhang H, Chen X, Gao E, MacDonnell SM, Wang W, Kolpakov M, Nakayama H, Zhang X, Jaleel N, Harris DM, Li Y, Tang M, Berretta R, Leri A, Kajstura J, Sabri A, Koch WJ, Molkentin JD, Houser SR (2010) Increasing cardiac contractility after myocardial infarction exacerbates cardiac injury and pump dysfunction. Circ Res 107(6):800–809

Zhang J, Chang L, Chen C, Zhang M, Luo Y, Hamblin M, Villacorta L, Xiong JW, Chen YE, Zhang J, Zhu X (2011) Rad GTPase inhibits cardiac fibrosis through connective tissue growth factor. Cardiovasc Res 91(1):90–98

Chapter 15
Epigenetic, Genetic, and Acquired Regulation of Ca_v3 T-Type Calcium Channel Expression and Function in Tumor Growth and Progression

Maria Beatrice Morelli, Sonia Liberati, Consuelo Amantini, Matteo Santoni, Massimo Nabissi, Valerio Farfariello and Giorgio Santoni

Abstract T-type Ca^{2+} channels are a specific channel family controlling proliferation, differentiation, angiogenesis, and invasion of tumor cells. Molecular biology has identified three main subfamilies of α_1 subunits, called Ca_v1, Ca_v2, and Ca_v3. The third subfamily contains three members, called T-types: $Ca_v3.1$ (α_{1G}), $Ca_v3.2$ (α_{1H}), and $Ca_v3.3$ (α_{1I}). The Ca_v3 channels are expressed in normal tissues throughout the body as well as in different types of tumors such as breast, glioma/neuroblastomas, colorectal, gastric, hepatic and prostate tumors, T cell leukemia, retinoblastoma (RB), and phaeochromocytoma. It has been shown that increased functional expression of Ca_v3 channels plays a role in abnormal proliferation of tumor cells. In addition, a crosstalk between the Rho-ROCK pathway and Ca_v3 channels in tumor cell migration and invasion has been demonstrated. Ca_v3 expression is strictly regulated during cell differentiation and tumor formation. Inactivation of Ca_v3 genes by aberrant methylation plays a major role in tumor development and progression. In this regard, since hypermethylation of CpG islands is thought to be the major epigenetic modification repressing gene transcription, $Ca_v3.1$ has been regarded as a candidate tumor suppressor gene. In contrast, CpG sites in the $Ca_v3.2$ gene were hypomethylated, and thus this gene is considered to be a candidate oncogene, regulated by methylation. Finally, Ca_v3 functional diversity may also result from the generation of splice variants.

M. B. Morelli · S. Liberati · C. Amantini · M. Nabissi · V. Farfariello · G. Santoni (✉)
School of Pharmacy, Section of Experimental Medicine, University of Camerino, Camerino, Italy
e-mail: giorgio.santoni@unicam.it

S. Liberati
Department of Molecular Medicine, Sapienza University, Rome, Italy

M. Santoni
Department of Medical Oncology, Polytechnic University of the Marche Region, Ancona, Italy

V. Farfariello
Department of Urology and Andrology, University of Perugia, Perugia, Italy

N. Weiss and A. Koschak (eds.), *Pathologies of Calcium Channels*,
DOI: 10.1007/978-3-642-40282-1_15,

Alterations in the expression of different splice variants, and their expression during tumor development and progression, may trigger variety in Ca^{2+} signaling, which may contribute to the generation of more aggressive tumor clones.

15.1 Involvement of T-Type Calcium Channels in Tumor Growth and Progression

15.1.1 Hypoxia Regulates T-Type Calcium Channel Gene Induction

Extreme hypoxia, a hallmark of solid tumors, leads to phenotypic alterations promoting tumor growth and progression (Harris 2002). In this regard, the hypoxia-inducible factors (HIFs) regulate the expression of a wide repertoire of oxygen-sensitive genes with roles in diverse cellular functions such as angiogenesis, glucose and energy metabolism, apoptosis, and proliferation (Bunn and Poyton 1996). Among these genes, the expression of T-type channel genes is influenced by hypoxia, and the respective protein channels are predominantly expressed in the G1/S transition stage of the cell cycle (Kuga et al. 1996).

Studies conducted in an oxygen-sensitive pheochromocytoma-derived PC12 cell line evidenced the expression of mRNA for both α_{1H} and α_{1G} subunits of T-type Ca^{2+} channels, with α_{1H} being the most abundant (Del Toro et al. 2003). Lowering oxygen tension in PC12 cells induces membrane depolarization, increase of extracellular Ca^{2+} influx, and catecholamine secretion (Zhu et al. 1996; Taylor and Peers 1999). Accumulation of α_{1H} mRNA in response to hypoxia paralleled the increase in the density of T-current. Moreover, functional inhibition of HIF-2α protein accumulation using antisense HIF-2α oligonucleotides reverses the effect of hypoxia on T-type Ca^{2+} channel expression. Importantly, regulation by oxygen tension is T-type-specific, since it is not observed with L-, N-, and P/Q-channel types (Del Toro et al. 2003). T-type channels may have a role in the modulation of cell excitability, and have been suspected of participating in cell cycle progression and proliferation, although the evidence is still being debated (Chemin et al. 2000). It is therefore possible that Ca^{2+} entry through T-type channels is implicated in the hypoxic environment of proliferating tumor cells. In this regard, the α_{1H} channels are found to be overexpressed in human prostate cancer (LNCaP) cells in their more aggressive and invasive stages (Mariot et al. 2002).

15.1.2 Role of T-Type Calcium Channels in Triggering Tumor Cell Proliferation

Calcium has an essential role in signal transduction and is involved in the regulation of many eukaryotic cellular functions including proliferation, differentiation, cell death, and apoptosis (Berridge et al. 1998). In tumor cell proliferation, Ca^{2+} fluxes regulate membrane potential, volume, secretion, as well as DNA synthesis and transcription (Berridge et al. 1998). It has been shown that increased T-type Ca^{2+} channel expression and function play a crucial role in abnormal proliferation of different types of tumors (Perez-Reyes 1999; Taylor et al. 2008; Monteil et al. 2000). T-type channels are voltage-gated Ca^{2+} channels with a unique low-voltage-dependent opening and slow-deactivation kinetics. These special electrophysiological characteristics allow these channels to open at low membrane potential and play a role in the regulation of intracellular Ca^{2+} homeostasis in non-excitable cells, including a variety of primary cancer cells (Panner and Wurster 2006; Lory et al. 2006; Díaz-Lezama et al. 2010; Santoni et al. 2012) and cancer cell lines (Table 15.1). Molecular biologists have expanded our knowledge of the repertoire of these Ca^{2+} channels, identifying three main subfamilies of the α_1 subunit, namely Ca_v1, Ca_v2, and Ca_v3. The third subfamily contains three members called T-types: $Ca_v3.1$ (α_{1G}), $Ca_v3.2$ (α_{1H}), and $Ca_v3.3$ (α_{1I}). The $Ca_v3.1$ and $Ca_v3.2$ T-types are broadly expressed in various tissues (Taylor et al. 2008; Li et al. 2011), while $Ca_v3.3$ expression is more restricted to the central nervous system, thyroid, and adrenal gland tissue (Bertolesi et al. 2002). T-type Ca^{2+} channels are expressed in numerous types of tumors or cell lines as well as in normal tissues throughout the body. The function of regulating Ca^{2+} homeostasis may allow T-type Ca^{2+} channels to play an important role in controlling cell proliferation, which is critical in understanding the mechanism of cancer progression. Therefore, loss of T-type Ca^{2+} channels and alterations in calcium signaling may lead to aberrant growth and tumor progression. T-type channel expression was found in breast (Bertolesi et al. 2002), neuroblastoma (Chemin et al. 2004), glioma (Latour et al. 2004), colorectal (Toyota et al. 1999a), gastric (Toyota et al. 1999b), renal (Panner and Wurster 2006), hepatic (Li et al. 2009) and prostate tumors (Mariot et al. 2002), leukaemic cells (Yoshida et al. 2004), RBs (Hirooka et al. 2002), and pheochromocytomas (Harkins et al. 2003).

The expression of α_{1G} and α_{1H} subunits in human MCF-7 breast carcinomas has been demonstrated (Taylor et al. 2008). In addition, the selective T-type inhibitor, NNC-55-0396, has been found to inhibit MCF-7 ($ER\alpha^+$) breast cancer cell proliferation. Knockdown of the expression of T-type Ca^{2+} channels through siRNA targeting of both α_{1G} and α_{1H} resulted in 45 % growth inhibition (Taylor et al. 2008). T-type channel current was elevated during the S phase of the cell cycle, resulting in a transient increase in $[Ca^{2+}]_i$ required for cell cycle progression. In addition, overexpression of $Ca_v3.1$ channels in MCF-7 cells suppressed cell proliferation and, by enhancing Ca^{2+} influx, increased the number of Annexin V^+ apoptotic cells. Moreover, knockdown of the $Ca_v3.1$ gene by siRNA, or treatment

Table 15.1 Cancerous cell lines that express Ca_v3 T-Type Ca^{2+} channels

Cell type	Cell line	T-type isoform	References
Adrenocarcinoma	H295R	α_{1H}	Lesouhaitier et al. (2001)
Breast carcinoma	MCF-7, MDA-435	α_{1G}, α_{1H}	Taylor et al. (2008), Gray et al. (2004), Toyota et al. (1999a), Latour et al. (2004)
	MDA-231, MDA-361 MB-468, MB-474, BT-20, CAMA1, SKBR-3	α_{1G}	Gray et al. (2004), Toyota et al. (1999a), Latour et al. (2004)
Colorectal carcinoma	Caco2, DLD-1, Lovo, SW837	α_{1G}	Toyota et al. (1999a)
Esophageal carcinoma	TE1, TE10, TE12, KYSE150, KYSE180, KYSE450	α_{1H}	Lu et al. (2008)
	SKGT4, TE3, TE7, KYSE70	α_{1G}, α_{1H}	Lu et al. (2008)
	COLO-680 N, SEG1, TE8, TE11, KYSE30, KYSE410, KYSE510	α_{1G}, α_{1H}, α_{1I}	Lu et al. (2008)
Fibrosarcoma	HT1080	α_{1G}	Huang et al. (2004)
Glioma	Primary (biopsy)	α_{1G}	Latour et al. (2004)
	U87	α_{1G}, α_{1H}	Panner et al. (2005)
Hepatocellular carcinoma	HuH-1, PLC/PRF5, SMMC7721, SNU182, SNU449, SNU475	α_{1G}, α_{1H}, α_{1I}	Li et al. (2009)
	Hep3B, HepG2, SKHep1	α_{1H}, α_{1I}	Li et al. (2009)
	SNU387	α_{1I}	Li et al. (2009)
Neuroblastoma	SK-N-SH, NG 108-15, SK-N-MC	α_{1G}	Gray et al. (2004), Panner et al. (2005), Leuranguer et al. (1998), Assandri et al. (1999)
	N1E−115	α_{1G}, α_{1H}	Panner et al. (2005)
Ovarian cancer	HO8910, A2780	α_{1G}, α_{1H}	Li et al. (2011)
Pheochromocytoma	MPC 9/3L	α_{1G}	Harkins et al. (2003)
	PC-12	α_{1H}	Del toro et al. (2003)
Prostate carcinoma	TSU-PRL, DUPRO	α_{1G}	Latour et al. (2004), Wang et al. (2002)
	LNCaP	α_{1H}	Gray et al. (2004), Toyota et al. (1999b)
	PC-3, DU-145	α_{1G}, α_{1H}	Gray et al. (2004), Toyota et al. (1999c)
Retinoblastoma	Y-79, WERI-Rb1	α_{1G}, α_{1H}, α_{1I}	Hirooka et al. (2002)

with ProTx-I, a relatively selective inhibitor for $Ca_v3.1$, promoted proliferation of MCF-7 cells. In addition, siRNA for $Ca_v3.1$ opposed cyclophosphamide-induced apoptosis (Ohkubo and Yamazaki 2012).

The expression of α_{1H} subunits was recently found in SNU499 hepatocarcinoma (HCC) cells (Li et al. 2009), and T-type current was recorded in SNU449 cells. In these cells, the growth inhibition caused by mibefradil, a mixed T/L channel blocker (Panner and Wurster 2006), was associated with persistent increase of phosphorylated ERK1/2 protein. Finally, cyclin E2 and cyclin-dependent kinase-like 3 were down-regulated in SNU449 cells by mibefradil (Li et al. 2009).

Recently, the expression of both $Ca_v3.1$ and $Ca_v3.2$ subunits in human ovarian cancer tissues and H08910 and A2780 ovarian cancer cell lines has been reported (Li et al. 2011). Increased expression of $Ca_v3.1$ and $Ca_v3.2$ was observed in the cytoplasm and membrane of cancerous ovarian tissues compared to normal ovarian tissue by immunocytochemistry, with α_{1H} channels mainly elevated in ovarian tumor tissues. Blockade of T-type Ca^{2+} channels with NNC55-0396, mibefradil, or with siRNA- $Ca_v3.1/3.2$ suppressed proliferation and increased G0/Gl phase distribution (Li et al. 2011).

Ghrelin, a multifunctional peptide hormone, significantly decreased proliferation and induced apoptosis in human PC3 prostate cancer cells; these processes were prevented by T-type channel antagonists. In addition, consistent with their role in apoptosis, an increase in intracellular free Ca^{2+} levels accompanied by up-regulation of $Ca_v3.1$ expression was observed in ghrelin-treated cells (Díaz-Lezama et al. 2010). Alternatively, an increase in both α_{1H} mRNA and T-current was found in LNCaP cells. $Ca_v3.2$ channels have been found to facilitate the neurite lengthening of LNCaP cells (Mariot et al. 2002), which is thought to be related to decrease in malignancy (Shimizu et al. 2000).

The expression of T-type channels can vary depending on tumor stage. For instance, differentiation of prostate cancer cells into more aggressive neuroendocrine cells expressing functional T-type Ca^{2+} channels triggers the release of growth factors stimulating the proliferation of neighboring prostate cancer cells (Gackière et al. 2008) (Fig. 15.1a–c).

Moreover, overexpression of α_{1H} alone or together with α_{1G} and α_{1I} has been demonstrated in human oesophageal tumors, as compared with the normal counterpart that shows lower $\alpha 1$ expression. In accordance with T-type channel functional expression, their blockade diminished the proliferation of TE8 oesophageal cancer cells through p53-dependent p21CIP1 up-regulation (Lu et al. 2008) (Fig. 15.2).

Panner and colleagues have demonstrated that a decrease of $Ca_v3.1$ (α_{1G}), $Ca_v3.2$ (α_{1H}) channel expression is associated with decreased U87 glioma cell proliferation, while stable overexpression of $Ca_v3.2$ resulted in an increase in proliferation of glioma and neuroblastoma cell lines (Panner et al. 2005).

Zhang and colleagues, using HEK-293 or CHO transfected cells, found that $Ca_v3.1$ and $Ca_v3.2$ currents were significantly inhibited by endostatin (ES) (Zhang et al. 2012). Moreover, pretreatment of U87 glioma cells with the T-type Ca^{2+} channel blocker, mibefradil, inhibited cell proliferation and migration. The data from Zhang and colleagues also demonstrate that the ES-mediated effects on the induced proliferation and migration of U87 glioma cells are not mediated by G-proteins or tyrosine-kinase signaling pathways, raising a crucial question regarding

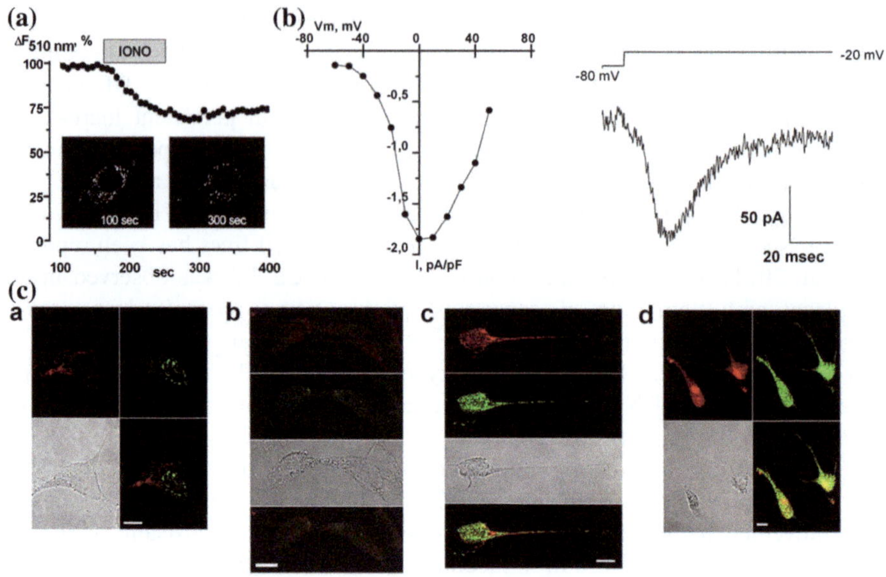

Fig. 15.1 Fluorescence imaging of FM1-43 and T-type calcium channel expression in human prostate neuroendocrine LNCaP cells. **a** Dynamic confocal measurement of FM1-43 fluorescence in an LNCaP cell preloaded with FM1-43 (5 μm) for 12 h. FM1-43 fluorescence appeared as a punctuated pattern inside the cell showing that FM1-43 had been trapped inside intracellular compartments. **b** Current/voltage (I/V) relationship (*left panel*) and example of T-type currents in a typical LNCaP-NE cell (*right panel*). Membrane potential was depolarized for 100 ms from −80 to −20 mV. **c** Labeling of LNCaP-CTL and LNCaP-NE cells with anti-$Ca_v3.2$ (*green* in all panels) and neuroendocrine markers (serotonin or chromogranin A labeled in *red*). **c(a)** LNCaP-CTL cells labeled with anti-chromogranin A (*red*) and anti-$Ca_v3.2$ (*green*). **c(b)** LNCaP-CTL cells labeled with anti-serotonin (*red*) and anti-$Ca_v3.2$ (*green*). **c(c)** LNCaP-NE cells labeled with anti-chromogranin A (*red*) and anti-$Ca_v3.2$ (*green*). **c(d)** LNCaP-NE cells labeled with anti-serotonin (*red*) and anti-$Ca_v3.2$ (*green*). Bar, 10 μm. *From:* (Gackière et al. 2008) $Ca_v3.2$ T-type calcium channels are involved in calcium-dependent secretion of neuroendocrine prostate cancer cells. J Biol Chem. 283(15):10162–73

the nature of the signaling pathways involved in ES inhibition of Ca^{2+} T-type channels and how they regulate $Ca_v3.1/2$ activity. T-channel activity can be modulated by hormones and neurotransmitters acting through signaling intermediates such as protein kinases A and C, calmodulin dependent protein kinase II, tyrosine kinase, G-proteins, and lipid derivatives such as arachidonic acid. Recent reports suggest a role for PKC and ERK pathways in T-type channel activation. Phorbol-12-myristate-13-acetate potently enhances, although to different extents, the current amplitude of $Ca_v3.1$, $Ca_v3.2$, and $Ca_v3.3$ channels, via PKC activation (Park et al. 2006). At present, it is not completely understood if the PKC-mediated stimulation might depend on direct phosphorylation of Ca_v3 channels or whether it is an indirect consequence of phosphorylation of associated targeting, anchoring, or signaling protein(s). Finally, T-type channels mediate release of growth factors

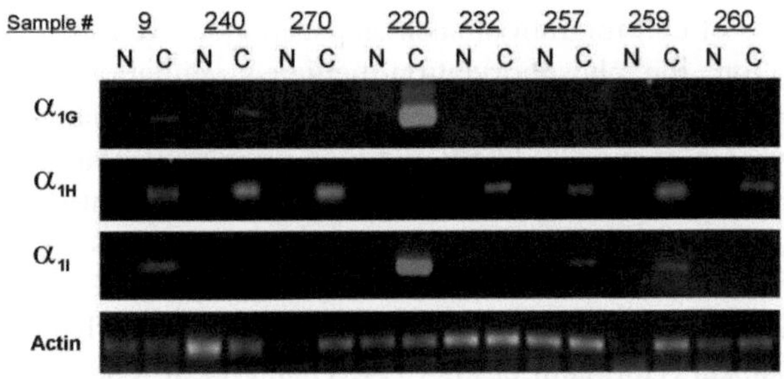

Fig. 15.2 Expression analysis of T-type Ca^{2+} channel α_1-subunits in in situ esophageal cancer foci by RT-PCR. α_{1G}-, α_{1H}-, and α_{1I}-subunits mRNA expression from the primary focus of esophageal cancer and normal mucosa adjacent to the focus from each of the eight patients pathologically diagnosed with esophageal squamous cell carcinoma. The mRNA for α_{1G}, α_{1H}, and α_{1I} was detected in different combinations in esophageal cancer foci but not normal mucosa. *From* Lu F, Chen H, Zhou C, Liu S, Guo M, Chen P, Zhuang H, Xie D, Wu S (Lu et al. 2008) T-type Ca^{2+} channel expression in human esophageal carcinomas: a functional role in proliferation. Cell calcium 43(1):49–58

in pheochromocytomas (Harkins et al. 2003); in RBs, the decreased proliferation was accompanied by a reduced expression of T-type channels mRNA and decreased T-currents (Hirooka et al. 2002).

The mechanisms underlying the in vivo anti-tumoural action of T-type Ca^{2+} channel antagonists are less well understood. In athymic nude mice implanted with MCF-7 breast cancer cells, injection of mibefradil (0.5 mg·100 mL-1, twice a week) at tumor sites resulted in marked tumor degeneration and necrosis (Taylor et al. 2008). Furthermore, local intra-cerebral micro-infusion of ES improved treatment efficiency and survival in an orthotopic xenograft model of human glioblastoma multiforme, and the specific inhibitor of T-type calcium channel, NNC 55-0396, suppressed the growth of $Ca_v3.1^+$ and $Ca_v3.2^+$ HO8910 ovarian cancer formation in nude mouse xenografts (Li et al. 2011). Finally, the TH-1177 blocker of T-type channels (Gray et al. 2004) was shown to extend the lifespan of SCID mice inoculated with human prostate cancer cells (Haverstick et al. 2000).

15.1.3 Role of Ca_v3 Channels in Tumor Invasion and Metastasis Formation

The invasion and metastasis of tumor cells require cell migration (Mareel and Leroy 2003). Cell migration is a cyclic process involving the repetitive extension of invadopodia/lamellipodia at the leading edge, the formation of adhesion sites, the contraction of cell body, and release of trailing adhesion sites. The cyclic morphological and adherence changes observed during cell migration are accompanied by repetitive Ca^{2+} signals, which take the form of Ca^{2+} spikes or oscillations. These oscillations are thought to participate in coordinating the cyclic temporal features of cell migration, such as pseudopod extension, actin assembly, integrin regulation, the phosphorylation-mediated regulation of focal adhesion formation, and pericellular proteolysis (Ronde et al. 2000; Giannone et al. 2002). Thus, repetitive Ca^{2+} signals likely play an important role in the repetitive structural and functional changes required for cell movement. Intracellular Ca^{2+} spikes may involve Ca^{2+} release from intracellular stores as well as Ca^{2+} entry from the extracellular environment that is mediated by Ca^{2+} permeable ion channels of the plasma membrane, the best characterized of which are the voltage-gated Ca^{2+} channels (Hofmann et al. 1999; Huang et al. 2004). Previous studies have demonstrated that mibefradil inhibits leukocyte adhesion and locomotion (Blaheta et al. 1998; Nebe et al. 2002).

There is a considerable overlap between Rho-ROCK activation and T-type channel activity during migration and invasion of tumor cells. The last step of invasion requires cytoskeletal rearrangements and formation of lamellipodia and fillopodia, for which the family of Rho GTPases plays an important role. Most Rho proteins cycle between the GTP-bound active and the GDP-bound inactive state. From the Rho GTPases family members, RhoA stimulates the formation of stress fibers and focal adhesion, whereas Rac is required for the formation of lamellipodia and Cdc42 regulates cell polarity and fillopodia formation (Teodorczyk and Martin-Villalba 2009). The Ca^{2+} T-type mediated entry may contribute to invasion by promoting actin-myosin interaction and the formation and disassembly of cell-substratum adhesions that are important for migration.

Activation of ROCK via the endogenous ligand lysophosphatidic acid (LPA) reversibly inhibited the peak current amplitudes of rat $Ca_v3.1$ and $Ca_v3.3$ channels; conversely, $Ca_v3.2$ currents showed a depolarizing shift. LPA-induced inhibition of $Ca_v3.1$ was dependent on intracellular GTP, and was antagonized by treatment with ROCK and RhoA inhibitors, LPA receptor antagonists or GDPβS. ROCK activation by LPA reduced native T-type currents in Y79 retinoblastoma and in neurons, and upregulated native $Ca_v3.2$ current in dorsal root ganglion neurons (Iftinca et al. 2007). Overall, these data suggest that ROCK is an important regulator of T-type calcium channels, with potential implications for multiple phases of cell locomotion.

15.1.4 Involvement of T-Type Channels in Tumor Cell Differentiation

Calcium channels, mostly T-types, are also regulated during cell differentiation and tumor formation (Chemin et al. 2002; Hirooka et al. 2002; Mariot et al. 2002; Bertolesi et al. 2003; Latour et al. 2004). For example, in human RB cells capable of differentiating into photoreceptors, neurons, and glia (Gallie et al. 1999; Seigel 1999), the current–voltage relation of undifferentiated cells is dominated by a transient inward current that disappears shortly after differentiation. RT-PCR analysis reveals the presence of α_{1G} and α_{1H} mRNA and T-currents in undifferentiated cells, but following differentiation, consistent with a role of T-type channels in embryonic tissue, a striking reduction of both α_{1G} and α_{1H} mRNA expression and loss of T-currents was found. The α_{1I} mRNA levels were low in undifferentiated and differentiated cells. Collectively, these results suggest that T-type channels play a role in undifferentiated RB cell physiology, since α_{1G} and α_{1H} channel expression are reduced in cells that have differentiated and exited the cell cycle (Hirooka et al. 2002).

In neuroblastoma NG108-15 cells, only the α_{1H} subunit or $Ca_v3.2$ displays nickel-sensitive T-currents (Perez-Reyes 1999), suggesting a specific role for this subunit in neuronal development. Calcium influx via low-voltage activated α_{1H} ($Ca_v3.2$) T-currents is involved in the onset of the morphological and electrical differentiation process, leading to the arrest of cell proliferation in neuroblastoma NG108-15 cells (Chemin et al. 2004).

Calcium channels are classically modulated by hormones and neurotransmitters. By RT-PCR, α_{1H} mRNA was evidenced in human adrenocarcinoma cells. In this regard, aldosterone brought about a 39 % increase in the density of α_{1H} currents in human H295R adrenocarcinoma cell lines, and induced qualitative change in calcium influx, thus favoring calcium flow through T-channels (Lesouhaitier et al. 2001).

The androgen-independent stage of prostate cancer has been shown to be associated with the development of neuroendocrine differentiation with over- (or neo-) expression of neuron-specific enolase, chromogranin A, neurotensin, parathyroid hormone-related peptide (Abrahamsson 1999). Neuroendocrine differentiation of prostate epithelial cells is usually associated with increased aggressivity and invasiveness of prostate tumors and poor prognosis (Abrahamsson 1999).

Voltage-dependent calcium channels are also expressed in normal rat prostate cells (Martin et al. 2000), and it has been shown that α_{1G} channels are underexpressed in prostate cancer because of CACNA1G gene hypermethylation (Burchardt et al. 1999). In addition, by RT-PCR, a small proportion of undifferentiated LNCaP cells has been found to express α_{1H} mRNA, and functional T-type calcium currents were found in LNCaP cells (Zelivianski et al. 2001). As evaluated by using the FMI-43 dye and confocal microscopy (Fig. 15.1a, b), neuroendocrine differentiation of LNCaP cells (LNCaP-NE) has been associated with neuritic extensions, secretory granules, and overexpression of the $Ca_v3.2$ (α_{1H}) subunit (Zelivianski et al. 2001). The $Ca_v3.2$ (α_{1H}) channel promotes the secretion of prostatic acid phosphatase (PAP) in LNCaP cells. The LNCaP-NE cells secrete more PAP than non-differentiated cells. Basal PAP secretion by LNCaP-NE cells depends on calcium entry through α_{1H}. In addition, in α_{1H} siRNA LNCaP-NE transfected cells, a decrease of basal release and PAP synthesis and secretion was observed (Fig. 15.1c). In addition, prostate cancer cells, obtained from surgical prostate biopsies, express α_{1H} channels which colocalize with serotonin and chromogranin A and cytokeratin 18 and are more abundant in LNCaP tissue samples than in hyperplasia (Gackière et al. 2008).

In tumor cells, T-type calcium channel functional diversity may result not only from different expression of the three α_1 subunits, but also from the generation of splice variants (Mittman et al. 1999; Chemin et al. 2002; Yunker et al. 2003; Bertolesi et al. 2006) that affect the differentiation in different tumor cells. For example, in undifferentiated RB cells, Ca^{2+} influx via $Ca_v3.1$ channels is essential for cell cycle progression and survival of mitogenic cell types (Hirooka et al. 2002; Bertolesi et al. 2006). A decrease in $Ca_v3.1$ mRNA and protein as well as T-current, and $Ca_v3.1$ alternative splicing have been reported during neuronal differentiation of RB cells (Bertolesi et al. 2006). In this study, Bertolesi and colleagues found that $Ca_v3.1$ alternative splicing induced variations that produce the isoforms e^+ and e^-, ac, bc, b, f, and d. $Ca_v3.1$ac, $Ca_v3.1$bc, and $Ca_v3.1$b were all identified in undifferentiated RB cells, while in differentiated cells, the expression was restricted to $Ca_v3.1$bc isoform. $Ca_v3.1$f was expressed independently of differentiation status of RB cells with or without $Ca_v3.1$d. The variant $Ca_v3.1$e was not expressed in either undifferentiated or differentiated RB cells (Table 15.2). Examination of the functional contribution of $Ca_v3.1$ protein to differentiation reveals that knockdown of $Ca_v3.1$ in RB cells transfected with $Ca_v3.1$ antisense oligodeoxynucleotides did not alter the time-course of differentiation or neuritogenesis, but contributed to the heterogeneity in Ca^{2+} signaling via differences in biophysical and pharmacological properties, protein expression and alterations in intracellular regulation, essential for the establishment of a mature differentiated phenotype (Bertolesi et al. 2006). In vivo, splice variants may give rise to different patterns of tissue distribution as well as different post-translational modifications (Klugbauer et al. 1999; Yunker et al. 2003; Bertolesi et al. 2006). Alterations in splice variant expression and distribution during development may lead to variety of Ca^{2+} signaling that may underlie alterations in cancer behavior. For example, differential electrophysiological properties result in reduction of the inactivation rates of $Ca_v3.1$ae and $Ca_v3.1$be versus $Ca_v3.1$a and $Ca_v3.1$b (Chemin et al. 2001; Bertolesi et al. 2006). An important observation in RB cells was the partial loss of $Ca_v3.1$b, with the $Ca_v3.1$bc remaining the principal splice variant in differentiated cells and the expression of $Ca_v3.1$ac splice variant in undifferentiated, but not in differentiated cells. The $Ca_v3.1$ac splice variant was recently evidenced in U251 N

Table 15.2 Specific Variants of Isoforms of Cav3.1 in Undifferentiated and Differentiated Y-79 Cells

Cytoplasmic region	Exon	Variants	Undifferentiated	Differentiated
II-III domains	Exon 14	e+	–	–
	Exon 14	e–	+	+
III-IV domains	Exon 25, Exon 26	ac	+	–
	Exon 25, Exon 26	bc	+	+
	Exon 25	b	+	–
C-terminal	Exon 34, Exon 35	f+ d+	+	+
	Exon 34	f+ d–	+	+

+ expressed, – not expressed

glioma cells and samples, but not in normal tissues (Latour et al. 2004; Bertolesi et al. 2006). Interestingly, the $Ca_v3.1ac$ splice variant was expressed alone in undifferentiated proliferating RB cells expressing glial markers, suggesting that ac isoforms have a dominant functional role in undifferentiated RB cycling cells with a glial-phenotype that is lost when RB cells undergo neuronal differentiation (Latour et al. 2004; Bertolesi et al. 2006).

15.2 Epigenetic Changes of T-Type Channel Genes in Cancers

15.2.1 DNA Hypermethylation

CpG islands are GC-rich regions of DNA that are coincident with the promoters of 60 % of human genes (Antequera and Bird 1993) and are normally unmethylated, regardless of gene expression. Methylation of CpG islands was detected in genes that are located on the inactive X chromosome and genes that are inactivated by imprinting (Razin and Cedar 1994). Aberrant cytosine methylation in the promoter region was also implicated as one mechanism of tumor suppressor gene inactivation in cancer. To date, various tumor suppressor genes, including Rb1, VHL, p16, BRCA1, hMLH1, and E-cad (Schmutte and Jones 1998), have been shown to be inactivated by hypermethylation in sporadic cancers. Moreover, CpG island hypermethylation of tumor suppressor genes and tissue specific genes seems to be crucial for immortalization of normal cells (Foster et al. 1998; Antequera et al. 1990; Ueki et al. 2002). Inactivation of calcium channel genes plays an important role in cancer development and progression. It has previously been shown that Ca^{2+} influx via T-type channels is an important factor during the initial stages of apoptotic cell death (Berridge et al. 1998). A high level of Ca^{2+} in mitochondria is essential to activate genes associated with programmed cell death (Trump and Berezesky 1995). Furthermore, calcium channel antagonists, which specifically block T-type channels, inhibit cell death by decreasing Ca^{2+} influx (Newsholme et al. 1993). Finally, T-type calcium channels are down-regulated in fibroblasts transformed by platelet-derived growth factor (Estacion and Mordan 1997).

The $Ca_v3.1$ (CACNA1G) gene has been reported to be a target of aberrant methylation and silencing in human tumors. Since the hypermethylation of CpG islands located in the promoter region is thought to be a major epigenetic modification that represses the transcription of tumor suppressor genes, the $Ca_v3.1$ (CACNATG) gene has been regarded as a candidate tumor suppressor gene (Toyota et al. 1999a). In this regard, hypermethylation of the CACNA1G gene on 17q21 was identified by methylated CpG island amplification in human colorectal and lung cancers (Castro et al. 2010). By RT-PCR, CACNA1G expression was detected in normal colon and bone marrow. Aberrant methylation of CACNA1G was examined in various human primary tumors and was detected in 17 of 49

(35 %) colorectal cancers, 4 of 16 (25 %) gastric cancers, and 3 of 23 (13 %) acute myelogenous leukemia cases. Inactivation of CACNA1G modulates calcium signaling, which potentially affects cell proliferation and apoptosis (Toyota et al. 1999a). These data were subsequently confirmed by methyLight assays on 920 and 889 colorectal cancers (Ogino et al. 2007; Kawasaki et al. 2008). Promoter CpG island methylation was shown to occur early in colorectal carcinogenesis (Jass 2005).

The causes of CACNA1G methylation remain to be determined. Methylation was not detected in normal colon mucosa, placenta, normal breast epithelium, and normal bone marrow, including samples from aged patients, suggesting that methylation of this region is cancer specific.

The CACNA1G gene has been used to identify specific clinicopathological features and determine the prognosis of patients with different cancers. Currently, the CACNAIG gene is regarded as a marker of the CpG island methylation phenotype (CIMP) which results in a form of epigenetic instability with simultaneous inactivation of multiple genes (Toyota et al. 1999a, b, c). A subsets of cancers (e.g., bladder, prostate cancer, colorectal carcinoma, leukemia, and gastric cancer) harbor this CIMP1 phenotype (Melki et al. 1999; Toyota et al. 1999a, b, c). The CIMP-positive colorectal tumors with extensive promoter methylation have a distinct clinical, pathological, and molecular profile, such as associations with proximal tumor location, female sex, mucinous and poor tumor differentiation, microsatellite instability, and high BRAF and low TP53 mutation rates (Toyota et al. 1999b; Hawkins et al. 2002; Ogino et al. 2007).

Similarly, Ueki and colleagues analyzed 45 pancreatic carcinomas and 14 normal pancreata (Ueki et al. 2000) as well as 10 pancreatic cancer cell lines (Ueki et al. 2002) for aberrant DNA methylation of CpG islands of multiple genes and clones using methylation-specific PCR. They found that hypermethylation was mainly confined to carcinomas (60 %), rather than normal pancreata. Their identification of genes targeted by hypermethylation provides insight into tumor-suppressive pathways that are inactivated in pancreatic cancer. In addition, in pancreatic carcinoma, aberrant promoter methylation was detected in the CACNA1G gene. Promoter methylation of these genes was shown to be associated with loss of their expression, suggesting that aberrant hypermethylation may be a common mechanism of tumor suppressor gene inactivation in pancreatic carcinoma (Ueki et al. 2002).

In addition to CACNA1G gene expression, also CACNA1H expression was found hypermethylated in MCF-7, MDA-MB-435, and T-47D cell lines, and in six breast cancer samples. Based on this single study, the rate of expression of the CACNA1H gene in breast cancer appears to be about 60–70 %, as expression occurred in 6 of the 9 tissue samples examined (Asaga et al. 2006).

15.2.2 DNA Hypomethylation

Hypomethylation has been reported in cancer cells (Feinberg and Vogelstein 1983) and demethylation due to decreased expression of DNA methyltransferase has been shown to be associated with oncogenesis (Gaudet et al. 2003). It has been associated with genetic instability and aberrant gene expression (Ehrlich 2002), and most hypomethylated genes have been identified as oncogenes and genes associated with drug resistance. Two mechanisms mainly link DNA hypomethylation with oncogenesis: (1) DNA hypomethylation directly induces genetic instability, and (2) aberrant transcription is linked with DNA hypomethylation. Undermethylation of the CpG sites is closely linked with organization and stabilization of chromatin structures. Although DNA hypomethylation in cancer cells was frequently observed in both highly and moderately repetitive DNA sequences (Ehrlich 2002), it was also detected in transcriptional units of single-copy genes such as MAGE-1,27, MDR1,28, and HOX11 (De Smet et al. 1996).

Since CpG sites in both the 5′ and 3′ regions in CACNA1H were hypomethylated in adult T cell leukemia (ATL-55T) cells compared with peripheral blood mononuclear cells (Yoshida et al. 2004), the CACNA1H gene is considered to be a candidate oncogenic gene, perhaps regulated by methylation. The CACNA1H mapping on human chromosome 16p13.3 was normally transcribed in kidney and heart tissues. Hypomethylation of the CACNA1H gene was associated with aberrant expression in ATL cells (2 of 6 ATL cell lines). Transcripts for the CACNA1H gene were detected in some HTLV-I–associated T- or non–HTLV-I–associated T-cell lines. In addition, antagonists of $Ca_v3.2$ channels inhibited cell proliferation (Bertolesi et al. 2002), suggesting that its aberrant expression is associated with cell proliferation.

15.3 Expression of T-Type Alternative Splice Variants in Tumor Cells

In tumor cells, alternative splicing generates additional variants of these channels with distinct biophysical properties such as the kinetics and voltage dependence of activation, inactivation, and deactivation (Monteil et al. 2000; Chemin et al. 2001; Latour et al. 2004); it is thus possible that these splice variants exhibit different functions in different tumors. In human gliomas, three $Ca_v3.1$ splice variants have been reported as a result of alternate splicing of exons 25 and 26 ($Ca_v3.1a$, $Ca_v3.1b$, or $Ca_v3.1bc$). The analysis of glioma samples and cell lines (U87, U563, and U251 N), primary cultures of human fetal astrocytes, as well as adult and fetal human brain cDNA, revealed that in normal adult brain, $Ca_v3.1a$ transcripts predominate, while Ca_v 3.1b is mostly fetal-specific. RT-PCR results on glioma tissue and cell lines showed that $Ca_v3.1$ expression in tumor cells resembles the pattern of fetal brain expression, as $Ca_v3.1bc$ is predominantly expressed. In addition, a glioma-specific novel T-type splice variant, Ca_v 3.1ac, was found to be expressed in some glioma biopsies and cell lines, but not in normal brain or fetal astrocytes. The presence of the Ca_v3.1ac splice variant showing a slower recovery from inactivation suggests that there is glioma-specific Ca_v3.1 gene regulation (Latour et al. 2004). Thus, the existence of a novel Ca_v3.1 splice variant, and the fact that voltage-gated calcium channel inhibition decreases glioma cell proliferation (Kunert-Radek et al. 1989; Latour et al. 2004), suggest that regulation of T-type gene expression may contribute to the pathophysiology of certain tumors.

In the NG 108-15 neuroblastoma/glioma cell line, the atypically slow kinetics displayed by T-type calcium channel α_{1I} (Ca_v3.3) compared to native T-channels (Chemin et al. 2001) suggest the expression of alternative splicing of the α_{1I} subunit. In this regard, two human α_{1I} and α_{1I}-a and α_{1I}-b splice variants that harbor distinct carboxy-terminal regions have been identified and functionally expressed in the NG 108-15 neuroblastoma/glioma cell line. Results indicate that the α_{1I}-b splice variant showed significantly slower deactivation kinetics both in NG 1080-15 and HEK-293 cells (Chemin et al. 2001). This study of two human α_{1I}-a and α_{1I}-b splice variants confirmed the importance of intracellular sequence variations in establishing the gating properties of T-type channels, (Chemin et al. 2001) and suggests that the expression of α_{1I} subunits, especially α_{1I}-b, in neuronal cells contributes to 'normalize' α_{1I} current kinetics to more typical fast-gating T-type Ca^{2+} currents.

Finally, a δ25B splice variant of Ca_v3.2 (α_{1H}) was found in PC3 prostate cancer cells, MDA,231, MDA-361, and MDA-435 breast cancer cell lines and in SK-N-SH neuroblastoma cells (Gray et al. 2004). The analysis of 320 base amplicons from SK-N-SH cells showing a 95 % identity with the δ25B splice variant, evidenced the presence of an unspliced intron containing a stop codon resulting in a truncated two domain protein. The δ25B splice variant participates in Ca^{2+} entry into cancer cells. In this regard, treatment of mice bearing xenografted human PC3 prostate cancer cells expressing the δ25B splice variant with the mibefradil-like compound, TH-1177, significantly extended their lifespan. (Tang et al. 1988).

15.4 Conclusion

T-type Ca^{2+} channels are a distinct channel family controlling tumor growth and progression. Three Ca_v3 α_1 subunits, called Ca_v3.1 (α_{1G}), Ca_v3.2 (α_{1H}), and Ca_v3.3 (α_{1I}), have been identified. Overexpression of the Ca_v3 channels seems to play an important role in abnormal proliferation of tumor cells. In addition, interaction between Rho GTPases and Ca_v3 channels has been demonstrated to regulate tumor migration and invasion. Ca_v3 channel also regulated tumor cell differentiation by the generation of Ca_v3 splice variants that trigger different Ca^{2+} signaling pathways. Finally, it has been demonstrated that inactivation of Ca_v3 genes by aberrant methylation (hypermethylation and hypomethylation) of CpG islands is involved in tumor development and progression.

References

Abrahamsson PA (1999) Neuroendocrine differentiation in prostatic carcinoma. Prostate 39(2):135–148

Antequera F, Bird A (1993) Number of CpG islands and genes in human and mouse. Proc Natl Acad Sci USA 90(24):11995–11999

Antequera F, Boyes J, Bird A (1990) High levels of de novo methylation and altered chromatin structure at CpG islands in cell lines. Cell 62(3):503–514

Asaga S, Ueda M, Jinno H, Kikuchi K, Itano O, Ikeda T, Kitajima M (2006) Identification of a new breast cancer-related gene by restriction landmark genomic scanning. Anticancer Res 26(1A):35–42

Assandri R, Egger M, Gassmann M, Niggli E, Bauer C, Forster I, Gorlach A (1999) Erythropoietin modulates intracellular calcium in a human neuroblastoma cell line. J Physiol 516(Pt 2):343–352

Berridge MJ, Bootman MD, Lipp P (1998) Calcium - a life and death signal. Nature 395(6703):645–648

Bertolesi GE, Jollimore CA, Shi C, Elbaum L, Denovan-Wright EM, Barnes S, Kelly ME (2003) Regulation of alpha1G T-type calcium channel gene (CACNA1G) expression during neuronal differentiation. Eur J Neurosci 17(9):1802–1810

Bertolesi GE, Shi C, Elbaum L, Jollimore C, Rozenberg G, Barnes S, Kelly ME (2002) The Ca(2+) channel antagonists mibefradil and pimozide inhibit cell growth via different cytotoxic mechanisms. Mol Pharmacol 62(2):210–219

Bertolesi GE, Walia Da Silva R, Jollimore CA, Shi C, Barnes S, Kelly ME (2006) Ca(v)3.1 splice variant expression during neuronal differentiation of Y-79 retinoblastoma cells. Neuroscience 141(1):259–268

Blaheta RA, Hailer NP, Brude N, Wittig B, Oppermann E, Leckel K, Harder S, Scholz M, Weber S, Encke A, Markus BH (1998) Novel mode of action of the calcium antagonist mibefradil (Ro 40–5967): potent immunosuppression by inhibition of T-cell infiltration through allogeneic endothelium. Immunology 94(2):213–220

Bunn HF, Poyton RO (1996) Oxygen sensing and molecular adaptation to hypoxia. Physiol Rev 76(3):839–885

Burchardt T, Burchardt M, Chen MW, Cao Y, de la Taille A, Shabsigh A, Hayek O, Dorai T, Buttyan R (1999) Transdifferentiation of prostate cancer cells to a neuroendocrine cell phenotype in vitro and in vivo. J Urol 162(5):1800–1805

Castro M, Grau L, Puerta P, Gimenez L, Venditti J, Quadrelli S, Sánchez-Carbayo M (2010) Multiplexed methylation profiles of tumor suppressor genes and clinical outcome in lung cancer. J Transl Med 8:86–97

Chemin J, Monteil A, Briquaire C, Richard S, Perez-Reyes E, Nargeot J, Lory P (2000) Overexpression of T-type calcium channels in HEK-293 cells increases intracellular calcium without affecting cellular proliferation. FEBS Lett 478(1–2):166–172

Chemin J, Monteil A, Dubel S, Nargeot J, Lory P (2001) The alpha1I T-type calcium channel exhibits faster gating properties when overexpressed in neuroblastoma/glioma NG 108-15 cells. Eur J Neurosci 14(10):1678–1686

Chemin J, Nargeot J, Lory P (2002) Neuronal T-type alpha 1H calcium channels induce neuritogenesis and expression of high-voltage-activated calcium channels in the NG108-15 cell line. J Neurosci 22(16):6856–6862

Chemin J, Nargeot J, Lory P (2004) Ca(v)3.2 calcium channels control an autocrine mechanism that promotes neuroblastoma cell differentiation. Neuroreport 15(4):671–675

De Smet C, De Backer O, Faraoni I, Lurquin C, Brasseur F, Boon T (1996) The activation of human gene MAGE-1 in tumor cells is correlated with genome-wide demethylation. Proc Natl Acad Sci U S A 93(14):7149–7153

Del Toro R, Levitsky KL, pez-Barneo JL, Chiara MD (2003) Induction of T-type Calcium Channel Gene Expression by Chronic Hypoxia. J Biol Chem 278(25):22316–22324

Díaz-Lezama N, Hernández-Elvira M, Sandoval A, Monroy A, Felix R, Monjaraz E (2010) Ghrelin inhibits proliferation and increases T-type Ca2+ channel expression in PC-3 human prostate carcinoma cells. Biochem Biophys Res Commun 403(1):24–29

Ehrlich M (2002) DNA methylation in cancer: too much, but also too little. Oncogene 21(35):5400–5413

Estacion M, Mordan LJ (1997) PDGF-stimulated calcium influx changes during in vitro cell transformation. Cell Signalling 9(5):363–366

Feinberg AP, Vogelstein B (1983) Hypomethylation distinguishes genes of some human cancers from their normal counterparts. Nature 301(5895):89–92

Foster SA, Wong DJ, Barrett MT, Galloway DA (1998) Inactivation of p16 in human mammary epithelial cells by CpG island methylation. Mol Cell Biol 18(4):1793–1801

Gackière F, Bidaux G, Delcourt P, Van Coppenolle F, Katsogiannou M, Dewailly E, Bavencoffe A, Van Chuoï-Mariot MT, Mauroy B, Prevarskaya N, Mariot P (2008) $Ca_v3.2$ T-type calcium channels are involved in calcium-dependent secretion of neuroendocrine prostate cancer cells. J Biol Chem 283(15):10162–10173

Gallie BL, Campbell C, Devlin H, Duckett A, Squire JA (1999) Developmental basis of retinal-specific induction of cancer by RB mutation. Cancer Res 59(7 Suppl):1731s–1735s

Gaudet F, Hodgson JG, Eden A, Jackson-Grusby L, Dausman J, Gray JW, Leonhardt H, Jaenisch R (2003) Induction of tumors in mice by genomic hypomethylation. Science 300(5618):489–492

Giannone G, Ronde P, Gaire M, Haiech J, Takeda K (2002) Calcium oscillations trigger focal adhesion disassembly in human U87 astrocytoma cells. J Biol Chem 277(29):26364–26371

Gray LS, Perez-Reyes E, Gomora JC, Haverstick DM, Shattock M, McLatchie L, Harper J, Brooks G, Heady T, Macdonald TL (2004) The role of voltage gated T-type Ca2+ channel isoforms in mediating "capacitative" Ca2+ entry in cancer cells. Cell Calcium 36(6):489–497

Harkins AB, Cahill AL, Powers JF, Tischler AS, Fox AP (2003) Expression of recombinant calcium channels support secretion in a mouse pheochromocytoma cell line. J Neurophysiol 90(4):2325–2333

Harris AL (2002) Hypoxia—a key regulatory factor in tumour growth. Nat Rev Cancer 2(1):38–47

Haverstick DM, Heady TN, Macdonald TL, Gray LS (2000) Inhibition of human prostate cancer proliferation in vitro and in a mouse model by a compound synthesized to block Ca2+ entry. Cancer Res 60(4):1002–1008

Hawkins N, Norrie M, Cheong K, Mokany E, Ku SL, Meagher A, O'Connor T, Ward R (2002) CpG island methylation in sporadic colorectal cancers and its relationship to microsatellite instability. Gastroenterology 122(5):1376–1387

Hirooka K, Bertolesi GE, Kelly ME, Denovan-Wright EM, Sun X, Hamid J, Zamponi GW, Juhasz AE, Haynes LW, Barnes S (2002) T-Type calcium channel alpha1G and alpha1H subunits in human retinoblastoma cells and their loss after differentiation. J Neurophysiol 88(1):196–205

Hofmann F, Lacinova L, Klugbauer N (1999) Voltage-dependent calcium channels: from structure to function. Rev Physiol Biochem Pharmacol 139:33–87

Huang JB, Kindzelskii AL, Clark AJ, Petty HR (2004) Identification of channels promoting calcium spikes and waves in HT1080 tumor cells: their apparent roles in cell motility and invasion. Cancer Res 64(7):2482–2489

Iftinca M, Hamid J, Chen L, Varela D, Tadayonnejad R, Altier C, Turner RW, Zamponi GW (2007) Regulation of T-type calcium channels by Rho-associated kinase. Nat Neurosci 10(7):854–860

Jass JR (2005) Serrated adenoma of the colorectum and the DNA-methylator phenotype. Nat Clin Pract Oncol 2(8):398–405

Kawasaki T, Ohnishi M, Nosho K, Suemoto Y, Kirkner GJ, Meyerhardt JA, Fuchs CS, Ogino S (2008) CpG island methylator phenotype-low (CIMP-low) colorectal cancer shows not only few methylated CIMP-high-specific CpG islands, but also low-level methylation at individual loci. Mod Pathol 21(3):245–255

Klugbauer N, Marais E, Lacinova L, Hofmann F (1999) A T-type calcium channel from mouse brain. Pflugers Arch 437(5):710–715

Kuga T, Kobayashi S, Hirakawa Y, Kanaide H, Takeshita A (1996) Cell cycle–dependent expression of L- and T-type Ca2+ currents in rat aortic smooth muscle cells in primary culture. Circ Res 79(1):14–19

Kunert-Radek J, Stepien H, Radek A, Lyson K, Pawlikowski M (1989) Inhibitory effect of calcium channel blockers on proliferation of human glioma cells in vitro. Acta Neurol Scand 79(2):166–169

Latour I, Louw DF, Beedle AM, Hamid J, Sutherland GR, Zamponi GW (2004) Expression of T-type calcium channel splice variants in human glioma. Glia 48:112–119

Lesouhaitier O, Chiappe A, Rossier MF (2001) Aldosterone increases T-type calcium currents in human adrenocarcinoma (H295R) cells by inducing channel expression. Endocrinology 142(10):4320–4330

Leuranguer V, Bourinet E, Lory P, Nargeot J (1998) Antisense depletion of beta-subunits fails to affect T-type calcium channels properties in a neuroblastoma cell line. Neuropharmacology 37(6):701–708

Li W, Zhang SL, Wang N, Zhang BB, Li M (2011) Blockade of T-type Ca(2+) channels inhibits human ovarian cancer cell proliferation. Cancer Invest 29(5):339–346

Li Y, Liu S, Lu F, Zhang T, Chen H, Wu S, Zhuang H (2009) A role of functional T-type Ca2+ channel in hepatocellular carcinoma cell proliferation. Oncol Rep 22(5):1229–1235

Lory P, Bidaud I, Chemin J (2006) T-type calcium channels in differentiation and proliferation. Cell Calcium 40(2):135–146

Lu F, Chen H, Zhou C, Liu S, Guo M, Chen P, Zhuang H, Xie D, Wu S (2008) T-type Ca2+ channel expression in human esophageal carcinomas: a functional role in proliferation. Cell Calcium 43(1):49–58

Mareel M, Leroy A (2003) Clinical, cellular, and molecular aspects of cancer invasion. Physiol Rev 83(2):337–376

Mariot P, Vanoverberghe K, Lalevee N, Rossier MF, Prevarskaya N (2002) Overexpression of an alpha 1H ($Ca_v3.2$) T-type calcium channel during neuroendocrine differentiation of human prostate cancer cells. J Biol Chem 277(13):10824–10833

Martin RL, Lee JH, Cribbs LL, Perez-Reyes E, Hanck DA (2000) mibefradil block of cloned T-type calcium channels. J Pharmacol Exp Ther 295(1):302–308

Melki JR, Vincent PC, Clark SJ (1999) Concurrent DNA hypermethylation of multiple genes in acute myeloid leukemia. Cancer Res 59(15):3730–3740

Mittman S, Guo J, Agnew WS (1999) Structure and alternative splicing of the gene encoding alpha1G, a human brain T calcium channel alpha1 subunit. Neurosci Lett 274(3):143–146

Monteil A, Chemin J, Leuranguer V, Altier C, Mennessier G, Bourinet E, Lory P, Nargeot J (2000) Specific properties of T-type calcium channels generated by the human alpha 1I subunit. J Biol Chem 275(22):16530–16535

Nebe B, Holzhausen C, Rychly J, Urbaszek W (2002) Impaired mechanisms of leukocyte adhesion in vitro by the calcium channel antagonist mibefradil. Cardiovasc Drugs Ther 16(3):183–193

Newsholme P, Adogu AA, Soos MA, Hales CN (1993) Complement induced Ca21 influx in cultured fibroblasts is decreased by the calcium-channel antagonist nifedipine or by some bivalent inorganic cations. Biochem J 295(Pt3):773–779

Ogino S, Kawasaki T, Kirkner GJ, Kraft P, Loda M, Fuchs CS (2007) Evaluation of markers for CpG island methylator phenotype (CIMP) in colorectal cancer by a large population-based sample. J Mol Diagn 9(3):305–314

Ohkubo T, Yamazaki Y (2012) T-type voltage-activated calcium channel $Ca_v3.1$, but not $Ca_v3.2$, is involved in the inhibition of proliferation and apoptosis in MCF-7 human breast cancer cells. Int J Oncol 41(1):267–275

Panner A, Cribbs LL, Zainelli GM, Origitano TC, Singh S, Wurster RD (2005) Variation of T-type calcium channel protein expression affects cell division of cultured tumor cells. Cell Calcium 37(2):105–119

Panner A, Wurster RD (2006) T-type calcium channels and tumor proliferation. Cell Calcium 40(2):253–259

Park JY, Kang HW, Moon HJ, Huh SU, Jeong SW, Soldatov NM, Lee JH (2006) Activation of protein kinase C augments T-type Ca2+ channel activity without changing channel surface density. J Physiol 577(Pt 2):513–523

Perez-Reyes E (1999) Three for T: molecular analysis of the low voltage-activated calcium channel family. Cell Mol Life Sci 56(7–8):660–669

Razin A, Cedar H (1994) DNA methylation and genomic imprinting. Cell 77(4):473–476

Ronde P, Giannone G, Gerasymova I, Stoeckel H, Takeda K, Haiech J (2000) Mechanism of calcium oscillations in migrating human astrocytoma cells. Biochim Biophys Acta 1498(2–3):273–280

Santoni G, Santoni M, Nabissi M (2012) Functional role of T-type calcium channels in tumour growth and progression: prospective in cancer therapy. Br J Pharmacol 166(4):1244–1246

Schmutte C, Jones PA (1998) Involvement of DNA methylation in human carcinogenesis. Biol Chem 379(10):377–388

Shimizu T, Ohta Y, Ozawa H, Matsushima H, Takeda K (2000) Papaverine combined with prostaglandin E2 synergistically induces neuron-like morphological changes and decrease of malignancy in human prostatic cancer LNCaP cells. Anticancer Res 20(2A):761–767

Seigel GM (1999) The golden age of retinal cell culture. Mol Vis 5:4

Tang C, Presser F, Morad M (1988) Amiloride selectively blocks the low threshold (T) calcium channel. Science 240(4849):213–215

Taylor JT, Huang L, Pottle JE, Liu K, Yang Y, Zeng X, Keyser BM, Agrawal KC, Hansen JB, Li M (2008) Selective blockade of T-type Ca2+ channels suppresses human breast cancer cell proliferation. Cancer Lett 267(1):116–124

Taylor SC, Peers C (1999) Chronic hypoxia enhances the secretory response of rat phaeochromocytoma cells to acute hypoxia. J Physiol 514(Pt 2):483–491

Teodorczyk M, Martin-Villalba A (2009) Sensing Invasion: Cell Surface Receptors. Driving Spreading of Glioblastoma. J Cell Physiol 222(1):1–10

Toyota M, Ahuja N, Ohe-Toyota M, Herman JG, Baylin SB, Issa J-PJ (1999a) CpG island methylator phenotype in colorectal cancer. Proc Natl Acad Sci USA 96(15):8681–8686

Toyota M, Ahuja N, Suzuki H, Itoh F, Ohe-Toyota M, Imai K, Baylin SB, Issa JP (1999b) Aberrant methylation in gastric cancer associated with the CpG island methylator phenotype. Cancer Res 59(21):5438–5442

Toyota M, Ho C, Ohe-Toyota M, Baylin SB, Issa JP (1999c) Inactivation of CACNA1G, a T-type calcium channel gene, by aberrant methylation of its 5' CpG island in human tumors. Cancer Res 59:4535–4541

Trump BF, Berezesky IK (1995) Calcium-mediated cell injury and cell death. FASEB J 9(2):219–228

Ueki T, Toyota M, Sohn T, Yeo CJ, Issa JP, Hruban RH, Goggins M (2000) Hypermethylation of multiple genes in pancreatic adenocarcinoma. Cancer Res 60(7):1835–1839

Ueki T, Walter KM, Skinner H, Jaffee E, Hruban RH, Goggins M (2002) Aberrant CpG island methylation in cancer cell lines arises in the primary cancers from which they were derived. Oncogene 21(13):2114–2117

Wang YQ, Brooks G, Zhu CB, Yuan WZ, Li YQ, Wu XS (2002) Functional analysis of the human T-type calcium channel alpha 1H subunit gene in cellular proliferation. Yi Chuan Xue Bao 29(8):659–665

Yoshida M, Nosaka K, Yasunaga J, Nishikata I, Morishita K, Matsuoka M (2004) Aberrant expression of the MEL1S gene identified in association with hypomethylation in adult T-cell leukemia cells. Blood 103(7):2753–2760

Yunker AM, Sharp AH, Sundarraj S, Ranganathan V, Copeland TD, McEnery MW (2003) Immunological characterization of T-type voltage-dependent calcium channel $Ca_v3.1$ (alpha 1G) and $Ca_v3.3$ (alpha 1I) isoforms reveal differences in their localization, expression, and neural development. Neuroscience 117(2):321–335

Zelivianski S, Verni M, Moore C, Kondrikov D, Taylor R, Lin MF (2001) Multipathways for transdifferentiation of human prostate cancer cells into neuroendocrine-like phenotype. Biochim Biophys Acta 1539(1–2):28–43

Zhang Y, Zhang J, Jiang D, Zhang D, Qian Z, Liu C, Tao J (2012) Inhibition of T-type Ca^{2+} channels by endostatin attenuates human glioblastoma cell proliferation and migration. Br J Pharmacol 166(4):1247–1260

Zhu WH, Conforti L, Czyzyk-Krzeska MF, Millhorn DE (1996) Membrane depolarization in PC-12 cells during hypoxia is regulated by an O2-sensitive K^+ current. Am J Physiol 271(2 Pt 1):C658–C665

Chapter 16
Pharmacology of Voltage-Gated Calcium Channels in Clinic

Lubica Lacinová and Lucia Lichvárová

Abstract The diversity of voltage-gated calcium channels (VGCC) allows precise time- and voltage-dependent regulation of calcium ions entry into cells. Individual calcium channel subtypes participate in the initiation of neuronal action potential and repetitive action potential firing; in cardiac pacemaking and signal conduction; in shaping an action potential plateau; and in action potential repolarization. Calcium entry through VGCC is necessary for cardiac and smooth muscle cell contraction. Since disrupted regulation of calcium homeostasis accompanies various pathologies, the complex pattern of regulation of cell physiology by VGCC makes them an attractive target for pharmaceutical agents. The most common pathologies treated with VGCC blockers are cardiovascular diseases, especially hypertension, angina, and tachycardia. In addition, calcium channels have been identified as targets of some antiepileptic and anesthetic drugs. More recently, specific VGCC blockers were introduced in the treatment of chronic pain, and these have also proved beneficial as support therapy in neurodegenerative diseases and cognitive impairment.

16.1 Introduction

Voltage-gated calcium channels (VGCC) are a family of voltage-gated ion channels which are far less diverse than the voltage-gated potassium channel family—in respect to the number of representatives and also to the diversity of the current waveform. It is, however, much more diverse than the voltage-gated sodium channel family in respect to its kinetics and voltage dependencies of channel gating. Since the calcium current is ionic, it flows through the VGCC and modulates the dynamics

L. Lacinová (✉) · L. Lichvárová
Institute of Molecular Physiology and Genetics Slovak Academy of Sciences, Vlarska 5, 833 34 Bratislava, Slovakia
e-mail: lubica.lacinova@savba.sk

N. Weiss and A. Koschak (eds.), *Pathologies of Calcium Channels*,
DOI: 10.1007/978-3-642-40282-1_16, © Springer-Verlag Berlin Heidelberg 2014

of transmembrane potential changes in excitable cells. Low-voltage-activated (LVA) calcium channels participate in the initiation of action potentials, and they are responsible for generation of 'low threshold calcium spikes' in specific neuronal tissues. They are a prerequisite for neuronal-burst firing, contribute to repetitive action potential firing and also participate in pacemaking during heart diastolic depolarization (Perez-Reyes 2003). High-voltage-activated calcium channels (HVA) participate in shaping an action potential plateau in cardiac myocytes and may also contribute to action potential repolarization in neurons. Furthermore, $Ca_v1.3$ calcium channels serve as pacemaker in the sinoatrial node (Mangoni et al. 2003) and in chromaffin cells (Comunanza et al. 2010). VGCC also contribute to rhythmic depolarization waves in smooth muscle tissue.

In contrast to potassium and sodium ions entry to cells by voltage-gated potassium and sodium channels, calcium ions also play an additional role as a second messenger. Calcium entry into cardiac and smooth muscle cells through L-type calcium channels is necessary for muscle contraction. Further, calcium ions participate in enzyme regulation, neurotransmitter release, regulation of gene transcription, cell growth and proliferation, and fertilization. It has been long established that disrupted regulation of calcium homeostasis accompanies various pathologies. The cloning era commencing in the1980s enabled identification of the specific contribution of individual calcium channel subtypes to cellular regulation. This was followed by identification of mutations in both the VGCC principal α_1 subunit and the auxiliary subunits responsible for hereditary diseases. The complex pattern of cell physiology regulation by VGCC makes them an attractive target for pharmaceutical agents, where cardiovascular diseases are the most common pathologies treated with VGCC blockers. While calcium channels were later identified as targets for some antiepileptic and anesthetic drugs, specific VGCC blockers have more recently been introduced in treatment of chronic pain.

16.2 L-Type Calcium Channels

L-type calcium channels (LTCC) are the most investigated and most ubiquitous VGCC in the human organism. In mammalian cells, the calcium current was first described in cardiac Purkinje fibers (Reuter 1967) and in cardiac ventricular myocytes (Reuter and Beeler 1969). An L-type calcium current was apparently reported in those works before such classification was introduced. In the 1960s, verapamil was introduced as the first clinically used calcium antagonist, and this was soon followed by Dihydropyridines such as nifedipine, niludipine, and nimodipine (Fleckenstein 1983). Initially, these drugs were not recognized as calcium channel blockers, but their binding to calcium channel complexes was soon appreciated, and this aided purification of the first VGCC α_1 subunit as a Dihydropyridine receptor (Tanabe et al. 1987). Currently, three major classes of LTCC blockers are used for treatment of cardiovascular disease: dihydropyridines (DHPs), phenylakylamines (PAAs), and benzothiazepines (BTZs).

Blockage of the two $Ca_v1.2$ and $Ca_v1.3$ LTCC subtypes by DHPs, PAAs, and BTZs is relevant in understanding their cardiovascular effects. These channels share up to 90 % sequence homology, yet they have distinct biophysical properties (Zuccotti et al. 2011). Different electrophysiological properties, together with different distribution of these channel subtypes in the cardiovascular system, form the basis for their different roles therein. While the $Ca_v1.2$ channel is the main channel responsible for initiation of both cardiac and smooth muscle contraction, the $Ca_v1.3$ channel is predominantly expressed in the sinoatrial node where it contributes to pacemaking (Mangoni et al. 2003), and also in the atrioventricular node where it controls conduction velocity (Marger et al. 2011).

Inhibition of calcium entry through LTCC has complex effects on cardiovascular system physiology. Attenuated contractility in vascular smooth muscle results in vasodilatation and consequently lower blood pressure. Contractility is also reduced in cardiac muscle as negative inotropy. These effects are based on the inhibition of $Ca_v1.2$ channels. The heart rate is reduced in the sinus node (negative chronotropy) and conduction velocity is slowed in the atrioventricular node (negative dromotropy), both mainly due to the inhibition of $Ca_v1.3$ channels.

16.2.1 1,4-Dihydropyridines

1,4-dihydropyridines (DHPs) are the best known group of calcium channel ligands. Although both agonist and antagonist DHPs are recognized, only antagonists are used in clinical work. DHP calcium channel blockers are easily identified by the suffix "-dipine." The dihydropyridine molecule is based on pyridine, but is semi-saturated with two substituents replacing one double bond. The presence of a C-4 phenyl group is a preferential requirement to optimize biological activity. The structure of several DHPs commonly used in clinical practice is shown in Fig. 16.1.

DHPs are still among the most widely used drugs in the treatment of cardiovascular disease. Four generations of DHPs are currently available. The first-generation drugs were effective in hypertension treatment, but their main disadvantage was the rapid onset and short duration of vasodilatory action. A typical representative of this class is nifedipine. Second-generation DHPs were characterized by improved control of the therapeutic effect due to either an intrinsically long-acting preparation or to the slow-release preparation of otherwise fast acting drugs. Amlodipine and nitrendipine are representatives of third-generation DHPs and possess more stable pharmacokinetics. Fourth-generation DHPs, including lercanidipine and lacidipine, are highly lipophilic. They are characterized by stable activity, reduced adverse effects, and a broad therapeutic spectrum.

The rapid onset of nifedipine (Procardia) action activated a sympathetic reflex and consequently significantly increased heart rate. This defect is alleviated in the long-acting formulation of Procardia XL. Due to its ability to suppress smooth muscle contraction, nifedipine is used to prevent preterm labor. Currently, nifedipine has 5 % share of the calcium channel blocker market (Elliott and Ram

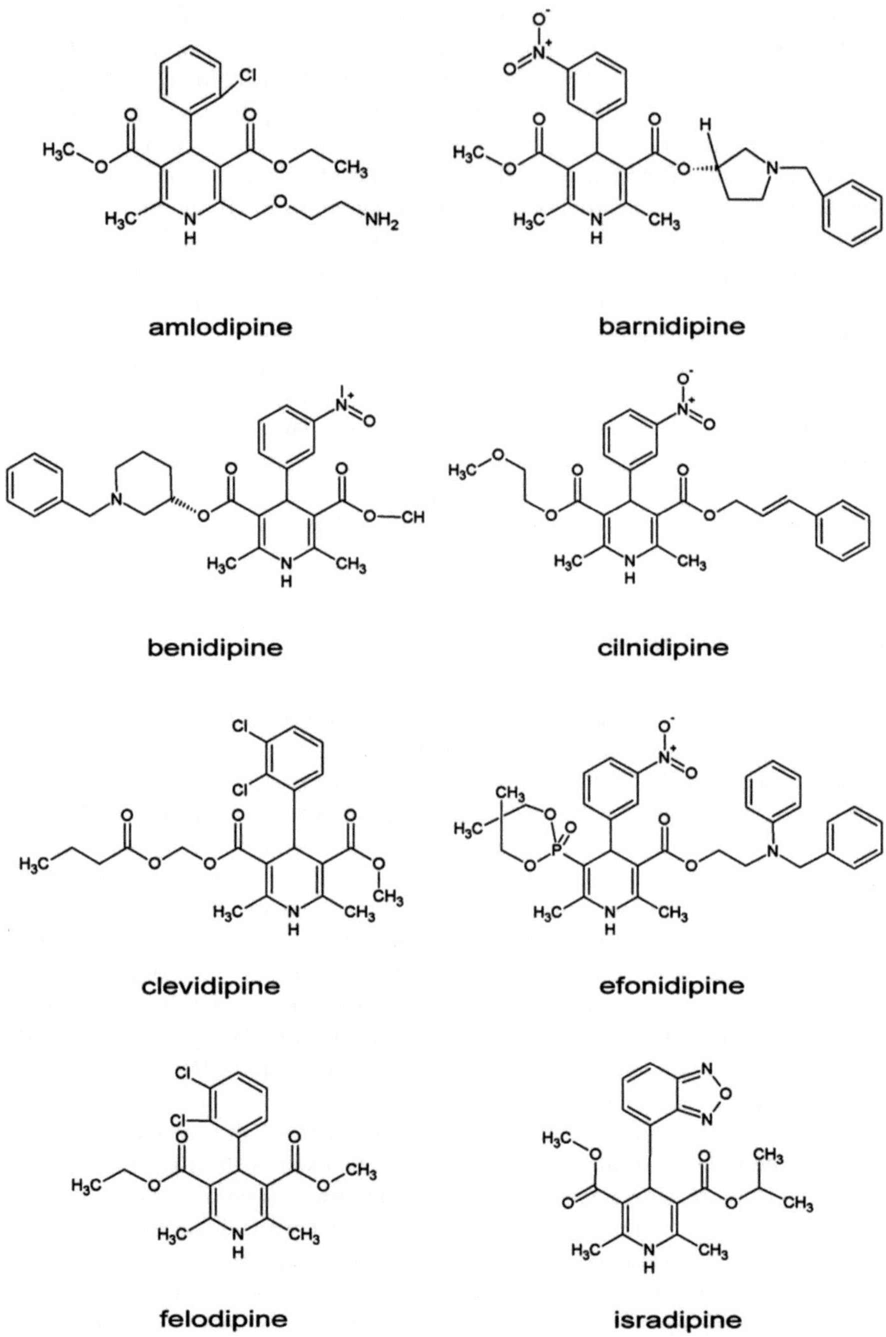

Fig. 16.1 Structural formulae of 1,4-dihydropyridines commonly used in clinical practice in the treatment of cardiovascular diseases. Compounds are listed in alphabetical order

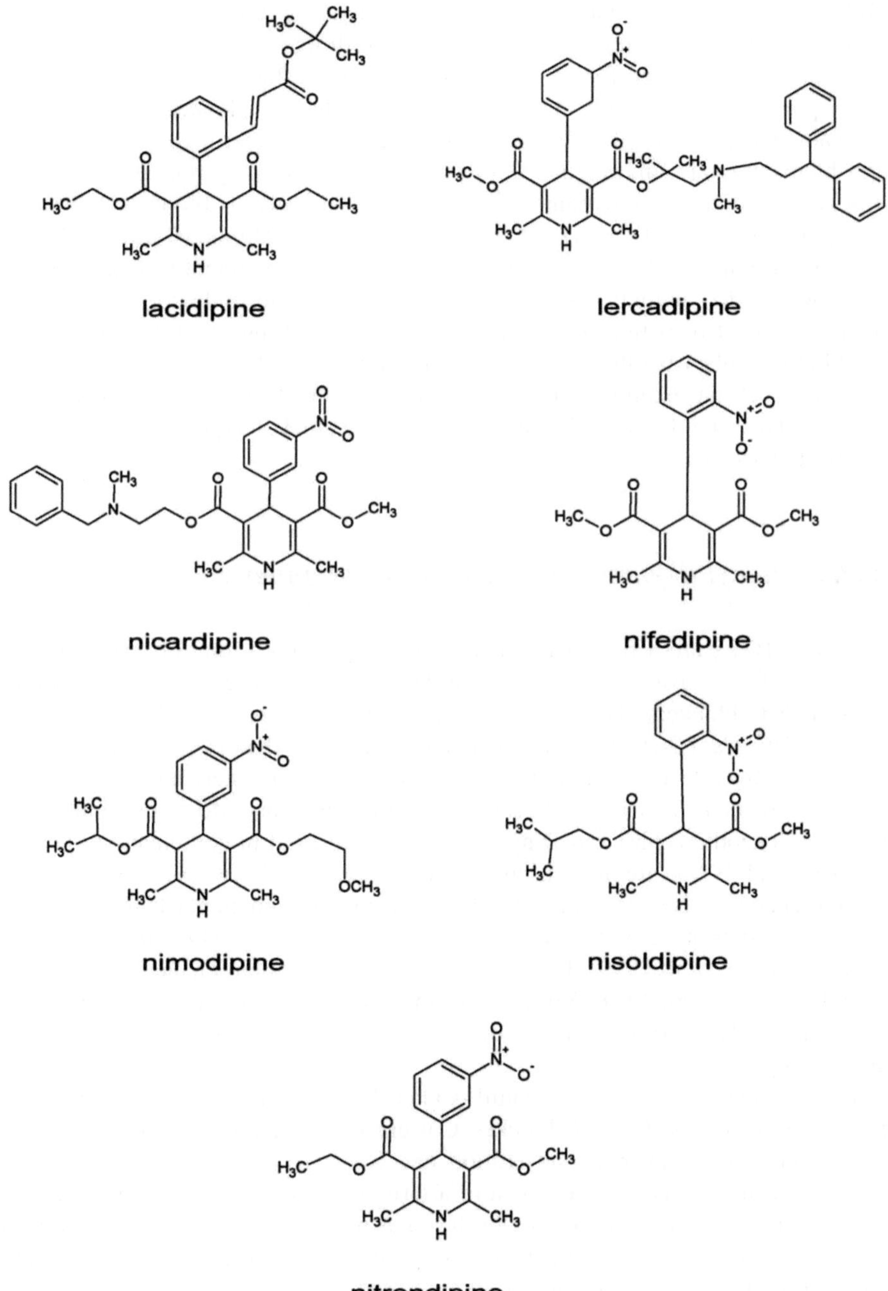

Fig. 16.1 (continued)

2011). Amlodipine (Norvasc) is a second-generation intrinsically long-acting agent. In contrast to neutral DHPs such as nisoldipine and isradipine, it is predominantly in a charged form at physiological pH. Amlodipine is the most commonly prescribed DHP with 71 % of the calcium channel blocker market (Elliott and Ram 2011). It is used alone or in combination with other medications to treat high blood pressure and angina. It lowers blood pressure by relaxing the blood vessels and controls chest pain by increasing the supply of blood to the heart. Isradipine (DynaCirc, Prescal), felodipine (Plendil), nicardipine (Cardene, Carden SR), and nifedipine (Procardia, Adalat) are used in the treatment of hypertension, and now also for angina. Felodipine (Plendil) is used not only for hypertension treatment but also to treat Raynaud's syndrome and congestive cardiac failure. Nisoldipine (Sular) is effective in treatment of essential hypertension, while Benidipe (Coniel) has exhibited significant beneficial effects compared to other DHP blockers. This may be related to its ability to also block T-type calcium channels (see Sect. 16.4.2).

16.2.2 Phenylalkylamines and Benzothiazepines

PAAs and BTZs are a further two classes of drugs with a long history of use in treatment of cardiovascular disease. Similar to DHPs, these drugs are predominantly LTCC blockers. The binding sites for DHPs, PAAs, and BTZs on the α_1 subunit of the $Ca_v1.2$ channel were mapped to the IIIS5, IIIS6, and IVS6 segments. Since they share several common amino acids and are allosterically linked (Lacinova 2005), there is a rationale for combination therapy. Most commonly, diltazem (a benzothiazepin) and Dihydropyridines are used in combination.

PAAs and BTZs are not strictly selective for LTCC. PAAs inhibit T-type calcium channels (Bergson et al. 2011), and also potassium channels including the delayed rectifier potassium channel (Catacuzzeno et al. 1999) and the hERG channel (Zhang et al. 1999). BTZs inhibit various potassium channels (Gao et al. 2005; Caballero et al. 2004). While DHPs mainly induce peripheral vasodilatation, PAAs and BTZs primary mode of action is negative cardiac inotropy and chronotropy.

The prototypical PAA is verapamil (Calan, Isoptin; Fig. 16.2), which was also the first clinically used VGCC blocker. Currently, verapamil has 8 % share of the calcium channel blocker market (Elliott and Ram 2011). Due to its vasodilator effect, verapamil is used in treatment of arterial hypertension and angina pectoris. In addition, it depresses the activity of the sinus node, decreases conduction rate at the atrioventricular node, and suppresses myocardial contractility. Intra-coronary administration of verapamil can safely and rapidly terminate reperfusion-induced ventricular tachyarrhythmias (Kato et al. 2004).

The clinically used BTZ diltiazem (Cardizem; Fig. 16.2) is a short-acting drug. This disadvantage was overcome by design of an extended-release drug delivery system (ditiazem XR, cardizem SR). Various forms of dilatiazem share

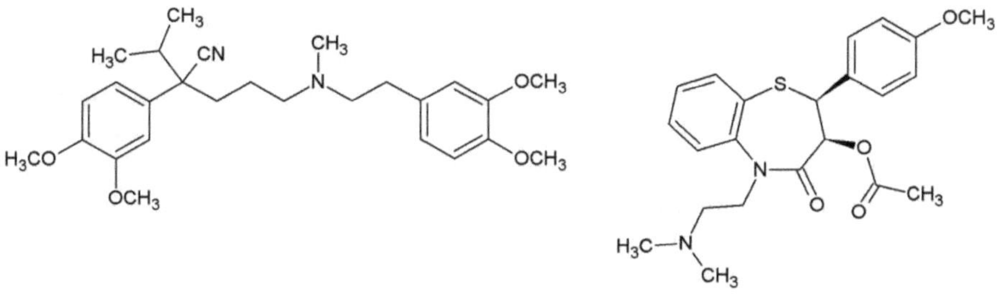

Fig. 16.2 Structural formulae of phenylalkylamine verapamil and benzothiazepine diltiazem

approximately 11 % of the calcium channel blocker market (Elliott and Ram 2011). Dihydropyridine calcium antagonists can increase heart rate due to reflex activation of the sympathetic nervous system. Since this is not the case with diltiazem, oral sustained-release diltiazem may be a preferable drug for arterial hypertension treatment (Kawano et al. 2000). In its injectable form, it is used in the treatment of tachycardia.

16.2.3 Cardiovascular Selectivity of DHPs, PAAs, and BTZs

While L-type calcium channels as such are ubiquitously expressed in the cardiovascular system, their subtypes and/or splice variants are selectively distributed. $Ca_v1.3$ channels were found in sinoatrial and atrioventricular node cells. The predominant LTCC in cardiac myocytes is the $Ca_v1.2a$ splice variant of a $Ca_v1.2$ channel, while $Ca_v1.2b$ is a splice variant expressed in smooth muscle, including vascular muscle. Little is known about the selectivity of LTCC blockers for the $Ca_v1.2$ channels compared to the $Ca_v1.3$ channels where in vitro experiments suggest that these are about one decimal order less sensitive to all three clinically used calcium channel blocker classes DHPs, PAAs, and BTZs (Tarabova et al. 2007; Michna et al. 2003).

Dihydropyridines, phenylalkylamines, and bezothiazepines interact with LTCC in a state-dependent manner. Individual blockers mainly differ in their affinity for inactivated and open channel states. Generally, DHPs have higher affinity for an inactivated channel state while PAAs and BTZs bind preferentially to the channel in an open state. Membrane depolarization favors channel inactivation and therefore facilitates current inhibition by DHPs. For a long time this was considered to be the sole cause of their relative vascular selectivity, since the resting membrane potential of vascular tissue is more positive than the resting membrane potential in cardiomycytes. Later, it was shown that the difference between the amino acid sequence of vascular ($Ca_v1.2b$) and cardiac ($Ca_v1.2a$) splice variants in transmembrane segment IS1 also contributes to this phenomenon (Welling et al. 1997).

Preferential binding to an open channel state is responsible for use-dependent current inhibition. This infers that an increase in the frequency of channel activation facilitates drug binding and enhances current inhibition. Increased heart rate increases the efficiency of a use-dependent drug which then attenuates the heart rate in a negative feedback mechanism. Frequency-dependent action is characteristic for current inhibition by verapamil and diltiazem. These effects may explain the anti-arrhythmic properties of verapamil and the general vascular selectivity of the 1,4-Dihydropyridines (Table 16.1).

16.2.4 L-Type Calcium Channels in Central Nervous System

LTCC are also ubiquitously expressed in the central nervous system. Although whether verapamil and diltiazem are present at therapeutic level concentrations in the brain is currently unknown, several neutral DHPs are known to penetrate the blood–brain barrier (felodipine, isradipine, nicardipine, nifedipine, nimodipine, nitrendipine, lacidipine, and lercandipine) while charged DHPs such as amlodipine can not cross it so readily. Therefore, DHP channel blockers have also been tested as support therapy in neuronal diseases. Nimodipine (Nimotop) which readily penetrates the blood–brain barrier was effective in pharmacotherapy of age-related mental deficits including primary degenerative dementia, multi-infarct dementia (Fischhof 1993) and also in HIV-related neurological disorders (Navia et al. 1998) and migraine headaches. Intra-arterial, intra-ventricular, or intra-cisternal application of nicardipine has proven effective in prophylaxis and treatment of cerebral vasospasm. In addition, patients with mild cognitive impairment treated with nivaldipine for hypertension demonstrated an improved score in memory tests (Hanyu et al. 2007).

Table 16.1 Selectivity of LTCC blockers in the cardiovascular system expressed as a ratio of IC_{50} for half-maximal contraction inhibition in vascular muscle to IC_{50} for half-maximal contraction inhibition in cardiac muscle; based on (Godfraind et al. 1992)

Drug	Selectivity
Diltiazem (BTZ)	1
Verapamil (PAA)	1
Nifedipine, amlodipine (DHPs)	10
Felodipine, isradipine, nitrendipine, nicardipine, lacidipine (DHPs)	100
Nisoldipine (DHP)	1,000

16.3 Neuronal Voltage-Gated Calcium Channels

Three neuronal calcium channels are currently recognized: $Ca_v2.1$, $Ca_v2.2$, and $Ca_v2.3$. With their quite restricted expression pattern in the central and peripheral nervous systems, they contribute to the pathology of neurodegenerative (Parkinson and Alzheimer diseases), neurological (chronic pain), and psychiatric (anxiety) disorders. Therefore, specific and effective blockers of these channels should have important therapeutic potential.

16.3.1 $Ca_v2.2$ Channel and Pain Therapy

Currently, one relatively selective $Ca_v2.2$ calcium channel blocker ziconotide (Prialt) has been approved for clinical use. It is a synthetic analogue of a conopeptide known as SNX-111 or ω-conotoxin MVIIA isolated from the venom of the *Conus magus* marine snail. Inhibition of the $Ca_v2.2$ channels in the spinal cord inhibits the release of pain-related neurotransmitters from the central terminals of primary afferent neurons. Ziconotide was shown to be effective for intrathecal treatment of severe refractory chronic cancer and non-cancer pain. It can be applied either as a monotherapy or in combination with opioids (Staats et al. 2004; Raffaeli et al. 2011; Smith and Deer 2009). However, because of its substantial side-effects, ziconotide has a narrow therapeutic window and the treatment is appropriate only in patients suffering severe chronic pain. Several other conotoxin calcium channel blockers are currently in preclinical and/or clinical testing (Essack et al. 2012).

16.3.2 Gabapentin and Pregabalin

The pathophysiology of neuropathic pain exhibits many similarities with the pathology of epilepsy. Therefore, several antiepileptic drugs, including those acting as VGCC blockers were tested for the treatment of neuropathic pain. Positive results have been achieved with gabapentin (1-(aminomethyl)cyclohexaneacetic acid) and also with its successor, pregabalin (S-(+)-3-isobutylgaba) which has improved bio-availability. Both drugs are lipophilic analogues of the inhibitory neurotransmitter γ-aminobutyric acid (GABA) (Fig. 16.3) designed to facilitate diffusion across the blood–brain barrier. Clinical studies demonstrated their effectiveness in the treatment of epilepsy (Taylor et al. 2007). Although the functional effects of gabapentin were attributed to activation of GABA-B receptors, direct evidence of such interaction has not been ascertained, and gabapentin does not mimic GABA physiologically. It was found that both gabapentin (Neurontin, Grimodin, Gabagamma) and pregabalin (Lyrica) bind with high affinity to

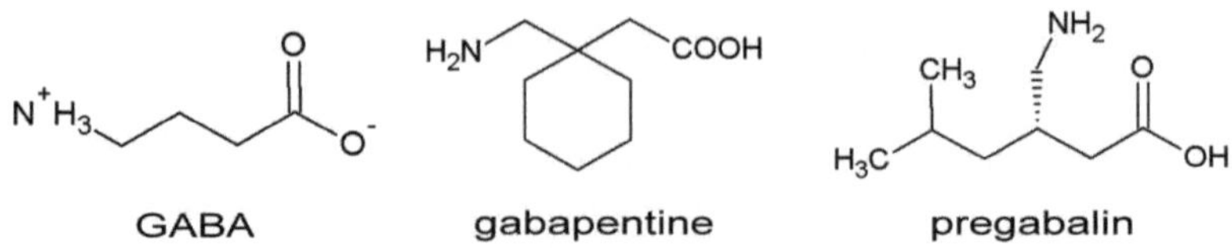

Fig. 16.3 Structural formulae of γ-amino butyric acid (GABA) and its synthetic analogues, gabapentin and pregabalin

the $\alpha_2\delta$-1 and $\alpha_2\delta$-2 subunits of VGCC. These drugs suppress neuronal excitability due to inhibition of presynaptic calcium entry through VGCC, thus attenuating neurotransmitter release. The decrease in voltage-dependent calcium current is due to disruption of the passage of $\alpha_2\delta$ subunits into the plasma membrane which causes a reduction in the total number of $\alpha_2\delta$ and α_1 subunits of VGCC, predominantly $Ca_v2.1$ and $Ca_v2.2$, at the cell surface (Bauer et al. 2010; Tran-Van-Minh and Dolphin 2010). Therefore, although no acute effects on current amplitude are observable in vitro, current inhibition was apparent following long-term application.

Interaction of both gabapentin and pregabalin with the $\alpha_2\delta$-1 subunit is essential for their therapeutic effect in the treatment of neuropathic pain. It is tempting to hypothesize that their antiepileptic action is mediated by an analogous interaction with the $\alpha_2\delta$-2 subunit, however, this mechanism remains to be fully investigated.

Both drugs have a favorable adverse effect profile with no negative impact on cardiac function and few interactions with other drugs. They are negligibly metabolized in the liver (<2 %) and do not interact with hepatic enzymes such as the cytochrome P450 system (Ben-Menachem 2004). Therefore, they can be considered the first choice therapy for neuropathic pain; specifically for treatment of painful diabetic neuropathy, post-herpetic neuralgia, and spinal cord injury.

16.4 T-Type Calcium Channels

T-type calcium channels (TTCC), with their lower voltage activation threshold, contribute to cellular excitability differently to HVA channels. Disrupted regulation of $Ca_v3.x$ channels forms part of several neuronal and cardiovascular diseases. The TTCC or $Ca_v3.x$ or LVA calcium channels were the last to be discovered and last cloned (Perez-Reyes 2003), and concurrent lack of detailed knowledge of the structure and function of these channels delayed the development of drugs specifically targeted at TTCC. When three $Ca_v3.x$ channels were cloned, a number of drugs used in the treatment of cardiovascular and/or neuronal diseases mainly as LTCC or sodium channel blockers were shown to also inhibit TTCC. Apart from mibefradil no drug aimed as a specific T-type current blocker has been developed, nevertheless TTCC block should be considered a contributing factor to the action of many drugs currently used in clinics.

16.4.1 Antispasmodic Drugs

Antispamodic drugs are used clinically to treat intestinal disorders manifested by increased contractility. While calcium entry through LTCC initiates gastrointestinal contraction, the propagation of intestinal slow wave can be mediated by LVA calcium channels. At least one antispasmodic agent, otilonium bromide, was shown to inhibit LVA channels in clinically relevant concentration (Strege et al. 2010). Part of the mechanism by which these drugs exert their effect can be attributed to blocking Ca_v3 channels, which attenuates propagation of the intestinal slow wave. The other part of this mechanism is inhibition of LTCC which hampers contraction initiation (Gallego et al. 2010; Strege et al. 2004).

16.4.2 Dihydropyridines

Dihydropyridines were originally accepted as specific LTCC blockers. This was reflected in the naming of LTCC "Dihydropyridine receptors." However, some DHPs are also relatively effective Ca_v3 channel blockers (Lacinova 2011). These blockers may be beneficial in patients treated for hypertension. Vascular contractility depends on both LTCC and TTCC, but while vascular endothelium expresses TTCC, it lacks LTCC (Wu et al. 2003). Therefore, efonidipine (Landel), a blocker of both TTCC and LTCC, was able to improve vascular endothelial dysfunction in patients with essential hypertension, whereas the LTCC blocker nifedipine was ineffective in this respect (Oshima et al. 2005). Efonidipine was originally considered to be the first DHP blocker specific for T-type calcium channels, but according to more recent reports it is an equally effective blocker of both LTCC and TTCC (Lacinova 2011).

Another Dihydropyridine benidipine (Coniel), which also blocks both LTCC and TTCC, had a similar beneficial effect on vascular endothelium (Miwa et al. 2009), and benidipine hydrochloride also appears the most preferable antihypertensive for patients with diabetes. Here, it is more effective in lowering blood pressure than amlodipine or cilnidipine and it also has an additional renal-protective effect (Seino et al. 2007; Ohishi et al. 2007). Although benidipine registered a two decimal order greater effectiveness in blocking the $Ca_v3.2$ channel than the $Ca_v1.2$ channel, these in vitro results should be taken with caution as they were obtained in different recombinant systems (Lacinova 2011).

16.4.3 Antiepileptics

Due to their negative activation, voltage T-type calcium channels are a key mechanism in the generation of neuronal burst-firing. Abnormally increased firing in a group of neurons in the central nervous system is related to various epileptiform activities. Although no antiepileptic drug dependent on specific TTCC blocking action has been developed so far, inhibition of T-type calcium current at least partly underlies the therapeutic mechanism of several drugs used to combat epileptiform seizures.

Zonisamide (Zonegran) is a synthetic 1,2-benzisoxazole-3-methanesulfonamide developed as an antiepileptic drug (Masuda et al. 1980). It is effective in patients resistant to other antiepileptic drugs in the treatment of both generalized and partial seizures. Since it has a long plasma elimination half-life, twice-daily, or even once-daily dosing is effective. Zonisamide was expected to suppress seizures through inhibition of carbonic anhydrase, as does the sulfonamide analogue acetazolamide. However, this mechanism was excluded in in vitro experiments, and inhibition of a T-type calcium current together with inhibition of the voltage-gated sodium current was suggested as the primary mechanism action (Leppik 2004). Blocking TTCC may also be a primary mechanism underlying the anti-tremor efficacy of zonisamide in the treatment of essential tremor and the tremor associated with Parkinson's disease. Although zonisamide is also used in prevention of migraine, this action is most likely independent of its ability to block T-type calcium channels (Bermejo and Dorado 2009).

Zonisamide shares its pharmacological properties with the other clinically used antiepileptics phenytoin, carbamazepine, and sodium valproate. Although these drugs were introduced as inhibitors of voltage-gated sodium channels, they may actually act as T-type calcium channel blockers. Phenytoin (Dilantin) is commonly used in the treatment of epilepsy, in the management of post-stroke seizures and it can also be used in prevention of seizures following brain injury. Its half-maximal inhibitory concentration of the $Ca_v3.1$ channel in vitro is between 70 and 140 μM and, since the therapeutic plasma concentration is approximately 80 μM, this interaction may significantly contribute to phenytoin's therapeutic action (Lacinova 2004). Other antiepileptics such as carbamazepine, valproate, ethosuximide, lamotrigine, and leviteracetam in therapeutically relevant concentrations only inhibit TTCC negligibly.

16.4.4 Neuroleptics and Anesthetics

Many clinically used neuroleptics are potent blockers of T-type VGCC. Two diphenylbutylpiperidines, pimozide (Orap) and penfluridol (Semap, Micefal), are particularly potent blockers of recombinant T-channels with estimated IC_{50} values of approximately 30–50 and 70–100 nM, respectively (Santi et al. 2002). Since this value is close to their therapeutic plasma concentration, inhibition of the T-type calcium current may form a part of the mechanism underlying their therapeutic action. Pimozide is used in the treatment of schizophrenia, chronic psychosis, and tics in Tourette syndrome, while penfluoridol is used in the treatment of schizophrenia. Therapeutic plasma concentrations of the antipsychotics haloperidol and flunarizine are also sufficient to inhibit T-type calcium current, and the general anesthetics, propofol, octanol, and isofluran block TTCC in clinically relevant concentrations (Lacinova 2004).

16.4.5 Mibefradil

Mibefradil (Fig. 16.4) was the first member of the new class of tetralol derivative calcium antagonists. These are presumed to selectively block T-type VGCC and they are expected to establish a new paradigm in the treatment of cardiovascular disorders. T-type calcium channels are expressed in vascular smooth muscle and in the myocardial conduction system, but not in ventricles. Therefore, a selective T-type channel blocker should be able to relax vascular muscle and slow the heart beat without decreasing heart contractility. Mibefradil proved more efficient than nifedipine in treating moderate to severe hypertension (Lacourciere et al. 1997), and it was also effective in treating angina pectoris (Frishman et al. 1998; Charlon and Kobrin 1998). It was more effective than amlodipine and comparable to diltiazem as an anti-anginal and anti-ischaemic drug (Kobrin et al. 1998; Davies et al. 1997). Nevertheless, the in vitro selectivity of mibefradil for T-type over the L-type VGCC was only 10- to 30-fold (Bezprozvanny and Tsien 1995). While mibefradil was well tolerated by patients in initial clinical studies, it was withdrawn just 1 year after its introduction to the market in 1997 because of multiple drug interactions related to its inhibition of the cytochrome P450 system (SoRelle 1998). More recently, it has been suggested that mibefradil's antihypertensive and vasodilatory effects are mediated by inhibition of L-type VGCC (Moosmang et al. 2006). Mibefradil is still the subject of clinical trials where it demonstrates beneficial effects compared to LTCC channel blockers due to its lack of negative chronotropic effects (Lindqvist et al. 2007). Although the role of T-type VGCC in regulating blood pressure remains controversial, research into selective T-type channel blockers with potential use in hypertension treatment still continues (Bui et al. 2008).

Fig. 16.4 Structural formula of mibefradil

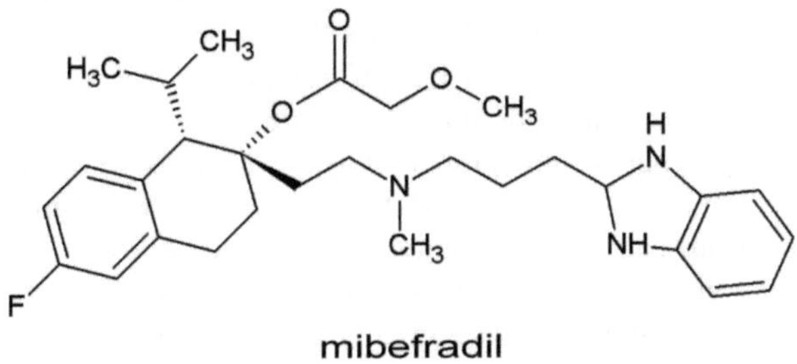

16.5 Calcium Channel Blockers in the Treatment of Heritable Channelopathies

Most chapters in the Part I of this book are devoted to inherited mutations in individual VGCC involved in development of various muscular and neuronal pathologies and are described in a detail in preceding chapters. Effects of some of these mutations can be equivocally classified as "gain of function." To such mutation belong mutation in the CACNA1S gene participates in pathology of hypokalemic periodic paralysis (see Chap. 7), mutations in the CACNA1C gene causing Timothy syndrome (see Chap. 11), or mutations in the CACNA1F gene leading to the incomplete congenital stationary night blindness type 2 (see Chap. 3). Currently known mutations in the CACNA1A gene are both of "gain of function" leading to familial hemiplegic migraine (see Chap. 1) or "loss of function" leading to episodic ataxia type 2 (see Chap. 2).

Precise mechanism leading to symptoms of VGCC channelopathies is subject of ongoing investigations. While no channels-specific therapy is available at the moment, certainly they represent attractive target for future pharmaceuticals.

16.6 Conclusion

While L-type calcium channels retain central focus in hypertension treatment, there is growing evidence for the possible role of T-type calcium channels in both vascular endothelium and vascular smooth muscle cells. Since clinically used LTCC blockers currently fail to prevent vasoconstriction of a number of small blood vessels and fail to reverse cerebral vasospasm, there is room for development of more specific T-type calcium channel blockers. The continued testing of mibefradil long after it was retracted from clinical use suggests that development of specific T-type calcium channel blockers as therapy in cardiovascular pathologies may be a hope for the future.

Since disrupted regulation of VGCC has been implied in neurodegenerative diseases such as Parkinson's and Alzheimer's, targeted regulation of VGCC could play a successful role in future therapy. A population-based Danish study showed that intake of DHPs that cross the blood–brain barrier is associated with approximately 30 % reduced risk of contracting Parkinson's disease (Ritz et al. 2010). This is such a surprisingly strong association that blood–brain barrier-permeable L-type channel antagonists may well represent a viable neuroprotective strategy.

Neuronal ($Ca_v2.x$) and T-type ($Ca_v3.x$) calcium channels play a role in nociception. Current treatments for acute and chronic pain lack sufficient efficacy and they are associated with serious side effects. Therefore, new selective and potent channel inhibitors may greatly benefit patients. Meta-analysis of published literature has revealed almost 100 conotoxins with therapeutic potential in chronic pain and neurodegenerative disorders (Essack et al. 2012). The $Ca_v3.2$ channel presents a further prospective target for specific analgesics in chronic pain treatment (Todorovic and Jevtovic-Todorovic 2011).

Activity of T-type calcium channels is linked to conditions such as autism, psychosis, cancer, and fertility, and the anti-proliferative efficacy of T-type channel blockers has already been demonstrated in vitro (Lee et al. 2006). In conclusion, while calcium channel antagonists related to the cardiovascular system have dominated clinical practice up until now, new areas for their application are emerging.

References

Bauer CS, Rahman W, Tran-van-Minh A, Lujan R, Dickenson AH, Dolphin AC (2010) The anti-allodynic $\alpha_2\delta$ ligand pregabalin inhibits the trafficking of the calcium channel $\alpha_2\delta$-1 subunit to presynaptic terminals in vivo. Biochem Soc Trans 38(2):525–528

Ben-Menachem E (2004) Pregabalin pharmacology and its relevance to clinical practice. Epilepsia 45(Suppl 6):13–18

Bergson P, Lipkind G, Lee SP, Duban ME, Hanck DA (2011) Verapamil block of T-type calcium channels. Mol Pharmacol 79(3):411–419

Bermejo PE, Dorado R (2009) Zonisamide for migraine prophylaxis in patients refractory to topiramate. Clin Neuropharmacol 32(2):103–106

Bezprozvanny I, Tsien RW (1995) Voltage-dependent blockade of diverse types of voltage-gated Ca_{2+} channels expressed in Xenopus oocytes by the Ca_{2+} channel antagonist mibefradil (Ro 40-5967). Mol Pharmacol 48(3):540–549

Bui PH, Quesada A, Handforth A, Hankinson O (2008) The mibefradil derivative NNC55-0396, a specific T-type calcium channel antagonist, exhibits less CYP3A4 inhibition than mibefradil. Drug Metab Dispos 36(7):1291–1299

Caballero R, Gomez R, Nunez L, Moreno I, Tamargo J, Delpon E (2004) Diltiazem inhibits hKv1.5 and Kv4.3 currents at therapeutic concentrations. Cardiovasc Res 64(3):457–466

Catacuzzeno L, Trequattrini C, Petris A, Franciolini F (1999) Mechanism of verapamil block of a neuronal delayed rectifier K channel: active form of the blocker and location of its binding domain. Br J Pharmacol 126(8):1699–1706

Charlon V, Kobrin I (1998) The efficacy and safety of mibefradil in subgroups of patients with chronic stable angina pectoris. Int J Clin Pract 52(4):257–264

Comunanza V, Marcantoni A, Vandael DH, Mahapatra S, Gavello D, Carabelli V, Carbone E (2010) $Ca_v1.3$ as pacemaker channels in adrenal chromaffin cells: specific role on exo- and endocytosis? Channels (Austin) 4 (6):440−446

Davies GJ, Kobrin I, Caspi A, Reisin LH, de Albuquerque DC, Armagnijan D, Coelho OR, Schneeweiss A (1997) Long-term antianginal and antiischemic effects of mibefradil, the novel T-type calcium channel blocker: a multicenter, double-blind, placebo-controlled, randomized comparison with sustained-release diltiazem. Am Heart J 134(2 Pt 1):220–228

Elliott WJ, Ram CV (2011) Calcium channel blockers. J Clin Hypertens (Greenwich) 13(9):687–689

Essack M, Bajic VB, Archer JA (2012) Conotoxins that confer therapeutic possibilities. Mar Drugs 10(6):1244–1265

Fischhof PK (1993) Divergent neuroprotective effects of nimodipine in PDD and MID provide indirect evidence of disturbances in Ca_{2+} homeostasis in dementia. Methods Find Exp Clin Pharmacol 15(8):549–555

Fleckenstein A (1983) History of calcium antagonists. Circ Res 52(2 Pt 2):I3–I16

Frishman WH, Bittar N, Glasser S, Habib G, Smith W, Pordy R (1998) Additional antianginal and anti-ischemic efficacy of mibefradil in patients concomitantly treated with long-acting nitrates for chronic stable angina pectoris. Clin Cardiol 21(7):483–490

Gallego D, Auli M, Aleu J, Martinez E, Rofes L, Marti-Rague J, Jimenez M, Clave P (2010) Effect of otilonium bromide on contractile patterns in the human sigmoid colon. Neurogastroenterol Motil 22(6):e180–e191

Gao Z, Sun H, Chiu SW, Lau CP, Li GR (2005) Effects of diltiazem and nifedipine on transient outward and ultra-rapid delayed rectifier potassium currents in human atrial myocytes. Br J Pharmacol 144(4):595–604

Godfraind T, Salomone S, Dessy C, Verhelst B, Dion R, Schoevaerts JC (1992) Selectivity scale of calcium antagonists in the human cardiovascular system based on in vitro studies. J Cardiovasc Pharmacol 20(Suppl 5):S34–S41

Hanyu H, Hirao K, Shimizu S, Sato T, Kiuchi A, Iwamoto T (2007) Nilvadipine prevents cognitive decline of patients with mild cognitive impairment. Int J Geriatr Psychiatry 22(12):1264–1266

Kato M, Dote K, Sasaki S, Takemoto H, Habara S, Hasegawa D (2004) Intracoronary verapamil rapidly terminates reperfusion tachyarrhythmias in acute myocardial infarction. Chest 126(3):702–708

Kawano Y, Makino Y, Okuda N, Takishita S, Omae T (2000) Effects of diltiazem retard on ambulatory blood pressure and heart rate variability in patients with essential hypertension. Blood Press Monit 5(3):181–185

Kobrin I, Bieska G, Charlon V, Lindberg E, Pordy R (1998) Anti-anginal and anti-ischemic effects of mibefradil, a new T-type calcium channel antagonist. Cardiology 89(Suppl 1):23–32

Lacinova L (2004) Pharmacology of recombinant low-voltage activated calcium channels. Curr Drug Targets CNS Neurol Disord 3(2):105–111

Lacinova L (2005) Voltage-dependent calcium channels. Gen Physiol Biophys 24(Suppl 1):1–78

Lacinova L (2011) T-type calcium channel blockers—new and notable. Gen Physiol Biophys 30(4):403–409

Lacourciere Y, Poirier L, Lefebvre J, Archambault F, Dalle Ave S, Ward C, Lindberg E (1997) The antihypertensive efficacy of the novel calcium antagonist mibefradil in comparison with nifedipine GITS in moderate to severe hypertensives with ambulatory hypertension. Am J Hypertens 10(2):189–196

Lee JY, Park SJ, Lee MJ, Rhim H, Seo SH, Kim KS (2006) Growth inhibition of human cancer cells in vitro by T-type calcium channel blockers. Bioorg Med Chem Lett 16(19):5014–5017

Leppik IE (2004) Zonisamide: chemistry, mechanism of action, and pharmacokinetics. Seizure 13 Suppl 1:S5−9; (discussion S10)

Lindqvist M, Kahan T, Melcher A, Ekholm M, Hjemdahl P (2007) Long-term calcium antagonist treatment of human hypertension with mibefradil or amlodipine increases sympathetic nerve activity. J Hypertens 25(1):169–175

Mangoni ME, Couette B, Bourinet E, Platzer J, Reimer D, Striessnig J, Nargeot J (2003) Functional role of L-type $Ca_v1.3$ Ca_{2+} channels in cardiac pacemaker activity. Proc Natl Acad Sci U S A 100(9):5543–5548

Marger L, Mesirca P, Alig J, Torrente A, Dubel S, Engeland B, Kanani S, Fontanaud P, Striessnig J, Shin HS, Isbrandt D, Ehmke H, Nargeot J, Mangoni ME (2011) Pacemaker activity and ionic currents in mouse atrioventricular node cells. Channels (Austin) 5(3):241–250

Masuda Y, Karasawa T, Shiraishi Y, Hori M, Yoshida K, Shimizu M (1980) 3-Sulfamoylmethyl-1,2-benzisoxazole, a new type of anticonvulsant drug. Pharmacological profile. Arzneimittelforschung 30(3):477–483

Michna M, Knirsch M, Hoda JC, Muenkner S, Langer P, Platzer J, Striessnig J, Engel J (2003) $Ca_v1.3$ (α1D) Ca_{2+} currents in neonatal outer hair cells of mice. J Physiol 553 (Pt 3):747−758

Miwa Y, Masai H, Shimizu M (2009) Differential effects of calcium-channel blockers on vascular endothelial function in patients with coronary spastic angina. Circ J 73(4):713–717

Moosmang S, Haider N, Bruderl B, Welling A, Hofmann F (2006) Antihypertensive effects of the putative T-type calcium channel antagonist mibefradil are mediated by the L-type calcium channel $Ca_v1.2$. Circ Res 98(1):105–110

Navia BA, Dafni U, Simpson D, Tucker T, Singer E, McArthur JC, Yiannoutsos C, Zaborski L, Lipton SA (1998) A phase I/II trial of nimodipine for HIV-related neurologic complications. Neurology 51(1):221–228

Ohishi M, Takagi T, Ito N, Terai M, Tatara Y, Hayashi N, Shiota A, Katsuya T, Rakugi H, Ogihara T (2007) Renal-protective effect of T-and L-type calcium channel blockers in hypertensive patients: an amlodipine-to-benidipine changeover (ABC) study. Hypertens Res 30(9):797–806

Oshima T, Ozono R, Yano Y, Higashi Y, Teragawa H, Miho N, Ishida T, Ishida M, Yoshizumi M, Kambe M (2005) Beneficial effect of T-type calcium channel blockers on endothelial function in patients with essential hypertension. Hypertens Res 28(11):889–894

Perez-Reyes E (2003) Molecular physiology of low-voltage-activated t-type calcium channels. Physiol Rev 83(1):117–161

Raffaeli W, Righetti D, Sarti D, Balestri M, Ferioli I, Monterubbianesi MC, Caminiti A (2011) Ziconotide: a rapid detoxification protocol for the conversion from intrathecal morphine–the raffaeli detoxification model. J Opioid Manag 7(1):21–26

Reuter H (1967) The dependence of slow inward current in Purkinje fibres on the extracellular calcium-concentration. J Physiol 192(2):479–492

Reuter H, Beeler GW Jr (1969) Calcium current and activation of contraction in ventricular myocardial fibers. Science 163(3865):399–401

Ritz B, Rhodes SL, Qian L, Schernhammer E, Olsen JH, Friis S (2010) L-type calcium channel blockers and Parkinson disease in Denmark. Ann Neurol 67(5):600–606

Santi CM, Cayabyab FS, Sutton KG, McRory JE, Mezeyova J, Hamming KS, Parker D, Stea A, Snutch TP (2002) Differential inhibition of T-type calcium channels by neuroleptics. J Neurosci 22(2):396–403

Seino H, Miyaguchi S, Yamazaki T, Ota S, Yabe R, Suzuki S (2007) Effect of benidipine hydrochloride, a long-acting T-type calcium channel blocker, on blood pressure and renal function in hypertensive patients with diabetes mellitus. Analysis after switching from cilnidipine to benidipine. Arzneimittelforschung 57(8):526–531

Smith HS, Deer TR (2009) Safety and efficacy of intrathecal ziconotide in the management of severe chronic pain. Ther Clin Risk Manag 5(3):521–534

SoRelle R (1998) Withdrawal of Posicor from market. Circulation 98(9):831–832

Staats PS, Yearwood T, Charapata SG, Presley RW, Wallace MS, Byas-Smith M, Fisher R, Bryce DA, Mangieri EA, Luther RR, Mayo M, McGuire D, Ellis D (2004) Intrathecal ziconotide in the treatment of refractory pain in patients with cancer or AIDS: a randomized controlled trial. Jama 291(1):63–70

Strege PR, Evangelista S, Lyford GL, Sarr MG, Farrugia G (2004) Otilonium bromide inhibits calcium entry through L-type calcium channels in human intestinal smooth muscle. Neurogastroenterol Motil 16(2):167–173

Strege PR, Sha L, Beyder A, Bernard CE, Perez-Reyes E, Evangelista S, Gibbons SJ, Szurszewski JH, Farrugia G (2010) T-type Ca_{2+} channel modulation by otilonium bromide. Am J Physiol Gastrointest Liver Physiol 298(5):G706–G713

Tanabe T, Takeshima H, Mikami A, Flockerzi V, Takahashi H, Kangawa K, Kojima M, Matsuo H, Hirose T, Numa S (1987) Primary structure of the receptor for calcium channel blockers from skeletal muscle. Nature 328(6128):313–318

Tarabova B, Lacinova L, Engel J (2007) Effects of phenylalkylamines and benzothiazepines on Ca(v)1.3-mediated Ca_{2+} currents in neonatal mouse inner hair cells. Eur J Pharmacol 573(1–3):39–48

Taylor CP, Angelotti T, Fauman E (2007) Pharmacology and mechanism of action of pregabalin: the calcium channel α_2-δ (α_2-δ) subunit as a target for antiepileptic drug discovery. Epilepsy Res 73(2):137–150

Todorovic SM, Jevtovic-Todorovic V (2011) T-type voltage-gated calcium channels as targets for the development of novel pain therapies. Br J Pharmacol 163(3):484–495

Tran-Van-Minh A, Dolphin AC (2010) The $\alpha_2\delta$ ligand gabapentin inhibits the Rab11-dependent recycling of the calcium channel subunit $\alpha_2\delta$-2. J Neurosci 30(38):12856–12867

Welling A, Ludwig A, Zimmer S, Klugbauer N, Flockerzi V, Hofmann F (1997) Alternatively spliced IS6 segments of the α 1C gene determine the tissue-specific Dihydropyridine sensitivity of cardiac and vascular smooth muscle L-type Ca_{2+} channels. Circ Res 81(4):526–532

Wu S, Haynes J, Jr, Taylor JT, Obiako BO, Stubbs JR, Li M, Stevens T (2003) $Ca_v3.1$ (α1G) T-type Ca_{2+} channels mediate vaso-occlusion of sickled erythrocytes in lung microcirculation. Circ Res 93 (4):346−353

Zhang S, Zhou Z, Gong Q, Makielski JC, January CT (1999) Mechanism of block and identification of the verapamil binding domain to HERG potassium channels. Circ Res 84(9):989–998

Zuccotti A, Clementi S, Reinbothe T, Torrente A, Vandael DH, Pirone A (2011) Structural and functional differences between L-type calcium channels: crucial issues for future selective targeting. Trends Pharmacol Sci 32(6):366–375

Part II
Pathologies of Transient Receptor Potential Channels

Chapter 17
TRPM1 and Congenital Stationary Night Blindness

Shoichi Irie and Takahisa Furukawa

Abstract The transient receptor potential (TRP) channels play a wide variety of essential roles in the sensory systems of various species, both invertebrates and vertebrates. The TRP channel was first identified as a molecule required for proper light response in *Drosophila melanogaster*. We and another group recently revealed that TRPM1, the founding member of the melanoma-related transient receptor potential (TRPM) subfamily, is required for the photoresponse in mouse retinal ON bipolar cells. We further demonstrated that TRPM1 is a component of the transduction cation channel negatively regulated by the metabotropic glutamate receptor 6 (mGulR6) cascade in ON bipolar cells through a reconstitution experiment using CHO cells expressing TRPM1, mGluR6, and Goα. Furthermore, human TRPM1 mutations are associated with congenital stationary night blindness (CSNB), whose patients lack rod function and suffer from night blindness starting in early childhood. In this chapter, we describe the physiological functions and regulatory mechanisms of the TRPM1 channel in retinal ON bipolar cells, and the association of human TRPM1 mutations with CSNB.

17.1 TRP Channels

The transient receptor potential channel (TRP) gene family, which encodes non-voltage-gated Ca^{2+}-permeable cation channels, crucially and broadly participates in sensory reception in a variety of living organisms. Based on the analysis of

S. Irie · T. Furukawa
Laboratory for Molecular and Developmental Biology, Institute for Protein Research, Osaka University, Osaka, Japan

S. Irie · T. Furukawa (✉)
JST, CREST, 3–2 Yamadaoka, Suita, Osaka 565–0871, Japan
e-mail: takahisa.furukawa@protein.osaka-u.ac.jp

N. Weiss and A. Koschak (eds.), *Pathologies of Calcium Channels*,
DOI: 10.1007/978-3-642-40282-1_17,

amino acid sequence homology, TRP superfamily members are categorized into seven subfamilies; TRPC, TRPM, TRPV, TRPA, TRPP, TRPML, and TRPN (Clapham 2003; Corey 2003; Delmas 2004; Minke 2010; Montell et al. 2002), although TRPN family genes have not yet been identified in mammals. TRP channel proteins share a common structure of six predicted transmembrane domains and a reentrant P-loop, which is involved in forming the pore of the channel, as it does in voltage-gated ion channels (Morgans et al. 2010). In addition to these features, a highly conserved 23–25 amino acid residue TRP domain containing an 'EWKFAR' TRP-box is located near the sixth transmembrane domain in the cytoplasmic region of TRPC, TRPM, and TRPN channels (Morgans et al. 2010; Venkatachalam and Montell 2007). A large body of research has shown that TRP channels function mainly in sensing activities, including taste, olfaction, hearing, and touch, in addition to thermosensation and osmosensation. In addition, some TRP channels function as effector channels in neurons. TRPC3 increases the firing rate of GABAergic neurons in mouse substantia nigra pars reticulate (SNr) upon dopamine or serotonin stimulation (Zhou and Lee 2011; Zhou et al. 2009). The TRPM2 channel increases burst firing upon glutamatergic input and modulates burst activity by reactive oxygen H_2O_2 in SNr GABAergic neurons (Lee et al. 2013).

A TRP channel was first cloned in *Drosophila melanogaster* as a necessary molecule for photoreceptor response to light (Montell et al. 1985). However, whether or not a TRP channel plays a crucial role in vertebrate vision had not been determined until a recent finding that TRPM1 is a transduction channel negatively regulated by the mGluR6 pathway in retinal ON bipolar cells (Koike et al. 2010b).

17.2 TRPM Family Genes

TRPM family genes are well conserved through evolution both in invertebrates and vertebrates, and encode non-selective cation channels involved in many biological functions. There are eight TRPM genes in human and mouse, four genes in *C. elegans*, and a single gene in *Drosophila melanogaster*. TRPM channels have intracellular N- and C-terminals, six transmembrane regions, and a hydrophobic pore loop between transmembrane regions five and six. This family possesses a TRPM family homology region of about 700 amino acid residues along with a CC domain in the N-terminal region, and a TRP-box and CC domain in the C-terminal region. The CC domain in the C-terminal region is involved in the assembly of subunits and the regulation of cellular localization of channel proteins (Erler et al. 2006; Fujiwara and Minor 2008; Tsuruda et al. 2006). Phylogenetic analysis of TRPM channels classified the TRPM family into four groups: TRPM1 and TRPM3, TRPM6 and TRPM7, TRPM4 and TRPM5, TRPM2 and TRPM8 (Harteneck 2005). TRPM1 is expressed in skin melanocytes, but its expression level is reduced in melanoma cells, suggesting a possible role for TRPM1 in body color formation and melanoma metastasis (Duncan et al. 1998; Bellone et al.

2008). TRPM2 is a redox status-dependent Ca^{2+}-permeable cation channel whose intrinsic sensitivity to an activation trigger, β-nicotinamide adenine dinucleotide, may underlie susceptibility to H_2O_2-induced cell death (Hara et al. 2002). TRPM2 can be activated by co-application of warm temperatures (>35 °C). Intracellular cyclic ADP-ribose dramatically potentiates TRPM2 activity. In pancreatic islets, TRPM2 is co-expressed with insulin, and mild heating of these cells increases both Ca^{2+} entry and insulin release (Perraud et al. 2001; Togashi et al. 2006). TRPM3 plays an essential role as a component of ionotropic steroid receptors in pancreatic beta cells (Wagner et al. 2008). TRPM4 is involved in the action potential generation in excitatory neurons. Absence of TRPM4 function can cause cerebral edema and arrhythmia (Simard et al. 2006; Guinamard et al. 2004). In non-excitatory neurons, TRPM4 is activated by the elevation of the intracellular Ca^{2+} concentration. Furthermore, TRPM4 is involved in the Ca^{2+} oscillation in T-lymphocytes and the migration of mast cells (Nilius et al. 2003; Hofmann et al. 2003; Launay et al. 2004; Vennekens et al. 2007). TRPM5 is activated by warm temperature (15–35 °C) and plays a role in taste sensation (Talavera et al. 2005; Zhang et al. 2003). TRPM5 is expressed in taste bud cells for detection of sweetness, bitterness, and umami. Mutations of human TRPM6 cause a defect of Mg^{2+} absorption and result in hypomagnesemia, indicating that TRPM6 is a Mg^{2+} absorber (Walder et al. 2002; Schlingmann et al. 2002). TRPM7 is activated by various stimuli, including MgATP concentration, reactive oxygen, mechanical stimulation, and pH. TRPM7 appears to play a role in the regulation of cell volume, cell adhesion, and cell growth (Numata et al. 2007; Su et al. 2006; Hanano et al. 2004). TRPM8 functions as a cold sensor. TRPM8 is activated by low temperature (<25 °C), menthol, incilin, and frescolar ML (Peier et al. 2002; Bautista et al. 2007; McKemy et al. 2002).

17.3 Identification of TRPM1 Gene

TRPM1, also known as melastatin, was the first cloned member of the TRPM subfamily (Duncan et al. 1998). A mouse *melastatin* cDNA, later named *TRPM1-S*, was differentially expressed between poorly and highly metastatic B16 melanoma cell lines (Duncan et al. 1998). *Melastatin* expression was correlated with pigmentation in melanoma cell lines, while reversely correlated with the metastatic potential of melanoma cells (Duncan et al. 1998; Fang and Setaluri 2000). *TRPM1* transcript is alternatively spliced to produce a long form (*TRPM1-L*) or a short N-terminal form devoid of transmembrane segments (*TRPM1-S*) in humans (Duncan et al. 1998; Hunter et al. 1998; Fang and Setaluri 2000; Xu et al. 2001). It was reported that TRPM1-S interacts with TRPM1-L to suppress TRPM1-L channel activity by inhibiting its translocation to the plasma membrane (Xu et al. 2001). Although mouse *TRPM1-S* was previously identified as *melastatin*, mouse *TRPM1-L* had not been identified until recently (Duncan et al. 1998).

17.4 Transduction Channel of Retinal on Bipolar Cells

Functional separation of neuronal signaling into the ON and OFF pathways that contribute to visual contrast recognition is a fundamental feature of the vertebrate visual system. Photoreceptors transmit signals to depolarizing ON bipolar cells and hyperpolarizing OFF cells by glutamate release, which is reduced by light-evoked hyperpolarization (DeVries and Baylor 1993; Dowling 1978). In mammalian retinas, rod photoreceptors form synapses with ON type rod bipolar cells, and cone photoreceptors connect with cone bipolar cells, which are subdivided into ON and OFF types. The expression pattern of different glutamate receptors (GluRs) at the postsynaptic region underlies the functional diversity of bipolar cells (Nakanishi et al. 1998). ON bipolar cells express a metabotropic GluR, mGluR6, on dendrites. The ON bipolar dendrite forms an invaginating single contact with a photoreceptor terminal. On the other hand, OFF bipolar cells express ionotropic GluRs (AMPA/Kainate receptors), glutamate-gated cation channels, on dendrites. The OFF bipolar dendrite makes multiple flat contacts with a photoreceptor terminal (Morigiwa and Vardi 1999; Haverkamp et al. 2001; Nomura et al. 1994). In the dark, the neurotransmitter glutamate is released at a high rate from rod terminals and depolarizes OFF bipolar cells through activation of ionotropic glutamate receptor, whereas glutamate hyperpolarizes ON bipolar cells through mGluR6 activation leading to the closure of cation channels and a decrease in cationic conductance (Euler et al. 1996; Masu et al. 1995; de la Villa et al. 1995; Slaughter and Miller 1981; Shiells et al. 1981). The transduction cation channel of retinal ON bipolars was hypothesized to be a cGMP-gated cation channel that is closed by increasing the rate of cGMP hydrolysis through phosphodiesterases (PDEs) through a G-protein-mediated process, based on analogous light transduction machinery in photoreceptors. Several studies reported that the response of the ON pathway is mediated by turning off the G-protein cascade that is activated in the dark when glutamate binds to mGluR6 (Nomura et al. 1994; Vardi and Morigiwa 1997; Vardi et al. 2000). In ON bipolar cells, mGluR6 couples to a heterotrimeric G-protein complex Go (Dhingra et al. 2000, 2002; Nawy 1999; Vardi 1998; Weng et al. 1997). Signals require Goα, which closes a downstream cation channel upon the activation of ON bipolar cells (Shiells and Falk 1990, 1992a, b, c; Dhingra et al. 2000, 2002; Nawy 1999; Weng et al. 1997; Nawy and Jahr 1990; Euler et al. 1996; de la Villa et al. 1995). From these results, it appears that Goα is required for the closure of the cation channel, but there is no evidence that PDEs are involved in the signal transduction of ON bipolar cells (Nawy 1999). Although a regulator of G-protein signaling (RGS), Ret-RGS1, which interacts with Goα in transfected cells and the retina, was identified (Dhingra et al. 2004), a transduction cation channel in the downstream of the mGluR6 pathway in retinal bipolar cells remains elusive.

17.5 The Function of the TRPM1 Channel in the Retina

A mouse *TRPM1-L* cDNA that corresponds to the human *TRPM1* long form was identified as being a highly expressed gene in the retina (GenBank Accession Number #AY180104) (Koike et al. 2010b). Northern blot analysis revealed the presence of both *TRPM1-L* and *-S* transcripts in the retina; however, only the latter was detected in the skin (Koike et al. 2010b). *In situ* hybridization (ISH) detected the presence of substantial *TRPM1-L* transcripts, specifically in the inner nuclear layer (INL), at postnatal stages (Koike et al. 2010b). An antibody against TRPM1-L was raised and the localization of TRPM1-L was examined. At postnatal day 14 (P14), TRPM1-L proteins were observed diffusely in bipolar cell somata, and at 1 month after birth (1 M), the proteins localized at the tips of Goα-expressing and mGluR6-expressing dendrites in the outer plexiform layer (OPL) (Koike et al. 2010b). Koike et al. generated *TRPM1* null mutant ($TRPM1^{-/-}$) mice by targeted gene disruption, and revealed that neither their rod bipolar cells nor their cone ON bipolar cells showed photoresponses by using whole-cell patch-clamp analysis (Koike et al. 2010b). On the other hand, light stimulation on both wild-type (WT) and $TRPM1^{-/-}$ mice cone OFF bipolar cells evoked photoresponses, and no significant differences were detected in either the amplitude of the light responses or the time for half-maximal amplitude after the light was turned off (Koike et al. 2010b). Examination of the optokinetic responses (OKRs) and electroretinograms (ERGs) of 2-month-old WT and $TRPM1^{-/-}$ mice revealed optokinetic deficiencies similar to those of mice lacking mGluR6 (Iwakabe et al. 1997; Koike et al. 2010b). The ERGs evoked by light stimuli in WT mice show normal a-waves and b-waves, originating mainly from photoreceptors and rod bipolar cells, respectively. The ERG b-wave in $TRPM1^{-/-}$ mice was absent and their ERG waveforms were very similar to those of $mGluR6^{-/-}$ mice (Masu et al. 1995). In the light-adapted state, the ERG b-wave was severely attenuated or absent leaving only the ERG a-wave in the $TRPM1^{-/-}$ mice (Koike et al. 2007). These ERG results suggested that the function of both rod and cone bipolar cells was severely impaired in $TRPM1^{-/-}$ mice.

Križaj et al. examined *TRPM1* expression in $Pde6b^{rd1}$ (*rd1*) and DBA/2 J retinas in which rod photoreceptors and retinal ganglion cells undergo degeneration, respectively (Križaj et al. 2010). In their DBA/2 J cohort, no significant changes were detected in the *TRPM1* mRNA content in the retinas compared to the control retinas (Križaj et al. 2010). In $Pde6b^{rd1}$ mice, prominent *TRPM1* mRNA signatures were observed in the surviving *rd1* bipolar cells, and adult *rd1* retinas with severely degenerated ONL express robust TRPM1 immunoreactivity in the distal retina (Križaj et al. 2010). These results suggest that TRPM1 is specifically expressed in bipolar cells, and unaffected by either photoreceptor or ganglion cell degeneration. In the human retina, TRPM1 is also expressed on ON-bipolar cell dendrites (Klooster et al. 2011; van Genderen et al. 2009). Two groups reported that nyctalopin, a small leucine-rich repeat protein, directly interacts and forms complexes with TRPM1 and mGluR6, and is required for the proper localization of TRPM1 to the dendritic tips of depolarizing bipolar cells (Cao et al. 2011; Pearring et al. 2011).

17.6 Functional Mechanism of the TRPM1 Channel in the Retina

To verify that TRPM1-L satisfies the properties of a non-selective cation channel regulated by the mGluR6 cascade, measurements were made of ionic currents under a whole-cell voltage clamp in a reconstitution system that mimicked the postsynaptic membrane of retinal ON bipolar cells by transfection of CHO cells with *TRPM1-L, mGluR6,* and *Goα* (Koike et al. 2010b). In CHO cells expressing mGluR6, Goα, and TRPM1-L, constitutively active inward currents were observed. Constitutively active currents were detected in TRPM1-L expressing cells even after replacing extracellular cations with Na^+, K^+, Ca^{2+}, or Mg^{2+} (Koike et al. 2010b). Furthermore, in CHO cells expressing mGluR6, Goα and TRPM1-L, constitutively active cationic currents were suppressed by the administration of glutamate to the bath solution, while subsequent washout of glutamate restored the suppressed currents to levels comparable to those prior to glutamate administration (Koike et al. 2010b). The effect of Go activation on the TRPM1-L current was further investigated by measuring whole-cell current in TRPM1-L expressing CHO cells transfected either with wild-type of *Goα* or a constitutively active mutant of *Goα* (*Goα-Q205L*) (Koike et al. 2010b). The current density obtained in *TRPM1-L*-transfected cells was significantly larger than that in *TRPM1-L* and *Goα*-co-transfected cells, while suppression of the current density in *TRPM1-L* and *Goα-Q205L*-co-transfected cells was observed at a level comparable to that in *TRPM1-L-* and *Goα*-co-transfected cells intracellularly perfused with GMP-PNP, an unhydrolyzable analog of GTP (Koike et al. 2010b). To show more directly that TRPM1-L is regulated by Go protein, the effect of addition of the purified Go protein from the intracellular side on TRPM1-L activity was examined in the excised inside-out patch recordings (Koike et al. 2010b). Application of the purified Goα protein gradually, but strongly, suppressed open probability, whereas administration of heat-denatured Goα protein failed to suppress it with GMP-PNP. Application of Gβγ did not suppress the open probability of TRPM1-L (Koike et al. 2010b). Western blot analysis distinctly showed that the purified Goα protein fraction from the bovine brain did not contain Gβγ (Koike et al. 2010a). Patch-clamp analysis on the *TRPM1-L*-transfected CHO cells effectively demonstrated that TRPM1-L is a non-selective cation channel and that TRPM1-L activity is negatively regulated by the glutamate-activated mGluR6-Goα signaling cascade (Koike et al. 2010b) (Fig. 17.1).

17.7 Congenital Stationary Night Blindness

Congenital Stationary Night Blindness (CSNB) is a clinically and genetically heterogeneous group of retinal disorders. CSNB patients exhibit defects in night vision while their day vision is normal. Two types of CSNB can be distinguished

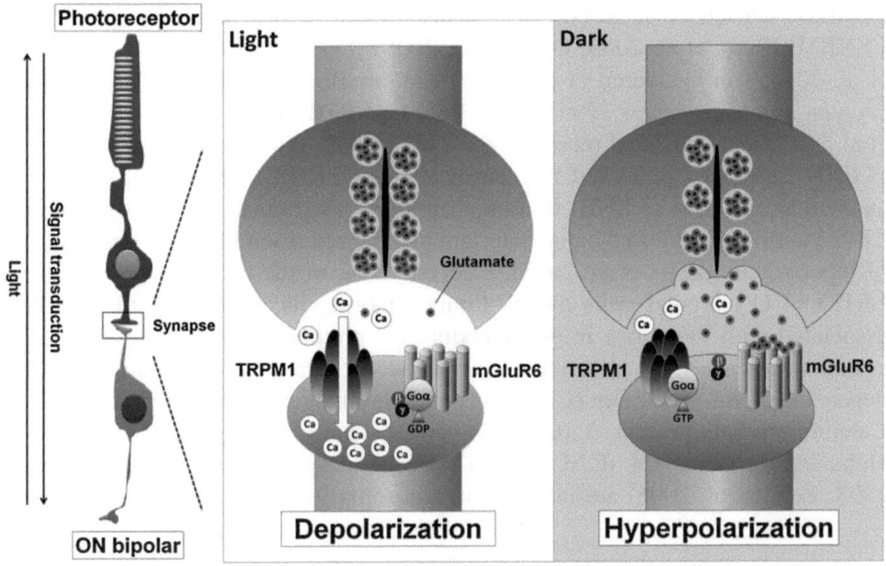

Fig. 17.1 Schematic representation of the mechanism of the ON bipolar cell response to the light. Schematic diagram of a photoreceptor and an ON bipolar cell which form a ribbon synapse between them (*left*). Upon light stimulation, the conversion of photons to neural signals in the photoreceptor cell decreases the concentration of glutamate released around the synaptic ribbon area of a photoreceptor cell as shown in the *middle* diagram. The decrease in the synaptic glutamate concentration inactivates an mGluR6 channel, leading to an open state of TRPM1 which is a constitutively active, non-selective cation channel. This leads to depolarization of the ON bipolar cells (*middle*). In contrast, in the dark, a decrease in the rate of photon absorption by photoreceptor cell increases the concentration of glutamate. An increase of glutamate activates a mGluR6 channel, results in activation of Goα, and finally inactivates the TRPM1 channel with a decrease in cationic conductance (*right*)

by using the standard flash ERG: the complete form (cCSNB) and the incomplete form (icCSNB). The complete form, also known as type1 CSNB or CSNB1, is characterized by a complete loss of the b-wave in response to a dim flash and an electronegative maximum response with a normal a-wave under scotopic conditions (Audo et al. 2008). The photopic single flash and 30 Hz flicker ERG has a normal a-wave amplitude with a broadened trough and a sharply rising peak with no photopic oscillatory potentials and a reduced b/a ratio (Audo et al. 2008). The incomplete form (icCSNB), also known as type 2 CSNB or CSNB2, is characterized by a reduction of b-wave, and that photopic responses are severely reduced and delayed both in response to a single flash or 30 Hz flicker compared with CSNB1 (Miyake et al. 1986; Audo et al. 2008). The primary cause of complete CSNB is postsynaptic defects in ON bipolar cell signaling, whereas the OFF bipolar cell pathway is unaffected. X-linked cCSNB has been associated with mutations in genes expressed and localized in ON bipolar cells including the leucine-rich proteoglycan *nyctalopin* (*NYX*) on chromosome X for X-linked

CSNB1A (Pusch et al. 2000; Bech-Hansen et al. 1998). *mGluR6* (*GRM6*) for CSNB1B (Zeitz et al. 2005; Dryja et al. 2005), *TRPM1* for CSNB1C (Nakamura et al. 2010; van Genderen et al. 2009; Audo et al. 2009; Li et al. 2009), and *G protein-coupled receptor 179* (*GPR179*) for CSNB1E (Peachey et al. 2012b; Audo et al. 2012) are autosomal recessive form of cCSNB. Whole-exome sequencing on cCSNB patient genomes led to identification of *LRIT3* mutations as a cause of autosomal recessive cCSNB. Immunostaining of the human retina using an anti-LRIT3 antibody exhibited a punctate-labeling pattern resembling the dendritic tips of bipolar cells (Zeitz et al. 2013). Understanding of the exact functional role of LRIT3 in ON bipolar cells awaits future analysis. Incomplete CSNB has been associated with mutations in genes coding for proteins important for continuous glutamate release at the photoreceptor synapse, including calcium channel voltage-dependent alpha-1F subunit (CACNA1F) on chromosome X for X-linked icCSNB. Calcium-binding protein 4 (CABP4), and calcium channel voltage-dependent alpha-2/delta subunit 4 (CACNA2D4) (Bech-Hansen et al. 1998; Strom et al. 1998; Zeitz et al. 2006) are involved in autosomal recessive form of icCSNB.

17.8 TRPM1 and CSNB

A single incomplete dominant gene, leopard complex (*LP*), causes the Appaloosa coat-spotting pattern in horses. Homozygous *LP* (*LP/LP*) causes CSNB in Appaloosa horses, characterized by a congenital and non-progressive scotopic visual deficit (Bellone et al. 2008). Furthermore, Bellone et al. mapped a 6-cM *LP* candidate region where the *TRPM1* gene is located. They found that in the retina of CSNB (*LP/LP*) horses, *TRPM1* expression was downregulated, suggesting that *TRPM1* is a strong candidate for *LP*, however, no *TRPM1* mutation in *LP/LP* horses was reported (Bellone et al. 2008). In an analysis of a large south Asian family with CSNB, Li et al. identified a large region of homozygosity on chromosome 15q which contains *TRPM1* (Li et al. 2009). Screening identified a single homozygous mutation in the affected moth (IVS16 + 2T > C) (Li et al. 2009). In a caucasian non-consanguineous family, the affected proband was found to have two likely disease-causing missense mutations in the TRPM1 gene (G138 fs and Y1035X) (Li et al. 2009). The proband of non-consanguineous Caucasian family was found to harbor heterozygote for two missense mutations in TRPM1 (R74C and I1002F) (Li et al. 2009). Furthermore, two groups have identified other homozygous mutations, including a premature truncation mutation (W856X), and a homozygous mutation for a 36,445 bp deletion of exons two to seven (Y72-K365del) (Audo et al. 2009; van Genderen et al. 2009). Nakamura et al. also identified five different novel mutations in the human *TRPM1* gene, IVS2-3C > G, IVS8 + 3_6felAAGT, R624C (c.1870C > T), S882X (c.2645C > A), and F1075S (c.3224T > C)) in three unrelated patients (Nakamura et al. 2010). All three patients were compound heterozygous. Biochemical and cell biological analyses revealed that the two intron mutations (IVS2-3C> and

IVS8 + 3_6delAAGT) were likely to result in abnormal protein production from abnormal splicing, and the two missense mutations (R624C and F1075S) lead to the mislocalization of the TRPM1 protein in ON bipolar cells (Nakamura et al. 2010). Peachy et al. reported a mouse mutant of *TRPM1*, *tvrm27*, which was identified through an ERG screen of chemically mutagenized mice (Won et al. 2011; Peachey et al. 2012a). The *tvrm27* mutant allele is caused from a point mutation (A1068T) in the *TRPM1* gene that results in a missense mutation in the pore domain of the TRPM1 protein (Peachey et al. 2012a). $TRPM1^{tvrm27/tvrm27}$ retinal histology is normal, whereas mice heterozygous for the $TRPM1^{tvrm27}$ allele showed decreased ERG b-wave (Peachey et al. 2012a).

17.9 TRPM1 and Paraneoplastic Retinopathy

Paraneoplastic retinopathy (PR), including melanoma-associated retinopathy (MAR) and cancer-associated retinopathy (CAR) is a progressive retinal disorder caused by antibodies generated against neoplasms not associated with the eye (Adamus 2009; Heckenlively and Ferreyra 2008; Chan 2003; Thirkill et al. 1989). Patients with PR can suffer from night blindness, photopsia, ring scotoma, attenuated retinal arteriole, and abnormal electroretinograms (ERGs). Recent reports showed that TRPM1 is an autoantigen targeted by autoantibodies in some patients with MAR or CAR (Dhingra et al. 2011; Kondo et al. 2011). The ERGs of a patient with lung CAR showed a severely reduced ON response with normal OFF response, indicating that the defect is in the signal transmission between photoreceptors and ON bipolar cells (Kondo et al. 2011). Western blot analysis of human TRPM1 using sera from MAR and CAR patients exhibited a significant immunoreactive band against *TRPM1*-transfected cell lysates (Kondo et al. 2011). These observations, the expression of TRPM1 in melanocytes, and its downregulation in melanoma cells suggest that TRPM1 is one of the retinal autoantigens in CAR or MAR associated with retinal ON bipolar cell dysfunction (Duncan et al. 1998).

There is evidently a high level of allelic heterogeneity in *TRPM1*, without any clear pattern with regard to location of mutations. These mutations affect residues throughout TRPM1, the intracellular N-terminus, the transmembrane domains and C-terminus. There is no noticeable genotype–phenotype relationship. However, TRPM1 is essential for the depolarizing ON bipolar cells in humans as well as in mice.

17.10 Conclusion

After decades of sizeable efforts on the ON bipolar transduction channel, TRPM1 was finally identified as a cation channel negatively regulated by mGluR6 in retinal ON-bipolar cells, based on TRPM1 mutant mouse phenotypes and a reconstitution system with TRPM1, mGluR6, and Goα, using cultured cells (Koike et al. 2010b). TRPM1 is specifically expressed in ON bipolar cells in human retinas as well as in mouse retinas. Furthermore, molecular genetic analysis on CSNB patients support the notion that TRPM1 plays an essential role in mediating the photoresponse in retinal ON bipolar cells. It should be noted that recent studies reported that other components or modulators, including GPR179 and LRIT3, function in the ON pathway in bipolar cells. Extensive future studies on both mouse retinas and human CSNB patients will greatly develop our understanding of the detailed molecular mechanisms that are the foundation of ON bipolar transduction mechanisms.

References

Adamus G (2009) Autoantibody targets and their cancer relationship in the pathogenicity of paraneoplastic retinopathy. Autoimmun Rev 8(5):410–414

Audo I, Bujakowska K, Orhan E, Poloschek CM, Defoort-Dhellemmes S, Drumare I, Kohl S, Luu TD, Lecompte O, Zrenner E, Lancelot ME, Antonio A, Germain A, Michiels C, Audier C, Letexier M, Saraiva JP, Leroy BP, Munier FL, Mohand-Saïd S (2012) Whole-exome sequencing identifies mutations in GPR179 leading to autosomal-recessive complete congenital stationary night blindness. Am J Hum Genet 90(2):321–330

Audo I, Kohl S, Leroy BP, Munier FL, Guillonneau X, Mohand-Saïdd S, Bujakowska K, Nandrot EF, Lorenz B, Preising M, Kellner U, Renner AB, Bernd A, Antonio A, Moskova-Doumanova V, Lancelot ME, Poloschek CM, Drumare I, Defoort-Dhellemmes S (2009) TRPM1 is mutated in patients with autosomal-recessive complete congenital stationary night blindness. Am J Hum Genet 85(5):720–729

Audo I, Robson AG, Holder GE, Moore AT (2008) The negative ERG: clinical phenotypes and disease mechanisms of inner retinal dysfunction. Surv Ophthalmol 53(1):16–40

Bautista DM, Siemens J, Glazer JM, Tsuruda PR, Basbaum AI, Stucky CL, Jordt SE, Julius D (2007) The menthol receptor TRPM8 is the principal detector of environmental cold. Nature 448(7150):204–208

Bech-Hansen NT, Naylor MJ, Maybaum TA, Pearce WG, Koop B, Fishman GA, Mets M, Musarella MA, Boycott KM (1998) Loss-of-function mutations in a calcium-channel alpha1-subunit gene in Xp11.23 cause incomplete X-linked congenital stationary night blindness. Nat Genet 19(3):264–267

Bellone RR, Brooks SA, Sandmeyer L, Murphy BA, Forsyth G, Archer S, Bailey E, Grahn B (2008) Differential gene expression of TRPM1, the potential cause of congenital stationary night blindness and coat spotting patterns (LP) in the Appaloosa horse (Equus caballus). Genetics 179(4):1861–1870

Cao Y, Posokhova E, Martemyanov KA (2011) TRPM1 forms complexes with nyctalopin in vivo and accumulates in postsynaptic compartment of ON-bipolar neurons in mGluR6-dependent manner. J Neurosci 31(32):11521–11526

Chan JW (2003) Paraneoplastic retinopathies and optic neuropathies. Surv Ophthalmol 48(1):12–38

Clapham DE (2003) TRP channels as cellular sensors. Nature 426(6966):517–524

Corey DP (2003) New TRP channels in hearing and mechanosensation. Neuron 39(4):585–588

de la Villa P, Kurahashi T, Kaneko A (1995) L-glutamate-induced responses and cGMP-activated channels in three subtypes of retinal bipolar cells dissociated from the cat. J Neurosci 15(5 Pt 1):3571–3582

Delmas P (2004) Assembly and gating of TRPC channels in signalling microdomains. Novartis Found Symp 258:75–89; discussion 89–102, 263–106

DeVries SH, Baylor DA (1993) Synaptic circuitry of the retina and olfactory bulb. Cell 72(Suppl):139–149

Dhingra A, Faurobert E, Dascal N, Sterling P, Vardi N (2004) A retinal-specific regulator of G-protein signaling interacts with Galpha(o) and accelerates an expressed metabotropic glutamate receptor 6 cascade. J Neurosci 24(25):5684–5693

Dhingra A, Fina ME, Neinstein A, Ramsey DJ, Xu Y, Fishman GA, Alexander KR, Qian H, Peachey NS, Gregg RG, Vardi N (2011) Autoantibodies in melanoma-associated retinopathy target TRPM1 cation channels of retinal ON bipolar cells. J Neurosci 31(11):3962–3967

Dhingra A, Jiang M, Wang TL, Lyubarsky A, Savchenko A, Bar-Yehuda T, Sterling P, Birnbaumer L, Vardi N (2002) Light response of retinal ON bipolar cells requires a specific splice variant of Galpha(o). J Neurosci 22(12):4878–4884

Dhingra A, Lyubarsky A, Jiang M, Pugh EN, Birnbaumer L, Sterling P, Vardi N (2000) The light response of ON bipolar neurons requires G[alpha]o. J Neurosci 20(24):9053–9058

Dowling JE (1978) How the retina "sees.". Invest Ophthalmol Vis Sci 17(9):832–834

Dryja TP, McGee TL, Berson EL, Fishman GA, Sandberg MA, Alexander KR, Derlacki DJ, Rajagopalan AS (2005) Night blindness and abnormal cone electroretinogram ON responses in patients with mutations in the GRM6 gene encoding mGluR6. Proc Natl Acad Sci USA 102(13):4884–4889

Duncan LM, Deeds J, Hunter J, Shao J, Holmgren LM, Woolf EA, Tepper RI, Shyjan AW (1998) Down-regulation of the novel gene melastatin correlates with potential for melanoma metastasis. Cancer Res 58(7):1515–1520

Erler I, Al-Ansary DM, Wissenbach U, Wagner TF, Flockerzi V, Niemeyer BA (2006) Trafficking and assembly of the cold-sensitive TRPM8 channel. J Biol Chem 281(50):38396–38404

Euler T, Schneider H, Wässle H (1996) Glutamate responses of bipolar cells in a slice preparation of the rat retina. J Neurosci 16(9):2934–2944

Fang D, Setaluri V (2000) Expression and up-regulation of alternatively spliced transcripts of melastatin, a melanoma metastasis-related gene, in human melanoma cells. Biochem Biophys Res Commun 279(1):53–61

Fujiwara Y, Minor DL (2008) X-ray crystal structure of a TRPM assembly domain reveals an antiparallel four-stranded coiled-coil. J Mol Biol 383(4):854–870

van Genderen MM, Bijveld MM, Claassen YB, Florijn RJ, Pearring JN, Meire FM, McCall MA, Riemslag FC, Gregg RG, Bergen AA, Kamermans M (2009) Mutations in TRPM1 are a common cause of complete congenital stationary night blindness. Am J Hum Genet 85(5):730–736

Guinamard R, Chatelier A, Demion M, Potreau D, Patri S, Rahmati M, Bois P (2004) Functional characterization of a Ca(2 +)-activated non-selective cation channel in human atrial cardiomyocytes. J Physiol 558(Pt 1):75–83

Hanano T, Hara Y, Shi J, Morita H, Umebayashi C, Mori E, Sumimoto H, Ito Y, Mori Y, Inoue R (2004) Involvement of TRPM7 in cell growth as a spontaneously activated Ca2 + entry pathway in human retinoblastoma cells. J Pharmacol Sci 95(4):403–419

Hara Y, Wakamori M, Ishii M, Maeno E, Nishida M, Yoshida T, Yamada H, Shimizu S, Mori E, Kudoh J, Shimizu N, Kurose H, Okada Y, Imoto K, Mori Y (2002) LTRPC2 Ca2 + -permeable channel activated by changes in redox status confers susceptibility to cell death. Mol Cell 9(1):163–173

Harteneck C (2005) Function and pharmacology of TRPM cation channels. Naunyn Schmiedebergs Arch Pharmacol 371(4):307–314

Haverkamp S, Grünert U, Wässle H (2001) The synaptic architecture of AMPA receptors at the cone pedicle of the primate retina. J Neurosci 21(7):2488–2500

Heckenlively JR, Ferreyra HA (2008) Autoimmune retinopathy: a review and summary. Semin Immunopathol 30(2):127–134

Hofmann T, Chubanov V, Gudermann T, Montell C (2003) TRPM5 is a voltage-modulated and Ca(2 +)-activated monovalent selective cation channel. Curr Biol 13(13):1153–1158

Hunter JJ, Shao J, Smutko JS, Dussault BJ, Nagle DL, Woolf EA, Holmgren LM, Moore KJ, Shyjan AW (1998) Chromosomal localization and genomic characterization of the mouse melastatin gene (Mlsn1). Genomics 54(1):116–123

Iwakabe H, Katsuura G, Ishibashi C, Nakanishi S (1997) Impairment of pupillary responses and optokinetic nystagmus in the mGluR6-deficient mouse. Neuropharmacology 36(2):135–143

Klooster J, Blokker J, Ten Brink JB, Unmehopa U, Fluiter K, Bergen AA, Kamermans M (2011) Ultrastructural localization and expression of TRPM1 in the human retina. Invest Ophthalmol Vis Sci 52(11):8356–8362

Koike C, Sanuki R, Miyata K, Koyasu T, Miyoshi T, Sawai H, Konda M, Usukura J, Furukawa T (2007) The functional analysis of TRPM1 in retinal bipolar cells. Neurosci Res 58, S41

Koike C, Numata T, Ueda H, Mori Y, Furukawa T (2010a) TRPM1: a vertebrate TRP channel responsible for retinal ON bipolar function. Cell Calcium 48(2–3):95–101

Koike C, Obara T, Uriu Y, Numata T, Sanuki R, Miyata K, Koyasu T, Ueno S, Funabiki K, Tani A, Ueda H, Kondo M, Mori Y, Tachibana M, Furukawa T (2010b) TRPM1 is a component of the retinal ON bipolar cell transduction channel in the mGluR6 cascade. Proc Natl Acad Sci USA 107(1):332–337

Kondo M, Sanuki R, Ueno S, Nishizawa Y, Hashimoto N, Ohguro H, Yamamoto S, Machida S, Terasaki H, Adamus G, Furukawa T (2011) Identification of autoantibodies against TRPM1 in patients with paraneoplastic retinopathy associated with ON bipolar cell dysfunction. PLoS ONE 6(5):e19911

Križaj D, Huang W, Furukawa T, Punzo C, Xing W (2010) Plasticity of TRPM1 expression and localization in the wild type and degenerating mouse retina. Vision Res 50(23):2460–2465

Launay P, Cheng H, Srivatsan S, Penner R, Fleig A, Kinet JP (2004) TRPM4 regulates calcium oscillations after T cell activation. Science 306(5700):1374–1377

Lee CR, Machold RP, Witkovsky P, Rice ME (2013) TRPM2 channels are required for NMDA-induced burst firing and contribute to H2O2-dependent modulation in substantia nigra pars reticulata GABAergic neurons. J Neurosci 33(3):1157–1168

Li Z, Sergouniotis PI, Michaelides M, Mackay DS, Wright GA, Devery S, Moore AT, Holder GE, Robson AG, Webster AR (2009) Recessive mutations of the gene TRPM1 abrogate ON bipolar cell function and cause complete congenital stationary night blindness in humans. Am J Hum Genet 85(5):711–719

Masu M, Iwakabe H, Tagawa Y, Miyoshi T, Yamashita M, Fukuda Y, Sasaki H, Hiroi K, Nakamura Y, Shigemoto R (1995) Specific deficit of the ON response in visual transmission by targeted disruption of the mGluR6 gene. Cell 80(5):757–765

McKemy DD, Neuhausser WM, Julius D (2002) Identification of a cold receptor reveals a general role for TRP channels in thermosensation. Nature 416(6876):52–58

Minke B (2010) The history of the Drosophila TRP channel: the birth of a new channel superfamily. J Neurogenet 24(4):216–233

Miyake Y, Yagasaki K, Horiguchi M, Kawase Y, Kanda T (1986) Congenital stationary night blindness with negative electroretinogram. a new classification. Arch Ophthalmol 104(7):1013–1020

Montell C, Birnbaumer L, Flockerzi V (2002) The TRP channels, a remarkably functional family. Cell 108(5):595–598

Montell C, Jones K, Hafen E, Rubin G (1985) Rescue of the Drosophila phototransduction mutation trp by germline transformation. Science 230(4729):1040–1043

Morgans CW, Brown RL, Duvoisin RM (2010) TRPM1: the endpoint of the mGluR6 signal transduction cascade in retinal ON-bipolar cells. BioEssays 32(7):609–614

Morigiwa K, Vardi N (1999) Differential expression of ionotropic glutamate receptor subunits in the outer retina. J Comp Neurol 405(2):173–184

Nakamura M, Sanuki R, Yasuma TR, Onishi A, Nishiguchi KM, Koike C, Kadowaki M, Kondo M, Miyake Y, Furukawa T (2010) TRPM1 mutations are associated with the complete form of congenital stationary night blindness. Mol Vis 16:425–437

Nakanishi S, Nakajima Y, Masu M, Ueda Y, Nakahara K, Watanabe D, Yamaguchi S, Kawabata S, Okada M (1998) Glutamate receptors: brain function and signal transduction. Brain Res Brain Res Rev 26(2–3):230–235

Nawy S (1999) The metabotropic receptor mGluR6 may signal through G(o), but not phosphodiesterase, in retinal bipolar cells. J Neurosci 19(8):2938–2944

Nawy S, Jahr CE (1990) Suppression by glutamate of cGMP-activated conductance in retinal bipolar cells. Nature 346(6281):269–271

Nilius B, Prenen J, Droogmans G, Voets T, Vennekens R, Freichel M, Wissenbach U, Flockerzi V (2003) Voltage dependence of the Ca2 + -activated cation channel TRPM4. J Biol Chem 278(33):30813–30820

Nomura A, Shigemoto R, Nakamura Y, Okamoto N, Mizuno N, Nakanishi S (1994) Developmentally regulated postsynaptic localization of a metabotropic glutamate receptor in rat rod bipolar cells. Cell 77(3):361–369

Numata T, Shimizu T, Okada Y (2007) Direct mechano-stress sensitivity of TRPM7 channel. Cell Physiol Biochem 19(1–4):1–8

Peachey NS, Pearring JN, Bojang P, Hirschtritt ME, Sturgill-Short G, Ray TA, Furukawa T, Koike C, Goldberg AF, Shen Y, McCall MA, Nawy S, Nishina PM, Gregg RG (2012a) Depolarizing bipolar cell dysfunction due to a Trpm1 point mutation. J Neurophysiol 108(9):2442–2451

Peachey NS, Ray TA, Florijn R, Rowe LB, Sjoerdsma T, Contreras-Alcantara S, Baba K, Tosini G, Pozdeyev N, Iuvone PM, Bojang P, Pearring JN, Simonsz HJ, van Genderen M, Birch DG, Traboulsi EI, Dorfman A, Lopez I, Ren H, Goldberg AF, Nis (2012b) GPR179 is required for depolarizing bipolar cell function and is mutated in autosomal-recessive complete congenital stationary night blindness. Am J Hum Genet 90(2):331–339

Pearring JN, Bojang P, Shen Y, Koike C, Furukawa T, Nawy S, Gregg RG (2011) A role for nyctalopin, a small leucine-rich repeat protein, in localizing the TRP melastatin 1 channel to retinal depolarizing bipolar cell dendrites. J Neurosci 31(27):10060–10066

Peier AM, Reeve AJ, Andersson DA, Moqrich A, Earley TJ, Hergarden AC, Story GM, Colley S, Hogenesch JB, McIntyre P, Bevan S, Patapoutian A (2002) A heat-sensitive TRP channel expressed in keratinocytes. Science 296(5575):2046–2049

Perraud AL, Fleig A, Dunn CA, Bagley LA, Launay P, Schmitz C, Stokes AJ, Zhu Q, Bessman MJ, Penner R, Kinet JP, Scharenberg AM (2001) ADP-ribose gating of the calcium-permeable LTRPC2 channel revealed by Nudix motif homology. Nature 411(6837):595–599

Pusch CM, Zeitz C, Brandau O, Pesch K, Achatz H, Feil S, Scharfe C, Maurer J, Jacobi FK, Pinckers A, Andreasson S, Hardcastle A, Wissinger B, Berger W, Meindl A (2000) The complete form of X-linked congenital stationary night blindness is caused by mutations in a gene encoding a leucine-rich repeat protein. Nat Genet 26(3):324–327

Schlingmann KP, Weber S, Peters M, Niemann Nejsum L, Vitzthum H, Klingel K, Kratz M, Haddad E, Ristoff E, Dinour D, Syrrou M, Nielsen S, Sassen M, Waldegger S, Seyberth HW, Konrad M (2002) Hypomagnesemia with secondary hypocalcemia is caused by mutations in TRPM6, a new member of the TRPM gene family. Nat Genet 31(2):166–170

Shiells RA, Falk G (1990) Glutamate receptors of rod bipolar cells are linked to a cyclic GMP cascade via a G-protein. Proc Biol Sci 242(1304):91–94

Shiells R, Falk G (1992a) Retinal on-bipolar cells contain a nitric oxide-sensitive guanylate cyclase. NeuroReport 3(10):845–848

Shiells RA, Falk G (1992b) The glutamate-receptor linked cGMP cascade of retinal on-bipolar cells is pertussis and cholera toxin-sensitive. Proc Biol Sci 247(1318):17–20

Shiells RA, Falk G (1992c) Properties of the cGMP-activated channel of retinal on-bipolar cells. Proc Biol Sci 247(1318):21–25

Shiells RA, Falk G, Naghshineh S (1981) Action of glutamate and aspartate analogues on rod horizontal and bipolar cells. Nature 294(5841):592–594

Simard JM, Chen M, Tarasov KV, Bhatta S, Ivanova S, Melnitchenko L, Tsymbalyuk N, West GA, Gerzanich V (2006) Newly expressed SUR1-regulated NC(Ca-ATP) channel mediates cerebral edema after ischemic stroke. Nat Med 12(4):433–440

Slaughter MM, Miller RF (1981) 2-amino-4-phosphonobutyric acid: a new pharmacological tool for retina research. Science 211(4478):182–185

Strom TM, Nyakatura G, Apfelstedt-Sylla E, Hellebrand H, Lorenz B, Weber BH, Wutz K, Gutwillinger N, Ruther K, Drescher B, Sauer C, Zrenner E, Meitinger T, Rosenthal A, Meindl A (1998) An L-type calcium-channel gene mutated in incomplete X-linked congenital stationary night blindness. Nat Genet 19(3):260–263

Su LT, Agapito MA, Li M, Simonson WT, Huttenlocher A, Habas R, Yue L, Runnels LW (2006) TRPM7 regulates cell adhesion by controlling the calcium-dependent protease calpain. J Biol Chem 281(16):11260–11270

Talavera K, Yasumatsu K, Voets T, Droogmans G, Shigemura N, Ninomiya Y, Margolskee RF, Nilius B (2005) Heat activation of TRPM5 underlies thermal sensitivity of sweet taste. Nature 438(7070):1022–1025

Thirkill CE, FitzGerald P, Sergott RC, Roth AM, Tyler NK, Keltner JL (1989) Cancer-associated retinopathy (CAR syndrome) with antibodies reacting with retinal, optic-nerve, and cancer cells. N Engl J Med 321(23):1589–1594

Togashi K, Hara Y, Tominaga T, Higashi T, Konishi Y, Mori Y, Tominaga M (2006) TRPM2 activation by cyclic ADP-ribose at body temperature is involved in insulin secretion. EMBO J 25(9):1804–1815

Tsuruda PR, Julius D, Minor DL (2006) Coiled coils direct assembly of a cold-activated TRP channel. Neuron 51(2):201–212

Vardi N (1998) Alpha subunit of Go localizes in the dendritic tips of ON bipolar cells. J Comp Neurol 395(1):43–52

Vardi N, Duvoisin R, Wu G, Sterling P (2000) Localization of mGluR6 to dendrites of ON bipolar cells in primate retina. J Comp Neurol 423(3):402–412

Vardi N, Morigiwa K (1997) ON cone bipolar cells in rat express the metabotropic receptor mGluR6. Vis Neurosci 14(4):789–794

Venkatachalam K, Montell C (2007) TRP channels. Annu Rev Biochem 76:387–417

Vennekens R, Olausson J, Meissner M, Bloch W, Mathar I, Philipp SE, Schmitz F, Weissgerber P, Nilius B, Flockerzi V, Freichel M (2007) Increased IgE-dependent mast cell activation and anaphylactic responses in mice lacking the calcium-activated nonselective cation channel TRPM4. Nat Immunol 8(3):312–320

Wagner TF, Loch S, Lambert S, Straub I, Mannebach S, Mathar I, Düfer M, Lis A, Flockerzi V, Philipp SE, Oberwinkler J (2008) Transient receptor potential M3 channels are ionotropic steroid receptors in pancreatic beta cells. Nat Cell Biol 10(12):1421–1430

Walder RY, Landau D, Meyer P, Shalev H, Tsolia M, Borochowitz Z, Boettger MB, Beck GE, Englehardt RK, Carmi R, Sheffield VC (2002) Mutation of TRPM6 causes familial hypomagnesemia with secondary hypocalcemia. Nat Genet 31(2):171–174

Weng K, Lu C, Daggett LP, Kuhn R, Flor PJ, Johnson EC, Robinson PR (1997) Functional coupling of a human retinal metabotropic glutamate receptor (hmGluR6) to bovine rod transducin and rat Go in an in vitro reconstitution system. J Biol Chem 272(52):33100–33104

Won J, Shi LY, Hicks W, Wang J, Hurd R, Naggert JK, Chang B, Nishina PM (2011) Mouse model resources for vision research. J Ophthalmol 2011:391384

Xu XZ, Moebius F, Gill DL, Montell C (2001) Regulation of melastatin, a TRP-related protein, through interaction with a cytoplasmic isoform. Proc Natl Acad Sci USA 98(19):10692–10697

Zeitz C, Jacobson SG, Hamel CP, Bujakowska K, Neuill M, Orhan E, Zanlonghi X, Lancelot ME, Michiels C, Schwartz SB, Bocquet B, Antonio A, Audier C, Letexier M, Saraiva JP, Luu TD, Sennlaub F, Nguyen H, Poch O, Dollfus H, Lecompte O (2013) Whole-Exome Sequencing Identifies LRIT3 Mutations as a Cause of Autosomal-Recessive Complete Congenital Stationary Night Blindness. Am J Hum Genet 92:1–9

Zeitz C, Kloeckener-Gruissem B, Forster U, Kohl S, Magyar I, Wissinger B, Mátyás G, Borruat FX, Schorderet DF, Zrenner E, Munier FL, Berger W (2006) Mutations in CABP4, the gene encoding the Ca2 + -binding protein 4, cause autosomal recessive night blindness. Am J Hum Genet 79(4):657–667

Zeitz C, van Genderen M, Neidhardt J, Luhmann UF, Hoeben F, Forster U, Wycisk K, Mátyás G, Hoyng CB, Riemslag F, Meire F, Cremers FP, Berger W (2005) Mutations in GRM6 cause autosomal recessive congenital stationary night blindness with a distinctive scotopic 15-Hz flicker electroretinogram. Invest Ophthalmol Vis Sci 46(11):4328–4335

Zhang Y, Hoon MA, Chandrashekar J, Mueller KL, Cook B, Wu D, Zuker CS, Ryba NJ (2003) Coding of sweet, bitter, and umami tastes: different receptor cells sharing similar signaling pathways. Cell 112(3):293–301

Zhou FM, Lee CR (2011) Intrinsic and integrative properties of substantia nigra pars reticulata neurons. Neuroscience 198:69–94

Zhou FW, Jin Y, Matta SG, Xu M, Zhou FM (2009) An ultra-short dopamine pathway regulates basal ganglia output. J Neurosci 29(33):10424–10435

Chapter 18
Alterations in TRPM2 and TRPM7 Functions in the Immune System Could Confer Susceptibility to Neurodegeneration

Meredith C. Hermosura

Abstract Transient Receptor Potential Melastatin 2 (TRPM2) and Transient Receptor Potential Melastatin 7 (TRPM7) are Ca^{2+} permeable channels found in neuronal and immune cells that have been linked to neuropathological disorders including neurodegeneration. Genetic variants of these channels, *Trpm2*^{P1018L} and *Trpm7*^{T1482I}, were identified in Guamanian cases of Western Pacific Amyotrophic Lateral Sclerosis and Parkinsonism Dementia Complex (WP ALS-PDC). This chapter will summarize current knowledge of the biophysical and physiological properties of these channels with special focus on their roles in the innate and adaptive immune responses. Attempts will be made to describe how dysfunctions in normal channel expression, regulation and function in the innate and adaptive immune system could potentially contribute to neurodegenerative pathophysiology, particularly in connection with WP ALS-PDC.

18.1 Introduction

Variants of two Transient Receptor Potential (TRP) channel genes, *Trpm2*^{P1018L} and *Trpm7*^{T1482I}, have been identified in Guamanian cases of Western Pacific Amyotrophic Lateral Sclerosis and Parkinsonism Dementia Complex (WP ALS-PDC) cases (Hermosura et al. 2005, 2008). WP ALS-PDC is a fatal progressive neurodegenerative disorder that occurred at high incidence in three separate disease foci among three genetically distinct groups of people more than 50 years ago: Chamorro residents of the Mariana Islands of Guam and Rota, Japanese living in the Kii Peninsula of Japan, and Auyu and Jaqai linguistic groups in Papua New Guinea (Kimura et al. 1963; Elizan et al. 1966; Gajducek and Salazar 1982).

M. C. Hermosura (✉)
Neuroscience and MRI Research Program, John A. Burns School of Medicine,
University of Hawaii, Honolulu, HI, USA
e-mail: hermosur@hawaii.edu

N. Weiss and A. Koschak (eds.), *Pathologies of Calcium Channels*,
DOI: 10.1007/978-3-642-40282-1_18, © Springer-Verlag Berlin Heidelberg 2014

Because WP ALS-PDC exhibits clinical and histological features of the more common forms of amyotrophic lateral sclerosis (ALS), Parkinson's Disease (PD), and Alzheimer's disease (AD), it is believed that the cellular and molecular pathogenic mechanisms underlying these disorders are similar and that studies on WP ALS-PDC would provide important mechanistic insights relevant to the disease processes in ALS, PD, and AD (Hermosura and Garruto 2007; Kisby and Spencer 2011). This view is supported by findings of protein aggregates characteristic of ALS, PD, and AD in WP ALS-PDC, i.e., phosphorylated and ubiquinylated deposits of tau, β-amyloid, α-synuclein (α-syn), and TDP-43 (Geser et al. 2008; Guiroy et al. 1987; Hirano 1992; Miklossy et al. 2008).

Both TRP Melastatin 2 (TRPM2) and TRP Melastatin 7 (TRPM7) are expressed in neurons and immune cells, although their biological roles are still under intense investigation. Both have been linked to neuropathological disorders, especially those associated with neuroinflammation and high oxidative stress (see Sects. 18.3 and 18.5). Notably, TRPM7 is one of only 18 genes demonstrated to be commonly regulated in AD, multiple sclerosis (MS), and cerebral stroke mouse models (Tseveleki et al. 2010). This review will summarize current knowledge on some of the biophysical and physiological properties of these channels. Attempts will be made to describe how dysfunctions in normal channel regulation and function, including the possible effects of the aforementioned genetic variants, could contribute to the pathophysiology of the neurodegenerative process. Because cumulative evidence has now established inflammatory responses as principal players in neurodegenerative pathophysiology (Appel et al. 2010, 2011; Evans et al. 2012; Glass et al. 2010; Mosley et al. 2012), special focus will be given on the roles of TRPM2 and TRPM7 in innate and adaptive immunity. Lastly, suggestions will be made for future studies that might provide information needed to achieve a better mechanistic understanding of the biological roles of these channels, knowledge that is vital toward future efforts at therapeutic intervention.

18.2 General Overview of TRPM2 and TRPM7

TRP proteins are cationic channels that are expressed in virtually all tissues and cell types, and which are increasingly thought to play crucial roles in how cells sense and respond to chemical and physical changes in their environment (reviewed in Clapham 2003; Damann et al. 2008). Similar to voltage-gated channels, TRP channels have a tetrameric architecture consisting of four subunits each with six transmembrane segments, with the loop between segments 5 and 6 forming the channel pore and selectivity filter (Fig. 18.1). In TRPM2, proline 1018 (P1018) is localized at the C-terminal end of the outer pore loop (Fig. 18.1a). The substitution of proline with leucine in the WP ALS-PDC variant $Trpm2^{P1018L}$ causes the channel to inactivate (Hermosura et al. 2008).

There are currently 28 known mammalian TRP genes classified into six families based on amino acid sequence. The broad cell and tissue distribution, the

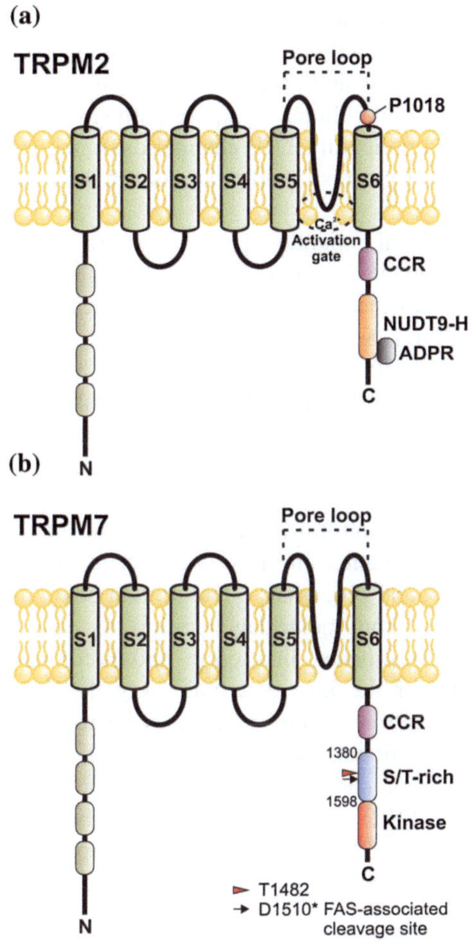

Fig. 18.1 Schematic diagrams of a TRPM2 and a TRPM7 subunit. In both channels, a subunit consists of a cytoplasmic N-terminal region containing TRPM homology domains, six transmembrane segments (S1-S6), a pore loop region between S5-S6 and a cytoplasmic C-terminal domain. The coiled-coil region (CCR) participates in subunit assembly to form the tetrameric channel. **a** In TRPM2, a Nudix homology domain (NUDT9-H) primarily serves as the binding site for one of the activating signals, ADPR. The other requisite activator, Ca^{2+}, is believed to bind somewhere in the intracellular mouth of the pore (region marked by dashed oval, 'activation gate'). Pro^{1018} is located near the C-terminal end of the outer pore vestibule. The presence of Leu in place of Pro in $TRPM2^{P1018L}$ destabilizes the outer pore conformation of the open channel causing channel inactivation. A similar effect was induced by Zn^{2+} (see text). **b** In TRPM7, a functional α-kinase is found at the C-terminal domain. The kinase is immediately preceded by a Ser/Thr-rich region (S/T-rich). The S/T-rich domain is autophosphorylated in vivo and is proposed to play a role in substrate recognition and access of substrates to the kinase catalytic domain. Thr^{1482} (solid red triangle) is in this region and is one of the residues found to be autophosphorylated. Asp^{1510} (D1510) marked by arrow is the site for Fas receptor activation-induced cleavage, an event that separates the kinase from the rest of the channel, causing increase in I_{TRPM7} that was shown to be necessary for effective FAS-induced apoptosis (see text)

existence of splice variants, and the ability of subunits within the same family to form heterotetramers collectively produce a group of channels with a wide variety of ionic permeability profiles and gating mechanisms that affect a broad range of physiological processes and cellular signaling pathways. Importantly, an increasing number of reports implicating TRP channel dysfunctions in various disorders and making TRP proteins promising drug targets, are being published (reviewed in Wu et al. 2010).

TRPM2 and TRPM7 are unique in that they are two of only three known channels that have both an ion-permeating pore and an enzymatic domain: a NudT9 adenosine diphosphate ribose (ADPR) hydrolase for TRMP2 (NudT9-Homology or NUDT9-H), and a serine-threonine α-kinase for TRPM7 (Perraud et al. 2001; Nadler et al. 2001; Runnels et al. 2001). Given the very weak enzymatic activity of NudT9-H, it not likely that the catalytic property has a role in TRPM2 gating. The prevailing view is that it provides a binding site for free ADPR, one of the two essential channel activators (Fig. 18.1a). The kinase moiety in TRPM7, on the other hand, is fully functional. As will be seen in later discussions, the dual channel and kinase functionality of the ubiquitously expressed TRPM7 confers on this channel an unusual ability to influence diverse signaling pathways and physiological processes, a feature that could underlie its association with disorders ranging from neurodegeneration to cancer (see Sect. 18.6.4). Threonine 1482 (T1482) is located in a serine- and threonine-rich (S/T-rich) region immediately upstream of the kinase domain (Fig. 18.1b). Studies suggest that massive autophosphorylation of this region plays a role in substrate recognition (Clark et al. 2008b). Additionally, an aspartate residue within this S/T-rich region, D1510, was identified as a caspase cleavage site following Fas receptor-induced activation (Desai et al. 2012). A phosphorylation site is lost in the WP ALS-PDC-associated variant, $Trpm7^{T1482I}$, where threonine is replaced with isoleucine. This change could impose subtle effects on substrate recognition and even on caspase cleavage at D1510.

18.3 TRM2 is a Metabolic and Oxidative Stress Sensor

TRPM2 functions as metabolic and oxidative stress sensor that senses intracellular levels of calcium (Ca^{2+}), ADPR, reactive oxygen species (ROS), glutathione (GSH) and protons, integrating the acquired information with changes in membrane potential and downstream cell type-specific Ca^{2+}-activated signaling networks. TRPM2 is also thermosensitive, its activity increasing at higher physiological temperatures (Togashi 2006). The channel is expressed in a wide range of cells and tissues including endothelial cells, cardiac myocytes, bone marrow, spleen, liver, heart, lung, pancreatic β-cells, but is particularly abundant in the brain and immune cells, particularly those of monocytic lineage (reviewed by Takahashi et al. 2011; Knowles et al. 2012).

18.3.1 Gating: TRPM2 is Activated by the Simultaneous Binding of ADPR and Ca^{2+} at Intracellular Sites on the Channel

Two coactivators are absolutely required for TRPM2 channel gating: the nicotinamide adenine dinucleotide $(NAD)^+$ metabolite, **ADPR**, bound to the intracellular NudT9-H domain, and $\mathbf{Ca^{2+}}$ bound to still undefined 'activation sites' close to the intracellular mouth of the pore, see Fig. 18.1a (Perraud et al. 2001; McHugh et al. 2003; Starkus et al. 2007; Csanady and Torocsik 2009; Toth and Csanady 2010). Neither ADPR nor Ca^{2+} alone is able to initiate channel opening. Because of the tetrameric nature of TRPM2, it was proposed that four Ca^{2+} ions are required; each binding event increases the open-closed equilibrium constant in a manner well predicted by the Monod-Wymann-Changeux (MWC) model (Csanady and Torocsik 2009). Closed channels do not sense extracellular Ca^{2+}, but once channels have opened, Ca^{2+} ions permeating through the pore keep the intracellular activating gate saturated. The opening rate is a property of the closed channel and is dependent on the levels of intracellular free Ca^{2+}, $[Ca^{2+}]_i$. What this implies is that in cells where the metabolic state is such that there is sufficient (micromolar) ADPR, sudden local increases in $[Ca^{2+}]_i$ can initiate channel opening. Ca^{2+} entering from the outside will keep the activating sites saturated, maintaining the open state until the channel closes normally (Csanady and Torocsik 2009). Readers who are interested in more detailed discussions of the biophysics of channel gating are referred to the excellent original reports cited.

Although exposure to ROS (H_2O_2) leads to channel activation in intact cells, direct gating could not be demonstrated in cell-free studies even when a high concentration (1 mM) of H_2O_2 was used (Toth and Csanady 2010). ROS could induce TRPM2 activation in three ways, the effects likely occurring synergistically: (i) by increasing ADPR production (Perraud et al. 2005); (ii) by elevating cytosolic Ca^{2+} levels (Bogeski et al. 2011; Gonzales et al. 2006); and (iii) by lowering the temperature activation threshold (Kashio et al. 2012). Increases in the cytosolic levels of ADPR occur primarily through the actions of the plasma membrane-bound CD38 NAD^+ glycohydrolase (CD38), and ROS-induced leakage from mitochondria which contains the largest pool of intracellular NAD^+ and thus, ADPR (Perraud et al. 2005). In addition, oxidative stress activates the polyADPR polymerase/glycohydrolase (PARP/PARG) pathway in the nucleus that also produces ADPR (Buelow et al. 2008). Questions that were initially raised as to how the ectoenzyme CD38 generates cytosolic ADPR are now answered by the finding that the enzyme exists in two opposing orientations: one with the catalytic domain facing outside, the other with the catalytic domain flipped, facing the cytosol (Zhao Y et al. 2012). Support for the notion that CD38 activity cross-talks with TRPM2 is provided by observations that genetic disruption of either gene affects dendritic cell migration and pulmonary inflammation (Gally et al. 2009; Sumoza-Toledo 2011). In addition to increasing cytosolic ADPR, ROS elevates intracellular Ca^{2+} levels through its effects on multiple Ca^{2+}-permeable channels located in the plasma membrane and in intracellular membrane systems (Bogeski et al. 2011; Gonzales et al. 2006). In sum, oxidative stress promotes TRPM2 activation by increasing the cytosolic levels of the two factors absolutely required for channel gating, ADPR and Ca^{2+}, and by lowering the temperature threshold for channel activation (Kashio et al. 2012).

Channel opening at physiological hyperpolarized membrane potentials will cause a large influx of Ca^{2+} and sodium (Na^{+}) because both ions will be moving down their electrochemical gradients. Ca^{2+} entering the cell will activate cell-type specific Ca^{2+}-dependent signal transduction pathways, whereas the movement of both Na^{+} and Ca^{2+} will depolarize the membrane. This membrane potential change will affect the driving force for Ca^{2+} through other Ca^{2+}-permeant channels such as the Ca^{2+} release-activated Ca^{2+} channels (CRAC) (Hoth and Penner 1992). Furthermore, it will affect the function of other membrane potential-sensitive proteins such as NADPH oxidase, the enzyme primarily responsible for ROS production in microglia, monocytes, and phagocytes. The significance of TRPM2-induced membrane depolarization on NADPH oxidase activity and ROS production is discussed in Sect. 18.3.3.

18.3.2 Modulation of TRPM2 Activity by Protons, Zn^{2+}, and GSH

18.3.2.1 Protons Suppress Channel Activity

Significant TRPM2 activity occurs in conditions associated with acidic pH such as during immunocyte respiratory burst, sites of neuronal injury following ischaemia, and sites of inflammation. Separate studies conducted by three research groups collectively found that protons inhibit TRPM2 activity (Du et al. 2009; Starkus et al. 2010; Yang et al. 2010). There were important similarities and differences.

All three studies reported that intracellular and extracellular protons suppress channel activity, but the proposed molecular mechanisms especially pertaining to the effects of extracellular protons were diverse. There were also differences found in terms of reversibility, effective pH, and voltage-dependence of inhibition; on whether protons affect single channel conductance; and the underlying mechanism of extracellular proton actions: do they bind to residues in the outer vestibule thereby interfering with the selectivity filter, or do they compete with Na^{+} and Ca^{2+} for permeation? These issues are beyond the scope of this review and interested readers are again referred to the original articles cited. However, the following generalizations are supported by existing evidence: (i) intracellular protons suppress TRPM2 gating by competing with Ca^{2+} for activating sites near the cytoplasmic mouth of the pore and (ii) inhibition by external protons is complex. When arising as a result of competition with Na^{+} and Ca^{2+} for permeation, inhibition by external protons is reversible; however, if it involves interaction with residues in the outer vestibule of the pore, interference with the selectivity filter and channel inactivation, external proton effect is irreversible (Yang et al. 2010). Irrespective of the actual mechanism, suppression of TRPM2 activity by protons is likely a physiological way of controlling Ca^{2+} influx and membrane potential changes in conditions of high oxidative stress.

18.3.2.2 Destabilization of the Outer Pore Vestibule of the Open Channel Causes TRPM2 Inactivation

In explaining $TRPM2^{P1018L}$-induced inactivation, we hypothesized that the mutation could be destabilizing the outer pore loop structure (Hermosura et al. 2008). Indeed, emerging evidence suggests that certain residues in the pore loop domain between S5 and S6 are important in maintaining a stable conformation for the selectivity filter and open channel (Yang et al. 2010; Toth and Csanady 2012). As noted in the previous section, channel inactivation occurs when protons interact with certain residues in the outer pore loop. A similar effect was reported for Zn^{2+} (Yang et al. 2011). Mutational analysis shows that while there are overlaps, the structural and mechanistic determinants of proton- and Zn^{2+}-induced inactivation are not the same (Yang et al. 2011). Nonetheless, collective evidence provided by studies on protons, Zn^{2+} and the P1018L mutation indicate that disrupting the stable conformation of the outer pore vestibule destabilizes the open channel and causes channel inactivation. A similar mechanism was recently proposed for TRPM2 channel rundown that is often observed in inside-out patches: prolonged stimulation causes the selectivity filter to undergo a conformational rearrangement similar to what happens during C-type inactivation of voltage-gated K^+ channels (Toth and Csanady 2012).

It is noteworthy that both Zn^{2+} and the P1018L mutation have been associated with neurodegeneration. Zn^{2+} functions as an important signaling molecule in neuronal synapses, but accumulates at abnormally high levels in pathological conditions characterized by high oxidative stress such as cerebral ischemia, epilepsy, and traumatic brain injury (Capasso et al. 2005). TRPM2 has similarly been linked to these disorders (Cook et al. 2010; Jia et al. 2011; Katano et al. 2012).

18.3.2.3 Glutathione (GSH) Modulates TRPM2 by Inhibiting its Activation

GSH, a reducing agent that functions as one of the cell's main defenses against oxidative stress, acts as a physiologically relevant regulator of TRPM2 (Lee et al. 2010; Naziroglu et al. 2011; Ozgul et al. 2012; Belrose et al. 2012). It inhibits TRPM2 activation, causing a rightward shift in the ADPR dose–response curve and decreasing the Hill coefficient (Belrose et al. 2012). These effects were determined not to be due to the reducing actions of GSH, but could arise as result of GSH acting as a channel blocker, competing with ADPR for binding sites, or even binding directly to ADPR. Thus, existing intracellular levels of GSH could influence the degree of TRPM2 activation at any one time.

18.3.3 TRPM2 Functions

18.3.3.1 In Oxidative Stress-Induced Cell Death

TRPM2's role in oxidative stress-induced cell death is complex (Hara et al. 2002). Using HEK-293 cells transfected with human TRPM2, Wilkinson et al. (2008) showed that while H_2O_2 did indeed cause loss of viability in TRPM2-expressing cells, cell death was not solely due to Ca^{2+} influx through this channel and that H_2O_2-induced cell death has a Ca^{2+}-independent component unrelated to TRPM2. The authors concluded that while Ca^{2+} influx via TRPM2 could contribute to, even potentiate, loss of cell viability as a result of oxidative stress, there are other cellular mechanisms involved. Indeed, it is more than likely that it is influenced by cell type, metabolic state, as well as other intrinsic and extrinsic factors. For instance, TRPM2 is abundant in monocytes and phagocytes, cells that are professional oxidant generators and therefore exposed to ROS in the course of their activities. However, when monocytes were exposed to lipopolysaccharide (LPS), an agent that initiates an inflammatory response characterized by ROS production, although TRPM2 activation occurred as expected, the outcome was cytokine production, not cell death (Wehrhahn et al. 2010). Providing further support is a recent study which established that $TRPM2^{-/-}$ T cells were not particularly resistant to oxidative stress-induced cell death (Melzer et al. 2012). Thus, at least in immune cells, TRPM2 activation has biological functions more pertinent to the immune response, and that ROS-induced activation of TRPM2 promotes these cell-type specific functions.

18.3.3.2 In Neuronal Cells

TRPM2 is present in various neuronal populations in the CNS and in the periphery. There is a paucity of information on TRPM2's biological functions in these cells because much of the research done so far investigated TRPM2 activity in connection with oxidative stress. It is anticipated, however, that the availability of the $TRPM2^{-/-}$ mouse model should lead to more studies reporting the consequences of not having the channel, and would therefore provide clues to its biological functions such as its recently described role in synaptic plasticity (Xie et al. 2011). $TRPM2^{-/-}$ mice do not show overt behavioral or neurological phenotypes (Yamamoto et al. 2008), but there could be subtle behavioral changes especially in the presence of stressors. Genetic association studies have linked TRPM2 with bipolar disorder (McQuillin et al. 2006; Xu et al. 2006, 2009).

Synaptic Plasticity
N-Methyl-D-aspartate (NMDA)-dependent long-term depression (LTD) in $TRPM2^{-/-}$ mice is reduced, whereas the threshold for long-term potentiation (LTP) is lowered, suggesting a role for TRPM2 in synaptic plasticity (Xie et al. 2011). TRPM2 appears to be necessary in keeping glycogen synthase kinase-3β (GSK-3β) activity, as well as postsynaptic density protein 95 (PSD95) and α-amino-3-hydroxyl-5-methyl-4-isoxazole-propionate receptor ((AMPAR) expression at appropriate levels. In the absence of TRPM2, there is increased phosphorylation and inhibition of GSK-3β, decreased expression of PSD95 and AMPAR, and reduced AMPA receptor-mediated fast excitatory transmission at CA1 synapses. Interestingly, GSK-3β has been implicated in the activity of cycad toxins linked to WP ALS-PDC (see Sect. 18.6).

Traumatic Brain Injury
TRPM2 mRNA and protein expression in the cerebral cortex and hippocampus were significantly increased in rats following experimental TBI indicating that the channel plays a role in the pathologic changes associated with this disorder: compromised blood brain barrier (BBB) integrity, glial activation, accumulation of inflammatory mediators and leukocytes in affected areas, and neuronal death (Cook et al. 2010). Further support for TRPM2 involvement was provided by a study using a mouse ischemic model (Jia et al. 2011), except that in this case, there is gender-specificity to the effects of TRPM2: it was found to be limited to male mice, an effect that was proposed to have been due to PARP hyperactivation in males.

Neurodegenerative Diseases
Freestone et al. (2009) reported TRPM2 participation in the response of rat dopaminergic substantia nigra pars compacta (SNc) neurons to rotenone, a neurotoxin used to model PD. Subsequently, the same group showed that TRPM2 mediates oxidative stress-induced changes in neuronal excitability and Ca^{2+} homeostasis in these neurons, suggesting that these channels could integrate Ca^{2+} dysregulation, oxidative stress, and mitochondrial dysfunctions in PD (Chung et al. 2011).

18.3.3.3 In Immune and Inflammatory Responses

The most studied function of TRPM2 is its role in immune and inflammatory processes. However, despite (or perhaps because of) the intense research conducted in a multitude of laboratories using different model systems, the best that can be said is that the role of TRPM2 in the immune system is complex. The growing number of studies being done using TRPM2-deficient mouse models will hopefully provide a better understanding of this channel's involvement in innate and adaptive immune responses (reviewed in Knowles et al. 2012).

TRPM2 in Cells of Monocytic Lineage and Neutrophils

As mentioned earlier, mice deficient in TRPM2 do not exhibit obvious phenotypic or behavioral deficiencies (Yamamoto et al. 2008). Based on studies published so far, the effects of TRPM2 ablation appear to be manifested only upon exposure to stressors. Thus, in the first published work on TRPM2$^{-/-}$ mice (C57BL/6 background) created by Y. Mori's group in Kyoto, a chemically induced inflammation model, dextran sodium sulfate (DSS)-induced colitis was used to provide insight on the role of TRPM2 in inflammation (Yamamoto et al. 2008). In the DSS model, TRPM2$^{-/-}$ mice exhibited diminished neutrophil influx into the colonic epithelium, and consequently less inflammation, compared to wildtype (wt). TRPM2$^{-/-}$ neutrophils are fully capable of migrating. The reduced neutrophil influx at inflammation sites resulted from decreased production of the neutrophil chemoattractant, CXCL2, by TRPM2$^{-/-}$ monocytes/macrophages.

Another example of a stressor uncovering the phenotypic manifestation of *Trpm2* genetic knockout was described in a study that used a TRPM2$^{-/-}$ mouse strain created by scientists at Glaxo Smith Kline (GSK), in the same C57BL/6 background (Knowles et al. 2011). Here, the stressor, *Listeria monocytogenes* (*Lm*), revealed TRPM2's role in innate immunity. This GSK TRPM2$^{-/-}$ strain was found to be extremely susceptible to bacterial infection. The susceptibility was shown to be caused, in part, by diminished and inefficient interferonγ (IFNγ)-mediated innate immune response. Compared to wt, TRPM2$^{-/-}$ mice infected with *Lm* had lower numbers of IFNγ^+CD8$^+$T cells, IFNγ^+ NK cells, and showed significantly reduced production of IL-12 by CD8$^+$ splenic dendritic cells.

Using the same GSK TRPM2$^{-/-}$ strain, a separate research group uncovered a protective role for TRPM2 in endotoxin (LPS)-induced pulmonary inflammation (Di et al. 2012). Compared to wt, TRPM2$^{-/-}$ phagocytes had higher ROS levels and released more proinflammatory mediators upon LPS challenge. These effects were ascribed to TRPM2 activity exerting negative feedback on NADPH oxidase, the enzyme responsible for generating ROS in phagocytes and other cells of monocytic lineage. NADPH oxidase produces superoxide by extracting electrons from intracellular NADPH and translocating the electrons across the membrane to the extracellular or intraphagosomal compartment where they reduce molecular oxygen (O_2). Thus, NADPH oxidase activity is electrogenic and the movement of electrons across the membrane is measurable as an electron current, I_e (Schrenzel et al. 1998; DeCoursey et al. 2000). This electrogenic activity sharply depolarizes the membrane, by $\sim$100 mV above resting potential in neutrophils (DeCoursey 2003). Because NADPH oxidase functions best at negative potentials (-100 to 0 mV) and neutral pH (Morgan et al. 2009; DeCoursey et al. 2003), the ensuing depolarization and acidification caused by the translocation of electrons would inhibit enzyme activity unless corrected. Voltage-gated proton channels and Na^+/H^+ transporters act in concert to correct the pH and membrane potential changes thereby allowing continued NADPH oxidase activity under normal conditions (DeCoursey et al. 2000; DeCoursey et al. 2003; Morgan et al. 2009). However, ROS-induced TRPM2 activation and the consequent influx of Na^+ and Ca^{2+} depolarizes the membrane sufficiently to interfere with ROS production by

NADPH oxidase. This is an elegant demonstration of a biological truism: in many cases, an initiating signal controls its own termination. It must be noted, however, that another study, using TRPM2$^{-/-}$ mice in a different background (Balb/c), did not find a role for TRPM2 in inflammatory chronic obstructive pulmonary disease, COPD (Hardaker et al. 2012). Whether the difference is due the difference in mouse strain, route of dosing (intranasal vs. intraperitoneal), and/or methods used to monitor cellular inflammation, remains to be clarified.

TRPM2 in Astrocytes and Microglia

In addition to controlling NADPH oxidase activity by depolarizing the membrane, TRPM2 activation leads to Ca^{2+} influx and the activation of downstream Ca^{2+}-dependent signaling pathways. In astrocytes and microglia, these were reported to include the p38 MAP kinase (p38MAPK), the Jun-*N*-terminal kinase (JNK), and the NFκB inflammatory pathways (Lee et al. 2010). In this study, oxidative stress was induced by exposing cells to buthionine sulfoximine (BSO), an inhibitor of γ–glutamylcysteine synthase, causing depletion of intracellular GSH. Astrocytes and microglia exposed to BSO secreted tumor necrosis factor-α (TNF-α), interleukin-6 (IL-6), nitrite ions, and other factors toxic to neuronal cells. The effects of BSO were reduced by TRPM2 siRNA knockdown, pharmacological inhibition of TRPM2, or addition of GSH to the medium. This study is a valuable first step in delineating the role of TRPM2 in glial-induced neurotoxicity. It would be interesting to study the extent of TRPM2 involvement using TRPM2$^{-/-}$ mice or glial cells from these mice.

Importantly, in the light of existing evidence that the initial astrocytic and microglial responses in neurodegenerative disorders are likely to be protective in nature, transforming to a chronic, inflammatory and neurotoxic type as the disease progresses (reviewed by Barres 2008; Liao et al. 2012), it would be valuable to know if TRPM2 is associated with either protective or neurotoxic glial activities. One study reported that astrocytes express TRPM2 only when exposed to oxidative stress, the expression occurring in the context of a more adaptive stress response that is geared toward neuronal survival instead of apoptotic (Bond and Greenfield 2007). If this is the case, then blocking TRPM2 is not going to be a good therapeutic option. Similarly, existing evidence indicate that there are at least two general subpopulations of microglia: those that are classically activated and neurotoxic (M1), and those that are alternatively activated and neuroprotective (M2) (Gordon 2003; Henkel et al. 2009; Liao et al. 2012). Although it has been established that microglia express TRPM2, studies that seek to determine if channel expression and activity differ between the M1 and M2 types would provide critical information needed to guide efforts targeting TRPM2 for therapeutic purposes.

TRPM2 in T Cells and the Adaptive Immune Response

Initial studies on TRPM2 in T cells were performed using the well-studied Jurkat cell line (Beck et al. 2006; Gasser et al. 2006). In these cells, crosslinking of surface receptors by ConA induced a rise of ADPR endogenously generated from

NAD^+, followed by Ca^{2+} rise and TRPM2-mediated current (Gasser et al. 2006). TRPM2 expression in T cells was further confirmed in a more recent study which sought to compare TRP channel expression profiles between the Jurkat T-cell line and primary human T cells (Wenning et al. 2011). TRP channel expression profiles in primary T cells were found to be significantly different from the cell line. Only five—TRPC1, TRPC3, TRPV1, TRPM2 and TRPM7—were consistently found in primary human CD4+T-cells of all blood donors analyzed. In this study, the cells were stimulated with antibody-coated beads for several days (in an attempt to mimic physiological stimulation including the formation of an immunological synapse). TRPM2 exhibited a complicated mRNA expression profile in primary T cells: it was first down-regulated 7 h following stimulation, slightly increased until day 3, then up-regulated until day 6. The significance of the delayed TRPM2 upregulation in T cells is not known, but could be related to a recently proposed role for this channel in mediating NAD^+-dependent control of the state of replenishment of ER Ca^{2+} stores (Magnone et al. 2012). The filling state of ER stores is a major factor in intracellular Ca^{2+} homeostasis, which is in turn crucial for various T-cell functions. Another study used cells from the Kyoto $TRPM2^{-/-}$ mice (made by Y. Mori's group) to demonstrate TRPM2 involvement in $CD4^+$ T-cell proliferation and secretion of effector cytokines such as IL-2, IFNγ, and IL-17 following polyclonal T-cell receptor (TCR) stimulation (Melzer et al. 2012). The important role of TRPM2 in the adaptive immune response was underscored when the same group showed that $TRPM2^{-/-}$ mice exhibited an attenuated clinical phenotype of experimental autoimmune encephalomyelitis (EAE), with reduced inflammatory infiltrates in the CNS and demyelination in the spinal cord. Finally, TRPM2's role in increasing microvessel endothelial permeability (Hecquet et al. 2008), while not strictly part of the adaptive immune response, is peripherally involved because endothelial cells participate in the process of lymphocyte migration to sites of immune activity and inflammation.

18.3.3.4 Potential Links Between TRPM2 (and $TRPM2^{P1018L}$) in the Innate and Adaptive Immune Systems to Neurodegenerative Pathophysiology

In this section, hypotheses as to how having the *Trpm2*^{P1018L} variant could affect TRPM2-dependent cellular functions and how these effects could potentially contribute to the neurodegenerative process, are presented and discussed. Suggestions for future studies are also provided.

Negative Feedback on NADPH Oxidase Activity

The proper control of NADPH oxidase activity is relevant to neurodegenerative pathology, especially in ALS where enzyme activity was found to be upregulated in familial and sporadic cases, and in mutant SOD1 mouse models (Wu et al. 2006; Lobsiger and Cleveland 2007; Harraz et al. 2008). The finding that TRPM2 activation is needed to control NADPH oxidase activity suggests an intriguing mechanism as to how the *Trpm2*^{P1018L} variant could contribute to the pathophysiology of WP ALS-PDC: because it inactivates quickly, TRPM2^{P1018L} might not be able to induce sufficient membrane depolarization to negatively modulate NADPH oxidase in microglia and macrophages thus keeping ROS levels high. One can imagine this situation eventually leading to a state of chronic inflammation which could be damaging if prolonged (as occurs when one has a genetic variant), especially when combined with environmental stressors such as cycad toxins or prolonged exposure to low Mg^{2+}, high metal conditions. Measurements of ROS levels and membrane potentials in TRPM2$^{-/-}$ phagocytes transfected with TRPM2^{P1018L} would be one way to test this hypothesis.

The Inability of TRPM2^{P1018L} to Sustain Normal Levels of Ca^{2+}Influx Could Attenuate T-cell Effector Functions

As discussed in the previous section, TRPM2 contributes to the proliferation and proinflammatory cytokine secretion of CD4^{+} effector T cells. Because it cannot sustain the normal level of Ca^{2+} influx, the inactivating TRPM2^{P1018L} variant would be expected to exhibit reduced T-cell effector functions and attenuated EAE, albeit not to the same degree seen in TRPM2$^{-/-}$mice (Melzer et al. 2012). Similar studies on mouse models expressing TRPM2^{P1018L} would be useful. Intriguingly, the *opposing effects of impaired T-cell activity in EAE and ALS* were recently described (Vaknin et al. 2011). Adoptive transfer of IL-4 activated bone marrow derived myeloid cells (M2 BMDMs) exacerbated ALS symptoms in the SOD^{G93A} mice, but mitigated EAE. M2 BMDMs suppress peripheral CD4+ T-cell activity indicating that T-cell deficiency promotes the neurodegenerative process in ALS as was shown by others, and highlighting the opposing cellular mechanistic nature of inflammation in EAE and ALS (Banerjee et al. 2008; Beers et al. 2008; Chiu et al. 2008).

TRPM2 Activity in the Neuroprotective and Neurotoxic Innate and Adaptive Immune Responses

There is compelling evidence for the important roles played by both innate and adaptive immune responses in neurodegeneration (Appel et al. 2010; Appel et al. 2011; Banerjee et al. 2008; Beers et al. 2008, 2011; Henkel et al. 2013; Mosley et al. 2012; Reynolds et al. 2010; Saunders et al. 2012; Zhao W et al. 2012). In both cases, neurotoxic and neuroprotective responses have been identified. Thus, there are neurotoxic (M1) and neuroprotective (M2) microglia (Gordon 2003; Liao et al. 2012). However, the mechanisms and molecular determinants that decide whether microglia will be neurotoxic or neuroprotective remain unknown. Similarly, there are neurotoxic and neuroprotective T-cell subtypes (see Sect 18.4.1). Findings of increased T-cell infiltration in the CNS, coupled with recent reports that numbers of activated T cells are elevated in the peripheral blood strongly suggest that the various T-cell subtypes play important roles in neurodegenerative pathophysiology (Lincecum et al. 2010; Henkel et al. 2013; Saunders et al. 2012; Lewis et al. 2012). Studies to investigate TRPM2 expression and function in M1 and M2 microglia and in the different T-cell subtypes could provide valuable information on the role of the immune response in neurodegeneration, as well as useful knowledge to guide therapeutic intervention efforts.

18.4 T-Cell Subtypes and T-Cell Activation

18.4.1 CD4+ T-Cell Subtypes

Due in part to more sophisticated methods of analyzing cytokine profiles and cell surface expression markers, the classification of CD4+ T-cells has evolved over the years. Briefly, there are effector (Teffs), memory (Tems), and regulatory CD4+ T cells (Tregs). Based on cytokine profiles, Teffs are classified into Th1 (IL-2, IFN-γ, TNF-α), Th2 (IL-4, IL-5, IL-13), and Th17 (IL-17, IL-22) (reviewed in Annunziato and Romagnini 2009; Huang et al. 2009). Tregs are CD4+, CD25+, and Fox-P3+. Th1 and Th17 produce proinflammatory cytokines, while Th2 produce anti-inflammatory cytokines thereby promoting neuroprotection. The role of Tregs is complex, but in the context of neurodegeneration, Tregs promote neuroprotection by attenuating inflammation (reviewed in Huang et al. 2009; Gendelman and Appel 2011). Tilting differentiation toward the neuroprotective subtypes is a logical therapeutic intervention plan, but that would require an understanding of the processes involved in subtype determination. It is more than likely that Ca^{2+} influx pathways are involved because the magnitude, shape, and pattern of cellular Ca^{2+} signals initiated by TCR engagement control the activation of transcription factors, and consequent gene expression that define the overall T-cell response whether it be proliferation, apoptosis, differentiation into subtypes and clonal expansion (Feske 2007; Qu et al. 2011).

18.4.2 T-Cell Activation

When a T cell encounters, a target antigen on another cell, a well-orchestrated series of events involving reorganization of surface receptors (i.e., TCR and associated costimulatory/coinhibitory molecules), cytoskeletal components, adhesion molecules, and even organelles such as mitochondria, endoplasmic reticulum (ER), centrioles, and Golgi complex, occurs in the process of forming the immunological synapse (IS) (Quintana et al. 2011; Ueda et al. 2011; Kumari et al. 2012). The IS serves as the site where signals delivered by the TCR complex are decoded and integrated before a decision is made on whether to initiate signaling cascades that could lead to either activation or tolerance. Three supramolecular activation clusters (SMACs) are sequentially formed during IS maturation: (1) the distal SMAC (dSMAC) forms first emerging as an F-actin-rich ring of close contact as the T cell spreads on the antigen surface, using invasive pseudopodia to increase the point of contact (Ueda et al. 2011). TCR and leukocyte function-associated antigen-1 (LFA-1) microclusters are formed at this stage. (2) The peripheral SMAC (pSMAC) is then assembled from distal microclusters that undergo centripetal transport toward the center. It is at this stage when Ca^{2+} influx pathways, mitochondria and ER interactions occur prominently (Quintana et al. 2011). It is critical that mitochondria accumulate near the IS for two primary reasons: (i) to provide ATP for the energy demanding events at the IS including motors such as myosin IIA which controls actin flux at the actomyosin contractile ring formed beneath pSMAC (Kumari et al. 2012) and (ii) to buffer Ca^{2+} entering through CRAC channels (reviewed in Junker and Hoth 2011). If allowed to accumulate, Ca^{2+} will immediately inactivate CRAC channels resulting in inefficient T-cell activation. This will have downstream consequences on gene expression and cytokine secretion, particularly those controlled by the transcription factor NFAT (Quintana et al. 2011). (3) In the last stage, the central SMAC (cSMAC) is formed when large TCR/CD3 microclusters reach the center. It is at the cSMAC where the activated TCR/CD3 signaling complexes are switched off, internalized and degraded (Lee et al. 2003). Activated T cells ultimately undergo Fas-induced apoptosis, a process that requires TRPM7 cleavage and increased I_{TRPM7} (see Sect. 18.5.2). Disruption of this process could lead to chronic inflammation.

18.5 TRPM7 is a Ubiquitously Expressed Ca^{2+}-Permeant Channel with a Functional Kinase Domain

TRPM7 is a ubiquitously expressed protein that can function both as a divalent cation channel at physiological membrane potentials, as well as a serine (Ser)/ threonine (Thr) kinase (Nadler et al. 2001; Runnels et al. 2001). Aside from the finding that intracellular Mg^{2+} blocks the channel, the gating mechanism remains largely unknown (Nadler et al. 2001; Hermosura et al. 2002; Kozak and Cahalan 2003). It now appears that there are two intracellular inhibition sites: one with low and the other with high affinity for Mg^{2+} (Chokshi et al. 2012a). Current understanding is that there will always be open channels at any time, depending on the local levels of free Mg^{2+} within the channel vicinity. This view is supported by experimental demonstrations of preactivated currents upon whole cell break-in (Hermosura et al. 2002; Kozak and Cahalan 2003; Chokshi et al. 2012b).

Perhaps due in part to its ubiquitous expression, its permeability to divalents including physiologically relevant ones such as Ca^{2+}, Mg^{2+}, Zn^{2+}, and Mn^{2+} (Monteihl-Zohler et al. 2003), and to its functional cytosolic kinase domain, TRPM7 has been implicated in a wide array of physiological processes including magnesium homeostasis, mechanotransduction, sensing of osmolarity and sheer stress, cell volume regulation, cytoskeletal rearrangements during cell adhesion and motility, synaptic vesicle fusion, melanopore maturation, skeletogenesis, ischaemic cell death, divalent sensing, coupling nutrient status with protein synthesis, Fas-induced apoptosis, vertebrate gastrulation, and thymopoiesis (Schmitz et al. 2003; Elizondo et al. 2005; Clark et al. 2006; Krapivinsky et al. 2006; Oancea et al. 2006; Su et al. 2006; Wei et al. 2007; Jin et al. 2008; Liu et al. 2011; Sun et al. 2009; Perraud et al. 2011; Desai et al. 2012). Importantly, global disruption is lethal before embryonic day 7 indicating that TRPM7 is required for early development (Jin et al. 2008). Subsequent studies showed that there is a temporal and spatial aspect to its role in development, and the absolute requirement for TRPM7 varies in each organ system. Pluripotent stem cells require TRPM7 (Jin et al. 2012). This review will only touch on aspects of TRPM7 function that might have a relevance to neurodegenerative pathophysiology, starting with its role in T cells.

As mentioned earlier, TRPM7 is one of only 5 TRP channels consistently expressed in activated primary human CD4+ T cells. It is also the most abundant TRP channel in these cells (Wenning et al. 2011). Unlike TRPM2 expression which varied over the 14-day experimental period, TRPM7 expression remained stable throughout, similar to ORAI1, the plasma membrane component of CRAC channels and considered to be a dominant Ca^{2+} influx pathway in lymphocytes (Feske 2007; Qu et al. 2011). Overall, these findings indicate that TRPM7 has an essential role in T cells.

18.5.1 TRPM7 is Required For T-Cell Development and Differentiation

To date, TRPM7 is the only ion channel known to be absolutely required for normal T-cell development. When specifically disrupted in thymocytes, a developmental block occurs, leaving a large fraction at the double negative DN3 stage, which are CD4-CD8-CD44-CD25+ (Jin et al. 2008). Failure to make the DN3 to DN4 transition reflects an inability to downregulate CD25, the IL-2 receptor (Koch and Radtke 2011). There are some thymocytes, however, that manage to escape this developmental block. Those could survive and populate peripheral lymphoid organs, but in low numbers possibly because of dysregulated synthesis of many relevant growth factors and the Stat3 transcription factor (Jin et al. 2008). Moreover, TRPM7$^{-/-}$ thymus has an abnormal architecture. Interestingly, a similar pattern of significantly low numbers of thymic progenitor-cells and abnormal thymic histology was observed in mSOD1^{G93A} mice (Seksenyan 2010). The same study also reported that ALS patients exhibited thymic dysfunction. These findings provide intriguing support for a connection between TRPM7, the only channel known to date to be required for thymic development, and ALS pathophysiology.

18.5.2 Increased TRPM7 Channel Activity is Required For Fas Receptor-Induced Apoptosis in T Cells

Depending on the initiating signals and cellular context, Fas receptor stimulation could either lead to apoptosis or nonapoptotic survival pathways. Fas receptor-induced apoptosis imposes control on the T-cell adaptive immune response, preventing chronic inflammation. Desai et al. (2012) found that TRPM7$^{-/-}$ T cells are deficient in Fas receptor-induced apoptosis. Further investigation revealed that upon Fas stimulation, TRPM7 is cleaved by caspases 3 and 8 at D1510, dissociating the kinase from the ion-conducting pore, an event that produces significantly larger I_{TRPM7}. They showed that increase in I_{TRPM7} is needed to potentiate Fas receptor signaling, endocytic internalization, and subsequent cell apoptosis. The cleaved kinase domain does not participate in Fas-induced apoptosis. In the absence of TRPM7, Fas internalization is disrupted, the activated T cell proceeds to a nonapoptotic Fas signaling pathway. The consequence of the latter route is evident in the presence of significant inflammatory infiltrates (T and B cells, neutrophils) in the lungs of aged TRPM7$^{-/-}$ mice (Desai et al. 2012), a phenotype that is remarkably similar to mice with T-cell-specific deletion of Fas (Hao et al. 2004). This age-dependent increase in inflammation associated with disrupted TRPM7 channel functions might contribute, along with other factors, to neurodegenerative pathophysiology.

The cleavage of TRPM7 at D1510 might have some phosphorylation dependence (Desai et al. 2012). It is within the Ser/Thr-rich region immediately N-terminal of the kinase domain that was reported to be massively autophosphorylated for proper substrate recognition (Clark et al. 2008b). The WP ALS-PDC-associated variant, TRPM7^{T1482I}, is part of this region, and although it is only one residue among many, phosphorylated T1482 could be an important determinant in interactions with a specific cellular substrate. The absence of a phosphorylated T1482 in TRPM7^{T1482I} could interfere with the caspase-mediated cleavage of TRPM7 at D1510. Although D1510 and T1482I are 28 residues apart, there is no way of knowing how the protein folds in this region, especially when the Ser and Thr residues are phosphorylated.

18.5.3 TRPM7 is a Mechanosensor That Can Interact with and Phosphorylate Myosin IIA

Studies have shown that TRPM7 kinase phosphorylates a short stretch of amino acids in the positively charged C-terminal end of myosin IIA heavy chain (MHCIIA) in a Ca^{2+}-dependent manner and in so doing, regulates myosin IIA filament stability and its incorporation into the actomyosin cytoskeleton (Clark et al. 2006, 2008a). Phosphorylation of MHCIIA by TRPM7 promotes local relaxation of cortical actomyosin cytoskeleton and loss of cortical tension that leads to the reorganization of focal adhesions and podosomes, to cell spreading and adhesion (Clark et al. 2008a). While phosphorylation of the regulatory myosin IIA light chains is the predominant mechanism regulating myosin II activity in mammalian cells, regulation of MHCIIA by TRPM7 could be important locally, especially given that the channel is enriched in cell adhesion structures and been shown to function as a mechanosensor (Oancea et al. 2006; Numata et al. 2007a,b). Both TRPM7 and MHCIIA are expressed in T cells where both could work cooperatively to control mechanosensation during T-cell activation (Kumari et al. 2012; Yu et al. 2012). The importance of mechanosensation in T cells is highlighted by a recent study reporting that substrate stiffness influences human T-cell proliferation, differentiation, and expansion (O'Connor et al. 2012). In the light of increasing attempts to manipulate T-cell subtype differentiation in vitro for therapeutic applications, information on how the interaction between TRPM7 and myosin IIA contributes to mechanosensation in T cells would be very useful.

18.5.4 TRPM7, Calpain 2, and T-Cell Migration

The channel activity of TRPM7 reportedly regulates cell adhesion by controlling the activity of calpain 2 (calpain-m), a protease implicated in the control of cell adhesion through focal adhesion disassembly (Su et al. 2006). The same group further reported that excessive TRPM7 channel activity causes oxidative stress, ultimately causing the activation of calpain 2 via MAPK/JNK (Su et al. 2010).

TRPM7 and calpain 2 might also be involved in lymphocyte migration (Svensson et al. 2010). T cells are highly motile because inherent to their role in immunosurveillance is the need to migrate from the peripheral blood circulation, to tissues and lymph nodes. The integrin LFA-1 on T cells recognize, attach, and migrate on substrates expressing the LFA-1 ligand ICAM-1. Calpain 2 was shown to play a crucial role in the turnover of LFA-1 adhesions on migrating T cells (Svensson et al. 2010). When calpain activity is blocked, the posterior end of the cell could not detach easily so as the front of the cell moves forward, trailing edge elongation and shedding of LFA-1 clusters occur. These events slow down cell migration. Ca^{2+} influx via a plasma membrane-bound constitutive channel was found to be an essential requirement to maintain calpain 2 activity (Svensson et al. 2010). The identity of this channel is not known, but Orai1/CRAC was ruled out and the strongest candidate is TRPM7, especially given that it colocalizes with calpain 2 in peripheral adhesion complexes and was further shown to regulate directional cell motility (Su et al. 2006, 2011). Supporting this view are findings that TRPM7 participates in high-calcium microdomains ('calcium flickers') in migrating human embryonic lung fibroblasts (Wei et al. 2009).

18.6 The Environmental Triggers in WP ALS-PDC, Altered Mineral Homeostasis and Cycad Toxins, Affect TRPM2- and TRPM7-Associated Cellular Processes

The three WP ALS-PDC foci - Guam, Kii Peninsula of Japan, and southern West New Guinea - shared two features: (1) use of the cycad plant for food and medicine (Whiting 1964) and (2) a unique mineral environment characterized by severely low levels of magnesium (Mg^{2+}) and high levels of bioavailable transition metals in the soil and drinking water, conditions that promote oxidative stress (Yase 1972; Garruto et al. 1984; Kihira et al. 2002). This led to the hypothesis that the *prolonged* exposure to these environmental triggers combined with genetic susceptibility could be involved in the pathogenesis of WP ALS and PDC (Hermosura et al. 2005, 2008; Hermosura and Garruto 2007). Disease incidence is declining in Guam for the ALS disorder, but not PDC, and importantly, a worrisome increase has been observed on Kii, Japan (Galasko et al. 2007; Kihira et al. 2008, 2010, 2012; Kuzuhara 2011). While the $Trpm7^{T1482I}$ variant was not found in Kii cases (Hara et al. 2010) and to date, no report is made regarding the $Trpm2^{P1018L}$ in this population, it is still possible that the signal transduction pathways and cellular processes associated with TRPM2 and TRPM7 could be involved in the disease process, especially those associated with chronic inflammation. This section will summarize relevant effects of the cycad toxins, β-N-methylamino-L-alanine (BMAA) and methylazoxymethanol (MAM) on cellular processes and signaling networks that interconnect with TRPM2 and TRPM7 activities.

18.6.1 Nonexcitotoxic Levels of BMAA Affect Protein Synthesis and Cause ER Stress

BMAA is a nonprotein amino acid implicated in WP ALS-PDC (Spencer et al. 1987). The controversy surrounding BMAA's role in WP ALS-PDC, and perhaps AD, is well documented in the literature and will not be covered in this review. Two very recent studies reported promising results that might shed light on some of the effects of BMAA that could have a bearing on its role in neurodegenerative pathophysiology (Okle et al. 2013; Karlsson et al. 2012). The first study used human SH-SY5Y neuroblastoma cells, a cell line lacking NMDA receptors (thus controlling for excitotoxic effects), and showed that at low nonexcitotoxic levels, BMAA became tightly associated with proteins, interfered with protein synthesis by affecting the phosphorylation state of the translation initiator protein, elf2-α, increased protein ubiquitinylation, 20S proteasomal and caspase 12 activities, and induced the expression of ER stress markers such as C/EBP homologous protein

(CHOP), a protein involved in ER stress-mediated apoptosis (Okle et al. 2013). BMAA-induced ER stress was also observed in the cell lines HEK-293 and THP-1, the latter commonly used to study monocyte and macrophage functions. In these two cell lines, BMAA promoted increased phosphorylation of PERK (PKR-like ER-localized eIF2-α kinase), as well as increased expression of CHOP and the molecular chaperone BAG1. This group also found that high (>2 mM) levels of BMAA increased ROS production, protein oxidation, and significantly upregulated the activities of caspases 3 and 7. It would be interesting to know if exposure to BMAA affects astrocytes similarly.

18.6.2 Neonatal Exposure to BMAA Causes Long-Term Brain Changes

The second study examined the effects of neonatal BMAA exposure and reported that at low doses, BMAA induced long-term changes in the adult hippocampus including reduced expression of proteins involved in energy metabolism and intracellular signaling, but not observable brain lesions (Karlsson et al. 2012). High doses of BMAA, however, caused marked neuronal loss and astrogliosis especially in the CA1 region, which was most susceptible and where positive staining for α-syn and ubiquitin was observed. Astrocytes, in particular, were strongly positive for ubiquitin. High doses of BMAA also induced changes in S100β, a glial specific Ca^{2+}-binding protein, and in proteins involved in Ca^{2+}, as well as guanine nucleotide signaling. Because of the critical roles of components of the metabolic, Ca^{2+}, and guanine nucleotide signaling pathways in cell, tissue, and organ system development, these results suggest that neonatal exposure to BMAA could disrupt many developmental processes, with the effects manifested in adulthood. Indeed, in a previous study, the same group reported that exposure of mice to BMAA during the postnatal period led to spatial learning and memory impairments in adulthood, perhaps due in part to easier transport of BMAA across the BBB in neonates compared to adults (Karlsson et al. 2009a). Similar age-dependent susceptibility to cycad exposure was also seen among Guam Chamorros; young adults exhibited increased risk for WP ALS-PDC (Borenstein et al. 2007).

Overall, these two studies suggest that low nonexcitotoxic levels of BMAA could affect protein synthesis, causing increased ubiquitinylation and misfolding of proteins that could promote ER stress. These observations support the prevailing view that protein misfolding and accumulation are early events in the onset of neurodegeneration (Ido et al. 2011; Bendotti et al. 2012). Certainly, accumulation of various forms of aggregates—tau, α-syn, β-amyloid, TDP-43—are seen in WP ALS-PDC (Geser et al. 2008; Miklossy et al. 2008). The finding that BMAA induced ubiquitin-positive deposits in astrocytes (Karlsson et al. 2012) is particularly noteworthy because of a possible connection to neuroinflammation.

Astrocytes generally support neurons by secreting growth promoting mediators; however, transcriptome analysis of astrocytes containing α-syn and ubiquitin inclusions revealed gene expression changes: proinflammatory mediators are upregulated, so are genes involved in antigen processing and presentation, toll-like receptor signaling, MAPK signaling, complement cascades, leukocyte trans-endothelial migration, and the Janus kinase-Signal Transducer and Activator of Transcription (JAK-STAT) signaling pathway (Lee et al. 2010). Thus, BMAA-induced ubiquitin deposits in astrocytes could promote proinflammatory cytokine release from these cells.

18.6.3 BMAA and TRPM7 Intersect in Protein Synthesis and in Melanocytes

TRPM7 has recently been shown to play a role in coupling the rate of protein synthesis to the availability of nutrients, particularly Mg^{2+} (Perraud et al. 2011). When Mg^{2+} levels are low, the kinase activity of TRPM7 influences the phosphorylation of eukaryotic elongation factor 2 (eeF2), inhibiting it, thereby down-regulating protein synthesis until Mg^{2+} becomes available. Thus, protein synthesis and handling might be a common interaction point for cellular signaling networks affected by low Mg^{2+}, BMAA, and TRPM7/TRPM7^{T1482I}.

Another example might be in melanocytes. In addition to being temporarily paralyzed, homozygous *Trpm7* zebrafish mutants exhibited melanophore death indicating that TRPM7 is essential for melanocyte survival (McNeill et al. 2007). It was determined that TRPM7 is required to detoxify intermediates of melanin synthesis. BMAA reportedly interacts with melanin, especially during its synthesis, accumulating in melanin and neuromelanin-containing cells over time (Karlsson et al. 2009b). BMAA could accumulate to such an extent as to overcome TRPM7-mediated detoxification of melanin intermediates. Similarly, any disruption in TRPM7 activity will impact the detoxification process, especially in the presence of BMAA.

18.6.4 Signaling Pathways in Neurodegeneration and Cancer: Intersection Points for Cycad Metabolites and TRPM2, TRPM7?

Cycasin, and It has been proposed that the cycad metabolite, MAM, affects cellular signaling pathways common to cancer and neurodegeneration, with the outcome determined by whether the cells can divide or not (Kisby and Spencer 2011; Kisby et al. 2011). MAM was shown to disrupt the expression of genes that regulate the neuronal cytoskeleton, protein degradation, mitochondrial metabolism, cytokine and Ca^{2+} signaling (Kisby et al. 2011). Many of these intersect with the signaling pathways affected by TRPM2 and TRPM7. For instance, the protein GSK-3β and the p38-MAPK pathway are also associated with TRPM2 (Yamamoto et al. 2008; Xie et al. 2011). Moreover, TRPM2 and TRPM7 are two proteins that have been linked to both neurodegenerative diseases and cancer: TRPM2 with WP ALS-PDC and prostate cancer (Hermosura et al. 2008; Zeng et al. 2010); TRPM7 with WP ALS-PDC, AD, head and neck cancers, and migration of human nasopharyngeal carcinoma (Chen et al. 2010; Hermosura et al. 2005; Jiang et al. 2007; Landman et al. 2006; Tseveleki et al. 2010). Furthermore, TRPM7T^{1482I} was associated with increased risk for adenomatous and hyperplastic polyps (Dai et al. 2007).

Studies that seek to examine if/how MAM affects TRPM2 and TRPM7 functions are especially important at this time because there are environmental agents and drugs that are in current use (e.g., nitrosamine, hydrazines, and streptozotocin) that have MAM-like genotoxic properties, and could thus be potential environmental triggers for neurodegeneration, cancers, and even neurodevelopmental disorders (Kisby et al. 2011).

18.7 Conclusion

Two Ca^{2+}-permeable, cation conducting channels expressed in neuronal and immune cells were identified as candidate susceptibility genes in WP ALS-PDC (Hermosura and Garruto 2007). Based on accumulating evidence, their roles in innate and adaptive immunity interacting with prolonged exposure to cycad toxins and low Mg^{2+} conditions, and even mitochondrial mutations (Lynch et al. 2008; Reiff et al. 2011) might contribute to the pathophysiology of this disease. Certainly, the connections described in this review are circumstantial at this point, and a lot of work is needed to confirm or reject those. However, it is not surprising that channels involved in inflammation and immunity could confer susceptibility to WP ALS-PDC. Even back in 1978 (more than 30 years ago), lymphopenia and defective T cell-mediated immunity were reported among Guam ALS-PDC cases (Hoffmann et al. 1978). That study was followed a few years later by a report describing age-dependent immune dysfunctions even among normal Guamanians suggesting that as a group, these people might be genetically predisposed to immune dysfunctions and age-associated neurodegeneration (Hoffmann et al. 1983). Therefore, identifying and characterizing genetic variants of other channels involved in innate and adaptive immune Ca^{2+} signaling within this population might be a promising avenue to pursue as we seek to understand the pathophysiology of neurodegenerative disorders.

Acknowledgments Support and encouragement from faculty and staff of the Bekesy Laboratory of Biology, PBRC, UH Manoa is gratefully acknowledged.

References

Annunziato F, Romagnani S (2009) Heterogeneity of human effector CD4 + T cells. Arthritis Res Ther 11:257

Appel SH, Beers DR, Henkel JS (2010) T cell-microglial dialogue in Parkinson's disease and amyotrophic lateral sclerosis: are we listening? Trends Immunol 31:7–17

Appel SH, Zhao W, Beers DR, Henkel JS (2011) The microglial-motoneuron dialogue in ALS. Acta Myol 30:4–8

Banerjee R, Mosley RL, Reynolds A, Dhar V, Jackson-Lewis PH, Gordon S Przedborski, Gendelman HE (2008) Adaptive immune neuroprotection in G93A-SOD1 amyotrophic lateral sclerosis mice. PLoS ONE 3:e2740

Barres BA (2008) The mystery and magic of glia: a perspective on their roles in health and disease. Neuron 60:430–440

Beck A, Kolisek M, Bagley LA, Fleig A, Penner R (2006) Nicotinic acid adenine dinucleotide phosphate and cyclic ADP-ribose regulate TRPM2 channels in T lymphocytes. FASEB J 20:962–964

Beers DR, Henkel JS, Zhao W, Wang J, Appel SH (2008) CD4 + T cells support glial neuroprotection, slow disease progression, and modify glial morphology in an animal model of inherited ALS. Proc Natl Acad Sci U S A 105:15558–15563

Beers DR, Henkel JS, Zhao W, Wang J, Huang A, Wen S, Liao B, Appel SH (2011) Endogenous regulatory T lymphocytes ameliorate amyotrophic lateral sclerosis in mice and correlate with disease progression in patients with amyotrophic lateral sclerosis. Brain 134:1293–1314

Belrose JC, Xie YF, Gierszewski LJ, MacDonald JF, Jackson MF (2012) Loss of glutathione homeostasis associated with neuronal senescence facilitates TRPM2 channel activation in cultured hippocampal pyramidal neurons. Mol Brain 5:11

Bendotti C, Marino M, Cheroni C, Fontana E, Crippa V, Poletti A, De Biasi S (2012) Dysfunction of constitutive and inducible ubiquitin-proteasome system in amyotrophic lateral sclerosis: implication for protein aggregation and immune response. Prog Neurobiol 97:101–126

Bogeski I, Kappl R, Kummerow C, Gulaboski R, Hoth M, Niemeyer BA (2011) Redox regulation of calcium ion channels: chemical and physiological aspects. Cell Calcium 50:407–423

Bond CE, Greenfield SA (2007) Multiple cascade effects of oxidative stress on astroglia. Glia 55:1348–1361

Borenstein AR, Mortimer JA, Schofield E, Wu Y, Salmon DP, Gamst A, Olichney J, Thal LJ, Silbert L, Kaye J, Craig UL, Schellenberg GD, Galasko DR (2007) Cycad exposure and risk of dementia, MCI, and PDC in the Chamorro population of Guam. Neurology 68:1764–1771

Buelow B, Song Y, Scharenberg AM (2008) The Poly(ADP-ribose) polymerase PARP-1 is required for oxidative stress-induced TRPM2 activation in lymphocytes. J Biol Chem 283:24571–24583

Capasso M, Jeng JM, Malavolta M, Mocchegiani E, Sensi SL (2005) Zinc dyshomeostasis: a key modulator of neuronal injury. J Alzheimers Dis 8:93–108

Chen JP, Luan Y, You CX, Chen XH, Luo RC, Li R (2010) TRPM7 regulates the migration of human nasopharyngeal carcinoma cell by mediating Ca(2 +) influx. Cell Calcium 47:425–432

Chiu IM, Chen A, Zheng Y, Kosaras B, Tsiftsoglou SA, Vartanian TK, Brown RH Jr, Carroll MC (2008) T lymphocytes potentiate endogenous neuroprotective inflammation in a mouse model of ALS. Proc Natl Acad Sci U S A 105:17913–17918

Chokshi R, Matsushita M, Kozak JA (2012a) Sensitivity of TRPM7 channels to Mg2 + characterized in cell-free patches of Jurkat T lymphocytes. Am J Physiol Cell Physiol 302:C1642–C1651

Chokshi R, Matsushita M, Kozak JA (2012b) Detailed examination of Mg2 + and pH sensitivity of human TRPM7 channels. Am J Physiol Cell Physiol 302:C1004–C1011

Chung KK, Freestone PS, Lipski J (2011) Expression and functional properties of TRPM2 channels in dopaminergic neurons of the substantia nigra of the rat. J Neurophysiol 106:2865–2875

Clapham DE (2003) TRP channels as cellular sensors. Nature 426:517–524

Clark K, Langeslag M, van Leeuwen B, Ran L, Ryazanov AG, Figdor CG, Moolenaar WH, Jalink K, van Leeuwen FN (2006) TRPM7, a novel regulator of actomyosin contractility and cell adhesion. EMBO J 25:290–301

Clark K, Middelbeek J, Lasonder E, Dulyaninova NG, Morrice NA, Ryazanov AG, Bresnick AR, Figdor CG, van Leeuwen FN (2008a) TRPM7 regulates myosin IIA filament stability and protein localization by heavy chain phosphorylation. J Mol Biol 378:790–803

Clark K, Middelbeek J, Morrice NA, Figdor CG, Lasonder E, van Leeuwen FN (2008b) Massive autophosphorylation of the Ser/Thr-rich domain controls protein kinase activity of TRPM6 and TRPM7. PLoS ONE 3:e1876

Cook NL, Vink R, Helps SC, Manavis J, van den Heuvel C (2010) Transient receptor potential melastatin 2 expression is increased following experimental traumatic brain injury in rats. J Mol Neurosci 42:192–199

Csanady L, Torocsik B (2009) Four Ca2 + ions activate TRPM2 channels by binding in deep crevices near the pore but intracellularly of the gate. J Gen Physiol 133:189–203

Dai Q, Shrubsole MJ, Ness RM, Schlundt D, Cai Q, Smalley WE, Li M, Shyr Y, Zheng W (2007) The relation of magnesium and calcium intakes and a genetic polymorphism in the magnesium transporter to colorectal neoplasia risk. Am J Clin Nutr 86:743–751

Damann N, Voets T, Nilius B (2008) TRPs in our senses. Curr Biol 18(18):R880–R889

DeCoursey TE, Cherny VV, Zhou W, Thomas LL (2000) Simultaneous activation of NADPH oxidase-related proton and electron currents in human neutrophils. Proc Natl Acad Sci U S A 97:6885–6889

DeCoursey TE, Morgan D, Cherny VV (2003) The voltage dependence of NADPH oxidase reveals why phagocytes need proton channels. Nature 422:531–534

Desai BN, Krapivinsky G, Navarro B, Krapivinsky L, Carter BC, Febvay S, Delling M, Penumaka A, Ramsey IS, Manasian Y, Clapham DE (2012) Cleavage of TRPM7 releases the kinase domain from the ion channel and regulates its participation in Fas-induced apoptosis. Dev Cell 22:1149–1162

Di A, Gao XP, Qian F, Kawamura T, Han J, Hecquet C, Ye RD, Vogel SM, Malik AB (2012) The redox-sensitive cation channel TRPM2 modulates phagocyte ROS production and inflammation. Nat Immunol 13:29–34

Du J, Xie J, Yue L (2009) Modulation of TRPM2 by acidic pH and the underlying mechanisms for pH sensitivity. J Gen Physiol 134:471–488

Elizan TS, Hirano A, Abrams BM, Need RL, Van Nuis C, Kurland LT (1966) Amyotrophic lateral sclerosis and parkinsonism-dementia complex of Guam. Neurological reevaluation, Arch Neurol 14:356–368

Elizondo MR, Arduini BL, Paulsen J, MacDonald EL, Sabel JL, Henion PD, Cornell RA, Parichy DM (2005) Defective skeletogenesis with kidney stone formation in dwarf zebrafish mutant for trpm7. Curr Biol 15(667–71):S1–S7

Evans MC, Couch Y, Sibson N, Turner MR (2012) Inflammation and neurovascular changes in amyotrophic lateral sclerosis. Mol Cell Neurosci. doi:10.1016/j.mcn.2012.10.008

Feske S (2007) Calcium signalling in lymphocyte activation and disease. Nat Rev Immunol 7:690–702

Freestone PS, Chung KK, Guatteo E, Mercuri NB, Nicholson LF, Lipski J (2009) Acute action of rotenone on nigral dopaminergic neurons–involvement of reactive oxygen species and disruption of Ca2 + homeostasis. Eur J Neurosci 30:1849–1859

Galasko D, Salmon D, Gamst A, Olichney J, Thal LJ, Silbert L, Kaye J, Brooks P, Adonay R, Craig UK, Schellenberg G, Borenstein AR (2007) Prevalence of dementia in Chamorros on Guam: relationship to age, gender, education, and APOE. Neurology 68:1772–1781

Gally F, Hartney JM, Janssen WJ, Perraud AL (2009) CD38 plays a dual role in allergen-induced airway hyperresponsiveness. Am J Respir Cell Mol Biol 40:433–442

Gajdusek DC, Salazar AM (1982) Amyotrophic lateral sclerosis and parkinsonian syndromes in high incidence among the Auyu and Jakai people of West New Guinea. Neurology 32:107–126

Garruto R.M., R. Yanagihara, D.C. Gajdusek, D.M. Arion (1984) Concentrations of Heavy Metals and Essential Minerals in Garden Soil and Drinking Water in the Western Pacific, Amyotrophic Lateral Sclerosis in Asia and Oceania. 265-330

Gasser A, Glassmeier G, Fliegert R, Langhorst MF, Meinke S, Hein D, Kruger S, Weber K, Heiner I, Oppenheimer N, Schwarz JR, Guse AH (2006) Activation of T cell calcium influx by the second messenger ADP-ribose. J Biol Chem 281:2489–2496

Gendelman HE, Appel SH (2011) Neuroprotective activities of regulatory T cells. Trends Mol Med 17:687–688

Geser F, Winton MJ, Kwong LK, Xu Y, Xie SX, Igaz LM, Garruto RM, Perl DP, Galasko D, Lee VM, Trojanowski JQ (2008) Pathological TDP-43 in parkinsonism-dementia complex and amyotrophic lateral sclerosis of Guam. Acta Neuropathol 115:133–145

Glass CK, Saijo K, Winner B, Marchetto MC, Gage FH (2010) Mechanisms underlying inflammation in neurodegeneration. Cell 140:918–934

Gonzalez A, Granados MP et al (2006) H2O2 mobilizes Ca2 + from agonist- and thapsigargin-sensitive and insensitive intracellular stores and stimulates glutamate secretion in rat hippocampal astrocytes. Neurochem Res 31(6):741–750

Gordon S (2003) Alternative activation of macrophages. Nat Rev Immunol 3:23–35

Guiroy DC, Miyazaki M, Multhaup G, Fischer P, Garruto RM, Beyreuther K, Masters CL, Simms G, Gibbs CJ Jr, Gajdusek DC (1987) Amyloid of neurofibrillary tangles of Guamanian parkinsonism-dementia and Alzheimer disease share identical amino acid sequence. Proc Natl Acad Sci U S A 84:2073–2077

Hao Z, Hampel B, Yagita H, Rajewsky K (2004) T cell-specific ablation of Fas leads to Fas ligand-mediated lymphocyte depletion and inflammatory pulmonary fibrosis. J Exp Med 199:1355–1365

Hara Y, Wakamori M, Ishii M, Maeno E, Nishida M, Yoshida T, Yamada H, Shimizu S, Mori E, Kudoh J, Shimizu N, Kurose H, Okada Y, Imoto K, Mori Y (2002) LTRPC2 Ca2 + -permeable channel activated by changes in redox status confers susceptibility to cell death. Mol Cell 9:163–173

Hara K, Kokubo Y, Ishiura H, Fukuda Y, Miyashita A, Kuwano R, Sasaki R, Goto J, Nishizawa M, Kuzuhara S, Tsuji S (2010) TRPM7 is not associated with amyotrophic lateral sclerosis-parkinsonism dementia complex in the Kii peninsula of Japan. Am J Med Genet B Neuropsychiatr Genet 153B:310–313

Hardaker L, Bahra P, de Billy BC, Freeman M, Kupfer N, Wyss D, Trifilieff A (2012) The ion channel transient receptor potential melastatin-2 does not play a role in inflammatory mouse models of chronic obstructive pulmonary diseases. Respir Res 13:30

Harraz MM, Marden JJ, Zhou W, Zhang Y, Williams A, Sharov VS, Nelson K, Luo M, Paulson H, Schoneich C, Engelhardt JF (2008) SOD1 mutations disrupt redox-sensitive Rac regulation of NADPH oxidase in a familial ALS model. J Clin Invest. 118:659–670

Hecquet CM, Ahmmed GU, Vogel SM, Malik AB (2008) Role of TRPM2 channel in mediating H2O2-induced Ca2 + entry and endothelial hyperpermeability. Circ Res 102:347–355

Henkel JS, Beers DR, Zhao W, Appel SH (2009) Microglia in ALS: the good, the bad, and the resting. J Neuroimmune Pharmacol 4:389–398

Henkel JS, Beers DR, Wen S, Rivera AL, Toennis KM, Appel JE, Zhao W, Moore DH, Powell SZ, Appel SH (2013) Regulatory T-lymphocytes mediate amyotrophic lateral sclerosis progression and survival. EMBO Mol Med 5:64–79

Hermosura MC, Monteilh-Zoller MK, Scharenberg AM, Penner R, Fleig A (2002) Dissociation of the store-operated calcium current I(CRAC) and the Mg-nucleotide-regulated metal ion current MagNuM. J Physiol 539:445–458

Hermosura MC, Nayakanti H, Dorovkov MV, Calderon FR, Ryazanov AG, Haymer DS, Garruto RM (2005) A TRPM7 variant shows altered sensitivity to magnesium that may contribute to the pathogenesis of two Guamanian neurodegenerative disorders. Proc Natl Acad Sci U S A 102:11510–11515

Hermosura MC, Garruto RM (2007) TRPM7 and TRPM2-Candidate susceptibility genes for Western Pacific ALS and PD? Biochim Biophys Acta 1772:822–835

Hermosura MC, Cui AM, Go RC, Davenport B, Shetler CM, Heizer JW, Schmitz C, Mocz G, Garruto RM, Perraud AL (2008) Altered functional properties of a TRPM2 variant in Guamanian ALS and PD. Proc Natl Acad Sci U S A 105:18029–18034

Hirano A (1992) Amyotrophic lateral sclerosis and parkinsonism-dementia complex on Guam: immunohistochemical studies. Keio J Med 41:6–9

Hoffman PM, Robbins DS, Gibbs CJ Jr, Gajdusek DC (1983) Immune function among normal Guamanians of different age. J Gerontol 38:414–419

Hoffman PM, Robbins DS, Nolte MT, Gibbs CJ Jr, Gajdusek DC (1978) Cellular immunity in Guamanians with amyotrophic lateral sclerosis and Parkinsonism-dementia. N Engl J Med 299:680–685

Hoth M, Penner R (1992) Depletion of intracellular calcium stores activates a calcium current in mast cells. Nature 355:353–356

Huang X, Reynolds AD, Mosley RL, Gendelman HE (2009) CD 4 + T cells in the pathobiology of neurodegenerative disorders. J Neuroimmunol 211:3–15

Ido A, Fukuyama H, Urushitani M (2011) Protein Misdirection Inside and Outside Motor Neurons in Amyotrophic Lateral Sclerosis (ALS): A Possible Clue for Therapeutic Strategies. Int J Mol Sci 12:6980–7003

Jia J, Verma S, Nakayama S, Quillinan N, Grafe MR, Hurn PD, Herson PS (2011) Sex differences in neuroprotection provided by inhibition of TRPM2 channels following experimental stroke. J Cereb Blood Flow Metab 31:2160–2168

Jiang J, Li MH, Inoue K, Chu XP, Seeds J, Xiong ZG (2007) Transient receptor potential melastatin 7-like current in human head and neck carcinoma cells: role in cell proliferation. Cancer Res 67:10929–10938

Jin J, Desai BN, Navarro B, Donovan A, Andrews NC, Clapham DE (2008) Deletion of Trpm7 disrupts embryonic development and thymopoiesis without altering Mg2 + homeostasis. Science 322:756–760

Jin J, Wu LJ, Jun J, Cheng X, Xu H, Andrews NC, Clapham DE (2012) The channel kinase, TRPM7, is required for early embryonic development. Proc Natl Acad Sci U S A 109:E225–E233

Junker C, Hoth M (2011) Immune synapses: mitochondrial morphology matters. EMBO J 30:1187–1189

Karlsson O, Lindquist NG, Brittebo EB, Roman E (2009a) Selective brain uptake and behavioral effects of the cyanobacterial toxin BMAA (beta-N-methylamino-L-alanine) following neonatal administration to rodents. Toxicol Sci 109:286–295

Karlsson O, Berg C, Brittebo EB, Lindquist NG (2009b) Retention of the cyanobacterial neurotoxin beta-N-methylamino-L-alanine in melanin and neuromelanin-containing cells–a possible link between Parkinson-dementia complex and pigmentary retinopathy. Pigment Cell Melanoma Res 22:120–130

Karlsson O, Berg AL, Lindstrom AK, Hanrieder J, Arnerup G, Roman E, Bergquist J, Lindquist NG, Brittebo EB, Andersson M (2012) Neonatal Exposure to the Cyanobacterial Toxin BMAA Induces Changes in Protein Expression and Neurodegeneration in Adult Hippocampus. Toxicol Sci 130:391–404

Kashio M, Sokabe T, Shintaku K, Uematsu T, Fukuta N, Kobayashi N, Mori Y, Tominaga M (2012) Redox signal-mediated sensitization of transient receptor potential melastatin 2 (TRPM2) to temperature affects macrophage functions. Proc Natl Acad Sci U S A 109:6745–6750

Katano M, Numata T, Aguan K, Hara Y, Kiyonaka S, Yamamoto S, Miki T, Sawamura S, Suzuki T, Yamakawa K, Mori Y (2012) The juvenile myoclonic epilepsy-related protein EFHC1 interacts with the redox-sensitive TRPM2 channel linked to cell death. Cell Calcium 51:179–185

Kihira T, Yoshida S, Yase Y, Ono S, Kondo T (2002) Chronic low-Ca/Mg high-Al diet induces neuronal loss. Neuropathology 22:171–179

Kihira T, Yoshida S, Okamoto K, Kazimoto Y, Ookawa M, Hama K, Miwa H, Kondo T (2008) Survival rate of patients with amyotrophic lateral sclerosis in Wakayama Prefecture, Japan, 1966 to 2005. J Neurol Sci 268:95–101

Kihira T, Yoshida S, Murata K, Ishiguti H, Kondo T, Kohmoto J, Okamoto K, Kokubo Y, Kuzuhara S (2010) Changes in the incidence and clinical features of ALS in the Koza, Kozagawa, and Kushimoto area of the Kii Peninsula–from the 1960s to the 2000s (follow-up study). Brain Nerve 62:72–80

Kihira T, Yoshida S, Kondo T, Iwai K, Wada S, Morinaga S, Kazimoto Y, Okamoto K, Kokubo Y, Kuzuhara S (2012) An increase in ALS incidence on the Kii Peninsula, 1960-2009: a possible link to change in drinking water source. Amyotroph Lateral Scler. 13:347–350

Kimura K, Yase Y, Higashi Y, Une S, Yamamoto K, Iwasaki M, Tsumoto I, Sugiura M, Yoshimura S, Namikawa K, Kumura J, Iwamoto S, Yamamoto I, Handa Y, Yata M, Yata Y (1963) Epidemiological and geomedical studies on amytrophic lateral sclerosis. Dis Nerv Syst 14:155–159

Kisby GE, Fry RC, Lasarev MR, Bammler TK, Beyer RP, Churchwell M, Doerge DR, Meira LB, Palmer VS, Ramos-Crawford AL, Ren X, Sullivan RC, Kavanagh TJ, Samson LD, Zarbl H, Spencer PS (2011) The cycad genotoxin MAM modulates brain cellular pathways involved in neurodegenerative disease and cancer in a DNA damage-linked manner. PLoS ONE 6:e20911

Kisby GE, Spencer PS (2011) Is neurodegenerative disease a long-latency response to early-life genotoxin exposure? Int J Environ Res Public Health 8:3889–3921

Knowles H, Heizer JW, Li Y, Chapman K, Ogden CA, Andreasen K, Shapland E, Kucera G, Mogan J, Humann J, Lenz LL, Morrison AD, Perraud AL (2011) Transient Receptor Potential Melastatin 2 (TRPM2) ion channel is required for innate immunity against Listeria monocytogenes. Proc Natl Acad Sci U S A. 108:11578–11583

Knowles H, Li Y, Perraud AL (2012) The TRPM2 ion channel, an oxidative stress and metabolic sensor regulating innate immunity and inflammation. Immunol Res. doi:10.1007/s12026-012-8373-8

Koch U, Radtke F (2011) Mechanisms of T cell development and transformation. Annu Rev Cell Dev Biol 27:539–562

Kozak JA, Cahalan MD (2003) MIC channels are inhibited by internal divalent cations but not ATP. Biophys J 84(2 Pt 1):922–927

Krapivinsky G, Mochida S, Krapivinsky L, Cibulsky SM, Clapham DE (2006) The TRPM7 ion channel functions in cholinergic synaptic vesicles and affects transmitter release. Neuron 52:485–496

Kumari S, Vardhana S, Cammer M, Curado S, Santos L, Sheetz MP, Dustin ML (2012) T Lymphocyte Myosin IIA is Required for Maturation of the Immunological Synapse. Front Immunol. 3:230

Kuzuhara S (2011) Muro disease: amyotrophic lateral sclerosis/parkinsonism-dementia complex in Kii peninsula of Japan. Brain Nerve 63:119–129

Landman N, Jeong SY, Shin SY, Voronov SV, Serban G, Kang MS, Park MK, Di Paolo G, Chung S, Kim TW (2006) Presenilin mutations linked to familial Alzheimer's disease cause an imbalance in phosphatidylinositol 4,5-bisphosphate metabolism. Proc Natl Acad Sci U S A 103:19524–19529

Lee KH, Dinner AR, Tu C, Campi G, Raychaudhuri S, Varma R, Sims TN, Burack WR, Wu H, Wang J, Kanagawa O, Markiewicz M, Allen PM, Dustin ML, Chakraborty AK, Shaw AS (2003) The immunological synapse balances T cell receptor signaling and degradation. Science 302:1218–1222

Lee M, Cho T, Jantaratnotai N, Wang YT, McGeer E, McGeer PL (2010) Depletion of GSH in glial cells induces neurotoxicity: relevance to aging and degenerative neurological diseases. FASEB J 24:2533–2545

Lewis CA, Manning J, Rossi F, Krieger C (2012) The Neuroinflammatory Response in ALS: The Roles of Microglia and T Cells. Neurol Res Int 2012:803701

Liao B, Zhao W, Beers DR, Henkel JS, Appel SH (2012) Transformation from a neuroprotective to a neurotoxic microglial phenotype in a mouse model of ALS. Exp Neurol 237:147–152

Lincecum JM, Vieira FG, Wang MZ, Thompson K, De Zutter GS, Kidd J, Moreno A, Sanchez R, Carrion IJ, Levine BA, Al-Nakhala BM, Sullivan SM, Gill A, Perrin S (2010) From transcriptome analysis to therapeutic anti-CD40L treatment in the SOD1 model of amyotrophic lateral sclerosis. Nat Genet 42:392–399

Liu W, Su LT, Khadka DK, Mezzacappa C, Komiya Y, Sato A, Habas R, Runnels LW (2011) TRPM7 regulates gastrulation during vertebrate embryogenesis. Dev Biol 350:348–357

Lobsiger CS, Cleveland DW (2007) Glial cells as intrinsic components of non-cell-autonomous neurodegenerative disease. Nat Neurosci 10:1355–1360

Lynch D, Wanglund C, Spathis R, Chan CW, Reiff DM, Lum JK, Garruto RM (2008) The contribution of mitochondrial dysfunction to a gene-environment model of Guamanian ALS and PD. Mitochondrion 8:109–116

Magnone M, Bauer I, Poggi A, Mannino E, Sturla L, Brini M, Zocchi E, De Flora A, Nencioni A, Bruzzone S (2012) NAD + levels control Ca2 + store replenishment and mitogen-induced increase of cytosolic Ca2 + by Cyclic ADP-ribose-dependent TRPM2 channel gating in human T lymphocytes. J Biol Chem 287:21067–21081

McHugh D, Flemming R, Xu SZ, Perraud AL, Beech DJ (2003) Critical intracellular Ca2 + dependence of transient receptor potential melastatin 2 (TRPM2) cation channel activation. J Biol Chem 278:11002–11006

McNeill MS, Paulsen J, Bonde G, Burnight E, Hsu MY, Cornell RA (2007) Cell death of melanophores in zebrafish trpm7 mutant embryos depends on melanin synthesis. J Invest Dermatol 127:2020–2030

McQuillin A, Bass NJ, Kalsi G, Lawrence J, Puri V, Choudhury K, Detera-Wadleigh SD, Curtis D, Gurling HM (2006) Fine mapping of a susceptibility locus for bipolar and genetically related unipolar affective disorders, to a region containing the C21ORF29 and TRPM2 genes on chromosome 21q22.3. Mol Psychiatry 11:134–142

Melzer N, Hicking G, Gobel K, Wiendl H (2012) TRPM2 cation channels modulate T cell effector functions and contribute to autoimmune CNS inflammation. PLoS ONE 7:e47617

Miklossy J, Steele JC, Yu S, McCall S, Sandberg G, McGeer EG, McGeer PL (2008) Enduring involvement of tau, beta-amyloid, alpha-synuclein, ubiquitin and TDP-43 pathology in the amyotrophic lateral sclerosis/parkinsonism-dementia complex of Guam (ALS/PDC). Acta Neuropathol 116:625–637

Monteilh-Zoller MK, Hermosura MC, Nadler MJ, Scharenberg AM, Penner R, Fleig A (2003) TRPM7 provides an ion channel mechanism for cellular entry of trace metal ions. J Gen Physiol 121:49–60

Morgan D, Capasso M, Musset B, Cherny VV, Rios E, Dyer MJ, DeCoursey TE (2009) Voltage-gated proton channels maintain pH in human neutrophils during phagocytosis. Proc Natl Acad Sci U S A 106:18022–18027

Mosley RL, Hutter-Saunders JA, Stone DK, Gendelman HE (2012) Inflammation and adaptive immunity in Parkinson's disease. Cold Spring Harb Perspect Med 2:a009381

Nadler MJ, Hermosura MC, Inabe K, Perraud AL, Zhu Q, Stokes AJ, Kurosaki T, Kinet JP, Penner R, Scharenberg AM, Fleig A (2001) LTRPC7 is a Mg.ATP-regulated divalent cation channel required for cell viability. Nature 411:590–595

Naziroglu M, Ozgul C, Cig B, Dogan S, Uguz AC (2011) Glutathione modulates Ca(2 +) influx and oxidative toxicity through TRPM2 channel in rat dorsal root ganglion neurons. J Membr Biol 242:109–118

Numata T, Shimizu T, Okada Y (2007a) TRPM7 is a stretch- and swelling-activated cation channel involved in volume regulation in human epithelial cells. Am J Physiol Cell Physiol 292:C460–C467

Numata T, Shimizu T, Okada Y (2007b) Direct mechano-stress sensitivity of TRPM7 channel. Cell Physiol Biochem 19:1–8

O'Connor RS, Hao X, Shen K, Bashour K, Akimova T, Hancock WW, Kam LC, Milone MC (2012) Substrate rigidity regulates human T cell activation and proliferation. J Immunol. 189:1330–1339

Oancea E, Wolfe JT, Clapham DE (2006) Functional TRPM7 channels accumulate at the plasma membrane in response to fluid flow. Circ Res 98:245–253

Okle O, Stemmer K, Deschl U, Dietrich DR (2013) L-BMAA induced ER-stress and enhanced caspase 12 cleavage in human neuroblastoma SH-SY5Y cells at low non-excitotoxic concentrations. Toxicol Sci 131:217–224

Ozgul C, Naziroglu M (2012) TRPM2 channel protective properties of N-acetylcysteine on cytosolic glutathione depletion dependent oxidative stress and Ca2 + influx in rat dorsal root ganglion. Physiol Behav 106(2):122–128

Partida-Sanchez S, Gasser A, Fliegert R, Siebrands CC, Dammermann W, Shi G, Mousseau BJ, Sumoza-Toledo A, Bhagat H, Walseth TF, Guse AH, Lund FE (2007) Chemotaxis of mouse bone marrow neutrophils and dendritic cells is controlled by adp-ribose, the major product generated by the CD38 enzyme reaction. J Immunol. 179:7827–7839

Perraud AL, Fleig A, Dunn CA, Bagley LA, Launay P, Schmitz C, Stokes AJ, Zhu Q, Bessman MJ, Penner R, Kinet JP, Scharenberg AM (2001) ADP-ribose gating of the calcium-permeable LTRPC2 channel revealed by Nudix motif homology. Nature 411:595–599

Perraud AL, Takanishi CL, Shen B, Kang S, Smith MK, Schmitz C, Knowles HM, Ferraris D, Li W, Zhang J, Stoddard BL, Scharenberg AM (2005) Accumulation of free ADP-ribose from mitochondria mediates oxidative stress-induced gating of TRPM2 cation channels. J Biol Chem 280:6138–6148

Perraud AL, Zhao X, Ryazanov AG, Schmitz C (2011) The channel-kinase TRPM7 regulates phosphorylation of the translational factor eEF2 via eEF2-k. Cell Signal 23:586–593

Qu B, Al-Ansary D, Kummerow C, Hoth M, Schwarz EC (2011) ORAI-mediated calcium influx in T cell proliferation, apoptosis and tolerance. Cell Calcium 50:261–269

Quintana A, Pasche M, Junker C, Al-Ansary D, Rieger H, Kummerow C, Nunez L, Villalobos C, Meraner P, Becherer U, Rettig J, Niemeyer BA, Hoth M (2011) Calcium microdomains at the immunological synapse: how ORAI channels, mitochondria and calcium pumps generate local calcium signals for efficient T-cell activation. EMBO J 30:3895–3912

Reiff DM, Spathis R, Chan CW, Vilar MG, Sankaranarayanan K, Lynch D, Ehrlich E, Kerath S, Chowdhury R, Robinowitz L, Koji Lum J, Garruto RM (2011) Inherited and somatic mitochondrial DNA mutations in Guam amyotrophic lateral sclerosis and parkinsonism-dementia. Neurol Sci 32:883–892

Reynolds AD, Stone DK, Hutter JA, Benner EJ, Mosley RL, Gendelman HE (2010) Regulatory T cells attenuate Th17 cell-mediated nigrostriatal dopaminergic neurodegeneration in a model of Parkinson's disease. J Immunol 184:2261–2271

Runnels LW, Yue L, Clapham DE (2001) TRP-PLIK, a bifunctional protein with kinase and ion channel activities. Science 291(5506):1043–1047

Saunders JA, Estes KA, Kosloski LM, Allen HE, Dempsey KM, Torres-Russotto DR, Meza JL, Santamaria PM, Bertoni JM, Murman DL, Ali HH, Standaert DG, Mosley RL, Gendelman HE (2012) CD4 + Regulatory and Effector/Memory T Cell Subsets Profile Motor Dysfunction in Parkinson's Disease. J Neuroimmune Pharmacol. 7:927–938

Schmitz C, Perraud AL, Johnson CO, Inabe K, Smith MK, Penner R, Kurosaki T, Fleig A, Scharenberg AM (2003) Regulation of vertebrate cellular Mg2 + homeostasis by TRPM7. Cell 114:191–200

Schrenzel J, Serrander L, Banfi B, Nusse O, Fouyouzi R, Lew DP, Demaurex N, Krause KH (1998) Electron currents generated by the human phagocyte NADPH oxidase. Nature 392:734–737

Seksenyan A, Ron-Harel N, Azoulay D, Cahalon L, Cardon M, Rogeri P, Ko MK, Weil M, Bulvik S, Rechavi G, Amariglio N, Konen E, Koronyo-Hamaoui M, Somech R, Schwartz M (2010) Thymic involution, a co-morbidity factor in amyotrophic lateral sclerosis. J Cell Mol Med 14:2470–2482

Spencer PS, Nunn PB, Hugon J, Ludolph AC, Ross SM, Roy DN, Robertson RC (1987) Guam amyotrophic lateral sclerosis-parkinsonism-dementia linked to a plant excitant neurotoxin. Science 237:517–522

Starkus J, Beck A, Fleig A, Penner R (2007) Regulation of TRPM2 by Extra- and Intracellular Calcium. J Gen Physiol 130:427–440

Starkus JG, Fleig A, Penner R (2010) The calcium-permeable non-selective cation channel TRPM2 is modulated by cellular acidification. J Physiol 588:1227–1240

Su LT, Agapito MA, Li M, Simonson WT, Huttenlocher A, Habas R, Yue L, Runnels LW (2006) TRPM7 regulates cell adhesion by controlling the calcium-dependent protease calpain. J Biol Chem 281:11260–11270

Su LT, Chen HC, Gonzalez-Pagan O, Overton JD, Xie J, Yue L, Runnels LW (2010) TRPM7 activates m-calpain by stress-dependent stimulation of p38 MAPK and c-Jun N-terminal kinase. J Mol Biol 396:858–869

Su LT, Liu W, Chen HC, Gonzalez-Pagan O, Habas R, Runnels LW (2011) TRPM7 regulates polarized cell movements. Biochem J 434:513–521

Sumoza-Toledo A, Lange I, Cortado H, Bhagat H, Mori Y, Fleig A, Penner R, Partida-Sanchez S (2011) Dendritic cell maturation and chemotaxis is regulated by TRPM2-mediated lysosomal Ca2 + release. FASEB J 25:3529–3542

Sun HS, Jackson MF, Martin LJ, Jansen K, Teves L, Cui H, Kiyonaka S, Mori Y, Jones M, Forder JP, Golde TE, Orser BA, Macdonald JF, Tymianski M (2009) Suppression of hippocampal TRPM7 protein prevents delayed neuronal death in brain ischemia. Nat Neurosci 12:1300–1307

Svensson L, McDowall A, Giles KM, Stanley P, Feske S, Hogg N (2010) Calpain 2 controls turnover of LFA-1 adhesions on migrating T lymphocytes. PLoS ONE 5:e15090

Takahashi N, Kozai D, Kobayashi R, Ebert M, Mori Y (2011) Roles of TRPM2 in oxidative stress. Cell Calcium 50:279–287

Togashi K, Hara Y, Tominaga T, Higashi T, Konishi Y, Mori Y, Tominaga M (2006) TRPM2 activation by cyclic ADP-ribose at body temperature is involved in insulin secretion. EMBO J 25:1804–1815

Toth B, Csanady L (2010) Identification of direct and indirect effectors of the transient receptor potential melastatin 2 (TRPM2) cation channel. J Biol Chem 285:30091–30102

Toth B, Csanady L (2012) Pore collapse underlies irreversible inactivation of TRPM2 cation channel currents. Proc Natl Acad Sci U S A 109:13440–13445

Tseveleki V, Rubio R, Vamvakas SS, White J, Taoufik E, Petit E, Quackenbush J, Probert L (2010) Comparative gene expression analysis in mouse models for multiple sclerosis, Alzheimer's disease and stroke for identifying commonly regulated and disease-specific gene changes. Genomics 96:82–91

Ueda H, Morphew MK, McIntosh JR, Davis MM (2011) CD4 + T-cell synapses involve multiple distinct stages. Proc Natl Acad Sci U S A 108:17099–17104

Vaknin I, Kunis G, Miller O, Butovsky O, Bukshpan S, Beers DR, Henkel JS, Yoles E, Appel SH, Schwartz M (2011) Excess circulating alternatively activated myeloid (M2) cells accelerate ALS progression while inhibiting experimental autoimmune encephalomyelitis. PLoS ONE 6:e26921

Wehrhahn J, Kraft R, Harteneck C, Hauschildt S (2010) Transient receptor potential melastatin 2 is required for lipopolysaccharide-induced cytokine production in human monocytes. J Immunol 184:2386–2393

Wei WL, Sun HS, Olah ME, Sun X, Czerwinska E, Czerwinski W, Mori Y, Orser BA, Xiong ZG, Jackson MF, Tymianski M, MacDonald JF (2007) TRPM7 channels in hippocampal neurons detect levels of extracellular divalent cations. Proc Natl Acad Sci U S A 104:16323–16328

Wei C, Wang X, Chen M, Ouyang K, Song LS, Cheng H (2009) Calcium flickers steer cell migration. Nature 457:901–905

Wenning AS, Neblung K, Strauss B, Wolfs MJ, Sappok A, Hoth M, Schwarz EC (2011) TRP expression pattern and the functional importance of TRPC3 in primary human T-cells. Biochim Biophys Acta 1813:412–423

Whiting MG (1964) Food practices in ALS foci in Japan, the Marianas and New Guinea. Fed Proc 23:1343–1345

Wilkinson JA, Scragg JL, Boyle JP, Nilius B, Peers C (2008) H2O2-stimulated Ca2 + influx via TRPM2 is not the sole determinant of subsequent cell death. Pflugers Arch 455:1141–1151

Wu DC, Re DB, Nagai M, Ischiropoulos H, Przedborski S (2006) The inflammatory NADPH oxidase enzyme modulates motor neuron degeneration in amyotrophic lateral sclerosis mice. Proc Natl Acad Sci U S A 103:12132–12137

Wu LJ, Sweet TB, Clapham DE (2010) International Union of Basic and Clinical Pharmacology. LXXVI. Current progress in the mammalian TRP ion channel family. Pharmacol Rev 62:381–404

Xie YF, Belrose JC, Lei G, Tymianski M, Mori Y, Macdonald JF, Jackson MF (2011) Dependence of NMDA/GSK-3beta mediated metaplasticity on TRPM2 channels at hippocampal CA3-CA1 synapses. Mol Brain 4:44

Xu C, Macciardi F, Li PP, Yoon IS, Cooke RG, Hughes B, Parikh SV, McIntyre RS, Kennedy JL, Warsh JJ (2006) Association of the putative susceptibility gene, transient receptor potential protein melastatin type 2, with bipolar disorder. Am J Med Genet B Neuropsychiatr Genet 141B:36–43

Xu C, Li PP, Cooke RG, Parikh SV, Wang K, Kennedy JL, Warsh JJ (2009) TRPM2 variants and bipolar disorder risk: confirmation in a family-based association study. Bipolar Disord 11:1–10

Yamamoto S, Shimizu S, Kiyonaka S, Takahashi N, Wajima T, Hara Y, Negoro T, Hiroi T, Kiuchi Y, Okada T, Kaneko S, Lange I, Fleig A, Penner R, Nishi M, Takeshima H, Mori Y (2008) TRPM2-mediated Ca2 + influx induces chemokine production in monocytes that aggravates inflammatory neutrophil infiltration. Nat Med 14:738–747

Yang W, Zou J, Xia R, Vaal ML, Seymour VA, Luo J, Beech DJ, Jiang LH (2010) State-dependent inhibition of TRPM2 channel by acidic pH. J Biol Chem 285:30411–30418

Yang W, Manna PT, Zou J, Luo J, Beech DJ, Sivaprasadarao A, Jiang LH (2011) Zinc inactivates melastatin transient receptor potential 2 channels via the outer pore. J Biol Chem 286:23789–23798

Yase Y (1972) The pathogenesis of amyotrophic lateral sclerosis. Lancet 2:292–296

Yu Y, Fay NC, Smoligovets AA, Wu HJ, Groves JT (2012) Myosin IIA modulates T cell receptor transport and CasL phosphorylation during early immunological synapse formation. PLoS ONE 7:e30704

Zeng X, Sikka SC, Huang L, Sun C, Xu C, Jia D, Abdel-Mageed AB, Pottle JE, Taylor JT, Li M (2010) Novel role for the transient receptor potential channel TRPM2 in prostate cancer cell proliferation. Prostate Cancer Prostatic Dis. 13:195–201

Zhao W, Beers DR, Liao B, Henkel JS, Appel SH (2012) Regulatory T lymphocytes from ALS mice suppress microglia and effector T lymphocytes through different cytokine-mediated mechanisms. Neurobiol Dis 48:418–428

Zhao, Y.J., C.M. Lam, and H.C. Lee (2012) The membrane-bound enzyme CD38 exists in two opposing orientations. Sci Signal 5: ra67

Chapter 19
TRPML Channels and Mucolipidosis Type IV

Christian Grimm and Math P. Cuajungco

Abstract The transient receptor potential (TRP) mucolipin channels TRPML1, TRPML2, and TRPML3 are non-selective cation channels predominantly found in the endolysosomal system. Mutations in the *TRPML1* gene (also known as *MCOLN1*) cause mucolipidosis type IV in humans manifested by psychomotor abnormalities, corneal clouding, retinal degeneration, and progressive neurodegeneration. The recent identification of small compound chemical activators for TRPML channels has opened up the possibility to study TRPML mutant and wild-type isoforms both in vitro and in vivo in more detail. These compounds will permit further investigation on the functional roles of TRPML channels in the endolysosomal system. Ultimately, these drugs or their derivatives could be used to design selective pharmacological tools to gate and rescue specific loss-of-function point mutations in TRPML1 that cause mucolipidosis type IV. Recently discovered TRPML channel interaction partners may serve as alternative pharmacological targets for the treatment of mucolipidosis type IV.

C. Grimm (✉)
Department of Pharmacy—Center for Drug Research and Center for Integrated Protein Science Munich (CIPSM), Ludwig-Maximilians-Universität, Munich, Germany
e-mail: chgrph@cup.uni-muenchen.de

M. P. Cuajungco
Department of Biological Science and Center for Applied Biotechnology Studies, California State University, Fullerton, CA, USA
e-mail: mcuajungco@fullerton.edu

M. P. Cuajungco
Mental Health Research Institute, Melbourne Brain Centre, Parkville, VIC, Australia

N. Weiss and A. Koschak (eds.), *Pathologies of Calcium Channels*,
DOI: 10.1007/978-3-642-40282-1_19,

19.1 Introduction

Lysosomes are important cell organelles involved in the breakdown of proteins, lipids, and other molecules. These intracellular vesicles have been implicated not only in endolysosomal storage disorders (LSDs) such as mucolipidoses or sphingolipidoses but also in several neurodegenerative disorders such as Alzheimer's and Parkinson's disease. Endolysosomal organelles contain a plethora of different receptors, ion channels, and transporters that are crucial for their normal functions. Ion homeostasis, regulation of distinct proton concentrations in different organelles and regulation of heavy metal concentrations such as copper, zinc, or iron are essential not only for lysosomes and endosomes but also for lysosome-related organelles (LROs) such as melanosomes, lytic granules, or platelet-dense granules. It has been known for a while that acidification of lysosomes and LROs is primarily regulated by ATP-driven proton pumps (vacuolar H^+-ATPases). However, it is still largely unclear how the fine-tuning of luminal pH in early endosomes (pH 6–7), late endosomes (pH 5–6) and lysosomes (pH 4–5) or different LROs is achieved. Likewise, the mechanisms underlying the regulation of typical intraluminal cation concentrations of sodium, potassium or calcium, and different heavy metal cation concentrations are largely unexplored. For example, luminal pH as well as cation concentrations need to be readjusted after fusion events between intracellular vesicles and after endocytosis or phagocytosis. Similarly, the luminal pH needs to be controlled during maturation processes of endolysosomal organelles or LROs like melanosomes (Gerasimenko et al. 1998; Lelouvier and Puertollano 2011; Morgan et al. 2011; Ancans et al. 2001). Calcium levels in vesicles after endocytic or phagocytic events reach millimolar values; however, micromolar concentrations are re-established in a matter of minutes. The proteins which control these processes remain unknown.

Likewise, the role of calcium on endolysosomal fusion processes and how endolysosomal membrane fusion is regulated remain to be elucidated (Pryor and Luzio 2009; Luzio et al. 2007, 2010; Mindell 2012). In analogy to fusion processes between the plasma membrane and synaptic vesicles which use voltage-gated calcium channels (N- or P/Q-type calcium channels) as their calcium source, endolysosomal fusion processes (Südhof and Rizo 2011; Südhof 2012) appear to be calcium dependent as well and require the presence and interaction of distinct SNARE (soluble N-ethylmaleimide-sensitive-factor attachment receptor) proteins (Jahn and Scheller 2006; Luzio et al. 2010). For example, homotypic fusion processes, between late endosomes, or heterotypic fusion processes, between endosomes and lysosomes, or phagosomes and lysosomes involve different SNARE proteins such as syntaxins, Vamp, or vti1 isoforms (Prekeris et al. 1998, 1999; Brandhorst et al. 2006; Luzio et al. 2007, 2010; Pattu et al. 2011; Fig. 19.1). Finally, the participation of specific ion channels in endolysosomal trafficking processes also needs further investigation. Members of the TRP family of nonselective cation channels, in particular, the mucolipins or TRPML channels (TRPML1-3) are expressed in the endolysosomal system. Essential contributions

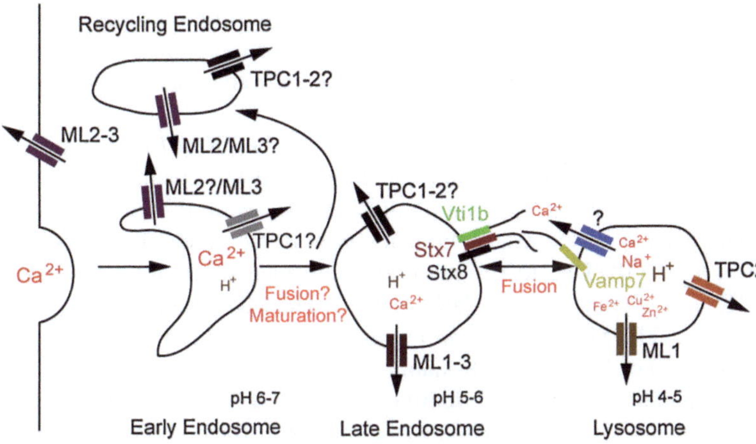

Fig. 19.1 Cartoon displaying putative subcellular localizations of TRPMLs and TPCs. While TPC2 and TRPML1 are predominantly found in lysosomes and possibly late endosomes, TRPML2, TRPML3, and TPC1 appear to be predominant in early endosomes, late endosomes, or recycling endosomes as well as the plasma membrane (TRPML2, TRPML3). Stx7 (Syntaxin7), Stx8 (Syntaxin8), Vti1b, and Vamp7 form a SNARE complex during fusion processes between late endosomes and lysosomes

of these ion channels to some of the aforementioned processes have recently been postulated (Dong et al. 2010b; Lelouvier and Puertollano 2011; Abe and Puertollano 2011; Scott and Gruenberg 2011; Shen et al. 2012; Grimm et al. 2012a).

Since their discovery by Christian de Duve in the 1950s (de Duve et al. 1955), it was found that lysosomes contain more than 60 different degradative enzymes that can hydrolyze proteins, DNA, RNA, polysaccharides, and lipids (Saftig and Klumperman 2009; Schröder et al. 2010). Mutations in the genes that encode lysosomal enzymes are responsible for more than 30 different human genetic diseases, which are called lysosomal storage diseases or disorders (LSD) because undegraded material accumulates within the endolysosomal system of affected individuals.

19.2 Mucolipidosis Type IV: A Lysosomal Storage Disorder Caused by Mutations in TRPML1

Mucolipidosis type IV (ML IV) is an autosomal recessive LSD characterized by severe neurological problems, progressive neurodegeneration, and ophthalmologic abnormalities such as corneal opacity, retinal degeneration, and strabismus. Most patients are unable to speak or walk independently. Motor and mental retardation can however also be mild in some cases. All patients have constitutive achlorhydria associated with a secondary elevation of serum gastrin levels. Storage bodies of lipids and water-soluble substances are seen by electron microscopy in almost every cell type of the patients. Iron deficiency anemia is another phenotypic feature of this condition.

Eventually, it was discovered that mutations in the *TRPML1* gene are causative for ML IV (Sun et al. 2000; Bargal et al. 2000). TRPML channels were originally called mucolipins (MCOLN). Cloning of *MCOLN1* led to the identification of two additional genes, both located on human chromosome 1, *MCOLN2* and *MCOLN3* (Bargal et al. 2000). Due to structural and sequence similarities with TRP channels, MCOLN channels were later named TRPML channels, i.e., TRPML1, TRPML2, and TRPML3 (for recent reviews on the TRP channel family see, e.g., Gees et al. 2010 or Nilius and Owsianik 2011). Today, a plethora of different ML IV causing mutations have been identified throughout the *TRPML1* gene (reviewed in more detail elsewhere, e.g., Altarescu et al. 2002; Bach 2005; Bach et al. 2005; Grimm et al. 2012a). There are ML IV-causing mutations which lead to non-functional protein, mislocalized protein, or short versions of the protein due to early stop codons. Besides those, there are also ML IV causing mutations found within intron regions of *TRPML1* which occur, as most ML IV mutations with very high prevalence in people with Ashkenazi Jewish ancestry. One such mutation, c.406-2A > G, changes a nucleotide in a region of the gene known as intron 3. This mutation, which is a splice-site mutation, introduces a premature stop signal. Another example with high prevalence is g.511_6943del, a mutation which deletes a large amount of DNA near the beginning of the *TRPML1* gene. Both of these mutations result in the production of abnormally short, non-functional proteins.

19.3 Mucolipidosis Type IV Mouse Models

Similar to the human phenotype, *Trpml1*$^{-/-}$ knockout mice display inclusion bodies, enlarged vacuoles, psychomotor defects, and retinal degeneration (Venugopal et al. 2007). Venugopal et al. also described a gastric phenotype in *Trpml1*$^{-/-}$ knockout mice with remarkable similarities to what has been described in ML IV patients, including elevated gastrin. Chandra et al. (2011) also reported that *Trpml1*$^{-/-}$ knockout mice have significant impairments in basal and histamine-stimulated gastric acid secretion. Specifically, the loss of *Trpml1* causes reduced levels and mislocalization of the gastric proton pump and alters the secretory canaliculi, causing hypochlorhydria and hypergastrinemia (Chandra et al. 2011). In histologic and ultrastructural analyses, Chandra et al. found that *Trpml1*$^{-/-}$ parietal cells are enlarged and have multivesicular and multilamellar lysosomes.

Mostly normal at birth, it was found that 3-month-old *Trpml1*$^{-/-}$ knockout mice start to show weakness and progressive neurological deficits. The lifespan of *Trpml1*$^{-/-}$ knockout mice is decreased (Venugopal et al. 2007). Retinal degeneration was found to be dramatic and similar to that described in patients. Mice show defects in the outer nuclear layer, the outer plexiform layer, and the inner nuclear layer (Venugopal et al. 2007). Micsenyi et al. (2009) characterized the neuropathology of *Trpml1*$^{-/-}$ knockout mice in more detail and found evidence of ganglioside accumulation, including increases in GM2, GM3, and GD3, and redistribution of GM1, found throughout the central nervous system (CNS). In contrast, cholesterol accumulation in the CNS was not as prominent as it is in other LSDs such as NPC1. Furthermore, evidence of reduced myelination in cerebral and cerebellar white matter tracts and accumulation of autofluorescent material throughout the brain were found (Micsenyi et al. 2009). Curcio-Morelli and colleagues reported that macroautophagy is defective in *Trpml1*$^{-/-}$ knockout mice and that LC3-II, the lipidated form of the microtubule-associated protein (LC3), which associates with autophagosomal membranes is strongly increased in neuronal cultures of *Trpml1*$^{-/-}$ knockout mice (Venugopal et al. 2009).

Thus, in summary, *Trpml1*$^{-/-}$ knockout mice appear to exhibit many clinical and cellular features of the human disease.

19.4 Electrophysiological and Functional Properties of TRPML Wild-Type and Mutant Channels

Functional characterization of TRPML channels has gained momentum with the electrophysiological characterization of the constitutively active inwardly rectifying TRPML3 varitint-waddler mutant isoforms Va (A419P) and Va^{J} (A419P + I362T) and their TRPML1 and TRPML2 counterparts (Grimm et al. 2007, 2010; Kim et al. 2007, 2010; Nagata et al. 2008; Xu et al. 2007; Dong et al. 2009; Samie et al. 2009; Lev et al. 2010). TRPML3 Va and Va^{J} mutations cause deafness, circling behavior, and coat color dilution in mice due to severe calcium overload in cells natively expressing the channel, e.g., hair cells of the inner ear and melanocytes. Equivalent mutations in TRPML1 and TRPML2 showed similar calcium overload effects upon heterologous expression in cultured cells (Grimm et al. 2007, 2009; Dong et al. 2009; Lev et al. 2010).

While the introduction of proline residues in TMD5 of TRPML channels causes a gain-of-function (constitutive activity), ML IV-causing mutations, generally, appear to render TRPML1 non-functional. Dong et al. (2008) demonstrated that ML IV mutant isoforms such as T232P, D362Y, F465L, or R403C in combination with the Va equivalent mutation in TRPML1, known as TRPML1(V432P), abolish the constitutive activity of TRPML1(V432P) (Dong et al. 2008). One exception was the F408 Δ deletion mutant which still showed some constitutive activity in combination with TRPML1(V432P). In accordance with this result, patients with the F408 Δ deletion in TRPML1 reportedly show a very mild ML IV phenotype (Raychowdhury et al. 2004).

Dong et al. (2008) further demonstrated both TRPML1 and TRPML2 channels, but not TRPML3 to be permeable to iron. The authors concluded, by showing that cytosolic Fe^{2+} deficiency is coincident with intralysosomal iron overload in ML IV cells, that TRPML1 may play a critical role in cellular iron homeostasis. In addition, they found that the constitutive activity of TRPML1 and TRPML2 Va isoforms can be potentiated by low pH while the activity of TRPML3 Va is inhibited by low pH (Xu et al. 2007; Dong et al. 2008, 2009). The latter finding is in accordance with the results obtained by Kim et al. (2008) with wild-type TRPML3, i.e., block of TRPML3 channel activity by low extracellular pH (Kim et al. 2008).

19.5 Identification of TRPML Channel Activators: A Perspective for the Development of Pharmacological Tools for Therapeutic Use

Dong and colleagues (2010a) also found that wild-type TRPML channels can be activated with $PI(3,5)P_2$. These measurements were performed as whole-lysosome patch-clamp experiments using vacuolin-1 to increase the size of endolysosomes as described previously (Dong et al. 2010a; Schieder et al. 2010a, b). Currents elicited with $PI(3,5)P_2$ showed inward rectification. In accordance with these results, TRPML3 was found to be inwardly rectifying when activated by small chemical compounds, identified in a recent high-throughput screening (Grimm et al. 2010, 2012b; Yamaguchi and Muallem 2010; Saldanha et al. 2011). The identified candidate compounds were inactive against other TRP channels tested, including members of the TRPC, TRPV, TRPM, TRPA, and TRPN subfamilies such as TRPC3, TRPV2, TRPM2, TRPA1, and TRPN1. It was further found that the identified candidate compounds are inactive against a plethora of other targets (generally more than 500; for further information see http://pubchem.ncbi.nlm.nih.gov; AID: 1448, 1525, 1526, 1562, 2719, 1809, and 2694).

While some of the compounds were found to activate only TRPML3 (Grimm et al. 2010), others also activate TRPML1 and/or TRPML2. Compound SF-22 (5-chloro-*N*-(2-piperidin-1-ylphenyl)-thiophene-2-sulfonamide, EC_{50} for TRPML3 = 900 nM), for example, was found to activate TRPML3 and TRPML1 but not TRPML2 at concentrations up to 10 μM (Grimm et al. 2010). Compounds SF-21 (4-chloro-*N*-(2-morpholin-4-ylcyclohexyl)-benzenesulfonamide, EC_{50} for TRPML3 = 860 nM), SF-41 (1-(2,4-dimethylphenyl)-4-piperidin-1-ylsulfonyl-piperazine, EC_{50} for TRPML3 = 1.44 μM), and SF-81 (4,6-di-methyl-3-(2-methylphenyl)-sulfonyl-1-propan-2-ylpyridin-2-one, EC_{50} for TRPML3 = 2.45 μM) activate TRPML2 and TRPML3 at a concentration of 10 μM (Grimm et al. 2012b). Compound SF-51 (2-[2-oxo-2-(2,2,4-trimethylquinolin-1-yl)-ethyl]isoindole-1,3-dione, EC_{50} for TRPML3 = 1.47 μM) activates TRPML1 at concentrations beyond 30 μM. A chemically modified version of SF-51 (2-[2-oxo-2-(2,2,4-trimethyl-3,4-dihydroquinolin-1(2H)-yl)ethyl]-1H-isoindole-1,3(2H)-dione; ML-SA1) was shown to activate all three TRPML isoforms at a concentration of 10 μM (Shen et al. 2012).

In addition to these dose-dependent cross reactions within the TRPML subfamily, it should be noted that some of the compounds do show other off-target effects, e.g., ML-SA1 (but not SF-51) inhibits TRPC4 with an IC_{50} of 4.2 μM (confirmation dose–response assay), SF-11 may potentiate/activate the calcium-activated chloride channel TMEM16A (primary screening) and may also inhibit the two-pore domain potassium channel KCNK9 (primary screening); finally, SF-21 (EC_{50} for TRPML3 = 860 nM) activates the CRF-binding protein, albeit with a much higher EC_{50} (13.7 μM; confirmation dose–response assay) (for further information see http://pubchem.ncbi.nlm.nih.gov; AID: 1448, 1525, 1526, 1562, 2719, 1809, and 2694).

In summary, while there are selective TRPML3 compounds available, it will be necessary for the future characterization of endogenous TRPML1 and TRPML2 channel properties, and for potential therapeutic use, to further increase selectivity for these TRPML isoforms. This may be achieved by either chemical modifications of the existing lead compounds or by additional high-throughput initiatives.

Compounds with improved selectivity profile may ultimately be applied to restore TRPML1 channel function in the case of loss-of-function mutations causing ML IV. A good example for such a strategy is the cystic fibrosis transmembrane conductance regulator (CFTR) channel. Here, compounds (benzothiophenes, phenylglycines, and sulfonamides) have been developed which are able to correct the defective gating of ΔF508-CFTR (Amaral 2011; Ashlock and Olson 2011). Other small molecules have been developed that correct its defective cellular processing (van Goor et al. 2011). For example, VX-770 (Ivacaftor, N-(2,4-Di-tert-butyl-5-hydroxyphenyl)-4-oxo-1,4-dihydroquinoline-3-carboxamide), is now commercially available under the trade name Kalydeco® for the treatment of cystic fibrosis patients carrying the G551D mutation in the CFTR gene. Some of the ML IV-causing TRPML1 point mutations lead to early sequence termination resulting in protein variants that are lacking the pore domain, other mutations (e.g., in intron regions) result in abnormally short, non-functional protein or prevent the protein from being produced as outlined above. Hence, this strategy may not be applicable to those cases. Nevertheless, in cases of defective TRPML1 channel gating, the approach may be promising for the treatment of ML IV.

19.6 Gene Therapy: A Treatment Option for Retinal Degeneration

Gene therapy may represent an alternative approach especially for the treatment of ML IV mutations which result in a complete loss of the TRPML1 protein, as it has the potential to theoretically provide a permanent source of the deficient protein, either by direct injection of vectors or by transplantation of gene-corrected cells (for recent reviews on LSDs and gene therapy see, e.g., Byrne et al. 2012; Parenti et al. 2012; van Gelder et al. 2012). A gene therapy approach restricted, e.g., to the retina is something that has been successfully done with other ion channels (see, e.g., Michalakis et al. 2012; Koch et al. 2012) and thus may well be a promising approach to treat retinal degeneration in ML IV patients effectively.

19.7 TRPML1 and Other Lysosomal Storage Disorders

Interestingly, there appears to be a link between TRPML1 dysfunction and other LSDs. Shen et al. (2012) demonstrated that increasing TRPML1 activity in Niemann-Pick type C cells is sufficient to restore normal lysosomal trafficking and prevent cholesterol accumulation. It thus appears that TRPML1 channel, e.g., by controlling calcium-dependent lysosomal trafficking, is the entry point of various lysosomal storage diseases, and may therefore represent a promising pharmacological target for the treatment of several lysosomal storage disorders (Weiss 2012).

A potentially more global role for TRPML1 in cellular clearance is supported by a study by Medina et al. (2011), demonstrating that the transcription factor EB (TFEB) regulates lysosomal exocytosis both by inducing the release of intracellular Ca^{2+} through its target gene *TRPML1* and by increasing the population of lysosomes ready to fuse with the plasma membrane. Moreover, Medina et al. (2011) demonstrated that the induction of lysosomal exocytosis by TFEB promotes cellular clearance in pathological conditions such as lysosomal storage diseases. TFEB-mediated increase of intracellular Ca^{2+} was blocked by transient silencing of TRPML1. Consistently, Ca^{2+} levels were not affected by TFEB overexpression in human ML IV cells that carry loss-of-function mutations of TRPML1 (Medina et al. 2011). Thus, pharmacological induction of TFEB may be a promising approach for the treatment of LSDs other than ML IV.

19.8 Heteromultimerization of TRPML1 with Related Ion Channels

TRPML channels are able to heteromultimerize with each other. This has been shown by several groups (Venkatachalam et al. 2006; Zeevi et al. 2009; Curcio-Morelli et al. 2010; Grimm et al. 2010; Zeevi et al. 2010). To what extent such heteromultimerizations occur in vivo and how they may affect physiological processes is not fully understood. Zeevi et al. (2010) have recently suggested that TRPMLs heteromultimerize with each other to regulate cell viability and starvation-induced autophagy, a process that mediates macromolecular and organellar turnover under cell starvation conditions. Although TRPMLs have been shown to reside mostly within different intracellular organelles (Karacsonyi et al. 2007; Martina et al. 2009; Vergarajauregui and Puertollano 2006; Fig. 19.1), Zeevi et al. (2009) found that endogenous TRPML channels partially colocalize with each other, in a subset of intracellular vesicles. In addition, TRPML2 and TRPML3 traffic, in part, to lysosomes (Karacsonyi et al. 2007; Kim et al. 2009; Martina et al. 2009; Zeevi et al. 2009) where they may like TRPML1 regulate lysosomal function(s). Indeed, it was found that gene-specific knockdown of TRPML2 or TRPML3 leads to lysosomal inclusions reminiscent of those found in MLIV patient TRPML1$^{-/-}$ cells (Zeevi et al. 2009).

Recent evidence also suggests that TRPML channels can interact with two-pore channels (TPCs), a family of novel endolysosomal ion channels containing 12 transmembrane domains and predicted to form dimers (Yamaguchi et al. 2011). TRPML1 and TRPML3 (albeit to a lesser extent) were shown to co-immunoprecipitate with TPC1 and TPC2. In addition, a comparison of the amino acid sequences of the TRP channels with TPC1 and TPC2 (Fig. 19.2) suggests some homology between TRPML channels and TPCs when the pore regions only are compared (TMD5-pore-TMD6). However, so far no functional consequences of TRPML-TPC interactions have been found (Yamaguchi et al. 2011) and thus it was concluded that although TRPMLs and TPCs are present in the same organelles

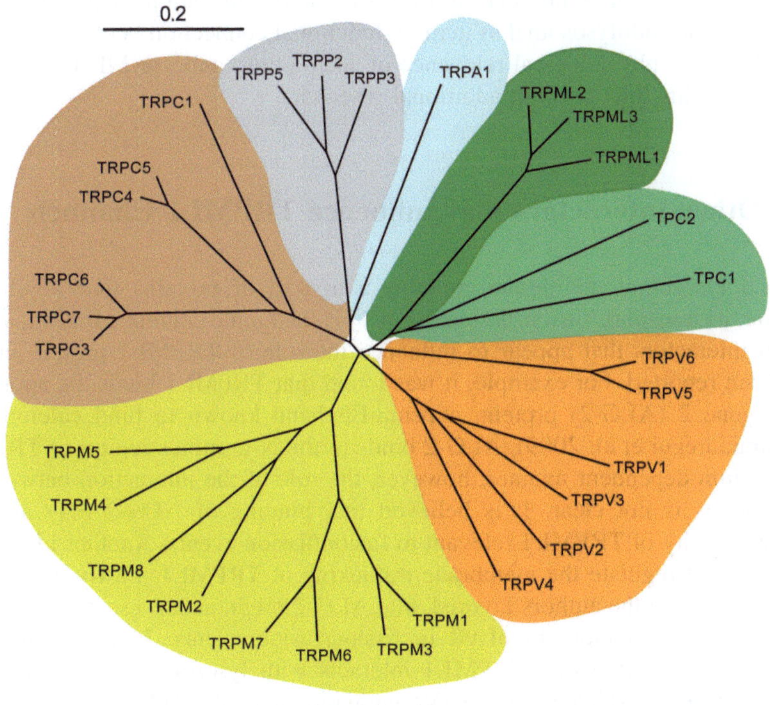

Fig. 19.2 Phylogenetic analysis of the pore regions of TRPs und TPCs. Phylogenetic analysis of human TRP channels and human TPCs based on amino acid alignments of the respective pore regions (TMD5-pore-TMD6 for TRPs; for TPCs the respective regions in domain II were used). The alignment was done using DNAMAN software (Lynnon Corporation, Pointe-Claire, Quebec, Canada); the phylogenetic tree was plotted using NJPlot software (http://pbil.univ-lyon1.fr/software/njplot.html)

and can physically interact with each other, they may function as independent organellar ion channels (Yamaguchi et al. 2011).

TPCs have initially been put forward as Nicotinic acid adenine dinuclecotide phosphate (NAADP)-stimulated endolysosomal calcium release channels by several groups (Brailoiu et al. 2009, 2010; Calcraft et al. 2009; Zong et al. 2009; Pitt et al. 2010; Ruas et al. 2010; Schieder et al. 2010a, b; Tugba Durlu-Zhu et al. 2010; Ogunbayo et al. 2011; Rybalchenko et al. 2012; Lin-Moshier et al. 2012; Walseth et al. 2012). Quite in contrast to these findings, Wang et al. have recently provided whole-lysosome patch-clamp data suggesting that NAADP does not activate TPCs, but is activated by $PI(3,5)P_2$ (phosphatidylinositol 3,5-bisphosphate) instead (Wang et al. 2012). The same group has shown previously that TRPML channels are also activated by $PI(3,5)P_2$ (Dong et al. 2010a; Shen et al. 2012; Zhang et al. 2012). The major functions of $PI(3,5)P_2$ are in membrane and protein trafficking, and in pH control of the endosome-lysosome axis (Michell et al. 2006). It remains to be further established under what physiological circumstances TPCs and TRPMLs are activated by $PI(3,5)P_2$, whether they are activated simultaneously within the same organelle or independently of each other, and why such a large number of $PI(3,5)P_2$ activated cation channels may be required in the endolysosomal system. A functional connection between TPCs and TRPMLs and a physiological relevance of such a cross talk for ML IV cannot be ruled out at this time without additional research.

19.9 Other Interactors that Influence TRPML1 Channels

The endolysosomal membrane contains a myriad of proteins that regulate its function, its transport, fusion/fission events, pH, and ionic contents. Indeed, several protein interactors that appear to influence the role of TRPML1 protein in cells have been reported. For example, it was found that TRPML1 binds the apoptosis-linked gene 2 (ALG-2) protein, a penta-EF hand known to bind calcium ions (Vergarajauregui et al. 2009). ALG-2 binds to the N-terminus region of TRPML1 in a calcium-dependent manner; however, the role of the interaction between the two proteins is not clear. It is believed that binding of ALG-2 may alter the channel activity of TRPML1 relevant to fusion/fission events. Another idea is that ALG-2 might regulate the membrane trafficking of TRPML1. Lastly, it was suggested, for which the authors favored, that ALG-2 might act as a scaffold to recruit other protein interactors involved in fission/fusion events. In 2011, the same authors have reported that TRPML1 interacts with lysosome-associated protein transmembrane (LAPTM) proteins (Vergarajauregui et al. 2011). LAPTM proteins are associated with the transport of molecules in the endolysosomal compartments. It was reported that overexpression of LAPTM proteins results in enlarged lysosomes, but their depletion via RNA interference produces a similar phenotypic defect observed in ML IV (Vergarajauregui et al. 2011). The authors suggested that the binding of TRPML1 with LAPTM proteins might aid the function of lysosomes in transporting specific molecules (e.g., provide energy for LAPTM-dependent transport via TRPML1-induced ionic gradients). With the knowledge that TRPML proteins form functional heteromeric channels, it would be interesting to study if any of these newly discovered binding partners directly affect the gating properties of heteromers compared to homomers.

19.10 Conclusion

Research on ML IV has come a long way from its first description in the 1970s (Berman et al. 1974) to the discovery of its genetic cause (Sun et al. 2000; Bargal et al. 2000), the generation of ML IV mouse models (Venugopal et al. 2007; Curcio-Morelli et al. 2010; Micsenyi et al. 2009; Chandra et al. 2011), the electrophysiological characterization of TRPML1 as a non-selective cation channel, its gating and permeation properties, and finally the availability of first chemical tools to modulate its gating (Kim et al. 2008; Dong et al. 2010a; Grimm et al. 2010, 2012b; Shen et al. 2012; Zhang et al. 2012). After more than 30 years of research, it seems about time to move to the next level and start to develop effective treatments for ML IV.

References

Abe K, Puertollano R (2011) Role of TRP channels in the regulation of the endosomalpathway. Physiology (Bethesda) 26:14–22

Altarescu G, Sun M, Moore DF, Smith JA, Wiggs EA, Solomon BI, Patronas NJ, Frei KP, Gupta S, Kaneski CR, Quarrell OW, Slaugenhaupt SA, Goldin E, Schiffmann R (2002) The neurogenetics of mucolipidosis type IV. Neurology 59:306–313

Amaral MD (2011) Targeting CFTR: how to treat cystic fibrosis by CFTR-repairing therapies. Curr Drug Targets 12:683–693

Ancans J, Tobin DJ, Hoogduijn MJ, Smit NP, Wakamatsu K, Thody AJ (2001) Melanosomal pH controls rate of melanogenesis, eumelanin/phaeomelanin ratio and melanosome maturation in melanocytes and melanoma cells. Exp Cell Res 268:26–35

Ashlock MA, Olson ER (2011) Therapeutics development for cystic fibrosis: a successful model for a multisystem genetic disease. Annu Rev Med 62:107–125

Bach G (2005) Mucolipin 1: endocytosis and cation channel—a review. Pflugers Arch 451:313–317

Bach G, Webb MB, Bargal R, Zeigler M, Ekstein J (2005) The frequency of mucolipidosis type IV in the Ashkenazi Jewish population and the identification of 3 novel MCOLN1 mutations. Hum Mutat 26:591

Bargal R, Avidan N, Ben-Asher E, Olender Z, Zeigler M, Frumkin A, Raas-Rothschild A, Glusman G, Lancet D, Bach G (2000) Identification of the gene causing mucolipidosis type IV. Nat Genet 26:118–123

Brailoiu E, Churamani D, Cai X, Schrlau MG, Brailoiu GC, Gao X, Hooper R, Boulware MJ, Dun NJ, Marchant JS, Patel S (2009) Essential requirement for two-pore channel 1 in NAADP-mediated calcium signaling. J Cell Biol 186:201–209

Brailoiu E, Rahman T, Churamani D, Prole DL, Brailoiu GC, Hooper R, Taylor CW, Patel S (2010) An NAADP-gated two-pore channel targeted to the plasma membrane uncouples triggering from amplifying Ca2+ signals. J Biol Chem 49:38511–38516

Brandhorst D, Zwilling D, Rizzoli SO, Lippert U, Lang T, Jahn R (2006) Homotypic fusion of early endosomes: SNAREs do not determine fusion specificity. Proc Natl Acad Sci USA 103:2701–2706

Byrne BJ, Falk DJ, Clément N, Mah CS (2012) Gene therapy approaches for lysosomalstorage disease: next-generation treatment. Hum Gene Ther 23:808–815

Calcraft PJ, Ruas M, Pan Z, Cheng X, Arredouani A, Hao X, Tang J, Rietdorf K, Teboul L, Chuang KT, Lin P, Xiao R, Wang C, Zhu Y, Lin Y, Wyatt CN, Parrington J, Ma J, Evans AM, Galione A, Zhu MX (2009) NAADP mobilizes calcium from acidic organelles through two-pore channels. Nature 459:596–600

Chandra M, Zhou H, Li Q, Muallem S, Hofmann SL, Soyombo AA (2011) A role for the Ca^{2+} channel TRPML1 in gastric acid secretion, based on analysis of knockout mice. Gastroenterology 140:857–867

Curcio-Morelli C, Zhang P, Venugopal B, Charles FA, Browning MF, Cantiello HF, Slaugenhaupt SA (2010) Functional multimerization of mucolipin channel proteins. J Cell Physiol 222:328–335

de Duve C, Pressman BC, Gianetto R, Wattiaux R, Appelmans F (1955) Tissue fractionation studies. 6. Intracellular distribution patterns of enzymes in rat-liver tissue. Biochem. J. 60:604–617

Dong XP, Cheng X, Mills E, Delling M, Wang F, Kurz T, Xu H (2008) The type IV mucolipidosis-associated protein TRPML1 is an endolysosomal iron release channel. Nature 455:992–996

Dong XP, Wang X, Shen D, Chen S, Liu M, Wang Y, Mills E, Cheng X, Delling M, Xu H (2009) Activating mutations of the TRPML1 channel revealed by proline-scanning mutagenesis. J Biol Chem 284:32040–32052

Dong XP, Shen D, Wang X, Dawson T, Li X, Zhang Q, Cheng X, Zhang Y, Weisman LS, Delling M, Xu H (2010a) PI(3,5)P(2) controls membrane trafficking by direct activation of mucolipin Ca(2+) release channels in the endolysosome. Nat Commun 1:38. doi: 10.1038/ncomms1037

Dong XP, Wang X, Xu H (2010b) TRP channels of intracellular membranes. J Neurochem 113:313–328

Gees M, Colsoul B, Nilius B (2010) The role of transient receptor potential cation channels in Ca^{2+} signaling. Cold Spring Harb Perspect Biol 2:a003962

Gerasimenko JV, Tepikin AV, Petersen OH, Gerasimenko OV (1998) Calcium uptake via endocytosis with rapid release from acidifying endosomes. Curr Biol 8:1335–1338

Grimm C, Cuajungco MP, van Aken AF, Schnee M, Jörs S, Kros CJ, Ricci AJ, Heller S (2007) A helix-breaking mutation in TRPML3 leads to constitutive activity underlying deafness in the varitint-waddler mouse. Proc Natl Acad Sci USA 104:19583–19588

Grimm C, Jörs S, Heller S (2009) Life and death of sensory hair cells expressing constitutively active TRPML3. J Biol Chem 284:13823–13831

Grimm C, Jörs S, Saldanha SA, Obukhov AG, Pan B, Oshima K, Cuajungco MP, Chase P, Hodder P, Heller S (2010) Small molecule activators of TRPML3. Chem Biol 17:135–148

Grimm C, Hassan S, Wahl-Schott C, Biel M (2012a) Role of TRPML and two-pore channels in endolysosomalcation homeostasis. J Pharmacol Exp Ther 342:236–244

Grimm C, Jörs S, Guo Z, Obukhov AG, Heller S (2012b) Constitutive activity of TRPML2 and TRPML3 channels versus activation by low extracellular sodium and small molecules. J Biol Chem 287:22701–22708

Jahn R, Scheller RH (2006) SNAREs -engines for membrane fusion. Nat Rev Mol Cell Biol 7:631–643

Karacsonyi C, Miguel AS, Puertollano R (2007) Mucolipin-2 localizes to the Arf6-associated pathway and regulates recycling of GPI-APs. Traffic 8:1404–1414

Kim HJ, Li Q, Tjon-Kon-Sang S, So I, Kiselyov K, Muallem S (2007) Gain-of-function mutation in TRPML3 causes the mouse Varitint-Waddler phenotype. J Biol Chem 282:36138–36142

Kim HJ, Li Q, Tjon-Kon-Sang S, So I, Kiselyov K, Soyombo AA, Muallem S (2008) A novel mode of TRPML3 regulation by extracytosolic pH absent in the varitint-waddler phenotype. EMBO J 27:1197–1205

Kim HJ, Soyombo AA, Tjon-Kon-Sang S, So I, Muallem S (2009) The Ca(2+) channel TRPML3 regulates membrane trafficking and autophagy.Traffic 10:1157–1167

Kim HJ, Yamaguchi S, Li Q, So I, Muallem S (2010) Properties of the TRPML3 channel pore and its stable expansion by the Varitint-Waddler-causing mutation. J Biol Chem 285:16513–16520

Koch S, Sothilingam V, Garcia Garrido M, Tanimoto N, Becirovic E, Koch F, Seide C, Beck SC, Seeliger MW, Biel M, Mühlfriedel R, Michalakis S (2012) Gene therapy restores vision and delays degeneration in the CNGB1(-/-) mouse model of retinitis pigmentosa. Hum Mol Genet 21:4486–4496

Lelouvier B, Puertollano R (2011) Mucolipin-3 regulates luminal calcium, acidification, and membrane fusion in the endosomal pathway. J Biol Chem 286:9826–9832

Lev S, Zeevi DA, Frumkin A, Offen-Glasner V, Bach G, Minke B (2010) Constitutive activity of the human TRPML2 channel induces cell degeneration. J Biol Chem 285:2771–2782

Lin-Moshier Y, Walseth TF, Churamani D, Davidson SM, Slama JT, Hooper R, Brailoiu E, Patel S, Marchant JS (2012) Photoaffinity labeling of nicotinic acid adenine dinucleotide phosphate (NAADP) targets in mammalian cells. J Biol Chem 287:2296–2307

Luzio JP, Bright NA, Pryor PR (2007) The role of calcium and other ions in sorting and delivery in the late endocytic pathway. Biochem Soc Trans 35:1088–1091

Luzio JP, Gray SR, Bright NA (2010) Endosome-lysosome fusion. Biochem Soc Trans 38:1413–1416

Martina JA, Lelouvier B, Puertollano R (2009) The calcium channel mucolipin-3 is a novel regulator of trafficking along the endosomal pathway. Traffic 10:1143–1156

Medina DL, Fraldi A, Bouche V, Annunziata F, Mansueto G, Spampanato C, Puri C, Pignata A, Martina JA, Sardiello M, Palmieri M, Polishchuk R, Puertollano R, Ballabio A (2011) Transcriptional activation of lysosomal exocytosis promotes cellular clearance. Dev Cell 21:421–430

Michalakis S, Mühlfriedel R, Tanimoto N, Krishnamoorthy V, Koch S, Fischer MD, Becirovic E, Bai L, Huber G, Beck SC, Fahl E, Büning H, Schmidt J, Zong X, Gollisch T, Biel M, Seeliger MW (2012) Gene therapy restores missing cone-mediated vision in the CNGA3-/- mouse model of achromatopsia. Adv Exp Med Biol 723:183–189

Michell RH, Heath VL, Lemmon MA, Dove SK (2006) Phosphatidylinositol 3,5-bisphosphate: metabolism and cellularfunctions. Trends Biochem Sci 31:52–63

Micsenyi MC, Dobrenis K, Stephney G, Pickel J, Vanier MT, Slaugenhaupt SA, Walkley SU (2009) Neuropathology of the Mcoln1(-/-) knockout mouse model of mucolipidosis type IV. J Neuropathol Exp Neurol 68:125–135

Mindell JA (2012) Lysosomal acidification mechanisms. Annu Rev Physiol 74:69–86

Morgan AJ, Platt FM, Lloyd-Evans E, Galione A (2011) Molecular mechanisms of endolysosomal Ca2+ signalling in health and disease. Biochem J 439:349–374

Nagata K, Zheng L, Madathany T, Castiglioni AJ, Bartles JR, Garcia-Anoveros J (2008) The varitint-waddler (Va) deafness mutation in TRPML3 generates constitutive, inward rectifying currents and causes cell degeneration. Proc Natl Acad Sci USA 105:353–358

Nilius B, Owsianik G (2011) The transient receptor potential family of ion channels. Genome Biol 12:218

Ogunbayo OA, Zhu Y, Rossi D, Sorrentino V, Ma J, Zhu MX, Evans AM (2011) Cyclic adenosine diphosphate ribose activates ryanodine receptors, whereas NAADP activates two-pore domain channels. J Biol Chem 286:9136–9140

Parenti G, Pignata C, Vajro P, Salerno M (2012) New strategies for the treatment of lysosomal storage diseases. Int J Mol Med 10:3892

Pattu V, Qu B, Marshall M, Becherer U, Junker C, Matti U, Schwarz EC, Krause E, Hoth M, Rettig J (2011) Syntaxin7 is required for lytic granule release from cytotoxic T lymphocytes. Traffic 12:890–901

Pitt SJ, Funnell TM, Sitsapesan M, Venturi E, Rietdorf K, Ruas M, Ganesan A, Gosain R, Churchill GC, Zhu MX, Parrington J, Galione A, Sitsapesan R (2010) TPC2 is a novel NAADP-sensitive Ca2 + release channel, operating as a dual sensor of luminal pH and Ca2+. J Biol Chem 285:35039–35046

Prekeris R, Klumperman J, Chen YA, Scheller RH (1998) Syntaxin 13mediates cycling of plasma membrane proteins via tubulovesicular recycling endosomes. J Cell Biol 143:957–971

Prekeris R, Yang B, Oorschot V, Klumperman J, Scheller RH (1999) Differential roles of syntaxin 7 and syntaxin 8 in endosomal trafficking. Mol Biol Cell 10:3891–3908

Pryor PR, Luzio JP (2009) Delivery of endocytosed membrane proteins to the lysosome. Biochim Biophys Acta 1793:615–624

Raychowdhury MK, Gonzalez-Perrett S, Montalbetti N, Timpanaro GA, Chasan B, Goldmann WH, Stahl S, Cooney A, Goldin E, Cantiello HF (2004) Molecular pathophysiology of mucolipidosis type IV: pH dysregulation of the mucolipin-1 cationchannel. Hum Mol Genet 13:617–627

Ruas M, Rietdorf K, Arredouani A, Davis LC, Lloyd-Evans E, Koegel H, Funnell TM, Morgan AJ, Ward JA, Watanabe K, Cheng X, Churchill GC, Zhu MX, Platt FM, Wessel GM, Parrington J, Galione A (2010) Purified TPC isoforms form NAADP receptors with distinct roles for Ca(2+) signaling and endolysosomal trafficking. Curr Biol 20:703–709

Rybalchenko V, Ahuja M, Coblentz J, Churamani D, Patel S, Kiselyov K, Muallem S (2012) Membrane potential regulates nicotinic acid adenine dinucleotide phosphate (NAADP) dependence of the pH- and Ca2+ -sensitive organellar two-pore channel TPC1. J Biol Chem 287:20407–20416

Saftig P, Klumperman J (2009) Lysosome biogenesis and lysosomal membrane proteins: trafficking meets function. Nat Rev Mol Cell Biol 10:623–635

Saldanha SA, Grimm C, Mercer BA, Choi JY, Allais C, Roush WR, Heller S, Hodder P (2011) Campaign to identify agonists of transient receptor potential channels 3 and 2 (TRPML3 & TRPML2), Probe reports from the NIH molecular libraries program, National Center for Biotechnology Information, Bethesda (MD) 2009 Nov 13 (updated 2011 May 5)

Samie MA, Grimm C, Evans JA, Curcio-Morelli C, Heller S, Slaugenhaupt SA, Cuajungco MP (2009) The tissue-specific expression of TRPML2 (MCOLN-2) gene is influenced by the presence of TRPML1. Pflugers Arch 459:79–91

Schieder M, Rötzer K, Brüggemann A, Biel M, Wahl-Schott CA (2010a) Characterization of two-pore channel 2 (TPCN2)-mediated Ca^{2+}currents in isolated lysosomes. J Biol Chem 285:21219–21222

Schieder M, Rötzer K, Brüggemann A, Biel M, Wahl-Schott C (2010b) Planar patch clamp approach to characterize ionic currents from intact lysosomes. Sci Signal 3:pl3

Schröder BA, Wrocklage C, Hasilik A, Saftig P (2010) The proteome of lysosomes. Proteomics 10:4053–4076

Scott CC, Gruenberg J (2011) Ion flux and the function of endosomes and lysosomes: pH is just the start: the flux of ions across endosomal membranes influences endosome function not only through regulation of the luminal pH. BioEssays 33:103–110

Shen D, Wang X, Li X, Zhang X, Yao Z, Dibble S, Dong XP, Yu T, Lieberman AP, Showalter HD, Xu H (2012) Lipid storage disorders block lysosomal trafficking by inhibiting a TRP channel and lysosomal calcium release. Nat Commun 3:731

Südhof TC (2012) Calcium control of neurotransmitter release. Cold Spring Harb Perspect Biol 4:a011353

Südhof TC, Rizo J (2011) Synaptic vesicle exocytosis. Cold Spring Harb Perspect Biol 3:a005637

Sun M, Goldin E, Stahl S, Falardeau JL, Kennedy JC, Acierno JS Jr, Bove C, Kaneski CR, Nagle J, Bromley MC, Colman M, Schiffmann R, Slaugenhaupt SA (2000) Mucolipidosis type IV is caused by mutations in a gene encoding a novel transient receptor potential channel. Hum Mol Genet 9:2471–2478

Tugba Durlu-Kandilci N, Ruas M, Chuang KT, Brading A, Parrington J, Galione A (2010) TPC2 proteins mediate nicotinic acid adenine dinucleotide phosphate (NAADP)- and agonist-evoked contractions of smooth muscle. J Biol Chem 285:24925–24932

van Gelder CM, Vollebregt AA, Plug I, van der Ploeg AT, Reuser AJ (2012) Treatment options for lysosomalstorage disorders: developing insights. Expert Opin Pharmacother 13:2281–2299
vanGoor F, Hadida S, Grootenhuis PD, Burton B, Stack JH, Straley KS, Decker CJ, Miller M, McCartney J, Olson ER, Wine JJ, Frizzell RA, Ashlock M, Negulescu PA (2011) Correction of the F508del-CFTR protein processing defect in vitro by the investigational drug VX-809. Proc Natl Acad Sci USA 108:18843–18848
Vergarajauregui S, Puertollano R (2006) Two di-leucine motifs regulate trafficking of mucolipin-1 to lysosomes. Traffic 7:337–353
Vergarajauregui S, Martina JA, Puertollano R (2009) Identification of the penta-EF-hand protein ALG-2 as a Ca2 + -dependent interactor of mucolipin-1. J Biol Chem 284:36357–36366
Vergarajauregui S, Martina JA, Puertollano R (2011) LAPTMs regulate lysosomal function and interact with mucolipin 1: new clues for understanding mucolipidosis type IV. J Cell Sci 124:459–468
Venkatachalam K, Hofmann T, Montell C (2006) Lysosomal localization of TRPML3 depends on TRPML2 and the mucolipidosis-associated protein TRPML1. J Biol Chem 281:17517–17527
Venugopal B, Browning MF, Curcio-Morelli C, Varro A, Michaud N, Nanthakumar N, Walkley SU, Pickel J, Slaugenhaupt SA (2007) Neurologic, gastric, and opthalmologic pathologies in a murine model of mucolipidosis type IV. Am J Hum Genet 81:1070–1083
Venugopal B, Mesires NT, Kennedy JC, Curcio-Morelli C, Laplante JM, Dice JF, Slaugenhaupt SA (2009) Chaperone-mediated autophagy is defective in mucolipidosis type IV. J Cell Physiol 219:344–353
Walseth TF, Lin-Moshier Y, Jain P, Ruas M, Parrington J, Galione A, Marchant JS, Slama JT (2012) Photoaffinity labeling of high affinity nicotinic acid adenine dinucleotide phosphate (NAADP)-binding proteins in sea urchin egg. J Biol Chem 287:2308–2315
Wang X, Zhang X, Dong XP, Samie M, Li X, Cheng X, Goschka A, Shen D, Zhou Y, Harlow J, Zhu MX, Clapham DE, Ren D, Xu H (2012) TPC proteins are phosphoinositide- activated sodium-selective ion channels in endosomes and lysosomes. Cell 151:372–383
Weiss N (2012) Cross-talk between TRPML1 channel, lipids and lysosomal storage diseases. Commun Integr Biol 5:111–113
Xu H, Delling M, Li L, Dong X, Clapham DE (2007) Activating mutationin a mucolipin transient receptor potential channel leads to melanocyteloss in varitint-waddler mice. Proc Natl Acad Sci USA 104:18321–18326
Yamaguchi S, Muallem S (2010) Opening the TRPML gates. Chem Biol 17:209–210
Yamaguchi S, Jha A, Li Q, Soyombo AA, Dickinson GD, Churamani D, Brailoiu E, Patel S, Muallem S (2011) Transient receptor potential mucolipin 1 (TRPML1) and two-pore channels are functionally independent organellar ion channels. J Biol Chem 286:22934–22942
Zeevi DA, Frumkin A, Offen-Glasner V, Kogot-Levin A, Bach G (2009) A potentially dynamic lysosomal role for the endogenous TRPML proteins. J Pathol 219:153–162
Zeevi DA, Lev S, Frumkin A, Minke B, Bach G (2010) Heteromultimeric TRPML channel assemblies play a crucial role in the regulation of cell viability models and starvation-induced autophagy. J Cell Sci 123:3112–3124
Zhang X, Li X, Xu H (2012) Phosphoinositide isoforms determine compartment-specific ion channel activity. Proc Natl Acad Sci USA 109:11384–11389
Zong X, Schieder M, Cuny H, Fenske S, Gruner C, Rötzer K, Griesbeck O, Harz H, Biel M, Wahl-Schott C (2009) The two-pore channel TPCN2 mediates NAADP-dependent Ca(2+)-release from lysosomal stores. Pflugers Arch 458:891–899

[illegible]

Chapter 20
Transient Receptor Potential Channels and Pain

Alexandre Denadai-Souza and Nicolas Cenac

Abstract Over the past decade, a considerable amount of data gave support to the implication of TRP channels in pain. TRPV1, 2, 3, and 4, along with TRPA1 and TRPM8 have been shown to be central players of nociceptive hypersensitivity in models of somatic and visceral inflammatory pain as well as in neuropathic pain. TRP channels are regulated by post-translational modifications in response to inflammatory mediators by mechanisms involving the recruitment of a complex repertoire of intracellular signaling pathways, resulting in channel sensitization and increased activity. Additionally, recent studies based on heterologous expression systems have shown that lipidergic mediators are putative endogenous activators of TRP channels with relevance to human painful chronic diseases. In this chapter, the reader is offered an overview, relevant to chronic pain diseases, of the original as well as the most recent discoveries regarding the superfamily of TRP channels.

20.1 Introduction

The transient receptor potential (TRP) channel superfamily is one of the largest families of ion channels with representative members across the phylogenetic tree, from yeast to humans. This superfamily is composed of 28 cation channels which

A. Denadai-Souza
Federal University of São Paulo, São Paulo, SP, Brazil

A. Denadai-Souza · N. Cenac (✉)
Inserm, U1043, F-31300 Toulouse, France
e-mail: nicolas.cenac@inserm.fr

A. Denadai-Souza · N. Cenac
CNRS, U5282, F-31300 Toulouse, France

A. Denadai-Souza · N. Cenac
Université de Toulouse, UPS, Centre de Physiopathologie de Toulouse Purpan (CPTP), F-31300 Toulouse, France

N. Weiss and A. Koschak (eds.), *Pathologies of Calcium Channels*,
DOI: 10.1007/978-3-642-40282-1_20,

are implicated in several perceptions, including thermal, chemo, and pressure sensation. The TRP channel superfamily is classified into six subfamilies: TRP channel subfamily C (canonical; TRPC), TRP channel subfamily V (vanilloid; TRPV), TRP channel subfamily M (melastatin; TRPM), TRP channel subfamily A (ankyrin; TRPA), TRP channel polycystin subfamily (TRPP), and TRP channel mucolipin subfamily (TRPML) (Vriens et al. 2009). The TRPML as TRPP subfamilies were characterized after a comparative analysis of the kidney human diseases mucolipidosis and polycystic kidney disease, respectively (Vriens et al. 2009). The TRPA subfamily has only one described member (TRPA1), its name is coming from the high number of ankyrin domains at the amino terminus of the channel protein (Vriens et al. 2009). Mammalian TRP channels that are comparable to the product of the *Drosophila Trp* gene are the TRPC channels, also called classical. The TRPV subfamily was named after the identification of its first representative TRPV1, which is the receptor for the irritant vanilloid, capsaicin. Unlike other families of ion channels, most members of the TRP channel superfamily do not possess a structural similarity. TRP channels have six transmembrane domains and intracellular amino and carboxyl termini of variable length with a pore loop between transmembrane domains five and six. TRP channels do not function as monomer; they need to be assembled in homo- and/or heterotetramers to form an efficient channel. The majority of TRP channels are non-selective cation channels with minor selectivity for calcium. Nevertheless, calcium–selective (TRPV5 and TRPV6) and sodium-selective (TRPM4 and TRPM5) members of the TRP channel subfamilies exist. In addition, some TRP channels can also transport iron (TRPML1) or magnesium (TRPV6 and TRPM7).

20.2 TRP Channels Expression in Sensory Neurons

Even if members of TRPC or TRPM family have been characterized in sensory neurons, the major TRP channels described as sensory transducers are TRPV1, TRPV2, TRPV3, TRPV4, TRPA1, and TRPM8. TRPV1, TRPV2, TRPV3, and TRPM8 are usually mentioned as thermoreceptors, while TRPV4 and TRPA1 are mechanoreceptors. Nevertheless, TRP channels possess multiple stimulus modalities; TRPV4 and TRPA1 are also described as thermoreceptors and they are all described as chemoreceptors.

Consistent with its role in pain perception, TRPV1 has been found to be expressed within sites of peripheral and central nervous system involved in pain detection, transmission, and regulation. In rodent, TRPV1 is expressed primarily in small- to medium-diameter sensory neurons, characteristic of nociceptive Aδ- and C-fibers (Caterina et al. 1997). TRPV1 is expressed in dorsal root (DRG), trigeminal (TG) and vagal sensory ganglia containing substance P, calcitonin gene-related peptide (CGRP), and co-localize with the nerve growth factor (NGF) receptor TrkA (Chuang et al. 2001; Zhu et al. 2004; Zhu and Oxford 2011; Caterina et al. 1997). In human whole cervical DRG, TRPV1 expression was characterized in 70 % of cells with a diameter <50 μm (Facer et al. 2007; Smith et al. 2002; Anand et al. 2006). In cultured neurons from human cervical DRG TRPV1-positive staining was observed in small/medium-sized neurons with a maximum intensity for cells with a diameter from 40 to 49 μm (Anand et al. 2006). In contrast to TRPV1, TRPV2 is principally expressed in neurons with medium- to large-sized cell bodies and myelinated A-fibers in DRG and TG (Ma 2001). Moreover, TRPV2 is expressed predominantly in a subpopulation of TrkC-positive neurons in mouse DRGs (Tamura et al. 2005). The cell size and axonal morphology suggest that co-expression of TRPV1 and TRPV2 is rare in the sensory ganglia (Ma 2001; Tamura et al. 2005). In contrast to DRG and TG, in somatic sensory neurons of the vagus nerve located in the jugular ganglion and in visceral sensory neurons located in the nodose ganglion, TRPV1 and TRPV2 are co-expressed in 3.9 and 63.4 % of the neurons, respectively (Ichikawa and Sugimoto 2003). Whereas in skin of mouse, TRPV2 is expressed in a majority of myelinated high-threshold mechanoreceptors, whereas in human, TRPV2 is localized mainly in unmyelinated nerve fibers (Axelsson et al. 2009). Moreover, TRPV2 co-localizes with CGRP and SP, but not with VIP and TH, suggesting that these ion channels are present mainly in C-fiber primary afferents in human skin (Axelsson et al. 2009). This discrepancy between human and rodents for the expression of TRP channels is observed for the majority of channels, being the most important concern TRPV3 expression. In mice, TRPV3 is not expressed in sensory neurons (Peier et al. 2002b). In contrast, in monkey and human tissue TRPV3 is found in sensory neurons and co-express with TRPV1 (Smith et al. 2002; Xu et al. 2002). However, in another study, expression of TRPV3 is characterized in keratinocytes but not in skin fibers that innervate the human breast (Gopinath et al. 2005).

TRPV4 is expressed in sensory neurons; interestingly a massive difference of expression of this channel depending of the projection area of the sensory neurons has been observed. In mice, using retrolabeling of sensory neurons and in situ hybridization, it has been shown an intense enrichment of TRPV4 expression in splanchnic and pelvic colonic afferents traced to their cell bodies in the thoraco-lumbar and lumbo-sacral DRG (Brierley et al. 2008). This was evident both in terms of the proportions of neurons expressing TRPV4 mRNA and the relative quantities of TRPV4 mRNA in the colonic population of DRG neurons (Brierley et al. 2008). TRPV4 protein was co-localized almost exclusively with the sensory neuropeptide CGRP in colonic nerve fibers (Brierley et al. 2008). In rat DRG and TG, TRPV4 expression was also confirmed at mRNA and protein level, including the single-cell RT-PCR (Alessandri-Haber et al. 2009; Denadai-Souza et al. 2012). In this study, it was also demonstrated that TRPV4 was co-expressed with TRPC1 and TRPC6 (Alessandri-Haber et al. 2009). In human, TRPV4 is expressed in sensory neurons (Facer et al. 2007). TRPV4 expression was not specific for any neuronal subtype in DRG and TG (Facer et al. 2007; Denadai-Souza et al. 2012). In contrast, distribution of TRPM8 was confined to a small proportion of small/medium neurons in human (Facer et al. 2007). TRPM8 mRNA has been shown to be localized mainly in A-delta fibers/C-fibers in rat primary afferent neurons

(Kobayashi et al. 2005). In rodents, TRPM8 and the cold-activated receptor TRPA1 are also detected in subpopulations of small neurons. In mouse DRG, TRPM8 mRNA does not co-express with many of the classical markers of nociception including TRPV1 (Kobayashi et al. 2005). However, TRPA1 is found in nociceptive sensory neurons in DRG and co-localizes with TRPV1, CGRP, and SP but not with TRPM8 in rat (Kobayashi et al. 2005; Story et al. 2003). As in rodents, TRPA1 was co-localized with TRPV1 in human sensory neurons (Patapoutian et al. 2003; Anand et al. 2008). Expression of TRPA1 was observed in human small/medium-sized DRG neurons (Anand et al. 2008). Dorsal nerve roots were positive for TRPA1, indicating transport of the ion channel to central sensory terminals; however, its presence in ventral roots and motoneurons may indicate other roles in the peripheral nervous system (Anand et al. 2008).

20.3 Endogenous Agonist of TRP Channels

Most of the studies performed to demonstrate the implication of TRP channel in sensation and pain have used synthetic, semi-synthetic, or natural exogenous agonist of TRP channels. As TRP channels are expressed in an area, such as brain, where they are not submitted to important temperature variations or to natural exogenous agonists, there is a need to characterize endogenous agonist of TRP channels. Since the past few years, studies trying to characterize TRP channels endogenous agonists are emerging for TRPV1, 3, 4, TRPA1, and TRPM8 (Bang et al. 2010a). For TRPV2, there is no study describing an endogenous agonist of this receptor (Bang et al. 2010a). Interestingly, for others, many endogenous TRP channels agonists are lipidergic substances. As they are structurally close to capsaicin, fatty acid conjugates of amine constitute a group of molecules defined as endovanilloid. They are constituted by a catecholic or a guaiacolic aromatic ring linked with an aliphatic chain by an amide bond: anandamide (N-arachidonyl ethanolamine, AEA), N-oleoyl ethanolamine (OEA), N-arachidonoyl dopamine (NADA), N-oleoyl dopamine (OLDA) (Chu et al. 2003; Huang et al. 2002; Zygmunt et al. 1999; Movahed et al. 2005; Starowicz et al. 2007). Whereas NADA is similar in potency to capsaicin in a variety of assays of VR1 activity, anandamide, and 12(S)-hydroperoxyeicosatetraenoic acid are at least 20-fold less potent than capsaicin (Chu et al. 2003; Hwang et al. 2000; Smart et al. 2000). The order of potency of the compounds from the dopamine series is OLDA > NADA = capsaicin; two other dopamine metabolites (N-palmitoyl-dopamine, PALDA and N-stearoyldopamine, STEARDA) have no effect on TRPV1 activation (Chu et al. 2003). In addition to TRPC3, TRPC6, and TRPC7, which are activated by diacylglycerol (DAG) (Hofmann et al. 1999), TRPV1 is also activated by DAG and other DAG analogs such as 1-oleoyl-2-acetyl-sn-glycerol (OAG), 1-Stearyl-2-arachinonyl-sn-glycerol (SAG), and 1, 2-dioctanoyl-sn-glycerol (DOG) (Woo et al. 2008). OAG, a membrane-permeable DAG analog, is five times less potent than capsaicin to induce a calcium response in sensory

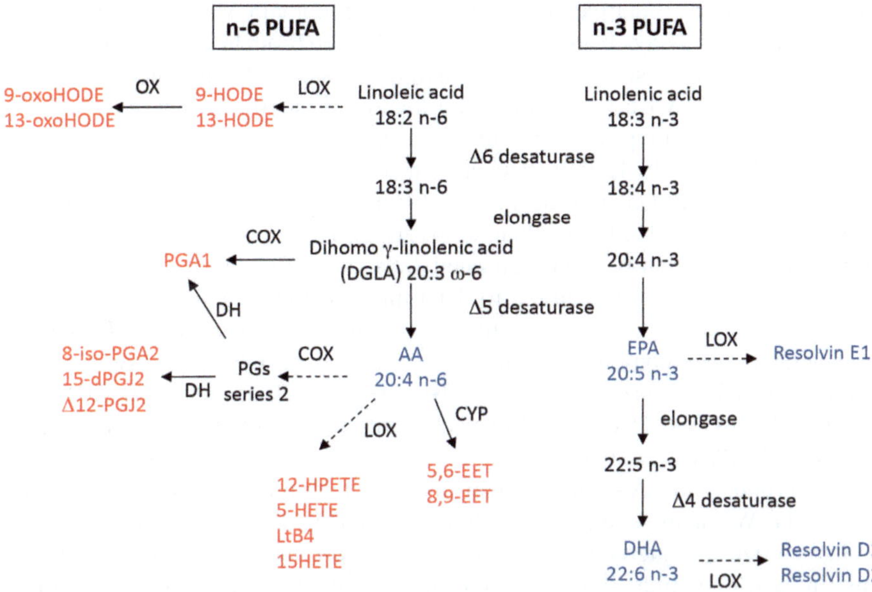

Fig. 20.1 Synthesis of TRP endogenous agonist by n-6 and n-3 polyunsaturated fatty acid metabolism. Metabolism of n-6 PUFA (AA, linoleic acid or DGLA) produced TRPV1 agonist by LOX pathway, TRPA1 agonist by COX, and TRPV4 agonist by CYP epoxygenase pathway. Metabolism of n-3 PUFA (EPA and DHA) produced TRPV1, TRPV4, and TRPA1 inhibitors. TRPM8 could be inhibited by n-6 and n-3 PUFA precursors (AA, EPA, and DHA). TRP agonists are represented in red and TRP inhibitors in blue. HODE: hydroxyoctadecadienoic acid; COX: cyclooxygenase; LOX: lipoxygenase; DH: dehydrogenase; CYP: cytochrome; EPA: eicosapentaenoic acid; DHA: docosahexaenoic acid; AA: arachidonic acid; PG: prostaglandin; HPETE: hydroperoxyeicosatetraenoic acid; HETE: hydroeicosatetraenoic acid; Lt: leukotriene; EET: epoxyeicosatrienoic acid; OX: oxidation; DGLA: Dihomo γ-linolenic acid

neurons suggesting that OAG is a partial agonist of TRPV1 (Woo et al. 2008). Capsaicin- and DAG-binding site is the same on the TRPV1 receptor (Woo et al. 2008). Other lipid substances which can activate TRPV1 are the products of arachidonic acid (AA), a n-6 polyunsaturated fatty acid, by lipoxygenase (LOX; Fig. 20.1). The 12-hydroperoxyeicosatetraenoic acid (12-(S)-HPETE) is able to activate TRPV1 in sensory neurons and in HEK transfected cells (Hwang et al. 2000). Moreover, the three-dimensional structure superimposed that of capsaicin (Hwang et al. 2000). Metabolite issued from LOX pathway other than 12-(S)-HPETE were also able to activate TRPV1 (Fig. 20.1). Among eicosanoids, 15-(S)-HPETE, 5-(S)-hydroxyeicosatetraenoic acid (5-(S)-HETE), leukotriene B4 (LTB4), and 15-(S)-HETE are the most effective in activating TRPV1 (Hwang et al. 2000). Other LO products such as hepoxilin B3, 5-(S)-HPETE, 8-(R)-15-(S)-dihydroxyeicosatetraenoic acid ((8R,15S)-diHETE), 12-(S)-HETE, hepoxilin A3, and LTC4 have no effect on TRPV1 activation (Hwang et al. 2000). Prostaglandins (PG) such as PGI_2, PGH_2, PGE_2, and PGD_2 failed to activate TRPV1 (Hwang et al.

2000). Even if the putative direct heat sensing domain was determined with domain-swapping assays using TRPV1 and TRPM8 (Brauchi et al. 2006; Brauchi et al. 2007) another study proposed that heat causes the production of lipid responsible of TRPV1 activation (Patwardhan et al. 2010). In fact, exposition of mouse and rat skin biopsies to noxious heat produces 9-hydroxyoctadecadienoic acids (9-HODE) and 13-HODE (Patwardhan et al. 2010). These two compounds as well as their dehydrogenated metabolites (9- and 13-oxoHODE) activate TRPV1 (Patwardhan et al. 2010; Patwardhan et al. 2009). HODE are LOX metabolites of linoleic acid, a C18 n-6 polyunsaturated fatty acid (Fig. 20.1) (Ramsden et al. 2012). Further studies are required to determine whether activation of TRPV1 by heat is entirely dependent on these linoleic metabolites. Moreover, as HODE can be generated also non-enzymatically via free radical-mediated oxidation, the implication of LOX enzymes needs to be confirmed (Ramsden et al. 2012). As TRPV1, TRPV4, and TRPA1 could be activated by metabolites of polyunsaturated fatty acid. TRPV4 is activated by the 5, 6-epoxyeicosatrienoic acid (EET) and 8, 9-EET, products of AA metabolism by a cytochrome epoxygenase (CYPe) (Fig. 20.1) (Watanabe et al. 2003; Vriens et al. 2005). Further, studies are needed to characterize the interaction of TRPV4 with those metabolites in sensory neurons. In fact, one study show that serosal and mesenteric colonic fibers showed increases in mechanosensitivity in response to the selective endogenous TRPV4 agonist 5, 6-EET with a loss of effect in TRPV4 deficient mice (Brierley et al. 2008). Another one shows that 5, 6-EET induce calcium signaling in sensory neurons through TRPA1 activation and not through TRPV4 activation (Sisignano et al. 2012). This discrepancy could be due to the relative high expression of TRPV4 in sensory neurons projecting from the colon compared to the whole population of sensory neurons. In other cell type than sensory neurons, 5, 6-EET has been well characterized as a TRPV4 agonist. Endogenously, dimethylallyl pyrophosphate (DMAPP) is an activator of TRPV4 (Bang et al. 2012b). DMAPP is an intermediate molecule in the mevalonate pathway: Isopentenyl pyrophosphate (IPP) isomerase generates DMAPP from IPP (Liang 2009). Condensation of DMAPP with IPP results in the formation of geranyl pyrophosphate (GPP), in turn used for farnesyl pyrophosphate (FPP) and cholesterol synthesis (Liang 2009). Interestingly, concerning the same metabolism pathway, FPP is the only endogenous agonist of TRPV3 described and IPP is presented as an endogenous inhibitor of TRPV3 and TRPA1 (Bang et al. 2010b, 2011). Endogenous agonists of TRPA1 described in the literature are for most of them cyclopentenone prostaglandins such as PGA1, 8-iso-PGA2, 15-deoxy-Δ12,14-PGJ2, and Δ12-PGJ2. These cylopentenone are issued from the peroxidation of AA by free radicals (Fig. 20.1) (Andersson et al. 2008; Cruz-Orengo et al. 2008; Materazzi et al. 2008; Taylor-Clark et al. 2009). PGA1 is issued from Dihomo γ-linolenic acid (DGLA, 20:3 n-6) metabolism by cyclooxygenase (COX; Fig. 20.1) (Xiao et al. 2011). These cyclopentenones could also be formed by the dehydration of the cylcopentane ring of other PGs such as PGE and PGD from the series-2 (Fukushima 1990). Their highly reactive α, β-unsaturated carbonyl residue in their structures was proposed to covalently react with lysines and cysteines in N-term of TRPA1 (Santoro 1997).

Cyclopentane PGs do not have this ability, while cyclopentenone PGs are not able to activate known G-protein coupled PG receptors. The same covalent interaction between α, β-unsaturated aldehyde or carbonyl group from nitrooleic acid has been shown (Taylor-Clark et al. 2009). Nitric oxide forms nitrated phospholipids, which in turn produce nitrated fatty acids including nitrooleic acid (Freeman et al. 2008). The nitrated carbon is essential to confer this reactivity with TRPA1 cysteines, as oleic acid failed to activate TRPA1 (Taylor-Clark et al. 2009). However, it seems that nitrooleic acid is not specific from TRPA1 as a study that demonstrates this substance was able to activate TRPV1 and TRPA1 channels in DRG neurons but may also target unidentified channels (Sculptoreanu et al. 2010). Reaction with TRPA1 cysteine is also implicated in the activation of this channel by radical oxygen species (ROS) (Andersson et al. 2008). Peroxynitrite, which results from the conversion of nitric oxide by superoxide ion, one of the ROS, was able to directly activate TRPV1 (Ito et al. 2013). Nevertheless, more studies are needed to confirm the implication of ROS as endogenous activators of TRP channels to detect pain stimuli.

Another class of PUFA metabolites, resolvins, has been described as TRP channel inhibitors. Resolvins are products of the metabolism of n-3 PUFA by LOX enzyme (Fig. 20.1). Resolvins E-series are metabolites of eicosapentaenoic acid (EPA) and D-series of docosahexaenoic acid (DHA) (Ji et al. 2011). Resolvin E1 has been described as an inhibitor of TRPV1, resolvin D1 as an inhibitor of TRPV4, TRPV3, and TRPA1, and resolvin D2 as an inhibitor of TRPA1 and TRPV1 (Park et al. 2011a; Bang et al. 2012a; Bang et al. 2010c). Nevertheless, for these compounds more studies are needed to determine if they directly bind to the receptor channel or if these effects are mediated through an activation of G-protein coupled receptor. In contrast to other TRP channels, TRPM8 did not distinguish the two types of PUFA metabolites, n-6 and n-3 (Fig. 20.1). Both the cold and menthol responses of TRPM8 are reduced by n-3 PUFA (EPA or DHA) or n-6 PUFA (AA) species (Andersson et al. 2007). These findings suggest that TRPM8 can be regulated by variations in the cellular concentrations of PUFAs. PUFA such as AA, DHA, and EPA are produced in the cells by PLA2 enzymes which hydrolyze the sn-2 ester of glycerophospholipids to release a free polyunsaturated fatty acid (PUFA) and a lysophospholipid (LPL) (Balsinde and Balboa 2005). Interestingly, in contrast to expected results, treatment of cells by PLA2 inhibitor blocks the activation of TRPM8 by cold, icillin, and menthol (Andersson et al. 2007). Authors demonstrated that LPL is able to activate TRPM8 (Andersson et al. 2007). From the same family of molecules, lysophosphatidylcholine, lysophosphatidylinositol, and lysophosphatidylserine activate TRPM8 (Andersson et al. 2007; Vanden Abeele et al. 2006). The LPLs also potentiate the cold responses of TRPM8. However, evidence of a direct binding of these molecules to TRPM8 has not been characterized. They may alter the local plasma membrane microcurvature thereby physically affecting channel gating (Vanden Abeele et al. 2006).

20.4 Participation of TRP Channels in the Physiopathology of Pain

How different animals perceive pain, mainly its emotional component, is still a matter of intense debate. Nevertheless, it is well accepted that the general neuroanatomical and biochemical nociceptive pathways are well-conserved among species (Jordt and Julius 2002). Nociception is the neural process of encoding a noxious stimulus, which plays a major role in survival, by triggering avoidance to harmful situations or by signaling a disease state (Gold and Gebhart 2010; Schaible et al. 2011). This process is performed by nociceptors, which are high-threshold sensory receptors of the peripheral somatosensory nervous system that are capable of transducing a noxious stimulus. On the other hand, exacerbated nociceptor activity can be, by itself, the physiopathological basis to a whole class of painful chronic diseases, an emerging public health concern due to the aging of population, wherein evidence for the participation of TRP channels has increased over the past decade.

In adults, chronic pain has a worldwide incidence of 20–25 % (Breivik et al. 2006; Gold and Gebhart 2010) and is most frequently associated with disorders of the musculoskeletal system rather than with skin diseases (Breivik et al. 2006). Likewise, nociceptors in the musculoskeletal system are sensitized to mechanical stimuli during inflammation, while cutaneous nociceptors are more usually sensitized to thermal stimuli (McDougall 2006; Schaible et al. 2009). Besides somatic tissues such as skin and the musculoskeletal system, in which nociceptors have high mechanical thresholds that differentiate them from non-nociceptive low threshold nerve terminals (Schaible et al. 2011), most putative nociceptors in viscera have their excitation threshold in the innocuous range (Schaible et al. 2011; Blackshaw et al. 2010). The aforementioned dichotomy on how pain processing differs among tissues in regard to the nature of stimuli, brilliantly raised and reviewed by (Schaible et al. 2011), along with the differential molecular sensory properties of each TRP channel described below, bring strong elements that should be taken into account when exploring the involvement of molecular transduction nociceptors and their potential as drug targets for the management of chronic pain diseases.

Physiological pain signals are transmitted when molecular sensors within nociceptors are activated with enough intensity to transduce the incoming stimulus into a depolarization, that once above the nerve terminal threshold, generates an action potential by opening voltage-gated Na^+ channels, and then the signal is conducted upstream within the nociceptive pathway (McDougall 2006; Schaible et al. 2009, 2011). On a very concise way, the nociceptive pathway consists of peripheral nerve terminals originating from primary sensory ganglion neurons, which in turn project their proximal fibers to spinal and trigeminal nociceptive centers, wherein the upcoming signals are modulated by interneurons and transmitted by second order neurons projecting to higher levels of the sensory nervous system. On the other hand, physiopathological pain can occur when neurons along with the nociceptive pathway are sensitized and the threshold is lowered. This process prompts a nerve fiber, whereas in the periphery or at a central level in the nociceptive pathway, to fire an action potential in response to a stimulus that once, was innocuous and under its threshold (allodynia), or leads to an exaggerated response to a noxious stimulation (hyperalgesia). Although the terms allodynia and hyperalgesia are widely employed in the literature, herein the term hypersensitivity will be adopted throughout the text in order to simplify the reading. TRPV1, 2, 3, and 4, along with TRPA1 and TRPM8, are specialized molecular transduction channels expressed in nociceptors that can be sensitized by a plethora of inflammatory mediators such as bradykinin, prostaglandin E_2, extracellular ATP, glutamate, proteases, and neurotrophins (Stein et al. 2009; Levine and Alessandri-Haber 2007). Aberrant channel activity can be the net product of several intracellular changes, such as transcriptional, translational, or post-translational regulation. Indeed, the implication of post-translational changes in TRP channels has been widely studied over the past years, enlightening some primary neurobiological processes that might be central to the physiopathology of chronic pain.

20.4.1 Regulation of TRP Channels

On the whole, the mechanism by which TRP channels activity is modulated involves fast and slow regulatory processes. Likewise, TRP channels can suffer post-translational modifications within a few minutes after activation of GPCRs by inflammatory mediators, such as downstream phosphorylation, leading to an increased gating activity in response to agonists or even a reduced desensitization to continued or repeated stimulation. Additionally, a mechanism involving enhanced translocation from the cytosol to the plasma membrane has also been proposed. On the other hand, slower process can be the result of changes induced by cytokines and growth factors on transcriptional regulation, RNA processing, or even enhanced translation. Whereas these different processes may be differentially regulated depending on target tissue, inflammatory mediators, and even at an intracellular microdomain-dependent way, the net result in most cases is an enhancement of TRP channel-mediated ionic currents and nociceptor sensitization.

20.4.2 Sensitization of TRP Channels by Inflammatory Mediators and Downstream Second Messenger Pathways

Early in vitro studies have shown that protons, within a pH range commonly observed during inflammation ($pH \leq 5.9$), besides directly interacting with TRPV1, could lead to a lower thermal threshold, turning this channel into an active state in the range below normal human body temperature (Tominaga et al. 1998; Jordt et al. 2000). Indeed, by now it is well established that TRPV1 and other TRP channels family members can be sensitized by a plethora of mechanisms, most likely involving the second messenger signaling pathways activation by inflammatory mediators and growth factors (Fig. 20.2). Clear evidences address TRPV1 as a substrate for multiple protein kinases such as cyclic AMP-dependent protein kinase (PKA) (Bhave et al. 2002), protein kinase C (PKC) (Premkumar and Ahern 2000), Ca^{2+}/calmodulin-dependent kinase II (CaMKII) (Jung et al. 2004), phosphatidylinositol-3 kinase (PI3 K), and mitogen-activated protein kinase (MAPK) (Zhuang et al. 2004).

Bradykinin is produced by the proteolytic processing of its precursor by kallikreins in response to inflammatory stimuli such as infection and tissue injury. Its effects are mediated by the activation of two distinct receptors named B1 and B2 (Marceau and Regoli 2004). The receptor B2 is constitutively expressed on a wide array of tissues, including sensory neurons, and appears to be important on acute pain, whereas B1 expression is induced by inflammation and is more related to chronic pain (Calixto et al. 2000; Campos et al. 2006). Interestingly, it has been demonstrated that TRPV1 can be sensitized by bradykinin via B2 receptors in a heterologous expression system and in nerve fiber preparations (Chuang et al. 2001; Liang et al. 2001). Bradykinin can activate capsaicin receptors via intracellular second messenger pathway involving mobilization of arachidonic acid by PLA_2 and the generation of 12-lipoxygenase (12-LOX) products, which are required for bradykinin-induced thermal hypersensitivity (Shin et al. 2002; Ferreira et al. 2004). So far, several metabolites of lipoxygenases such as 12-(S)-HPETE, 15-(S)-HPETE, 5-(S)-HETE, and LTB_4 were demonstrated to activate TRPV1 due to the structural similarity with capsaicin (Hwang et al. 2000). Additionally, it has been shown that TRPV1 phosphorylation by the cAMP-PKA pathway delays its Ca^{2+}-dependent desensitization (Chuang et al. 2001; Mohapatra and Nau 2003). The arachidonic acid metabolite PGE_2, abundant during inflammation (Julius and Basbaum 2001), binds to its $G\alpha_s$-coupled receptors EP3C and EP4 (Southall and Vasko 2001), activates adenylyl cyclase and promotes cAMP-mediated activation of PKA, which in turn transiently sensitizes TRPV1 via phosphorylation, increasing Ca^{2+} currents to the above channel activators (Lopshire and Nicol 1998; Rathee et al. 2002). This phosphorylation-mediated gain of function, which seems to be dependent on Ser-166 and putatively Thr-370, is thought to be due to an inhibition of TRPV1 desensitization by the Ca^{2+}-and calmodulin-dependent protein phosphatase B (calcineurin) (Docherty et al. 1996; Mohapatra and Nau 2003).

TRPV3 is thought to mediate thermal hypersensitivity by a process involving intercellular communication, where the activation of TRPV3 in keratinocytes may signal to sensory nerve terminals within the epidermis on a paracrine way (Caterina 2007; Chung et al. 2004; Guler et al. 2002). In support to this hypothesis, it was shown that keratinocytes from transgenic mice overexpressing TRPV3 exhibited an augmented release of PGE_2 and thermal hypersensitivity in response to its agonists (Huang et al. 2008). It is possible that the nociceptive effect induced by TRPV3 activation is mediated by the sensitization of TRP channels expressed in neurons by a PGE_2 and PKA-dependent mechanism. PKA activation has also been shown to be a downstream effector event on the sensitization of TRPV4 after the intradermal administration of an inflammatory soup composed of PGE_2 and 5-HT, leading to a TRPV4-mediated mechanical and osmotic hypersensitivity in mice (Alessandri-Haber et al. 2006). Furthermore, histamine and 5-HT induces the translocation of TRPV4 from cytosol to the plasma membrane of DRG sensory neurons, by a MAPK-dependent mechanism (Cenac et al. 2010b).

Besides PKA, PKCε plays a prominent role in thermal hypersensitivity induced by inflammation (Cesare et al. 1999) and this effect is mediated by TRPV1 sensitization (Premkumar and Ahern 2000). Likewise, mice lacking a functional *Trpv1* gene (*Trpv1−/−*) did not display nocifensive behavior following intraplantar injection of phorbol 12-myristate 13-acetate (PMA), a PKC activator, suggesting that PMA-induced nociceptive behavior was exclusively dependent on TRPV1 (Bolcskei et al. 2005). Additionally, it has been shown that PKCε-mediated TRPV1 sensitization involves the phosphorylation of the amino acid residues S502 and S800, a mechanism that has been hypothesized to be central to bradykinin and ATP-induced TRPV1 sensitization (Cesare et al. 1999; Numazaki et al. 2002). It was recently shown that the sensitization of TRPV4 is mediated by phosphorylation of S824 by PKC (Peng et al. 2010). Contrary to the aforementioned TRP channels family members, TRPM8 activity is downregulated by the activation of PKC (Premkumar et al. 2005) and inhibited by a disruption of its interaction with the membrane phospholipid PtdIns(4,5)P_2 (PIP2) (Weil et al. 2005; Benedikt et al. 2007). Additionally, PIP2 has been shown to upregulate TRPV1 and TRPV2 activity by a putative interaction with their proximal C-terminus (Mercado et al. 2010; Ufret-Vincenty et al. 2011; Grycova et al. 2012).

Proteinase-activated receptors (PARs) have been shown to play an important role on inflammatory pain. Within the four members of this family, the prototypic agonist of PAR_1, PAR_3, and PAR_4 is thrombin, whereas PAR_2 is activated by trypsin-like proteinases (Vergnolle 2009) and was the first member of this family to be implicated in nociception (Vergnolle et al. 2001). Interestingly, PAR_2 has been shown to sensitize TRPV1, TRPV4, and TRPA1 in several models of inflammatory pain (Amadesi et al. 2004, 2006; Dai et al. 2004); (Grant et al. 2007; Cenac et al. 2008b; Sipe et al. 2008; Helyes et al. 2010; Chen et al. 2011b; Denadai-Souza et al. 2012). Likewise, the activation of this receptor sensitizes TRPV1 to induce thermal hypersensitivity through the PKA and PKCε signaling pathways (Amadesi et al. 2004, 2006; Dai et al. 2004). In addition to other inflammatory mediators, PAR_2 agonists were demonstrated to sensitize TRPV4, leading to somatic (Grant et al. 2007; Chen et al. 2011b; Denadai-Souza et al. 2012) and visceral nociceptive hypersensitivity (Cenac et al. 2008b; Sipe et al. 2008), by a mechanism involving the activation of PLCβ, PKA, PKC, and PKD (Grant et al. 2007). On the contrary, the thrombin-activated receptor PAR_4 inhibits TRPV4-induced Ca^{2+} mobilization in dorsal root ganglion neurons and visceral hypersensitivity, by a mechanism yet to be further investigated (Auge et al. 2009; Molliver and Snider 1997).

The neurotrophin nerve growth factor (NGF) is essential for the survival of sensory nociceptive neurons during embryonic development via activation of the receptor tyrosine kinase (TrkA). At early postnatal stage, nociceptive neurons differentiate into two roughly proportional subpopulations where one is positive for neuropeptides and dependent on NGF/TrkA for phenotype maintenance, whereas the other downregulates TrkA and gives place to the receptor tyrosine kinase Ret, which is responsive to glial cell-derived neurotrophic factors (GDNF) family of ligands (GLF), and most of these neurons are characterized by the IB4-binding capacity (Molliver and Snider 1997). Then, from neonatal to adulthood transition, major changes occur within the downstream signaling pathways for these neurotrophic factors in such a way that its activation couples mostly to nociceptor sensitization during inflammatory responses rather than to neuronal survival (Zhu and Oxford 2011). Indeed, NGF and GDNF release increases during inflammation and have been implicated in thermal hypersensitivity via TRPV1 and TRPA1 (Amaya et al. 2004; Petruska and Mendell 2004); (Malin et al. 2006, 2011), wherein the activation of several intracellular signaling pathways have been proposed so far. Furthermore, it was suggested that NGF can induce sensitization of TRPV1 to protons or heat by the formation of a ternary complex with TrkA/PLCγ (Chuang et al. 2001; Bhave and Gereau 2004). In addition, it was proposed that NGF sensitizes TRPV1 by a mechanism dependent on PI3 K and MAPK (Zhuang et al. 2004), but not Src (Zhu and Oxford 2007). On the contrary, another study has shown that TRPV1 is rapidly addressed to the plasma membrane in the presence of NGF, via TrkA/PI3 K/PKCδ/Src, being the former the downstream element directly responsible for phosphorylating TRPV1 at Y200, an amino acid residue highly conserved in this channel among species and within other TRPV family members (Zhang et al. 2005). Moreover, NGF induces an increased protein expression of TRPV1 in peripheral nerve terminals, by a mechanism dependent on generation of reactive oxygen species (ROS) and MAPK p38 (Ji et al. 2002). Both NGF and GDNF have been implicated in the upregulation of TRPV1 in rat dorsal root ganglia, respectively in TrkA- and IB4-positive neurons, and the neutralization of each neurotrophin with antibodies alleviate thermal hypersensitivity (Amaya et al. 2004), while NGF seems to have no effect on TRPV2 expression (Shimosato et al. 2005). Otherwise, the sensitization of TRPV1 and TRPA1 by neurotrophic factors is dependent on the target tissue and seems to be more relevant in muscle and visceral afferents rather than in skin (Malin et al. 2011). The injection of NGF into the human muscle fascia induces nociceptor sensitization to mechanical and chemical (protons) stimulation, while the thermal threshold is not altered (Deising et al. 2012). The regulation of other TRP channels family members by neurotrophic factors in nociceptors has not been addressed yet, but it is reasonable to hypothesize that most of features shown by TRPV1 and TRPA1 may be conserved, with varying degrees of complexity.

20.4.3 Participation of TRP Channels in Inflammatory Acute Pain

Although the incidence of pain in humans is not as much frequent on skin diseases as in other systems such as the musculoskeletal and visceral systems (Breivik et al. 2006), as soon as transgenic TRP channels mice became available, most of the primordial studies were carried out on well-established models of cutaneous pain. Notwithstanding to the relevance to human chronic inflammatory diseases, these studies generated much of what is known about the role of TRP channels in physiological pain and inflammatory diseases. Thus far, the best-explored channel implicated in pain within the TRP channels superfamily is TRPV1. *Trpv1−/−* mice have an increased thermal threshold but a normal physiological response to noxious mechanical stimuli. Additionally, whereas mechanical hypersensitivity is able to develop in these animals the setting of an acute inflammation, they are devoid of thermal hypersensitivity induced by several pro-inflammatory mediators such as bradykinin, nerve growth factor (Chuang et al. 2001), and extracellular ATP (Moriyama et al. 2003), or to irritant substances such as mustard oil, complete Freund's adjuvant (Caterina et al. 2000) and carrageenan (Davis et al. 2000). Similarly to TRPV1, a recent study suggested that TRPM3 is also involved in the detection of heat noxious stimulation, *albeit* with a shift to higher temperatures. Additionally, heat hypersensitivity induced by skin inflammation was abrogated in *Trpm3−/−* animals (Vriens et al. 2011). Further studies are needed to clarify whether this TRP channel is also relevant to nociception in other systems, such as the musculoskeletal and in viscera.

TRPV2, which can be activated by extreme heat (>52 °C), hypoosmotic stimulus or cell stretch, was hypothesized as a potential thermal and mechanical transduction molecule. Intriguingly, whereas *Trpv2−/−* mice showed reduced perinatal viability and weight, the response to noxious heat over a broad range of temperatures as well as to mechanical stimuli is not altered, either in basal conditions or under inflammation or nerve injury (Park et al. 2011b). As stated before, TRPV3 has been implicated in skin thermal hypersensitivity by a mechanism involving intercellular communication between keratinocytes and sensory neurons (Levine and Alessandri-Haber 2007; Caterina 2007); (Huang et al. 2008); (Chung et al. 2004, 2005); (Bang et al. 2010b).

Trpv4−/− mice have impaired sensitivity to acid, an increase in mechanical nociceptive threshold, and altered thermal selection behavior (Suzuki et al. 2003; Liedtke et al. 2003; Liedtke and Friedman 2003; Lee et al. 2005). Otherwise, responses to noxious heat and low-threshold mechanical stimuli are conserved (Suzuki et al. 2003) (Liedtke et al. 2003; Liedtke and Friedman 2003). These initial studies not only proposed a role of TRPV4 in nociception, as well as suggested that this channel could be a molecular mechanotransducer of major relevance to human painful diseases. The contribution of TRPV4 for the detection of warm temperatures and thermal hypersensitivity has also been investigated (Todaka et al. 2004). Likewise, inflammatory and thermal hypersensitivity induced by carrageenan injection was markedly reduced in *Trpv4−/−* mice, which displayed a longer latency to escape from a hot-plate stimulus set at 35–45 °C. Additionally, in mice lacking functional TRPV4 or in rats submitted to a transient downregulation of the level of TRPV4, the nocifensive behaviors to small increases or decreases in osmolarity in the presence of PGE_2 is impaired (Alessandri-Haber et al. 2003, 2005). This finding is relevant for a role of TRPV4 in pathological pain states because changes in osmolarity and pH have been described in various diseases and are believed to contribute to inflammatory pain (Hamamoto et al. 2000).

TRPA1 was initially implicated in sensing cold and mechanical stimuli. In *Trpa1−/−* mice, this channel is not required to acute cold transduction in normal skin whereas it seems to modulate the mechanically evoked firing rate of multiple classes of cutaneous sensory neurons (Kwan et al. 2009). Otherwise, the role of TRPA1 in sensation of painful mechanical stimuli is still controversial (Bautista et al. 2006; Kwan et al. 2006) and seems to be more restricted to intense mechanical force applied to skin (Kwan et al. 2006). As discussed in the following section, TRPA1 seems to have a more pronounced role in musculoskeletal nociception.

TRPM8 is activated by cold and menthol (McKemy et al. 2002; Peier et al. 2002a) and is essential for the neuronal sensing of innocuous cool temperatures and noxious cold (Bautista et al. 2007; Colburn et al. 2007; Dhaka et al. 2007; Knowlton et al. 2010; McCoy et al. 2011). *Trpm8−/−* mice have an impaired discrimination of mildly warm or cool temperatures as well as a lack of aversion to noxious cold (Knowlton et al. 2010). Recent evidences suggest that TRPM8 is essential for cold hypersensitivity associated with inflammation. Likewise, the nocifensive response of *Trpm8−/−* mice to cold stimulation in animals submitted to an intraplantar injection of CFA is reduced (Colburn et al. 2007; Knowlton et al. 2011). Surprisingly, the channel is also required for the antinociceptive effect of mild cooling or cooling compounds (Dhaka et al. 2007; Proudfoot et al. 2006).

20.4.4 Implication of TRP Channels in Neuropathic Pain

Neuropathic pain is caused by a lesion or dysfunction of the somatosensory system and is characterized by symptoms such as ongoing, paroxysmal, or evoked pain and can be associated with underlining disorders such as traumatic nerve lesions, painful diabetic, and nondiabetic neuropathies, postherpetic neuralgia, central pain, and others (Attal 2012). The implication of TRP channels in the physiopathology of neuropathic pain has been the focus of several studies in the past decade. For example, in a study on *Trpv1−/−*mice, it was shown that the chronic mechanical hypersensitivity induced by streptozotocin-diabetic or cisplatin-induced toxic polyneuropathy occurred earlier than in wild-type animals, suggesting a protective role for TRPV1 (Bolcskei et al. 2005). Interestingly, it has been shown that on streptozotocin-induced diabetic rats, TRPV1 undergoes several modifications on neurons including an increased oligomerization, phosphorylation, and addressing to the neuronal plasma membrane, alongside with a shift on TRPV1 expression predominance from C- toward A-fibers (Hong and Wiley 2005). On the same way, an increase in TRPV1 expression has been reported following peripheral nerve injury (Fukuoka et al. 2002). The involvement of TRPV1 has also been evaluated in functional assays by using models of neuropathic pain associated with nerve lesion. Likewise, after a traumatic neuropathy caused by the ligation of the sciatic nerve, the cold hypersensitivity was reduced by *Trpv1* knockdown (Christoph et al. 2006, 2007). Similarly, *Trpv1* antisense oligodeoxynucleotides reduce the mechanical hypersensitivity associated with spinal nerve ligation (Christoph et al. 2007). TRPA1 seems to be implicated in neuropathic pain as well. The TRPA1 antagonist A-967079 attenuates the cold hypersensitivity produced by nerve injury but without changing noxious cold sensation in naïve animals, suggesting distinct roles of TRPA1 in physiological and pathological states (Chen et al. 2011a).

The contribution of TRPV4 in neuropathic pain has been investigated in models of small-fiber peripheral neuropathy induced by taxol (Alessandri-Haber et al. 2004) paclitaxel- and nerve chronic compression-induced neuropathic pain (Chen 2011b; Wang et al. 2011). Likewise, taxol treatment enhanced nociceptive behavioral responses to mechanical and hypotonic stimulation of the rat hind paw and the treatment with *Trpv4* antisense oligodeoxynucleotides reversed these pronociceptive effects. It has been suggested that TRPV4 may be engaged in neuropathic pain induced by taxol via a second messenger pathway involving integrin/Src tyrosine kinase/PKA/PKCε signaling pathways (Alessandri-Haber et al. 2006, 2007, 2008).

Nevertheless, in recent years, TRPM8 has been pointed out as a putative new target for the treatment of neuropathic pain. The expression of TRPM8 increased in the corresponding DRG and TG ipsilateral to the nerve injury with an apparent correlation to the development of hypersensitivity (Knowlton et al. 2011; Rossi et al. 2012). Additionally, the symptoms of cold hypersensitivity have been reported to be TRPM8-dependent on models of neuropathic pain induced by chronic constriction injury (CCI) of the sciatic nerve in *Trpm8−/−* mice (Colburn et al. 2007; Knowlton et al. 2011). Similarly, a recently developed TRPM8 antagonist presented strong inhibitory activity of cold hypersensitivity in a model of neuropathic pain (Calvo et al. 2012). On the other hand, TRPM8 may play different roles in mechanical and thermal hypersensitivity that develop after nerve injury. For example, while intrathecal application of menthol resulted in inhibition of mechanical and thermal hypersensitivity, responsiveness to cold was enhanced. In contrast, downregulation of TRPM8 by antisense oligonucleotides attenuated cold hypersensitivity, but it had no effect on CCI-induced mechanical and thermal hypersensitivity (Su et al. 2011).

20.4.5 Implication of TRP Channels in Pain Associated with Arthritis

Arthritis refers to a class of chronic degenerative diseases associated with painful episodes during peaks of acute reincidence and a marked baseline chronic discomfort with impact on day life activities. One of the clinical hallmarks of rheumatoid arthritis, a chronic and painful autoimmune disease remaining to this class of disorders, is the bilateral symmetry by which this disease affects most joints. For several years, this feature has been suggested to be mediated by neurogenic mechanisms (Levine et al. 1987). The first solid evidence that TRPV1 had a role on joint pain during arthritis came from studies based on knockout animals. Indeed, it was shown that in *Trpv1−/−* mice, the development of thermal (Keeble et al. 2005) and mechanical hypersensitivity (Szabo et al. 2005) associated with adjuvant-induced chronic arthritis was abrogated. The discrepancy present on the fact that TRPV1 may mediate mechanical hypersensitivity during arthritis but not in cutaneous pain processes suggest that the TRPV1 activity and biological function may be regulated on a complex way, depending on the molecules present in an inflammatory context (Caterina and Julius 2001; Szabo et al. 2005). Chronic inflammatory diseases may involve the recruitment of a different pool of inflammatory mediators, leading to a bias in TRP channels sensitization/activation and pain processing. TRPV1 has also been suggested to be involved in osteoarthritis pain on a genetic study, wherein the I585 V TRPV1 variant encoded by rs8065080 was associated with symptomatic knee osteoarthritis in humans (Valdes et al. 2011). On the other hand, it was recently shown that while the peripheral administration of SB-366791, a TRPV1 antagonist, had no effect on bilateral mechanical hypersensitivity induced by TNF on mice, the intrathecal administration of this drug was effective. In the same study, local or central blockade of TRPA1 with AP-18 inhibited the bilateral mechanical hypersensitivity induced by TNF (Fernandes et al. 2011). On a model of monosodium iodoacetate-induced osteoarthritis, a TRPV1 antagonist (AMG9810), but not a TRPA1 antagonist (HC030031), inhibited the thermal hypersensitivity, while neither antagonists affected ongoing pain (Okun et al. 2012). On an experimental model of gout, one of the most painful forms of arthritis, the TRPV1 antagonists AMG9810 or SB366791 inhibited the nocifensive behavior displayed after the intraplantar injection of monosodium urate crystals (Hoffmeister et al. 2011). Besides TRPV1, TRPV4 activation by the agonists 4αPDD or GSK1016790A induces joint mechanical hypersensitivity (Denadai-Souza et al. 2012), while the nociceptive effect of both channels are potentiated by PAR_2 agonists (Helyes et al. 2010; Denadai-Souza et al. 2012).

20.4.6 Implication of TRP Channels in Visceral Pain

Only the implication of TRPV1, TRPV4, TRPM8, and TRPA1 has been studied in visceral pain. For TRPV2 and TRPV3, there is no study showing a role of these channels in visceral pain. A large number of studies have characterized the role of the activation of TRPV1 on primary afferent neurons innervating the gut evoking pain-related behavior in rodents (Holzer 2011). Even if TRPV1 has been characterized as a nociceptor implicated in thermal and chemical pain, there is evidence that TRPV1 contributes to mechanical hypersensitivity (Holzer 2011). In pelvis afferent innervating the murine colon, TRPV1 is functional and expressed by mechanoreceptors that respond to distension with a low frequency of firing (Malin et al. 2009). Therefore, capsaicin induces hypersensitivity in response to colorectal distension (van den Wijngaard et al. 2009a). Similarly, the hypersensitivity to colorectal distension, which is seen after acute stress exposure of adult rats that have been subjected to maternal separation as neonates, is reversed by TRPV1 blockade (van den Wijngaard et al. 2009b). In contrast, for high-threshold splanchnic afferents innervating the mesentery and serosa of the colon, TRPV1 blocker decreases mechanical probing in inflammation but not in control condition (Phillis et al. 2009). The role of TRPV1 in mechanical pain seems to be that of a secondary transducer, because a mechanodetector role of TRPV1 has not been proven. Instillation of capsaicin at different level of the gut, in human volunteers, evokes symptoms such as heartburn, cramps, pressure, warmth, and nausea and increase the sensibility to balloon distension (Kindt et al. 2009; Chen et al. 2010; Lee et al. 2004; Hammer and Vogelsang 2007; Schmidt et al. 2004). In patients, local capsaicin instillation induces painful sensations after an ileostomy or a colostomy (Drewes et al. 2003; Arendt-Nielsen et al. 2008; Akbar et al. 2010). In irritable bowel syndrome (IBS) and quiescent inflammatory bowel disease patients a correlation between pain intensity and number of mucosal TRPV1-positive nerve fibers (Akbar et al. 2010, 2008). For diarrhea-predominant IBS this increase in expression was linked to an increase of function. In fact, capsaicin-containing capsule induces hypersensitivity in these patients compared to healthy control (Hammer et al. 2008; Gonlachanvit et al. 2009). Increase of TRPV1 expression in the absence of inflammation has been also observed in idiopathic rectal hypersensitivity and fecal urgency (Chan et al. 2003) and Hirschsprung's disease (Facer et al. 2001). Patients with uninvestigated dyspepsia have been found hypersensitive to intrajejunal capsaicin infusion (Hammer 2006), and a proportion of patients with functional dyspepsia are more responsive to ingestion of capsaicin capsules than healthy controls (Hammer et al. 2008). In upper gastrointestinal (GI) tract, prolonged treatment with capsaicin-containing capsules induces an antinociceptive effect in healthy volunteers and patients with functional dyspepsia (Bortolotti et al. 2002; Fuhrer and Hammer 2009). Furthermore, the homozygous G315C polymorphism of the TRPV1 gene has been found to be inversely related to symptom severity in functional dyspepsia patients (Tahara et al. 2010).

In contrast to TRPV1, TRPV4 has turned out to play a major role in mechanical pain and hypersensitivity in the colon. TRPV4 agonists enhance the mechanosensory responses of colonic serosal and mesenteric afferent nerve fibers, while the mechanosensitivity of these high-threshold afferent nerve fibers is substantially attenuated in TRPV4 knockout mice (Brierley et al. 2008). The TRPV4 agonist induces mechanical hypersensitivity in response to colorectal distension (Cenac et al. 2008a). More importantly, in TRPV4 knockout mice or following intravertebral pretreatment of mice with TRPV4-directed small interfering RNA (siRNA) the sensitivity to colorectal distension is reduced in the noxious range (Brierley et al. 2008; Cenac et al. 2008a). Taken these findings together, the specific contribution to colonic high-threshold mechanosensory function makes TRPV4 currently the only nociceptor-specific TRP channel in the gut (Blackshaw et al. 2010). Under basal conditions, TRPV4 is involved in mechanical nociception and function as primary mechanosensor. In addition, TRPV4 mediates the mechanical hypersensitivity induced by PAR_2 activation, serotonin, and histamine (Cenac et al. 2008a, 2010a). Thus, TRPV4 seems to be a common mediator of visceral hypersensitivity. Importantly, these studies suggest that TRPV4 per se or mediators involved in its regulation could constitute new potential therapeutic options for the treatment of abdominal pain, particularly in patients with IBS. TRPV4 contribute to the pain associated with pancreatitis. Intraductal administration of a TRPV4 agonist to the murine pancreas causes expression of c-fos in the spinal cord (Ceppa et al. 2010). Moreover, pain behavior and c-fos expression in experimental pancreatitis due to caerulein is inhibited in TRPV4 knockout mousse (Ceppa et al. 2010).

Even if there is strong evidences that TRPA1 contributes to mechanosensation in the gut (Brierley et al. 2009), it appears that TRPA1 as TRPV1 does not operate as primary mechanosensor. In fact, analysis of TRPA1-transfected HEK293 cells shows that TRPA1 alone is not sufficient to confer mechanical sensitivity (Vilceanu and Stucky 2010). It would appear, therefore, that TRPA1 functions as a secondary transducer of primary mechanosensors. In fact, treatment of afferent neurons by TRPA1 agonists sensitizes colonic sensory neurons to mechanical stimulation (Brierley et al. 2009) and enhances the visceromotor response to colorectal distension (Cattaruzza et al. 2010). The sensitization of mechanosensitive afferent by TRPA1 agonist could be due to its effect to increase the level of TRPA1 in the colon (Kimball et al. 2007). This hypothesis is reinforced by intracolonic treatment of newborn mouse pups provoking a long-lasting hypersensitivity to colorectal distension in adult animals (Christianson et al. 2010). TRPA1 functions as a secondary transducer is also observed in GI inflammation and stress. In TNBS-induced colitis model, colitis expression of TRPA1 in DRG neurons and hypersensitivity to colorectal distension are increased (Yang et al. 2008). Moreover, TNBS increases the ability of TRPA1 agonists to sensitize mechanosensitive afferents in the splanchnic and pelvic nerves (Brierley et al. 2009). In this colitis model, the hypersensitivity to colorectal distension is inhibited in TRPA1 knockout mice (Cattaruzza et al. 2010). Same observations have been performed in stress-induced withdrawal reflex to colorectal distension in rat (Yu et al. 2010).

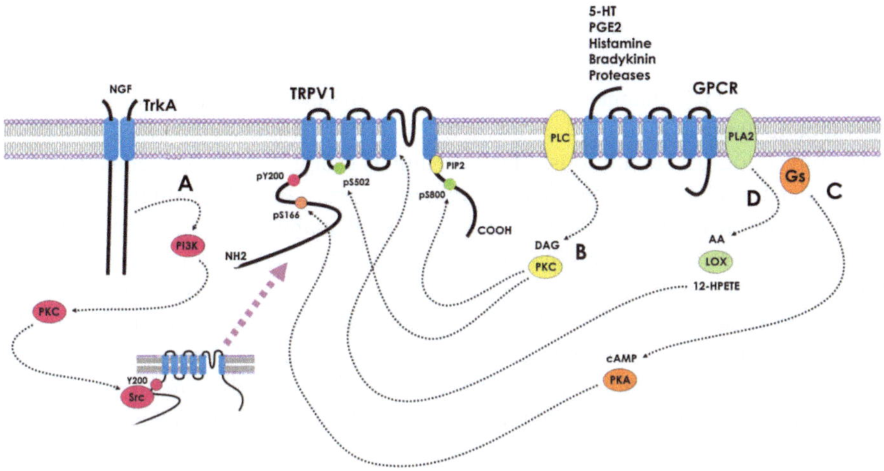

Fig. 20.2 Simplified schematization of the fast regulatory mechanisms that leads to TRPV1 sensitization. **a** The binding of NGF to TrkA activates PI3 K, leading to activation of PKC and ultimately Src, which phosphorylates Y200, translocating the TRPV1 channel to the membrane. **b** Activation of GPCRs may activate PLC, which releases DAG and activates PKC, which subsequently phosphorylates TRPV1 at S502 and S800, increasing its gate activity. **c** Inflammatory mediators like PGE_2 activates Gs Protein-Coupled Receptor, transactivating adenylyl cyclase and induces the production of cAMP. This second messenger activates PKA, which phosphorylates S166 of TRPV1, inhibiting its desensitization. **d** The activation of GPCRs by inflammatory mediators can also activate PLA2, promoting the conversion of arachidonic acid, which once available is converted into active metabolites by cyclooxygenases or lipoxygenases, such as 12-HPETE, which can activate TRPV1 due to its structural similarity with capsaicin

TRPA1 knockout or TRPA1 antagonist inhibits bradykinin-induced mechanical sensitization of splanchnic afferents innervating the colonic serosa of the mouse (Brierley et al. 2009) and mechanical hypersensitivity of vagal afferent neurons in the guinea-pig esophagus (Yu and Ouyang 2009). Importantly, the activation of vagal and splanchnic afferents by bradykinin itself is independent of TRPA1 (Brierley et al. 2009; Yu and Ouyang 2009). Similarly to TRPV4, TRPA1 expression is essential to PAR_2 agonist-induced hypersensitivity in response to colorectal distension (Cattaruzza et al. 2010), but not to the effect of a PAR-2 agonist to stimulate colonic afferents in the splanchnic nerves (Brierley et al. 2009). Interestingly, the severity of abdominal pain is significantly reduced after peppermint administration (Cappello et al. 2007; Merat et al. 2010). Based on these results, TRPM8 functions in visceral pain have been studied. Authors demonstrate that TRPM8 is present on a selected population of colonic high threshold sensory neurons, which may also co-express TRPV1 (Harrington et al. 2011). Moreover, TRPM8 couples to TRPV1 and TRPA1 to inhibit their downstream chemosensory and mechanosensory actions (Harrington et al. 2011).

20.5 Conclusion

The ability of TRP channels to regulate pain demonstrates their value as potential therapeutic targets in this phenomenon. Based on studies performed mostly in pathophysiological conditions, TRP channels should also be considered as molecular targets in the treatment of inflammatory pain. Most studies have concentrated so far at understanding the role of TRP channels as signaling molecules in sensory neurons, however, the role of TRP channels has now to be considered in the context of human painful diseases. Moreover, even if basic science results point to a crucial role for TRP channels in pain, there is a severe lack of knowledge concerning the nature of their endogenous agonist released upon human pain-associated diseases. Future directions in this field would have to include the identification of endogenous regulators of TRP channels in human pathologies associated with pain.

References

Akbar A, Yiangou Y, Facer P, Brydon WG, Walters JR, Anand P, Ghosh S (2010) Expression of the TRPV1 receptor differs in quiescent inflammatory bowel disease with or without abdominal pain. Gut 59(6):767–774

Akbar A, Yiangou Y, Facer P, Walters JR, Anand P, Ghosh S (2008) Increased capsaicin receptor TRPV1-expressing sensory fibres in irritable bowel syndrome and their correlation with abdominal pain. Gut 57(7):923–929

Alessandri-Haber N, Dina OA, Chen X, Levine JD (2009) TRPC1 and TRPC6 channels cooperate with TRPV4 to mediate mechanical hyperalgesia and nociceptor sensitization. J Neurosci 29(19):6217–6228

Alessandri-Haber N, Dina OA, Joseph EK, Reichling D, Levine JD (2006) A transient receptor potential vanilloid 4-dependent mechanism of hyperalgesia is engaged by concerted action of inflammatory mediators. J Neurosci 26(14):3864–3874

Alessandri-Haber N, Dina OA, Joseph EK, Reichling DB, Levine JD (2008) Interaction of transient receptor potential vanilloid 4, integrin, and SRC tyrosine kinase in mechanical hyperalgesia. J Neurosci 28(5):1046–1057

Alessandri-Haber N, Dina OA, Yeh JJ, Parada CA, Reichling DB, Levine JD (2004) Transient receptor potential vanilloid 4 is essential in chemotherapy-induced neuropathic pain in the rat. J Neurosci 24(18):4444–4452

Alessandri-Haber N, Joseph E, Dina OA, Liedtke W, Levine JD (2005) TRPV4 mediates pain-related behavior induced by mild hypertonic stimuli in the presence of inflammatory mediator. Pain 118(1–2):70–79

Alessandri-Haber N, Yeh JJ, Boyd AE, Parada CA, Chen X, Reichling DB, Levine JD (2003) Hypotonicity induces TRPV4-mediated nociception in rat. Neuron 39(3):497–511

Amadesi S, Cottrell GS, Divino L, Chapman K, Grady EF, Bautista F, Karanjia R, Barajas-Lopez C, Vanner S, Vergnolle N, Bunnett NW (2006) Protease-activated receptor 2 sensitizes TRPV1 by protein kinase Cepsilon- and A-dependent mechanisms in rats and mice. J Physiol 575(Pt 2):555–571

Amadesi S, Nie J, Vergnolle N, Cottrell GS, Grady EF, Trevisani M, Manni C, Geppetti P, McRoberts JA, Ennes H, Davis JB, Mayer EA, Bunnett NW (2004) Protease-activated

receptor 2 sensitizes the capsaicin receptor transient receptor potential vanilloid receptor 1 to induce hyperalgesia. J Neurosci 24(18):4300–4312
Amaya F, Shimosato G, Nagano M, Ueda M, Hashimoto S, Tanaka Y, Suzuki H, Tanaka M (2004) NGF and GDNF differentially regulate TRPV1 expression that contributes to development of inflammatory thermal hyperalgesia. Eur J Neurosci 20(9):2303–2310
Anand U, Otto WR, Casula MA, Day NC, Davis JB, Bountra C, Birch R, Anand P (2006) The effect of neurotrophic factors on morphology, TRPV1 expression and capsaicin responses of cultured human DRG sensory neurons. Neurosci Lett 399(1–2):51–56
Anand U, Otto WR, Facer P, Zebda N, Selmer I, Gunthorpe MJ, Chessell IP, Sinisi M, Birch R, Anand P (2008) TRPA1 receptor localisation in the human peripheral nervous system and functional studies in cultured human and rat sensory neurons. Neurosci Lett 438(2):221–227
Andersson DA, Gentry C, Moss S, Bevan S (2008) Transient receptor potential A1 is a sensory receptor for multiple products of oxidative stress. J Neurosci 28(10):2485–2494
Andersson DA, Nash M, Bevan S (2007) Modulation of the cold-activated channel TRPM8 by lysophospholipids and polyunsaturated fatty acids. J Neurosci 27(12):3347–3355
Arendt-Nielsen L, Schipper KP, Dimcevski G, Sumikura H, Krarup AL, Giamberardino MA, Drewes AM (2008) Viscero-somatic reflexes in referred pain areas evoked by capsaicin stimulation of the human gut. Eur J Pain 12(5):544–551
Attal N (2012) Neuropathic pain: mechanisms, therapeutic approach, and interpretation of clinical trials. Continuum (Minneap Minn) 18(1):161–175
Auge C, Balz-Hara D, Steinhoff M, Vergnolle N, Cenac N (2009) Protease-activated receptor-4 (PAR 4): a role as inhibitor of visceral pain and hypersensitivity. Neurogastroenterol Motil 21(11):1189–e1107
Axelsson HE, Minde JK, Sonesson A, Toolanen G, Hogestatt ED, Zygmunt PM (2009) Transient receptor potential vanilloid 1, vanilloid 2 and melastatin 8 immunoreactive nerve fibers in human skin from individuals with and without Norrbottnian congenital insensitivity to pain. Neuroscience 162(4):1322–1332
Balsinde J, Balboa MA (2005) Cellular regulation and proposed biological functions of group VIA calcium-independent phospholipase A2 in activated cells. Cell Signal 17(9):1052–1062
Bang S, Yoo S, Oh U, Hwang SW (2010a) Endogenous lipid-derived ligands for sensory TRP ion channels and their pain modulation. Arch Pharm Res 33(10):1509–1520
Bang S, Yoo S, Yang TJ, Cho H, Hwang SW (2010b) Farnesyl pyrophosphate is a novel pain-producing molecule via specific activation of TRPV3. J Biol Chem 285(25):19362–19371
Bang S, Yoo S, Yang TJ, Cho H, Hwang SW (2011) Isopentenyl pyrophosphate is a novel antinociceptive substance that inhibits TRPV3 and TRPA1 ion channels. Pain 152(5):1156–1164
Bang S, Yoo S, Yang TJ, Cho H, Hwang SW (2012a) 17(R)-resolvin D1 specifically inhibits transient receptor potential ion channel vanilloid 3 leading to peripheral antinociception. Br J Pharmacol 165(3):683–692
Bang S, Yoo S, Yang TJ, Cho H, Hwang SW (2012b) Nociceptive and pro-inflammatory effects of dimethylallyl pyrophosphate via TRPV4 activation. Br J Pharmacol 166(4):1433–1443
Bang S, Yoo S, Yang TJ, Cho H, Kim YG, Hwang SW (2010c) Resolvin D1 attenuates activation of sensory transient receptor potential channels leading to multiple anti-nociception. Br J Pharmacol 161(3):707–720
Bautista DM, Jordt SE, Nikai T, Tsuruda PR, Read AJ, Poblete J, Yamoah EN, Basbaum AI, Julius D (2006) TRPA1 mediates the inflammatory actions of environmental irritants and proalgesic agents. Cell 124(6):1269–1282
Bautista DM, Siemens J, Glazer JM, Tsuruda PR, Basbaum AI, Stucky CL, Jordt SE, Julius D (2007) The menthol receptor TRPM8 is the principal detector of environmental cold. Nature 448(7150):204–208
Benedikt J, Teisinger J, Vyklicky L, Vlachova V (2007) Ethanol inhibits cold-menthol receptor TRPM8 by modulating its interaction with membrane phosphatidylinositol 4,5-bisphosphate. J Neurochem 100(1):211–224

Bhave G, Gereau RWt (2004) Posttranslational mechanisms of peripheral sensitization. J Neurobiol 61(1):88–106

Bhave G, Zhu W, Wang H, Brasier DJ, Oxford GS, Gereau RWt (2002) cAMP-dependent protein kinase regulates desensitization of the capsaicin receptor (VR1) by direct phosphorylation. Neuron 35(4):721–731

Blackshaw LA, Brierley SM, Hughes PA (2010) TRP channels: new targets for visceral pain. Gut 59(1):126–135

Bolcskei K, Helyes Z, Szabo A, Sandor K, Elekes K, Nemeth J, Almasi R, Pinter E, Petho G, Szolcsanyi J (2005) Investigation of the role of TRPV1 receptors in acute and chronic nociceptive processes using gene-deficient mice. Pain 117(3):368–376

Bortolotti M, Coccia G, Grossi G, Miglioli M (2002) The treatment of functional dyspepsia with red pepper. Aliment Pharmacol Ther 16(6):1075–1082

Brauchi S, Orta G, Mascayano C, Salazar M, Raddatz N, Urbina H, Rosenmann E, Gonzalez-Nilo F, Latorre R (2007) Dissection of the components for PIP2 activation and thermosensation in TRP channels. Proc Natl Acad Sci U S A 104(24):10246–10251

Brauchi S, Orta G, Salazar M, Rosenmann E, Latorre R (2006) A hot-sensing cold receptor: C-terminal domain determines thermosensation in transient receptor potential channels. J Neurosci 26(18):4835–4840

Breivik H, Collett B, Ventafridda V, Cohen R, Gallacher D (2006) Survey of chronic pain in Europe: prevalence, impact on daily life, and treatment. Eur J Pain 10(4):287–333

Brierley SM, Hughes PA, Page AJ, Kwan KY, Martin CM, O'Donnell TA, Cooper NJ, Harrington AM, Adam B, Liebregts T, Holtmann G, Corey DP, Rychkov GY, Blackshaw LA (2009) The ion channel TRPA1 is required for normal mechanosensation and is modulated by algesic stimuli. Gastroenterology 137 (6):2084–2095 e2083

Brierley SM, Page AJ, Hughes PA, Adam B, Liebregts T, Cooper NJ, Holtmann G, Liedtke W, Blackshaw LA (2008) Selective role for TRPV4 ion channels in visceral sensory pathways. Gastroenterology 134(7):2059–2069

Calixto JB, Cabrini DA, Ferreira J, Campos MM (2000) Kinins in pain and inflammation. Pain 87(1):1–5

Calvo RR, Meegalla SK, Parks DJ, Parsons WH, Ballentine SK, Lubin ML, Schneider C, Colburn RW, Flores CM, Player MR (2012) Discovery of vinylcycloalkyl-substituted benzimidazole TRPM8 antagonists effective in the treatment of cold allodynia. Bioorg Med Chem Lett 22(5):1903–1907

Campos MM, Leal PC, Yunes RA, Calixto JB (2006) Non-peptide antagonists for kinin B1 receptors: new insights into their therapeutic potential for the management of inflammation and pain. Trends Pharmacol Sci 27(12):646–651

Cappello G, Spezzaferro M, Grossi L, Manzoli L, Marzio L (2007) Peppermint oil (Mintoil) in the treatment of irritable bowel syndrome: a prospective double blind placebo-controlled randomized trial. Dig Liver Dis 39(6):530–536

Caterina MJ (2007) Transient receptor potential ion channels as participants in thermosensation and thermoregulation. Am J Physiol Regul Integr Comp Physiol 292(1):R64–R76

Caterina MJ, Julius D (2001) The vanilloid receptor: a molecular gateway to the pain pathway. Annu Rev Neurosci 24:487–517

Caterina MJ, Leffler A, Malmberg AB, Martin WJ, Trafton J, Petersen-Zeitz KR, Koltzenburg M, Basbaum AI, Julius D (2000) Impaired nociception and pain sensation in mice lacking the capsaicin receptor. Science 288(5464):306–313

Caterina MJ, Schumacher MA, Tominaga M, Rosen TA, Levine JD, Julius D (1997) The capsaicin receptor: a heat-activated ion channel in the pain pathway. Nature 389(6653):816–824

Cattaruzza F, Spreadbury I, Miranda-Morales M, Grady EF, Vanner S, Bunnett NW (2010) Transient receptor potential ankyrin-1 has a major role in mediating visceral pain in mice. Am J Physiol Gastrointest Liver Physiol 298(1):G81–G91

Cenac N, Altier C, Chapman K, Liedtke W, Zampon G, Vergnolle N (2008a) Transient receptor potential vanilloid-4 has a major role in visceral hypersensitivity symptoms. Gastroenterology 135(3):937–946

Cenac N, Altier C, Chapman K, Liedtke W, Zamponi G, Vergnolle N (2008b) Transient receptor potential vanilloid-4 has a major role in visceral hypersensitivity symptoms. Gastroenterology 135(3):937–946, 946 e931–932

Cenac N, Altier C, Motta JP, d'Aldebert E, Galeano S, Zamponi GW, Vergnolle N (2010a) Potentiation of TRPV4 signalling by histamine and serotonin: an important mechanism for visceral hypersensitivity Gut 59(4):481–488

Cenac N, Altier C, Motta JP, d'Aldebert E, Galeano S, Zamponi GW, Vergnolle N (2010b) Potentiation of TRPV4 signalling by histamine and serotonin: an important mechanism for visceral hypersensitivity. Gut 59(4):481–488

Ceppa E, Cattaruzza F, Lyo V, Amadesi S, Pelayo JC, Poole DP, Vaksman N, Liedtke W, Cohen DM, Grady EF, Bunnett NW, Kirkwood KS (2010) Transient receptor potential ion channels V4 and A1 contribute to pancreatitis pain in mice. Am J Physiol Gastrointest Liver Physiol 299(3):G556–G571

Cesare P, Dekker LV, Sardini A, Parker PJ, McNaughton PA (1999) Specific involvement of PKC-epsilon in sensitization of the neuronal response to painful heat. Neuron 23(3):617–624

Chan CL, Facer P, Davis JB, Smith GD, Egerton J, Bountra C, Williams NS, Anand P (2003) Sensory fibres expressing capsaicin receptor TRPV1 in patients with rectal hypersensitivity and faecal urgency. Lancet 361(9355):385–391

Chen CL, Liu TT, Yi CH, Orr WC (2010) Effects of capsaicin-containing red pepper sauce suspension on esophageal secondary peristalsis in humans. Neurogastroenterol Motil 22(11):1177–1182, e1312–1173

Chen J, Joshi SK, DiDomenico S, Perner RJ, Mikusa JP, Gauvin DM, Segreti JA, Han P, Zhang XF, Niforatos W, Bianchi BR, Baker SJ, Zhong C, Simler GH, McDonald HA, Schmidt RG, McGaraughty SP, Chu KL, Faltynek CR, Kort ME, Reilly RM, Kym PR (2011a) Selective blockade of TRPA1 channel attenuates pathological pain without altering noxious cold sensation or body temperature regulation. Pain 152(5):1165–1172

Chen Y, Yang C, Wang ZJ (2011b) Proteinase-activated receptor 2 sensitizes transient receptor potential vanilloid 1, transient receptor potential vanilloid 4, and transient receptor potential ankyrin 1 in paclitaxel-induced neuropathic pain. Neuroscience 193:440–451

Christianson JA, Bielefeldt K, Malin SA, Davis BM (2010) Neonatal colon insult alters growth factor expression and TRPA1 responses in adult mice. Pain 151(2):540–549

Christoph T, Gillen C, Mika J, Grunweller A, Schafer MK, Schiene K, Frank R, Jostock R, Bahrenberg G, Weihe E, Erdmann VA, Kurreck J (2007) Antinociceptive effect of antisense oligonucleotides against the vanilloid receptor VR1/TRPV1. Neurochem Int 50(1):281–290

Christoph T, Grunweller A, Mika J, Schafer MK, Wade EJ, Weihe E, Erdmann VA, Frank R, Gillen C, Kurreck J (2006) Silencing of vanilloid receptor TRPV1 by RNAi reduces neuropathic and visceral pain in vivo. Biochem Biophys Res Commun 350(1):238–243

Chu CJ, Huang SM, De Petrocellis L, Bisogno T, Ewing SA, Miller JD, Zipkin RE, Daddario N, Appendino G, Di Marzo V, Walker JM (2003) N-oleoyldopamine, a novel endogenous capsaicin-like lipid that produces hyperalgesia. J Biol Chem 278(16):13633–13639

Chuang HH, Prescott ED, Kong H, Shields S, Jordt SE, Basbaum AI, Chao MV, Julius D (2001) Bradykinin and nerve growth factor release the capsaicin receptor from PtdIns(4,5)P2-mediated inhibition. Nature 411(6840):957–962

Chung MK, Guler AD, Caterina MJ (2005) Biphasic currents evoked by chemical or thermal activation of the heat-gated ion channel, TRPV3. J Biol Chem 280(16):15928–15941

Chung MK, Lee H, Mizuno A, Suzuki M, Caterina MJ (2004) TRPV3 and TRPV4 mediate warmth-evoked currents in primary mouse keratinocytes. J Biol Chem 279(20):21569–21575

Colburn RW, Lubin ML, Stone DJ Jr, Wang Y, Lawrence D, D'Andrea MR, Brandt MR, Liu Y, Flores CM, Qin N (2007) Attenuated cold sensitivity in TRPM8 null mice. Neuron 54(3):379–386

Cruz-Orengo L, Dhaka A, Heuermann RJ, Young TJ, Montana MC, Cavanaugh EJ, Kim D, Story GM (2008) Cutaneous nociception evoked by 15-delta PGJ2 via activation of ion channel TRPA1. Mol Pain 4:30

Dai Y, Moriyama T, Higashi T, Togashi K, Kobayashi K, Yamanaka H, Tominaga M, Noguchi K (2004) Proteinase-activated receptor 2-mediated potentiation of transient receptor potential vanilloid subfamily 1 activity reveals a mechanism for proteinase-induced inflammatory pain. J Neurosci 24(18):4293–4299

Davis JB, Gray J, Gunthorpe MJ, Hatcher JP, Davey PT, Overend P, Harries MH, Latcham J, Clapham C, Atkinson K, Hughes SA, Rance K, Grau E, Harper AJ, Pugh PL, Rogers DC, Bingham S, Randall A, Sheardown SA (2000) Vanilloid receptor-1 is essential for inflammatory thermal hyperalgesia. Nature 405(6783):183–187

Deising S, Weinkauf B, Blunk J, Obreja O, Schmelz M, Rukwied R (2012) NGF-evoked sensitization of muscle fascia nociceptors in humans. Pain 153(8):1673–1679

Denadai-Souza A, Martin L, de Paula MA, de Avellar MC, Muscara MN, Vergnolle N, Cenac N (2012) Role of transient receptor potential vanilloid 4 in rat joint inflammation. Arthritis Rheum 64(6):1848–1858

Dhaka A, Murray AN, Mathur J, Earley TJ, Petrus MJ, Patapoutian A (2007) TRPM8 is required for cold sensation in mice. Neuron 54(3):371–378

Docherty RJ, Yeats JC, Bevan S, Boddeke HW (1996) Inhibition of calcineurin inhibits the desensitization of capsaicin-evoked currents in cultured dorsal root ganglion neurones from adult rats. Pflugers Arch 431(6):828–837

Drewes AM, Schipper KP, Dimcevski G, Petersen P, Gregersen H, Funch-Jensen P, Arendt-Nielsen L (2003) Gut pain and hyperalgesia induced by capsaicin: a human experimental model. Pain 104(1–2):333–341

Facer P, Casula MA, Smith GD, Benham CD, Chessell IP, Bountra C, Sinisi M, Birch R, Anand P (2007) Differential expression of the capsaicin receptor TRPV1 and related novel receptors TRPV3, TRPV4 and TRPM8 in normal human tissues and changes in traumatic and diabetic neuropathy. BMC Neurol 7:11

Facer P, Knowles CH, Tam PK, Ford AP, Dyer N, Baecker PA, Anand P (2001) Novel capsaicin (VR1) and purinergic (P2X3) receptors in Hirschsprung's intestine. J Pediatr Surg 36(11):1679–1684

Fernandes ES, Russell FA, Spina D, McDougall JJ, Graepel R, Gentry C, Staniland AA, Mountford DM, Keeble JE, Malcangio M, Bevan S, Brain SD (2011) A distinct role for transient receptor potential ankyrin 1, in addition to transient receptor potential vanilloid 1, in tumor necrosis factor alpha-induced inflammatory hyperalgesia and Freund's complete adjuvant-induced monarthritis. Arthritis Rheum 63(3):819–829

Ferreira J, da Silva GL, Calixto JB (2004) Contribution of vanilloid receptors to the overt nociception induced by B2 kinin receptor activation in mice. Br J pharmacol 141(5):787–794

Freeman BA, Baker PR, Schopfer FJ, Woodcock SR, Napolitano A, d'Ischia M (2008) Nitro-fatty acid formation and signaling. J Biol Chem 283(23):15515–15519

Fuhrer M, Hammer J (2009) Effect of repeated, long term capsaicin ingestion on intestinal chemo- and mechanosensation in healthy volunteers. Neurogastroenterol Motil 21(5):521–527, e527

Fukuoka T, Tokunaga A, Tachibana T, Dai Y, Yamanaka H, Noguchi K (2002) VR1, but not P2X(3), increases in the spared L4 DRG in rats with L5 spinal nerve ligation. Pain 99(1–2):111–120

Fukushima M (1990) Prostaglandin J2–anti-tumour and anti-viral activities and the mechanisms involved. Eicosanoids 3(4):189–199

Gold MS, Gebhart GF (2010) Nociceptor sensitization in pain pathogenesis. Nat Med 16(11):1248–1257

Gonlachanvit S, Mahayosnond A, Kullavanijaya P (2009) Effects of chili on postprandial gastrointestinal symptoms in diarrhoea predominant irritable bowel syndrome: evidence for capsaicin-sensitive visceral nociception hypersensitivity. Neurogastroenterol Motil 21(1):23–32

Gopinath P, Wan E, Holdcroft A, Facer P, Davis JB, Smith GD, Bountra C, Anand P (2005) Increased capsaicin receptor TRPV1 in skin nerve fibres and related vanilloid receptors TRPV3 and TRPV4 in keratinocytes in human breast pain. BMC Womens Health 5(1):2

Grant AD, Cottrell GS, Amadesi S, Trevisani M, Nicoletti P, Materazzi S, Altier C, Cenac N, Zamponi GW, Bautista-Cruz F, Lopez CB, Joseph EK, Levine JD, Liedtke W, Vanner S, Vergnolle N, Geppetti P, Bunnett NW (2007) Protease-activated receptor 2 sensitizes the transient receptor potential vanilloid 4 ion channel to cause mechanical hyperalgesia in mice. J Physiol 578(Pt 3):715–733

Grycova L, Holendova B, Bumba L, Bily J, Jirku M, Lansky Z, Teisinger J (2012) Integrative Binding Sites within Intracellular Termini of TRPV1 Receptor. PLoS One 7(10):e48437

Guler AD, Lee H, Iida T, Shimizu I, Tominaga M, Caterina M (2002) Heat-evoked activation of the ion channel, TRPV4. J Neurosci 22(15):6408–6414

Hamamoto DT, Forkey MW, Davis WL, Kajander KC, Simone DA (2000) The role of pH and osmolarity in evoking the acetic acid-induced wiping response in a model of nociception in frogs. Brain Res 862(1–2):217–229

Hammer J (2006) Hyperalgesia against capsaicin in persons with un-investigated dyspepsia: potential as a new diagnostic test. Wien Klin Wochenschr 118(1–2):43–48

Hammer J, Fuhrer M, Pipal L, Matiasek J (2008) Hypersensitivity for capsaicin in patients with functional dyspepsia. Neurogastroenterol Motil 20(2):125–133

Hammer J, Vogelsang H (2007) Characterization of sensations induced by capsaicin in the upper gastrointestinal tract. Neurogastroenterol Motil 19(4):279–287

Harrington AM, Hughes PA, Martin CM, Yang J, Castro J, Isaacs NJ, Blackshaw LA, Brierley SM (2011) A novel role for TRPM8 in visceral afferent function. Pain 152(7):1459–1468

Helyes Z, Sandor K, Borbely E, Tekus V, Pinter E, Elekes K, Toth DM, Szolcsanyi J, McDougall JJ (2010) Involvement of transient receptor potential vanilloid 1 receptors in protease-activated receptor-2-induced joint inflammation and nociception. Eur J Pain 14(4):351–358

Hoffmeister C, Trevisan G, Rossato MF, de Oliveira SM, Gomez MV, Ferreira J (2011) Role of TRPV1 in nociception and edema induced by monosodium urate crystals in rats. Pain 152(8):1777–1788

Hofmann T, Obukhov AG, Schaefer M, Harteneck C, Gudermann T, Schultz G (1999) Direct activation of human TRPC6 and TRPC3 channels by diacylglycerol. Nature 397(6716):259–263

Holzer P (2011) Transient receptor potential (TRP) channels as drug targets for diseases of the digestive system. Pharmacol Ther 131(1):142–170

Hong S, Wiley JW (2005) Early painful diabetic neuropathy is associated with differential changes in the expression and function of vanilloid receptor 1. J Biol Chem 280(1):618–627

Huang SM, Bisogno T, Trevisani M, Al-Hayani A, De Petrocellis L, Fezza F, Tognetto M, Petros TJ, Krey JF, Chu CJ, Miller JD, Davies SN, Geppetti P, Walker JM, Di Marzo V (2002) An endogenous capsaicin-like substance with high potency at recombinant and native vanilloid VR1 receptors. Proc Natl Acad Sci U S A 99(12):8400–8405

Huang SM, Lee H, Chung MK, Park U, Yu YY, Bradshaw HB, Coulombe PA, Walker JM, Caterina MJ (2008) Overexpressed transient receptor potential vanilloid 3 ion channels in skin keratinocytes modulate pain sensitivity via prostaglandin E2. J Neurosci 28(51):13727–13737

Hwang SW, Cho H, Kwak J, Lee SY, Kang CJ, Jung J, Cho S, Min KH, Suh YG, Kim D, Oh U (2000) Direct activation of capsaicin receptors by products of lipoxygenases: endogenous capsaicin-like substances. Proc Natl Acad Sci U S A 97(11):6155–6160

Ichikawa H, Sugimoto T (2003) The co-expression of VR1 and VRL-1 in the rat vagal sensory ganglia. Brain Res 980(2):293–296

Ito N, Ruegg UT, Kudo A, Miyagoe-Suzuki Y, Takeda S (2013) Activation of calcium signaling through Trpv1 by nNOS and peroxynitrite as a key trigger of skeletal muscle hypertrophy. Nat Med 19(1):101–106

Ji RR, Samad TA, Jin SX, Schmoll R, Woolf CJ (2002) p38 MAPK activation by NGF in primary sensory neurons after inflammation increases TRPV1 levels and maintains heat hyperalgesia. Neuron 36(1):57–68

Ji RR, Xu ZZ, Strichartz G, Serhan CN (2011) Emerging roles of resolvins in the resolution of inflammation and pain. Trends Neurosci 34(11):599–609

Jordt SE, Julius D (2002) Molecular basis for species-specific sensitivity to "hot" chili peppers. Cell 108(3):421–430

Jordt SE, Tominaga M, Julius D (2000) Acid potentiation of the capsaicin receptor determined by a key extracellular site. Proc Nat Acad Sci U S A 97(14):8134–8139

Julius D, Basbaum AI (2001) Molecular mechanisms of nociception. Nature 413(6852):203–210

Jung J, Shin JS, Lee SY, Hwang SW, Koo J, Cho H, Oh U (2004) Phosphorylation of vanilloid receptor 1 by Ca2 +/calmodulin-dependent kinase II regulates its vanilloid binding. J Biol Chem 279(8):7048–7054

Keeble J, Russell F, Curtis B, Starr A, Pinter E, Brain SD (2005) Involvement of transient receptor potential vanilloid 1 in the vascular and hyperalgesic components of joint inflammation. Arthritis Rheum 52(10):3248–3256

Kimball ES, Prouty SP, Pavlick KP, Wallace NH, Schneider CR, Hornby PJ (2007) Stimulation of neuronal receptors, neuropeptides and cytokines during experimental oil of mustard colitis. Neurogastroenterol Motil 19(5):390–400

Kindt S, Vos R, Blondeau K, Tack J (2009) Influence of intra-oesophageal capsaicin instillation on heartburn induction and oesophageal sensitivity in man. Neurogastroenterol Motil 21(10):1032–e1082

Knowlton WM, Bifolck-Fisher A, Bautista DM, McKemy DD (2010) TRPM8, but not TRPA1, is required for neural and behavioral responses to acute noxious cold temperatures and cold-mimetics in vivo. Pain 150(2):340–350

Knowlton WM, Daniels RL, Palkar R, McCoy DD, McKemy DD (2011) Pharmacological blockade of TRPM8 ion channels alters cold and cold pain responses in mice. PLoS One 6(9):e25894

Kobayashi K, Fukuoka T, Obata K, Yamanaka H, Dai Y, Tokunaga A, Noguchi K (2005) Distinct expression of TRPM8, TRPA1, and TRPV1 mRNAs in rat primary afferent neurons with adelta/c-fibers and colocalization with trk receptors. J Comp Neurol 493(4):596–606

Kwan KY, Allchorne AJ, Vollrath MA, Christensen AP, Zhang DS, Woolf CJ, Corey DP (2006) TRPA1 contributes to cold, mechanical, and chemical nociception but is not essential for hair-cell transduction. Neuron 50(2):277–289

Kwan KY, Glazer JM, Corey DP, Rice FL, Stucky CL (2009) TRPA1 modulates mechano-transduction in cutaneous sensory neurons. J Neurosci 29(15):4808–4819

Lee H, Iida T, Mizuno A, Suzuki M, Caterina MJ (2005) Altered thermal selection behavior in mice lacking transient receptor potential vanilloid 4. J Neurosci 25(5):1304–1310

Lee KJ, Vos R, Tack J (2004) Effects of capsaicin on the sensorimotor function of the proximal stomach in humans. Aliment Pharmacol Ther 19(4):415–425

Levine JD, Alessandri-Haber N (2007) TRP channels: targets for the relief of pain. Biochim Biophys Acta 1772(8):989–1003

Levine JD, Goetzl EJ, Basbaum AI (1987) Contribution of the nervous system to the pathophysiology of rheumatoid arthritis and other polyarthritides. Rheum Dis Clin North Am 13(2):369–383

Liang PH (2009) Reaction kinetics, catalytic mechanisms, conformational changes, and inhibitor design for prenyltransferases. Biochemistry 48(28):6562–6570

Liang YF, Haake B, Reeh PW (2001) Sustained sensitization and recruitment of rat cutaneous nociceptors by bradykinin and a novel theory of its excitatory action. J Physiol 532(Pt 1):229–239

Liedtke W, Friedman JM (2003) Abnormal osmotic regulation in trpv4-/- mice. Proc Natl Acad Sci U S A 100(23):13698–13703

Liedtke W, Tobin DM, Bargmann CI, Friedman JM (2003) Mammalian TRPV4 (VR-OAC) directs behavioral responses to osmotic and mechanical stimuli in Caenorhabditis elegans. Proc Natl Acad Sci U S A 100 Suppl 2:14531–14536

Lopshire JC, Nicol GD (1998) The cAMP transduction cascade mediates the prostaglandin E2 enhancement of the capsaicin-elicited current in rat sensory neurons: whole-cell and single-channel studies. J Neurosci : Official J Soc Neurosci 18(16):6081–6092

Ma QP (2001) Vanilloid receptor homologue, VRL1, is expressed by both A- and C-fiber sensory neurons. Neuroreport 12(17):3693–3695

Malin S, Molliver D, Christianson JA, Schwartz ES, Cornuet P, Albers KM, Davis BM (2011) TRPV1 and TRPA1 function and modulation are target tissue dependent. J Neurosci 31(29):10516–10528

Malin SA, Christianson JA, Bielefeldt K, Davis BM (2009) TPRV1 expression defines functionally distinct pelvic colon afferents. J Neurosci 29(3):743–752

Malin SA, Molliver DC, Koerber HR, Cornuet P, Frye R, Albers KM, Davis BM (2006) Glial cell line-derived neurotrophic factor family members sensitize nociceptors in vitro and produce thermal hyperalgesia in vivo. J Neurosci 26(33):8588–8599

Marceau F, Regoli D (2004) Bradykinin receptor ligands: therapeutic perspectives. Nat Rev Drug Discovery 3(10):845–852

Materazzi S, Nassini R, Andre E, Campi B, Amadesi S, Trevisani M, Bunnett NW, Patacchini R, Geppetti P (2008) Cox-dependent fatty acid metabolites cause pain through activation of the irritant receptor TRPA1. Proc Natl Acad Sci U S A 105(33):12045–12050

McCoy DD, Knowlton WM, McKemy DD (2011) Scraping through the ice: uncovering the role of TRPM8 in cold transduction. Am J Physiol Regul Integr Comp Physiol 300(6):R1278–R1287

McDougall JJ (2006) Arthritis and pain. Neurogenic origin of joint pain. Arthritis Res Ther 8(6):220

McKemy DD, Neuhausser WM, Julius D (2002) Identification of a cold receptor reveals a general role for TRP channels in thermosensation. Nature 416(6876):52–58

Merat S, Khalili S, Mostajabi P, Ghorbani A, Ansari R, Malekzadeh R (2010) The effect of enteric-coated, delayed-release peppermint oil on irritable bowel syndrome. Dig Dis Sci 55(5):1385–1390

Mercado J, Gordon-Shaag A, Zagotta WN, Gordon SE (2010) Ca2 + -dependent desensitization of TRPV2 channels is mediated by hydrolysis of phosphatidylinositol 4,5-bisphosphate. J Neurosci 30(40):13338–13347

Mohapatra DP, Nau C (2003) Desensitization of capsaicin-activated currents in the vanilloid receptor TRPV1 is decreased by the cyclic AMP-dependent protein kinase pathway. J Biol Chem 278(50):50080–50090

Molliver DC, Snider WD (1997) Nerve growth factor receptor TrkA is down-regulated during postnatal development by a subset of dorsal root ganglion neurons. J Comp Neurol 381(4):428–438

Moriyama T, Iida T, Kobayashi K, Higashi T, Fukuoka T, Tsumura H, Leon C, Suzuki N, Inoue K, Gachet C, Noguchi K, Tominaga M (2003) Possible involvement of P2Y2 metabotropic receptors in ATP-induced transient receptor potential vanilloid receptor 1-mediated thermal hypersensitivity. J Neurosci 23(14):6058–6062

Movahed P, Jonsson BA, Birnir B, Wingstrand JA, Jorgensen TD, Ermund A, Sterner O, Zygmunt PM, Hogestatt ED (2005) Endogenous unsaturated C18 N-acylethanolamines are vanilloid receptor (TRPV1) agonists. J Biol Chem 280(46):38496–38504

Numazaki M, Tominaga T, Toyooka H, Tominaga M (2002) Direct phosphorylation of capsaicin receptor VR1 by protein kinase Cepsilon and identification of two target serine residues. J Biol Chem 277(16):13375–13378

Okun A, Liu P, Davis P, Ren J, Remeniuk B, Brion T, Ossipov MH, Xie J, Dussor GO, King T, Porreca F (2012) Afferent drive elicits ongoing pain in a model of advanced osteoarthritis. Pain 153(4):924–933

Park CK, Xu ZZ, Liu T, Lu N, Serhan CN, Ji RR (2011a) Resolvin D2 is a potent endogenous inhibitor for transient receptor potential subtype V1/A1, inflammatory pain, and spinal cord synaptic plasticity in mice: distinct roles of resolvin D1, D2, and E1. J Neurosci 31(50):18433–18438

Park U, Vastani N, Guan Y, Raja SN, Koltzenburg M, Caterina MJ (2011b) TRP vanilloid 2 knock-out mice are susceptible to perinatal lethality but display normal thermal and mechanical nociception. J Neurosci 31(32):11425–11436

Patapoutian A, Peier AM, Story GM, Viswanath V (2003) ThermoTRP channels and beyond: mechanisms of temperature sensation. Nat Rev Neurosci 4(7):529–539

Patwardhan AM, Akopian AN, Ruparel NB, Diogenes A, Weintraub ST, Uhlson C, Murphy RC, Hargreaves KM (2010) Heat generates oxidized linoleic acid metabolites that activate TRPV1 and produce pain in rodents. J Clin Invest 120(5):1617–1626

Patwardhan AM, Scotland PE, Akopian AN, Hargreaves KM (2009) Activation of TRPV1 in the spinal cord by oxidized linoleic acid metabolites contributes to inflammatory hyperalgesia. Proc Natl Acad Sci U S A 106(44):18820–18824

Peier AM, Moqrich A, Hergarden AC, Reeve AJ, Andersson DA, Story GM, Earley TJ, Dragoni I, McIntyre P, Bevan S, Patapoutian A (2002a) A TRP channel that senses cold stimuli and menthol. Cell 108(5):705–715

Peier AM, Reeve AJ, Andersson DA, Moqrich A, Earley TJ, Hergarden AC, Story GM, Colley S, Hogenesch JB, McIntyre P, Bevan S, Patapoutian A (2002b) A heat-sensitive TRP channel expressed in keratinocytes. Science 296(5575):2046–2049

Peng H, Lewandrowski U, Muller B, Sickmann A, Walz G, Wegierski T (2010) Identification of a Protein Kinase C-dependent phosphorylation site involved in sensitization of TRPV4 channel. Biochemical and biophysical research communications 391(4):1721–1725

Petruska JC, Mendell LM (2004) The many functions of nerve growth factor: multiple actions on nociceptors. Neurosci Lett 361(1–3):168–171

Phillis BD, Martin CM, Kang D, Larsson H, Lindstrom EA, Martinez V, Blackshaw LA (2009) Role of TRPV1 in high-threshold rat colonic splanchnic afferents is revealed by inflammation. Neurosci Lett 459(2):57–61

Premkumar LS, Ahern GP (2000) Induction of vanilloid receptor channel activity by protein kinase C. Nature 408(6815):985–990

Premkumar LS, Raisinghani M, Pingle SC, Long C, Pimentel F (2005) Downregulation of transient receptor potential melastatin 8 by protein kinase C-mediated dephosphorylation. J Neurosci : Official J Soc Neurosci 25(49):11322–11329

Proudfoot CJ, Garry EM, Cottrell DF, Rosie R, Anderson H, Robertson DC, Fleetwood-Walker SM, Mitchell R (2006) Analgesia mediated by the TRPM8 cold receptor in chronic neuropathic pain. Curr Biol 16(16):1591–1605

Ramsden CE, Ringel A, Feldstein AE, Taha AY, Macintosh BA, Hibbeln JR, Majchrzak-Hong SF, Faurot KR, Rapoport SI, Cheon Y, Chung YM, Berk M, Douglas Mann J (2012) Lowering dietary linoleic acid reduces bioactive oxidized linoleic acid metabolites in humans. Prostaglandins Leukot Essent Fatty Acids 87(4–5):135–141

Rathee PK, Distler C, Obreja O, Neuhuber W, Wang GK, Wang SY, Nau C, Kress M (2002) PKA/AKAP/VR-1 module: a common link of Gs-mediated signaling to thermal hyperalgesia. J Neurosci : Official J Soc Neurosci 22(11):4740–4745

Rossi HL, Jenkins AC, Kaufman J, Bhattacharyya I, Caudle RM, Neubert JK (2012) Characterization of bilateral trigeminal constriction injury using an operant facial pain assay. Neuroscience 224:294–306

Santoro MG (1997) Antiviral activity of cyclopentenone prostanoids. Trends Microbiol 5(7):276–281

Schaible HG, Ebersberger A, Natura G (2011) Update on peripheral mechanisms of pain: beyond prostaglandins and cytokines. Arthritis Res Ther 13(2):210

Schaible HG, Richter F, Ebersberger A, Boettger MK, Vanegas H, Natura G, Vazquez E, Segond von Banchet G (2009) Joint pain. Exp Brain Res 196(1):153–162

Schmidt B, Hammer J, Holzer P, Hammer HF (2004) Chemical nociception in the jejunum induced by capsaicin. Gut 53(8):1109–1116

Sculptoreanu A, Kullmann FA, Artim DE, Bazley FA, Schopfer F, Woodcock S, Freeman BA, de Groat WC (2010) Nitro-oleic acid inhibits firing and activates TRPV1- and TRPA1-mediated inward currents in dorsal root ganglion neurons from adult male rats. J Pharmacol Exp Ther 333(3):883–895

Shimosato G, Amaya F, Ueda M, Tanaka Y, Decosterd I, Tanaka M (2005) Peripheral inflammation induces up-regulation of TRPV2 expression in rat DRG. Pain 119(1–3):225–232

Shin J, Cho H, Hwang SW, Jung J, Shin CY, Lee SY, Kim SH, Lee MG, Choi YH, Kim J, Haber NA, Reichling DB, Khasar S, Levine JD, Oh U (2002) Bradykinin-12-lipoxygenase-VR1 signaling pathway for inflammatory hyperalgesia. Proc Nat Acad Sci U S A 99(15):10150–10155

Sipe WE, Brierley SM, Martin CM, Phillis BD, Cruz FB, Grady EF, Liedtke W, Cohen DM, Vanner S, Blackshaw LA, Bunnett NW (2008) Transient receptor potential vanilloid 4 mediates protease activated receptor 2-induced sensitization of colonic afferent nerves and visceral hyperalgesia. Am J Physiol Gastrointest Liver Physiol 294(5):G1288–G1298

Sisignano M, Park CK, Angioni C, Zhang DD, von Hehn C, Cobos EJ, Ghasemlou N, Xu ZZ, Kumaran V, Lu R, Grant A, Fischer MJ, Schmidtko A, Reeh P, Ji RR, Woolf CJ, Geisslinger G, Scholich K, Brenneis C (2012) 5,6-EET is released upon neuronal activity and induces mechanical pain hypersensitivity via TRPA1 on central afferent terminals. J Neurosci 32(18):6364–6372

Smart D, Gunthorpe MJ, Jerman JC, Nasir S, Gray J, Muir AI, Chambers JK, Randall AD, Davis JB (2000) The endogenous lipid anandamide is a full agonist at the human vanilloid receptor (hVR1). Br J Pharmacol 129(2):227–230

Smith GD, Gunthorpe MJ, Kelsell RE, Hayes PD, Reilly P, Facer P, Wright JE, Jerman JC, Walhin JP, Ooi L, Egerton J, Charles KJ, Smart D, Randall AD, Anand P, Davis JB (2002) TRPV3 is a temperature-sensitive vanilloid receptor-like protein. Nature 418(6894):186–190

Southall MD, Vasko MR (2001) Prostaglandin receptor subtypes, EP3C and EP4, mediate the prostaglandin E2-induced cAMP production and sensitization of sensory neurons. J Biol Chem 276(19):16083–16091. doi:10.1074/jbc.M011408200

Starowicz K, Maione S, Cristino L, Palazzo E, Marabese I, Rossi F, de Novellis V, Di Marzo V (2007) Tonic endovanilloid facilitation of glutamate release in brainstem descending antinociceptive pathways. J Neurosci 27(50):13739–13749

Stein C, Clark JD, Oh U, Vasko MR, Wilcox GL, Overland AC, Vanderah TW, Spencer RH (2009) Peripheral mechanisms of pain and analgesia. Brain Res Rev 60(1):90–113

Story GM, Peier AM, Reeve AJ, Eid SR, Mosbacher J, Hricik TR, Earley TJ, Hergarden AC, Andersson DA, Hwang SW, McIntyre P, Jegla T, Bevan S, Patapoutian A (2003) ANKTM1, a TRP-like channel expressed in nociceptive neurons, is activated by cold temperatures. Cell 112(6):819–829

Su L, Wang C, Yu YH, Ren YY, Xie KL, Wang GL (2011) Role of TRPM8 in dorsal root ganglion in nerve injury-induced chronic pain. BMC Neurosci 12:120

Suzuki M, Mizuno A, Kodaira K, Imai M (2003) Impaired pressure sensation in mice lacking TRPV4. J Biol Chem 278(25):22664–22668

Szabo A, Helyes Z, Sandor K, Bite A, Pinter E, Nemeth J, Banvolgyi A, Bolcskei K, Elekes K, Szolcsanyi J (2005) Role of transient receptor potential vanilloid 1 receptors in adjuvant-induced chronic arthritis: in vivo study using gene-deficient mice. J Pharmacol Exp Ther 314(1):111–119

Tahara T, Shibata T, Nakamura M, Yamashita H, Yoshioka D, Hirata I, Arisawa T (2010) Homozygous TRPV1 315C influences the susceptibility to functional dyspepsia. J Clin Gastroenterol 44(1):e1–e7

Tamura S, Morikawa Y, Senba E (2005) TRPV2, a capsaicin receptor homologue, is expressed predominantly in the neurotrophin-3-dependent subpopulation of primary sensory neurons. Neuroscience 130(1):223–228

Taylor-Clark TE, Ghatta S, Bettner W, Undem BJ (2009) Nitrooleic acid, an endogenous product of nitrative stress, activates nociceptive sensory nerves via the direct activation of TRPA1. Mol Pharmacol 75(4):820–829

Todaka H, Taniguchi J, Satoh J, Mizuno A, Suzuki M (2004) Warm temperature-sensitive transient receptor potential vanilloid 4 (TRPV4) plays an essential role in thermal hyperalgesia. J Biol Chem 279(34):35133–35138

Tominaga M, Caterina MJ, Malmberg AB, Rosen TA, Gilbert H, Skinner K, Raumann BE, Basbaum AI, Julius D (1998) The cloned capsaicin receptor integrates multiple pain-producing stimuli. Neuron 21(3):531–543

Ufret-Vincenty CA, Klein RM, Hua L, Angueyra J, Gordon SE (2011) Localization of the PIP2 sensor of TRPV1 ion channels. J Biol Chem 286(11):9688–9698

Valdes AM, De Wilde G, Doherty SA, Lories RJ, Vaughn FL, Laslett LL, Maciewicz RA, Soni A, Hart DJ, Zhang W, Muir KR, Dennison EM, Wheeler M, Leaverton P, Cooper C, Spector TD, Cicuttini FM, Chapman V, Jones G, Arden NK, Doherty M (2011) The Ile585Val TRPV1 variant is involved in risk of painful knee osteoarthritis. Ann Rheum Dis 70(9):1556–1561

van den Wijngaard RM, Klooker TK, Welting O, Stanisor OI, Wouters MM, van der Coelen D, Bulmer DC, Peeters PJ, Aerssens J, de Hoogt R, Lee K, de Jonge WJ, Boeckxstaens GE (2009a) Essential role for TRPV1 in stress-induced (mast cell-dependent) colonic hypersensitivity in maternally separated rats. Neurogastroenterol Motil 21(10):1107–e1194

van den Wijngaard RM, Welting O, Bulmer DC, Wouters MM, Lee K, de Jonge WJ, Boeckxstaens GE (2009b) Possible role for TRPV1 in neomycin-induced inhibition of visceral hypersensitivity in rat. Neurogastroenterol Motil 21(8):863–e860

Vanden Abeele F, Zholos A, Bidaux G, Shuba Y, Thebault S, Beck B, Flourakis M, Panchin Y, Skryma R, Prevarskaya N (2006) Ca2 + -independent phospholipase A2-dependent gating of TRPM8 by lysophospholipids. J Biol Chem 281(52):40174–40182

Vergnolle N (2009) Protease-activated receptors as drug targets in inflammation and pain. Pharmacol Ther 123(3):292–309

Vergnolle N, Bunnett NW, Sharkey KA, Brussee V, Compton SJ, Grady EF, Cirino G, Gerard N, Basbaum AI, Andrade-Gordon P, Hollenberg MD, Wallace JL (2001) Proteinase-activated receptor-2 and hyperalgesia: A novel pain pathway. Nat Med 7(7):821–826

Vilceanu D, Stucky CL (2010) TRPA1 mediates mechanical currents in the plasma membrane of mouse sensory neurons. PLoS One 5(8):e12177

Vriens J, Appendino G, Nilius B (2009) Pharmacology of vanilloid transient receptor potential cation channels. Mol Pharmacol 75(6):1262–1279

Vriens J, Owsianik G, Fisslthaler B, Suzuki M, Janssens A, Voets T, Morisseau C, Hammock BD, Fleming I, Busse R, Nilius B (2005) Modulation of the Ca2 permeable cation channel TRPV4 by cytochrome P450 epoxygenases in vascular endothelium. CircRes 97(9):908–915

Vriens J, Owsianik G, Hofmann T, Philipp SE, Stab J, Chen X, Benoit M, Xue F, Janssens A, Kerselaers S, Oberwinkler J, Vennekens R, Gudermann T, Nilius B, Voets T (2011) TRPM3 is a nociceptor channel involved in the detection of noxious heat. Neuron 70(3):482–494

Wang C, Ning LP, Wang YH, Zhang Y, Ding XL, Ge HY, Arendt-Nielsen L, Yue SW (2011) Nuclear factor-kappa B mediates TRPV4-NO pathway involved in thermal hyperalgesia following chronic compression of the dorsal root ganglion in rats. Behav Brain Res 221(1):19–24

Watanabe H, Vriens J, Prenen J, Droogmans G, Voets T, Nilius B (2003) Anandamide and arachidonic acid use epoxyeicosatrienoic acids to activate TRPV4 channels. Nature 424(6947):434–438

Weil A, Moore SE, Waite NJ, Randall A, Gunthorpe MJ (2005) Conservation of functional and pharmacological properties in the distantly related temperature sensors TRVP1 and TRPM8. Mol pharmacol 68(2):518–527. doi:10.1124/mol.105.012146

Woo DH, Jung SJ, Zhu MH, Park CK, Kim YH, Oh SB, Lee CJ (2008) Direct activation of transient receptor potential vanilloid 1(TRPV1) by diacylglycerol (DAG). Mol Pain 4:42

Xiao Y, Gu Y, Purwaha P, Ni K, Law B, Mallik S, Qian SY (2011) Characterization of free radicals formed from COX-catalyzed DGLA peroxidation. Free Radic Biol Med 50(9):1163–1170

Xu H, Ramsey IS, Kotecha SA, Moran MM, Chong JA, Lawson D, Ge P, Lilly J, Silos-Santiago I, Xie Y, DiStefano PS, Curtis R, Clapham DE (2002) TRPV3 is a calcium-permeable temperature-sensitive cation channel. Nature 418(6894):181–186

Yang J, Li Y, Zuo X, Zhen Y, Yu Y, Gao L (2008) Transient receptor potential ankyrin-1 participates in visceral hyperalgesia following experimental colitis. Neurosci Lett 440(3):237–241

Yu S, Ouyang A (2009) TRPA1 in bradykinin-induced mechanical hypersensitivity of vagal C fibers in guinea pig esophagus. Am J Physiol Gastrointest Liver Physiol 296(2):G255–G265

Yu YB, Yang J, Zuo XL, Gao LJ, Wang P, Li YQ (2010) Transient receptor potential vanilloid-1 (TRPV1) and ankyrin-1 (TRPA1) participate in visceral hyperalgesia in chronic water avoidance stress rat model. Neurochem Res 35(5):797–803

Zhang X, Huang J, McNaughton PA (2005) NGF rapidly increases membrane expression of TRPV1 heat-gated ion channels. EMBO J 24(24):4211–4223

Zhu W, Galoyan SM, Petruska JC, Oxford GS, Mendell LM (2004) A developmental switch in acute sensitization of small dorsal root ganglion (DRG) neurons to capsaicin or noxious heating by NGF. J Neurophysiol 92(5):3148–3152

Zhu W, Oxford GS (2007) Phosphoinositide-3-kinase and mitogen activated protein kinase signaling pathways mediate acute NGF sensitization of TRPV1. Mol Cell Neurosci 34(4):689–700

Zhu W, Oxford GS (2011) Differential gene expression of neonatal and adult DRG neurons correlates with the differential sensitization of TRPV1 responses to nerve growth factor. Neurosci Lett 500(3):192–196

Zhuang ZY, Xu H, Clapham DE, Ji RR (2004) Phosphatidylinositol 3-kinase activates ERK in primary sensory neurons and mediates inflammatory heat hyperalgesia through TRPV1 sensitization. J Neurosci 24(38):8300–8309

Zygmunt PM, Petersson J, Andersson DA, Chuang H, Sorgard M, Di Marzo V, Julius D, Hogestatt ED (1999) Vanilloid receptors on sensory nerves mediate the vasodilator action of anandamide. Nature 400(6743):452–457

[illegible]

Chapter 21
Hereditary Channelopathies Caused by *TRPV4* Mutations

Jeremy M. Sullivan, Thomas E. Lloyd and Charlotte J. Sumner

Abstract Mutations in the *transient receptor potential vanilloid 4* gene (*TRPV4*), encoding a widely expressed Ca^{2+}-permeable ion channel, result in autosomal dominant diseases of peripheral nerve (Charcot-Marie-Tooth disease type 2C, congenital distal spinal muscular atrophy, scapuloperoneal spinal muscular atrophy) or the skeletal system (metatropic dysplasia, spondylometaphyseal dysplasia Kozlowski type, spondyloepiphyseal dysplasia Maroteaux type, autosomal dominant brachyolmia, parastremmatic dysplasia, familial digital arthropathy-brachydactyly). TRPV4-mediated nerve and bone disorders are characterised by considerable phenotypic variability, including severe congenital-onset disease to complete nonpenetrance in the case of nerve disease-associated mutations. Most often, individual TRPV4 mutations have been associated with distinct organ system diseases; although overlap syndromes may occur. Heterologous expression studies have identified a number of functional changes in the TRPV4 ion channel resulting from disease-causing mutations, the most prominent being increased constitutive channel activity. As many of these changes are shared by TRPV4 mutants causing different diseases, however, it remains unclear how particular mutations in TRPV4 result in such diverse disease phenotypes. In this chapter, we outline the clinical and pathological features of the known forms of TRPV4-mediated disease, as well as the effects of disease-causing mutations on TRPV4 assembly, expression and channel activity. We also explore current thinking on the pathogenenic mechanisms contributing to different forms of TRPV4 channelopathy. Elucidation of these mechanisms will be an important step in the development of new therapeutic interventions for these diseases.

J. M. Sullivan (✉) · T. E. Lloyd · C. J. Sumner
Department of Neurology, Johns Hopkins University, Baltimore, MD 21205, USA
e-mail: jsulli45@jhmi.edu

T. E. Lloyd · C. J. Sumner
Department of Neuroscience, Johns Hopkins University, Baltimore, MD 21205, USA

N. Weiss and A. Koschak (eds.), *Pathologies of Calcium Channels*,
DOI: 10.1007/978-3-642-40282-1_21, © Springer-Verlag Berlin Heidelberg 2014

21.1 Introduction

Mutations in TRPV4 are associated with three distinct categories of human disease: neuromuscular disease (affecting peripheral nerves), skeletal dysplasia (affecting bone development) and osteoarthropathy (affecting joints). Over 40 disease-causing mutations in TRPV4 have now been described. TRPV4 encodes a non-selective, Ca^{2+}-permeable cation channel in the transient receptor potential (TRP) superfamily of ion channels. Most disease-causing mutations in TRPV4 result in alterations of TRPV4 channel assembly and/or function in vitro. A clear understanding of how these changes engender the diverse disease phenotypes, however, is only slowly emerging (Nilius and Voets 2013; McEntagart 2012).

TRPV4 is a polymodal channel that can be activated by a variety of environmental stimuli including low osmolarity, warm temperatures (25–34 °C), and mechanical shear stress, as well as a number of endogenous agonists (e.g., arachidonic acid and its metabolites) (reviewed in Everaerts et al. 2010a). The channel is also activated by several exogenous ligands, including the botanical compounds apigenin, bisandrographolide A and cannabidivarin (De Petrocellis et al. 2012; Smith et al. 2006; Ma et al. 2012) and the synthetic agonists 4α-phorbol 12,13-didecanoate (4αPDD) and GSK1016790A (Watanabe et al. 2002; Thorneloe et al. 2008; Jin et al. 2011; Klausen et al. 2009). Several TRPV4 antagonists have also been identified, including the non-selective TRP channel blocker ruthenium red and the selective antagonists HC-067047, GSK205, GSK2193874 and RN-1734 (Phan et al. 2009; Ye et al. 2012; Vincent and Duncton 2011; Thorneloe et al. 2012).

21.1.1 TRPV4 Biogenesis

Functional TRPV4 channels are tetrameric in structure, and are thought to occur principally as homotetramers (Shigematsu et al. 2010; Stewart et al. 2010; Hellwig et al. 2005; Fernandez-Fernandez et al. 2008; Arniges et al. 2006). During protein synthesis, TRPV4 protomers are processed via the endoplasmic reticulum where they undergo core glycosylation and assemble into tetramers (Arniges et al. 2006). Further maturation of the attached oligosaccharides occurs upon the transfer of TRPV4 to the Golgi apparatus, and the mature channel is then trafficked to the plasma membrane (Arniges et al. 2006; Becker et al. 2008; Lamandé et al. 2011). Within the plasma membrane TRPV4 appears to operate in supramolecular complexes comprising plasma membrane components, cytoskeletal elements and signalling proteins (Lanciotti et al. 2012; Fernandez-Fernandez et al. 2008; Goswami et al. 2010; Clark et al. 2008; Fan et al. 2009). Homomeric TRPV4 channels exhibit a relative permeability (P_x/P_{Na}) to Ca^{2+} and Mg^{2+} of 6–10 and 2–3, respectively (Everaerts et al. 2010a). TRPV4 has also been reported to interact with other TRP channel family members, including TRPC1 and TRPP2, forming heterotetrameric channels with distinct functional properties (TRPC1: Ma et al. 2010; Ma et al. 2011a, b; TRPP2: Köttgen et al. 2008; Stewart et al. 2010; Du et al. 2012). Little is known currently about the extent to which these heteromeric channels might contribute to TRPV4-mediated disease.

21.1.2 TRPV4 Structure

Structurally, each TRPV4 protomer is comprised of six transmembrane segments with a putative pore loop between the fifth and sixth segments and intracellular N- and C-termini, both containing several function domains (Fig. 21.1aii). The cytoplasmic N-terminus represents more than half the TRPV4 protein and contains a prominent ankyrin repeat domain (ARD; residues 149–396) comprising six ankyrin repeats, a ubiquitous motif often involved in protein–protein and protein–ligand interactions (Gaudet 2008). The ARD currently represents the only TRPV4 domain for which the crystal structure has been solved (Landouré et al. 2010; Inada et al. 2012). Each ankyrin repeat comprises inner and outer helices followed by a connecting finger loop and the six repeats stack together forming a hand-shaped ARD with a concave palm surface, and a convex surface analogous to the back of the hand (Gaudet 2008). Interestingly, as will be discussed below, ARD mutations that cause neuromuscular disease occur primarily on the opposite surface (convex) to those causing skeletal dysplasia (concave). The TRPV4-ARD has been shown to have important roles in TRPV4 multimerisation, trafficking to the plasma membrane and binding to regulatory factors, including calmodulin and ATP (Phelps et al. 2010; Gaudet 2008; Inada et al. 2012; Everaerts et al. 2010a; Arniges et al. 2006). The cytoplasmic termini of TRPV4 also contain a number of additional functional domains (Fig. 21.1aii), including a proline-rich domain (residues 132–144) and a calmodulin-binding domain (residues 812–831). The proline-rich domain interacts with the intracellular adaptor protein PACSIN3 which influences both TRPV4 localisation and activation (Cuajungco et al. 2006; D'Hoedt et al. 2008); the calmodulin-binding domain plays roles in the calcium-dependent potentiation and inositol trisphosphate (IP3)-mediated sensitisation of TRPV4 (Strotmann et al. 2003; Strotmann et al. 2010; Garcia-Elias et al. 2008).

21.1.3 TRPV4 Regulation and Function

Multiple gating mechanisms regulate activation of the plasma membrane-bound TRPV4 channel by different stimuli (Everaerts et al. 2010a; Gao et al. 2003; Vriens et al. 2004; Loukin et al. 2010a). Hypotonic stress, heat and 4αPDD, for example, all promote TRPV4 channel opening via distinct molecular pathways (Vriens et al. 2004; Gao et al. 2003). TRPV4 responses are also modulated by diverse mechanisms including phosphorylation, glycosylation, S-nitrosylation, intracellular signalling

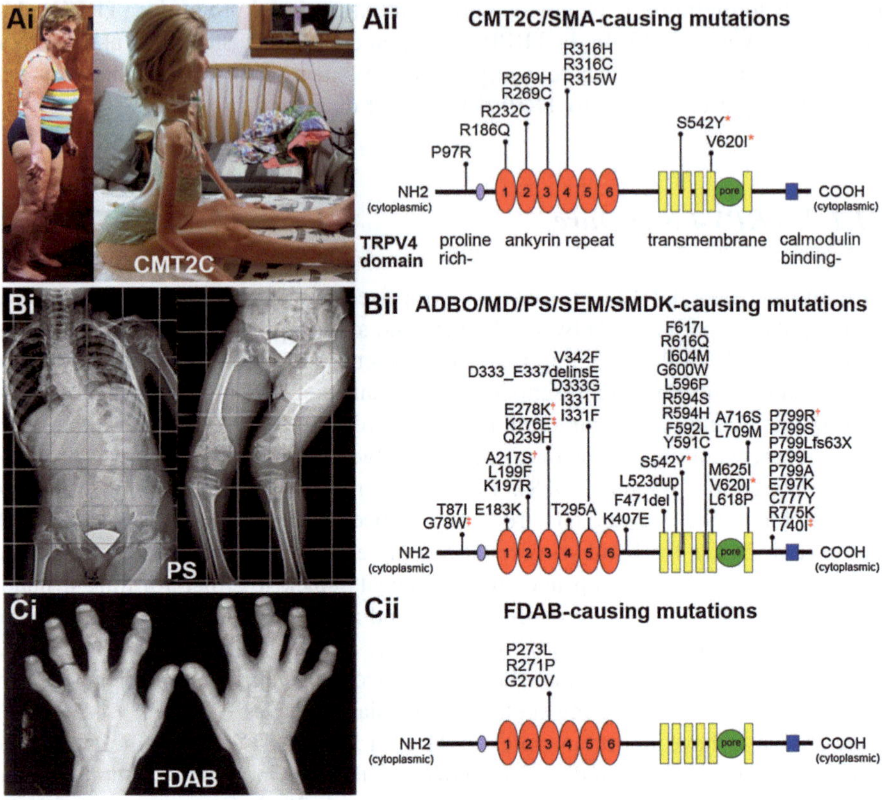

Fig. 21.1 TRPV4-mediated neuromuscular and skeletal disease. **a** TRPV4-mediated neuromuscular disease. **ai** The marked variability characteristic of CMT2C is demonstrated by mild, late-onset weakness in the individual shown at left (age 71 years), but severe quadriparesis and respiratory failure in her daughter shown at right (age 44 years; from Landouré et al. 2010, with permission from Nature Publishing Group and the written consent of the subject). Such phenotypic variability is also characteristic of other forms of TRPV4-mediated neuromuscular disease. **aii** Schematic illustrating the principal domains comprising TRPV4 (at *bottom*) and the locations of CMT2C/SMA-causing mutations. *Asterisks* denote mutations associated with both CMT2C and skeletal abnormalities. **b** TRPV4-mediated skeletal dysplasia. **bi** Radiographs of an individual (age 7 years) with parastremmatic dysplasia (PS) exhibiting platyspondyly with scoliosis (left panel) and marked metaphyseal changes in the long bones (centre and right panels) (from Nishimura et al. 2010, with permission from John Wiley & Sons). **bii** Schematic illustrating the locations of skeletal dysplasia-causing TRPV4 mutations. *Asterisks* highlight mutations associated with both CMT2C and skeletal dysplasia. *Daggers* indicate mutations found in individuals manifesting signs of both skeletal dysplasia and peripheral neuropathy. *Double daggers* denote mutations described in patients with MD and foetal akinesia. TRPV4 domains as in Fig. 21.1aii. **c** TRPV4-mediated FDAB. **ci** Hands of an individual (age 35 years) with FDAB exhibiting marked osteoarthropathy and distal brachydactyly with finger deformities (from Amor et al. 2002, with permission from John Wiley & Sons). **cii** Schematic illustrating the locations of FDAB-causing TRPV4 mutations. TRPV4 domains as in Fig. 21.1aii

pathways and inter- and intramolecular interactions. Consequently, TRPV4 responses are influenced strongly by both environmental conditions and the cellular context of the channel (Liedtke 2008; Güler et al. 2002; Chung et al. 2003). TRPV4 is expressed broadly and has diverse physiological roles, including in osmosensation and flow sensing in the kidney (Pochynyuk et al. 2013), mechanosensation and nociception in the sensory nervous system (Suzuki et al. 2003; Alessandri-Haber et al. 2009), stretch sensation in the bladder (Gevaert et al. 2007), skin barrier function (Kida et al. 2012), chondrogenesis (Muramatsu et al. 2007), bone homeostasis (Masuyama et al. 2012) and regulation of adipose oxidative metabolism (Ye et al. 2012). Interestingly, TRPV4-null mice have only mild impairments of these functions suggesting primarily regulatory rather than essential roles (Everaerts et al. 2010a). Of relevance to TRPV4-mediated disease, pathological features characteristic of neuromuscular disease (e.g., peripheral axonal degeneration, muscle atrophy) or skeletal dysplasia (e.g., congenital pathologies of the vertebral bodies) have not been described in TRPV4-null mice. These findings suggest that these diseases are not due to a loss-of-channel function. TRPV4-null mice do, however, exhibit an increased susceptibility to age- and diet-related osteoarthritis (Clark et al. 2010; O'Conor et al. 2013), consistent with the hypothesis that TRPV4-mediated osteoarthropathy (familial digital arthropathy-brachydactyly) is caused by a loss-of-TRPV4 function (see Sects. 21.2.3 and 21.4.3).

21.2 TRPV4 Mutations and Human Disease

Despite its broad expression, TRPV4 mutations appear to result primarily in peripheral nerve (see Sect. 21.2.1) or skeletal disorders (see Sects. 21.2.2 and 21.2.3). Genetic association analyses have also linked the TRPV4 polymorphism rs3742030, giving rise to a proline to serine substitution at residue 19 (P19S), to both hyponatremia (Tian et al. 2009) and chronic obstructive pulmonary disease (Zhu et al. 2009; Li et al. 2011; Obeidat et al. 2011). As the extent to which TRPV4 contributes to these disorders remains unclear, this variant is not discussed further herein, and we will focus on mutations that have been clearly shown to cause disease.

21.2.1 Neuromuscular Disease

The hereditary axonal neuropathies represent a heterogeneous group of disorders characterised by degeneration of peripheral nerves, resulting in disabling muscle weakness and/or sensory loss. Muscle weakness is also characteristic of spinal muscular atrophies (SMAs), but these disorders are thought to result primarily from degeneration of anterior horn cells (motor neurons innervating skeletal muscle) in the spinal cord, without sensory neuron involvement. Axonal neuropathies and distal

SMAs can be difficult to distinguish from one another clinically and electrophysiologically, and histopathological data for many of these disorders is limited. In 2010, dominant missense mutations in TRPV4 were shown to cause three forms of hereditary neuromuscular disease with autosomal dominant inheritance: Charcot-Marie-Tooth disease type 2C (also known as hereditary motor and sensory neuropathy 2C), congenital distal SMA, and scapuloperoneal SMA (Auer-Grumbach et al. 2010; Chen et al. 2010; Deng et al. 2010; Landouré et al. 2010). While these diseases are clinically heterogeneous, involving distinct muscles and nerves, the majority of patients with TRPV4 mutations share the feature of displaying pure or predominantly motor axonal neuropathy with frequent involvement of the vocal fold muscles (Yiu and Ryan 2012; McEntagart 2012).

To date, eight CMT2C/SMA-causing mutations have been described in TRPV4, occurring across six residues in the N-terminal cytoplasmic domain (Fig. 21.1aii). In addition, two TRPV4 mutations causing both CMT2C and skeletal abnormalities have been identified in the transmembrane domain (Fig. 21.1aii). Most of the pure CMT2C/SMA-causing mutations in TRPV4 occur at exposed arginine residues situated on the convex face of the ARD (Zimoń et al. 2010; Landouré et al. 2010; Inada et al. 2012). These arginine residues are highly conserved across vertebrates, but are not characteristic of the ARDs of other TRPV channel family members (Landouré et al. 2010; Auer-Grumbach et al. 2010; Zimoń et al. 2010; Landouré et al. 2012; Deng et al. 2010).

Owing to marked variability in clinical phenotype and severity, no clear correlation between individual mutations and disease type has been established. A given mutation, for example, can sometimes cause different diseases even within the same family. The R269C, R269H and R315W mutations are each associated with all three forms of neuromuscular disease. In addition, age of disease onset can vary from congenital/early childhood to the sixth decade. TRPV4-mediated nerve disease is also characterised by reduced penetrance, with unaffected mutation carriers having been identified in several families with different TRPV4 mutations (Auer-Grumbach et al. 2010; Landouré et al. 2010; Zimoń et al. 2010; Aharoni et al. 2011; Berciano et al. 2011).

In addition to the above mutations, three potentially pathogenic TRPV4 variants (G20R, N302Y and T701I) have been identified in patients with inherited neuropathy; these variants await more detailed characterisation (Fawcett et al. 2012). TRPV4 mutations have also been described in several individuals manifesting both skeletal dysplasia and either peripheral neuropathy or foetal akinesia, and these are outlined in Sect. 21.2.4.

21.2.1.1 Charcot-Marie-Tooth Disease Type 2C (CMT2C; OMIM 606071)

Charcot-Marie-Tooth disease is a heterogeneous group of hereditary sensorimotor neuropathies that together constitute the most common inherited neurological disease, with an incidence of 1 in 2,500 (Reilly et al. 2011). CMT2C is characterised clinically by distal muscle weakness and wasting in the limbs, distal sensory loss, vocal fold paresis, and weakness of the diaphragm muscles (Santoro et al. 2002; McEntagart et al. 2005; Donaghy and Kennett 1999; Dyck et al. 1994; Zimoń et al. 2010; Landouré et al. 2010; Auer-Grumbach et al. 2010; Klein et al. 2003, 2011; Deng et al. 2010). Sensorineuronal hearing loss, scoliosis and bladder urgency and incontinence may also be present. Muscle and nerve biopsies from a severely affected individual with CMT2C revealed marked denervation atrophy of the gastrocnemius muscle, indicating severe motor axonal degeneration, as well as a modest loss of sensory axons in the sural nerve (Landouré et al. 2010). To date, seven TRPV4 mutations (R186Q, R232C, R269C, R269H, R315W, R316C, R316H) have been associated with pure CMT2C (Zimoń et al. 2010; Landouré et al. 2010, 2012; Chen et al. 2010; Deng et al. 2010; Klein et al. 2011; Auer-Grumbach et al. 2010), all occurring at arginine residues located on the convex face of the ARD. In addition, TRPV4 mutations have also been described in both a multigenerational family (S542Y, Chen et al. 2010) and an isolated individual (V620I, Zimoń et al. 2010) exhibiting both CMT2C and skeletal abnormalities (Fig. 21.1aii; see also Sects. 21.2.2.3 and 21.2.4).

21.2.1.2 Congenital Distal Spinal Muscular Atrophy (CDSMA; OMIM 600175)

CDSMA is a non- or slowly progressive disease characterised by congenital muscle weakness predominantly affecting the lower limbs, as well as contractures of the hip, knee and ankle (Oates et al. 2012; Zimoń et al. 2010; Berciano et al. 2011; Auer-Grumbach et al. 2010; Astrea et al. 2012; Fleury and Hageman 1985; Adams et al. 1998; van der Vleuten et al. 1998; Reddel et al. 2008; Fiorillo et al. 2012). A number of CDSMA patients with TRPV4 mutations also display vocal fold paresis (Fiorillo et al. 2012; Zimoń et al. 2010; Berciano et al. 2011). Clinical and pathological findings are consistent with CDSMA arising due to a congenital deficiency of spinal cord anterior horn cells (Fleury and Hageman 1985; Reddel et al. 2008; Oates et al. 2012). Congenital anterior horn cell loss has not yet been demonstrated, however, in patients with TRPV4 mutations. Five mutations in TRPV4 have been shown to cause CDSMA. While four of these mutations (R232C, R269C, R269H, R315W) are situated in the ARD, the fifth (P97R) occurs upstream of the ARD in the N-terminal cytoplasmic region, outside any known functional domain (Auer-Grumbach et al. 2010; Zimoń et al. 2010; Fiorillo et al. 2012; Berciano et al. 2011). A recent study of a four-generation Australian pedigree with CDSMA did not find evidence of TRPV4 mutations, indicating genetic heterogeneity of this disease (Oates et al. 2012).

21.2.1.3 Scapuloperoneal Spinal Muscular Atrophy (SPSMA; OMIM 181405)

SPSMA is characterised by slowly progressive weakness and atrophy of shoulder girdle and peroneal muscles, vocal fold paresis, congenital absence of muscle and progressive distal muscle weakness and wasting (Zimoń et al. 2010; Deng et al. 2010; Isozumi et al. 1996; DeLong and Siddique 1992). Scoliosis and diminished vibratory sense may also be present (DeLong and Siddique 1992; Zimoń et al. 2010). Pathological data from biopsy and autopsy samples suggest a process of denervation-induced muscle atrophy with preservation of anterior horn cell number (DeLong and Siddique 1992; Deng et al. 2010), consistent with a peripheral motor neuropathy. Four TRPV4 mutations mediating SPSMA have been described (R269C, R269H, R315W, R316C) (Deng et al. 2010; Berciano et al. 2011; Auer-Grumbach et al. 2010; Zimoń et al. 2010), each of which can also cause CMT2C.

21.2.2 Skeletal Dysplasias

The skeletal dysplasias represent a large, heterogeneous group of developmental bone and cartilage disorders. Autosomal dominant mutations in TRPV4 have been associated with five forms of skeletal dysplasia sharing a number of clinical features, including platyspondyly (flattened vertebral bodies), scoliosis and short stature (Fig. 21.1bi; Sects. 21.2.2.1–21.2.2.5). Thirty-nine TRPV4 mutations have now been identified in individuals with these disorders (Fig. 21.1bii). As in TRPV4-mediated neuromuscular disease, several of these TRPV4 mutants (e.g., Y591C, R594H, E797K, P799L) are associated with multiple disease phenotypes. In contrast to the pure CMT2C/SMA-causing mutations, which are largely restricted to the convex face of the ARD, TRPV4 mutations associated with skeletal dysplasia are scattered throughout the protein (Fig. 21.1bii). Interestingly, those skeletal dysplasia-causing mutations located within the ARD occur primarily on the concave face (Inada et al. 2012). No correlation has yet been established between individual forms of skeletal dysplasia and the locations of mutated TRPV4 residues. In contrast to TRPV4-mediated neuromuscular disease, there is no evidence currently of reduced penetrance associated with any of the TRPV4 mutations causing skeletal dysplasia.

21.2.2.1 Metatropic Dysplasia (MD; OMIM 156530)

MD is a severe congenital bone dysplasia characterised by short stature, severe progressive kyphoscoliosis (backward and lateral curvature of the spinal column) and short limbs with enlarged joints. Disease severity in MD ranges from perinatal lethal to mild, with mild cases resembling spondylometaphyseal dysplasia

Kozlowski type (see Sect. 21.2.2.2). The principal abnormalities in MD include platyspondyly, shortening of the longs bones and severe enlargement of the metaphyseal portions of the long bones (Kannu et al. 2007; Geneviève et al. 2008). Histopathological data from two cases of perinatal lethal MD suggest that these abnormalities result primarily from altered chondrocyte differentiation during bone formation (Camacho et al. 2010; see also Sect. 21.4.2). Both mild and severe (lethal) forms of MD are associated with dominant mutations in TRPV4 (Camacho et al. 2010). To date, 21 missense mutations have been described in patients with MD (T89I, K197R, L199F, Q239H, T295A, I331F, I331T, V342F, K407E, F592L, R594H, R594S, I604M, F617L, L618P, R775K, E797K, P799A, P799L, P799S, P799R), as well as one deletion (F471del) and an in-frame deletion–insertion (D333_E337delinsE) (Fig. 21.1bii; (Camacho et al. 2010; Krakow et al. 2009; Dai et al. 2010; Andreucci et al. 2011). Most TRPV4 mutations causing MD appear to have arisen de novo, based on the absence of the changes in both parents. In addition, three de novo heterozygous missense TRPV4 mutations (G78W, K276E, T740I) have been described in four individuals with severe MD that also displayed foetal akinesia (Unger et al. 2011).

21.2.2.2 Spondylometaphyseal Dysplasia Kozlowski Type (SMDK, OMIM 1842522)

SMDK is an autosomal dominant disease characterised primarily by postnatal onset short stature, platyspondyly, and progressive kyphoscoliosis. Additional features include odontoid hypoplasia, flat acetabula and short square ilia, metaphyseal irregularities of the long bones, and delayed ossification of carpal bone (Dai et al. 2010; Krakow et al. 2009; Andreucci et al. 2011; Nural et al. 2006). Thirteen heterozygous missense mutations in TRPV4 have been associated with SMDK (E278K, D333G, Y591C, R594H, L596P, G600W, F617L, M625I, L709M, A716S, C777Y, E797K), as well as one duplication (L523dup) (Krakow et al. 2009; Andreucci et al. 2011; Dai et al. 2010). As in MD, many SMDK-causing TRPV4 mutations appear to have arisen de novo.

21.2.2.3 Autosomal Dominant Brachyolmia (ADBO, OMIM 113500)

The brachyolymias constitute a group of disorders characterised by mild short stature, platyspondyly, and mild shortening of the limbs without significant epiphyseal and metaphyseal abnormalities (Rock et al. 2008; Shohat et al. 1989). ADBO is distinguished from other forms of brachyolmia by severe, progressive kyphoscoliosis, and abnormalities of the cervical vertebrae (Rock et al. 2008; Shohat et al. 1989). Rock et al. (2008) identified TRPV4 mutations in affected individuals in two families with ADBO: R616Q and V620I, respectively (Fig. 21.1bii). The V620I mutation was also described subsequently in an isolated individual with both skeletal abnormalities and CMT2C (Zimoń et al. 2010; see also Sect. 21.2.4). More recently, Andreucci et al. (2011) reported a heterozygous Y591C mutation in a patient with ADBO (Fig. 21.1bii). This mutation was also identified in the patient's sister and in a half-brother who had been diagnosed with SMDK (Andreucci et al. 2011), suggesting a phenotypic continuum between the disorders (Andreucci et al. 2011; Dai et al. 2010).

21.2.2.4 Spondyloepiphyseal Dysplasia, Maroteaux Type (SEM, OMIM 184095)

SEM, also known as pseudo-Morquio syndrome type 2, is an autosomal dominant condition characterised by short stature with progressive truncal shortening in the absence of significant scoliosis, uniform platyspondyly, and shortening of metacarpals and phalanges (brachydactyly). Nishimura et al. (2010) identified four heterozygous missense TRPV4 mutations in patients with SEM (E183K, Y602C, E797K, P799L), only one of which (E183K) has not been associated with other forms of skeletal dysplasia, as well as a heterozygous deletion (P799Lfs63X) (Nishimura et al. 2010).

21.2.2.5 Parastremmatic Dysplasia (PD, OMIM 168400)

PD is a rare skeletal dysplasia characterised by severe platyspondyly, scoliosis, deformed limbs and a peculiar 'flaky' appearance of endochondral bone (Nishimura et al. 2010; Langer et al. 1970; Sensenbrenner et al. 1974). Nishimura et al. (2010) identified a heterozygous missense mutation in TRPV4 (R594H) in one individual with PD. This mutation has also been associated with both MD and SMDK (Krakow et al. 2009; Andreucci et al. 2011; Dai et al. 2010).

21.2.3 Inherited Osteoarthropathy

TRPV4 mutations have also been shown to cause a form of inherited osteoarthropathy, known as familial digital arthropathy-brachydactyly (FDAB); a skeletal condition clinically distinct from the skeletal dysplasias outlined above in Sect. 21.2.2. FDAB is characterised by aggressive osteoarthropathy of the fingers and toes and shortening of the middle and distal phalanges (Fig. 21.1ci), with no evidence of developmental skeletal dysplasia. The three known FDAB-causing TRPV4 mutations reside in close proximity to one another within the ARD (Fig. 21.1cii). There is no evidence currently of reduced penetrance associated with FDAB-causing TRPV4 mutations.

21.2.3.1 Familial Digital Arthropathy-Brachydactyly (FDAB, OMIM 606835)

FDAB is an autosomal dominant condition with onset in the first decade characterised by severe, deforming osteoarthropathy of the interphalangeal, metacarpophalangeal and metatarsophalangeal joints, and progressive shortening of the middle and distal phalanges (Lamandé et al. 2011; Amor et al. 2002). Bones and joints outside the hands and feet remain unaffected. Examination of two multigenerational families with FDAB identified two TRPV4 mutations (R271P, F273L) which segregated with disease (Lamandé et al. 2011). A heterozygous missense mutation in TRPV4 (G270V) was also identified in an isolated individual with sporadic FDAB (Lamandé et al. 2011). All three FDAB-causing TRPV4 mutations occur in the ARD at residues comprising the finger loop of the third ankyrin repeat (Lamandé et al. 2011).

21.2.4 Co-occurrence of TRPV4-Mediated Neuromuscular Disease and Skeletal Dysplasia

While most patients with TRPV4 mutations exhibit either neuromuscular or skeletal disorders, increasing evidence suggests that nerve and bone disease can co-occur in individuals with certain TRPV4 mutations. Zimoń et al. (2010) reported an isolated individual exhibiting both CMT2C and skeletal abnormalities. This patient was found to harbour a de novo heterozygous V620I mutation in TRPV4, a mutation identified originally in a family with ADBO (Rock et al. 2008). Chen et al. (2010) subsequently described a multigenerational family exhibiting CMT2C, platyspondyly, and pronounced short stature, in which a heterozygous S542Y mutation in TRPV4 segregated with disease. Platyspondyly, brachydactyly and dysmorphic femoral heads were also reported in a CDSMA patient with the R232C mutation (Fiorillo et al. 2012). It should be noted that although scoliosis is a common feature of both TRPV4-mediated neuromuscular disease and skeletal dysplasia, the predominant aetiology underlying this condition is likely distinct between the two disease groups. Clinical and radiological findings suggest that scoliosis in most patients with TRPV4-mediated CMT2C/SMA is secondary to the underlying neuromuscular dysfunction, rather than as a result of bone pathology (Auer-Grumbach et al. 2010; Fleury and Hageman 1985; Aharoni et al. 2011; McEntagart 2012).

Cho et al. (2012) described three individuals with heterozygous TRPV4 mutations (A217S, E278K, P977R; Fig. 21.1bii) manifesting signs of both nerve and bone disease. While each of these patients exhibited clinical features consistent with a known form of TRPV4-mediated skeletal dysplasia (A217S & E278K, SMDK; P977R, SEM) the neurological findings differed substantially from those of patients with CMT2C, CDSMA or SPSMA (Cho et al. 2012).

Consequently, in the absence of further clinical and genetic analyses it remains unclear whether the neuropathy described in these individuals can be directly attributed to the TRPV4 mutations.

As noted in Sect. 21.2.2.1, Unger et al. (2011) described four individuals exhibiting both severe MD and foetal akinesia. Each of these patients harboured one of three novel heterozygous missense mutations in TRPV4 (G78W, K276E, T740I; Fig. 21.1bii). As foetal akinesia is not a common feature of MD, the authors of this study proposed that the absence of movement in these individuals may have been the result of TRPV4-mediated neuropathy; suggesting that these particular TRPV4 mutations might cause both nerve and bone disease. In support of this idea, an EMG recording performed on the deltoid muscle of one individual at 3 months of age was reported to demonstrate signs of chronic axonal degeneration (Unger et al. 2011). Conclusive evidence of peripheral axonal neuropathy in individuals with these mutations, however, is still lacking.

21.3 Effects of Disease-Causing Mutations on TRPV4 Expression and Function

Several transfection-based studies have examined the effects of disease-causing mutations on TRPV4 assembly, trafficking, localisation and function (Table 21.1, and references therein). One comprehensive study has also examined the structural and biochemical properties of thirteen CMT2C/SMA- and skeletal dysplasia-causing mutations located in the TRPV4-ARD (Inada et al. 2012). While these investigations have identified a number of differences between the wild-type (WT) and mutant channels, they have yet to generate significant insights into the mechanisms mediating the tissue-specific pathologies observed in most forms of TRPV4-mediated disease. Most CMT2C/SMA- and skeletal dysplasia-causing mutants, for example, have been found to traffic and localise normally but to exhibit increased constitutive activity relative to $TRPV4^{WT}$. Consequently, we currently have little understanding of how the diverse disease phenotypes arising from mutations in TRPV4 correlate with changes in TRPV4 function.

21.3.1 Neuromuscular Disease-Causing TRPV4 Mutants

Studies in heterologous expression systems have generated conflicting results on the effects of CMT2C/SMA-causing TRPV4 mutations on channel trafficking and function (Table 21.1). Auer-Grumbach et al. (2010), in their study of the R269H, R315W and R316C mutations, observed reduced plasma membrane expression of these mutant channels as well as decreased responses to activating stimuli, suggesting a loss-of-function. In contrast, several studies across each of the seven

Table 21.1 Findings of transfection-based studies of disease-causing TRPV4 mutants

Disease	Mutation	Cell surface expression	Basal $[Ca^{2+}]_i$	Constit. activity	4αPDD	Temp.	Hyposm.	Arach. acid	GSK101	Reference
Neuromuscular Disease	P97L	–	WT	–	–	–	↓	–	–	Fiorillo et al. (2012)
CMT2C	R186Q	–	WT	–	–	–	–	–	–	Landouré et al. (2012)
CDSMA										
SPSMA	R232C	–	↑	–	↑	–	–	–	–	Klein et al. (2011)
	R269C	WT	↑	↑	WT	↑	↑	–	–	Landouré et al. (2010)
	R269H	↓	WT	–	↓	NR	↓	–	–	Auer-Grumbach et al. (2010)
		WT	↑	↑	↑	↑	↑	↑	–	Deng et al. (2010)
		WT	↑	↑	↑	↑	↑	–	–	Landouré et al. (2010)
		–	↑	↑	↑	–	–	–	–	Fecto et al. (2011)
		–	↑	–	↑	–	–	–	–	Klein et al. (2011)
	R315W	↓	WT	–	↓	NR	↓	–	–	Auer-Grumbach et al. (2010)
		–	↑	↑	↑	–	–	–	–	Fecto et al. (2011)
	R316C	↓	WT	–	↓	NR	↓	–	–	Auer-Grumbach et al. (2010)
		WT	↑	↑	↑	↑	↑	↑	–	Deng et al. (2010)
		–	↑	↑	↑†	–	–	–	–	Fecto et al. (2011)
	R316H	–	↑	–	WT	–	–	–	–	Klein et al. (2011)
Skeletal dysplasia	T89I	–	–	–	–	–	↓	–	↓	Loukin et al. (2011)
	I331F	–	↑	–	↑	–	↑	↑‡	–	Camacho et al. (2010)
		–	–	–	–	–	↓	–	↓	Loukin et al. (2011)
	D333G	WT	↑	–	↑	–	↓	↓	–	Krakow et al. (2009)
		–	–	–	–	–	↓	–	↓	Loukin et al. (2011)

(continued)

Table 21.1 (continued)

Disease	Mutation	Cell surface expression	Basal $[Ca^{2+}]_i$	Constit. activity	4αPDD	Temp.	Hyposm.	Arach. acid	GSK101	Reference
ADBO	Δ333-7	–	–	–	–	–	↓	–	↓	Loukin et al. (2011)
MD										
PS										
SEM	F471Δ	–	–	–	–	–	NR	–	↓	Loukin et al. (2011)
SMDK	R594H	WT	↑	–	NR	–	↓	↓	–	Krakow et al. (2009)
		–	–	–	–	–	↓	–	↓	Loukin et al. (2011)
	I604M	–	–	↑	–	–	NR	–	↓	Loukin et al. (2011)
	R616Q	WT	↑	–	↑	–	↑	↑	–	Rock et al. (2008)
		–	–	↑	–	–	–	–	–	Loukin et al. (2010)
		–	–	–	–	–	↓	–	↓	Loukin et al. (2011)
	F617L	–	–	–	–	–	↓	–	↓	Loukin et al. (2011)
	L618P	–	–	–	–	–	NR	–	↓	Loukin et al. (2011)
	V620I*	WT	↑	–	↑	–	↑	↑	–	Rock et al. (2008)
		–	–	↑	↑	–	↑	–	–	Auer-Grumbach et al. (2010)
		–		–	–	–	↓	–	↓	Loukin et al. (2011)
	A716S	WT	WT	–	NR	–	NR	NR	–	Krakow et al. (2009)
		–	–	–	–	–	↓	–	↓	Loukin et al. (2011)
	E797K	–	–	–	–	–	↓	–	↓	Loukin et al. (2011)
	P799L	–	↑	–	↑	–	↑	↑‡	–	Camacho et al. (2010)
		–	–	–	–	–	↓	–	↓	Loukin et al. (2011)
Osteoarthropathy	G270V	↓	↑	–	↓	–	NR	–	↓	Lamandé et al. (2011)
FDAB	R271P	↓	↑	–	↓	–	NR	–	↓	Lamandé et al. (2011)
	F273L	↓	↑	–	↓	–	NR	–	↓	Lamandé et al. (2011)

4αPDD, responses to 4α-phorbol 12,13-didecanoate; Arach. acid, responses to arachidonic acid; $[Ca^{2+}]_i$, cytosolic free calcium levels; Constit. activity, constitutive channel activity, as assessed electrophysiologically; GSK101, responses to GSK101679A; Hyposm., responses to hyposmotic stress; NR no response; Temp., responses to temperature ramps; WT not different from WT; –, not yet determined; ↑, increased relative to WT; ↑†, increase responses not observed in all cell lines examined; ↑‡, increased responses observed with electrophysiology but not with Ca^{2+} imaging; ↓, decreased relative to WT. *Grey shading* indicates channel properties differing between studies. *Asterisk* indicates mutation associated with both skeletal dysplasia and CMT2C

CMT2C/SMA-causing mutations in the ARD have revealed normal plasma membrane expression of the mutant channels with increased channel activity in both the basal and activated states (Landouré et al. 2010; Fecto et al. 2011; Deng et al. 2010; Klein et al. 2011; Landouré et al. 2012). Although the specific factors contributing to these disparate results remain unclear, it is becoming increasingly apparent that the properties of individual TRP channels can vary markedly in different expression systems (Lev et al. 2012). TRPV4 responses are also likely to be influenced by expression levels and the stoichiometry of the channel with its binding partners. While these factors demonstrate the complexity of examining mutant TRPV4 in transfection-based experiments, the bulk of evidence obtained to date is consistent with these CMT2C/SMA-causing mutations conferring a gain-of-channel function. The single functional study of the CDSMA-causing P97R mutation, located upstream of the ARD, observed no evidence of increased constitutive channel activity and reduced responses to hyposmotic saline (Fiorillo et al. 2012), consistent with a loss-of-function. Given the above caveats, however, further study of this mutant will be required to firmly establish the extent to which its properties differ from those of other CMT2C/SMA-causing mutants. Examination of the structural and biochemical properties of five CMT2C/SMA-causing mutants revealed that the R232C and R269H mutants exhibit significantly decreased ATP binding compared to $TRPV4^{WT}$ (Inada et al. 2012). Consistent differences in ATP binding or thermal stability, however, were not observed across the mutants (Inada et al. 2012).

21.3.2 Skeletal Dysplasia-Causing TRPV4 Mutants

Cell surface biotinylation assays suggest that skeletal dysplasia-causing mutations do not appreciably affect TRPV4 plasma membrane expression levels (Rock et al. 2008; Krakow et al. 2009). Calcium imaging studies in heterologous expression systems, however, have consistently shown that expression of most skeletal dysplasia mutants results in elevations of basal cytosolic free Ca^{2+} levels ($[Ca^{2+}]_i$), suggesting increased constitutive calcium channel activity. Similarly, increased constitutive activity has also been observed in patch clamp recordings of transfected cells (Loukin et al. 2010b, 2011; Auer-Grumbach et al. 2010). Investigations of mutant channel Ca^{2+} responses to activating stimuli have generated a complex picture, with examples of increased, decreased, normal and absent responses relative to the WT channel. In an extensive electrophysiological study of 14 skeletal dysplasia-causing mutants, however, Loukin et al. (2011) found that all of the mutants examined exhibited higher basal open probabilities than $TRPV4^{WT}$, as well as decreased responses to hypotonic saline and GSK1016790A (Fig. 21.2b). Furthermore, they determined a negative correlation between the size of the response to either stimuli and disease severity, with ADBO and neonatal lethal MD representing opposite ends of the clinical severity spectrum (Fig. 21.2b, inset). The decreased responses of the mutant channels appear to result primarily

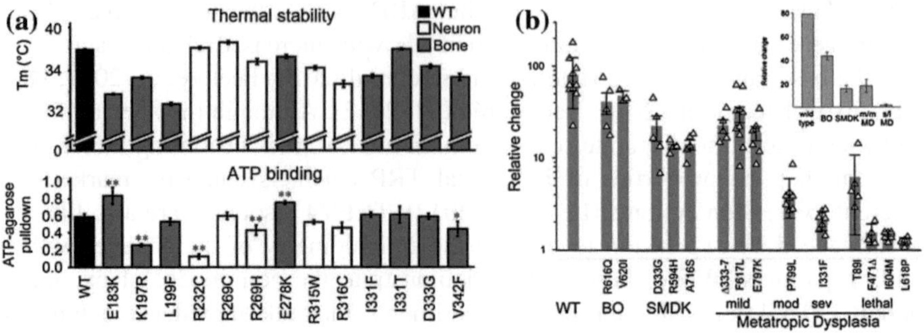

Fig. 21.2 Structural, biochemical and functional properties of disease-causing TRPV4 mutants. **a** Thermal stability and ATP binding of wild-type (WT) TRPV4 and disease-causing forms of TRPV4 with mutations in the ARD (mean ± SD; $p < 0.05$ and $p < 0.01$ indicated by one *asterisk* and two *asterisks*, respectively; from Inada et al. 2012, with permission from ACS Publications). **b** Relative increases in peak currents in *Xenopus* oocytes expressing WT or skeletal dysplasia-causing mutant TRPV4 channels in response to application of 1 μM GSK1016790A (mean ± SEM). Inset shows the average relative increase grouped by clinical severity (mean ± SEM). m/m, mild and moderate; s/l, severe and lethal (modified from Loukin et al. 2011)

from their higher basal open probability, and likely represent a "ceiling effect" rather than reduced sensitivity to the stimuli (Loukin et al. 2011). These results suggest that increased constitutive TRPV4 activity may represent a primary cause of pathogenesis in TRPV4-mediated skeletal dysplasia. That increased constitutive activity is also characteristic of CMT2C/SMA-causing TRPV4 mutants (see Sect. 21.3.1), however, highlights the complexity of identifying pathogenic changes in TRPV4 channel function. As with the CMT2C/SMA-causing mutants, no consistent differences in thermal stability or ATP binding have been observed across TRPV4 mutants associated with skeletal dysplasia (Inada et al. 2012).

21.3.3 FDAB-Causing TRPV4 Mutants

TRPV4 mutations mediating FDAB appear to have a marked influence on channel assembly (Lamandé et al. 2011). Non-reducing western blots of lysates from cell lines stably transfected with FDAB-causing TRPV4 mutants demonstrate reduced amounts of tetrameric channel compared to $TRPV4^{WT}$-expressing lines (Lamandé et al. 2011). Similarly, cell surface biotinylation assays reveal diminished plasma membrane expression levels of these mutants. Calcium imaging studies demonstrate reduced TRPV4 responses to 4αPDD and GSK1016790A and an absence of responsiveness to hypotonic saline (Lamandé et al. 2011). Interestingly, expression of each of the three FDAB mutants results in elevations in basal $[Ca^{2+}]_i$, suggesting they exhibit increased constitutive calcium channel activity (Lamandé et al. 2011). Taken together, however, these results are consistent with FDAB-causing mutants resulting in a loss-of-channel function. The structural and biochemical properties of FDAB mutants have not yet been examined.

21.3.4 Interactions Between WT and Mutant TRPV4

Owing to the autosomal dominant inheritance pattern of all known forms of TRPV4-mediated disease, it is likely that patients express both WT and mutant TRPV4 protomers. The ability of these two protomer types to oligomerize and form heteromeric TRPV4 channels has not yet been examined. It remains unclear, therefore, what proportion of tetrameric TRPV4 channels in patients are comprised entirely of mutant TRPV4 protomers. Only one study to date has examined TRPV4 function in cells co-transfected with both WT and mutant TRPV4. This investigation of three CMT2C/SMA mutants (R269H, R315W, R316C) found that whereas TRPV4 responses to hypotonic saline differed between cells transfected with either WT or mutant TRPV4 alone, co-transfection of both protomers resulted in WT-like TRPV4 responses to this stimulus (Auer-Grumbach et al. 2010). Further study will be required to determine the exact extent to which WT and mutant TRPV4 protomers interact and how these interactions influence channel function.

21.4 Putative Pathological Mechanisms Underlying TRPV4-Mediated Disease

No animal models of TRPV4-mediated disease have yet been generated. Insights into potential pathomechanisms underlying these disorders have therefore been drawn primarily from the limited histopathological data available from patients and the results of transfection-based studies (see Sect. 21.3). Here, we outline current thinking on the possible cellular and molecular mechanisms mediating TRPV4 channelopathies (Fig. 21.3). It is important to stress that clear experimental evidence linking any of these proposed mechanisms to disease pathogenesis in vivo is still lacking. Development of animal models of TRPV4-mediated diseases will therefore play a central role in the elucidation of pathological mechanisms underlying these disorders. The evidence that neuromuscular disease- and skeletal dysplasia-causing TRPV4 mutations result in a gain-of-channel function suggests that inhibition of TRPV4 channel activity may represent a therapeutic target. Efforts in the pharmaceutical industry have led to the identification of several TRPV4-specific antagonists, three of which have shown efficacy in vivo without prohibitive toxicity in mouse models of cystitis-induced bladder dysfunction (HC-067047, Everaerts et al. 2010b), diet-induced glucose intolerance (GSK205, Ye et al. 2012) and heart failure-induced pulmonary oedema

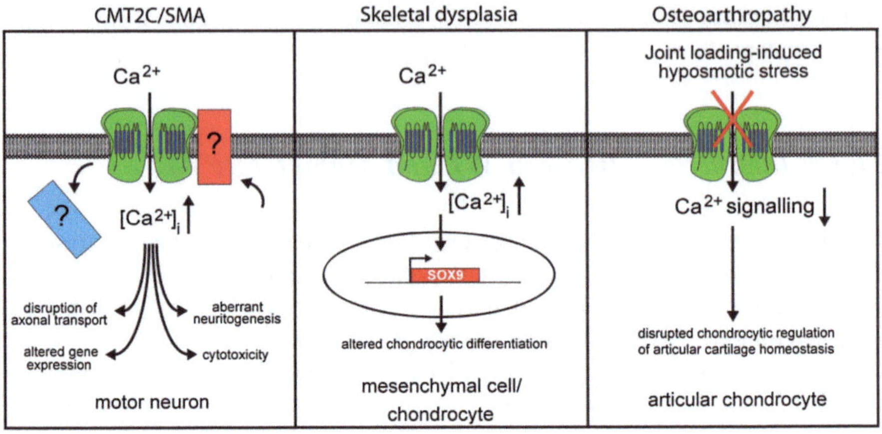

Fig. 21.3 Putative cellular and molecular mechanisms contributing to TRPV4-mediated disease. See Sect. 21.4 for detailed explanations. As outlined in the text, experimental evidence supporting pathogenic roles in vivo is still lacking for each of the proposed mechanisms

(GSK2193874, Thorneloe et al. 2012). The design of more complex model systems will therefore enable proof-of-principle studies examining the therapeutic potential of TRPV4 antagonism in the treatment of TRPV4-mediated diseases, as well as provide important insights into the pathological mechanisms underlying each of these disorders.

21.4.1 Neuromuscular Disease (CMT2C, CDSMA, SPSMA)

The localisation of most CMT2C/SMA-causing TRPV4 mutations to exposed arginine residues on the convex face of the ARD suggests that the pathomechanisms mediating these diseases are intimately linked to the function of this region. While the TRPV4-ARD has been ascribed roles in channel assembly, trafficking and binding to regulatory factors (Arniges et al. 2006; Phelps et al. 2010; Everaerts et al. 2010a; Gaudet 2008; Inada et al. 2012), little is known about the specific functional roles of the convex face. CMT2C/SMA-causing mutations do not alter TRPV4 binding to calmodulin, which interacts with the ARD concave surface, suggesting retained folding and structure of the ARD (Landouré et al. 2010). Given their exposed location, it has been postulated that the CMT2C/SMA-causing mutations may alter intramolecular or intermolecular (protein–protein or protein–phospholipid) interactions critical for TRPV4 function in the peripheral nervous system (Landouré et al. 2010; Klein et al. 2011). These interactions remain to be defined. The extent to which pathomechanisms associated with the P97R mutation overlap with those of CMT2C/SMA-causing mutations in the ARD also remains unclear (Fiorillo et al. 2012).

Several transfection-based studies have demonstrated that expression of CMT2C/SMA-causing TRPV4 mutants results in extensive cytotoxicity (Landouré et al. 2010, 2012; Klein et al. 2011; Deng et al. 2010; Fecto et al. 2011). This toxicity is abrogated both by application of the TRP channel antagonist ruthenium red or by the introduction of an additional mutation (M680K) that blocks the channel pore (Fecto et al. 2011; Klein et al. 2011; Landouré et al. 2010, 2012). Together, these results suggest that intracellular hypercalcaemia may represent an important pathological factor underlying TRPV4-mediated neuromuscular disease (Fig. 21.3).

TRPV4 is widely expressed in the nervous system, including in neurons of the dorsal root and trigeminal ganglia (Lechner et al. 2011; Vandewauw et al. 2013; Alessandri-Haber et al. 2004) and anterior horn cells (Landouré et al. 2010; Facer et al. 2007; Jang et al. 2012). The limited pathological data available from patients with CMT2C and SPSMA indicate that mutant TRPV4 causes distal axonal degeneration with preservation of anterior horn cell number (Landouré et al. 2010; Deng et al. 2010). How mutant TRPV4 may trigger axonal degeneration is unknown. Many spinal motor neurons have low Ca^{2+} buffering capacities, resulting from low concentrations of cytosolic Ca^{2+} binding proteins (Grosskreutz et al. 2010). As mitochondria play a prominent role in Ca^{2+} buffering in these neurons, excessive Ca^{2+} influx could lead to mitochondrial Ca^{2+} overload, and the many deleterious consequences thereof (Grosskreutz et al. 2010). Ca^{2+} influx can also regulate associations between molecular motors and their cargo, including mitochondria (Wang and Schwarz 2009; Macaskill et al. 2009; Saotome et al. 2008), and disruption of axonal transport has been proposed to be a common pathogenic mechanism in Charcot-Marie-Tooth disease (D'Ydewalle et al. 2011; Almeida-Souza et al. 2011; Pareyson et al. 2009). Furthermore, TRPV4 has reciprocal interactions with a number of microtubule and cytoskeletal proteins involved in axonal transport, including tubulin and actin (Masuyama et al. 2012; Jang et al. 2012; Fiorio Pla et al. 2012; Goswami et al. 2010; Shin et al. 2012). As Ca^{2+} signalling resulting from TRPV4 channel activity can regulate gene expression (Muramatsu et al. 2007; Ye et al. 2012), a gain-of-TRPV4 channel function could result in pathological changes in gene expression patterns in anterior horn cells (Fig. 21.3). On the basis of recent work examining effects of TRPV4 on neuritogenesis in vitro, it has also been postulated that developmental abnormalities of axonal outgrowth may mediate some forms of TRPV4-mediated neuromuscular disease (Jang et al. 2012).

21.4.2 Skeletal Dysplasia (MD, SMDK, ADBS, SEM, PS)

Skeletal development outside the flat bones of the skull is mediated by the process of endochondral ossification through which bone is generated via a cartilage intermediate (Long and Ornitz 2013). Formation of the initial cartilage template of a prospective bone begins with the condensation of mesenchymal cells and their subsequent differentiation into chondrocytes. Chondrocyte proliferation and secretion of extracellular matrix components result in the expansion of this template, while its ossification is initiated and regulated by chondrocytes that have undergone a programme of hypertrophic differentiation. Histopathological examination of the developing bones of two individuals with perinatal lethal MD provided evidence of disrupted endochondral ossification, primarily due to alterations in mesenchymal cell differentiation, and chondrocyte proliferation and differentiation (Camacho et al. 2010). The transcription factor SOX9 plays a critical role in all three of these processes (Long and Ornitz 2013; Akiyama 2008). The finding that TRPV4 activates SOX9-dependent transcription in a chondrocyte cell line (ATDC5), likely through a Ca^{2+}/calmodulin-dependent pathway (Muramatsu et al. 2007), has led to the hypothesis that dysregulation of SOX9 activity may underlie TRPV4-mediated skeletal dysplasia (Camacho et al. 2010; Rock et al. 2008). This idea is supported by the observation in embryonic mice that TRPV4 mRNA is highly expressed in the cartilage templates of developing hindlimb bones (Cameron et al. 2007).

21.4.3 FDAB

In contrast to TRPV4-mediated skeletal dysplasia, there is no evidence of developmental abnormalities in FDAB. On the basis of radiographic findings, it was hypothesised initially that FDAB primarily involves pathology of subchondral bone of the phalanges, metacarpals and metatarsals, with arthropathy and brachydactyly arising secondarily (Amor et al. 2002). It has more recently been proposed, however, that the primary pathology occurs in articular chondrocytes (Lamandé et al. 2011). TRPV4 is expressed abundantly in porcine and murine articular chondrocytes (Lamandé et al. 2011; Muramatsu et al. 2007; Hdud et al. 2012; Phan et al. 2009) where it is thought to play a role in cartilage homeostasis (Clark et al. 2010), primarily by initiating chondrocyte responses to decreases in pericellular osmolarity resulting from mechanical loading of joints (Clark et al. 2010; Phan et al. 2009; Liedtke and Guilak 2010). TRPV4 may also modulate inflammatory responses to osmotic stress in articular cartilage (Phan et al. 2009). The absence of TRPV4-mediated responses to hyposmotic saline in cells transfected with FDAB-causing mutants suggests that an inability of articular chondrocytes to respond to osmotic stress may contribute to disease pathogenesis in FDAB (Lamandé et al. 2011). In support of this hypothesis, TRPV4-null mice exhibit increased susceptibility to age- and obesity-dependent osteoarthritic degeneration in the knee (Clark et al. 2010; O'Conor et al. 2013). Phalangeal joints were not examined in these studies. Immunohistochemical investigations have not yet identified differences in TRPV4 expression between articular cartilage of murine knee and phalangeal joints (Lamandé et al. 2011). Further study will be required, therefore, to elucidate the mechanisms restricting FDAB to the joints of the fingers and toes.

21.5 Conclusion

Mutations in the TRPV4 gene cause several forms of autosomal dominant neuromuscular or skeletal disease. Increasing evidence also indicates that some TRPV4 mutations can result in co-occurrence of both nerve and bone disease. TRPV4-mediated neuromuscular disorders and skeletal dysplasias are characterised by considerable variability in disease phenotype and severity. Studies in heterologous expression systems suggest that both CMT2C/SMA- and skeletal dysplasia-causing TRPV4 mutations result in a gain-of-channel function. How such apparently similar changes in channel function result in such diverse disease phenotypes remains essentially unknown. Resolution of this question will likely require the generation of more sophisticated models of TRPV4-mediated disease, including animal models. Such models will also contribute to our understanding of the pathological mechanisms underlying FDAB, which appears to result from a loss-of-TRPV4 channel function. Definition of the molecular and cellular pathomechanisms specific to each form of TRPV4 channelopathy will represent an important step in the development of new therapeutic interventions for these conditions.

References

Adams C, Suchowersky O, Lowry RB (1998) Congenital autosomal dominant distal spinal muscular atrophy. Neuromuscul Disord 8:405–408

Aharoni S, Harlalka G, Offiah A, Shuper A, Crosby AH, McEntagart M (2011) Striking phenotypic variability in familial TRPV4-axonal neuropathy spectrum disorder. Am J Med Genet Part A 155:3153–3156

Akiyama H (2008) Control of chondrogenesis by the transcription factor Sox9. Mod Rheumatol 18:213–219

Alessandri-Haber N, Dina OA, Chen X, Levine JD (2009) TRPC1 and TRPC6 channels cooperate with TRPV4 to mediate mechanical hyperalgesia and nociceptor sensitization. J Neurosci 29(19):6217–6228

Alessandri-Haber N, Dina OA, Yeh JJ, Parada CA, Reichling DB, Levine JD (2004) Transient receptor potential vanilloid 4 is essential in chemotherapy-induced neuropathic pain in the rat. J Neurosci 24(18):4444–4452

Almeida-Souza L, Asselbergh B, d'Ydewalle C, Moonens K, Goethals S, de Winter V, Azmi A, Irobi J, Timmermans JP, Gevaert K, Remaut H, Van Den Bosch L, Timmerman V, Janssens S (2011) Small heat-shock protein HSPB1 mutants stabilize microtubules in Charcot-Marie-Tooth neuropathy. J Neurosci 31(43):15320–15328

Amor DJ, Tudball C, Gardner RJ, Lamandé SR, Bateman JF, Savarirayan R (2002) Familial digital arthropathy-brachydactyly. Am J Med Genet 108:235–240

Andreucci E, Aftimos S, Alcausin M, Haan E, Hunter W, Kannu P, Kerr B, McGillivray G, McKinlay Gardner RJ, Patricelli MG, Sillence D, Thompson E, Zacharin M, Zankl A, Lamande SR, Savarirayan R (2011) TRPV4 related skeletal dysplasias: a phenotypic spectrum highlighted byclinical, radiographic, and molecular studies in 21 new families. Orphanet J Rare Dis 6:37

Arniges M, Fernandez-Fernandez JM, Albrecht N, Schaefer M, Valverde MA (2006) Human TRPV4 channel splice variants revealed a key role of ankyrin domains in multimerization and trafficking. J Biol Chem 281(3):1580–1586

Astrea G, Brisca G, Fiorillo C, Valle M, Tosetti M, Bruno C, Santorelli FM, Battini R (2012) Muscle MRI in TRPV4-related congenital distal SMA. Neurology 78:364–365

Auer-Grumbach M, Olschewski A, Papić L, Kremer H, McEntagart ME, Uhrig S, Fischer C, Fröhlich E, Bálint Z, Tang B, Strohmaier H, Lochmüller H, Schlotter-Weigel B, Senderek J, Krebs A, Dick KJ, Petty R, Longman C, Anderson NE, Padberg GW, Schelhaas HJ, van Ravenswaaij-Arts CM, Pieber TR, Crosby AH, Guelly C (2010) Alterations in the ankyrin domain of TRPV4 cause congenital distal SMA, scapuloperoneal SMA and HMSN2C. Nat Genet 42(2):160–164

Becker D, Muller M, Leuner K, Jendrach M (2008) The C-terminal domain of TRPV4 is essential for plasma membrane localization. Mol Membr Biol 25(2):139–151

Berciano J, Baets J, Gallardo E, Zimoń M, Garcia A, López-Laso E, Combarros O, Infante J, Timmerman V, Jordanova A, De Jonghe P (2011) Reduced penetrance in hereditary motor neuropathy caused by TRPV4 Arg269Cys mutation. J Neurol 258:1413–1421

Camacho N, Krakow D, Johnykutty S, Katzman PJ, Pepkowitz S, Vriens J, Nilius B, Boyce BF, Cohn DH (2010) Dominant TRPV4 mutations in nonlethal and lethal metatropic dysplasia. Am J Med Genet Part A 152(5):1169–1177

Cameron TL, Belluoccio D, Farlie PG, Brachvogel B, Bateman JF (2007) Global transcriptome analysis of cartilage formation in vivo. BMC Dev Biol 9:20

Chen DH, Sul Y, Weiss M, Hillel A, Lipe H, Wolff J, Matsushita M, Raskind W, Bird T (2010) CMT2C with vocal cord paresis associated with short stature and mutations in the TRPV4 gene. Neurology 75(22):1968–1975

Cho TJ, Matsumoto K, Fano V, Dai J, Kim OH, Chae JH, Yoo WJ, Tanaka Y, Matsui Y, Takigami I, Monges S, Zabel B, Shimizu K, Nishimura G, Lausch E, Ikegawa S (2012) TRPV4-pathy manifesting both skeletal dysplasia and peripheral neuropathy: a report of three patients. Am J Med Genet Part A 158:795–802

Chung MK, Lee H, Caterina MJ (2003) Warm temperatures activate TRPV4 in mouse 308 keratinocytes. J Biol Chem 278:32037–32046

Clark AL, Votta BJ, Kumar S, Liedtke W, Guilak F (2010) Chondroprotective role of the osmotically sensitive ion channel transient receptor potential vanilloid 4: age- and sex-dependent progression of osteoarthritis in *Trpv4*-deficient mice. Arthritis Rheum 62:2973–2983

Clark K, Middelbeek J, van Leeuwen FN (2008) Interplay between TRP channels and the cytoskeleton in health and disease. Eur J Cell Biol 87:631–640

Cuajungco MP, Grimm C, Oshima K, D'Hoedt D, Nilius B, Mensenkamp AR, Bindels RJ, Plomann M, Heller S (2006) PACSINs bind to the TRPV4 cation channel. PACSIN 3 modulates the subcellular localization of TRPV4. J Biol Chem 281(27):18753–18762

D'Hoedt D, Owsianik G, Prenen J, Cuajungco MP, Grimm C, Heller S, Voets T, Nilius B (2008) Stimulus-specific modulation of the cation channel TRPV4 by PACSIN 3. J Biol Chem 283(10):6272–6280

D'Ydewalle C, Krishnan J, Chiheb DM, Van Damme P, Irobi J, Kozikowski AP, Vanden Berghe P, Timmerman V, Robberecht W, Van Den Bosch L (2011) HDAC6 inhibitors reverse axonal loss in a mouse model of mutant HSPB1-induced Charcot-Marie-Tooth disease. Nat Med 17(8):968–974

Dai J, Kim OH, Cho TJ, Schmidt-Rimpler M, Tonoki H, Takikawa K, Haga N, Miyoshi K, Kitoh H, Yoo WJ, Choi IH, Song HR, Jin DK, Kim HT, Kamasaki H, Bianchi P, Grigelioniene G, Nampoothiri S, Minagawa M, Miyagawa SI, Fukao T, Marcelis C, Jansweijer MC, Hennekam RC, Bedeschi F, Mustonen A, Jiang Q, Ohashi H, Furuichi T, Unger S, Zabel B, Lausch E, Superti-Furga A, Nishimura G, Ikegawa S (2010) Novel and recurrent TRPV4 mutations and their association with distinct phenotypes within the TRPV4 dysplasia family. J Med Genet 47(10):704–709

De Petrocellis L, Orlando P, Moriello AS, Aviello G, Stott C, Izzo AA, Di Marzo V (2012) Cannabinoid actions at TRPV4 channels: effects on TRPV3 and TRPV4 and their potential relevance to gastrointestinal inflammation. Acta Physiol 204:255–266

DeLong R, Siddique T (1992) A large New England kindred with autosomal dominant neurogenic scapuloperoneal amyotrophy with unique features. Arch Neurol 49:905–908

Deng HX, Klein CJ, Yan J, Shi Y, Wu Y, Fecto F, Yau HJ, Yang Y, Zhai H, Siddique N, Hedley-Whyte ET, Delong R, Martina M, Dyck PJ, Siddique T (2010) Scapuloperoneal spinal muscular atrophy and CMT2C are allelic disorders caused by alterations in TRPV4. Nat Genet 42 (2):165-169

Donaghy M, Kennett R (1999) Varying occurrence of vocal cord paralysis in a family with autosomal dominant hereditary motor and sensory neuropathy. J Neurol 246:552–555

Du J, Wong WY, Sun L, Huang Y, Yao X (2012) Protein kinase G inhibits flow-induced Ca^{2+} entry into collecting duct cells. J Am Soc Nephrol 23:1172–1180

Dyck PJ, Litchy WJ, Minnerath S, Bird TD, Chance PF, Schaid DJ, Aronson AE (1994) Hereditary motor and sensory neuropathy with diaphragm and vocal cord paresis. Ann Neurol 35:608–615

Everaerts W, Nilius B, Owsianik G (2010a) The vanilloid transient receptor potential channel TRPV4: from structure to disease. Prog Biophys Mol Biol 103(1):2–17

Everaerts W, Zhen X, Ghosh D, Vriens J, Gevaert T, Gilbert JP, Hayward NJ, McNamara CR, Xue F, Moran MM, Strassmaier T, Uykal E, Owsianik G, Vennekens R, De Ridder D, Nilius B, Fanger CM, Voets T (2010b) Inhibition of the cation channel TRPV4 improves bladder function in mice and rats with cyclophosphamide-induced cystitis. Proc Nat Acad Sci USA 107(44):19084–19089

Facer P, Casula MA, Smith GD, Benham CD, Chessell IP, Bountra C, Sinisi M, Birch R, Anand P (2007) Differential expression of the capsaicin receptor TRPV1 and related novel receptors TRPV3, TRPV4 and TRPM8 in normal human tissues and changes in traumatic and diabetic neuropathy. BMC Neurol 7:11

Fan HC, Zhang X, McNaughton PA (2009) Activation of the TRPV4 ion channel is enhanced by phosphorylation. J Biol Chem 284:27884-27891

Fawcett KA, Murphy SM, Polke JM, Wray S, Burchell VS, Marji H, Quinlivan RM, Zdebik AA, Reilly MM, Houlden H (2012) Comprehensive analysis of the TRPV4 gene in a large series of inherited neuropathies and controls. J Neurol Neurosurg Psychiatry 83:1204–1209

Fecto F, Shi Y, Huda R, Martina M, Siddique T, Deng HX (2011) Mutant TRPV4-mediated toxicity is linked to increased constitutive function in axonal neuropathies. J Biol Chem 286(19):17281–17291

Fernandez–Fernandez JM, Andrade YN, Arniges M, Fernandes J, Plata C, Rubio-Moscardo F, Vazquez E, Valverde MA (2008) Functional coupling of TRPV4 cationic channel and large conductance, calcium-dependent potassium channel in human bronchial epithelial cell lines. Pflügers Arch 457(1):149–159

Fiorillo C, Moro F, Brisca G, Astrea G, Nesti C, Balint Z, Olschewski A, Meschini MC, Guelly C, Auer-Grumbach M, Battini R, Pedemonte M, Romano A, Menchise V, Biancheri R, Santorelli FM, Bruno C (2012) TRPV4 mutations in children with congenital distal spinal muscular atrophy. Neurogenetics 13:195–203

Fiorio Pla A, Ong HL, Cheng KT, Brossa A, Bussolati B, Lockwich T, Paria B, Munaron L, Ambudkar IS (2012) TRPV4 mediates tumor-derived endothelial cell migration via arachidonic acid-activated actin remodeling. Oncogene 31(2):200–212

Fleury P, Hageman G (1985) A dominantly inherited lower motor neuron disorder presenting at birth with associated arthrogryposis. J Neurol Neurosurg Psychiatry 48:1037–1048

Gao X, Wu L, O'Neil RG (2003) Temperature-modulated diversity of TRPV4 channel gating: activation by physical stresses and phorbol ester derivatives through protein kinase C-dependent and -independent pathways. J Biol Chem 278(29):27129–27137

Garcia-Elias A, Lorenzo IM, Vicente R, Valverde MA (2008) IP_3 receptor binds to and sensitizes TRPV4 channel to osmotic stimuli via a calmodulin-binding site. J Biol Chem 283:31284–31288

Gaudet R (2008) A primer on ankyrin repeat function in TRP channels and beyond. Mol BioSyst 4(5):372–379

Geneviève D, Le Merrer M, Feingold J, Munnich A, Maroteaux P, Cormier-Daire V (2008) Revisiting metatropic dysplasia: presentation of a series of 19 novel patients and review of the literature. Am J Med Genet Part A 146:992–996

Gevaert T, Vriens J, Segal A, Everaerts W, Roskams T, Talavera K, Owsianik G, Liedtke W, Daelemans D, Dewachter I, Van Leuven F, Voets T, De Ridder D, Nilius B (2007) Deletion of the transient receptor potential cation channel TRPV4 impairs murine bladder voiding. J Clin Invest 117(11):3453–3462

Goswami C, Kuhn J, Heppenstall PA, Hucho T (2010) Importance of non-selective cation channel TRPV4 interaction with cytoskeleton and their reciprocal regulations in cultured cells. PLoS ONE 5(7):e11654

Grosskreutz J, Van Den Bosch L, Keller BU (2010) Calcium dysregulation in amyotrophic lateral sclerosis. Cell Calcium 47:165–174

Güler AD, Lee H, Iida T, Shimizu I, Tominaga M, Caterina M (2002) Heat-evoked activation of the ion channel, TRPV4. J Neurosci 22:6408–6414

Hdud IM, El-Shafei AA, Loughna P, Barrett-Jolley R, Mobasheri A (2012) Expression of transient receptor potential vanilloid (TRPV) channels in different passages of articular chondrocytes. Int J Mol Sci 13:4433–4445

Hellwig N, Albrecht N, Harteneck C, Schultz G, Schaefer M (2005) Homo- and heteromeric assembly of TRPV channel subunits. J Cell Sci 118(Pt 5):917–928

Inada H, Procko E, Sotomayor M, Gaudet R (2012) Structural and biochemical consequences of disease-causing mutations in the ankyrin repeat domain of the human TRPV4 channel. Biochemistry 51:6195–6206

Isozumi K, DeLong R, Kaplan J, Deng HX, Iqbal Z, Hung WY, Wilhelmsen KC, Hentati A, Pericak-Vance MA, Siddique T (1996) Linkage of scapuloperoneal spinal muscular atrophy to chromosome 12q24.1-q24.31. Hum Mol Genet 5:1377–1382

Jang Y, Jung J, Kim H, Oh J, Jeon JH, Jung S, Kim KT, Cho H, Yang DJ, Kim SM, Kim IB, Song MR, Oh U (2012) Axonal neuropathy-associated TRPV4 regulates neurotrophic factor-derived axonal growth. J Biol Chem 287(8):6014–6024

Jin M, Wu Z, Chen L, Jaimes J, Collins D, Walters ET, O'Neil RG (2011) Determinants of TRPV4 activity following selective activation by small molecule agonist GSK1016790A. PLoS ONE 6:e16713

Kannu P, Aftimos S, Mayne V, Donnan L, Savarirayan R (2007) Metatropic dysplasia: clinical and radiographic findings in 11 patients demonstrating long-term natural history. Am J Med Genet Part A 143:2512–2522

Kida N, Sokabe T, Kashio M, Haruna K, Mizuno Y, Suga Y, Nishikawa K, Kanamaru A, Hongo M, Oba A, Tominaga M (2012) Importance of transient receptor potential vanilloid 4 (TRPV4) in epidermal barrier function in human skin keratinocytes. Pflügers Arch 463:715–725

Klausen TK, Pagani A, Minassi A, Ech-Chahad A, Prenen J, Owsianik G, Hoffmann EK, Pedersen SF, Appendino G, Nilius B (2009) Modulation of the transient receptor potential vanilloid channel TRPV4 by 4α-phorbol esters: a structure-activity study. J Med Chem 52:2933–2939

Klein CJ, Cunningham JM, Atkinson EJ, Schaid DJ, Hebbring SJ, Anderson SA, Klein DM, Dyck PJ, Litchy WJ, Thibodeau SN, Dyck PJ (2003) The gene for HMSN2C maps to 12q23–24: a region of neuromuscular disorders. Neurology 60:1151–1156

Klein CJ, Shi Y, Fecto F, Donaghy M, Nicholson G, McEntagart ME, Crosby AH, Wu Y, Lou H, McEvoy KM, Siddique T, Deng HX, Dyck PJ (2011) *TRPV4* mutations and cytotoxic hypercalcemia in axonal Charcot-Marie-Tooth neuropathies. Neurology 76(10):887–894

Köttgen M, Buchholz B, Garcia-Gonzalez MA, Kotsis F, Fu X, Doerken M, Boehlke C, Steffl D, Tauber R, Wegierski T, Nitschke R, Suzuki M, Kramer-Zucker A, Germino GG, Watnick T, Prenen J, Nilius B, Kuehn EW, Walz G (2008) TRPP2 and TRPV4 form a polymodal sensory channel complex. J Cell Biol 182:437–447

Krakow D, Vriens J, Camacho N, Luong P, Deixler H, Funari TL, Bacino CA, Irons MB, Holm IA, Sadler L, Okenfuss EB, Janssens A, Voets T, Rimoin DL, Lachman RS, Nilius B, Cohn DH (2009) Mutations in the gene encoding the calcium-permeable ion channel TRPV4 produce spondylometaphyseal dysplasia, Kozlowski type and metatropic dysplasia. Am J Hum Genet 84(3):307–315

Lamandé SR, Yuan Y, Gresshoff IL, Rowley L, Belluoccio D, Kaluarachchi K, Little CB, Botzenhart E, Zerres K, Amor DJ, Cole WG, Savarirayan R, McIntyre P, Bateman JF (2011) Mutations in TRPV4 cause an inherited arthropathy of hands and feet. Nat Genet 43(11):1142–1146

Lanciotti A, Brignone MS, Molinari P, Visentin S, De Nuccio C, Macchia G, Aiello C, Bertini E, Aloisi F, Petrucci TC, Ambrosini E (2012) Megalencephalic leukoencephalopathy with subcortical cysts protein 1 functionally cooperates with the TRPV4 cation channel to activate the response of astrocytes to osmotic stress: dysregulation by pathological mutations. Hum Mol Genet 21(10):2166–2180

Landouré G, Sullivan JM, Johnson JO, Munns CH, Shi Y, Diallo O, Gibbs JR, Gaudet R, Ludlow CL, Fischbeck KH, Traynor BJ, Burnett BG, Sumner CJ (2012) Exome sequencing identifies a novel TRPV4 mutation in a CMT2C family. Neurology 79:192–194

Landouré G, Zdebik AA, Martinez TL, Burnett BG, Stanescu HC, Inada H, Shi Y, Taye AA, Kong L, Munns CH, Choo SS, Phelps CB, Paudel R, Houlden H, Ludlow CL, Caterina MJ, Gaudet R, Kleta R, Fischbeck KH, Sumner CJ (2010) Mutations in TRPV4 cause Charcot-Marie-Tooth disease type 2C. Nat Genet 42(2):170–174

Langer LO, Petersen D, Spranger J (1970) An unusual bone dysplasia: parastremmatic dwarfism. Am J Roentgenol Radium Ther Nucl Med 110:550–560

Lechner SG, Markworth S, Poole K, Smith ES, Lapatsina L, Frahm S, May M, Pischke S, Suzuki M, Ibañez-Tallon I, Luft FC, Jordan J, Lewin GR (2011) The molecular and cellular identity of peripheral osmoreceptors. Neuron 69(2):332–344

Lev S, Katz B, Minke B (2012) The activity of the TRP-like channel depends on its expression system. Channels 6:86–93

Li J, Kanju P, Patterson M, Chew WL, Cho SH, Gilmour I, Oliver T, Yasuda R, Ghio A, Simon SA, Liedtke W (2011) TRPV4-mediated calcium influx into human bronchial epithelia upon exposure to diesel exhaust particles. Environ Health Perspect 119(6):784–793

Liedtke W (2008) Molecular mechanisms of TRPV4-mediated neural signaling. Ann NY Acad Sci 1144:42–52

Liedtke W, Guilak F (2010) Transient receptor potential vanilloid 4: the sixth sense of the musculoskeletal system? Ann N Y Acad Sci 1192:404–409

Long F, Ornitz DM (2013) Development of the endochondral skeleton. Cold Spring Harb Perspect Biol 5(1):a008334

Loukin S, Su Z, Kung C (2011) Increased basal activity is a key determinant in the severity of human skeletal dysplasia caused by TRPV4 mutations. PLoS ONE 6(5):e19533

Loukin S, Su Z, Zhou X, Kung C (2010a) Forward genetic analysis reveals multiple gating mechanisms of TRPV4. J Biol Chem 285(26):19884–19890

Loukin S, Zhou X, Su Z, Saimi Y, Kung C (2010b) Wild-type and brachyolmia-causing mutant TRPV4 channels respond directly to stretch force. J Biol Chem 285:27176–27181

Ma X, Cao J, Luo J, Nilius B, Huang Y, Ambukar IS, Yao X (2010) Depletion of intracellular Ca^{2+} store stimulates the translocation of vanilloid transient receptor potential 4-C1 heteromeric channels to the plasma membrane. Arterioscler Thromb Vasc Biol 30:2249–2255

Ma X, Cheng KT, Wong CO, O'Neil RG, Birnbaumer L, Ambudkar IS, Yao X (2011a) Heteromeric TRPV4-C1 channels contribute to store-operated $Ca^{(2+)}$ entry in vascular endothelial cells. Cell Calcium 50:502–509

Ma X, Nilius B, Wong JW, Huang Y, Yao X (2011b) Electrophysiological properties of heteromeric TRPV4-C1 channels. Biochem Biophys Acta 1808:2789–2797

Ma X, He D, Ru X, Chen Y, Cai Y, Bruce IC, Xia Q, Yao X, Jin J (2012) Apigenin, a plant-derive flavone, activates transient receptor potential vanilloid 4 cation channel. Br J Pharmacol 166:349–358

Macaskill AF, Rinholm JE, Twelvetrees AE, Arancibia-Carcamo IL, Muir J, Fransson A, Aspenstrom P, Attwell D, Kittler JT (2009) Miro1 is a calcium sensor for glutamate receptor-dependent localization of mitochondria at synapses. Neuron 61(4):541–555
Masuyama R, Mizuno A, Komori H, Kajiya H, Uekawa A, Kitaura H, Okabe K, Ohyama K, Komori T (2012) Calcium/calmodulin-signaling supports TRPV4 activation in osteoclasts and regulates bone mass. J Bone Mineral Res 27:1708–1721
McEntagart M (2012) TRPV4 axonal neuropathy spectrum disorder. J Clin Neurosci 19:927–933
McEntagart ME, Reid SL, Irrthum A, Douglas JB, Eyre KE, Donaghy MJ, Anderson NE, Rahman N (2005) Confirmation of a hereditary motor and sensory neuropathy IIC locus at chromosome 12q23-q24. Ann Neurol 57:293–297
Muramatsu S, Wakabayashi M, Ohno T, Amano K, Ooishi R, Sugahara T, Shiojiri S, Tashiro K, Suzuki Y, Nishimura R, Kuhara S, Sugano S, Yoneda T, Matsuda A (2007) Functional gene screening system identified TRPV4 as a regulator of chondrogenic differentiation. J Biol Chem 282:32158–32167
Nilius B, Voets T (2013) The puzzle of TRPV4 channelopathies. EMBO Rep 14(2):152–163
Nishimura G, Dai J, Lausch E, Unger S, Megarbané A, Kitoh H, Kim OH, Cho TJ, Bedeschi F, Benedicenti F, Mendoza-Londono R, Silengo M, Schmidt-Rimpler M, Spranger J, Zabel B, Ikegawa S, Superti-Furga A (2010) Spondylo-epiphyseal dysplasia, Maroteaux type (pseudo-Morquio syndrome type 2), and parastremmatic dysplasia are caused by TRPV4 mutations. Am J Med Genet 152A(6):1443–1449
Nural MS, Diren HB, Sakarya O, Yalin T, Dağdemir A (2006) Kozlowski type spondylometaphyseal dysplasia: a case report with literature review. Diagn Interv Radiol 12:70–73
O'Conor CJ, Griffin TM, Liedtke W, Guilak F (2013) Increased susceptibility of *TRPV4*-deficient mice to obesity and obesity-induced osteoarthritis with very high-fat diet. Ann Rheum Dis 72:300–304
Oates EC, Reddel S, Rodriguez ML, Gandolfo LC, Bahlo M, Hawke SH, Lamande SR, Clarke NF, North KN (2012) Autosomal dominant congenital spinal muscular atrophy: a true form of spinal muscular atrophy caused by early loss of anterior horn cells. Brain 135:1714–1723
Obeidat M, Wain LV, Shrine N, Kalsheker N, Soler Artigas M, Repapi E, Burton PR, Johnson T, Ramasamy A, Zhao JH, Zhai G, Huffman JE, Vitart V, Albrecht E, Igl W, Hartikainen AL, Pouta A, Cadby G, Hui J, Palmer LJ, Hadley D, McArdle WL, Rudnicka AR, Barroso I, Loos RJ, Wareham NJ, Mangino M, Soranzo N, Spector TD, Gläser S, Homuth G, Völzke H, Deloukas P, Granell R, Henderson J, Grkovic I, Jankovic S, Zgaga L, Polašek O, Rudan I, Wright AF, Campbell H, Wild SH, Wilson JF, Heinrich J, Imboden M, Probst-Hensch NM, Gyllensten U, Johansson A, Zaboli G, Mustelin L, Rantanen T, Surakka I, Kaprio J, Jarvelin MR, Hayward C, Evans DM, Koch B, Musk AW, Elliott P, Strachan DP, Tobin MD, Sayers I, Hall IP, Consortium S (2011) A comprehensive evaluation of potential lung function associated genes in the SpiroMeta general population sample. PLoS ONE 6(5):e19382
Pareyson D, Marchesi C, Salsano E (2009) Hereditary predominantly motor neuropathies. Curr Opin Neurol 22(5):451–459
Phan MN, Leddy HA, Votta BJ, Kumar S, Levy DS, Lipshutz DB, Lee SH, Liedtke W, Guilak F (2009) Functional characterization of TRPV4 as an osmotically sensitive ion channel in porcine articular chondrocytes. Arthritis Rheum 60:3028–3037
Phelps CB, Wang RR, Choo SS, Gaudet R (2010) Differential regulation of TRPV1, TRPV3, and TRPV4 sensitivity through a conserved binding site on the ankyrin repeat domain. J Biol Chem 285(1):731–740
Pochynyuk O, Zaika O, O'Neil RG, Mamenko M (2013) Novel insights into TRPV4 function in the kidney. Pflügers Arch 465:177–186
Reddel S, Ouvrier RA, Nicholson G, Dierick I, Irobi J, Timmerman V, Ryan MM (2008) Autosomal dominant congenital spinal muscular atrophy—a possible developmental deficiency of motor neurones? Neuromuscul Disord 18:530–555
Reilly MM, Murphy SM, Laurá M (2011) Charcot-Marie-Tooth disease. J Peripher Nerv Syst 16:1–14

Rock MJ, Prenen J, Funari VA, Funari TL, Merriman B, Nelson SF, Lachman RS, Wilcox WR, Reyno S, Quadrelli R, Vaglio A, Owsianik G, Janssens A, Voets T, Ikegawa S, Nagai T, Rimoin DL, Nilius B, Cohn DH (2008) Gain-of-function mutations in TRPV4 cause autosomal dominant brachyolmia. Nat Genet 40(8):999–1003

Santoro L, Manganelli F, Di Maio L, Barbieri F, Carella M, D'Adamo P, Casari G (2002) Charcot-Marie-Tooth disease type 2C: a distinct genetic identity. Clinical and molecular characterization of the first European family. Neuromuscul Disord 12:399–404

Saotome M, Safiulina D, Szabadkai G, Das S, Fransson A, Aspenstrom P, Rizzuto R, Hajnóczky G (2008) Bidirectional Ca^{2+}-dependent control of mitochondrial dynamics by the Miro GTPase. Proc Natl Acad Sci USA 105(52):20728–20733

Sensenbrenner JA, Dorst JP, Hungerford DS (1974) Parastremmatic dwarfism. Birth Defects Orig Artic Ser 10:424–429

Shigematsu H, Sokabe T, Danev R, Tominaga M, Nagayama K (2010) A 3.5-nm structure of rat TRPV4 cation channel revealed by Zernike phase-contrast cryoelectron microscopy. J Biol Chem 285(15):11210–11218

Shin SH, Lee EJ, Hyun S, Chun J, Kim Y, Kang SS (2012) Phosphorylation on the Ser 824 residue of TRPV4 prefers to bind with F-actin than with microtubules to expand the cell surface area. Cell Signal 24(3):641–651

Shohat M, Lachman R, Gruber HE, Rimoin DL (1989) Brachyolmia: radiographic and genetic evidence of heterogeneity. Am J Med Genet 33:209–219

Smith PL, Maloney KN, Pothen RG, Clardy J, Clapham DE (2006) Bisandrographolide from *Andrographis paniculata* activates TRPV4 channels. J Biol Chem 281:29897–29904

Stewart AP, Smith GD, Sandford RN, Edwardson JM (2010) Atomic force microscopy reveals the alternating subunit arrangement of the TRPP2-TRPV4 heterotetramer. Biophys J 99(3):790–797

Strotmann R, Schultz G, Plant TD (2003) Ca^{2+}-dependent potentiation of the nonselective cation channel TRPV4 is mediated by a C-terminal calmodulin binding site. J Biol Chem 278:26541–26549

Strotmann R, Semtner M, Kepura F, Plant TD, Schöneberg T (2010) Interdomain interactions control Ca^{2+}-dependent potentiation in the cation channel TRPV4. PLoS ONE 5(5):e10580

Suzuki M, Mizuno A, Kodaira K, Imai M (2003) Impaired pressure sensation in mice lacking TRPV4. J Biol Chem 278:22664–22668

Thorneloe KS, Cheung M, Bao W, Alsaid H, Lenhard S, Jian MY, Costell M, Maniscalco-Hauk K, Krawiec JA, Olzinski A, Gordon E, Lozinskaya I, Elefante L, Qin P, Matasic DS, James C, Tunstead J, Donovan B, Kallal L, Waszkiewicz A, Vaidya K, Davenport EA, Larkin J, Burgert M, Casillas LN, Marquis RW, Ye G, Eidam HS, Goodman KB, Toomey JR, Roethke TJ, Jucker BM, Schnackenberg CG, Townsley MI, Lepore JJ, Willette RN (2012) An orally active TRPV4 channel blocker prevents and resolves pulmonary edema induced by heart failure. Sci Transl Med 4(159):159ra148

Thorneloe KS, Sulpizio AC, Lin Z, Figueroa DJ, Clouse AK, McCafferty GP, Chendrimada TP, Lashinger ESR, Gordon E, Evans L, Misajet BA, DeMarini DJ, Nation JH, Casillas LN, Marquis RW, Votta BJ, Sheardown SA, Xu X, Brooks DP, Laping NJ, Westfall TD (2008) N-((1S)-1-{[4-((2S)-2-{[2,4-dichlorophenyl)sulfonyl]amino}-3-hydroxypropanol)-1-piperazinyl]carbonyl}-3-methylbutyl)-1-benzothiophene-2-carboxamide (GSK1016790A), a novel and potent transient receptor potential vanilloid 4 channel agonist induces urinary bladder contraction and hyperactivity: Part 1. J Pharmacol Exp Ther 326:432–442

Tian W, Fu Y, Garcia-Elias A, Fernández-Fernández JM, Vicente R, Kramer PL, Klein RF, Hitzemann R, Orwoll ES, Wilmot B, McWeeney S, Valverde MA, Cohen DM (2009) A loss-of-function nonsynonymous polymorphism in the osmoregulatory *TRPV4* gene is associated with human hyponatremia. Proc Natl Acad Sci USA 106(33):14034–14039

Unger S, Lausch E, Stanzial F, Gillessen-Kaesbach G, Stefanova I, Di Stefano I, Bertini E, Dionisi-Vici C, Nilius B, Zabel B, Superti-Furga A (2011) Fetal akinesia in metatropic dysplasia: the combined phenotype of chondrodysplasia and neuropathy? Am J Med Genet Part A 155:2860–2864 van der Vleuten AJW, van Ravenswaaij-Arts CMA, Frijns CJM, Smits APT, Hageman G, Padberg GW, Kremer H (1998) Localisation of the gene for a dominant congenital spinal muscular atrophy predominantly affecting the lower limbs to chromosome 12q23-q24. Eur J Hum Genet 6:376–382

Vandewauw I, Owsianik G, Voets T (2013) Systematic and quantitative mRNA expression analysis of TRP channel genes at the single trigeminal and dorsal root ganglion level in mouse. BMC Neurosci 14:21

Vincent F, Duncton MA (2011) TRPV4 agonists and antagonists. Curr Top Med Chem 11(17):2216–2226

Vriens J, Watanabe H, Janssens A, Droogmans G, Voets T, Nilius B (2004) Cell swelling, heat, and chemical agonists use distinct pathways for the activation of the cation channel TRPV4. Proc Natl Acad Sci USA 101(1):396–401

Wang X, Schwarz TL (2009) The mechanism of Ca^{2+}-dependent regulation of kinesin-mediated mitochondrial motility. Cell 136:163–174

Watanabe H, Davis JB, Smart D, Jerman JC, Smith GD, Hayes P, Vriens J, Cairns W, Wissenbach U, Prenen J, Flockerzi V, Droogmans G, Benham CD, Nilius B (2002) Activation of TRPV4 channels (hVRL-2/mTRP12) by phorbol derivatives. J Biol Chem 277:13569–13577

Ye L, Kleiner S, Wu J, Sah R, Gupta RK, Banks AS, Cohen P, Khandekar MJ, Boström P, Mepani RJ, Laznik D, Kamenecka TM, Song X, Liedtke W, Mootha VK, Puigserver P, Griffin PR, Clapham DE, Spiegelman BM (2012) TRPV4 is a regulator of adipose oxidative metabolism, inflammation, and energy homeostasis. Cell 151:96–110

Yiu EM, Ryan MM (2012) Genetic axonal neuropathies and neuronopathies of pre-natal and infantile onset. J Peripher Nerv Syst 17(3):285–300

Zhu G, Investigators I, Gulsvik A, Bakke P, Ghatta S, Anderson W, Lomas DA, Silverman EK, Pillai SG (2009) Association of *TRPV4* gene polymorphisms with chronic obstructive pulmonary disease. Hum Mol Genet 18(11):2053–2062

Zimoń M, Baets J, Auer-Grumbach M, Berciano J, Garcia A, Lopez-Laso E, Merlini L, Hilton-Jones D, McEntagart M, Crosby AH, Barisic N, Boltshauser E, Shaw CE, Landouré G, Ludlow CL, Gaudet R, Houlden H, Reilly MM, Fischbeck KH, Sumner CJ, Timmerman V, Jordanova A, Jonghe PD (2010) Dominant mutations in the cation channel gene transient receptor potential vanilloid 4 cause an unusual spectrum of neuropathies. Brain 133(Pt 6):1798–1809

Chapter 22
TRPC Channels in Cardiac Hypertrophy

Nadine Kirschmer, Kristina Lorenz and Petra Eder-Negrin

Abstract Impaired cardiomyocyte Ca^{2+} handling is considered as crucial factor programming cardiac hypertrophy and heart failure. It has, however, not been entirely resolved what type of Ca^{2+} signals is involved in mechanisms inducing hypertrophic remodelling. Ca^{2+} microdomains, which are restricted from the fluctuating Ca^{2+} during excitation–contraction coupling, could generate local Ca^{2+} signals and activate downstream Ca^{2+} dependent signalling pathways. With the non-selective ion channels of the TRPC family, attractive entities in microdomain Ca^{2+} signalling have been found. Within a short period of time important aspects of TRPC dependent signalling have been revealed. TRPC channels contribute to a store-operated Ca^{2+} entry, are functionally cross-linked with L-type Ca^{2+} channels and can indirectly sense stress stimuli during cardiac hypertrophy. Pressure-overload or agonist stimulation of the heart results in an increased TRPC channel activity which is mechanistically coupled to Ca^{2+} dependent signalling pathways. In particular, the calcineurin/nuclear factor of activated T-cells signalling pathway has been elucidated as central mediator translating TRPC dependent Ca^{2+} signals to hypertrophic cardiac growth. Interference with the TRPC/calcineurin/nuclear factor of activated T-cells circuit or TRPC blockade has been found beneficial in opposing cardiac remodelling. TRPC channels can therefore be regarded as promising targets in the therapy of heart failure

N. Kirschmer · P. Eder-Negrin
Department of Internal Medicine I—Cardiology, University Hospital Würzburg, Würzburg, Germany

N. Kirschmer · P. Eder-Negrin (✉)
Comprehensive Heart Failure Center, University Hospital Würzburg, Würzburg, Germany
e-mail: Eder_P@medizin.uni-wuerzburg.de

K. Lorenz
Institute of Pharmacology and Toxicology, University of Würzburg, Würzburg, Germany

N. Weiss and A. Koschak (eds.), *Pathologies of Calcium Channels*,
DOI: 10.1007/978-3-642-40282-1_22, © Springer-Verlag Berlin Heidelberg 2014

22.1 Ca^{2+}-Dependent Signalling Pathways in Pathologic Cardiac Hypertrophy

The remodelling process in response to stress stimulation or injury is referred to as pathological cardiac hypertrophy. Although primarily adaptive, this type of hypertrophy eventually leads to heart failure and dilation and is therefore considered as a predictor of cardiac deterioration. Characteristic of pathological hypertrophy is one the too much re-activation of a foetal transcriptional programme. It has been found that the ß-isoform of the myosin heavy chain (*β*-MHC) and skeletal *α*-actin are up-regulated, whereas the sarcoplasmic/endoplasmic reticulum calcium ATPase (SERCA), the glucose transporter 4 (GLUT4) and the mitochondrial creatine kinase (mCK) are down-regulated which may ultimately affect excitation contraction coupling (ECC), total cardiac energetic and metabolism (Rajabi et al. 2007). The development of pathologic hypertrophy is dependent on a series of intracellular signalling pathways and proteins. In particular, the serine-threonine phosphatase calcineurin and the serine/threonine kinase Ca^{2+}/calmodulin (CaM)-dependent protein kinase II (CamKII) have emerged as central proteins in hypertrophic signalling. Activation of both proteins is dependent on a sustained rise in intracellular Ca^{2+} and the presence of calmodulin (Kehat and Molkentin 2010).

Calcineurin exists as a heterotrimer consisting of the catalytic A subunit (CnA) and the Ca^{2+} binding proteins calcineurin B (CnB) and CaM. At low Ca^{2+} concentrations, CnA activity is blocked by an auto-inhibitory domain. A rise of Ca^{2+}, however, strengthens the interaction between CaM and CnA resulting in the release of the auto-inhibitory domain and activation of calcineurin (Klee et al. 1998). The bona fide target of calcineurin is the transcription factor nuclear factor of activated T-cells (NFAT) which upon its dephosphorylation is translocated to the nucleus (Molkentin et al. 1998). As a co-factor of the cardiac zinc-finger transcription factor GATA4, NFAT regulates the expression of pro-hypertrophic genes. As upstream signalling pathway the G*α*q-protein coupled receptor pathway (GPCR) has been repeatedly described to increase calcineurin activity (Eder and Molkentin 2011).

Comparable to calcineurin, the signalling protein CamKII playsignalling cascade in pathologichronic pathologic situations. CamKII is activated by local and high Ca^{2+} waves to exert its function (Anderson et al. 2011). Intriguingly, a series of proteins of the ECC machinery, such as the ryanodine receptor (RyR), the L-type Ca^{2+} channel (LTCC) and the SERCA are targets of CamKII dependent phosphorylation. These modifications are, however, associated with a chain of maladaptive reactions comprising an increased sarcoplasmic reticulum (SR) Ca^{2+} leak, elevation of diastolic Ca^{2+}, arrhythmias and apoptosis (Anderson et al. 2011). Also, the hypertrophic programme is regulated by CamKII. In particular, CamKII dependent phosphorylation of the histone deacetylases HDAC4 and HDAC5 relieves the transcriptional activity of the myocytes enhancer factor-2 (MEF-2) which is a crucial regulator of cardiac growth (Passier et al. 2000).

Regulation of these two particular signalling pathways has evoked great research interest during the last couple of years because both calcineurin and CamKII are strictly Ca^{2+} dependent and their activity is exclusively associated with the pathologic cardiac hypertrophic response (Kehat and Molkentin 2010). Hence, it has been a matter of speculation as to what Ca^{2+} entry pathways are selectively activated in chronic pathologic situations and as to how Ca^{2+} signals can be specifically integrated in signalling mechanisms. Cardiomyocytes are probably the most difficult cell type to answer these questions given the fact that they are continuously bathed in Ca^{2+} during ECC.

22.2 Ca^{2+} Sources Linked to Pro-Hypertrophic Signalling Pathways

The ECC is initiated by the opening of voltage-gated Na^{+} channels depolarizing the sarcolemma. LTCCs allow a Ca^{2+} entry in the cleft region of cardiomyocytes which triggers the opening of the juxta-positioned Ca^{2+}-sensitive RyRs. Ca^{2+} from intracellular stores is released via RyRs causing myocyte contraction. During diastole, Ca^{2+} is transported back into intracellular stores via the SERCA and out of the cells through the Na^{+}/Ca^{2+} exchanger (NCX) (Bers 2008). Almost all observable Ca^{2+} changes are reserved for ECC which raises the question how Ca^{2+} signals in cardiac myocytes can be separated and used for signalling purposes within the cell.

22.2.1 *Contractile Ca^{2+} Signals*

It has been speculated that the signalling information could be integrated in the Ca^{2+} signals related to the ECC machinery as alterations in amplitude or/and frequency of transients or changes in diastolic Ca^{2+}. In a study by Collella et al. (2008) a positive correlation between the frequency of Ca^{2+} transients and the activation of hypertrophic signalling pathways was described. Cardiomyocytes when stimulated with angiotensin II (Ang II) or when depolarized by KCL respond with an increased frequency of Ca^{2+} oscillations which are sufficient to activate the calcineurin/NFAT pathway and finally the hypertrophic program. With a shortened interval between the Ca^{2+} transients the balance of the phosphorylation status of NFAT could shift from phosphorylated to dephosphorylated NFAT which would accumulate over time. Similarly, an increased LTCC-mediated peak amplitude could activate a higher percentage of calcineurin molecules which in turn could de-phosphorylate NFAT. In line with this hypothesis, a recent study shows that overexpression of the L-type channel $\beta 2a$ subunit in the heart is associated with a pronounced hypertrophic response (Chen et al. 2011). The authors suggest that apart from the increased LTCC mediated Ca^{2+} influx an enhanced Ca^{2+} SR load could be a critical determinant controlling hypertrophic signalling pathways. Mechanistically, overexpression of β2a is coupled to the activation of both the calcineurin/NFAT as well as the CamKII/HDAC signalling pathways. It seems, however, that the ultimate Ca^{2+} source necessary for their activation differs. While a decrease in the SR Ca^{2+} load which was experimentally achieved by a SERCA inhibition impairs CamKII/HDAC signalling in β2a overexpressing cardiomyocytes it leaves the calcineurin/NFAT signalling pathway unaffected. It is possible that inhibition of SERCA and the reduced SR Ca^{2+} load causes nuclear Ca^{2+} to decrease which negatively affects CamKII activation and the hypertrophic response. Based on the fact that the SR Ca^{2+} load is most critical for cardiomyocyte Ca^{2+} cycling, several studies have examined its effect on the progression of hypertrophy towards heart failure. Cardiomyocytes from conditional SERCA2 knock-out mice, for example, exhibit a reduced SR Ca^{2+} load, which, however, is not associated with any functional deterioration at early stages of SERCA gene excision (Andersson et al. 2009). Apparently, non-SR Ca^{2+} components, such as the LTCC and the NCX can compensate for the impaired SR function. Signs of hypertrophic remodelling are not noticeable in the absence of SERCA2a which is a rather puzzling result assuming that the absence of SERCA2 per se would be sufficient to increase diastolic Ca^{2+} or certain Ca^{2+} microdomains in the cytosol which could activate Ca^{2+} dependent hypertrophic pathways. Instead it supports the hypothesis of an enhanced SR Ca^{2+} load as critical factor modulating hypertrophic signalling. Accordingly, a recent study shows that facilitated SERCA activity achieved by the genetic ablation of phospholamban (PLB) promotes cardiac hypertrophy and cardiac dysfunction in combination with a dysregulation of the RyR (Kalyanasundaram et al. 2012). It is well known that hyperphosphorylation of the RyR upon β-adrenergic receptor stimulation and protein kinase A (PKA) and CaMKII activation is associated with greater cardiac hypertrophy. For example, mice heterozygous for the gain-of-function mutation R176Q in *RyR* which predisposes the heart to a catecholamine-induced oscillatory Ca^{2+} release respond with an exaggerated cardiac hypertrophic response upon pressure overload induction (van Oort et al. 2010). In such a situation, attempts to facilitate SR Ca^{2+} uptake, for example via ablation of PLB or overexpression of SERCA, would be rather detrimental than therapeutic.

In the end, the extent of cardiac hypertrophy largely depends on the expressional balance and phosphorylation status of the Ca^{2+} handling proteins, the neurohumoral input and the aetiology of disease. In particular, the plasticity of the myocardium becomes evident when alterations of the LTCC occur. As explained above, overexpression of the LTCC accelerates cardiac hypertrophy. Interestingly, less LTCC does not reverse this effect but even accelerates hypertrophy towards heart failure (Goonasekera et al. 2012). It seems that a reduced LTCC activity leads to a neuro-endocrinic stress situation resulting in a compensatory SR Ca^{2+} leak and elevation of diastolic Ca^{2+}. In this scenario calcineurin which is known to be located at the z discs in the cleft microenvironment samples Ca^{2+} to further promote the hypertrophic response.

22.2.2 Ca^{2+} Microdomains

Cardiomyocytes also harbour Ca^{2+} sources which do not interfere with the contractile Ca^{2+} generated in the cleft region. They are referred to as Ca^{2+} microdomains. Despite the lack of a uniform definition, it has been suggested that Ca^{2+} in microdomains exceeds global cytosolic Ca^{2+} levels and functions as signalling messenger that especially becomes important in pathology. In particular, pathologic hypertrophy could result in structural, expressional alterations yielding new Ca^{2+} platforms and Ca^{2+} entities in cardiomyocytes. Within the sarcolemma, lipid raft domains, for example, could become more important as signalling platform. Interestingly, a recent study has found indications for an LTCC—calcineurin interaction in caveolae (Makarewich et al. 2012). Here, a genetically modified LTCC inhibitor diminished LTTC activity in caveolae without affecting contractility and global Ca^{2+} transients. In contrast, calcineurin-NFAT signalling was reduced. It will be intriguing to find out whether a specific inhibition of LTCCs in caveolae also blunts the pathological hypertrophic response in vivo.

Up-regulation of Ca^{2+} handling proteins during the remodelling process could also generate new Ca^{2+} pools with a specific signalling function. For example, the T-type Ca^{2+} channels (TTCCs) $Ca_v3.1$ (α1G) and $Ca_v3.2$ (α1H) are expressed during cardiac development but not in healthy adult cardiac tissue. There is, however, re-expression during pathological cardiac hypertrophy and in the post-infarction heart suggesting that they are part of the foetal gene programme during pathological hypertrophy (Nuss and Houser 1993; Huang et al. 2000). The functional role of TTCCs in cardiac remodelling has been evaluated using gene targeted mouse models. Interestingly, pathological hypertrophy was suppressed in mice deficient for α1H ($Ca_v3.2$−/−) but not in mice deficient for α1G ($Ca_v3.1$−/−). This phenotype was mechanistically coupled with less activation of the calcineurin/NFAT signalling pathway in α1H deficient mice (Chiang et al. 2009). Since Ca^{2+} cycling and contractility in these mice were not impaired, the authors suggested that α1H is part of a membrane complex that allows local Ca^{2+} changes instead of bulk Ca^{2+}. While re-expression of α1H has been suggested to impair cardiac function during cardiac hypertrophy, there are controversial findings regarding α1G ($Ca_v3.1$). $Ca_v3.1$ −/− mice either do not develop a cardiac phenotype (Chiang et al. 2009) or respond with an increased cardiac hypertrophy (Nakayama et al. 2009). The latter finding even suggests that α1G and α1H have diametric functional effects with α1G protecting against pathological cardiac hypertrophy. Indeed, overexpression of α1G in transgenic mice results in less cardiac hypertrophy after pressure overload induction despite an increased α1G Ca^{2+} current in cardiomyocytes. In this case, the α1G dependent current antagonises cardiac hypertrophy through an endothelial nitric oxide synthase (eNOS)- cyclic guanosine monophosphate (cGMP) dependent mechanism which could also be locally restricted in caveolae.

With the ion channels of the TRPC (Transient Receptor Potential, Canonical) family, further promising Ca^{2+} entities in cardiac microdomain Ca^{2+} signalling have been found. Essentially, three characteristics make them ideal candidates in hypertrophic signalling: They are preferably activated by hypertrophic agonists, they are expressionally regulated by the transcription factor NFAT and are up-regulated in heart failure.

22.3 TRPC Channels are Implicated in the Hypertrophic Remodelling Process

22.3.1 Signalling Pathways Regulating TRPC Channel Activity

TRPC channels are tetrameric ion channels in the plasma membrane mediating a non-selective Na^+ and Ca^{2+} entry. There are 7 subunits (TRPC1-7) which can either assemble to homo- or heteromers. According to their sequence homologies, the subunits TRPC1/4/5 and TRPC3/6/7 have been suggested as preferable heteromers (Hofmann et al. 2002) although heteromerization can also occur across TRPC subfamilies (Poteser et al. 2006) and even TRP families (Schindl et al. 2012; Roedding et al. 2012). TRPC channels are typically activated upon stimulation of GPCRs or receptor tyrosine kinases and the subsequent activation of phospholipase C (PLC ß/γ; Fig. 22.1) (Trebak et al. 2007). The precise activation mechanism in the heart has not been entirely resolved. It could comprise the PLC mediated generation of diacylglycerol (DAG) or could take place in the course of the inositol triphosphate (IP_3) dependent SR Ca^{2+} store depletion.

Activation of TRPC channels in cardiac myocytes can be facilitated with Ang II or endothelin 1 (ET-1) and α adrenergic agonists (Ohba et al. 2007). Stimulation with ET-1 for example results in an increased TRPC6 current in cardiac myocytes from TRPC6 transgenic mice (TG) (Kuwahara et al. 2006). In HL-1 cells, a cardiac cell model, a TRPC3 specific current can be increased when ET-1 is acutely applied (Poteser et al. 2011). More often Ang II is used to activate TRPC channels in mouse or rat cardiac myocytes, as shown for TRPC3 in myocytes from TRPC3 TG mice (Nakayama et al. 2006) or in rat cardiomyocytes (Eder et al. 2007). TRPC3/TRPC7 heteromers in rat cardiomyocytes can also be activated downstream of the ATP/UTP-P2Y2 receptor signalling pathway (Alvarez et al. 2008). ATP/UTP release is enhanced in early ischemia reperfusion and could favour arrhythmias through the activation of TRPC channels. The activity of TRPC1 as shown in mouse cardiac myocytes can also be increased with Ang II stimulation (Seth et al. 2009). In this case it seems that the Ang II receptor confers its stretch sensitivity to TRPC1. Accordingly, blockade of the Ang II receptor type 1 by losartan reduces a TRPC1 specific current elicited by myocyte stretching.

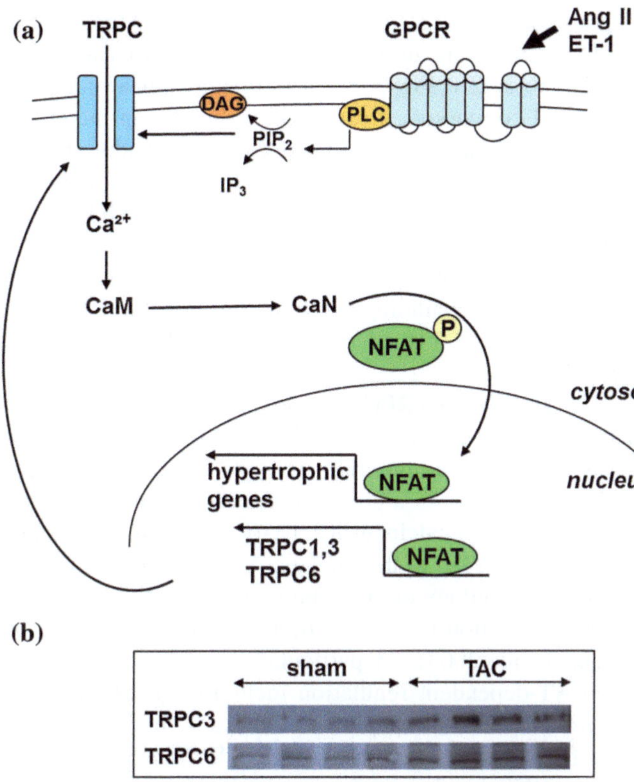

Fig. 22.1 Expressional regulation of TRPC channels in cardiomyocytes during cardiac pathological stress stimulation. **a** Stimulation of GPCRs generates DAG and IP_3 through the activation of PLC which results in activation of TRPC channels. The TRPC mediated Ca^{2+} influx is sensed by CaM which results in the activation of CaN and the dephosphorylation of NFAT. Translocation of dephosphorylated NFAT to the nucleus results in an increased expression of hypertrophic genes and genes encoding TRPC1, TRPC3 and TRPC6. **b** Up-regulation of TRPC3 and TRPC6 in response to pressure overload induction. To induce cardiac hypertrophy, mice were subjected to transverse aortic constriction (TAC) treatment for 8 weeks. Western blotting was performed on whole heart homogenates. TRPC3 and TRPC6 expression is increased compared to sham treated animals, Ang II, angiotensin II; CaM, calmodulin; CaN, Calcineurin; DAG, diacylglycerol; ET-1, endothelin-1; GPCR, Gα protein coupled receptor; IP_3, inositol 3-phosphate; NFAT, nuclear factor of activated T cells; PIP_2, Phosphatidylinositol 4,5-bisphosphate; PLC, phospholipase C; TRPC, transient receptor potential canonical

22.3.2 TRPC Channel Expression and Transcriptional Regulation in the Heart

All of the 7 TRPC isoforms have been detected in cardiac tissue on the mRNA level at baseline (Garcia and Schilling 1997; Ohba et al. 2007; Riccio et al. 2002). Regarding the protein expression there are divergent findings but it seems that all of the isoforms are expressed at a low level (Table 22.1). Most interesting is the finding that TRPC channels are up-regulated in models of cardiac hypertrophy (TRPC1, 3, 6 and 7; Table 22.1; Fig. 22.1a) and in human heart failure (TRPC5, 6; Table 22.1). Olson and co-workers, for example, showed an increased transcription of TRPC6 in heart samples with dilated cardiomyopathy (Kuwahara et al. 2006). Up-regulation was also detected in calcineurin TG mice, a model of cardiac hypertrophy, and in response to hypertrophic stimulation with ET-1. Another study shows that only TRPC5 is up-regulated in human heart failure (Bush et al. 2006). It is probably the heterogeneity of human tissue that causes the discrepancy of the results. In the experimental pathological hypertrophic TAC (transverse aortic constriction) model, also TRPC3 (Bush et al. 2006) and TRPC1 (Ohba et al. 2007) are up-regulated.

Intriguingly, up-regulation of TRPC channels is enhanced by their own activity and can be described as feed-forward loop (Fig. 22.1a). During pathologic cardiac hypertrophy the increased neuro-humoral activation results in activation of TRPC channels in cardiac myocytes. As a consequence increased Ca^{2+} signals activate CaM which is required for full calcineurin activity and subsequent translocation of the transcription factor NFAT into the nucleus. Since NFAT transcription factors share an imperfect Rel homology domain their capability to bind DNA promoters is dependent on the interaction with other transcription factors, such as GATA-4 and MEF-2 (Hogan et al. 2003). In particular, hypertrophic marker genes are subjected to an NFAT-dependent regulation including genes encoding the natriuretic peptides, atrial natriuretic peptide (ANP) and brain natriuretic peptide (BNP), as well as foetal genes, such as β- MHC (Wilkins and Molkentin 2002). Increased ANP and BNP levels activate compensatory mechanisms protecting against hypertrophic progression (see also paragraph 5.). Also, the NFAT mediated expression of the calcineurin inhibitor, regulator of calcineurin1 (Rcan1) might initiate a negative feedback loop on calcineurin signalling and hypertrophic remodelling. On the other hand, increased β-MHC expression might positively contribute to the remodelling programming and functional weakening of the heart. With the up-regulation of ET-1 (Wilkins and Molkentin 2002) and TRPC channels (Kuwahara et al. 2006; Ohba et al. 2007), calcineurin/NFAT signalling could be further promoted. Conserved NFAT-binding sites have been found in the promotors of TRPC6 and TRPC1 genes (Kuwahara et al. 2006; Ohba et al. 2007). This remodelling process could generate new Ca^{2+} pools further maintaining the process of maladaptive hypertrophy in a select interaction with the pro-hypertrophic calcineurin–NFAT signalling pathway. The subcellular distribution of these signalling microdomains has, however, not entirely been resolved. So far, TRPC channels have been described to localise to the outer membrane of cardiomyocytes (Seth et al. 2009; Kuwahara et al. 2006), to intercalated discs regions (Eder et al. 2007) and to T-tubules (Huang et al. 2009) or to a vast network of intracellular membranes (Goel et al. 2007). Immunocytochemistry studies also showed that TRPC3 does not co-localise with the RyR and SERCA suggesting that TRPC channels are excluded from the cleft region and thus from ECC (Goel et al. 2007).

Table 22.1 Expressional profiling of TRPC channels in the heart. (f) indicates expression in cardiac fibroblasts, n.d.: not determined; TRPC2 is not mentioned as it is not expressed in human

	C1	C3	C4	C5	C6	C7
mRNA						
• Human	+ (Riccio et al. 2002)	– (Bush et al. 2006)	+ (Riccio et al. 2002)	+ (Riccio et al. 2002)	+ (Kuwahara et al. 2006)	– (Riccio et al. 2002)
• Rat	+ (f) (Nishida et al. 2007) (Ohba et al. 2007)	+ (f) (Nishida et al. 2007) (Ohba et al. 2007)	+ (Garcia and Schilling 1997)	+ (Ohba et al. 2007)	+ (f) (Nishida et al. 2007) (Ohba et al. 2007)	+ (f) (Nishida et al. 2007) (Alvarez et al. 2008)
• Mouse	+ (Kuwahara et al. 2006)	+ (Kuwahara et al. 2006)	+ (Kuwahara et al. 2006)	n.d.	+ (Kuwahara et al. 2006)	n.d.
Protein						
• Human	n.d.	– (Bush et al. 2006)	n.d.	+ (Bush et al. 2006)	n.d.	n.d.
• Rat	+ (Ohba et al. 2007)	+ (f) (Nishida et al. 2007)	n.d.	n.d.	n.d.	+ (Alvarez et al. 2008)
• Mouse	+ (Shan et al. 2008)	+ (Bush et al. 2006)	+ (Wu et al. 2010)	n.d.	+ (f) (Nishida et al. 2007) (Kuwahara et al. 2006)	+ (f) (Nishida et al. 2007)
Immunocyto-Chemistry						
• Human	n.d.	n.d.	n.d.	n.d.	n.d.	n.d.
• Rat	+ (Huang et al. 2009)	+ (Eder et al. 2007; Goel et al. 2007)	n.d.	n.d.	n.d.	n.d.
• Mouse	+ (Seth et al. 2009)	n.d.	n.d.	n.d.	+ (Kuwaharaet al. 2006)	n.d.

(continued)

Table 22.1 (continued)

	C1	C3	C4	C5	C6	C7
Up-regulation/heart failure						
• Human	– (Bush et al. 2006)	– (Bush et al. 2006)	– (Bush et al. 2006)	+ (Bush et al. 2006)	+ (Kuwahara et al. 2006)	n.d.
• Rat	+ (Ohba et al. 2007)	+ (Bush et al. 2006)	n.d.	n.d.	+ (Kuwahara et al. 2006)	+ (Satoh et al. 2007)
• Mouse	n.d.	+ (Fig. 22.1)	n.d.	n.d.	+ (Kuwahara et al. 2006, Fig. 22.1)	n.d.

TRPC channels are not only expressed in cardiomyocytes but also in cardiac fibroblasts (Table 22.1) (Nishida et al. 2007). A recent study suggests a functional link between myofibroblast differentiation and TRPC6- calcineurin-NFAT signalling (Davis et al. 2012). Cardiac injury, for example after myocardial infarction, initiates transforming growth factor (TGFß) and angiotensin/endothelin signalling in fibroblasts resulting in the activation of the downstream effector p38 MAPK. This pathway mobilises the transcriptional activity of the serum response factor (SRF) which specifically up-regulates TRPC6. The TRPC6 dependent Ca^{2+} influx is finally linked to enhanced calcineurin NFAT signalling which induces myofibroblast trans-differentiation (Fig. 22.2).

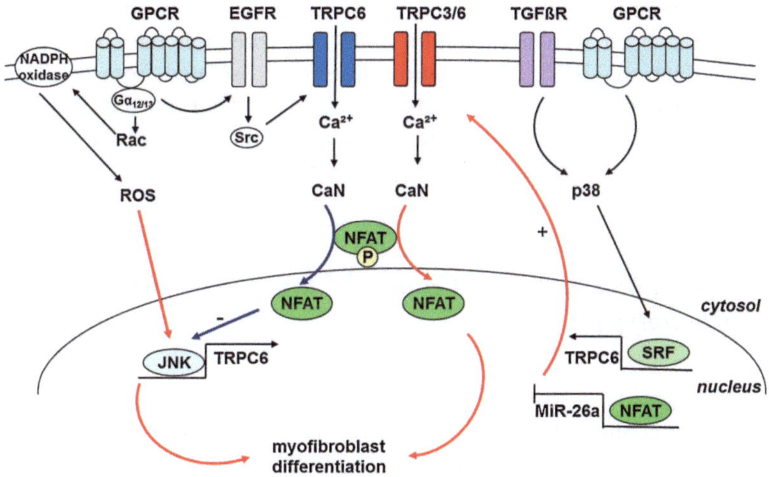

Fig. 22.2 Expressional regulation of TRPC channels during cardiac myofibroblast formation. Stimulation of GPCRs or TGFßRs results in an increased TRPC6 gene expression through p38 signalling and SRF dependent transcription. The enhanced TRPC6 mediated Ca^{2+} influx leads to CaN activation resulting in NFAT dephosphorylation and translocation. Nuclear NFAT drives myofibroblast differentiation. On the other hand, GPCRs/$G\alpha_{12/13}$ activates the NAPDH oxidase via Rac which results in ROS generation and subsequent JNK mediated myofibroblast formation and TRPC6 expression. In parallel, GPCRs/$G\alpha_{12/13}$ signalling might induce EGFR and Src kinase activation which enhances a TRPC6 mediated Ca^{2+} entry. This Ca^{2+} influx and the activation of CaN/NFAT negatively regulate JNK induced myofibroblast formation. In atrial fibroblasts TRPC3 activation is reduced by the micro RNA miR-26a. During fibroblast remodelling NFAT negatively regulates miR-26 transcription which gives rise to an increased TRPC3 expression and function. *Red arrows* indicate stimulatory and *blue arrows* inhibitory effects on myofibroblast formation, CaN, calcineurin; DAG, diacylglycerol; EGFR, epidermal growth factor receptor; ET-1, endothelin-1; GPCR, $G\alpha$ protein coupled receptor; $G\alpha_{12/13}$ heterotrimeric G protein 12/13; JNK, c-Jun NH_2-terminal kinase; NADPH, nicotinamide adenine dinucleotide phosphate-oxidase; NFAT, nuclear factor of activated T cells; p38, p38 MAPK (mitogen-activated protein kinase); Rac, member of the Rho family of small GTPases; ROS, reactive oxygen species; SRF, serum response factor; Src kinase, tyrosine kinase; TGFßR, transforming growth factor ß receptor; TRPC, transient receptor potential canonical

Cardiac myofibroblast formation might also be initiated by the production of reactive oxygen species (ROS). Nishida et al. (2007) have found evidence that TRPC6 interferes with GPCR$\alpha_{12/13}$ signalling which precedes fibroblast remodelling through the activation of the c-Jun NH_2-terminal kinase (JNK). In this case, JNK up-regulates TRPC6 which, when activated via epidermal growth factor receptor (EGFR) signalling and Src kinase phosphorylation, mediates a Ca^{2+} entry and calcineurin/NFAT activation. In this case, TRPC6 and calcineurin/NFAT act as negative feedback regulators of ET-1 induced myofibroblast formation (Fig. 22.2). Myofibroblast differentiation has also been associated with functional abnormalities which could implicate TRPC dependent mechanisms. In atrial tissue from patients with atrial fibrillation (AF) and AF goats, for example, TRPC3 has been found to be up-regulated (Harada et al. 2012). Here, NFAT transcription factors might indirectly control TRPC expression through the regulation of microRNAs. Accordingly, an enhanced transcriptional activity of NFATc3 in AF fibroblasts has been shown to inhibit miR-26 expression thereby relieving its inhibitory effect on TRPC3. In addition, fibroblasts isolated from dog AF tissue have revealed a Pyr3 (TRPC3 inhibitor) sensitive cation current which causes fibroblast proliferation, differentiation and activation through an extracellular signal-regulated kinase (ERK) dependent mechanism.

22.3.3 The Effect of TRPC Channel Function on the Cardiac Phenotype

The cardiac specific detection of TRPC channels was the trigger for a series of studies characterising their role in cardiac function and dysfunction. Genetic mouse models were generated selectively overexpressing different TRPC isoforms in cardiomyocytes. In addition, already existing TRPC deleted mice were examined according to their cardiac phenotype. Due to their low expression and therefore low activity TRPC channels do not affect basal cardiac function. Also cardiac remodelling during physiological stress as initiated in swimming experiments does not seem to be affected by a TRPC mediated Na/Ca influx (Nakayama et al. 2006; Wu et al. 2010). Up-regulation of different TRPC isoforms in different heart failure models (Table 22.1) was already an indication of TRPC channels as mediators of pathological stress stimuli. This increase was simulated in genetic mouse models overexpressing TRPC3 and TRPC6 (Nakayama et al. 2006; Kuwahara et al. 2006). Even at baseline, increased TRPC expression results in cardiac myopathy which is exaggerated after TAC treatment or neuro-humoral stimulation. Accordingly, TRPC deletion has the opposite effect. While wildtype (WT) mice hypertrophy normally with the characteristic functional and histological alterations, TRPC1 knock-out mice are almost completely preserved from cardiac damage after pressure over load induction (Seth et al. 2009). It is an intriguing finding given the existence of several other functionally important

TRPC isoforms in the heart (Table 22.1). It seems that TRPC1 function cannot be compensated by its TRPC family. It might be speculated that TRPC channels containing TRPC1 as dominant pore forming subunit or TRPC channels which are associated in a complex with TRPC1 channels are tethered to signalling pathways or membrane microdomains crucial for the hypertrophic response. For example, TRPC channels could be harboured in signalling platforms, such as lipid raft domains or caveolae (Sundivakkam et al. 2009). Correct insertion of the channels and their localisation could be mediated by a select interaction between caveolin3 and TRPC1. A loss of interaction, due to the down-regulation of TRPC1, for example, could result in a misregulation and decreased functionality of ion channels associated with TRPC1. It might also be speculated that the heteromerization profile of TRPC channels in cardiac myocytes is more promiscuous than expected. Indeed, TRPC1 not only associates with TRPC4 and 5 (Strubing et al. 2003) but also with members of other TRP families (Lintschinger et al. 2000; Schindl et al. 2012). The weak basal expression in the heart might, however, be a challenge to prove the actual preference of individual TRPC subunits for tetramerization. A recent study might have found some indications in this direction. In an approach to characterise the cardiac hypertrophic response on the background of inactive TRPC channels, genetic mouse models were generated overexpressing dominant negative (dn) TRPC mutants (Wu et al. 2010). Three dn-TRPC mutants belonging to the TRPC subfamilies 1/4/5 and 3/6/7 were selected for transgenesis (dn-TRPC6, dn-TRPC3 and dn-TRPC4) to achieve a broad spectrum of TRPC channel inhibition. Similar to the TRPC1 knock out, cardiac specific expression of any dn-TRPC mutant resulted in an attenuated cardiac hypertrophic response after TAC treatment which was associated with a reduced Ca^{2+} influx and less activation of the calcineurin/NFAT signalling pathway. It seems that the inhibition of each TRPC subfamily is equally potent in the suppression of pathological cardiac hypertrophy and that cross-inhibition among dn-TRPC isoforms occurs. Indeed, functional experiments in cardiac myocytes revealed a profound inhibition of a TRPC3 mediated Ca^{2+} influx by dn-TRPC4 and a complex formation between these TRPC isoforms in cardiac myocytes.

22.4 Modes of TRPC Channel Activation in Cardiac Hypertrophy

22.4.1 *Activation By Ca^{2+} Store Depletion*

Store-operated Ca^{2+} entry (SOCE) or capacitative Ca^{2+} entry is known as important step in the immune response (Lewis 2001). Typically, it comprises the release of IP_3 (downstream of the GPCR-PLC signalling pathway) its binding to the IP_3 receptor (IP3R) and the release of Ca^{2+} from intracellular Ca^{2+} stores. Ca^{2+} store depletion activates ion channels in the plasma membrane, the store-operated

Ca^{2+} channels (SOCs), mediating a sustained Ca^{2+} entry which is essential for driving the programme of gene expression underlying T cell activation. TRPC channels were handled as a priori components of SOCs until the small plasma membrane protein ORAI was identified as the Ca^{2+} release-activated Ca^{2+} channel pore forming subunit (CRAC) in immune cells (Frischauf et al. 2008).

That excitable cells such as cardiac myocytes could exhibit SOCE was rather un-realistic given the fact that Ca^{2+} cycling during ECC can be fully maintained by the classical Ca^{2+} handling proteins, the LTCC, RyR and SERCA. It was therefore all the more intriguing to learn about its existence in neonatal cardiomyocytes which indeed react with a sustained Ca^{2+} influx upon passive Ca^{2+} store depletion using the SERCA inhibitor thapsigargin (Hunton et al. 2002). This sustained Ca^{2+} increase is not inhibited by verapamil (LTCC) or KB-R7943 (reverse mode NCX inhibition) suggesting additional Ca^{2+} pools in the sarcolemma. The physiological purpose of a SOC in neonatal cardiomyocytes could be necessary for the regulation of transcriptional programmes regulating cardiac differentiation and growth. Once fully developed healthy adult cardiomyocytes lose their ability to generate a SOCE (Uehara et al. 2002). It might, however, be retained in the course of cardiac hypertrophy as consequence of the re-activated foetal gene programme (Fig. 22.1). Indeed, by comparing Ca^{2+} characteristics in hypertrophic myocytes from WT and TG mice with an overexpression of dn- TRPC mutants. Wu et al. (2010) demonstrated a reduction of SOCE when TRPC channels were blocked. Interestingly, inhibition of any potential TRPC heteromer showed the same tendency of a reduced SOCE, attenuated calcineurin signalling and reduced hypertrophic response after TAC treatment (see also 21.3.3).

Despite this finding it is, however, uncertain to what extent TRPC channels contribute to SOCE in cardiomyocytes, given their non-selectivity for Na^{+} and Ca^{2+} and the recent discovery of the two key players in SOCE, the stromal interaction molecule 1 (STIM1) and ORAI1. Endoplasmic reticulum (ER)—Ca^{2+} release by IP3R causes STIM1 to re-localise to ER sites adjacent to the plasma membrane. Subsequently, ORAI1 is activated and allows a Ca^{2+} entry into the cytosol (Fig. 22.3a) (Frischauf et al. 2008). Both ORAI1 and STIM1 are expressed in the heart where they might fulfil a similar function. For example, down regulation of ORAI in neonatal cardiomyocytes results in reduced SOCE elicited by passive Ca^{2+} store depletion, less calcineurin activation and reduced agonist induced hypertrophy (Voelkers et al. 2010). The absence of STIM1, on the other hand, is associated with less SR Ca^{2+} load indicating its role as a critical Ca^{2+} sensor in the SR in pathological conditions. Indeed, down-regulation and inhibition of STIM1 result in less pressure-overload-induced cardiac hypertrophy which is functionally associated with a reduced SOCE (Hulot et al. 2011; Luo et al. 2012).

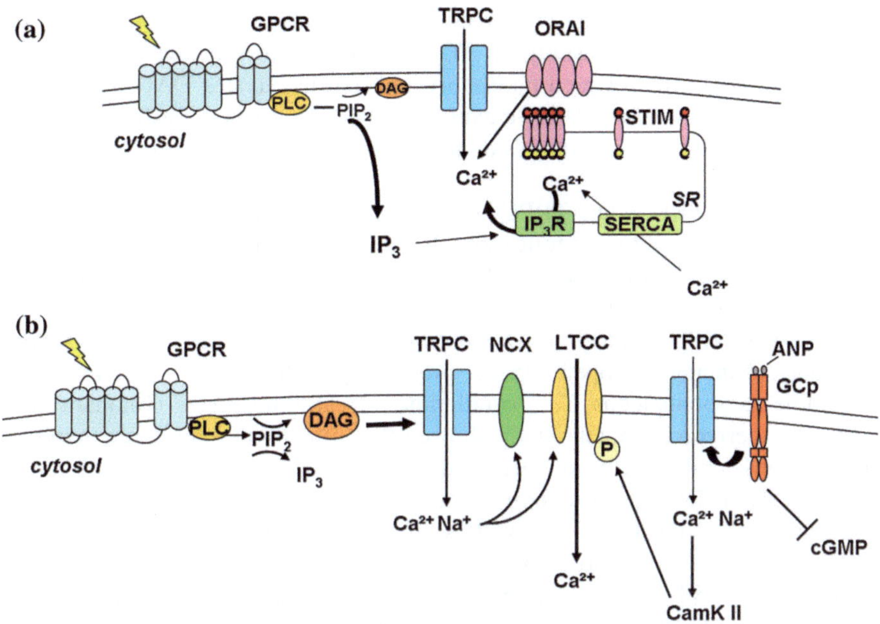

Fig. 22.3 **a** Schematic of store-operated TRPC channel activation. GPCR stimulation activates PLC generating DAG and IP_3 which leads to an IP_3-induced Ca^{2+} store depletion activating SOCs in the plasma membrane. Ca^{2+} store depletion is sensed by STIM1 activating ORAI to mediate a Ca^{2+} entry. STIM1 might also activate TRPC channels. **b** Schematic of receptor-operated TRPC channel activation. The GPCR- PLC dependent formation of DAG activates TRPC channels in the plasma membrane. The TRPC mediated non-selective cation entry causes a local membrane depolarisation affecting LTCCs and the NCX. Desensitisation of GCps inhibits the formation of cGMP resulting in an increased TRPC channel mediated cation entry which further activates LTCCs probably through a CamKII dependent mechanism, DAG, diacylglycerol; GPCR, $G\alpha$ **a** or $G\alpha q$ **b** protein coupled receptor; IP_3, inositol 3-phosphate; IP_3R, inositol 3-phosphate receptor; ORAI, Ca^{2+} release-activated channel protein; PIP_2, Phosphatidylinositol 4,5-bisphosphate; PLC, phospholipase C; SR, sarcoplasmic reticulum; STIM, stromal interaction molecule; TRPC, transient receptor potential canonical; ANP, atrial natriuretic peptide; CamK, Ca^{2+}/calmodulin-dependent protein kinase; cGMP, cyclic guanosine monophosphate; GCp, guanylyl cyclase particulate; LTCC, L-type Ca^{2+} channel; NCX, Na^{+}/Ca^{2+} exchanger

22.4.2 Receptor Activation

TRPC channels have often been described as receptor-operated channels (ROCs) whose activation is independent of Ca^{2+} store depletion (Hofmann et al. 1999). Although the identity of the molecule mediating ROC activation has not been entirely resolved, diacylglycerol (DAG) has been suggested to increase TRPC channel activity. Also, in cardiac myocytes TRPC channels can be directly activated by the DAG analogue 1-oleoyl-2-acetyl-sn-glycerol (OAG) (Onohara et al. 2006), which is sufficient to induce the calcineurin-NFAT signalling pathway (Bush et al. 2006). Even more it seems that DAG activation positively regulates the calcineurin/NFAT/TRPC circuit as depicted in Fig. 22.1a. Support for this assumption comes from a study on TG mice overexpressing the diacylglycerol kinase epsilon (DGK-ε) which phosphorylates and inactivates DAG in the heart. Interestingly, an increased DGK-ε expression restores cardiac dysfunction under pressure overload by controlling DAG levels and TRPC6 expression (Niizeki et al. 2008).

In particular, TRPC3 and TRPC6 channel activity have been repeatedly shown to be increased by OAG application. It was further assumed that the TRPC mediated Na^+/Ca^{2+} influx does not directly affect the calcineurin/NFAT signalling pathway but indirectly through a crosstalk with the LTCC (Fig. 22.3b). It is possible that TRPC channel activation results in low increases in the membrane potential enhancing the open channel probability of the LTCC, which becomes noticeable as spontaneous Ca^{2+} oscillations in cardiomyocytes (Onohara et al. 2006). Furthermore, a recent study showed that nifedipine inhibits the hypertrophic response induced by an increased TRPC3 expression in neonatal cardiac myocytes. Here TRPC3 expression was associated with a greater LTCC abundance and amplitude of the LTCC, as well as alterations of the Ca^{2+} transients with a prolonged Ca^{2+} decay (Gao et al. 2012). A similar crosstalk could exist with the electrogenic NCX1. It was for example shown that a PLC dependent activation of TRPC3 favours the NCX reverse mode activity in rat cardiac myocytes (Eder et al. 2007). These findings suggest a close proximity between TRPC channels, the LTCC and/or NCX, probably as signalplexes in Ca^{2+} microdomains. The precise protein composition and cellular localisation, in particular in the progression of pathological cardiac hypertrophy, however, needs to be further elucidated.

Stretch dependent activation is another aspect to be considered when analysing TRPC channel function in cardiac hypertrophy. TRPC1 function is increased when cardiac myocytes are subjected to mechanical stretch. It has been observed that TRPC channels can sense membrane lipid formation either directly (Spassova et al. 2006) or indirectly through stretch sensitive GPCRs. In the heart the Ang II receptor is functionally coupled to TRPC1 conferring membrane lipid deformation to an increased TRPC1 mediated cation entry (Seth et al. 2009). To what extent other TRPC isoforms contribute to stretch dependent signalling in the heart needs to be determined.

22.5 TRPC Channel Inhibition and Treatment Strategies

It is well accepted that the secretion of natriuretic peptides initiates a counteracting signalling cascade in pathological cardiac hypertrophy (Tamura et al. 2000; Klaiber et al. 2010). Once released the atrial natriuretic peptide ANP and the brain natriuretic peptide BNP bind to their common receptor, the guanylyl cyclase A (GC-A) which synthesises the second messenger cGMP further activating the protein kinase G 1 (PKG1). Among several pro-hypertrophic pathways also the calcineurin/NFAT signalling pathway has been described to be inhibited by

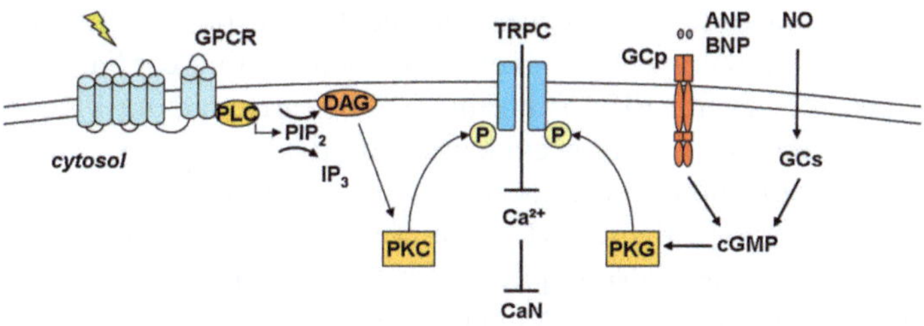

Fig. 22.4 TRPC channel inhibition in the heart. TRPC channels are inhibited by PKG phosphorylation, which is activated by cGMP preceded either by GCp stimulation with ANP/BNP or stimulation of GCs by NO. TRPC channel inhibition by PKC results in an un-coupling of TRPC mediated Ca signals and the CaN signalling pathway, ANP, atrial natriuretic peptide; BNP, brain natriuretic peptide; CaN, Calcineurin; cGMP, cyclic guanosine monophosphate; DAG, diacylglycerol; GCp, guanylyl cyclase particulate; GCs guanylyl cyclase soluble; GPCR, G protein coupled receptor; IP_3, inositol 3-phosphate; NO, Nitric oxide; PIP_2, Phosphatidylinositol 4,5-bisphosphate; PKC, protein kinase C; PKG, protein kinase G; PLC, phospholipase C; TRPC, transient receptor potential canonical

ANP/BNP signalling. For example, PKG1 inhibits LTCCs resulting in an attenuated GPCR mediated Ca^{2+} influx and calcineurin/NFAT activation (Fiedler et al. 2002). More recently, it was shown that the anti-hypertrophic effect of PKG1 also involves TRPC channel inhibition. According to heterologous cell culture experiments TRPC3 and TRPC6 can be phosphorylated by PKG I (phosphorylation sites at Thr11 and Ser263 in TRPC3 and Thr69 in TRPC6), which greatly reduces channel activity (Kwan et al. 2004; Takahashi et al. 2008) (Fig. 22.4). This negative feedback regulation has also turned out important in vivo, as shown by a study on TRPC6 TG mice crossed with GC-A knock-out mice (Kinoshita et al. 2010). Although TRPC6 TG mice are prone to cardiac hypertrophy even at baseline, disease is accelerated due to the loss of GC-A. Mechanistically, cGMP/PKG mediated inhibition results in an interruption of the TRPC6/calcineurin/NFAT feed–forward loop. Therapeutically, a blocking effect might be achieved by the application of phosphodiesterase inhibitors such as sildenafil or the addition of nitric oxide (NO) donors, as TRPC channels (TRPC1, 3 and 6) are sensitive to both treatment strategies (Chen et al. 2009; Koitabashi et al. 2010) (Fig. 22.4).

Intact cGMP/PKG signalling not only counteracts an overshoot of TRPC mediated Ca^{2+} entry via PKG phosphorylation but is also important for the conformation and function of TRPC channels per se. According to a recent study (Klaiber et al. 2011), reason for impaired cGMP/PKG signalling could be a desensitisation and therefore impairment of GC-A when chronically exposed to ANP/BNP. In such a scenario, pre-existing complexes of GC-A and TRPC3/6 undergo a steric alteration which results in an increased TRPC channel activity which further activates LTCC currents and finally increases cardiomyocyte Ca^{2+} levels. Here, the TRPC- LTCC crosstalk could be mediated by CamKII which upon activation by TRPC mediated Ca signals could phosphorylate and increase LTCC activity (Fig. 22.3b).

TRPC channel activity can also modulated by the protein kinase C (PKC), which could become therapeutically relevant when treating cardiac diseases. Due to the existence of several PKC isoforms in the heart targeting PKC has been a critical goal. Nevertheless, there is evidence that broad-acting PKC inhibitors such as ruboxistaurin can improve cardiac function and ameliorate disease symptoms during chronic pathological stress (Ladage et al. 2011). It is still too early to tell that this therapeutic benefit also implicates an effect on TRPC channel function. A recent study (Poteser et al. 2011) demonstrates that a conventional PKC inhibitor (GF109203X) prevents complex formation between CnA and TRPC3 which is associated with reduced phosphorylation of the channel protein at Thr 573. As a result, TRPC3 dependent Ca^{2+} signals are un-coupled form CaN/NFAT signalling (Fig. 22.4) but are rather translated into an increased LTCC activity. Interestingly, in this situation LTCC activity does only have negligible effects on NFAT activation.

While these studies have advanced our understanding of cardiac TRPC signalling enormously, therapeutic strategies that interfere with multiple signalling cascades need to be carefully balanced. The ultimate achievement would definitely be the development of selective TRPC channels blockers. To date the pyrazole compound, Pyr3, seems to be most promising. It is directed against TRPC3 and when administered in vivo inhibits pressure overload induced cardiac remodelling and cardiac hypertrophy (Kiyonaka et al. 2009).

References

Alvarez J, Coulombe A, Cazorla O, Ugur M, Rauzier JM, Magyar J, Mathieu EL, Boulay G, Souto R, Bideaux P, Salazar G, Rassendren F, Lacampagne A, Fauconnier J, Vassort G (2008) ATP/UTP activate cation-permeable channels with TRPC3/7 properties in rat cardiomyocytes. Am J Physiol Heart Circ Physiol 295(1):H21–H28

Anderson ME, Brown JH, Bers DM (2011) CaMKII in myocardial hypertrophy and heart failure. J Mol Cell Cardiol 51(4):468–473

Andersson KB, Birkeland JA, Finsen AV, Louch WE, Sjaastad I, Wang Y, Chen J, Molkentin JD, Chien KR, Sejersted OM, Christensen G (2009) Moderate heart dysfunction in mice with inducible cardiomyocyte-specific excision of the Serca2 gene. J Mol Cell Cardiol 47(2):180–187

Bers DM (2008) Calcium cycling and signaling in cardiac myocytes. Annu Rev Physiol 70:23–49

Bush EW, Hood DB, Papst PJ, Chapo JA, Minobe W, Bristow MR, Olson EN, McKinsey TA (2006) Canonical transient receptor potential channels promote cardiomyocyte hypertrophy through activation of calcineurin signaling. J Biol Chem 281(44):33487–33496

Chen J, Crossland RF, Noorani MM, Marrelli SP (2009) Inhibition of TRPC1/TRPC3 by PKG contributes to NO-mediated vasorelaxation. Am J Physiol Heart Circ Physiol 297(1):H417–H424

Chen X, Nakayama H, Zhang X, Ai X, Harris DM, Tang M, Zhang H, Szeto C, Stockbower K, Berretta RM, Eckhart AD, Koch WJ, Molkentin JD, Houser SR (2011) Calcium influx through $Ca_v1.2$ is a proximal signal for pathological cardiomyocyte hypertrophy. J Mol Cell Cardiol 50(3):460–470

Chiang CS, Huang CH, Chieng H, Chang YT, Chang D, Chen JJ, Chen YC, Chen YH, Shin HS, Campbell KP, Chen CC (2009) The Ca(v)3.2 T-type Ca(2 +) channel is required for pressure overload-induced cardiac hypertrophy in mice. Circ Res 104(4):522–530

Colella M, Grisan F, Robert V, Turner JD, Thomas AP, Pozzan T (2008) Ca^{2+} oscillation frequency decoding in cardiac cell hypertrophy: role of calcineurin/NFAT as Ca^{2+} signal integrators. Proc Natl Acad Sci USA 105(8):2859–2864

Davis J, Burr AR, Davis GF, Birnbaumer L, Molkentin JD (2012) A TRPC6-dependent pathway for myofibroblast transdifferentiation and wound healing in vivo. Dev Cell 23(4):705–715

Eder P, Molkentin JD (2011) TRPC channels as effectors of cardiac hypertrophy. Circ Res 108(2):265–272

Eder P, Probst D, Rosker C, Poteser M, Wolinski H, Kohlwein SD, Romanin C, Groschner K (2007) Phospholipase C-dependent control of cardiac calcium homeostasis involves a TRPC3-NCX1 signaling complex. Cardiovasc Res 73(1):111–119

Fiedler B, Lohmann SM, Smolenski A, Linnemuller S, Pieske B, Schroder F, Molkentin JD, Drexler H, Wollert KC (2002) Inhibition of calcineurin-NFAT hypertrophy signaling by cGMP-dependent protein kinase type I in cardiac myocytes. Proc Natl Acad Sci USA 99(17):11363–11368

Frischauf I, Schindl R, Derler I, Bergsmann J, Fahrner M, Romanin C (2008) The STIM/Orai coupling machinery. Channels (Austin) 2(4):261–268

Gao H, Wang F, Wang W, Makarewich CA, Zhang H, Kubo H, Berretta RM, Barr LA, Molkentin JD, Houser SR (2012) Ca(2 +) influx through L-type Ca(2 +) channels and transient receptor potential channels activates pathological hypertrophy signaling. J Mol Cell Cardiol 53(5):657–667

Garcia RL, Schilling WP (1997) Differential expression of mammalian TRP homologues across tissues and cell lines. Biochem Biophys Res Commun 239(1):279–283

Goel M, Zuo CD, Sinkins WG, Schilling WP (2007) TRPC3 channels colocalize with Na^{+}/Ca^{2+} exchanger and Na^{+} pump in axial component of transverse-axial tubular system of rat ventricle. Am J Physiol Heart Circ Physiol 292(2):H874–H883

Goonasekera SA, Hammer K, Auger-Messier M, Bodi I, Chen X, Zhang H, Reiken S, Elrod JW, Correll RN, York AJ, Sargent MA, Hofmann F, Moosmang S, Marks AR, Houser SR, Bers DM, Molkentin JD (2012) Decreased cardiac L-type Ca(2)(+) channel activity induces hypertrophy and heart failure in mice. J Clin Invest 122(1):280–290

Harada M, Luo X, Qi XY, Tadevosyan A, Maguy A, Ordog B, Ledoux J, Kato T, Naud P, Voigt N, Shi Y, Kamiya K, Murohara T, Kodama I, Tardif JC, Schotten U, Van Wagoner DR, Dobrev D, Nattel S (2012) Transient receptor potential canonical-3 channel-dependent fibroblast regulation in atrial fibrillation. Circulation 126(17):2051–2064

Hofmann T, Obukhov AG, Schaefer M, Harteneck C, Gudermann T, Schultz G (1999) Direct activation of human TRPC6 and TRPC3 channels by diacylglycerol. Nature 397(6716):259–263

Hofmann T, Schaefer M, Schultz G, Gudermann T (2002) Subunit composition of mammalian transient receptor potential channels in living cells. Proc Natl Acad Sci USA 99(11):7461–7466

Hogan PG, Chen L, Nardone J, Rao A (2003) Transcriptional regulation by calcium, calcineurin, and NFAT. Genes Dev 17(18):2205–2232

Huang B, Qin D, Deng L, Boutjdir M, El-Sherif N (2000) Reexpression of T-type Ca^{2+} channel gene and current in post-infarction remodeled rat left ventricle. Cardiovasc Res 46(3):442–449

Huang H, Wang W, Liu P, Jiang Y, Zhao Y, Wei H, Niu W (2009) TRPC1 expression and distribution in rat hearts. Eur J Histochem 53(4):217–223

Hulot JS, Fauconnier J, Ramanujam D, Chaanine A, Aubart F, Sassi Y, Merkle S, Cazorla O, Ouille A, Dupuis M, Hadri L, Jeong D, Muhlstedt S, Schmitt J, Braun A, Benard L, Saliba Y, Laggerbauer B, Nieswandt B, Lacampagne A, Hajjar RJ, Lompre AM, Engelhardt S (2011) Critical role for stromal interaction molecule 1 in cardiac hypertrophy. Circulation 124(7):796–805

Hunton DL, Lucchesi PA, Pang Y, Cheng X, Dell'Italia LJ, Marchase RB (2002) Capacitative calcium entry contributes to nuclear factor of activated T-cells nuclear translocation and hypertrophy in cardiomyocytes. J Biol Chem 277(16):14266–14273

Kalyanasundaram A, Lacombe VA, Belevych AE, Brunello L, Carnes CA, Janssen PM, Knollmann BC, Periasamy M, Gyorke S (2012) Up-regulation of sarcoplasmic reticulum Ca^{2+} uptake leads to cardiac hypertrophy, contractile dysfunction and early mortality in mice deficient in CASQ2. Cardiovasc Res. doi:10.1093/cvr/cvs334

Kehat I, Molkentin JD (2010) Molecular pathways underlying cardiac remodeling during pathophysiological stimulation. Circulation 122(25):2727–2735

Kinoshita H, Kuwahara K, Nishida M, Jian Z, Rong X, Kiyonaka S, Kuwabara Y, Kurose H, Inoue R, Mori Y, Li Y, Nakagawa Y, Usami S, Fujiwara M, Yamada Y, Minami T, Ueshima K, Nakao K (2010) Inhibition of TRPC6 channel activity contributes to the antihypertrophic effects of natriuretic peptides-guanylyl cyclase-A signaling in the heart. Circ Res 106(12):1849–1860

Kiyonaka S, Kato K, Nishida M, Mio K, Numaga T, Sawaguchi Y, Yoshida T, Wakamori M, Mori E, Numata T, Ishii M, Takemoto H, Ojida A, Watanabe K, Uemura A, Kurose H, Morii T, Kobayashi T, Sato Y, Sato C, Hamachi I, Mori Y (2009) Selective and direct inhibition of TRPC3 channels underlies biological activities of a pyrazole compound. Proc Natl Acad Sci USA 106(13):5400–5405

Klaiber M, Dankworth B, Kruse M, Hartmann M, Nikolaev VO, Yang RB, Volker K, Gassner B, Oberwinkler H, Feil R, Freichel M, Groschner K, Skryabin BV, Frantz S, Birnbaumer L, Pongs O, Kuhn M (2011) A cardiac pathway of cyclic GMP-independent signaling of guanylyl cyclase A, the receptor for atrial natriuretic peptide. Proc Natl Acad Sci USA 108(45):18500–18505

Klaiber M, Kruse M, Volker K, Schroter J, Feil R, Freichel M, Gerling A, Feil S, Dietrich A, Londono JE, Baba HA, Abramowitz J, Birnbaumer L, Penninger JM, Pongs O, Kuhn M (2010) Novel insights into the mechanisms mediating the local antihypertrophic effects of cardiac atrial natriuretic peptide: role of cGMP-dependent protein kinase and RGS2. Basic Res Cardiol 105(5):583–595

Klee CB, Ren H, Wang X (1998) Regulation of the calmodulin-stimulated protein phosphatase, calcineurin. J biol chem 273(22):13367–13370

Koitabashi N, Aiba T, Hesketh GG, Rowell J, Zhang M, Takimoto E, Tomaselli GF, Kass DA (2010) Cyclic GMP/PKG-dependent inhibition of TRPC6 channel activity and expression negatively regulates cardiomyocyte NFAT activation Novel mechanism of cardiac stress modulation by PDE5 inhibition. J Mol Cell Cardiol 48(4):713–724

Kuwahara K, Wang Y, McAnally J, Richardson JA, Bassel-Duby R, Hill JA, Olson EN (2006) TRPC6 fulfills a calcineurin signaling circuit during pathologic cardiac remodeling. J Clin Invest 116(12):3114–3126

Kwan HY, Huang Y, Yao X (2004) Regulation of canonical transient receptor potential isoform 3 (TRPC3) channel by protein kinase G. Proc Natl Acad Sci USA 101(8):2625–2630

Ladage D, Tilemann L, Ishikawa K, Correll RN, Kawase Y, Houser SR, Molkentin JD, Hajjar RJ (2011) Inhibition of PKCalpha/beta with ruboxistaurin antagonizes heart failure in pigs after myocardial infarction injury. Circ Res 109(12):1396–1400

Lewis RS (2001) Calcium signaling mechanisms in T lymphocytes. Annu Rev Immunol 19:497–521

Lintschinger B, Balzer-Geldsetzer M, Baskaran T, Graier WF, Romanin C, Zhu MX, Groschner K (2000) Coassembly of Trp1 and Trp3 proteins generates diacylglycerol- and Ca^{2+}-sensitive cation channels. J Biol Chem 275(36):27799–27805

Luo X, Hojayev B, Jiang N, Wang ZV, Tandan S, Rakalin A, Rothermel BA, Gillette TG, Hill JA (2012) STIM1-dependent store-operated Ca(2)(+) entry is required for pathological cardiac hypertrophy. J Mol Cell Cardiol 52(1):136–147

Makarewich CA, Correll RN, Gao H, Zhang H, Yang B, Berretta RM, Rizzo V, Molkentin JD, Houser SR (2012) A caveolae-targeted L-type Ca(2) + channel antagonist inhibits hypertrophic signaling without reducing cardiac contractility. Circ Res 110(5):669–674

Molkentin JD, Lu JR, Antos CL, Markham B, Richardson J, Robbins J, Grant SR, Olson EN (1998) A calcineurin-dependent transcriptional pathway for cardiac hypertrophy. Cell 93(2):215–228

Nakayama H, Bodi I, Correll RN, Chen X, Lorenz J, Houser SR, Robbins J, Schwartz A, Molkentin JD (2009) Alpha1G-dependent T-type Ca^{2+}current antagonizes cardiac hypertrophy through a NOS3-dependent mechanism in mice. J Clin Invest 119(12):3787–3796

Nakayama H, Wilkin BJ, Bodi I, Molkentin JD (2006) Calcineurin-dependent cardiomyopathy is activated by TRPC in the adult mouse heart. Faseb J 20(10):1660–1670

Niizeki T, Takeishi Y, Kitahara T, Arimoto T, Ishino M, Bilim O, Suzuki S, Sasaki T, Nakajima O, Walsh RA, Goto K, Kubota I (2008) Diacylglycerol kinase-epsilon restores cardiac dysfunction under chronic pressure overload: a new specific regulator of Galpha(q) signaling cascade. Am J Physiol Heart Circ Physiol 295(1):H245–H255

Nishida M, Onohara N, Sato Y, Suda R, Ogushi M, Tanabe S, Inoue R, Mori Y, Kurose H (2007) Galpha12/13-mediated up-regulation of TRPC6 negatively regulates endothelin-1-induced cardiac myofibroblast formation and collagen synthesis through nuclear factor of activated T cells activation. J Biol Chem 282(32):23117–23128

Nuss HB, Houser SR (1993) T-type Ca^{2+}current is expressed in hypertrophied adult feline left ventricular myocytes. Circ Res 73(4):777–782

Ohba T, Watanabe H, Murakami M, Takahashi Y, Iino K, Kuromitsu S, Mori Y, Ono K, Iijima T, Ito H (2007) Upregulation of TRPC1 in the development of cardiac hypertrophy. J Mol Cell Cardiol 42(3):498–507

Onohara N, Nishida M, Inoue R, Kobayashi H, Sumimoto H, Sato Y, Mori Y, Nagao T, Kurose H (2006) TRPC3 and TRPC6 are essential for angiotensin II-induced cardiac hypertrophy. EMBO J 25(22):5305–5316

Passier R, Zeng H, Frey N, Naya FJ, Nicol RL, McKinsey TA, Overbeek P, Richardson JA, Grant SR, Olson EN (2000) CaM kinase signaling induces cardiac hypertrophy and activates the MEF2 transcription factor in vivo. J Clin Investig 105(10):1395–1406

Poteser M, Graziani A, Rosker C, Eder P, Derler I, Kahr H, Zhu MX, Romanin C, Groschner K (2006) TRPC3 and TRPC4 associate to form a redox-sensitive cation channel. Evidence for expression of native TRPC3-TRPC4 heteromeric channels in endothelial cells. J Biol Chem 281(19):13588–13595

Poteser M, Schleifer H, Lichtenegger M, Schernthaner M, Stockner T, Kappe CO, Glasnov TN, Romanin C, Groschner K (2011) PKC-dependent coupling of calcium permeation through transient receptor potential canonical 3 (TRPC3) to calcineurin signaling in HL-1 myocytes. Proc Natl Acad Sci USA 108(26):10556–10561

Rajabi M, Kassiotis C, Razeghi P, Taegtmeyer H (2007) Return to the fetal gene program protects the stressed heart: a strong hypothesis. Heart Fail Rev 12(3–4):331–343

Riccio A, Medhurst AD, Mattei C, Kelsell RE, Calver AR, Randall AD, Benham CD, Pangalos MN (2002) mRNA distribution analysis of human TRPC family in CNS and peripheral tissues. Brain Res Mol Brain Res 109(1–2):95–104

Roedding AS, Gao AF, Au-Yeung W, Scarcelli T, Li PP, Warsh JJ (2012) Effect of oxidative stress on TRPM2 and TRPC3 channels in B lymphoblast cells in bipolar disorder. Bipolar Disord 14(2):151–161

Satoh S, Tanaka H, Ueda Y, Oyama J, Sugano M, Sumimoto H, Mori Y, Makino N (2007) Transient receptor potential (TRP) protein 7 acts as a G protein-activated Ca^{2+} channel mediating angiotensin II-induced myocardial apoptosis. Mol Cell Biochem 294(1–2):205–215

Schindl R, Fritsch R, Jardin I, Frischauf I, Kahr H, Muik M, Riedl MC, Groschner K, Romanin C (2012) Canonical transient receptor potential (TRPC) 1 acts as a negative regulator for vanilloid TRPV6-mediated Ca^{2+} influx. J Biol Chem 287(42):35612–35620

Seth M, Zhang ZS, Mao L, Graham V, Burch J, Stiber J, Tsiokas L, Winn M, Abramowitz J, Rockman HA, Birnbaumer L, Rosenberg P (2009) TRPC1 Channels are Critical for Hypertrophic Signaling in the Heart. Circ Res 105:1023–1030

Shan D, Marchase RB, Chatham JC (2008) Overexpression of TRPC3 increases apoptosis but not necrosis in response to ischemia-reperfusion in adult mouse cardiomyocytes. Am J Physiol Cell Physiol 294(3):C833–C841

Spassova MA, Hewavitharana T, Xu W, Soboloff J, Gill DL (2006) A common mechanism underlies stretch activation and receptor activation of TRPC6 channels. Proc Natl Acad Sci USA 103(44):16586–16591

Strubing C, Krapivinsky G, Krapivinsky L, Clapham DE (2003) Formation of novel TRPC channels by complex subunit interactions in embryonic brain. J Biol Chem 278(40):39014–39019

Sundivakkam PC, Kwiatek AM, Sharma TT, Minshall RD, Malik AB, Tiruppathi C (2009) Caveolin-1 scaffold domain interacts with TRPC1 and IP3R3 to regulate Ca^{2+} store release-induced Ca^{2+} entry in endothelial cells. Am J Physiol Cell Physiol 296(3):C403–C413

Takahashi S, Lin H, Geshi N, Mori Y, Kawarabayashi Y, Takami N, Mori MX, Honda A, Inoue R (2008) Nitric oxide-cGMP-protein kinase G pathway negatively regulates vascular transient receptor potential channel TRPC6. J Physiol 586(Pt 17):4209–4223

Tamura N, Ogawa Y, Chusho H, Nakamura K, Nakao K, Suda M, Kasahara M, Hashimoto R, Katsuura G, Mukoyama M, Itoh H, Saito Y, Tanaka I, Otani H, Katsuki M (2000) Cardiac fibrosis in mice lacking brain natriuretic peptide. Proc Natl Acad Sci USA 97(8):4239–4244

Trebak M, Lemonnier L, Smyth JT, Vazquez G, Putney JW Jr (2007) Phospholipase C-coupled receptors and activation of TRPC channels. Handb Exp Pharmacol 179:593–614

Uehara A, Yasukochi M, Imanaga I, Nishi M, Takeshima H (2002) Store-operated Ca^{2+} entry uncoupled with ryanodine receptor and junctional membrane complex in heart muscle cells. Cell Calcium 31(2):89–96

van Oort RJ, Respress JL, Li N, Reynolds C, De Almeida AC, Skapura DG, De Windt LJ, Wehrens XH (2010) Accelerated development of pressure overload-induced cardiac hypertrophy and dysfunction in an RyR2-R176Q knockin mouse model. Hypertension 55(4):932–938

Voelkers M, Salz M, Herzog N, Frank D, Dolatabadi N, Frey N, Gude N, Friedrich O, Koch WJ, Katus HA, Sussman MA, Most P (2010) Orai1 and Stim1 regulate normal and hypertrophic growth in cardiomyocytes. J Mol Cell Cardiol 48(6):1329–1334

Wilkins BJ, Molkentin JD (2002) Calcineurin and cardiac hypertrophy: where have we been? Where are we going? J Physiol 541(Pt 1):1–8

Wu X, Eder P, Chang B, Molkentin JD (2010) TRPC channels are necessary mediators of pathologic cardiac hypertrophy. Proc Natl Acad Sci USA 107(15):7000–7005

Chapter 23
Role of TRPC and Orai Channels in Vascular Remodeling

Michael Poteser, Sarah Krenn and Klaus Groschner

Abstract Calcium permeable channels that are barely sensitive to membrane voltage but controlled by a complex array of external stimuli to generate coordinated cellular Ca^{2+} signals have been identified and extensively characterized in the past decades. Vascular cells express Ca^{2+} conductances, which lack primary voltage sensitivity and are formed by proteins of the TRP (transient receptor potential) and Orai families. Channel complexes composed of TRPC (canonical TRP) or Orai proteins operate in concert with other Ca^{2+} transporters to govern both acute functions, such as contraction and migration and long-term fate of vascular cells due to a pivotal role in Ca^{2+} transcription coupling. Both TRPC and Orai channels have been recently suggested to control phenotype of vascular cells and therefore to represent attractive novel targets for pharmacological prevention of vascular remodeling and aging. With this chapter, we aim to provide an overview on current knowledge regarding the role of TRPC and Orai channels in vascular endothelium and smooth muscle. The significance of these channels for maladaptive remodeling processes in the vascular system and potential therapeutic strategies based on modulation or block of these Ca^{2+} channels are discussed.

23.1 Introduction

The vascular system is a highly dynamic structure that undergoes constant remodeling. Physiological as well as maladaptive restructuring processes are associated with altered Ca^{2+} homeostasis in both vascular endothelium and smooth muscle. To enable coordinated control of acute vascular functions and long-term adaptation, Ca^{2+} signals require efficient temporal and spatial segregation.

M. Poteser (✉) · S. Krenn · K. Groschner
Institute of Biophysics, Medical University of Graz, Harrachgasse 21/4,
A-8010 Graz, Austria
e-mail: michael.poteser@medunigraz.at

N. Weiss and A. Koschak (eds.), *Pathologies of Calcium Channels*,
DOI: 10.1007/978-3-642-40282-1_23,

Specifically, Ca^{2+} transcription coupling involves Ca^{2+} entry channels that are organized in dynamic signalplexes targeted into specialized membrane domains. Membrane proteins of the TRPC and Orai family have been recognized as pivotal elements of Ca^{2+} transcription coupling (Beech 2013; Trebak 2012; Dietrich et al. 2010) and demonstrated to control phenotype transitions, proliferation, and differentiation in smooth muscle and endothelium. Exploiting TRPC and Orai channel proteins as therapeutic targets requires in-depth understanding of their signaling function. Here we provide an overview on current evidence demonstrating a role of TRPC and Orai channels in vascular remodeling.

23.2 Receptor and Store-Operated Ca^{2+} Entry in the Vasculature

Despite a long history of investigations on TRPC channel function, key issues, such as the cellular events leading to channel opening, are still ill defined and remain a matter of ongoing dispute. Widely accepted and well documented is the view of TRPC3, TRPC6, and TRPC7 as ion channels that are, in a physiological context, activated by signaling mechanisms initiated in response to stimulation of G protein coupled receptors. These pathways typically involve phospholipase C (PLC) and generation of diacylglycerol (DAG) along with inositol 1, 4, 5-tris-phosphate (IP3). Calcium entry initiated by this process is referred to as receptor-activated (operated) calcium entry (ROCE). There is good evidence, that TRPC3, TRPC6, and TRPC7 are directly activated by diacylglycerols (Okada et al. 1999; Trebak et al. 2003; Gudermann et al. 2004; Hofmann et al. 1999) thereby fitting into this category. In contrast, TRPC1, 4, and 5 were found insensitive to lipid mediators in heterologous expression experiments (Hofmann et al. 1999) and were initially suggested as store-operated channel subunits (Beech et al. 2003; Wang et al. 2004; Worley et al. 2007; Yuan et al. 2007; Philipp et al. 1996, 1998). These findings are contradicted by a number of studies suggesting receptor- and second messenger-dependent regulation of TRPC1/4/5, which may assemble into heteromers (Plant and Schaefer 2003; Schaefer et al. 2000; Okada et al. 1998; Strubing et al. 2001). In line with this unclear situation, evidence for a role of vascular TRPC channels in both ROCE and store-operated Ca^{2+} entry (SOCE) has been presented during the past decades. Several reports demonstrate TRPC3, TRPC6, and TRPC7 as receptor-activated vascular channels (Inoue et al. 2001; Imai et al. 2012; Lemos et al. 2007; Soboloff et al. 2005; Ju et al. 2010; Maruyama et al. 2006). Nonetheless, TRPC channels have repeatedly been proposed to form SOCE channels in various tissues including the endothelium (Freichel et al. 2001) and smooth muscle (Xu and Beech 2001). Using a dominant negative fragment of TRPC3, Groschner et al. obtained evidence for a contribution of TRPC channels to SOCE in HUVEC (human umbilical vein endothelial cells) (Groschner et al. 1998) and Shi et al. demonstrated suppression of SOCE in vascular smooth muscle of

TRPC1−/− mice (Shi et al. 2012). By contrast, knockdown of TRPC1 and TRPC4 did not reduce SOCE in HUVEC (Abdullaev et al. 2008). Similarly, down-regulation of TRPC1, 4, and 6 by RNA interference did not affect SOCE in primary rat aortic smooth muscle cells (Potier et al. 2009) and refilling of internal Ca^{2+} stores was found unchanged in neuronal cells of TRPC1, 4, and 6 triple knockout mice (Hartmann et al. 2008). It appears important to note that clear evidence for TRPC-mediated SOCE by reconstitution of a store-operated Ca^{2+} conductance by heterologous expression is sparse (Ong et al. 2007; Worley et al. 2007) and further analysis of the molecular mechanism of gating of these channels is required. Specific conditions in terms of signaling partners may be required for TRPC proteins to generate classical store-operated Ca^{2+} conductances.

Another type of Ca^{2+} channel, that was repeatedly suggested to be of importance for vascular physiology and as functionally related to TRPC signaling (Liao et al. 2009), is Orai, the protein underlying calcium release activated Ca^{2+} (CRAC) conductances in immune cells (Li et al. 2011a; Prakriya et al. 2006; Gwack et al. 2007). Orai has been identified as the molecular basis of SOCE in many cellular systems (Peel et al. 2008; Schaff et al. 2010; Tolhurst et al. 2008; Voelkers et al. 2010; Lyfenko and Dirksen 2008; Jones et al. 2008). Orai channels generate distinct Ca^{2+} selective conductances that are activated by physical communication of plasmalemmal Orai hexamers (Hou et al. 2012) with stromal interaction molecules (STIM), located in the membrane of internal Ca^{2+} stores (Muik et al. 2012; Prakriya et al. 2006). The N-terminal segment of STIM holds a Ca^{2+} sensing domain that mediates conformational changes of the protein across the ER-membrane upon emptying of the luminal calcium content. A basic scheme of the Orai activation mechanism is shown in Fig. 23.1. Reduction in ER Ca^{2+} content promotes dimerization and clustering of STIM1 into punctae at the subplasmalemmal part of the ER surface and enables physical interaction with plasmalemmal Orai channels (Frischauf et al. 2008). The conformational changes initiated by STIM/Orai coupling activate the Ca^{2+} entry channel and are apparently essential for generating the typically observed high Ca^{2+} selectivity (McNally and Prakriya 2012). Thereby, protein–protein interactions in STIM/Orai complexes enable Ca^{2+} entry through the Orai pore, generating the inwardly rectifying CRAC conductance.

Upon discovery of the STIM/Orai signaling machinery, which explains the SOCE phenomenon in many cell types, some evidence has been obtained for functional and may be even physical association of TRPC and ORAI channels (Liao et al. 2008, 2009; Salido et al. 2011; Lee et al. 2010a). TRPC1 channels were shown to interact via an electrostatic interaction with STIM1, involving a polybasic domain (K684, K685) in STIM and two negatively charged amino acids in TRPC1 (D639, D640) or TRPC3 (D697, D698) (Yuan et al. 2009). Some studies even indicate the existence of ternary STIM/Orai/TRPC complexes (Ambudkar et al. 2007; Lee et al. 2010a), as suggested by cellular co-localization of the proteins. In pulmonary artery smooth muscle cells, TRPC1 co-immunoprecipitated with STIM1 and Orai1 and was suggested to contribute to SOCE, as demonstrated by siRNA and anti TRPC1 antibodies (Yuan et al. 2012). Sundivakkam et al.

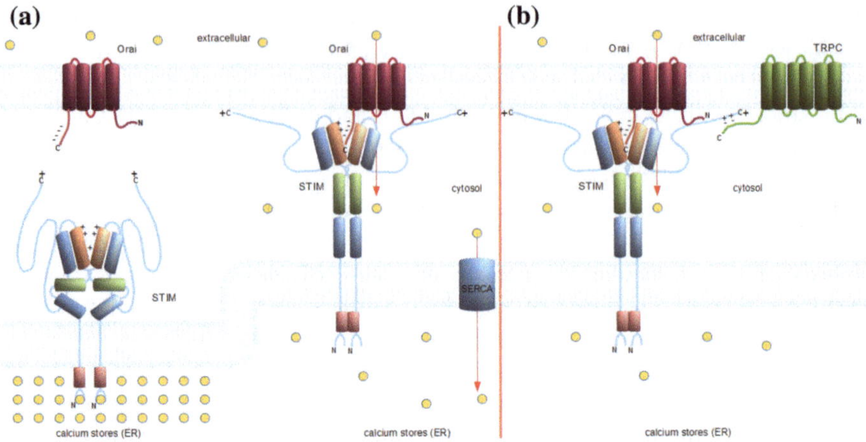

Fig. 23.1 STIM/Orai and Orai/TRPC coupling. **a** Scheme of STIM/Orai coupling. STIM remains in a folded state as long as intracellular Ca^{2+} stores are filled and Ca^{2+} is bound to the intraluminal EF-hand-domain (*left*). Store depletion leads to change of STIM to an extended conformation that enables oligomerization and formation in punctae (not shown). Orai (only one of six subunits depicted) is targeted to STIM punctae by interaction of the STIM carboxyl-terminus with the Orai N-terminus. Additional binding of the C-terminal part of Orai triggers Ca^{2+} entry via the Orai pore. The rise of cytosolic [Ca^{2}] allows for refilling of the stores by sarcoplasmatic endoplasmatic Ca^{2+}-ATPase (SERCA). **b** Possible interaction between Orai and TRPC channels. C-terminal aspartate residues of the TRPC channels (only one of four subunits depicted) enable an electrostatic interaction with positively charged sites of the STIM carboxyl-terminus, providing the prerequisites for the formation of a possible ternary STIM/Orai/TRPC complex. Modified according to Muik et al. (2012)

reported that interaction of STIM1 with TRPC4 was required and sufficient for SOCE function in mouse lung endothelial cells (Sundivakkam et al. 2012). However, typical receptor-operated activity of TRPC channels apparently lacked the requirement of association with STIM1 (DeHaven et al. 2009) and SOCE proved insensitive to simultaneous knockout of multiple TRPC isoforms, at least in some cellular systems (Potier et al. 2009; Hartmann et al. 2008). Thus, the physiological significance of TRPC/Orai/STIM interactions still remains unclear. Collectively, current evidence favors the view that some TRPC channels may be activated in either a STIM-dependent or a STIM-independent fashion, adding versatility to TRPC channel function (Lee et al. 2010b) while the Orai/STIM interaction is seen as obligatory.

Diversification of Ca^{2+} entry into endothelial cells may be based on the general ability of TRPCs to form heteromeric complexes with other isoforms of the same subfamily (Poteser et al. 2006; Schaefer 2005), but also to communicate with members of other TRP channel families and potentially with Orai1 (Cioffi et al. 2012). Moreover, a heteromeric channel comprised of TRPC1 and TRPV4 has been reported to contribute to endothelial SOCE (Ma et al. 2011). Already before the discovery of Orai proteins, SOCE was proposed essential and sufficient to induce endothelial remodeling. Store depletion induced changes in endothelial shape and rearrangement of cytoskeletal elements in pulmonary arterial endothelial cells, natively expressing TRPC1 (Moore et al. 1998). Accordingly, Orai, STIM, and TRPC channel functions have been linked to physiological downstream pathways that are essential for vascular remodeling (Baryshnikov et al. 2009; Aubart et al. 2009; Wang et al. 2009). TRPC channels, for example, have been demonstrated as crucial for the activation of transcription factors like nuclear factor of activated T-cells (NFAT) (Schlondorff et al. 2009; Wamhoff et al. 2006; Poteser et al. 2011; Onohara et al. 2006) or cAMP-responsive element binding protein (CREB) (Yao et al. 2009; Tai et al. 2008). Both transcription factors were identified as key mediators of vascular proliferation (Tokunou et al. 2001; Fukuyama et al. 2006; Zhang et al. 2004; de Frutos et al. 2010; Pang and Sun 2009).

23.2.1 Endothelial Expression of TRPC and Orai Channels

Within the spectrum of ion channels expressed in endothelial cells, the TRP channel superfamily has been recognized as a prominent group of molecules that govern Ca^{2+} signaling and cellular functions. As reviewed by Yao et al. (Yao and Garland 2005), the canonical subfamily of TRP channels (TRPC) was found abundantly expressed in endothelial cells of many vertebrates and expression of most of the seven known TRPC isoforms was demonstrated at the mRNA level. The presence of all TRPC isoforms, except the pseudogene TRPC2, in human endothelial cells was confirmed at the protein level for coronary and pulmonary artery cells (Yip et al. 2004; Smedlund and Vazquez 2008; Paria et al. 2004). However, as expected from variability in mRNA expression pattern, TRPC protein levels in endothelial cells vary significantly among different vascular beds/species and generalized conclusions about the function of these channel proteins are difficult (Antoniotti et al. 2006; Garcia and Schilling 1997; Kamouchi et al. 1999). In addition to the canonical TRP proteins, members like vanilloid (TRPV), melastatin (TRPM), and polycystin-related (TRPP) channel subfamilies were found expressed in endothelial cells and have been assigned to specific cellular functions (Nilius et al. 2003; Dietrich and Gudermann 2008; Hecquet et al. 2008; Kim et al. 2000; AbouAlaiwi et al. 2009). The presence of Orai channels was repeatedly demonstrated in endothelial cells, by studies providing evidence for expression of all three Orai isoforms at the mRNA level (Abdullaev et al. 2008; Shimoda and Laurie 2013; Watanabe et al. 2008). Orai1 protein expression was detected in rat pulmonary endothelial cells, human pulmonary microvascular endothelial cells (Sundivakkam et al. 2012), and HUVEC (Li et al. 2011a).

23.2.2 *Smooth Muscle Expression of TRPC and Orai Channels*

Initiation and regulation of vascular smooth muscle contraction involves a unique repertoire of Ca^{2+} entry mechanisms (Wang et al. 2008a; Wamhoff et al. 2006). Importantly, the underlying Ca^{2+} entry channels comprise both voltage-gated, highly selective channels as well as nonselective cation channels, which control ion homeostasis in a more complex manner by conducting Ca^{2+}, Mg^{2+}, and Na^{+} (Wang et al. 2008a; Berra-Romani et al. 2008). The TRP superfamily contributes to a large group of nonselective, nonvoltage-gated ion channels that appear to be essential for signal transduction in vascular smooth muscle cells (VSMC), including TRPV, TRPM, and TRPP members (Wang et al. 2008b; Yang et al. 2006; Dietrich et al. 2006). Similarly to the endothelium, all members, except TRPC2, of the TRPC subfamily have been detected in human vasculature by PCR, Western Blot, and/or immunofluorescence analysis (Guibert et al. 2011; Dietrich et al. 2006). TRPC1, TRPC4, and TRPC5 appear to prevail in many vascular beds (Yip et al. 2004; Watanabe et al. 2008; Xu and Beech 2001), TRPC3 expression in vascular smooth muscle appears rather low as compared to tissues with prominent expression such as myocardium and brain (Dietrich et al. 2006). Nonetheless, TRPC3 may be a significant player in certain vascular beds or (patho)physiological conditions. TRPC6 is an ubiquitous TRPC isoform found in the whole vasculature, but is most prominently expressed in pulmonary vascular smooth muscle as indicated by northern blot analysis (Dietrich et al. 2005). Ulloa et al. demonstrated TRPC1, TRPC4, and TRPC6 mRNA expression in smooth muscle of the human myometrium (Ulloa et al. 2009). TRPC7 had only been identified in vascular endothelial cells but appears sparse in VSMC (Yip et al. 2004). Interestingly, Orai expression levels were found relatively low in native contractile VSMC compared to smooth muscle cells residing in a proliferating and migrating state (Beech 2012). Nevertheless, immunostainings detected Orai1 in VSMC of arterial sections, demonstrating Orai1 protein expression in pig coronary artery and rat carotid artery (Zhang et al. 2011; Edwards et al. 2010) and Orai2 and Orai3 mRNA (Li et al. 2011b) as well as protein (Berra-Romani et al. 2008) have as well been identified in VSMCs.

23.3 Endothelial Remodeling

The luminal border of blood vessels and the foremost barrier of the vessel wall against blood born factors are formed by the endothelium. A thin single layer of endothelial cells governs pivotal vascular functions including the exchange of compounds between blood and tissue (Irie and Tavassoli 1991; Rippe et al. 2002), regulation of blood pressure (Furchgott and Zawadzki 1980), haemostasis, and fibrinolysis (Stalcup et al. 1982) as well as control of permeability for circulating cells (Vestweber 2012) and enables initial steps in angiogenesis (Herbert and Stainier 2011). Accordingly, endothelial cells take a central role in vascular remodeling and are generally considered to pass on proliferative signals to adjacent smooth muscle cell layers (Burnstock 2008; Zhang et al. 2012; Wedgwood et al. 2001). Endothelial cells can generate vasoactive substances that influence cellular growth (Damon 2000; Peiro et al. 1995), migration (Meoli and White 2010; Kernochan et al. 2002), and death (Nakahashi et al. 1998) as well as the architecture of the extracellular matrix (Shi-Wen et al. 2001). In turn, endothelial cells are equipped with receptors enabling them to sense vascular mediators and signaling molecules, of which many trigger intracellular Ca^{2+} signals (Ryan 1989; Neufeld et al. 1994). Adverse remodeling of the vascular system is typically initiated by a critical level of physical or biochemical factors that are either limited to distinct localizations within blood vessels or affect vascular branches and large parts of the vascular bed. The latter situation is common for haemodynamic, physical factors, predominantly generating mechanical stress that is induced by changes in blood pressure and flow rate. Local vascular remodeling may also be triggered by injury and/or oxidative stress, as typically observed in atherosclerosis, or due to the aberrant cellular signaling associated with tumor formation. All these initiating conditions are potentially linked to activation of either TRP or Orai signaling.

23.3.1 The Role of Endothelial TRPC Channels in Growth Factor-Mediated Vascular Remodeling

A variety of vasoactive factors regulate endothelial permeability, proliferation, and angiogenesis as typically measured by tube formation (Gerwins et al. 2000). Numerous studies investigating and proposing a potential role of TRPC channels in endothelial responses to vasoactive factors were focusing on barrier function, while little direct evidence is available for a contribution of these channels to proliferation or remodeling. Nonetheless, the existence of overlapping pathways involved in these processes supports the assumption of a common set of ion channels which may be required for these functions (Qin et al. 2013; Shibuya 2013). Angiogenic factors, which stimulate proliferation and migration of endothelial cells, include basic fibroblast growth factor (bFGF), platelet-derived growth factor (PDGF), and vascular endothelial growth factor (VEGF). VEGF appears as an activator of endothelial TRPC channels of the "receptor-operated" type, such as TRPC6. TRPC6 was shown as a prerequisite for VEGF-induced proliferation and tube formation in HUVEC (Ge et al. 2009; Hamdollah Zadeh et al. 2008) and TRPC3 and TRPC6 were found involved in VEGF receptor signaling in human microvascular endothelial cells (HMVEC) (Cheng et al. 2006). In addition to a functional linkage of VEGF signaling to TRPC channel function, these receptors and channels may be even elements of a common signaling complex as endothelial

TRPC1 was found associated with VEGF receptor-2 (Kusaba et al. 2010). The sensitivity of TRPC1 to VEGF is further supported by the finding that VEGF-induced activity of endothelial TRPC1 channels is reduced by angiopoietin-1 which disturbs the interaction of TRPC1 and the IP3-receptor (Jho et al. 2005).

In addition to its direct effect, VEGF may also be responsible for the release of other vasoactive factors, which as well act at TRPC channels as target. VEGF promotes the synthesis of endothelial inflammatory mediators such as platelet activating factor (PAF), representing a molecular link between vascular remodeling and inflammatory stimuli (Bernatchez et al. 2002). PAF was shown to induce nonselective currents in lung endothelial cells that were attributed to TRPC6 activity by pharmacological strategies (Samapati et al. 2012). Exposure to PAF resulted in increased microvascular wall thickness, perivascular cell proliferation, and rearrangement of F-actin filaments (Yuan and He 2012), typically indicating vascular remodeling. In addition to VEGF, bFGF has also been shown to induce TRPC channel activity in bovine aortic endothelial cells (Antoniotti et al. 2002, 2006). In summary, there is relatively good evidence for a role of endothelial TRPC 1, 3, and 6 channels in growth factor-induced signaling, a set of isoforms that is potentially able to form heteromeric channels (Liu et al. 2005; Storch et al. 2012) and may thus confer the observed effects to a single pore.

Growth factor-activated signaling is of particular significance for development and proliferation of vascular progenitor cells. Proliferation and migration of endothelial progenitor cells (EPCs) are considered critical for endothelial repair after injury and the initial steps of neovascularization. EPCs are a rare species of circulating blood cells that adhere to vascular spots of injury or a site that requires *de novo* vessel formation (Asahara et al. 1999). TRPC3 was identified as a VEGF-regulated Ca^{2+} entry pathway in adipose tissue-resident endothelial progenitor cells (Poteser et al. 2008) and silencing of TRPC1 was shown to reduce proliferation and migration of bone marrow derived EPCs (Kuang et al. 2012). The latter group reported the requirement of STIM1 for TRPC1-mediated proliferation and migration of EPCs (Kuang et al. 2010). Endothelial tube formation as an experimental read-out of angiogenesis was recently demonstrated to depend on TRPC channels (Antigny et al. 2012). Using a siRNA-knockdown strategy in an in vitro matrigel assay, the authors demonstrated the significance of TRPC3, TRPC4, and TRPC5 channels in EA.hy926 cells, while silencing of STIM1 and Orai1 failed to prevent tubulogenesis. By contrast, a paper by Li et al. communicated that endothelial (in vitro) tube formation by HUVEC depends on Orai1 and CRAC channel activity (Li et al. 2011a). As indicated by these discrepancies, the relative contribution of TRPC and Orai signaling to angiogenesis may again markedly depend on the vascular bed and environmental situation. Nonetheless, taken together, evidence has been accumulated in favor of a key role of TRPC channels in growth factor-induced vascular remodeling and therefore potentially in tumor angiogenesis and neovascularization. TRPC channels thus deserve consideration as targets for future anti-tumor therapies (He et al. 2012). VEGF and its receptors have already been recognized as targets in the fight against cancer and led to the development of promising tools like bevacizumab (Avastin) (Awazu et al. 2013;

Kabbinavar et al. 2003; Yang et al. 2003). Currently, pharmacological means to directly target TRPC channels in a selective way are limited, but the recent development of pyrazole derivatives (Kiyonaka et al. 2009; Glasnov et al. 2009; Schleifer et al. 2012), some of which exhibiting selectivity between TRPC and Orai channels, gives reason to anticipate fairly rapid progress in this field.

23.3.2 The Role of Endothelial TRPC Channels in Oxidative Stress-Mediated Vascular Remodeling

Free oxygen radicals are important mediators of endothelial injury and pathological impairment of endothelial function, a typical scenario in atherosclerosis, reperfusion-injury, or respiratory distress syndrome. Accordingly, oxidative stress has been shown to disrupt endothelial barrier (Usatyuk and Natarajan 2004) as well as blood–brain barrier function (Huppert et al. 2010). Cellular Ca^{2+} signals take a central role in the pathways leading to redox-mediated injury and remodeling of the affected tissue (Geeraerts et al. 1991) and several members of the TRPC channel family were found involved in oxidative stress responses (Miller and Zhang 2011; Cioffi 2011). A membrane conductance triggered by external oxidative stress, was characterized in porcine aortic endothelial cells and reported to display typical TRPC properties (Balzer et al. 1999), suggesting that TRPC proteins might serve as vascular redox sensors. In a heterologous expression system, TRPC6 was found to be regulated by externally applied H_2O_2, an effect that was attributed to redox- sensitive thiol groups of (not identified) intracellular proteins, as H_2O_2-induced activation was not observed in electrophysiological recordings from inside-out patches (Graham et al. 2010). Thus, TRPC6 may be an indirect target of oxidative stress, potentially conferred by means of altered phosphatase activity (Hisatsune et al. 2004). TRPC6 was recently suggested as an endothelial mediator of oxidative injury. TRPC6 knockout mice were shown to be protected from lung ischemia–reperfusion-induced oedema and explanted lung endothelial cells of these mice showed attenuated Ca^{2+} influx, modified cellular shape as well as defective barrier function which are typical cellular indicators of endothelial remodeling (Weissmann et al. 2012).

Oxidative stress does not necessarily depend on external factors, but might also be generated within the endothelial cells. One primary source of internal redox stress, in addition to NADPH oxidase (nicotinamide adenine dinucleotide phosphate oxidase, NOX) (Gray et al. 2013) and xanthine oxidase (Rodell et al. 1987) is the activity of nitric oxide synthase (NOS). Oxidation of tetrahydrobiopterin (BH4), an essential cofactor of NO synthase by NOX leads to uncoupling of NOS, subsequent reduced synthesis of NO and enhanced production of superoxide (Crabtree et al. 2013). NO, as generated by eNOS activity, controls vascular tone (Lefroy et al. 1993) and arterial pressure (Rees et al. 1989), but also represents a highly reactive radical. TRPC5 has been proposed as a potential target of NO or nitric oxide derived species, such as N_2O_3, via S-nitrosylation of cytoplasmic accessible cysteine residues (cys553 and cys558) (Yoshida et al. 2006). These authors proposed that oxidation-induced channel activation causes Ca^{2+} entry in endothelial cells after G-protein coupled receptor activation. This finding was challenged by Wong et al., who demonstrated that homomeric TRPC5 was not activated by S-nitroso-N-actetylpenicillamine (SNAP) and diethylamine NONOate (DEA-NONOate) (Wong et al. 2010). As stressed by the authors, a possible explanation for this discrepancy may be the ability of TRPC5 to form heteromeric channels with other native, redox-sensitive endothelial channels. Heteromeric assembly of TRPC channels was demonstrated in a HEK293 cell-based heterologous expression system, where heteromeric TRPC channels comprised of TRPC3 and TRPC4 isoforms were found to be redox-sensitive and the IV-signature of the current resembled those observed in endothelial cells (Poteser et al. 2006).

Interestingly, a rather protective role of TRPC3 was reported in porcine coronary artery cells exposed to hypoxia-reoxigenation (Huang et al. 2011). Such reoxygenation-mediated vascular injury is typically associated with decreased production of NO and impaired endothelium-derived hyperpolarizing factor (EDHF) signaling (Carden and Granger 2000; Dong et al. 2005). TRPC3 was shown to contribute to endothelial NO release in porcine endothelial cells and surface expression of TRPC3 was consistently found to enhance NO production. TRPC3 activity was found inhibited by hypoxia-reoxigenation (Huang et al. 2011). Considering the findings showing TRPC channels as contributors or enhancers of redox injury, this indicates that both, inhibition and moderate enhancement of TRPC channel activity might have beneficial effects on endothelial functions.

Collectively, these studies support the view of a pivotal involvement of TRPC channels in endothelial redox injury, while their role as direct targets of oxidative reactions still remains elusive. However, TRPC channels again qualify as potential targets for therapeutic strategies aimed to prevent endothelial injury, a process that is tightly linked to vascular aging. In line with this concept, Naylor et al. reported sensitivity of TRPC5 to antioxidants (Naylor et al. 2011). TRPC5 was found to be inhibited by redox scavengers, indicating a mode of activation that depends on endogenous redox stress, while hydroxylated stilbenes and reservatrol inhibited the channel independently of the antioxidant effect. Thus, endothelial TRPC channels might be a target of common antioxidative therapies. The therapeutic consequences of TRPC modulation or block in endothelial cells definitely require further studies to elucidate basic principles.

23.3.3 The Role of Endothelial TRPC Channels in Mechanical Stress-Induced Vascular Remodeling

The vasculature is under permanent stress by alternating mechanical forces as generated by luminal blood flow and pulse wave. Vascular remodeling is hence not initiated by this mechanical stimulation per se, but is rather triggered by pathophysiological deviations from a "normal" level (Lu and Kassab 2011). Experimental approaches to study responses to mechanical stimulation in the vascular endothelium are challenging and a generally accepted picture of the role of TRPC channels mechanosensitivity has not been obtained so far (Gottlieb et al. 2008). Consequently, the role of endothelial TRPC channels in blood pressure- or flow-induced vascular remodeling is still unclear. Several studies indicate a potential involvement of TRPC channels in mechanically triggered Ca^{2+} entry (Maroto et al. 2005; Berrout et al. 2012; Spassova et al. 2006). In a CHO-cell based heterologous expression system, overexpression of TRPC1 resulted in a mechanically triggered conductance, while genetic knockdown of TRPC1 eliminated this mechanosensitivity (Maroto et al. 2005). The authors discuss a possible interaction of TRPC1 with dystrophin as a possible base of the mechanosensitivity, which could also explain the reduction of the abnormal mechanosensive channel activity in dystrophic muscle, as induced by TRPC1 antisense oligonucleotides (Vandebrouck et al. 2002). Stretch and receptor activation of TRPC6 channels have been proposed to involve a common mechanism induced by lateral-lipid tension. Strikingly, the responses to both stimuli were clearly suppressed by the tarantula toxin GsMTx-4, an established inhibitor of mechanosensitve conductances (Spassova et al. 2006). Importantly, a solid investigation in expression systems argued against a direct mechanosensitivity of TRPC1 and TRPC6 (Gottlieb 2008). Nonetheless, TRPC channels might represent target proteins in mechanically triggered signaling cascades. One potential mechanism linking mechanical stress indirectly to TRPC channel activity is mechanosensitivity of G-protein coupled receptors as upstream signaling elements (Mederos y Schnitzler et al. 2008). So far, only little information about involvement of TRPC channels in mechanically induced endothelial conductances is available. In blood–brain barrier endothelial cells, stretch injury-associated Ca^{2+} elevation was reduced by silencing of TRPC1 and TRPP2 (Berrout et al. 2012). Interestingly, pulsatile shear stress but not constant laminar flow was demonstrated to promote expression of TRPC6 and TRPV1 in human umbilical vascular endothelial cells (Thilo et al. 2012). Thus, certain states of mechanical stimulation might initiate a positive feed-forward mechanism involving up-regulation of certain TRP molecules thereby leading to altered Ca^{2+} signaling.

In summary, TRPC channels are suggested to play a role in mechanosensitivity of vascular cells, most likely due to cellular mediators that are generated downstream of primary mechanosensors, such as receptors or enzymes, and to contribute to vascular mechanoregulation (Ungvari et al. 2006; Zhu et al. 2011).

23.3.4 The Role of Endothelial Orai Channels in Vascular Remodeling

The current available information about the potential involvement of Orai in endothelial remodeling is limited. In the human umbilical vein derived endothelial cell line EA.hy926, knockdown of STIM1 and Orai1 did not prevent in vitro tubulogenesis, whereas knockdown of TRPC3, TRPC4, or TRPC5 were found to be essential. In the same system, Orai1, but not STIM1 knockdown was found to prevent cell proliferation (Antigny et al. 2012).

In contrast, SOCE and CRAC in HUVECs were demonstrated to be mediated by Stim1 and Orai1 and TRPC1 and TRPC4 were found not involved in SOCE after siRNA induced silencing. Accordingly, knockdown of Orai1, STIM1 and STIM2 inhibited endothelial proliferation (Abdullaev et al. 2008). Also in HUVEC, Orai1 silencing inhibited VEGF-induced cell migration and in vitro tube formation (Li et al. 2011a). The discrepancies observed in similar endothelial cell systems have been attributed to different levels of VEGF used in the experiments (Antigny et al. 2012).

Generally, a prominent role for Orai1 and SOCE in VEGF-induced endothelial proliferation, migration, and angiogenesis is very likely. In contrast, thrombin-mediated signaling of endothelial cells might involve STIM1, but not Orai1 and SOCE (Shinde et al. 2013). This finding is supported by the observation that dominant negative Orai1 (R91 W) and Orai3 (E81 W) mutants failed to prevent thrombin induced Ca^{2+} entry in murine lung endothelial cells (Sundivakkam et al. 2012). Thus, the relative contribution of Orai and TRPC channels to endothelial remodeling might depend on involved types of receptors. However, acknowledging the important role for VEGF, Orai1 has to be considered as a key element of endothelial remodeling.

23.4 Smooth Muscle Remodeling

Vascular smooth muscle cells (VSMC) are essential for vessel integrity and blood pressure control (Yoshida and Owens 2005). Compared to skeletal and cardiac muscle, which are terminally differentiated, VSMC retain a high degree of plasticity and are capable to switch from a contractile excitable to a proliferative nonexcitable phenotype (Owens et al. 2004). The VSMC phenotype switch from a contractile to a synthetic morphology is an essential process of vascular growth, remodeling, and repair. Moreover, the process is critical for maladaptive responses in pathological states such as atherosclerosis, neointimal hyperplasia after percutaneous transluminal coronary angioplast or in-stent restenosis (Alvaro-Gonzalez et al. 2002; Ailawadi et al. 2009; Hao et al. 2003). Under normal, physiological conditions, VSMC are characterized by an essentially low proliferation rate, a minor synthetic activity and a unique pool of ion channels, contractile proteins, and signaling molecules (Owens et al. 2004; House et al. 2008). The VSMC phenotype switch from the quiescent to proliferative state is a highly complex mechanism that is accompanied by down-regulation of smooth muscle differentiation markers such as smooth muscle α-actin, myosin heavy chain or smoothelin, as well as a loss of voltage-gated channels like the L-type Ca^{2+} channel $Ca_v1.2$ and Ca^{2+} activated K^+ channel $K_{Ca}1.1$. Typically, dedifferentiation is associated with increased expression of components assigned to SOCE (TRPC, STIM1 and Orai1) and a *de novo* expression of the K^+ channel $K_{Ca}3.1$ (Golovina et al. 2001; Beech 2007; Lang et al. 2005). VSMC additionally are stocked with appropriate pro-proliferative and pro-migratory signaling pathways targeted by growth factors and cytokines to promote the transcriptional up-regulation of remodeling genes (Kawai-Kowase and Owens 2007), that are associated with altered intracellular Ca^{2+} signaling. These alterations in Ca^{2+} homeostasis typically involve changes in the sarcoplasmatic reticulum (SR) Ca^{2+} and in Ca^{2+} entry pathways. The intracellular Ca^{2+} signal that drives VSMC proliferation and remodeling has been attributed to SOCE as well as ROCE (Berra-Romani et al. 2008; Bergdahl et al. 2005; Yu et al. 2003; Baryshnikov et al. 2009). Hence, ion channels involved in these Ca^{2+} entry mechanisms are considered as pharmacological targets to prevent vascular remodeling rather than therapeutically well-established blockers of L-type voltage-gated Ca^{2+} channels, which proofed inefficient to prevent smooth muscle proliferation. This is consistent with results obtained by down-regulation of voltage-gated channels (Li et al. 2004) in proliferating states of smooth muscle. Indeed, targeting of nonvoltage-gated ion channels may represent a promising therapeutic approach (Nilius et al. 2007; Courjaret and Machaca 2012).

23.4.1 The Role of Smooth Muscle TRPC Channels in Vascular Remodeling

Phenotype transition to proliferative VSMC is associated with decreased density of functional L-type channels, while Ca^{2+} entry is still prominent but based on other pathways. Besides the Ca^{2+} selective Orai1 pore complex, less Ca^{2+} selective SOCE conductances have been proposed to induce smooth muscle proliferation. These conductances display distinctly different properties as CRAC channels (Lee et al. 2010a; Beech 2007). Several studies indicated up-regulation of TRPC expression associated with vascular injury and involvement of TRPCs in the invasive migratory state of synthetic VSMC (Rowell et al. 2010; Watanabe et al. 2008; Inoue et al. 2006; Kumar et al. 2006; Spassova et al. 2006). TRPC1 was identified as a molecular component of this nonCRAC, nonselective SOC channel in proliferating smooth muscle cells (Bergdahl et al. 2005; Sweeney et al. 2002). Kumar et al. reported that TRPC1 is up-regulated in response to vascular injury and measurements of neointimal hyperplasia in human saphenous vein organ culture revealed, that TRPC1 expression is enhanced in intimal layers containing

SMC as compared to the media. The use of an antibody, recognizing the pore forming region of TRPC1 and presumably blocking the channel, inhibited neo-intimal hyperplasia, suggesting a pivotal role of TRPC1 in VSMC proliferation. Moreover, enhanced TRPC1 expression has been detected after vascular injury elicited by balloon dilation in the mammary artery. TRPC1 protein expression as well as SOCE was found increased in proliferative human pulmonary artery SMC as compared to nonproliferative cells (Kumar et al. 2006). Interestingly, it has been demonstrated that TRPC expression is altered upon exposure of VSMC to mechanical stretch. A 5-fold increase in TRPC1 and TRPC6 mRNA and protein expression levels has been observed after exposure of human internal mammary artery to static stretch via balloon dilatation or stenting (Bergdahl et al. 2005). TRPC1 appears to represent an important component of voltage-independent Ca^{2+} conductances in proliferating VSMC, but by itself is barely able to mediate SOCE in VSMC (Beech 2005). Strubing et al. were the first to suggest that TRPC1 is likely to function in a heteromultimer with TRPC5. Heteromultimeric TRPC1/5 complexes were tested for lipid-sensing, as lipid mediators are critical players in progression of atherosclerosis. Both, LPC (lysophosphatidylcholine), a major component of oxidized LDL in atherosclerotic plaques and sphingosine-1-phosphate (S1P) were found to activate the TRPC1/5 hetermultimeric channel (Strubing et al. 2001; Flemming et al. 2006). Xu et al. investigated the S1P sensing by TRPC5 and its impact on SMC motility by inhibiting channel functionality. Human SMCs of saphenous veins lack the S1P-induced motility when TRPC5 channels are inhibited by specific TRPC5-blocking antibodies or by the expression of a dominant negative ion-pore mutant of TRPC5 (Xu et al. 2006). Another TRPC isoform potentially involved in initiation of VSMC proliferation, after vascular injury, is TRPC4. TRPC4 was suggested to participate in phenotype transitions of human pulmonary arterial smooth muscle cells as increased TRPC4 protein expression was detected along with enhanced proliferation (Rowell et al. 2010). This phenomenon is associated with increased CREB phosphorylation, which in turn activates gene transcription of growth factors and cytokines (Zhang et al. 2004).

Stimuli inducing activation of genes in VSMCs, as associated with neointima hyperplasia, may be provided by the local mechanical disturbances that are produced by stent implantation (Antigny et al. 2012). As shortly outlined for endothelial cells above, TRPC channels are in discussion to be, at least indirectly, mechanosensitive. Interestingly, mechanical stretch appears to modulate the expression of TRPC3 and TRPC4 in myometrial smooth muscle cells. Exposure of cells to prolonged mechanical stretch in vitro resulted in increased TRPC3 and TRPC4 expression along with altered Ca^{2+} signaling (Dalrymple et al. 2007). Our group recently demonstrated changes in TRPC3 expression in a human aortic ex vivo model of smooth muscle hyperplasia (Koenig et al. 2013). Smooth muscle cell proliferation was initiated by overexpansion of the human aorta via stent implantation. The pyrazole derived TRPC3 blocker Pyr3 (ethyl-1-4-2,3,3-trichloroacrylamide phenyl-5-trifluoromethyl-1H-pyrazole-4-carboxylate) inhibited proliferation of VSMC, but not of endothelial cells. This suggests that TRPC3 is a critical player in smooth muscle remodeling but might be barely essential for endothelial proliferation. Stent-induced injury of aortic constructs was accompanied by a significantly enhanced expression of TRPC3 and Pyr3-coating of stents was efficient in suppression of injury-induced proliferation. This points toward a potential suitability of TRPC inhibitors in drug-eluting stent strategies that target VSMC proliferation without impeding re-endothelialization.

TRPC3 is generally considered as receptor operated in a physiological context and Berra-Romani et al. provided evidence for enhanced ROCE as well as TRPC3 and TRPC6 expression in proliferating myocytes. Cultured rat arterial smooth muscle cells were stimulated with the diacylglycerol analog OAG, a direct activator of the TRPC3 and TRPC6 channels, resulting in augmented ROCE of proliferating cells (Berra-Romani et al. 2008). To distinguish between SOCE and ROCE, the authors measured OAG-induced Ba^{2+} entry, which was significantly higher in proliferating arterial smooth muscle cells, as compared to control cells. Thus, it has been suggested that increased TRPC3 and TRPC6 protein expression is the basis of augmented ROCE in arterial myocytes, residing in a synthetic phenotype (Berra-Romani et al. 2008). Another line of evidence that favors a key role of TRPCs in vascular smooth muscle phenotype transitions was provided by the observation of TRPC6 up-regulation in platelet-derived growth factor (PDGF)-induced proliferation of pulmonary vascular smooth muscle cells (PASMC). Inhibition of the channel′s expression by antisense oligonucleotide knock-down significantly reduced PDGF-induced cell proliferation (Yu et al. 2003).

In conclusion, SOCE as well as ROCE appear to be involved in molecular pathways linked to smooth muscle proliferation, and TRPC channels were suggested to contribute to both Ca^{2+} entry mechanisms depending on isoform/heteromerization and stimulus. Most studies focussing on TRPC function and distribution in terms of VSMC remodeling revealed an increased expression of these ion channels in proliferating VSMC of diverse disease models, suggesting enhanced TRPC expression as part of the remodeling process and these molecules as promising targets for pharmacological or genetic approaches in the therapy of cardiovascular diseases. These approaches might not only be based on inhibition of channel activity by selective blockers but also on suppression of specific triggers for TRPC activation, such as growth factors (Yu et al. 2003), lipids (Xu et al. 2006; Flemming et al. 2006) or mechanical stimuli (Kumar et al. 2006; Spassova et al. 2006) as well as inhibition of signaling pathways involved in feed-forward promotion of TRPC expression.

23.4.2 The Role of Smooth Muscle Orai Channels in Vascular Remodeling

The best characterized SOCE mechanism is the highly Ca^{2+} selective I_{CRAC} (Ca^{2+} released activated Ca^{2+}) (Cahalan et al. 2007). Several studies demonstrated that transition from a contractile to a synthetic phenotype in cultured VSMC is tightly linked to alterations in Ca^{2+} handling by intracellular Ca^{2+} stores and to SOCE (Yang et al. 2012; Berra-Romani et al. 2008). In proliferating myocytes, Orai1 and STIM1 expression levels were found enhanced, suggesting involvement of Orai1 and STIM1 in vascular remodeling. Results obtained by siRNAs down-regulation of Orai1 expression corroborated the importance of the Orai1-mediated CRAC conductance in VSMC proliferation (Baryshnikov et al. 2009). Orai1 and STIM1 expression was found significantly up-regulated in freshly isolated arterial myocytes undergoing in vitro transition to the synthetic phenotype. Again, knock-down of Orai1 and STIM1, respectively, resulted in inhibition of VSMC proliferation and migration (Potier et al. 2009; Baryshnikov et al. 2009). The role of STIM1 and Orai1 in VSMC proliferation and neointima formation in vivo was elucidated by Zhang et al., using a rat carotid artery balloon-injury model. Vessel injury triggered neointima formation with a peak at two weeks after injury. Measuring cell proliferation markers along with Orai1- and STIM1-distribution in neointimal VSMC layers, the study demonstrated augmented expression levels of Orai1, STIM1 as well as proliferation markers. Moreover, the authors reported that knockdown of Orai1 and STIM1 by use of shRNAs abrogated store operated Ca^{2+} entry and I_{CRAC} in cultured rat aortic VSMC. Neointima formation in injured vessels was markedly reduced by shOrai1 and shSTIM1 knock-down. Furthermore, this study revealed up regulation of Orai1 und STIM1 in a model of vascular remodeling based on enhanced shear stress in carotid artery sections due to carotid ligation. Orai1 and STIM1 protein expression was significantly enhanced in the media and neointima of ligation-injured tissue (Zhang et al. 2011). Recently, the same group addressed the question whether the Orai1 homolog Orai3 is as well involved in neointima formation in response to vessel injury and reported that different pathophysiologically relevant agonists activate two distinct Orai-mediated conductances and Ca^{2+} entry pathways. Platelet-derived growth factor activates the classical Stim1/Orai1 pathway, which was found to promote VSMC migration in vitro (Bisaillon et al. 2010). In contrast, thrombin induces Ca^{2+} entry that appears mediated by Orai1/3 heteromeric channels and depends on STIM1. Both cultured synthetic VSMC and freshly isolated medial and neointimal VSMC from a rat in vivo model of vascular remodeling displayed elevated Orai3 expression. Moreover, in vivo knockdown of Orai3 by shRNA (shOrai3) clearly affected these vascular remodeling processes, in that both neointimal size and the ratio of neointima/media were markedly reduced (Gonzalez-Cobos and Trebak 2010; Trebak 2012). Recently, Gonzalez-Cobos et al. presented results showing that thrombin, but not PDGF-induced activation of Orai1/Orai3 in VSMCs required the intracellular production of leukotriene C_4, indicating that thrombin acts on these channels via an unidentified store-independent mechanism (Sundivakkam et al. 2012). Accordingly, Orai3 has been shown essential for an arachidonic acid-induced conductance observed in a HEK293 expression system, suggesting the existence of an arachidonate-regulated Ca^{2+} (ARC) entry based on Orai1/Orai3 heteropentamers (Shinde et al. 2013; Abdullaev et al. 2008; Li et al. 2011a). Thus, in addition to known pathways of SOCE, Orai channels may, depending on the heteromeric pore composition, also mediate receptor-operated Ca^{2+} entry that stimulates smooth muscle proliferation, and vascular remodeling.

These results suggest that nonvoltage-dependent Ca^{2+} entry channels such as Orai or TRPC complexes may respond to particular trigger stimuli to provide Ca^{2+} signals that initiate Ca^{2+} transcription coupling associated with feed-forward changes of Ca^{2+} channel expression and cellular phenotype transition.

With the discovery of STIM1 and Orai1 as the mediators for SOCE in VSMC (Beech 2012; Zhang et al. 2011) as well as other cell types (Abdullaev et al. 2008; Zhao et al. 2012; Frischauf et al. 2008), attention has also been directed to their contribution in VSMC remodeling in cardiovascular disease. The reports outlined above demonstrate an involvement of Orai1, alone or in heteromeric arrangements with Orai3, in VSMC proliferation.

23.5 Conclusion

In summary, Ca^{2+} signaling mediated by TRPC and Orai channels determines initiation and progress of vascular remodeling via signaling processes that govern transcriptional programs and phenotype transitions. Ca^{2+} channels formed by these proteins appear as the basis of Ca^{2+} conductances, which share the common feature of lacking primary voltage sensitivity but display variable selectivity for divalents. These channels are activated in response to a variety of external stimuli such as growth factors or oxidative injury. Their role in adaptive responses of blood vessels to mechanical stimuli remains elusive. Currently available evidence suggests a high degree of heterogeneity in TRPC and Orai signaling in vascular cells most likely due to variable composition of channel complexes among different vascular beds and an accordingly variable contribution of these channels to ROCE and SOCE. A pivotal role of STIM1/Orai1 as well as TRPC channels in the VSMC phenotype switch was repeatedly suggested and is supported by the observation that knockdown of these Ca^{2+} signaling molecules suppresses cell proliferation. Importantly, phenotype transitions in VSMC are accompanied by significant changes in the ion channel expression pattern in the cells. A crucial remodeling-associated change in the pattern of Ca^{2+} entry channels seems to be replacement of L-type voltage-gated channels by members of the TRPC family and components of the I_{CRAC}, STIM1 and Orai1. This reorganization in Ca^{2+} entry mechanisms results in profound changes in Ca^{2+}-transcription coupling and appears as a potential feed-forward mechanism in maladaptive restructuring of the vascular system. Our increasing knowledge about the molecular players in the vascular pathophysiology prompts the development of novel therapeutic strategies based on molecular targets related to TRP and Orai signaling.

References

Abdullaev IF, Bisaillon JM, Potier M, Gonzalez JC, Motiani RK, Trebak M (2008) Stim1 and Orai1 mediate CRAC currents and store-operated calcium entry important for endothelial cell proliferation. Circ Res 103(11):1289–1299

AbouAlaiwi WA, Takahashi M, Mell BR, Jones TJ, Ratnam S, Kolb RJ, Nauli SM (2009) Ciliary polycystin-2 is a mechanosensitive calcium channel involved in nitric oxide signaling cascades. Circ Res 104(7):860–869

Ailawadi G, Moehle CW, Pei H, Walton SP, Yang Z, Kron IL, Lau CL, Owens GK (2009) Smooth muscle phenotypic modulation is an early event in aortic aneurysms. J Thorac Cardiovasc Surg 138(6):1392–1399

Alvaro-Gonzalez LC, Freijo-Guerrero MM, Sadaba-Garay F (2002) Inflammatory mechanisms, arteriosclerosis and ischemic stroke: clinical data and perspectives. Revista de neurologia 35(5):452–462

Ambudkar IS, Ong HL, Liu X, Bandyopadhyay BC, Cheng KT (2007) TRPC1: the link between functionally distinct store-operated calcium channels. Cell Calcium 42(2):213–223

Antigny F, Girardin N, Frieden M (2012) Transient receptor potential canonical channels are required for in vitro endothelial tube formation. J Biol Chem 287(8):5917–5927

Antoniotti S, Fiorio Pla A, Barral S, Scalabrino O, Munaron L, Lovisolo D (2006) Interaction between TRPC channel subunits in endothelial cells. J Recept Signal Transduct Res 26(4):225–240

Antoniotti S, Lovisolo D, Fiorio Pla A, Munaron L (2002) Expression and functional role of bTRPC1 channels in native endothelial cells. FEBS Lett 510(3):189–195

Asahara T, Masuda H, Takahashi T, Kalka C, Pastore C, Silver M, Kearne M, Magner M, Isner JM (1999) Bone marrow origin of endothelial progenitor cells responsible for postnatal vasculogenesis in physiological and pathological neovascularization. Circ Res 85(3):221–228

Aubart FC, Sassi Y, Coulombe A, Mougenot N, Vrignaud C, Leprince P, Lechat P, Lompre AM, Hulot JS (2009) RNA interference targeting STIM1 suppresses vascular smooth muscle cell proliferation and neointima formation in the rat. Mol Ther 17(3):455–462

Awazu Y, Nakamura K, Mizutani A, Kakoi Y, Iwata H, Yamasaki S, Miyamoto N, Imamura S, Miki H, Hori A (2013) A Novel Inhibitor of c-Met and VEGF Receptor Tyrosine Kinases with a Broad Spectrum of In Vivo Antitumor Activities. Mol Cancer Ther 12(6):913–924

Balzer M, Lintschinger B, Groschner K (1999) Evidence for a role of Trp proteins in the oxidative stress-induced membrane conductances of porcine aortic endothelial cells. Cardiovasc Res 42(2):543–549

Baryshnikov SG, Pulina MV, Zulian A, Linde CI, Golovina VA (2009) Orai1, a critical component of store-operated Ca^{2+} entry, is functionally associated with Na^{+}/Ca^{2+} exchanger and plasma membrane Ca^{2+} pump in proliferating human arterial myocytes. Am J Physiol Cell Physiol 297(5):C1103–C1112

Beech DJ (2005) TRPC1: store-operated channel and more. Pflugers Arch 451(1):53–60

Beech DJ (2007) Ion channel switching and activation in smooth-muscle cells of occlusive vascular diseases. Biochem Soc Trans 35(Pt 5):890–894

Beech DJ (2012) Orai1 calcium channels in the vasculature. Pflugers Arch 463(5):635–647

Beech DJ (2013) Characteristics of transient receptor potential canonical calcium-permeable channels and their relevance to vascular physiology and disease. Circulation journal : official journal of the Japanese Circulation Society 77(3):570–579

Beech DJ, Xu SZ, McHugh D, Flemming R (2003) TRPC1 store-operated cationic channel subunit. Cell Calcium 33(5–6):433–440

Bergdahl A, Gomez MF, Wihlborg AK, Erlinge D, Eyjolfson A, Xu SZ, Beech DJ, Dreja K, Hellstrand P (2005) Plasticity of TRPC expression in arterial smooth muscle: correlation with store-operated Ca^{2+} entry. Am J Physiol Cell Physiol 288(4):C872–C880

Bernatchez PN, Rollin S, Soker S, Sirois MG (2002) Relative effects of VEGF-A and VEGF-C on endothelial cell proliferation, migration and PAF synthesis: Role of neuropilin-1. J Cell Biochem 85(3):629–639
Berra-Romani R, Mazzocco-Spezzia A, Pulina MV, Golovina VA (2008) Ca^{2+} handling is altered when arterial myocytes progress from a contractile to a proliferative phenotype in culture. Am J Physiol Cell Physiol 295(3):C779–C790
Berrout J, Jin M, O'Neil RG (2012) Critical role of TRPP2 and TRPC1 channels in stretch-induced injury of blood-brain barrier endothelial cells. Brain Res 1436:1–12
Bisaillon JM, Motiani RK, Gonzalez-Cobos JC, Potier M, Halligan KE, Alzawahra WF, Barroso M, Singer HA, Jourd'heuil D, Trebak M (2010) Essential role for STIM1/Orai1-mediated calcium influx in PDGF-induced smooth muscle migration. Am J Physiol Cell Physiol 298(5):C993–C1005
Burnstock G (2008) Dual control of vascular tone and remodelling by ATP released from nerves and endothelial cells. Pharmacol Rep 60(1):12–20
Cahalan MD, Zhang SL, Yeromin AV, Ohlsen K, Roos J, Stauderman KA (2007) Molecular basis of the CRAC channel. Cell Calcium 42(2):133–144
Carden DL, Granger DN (2000) Pathophysiology of ischaemia-reperfusion injury. J Pathol 190(3):255–266
Cheng HW, James AF, Foster RR, Hancox JC, Bates DO (2006) VEGF activates receptor-operated cation channels in human microvascular endothelial cells. Arterioscler Thromb Vasc Biol 26(8):1768–1776
Cioffi DL (2011) Redox regulation of endothelial canonical transient receptor potential channels. Antioxid Redox Signal 15(6):1567–1582
Cioffi DL, Wu S, Chen H, Alexeyev M, St Croix CM, Pitt BR, Uhlig S, Stevens T (2012) Orai1 determines calcium selectivity of an endogenous TRPC heterotetramer channel. Circ Res 110(11):1435–1444
Courjaret R, Machaca K (2012) STIM and Orai in cellular proliferation and division. Frontiers in bioscience 4:331–341
Crabtree MJ, Brixey R, Batchelor H, Hale AB, Channon KM (2013) Integrated redox sensor and effector functions for tetrahydrobiopterin- and glutathionylation-dependent endothelial nitric-oxide synthase uncoupling. J Biol Chem 288(1):561–569
Dalrymple A, Mahn K, Poston L, Songu-Mize E, Tribe RM (2007) Mechanical stretch regulates TRPC expression and calcium entry in human myometrial smooth muscle cells. Mol Hum Reprod 13(3):171–179
Damon DH (2000) Adrenoceptor-mediated modulation of endothelial-dependent vascular smooth muscle growth. J Auton Pharmacol 20(1):47–54
de Frutos S, Caldwell E, Nitta CH, Kanagy NL, Wang J, Wang W, Walker MK, Gonzalez Bosc LV (2010) NFATc3 contributes to intermittent hypoxia-induced arterial remodeling in mice. Am J Physiol Heart Circ Physiol 299(2):H356–H363
DeHaven WI, Jones BF, Petranka JG, Smyth JT, Tomita T, Bird GS, Putney JW Jr (2009) TRPC channels function independently of STIM1 and Orai1. J Physiol 587(Pt 10):2275–2298
Dietrich A, Chubanov V, Kalwa H, Rost BR, Gudermann T (2006) Cation channels of the transient receptor potential superfamily: their role in physiological and pathophysiological processes of smooth muscle cells. Pharmacol Ther 112(3):744–760
Dietrich A, Gudermann T (2008) Another TRP to endothelial dysfunction: TRPM2 and endothelial permeability. Circ Res 102(3):275–277
Dietrich A, Kalwa H, Gudermann T (2010) TRPC channels in vascular cell function. Thromb Haemost 103(2):262–270
Dietrich A, Kalwa H, Rost BR, Gudermann T (2005) The diacylgylcerol-sensitive TRPC3/6/7 subfamily of cation channels: functional characterization and physiological relevance. Pflugers Arch 451(1):72–80
Dong YY, Wu M, Yim AP, He GW (2005) Hypoxia-reoxygenation, St. Thomas cardioplegic solution, and nicorandil on endothelium-derived hyperpolarizing factor in coronary microarteries. The Annals of thoracic surgery 80(5):1803–1811

Edwards JM, Neeb ZP, Alloosh MA, Long X, Bratz IN, Peller CR, Byrd JP, Kumar S, Obukhov AG, Sturek M (2010) Exercise training decreases store-operated Ca^{2+} entry associated with metabolic syndrome and coronary atherosclerosis. Cardiovasc Res 85(3):631–640

Flemming PK, Dedman AM, Xu SZ, Li J, Zeng F, Naylor J, Benham CD, Bateson AN, Muraki K, Beech DJ (2006) Sensing of lysophospholipids by TRPC5 calcium channel. J Biol Chem 281(8):4977–4982

Freichel M, Suh SH, Pfeifer A, Schweig U, Trost C, Weissgerber P, Biel M, Philipp S, Freise D, Droogmans G, Hofmann F, Flockerzi V, Nilius B (2001) Lack of an endothelial store-operated Ca^{2+} current impairs agonist-dependent vasorelaxation in TRP4-/- mice. Nat Cell Biol 3(2):121–127

Frischauf I, Schindl R, Derler I, Bergsmann J, Fahrner M, Romanin C (2008) The STIM/Orai coupling machinery. Channels (Austin) 2(4):261–268

Fukuyama K, Ichiki T, Imayama I, Ohtsubo H, Ono H, Hashiguchi Y, Takeshita A, Sunagawa K (2006) Thyroid hormone inhibits vascular remodeling through suppression of cAMP response element binding protein activity. Arterioscler Thromb Vasc Biol 26(9):2049–2055

Furchgott RF, Zawadzki JV (1980) The obligatory role of endothelial cells in the relaxation of arterial smooth muscle by acetylcholine. Nature 288(5789):373–376

Garcia RL, Schilling WP (1997) Differential expression of mammalian TRP homologues across tissues and cell lines. Biochem Biophys Res Commun 239(1):279–283

Ge R, Tai Y, Sun Y, Zhou K, Yang S, Cheng T, Zou Q, Shen F, Wang Y (2009) Critical role of TRPC6 channels in VEGF-mediated angiogenesis. Cancer Lett 283(1):43–51

Geeraerts MD, Ronveaux-Dupal MF, Lemasters JJ, Herman B (1991) Cytosolic free Ca^{2+} and proteolysis in lethal oxidative injury in endothelial cells. Am J Physiol 261(5 Pt 1):C889–C896

Gerwins P, Skoldenberg E, Claesson-Welsh L (2000) Function of fibroblast growth factors and vascular endothelial growth factors and their receptors in angiogenesis. Crit Rev Oncol Hematol 34(3):185–194

Glasnov TN, Groschner K, Kappe CO (2009) High-speed microwave-assisted synthesis of the trifluoromethylpyrazol-derived canonical transient receptor potential (TRPC) channel inhibitor Pyr3. ChemMedChem 4(11):1816–1818

Golovina VA, Platoshyn O, Bailey CL, Wang J, Limsuwan A, Sweeney M, Rubin LJ, Yuan JX (2001) Upregulated TRP and enhanced capacitative Ca(2 +) entry in human pulmonary artery myocytes during proliferation. Am J Physiol Heart Circ Physiol 280(2):H746–H755

Gonzalez-Cobos JC, Trebak M (2010) TRPC channels in smooth muscle cells. Front Biosci j Virtual Libr 15:1023–1039

Gottlieb P, Folgering J, Maroto R, Raso A, Wood TG, Kurosky A, Bowman C, Bichet D, Patel A, Sachs F, Martinac B, Hamill OP, Honore E (2008) Revisiting TRPC1 and TRPC6 mechanosensitivity. Pflugers Arch 455(6):1097–1103

Graham S, Ding M, Ding Y, Sours-Brothers S, Luchowski R, Gryczynski Z, Yorio T, Ma H, Ma R (2010) Canonical transient receptor potential 6 (TRPC6), a redox-regulated cation channel. J Biol Chem 285(30):23466–23476

Gray SP, Di Marco E, Okabe J, Szyndralewiez C, Heitz F, Montezano AC, de Haan JB, Koulis C, El-Osta A, Andrews KL, Chin-Dusting JP, Touyz RM, Wingler K, Cooper ME, Schmidt HH, Jandeleit-Dahm KA (2013) Nox1 Plays a Key Role in Diabetes Accelerated Atherosclerosis. Circulation 127:1888–1902

Groschner K, Hingel S, Lintschinger B, Balzer M, Romanin C, Zhu X, Schreibmayer W (1998) Trp proteins form store-operated cation channels in human vascular endothelial cells. FEBS Lett 437(1–2):101–106

Gudermann T, Hofmann T, Mederos y Schnitzler M, Dietrich A (2004) Activation, subunit composition and physiological relevance of DAG-sensitive TRPC proteins. Novartis Found Symp 258:103–118; discussion 118–122, 155–109, 263–106

Guibert C, Ducret T, Savineau JP (2011) Expression and physiological roles of TRP channels in smooth muscle cells. Adv Exp Med Biol 704:687–706

Gwack Y, Srikanth S, Feske S, Cruz-Guilloty F, Oh-hora M, Neems DS, Hogan PG, Rao A (2007) Biochemical and functional characterization of Orai proteins. J Biol Chem 282(22):16232–16243

Hamdollah Zadeh MA, Glass CA, Magnussen A, Hancox JC, Bates DO (2008) VEGF-mediated elevated intracellular calcium and angiogenesis in human microvascular endothelial cells in vitro are inhibited by dominant negative TRPC6. Microcirculation 15(7):605–614

Hao H, Gabbiani G, Bochaton-Piallat ML (2003) Arterial smooth muscle cell heterogeneity: implications for atherosclerosis and restenosis development. Arterioscler Thromb Vasc Biol 23(9):1510–1520

Hartmann J, Dragicevic E, Adelsberger H, Henning HA, Sumser M, Abramowitz J, Blum R, Dietrich A, Freichel M, Flockerzi V, Birnbaumer L, Konnerth A (2008) TRPC3 channels are required for synaptic transmission and motor coordination. Neuron 59(3):392–398

He B, Liu F, Ruan J, Li A, Chen J, Li R, Shen J, Zheng D, Luo R (2012) Silencing TRPC1 expression inhibits invasion of CNE2 nasopharyngeal tumor cells. Oncol Rep 27(5):1548–1554

Hecquet CM, Ahmmed GU, Vogel SM, Malik AB (2008) Role of TRPM2 channel in mediating H2O2-induced Ca^{2+} entry and endothelial hyperpermeability. Circ Res 102(3):347–355

Herbert SP, Stainier DY (2011) Molecular control of endothelial cell behaviour during blood vessel morphogenesis. Nat Rev Mol Cell Biol 12(9):551–564

Hisatsune C, Kuroda Y, Nakamura K, Inoue T, Nakamura T, Michikawa T, Mizutani A, Mikoshiba K (2004) Regulation of TRPC6 channel activity by tyrosine phosphorylation. J Biol Chem 279(18):18887–18894

Hofmann T, Obukhov AG, Schaefer M, Harteneck C, Gudermann T, Schultz G (1999) Direct activation of human TRPC6 and TRPC3 channels by diacylglycerol. Nature 397(6716):259–263

Hou X, Pedi L, Diver MM, Long SB (2012) Crystal structure of the calcium release-activated calcium channel Orai. Science 338(6112):1308–1313

House SJ, Potier M, Bisaillon J, Singer HA, Trebak M (2008) The non-excitable smooth muscle: calcium signaling and phenotypic switching during vascular disease. Pflugers Arch 456(5):769–785

Huang JH, He GW, Xue HM, Yao XQ, Liu XC, Underwood MJ, Yang Q (2011) TRPC3 channel contributes to nitric oxide release: significance during normoxia and hypoxia-reoxygenation. Cardiovasc Res 91(3):472–482

Huppert J, Closhen D, Croxford A, White R, Kulig P, Pietrowski E, Bechmann I, Becher B, Luhmann HJ, Waisman A, Kuhlmann CR (2010) Cellular mechanisms of IL-17-induced blood-brain barrier disruption. Faseb J 24(4):1023–1034

Imai Y, Itsuki K, Okamura Y, Inoue R, Mori MX (2012) A self-limiting regulation of vasoconstrictor-activated TRPC3/C6/C7 channels coupled to PI(4,5)P(2)-diacylglycerol signalling. J Physiol 590(Pt 5):1101–1119

Inoue R, Jensen LJ, Shi J, Morita H, Nishida M, Honda A, Ito Y (2006) Transient receptor potential channels in cardiovascular function and disease. Circ Res 99(2):119–131

Inoue R, Okada T, Onoue H, Hara Y, Shimizu S, Naitoh S, Ito Y, Mori Y (2001) The transient receptor potential protein homologue TRP6 is the essential component of vascular alpha(1)-adrenoceptor-activated Ca(2 +)-permeable cation channel. Circ Res 88(3):325–332

Irie S, Tavassoli M (1991) Transendothelial transport of macromolecules: the concept of tissue-blood barriers. Cell Biol Rev 25 (4):317–333, 340–311

Jho D, Mehta D, Ahmmed G, Gao XP, Tiruppathi C, Broman M, Malik AB (2005) Angiopoietin-1 opposes VEGF-induced increase in endothelial permeability by inhibiting TRPC1-dependent Ca2 influx. Circ Res 96(12):1282–1290

Jones BF, Boyles RR, Hwang SY, Bird GS, Putney JW (2008) Calcium influx mechanisms underlying calcium oscillations in rat hepatocytes. Hepatology 48(4):1273–1281

Ju M, Shi J, Saleh SN, Albert AP, Large WA (2010) Ins(1,4,5)P3 interacts with PIP2 to regulate activation of TRPC6/C7 channels by diacylglycerol in native vascular myocytes. J Physiol 588(Pt 9):1419–1433

Kabbinavar F, Hurwitz HI, Fehrenbacher L, Meropol NJ, Novotny WF, Lieberman G, Griffing S, Bergsland E (2003) Phase II, randomized trial comparing bevacizumab plus fluorouracil (FU)/ leucovorin (LV) with FU/LV alone in patients with metastatic colorectal cancer. J Clin Oncol 21(1):60–65

Kamouchi M, Philipp S, Flockerzi V, Wissenbach U, Mamin A, Raeymaekers L, Eggermont J, Droogmans G, Nilius B (1999) Properties of heterologously expressed hTRP3 channels in bovine pulmonary artery endothelial cells. J Physiol 518(Pt 2):345–358

Kawai-Kowase K, Owens GK (2007) Multiple repressor pathways contribute to phenotypic switching of vascular smooth muscle cells. Am J Physiol Cell Physiol 292(1):C59–C69

Kernochan LE, Tran BN, Tangkijvanich P, Melton AC, Tam SP, Yee HF Jr (2002) Endothelin-1 stimulates human colonic myofibroblast contraction and migration. Gut 50(1):65–70

Kim K, Drummond I, Ibraghimov-Beskrovnaya O, Klinger K, Arnaout MA (2000) Polycystin 1 is required for the structural integrity of blood vessels. Proc Natl Acad Sci U S A 97(4):1731–1736

Kiyonaka S, Kato K, Nishida M, Mio K, Numaga T, Sawaguchi Y, Yoshida T, Wakamori M, Mori E, Numata T, Ishii M, Takemoto H, Ojida A, Watanabe K, Uemura A, Kurose H, Morii T, Kobayashi T, Sato Y, Sato C, Hamachi I, Mori Y (2009) Selective and direct inhibition of TRPC3 channels underlies biological activities of a pyrazole compound. Proc Natl Acad Sci U S A 106(13):5400–5405

Koenig S, Schernthaner M, Maechler H, Kappe CO, Glasnov TN, Hoefler G, Braune M, Wittchow E, Groschner K (2013) A TRPC3 Blocker, Ethyl-1-(4-(2,3,3-Trichloroacrylamide)Phenyl)-5-(Trifluoromethyl)-1H-Pyrazole-4-C arboxylate (Pyr3), Prevents Stent-Induced Arterial Remodeling. J Pharmacol Exp Ther 344(1):33–40

Kuang CY, Yu Y, Guo RW, Qian DH, Wang K, Den MY, Shi YK, Huang L (2010) Silencing stromal interaction molecule 1 by RNA interference inhibits the proliferation and migration of endothelial progenitor cells. Biochem Biophys Res Commun 398(2):315–320

Kuang CY, Yu Y, Wang K, Qian DH, Den MY, Huang L (2012) Knockdown of transient receptor potential canonical-1 reduces the proliferation and migration of endothelial progenitor cells. Stem Cells Devel 21(3):487–496

Kumar B, Dreja K, Shah SS, Cheong A, Xu SZ, Sukumar P, Naylor J, Forte A, Cipollaro M, McHugh D, Kingston PA, Heagerty AM, Munsch CM, Bergdahl A, Hultgardh-Nilsson A, Gomez MF, Porter KE, Hellstrand P, Beech DJ (2006) Upregulated TRPC1 channel in vascular injury in vivo and its role in human neointimal hyperplasia. Circ Res 98(4):557–563

Kusaba T, Okigaki M, Matui A, Murakami M, Ishikawa K, Kimura T, Sonomura K, Adachi Y, Shibuya M, Shirayama T, Tanda S, Hatta T, Sasaki S, Mori Y, Matsubara H (2010) Klotho is associated with VEGF receptor-2 and the transient receptor potential canonical-1 Ca^{2+} channel to maintain endothelial integrity. Proc Natl Acad Sci U S A 107(45):19308–19313

Lang F, Foller M, Lang KS, Lang PA, Ritter M, Gulbins E, Vereninov A, Huber SM (2005) Ion channels in cell proliferation and apoptotic cell death. J Membr Biol 205(3):147–157

Lee KP, Yuan JP, Hong JH, So I, Worley PF, Muallem S (2010a) An endoplasmic reticulum/ plasma membrane junction: STIM1/Orai1/TRPCs. FEBS Lett 584(10):2022–2027

Lee KP, Yuan JP, So I, Worley PF, Muallem S (2010b) STIM1-dependent and STIM1-independent function of transient receptor potential canonical (TRPC) channels tunes their store-operated mode. J Biol Chem 285(49):38666–38673

Lefroy DC, Crake T, Uren NG, Davies GJ, Maseri A (1993) Effect of inhibition of nitric oxide synthesis on epicardial coronary artery caliber and coronary blood flow in humans. Circulation 88(1):43–54

Lemos VS, Poburko D, Liao CH, Cole WC, van Breemen C (2007) Na^{+} entry via TRPC6 causes Ca^{2+} entry via NCX reversal in ATP stimulated smooth muscle cells. Biochem Biophys Res Commun 352(1):130–134

Li J, Cubbon RM, Wilson LA, Amer MS, McKeown L, Hou B, Majeed Y, Tumova S, Seymour VA, Taylor H, Stacey M, O'Regan D, Foster R, Porter KE, Kearney MT, Beech DJ (2011a) Orai1 and CRAC channel dependence of VEGF-activated Ca^{2+} entry and endothelial tube formation. Circ Res 108(10):1190–1198

Li J, McKeown L, Ojelabi O, Stacey M, Foster R, O'Regan D, Porter KE, Beech DJ (2011b) Nanomolar potency and selectivity of a Ca(2)(+) release-activated Ca(2)(+) channel inhibitor against store-operated Ca(2)(+) entry and migration of vascular smooth muscle cells. Br J Pharmacol 164(2):382–393

Li S, Westwick J, Cox B, Poll CT (2004) TRP channels as drug targets. Novartis Found Symp 258:204–213; discussion 213–221, 263–206

Liao Y, Erxleben C, Abramowitz J, Flockerzi V, Zhu MX, Armstrong DL, Birnbaumer L (2008) Functional interactions among Orai1, TRPCs, and STIM1 suggest a STIM-regulated heteromeric Orai/TRPC model for SOCE/Icrac channels. Proc Natl Acad Sci USA 105(8):2895–2900

Liao Y, Plummer NW, George MD, Abramowitz J, Zhu MX, Birnbaumer L (2009) A role for Orai in TRPC-mediated Ca^{2+} entry suggests that a TRPC:Orai complex may mediate store and receptor operated Ca^{2+} entry. Proc Natl Acad Sci U S A 106(9):3202–3206

Liu X, Bandyopadhyay BC, Singh BB, Groschner K, Ambudkar IS (2005) Molecular analysis of a store-operated and 2-acetyl-sn-glycerol-sensitive non-selective cation channel. Heteromeric assembly of TRPC1-TRPC3. J Biol Chem 280(22):21600–21606

Lu D, Kassab GS (2011) Role of shear stress and stretch in vascular mechanobiology. J Roy Soc Interface/Roy Soc 8(63):1379–1385

Lyfenko AD, Dirksen RT (2008) Differential dependence of store-operated and excitation-coupled Ca^{2+} entry in skeletal muscle on STIM1 and Orai1. J Physiol 586(Pt 20):4815–4824

Ma X, Cheng KT, Wong CO, O'Neil RG, Birnbaumer L, Ambudkar IS, Yao X (2011) Heteromeric TRPV4-C1 channels contribute to store-operated Ca(2 +) entry in vascular endothelial cells. Cell Calcium 50(6):502–509

Maroto R, Raso A, Wood TG, Kurosky A, Martinac B, Hamill OP (2005) TRPC1 forms the stretch-activated cation channel in vertebrate cells. Nat Cell Biol 7(2):179–185

Maruyama Y, Nakanishi Y, Walsh EJ, Wilson DP, Welsh DG, Cole WC (2006) Heteromultimeric TRPC6-TRPC7 channels contribute to arginine vasopressin-induced cation current of A7r5 vascular smooth muscle cells. Circ Res 98(12):1520–1527

McNally BA, Prakriya M (2012) Permeation, selectivity and gating in store-operated CRAC channels. J Physiol 590(Pt 17):4179–4191

Mederos y Schnitzler M, Storch U, Meibers S, Nurwakagari P, Breit A, Essin K, Gollasch M, Gudermann T (2008) Gq-coupled receptors as mechanosensors mediating myogenic vasoconstriction. EMBO J 27(23):3092–3103

Meoli DF, White RJ (2010) Endothelin-1 induces pulmonary but not aortic smooth muscle cell migration by activating ERK1/2 MAP kinase. Can J Physiol Pharmacol 88(8):830–839

Miller BA, Zhang W (2011) TRP channels as mediators of oxidative stress. Adv Exp Med Biol 704:531–544

Moore TM, Brough GH, Babal P, Kelly JJ, Li M, Stevens T (1998) Store-operated calcium entry promotes shape change in pulmonary endothelial cells expressing Trp1. Am J Physiol 275(3 Pt 1):L574–L582

Muik M, Schindl R, Fahrner M, Romanin C (2012) Ca(2 +) release-activated Ca(2 +) (CRAC) current, structure, and function. Cell Mol Life Sci 69(24):4163–4176

Nakahashi T, Fukuo K, Nishimaki H, Hata S, Shimizu M, Suhara T, Takimoto M, Morimoto S, Ogihara T (1998) Endothelin-1 enhances nitric oxide-induced cell death in cultured vascular smooth-muscle cells. J Cardiovasc Pharmacol 31(Suppl 1):S351–S353

Naylor J, Al-Shawaf E, McKeown L, Manna PT, Porter KE, O'Regan D, Muraki K, Beech DJ (2011) TRPC5 channel sensitivities to antioxidants and hydroxylated stilbenes. J Biol Chem 286(7):5078–5086

Neufeld G, Tessler S, Gitay-Goren H, Cohen T, Levi BZ (1994) Vascular endothelial growth factor and its receptors. Prog Growth Factor Res 5(1):89–97

Nilius B, Droogmans G, Wondergem R (2003) Transient receptor potential channels in endothelium: solving the calcium entry puzzle? Endothelium J Endothelial Cell Res 10(1):5–15

Nilius B, Owsianik G, Voets T, Peters JA (2007) Transient receptor potential cation channels in disease. Physiol Rev 87(1):165–217

Okada T, Inoue R, Yamazaki K, Maeda A, Kurosaki T, Yamakuni T, Tanaka I, Shimizu S, Ikenaka K, Imoto K, Mori Y (1999) Molecular and functional characterization of a novel mouse transient receptor potential protein homologue TRP7. Ca(2 +)-permeable cation channel that is constitutively activated and enhanced by stimulation of G protein-coupled receptor. J Biol Chem 274(39):27359–27370

Okada T, Shimizu S, Wakamori M, Maeda A, Kurosaki T, Takada N, Imoto K, Mori Y (1998) Molecular cloning and functional characterization of a novel receptor-activated TRP Ca^{2+} channel from mouse brain. J Biol Chem 273(17):10279–10287

Ong HL, Cheng KT, Liu X, Bandyopadhyay BC, Paria BC, Soboloff J, Pani B, Gwack Y, Srikanth S, Singh BB, Gill DL, Ambudkar IS (2007) Dynamic assembly of TRPC1-STIM1-Orai1 ternary complex is involved in store-operated calcium influx. Evidence for similarities in store-operated and calcium release-activated calcium channel components. J Biol Chem 282(12):9105–9116

Onohara N, Nishida M, Inoue R, Kobayashi H, Sumimoto H, Sato Y, Mori Y, Nagao T, Kurose H (2006) TRPC3 and TRPC6 are essential for angiotensin II-induced cardiac hypertrophy. EMBO J 25(22):5305–5316

Owens GK, Kumar MS, Wamhoff BR (2004) Molecular regulation of vascular smooth muscle cell differentiation in development and disease. Physiol Rev 84(3):767–801

Pang X, Sun NL (2009) Calcineurin-NFAT signaling is involved in phenylephrine-induced vascular smooth muscle cell proliferation. Acta Pharmacol Sin 30(5):537–544

Paria BC, Vogel SM, Ahmmed GU, Alamgir S, Shroff J, Malik AB, Tiruppathi C (2004) Tumor necrosis factor-alpha-induced TRPC1 expression amplifies store-operated Ca^{2+} influx and endothelial permeability. Am J Physiol Lung Cell Mol Physiol 287(6):L1303–L1313

Peel SE, Liu B, Hall IP (2008) ORAI and store-operated calcium influx in human airway smooth muscle cells. Am J Respir Cell Mol Biol 38(6):744–749

Peiro C, Redondo J, Rodriguez-Martinez MA, Angulo J, Marin J, Sanchez-Ferrer CF (1995) Influence of endothelium on cultured vascular smooth muscle cell proliferation. Hypertension 25(4 Pt 2):748–751

Philipp S, Cavalie A, Freichel M, Wissenbach U, Zimmer S, Trost C, Marquart A, Murakami M, Flockerzi V (1996) A mammalian capacitative calcium entry channel homologous to Drosophila TRP and TRPL. EMBO J 15(22):6166–6171

Philipp S, Hambrecht J, Braslavski L, Schroth G, Freichel M, Murakami M, Cavalie A, Flockerzi V (1998) A novel capacitative calcium entry channel expressed in excitable cells. EMBO J 17(15):4274–4282

Plant TD, Schaefer M (2003) TRPC4 and TRPC5: receptor-operated Ca^{2+}-permeable nonselective cation channels. Cell Calcium 33(5–6):441–450

Poteser M, Graziani A, Eder P, Yates A, Machler H, Romanin C, Groschner K (2008) Identification of a rare subset of adipose tissue-resident progenitor cells, which express CD133 and TRPC3 as a VEGF-regulated Ca^{2+}entry channel. FEBS Lett 582(18):2696–2702

Poteser M, Graziani A, Rosker C, Eder P, Derler I, Kahr H, Zhu MX, Romanin C, Groschner K (2006) TRPC3 and TRPC4 associate to form a redox-sensitive cation channel. Evidence for expression of native TRPC3-TRPC4 heteromeric channels in endothelial cells. J Biol Chem 281(19):13588–13595

Poteser M, Schleifer H, Lichtenegger M, Schernthaner M, Stockner T, Kappe CO, Glasnov TN, Romanin C, Groschner K (2011) PKC-dependent coupling of calcium permeation through transient receptor potential canonical 3 (TRPC3) to calcineurin signaling in HL-1 myocytes. Proc Natl Acad Sci USA 108(26):10556–10561

Potier M, Gonzalez JC, Motiani RK, Abdullaev IF, Bisaillon JM, Singer HA, Trebak M (2009) Evidence for STIM1- and Orai1-dependent store-operated calcium influx through ICRAC in vascular smooth muscle cells: role in proliferation and migration. Faseb J 23(8):2425–2437

Prakriya M, Feske S, Gwack Y, Srikanth S, Rao A, Hogan PG (2006) Orai1 is an essential pore subunit of the CRAC channel. Nature 443(7108):230–233

Qin L, Zhao D, Xu J, Ren X, Terwilliger EF, Parangi S, Lawler J, Dvorak HF, Zeng H (2013) The vascular permeabilizing factors histamine and serotonin induce angiogenesis through TR3/ Nur77 and subsequently truncate it through thrombospondin-1. Blood

Rees DD, Palmer RM, Moncada S (1989) Role of endothelium-derived nitric oxide in the regulation of blood pressure. Proc Natl Acad Sci USA 86(9):3375–3378

Rippe B, Rosengren BI, Carlsson O, Venturoli D (2002) Transendothelial transport: the vesicle controversy. J Vasc Res 39(5):375–390

Rodell TC, Cheronis JC, Ohnemus CL, Piermattei DJ, Repine JE (1987) Xanthine oxidase mediates elastase-induced injury to isolated lungs and endothelium. J Appl Physiol 63(5):2159–2163

Rowell J, Koitabashi N, Kass DA (2010) TRP-ing up heart and vessels: canonical transient receptor potential channels and cardiovascular disease. J cardiovasc transl res 3(5):516–524

Ryan US (1989) Endothelium as a transducing surface. J Mol Cell Cardiol 21(Suppl 1):85–90

Salido GM, Jardin I, Rosado JA (2011) The TRPC ion channels: association with Orai1 and STIM1 proteins and participation in capacitative and non-capacitative calcium entry. Adv Exp Med Biol 704:413–433

Samapati R, Yang Y, Yin J, Stoerger C, Arenz C, Dietrich A, Gudermann T, Adam D, Wu S, Freichel M, Flockerzi V, Uhlig S, Kuebler WM (2012) Lung endothelial $Ca^{2+}$and permeability response to platelet-activating factor is mediated by acid sphingomyelinase and transient receptor potential classical 6. Am J Respir Crit Care Med 185(2):160–170

Schaefer M (2005) Homo- and heteromeric assembly of TRP channel subunits. Pflugers Arch 451(1):35–42

Schaefer M, Plant TD, Obukhov AG, Hofmann T, Gudermann T, Schultz G (2000) Receptor-mediated regulation of the nonselective cation channels TRPC4 and TRPC5. J Biol Chem 275(23):17517–17526

Schaff UY, Dixit N, Procyk E, Yamayoshi I, Tse T, Simon SI (2010) Orai1 regulates intracellular calcium, arrest, and shape polarization during neutrophil recruitment in shear flow. Blood 115(3):657–666

Schleifer H, Doleschal B, Lichtenegger M, Oppenrieder R, Derler I, Frischauf I, Glasnov TN, Kappe CO, Romanin C, Groschner K (2012) Novel pyrazole compounds for pharmacological discrimination between receptor-operated and store-operated Ca(2 +) entry pathways. Br J Pharmacol 167(8):1712–1722

Schlondorff J, Del Camino D, Carrasquillo R, Lacey V, Pollak MR (2009) TRPC6 mutations associated with focal segmental glomerulosclerosis cause constitutive activation of NFAT-dependent transcription. Am J Physiol Cell Physiol 296(3):C558–C569

Shi-Wen X, Denton CP, Dashwood MR, Holmes AM, Bou-Gharios G, Pearson JD, Black CM, Abraham DJ (2001) Fibroblast matrix gene expression and connective tissue remodeling: role of endothelin-1. J Invest Dermatol 116(3):417–425

Shi J, Ju M, Abramowitz J, Large WA, Birnbaumer L, Albert AP (2012) TRPC1 proteins confer PKC and phosphoinositol activation on native heteromeric TRPC1/C5 channels in vascular smooth muscle: comparative study of wild-type and TRPC1-/- mice. Faseb J 26(1):409–419

Shibuya M (2013) Vascular endothelial growth factor and its receptor system: physiological functions in angiogenesis and pathological roles in various diseases. J Biochem 153(1):13–19

Shimoda LA, Laurie SS (2013) Vascular remodeling in pulmonary hypertension. J Mol Med 91(3):297–309

Shinde AV, Motiani RK, Zhang X, Abdullaev IF, Adam AP, Gonzalez-Cobos JC, Zhang W, Matrougui K, Vincent PA, Trebak M (2013) STIM1 controls endothelial barrier function independently of Orai1 and Ca^{2+} entry. Sci Signal 6 (267):ra18

Smedlund K, Vazquez G (2008) Involvement of native TRPC3 proteins in ATP-dependent expression of VCAM-1 and monocyte adherence in coronary artery endothelial cells. Arterioscler Thromb Vasc Biol 28(11):2049–2055

Soboloff J, Spassova M, Xu W, He LP, Cuesta N, Gill DL (2005) Role of endogenous TRPC6 channels in Ca^{2+} signal generation in A7r5 smooth muscle cells. J Biol Chem 280(48):39786–39794

Spassova MA, Hewavitharana T, Xu W, Soboloff J, Gill DL (2006) A common mechanism underlies stretch activation and receptor activation of TRPC6 channels. Proc Natl Acad Sci USA 103(44):16586–16591

Stalcup SA, Davidson D, Mellins RB (1982) Endothelial cell functions in the hemodynamic responses to stress. Ann N Y Acad Sci 401:117–131

Storch U, Forst AL, Philipp M, Gudermann T, Mederos y Schnitzler M (2012) Transient receptor potential channel 1 (TRPC1) reduces calcium permeability in heteromeric channel complexes. J Biol Chem 287(5):3530–3540

Strubing C, Krapivinsky G, Krapivinsky L, Clapham DE (2001) TRPC1 and TRPC5 form a novel cation channel in mammalian brain. Neuron 29(3):645–655

Sundivakkam PC, Freichel M, Singh V, Yuan JP, Vogel SM, Flockerzi V, Malik AB, Tiruppathi C (2012) The Ca(2 +) sensor stromal interaction molecule 1 (STIM1) is necessary and sufficient for the store-operated Ca(2 +) entry function of transient receptor potential canonical (TRPC) 1 and 4 channels in endothelial cells. Mol Pharmacol 81(4):510–526

Sweeney M, Yu Y, Platoshyn O, Zhang S, McDaniel SS, Yuan JX (2002) Inhibition of endogenous TRP1 decreases capacitative Ca^{2+} entry and attenuates pulmonary artery smooth muscle cell proliferation. Am J Physiol Lung Cell Mol Physiol 283(1):L144–L155

Tai Y, Feng S, Ge R, Du W, Zhang X, He Z, Wang Y (2008) TRPC6 channels promote dendritic growth via the CaMKIV-CREB pathway. J Cell Sci 121(Pt 14):2301–2307

Thilo F, Vorderwulbecke BJ, Marki A, Krueger K, Liu Y, Baumunk D, Zakrzewicz A, Tepel M (2012) Pulsatile atheroprone shear stress affects the expression of transient receptor potential channels in human endothelial cells. Hypertension 59(6):1232–1240

Tokunou T, Ichiki T, Takeda K, Funakoshi Y, Iino N, Takeshita A (2001) cAMP response element-binding protein mediates thrombin-induced proliferation of vascular smooth muscle cells. Arterioscler Thromb Vasc Biol 21(11):1764–1769

Tolhurst G, Carter RN, Amisten S, Holdich JP, Erlinge D, Mahaut-Smith MP (2008) Expression profiling and electrophysiological studies suggest a major role for Orai1 in the store-operated Ca^{2+} influx pathway of platelets and megakaryocytes. Platelets 19(4):308–313

Trebak M (2012) STIM/Orai signalling complexes in vascular smooth muscle. J Physiol 590(Pt 17):4201–4208

Trebak M, Vazquez G, Bird GS, Putney JW Jr (2003) The TRPC3/6/7 subfamily of cation channels. Cell Calcium 33(5–6):451–461

Ulloa A, Gonzales AL, Zhong M, Kim YS, Cantlon J, Clay C, Ku CY, Earley S, Sanborn BM (2009) Reduction in TRPC4 expression specifically attenuates G-protein coupled receptor-stimulated increases in intracellular calcium in human myometrial cells. Cell Calcium 46(1):73–84

Ungvari Z, Wolin MS, Csiszar A (2006) Mechanosensitive production of reactive oxygen species in endothelial and smooth muscle cells: role in microvascular remodeling? Antioxid Redox Signal 8(7–8):1121–1129

Usatyuk PV, Natarajan V (2004) Role of mitogen-activated protein kinases in 4-hydroxy-2-nonenal-induced actin remodeling and barrier function in endothelial cells. J Biol Chem 279(12):11789–11797

Vandebrouck C, Martin D, Colson-Van Schoor M, Debaix H, Gailly P (2002) Involvement of TRPC in the abnormal calcium influx observed in dystrophic (mdx) mouse skeletal muscle fibers. J Cell Biol 158(6):1089–1096

Vestweber D (2012) Relevance of endothelial junctions in leukocyte extravasation and vascular permeability. Ann N Y Acad Sci 1257:184–192

Voelkers M, Salz M, Herzog N, Frank D, Dolatabadi N, Frey N, Gude N, Friedrich O, Koch WJ, Katus HA, Sussman MA, Most P (2010) Orai1 and Stim1 regulate normal and hypertrophic growth in cardiomyocytes. J Mol Cell Cardiol 48(6):1329–1334

Wamhoff BR, Bowles DK, Owens GK (2006) Excitation-transcription coupling in arterial smooth muscle. Circ Res 98(7):868–878

Wang C, Li JF, Zhao L, Liu J, Wan J, Wang YX, Wang J (2009) Inhibition of SOC/Ca2 +/NFAT pathway is involved in the anti-proliferative effect of sildenafil on pulmonary artery smooth muscle cells. Respir Res 10:123

Wang X, Pluznick JL, Wei P, Padanilam BJ, Sansom SC (2004) TRPC4 forms store-operated Ca^{2+} channels in mouse mesangial cells. Am J Physiol Cell Physiol 287(2):C357–C364

Wang Y, Deng X, Hewavitharana T, Soboloff J, Gill DL (2008a) Stim, ORAI and TRPC channels in the control of calcium entry signals in smooth muscle. Clin Exp Pharmacol Physiol 35(9):1127–1133

Wang YX, Wang J, Wang C, Liu J, Shi LP, Xu M (2008b) Functional expression of transient receptor potential vanilloid-related channels in chronically hypoxic human pulmonary arterial smooth muscle cells. J Membr Biol 223(3):151–159

Watanabe H, Murakami M, Ohba T, Takahashi Y, Ito H (2008) TRP channel and cardiovascular disease. Pharmacol Ther 118(3):337–351

Wedgwood S, Dettman RW, Black SM (2001) ET-1 stimulates pulmonary arterial smooth muscle cell proliferation via induction of reactive oxygen species. Am J Physiol Lung Cell Mol Physiol 281(5):L1058–L1067

Weissmann N, Sydykov A, Kalwa H, Storch U, Fuchs B, Mederos y Schnitzler M, Brandes RP, Grimminger F, Meissner M, Freichel M, Offermanns S, Veit F, Pak O, Krause KH, Schermuly RT, Brewer AC, Schmidt HH, Seeger W, Shah AM, Gudermann T, Ghofrani HA, Dietrich A (2012) Activation of TRPC6 channels is essential for lung ischaemia-reperfusion induced oedema in mice. Nat commun 3:649

Wong CO, Sukumar P, Beech DJ, Yao X (2010) Nitric oxide lacks direct effect on TRPC5 channels but suppresses endogenous TRPC5-containing channels in endothelial cells. Pflugers Arch 460(1):121–130

Worley PF, Zeng W, Huang GN, Yuan JP, Kim JY, Lee MG, Muallem S (2007) TRPC channels as STIM1-regulated store-operated channels. Cell Calcium 42(2):205–211

Xu SZ, Beech DJ (2001) TrpC1 is a membrane-spanning subunit of store-operated Ca(2 +) channels in native vascular smooth muscle cells. Circ Res 88(1):84–87

Xu SZ, Muraki K, Zeng F, Li J, Sukumar P, Shah S, Dedman AM, Flemming PK, McHugh D, Naylor J, Cheong A, Bateson AN, Munsch CM, Porter KE, Beech DJ (2006) A sphingosine-1-phosphate-activated calcium channel controlling vascular smooth muscle cell motility. Circ Res 98(11):1381–1389

Yang B, Gwozdz T, Dutko-Gwozdz J, Bolotina VM (2012) Orai1 and Ca^{2+} -independent phospholipase A2 are required for store-operated Icat-SOC current, Ca^{2+} entry, and proliferation of primary vascular smooth muscle cells. Am J Physiol Cell Physiol 302(5):C748–C756

Yang JC, Haworth L, Sherry RM, Hwu P, Schwartzentruber DJ, Topalian SL, Steinberg SM, Chen HX, Rosenberg SA (2003) A randomized trial of bevacizumab, an anti-vascular endothelial growth factor antibody, for metastatic renal cancer. N Engl J Med 349(5):427–434

Yang XR, Lin MJ, McIntosh LS, Sham JS (2006) Functional expression of transient receptor potential melastatin- and vanilloid-related channels in pulmonary arterial and aortic smooth muscle. Am J Physiol Lung Cell Mol Physiol 290(6):L1267–L1276

Yao H, Peng F, Fan Y, Zhu X, Hu G, Buch SJ (2009) TRPC channel-mediated neuroprotection by PDGF involves Pyk2/ERK/CREB pathway. Cell Death Differ 16(12):1681–1693

Yao X, Garland CJ (2005) Recent developments in vascular endothelial cell transient receptor potential channels. Circ Res 97(9):853–863

Yip H, Chan WY, Leung PC, Kwan HY, Liu C, Huang Y, Michel V, Yew DT, Yao X (2004) Expression of TRPC homologs in endothelial cells and smooth muscle layers of human arteries. Histochem Cell Biol 122(6):553–561

Yoshida T, Inoue R, Morii T, Takahashi N, Yamamoto S, Hara Y, Tominaga M, Shimizu S, Sato Y, Mori Y (2006) Nitric oxide activates TRP channels by cysteine S-nitrosylation. Nat Chem Biol 2(11):596–607

Yoshida T, Owens GK (2005) Molecular determinants of vascular smooth muscle cell diversity. Circ Res 96(3):280–291

Yu Y, Sweeney M, Zhang S, Platoshyn O, Landsberg J, Rothman A, Yuan JX (2003) PDGF stimulates pulmonary vascular smooth muscle cell proliferation by upregulating TRPC6 expression. Am J Physiol Cell Physiol 284(2):C316–C330

Yuan D, He P (2012) Vascular remodeling alters adhesion protein and cytoskeleton reactions to inflammatory stimuli resulting in enhanced permeability increases in rat venules. J Appl Physiol 113(7):1110–1120

Yuan JP, Kim MS, Zeng W, Shin DM, Huang G, Worley PF, Muallem S (2009) TRPC channels as STIM1-regulated SOCs. Channels (Austin) 3(4):221–225

Yuan JP, Zeng W, Huang GN, Worley PF, Muallem S (2007) STIM1 heteromultimerizes TRPC channels to determine their function as store-operated channels. Nat Cell Biol 9(6):636–645

Yuan P, Leonetti MD, Hsiung Y, MacKinnon R (2012) Open structure of the Ca^{2+} gating ring in the high-conductance Ca^{2+} -activated K^{+} channel. Nature 481(7379):94–97

Zhang J, Hu H, Palma NL, Harrison JK, Mubarak KK, Carrie RD, Alnuaimat H, Shen X, Luo D, Patel JM (2012) Hypoxia-induced endothelial CX3CL1 triggers lung smooth muscle cell phenotypic switching and proliferative expansion. Am J Physiol Lung Cell Mol Physiol 303(10):L912–L922

Zhang S, Remillard CV, Fantozzi I, Yuan JX (2004) ATP-induced mitogenesis is mediated by cyclic AMP response element-binding protein-enhanced TRPC4 expression and activity in human pulmonary artery smooth muscle cells. Am J Physiol Cell Physiol 287(5):C1192–C1201

Zhang W, Halligan KE, Zhang X, Bisaillon JM, Gonzalez-Cobos JC, Motiani RK, Hu G, Vincent PA, Zhou J, Barroso M, Singer HA, Matrougui K, Trebak M (2011) Orai1-mediated I (CRAC) is essential for neointima formation after vascular injury. Circ Res 109(5):534–542

Zhao X, Moloughney JG, Zhang S, Komazaki S, Weisleder N (2012) Orai1 mediates exacerbated Ca(2 +) entry in dystrophic skeletal muscle. PLoS ONE 7(11):e49862

Zhu JH, Chen CL, Flavahan S, Harr J, Su B, Flavahan NA (2011) Cyclic stretch stimulates vascular smooth muscle cell alignment by redox-dependent activation of Notch3. Am J Physiol Heart Circ Physiol 300(5):H1770–H1780

Chapter 24
TRPP2 in Polycystic Kidney Disease

Andrew Streets and Albert Ong

Abstract Mutations in *PKD1* (85 %) or *PKD2* (15 %) account for almost all cases of autosomal dominant polycystic kidney disease (ADPKD). The ADPKD proteins, polycystin-1 and polycystin-2 interact to form a receptor-ion channel complex which regulates tubular structure in the developing and adult kidney. Polycystin-2 (TRPP2) is the founding member of the Transient Receptor Potential Polycystic (TRPP) subfamily of non-selective ion channels. Considerable progress has been made in recent years to understand the structure and function of TRPP2 and its relevance to the pathogenesis of ADPKD. TRPP2 channels have been shown to reside in a number of different subcellular compartments regulating intracellular calcium levels as part of a complex with polycystin-1 and other channel proteins. Disturbed calcium signalling mediated by loss of TRPP2 or its interacting partner polycystin-1 is clearly an early event in the development of a cystic phenotype. Understanding the functional significance of disrupted calcium signalling and identifying affected downstream pathways may lead to the development of new therapeutic approaches in the treatment of ADPKD.

24.1 Introduction

Polycystin-2 (TRPP2) is the founding member of the Transient Receptor Potential Polycystic (TRPP) subfamily of non-selective ion channels. TRPP2 is mutated in around 15 % of cases of Autosomal Dominant Polycystic Kidney Disease. TRPP2 has been shown in a large number of studies to be involved in mechanosensitive and ligand-activated Ca^{2+}-dependent cell signalling pathways mediating a diverse range of biological functions including proliferation, apoptosis, body axis

A. Streets (✉) · A. Ong
Kidney Genetics Group, Academic Nephrology Unit, The Henry Wellcome Laboratories for Medical Research, University of Sheffield Medical School, Sheffield, UK
e-mail: a.j.streets@sheffield.ac.uk

N. Weiss and A. Koschak (eds.), *Pathologies of Calcium Channels*,
DOI: 10.1007/978-3-642-40282-1_24, © Springer-Verlag Berlin Heidelberg 2014

patterning and heart loop morphogenesis. TRPP2 is found at a number of locations within the cell, at the plasma membrane, endoplasmic reticulum, primary cilia and mitotic spindle. This complex subcellular localisation and the precise role of TRPP2 have remained controversial. This chapter will summarise the current evidence regarding the structural and functional regulation of TRPP2 channels.

24.1.1 Autosomal Dominant Polycystic Kidney Disease

Autosomal dominant polycystic kidney disease (ADPKD) is the most common inherited human renal disease (incidence 1 in 1,000 live births). In almost all patients, it is due to germline mutations in one of two genes, *PKD1* (85 %) or *PKD2* (15 %) which are clinically indistinguishable (Ong and Harris 2005; Torres et al. 2007). ADPKD is a major cause of end stage renal disease (ESRD), accounting for ~10 % of all patients on renal replacement therapy. ADPKD has high penetrance and usually progresses to end stage renal disease within the fourth to fifth decades of life (Cuppage et al. 1980).

Its cardinal feature is the formation of multiple renal cysts, which over time expand and distort kidney structure ultimately resulting in ESRD. Extrarenal manifestations are found in both *PKD1* and *PKD2* patients which include cysts in the liver or pancreas, hypertension, cardiac valvular abnormalities and cerebral aneurysms (Torres et al. 2007).

Diagnosis and screening of ADPKD are frequently done using ultrasound, although this is less accurate than MRI for assessing renal volume and cyst volume, so it is not useful for follow up of disease progression (O'Neill et al. 2005). Ultrasound is also less effective than CT scanning for detecting other abnormalities such as renal calculi (Nishiura et al. 2009). Genetic testing is available to diagnose ADPKD if it cannot be definitively diagnosed through imaging studies (Garcia-Gonzalez et al. 2007; Rossetti et al. 2002).

Symptoms and signs of ADPKD include hypertension, abdominal masses, (Fick et al. 1993), lumbar and abdominal pain (Nishiura et al. 2009). Several complications can result from the disease and these may include diverticular disease (Scheff et al. 1980) liver cysts, cardiac valve abnormalities, hypertension and intracranial aneurysms (Patel et al. 2008), urinary tract infection (Fick et al. 1993) arachnoid cysts (Alehan et al. 2002) and stroke (Rivera et al. 1992). ADPKD patients have an increased incidence of nephrolithiasis, which often increases lower back pain. Larger kidneys are more likely to contain calculi. In patients without ADPKD, nephrolithiasis is frequently caused by hypercalciuria, but ADPKD patients with nephrolithiasis do not usually have hypercalciuria and are more likely to have hyperoxaluria. The reasons for this are not fully understood. (Nishiura et al. 2009).

ADPKD patients usually retain normal levels of renal function for many years, although the kidneys are enlarged. This is because non-cystic nephrons hyper-filter to compensate for the deficit caused by destruction of cystic nephrons. After 50 %

of the healthy renal tissue has been destroyed, serum creatinine begins to rise (Tokiwa et al. 2011), indicating renal failure. There is also a decrease in glomerular filtration rate (GFR), an increase in blood pressure and albuminuria. Patients who have larger kidneys and greater cyst volume are likely to have worse renal function and kidney size is inversely related to renal function (Chapman et al. 2003; Fick-Brosnahan et al. 2002). The ADPKD phenotype is generally worse in men than in women: men have higher blood pressure, a faster increase in renal volume and decline in GFR (Fick-Brosnahan et al. 2002).

24.1.2 The Genetics of ADPKD

ADPKD is a genetic condition, in which 85 % of cases are caused by a mutation in *PKD1* and 15 % of cases are caused by a mutation in *PKD2* (Tan et al. 2009). *PKD1* is located on the short arm of chromosome 16 in region 16p13.3 (Reeders et al. 1988) and *PKD2* is located on the long arm of chromosome 4 in region 4q21 (Kimberling et al. 1993; Peters et al. 1993).

Two-thirds of the *PKD1* sequence is duplicated at other loci within the same region of chromosome 16 (The polycystic kidney disease 1 gene encodes a 14 kb transcript and lies within a duplicated region on chromosome 16. The European Polycystic Kidney Disease Consortium 1994). These *PKD1-like* sequences share up to 97 % identity but are not translated and thus represent pseudogenes (Bogdanova et al. 2001). *PKD1* encodes an mRNA transcript which is 14 kb long (The polycystic kidney disease 1 gene encodes a 14 kb transcript and lies within a duplicated region on chromosome 16. The European Polycystic Kidney Disease Consortium 1994) and *PKD2* encodes a 5.4 kb transcript (Mochizuki et al. 1996).

Germline mutations are scattered throughout the genes with no apparent mutation ‘hot spot’. The mutations identified include splicing defects, small insertions or deletions, gross deletions or insertions, complex rearrangements and missense/nonsense mutations. 84 % of these mutations are nonsense/missense site mutations and small deletions/insertions resulting in truncated polycystins, suggesting that loss of polycystin function leads to ADPKD.

Mutations can occur in any region of each of the two genes (Rossetti et al. 2007). Most mutations are unique to the family in which they occur (Rossetti et al. 2001; Rossetti et al. 2002). *PKD2* has a milder phenotype than *PKD1*: *PKD1* patients have an earlier disease onset, reach end stage renal failure about 10 years sooner and at any age *PKD1* patients have worse renal function than *PKD2* patients. Hypertension is four times more prevalent in *PKD1* than *PKD2* and *PKD1* patients are also more likely to have haematuria. (Hateboer et al. 1999; Torra et al. 1996).

24.1.3 Molecular Pathogenesis of ADPKD

ADPKD is thought to occur by a loss of function mechanism, i.e. by 'two hit' or haploinsufficiency models (Ong and Harris 2005). In the 'two hit' model, cystogenesis occurs in cells carrying a germline mutation on one allele however a somatic mutation to the other allele is required to trigger cyst formation. Thousands of renal cysts can be found in the fifth decade of most ADPKD patients, while only a few are evident before 20 years of age (Pei 2001). The 'two-hit' model of cyst formation could help to explain the focal and sporadic nature of cyst development and the significant intra-familial variability of the cystic phenotype (Ong and Harris 2005; Wu and Somlo 2000; Pei 2001).

Evidence for a 'two hit' model has been obtained from both animal and human studies. In these studies, loss of heterozygosity (LOH) of *PKD1* was detected in 20 % of kidney cysts examined (Qian et al. 1996) and 30 % of liver cysts (Watnick et al. 1998). In other studies, somatic mutations of *PKD2* were also observed in some renal cysts isolated from *PKD2* patients (Koptides et al. 1999). Pei et al. found that somatic mutations all occurred in the *PKD2* allele from the unaffected parent in a non-random pattern, data which strongly supports a 'two hit' model (Pei 2001). Nauli et al. have also demonstrated that a somatic mutation in *PKD1* in addition to a germline *PKD1* mutation is necessary for disrupting the flow sensing function of primary cilia in renal epithelial cells, a feature which could lead to cyst formation (Nauli et al. 2006).

Heterozygous *pkd1* or *pkd2* knockout mice develop focal renal cysts in late adult life while homozygous inactivation of *pkd1* or *pkd2* results in massive polycystic kidney disease and embryonic lethality (reviewed in (Pei 2001)). Cyst development in the ADPKD mouse model $Pkd2^{-/ws25}$ is also consistent with the 'two hit' model. This mouse model contains a null allele $Pkd2^{-}$ and an unstable allele $Pkd2^{WS25}$, which could undergo rearrangement to form a null allele by random intra or intergenic somatic recombination, thus generating true null or wild-type alleles (Wu et al. 1998). $Pkd2^{-/ws25}$ mice developed more cysts than $pkd2^{-/+}$ mice, suggesting that a two hit model was necessary for cystogenesis in *PKD2* (Wu and Somlo 2000). Apart from the supportive evidence for 'two hit' model, there are other controversies which are not yet resolved. If the somatic mutation is a 'trigger step' for cyst development, all cyst cells should have the somatic mutation. However, neither *PKD1* nor *PKD2* have somatic mutations in all of the cysts examined. Only ~10–30 % of cysts examined have somatic mutations. In some studies, a trans-heterozygous 'two hit' phenomenon has been proposed to occur in ~10 % of the isolated cysts (Koptides et al. 2000; Watnick et al. 2000). The trans-heterozygous state, in which somatic *PKD2* mutations were found in *PKD1* cystic epithelia, represents a variation of the 'two hit' model.

The simple 'two hit' model does not seem to be the only mechanism underlying cyst formation because expression of PC1 and PC2 is still detectable in most ADPKD cysts (Ong et al. 1999) suggesting that somatic mutations are not required in all of the cysts. The alternative hypothesis of cyst formation is haploinsufficiency, which supposes that mutations in one allele may be sufficient to initiate cyst formation and other non-cystic features in ADPKD. This is a gene-dosage-dependent mechanism in which gene expression level may be more variable due to the presence of only one functional allele (Peters and Breuning 2001). Lowering the level of PC1 or/and PC2, below a tissue-specific threshold could predispose renal epithelial cells to stochastic events which trigger cyst initiation (Lantinga-van Leeuwen et al. 2004). Wu et al. have reported that haploinsufficient *pkd2* mice have a reduced life span (Qian et al. 2003; Wu et al. 2000). *pkd2* heterozygous vascular smooth muscle cells (VSC) have lower steady-state levels of intracellular Ca^{2+} (Qian et al. 2003; Wu et al. 2000). Using a *PKD1* mouse model in which only 13–20 % normally spliced *pkd1* transcripts were generated, Lantinga-van Leeuwen et al. demonstrated that lower expression of PC1 is sufficient to cause cyst formation (Lantinga-van Leeuwen et al. 2004). Finally, over-expression of wild-type human PC1 via a transgene can lead to age-related cyst formation in the mouse kidney and liver (Pritchard et al. 2000; Lantinga-van Leeuwen et al. 2004).

24.1.4 Cellular Phenotype of ADPKD

Disruption to normal calcium signalling is believed to be a primary event in the pathogenesis of ADPKD. A number of important signalling pathways are directly affected by defects in normal calcium signalling caused by mutations in polycystin-1 and TRPP2. These include cAMP (cyclic AMP) signalling, cell proliferation and apoptosis and mTOR signalling (Torres and Harris 2009). In cells derived from ADPKD cysts cytosolic calcium concentrations are typically 20 nM lower than normal. Lower cytosolic calcium levels resulting from loss of polycystin-1 or TRPP2 are thought to be at least partly responsible for the high levels of cAMP seen in cystic cells. High levels of cAMP and also of mitogen-activated protein kinase (MAPK) activity stimulate the proliferation of cystic cells whereas cAMP has an anti-proliferative effect in normal cells.

Normal renal epithelial cells are growth inhibited by cAMP, while cystic cells are growth stimulated. cAMP can stimulate or inhibit proliferation depending on the cell type. The anti-mitogenic role of cAMP occurs at the level of Raf-1 through inhibition of Ras/MEK/Erk (Yamaguchi et al. 2004). Reducing intracellular calcium levels through treatment of normal renal epithelial cell lines with calcium channel blockers, results in a proliferative response to cAMP. This mimics the response to cAMP seen in cystic cell lines. The PI3-kinase/Akt pathway is a well-established cell survival pathway. Under normal conditions, inhibitors of PI3-Kinase/Akt induce a growth-stimulated response to cAMP mediated via B-Raf (Yamaguchi et al. 2004). Conversely, treatment of cystic cells with Ca^{2+} channel activators or Ca^{2+} ionophores that increase intracellular Ca^{2+} levels reverses the mitogenic response by elevating levels of activated Akt (Yamaguchi et al. 2006). Vasopressin receptor V2 (VRV2) activation leads to cAMP release in renal collecting ducts (Koulen et al. 2002). An increased level of cAMP is seen in PKD kidneys and this may play an important role in cyst growth by stimulating fluid secretion and cell proliferation (Calvet and Grantham 2001). Delineating the pathways involved in mitogenic stimulation of cystic cells led to the testing of a VRV2 antagonist in pck rats, pcy and ws25/− mouse models of ADPKD (Wang et al. 2005). More recently, clinical trials of the VRV2 antagonist tolvaptan has shown some clinical benefits in human ADPKD patients (Torres et al. 2012).

Polycystin-1 has been shown to regulate a number of pathways associated with cell proliferation including G-protein signalling (Parnell et al. 1998), Wnt (Kim et al. 1999), AP-1 (Parnell et al. 2002) and JAK-STAT cascades (Bhunia et al. 2002). The evidence for a direct role of TRPP2 in cell proliferation is less clear although a number of recent studies have begun to address this issue. TRPP2 has been shown to directly regulate proliferation in kidney epithelial cells (Grimm et al. 2006). Loss of TRPP2 resulted in increased rates of basal and EGF stimulated proliferation. Cells expressing a channel dead TRPP2 mutant (D511 V) proliferated faster than cells expressing wild-type TRPP2 suggesting an important role for TRPP2 channel activity. The authors show some experimental evidence to suggest that TRPP2 may mediate its effect on cell proliferation by preventing phosphorylated ERK from entering the nucleus and stimulating proliferation. Further evidence from a heterozygous *Pkd2* mouse model shows that an increase in cell proliferation is an early event preceding cyst formation and can result from haploinsufficiency at *Pkd2* (Chang et al. 2006). TRPP2 has also been shown to act together with polycystin-1 to reduce cell proliferation via upregulation of p21 (Bhunia et al. 2002). TRPP2 has been shown to bind to the p21 repressing helix-loop-helix protein Id2 (Li et al. 2005b) preventing its nuclear localisation. TRPP2, but not pathogenic mutants E837X and R872X has been shown to repress cell proliferation by promoting the phosphorylation of eukaryotic translation initiation factor eIF2alpha by pancreatic ER-resident eIF2alpha kinase (PERK) (Liang et al. 2008). All these studies point to a direct role for TRPP2 in the regulation of cell proliferation.

Cystic epithelial cells show an increased cellular proliferation rate together with an associated increase in apoptosis. An imbalance between apoptosis and proliferation has been reported in ADPKD and may form the basis for its onset and progression (Lanoix et al. 1996). Whether apoptosis is a primary event in cystogenesis and disease progression or secondary to increased proliferation is not known. Apoptosis is seen in kidneys from ADPKD patients with apoptotic cells widely spread in interstitium, glomeruli, cyst wall as well as cystic and non-cystic tubules (Woo 1995; Winyard et al. 1996). Tubular cell apoptosis has also been reported in a number of animal models of PKD (Woo 1995). Evidence supporting a primary role of apoptosis in cyst formation comes from Bcl-2 knockout mice (Sorenson et al. 1996). Bcl-2 is developmentally expressed and its deficiency in mice results in increased lymphoid apoptosis and renal cystic hypoplasia/dysplasia. The Han: Sprague–Dawley (SPRD) rat model shows a significant increase in caspase-3 activity compared to control littermates. It has been reported that caspase inhibition not only decreased apoptosis, but also decreased proliferation in the renal tubules of Han: SPRD rats with a marked decrease in cyst volume and kidney size. No deleterious effects were observed in treated rats (Tao et al. 2005; Edelstein 2005). Inhibition of apoptosis may therefore have a protective role in ADPKD.

Defects in normal calcium signalling which is a hallmark of ADPKD may be responsible for the increased apoptotic rate. Work by Wegierski et al. (2009) put forward a model in which TRPP2 acts as a calcium leak pathway by increasing ER calcium permeability. This protects cells from apoptosis by lowering the calcium concentration in the ER thereby reducing calcium release from the ER in response to apoptotic stimuli. In this study, the loss of TRPP2 had the opposite effect leading to increased apoptosis in ADPKD.

TRPP2 was also shown to play a role in protecting renal epithelial cells from apoptosis induced by mechanical stress via opening of stretch-activated K_{2P} channels (Peyronnet et al. 2012). *Pkd1* knockout or expression of a TRPP2 pathogenic mutant, mimicking ADPKD, dramatically increased mechanical stress-induced tubular apoptotic cell death.

24.2 The TRPP Protein Family

The metazoan TRP superfamily has been divided into seven subfamilies sharing common features. TRP channels function as ion channels regulating a diverse range of cellular functions and responding to a wide range of sensory stimuli including phototransduction, thermosensation and mechanosensation. TRP channels are thought to form tetramers with each channel protein including six transmembrane spanning domains (S1–S6) with a pore loop between S5 and S6. Each TRP subfamily is named after the first described member: TRPC (canonical), TRPV (vanilloid), TRPM (melastatin), TRPN (NOMPC, from no mechanoreceptor potential-C), TRPA (ankyrin-like with transmembrane domain-1), TRPML (mucolipin) and TRPP (polycystin) (Nilius and Owsianik 2011; Montell 2005; Montell et al. 2002a, b).

The founder member of the TRPP subfamily is polycystin-2, the protein product of the *PKD2* gene shown to be mutated in ADPKD (Mochizuki et al. 1996). Currently there are eight proteins in the TRPP subfamily which has been further subdivided into two groups; The *PKD1* like proteins and the TRPP2 like proteins. Both groups have been shown to have a limited degree of similarity in their C-termini (Delmas 2005; Delmas et al. 2004b; Delmas 2004; Gallagher et al. 2010).

The *PKD1*-like proteins are not true members of the TRP superfamily as they show very little sequence similarity with TRP channels. However, they have clearly been shown to be physically and functionally linked to the TRPP channels. The PKD-1 like proteins include *PKD1* (polycystin-1) (Harris et al. 1995), PKD1L1 (Yuasa et al. 2002), PKD1L2 (Yuasa et al. 2004), PKD1L3 (Li et al. 2003a) and PKDREJ (receptor for egg jelly) (Sutton et al. 2006). The *PKD1*-like

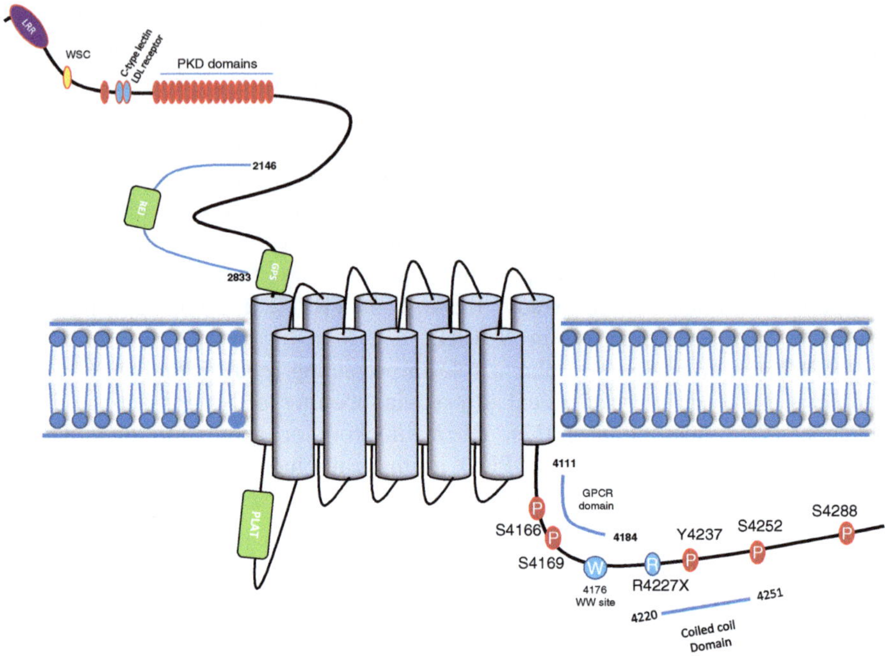

Fig. 24.1 Topological diagram of Polycystin-1 showing the location of published phosphorylation sites, key residues and major functional domains. The position of published phosphorylation sites in polycystin-1 identified from experimental studies and phosphoproteomic screens is depicted. Several functional motifs important for targeting and function are shown: Polycystin-1 is a membrane protein with 11 transmembrane domains, a large extracellular N-terminal domain and cytoplasmic C-terminal domain. The extracellular N-terminus has a complex domain structure containing 16 PKD repeats, two leucine-rich repeats, two cysteine rich domains, a WSC domain, a C-type lectin domain, an LDL-A receptor motif, a receptor for egg jelly (REJ) domain and a GPS cleavage site. The PLAT domain is found in the intracellular loop between transmembrane domains 1 and 2, while the C-terminal region contains a coiled-coil domain (aa 4220–4251) required for interaction with TRPP2

proteins all contain 11 transmembrane domains with a large extracellular N-terminal domain and a short cytoplasmic C-terminal domain. They all possess the combination of REJ (receptor for egg jelly), GPS (G-protein-coupled receptor proteolytic site) and PLAT/LH2 (lipoxygenase homology/polycystin, lipoxygenase, α-toxin) domains that uniquely define them as *PKD1* family members. The extracellular region of polycystin-1 (Fig. 24.1) spans more than ~3,000 amino acids and contains a number of adhesive domains that implicate polycystin-1 in cell–cell and cell-matrix interactions. Polycystin-1 is cleaved at its predicted GPS (Chapin et al. 2010; Qian et al. 2002), a feature common to members of the family-B (latrophilin) G-protein-coupled receptors and which may be important for receptor activation or localisation (Chapin et al. 2010). *PKD1*, PKD1L1 and PKD1L2 encompasses a G-protein-interacting site that may be used to regulate at least four different classes of heterotrimeric G-protein activity and multiple downstream effectors, including among others phospholipase C, protein kinase C, adenylyl cyclase, protein kinase A, Janus kinase 2 and nuclear factor of activated T cells (NFAT). The intracellular carboxyl terminus of *PKD1* also harbours a coiled-coil domain that is involved in physical interaction with TRPP2 and possibly TRPP3.

The TRPP2-like proteins is comprised of three members, *PKD2* (polycystin-2), PKD2L1 and PKD2L2. These are commonly referred to as TRPP2, TRPP3 and TRPP5 although recently a review on TRP channels by Wu et al. suggested that they can be renamed TRPP1, TRPP2 and TRPP3, respectively (Wu et al. 2010). The mammalian orthologues are highly conserved over the entire length (80–90 % identity) but show sequence divergence in their cytosolic N-terminal regions. All TRPP2-like proteins are predicted to possess a putative coiled-coil domain at their C-termini, but only TRPP2 and TRPP3 have a Ca^{2+}-binding EF-hand motif. TRPP2 (polycystin-2) is a 110 kDa membrane glycoprotein (Fig. 24.2). It has six transmembrane domains with cytoplasmic N- and C-termini and shows homology with transmembrane domains 6–11 of polycystin-1. The C-terminus contains a coiled-coil domain, thought to mediate the formation of homodimers, an ER retention motif and an EF-hand, which binds calcium. A pore homology region is found in the extracellular loop between transmembrane domains 5 and 6.

TRPP3 was the first *PKD2*-like protein to be identified as a functional TRP-like cation channel (Chen et al. 1999). TRPP2-related channels have large single-channel conductance (80–160 pS) and permeate a number of mono- and divalent cations, including Na^+, K^+, Ba^{2+} and Ca^{2+} (Delmas et al. 2004b). Like many TRP channels, TRPP2 activity is blocked by La^{3+} and reduced by the diuretic amiloride. TRPP3 is usually found to be more permeable to Ca^{2+} than to other monovalent cations such as Na^+ and K^+ ($P_{Ca}/P_{Na} = 4$), whereas TRPP2 has a P_{Ca}/P_{Na} selectivity ranging from 1 to 3. TRPP2-like cation channels therefore represent relevant routes for Ca2 + entry or release (Giamarchi and Delmas 2007; Delmas et al. 2004b).

Mutations in two TRPP family proteins, polycystin-1 and TRPP2 have been shown to cause ADPKD. Both proteins have been clearly shown to interact to form a heterodimeric complex in vivo and together regulate key signalling pathways regulating tubular morphogenesis (Newby et al. 2002).

24.3 TRPP2 Localisation: A Complex Picture

TRPP2 is expressed throughout renal development and unlike polycystin-1 strong expression levels persist in the adult kidney. Expression is also seen in other tissues including heart, vascular smooth muscle, pancreas, intestine, bile ducts and placenta. Analysis of TRPP2 has been complicated by its complex and dynamic subcellular localisation in epithelial cells which is dependent on cell type, developmental status, differentiation, confluence and quiescence (Ong 2000). In quiescent cells, TRPP2 can be localised in primary cilia and centrosomes, where a

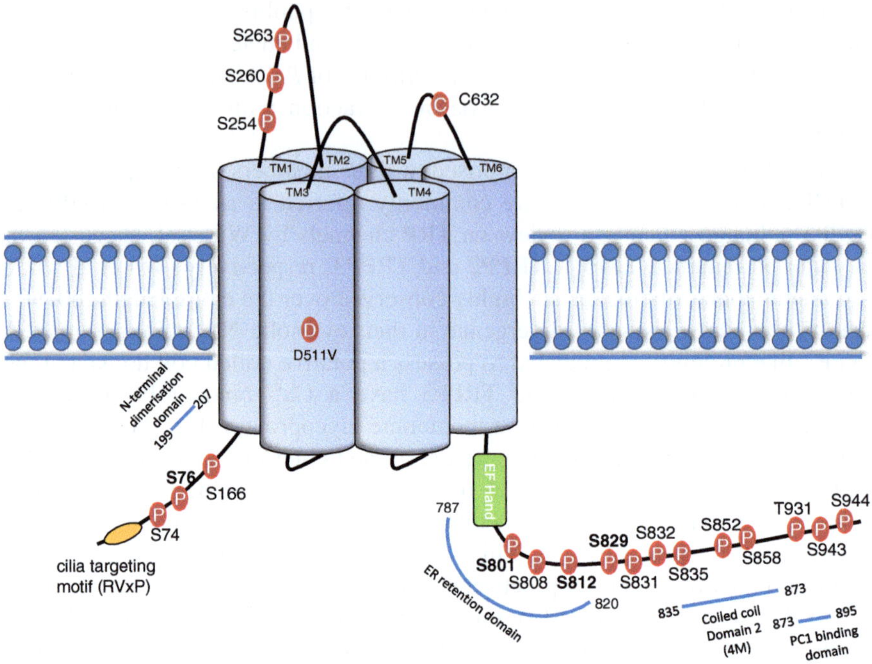

Fig. 24.2 Topological diagram of TRPP2 showing the location of published phosphorylation sites, key residues and major functional domains. The position of all known phosphorylation sites in TRPP2 is depicted. For references, please refer to Table 24.1. Sites with clear evidence of function are indicated in bold. Several functional motifs important for targeting and function are shown: a cilia targeting motif (RVxP, aa 6–9), an ER retention domain (aa 787–820) and an EF-hand (aa720–797) (Mochizuki et al. 1996; Geng et al. 2006). GSK3-mediated Ser^{76} phosphorylation can mediate basolateral targeting or retention (Streets et al. 2006). Three discrete dimerisation domains reported for TRPP2, i.e. the C-terminal coiled-coil domain (aa 835–873), an N-terminal dimerisation domain (aa 199–207) and a cysteine residue (C^{632}) in the third extracellular loop are shown (Feng et al. 2008, 2011; Giamarchi et al. 2010). An essential requirement for PC1 recognition is the prior C-terminal dimerisation of PC2 (Giamarchi et al. 2010). PC1 binding also requires a short heterodimerisation sequence (aa 873–895) distal to the coiled-coil domain. The position of the D511 V mutation which abolishes channel activity is also shown

role in mechanosensitive Ca^{2+} signalling has been reported (Nauli et al. 2003). In dividing cells, TRPP2 has been reported to localise at mitotic spindle poles in association with mDia-1 and pericentrin, where its function is thought to regulate intracellular Ca^{2+} during or after mitosis (Rundle et al. 2004; Jurczyk et al. 2004). A direct role in cell cycle regulation has also been proposed for TRPP2 through the regulation of the CDK inhibitor p21 (Bhunia et al. 2002; Li et al. 2005b). Both proteins are expressed at or close to the basolateral membrane in confluent cells where it is likely to mediate or regulate cell–cell and cell-matrix adhesion (Streets et al. 2003; Ibraghimov-Beskrovnaya et al. 2000; Streets et al. 2009). TRPP2 may also reconstitute an ER Ca^{2+} release channel together with the IP3 receptor

(Li et al. 2009; Giamarchi et al. 2010; Mekahli et al. 2012). Finally, TRPP2 has been detected in urinary exosomes where a role in urocrine signalling has been postulated (Hogan et al. 2009).

Recent studies have begun to make sense of this complex localisation with the realisation that TRPP2 contains a number of sequence motifs and binding domains regulating its trafficking between different subcellular compartments (Fig. 24.3). TRPP2 has been shown to contain an endoplasmic reticulum (ER) retention domain; deletion of amino acids Glu^{787}–Ser^{820} resulted in the relocation of a significant proportion of the TRPP2 pool from the ER to the lateral plasma membrane (Cai et al. 1999). Within this ER retention domain, phosphorylation at Ser^{812} has been shown to mediate trafficking of TRPP2 between the ER, golgi and plasma membrane via phosphorylation-dependent interactions with phosphofurin acidic cluster sorting protein (PACS)-1 and PACS-2 adaptor proteins (Kottgen et al. 2005). Trafficking of TRPP2 to the primary cilia is mediated via a 15 amino acid N-terminal domain containing a conserved RVxP motif (Geng et al. 2006). Expression of this conserved domain was sufficient to localise heterologous proteins to the primary cilia.

Expression of TRPP2 at the plasma membrane is regulated by GSK3 phosphorylation of Ser^{76}. In the presence of specific GSK3 inhibitors, the lateral plasma

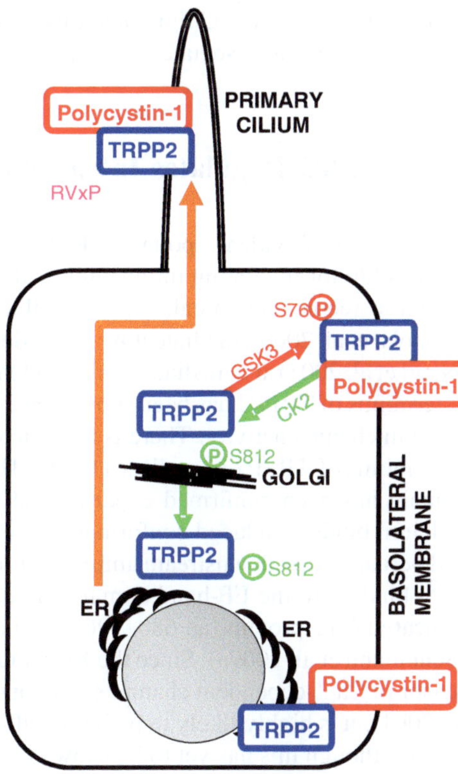

Fig. 24.3 Cellular localisation of TRPP2. Intracellular localisation of TRPP2 is highly regulated. Phosphorylation at Ser^{76} by GSK3 is required for trafficking to the plasma membrane. Retrograde trafficking between plasma membrane, golgi and ER is mediated by phosphorylation at Ser^{812} by CK2. Trafficking to the cilia is dependent on an N-terminal RVxP motif. TRPP2 transport to the ciliary membrane independent of Golgi trafficking has been recently demonstrated

membrane pool of endogenous TRPP2 redistributes into an intracellular compartment without any change in primary cilia localisation (Streets et al. 2006). Coexpression of *PKD1* and TRPP2 in CHO cells as well as in sympathetic neurons also has been shown to promote the translocation of TRPP2 to the plasma membrane (Hanaoka et al. 2000; Delmas et al. 2004a) resulting in calcium permeable non-selective cation currents. Disease-associated mutants of either *PKD1* or 2 which prevent their association do not result in new channel activity implying that the formation of a complex between *PKD1* and TRPP2 is required for the targeting/retention of TRPP2 to the plasma membrane. TRPP2 has also been shown to be required for polycystin-1 transport to the plasma and ciliary membrane. This was shown to be independent of TRPP2 channel activity (Chapin et al. 2010) or TRPP2's ability to bind polycystin-1. Polycystin-1 surface localisation was mediated by TRPP2 stimulated polycystin-1 GPS cleavage through an as yet unknown mechanism.

A recent study suggests that TRPP2 may take different routes to the somatic and ciliary plasma membrane (Hoffmeister et al. 2011). After studying a number of TRPP2 deletion constructs they determined that trafficking of TRPP2 to the plasma membrane was dependent on two amino acids (K^{572} and F^{576}) in loop 4. Trafficking of TRPP2 to the ciliary and to the somatic plasma membrane compartments originates in a COPII-dependent fashion at the ER and in both cases TRPP2 reaches the cis side of the Golgi apparatus. Trafficking to the somatic plasma membrane however goes through the Golgi apparatus, whereas transport vesicles to the cilium leave the Golgi apparatus at the cis compartment and is regulated by Rab8a, a BBSome associated monomeric GTPase.

24.4 TRPP2 Regulates Intracellular Calcium Homeostasis

Many pieces of evidence point to TRPP2 acting as a non-selective calcium channel responsible for regulating intracellular calcium homeostasis in the cell. It has been shown to reconstitute a calcium channel at the plasma membrane (Tsiokas 2009; Hanaoka et al. 2000), mediate flow-dependent calcium signalling in the primary cilia (Nauli et al. 2003) and mediate calcium release from the ER (Koulen et al. 2002). The C-terminus of TRPP2 has been shown to be critically important for the regulation of calcium channel activity. There is a predicted EF-hand domain (aa 750–785) in the C-terminus of TRPP2 as well as TRPP3. The calcium binding capacity of EF-hand domain has been confirmed experimentally (Celic et al. 2008; Petri et al. 2010). Calcium binding induced conformational changes of the globular EF-hand domain and perhaps other downstream amino acids (Petri et al. 2010). It has been shown that Ca^{2+} binding to the EF-hand domain can dissociate the protein dimer formed by a truncated CT2 protein (aa 680–796) lacking the second coiled-coil domain (CC2) (Schumann et al. 2009a). Since the EF-hand domain is commonly found as the Ca^{2+} sensor in Ca^{2+}-dependent channels (Kubota et al. 2009; Xiao et al. 2008; Braun and Sy 2001), it is highly likely to be involved in the regulation of polycystin-2 channel activity though this has yet to be shown experimentally.

24.4.1 TRPP2 Acts as a Flow-Induced Mechanosensitive Calcium Channel

Primary cilia are microtubule-based membrane-bound projections on the apical surface of almost all cells. In the kidney, each epithelial cell (except intercalated cells) has a single apical cilium projecting into the fluid-filled lumen (Deane and Ricardo 2007). Unlike motile cilia that have dynein arms and central microtubules ($9 + 2$ microtubule arrangement), primary cilia have a $9 + 0$ microtubule arrangement and are also referred to as non-motile cilia (Nauli and Zhou 2004). The role of primary cilia in PKD pathogenesis was first identified in 2000. Mutations in Tg737 caused a cystic phenotype in mice and were associated with abnormal shortened primary cilia in renal epithelial cells, suggesting that ciliary defects are associated with cytogenesis (Yoder et al. 2002; Pazour et al. 2000). Tg737 is the mouse homologue of the Chlamydomonas IFT88 gene. Mutation of IFT88 leads to defective flagella function in the algae Chlamydomonas (Yoder et al. 2002; Pazour et al. 2000). Direct evidence for the correlation between a defect renal primary cilia and the PKD phenotype was provided by the observation that a kidney-specific KIF3 (a ciliary motor subunit) knockout in tubular epithelial cells caused cyst formation (Lin et al. 2003).

When exposed to fluid flow, the flow-induced bending of the MDCK cell primary cilia increased the level of intracellular Ca^{2+}, which suggested a mechanosensory function for renal primary cilia (Praetorius and Spring 2001). PC1 and TRPP2 seem to be important for the Ca^{2+} response to fluid-flow stimulation and *pkd1* null collecting duct cells lost this response (Nauli et al. 2003). *Pkd2* knockout mice cells also did not respond to flow stimulation, which could be rescued by transfection of functional TRPP2 (Nauli and Zhou 2004). TRPP2 has been shown to function as a Ca^{2+} permeable channel. TRPP2 channel function is regulated by PC1 through specific interactions between PC1 and TRPP2 via their C-termini. Loss of the mechanosensory ciliary signal due to defective PC1 and TRPP2 results in the disruption of Ca^{2+} influx (Ong and Harris 2005). A low Ca^{2+} level could lead to increased levels of cAMP which stimulate cellular proliferation and apoptosis and thus promote cyst formation and enlargement (Deane and Ricardo 2007). It is possible that this cilia-mediated pathway controls a series of Ca^{2+}-dependent cellular processes, the disruption of which will result in development of PKD (Ong and Wheatley 2003).

24.4.2 TRPP2 Regulates Calcium Release from the ER

TRPP2 has been shown to act as a calcium-activated calcium release channel in ER membranes (Koulen et al. 2002; Cai et al. 1999) amplifying calcium transients initiated by IP_3R. TRPP2 has been shown to functionally interact with IP3R (Li et al. 2005c). A second study confirmed a direct interaction between the two proteins and showed that TRPP2 could function as an IP3-induced Ca^{2+} release channel (Sammels et al. 2010a). TRPP2 in the ER has also been shown to be involved in G-protein coupled receptor (GPCR) induced calcium signalling both in *PKD2* knock mice and ADPKD patients (Qian et al. 2003; Aguiari et al. 2004). The first evidence that TRPP2 may act as an intracellular calcium release channel came from the observation that TRPP2 exogenously expressed in LLCPK1 renal epithelial cells resulted in a significant increase in intracellular calcium release following vasopressin stimulation (Koulen et al. 2002). More recent work has shown that TRPP2-mediated calcium release from the ER requires activation of IP3R (Li et al. 2005c; Mekahli et al. 2012; Sammels et al. 2010b; Sammels et al. 2010a). Disruption of the IP3R-TRPP2 protein complex prevents the stimulation of IP3-induced calcium release by TRPP2. Further evidence for the requirement of TRPP2 channel activity was demonstrated when the channel defective TRPP2 D511 V mutation failed to induce stimulation of IP3-induced calcium release. The interaction between polycystin-1 and TRPP2 is believed to be important in the activation of TRPP2 channel activity (Hanaoka et al. 2000). It is not clear whether this requires co-localisation in the same membrane compartment and as such a model has been proposed in which polycystin-1 interaction with TRPP2 can occur across membrane compartments (Giamarchi et al. 2010; Delmas et al. 2004b). Studies using human proximal-tubule epithelial cells in which one of both of polycystin-1 and TRPP2 were knocked down demonstrated that both are required to mediate IP3R-induced calcium release indicating that these two proteins operate as a functional complex (Mekahli et al. 2012). TRPP2 has also been shown to form a complex with the cardiac ryanodine receptor (RyR2) from mouse heart (Anyatonwu et al. 2007). TRPP2 was shown to inhibit RyR2 channel activity. Loss of TRPP2 in *Pkd2* (*−/−*) cardiomyocytes led to reduced Ca^{2+} release from sarcoplasmic reticulum stores and reduced Ca^{2+} content compared with normal cells. This data suggest that TRPP2 is important for the regulation of RyR2 activity and that TRPP2 mutations result in altered Ca^{2+} signalling in the heart.

24.4.3 TRPP2 Functions as a Plasma Membrane Calcium Channel

As well as reconstituting calcium channel in the primary cilia and within the ER membrane, TRPP2 has been shown to form a calcium channel in the plasma membrane (Ma et al. 2005; Tsiokas 2009). Data from subcellular membrane fractionation experiments clearly show that the majority (>90 %) of TRPP2 is located in ER membranes however a small fraction (<10 %) can be seen in the plasma membrane (Newby et al. 2002; Scheffers et al. 2002). Further experimental evidence for TRPP2 plasma membrane localisation has come from membrane fractionation, surface biotinylation and live cell surface antibody labelling (Scheffers et al. 2002; Streets et al. 2006; Newby et al. 2002; Ma et al. 2005; Luo et al. 2003). The surface fraction of TRPP2 is reported to be increased by the use of proteosome inhibitors and/or chemical chaperones in some experimental systems (Vassilev et al. 2001). TRPP2 surface localisation has been the subject of controversy as a number of studies have been unable to detect an EndoH resistant TRPP2 fraction, thought to be a hallmark of plasma membrane trafficking (Koulen et al. 2002; Cai et al. 1999). Differences in cell type, differentiation status and culture conditions may account for some of the variability in surface expression.

Luo et al. (2003) found that endogenous TRPP2 was expressed in the plasma membrane and the primary cilium of mouse inner medullar collecting duct (IMCD) cells and Madin-Darby canine kidney (MDCK) cells, whereas heterologously expressed TRPP2 showed a predominant ER localisation. Patch-clamping of IMCD cells expressing endogenous or heterologous TRPP2 confirmed the presence of the channel on the plasma membrane. These results show that TRPP2 is able to function as a plasma membrane channel in renal epithelia. TRPP2 is also activated by cell surface receptor stimulation. Ma et al. (2005) showed that TRPP2 over expression increased EGF induced inward currents in LLCPK1 renal cells requiring the activation of phospholipase C and PI3 K. This action of TRPP2 did not require depletion of intracellular stores and activation of store operated calcium channels but was dependent on extracellular calcium levels.

24.5 TRPP2 Forms Complexes with Multiple Proteins in the Cell

An increasing number of proteins have been shown to form a complex with TRPP2 by yeast-2-hybrid screens or biochemical assays (Fig. 24.4). The C-terminus of TRPP2 mediates binding of the majority of these binding partners. These include other TRP channels such as TRPC1 (Kobori et al. 2009; Bai et al. 2008) and TRPV4 (Kottgen et al. 2008), calcium channels including IP_3R (Li et al. 2005c) and cytoskeletal components including Hax-1 (Gallagher et al. 2000) and Kif3A (Li et al. 2006).

Of most relevance for the pathogenesis of ADPKD is the formation of TRPP2 and polycystin-1 heterodimers and TRPP2 homodimers. Both proteins have been shown to interact forming a heterodimeric complex in vivo and together regulate key signalling pathways regulating tubular morphogenesis (Newby et al. 2002). Complex formation requires the obligate formation of TRPP2 dimers (or trimers) via a C-terminal coiled-coil domain (aa 835–873) followed by polycystin-1 binding through a coiled-coil domain in its C-terminus to a short heterodimerisation sequence on PC2 (aa 873–895) (Giamarchi et al. 2010; Yu et al. 2009). Two additional dimerisation motifs have been reported for TRPP2 in its N-terminus (aa199–207) and a cysteine (C^{632}) in the third extracellular loop (Feng et al. 2008, 2011). These mediate the likely tetramerisation of TRPP2 to generate non-selective Ca^{2+} channels. In view of the phenotypic similarity of both patients and

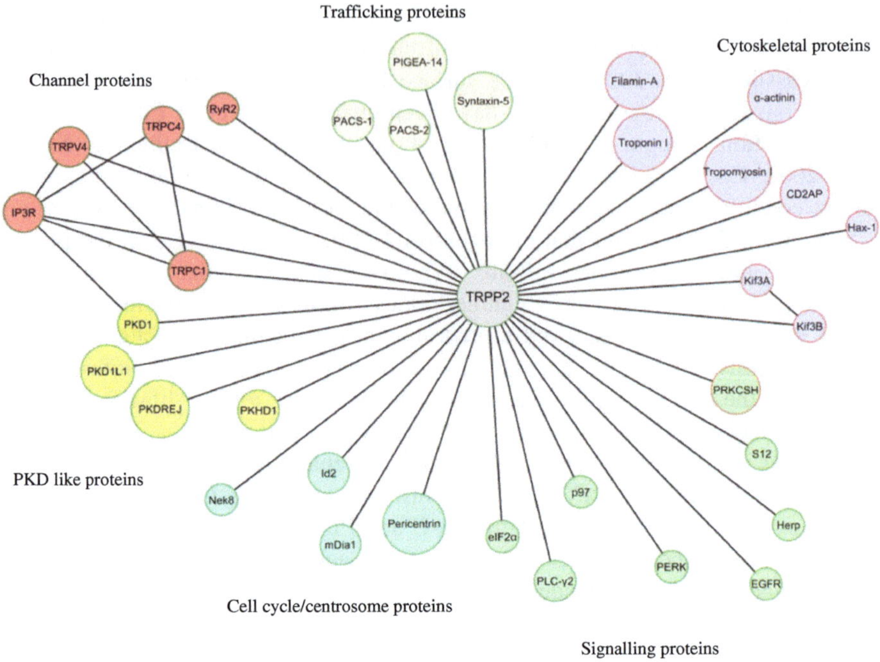

Fig. 24.4 TRPP2 interactome showing experimentally verified TRPP2 interacting partners clustered by function. TRPP2 has been shown to interact with a large number of proteins including channel proteins, ER trafficking proteins, cell cycle and centrosomal proteins, signal transduction proteins and cytoskeletal proteins. The functional significance of these protein interactions in the pathogenesis of ADPKD remains to be determined. Data were compiled from published reviews and the STRING 9.0 interaction database (Szklarczyk et al. 2011)

animal models, it is very likely that polycystin-1 and TRPP2 function together in regulating epithelial morphogenesis. However, a polycystin-1-independent function of TRPP2 has been shown in the embryonic node which determines left–right axis patterning during development. Here TRPP2 is thought to interact with the polycystin-1 homologue, PKD1L1, to mediate laterality (Pennekamp et al. 2002; Kamura et al. 2011; Field et al. 2011).

TRPP2 has been shown to interact with and form heteromeric channels with other TRP channel proteins including TRPC1 (Kobori et al. 2009; Zhang et al. 2009; Bai et al. 2008), TRPC4 (Du et al. 2008) and TRPV4 (Kottgen et al. 2008). TRPP2 was shown to interact with TRPC1 and TRPC4 in mesangial cells to form channel complexes mediating angiotensin II-induced calcium responses (Du et al. 2008). TRPP2 has been shown to interact with TRPC1 at the primary cilia and can be activated in response to G-protein-coupled receptor activation with distinct properties from TRPP2 or TRPC1 alone (Bai et al. 2008). TRPP2 and TRPV4 have also been shown to form a functional channel in primary cilia (Kottgen et al. 2008) and TRPV4 is an essential component of the ciliary flow sensor in MDCK cells. However, the relevance of these interactions to cystogenesis in ADPKD is unclear.

TRPV4 knockout mouse and zebrafish models do not show a cystic phenotype yet flow-induced calcium transients are absent in the distal nephrons of adult TRPV4-deficient mice. These results therefore question the importance of flow-mediated calcium signalling in cystogenesis.

Many of the TRPP2 interacting proteins are components of the cytoskeleton including tropomyosin-1, troponin I (cardiac), α-actinin, hax-1, kinesin-2, mDia1, pericentrin and filamin (Ong and Harris 2005). The TRPP2 C-terminus was found to interact with tropomyosin-1 in a yeast-2-hybrid screen; the interaction was confirmed using in vitro and in vivo biochemical techniques (Li et al. 2003b). The same group also identified an interaction between the C-terminal domain of polycystin-2 and cardiac troponin-I by yeast-2-hybrid (Schumann et al. 2009b). TRPP2 has also been shown to interact with: α-actinin—an actin-binding and actin-bundling protein (Li et al. 2005a), CD2AP—a scaffold that directly binds actin and regulates the actin cytoskeleton (Lehtonen et al. 2000); Hax-1 which interacts with cortactin (Gallagher et al. 2000), a scaffold involved in organisation of the cytoskeletal network; KIF3A and KIF3B which function as a heterodimer and are involved in cellular movement of membranous organelles including the primary cilium (Li et al. 2006; Wu et al. 2006) and filamin—an important actin cross-linking protein implicated in scaffolding, membrane stabilization and signal transduction, through interaction with ion channels, receptors and signalling proteins (Wang et al. 2012; Sharif-Naeini et al. 2009). It was shown that filamin-A substantially inhibits polycystin-2 channel activity indicating that filamins are important regulators of polycystin-2 channel function, and further links actin cytoskeletal dynamics to the regulation of TRPP2 (Wang et al. 2012). Therefore, TRPP2 is linked to the cytoskeleton through interactions with multiple actin-binding proteins. Aberrant interaction between TRPP2 and these proteins may be a consequence of pathogenic TRPP2 mutations; breaking the interaction between TRPP2 and the actin cytoskeleton could play a role in cystogenesis or other extra-renal manifestations of ADPKD.

TRPP2 has also been shown to interact with pericentrin and other centrosomal proteins controlling the assembly of primary cilia (Jurczyk et al. 2004). It has also been found in mitotic spindles (Rundle et al. 2004) where it forms a complex with mDia1, suggesting a role in cell division linked to intracellular Ca^{2+} release during mitosis. This hypothesis is supported by the finding that calcium is involved in the initiation of centrosome duplication in *Xenopus* egg extracts (Matsumoto and Maller 2002). A role for TRPP2 in mitosis and centrosome duplication was demonstrated by the observation that TRPP2 over expression in a transgenic mouse model was associated with mitotic instability and centrosome overduplication (Burtey et al. 2008).

24.6 TRPP2 is Regulated by Reversible Protein Phosphorylation

Protein phosphorylation is a reversible post-translational modification involved in regulating the function of many proteins. The addition of a phosphate group to specific serine, threonine, or tyrosine residues is catalysed by protein kinases and reversed by protein phosphatases. Phosphorylation generally results in a conformational change of the protein resulting in activation or deactivation.

Phosphorylation of TRPP2 has been shown on multiple residues. Phosphorylation at Ser^{812} by Casein Kinase 2 (CK2) is important for channel function and TRPP2 localisation. Phosphorylation at Ser^{76} by Glycogen Synthase Kinase 3 (GSK3) is also important for cellular localisation of TRPP2. Phosphorylation at Ser^{801} by Protein Kinase D (PrKD) is important for ER localised TRPP2 channel function in response to growth factor stimulation. Recent studies have shown that TRPP2 can be phosphorylated at Ser^{829} by AuroraA kinase acting to inhibit TRPP2 channel activity (Plotnikova et al. 2011). Phosphorylation at a number of other residues has been demonstrated in several large-scale phosphoproteomic screens (Yu et al. 2011; Huttlin et al. 2010) by mass spectrometry (Table 24.1). The majority of these residues are highly conserved in vertebrates suggesting that they may be necessary for TRPP2 function; however, their functional significance is yet to be elucidated.

24.6.1 Casein Kinase II Phosphorylation of TRPP2

Constitutive phosphorylation of TRPP2 was observed in vivo (Cai et al. 2004; Streets et al. 2006). Cai et al. (2004) identified a single phosphorylated site at Ser^{812} in the C-terminal of TRPP2 (Cai et al. 2004). It was demonstrated that casein kinase II (CK2), a serine/threonine protein kinase, phosphorylates at this site. The authors also demonstrated that CK2 phosphorylation was important for channel function; loss of phosphorylation resulted in reduced sensitivity to TRPP2 activation by intracellular calcium concentrations.

In this initial study, phosphorylation at Ser^{812} did not affect the localisation of TRPP2 nor its interaction with polycystin-1. A more recent study utilising a chimeric construct, with the C-terminal domain of TRPP2 fused to a transmembrane protein (CD16.7), did implicate a role for CK2 phosphorylation in TRPP2 trafficking (Kottgen et al. 2005). Ser^{812} is in a cluster of acidic amino acid residues which were found to mediate interaction between TRPP2 and phosphofurin acidic cluster sorting (PACS) proteins, PACS-1 and PACS-2 (Kottgen et al. 2005). Interaction with these PACS proteins proved essential for TRPP2 retrograde trafficking between the plasma membrane, golgi and ER; when binding of PACS proteins was blocked, TRPP2 localised to the plasma membrane. Binding of PACS to TRPP2 was found to be dependent on CK2 phosphorylation at Ser^{812}; mutation

Table 24.1 List of reported phosphorylation sites in TRPP2

Residue	Kinase	Species/ Cell Line	Methodology	Function	Reference
S74	Unknown	Mouse tissue	Proteomic screen	Unknown	Huttlin et al. (2010)
S76	GSK3	Human TRPP2	Biochemistry/site-directed mutagenesis	TRPP2 trafficking	Streets et al. (2006)
		Mouse tissue	Proteomic screen	Unknown	Huttlin et al. (2010)
S166	Unknown	HeLa cells	Proteomic screen	Unknown	Olsen et al. (2006)
S254	Unknown	HEK 293T cells	Proteomic screen	Unknown	Molina et al. (2007)
T260	Unknown	HEK 293T cells	Proteomic screen	Unknown	Molina et al. (2007)
S263	Unknown	HEK 293T cells	Proteomic screen	Unknown	Molina et al. (2007)
S801	PrKD1	Human TRPP2	Biochemistry/site-directed mutagenesis/ phosphoserine antibody mapping	Cell proliferation calcium channel regulation	Streets et al. (2010)
S808	Unknown	Mouse tissue	Proteomic screen	Unknown	Huttlin et al. (2010)
	Unknown	Mouse tissue	Proteomic screen	Unknown	
S812	CKII	Human TRPP2	Biochemistry/site-directed mutagenesis	Calcium channel regulation	Cai et al. (2004)
		Human TRPP2	Biochemistry	TRPP2 trafficking	Kottgen et al. (2005)
		Mouse tissue	Proteomic screen	Unknown	Huttlin et al. (2010)
		Mouse cells	Proteomic screen	Unknown	Yu et al. (2011)
		Mouse cells	Proteomic screen	Unknown	Li et al. (2009)
		Mouse tissue	Proteomic screen	Unknown	Zanivan et al. (2008)
		Mouse tissue	Proteomic screen	Unknown	
S829	AuroraA	HEK 293	Biochemistry/site-directed mutagenesis/ phosphoserine antibody mapping	Calcium channel regulation (indirect)	Plotnikova et al. (2011)
		Mouse tissue	Proteomic screen	Unknown	Huttlin et al. (2010)

(continued)

Table 24.1 (continued)

Residue	Kinase	Species/Cell Line	Methodology	Function	Reference
		Mouse cells	Proteomic screen	Unknown	Hsu et al. (2011)
		Mouse cells	Proteomic screen	Unknown	Yu et al. (2011)
	PKA	Human cells and tissue	Biochemistry/site-directed mutagenesis/phosphoserine antibody mapping	Calcium channel regulation/cell proliferation	Streets et al. (unpublished data)
S831	Unknown	Mouse tissue	Proteomic screen	Unknown	Huttlin et al. (2010)
		Mouse cells	Proteomic screen	Unknown	Hsu et al. (2011)
S832	Unknown	Mouse tissue	Proteomic screen	Unknown	Huttlin et al. (2010)
		Mouse cells	Proteomic screen	Unknown	Hsu et al. (2011)
S835	Unknown	Mouse tissue	Proteomic screen	Unknown	Huttlin et al. (2010)
S852	Unknown	HEK 293T	Proteomic screen	Unknown	Molina et al. (2007)
S858	Unknown	HEK 293T	Proteomic screen	Unknown	Molina et al. (2007)
T931	Unknown	Rat tissue	Proteomic screen	Unknown	Hoffert et al. (2006)
S943	Unknown	Rat tissue	Proteomic screen	Unknown	Hoffert et al. (2006)
S944	Unknown	Rat tissue	Proteomic screen	Unknown	Hoffert et al. (2006)

A summary of all reported phosphorylation sites in TRPP2, combining biochemical analysis and proteomic-driven mass spectrometry of cell lysates

of Ser^{812} to alanine or disruption of the acidic cluster weakened PACS binding resulting in translocation of TRPP2 to the lateral plasma membrane. Similar results were observed for full length TRPP2 though changes in localisation were less pronounced. A further study confirmed the interaction of TRPP2 with PACS proteins in a zebrafish model (Fu et al. 2008) mediated by phosphorylation of Ser^{812}. These findings support the hypothesis that TRPP2 assumes distinct subcellular localisations to exert tissue-specific functions mediated by changes in phosphorylation status.

Phosphorylation at Ser^{812} is also important for TRPP2 binding to ID2 (inhibitor of DNA binding 2)—a member of the ID family of helix-loop-helix (HLH) transcription factors (Li et al. 2005b). ID proteins do not bind DNA themselves—they form heteromers with other HLH transcription factors, preventing their binding to DNA and blocking transcription of their target genes. Co-immunoprecipitation studies revealed that the TRPP2-ID2 interaction was dependent on phosphorylation of TRPP2 at Ser^{812}; mutagenesis of this residue abolished the interaction. Binding TRPP2 appeared to prevent ID2 entering the nucleus; overexpression of TRPP2 reduced the nuclear fraction of ID2, while knockdown of TRPP2 resulted in an increase. TRPP2-mediated re-distribution of ID2 may contribute to increase cell proliferation and differentiation seen in ADPKD. Increased expression and nuclear localisation of ID2 has been detected in *PKD1* or PKD2 ADPKD cyst lining epithelia compared to non-cystic tubular epithelia in which ID2 expression is generally confined to the cytosol. This implied that polycystin-1 may also affect ID2 expression and localisation; indeed polycystin-1 overexpression saw relocalisation of a nuclear pool of ID2 to the cytosol. While TRPP2 expression was unchanged in polycystin-1 overexpressing cells, the Ser^{812} phosphorylated pool of TRPP2 was increased.

Studies on the C.elegans TRPP2 homolog demonstrated that phosphorylation of Ser^{534} by CKII was important in regulating TRPP2 ciliary localisation and its function during male mating behaviour (Hu et al. 2006). Ser^{534} is the only CKII site conserved in mouse (Ser^{681}) and human (Thr^{683}) TRPP2. Evolutionary conservation of this CKII site suggests it may have an important cellular function although that has yet to be shown in mammalian TRPP2.

24.6.2 GSK3 Phosphorylation of TRPP2

Although it had been suggested that TRPP2 was phosphorylated only at Ser^{812}, the possibility of other sites of phosphorylation has since been demonstrated. TRPP2 has been shown to be phosphorylated on its N-terminus on Ser^{76} by GSK3 with a potential CKI priming site at Ser^{80} (Streets et al. 2006). Phosphorylation at this site proved to be important for correct localisation of TRPP2; in MDCK cells, treatment with GSK3 inhibitors resulted in redistribution of lateral TRPP2 to an intracellular compartment (Streets et al. 2006). TRPP2 was still able to traffic the primary cilia, therefore GSK3 phosphorylation is specifically important for trafficking to the lateral plasma membrane and may not be required for ER export. The inability of a $Ser^{76/80}$ mutant of TRPP2 to rescue a cystic phenotype in the zebrafish pronephric kidney, induced by an antisense morpholino nucleotide to *PKD2*, provided further evidence of the functional importance of phosphorylation at this site (Streets et al. 2006).

24.6.3 PrKD Phosphorylation of TRPP2

Streets et al. (2006) had suggested that TRPP2 could be phosphorylated at sites other than Ser^{812} and Ser^{76}, as mutation of both these residues did not abolish TRPP2 phospholabelling (Streets et al. 2006). Phosphorylation at Ser^{801} in the

TRPP2 C-terminal domain by Protein Kinase D (PrKD) was later demonstrated (Streets et al. 2010). Phosphorylation at this site was shown to be increased following serum and EGF stimulation and was critical for TRPP2-mediated ER calcium release.

In addition, phosphorylation at Ser^{801} was important in the regulation of TRPP2's effects on proliferation; expression of S^{801G} TRPP2 in MDCK I cells had no effect on basal proliferation in contrast to suppressed proliferation observed following expression of wild-type or phospho-mimic S^{801D} TRPP2. The physiological importance of this site was further confirmed by findings that a reported missense mutation, S^{804N}, which lies within the PrKD consensus sequence (Rossetti et al. 2007), abolished Ser^{801} phosphorylation. Phosphorylation at Ser^{801} by PrKD thus appears to be essential for TRPP2 channel function in the ER in response to growth factor stimulation.

24.6.4 AuroraA Phosphorylation of TRPP2

A recent study demonstrated that TRPP2 could be phosphorylated by AuroraA kinase on Ser^{829} (Plotnikova et al. 2011). AuroraA kinase activity was shown to be upregulated in cyst lining epithelial cells from patients with ADPKD. Why AuroraA kinase activity should be increased in ADPKD is unclear. Data from the same group suggest that release of Ca^{2+} from intracellular endoplasmic reticulum stores rapidly and transiently activate AuroraA (Plotnikova et al. 2010) however calcium levels are normally lower in cystic cells. AuroraA was shown to bind, phosphorylate and reduce the activity of TRPP2 limiting the amplitude of calcium release from the endoplasmic reticulum in cultured kidney epithelial cells. Recent work from our group has demonstrated that TRPP2 can be phosphorylated on Ser^{829} by protein kinase A suggesting a more complex regulation of phosphorylation at this residue (Streets et al. 2013).

24.6.5 Other Kinases

Increased phosphorylation of TRPP2 on an unknown residue was seen in the *jck* mouse model of PKD (Sohara et al. 2008). This model contains a mutation in the kinase Nek8 which forms a complex with TRPP2 in tissues taken from *jck* mouse kidneys. As well as increased phosphorylation there was increased ciliary localisation of TRPP2 in *jck* mice suggest that protein phosphorylation may serve as a mechanism for the control of ciliary targeting of TRPP2. In addition, there are a large number of phosphorylation sites identified through large-scale phosphoproteomic screens (Table 24.1). The kinases responsible for phosphorylating these sites on TRPP2 are unknown as it is their functional significance.

24.7 Regulation of TRPP2 by MicroRNAs

MicroRNAs (miRNAs) are regulatory RNAs that act as post-transcriptional repressors by binding the 3' untranslated region of target genes (Bartel 2004). MicroRNAs have been shown to control the expression of TRPP2 mediating its functional effects. Two groups have shown that miR-17 directly targets the 3'UTR of the *PKD2* gene and represses its expression (Tran et al. 2010; Sun et al. 2010). Overexpression of miR-17 promoted cell proliferation in HEK-293 cells via post-transcriptional silencing of TRPP2 (Sun et al. 2010). The repressive activity of miR-17 was found to be antagonised by the RNA binding protein Bicaudal C (Bicc1). Bicc1 has been shown to modulate the expression of TRPP2 by regulating the stability of Pkd2 mRNA and its translational efficiency (Tran et al. 2010). Polycystic diseases also develop in mice and humans carrying mutations in Bicc1 (Cogswell et al. 2003; Kraus et al. 2012) and may be linked to the regulation of cAMP signalling in polycystic kidneys via microRNA-induced gene silencing of adenylate cyclase-6 (Piazzon et al. 2012). The regulation of TRPP2 and other proteins by microRNAs may potentially represent a novel mechanism for cystogenesis as well as a potential therapeutic target for ADPKD.

24.8 Conclusion

Considerable progress has been made in recent years to understand the structure and function of TRPP2 and its relevance to the pathogenesis of ADPKD. TRPP2 channels have been shown to reside in a number of different subcellular compartments. This spatial regulation enables it to perform specific functions in different areas of the cell. Disturbed calcium signalling mediated by loss of TRPP2 or its interacting partner polycystin-1 is clearly an early event in the development of a cystic phenotype. Further research is required to further understand the connection between altered calcium signalling and the pathogenesis of ADPKD. TRPP2 and polycystin-1 interact with many other proteins involved in regulating intracellular calcium levels and loss of either protein leads to changes in cellular calcium signalling within microdomains located in the primary cilium or at the ER membrane. Understanding the functional significance of this distinct calcium signalling microdomains and identifying the downstream signalling pathways affected will give a better understanding of their contribution to the pathogenesis of ADPKD.

References

Aguiari G, Banzi M, Gessi S, Cai Y, Zeggio E, Manzati E, Piva R, Lambertini E, Ferrari L, Peters DJ, Lanza F, Harris PC, Borea PA, Somlo S, Del Senno L (2004) Deficiency of polycystin-2 reduces Ca2 + channel activity and cell proliferation in ADPKD lymphoblastoid cells. Faseb J 18(7):884–886

Alehan FK, Gurakan B, Agildere M (2002) Familial arachnoid cysts in association with autosomal dominant polycystic kidney disease. Pediatrics 110(1):3

Anyatonwu GI, Estrada M, Tian X, Somlo S, Ehrlich BE (2007) Regulation of ryanodine receptor-dependent calcium signaling by polycystin-2. Proc Nat Acad Sci USA 104(15):6454–6459

Bai CX, Giamarchi A, Rodat-Despoix L, Padilla F, Downs T, Tsiokas L, Delmas P (2008) Formation of a new receptor-operated channel by heteromeric assembly of TRPP2 and TRPC1 subunits. EMBO Rep 9(5):472–479

Bartel DP (2004) MicroRNAs: genomics, biogenesis, mechanism, and function. Cell 116(2):281–297

Bhunia AK, Piontek K, Boletta A, Liu L, Qian F, Xu PN, Germino FJ, Germino GG (2002) PKD1 induces p21(waf1) and regulation of the cell cycle via direct activation of the JAK-STAT signaling pathway in a process requiring PKD2. Cell 109(2):157–168

Bogdanova N, Markoff A, Gerke V, McCluskey M, Horst J, Dworniczak B (2001) Homologues to the first gene for autosomal dominant polycystic kidney disease are pseudogenes. Genomics 74(3):333–341

Braun AP, Sy L (2001) Contribution of potential EF hand motifs to the calcium-dependent gating of a mouse brain large conductance, calcium-sensitive K(+) channel. J physiol 533(Pt 3):681–695

Burtey S, Riera M, Ribe E, Pennenkamp P, Rance R, Luciani J, Dworniczak B, Mattei MG, Fontes M (2008) Centrosome overduplication and mitotic instability in PKD2 transgenic lines. Cell Biol Int 32(10):1193–1198

Cai Y, Anyatonwu G, Okuhara D, Lee KB, Yu Z, Onoe T, Mei CL, Qian Q, Geng L, Wiztgall R, Ehrlich BE, Somlo S (2004) Calcium dependence of polycystin-2 channel activity is modulated by phosphorylation at Ser812. J Biol Chem 279(19):19987–19995

Cai Y, Maeda Y, Cedzich A, Torres VE, Wu G, Hayashi T, Mochizuki T, Park JH, Witzgall R, Somlo S (1999) Identification and characterization of polycystin-2, the PKD2 gene product. J Biol Chem 274(40):28557–28565

Calvet JP, Grantham JJ (2001) The genetics and physiology of polycystic kidney disease. Semin Nephrol 21(2):107–123

Celic A, Petri ET, Demeler B, Ehrlich BE, Boggon TJ (2008) Domain mapping of the polycystin-2 C-terminal tail using de novo molecular modeling and biophysical analysis. J Biol Chem 283(42):28305–28312

Chang MY, Parker E, Ibrahim S, Shortland JR, Nahas ME, Haylor JL, Ong AC (2006) Haploinsufficiency of Pkd2 is associated with increased tubular cell proliferation and interstitial fibrosis in two murine Pkd2 models. Nephrol Dial Transplant 21(8):2078–2084

Chapin HC, Rajendran V, Caplan MJ (2010) Polycystin-1 surface localization is stimulated by polycystin-2 and cleavage at the G protein-coupled receptor proteolytic site. Mol Biol Cell 21(24):4338–4348

Chapman AB, Guay-Woodford LM, Grantham JJ, Torres VE, Bae KT, Baumgarten DA, Kenney PJ, King BF, Glockner JF, Wetzel LH, Brummer ME, O'Neill WC, Robbin ML, Bennett WM, Klahr S, Hirschman GH, Kimmel PL, Thompson PA, Miller JP (2003) Renal structure in early autosomal-dominant polycystic kidney disease (ADPKD): The consortium for radiologic imaging studies of polycystic kidney disease (CRISP) cohort. Kidney Int 64(3):1035–1045

Chen XZ, Vassilev PM, Basora N, Peng JB, Nomura H, Segal Y, Brown EM, Reeders ST, Hediger MA, Zhou J (1999) Polycystin-L is a calcium-regulated cation channel permeable to calcium ions. Nature 401(6751):383–386

Cogswell C, Price SJ, Hou X, Guay-Woodford LM, Flaherty L, Bryda EC (2003) Positional cloning of jcpk/bpk locus of the mouse. Mamm Genome 14(4):242–249

Cuppage FE, Huseman RA, Chapman A, Grantham JJ (1980) Ultrastructure and function of cysts from human adult polycystic kidneys. Kidney Int 17(3):372–381

Deane JA, Ricardo SD (2007) Polycystic kidney disease and the renal cilium. Nephrology (Carlton) 12(6):559–564

Delmas P (2004) Polycystins: from mechanosensation to gene regulation. Cell 118(2):145–148

Delmas P (2005) Polycystins: polymodal receptor/ion-channel cellular sensors. Pflugers Arch 451(1):264–276

Delmas P, Nauli SM, Li X, Coste B, Osorio N, Crest M, Brown DA, Zhou J (2004a) Gating of the polycystin ion channel signaling complex in neurons and kidney cells. FASEB J 18(6):740–742

Delmas P, Padilla F, Osorio N, Coste B, Raoux M, Crest M (2004b) Polycystins, calcium signaling, and human diseases. Biochem Biophys Res Commun 322(4):1374–1383

Du J, Ding M, Sours-Brothers S, Graham S, Ma R (2008) Mediation of angiotensin II-induced Ca2 + signaling by polycystin 2 in glomerular mesangial cells. Am J Physiol Renal Physiol 294(4):F909–F918

Edelstein CL (2005) What is the role of tubular epithelial cell apoptosis in polycystic kidney disease (PKD). Cell Cycle 4(11):1550–1554

Feng S, Okenka GM, Bai CX, Streets AJ, Newby LJ, DeChant BT, Tsiokas L, Obara T, Ong AC (2008) Identification and functional characterization of an N-terminal oligomerization domain for polycystin-2. J Biol Chem 283(42):28471–28479

Feng S, Rodat-Despoix L, Delmas P, Ong AC (2011) A single amino acid residue constitutes the third dimerization domain essential for the assembly and function of the tetrameric polycystin-2 (TRPP2) channel. J Biol Chem 286(21):18994–19000

Fick-Brosnahan GM, Belz MM, McFann KK, Johnson AM, Schrier RW (2002) Relationship between renal volume growth and renal function in autosomal dominant polycystic kidney disease: A longitudinal study. Am J Kidney Dis 39(6):1127–1134

Fick GM, Johnson AM, Strain JD, Kimberling WJ, Kumar S, Mancojohnson ML, Duley IT, Gabow PA (1993) Characteristics of very early-onset autosomal-dominant polycystic kidney-disease. J Am Soc Nephrol 3(12):1863–1870

Field S, Riley KL, Grimes DT, Hilton H, Simon M, Powles-Glover N, Siggers P, Bogani D, Greenfield A, Norris DP (2011) Pkd1l1 establishes left-right asymmetry and physically interacts with Pkd2. Development 138(6):1131–1142

Fu X, Wang Y, Schetle N, Gao H, Putz M, von Gersdorff G, Walz G, Kramer-Zucker AG (2008) The subcellular localization of TRPP2 modulates its function. J Am Soc Nephrol 19(7):1342–1351

Gallagher AR, Cedzich A, Gretz N, Somlo S, Witzgall R (2000) The polycystic kidney disease protein PKD2 interacts with Hax-1, a protein associated with the actin cytoskeleton. Proc Natl Acad Sci USA 97(8):4017–4022

Gallagher AR, Germino GG, Somlo S (2010) Molecular advances in autosomal dominant polycystic kidney disease. Adv Chronic Kidney Dis 17(2):118–130

Garcia-Gonzalez MA, Jones JG, Allen SK, Palatucci CM, Batish SD, Seltzer WK, Lan Z, Allen E, Qian F, Lens XM, Pei Y, Germino GG, Watnick TJ (2007) Evaluating the clinical utility of a molecular genetic test for polycystic kidney disease. Mol Genet Metab 92(1–2):160–167

Geng L, Okuhara D, Yu Z, Tian X, Cai Y, Shibazaki S, Somlo S (2006) Polycystin-2 traffics to cilia independently of polycystin-1 by using an N-terminal RVxP motif. J Cell Sci 119(Pt 7):1383–1395

Giamarchi A, Delmas P (2007) Activation mechanisms and functional roles of TRPP2 cation channels. doi:NBK5258 [bookaccession]

Giamarchi A, Feng S, Rodat-Despoix L, Xu Y, Bubenshchikova E, Newby LJ, Hao J, Gaudioso C, Crest M, Lupas AN, Honore E, Williamson MP, Obara T, Ong AC, Delmas P (2010) A polycystin-2 (TRPP2) dimerization domain essential for the function of heteromeric polycystin complexes. EMBO J 29(7):1176–1191

Grimm DH, Karihaloo A, Cai Y, Somlo S, Cantley LG, Caplan MJ (2006) Polycystin-2 regulates proliferation and branching morphogenesis in kidney epithelial cells. J Biol Chem 281(1):137–144

Hanaoka K, Qian F, Boletta A, Bhunia AK, Piontek K, Tsiokas L, Sukhatme VP, Guggino WB, Germino GG (2000) Co-assembly of polycystin-1 and -2 produces unique cation-permeable currents. Nature 408(6815):990–994. doi:10.1038/35050128

Harris PC, Germino G, Klinger K, Landes G, van Adelsberg J (1995) The PKD1 gene product. Nat Med 1(6):493

Hateboer N, Dijk MAV, Bogdanova N, Coto E, Saggar-Malik AK, San Millan JL, Torra R, Breuning M, Ravine D (1999) Comparison of phenotypes of polycystic kidney disease types 1 and 2. European PKD1-PKD2 Study Group. Lancet 353(9147):103–107

Hoffert JD, Pisitkun T, Wang G, Shen RF, Knepper MA (2006) Quantitative phosphoproteomics of vasopressin-sensitive renal cells: regulation of aquaporin-2 phosphorylation at two sites. Proc Natl Acad Sci USA 103:7159–7164

Hoffmeister H, Babinger K, Gurster S, Cedzich A, Meese C, Schadendorf K, Osten L, de Vries U, Rascle A, Witzgall R (2011) Polycystin-2 takes different routes to the somatic and ciliary plasma membrane. J Cell Biol 192(4):631–645

Hogan MC, Manganelli L, Woollard JR, Masyuk AI, Masyuk TV, Tammachote R, Huang BQ, Leontovich AA, Beito TG, Madden BJ, Charlesworth MC, Torres VE, LaRusso NF, Harris PC, Ward CJ (2009) Characterization of PKD protein-positive exosome-like vesicles. J Am Soc Nephrol 20(2):278–288

Hsu PP, Kang SA, Rameseder J, Zhang Y, Ottina KA, Lim D, Peterson TR, Choi Y, Gray NS, Yaffe MB, Marto JA, Sabatini DM (2011) The mTOR-regulated phosphoproteome reveals a mechanism of mTORC1-mediated inhibition of growth factor signaling. Science 332:1317–1322

Hu J, Bae YK, Knobel KM, Barr MM (2006) Casein kinase II and calcineurin modulate TRPP function and ciliary localization. Mol Biol Cell 17(5):2200–2211

Huttlin EL, Jedrychowski MP, Elias JE, Goswami T, Rad R, Beausoleil SA, Villen J, Haas W, Sowa ME, Gygi SP (2010) A tissue-specific atlas of mouse protein phosphorylation and expression. Cell 143(7):1174–1189

Ibraghimov-Beskrovnaya O, Bukanov NO, Donohue LC, Dackowski WR, Klinger KW, Landes GM (2000) Strong homophilic interactions of the Ig-like domains of polycystin-1, the protein product of an autosomal dominant polycystic kidney disease gene, PKD1. Hum Mol Genet 9(11):1641–1649

Jurczyk A, Gromley A, Redick S, San Agustin J, Witman G, Pazour GJ, Peters DJ, Doxsey S (2004) Pericentrin forms a complex with intraflagellar transport proteins and polycystin-2 and is required for primary cilia assembly. J Cell Biol 166(5):637–643

Kamura K, Kobayashi D, Uehara Y, Koshida S, Iijima N, Kudo A, Yokoyama T, Takeda H (2011) Pkd1l1 complexes with Pkd2 on motile cilia and functions to establish the left-right axis. Development 138(6):1121–1129

Kim E, Arnould T, Sellin LK, Benzing T, Fan MJ, Gruning W, Sokol SY, Drummond I, Walz G (1999) The polycystic kidney disease 1 gene product modulates Wnt signaling. J Biol Chem 274(8):4947–4953

Kimberling WJ, Kumar S, Gabow PA, Kenyon JB, Connolly CJ, Somlo S (1993) Autosomal-dominant polycystic kidney-disease—localization of the 2nd gene to chromosome-4q13-q23. Genomics 18(3):467–472

Kobori T, Smith GD, Sandford R, Edwardson JM (2009) The transient receptor potential channels TRPP2 and TRPC1 form a heterotetramer with a 2:2 stoichiometry and an alternating subunit arrangement. J Biol Chem 284(51):35507–35513

Koptides M, Hadjimichael C, Koupepidou P, Pierides A, Constantinou Deltas C (1999) Germinal and somatic mutations in the PKD2 gene of renal cysts in autosomal dominant polycystic kidney disease. Hum Mol Genet 8(3):509–513

Koptides M, Mean R, Demetriou K, Pierides A, Deltas CC (2000) Genetic evidence for a trans-heterozygous model for cystogenesis in autosomal dominant polycystic kidney disease. Hum Mol Genet 9(3):447–452

Kottgen M, Benzing T, Simmen T, Tauber R, Buchholz B, Feliciangeli S, Huber TB, Schermer B, Kramer-Zucker A, Hopker K, Simmen KC, Tschucke CC, Sandford R, Kim E, Thomas G, Walz G (2005) Trafficking of TRPP2 by PACS proteins represents a novel mechanism of ion channel regulation. EMBO J 24(4):705–716

Kottgen M, Buchholz B, Garcia-Gonzalez MA, Kotsis F, Fu X, Doerken M, Boehlke C, Steffl D, Tauber R, Wegierski T, Nitschke R, Suzuki M, Kramer-Zucker A, Germino GG, Watnick T, Prenen J, Nilius B, Kuehn EW, Walz G (2008) TRPP2 and TRPV4 form a polymodal sensory channel complex. J Cell Biol 182(3):437–447

Koulen P, Cai Y, Geng L, Maeda Y, Nishimura S, Witzgall R, Ehrlich BE, Somlo S (2002) Polycystin-2 is an intracellular calcium release channel. Nat Cell Biol 4(3):191–197

Kraus MR, Clauin S, Pfister Y, Di Maio M, Ulinski T, Constam D, Bellanne-Chantelot C, Grapin-Botton A (2012) Two mutations in human BICC1 resulting in Wnt pathway hyperactivity associated with cystic renal dysplasia. Hum Mutat 33(1):86–90

Kubota T, Kinoshita M, Sasaki R, Aoike F, Takahashi MP, Sakoda S, Hirose K (2009) New mutation of the Na channel in the severe form of potassium-aggravated myotonia. Muscle Nerve 39(5):666–673

Lanoix J, D'Agati V, Szabolcs M, Trudel M (1996) Dysregulation of cellular proliferation and apoptosis mediates human autosomal dominant polycystic kidney disease (ADPKD). Oncogene 13(6):1153–1160

Lantinga-van Leeuwen IS, Dauwerse JG, Baelde HJ, Leonhard WN, van de Wal A, Ward CJ, Verbeek S, Deruiter MC, Breuning MH, de Heer E, Peters DJ (2004) Lowering of Pkd1 expression is sufficient to cause polycystic kidney disease. Hum Mol Genet 13(24):3069–3077

Lehtonen S, Ora A, Olkkonen VM, Geng L, Zerial M, Somlo S, Lehtonen E (2000) In vivo interaction of the adapter protein CD2-associated protein with the type 2 polycystic kidney disease protein, polycystin-2. J Biol Chem 275(42):32888–32893

Li A, Tian X, Sung SW, Somlo S (2003a) Identification of two novel polycystic kidney disease-1-like genes in human and mouse genomes. Genomics 81(6):596–608

Li Q, Dai Y, Guo L, Liu Y, Hao C, Wu G, Basora N, Michalak M, Chen XZ (2003b) Polycystin-2 associates with tropomyosin-1, an actin microfilament component. J Mol Biol 325(5):949–962

Li Q, Montalbetti N, Shen PY, Dai XQ, Cheeseman CI, Karpinski E, Wu G, Cantiello HF, Chen XZ (2005a) Alpha-actinin associates with polycystin-2 and regulates its channel activity. Hum Mol Genet 14(12):1587–1603

Li X, Luo Y, Starremans PG, McNamara CA, Pei Y, Zhou J (2005b) Polycystin-1 and polycystin-2 regulate the cell cycle through the helix-loop-helix inhibitor Id2. Nat Cell Biol 7(12):1202–1212

Li Y, Wright JM, Qian F, Germino GG, Guggino WB (2005c) Polycystin 2 interacts with type I inositol 1,4,5-trisphosphate receptor to modulate intracellular Ca2 + signaling. J Biol Chem 280(50):41298–41306

Li Q, Montalbetti N, Wu Y, Ramos AJ, Raychowdhury MK, Chen XZ, Cantiello HF (2006) Polycystin-2 cation channel function is under the control of microtubular structures in primary cilia of renal epithelial cells. J Biol Chem 281:37566–37575

Li Y, Santoso NG, Yu S, Woodward OM, Qian F, Guggino WB (2009) Polycystin-1 interacts with inositol 1,4,5-trisphosphate receptor to modulate intracellular Ca2 + signaling with implications for polycystic kidney disease. J Biol Chem 284(52):36431–36441

Liang G, Yang J, Wang Z, Li Q, Tang Y, Chen XZ (2008) Polycystin-2 down-regulates cell proliferation via promoting PERK-dependent phosphorylation of eIF2alpha. Hum Mol Genet 17(20):3254–3262

Lin F, Hiesberger T, Cordes K, Sinclair AM, Goldstein LS, Somlo S, Igarashi P (2003) Kidney-specific inactivation of the KIF3A subunit of kinesin-II inhibits renal ciliogenesis and produces polycystic kidney disease. Proc Natl Acad Sci USA 100(9):5286–5291

Luo Y, Vassilev PM, Li X, Kawanabe Y, Zhou J (2003) Native polycystin 2 functions as a plasma membrane Ca2 + -permeable cation channel in renal epithelia. Mol Cell Biol 23(7):2600–2607

Ma R, Li WP, Rundle D, Kong J, Akbarali HI, Tsiokas L (2005) PKD2 functions as an epidermal growth factor-activated plasma membrane channel. Mol Cell Biol 25(18):8285–8298

Matsumoto Y, Maller JL (2002) Calcium, calmodulin, and CaMKII requirement for initiation of centrosome duplication in Xenopus egg extracts. Science 295(5554):499–502

Mekahli D, Sammels E, Luyten T, Welkenhuyzen K, van den Heuvel LP, Levtchenko EN, Gijsbers R, Bultynck G, Parys JB, De Smedt H, Missiaen L (2012) Polycystin-1 and polycystin-2 are both required to amplify inositol-trisphosphate-induced Ca(2 +) release. Cell Calcium. doi:10.1016/j.ceca.2012.03.002

Mochizuki T, Wu G, Hayashi T, Xenophontos SL, Veldhuisen B, Saris JJ, Reynolds DM, Cai Y, Gabow PA, Pierides A, Kimberling WJ, Breuning MH, Deltas CC, Peters DJ, Somlo S (1996) PKD2, a gene for polycystic kidney disease that encodes an integral membrane protein. Science 272(5266):1339–1342

Molina H, Horn DM, Tang N, Mathivanan S, Pandey A (2007) Global proteomic profiling of phosphopeptides using electron transfer dissociation tandem mass spectrometry. Proc Natl Acad Sci USA 104:2199–2204

Montell C (2005) The TRP superfamily of cation channels. Sci STKE 2005(272):re3

Montell C, Birnbaumer L, Flockerzi V (2002a) The TRP channels, a remarkably functional family. Cell 108(5):595–598

Montell C, Birnbaumer L, Flockerzi V, Bindels RJ, Bruford EA, Caterina MJ, Clapham DE, Harteneck C, Heller S, Julius D, Kojima I, Mori Y, Penner R, Prawitt D, Scharenberg AM, Schultz G, Shimizu N, Zhu MX (2002b) A unified nomenclature for the superfamily of TRP cation channels. Mol Cell 9(2):229–231

Nauli SM, Alenghat FJ, Luo Y, Williams E, Vassilev P, Li X, Elia AE, Lu W, Brown EM, Quinn SJ, Ingber DE, Zhou J (2003) Polycystins 1 and 2 mediate mechanosensation in the primary cilium of kidney cells. Nat Genet 33(2):129–137

Nauli SM, Rossetti S, Kolb RJ, Alenghat FJ, Consugar MB, Harris PC, Ingber DE, Loghman-Adham M, Zhou J (2006) Loss of polycystin-1 in human cyst-lining epithelia leads to ciliary dysfunction. J Am Soc Nephrol 17(4):1015–1025

Nauli SM, Zhou J (2004) Polycystins and mechanosensation in renal and nodal cilia. BioEssays 26(8):844–856

Newby LJ, Streets AJ, Zhao Y, Harris PC, Ward CJ, Ong AC (2002) Identification, characterization, and localization of a novel kidney polycystin-1-polycystin-2 complex. J Biol Chem 277(23):20763–20773

Nilius B, Owsianik G (2011) The transient receptor potential family of ion channels. Genome Biol 12(3):218

Nishiura JL, Neves RFCA, Eloi SRM, Cintra SMLF, Ajzen SA, Heilberg IP (2009) Evaluation of nephrolithiasis in autosomal dominant polycystic kidney disease patients. Clin J Am Soc Nephrol 4(4):838–844

Olsen JV, Blagoev B, Gnad F, Macek B, Kumar C, Mortensen P, Mann M (2006) Global, in vivo, and site-specific phosphorylation dynamics in signaling networks. Cell 127:635–648

O'Neill WC, Robbin ML, Bae KT, Grantham JJ, Chapman AB, Guay-Woodford LM, Torres VE, King BF, Wetzel LH, Thompson PA, Miller JP (2005) Sonographic assessment of the severity and progression of autosomal dominant polycystic kidney disease: The consortium of renal imaging studies in polycystic kidney disease (CRISP). Am J Kidney Dis 46(6):1058–1064

Ong AC (2000) Polycystin expression in the kidney and other tissues: complexity, consensus and controversy. Exp Nephrol 8(4–5):208–214

Ong AC, Harris PC (2005) Molecular pathogenesis of ADPKD: the polycystin complex gets complex. Kidney Int 67(4):1234–1247

Ong AC, Ward CJ, Butler RJ, Biddolph S, Bowker C, Torra R, Pei Y, Harris PC (1999) Coordinate expression of the autosomal dominant polycystic kidney disease proteins, polycystin-2 and polycystin-1, in normal and cystic tissue. Am J Pathol 154(6):1721–1729

Ong AC, Wheatley DN (2003) Polycystic kidney disease–the ciliary connection. Lancet 361(9359):774–776

Parnell SC, Magenheimer BS, Maser RL, Rankin CA, Smine A, Okamoto T, Calvet JP (1998) The polycystic kidney disease-1 protein, polycystin-1, binds and activates heterotrimeric G-proteins in vitro. Biochem Biophys Res Commun 251(2):625–631

Parnell SC, Magenheimer BS, Maser RL, Zien CA, Frischauf AM, Calvet JP (2002) Polycystin-1 activation of c-Jun N-terminal kinase and AP-1 is mediated by heterotrimeric G proteins. J Biol Chem 277(22):19566–19572

Patel V, Li L, Cobo-Stark P, Shao X, Somlo S, Lin F, Igarashi P (2008) Acute kidney injury and aberrant planar cell polarity induce cyst formation in mice lacking renal cilia. Hum Mol Genet 17(11):1578–1590

Pazour GJ, Dickert BL, Vucica Y, Seeley ES, Rosenbaum JL, Witman GB, Cole DG (2000) Chlamydomonas IFT88 and its mouse homologue, polycystic kidney disease gene tg737, are required for assembly of cilia and flagella. J Cell Biol 151(3):709–718

Pei Y (2001) A "two-hit" model of cystogenesis in autosomal dominant polycystic kidney disease. Trends Mol Med 7(4):151–156

Pennekamp P, Karcher C, Fischer A, Schweickert A, Skryabin B, Horst J, Blum M, Dworniczak B (2002) The ion channel polycystin-2 is required for left-right axis determination in mice. Curr Biol 12(11):938–943

Peters DJ, Breuning MH (2001) Autosomal dominant polycystic kidney disease: modification of disease progression. Lancet 358(9291):1439–1444

Peters DJM, Spruit L, Saris JJ, Ravine D, Sandkuijl LA, Fossdal R, Boersma J, Vaneijk R, Norby S, Constantinoudeltas CD, Pierides A, Brissenden JE, Frants RR, Vanommen GJB, Breuning MH (1993) Chromosome 4 localization of a 2nd gene for autosomal-dominant polycystic kidney-disease. Nat Genet 5(4):359–362

Petri ET, Celic A, Kennedy SD, Ehrlich BE, Boggon TJ, Hodsdon ME (2010) Structure of the EF-hand domain of polycystin-2 suggests a mechanism for Ca2 + -dependent regulation of polycystin-2 channel activity. Proc Natl Acad Sci USA 107(20):9176–9181

Peyronnet R, Sharif-Naeini R, Folgering JH, Arhatte M, Jodar M, El Boustany C, Gallian C, Tauc M, Duranton C, Rubera I, Lesage F, Pei Y, Peters DJ, Somlo S, Sachs F, Patel A, Honore E, Duprat F (2012) Mechanoprotection by polycystins against apoptosis is mediated through the opening of stretch-activated K(2P) channels. Cell Rep 1(3):241–250

Piazzon N, Maisonneuve C, Guilleret I, Rotman S, Constam DB (2012) Bicc1 links the regulation of cAMP signaling in polycystic kidneys to microRNA-induced gene silencing. J Mol Cell Biol. doi:mjs027 [pii] 10.1093/jmcb/mjs027

Plotnikova OV, Pugacheva EN, Dunbrack RL, Golemis EA (2010) Rapid calcium-dependent activation of Aurora-A kinase. Nat Commun 1:64. doi:ncomms1061 [pii] 10.1038/ncomms1061

Plotnikova OV, Pugacheva EN, Golemis EA (2011) Aurora A kinase activity influences calcium signaling in kidney cells. J Cell Biol 193(6):1021–1032

Praetorius HA, Spring KR (2001) Bending the MDCK cell primary cilium increases intracellular calcium. J Membr Biol 184(1):71–79

Pritchard L, Sloane-Stanley JA, Sharpe JA, Aspinwall R, Lu W, Buckle V, Strmecki L, Walker D, Ward CJ, Alpers CE, Zhou J, Wood WG, Harris PC (2000) A human PKD1 transgene generates functional polycystin-1 in mice and is associated with a cystic phenotype. Hum Mol Genet 9(18):2617–2627

Qian F, Boletta A, Bhunia AK, Xu H, Liu L, Ahrabi AK, Watnick TJ, Zhou F, Germino GG (2002) Cleavage of polycystin-1 requires the receptor for egg jelly domain and is disrupted by human autosomal-dominant polycystic kidney disease 1-associated mutations. Proc Natl Acad Sci USA 99(26):16981–16986

Qian F, Watnick TJ, Onuchic LF, Germino GG (1996) The molecular basis of focal cyst formation in human autosomal dominant polycystic kidney disease type I. Cell 87(6):979–987

Qian Q, Hunter LW, Li M, Marin-Padilla M, Prakash YS, Somlo S, Harris PC, Torres VE, Sieck GC (2003) Pkd2 haploinsufficiency alters intracellular calcium regulation in vascular smooth muscle cells. Hum Mol Genet 12(15):1875–1880

Reeders ST, Keith T, Green P, Germino GG, Barton NJ, Lehmann OJ, Brown VA, Phipps P, Morgan J, Bear JC, Parfrey P (1988) Regional localization of the autosomal dominant polycystic kidney disease locus. Genomics 3(2):150–155

Rivera M, Gonzalo A, Gobernado JM, Orte L, Quereda C, Ortuno J (1992) Stroke in adult polycystic kidney-disease. Postgrad Med J 68(803):735–738

Rossetti S, Chauveau D, Walker D, Saggar-Malik A, Winearls CG, Torres VE, Harris PC (2002) A complete mutation screen of the ADPKD genes by DHPLC. Kidney Int 61(5):1588–1599

Rossetti S, Consugar MB, Chapman AB, Torres VE, Guay-Woodford LM, Grantham JJ, Bennett WM, Meyers CM, Walker DL, Bae K, Zhang QJ, Thompson PA, Miller JP, Harris PC (2007) CRISP consortium: Comprehensive molecular diagnostics in autosomal dominant polycystic kidney disease. J Am Soc Nephrol 18(7):2143–2160

Rossetti S, Strmecki L, Gamble V, Burton S, Sneddon V, Peral B, Roy S, Bakkaloglu A, Komel R, Winearls CG, Harris PC (2001) Mutation analysis of the entire PKD1 gene: Genetic and diagnostic implications. Am J Hum Genet 68(1):46–63

Rundle DR, Gorbsky G, Tsiokas L (2004) PKD2 interacts and co-localizes with mDia1 to mitotic spindles of dividing cells: role of mDia1 IN PKD2 localization to mitotic spindles. J Biol Chem 279(28):29728–29739

Sammels E, Devogelaere B, Mekahli D, Bultynck G, Missiaen L, Parys JB, Cai Y, Somlo S, De Smedt H (2010a) Polycystin-2 activation by inositol 1,4,5-trisphosphate-induced Ca2 + r-elease requires its direct association with the inositol 1,4,5-trisphosphate receptor in a signaling microdomain. J Biol Chem 285(24):18794–18805

Sammels E, Devogelaere B, Mekahli D, Bultynck G, Missiaen L, Parys JB, De Smedt H (2010b) Unraveling the role of polycystin-2/inositol 1,4,5-trisphosphate receptor interaction in Ca signaling. Commun Integr Biol 3(6):530–532

Scheff RT, Zuckerman G, Harter H, Delmez J, Koehler R (1980) Diverticular-disease in patients with chronic-renal-failure due to polycystic kidney-disease. Ann Intern Med 92(2):202–204

Scheffers MS, Le H, van der Bent P, Leonhard W, Prins F, Spruit L, Breuning MH, de Heer E, Peters DJ (2002) Distinct subcellular expression of endogenous polycystin-2 in the plasma membrane and Golgi apparatus of MDCK cells. Hum Mol Genet 11(1):59–67

Schumann F, Hoffmeister H, Bader R, Schmidt M, Witzgall R, Kalbitzer HR (2009a) Ca2 + -dependent conformational changes in a C-terminal cytosolic domain of polycystin-2. J Biol Chem 284(36):24372–24383

Schumann FH, Hoffmeister H, Schmidt M, Bader R, Besl E, Witzgall R, Kalbitzer HR (2009b) NMR-assignments of a cytosolic domain of the C-terminus of polycystin-2. Biomol NMR Assign 3(1):141–144

Sharif-Naeini R, Folgering JH, Bichet D, Duprat F, Lauritzen I, Arhatte M, Jodar M, Dedman A, Chatelain FC, Schulte U, Retailleau K, Loufrani L, Patel A, Sachs F, Delmas P, Peters DJ, Honore E (2009) Polycystin-1 and -2 dosage regulates pressure sensing. Cell 139(3):587–596

Sohara E, Luo Y, Zhang J, Manning DK, Beier DR, Zhou J (2008) Nek8 regulates the expression and localization of polycystin-1 and polycystin-2. J Am Soc Nephrol 19(3):469–476

Sorenson CM, Padanilam BJ, Hammerman MR (1996) Abnormal postpartum renal development and cystogenesis in the bcl-2 (-/-) mouse. Am J Physiol 271(1 Pt 2):F184–F193

Streets AJ, Moon DJ, Kane ME, Obara T, Ong AC (2006) Identification of an N-terminal glycogen synthase kinase 3 phosphorylation site which regulates the functional localization of polycystin-2 in vivo and in vitro. Hum Mol Genet 15(9):1465–1473

Streets AJ, Needham AJ, Gill SK, Ong AC (2010) Protein kinase D-mediated phosphorylation of polycystin-2 (TRPP2) is essential for its effects on cell growth and calcium channel activity. Mol Biol Cell 21(22):3853–3865

Streets AJ, Newby LJ, O'Hare MJ, Bukanov NO, Ibraghimov-Beskrovnaya O, Ong AC (2003) Functional analysis of PKD1 transgenic lines reveals a direct role for polycystin-1 in mediating cell-cell adhesion. J Am Soc Nephrol 14(7):1804–1815

Streets AJ, Wagner BE, Harris PC, Ward CJ, Ong AC (2009) Homophilic and heterophilic polycystin 1 interactions regulate E-cadherin recruitment and junction assembly in MDCK cells. J Cell Sci 122(Pt 9):1410–1417

Streets AJ, Wessely O, Peters DJM, Ong ACM (2013) Hyperphosphorylation of polycystin-2 at a critical residue in disease reveals an essential role for polycystin-1 mediated dephosphorylation. Hum Mol Genet. doi:10.1093/hmg/ddt031)

Sun H, Li QW, Lv XY, Ai JZ, Yang QT, Duan JJ, Bian GH, Xiao Y, Wang YD, Zhang Z, Liu YH, Tan RZ, Yang Y, Wei YQ, Zhou Q (2010) MicroRNA-17 post-transcriptionally regulates polycystic kidney disease-2 gene and promotes cell proliferation. Mol Biol Rep 37(6):2951–2958

Sutton KA, Jungnickel MK, Ward CJ, Harris PC, Florman HM (2006) Functional characterization of PKDREJ, a male germ cell-restricted polycystin. J Cell Physiol 209(2):493–500

Szklarczyk D, Franceschini A, Kuhn M, Simonovic M, Roth A, Minguez P, Doerks T, Stark M, Muller J, Bork P, Jensen LJ, von Mering C (2011) The STRING database in 2011: functional interaction networks of proteins, globally integrated and scored. Nucleic Acids Res 39 (Database issue):D561–568

Tan Y-C, Blumenfeld JD, Anghel R, Donahue S, Belenkaya R, Balina M, Parker T, Levine D, Leonard DGB, Rennert H (2009) Novel method for genomic analysis of PKD1 and PKD2 mutations in autosomal dominant polycystic kidney disease. Hum Mutat 30(2):264–273

Tao Y, Kim J, Faubel S, Wu JC, Falk SA, Schrier RW, Edelstein CL (2005) Caspase inhibition reduces tubular apoptosis and proliferation and slows disease progression in polycystic kidney disease. Proc Natl Acad Sci USA 102(19):6954–6959

The European Polycystic Kidney Disease Consortium (1994) The polycystic kidney disease 1 gene encodes a 14 kb transcript and lies within a duplicated region on chromosome 16. Cell 78 (4):725

Tokiwa S, Muto S, China T, Horie S (2011) The relationship between renal volume and renal function in autosomal dominant polycystic kidney disease. Clin Exp Nephrol 15(4):539–545

Torra R, Badenas C, Darnell A, Nicolau C, Volpini V, Revert L, Estivill X (1996) Linkage, clinical features, and prognosis of autosomal dominant polycystic kidney disease types 1 and 2. J Am Soc Nephrol 7(10):2142–2151

Torres VE, Chapman AB, Devuyst O, Gansevoort RT, Grantham JJ, Higashihara E, Perrone RD, Krasa HB, Ouyang J, Czerwiec FS (2012) Tolvaptan in patients with autosomal dominant polycystic kidney disease. N Engl J Med. doi:10.1056/NEJMoa1205511

Torres VE, Harris PC (2009) Autosomal dominant polycystic kidney disease: the last 3 years. Kidney Int 76(2):149–168

Torres VE, Harris PC, Pirson Y (2007) Autosomal dominant polycystic kidney disease. Lancet 369(9569):1287–1301

Tran U, Zakin L, Schweickert A, Agrawal R, Doger R, Blum M, De Robertis EM, Wessely O (2010) The RNA-binding protein bicaudal C regulates polycystin 2 in the kidney by antagonizing miR-17 activity. Development 137(7):1107–1116

Tsiokas L (2009) Function and regulation of TRPP2 at the plasma membrane. Am J Physiol Renal Physiol 297(1):F1–F9

Vassilev PM, Guo L, Chen XZ, Segal Y, Peng JB, Basora N, Babakhanlou H, Cruger G, Kanazirska M, Ye C, Brown EM, Hediger MA, Zhou J (2001) Polycystin-2 is a novel cation channel implicated in defective intracellular Ca(2 +) homeostasis in polycystic kidney disease. Biochem Biophys Res Commun 282(1):341–350

Wang Q, Dai XQ, Li Q, Wang Z, Cantero Mdel R, Li S, Shen J, Tu JC, Cantiello H, Chen XZ (2012) Structural interaction and functional regulation of polycystin-2 by filamin. PLoS ONE 7(7):e40448

Wang X, Gattone V 2nd, Harris PC, Torres VE (2005) Effectiveness of vasopressin V2 receptor antagonists OPC-31260 and OPC-41061 on polycystic kidney disease development in the PCK rat. J Am Soc Nephrol 16(4):846–851

Watnick T, He N, Wang K, Liang Y, Parfrey P, Hefferton D, St George-Hyslop P, Germino G, Pei Y (2000) Mutations of PKD1 in ADPKD2 cysts suggest a pathogenic effect of trans-heterozygous mutations. Nat Genet 25(2):143–144

Watnick TJ, Torres VE, Gandolph MA, Qian F, Onuchic LF, Klinger KW, Landes G, Germino GG (1998) Somatic mutation in individual liver cysts supports a two-hit model of cystogenesis in autosomal dominant polycystic kidney disease. Mol Cell 2(2):247–251

Wegierski T, Steffl D, Kopp C, Tauber R, Buchholz B, Nitschke R, Kuehn EW, Walz G, Kottgen M (2009) TRPP2 channels regulate apoptosis through the Ca2 + concentration in the endoplasmic reticulum. EMBO J 28(5):490–499

Winyard PJ, Nauta J, Lirenman DS, Hardman P, Sams VR, Risdon RA, Woolf AS (1996) Deregulation of cell survival in cystic and dysplastic renal development. Kidney Int 49(1):135–146

Woo D (1995) Apoptosis and loss of renal tissue in polycystic kidney diseases. N Engl J Med 333(1):18–25

Wu G, D'Agati V, Cai Y, Markowitz G, Park JH, Reynolds DM, Maeda Y, Le TC, Hou H Jr, Kucherlapati R, Edelmann W, Somlo S (1998) Somatic inactivation of Pkd2 results in polycystic kidney disease. Cell 93(2):177–188

Wu G, Markowitz GS, Li L, D'Agati VD, Factor SM, Geng L, Tibara S, Tuchman J, Cai Y, Park JH, van Adelsberg J, Hou H Jr, Kucherlapati R, Edelmann W, Somlo S (2000) Cardiac defects and renal failure in mice with targeted mutations in Pkd2. Nat Genet 24(1):75–78

Wu G, Somlo S (2000) Molecular genetics and mechanism of autosomal dominant polycystic kidney disease. Mol Genet Metab 69(1):1–15

Wu LJ, Sweet TB, Clapham DE (2010) International Union of Basic and Clinical Pharmacology. LXXVI. Current progress in the mammalian TRP ion channel family. Pharmacol Rev 62(3):381–404

Wu Y, Dai XQ, Li Q, Chen CX, Mai W, Hussain Z, Long W, Montalbetti N, Li G, Glynne R, Wang S, Cantiello HF, Wu G, Chen XZ (2006) Kinesin-2 mediates physical and functional interactions between polycystin-2 and fibrocystin. Hum Mol Genet 15(22):3280–3292

Xiao Q, Prussia A, Yu K, Cui YY, Hartzell HC (2008) Regulation of bestrophin Cl channels by calcium: role of the C terminus. J Gen Physiol 132(6):681–692

Yamaguchi T, Hempson SJ, Reif GA, Hedge AM, Wallace DP (2006) Calcium restores a normal proliferation phenotype in human polycystic kidney disease epithelial cells. J Am Soc Nephrol 17(1):178–187

Yamaguchi T, Wallace DP, Magenheimer BS, Hempson SJ, Grantham JJ, Calvet JP (2004) Calcium restriction allows cAMP activation of the B-Raf/ERK pathway, switching cells to a cAMP-dependent growth-stimulated phenotype. J Biol Chem 279(39):40419–40430

Yoder BK, Tousson A, Millican L, Wu JH, Bugg CE Jr, Schafer JA, Balkovetz DF (2002) Polaris, a protein disrupted in orpk mutant mice, is required for assembly of renal cilium. Am J Physiol Renal Physiol 282(3):F541–F552

Yu Y, Ulbrich MH, Li MH, Buraei Z, Chen XZ, Ong AC, Tong L, Isacoff EY, Yang J (2009) Structural and molecular basis of the assembly of the TRPP2/PKD1 complex. Proc Natl Acad Sci USA 106(28):11558–11563

Yu Y, Yoon SO, Poulogiannis G, Yang Q, Ma XM, Villen J, Kubica N, Hoffman GR, Cantley LC, Gygi SP, Blenis J (2011) Phosphoproteomic analysis identifies Grb10 as an mTORC1 substrate that negatively regulates insulin signaling. Science 332(6035):1322–1326

Yuasa T, Takakura A, Denker BM, Venugopal B, Zhou J (2004) Polycystin-1L2 is a novel G-protein-binding protein. Genomics 84(1):126–138

Yuasa T, Venugopal B, Weremowicz S, Morton CC, Guo L, Zhou J (2002) The sequence, expression, and chromosomal localization of a novel polycystic kidney disease 1-like gene, PKD1L1, in human. Genomics 79(3):376–386

Zanivan S, Gnad F, Wickstrom SA, Geiger T, Macek B, Cox J, Fassler R, Mann M (2008) Solid tumor proteome and phosphoproteome analysis by high resolution mass spectrometry. J Proteome Res 7:5314–5326

Zhang P, Luo Y, Chasan B, Gonzalez-Perrett S, Montalbetti N, Timpanaro GA, Cantero Mdel R, Ramos AJ, Goldmann WH, Zhou J, Cantiello HF (2009) The multimeric structure of polycystin-2 (TRPP2): structural-functional correlates of homo- and hetero-multimers with TRPC1. Hum Mol Genet 18(7):1238–1251

Chapter 25
TRPM6 and Hypomagnesaemia/ Hypocalcaemia

Daniel Landau and Hanna Shalev

Abstract Magnesium (Mg^{2+}) is an abundant intra- as well as extracellular divalent cation. It is essential for multiple intracellular processes. Its major reservoir is in the bone. Mg^{2+} homeostasis is regulated by its absorption from the gut and its secretion or reabsorption by the kidneys through specific transcellular and paracellular Mg^{2+} channels. Derangements in Mg^{2+} homeostasis may cause elevations (hypermagnesemia) or decrease (hypomagnesemia) in its blood levels, both potentially causing adverse effects, some of them directly ascribed to Mg^{2+} effects on the nervous as well as the cardiac electrical conduction systems. Most clinical conditions are associated with hypomagnesemia, whereas most cases of hypermagnesemia are asymptomatic. Subclinical hypomagnesemia may cause cardiovascular disease in adults. Other clinical adverse effects derive from the secondary hypocalcemia in cases of severe hypomagnesemia. Different clinical conditions may cause hypomagnesemia, including acquired (such as chronic diarrhea or medications) and genetic. Most affected genes that cause hypomagnesemia are expressed in the nephron's distal tubule, mainly the thick ascending limb (such as the paracellin/claudin 16 and claudin 19 genes) and distal convoluted tubule segments (such as the thiazide sensitive channel, TRPM6 and the Na–K-ATPase). Mutations in TRPM6, a channel expressed in both the distal tubule as well as in the small intestine cause early (usually in the first months of life) and severe hypomagnesemia with secondary hypocalcemia. Differentiation between the mutated different genes can be initially approached by several clinical as well as laboratory observations (for example: the presence of hypercalciuria, metabolic alkalosis, etc.). The aim of this chapter is to summarize Mg^{2+} physiology, its handling by kidney and intestine and the clinical syndromes of Mg^{2+} derangement, concentrating on a crucial protein involved in Mg^{2+} reabsorption in both kidney and intestine: TRPM6 channel.

D. Landau (✉) · H. Shalev
Department of Pediatrics, Ben Gurion University of the Negev, Beer Sheva, Israel
e-mail: ldaniel@bgu.ac.il

N. Weiss and A. Koschak (eds.), *Pathologies of Calcium Channels*,
DOI: 10.1007/978-3-642-40282-1_25,

25.1 The Physiology of Mg^{2+} Handling

Mg^{2+} is the second most common intracellular cation (after calcium). It is a co-factor in multiple enzymatic reactions and it participates in the regulation of other ion channels. The vast majority of body Mg^{2+} (99 %) is intracellular. Therefore, its plasma levels may not reflect total body Mg^{2+}. More than 50 % of body Mg^{2+} is stored in bones as part of the hydroxyapatite complex. Mg^{2+} plays major roles in overall cell function, including: DNA and protein synthesis, glucose and fat metabolism, oxidative phosphorylation, neuromuscular excitability, and enzyme activity. Its body balance depends on the interplay between its intestinal absorption, exchange with bone and renal excretion. Mg^{2+} is ingested along the entire gastrointestinal tract, either by passive paracellular route or the active transcellular route. Intestinal paracellular absorption can be inhibited by the commonly used antacid medication omeprazole (Thongon and Krishnamra 2011).

The main organ that regulates Mg^{2+} homeostasis, in cases of excess or deficiency, is the kidney (Fig. 25.1). Mg^{2+} flows in blood partly linked to albumin. 80 % of total plasma Mg^{2+} is filtered through the glomerulus and most of it is reabsorbed along the nephron. Under normal diet the usual fractional excretion of Mg^{2+} is 3–5 %, but it can be as low as 0.5 % in states of Mg^{2+} deficiency. 20 % of filtered Mg^{2+} is reabsorbed by the proximal tubule, 70 % at the loop of Henle (mostly paracellular), and the remaining fraction is absorbed by an active trans-cellular mechanism at the distal convoluted tubule. This latter segment can be upregulated or downregulated in states of deficiency of excess and is the major contributor of Mg^{2+} regulation.

Paracellular renal Mg^{2+} transport takes place at the thick ascending limb of the loop of Henle, through the tight junction proteins paracellin 1 (claudin 16) and claudin 19. These proteins participate also in calcium reabsorption. Therefore, defects in these proteins, either congenital or acquired can lead to a combination of hypomagnesemia and hypercalciuria. Transcellular Mg^{2+} reabsorption at the distal convoluted tubule starts with an apical channel called transient receptor potential melastatin 6 (TRPM6), which is the major inducible protein for Mg^{2+} reabsorption in the colon and kidney (see later) (Rondon et al. 2008). The mechanism of basolateral efflux of Mg^{2+} in this tubular segment is not well characterized yet.

25.2 Hypomagnesemia

Hypomagnesemia is a clinical state of plasma Mg^{2+} deficiency. Given its main intracellular location, the discovery of low plasma (extracellular) Mg^{2+} already reflects a state of severe intracellular deficiency. There is also a possibility of subclinical hypomagnesemia, when only the intracellular Mg^{2+} levels may be low (Ryzen et al. 1987; Lim and Jacob 1972). Since Mg^{2+} plays so many and crucial roles in cell function, its deficiency is associated with a variation of clinical

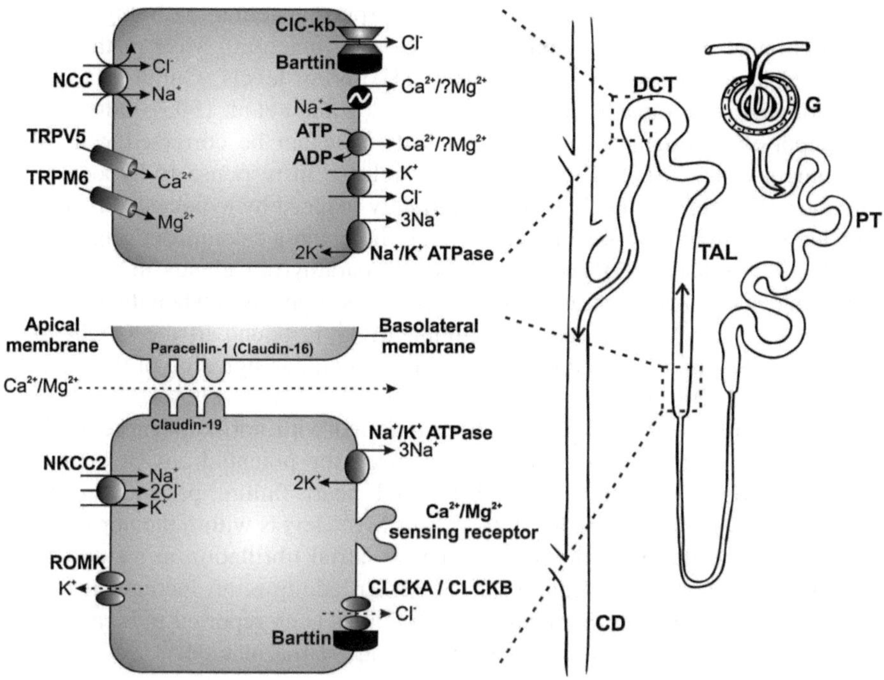

Fig. 25.1 Renal Mg^{2+} transport. 80 % of plasma Mg^{2+} (the fraction unbound to albumin) is filtered through the glomerulus (G) and most of it is reabsorbed along the nephron: 20 % is reabsorbed by the proximal tubule (PT), 70 % at the thick ascending limb of loop of Henle (TAL) and the remaining fraction is absorbed by an active transcellular mechanism at the distal convoluted tubule (DCT). Paracellular renal Mg^{2+} transport takes place at the TAL through the tight junction proteins paracellin 1 (claudin 16) and claudin 19. These proteins participate also in calcium reabsorption. This tubular segment is also involved in active NaCl reabsorption through the luminal NKCC2 (which can be inhibited by the popular diuretic drug furosemide) and ROMK channels and the basolateral Cl channel CLCKA and CLCKB. This function is regulated also by extracellular Ca^{++} and Mg^{2+} levels through the basolateral calcium sensing receptor. Transcellular Mg^{2+} reabsorption at the DCT starts with the apical channel TRPM6. The mechanism of basolateral efflux of Mg in this tubular segment is not well characterized yet. The DCT is also the site of NaCl reabsorption through the luminal NCC channel (affected by thiazide diuretics). *CD* medullary collecting duct

findings, including generalized weakness and anorexia. When secondary hypocalcemia appears, it may elicit signs of muscle hyperexcitability, including Chvostek and Trousseau signs, irritability, and even frank generalized seizures. Hypomagnesemia is also associated with ventricular arrhythmias. The electrocardiographic findings may include widening of the QRS interval, peaking of T waves and prolongation of the PR interval (Dyckner 1980). Associated laboratory findings of hypomagnesemia include hypokalemia and metabolic alkalosis.

The associated hypocalcemia seen in severe hypomagnesemia (i.e., serum Mg^{2+} levels below 0.5 mmol/L or 1.2 mg/dL) deserves an explanation. The clinical and laboratory findings in severe hypomagnesemia resemble severe hypoparathyroidism (hypocalcemia and hyperphosphatemia with high tubular reabsorption of phosphate in spite of normal renal function). Most of the studies performed have shown low parathyroid (PTH) serum levels (Guran et al. 2012) that normalize with the administration of Mg^{2+} (Mihara et al. 1995). Other studies have reported a resistance to PTH which could again be corrected with Mg^{2+} (Yamamoto et al. 2011). The lack of PTH release in response to hypocalcemia suggests a PTH release block, which is normally induced by hypocalcemia through the release of inhibition of signaling through the gland's calcium sensing receptor (CASR). The release of PTH from dispersed parathyroid glands in response to decreasing calcium concentration in the medium is suppressed when the reaction is performed at low Mg^{2+} levels in medium. Mg^{2+} is bound to the CASR at an alternative site, and is needed for the proper intracellular signaling of this receptor (Quitterer et al. 2001).

A small number of patients have been reported with normomagnesemic Mg^{2+} depletion. Other studies have raised concern on the potential adverse effects of subclinical Mg^{2+} deficiency in diabetic and heart failure patients, including arrhythmias (Dyckner 1980). Recently, lower Mg^{2+} levels within the normal range have been associated with the development of atrial fibrillation in a study of the Framingham cohort (May Khan et al. 2012). In addition, an increase in overall inflammation (Almoznino-Sarafian et al. 2007) has been reported to be associated with hypomagnesemia. The causes of hypomagnesemia include:

(1) Gastrointestinal losses, mainly in cases of chronic diarrhea and malabsorption, and previous small bowel surgery.
(2) Metabolic conditions such as uncontrolled diabetes, alcoholism and pancreatitis.
(3) Renal losses, due to specific hereditary channel defects (see below), the use of diuretics or the diuretic effect of hypercalcemia.
(4) Other drugs may also cause hypomagnesemia through unique mechanisms, including the effects of tacrolimus and cyclosporine on the paracellular renal Mg^{2+} absorption (Efrati E et al. 2010); the effects of the proton pump inhibitor omeprazole on the gastrointestinal paracellular Mg^{2+} absorption (Thongon and Krishnamra 2011).

25.3 Inherited Forms of Renal Hypomagnesemia

The main organ that regulates Mg^{2+} homeostasis, in cases of excess or deficiency, is the kidney. Therefore, most forms of inherited hypomagnesemia are due to mutations in renal tubular proteins, including (Table 25.1):

(1) In the thick ascending limb, Mg^{2+} and calcium are reabsorbed via the paracellular pathway. Mutations in the tight junction proteins, claudin 16 and 19, cause renal hypomagnesemia with hypercalciuria and nephrocalcinosis (Simon

Table 25.1 List of genes related to Mg^{2+} abnormalities

Disease	Inheritance	Gene	Protein	Main nephron segment	Calciuria	Hypokalemic alkalosis	Remarks
Gitelman syndrome	AR	SLC12A3	NCC	DCT	Hypo-	Yes	Age onset may be late
Classical Bartter syndrome	AR	CLCKb	Basolateral chloride channel	TAL	Hyper-	Yes	
Familial hypomagnesemia with hypercalciuria and nephrocalcinosis	AR	CLDN16 CLDN19	Claudin-16 Claudin-19	TAL	Hyper-	No	May lead to end stage renal failure
Autosomal dominant isolated renal Mg^{2+} loss	AD	FXYD2	Na–K-ATPase-γ-subunit	DCT	Hypo-	No	
Familial hypomagnesemia with secondary hypocalcemia	AR	TRPM6	Epithelial Mg^{2+} channel	DCT	Hypo-	No	Combined kidney and intestine defect
Autosomal recessive isolated renal Mg^{2+} loss	AR	EGF	Epidermal growth factor	DCT	Hypo-	No	
EAST syndrome	AR	KCNJ10	Basolateral potassium channel	DCT		Yes	Early onset Seizures, Ataxia, deafness and Tubulopathy

AD autosomal dominant; *AR* autosomal recessive; *DCT:* distal convoluted tubule; *TAL* thick ascending limb et al. 1999; Konrad et al. 2006). Symptoms begin during the first months of life, including: urinary tract infection, polyuria, and failure to thrive. In many cases progressive renal failure may also develop (Praga et al. 1995).

(2) In the distal convoluted tubule active transcellular Mg^{2+} absorption takes place through the luminal TRPM6 channel. The putative basolateral Mg^{2+} channel has not yet been identified. Mutations in TRPM6 cause the classical familial hypomagnesemia with secondary hypocalcemia (HSH) (which will be discussed in detail later). However, mutations in several other proteins influence Mg^{2+} reabsorption through TRPM6, including:

a. The thiazide sensitive NaCl channel (NCC). When this gene is mutated, hypokalemic metabolic alkalosis with hypocalciuria (Gitelman syndrome) develops. Hypomagnesemia may develop years after the onset of hypokalemia and alkalosis. This may lead to weakness, paresthesias and tetany (Calo et al. 2000). This clinical and laboratory observation was originally described by Gitelman et al. (1996) many years before the exact affected gene was found. The exact mechanism linking this defect with hypomagnesemia is poorly understood, but includes a decreased expression of TRPM6 (Nijenhuis et al. 2005). A similar combination of hypokalemic metabolic alkalosis and hypomagnesemia has been described in some patients with mutations in the basolateral chloride channel CLCKb. However, this Bartter-like syndrome is associated with hypercalciuria (Zelikovic et al. 2003).
b. A separate form of hypomagnesemia with hypocalciuria has been described, in association with autosomal dominant inheritance. Mutations in the γ-subunit of the Na–K-ATPase have been described for this disease. However, the exact explanation how such mutated protein (which is expressed in every cell) is not yet understood (Meij et al. 2000).
c. Another rare form of autosomal recessive hypomagnesemia with normal urinary calcium excretion has been found to be associated with mutations in the pro-epidermal growth factor (pro-EGF) gene. EGF binding to its receptor is essential for the function of TRPM6 (Groenestge et al. 2007). Moreover, treatment with EGFR antagonists may cause hypomagnesemia (Tejpar et al. 2007).
d. Recently, mutations in the basolateral potassium channel KCNJ10 has been shown to be associated with a unique combination of early onset seizures, ataxia, deafness, and tubulopathy. The latter includes mild hypokalemia alkalosis and mild hypomagnesemia (Bockenhauer et al. 2009).

25.4 Physiology of TRPM6

TRPM6 belongs to the family of transient receptor potential (TRP) channels and is the only one of this family associated with Mg^{2+} transport. It is abundantly expressed in small and large intestine, as well as the distal renal tubule. Specifically to the kidney tissue its exact cellular location is the luminal membrane of the DCT. It has a much higher affinity to Mg^{2+} than to calcium. The unique structure of TRP channels includes the existence of an intracellular α-kinase domain. The role of such kinase is not yet well established but seems not to be crucial for TRPM6's basal activity. Nevertheless, this channel is tightly regulated by different other signals, including hypo- and hypermagnesemia up- and downregulate TRPM6 transcription. In addition, metabolic acidosis, the use of tacrolimus and cyclosporine and the inhibition of the NCC channel (either genetic, as happens in Gitelman syndrome, see above, or by the chronic use of thiazide diuretics) cause a decreased expression of TRPM6. In contrast, metabolic alkalosis and the use of estrogen upregulate TRPM6 (Cao et al. 2008).

In addition to its carboxyterminal kinase domain, TRPM6 has a long amino terminal intraluminal chain. Most of the described TRPM6 human mutations that cause HSH are located in the N-terminal tail and fewer are located in the central transmembrane domain (Schlingmann et al. 2005), suggesting this tail to be of importance in TRPM6 activity. One explanation for that may be the putative quaternary structure of the Mg^{2+} channel which includes 2 TRPM6 and 2 similar molecules: TRPM7 (Runnels 2011).

25.5 Clinical Findings in TRPM6 Mutations

Mutations in the TRPM6 gene cause early onset severe hypomagnesemia with secondary hypocalcemia, as previously mentioned. The clinical and laboratory findings were described well before the genetic defect was described (Shalev et al. 1998). Disease onset is usually very early, as soon as 4 weeks of age. The common clinical presentation of the combination of hypocalcemia and hypomagnesemia is generalized seizures or tetany. Hypocalcemia can only be corrected when Mg^{2+} is provided, without the need to supplement with calcium. There is a major discrepancy between the enteral versus the parenteral doses of Mg^{2+} needed to correct this condition: the enteral dose needed is much higher than the parenteral one, hinting for a role of the gene product in intestinal Mg^{2+} absorption. Similar clinical findings were described in other series, involving more families of different ethnic origins (Schlingmann et al. 2005; Guran et al. 2012). Originally, renal handling of Mg^{2+} was thought to be intact, since the fractional excretion of Mg^{2+} during severe hypomagnesemia was very low. However, when serum Mg^{2+} levels were gradually increased a lower renal Mg^{2+} threshold was identified (Walder et al. 2002), supporting the co-expression of the identified responsible gene, TRPM6 in both kidney and intestine cells, as shown by us and others (Schlingmann et al. 2002). Treatment with high dose oral Mg^{2+} is usually able to correct serum Mg^{2+} levels to the subnormal values. However, achieved serum levels >1 mg/dL (or >0.4 mmol/L) is sufficient to prevent secondary hypocalcemia on the one hand and also to prevent gastrointestinal side effects from too much enteral Mg^{2+} dosage. If disease is diagnosed early on and seizures are prevented, then normal intellectual capacity is achieved. All genetic studies done so far have not found a major variability in genotype-phenotype correlation (Walder et al. 2002; Schlingmann et al. 2002, 2005; Guran et al. 2012).

References

Almoznino-Sarafian D, Berman S, Mor A, Shteinshnaider M, Gorelik O, Tzur I, Alon I, Modai D, Cohen N (2007) Magnesium and C-reactive protein in heart failure: an anti-inflammatory effect of magnesium administration? Eur J Nutr 46:230–237

Bockenhauer D, Feather S, Stanescu HC, Bandulik S, Zdebik AA, Reichold M, Tobin J, Lieberer E, Sterner C, Landoure G, Arora R, Sirimanna T, Thompson D, Cross JH, van't Hoff W, Al Masri O, Tullus K, Yeung S, Anikster Y, Klootwijk E, Hubank M, Dillon MJ, Heitzmann D, Arcos-Burgos M, Knepper MA, Dobbie A, Gahl WA, Warth R, Sheridan E, Kleta R (2009) Epilepsy, ataxia, sensorineural deafness,tubulopathy, and KCNJ10 mutations. N Engl J Med 360:1960–1970

Cao G, Hoenderop JG, Bindels RJ (2008) Insight into the molecular regulation of the epithelial magnesium channel TRPM6. Curr Opin Nephrol Hypertens 17:373–378

Calò L, Punzi L, Semplicini A (2000) Hypomagnesemia and chondrocalcinosis in Bartter's and Gitelman's syndrome: review of the pathogenetic mechanisms. Am J Nephrol 20:347–350

Dyckner T (1980) Serum magnesium in acute myocardial infarction. Relation to arrhythmias. Acta Med Scand 207:59–66

Efrati E, Hirsch A, Kladnitsky O, Rozenfeld J, Kaplan M, Zinder O, Zelikovic I (2010) Transcriptional regulation of the claudin-16 gene by Mg^{2+} availability. Cell Physiol Biochem 25:705–714

Gitelman HJ, Graham JB, Welt LG (1996) A new familial disorder characterized by hypokalemia and hypomagnesemia. Trans Assoc Am Physicians 79:221–235

Groenestege WM, Thébault S, van der Wijst J, van den Berg D, Janssen R, Tejpar S, van den Heuvel LP, van Cutsem E, Hoenderop JG, Knoers NV, Bindels RJ (2007) Impaired basolateral sorting of pro-EGF causes isolated recessive renal hypomagnesemia. J Clin Invest 117:2260–2267

Guran T, Akcay T, Bereket A, Atay Z, Turan S, Haisch L, Konrad M, Schlingmann KP (2012) Clinical and molecular characterization of Turkish patients with familial hypomagnesaemia: novel mutations in TRPM6 and CLDN16 genes. Nephrol Dial Transplant 27:667–673

Konrad M, Schaller A, Seelow D, Pandey AV, Waldegger S, Lesslauer A, Vitzthum H, Suzuki Y, Luk JM, Becker C, Schlingmann KP, Schmid M, Rodriguez-Soriano J, Ariceta G, Cano F, Enriquez R, Juppner H, Bakkaloglu SA, Hediger MA, Gallati S, Neuhauss SC, Nurnberg P, Weber S (2006) Mutations in the tight-junction gene claudin 19 (CLDN19) are associated with renal magnesium wasting, renal failure, and severe ocular involvement. Am J Hum Genet 79:949–957

Lim P, Jacob E (1972) Magnesium status of alcoholic patients. Metabolism 21:1045–1051

May Khan A, Lubitz SA, Sullivan LM, Sun JX, Levy D, Vasan RS, Magnani JW, Ellinor PT, Benjamin EJ, Wang TJ (2012) Low Serum Magnesium and the Development of Atrial Fibrillation in the Community: The Framingham Heart Study. Circulation. Nov 21. [Epub ahead of print]

Meij IC, Koenderink JB, van Bokhoven H, Assink KF, Groenestege WT, de Pont JJ, Bindels RJ, Monnens LA, van den Heuvel LP, Knoers NV (2000) Dominant isolated renal magnesium loss is caused by misrouting of the Na(+), K(+) ATPase gamma-subunit. Nat Genet 26:265–266

Mihara M, Kamikubo K, Hiramatsu K, Itaya S, Ogawa T, Sakata S (1995) Renal refractoriness to phosphaturic action of parathyroid hormone in a patient with hypomagnesemia. Intern Med 34:666–669

Nijenhuis T, Vallon V, van der Kemp AW, Loffing J, Hoenderop JG, Bindels RJ (2005) Enhanced passive Ca2 + reabsorption and reduced Mg^{2+} channel abundance explains thiazide-induced hypocalciuria and hypomagnesemia. J Clin Invest 115:1651–1658

Praga M, Vara J, González-Parra E, Andrés A, Alamo C, Araque A, Ortiz A, Rodicio JL (1995) Familial hypomagnesemia with hypercalciuria and nephrocalcinosis. Kidney Int 47:1419–1425

Quitterer U, Hoffmann M, Freichel M, Lohse MJ (2001) Paradoxical block of parathormone secretion is mediated by increased activity of G alpha subunits. J Biol Chem 276:6763–6769

Rondón LJ, Groenestege WM, Rayssiguier Y, Mazur A (2008) Relationship between low magnesium status and TRPM6 expression in the kidney and large intestine. Am J Physiol Regul Integr Comp Physiol 294:R2001–R2007

Runnels LW (2011) TRPM6 and TRPM7: A Mul-TRP-PLIK-cation of channel functions. Curr Pharm Biotechnol 12:42–53

Ryzen E, Nelson TA, Rude RK (1987) Low blood mononuclear cell magnesium content and hypocalcemia in normomagnesemic patients. West J 147:549–553

Schlingmann KP, Sassen MC, Weber S, Pechmann U, Kusch K, Pelken L, Lotan D, Syrrou M, Prebble JJ, Cole DE, Metzger DL, Rahman S, Tajima T, Shu SG, Waldegger S, Seyberth HW, Konrad M (2005) Novel TRPM6 mutations in 21 families with primary hypomagnesemia and secondary hypocalcemia. J Am Soc Nephrol 16:3061–3069

Schlingmann KP, Weber S, Peters M, Niemann Nejsum L, Vitzthum H, Klingel K, Kratz M, Haddad E, Ristoff E, Dinour D, Syrrou M, Nielsen S, Sassen M, Waldegge S, Seyberth HW, Konrad M (2002) Hypomagnesemia with secondary hypocalcemia is caused by mutations in TRPM6, a new member of the TRPM gene family. Nature Genet 31:166–170

Shalev H, Phillip M, Galil A, Carmi R, Landau D (1998) Clinical presentation and outcome in primary familial hypomagnesaemia. Arch Dis Child 78:127–130

Simon DB, Lu Y, Choate KA, Velazquez H, Al-Sabban E, Praga M, Casari Bettinelli A, Colussi G, Rodriguez-Soriano J, McCredie D, Milford D, Sanjad S, Lifton RP (1999) Paracellin-1, a renal tight junction protein required for paracellular Mg2 + resorption. Science 285:103–106

Tejpar S, Piessevaux H, Claes K, Piront P, Hoenderop JG, Verslype C, Van Cutsem E (2007) Magnesium wasting associated with epidermal-growth-factor receptor-targeting antibodies in colorectal cancer: a prospective study. Lancet Oncol 8:387–394

Thongon N, Krishnamra N (2011) Omeprazole decreases magnesium transport across Caco-2 monolayers. World J Gastroenterol 17:1574–1583

Walder RY, Landau D, Meyer P, Shalev H, Tsolia M, Borochowitz Z, Boettger MB, Beck GE, Englehardt RK, Carmi R, Sheffield VC (2002) Mutation of TRPM6 causes familial hypomagnesemia with secondary hypocalcemia. Nat Genet 2002:171–174

Yamamoto M, Yamaguchi T, Yamauchi M, Yano S, Sugimoto T (2011) Acute-onset hypomagnesemia-induced hypocalcemia caused by the refractoriness of bones and renal tubules to parathyroid hormone. J Bone Miner Metab 29:752–755

Zelikovic I, Szargel R, Hawash A, Labay V, Hatib I, Cohen N, Nakhoul FA (2003) novel mutation in the chloride channel gene, CLCNKB, as a cause of Gitelman and Bartter syndromes. Kidney Int 63:24–32

Chapter 26
TRP Channels in Prostate Cancer

George Shapovalov, Roman Skryma and Natalia Prevarskaya

Abstract TRP channels are often associated with sensory roles, but they also participate in such critical cellular processes as apoptosis, proliferation and differentiation. Disregulation of normal TRP channel activity and associated changes in calcium homeostasis often leads to disruption of normal cell cycle and, as a result, onset and progression of cancer. While early evidence specific to prostate cancer stemmed from studies involving blockers of voltage-gated calcium channels, later studies identified multiple TRP channels that appear to be at the center of prostate cancer progression. This chapter discusses the involvement of TRP channels in regulation of cell fate in prostate tissues, focusing on the prominent players belonging to TRPC, TRPM, and TRPV families as well as giving a general overview of the roles that TRP channels play in prostate cancer.

26.1 Introduction

Cancer is a serious widespread terminal illness responsible for approximately 13 % of deaths worldwide. One of the most prevalent types of cancer is a prostate cancer, diagnosed in, approximately, 1 out of 6 men. While majority of diagnosed patients develop a mild and slow-growing form of the prostate cancer, in a significant number of cases cancer develops metastases leading to eventual death in, approximately, 1 man in 36. Early diagnosis of prostate cancer became possible after identification of prostate-specific antigen (PSA) in early 1990s, which

G. Shapovalov (✉) · R. Skryma · N. Prevarskaya
Inserm U1003, Equipe Labellisee par la Ligue Nationale Contre le Cancer, Laboratory of Excellence Ion Channels Sciences and Therapeutics, Universite de Sciences et Technologies de Lille (USTL), 59655 Villeneuve dAscq, France
e-mail: george.shapovalov@inserm.fr

N. Prevarskaya
e-mail: natacha.prevarskaya@univ-lille1.fr

N. Weiss and A. Koschak (eds.), *Pathologies of Calcium Channels*,
DOI: 10.1007/978-3-642-40282-1_26,

allowed to address prostate cancer in its early form, significantly increasing chances of arresting disease in benign stage. However appropriate treatment of all stages of cancer, especially in high-risk cases, requires proper understanding of the mechanisms that lead to onset and advancement of prostate cancer.

Progression of various forms of cancer is associated with the dysregulation of cell cycle. Common impairments are usually focused around enhancement of proliferation pathways and suppression of mechanisms leading to cell death. Many aspects of cell cycle are highly sensitive to the fine details of $[Ca^{2+}]_i$ homeostasis in cells. For example, an increase in Ca^{2+} entry via the plasma membrane (PM) expressed ion channels often promotes Ca^{2+}-dependent proliferation and decreases susceptibility to programmed cell death—apoptosis of affected cells (Antoniotti et al. 2006; Monteith et al. 2007; Prevarskaya et al. 2010). On the other hand, the onset of apoptosis often involves Ca^{2+} influx through cytoplasmic, ER-mediated or mitochondrial mechanisms (Orrenius et al. 2003; Pinton et al. 2008). At the same time, the transformed cells as well as established tumor cell lines often become less sensitive to Ca^{2+} signals and even less dependent on external Ca^{2+} to survive and proliferate (Cook and Lockyer 2006; Whitfield 1992).

Changes in intracellular calcium trigger, in turn, a variety of Ca^{2+}-dependent pathways regulating cell cycle. Thus, CaMKII—a calcium/calmodulin-dependent kinase II plays an important role in calcium reuptake regulated by Ca^{2+}-calmodulin complex (Yamauchi 2005). Recently, it was shown to participate in androgen signaling in prostate cancer cells, where its overexpression led to induction of apoptosis resistance (Rokhlin et al. 2007). Similarly, NF-κB—a nuclear factor kappa-light-chain-enhancer of activated B cells is involved in regulation of immune and inflammatory responses and, in addition, cell survival and proliferation, by modulating gene expression (Gilmore 2006). It has been shown to be activated by tumor necrosis factor alpha (TNFα) and interleukin 1-beta (IL-1β), as well as by reactive oxygen species in the presence of sustained elevated Ca^{2+} (Sée et al. 2004; Sen et al. 1996).

Other involved factors include a member of a family of calcium-dependent cysteine proteases, calpain, which is activated by local and transient changes in Ca^{2+} concentration. The activated calpains then cleave various cytoplasmic proteins significantly affecting cellular development (Santella et al. 2000; Smith and Schnellmann 2012). Another common factor, calcium-dependent serine-threonine phosphatase calcineurin, activates nuclear transcription factors that are then translocated to the nucleus where they upregulate the expression of interleukins which, in turn, stimulate cellular growth and differentiation (Heit et al. 2006). Recent studies have even demonstrated the interaction between these two factors, describing an involved calpain-calcineurin signaling complex (Wu et al. 2007).

Calcium entry into the cell was traditionally associated with Ca^{2+} permeable ion channels present in the plasma membrane. Based on observations that classical blockers of these channels can influence cell death rates, the involvement of Ca^{2+}-permeable channels in regulation of cellular proliferation or apoptosis has been established as early as 1980s. However, later evidence pointed at other avenues that could supply calcium to the cells. Ca^{2+} release from internal stores was found to augment proliferation, aberrant differentiation, and impaired apoptosis and leading to hyperplasticity of cancer tissue (Bidaux et al. 2007; Xin et al. 2005; Xu et al. 2001). This emptying of internal Ca^{2+} stores leads, in turn, to store-operated calcium entry (SOCE), that first has been identified in nonexcitable and, later, excitable cells (Hoth and Penner 1992; Parekh et al. 1997; Stiber et al. 2008) and was shown to participate in regulation of gene transcription (Parekh and Putney 2005; Kahl and Means 2003) as well as mitochondrial and cytoplasmic apoptotic pathways (Prevarskaya et al. 2004; Vanden Abeele et al. 2002; Vanoverberghe et al. 2004). Moreover, calcium channels are located not only on PM, but also in a number of intracellular membranes, such as endoplasmic reticulum (ER), lysosomes, etc., where they provide pathways for Ca^{2+} release from cellular organelles (Pedersen et al. 2005). IP3 and RyR channels are the earliest identified examples of such channels that are known to be localized to the ER and have been implicated in regulating apoptosis (Høyer-Hansen et al. 2007; Høyer-Hansen and Jäättelä 2007; Johnson et al. 2004). Lately, more channels were found to be intracellularly localized leading to re-evaluation of the mechanisms of internal Ca^{2+} regulation. Incidentally, many of the calcium channels with recently identified subcellular localization belong to the superfamily of transient receptor potential (TRP) channels, which are the focus of this review.

26.2 Superfamily of TRP Channels

Initial identification of transient receptor potential channels was associated with investigating how Ca^{2+} entry regulates Drosophila phototransduction (Minke and Selinger 1996). Interestingly, this initial report presents TRPC1 ion channel in the role that is known as store-operated Ca^{2+} entry (SOCE) and is associated primarily with Orai1 protein. Numerous later studies identified many other TRP channels, sharing a high degree of structural homology. These studies revealed a superfamily of proteins presently consisting of 28 members grouped into smaller subfamilies that were denoted as TRPC (canonical), TRPV (vanilloid), TRPM (melastatin), TRPML (mucolipin), TRPA (ankyrin-like), and TRPP (polycystin) and share structural and, often functional similarity. TRP channels are tetrameric, composed of identical or closely homologous subunits belonging to the same subfamily (see, e.g., (Owsianik et al. 2006; Nilius et al. 2007) for review) and resemble structurally the superfamily of voltage-gated ion channels. The subunits have six transmembrane segments, a putative pore-forming loop between segments S5 and S6 with both C- and N-termini being intracellular. The reported so far changes to endogenous TRP activity that are associated with cancer progression include alteration of expression levels of TRP channels rather than mutagenic changes of TRP genes.

Activity of many TRP channels can be modulated by voltage or ligands, but they are also known to respond to a wide variety of chemical, mechanical, and even physical stimuli, such as temperature or mechanical stress. This naturally correlates with observation of many TRP channels being expressed in sensory neurons (Yao et al. 2005). However, as the majority of TRP channels are calcium permeable, they are often found to participate in many regulatory pathways involving Ca^{2+} (Roderick and Cook 2008). In addition to cell depolarization, their activity often leads to increase in intracellular Ca^{2+} and influences various cellular processes such as proliferation, apoptosis, or gene transcription (Pedersen et al. 2005). Indeed, increased expression of TRP channels in plasma membrane was shown to promote Ca^{2+}-dependent proliferative pathways leading to inhibition of apoptosis in affected cells and tumorigenesis (Berridge et al. 2003; Monteith et al. 2007; Roderick and Cook 2008). These observations naturally position the TRP channels as proteins of importance for regulating cell fate and explain the abundance of recent studies devoted to them.

26.2.1 TRPM Channels

We start the discussion of specific TRP channels with the channel that has the largest collection of observations tying it to prostate cancer in literature,—the TRPM8 channel. TRPM8 belongs to a melastatin subfamily of TRP channels. The channel possesses clear sensitivity to cold and menthol and plays a well-established role of temperature and chemical sensor in peripheral nervous system. It is interesting, therefore, to note that the initial identification of the channel happened in prostate, where it has been found to be expressed at high level (Tsavaler et al. 2001).

The localization of TRPM8 was initially thought to be limited to epithelial cells of the prostate (Tsavaler et al. 2001); however, later experiments identified its presence additionally in apical epithelial cells and smooth muscle cells of human prostate (Bidaux et al. 2005; Bidaux et al. 2007). Further studies clearly established the important role that the channel plays an in prostate cancer progression. For example, it was shown that the expression of TRPM8 is increased in cancerous cells (Bidaux et al. 2007; Prevarskaya et al. 2007), with whole-cell currents exhibiting responses resembling those of cold/menthol receptors. Additionally, a significant difference in the TRPM8 mRNA levels was reported between malignant and nonmalignant tissues (Fuessel et al. 2003). Recent evidence also points at the channel affecting the motility of the prostate epithelial cells. Thus, migration assays revealed that TRPM8 activation by prostate-specific antigen reduced motility of the PC-3 cell line (Gkika et al. 2010). Moreover, overexpression of functional TRPM8 channel on both ER and PM facilitated apoptosis induced by starvation and inhibited the migration of the PC-3-TRPM8 cells through the inactivation of focal-adhesion kinase (Yang et al. 2009).

Expression of TRPM8 is known to be directly regulated by androgen receptors (AR) in prostate (Zhang and Barritt 2004; Bidaux et al. 2005), making it one of the channels whose expression level is affected by the progression of prostate cancer. The channels can, primarily, be observed during the androgen-dependent phase of the cancer, with its expression levels decreasing in cells that are losing androgen sensitivity and regressing to the basal epithelial phenotype (Henshall et al. 2003; Bidaux et al. 2005). In more direct experiments, expression of TRPM8 protein could be induced in PNT1A cells that normally lack androgen receptor, when these cells were transfected with AR plasmid. Moreover, subsequent treatment of the cells with AR-siRNA reversed expression of TRPM8 channel (Bidaux et al. 2005). In general, the presence of TRPM8 is associated with the higher rate of growth of the cells in the androgen-dependent phase of prostate cancer (Kiessling et al. 2003; Berges et al. 1995), while the loss of TRPM8 with progression of the disease to androgen-independent phase (Fuessel et al. 2003) was associated with reduced doubling time of LNCaP cells resistant to antiandrogen bicalutamide treatment (Prevarskaya et al. 2007). Furthermore, it has been shown that pharmacological activation of TRPM8 as well as silencing of the channel with siRNA in LNCaP cells negatively influences cell survival, likely by perturbing Ca^{2+} homeostasis (Zhang and Barritt 2004).

One interesting property of the TRPM8 channel is the existence of multiple splice variants (Bidaux et al. 2007; Nealen et al. 2003; Bidaux et al. 2012). The classical, full-length isoform was shown to be expressed in the plasma membrane, while recent evidence indicates that shorter isoforms are localized to ER, where they serve as a Ca^{2+} release channels (Bidaux et al. 2007; Thebault et al. 2005; Zhang and Barritt 2004; Bidaux et al. 2012). It has been suggested that, depending on the balance in expression levels of PM or ER-localized isoforms, the TRPM8 may shift cellular Ca^{2+} homeostasis toward either proliferation or apoptosis in prostate tissue (Prevarskaya et al. 2007; Bidaux et al. 2012). More specifically, increased activity of shorter isoforms is believed to enhance proliferation and shift cell cycle toward malignant phenotype (Prevarskaya et al. 2007; Shapovalov et al. 2011). Additionally, it was suggested that expression of PM and ER-localized isoforms is sensitive to the stages of the prostate cancer. Thus full-length and PM-localized TRPM8 is expressed only in highly differentiated PCa epithelial cells, while its activity is abolished in dedifferentiated cells. The ER-localized isoforms, however, remain functional independently of cell differentiation (Bidaux et al. 2007).

Another channel of TRPM family that has an emerging evidence of its involvement in prostate cancer is TRPM2. Its expression in prostate, albeit at low levels, was identified during initial sweep of tissue distribution profile studies soon after mapping of the TRPM subfamily of the channels (Wang et al. 2007; Fonfria et al. 2006). Later, expression of the channel was found to be significantly increased in transformed cells. Moreover, selectively knocking down TRPM2 inhibited the growth of prostate cancer but not of the normal cells (Zeng et al. 2010). The channel was found to be translocated from PM in normal cells to the nuclei in a clustered pattern in PC-3 and DU-145 cells inhibiting nuclear ADP-ribosylation (Zeng et al. 2010). This regulatory function of TRPM2 in prostate cancer may be related to the role of a lysosomal Ca^{2+}-release channel it was found to play in beta cells (Lange et al. 2009).

Other channels belonging to TRPM subfamily also participate in regulation of cell fate. However, they are primarily involved in other types of cancer. Thus,

TRPM1 channels was shown to be regulated by the microphthalmia transcription factor (MIRF) (Zhiqi et al. 2004; Miller et al. 2004) and involved in regulating proliferation and survival of melanoma cells (Bödding 2007; Fang and Setaluri 2000). Alterations in TRPM4 and TRPM5 expression levels were reported in association with CD5+ B cell lymphomas and Wilms tumors and rhabdomyosarcomas (Beckwith-Wiedemann syndrome), respectively (Prawitt et al. 2000; Suguro et al. 2006). Interestingly, differential expression of TRPM4 channel in prostate transcripts of cogenital androgen-deficient mice was shown in a single study utilizing an array screening technique (Singh et al. 2006). However, no further evidence was presented in the literature since this initial observation. Finally, TRPM7 has been reported to be overexpressed in large breast tumors, with its suppression reducing proliferative potential of the cells (Guilbert et al. 2009). As can be seen, practically all the evidence concerning the prostate cancer that is available at present concentrates on TRPM8 channel.

26.2.2 TRPC Channels

TRPC, or the "canonical" TRP channels, comprise the subfamily closely related to the first identified channel in the superfamily, the Drosophila TRP, or TRPC1 in modern nomenclature. While numerous studies were devoted to the function and the role of this channel, its physiological significance is still far from being completely understood. Thus, early reports suggested its involvement in SOCE, implicating the channel in regulation of cell proliferation and apoptosis (Liu et al. 2003; Ambudkar 2007). Identification, at a later time, of Orai1 and STIM1 as proteins responsible for SOCE, however, questioned this role of TRPC1 channel. Certain models were proposed that involved a combination of TRPC1 with these proteins as principal players responsible for SOC (Ong et al. 2007). This, however, was questioned later by publications demonstrating that suppression of TRPC1 removes only a portion of SOCE response in some cells (Li et al. 2008; Abdullaev et al. 2008), while capacitative Ca^{2+} entry is completely independent of TRPC1 in other cells (Dietrich et al. 2007; Zanou et al. 2010). In prostate cancer cells, SOCE was even shown to be affected by disrupting actin cytoskeleton while entirely omitting TRPC1 as well as Orai1/STIM1 proteins (Vanoverberghe et al. 2012).

In spite of the unclear role that TRPC1 plays in SOCE, its involvement in regulating cell fate and motility was demonstrated in multiple knockout or overexpression studies, carried out in renal epithelial and skeletal muscle cells as well as immortalized GnRH neurons (Zanou et al. 2010; Ariano et al. 2011; Fabian et al. 2011; Fabian et al. 2008). Interestingly, while in normal tissue presence of TRPC1 promoted cell proliferation, in transformed tissues TRPC1 seemed to be involved in regulation of apoptosis. In prostate, the involvement of TRPC1 channel was found to be related to androgen receptor. The expression level of TRPC1 was found to decrease during the progression of the prostate cancer from androgen-dependent to androgen-independent phase (Nilius et al. 2007), suggesting that while in normal tissue presence of TRPC1 promotes cell proliferation, in transformed tissues TRPC1 seems to be shifting balance toward apoptosis. Similarly, the transient knockdown of TRPC1 and TRPC4 channels significantly reduced the ability of ATP to induce growth arrest in PCa cells (Thebault et al. 2006).

Following on, with the other members of TRPC subfamily, expression of TRPC3 channel was shown to increase in LNCaP cells after prolonged (24–48 h) depletion of calcium stores in a manner dependent on residual store content and involving the Ca^{2+}/calmodulin/calcineurin/NFAT pathway and suggesting that TRPC1 and/or TRPC3 proteins are involved in the response to alpha-adrenergic stimulation (Pigozzi et al. 2006). Additionally, TRPC3 was shown to colocalize with transcription factor TFII-I, suppression of which was shown to interfere with antiapoptotic signaling in 1-Ln and DU-145 human prostate cancer cells (Misra et al. 2011). TRPC4 ion channel, that was shown to form heteromers with TRPC1 protein (Antoniotti et al. 2006), was also implicated in early reports suggesting its involvement in SOCE in prostate cancer cell lines (Vanden Abeele et al. 2004; Pigozzi et al. 2006); however, no specific role in prostate cancer has been suggested for this channel.

26.2.3 TRPV Channels

The principal member of TRPV subfamily, involved in prostate cancer, is TRPV6 channel. Unlike many of TRP proteins that exhibit only mild selectivity for divalent cations TRPV6 has a high preference for Ca^{2+} over Na^{+} when heterologously expressed in mammalian cells (Yue et al. 2001). However, specifics of TRPV6 gating remain elusive, as whole-cell patch clamping experiments did not reveal a corresponding whole-cell current under physiological conditions. The majority of TRPV6 studies were, therefore, carried out via fura-2 measurements. The experiments suggested that the channel exhibits constitutive activity and that Ca^{2+} entry through TRPV6 seems to depend upon presence of intracellular Ca^{2+} and extracellular divalent cations (Bödding and Flockerzi 2004; Bödding et al. 2002). Activity of TRPV6 was also shown to be modulated by progesterone, estrogen, tamoxifen, and vitamin D, affecting proliferation and survival of cancer cells (Bolanz et al. 2008).

Initially, the expression of TRPV6 channel was identified in the epithelial cells belonging to kidney, intestine, placenta, and pancreas, where it is believed to be responsible for transcellular movement of Ca^{2+} (Peng et al. 1999; Wissenbach et al. 2001). Shortly thereafter, though, channel was found to be present in LNCaP and PC-3 cell lines, with its levels elevated with progression of the cancer toward aggressive forms (Peng et al. 2001). Similarly to TRPM8, the expression of the TRPV6 is known to be regulated by the androgen receptors, however in an agonist independent way. Thus, TRPV6 does not exhibit a clear change in expression levels with prostate cancer progressing from androgen-dependent to independent phases. Instead, expression level of the channel starts at a very low level in the healthy tissue (Fixemer et al. 2003; Wissenbach et al. 2001), with its mRNA levels increasing during the progression of cancer in correlation with cancer Gleeson score and development of metastases outside the prostate (Fixemer et al. 2003; Peng et al. 2001). However, expression of TRPV6 was not identified in androgen-insensitive prostate cancer cell lines (DU-145 and PC-3) (Fixemer et al. 2003). Additionally, TRPV6 expression levels were shown to be inhibited by dihydrotestosterone an androgen receptor agonist and stimulated by bicalutamide an androgen receptor antagonist (Bödding et al. 2003; Peng et al. 2001; Vanden Abeele et al. 2003).

Unlike the TRPM channels, the involvement of TRPV channels in prostate cancer is not limited to only one channel of the subfamily. While the majority of published so far results concentrate on the TRPV6 channel, other proteins belonging to TRPV subfamily have a fair share of studies devoted to them too. Thus, recent studies positioned TRPV1 channel as one of the players regulating progression of prostate cancer. First, the presence of TRPV1 was found in prostate epithelial cell lines PC-3 and LNCaP as well as in human prostate tissue (Sanchez et al. 2005), and later its expression was found to increase with increasing tumor grades in human prostate cancer and benign prostate hyperplasia (Czifra et al. 2009). Moreover, vanilloids through the interaction with TRPV1 were shown to induce apoptosis in prostate cancer cells via inducing $[Ca^{2+}]_i$ overloading as well as fast decrease of the transmembrane mitochondrial potential (Ziglioli et al. 2009). Interestingly, while TRPV1 was found to play a similar tumor-suppressing role in transitional urothelial cancer of human bladder, its expression was found to decrease with carcinoma progression (Santoni et al. 2012).

TRPV2 channel has also been recently found to be associated with progression of prostate cancer. Presence of TRPV2 has been detected in androgen-independent and metastatic phases in PC3 cells (Monet et al. 2010). Moreover, treatments that induce neuroendocrine differentiations were shown to increase expression of TRPV2, while siRNA suppression of the channel was shown to reduce growth and invasive properties of prostate tumors in nude mice xenografts (Monet et al. 2010). Additionally, TRPV2 can be activated by endogenous lysophospholipids, leading to an increase in cell migration of the PC3 prostate cancer cells (Monet et al. 2009). Finally, androgen receptor-mediated stimulation of TRPV3 gene was found in castrate resistant C4-2B PCa cells (Jariwala et al. 2007), however this initial finding was not followed up by further studies.

Table 26.1 TRP channels and prostate cancer

Channel	Changes in expression	Effect on survival	References
TRPC1	Decreases in androgen-independent phase	Pro-proliferation in normal cells. Pro-apoptotic in transformed cells. Possibly involved in SOCE	(Ariano et al. 2011; Fabian et al. 2008, 2001; Nilius et al. 2007; Thebault et al. 2006; Zanou et al. 2010)
TRPC3	Increases after prolonged store depletion	Possibly involved in the response to alpha-adrenergic stimulation of LNCaP cells	(Misra et al. 2011; Pigozzi et al. 2006)
TRPC4	No data	ATP induced growth arrest in PCa cells. Possibly involved in SOCE	(Vanden Abeele et al. 2004; Antoniotti et al. 2006; Pigozzi et al. 2006)
TRPM2	Elevated in prostate cancer. Relocates to nucleus	Selective knockdown inhibits growth of cancer cells	(Wang et al. 2007; Fonfria et al. 2006; Zeng et al. 2010; Lange et al. 2009)
TRPM8	Elevated in prostate cancer. Decreases with progression to androgen-independent phase	Cold and pain sensor, possibly participates in SOCE via iPLA2. Isoform-specific localization to PM or ER, leading to pro-proliferatory or pro-apoptotic effects	(Bidaux et al. 2005, 2012, 2007; Fuessel et al. 2003; Prevarskaya et al. 2007; Gkika et al. 2010; Henshall et al. 2003; Nealen et al. 2003; Thebault et al. 2005; Tsavaler et al. 2001; Yang et al. 2009; Zhang and Barritt 2004)
TRPV2	Elevated in androgen-independent and metastatic phases in PC3 cell	Stimulates cell migration	(Monet et al. 2009, 2010)
TRPV3	Found in castrate resistant C4-2B PCa cells	No data	(Jariwala et al. 2007)
TRPV6	Progressively increased in cancer, regulated by vitamin D	Pro-proliferative and antiapoptotic in normal and cancer tissues. Proposed role in SOCE	(Vanden Abeele et al. 2003; Bödding et al. 2003; Bolanz et al. 2008; Fixemer et al. 2003; Peng et al. 1999, 2001; Wissenbach et al. 2001)

26.3 Conclusion

Transient receptor potential channels exhibit a multitude of sensory properties, but also act as Ca^{2+} homeostasis regulators and, thus, play an important role regulating balance between cellular proliferation and apoptosis. In many cases, transformed cells rely on a disturbed cell cycle, commonly featuring a suppression of excessive Ca^{2+} entry that would normally lead to $[Ca^{2+}]_i$ overload and cell death. This often leads to an emergence of a more excitable phenotype that becomes self-sustaining in advanced stages of cancer. This review presented the accumulated evidence of the involvement of transient receptor potential channels in regulation of prostate cancer progression. Channels important for prostate cancer regulation belong to the TRPC, TRPM, and TRPV subfamilies, typically with the bulk of evidence concentrating on one prominent member (TRPM8, TRPV6, and TRPC1) and some sporadic reports suggesting participation of other channels of these subfamilies, as summarized in Table 26.1.

Acknowledgments This work was supported by grants from INSERM, la Ligue Nationale Contre le Cancer, le Ministere de l'Education Nationale, the Region Nord/Pas-de-Calais. George Shapovalov was supported by grant from FRM.

References

Abdullaev IF, Bisaillon JM, Potier M, Gonzalez JC, Motiani RK, Trebak M (2008) Stim1 and Orai1 mediate CRAC currents and store-operated calcium entry important for endothelial cell proliferation. Circ Res 103:1289–1299

Ambudkar IS (2007) TRPC1: a core component of store-operated calcium channels. Biochem Soc Trans 35:96–100

Antoniotti S, Fiorio Pla A, Barral S, Scalabrino O, Munaron L, Lovisolo D (2006) Interaction between TRPC channel subunits in endothelial cells. J Recept Signal Transduct Res 26:225–240

Ariano P, Dalmazzo S, Owsianik G, Nilius B, Lovisolo D (2011) TRPC channels are involved in calcium-dependent migration and proliferation in immortalized GnRH neurons. Cell Calcium 49:387–394

Berges RR, Vukanovic J, Epstein JI, CarMichel M, Cisek L, Johnson DE, Veltri RW, Walsh PC, Isaacs JT (1995) Implication of cell kinetic changes during the progression of human prostatic cancer. Clin Cancer Res 1:473–480

Berridge MJ, Bootman MD, Roderick HL (2003) Calcium signalling: dynamics, homeostasis and remodelling. Nat Rev Mol Cell Biol 4:517–529

Bidaux G, Roudbaraki M, Merle C, Crépin A, Delcourt P, Slomianny C, Thebault S, Bonnal JL, Benahmed M, Cabon F, Mauroy B, Prevarskaya N (2005) Evidence for specific TRPM8 expression in human prostate secretory epithelial cells: functional androgen receptor requirement. Endocr Relat Cancer 12:367–382

Bidaux G, Flourakis M, Thebault S, Zholos A, Beck B, Gkika D, Roudbaraki M, Bonnal JL, Mauroy B, Shuba Y, Skryma R, Prevarskaya N (2007) Prostate cell differentiation status determines transient receptor potential melastatin member 8 channel subcellular localization and function. J Clin Invest 117:1647–1657

Bidaux G, Beck B, Zholos A, Gordienko D, Lemonnier L, Flourakis M, Roudbaraki M, Borowiec AS, Fernández J, Delcourt P, Lepage G, Shuba Y, Skryma R, Prevarskaya N (2012) Regulation of activity of transient receptor potential melastatin 8 (TRPM8) channel by its short isoforms. J Biol Chem 287:2948–2962

Bödding M (2007) TRP proteins and cancer. Cell Signal 19:617–624

Bödding M, Flockerzi V (2004) Ca^{2+} dependence of the Ca^{2+} -selective TRPV6 channel. J Biol Chem 279:36546–36552

Bödding M, Wissenbach U, Flockerzi V (2002) The recombinant human TRPV6 channel functions as Ca^{2+} sensor in human embryonic kidney and rat basophilic leukemia cells. J Biol Chem 277:36656–36664

Bödding M, Fecher-Trost C, Flockerzi V (2003) Store-operated Ca^{2+} current and TRPV6 channels in lymph node prostate cancer cells. J Biol Chem 278:50872–50879

Bolanz KA, Hediger MA, Landowski CP (2008) The role of TRPV6 in breast carcinogenesis. Mol Cancer Ther 7:271–279

Cook SJ, Lockyer PJ (2006) Recent advances in Ca($^{2+}$)-dependent Ras regulation and cell proliferation. Cell Calcium 39:101–112

Czifra G, Varga A, Nyeste K, Marincsák R, Tóth BI, Kovács I, Kovács L, Bíró T (2009) Increased expressions of cannabinoid receptor-1 and transient receptor potential vanilloid-1 in human prostate carcinoma. J Cancer Res Clin Oncol 135:507–514

Dietrich A, Kalwa H, Storch U, Mederos y Schnitzler M, Salanova B, Pinkenburg O, Dubrovska G, Essin K, Gollasch M, Birnbaumer L, Gudermann T (2007) Pressure-induced and store-operated cation influx in vascular smooth muscle cells is independent of TRPC1. Pflugers Arch 455:465–477

Fabian A, Fortmann T, Bulk E, Bomben VC, Sontheimer H, Schwab A (2001) Chemotaxis of MDCK-F cells toward fibroblast growth factor-2 depends on transient receptor potential canonical channel 1. Pflugers Arch 461:295–306

Fabian A, Fortmann T, Dieterich P, Riethmüller C, Schön P, Mally S, Nilius B, Schwab A (2008) TRPC1 channels regulate directionality of migrating cells. Pflugers Arch 457:475–484

Fabian A, Fortmann T, Bulk E, Bomben VC, Sontheimer H, Schwab A (2011) Chemotaxis of MDCK-F cells toward fibroblast growth factor-2 depends on transient receptor potential canonical channel 1. Pflugers Arch 461(2):295–306

Fang D, Setaluri V (2000) Expression and Up-regulation of alternatively spliced transcripts of melastatin, a melanoma metastasis-related gene, in human melanoma cells. Biochem Biophys Res Commun 279:53–61

Fixemer T, Wissenbach U, Flockerzi V, Bonkhoff H (2003) Expression of the Ca^{2+} -selective cation channel TRPV6 in human prostate cancer: a novel prognostic marker for tumor progression. Oncogene 22:7858–7861

Fonfria E, Murdock PR, Cusdin FS, Benham CD, Kelsell RE, McNulty S (2006) Tissue distribution profiles of the human TRPM cation channel family. J Recept Signal Transduct Res 26:159–178

Fuessel S, Sickert D, Meye A, Klenk U, Schmidt U, Schmitz M, Rost AK, Weigle B, Kiessling A, Wirth MP (2003) Multiple tumor marker analyses (PSA, hK2, PSCA, trp-p8) in primary prostate cancers using quantitative RT-PCR. Int J Oncol 23:221–228

Gilmore TD (2006) Introduction to NF-kappaB: players, pathways, perspectives. Oncogene 25:6680–6684

Gkika D, Flourakis M, Lemonnier L, Prevarskaya N (2010) PSA reduces prostate cancer cell motility by stimulating TRPM8 activity and plasma membrane expression. Oncogene 29:4611–4616

Guilbert A, Gautier M, Dhennin-Duthille I, Haren N, Sevestre H, Ouadid-Ahidouch H (2009) Evidence that TRPM7 is required for breast cancer cell proliferation. Am J Physiol Cell Physiol 297:C493–C502

Heit JJ, Apelqvist AA, Gu X, Winslow MM, Neilson JR, Crabtree GR, Kim SK (2006) Calcineurin/NFAT signalling regulates pancreatic beta-cell growth and function. Nature 443:345–349

Henshall SM, Afar DE, Hiller J, Horvath LG, Quinn DI, Rasiah KK, Gish K, Willhite D, Kench JG, Gardiner-Garden M, Stricker PD, Scher HI, Grygiel JJ, Agus DB, Mack DH, Sutherland RL (2003) Survival analysis of genome-wide gene expression profiles of prostate cancers identifies new prognostic targets of disease relapse. Cancer Res 63:4196–4203

Hoth M, Penner R (1992) Depletion of intracellular calcium stores activates a calcium current in mast cells. Nature 355:353–356

Høyer-Hansen M, Jäättelä M (2007) Connecting endoplasmic reticulum stress to autophagy by unfolded protein response and calcium. Cell Death Differ 14:1576–1582

Høyer-Hansen M, Bastholm L, Szyniarowski P, Campanella M, Szabadkai G, Farkas T, Bianchi K, Fehrenbacher N, Elling F, Rizzuto R, Mathiasen IS, Jäättelä M (2007) Control of macroautophagy by calcium, calmodulin-dependent kinase kinase-beta, and Bcl-2. Mol Cell 25:193–205

Jariwala U, Prescott J, Jia L, Barski A, Pregizer S, Cogan JP, Arasheben A, Tilley WD, Scher HI, Gerald WL, Buchanan G, Coetzee GA, Frenkel B (2007) Identification of novel androgen receptor target genes in prostate cancer. Mol Cancer 6:39

Johnson JD, Han Z, Otani K, Ye H, Zhang Y, Wu H, Horikawa Y, Misler S, Bell GI, Polonsky KS (2004) RyR2 and calpain-10 delineate a novel apoptosis pathway in pancreatic islets. J Biol Chem 279:24794–24802

Kahl CR, Means AR (2003) Regulation of cell cycle progression by calcium/calmodulin-dependent pathways. Endocr Rev 24:719–736

Kiessling A, Füssel S, Schmitz M, Stevanovic S, Meye A, Weigle B, Klenk U, Wirth MP, Rieber EP (2003) Identification of an HLA-A*0201-restricted T-cell epitope derived from the prostate cancer-associated protein trp-p8. Prostate 56:270–279

Lange I, Yamamoto S, Partida-Sanchez S, Mori Y, Fleig A, Penner R (2009) TRPM2 functions as a lysosomal Ca^{2+} -release channel in beta cells. Sci Signal 2:ra23

Li J, Sukumar P, Milligan CJ, Kumar B, Ma ZY, Munsch CM, Jiang LH, Porter KE, Beech DJ (2008) Interactions, functions, and independence of plasma membrane STIM1 and TRPC1 in vascular smooth muscle cells. Circ Res 103:e97–e104

Liu X, Singh BB, Ambudkar IS (2003) TRPC1 is required for functional store-operated Ca^{2+} channels. Role of acidic amino acid residues in the S5-S6 region. J Biol Chem 278:11337–11343

Miller AJ, Du J, Rowan S, Hershey CL, Widlund HR, Fisher DE (2004) Transcriptional regulation of the melanoma prognostic marker melastatin (TRPM1) by MITF in melanocytes and melanoma. Cancer Res 64:509–516

Minke B, Selinger Z (1996) The roles of trp and calcium in regulating photoreceptor function in Drosophila. Curr Opin Neurobiol 6:459–466

Misra UK, Mowery YM, Gawdi G, Pizzo SV (2011) Loss of cell surface TFII-I promotes apoptosis in prostate cancer cells stimulated with activated α2-macroglobulin. J Cell Biochem 112:1685–1695

Monet M, Gkika D, Lehen'kyi V, Pourtier A, Vanden Abeele F, Bidaux G, Juvin V, Rassendren F, Humez S, Prevarsakaya N (2009) Lysophospholipids stimulate prostate cancer cell migration via TRPV2 channel activation. Biochim Biophys Acta 1793:528–539

Monet M, Lehen'kyi V, Gackiere F, Firlej V, Vandenberghe M, Roudbaraki M, Gkika D, Pourtier A, Bidaux G, Slomianny C, Delcourt P, Rassendren F, Bergerat JP, Ceraline J, Cabon F, Humez S, Prevarskaya N (2010) Role of cationic channel TRPV2 in promoting prostate cancer migration and progression to androgen resistance. Cancer Res 70:1225–1235

Monteith GR, McAndrew D, Faddy HM, Roberts-Thomson SJ (2007) Calcium and cancer: targeting Ca^{2+} transport. Nat Rev Cancer 7:519–530

Nealen ML, Gold MS, Thut PD, Caterina MJ (2003) TRPM8 mRNA is expressed in a subset of cold-responsive trigeminal neurons from rat. J Neurophysiol 90:515–520

Nilius B, Owsianik G, Voets T, Peters JA (2007) Transient receptor potential cation channels in disease. Physiol Rev 87:165–217

Ong HL, Cheng KT, Liu X, Bandyopadhyay BC, Paria BC, Soboloff J, Pani B, Gwack Y, Srikanth S, Singh BB, Gill DL, Ambudkar IS (2007) Dynamic assembly of TRPC1-STIM1-

Orai1 ternary complex is involved in store-operated calcium influx. Evidence for similarities in store-operated and calcium release-activated calcium channel components. J Biol Chem 282:9105–9116

Orrenius S, Zhivotovsky B, Nicotera P (2003) Regulation of cell death: the calcium-apoptosis link. Nat Rev Mol Cell Biol 4:552–565

Owsianik G, D'hoedt D, Voets T, Nilius B (2006) Structure-function relationship of the TRP channel superfamily. Rev Physiol Biochem Pharmacol 156:61–90

Parekh AB, Putney JW (2005) Store-operated calcium channels. Physiol Rev 85:757–810

Parekh AB, Fleig A, Penner R (1997) The store-operated calcium current I(CRAC): nonlinear activation by InsP3 and dissociation from calcium release. Cell 89:973–980

Pedersen SF, Owsianik G, Nilius B (2005) TRP channels: an overview. Cell Calcium 38:233–252

Peng JB, Chen XZ, Berger UV, Vassilev PM, Tsukaguchi H, Brown EM, Hediger MA (1999) Molecular cloning and characterization of a channel-like transporter mediating intestinal calcium absorption. J Biol Chem 274:22739–22746

Peng JB, Zhuang L, Berger UV, Adam RM, Williams BJ, Brown EM, Hediger MA, Freeman MR (2001) CaT1 expression correlates with tumor grade in prostate cancer. Biochem Biophys Res Commun 282:729–734

Pigozzi D, Ducret T, Tajeddine N, Gala JL, Tombal B, Gailly P (2006) Calcium store contents control the expression of TRPC1, TRPC3 and TRPV6 proteins in LNCaP prostate cancer cell line. Cell Calcium 39:401–415

Pinton P, Giorgi C, Siviero R, Zecchini E, Rizzuto R (2008) Calcium and apoptosis: ER-mitochondria Ca^{2+} transfer in the control of apoptosis. Oncogene 27:6407–6418

Prawitt D, Enklaar T, Klemm G, Gärtner B, Spangenberg C, Winterpacht A, Higgins M, Pelletier J, Zabel B (2000) Identification and characterization of MTR1, a novel gene with homology to melastatin (MLSN1) and the trp gene family located in the BWS-WT2 critical region on chromosome 11p15.5 and showing allele-specific expression. Hum Mol Genet 9:203–216

Prevarskaya N, Skryma R, Shuba Y (2004) Ca^{2+} homeostasis in apoptotic resistance of prostate cancer cells. Biochem Biophys Res Commun 322:1326–1335

Prevarskaya N, Zhang L, Barritt G (2007) TRP channels in cancer. Biochim Biophys Acta 1772:937–946

Prevarskaya N, Skryma R, Shuba Y (2010) Ion channels and the hallmarks of cancer. Trends Mol Med 16:107–121

Roderick HL, Cook SJ (2008) Ca^{2+} signalling checkpoints in cancer: remodelling Ca^{2+} for cancer cell proliferation and survival. Nat Rev Cancer 8:361–375

Rokhlin OW, Taghiyev AF, Bayer KU, Bumcrot D, Koteliansk VE, Glover RA, Cohen MB (2007) Calcium/calmodulin-dependent kinase II plays an important role in prostate cancer cell survival. Cancer Biol Ther 6:732–742

Sanchez MG, Sanchez AM, Collado B, Malagarie-Cazenave S, Olea N, Carmena MJ, Prieto JC, Diaz-Laviada II (2005) Expression of the transient receptor potential vanilloid 1 (TRPV1) in LNCaP and PC-3 prostate cancer cells and in human prostate tissue. Eur J Pharmacol 515:20–27

Santella L, Kyozuka K, Hoving S, Munchbach M, Quadroni M, Dainese P, Zamparelli C, James P, Carafoli E (2000) Breakdown of cytoskeletal proteins during meiosis of starfish oocytes and proteolysis induced by calpain. Exp Cell Res 259:117–126

Santoni G, Caprodossi S, Farfariello V, Liberati S, Gismondi A, Amantini C (2012) Antioncogenic effects of transient receptor potential vanilloid 1 in the progression of transitional urothelial cancer of human bladder. ISRN Urol 2012:458238

Sée V, Rajala NK, Spiller DG, White MR (2004) Calcium-dependent regulation of the cell cycle via a novel MAPK–NF-kappaB pathway in Swiss 3T3 cells. J Cell Biol 166:661–672

Sen CK, Roy S, Packer L (1996) Involvement of intracellular Ca^{2+} in oxidant-induced NF-kappa B activation. FEBS Lett 385:58–62

Shapovalov G, Lehen'kyi V, Skryma R, Prevarskaya N (2011) TRP channels in cell survival and cell death in normal and transformed cells. Cell Calcium 50:295–302

Singh J, Manickam P, Shmoish M, Natik S, Denyer G, Handelsman D, Gong DW, Dong Q (2006) Annotation of androgen dependence to human prostate cancer-associated genes by microarray analysis of mouse prostate. Cancer Lett 237:298–304

Smith MA, Schnellmann RG (2012) Calpains, mitochondria, and apoptosis. Cardiovasc Res 96:32–37

Stiber J, Hawkins A, Zhang ZS, Wang S, Burch J, Graham V, Ward CC, Seth M, Finch E, Malouf N, Williams RS, Eu JP, Rosenberg P (2008) STIM1 signalling controls store-operated calcium entry required for development and contractile function in skeletal muscle. Nat Cell Biol 10:688–697

Suguro M, Tagawa H, Kagami Y, Okamoto M, Ohshima K, Shiku H, Morishima Y, Nakamura S, Seto M (2006) Expression profiling analysis of the CD5+ diffuse large B-cell lymphoma subgroup: development of a CD5 signature. Cancer Sci 97:868–874

Thebault S, Lemonnier L, Bidaux G, Flourakis M, Bavencoffe A, Gordienko D, Roudbaraki M, Delcourt P, Panchin Y, Shuba Y, Skryma R, Prevarskaya N (2005) Novel role of cold/menthol-sensitive transient receptor potential melastatine family member 8 (TRPM8) in the activation of store-operated channels in LNCaP human prostate cancer epithelial cells. J Biol Chem 280:39423–39435

Thebault S, Flourakis M, Vanoverberghe K, Vandermoere F, Roudbaraki M, Lehen'kyi V, Slomianny C, Beck B, Mariot P, Bonnal JL, Mauroy B, Shuba Y, Capiod T, Skryma R, Prevarskaya N (2006) Differential role of transient receptor potential channels in Ca^{2+} entry and proliferation of prostate cancer epithelial cells. Cancer Res 66:2038–2047

Tsavaler L, Shapero MH, Morkowski S, Laus R (2001) Trp-p8, a novel prostate-specific gene, is up-regulated in prostate cancer and other malignancies and shares high homology with transient receptor potential calcium channel proteins. Cancer Res 61:3760–3769

Vanden Abeele F, Skryma R, Shuba Y, Van Coppenolle F, Slomianny C, Roudbaraki M, Mauroy B, Wuytack F, Prevarskaya N (2002) Bcl-2-dependent modulation of Ca^{2+} homeostasis and store-operated channels in prostate cancer cells. Cancer cell 1(2):169–179

Vanden Abeele F, Roudbaraki M, Shuba Y, Skryma R, Prevarskaya N (2003) Store-operated Ca^{2+} current in prostate cancer epithelial cells. Role of endogenous Ca^{2+} transporter type 1. J Biol Chem 278:15381–15389

Vanden Abeele F, Lemonnier L, Thébault S, Lepage G, Parys JB, Shuba Y, Skryma R, Prevarskaya N (2004) Two types of store-operated Ca^{2+} channels with different activation modes and molecular origin in LNCaP human prostate cancer epithelial cells. J Biol Chem 279:30326–30337

Vanden Abeele F, Skryma R, Shuba Y, Van Coppenolle F, Slomianny C, Roudbaraki M, Mauroy B, Wuytack F, Prevarskaya N (2007) Bcl-2-dependent modulation of $Ca(^{2+})$ homeostasis and store-operated channels in prostate cancer cells. Cancer Cell 1:169–179

Vanoverberghe K, Vanden Abeele F, Mariot P, Lepage G, Roudbaraki M, Bonnal JL, Mauroy B, Shuba Y, Skryma R, Prevarskaya N (2004) Ca^{2+} homeostasis and apoptotic resistance of neuroendocrine-differentiated prostate cancer cells. Cell Death Differ 11:321–330

Vanoverberghe K, Lehen'kyi V, Thébault S, Raphaël M, Vanden Abeele F, Slomianny C, Mariot P, Prevarskaya N (2012) Cytoskeleton reorganization as an alternative mechanism of store-operated calcium entry control in neuroendocrine-differentiated cells. PLoS ONE 7:e45615

Wang HP, Pu XY, Wang XH (2007) Distribution profiles of transient receptor potential melastatin-related and vanilloid-related channels in prostatic tissue in rat. Asian J Androl 9:634–640

Whitfield JF (1992) Calcium signals and cancer. Crit Rev Oncog 3:55–90

Wissenbach U, Niemeyer BA, Fixemer T, Schneidewind A, Trost C, Cavalie A, Reus K, Meese E, Bonkhoff H, Flockerzi V (2001) Expression of CaT-like, a novel calcium-selective channel, correlates with the malignancy of prostate cancer. J Biol Chem 276:19461–19468

Wu HY, Tomizawa K, Matsui H (2007) Calpain-calcineurin signaling in the pathogenesis of calcium-dependent disorder. Acta Med Okayama 61:123–137

Xin H, Tanaka H, Yamaguchi M, Takemori S, Nakamura A, Kohama K (2005) Vanilloid receptor expressed in the sarcoplasmic reticulum of rat skeletal muscle. Biochem Biophys Res Commun 332:756–762

Xu XZ, Moebius F, Gill DL, Montell C (2001) Regulation of melastatin, a TRP-related protein, through interaction with a cytoplasmic isoform. Proc Natl Acad Sci U S A 98:10692–10697

Yamauchi T (2005) Neuronal Ca2 +/calmodulin-dependent protein kinase II–discovery, progress in a quarter of a century, and perspective: implication for learning and memory. Biol Pharm Bull 28:1342–1354

Yang ZH, Wang XH, Wang HP, Hu LQ (2009) Effects of TRPM8 on the proliferation and motility of prostate cancer PC-3 cells. Asian J Androl 11:157–165

Yao X, Kwan HY, Huang Y (2005) Regulation of TRP channels by phosphorylation. Neurosignals 14:273–280

Yue L, Peng JB, Hediger MA, Clapham DE (2001) CaT1 manifests the pore properties of the calcium-release-activated calcium channel. Nature 410:705–709

Zanou N, Shapovalov G, Louis M, Tajeddine N, Gallo C, Van Schoor M, Anguish I, Cao ML, Schakman O, Dietrich A, Lebacq J, Ruegg U, Roulet E, Birnbaumer L, Gailly P (2010) Role of TRPC1 channel in skeletal muscle function. Am J Physiol Cell Physiol 298:C149–C162

Zeng X, Sikka SC, Huang L, Sun C, Xu C, Jia D, Abdel-Mageed AB, Pottle JE, Taylor JT, Li M (2010) Novel role for the transient receptor potential channel TRPM2 in prostate cancer cell proliferation. Prostate Cancer Prostatic Dis 13:195–201

Zhang L, Barritt GJ (2004) Evidence that TRPM8 is an androgen-dependent Ca^{2+} channel required for the survival of prostate cancer cells. Cancer Res 64:8365–8373

Zhiqi S, Soltani MH, Bhat KM, Sangha N, Fang D, Hunter JJ, Setaluri V (2004) Human melastatin 1 (TRPM1) is regulated by MITF and produces multiple polypeptide isoforms in melanocytes and melanoma. Melanoma Res 14:509–516

Ziglioli F, Frattini A, Maestroni U, Dinale F, Ciufifeda M, Cortellini P (2009) Vanilloid-mediated apoptosis in prostate cancer cells through a TRPV-1 dependent and a TRPV-1-independent mechanism. Acta Biomed 80:13–20

Xin H, Tanaka H, Yamaguchi M, Takemori S, Nakamura A, Kohama K (2005) Vanilloid receptor expressed in the sarcoplasmic reticulum of rat skeletal muscle. Biochem Biophys Res Commun 332:756–762

Xu XZ, Moebius F, Gill DL, Montell C (2001) Regulation of melastatin, a TRP-related protein, through interaction with a cytoplasmic isoform. [illegible]

[illegible]

Chapter 27
Pharmacology of TRPV Channels

Ulrich Wissenbach

Abstract From a pharmacological point of view, the transient receptor potential (TRPV) family of ion channels are interesting drug targets. After the identification of the TRPV1 channel, an attractive candidate for analgesic intervention, many agonists and antagonists were generated. Although TRPV1 modifying drugs partially induce serious side effects, several efforts are still under development. There is growing evidence that also the other thermo-gated TRPs are promising drug targets. Most interestingly, mutations within the TRPV4 gene are causing many skeletal dysplasias underlining the importance of the function of this ion channel. Another interesting channel, TRPV6, was shown to be up-regulated in a few malignancies including breast and prostate cancer. Several data show that the overexpression of TRPV6 contributes to an aggressive cancel cell phenotype.

27.1 Introduction

Transient receptor potentials (TRPs) are a heterogenic family of ion channels that primarily conduct cations. The common feature of the channels is unique structure composed of six transmembrane domains. It is believed that the pore region of all these channels is located between transmembrane domain 5 and 6. Although it has not been shown for all channels, it seems that functional TRP channels are composed of four subunits. Thereby the TRP channels can form homomers and heteromers with other TRP channel subunits. Based on sequence homology, TRP channels of mammals are categorized in six groups namely TRPC1-7, TRPV1-6, TRPM1-8, TRPP1-3 (polycystins), TRPML1-3 (mucolipins), and a single representative protein TRPA1. It is noteworthy that the number of TRP genes is highly

U. Wissenbach (✉)
Institut für Experimentelle und Klinische Pharmakologie und Toxikologie,
Universität des Saarlandes, 66421 Homburg/Saar, Germany
e-mail: ulrich.wissenbach@uks.eu

N. Weiss and A. Koschak (eds.), *Pathologies of Calcium Channels*,
DOI: 10.1007/978-3-642-40282-1_27,

variable among organisms. In humans, there are 27 TRP genes present, in mouse 28, in *Drosophila* 16 and one in yeast. The TRPC2 gene although present in humans, is interrupted by several stop codons and therefore believed to be a pseudogene. This review focuses on the pharmacology and disease relevance of TRPV channels.

27.2 TRPV1

27.2.1 Cloning and Expression

The most prominent candidate of the TRPV family for pharmacological intervention is the capsaicin receptor TRPV1. TRPV1 was cloned more than a decade ago, in outstanding work by Michael J. Caterina in the laboratory of David Julius (Caterina et al. 1997). In fact capsaicin, a potent agonist of TRPV1, has been used in medical practice for more than a century. The principle of the pungent of red pepper (cayenne pepper-derived from chilli) was discovered by Buchholz (1816). Later the isolation and purification was published by Thresh (1846) and Micko (1898). The correct formula of capsaicin was first published by Nelson (1919) and capsaicin was synthesized by Spath and Darling (1930). The so-called ABC plaster was introduced in 1928 by Raubenheimer (Beiersdorf Company) and contained plant extracts of arnika, belladonna and cayenne pepper for the treatment of backache.

TRPV1 is a nonselective cation channel expressed in nociceptive sensory neurons and neurons of the central nervous system (Caterina et al. 1997; Cui et al. 2006). TRPV1 can be activated by noxious heat, protons, ethanol, camphor, divalent cations, and polyamines and by the spicy compound capsaicin present in many plants as chili and pepper (Ahern et al. 2005, 2006; Caterina et al. 1997; Trevisani et al. 2002; Xu et al. 2005).

27.2.2 Channel Properties, Sensitization, and Desentization Mechanisms

The properties of TRPV1 predestine the channel as a target to treat several types of pain including chronic pain, inflammatory pain, cluster headache and migraine. The modulation of TRPV1 follows two principles. First, TRPV1 activation by agonists like capsaicin causes a depolarization of nociceptive neurons leading to an increased sensitivity to heat. In case of inflammatory pain, the sensitivity to heat is lowered by inflammatory compounds like protons endovanilloids as arachidonic acid, *N*-arachidonoyl-dopamine (NADA) and others (Premkumar et al. 2004;

Zygmunt et al. 1999). After sensitization, TRPV1 channels become insensitive to activation after prolonged application of capsaicin.

Sensitization and desensitization processes are matter of debate and were reviewed in detail by Peter Holzer (Holzer 1991, 2008). As suggested by Caterina and Julius (2001) sensitization may reflect enhanced open channel probability after TRPV1 exposure to the mentioned stimuli. Zhang and coworkers found that TRPV1 channels were released from an internal pool after exposure to NGF. In their model, NGF activates TrkA receptors and downstream PI3- and SRC-kinases which in turn phosphorylate the channel resulting in an enhanced surface localization of TRPV1 (Zhang et al. 2005; Huang et al. 2006; Stein et al. 2006).

The desensitized state may result from calcium influx that activates phospholipase C (PLC) followed by depletion of PIP2 as proposed by Prescott and Julius (2003). Lukacs and co-workers reported a dual regulation of PIP2, thus phosphoinositides have both inhibitory and activating effects on TRPV1 (Lukacs et al. 2007). However, others reported TRPV1 activation by PIP2 (Brauchi et al. 2007). Also Ufret-Vincenty et al. (2011) identified a PIP2 binding site localized within the TRPV1 C-terminal region and found that PIP2 activates it. Another inhibitory mechanism was presented by Mohapatra and Nau (2005) who demonstrated that desensitization involves the dephosphorylation of the channel by calcineurin in a calcium dependent manner. Jung et al. (2004) published that phoshorylation by Ca-calmodulin kinase II is critical for vanilloid binding. The authors presented mutations within TRPV1 which affect channel activation by vanilloids and resiniferatoxin (a naturally occurring agonist). In the case of ABC plasters, the initial capsaicin induced burning of the skin correlates very well with the direct activation of TRPV1 by capsaicin whereas the analgetic effect results from desensitization of the TRPV1 channel although the precise mechanism(s) are not fully understood. It has to be noted that capsaicin desensitization should be reversible, but administration of high doses can result in degeneration of nerve fibers (Holzer 1991; Nolano et al. 1999). Nerve degeneration is dose-dependent and affects myelinated and nonmyelinated nerve fibers. Administration of capsaicin doses up to $\sim$25 mg/kg selectively reduced only the number of unmyelinated fibers of dorsal root ganglia (Nagy et al. 1983; Holzer 1991). However, the long lasting analgetic effect of Qutenza, an injectable capsaicin formulation, is most likely based on neurotoxic modulation of nerve fibers.

The second principle of treating TRPV1 related pain is the selective blockade by an antagonist. As mentioned, TRPV1 can be activated by several stimuli including protons, arachidonic acid and other endogenous vanilloids. Thus, a TRPV1 antagonist should prevent activation by endogenous vanilloids. One major problem of blocking TRPV1 channels is that hyperthermia occurs as a severe side effect in humans and animals (Swanson et al. 2005). Several studies indicate that TRPV1 is involved in the body's thermoregulation, therefore the core temperature is affected (Gavva 2008; Gavva et al. 2007a, b, 2008; Tamayo et al. 2008). Another problem is that antagonists have species specific effects on TRPV1 for review (Szallasi and Sheta 2012). The hyperthermia as a side effect can be avoided by several means. First, local application (topic or local injection) of antagonists as well as application of antagonists, which do not penetrate the CNS had little effect on body temperature. Indeed, the antagonist AMG0347 developed by Amgen does not cause hyperthermia (Steiner et al. 2007). Second, several blockers showed stimulus specific action. Thus the antagonist AMG0610 blocked capsaicin activation of TRPV1 but not activation of protons. On the other hand, the antagonist AMG6880 blocked both mechanisms. The authors suggest independent binding sites for protons and capsaicin. Interestingly after repeated administration of the antagonist ABT-102 (Abott) hyperthermia in animals was attenuated (Othman et al. 2012). One may speculate that blocking the pore of TRPV1 may lead to hyperthermia whereas blocking binding sites of endogenous activators may not necessarily cause this side effect. According to Wong and Gavva (Gavva 2008; Wong and Gavva 2009), TRPV1 antagonists are grouped according to their blocking properties: Group A and B antagonists block (group A) or partially block (group B) TRPV1 activation of capsaicin, acid and heat. All blockers of these two groups cause hyperthermia. Antagonists of group C and D antagonize capsaicin effects and potentiate activation by acid. Group C blocker do not affect heat activation whereas group D blocker potentiate heat activation. However antagonists of group C and D have little or no effect on body core temperature.

Another problem of TRPV1 antagonists is that heat perception threshold may be affected. This was reported for the TRPV1 blockers AZD1386 and MK-2295 which altered heat threshold in healthy volunteers (Chizh and Sang 2009). Despite these problems, it is reasonable to screen both agonists and antagonists of the TRPV1 channel.

27.2.3 TRPV1 Antagonists

ABT-102 an urea derivate, was developed by Abbott. The substance blocks specifically TRPV1 and blocks a variety of stimuli including capsaicin, NADA, acid and heat (Gomtsyan et al. 2008; Surowy et al. 2008) Systemic administration affects core body temperature about 0.6–0.8 °C in humans (Othman et al. 2012). Very interestingly, this effect was attenuated after repeated exposure to the drug in humans and rodents (Honore et al. 2009; Othman et al. 2012). The drug entered phase I but study results are not published.

Amgen developed a set of blockers and came up with the most promising substance AMG-517. AMG-517 blocks all TRPV1 activating stimuli but induces hyperthermia in dogs and monkeys about 1.4 °C (Gavva et al. 2007a). The substance reveals a relatively long half life time depending on species. Half life times were found to be 31–64 h in dogs and monkeys, respectively (Gavva et al. 2007b). The drug entered phase I but was later found to increase body temperature up to 40 °C in humans (Gavva et al. 2008) and has a long half life time in humans (300 h) (Gavva et al. 2008). It was concluded that AMG-517 is not useful for systemic administration. Phase II studies were terminated. Other Amgen developed compounds did not increase core body temperature in rats as AMG8562 (Lehto et al. 2008).

AZD-1386 (AstraZeneca) was tested for efficiency in treating nociceptive chronic pain but was discontinued from development due to liver enzyme elevations (http://www.astrazenecaclinicaltrials.com). As mentioned above, the drug was found to alter heat sensitivity in healthy volunteers (Chizh and Sang 2009).

The blocker DWP-05195 was developed by Daewong Pharma and entered phase I but no study results were published.

GRC-6211 derived from Eli Lilly/Glenmark, was orally active and tested for efficacy against bladder overactivity in animals (Charrua et al. 2009). The authors concluded the drug may be useful for treating cystitis. However, GRC-6211 was tested treating dental pain, osteoarthritis and many other pain related indications but GRC-6211 was suspended from further clinical development (unavailable reason).

A new blocker was recently published by Japan Tobacco JTS-653 and was found to inhibit all modes of TRPV1 action including capsaicin, heat, acid activation, and binding of resiniferatoxin (Kitagawa et al. 2012a). JTS-653 was evaluated in treating bladder overactivity in rats with positive results (Kitagawa et al. 2012b). The drug entered clinical phase I.

MK-2295/NGD-8243 was published by Merck/Neurogen for the treatment of acute pain and dental pain but there are no data available since 5 years although the drug completed phase II (unpublished results).

A PharmEste derived blocker PHE377 shows a long lasting effect in different models of inflammatory and neuropathic pain and has entered phase I. Another TRPV1 blocker PHE575, is under preclinical development. A study of PHE377 on pain models in healthy volunteers is ongoing and current data are available on the company's homepage (see:http://www.pharmeste.com/home.asp?op=interna&id=2&id_pag=10&tit=Pipeline).

SB-705498 from GlaxoSmithKline (GSK) inhibits capsaicin-mediated activation. SB705498 produces full blockade of heat as well as pH activation of human TRPV1 (Lambert et al. 2009). The drug was tested for treating allergic rhinitis, pain and migraine and molar tooth extraction. In humans, heat-evoked pain and skin sensitisation induced by capsaicin or UV-light were tested and treatment was effective although an increased toleration of pain at the inflammation side evoked by UV-light was observed. The drug entered phase II and several indications were tested including acute migraine, allergic/nonallergic rhinitis, toothache, chronic cough, dermatitis, and more. A rectal pain study was terminated. Other studies are completed but results are not published.

The TRPV1 antagonist MR1817 (Mochida) completed phase I (unpublished), whereas others entered phase II as PF-03864086 (Pfizer), JNJ-39439335 (Johnson&Johnson) and XEN-D0501 (Provesica). The blocker XEN-D0501 increases modestly core body temperature in humans in a concentration-dependent manner (Round et al. 2011).

27.2.4 TRPV1 Agonists

Anesiva Inc. developed Adlea (ALRGX-4975) an injectable preparation of capsaicin which reduced pain levels in patients following total knee arthroplasty (Gullapalli) or bunionectomy, and reduced pain in patients with osteoarthritis or Morton's neuroma (Remadevi and Szallisi 2008). ALRGX-4975 completed successful phase 3 trial in 2007. The preparation has a short half life time (1–2 h) and reduced opioid medication consumption by the patients.

Civamide (WN-1001, cis-8-methyl-*N*-vanillyl-6-nonenamide) was introduced by Winston laboratories. Civamide cream was tolerated and efficient up to 1 year after continuous use. The compound is not systemically absorbed and lacks systemic toxicity (Schnitzer et al. 2012). A modest preventive effect was observed in patients with cluster headache using intranasal application of civamide (Saper et al. 2002). Other civamide containing formulations are under development (see http://www.winstonlabs.com/).

Qutenza (NGX-4010) from NeurogesX, contains a patch with 8 % capsaicin for treatment of peripheral neuropathic pain. Qutenza provides a long lasting therapy (up to 12 weeks) after a single application of 1 h (Wallace and Pappagallo 2011). The application of Qutenza reduces expression of TRPV1 and the density of epidermal nerve fibers in the patch area and the related pain can be managed by local cooling or analgetics (Jones et al. 2011; Peppin et al. 2011). Qutenza is used within the EU for treating neuropathic pain in nondiabetic patients. Side effects include redness, pain, small bumps, and itching (see http://www.qutenza.com/).

27.3 TRPV2

27.3.1 Cloning and Expression

TRPV2 was cloned by Caterina et al. (Caterina et al. 1999) and published as second heat sensitive channel with activation temperatures of 52 °C. TRPV2 was also published at the same time by Kanzaki and coworkers (Kanzaki et al. 1999) and was termed GRC (growth-factor-regulated channel). The authors showed that stimulation with insulin-like-growth factor translocates the channel to the plasma membrane. A similar mechanism was later described in neuronal/neuroendocrine cells (Boels et al. 2001) and in a primary culture of retinal pigment epithelium (Cordeiro et al. 2010). The TRPV2 was originally cloned from rat brain and from murine spleen cDNA libraries. Expression of TRPV2 was demonstrated in neurons of the CNS as well as in dorsal root, gastrointestinal tract and smooth muscle cell, heart, spleen and keratinocytes of the skin (Caterina et al. 1999; Ichikawa et al. 2004; Lewinter et al. 2004; Zhang et al. 2004; Muraki et al. 2003; Chung et al. 2003; Gu et al. 2005).

27.3.2 Channel Properties and Related Diseases

TRPV2 seems to be involved in several diseases. TRPV2 was found to play a role in developing a cardiomyopathy depending on muscular dystrophy (Iwata et al. 2003, 2009; Watanabe et al. 2009; Zanou et al. 2009) and elevated levels of TRPV2 were found in prostate cancer (Monet et al. 2010). Because TRPV2 is expressed in skin and is activated by heat, it was speculated that TRPV2 may also serve as a target to treat certain types of pain. This idea was underlined by the finding that in a Norrbottnian-based populations with congenital insensitivity to pain TRPV2 transcript and protein level was reduced (Axelsson et al. 2009). Despite this, there is almost no clinical relevant pharmacology of TRPV2 channels. Endogenous ligands are not identified but TRPV2 can be activated with probenecid, THC, and cannabidiol (Bang et al. 2007; Qin et al. 2008). TRPV2 antagonists are the unspecific channel blocker ruthenium red, trivalent cations such as La^{3+} and Gd^{3+} and SKF96365 (Boels et al. 2001; Leffler et al. 2007). To date, there is no agonist/antagonist published that is clinically relevant.

27.4 TRPV3

27.4.1 Cloning and Expression

A third temperature-gated channel was cloned in 2002 by Peier and co-workers from a murine skin cDNA library (Peier et al. 2002). The authors demonstrated expression of TRPV3 in keratinocytes and snout but not DRG neurons by northern analysis. The channel was found to be activated by warm temperatures (32–39 °C) and repeated heat steps sensitized the channel. At the same time, Xu cloned TRPV3 from human brain derived cDNA library and published similar functional results (Xu et al. 2002). In contrast, Xu found expression in monkey derived dorsal root ganglion, trigeminal ganglion, spinal cord and brain as demonstrated by In-Situ-Hybridization. Furthermore, a GlaxoSmithKline based work was published in 2002 describing the cloning of human TRPV3 from small intestine and colon and showed by co-immuno staining TRPV3 presence in dorsal root ganglia (Smith et al. 2002). Furthermore, the authors demonstrate interaction of TRPV1 with TRPV3. A specific role of TRPV3 in the detection of thermosensation in keratinocytes but not in sensory neurons was suggested by Moqrich concluded from genetic deletion (Moqrich et al. 2005). TRPV3 can be activated and blocked by a long list of compounds.

27.4.2 TRPV3 Antagonists

The super cooling compound icilin a potent TRPM8 activator, blocks TRPV3 (McKemy et al. 2002; Sherkheli et al. 2012). Other antagonists of TRPV3 namely HC-001403, GRC 15133 and GRC 17173 were published by Moran and Gullapalli (Moran et al. 2007; Gullapalli et al. 2008). These antagonists were reported to be active in animal models of pain. Glenmark developed the first TRPV3 antagonist GRC 15300 and completed clinical phase I in 2011 and as an outcome, the drug was well tolerated with a good pharmacokinetic profile (http://www.glenmarkpharma.cz/). But further clinical studies were not published.

27.4.3 TRPV3 Agonists

First, it was recognized by Chung and co-workers that the rather nonspecific TRP modulator 2-APB was found to activate TRPV3 (Hu et al. 2004; Chung et al. 2004a). Later plant derived compounds from oregano, savory, clove and thyme like camphor, carvacrol, eugenol and thymol were identified as activators (Moqrich et al. 2005; Xu et al. 2005, 2006). Also vanillin (from Vanilla plants) and a derivative thereof ethyl-vanillin was active (Xu et al. 2006). Interestingly, burning of the Boswellia tree often used in the context of religious rituals contains incensole acetate, a potent TRPV3 activator (Moussaieff et al. 2008). The monoterpene cresol activates TRPV3 (Vogt-Eisele et al. 2007) and most interestingly the peppermint derived compound menthol (Macpherson et al. 2006). Menthol as earlier reported is a potent agonist of the cold receptor TRPM8 (McKemy et al. 2002; Peier et al. 2002).

27.5 TRPV4

27.5.1 Cloning and Expression

TRPV4 was independently cloned from Strotmann et al. termed OTRPC4 (Strotmann et al. 2000), Liedtke et al. 2000 termed VR-OAC (Liedtke et al. 2000) and Wissenbach et al. 2000 termed Trp12 (Wissenbach et al. 2000). Liedtke and co-workers cloned the channel from rat, mouse, human, and chicken whereas the others presented the murine sequence. The human sequence was cloned by Delany et al. (2001). In all publications it was reported that the channel could be activated by physiologically relevant hypotonicity leading to a non selective inward current. Furthermore, all authors demonstrated by northern analysis strong expression within the kidney although transcripts were detectable in many organs such as lung, spleen, heart, and others. Furthermore, expression in the CNS as well as in the inner ear was demonstrated by Liedtke (Liedtke et al. 2000). Later several publications expanded the expression pattern to smooth muscle and vascular endothelium (Earley 2005; Fian et al. 2007; Tanaka et al. 2008; Yang et al. 2006; Yin and Kuebler 2008), keratinocytes (Chung et al. 2003, 2004b) pancreatic ß-cells (Casas et al. 2008) and bone derived osteoblasts and osteoclasts (Masuyama et al. 2008; Mizoguchi et al. 2008).

27.5.2 Channel Properties and Related Diseases

Despite the activation of TRPV4 by hypoosmotic stimuli, a broad spectrum of activation mechanisms was reported during the following years. Detailed work of the group of Bernd Nilius shows that TRPV4 could be activated by sheer stress or cell swelling but it is likely to depend on anandamide respectively arachidonic acid production in a phospholipase A2-dependent manner. Anandamide and its metabolite arachidonic acid in turn are metabolized to epoxyeicosatrienoic acid which activates the channel (Nilius et al. 2001b; Nilius and Voets 2004; Watanabe et al. 2003). Furthermore, TRPV4 can be stimulated by heat above room temperature; however, this seems to be an indirect mechanism because heat stimulus is ineffective in cell free inside-out-patches of TRPV4 containing membranes (Chung et al. 2004a; Guler et al. 2002; Vriens et al. 2004; Watanabe et al. 2003). In addition, several activators have been identified. The first synthetic activator of TRPV4 4a-PDD (4alpha-phorbol-didecanoate) a phorbolester characterized by Bernd Nilius activates the channel by binding to the transmembrane domains 3 and 4 of the channel (Vriens et al. 2007; Watanabe et al. 2002). The ester activates TRPV4 in the range of 200–600 nM. Other derivatives were found to be more potent as 4a-PDH (4a-phorbol 12,13-dihexanoate) which activates in the range of 70 nM (Klausen et al. 2009). A very potent activator GSK1016790A developed by GlaxoSmithKline was also found to be active on TRPV1 channels (Thorneloe et al. 2008; Willette et al. 2008). This compound induces bladder contraction as well as application of phorbolesters (Everaerts et al. 2010; Gevaert et al. 2007). Mutations critical for conducting calcium ions have been identified by Thomas Voets (Voets et al. 2002). Involvement of TRPV4 in hypotonicity induced nociception and taxol induced neuropathic pain was published by Alessandri-Haber et al. (Alessandri-Haber et al. 2003, 2004). TRPV4 knockout mice were published by Liedtke and Friedman and by Mizuno and co-workers (Liedtke and Friedman 2003; Mizuno et al. 2003). The Liedtke-TRPV4 knockout mice shows reduced water intake behavior, had increased serum osmolarity and reduced serum vasopressin level compared to TRPV4 +/+ littermates whereas the Mizuno-K.O. did not show altered drinking behavior, serum osmolarity, and serum vasopressin level. The role of mammalian TRPV4 in the nematode *Caenorhabditis elegans* was illuminated by Liedtke and co-workers (Liedtke 2005; Liedtke et al. 2003). In a *C. elegans* osm-9 mutant (TRPV4 orthologue), the murine TRPV4 was expressed and restored behavior to avoid mechanical stimuli (nose touch) and hypertonicity. However

TRPV4 is strongly expressed in the mammalian kidney and Wu and co-workers demonstrated TRPV4 channels in the cortical collecting duct (Wu et al. 2007). Calcium influx was induced by application of by 4alpha-PDD and hypotonic medium. Activation of TRPV4 by an increase of the luminal flow and shear stress increased sodium uptake and potassium secretion in renal distal tubules of kidney (Taniguchi et al. 2007). These data show convincingly that TRPV4 is involved in the regulation of osmosensation. As mentioned, TRPV4 is expressed in mammalian lung and its role was coupled to expression of the water channel AQP5 (Sidhaye et al. 2006, 2008). Thus, hypotonicity or 4a-PDD application decreased AQP5 abundance in airway epithelia whereas ruthenium red blocked this effect. These data show that TRPV4 is regulated by multiple stimuli including nociceptive, hyposmolar and mechanical stimulation, and stimulation by heat. Interestingly many mutations within the TRPV4 cDNA have been identified in humans and seem to be associated with several diseases reviewed by Verma et al. (2010).

Brachyolmia is a skeletal dysplasia which can be characterized by childhood onset, short trunk and short stature and a generalized platyspondyly. Brachyolmia associated mutations were found to be located nearby the pore of TRPV4 resulting in an increased channel activity after mechanical or chemical stimulation by 4a-PDD and arachidonic acid (Rock et al. 2008).

In 2009, Krakow and co-workers found TRPV4 mutations in spondylometaphyseal dysplasias (SMDs) patients from the Kozlowski subtype (SMDK). This disorder belongs to a group of short stature disorders with abnormalities in the vertebrae and the metaphyses of the tubular bones. In addition, mutations were also found in metatropic dysplasia patients (MD). These patients suffer from shortened limbs, a long, narrow trunk and develop progressive kyphoscoliosis. Expression of the mutated channels in HEK cells revealed that some of the mutations showed an increased basal channel activity (D333G and R594H). More TRPV4 mutations were analyzed in MD patients by Camacho et al. (2010). The authors found that autosomal-dominant brachyolmia, SMDK, and both nonlethal and lethal forms of metatropic dysplasia result from independent TRPV4 mutations. Unfortunately, a clear relationship of particular domains of the channel and disease severity could not be observed.

Hereditary motor and sensory neuropathy type 2 (HMSN2C) include the Charcot-Maries-Tooth disease type 2c (CMT2C) and scapuloperoneal spinal muscular atrophy (SPSMA). CMTC2 patients are characterized by distal muscle weakness and atrophy, mild sensory loss, and normal or near-normal nerve conduction velocities. Hearing, vision and vocal cords can be affected. SPSMA patients show bone abnormalities and weakness of the scapular muscle. Mutations were reported to occur within the N-terminal ankyrin repeat region of TRPV4 (Auer-Grumbach et al. 2010). The author showed reduced channel activity as a result of decreased surface localization in HeLa cells.

Deng et al. (2010) published two mutations within TRPV4 being present in SPSMA and CMTC2 families which are also located within the ankyrin repeats. These mutations did not affect trafficking of the channel as shown by biotinylation experiments. The mutant channels were overexpressed in HEK cells and showed more conductivity compared to wild type TRPV6 after stimulation with 4a-PDD. Interestingly one TRPV4 mutation R269H was found by both studies (Auer-Grumbach et al. 2010; Deng et al. 2010). Another study mapping mutations in two unrelated CMTC2 families found the mentioned mutation R269H in one family and R269C mutation in a second family (Landoure et al. 2010). Both mutant TRPV4 channels were expressed in *Xenopus* oocytes and showed enlarged currents after stimulation with heat or 4a-PDD. More TRPV4 mutations were published by Zimon et al. (2010). They found that TRPV4 mutations caused an unusual spectrum of neuropathies. Another dysplasia from the Maroteaux type was directly tested for association with TRPV4 and several mutations were found (Nishimura et al. 2010). From these data, one would expect an increasing number of studies showing involvement of TRPV4 and its relation to several skeletal dysplasias and neuromuscular diseases. Interestingly, Cho and co-workers reported TRPV4 mutations in three unrelated patients exhibiting both skeletal dysplasia and peripheral neuropathy (Cho et al. 2012).

Because TRPV4 is highly expressed in the kidney one would expect to find also TRPV4 mutations in kidney associated dysfunction. This is the case a mutation within the TRPV4 N-terminus (P19S) is found in patients developing hyponatremia and relative water excess (Tian et al. 2009). This mutation exhibits decreased response to hypotonic stress after overexpression in HEK cells. Surprisingly hyponatremia was pronounced only in male carriers of the mutation. Results of the last few years clearly show that it is expectable to find many more TRPV4 related diseases. Therefore, modulation of TRPV4 channel function with respect to the activating mechanisms may be interesting tools. On the other hand, one would expect side effects of TRPV4 modulators because of the wide tissue distribution of TRPV4.

27.5.3 TRPV4 Antagonists

The classic TRP-inhibitor ruthenium red inhibits TRPV4 inward but not outward currents (Watanabe et al. 2002). Also trivalent cations Gd^{3+} and La^{3+} block the channel in the upper micromolar range (Strotmann et al. 2000). A sulphonamide RN-1734 blocks human rat and mouse TRPV4 in the range of 100 μM (Vincent et al. 2009) and another blocker of unpublished structure RN-9893 seems to block specifically in the nanomolar range (Vincent and Duncton 2011). The blocker inhibits 4a-PDD activation of TRPV4. A natural isolate from lemongrass, citral, binds TRPV4 in a reversible manner but lacked specificity within the TRP family (Stotz et al. 2008). A company based screen identified a novel TRPV4 blocker termed HC-067047. The blocker was described by Cain and Writer (2010, see http://www.nature.com/scibx/journal/v3/n44/full/scibx.2010.1311.html) but on the homepage of the company there is no more information available (www.hydrabiosciences.com). A long list of antagonists related to the agonist GSK1016790A (GlaxoSmithKline) was patented including GSK205 which was claimed to be selective for TRPV4. Also Kobayashi Pharmaceutical developed a few antagonists which have been patented in 2007 but none of the compounds entered a clinical phase.

27.5.4 TRPV4 Agonists

Besides 4a-PDD and derivatives thereof, TRPV4 can be activated by a naturally occurring compound bisandrographolide from *Andrographis paniculata* (Smith et al. 2006). The compound activates the channel in the upper nanomolar range and does not affect other TRPV channels. The compound is used in traditional Chinese medicine as an anti-inflammatory drug. As mentioned epoxyeicosatrienoic acid an endogenous metabolite of arachidonic acid and anadamide also activate the channel (Watanabe et al. 2003). Furthermore, a GlaxoSmithKline-screened compound GSK1016790A turned out to be active also on TRPV1. However, systemic application of GSK1016790A leads to circulatory collapse in wild type but not in TRPV4 K.O. (Willette et al. 2008). Later it was demonstrated that the drug induced relaxation of small mesenteric arteries in a NO and EDHF (endothelium-derived hyperpolarizing factor) dependent manner (Mendoza et al. 2010). A sulfonamide-like compound RN-1747 activates human, mouse and rat TRPV4 in the low micromolar range (Vincent et al. 2009).

27.6 TRPV5/6

27.6.1 Cloning and Expression

TRPV5 and TRPV6 are very similar genes and the tandem localization on the human chromosome 7 indicates a duplication event from an ancestral gene. TRPV5 was the first of this subgroup that was cloned from rabbit kidney by Jost Hoenderop in the laboratory of Rene Bindels (Hoenderop et al. 1999). TRPV6 was cloned by Peng and co-workers from rat intestine (Peng et al. 1999), the human orthologue from Peng and Wissenbach and the murine one by Hirnet and co-workers (Hirnet et al. 2003; Peng et al. 2000; Wissenbach et al. 2001). TRPV5 and TRPV6 were overexpressed in several cell types and *Xenopus* oocytes and result in constitutively active channels with high prevalence of calcium ions (Hoenderop et al. 1999; Peng et al. 1999; Vennekens et al. 2000; Wissenbach et al. 2001). Thus reducing the extracellular calcium activates the channel and increasing intracellular calcium blocks it (Vennekens et al. 2000). Like most calcium selective channels, TRPV5/6 conduct monovalent ions under divalent free conditions. In the absence of calcium, it was found that TRPV5 conducts $Na^+ > Li^+ > K^+ > Cs^+ > NMDG^+$ whereas Mg^{2+} blocks TRPV5/6 channels from the inside (Nilius et al. 2000; Voets et al. 2003). Recently, permeability of heavy metal ions was presented by Kovacs et al. (2010). Interestingly conductance of La^{3+} and Gd^{3+} was shown by overexpresssed TRPV6 channels although a higher concentration of trivalent ions (micromolar range) blocks the channel. In addition Ba^{2+}, Sr^{2+}, Mn^{2+}, Zn^{2+}, Cd^{2+} ions can penetrate through TRPV6 channels. TRPV5/6 are expressed in epithelial tissues, which exhibit calcium uptake properties. TRPV5 is expressed in mammalian kidneys whereas TRPV6 has a broader expression pattern. TRPV5 is expressed in the late part of the distal convolute and regulates re-uptake of calcium ions in the kidney from the primary urine as shown by many papers of the Bindels laboratory (Dimke et al. 2011; Hoenderop et al. 2001). TRPV5 is thought to be involved in the re-uptake of calcium in the kidney where calcium is subsequently bound to calbindins (28k and 9k) and exported from the epithelial cells by a Ca-ATPase (PMCA1b) and/or calcium exchanger (NCX1). Thereby expression of TRPV5 is regulated by vitamin D3 and dietary calcium (Hoenderop et al. 2002). Accordingly TRPV5 K.O. mice exhibit exaggerated renal calcium waste and as a consequence, reduced thickness of the bones (Hoenderop et al. 2003).

In humans, TRPV6 is predominantly expressed in placenta and some exocrine tissues as salivary and lacrimal gland as well as in the exocrine pancreas (Peng et al. 2000; Wissenbach et al. 2001; Stoerger and Flockerzi, unpublished data). TRPV6 was first cloned from rat small intestine (Peng et al. 1999) and the authors presented data showing expression within the small intestine (duodenum) in human tissue by northern blot analysis (Peng et al. 2000). From the expression pattern and the function of the channel, it was concluded that TRPV6 takes part of the transplacental calcium transport and calcium uptake in the small intestine. However, TRPV6 deficient mice did not show decreased calcium serum level even under low calcium diet indicating that TRPV6 contribution to intestinal calcium uptake is not too dramatic although intestinal uptake was clearly reduced (Woudenberg-Vrenken et al. 2012). The most pronounced phenotype of the TRPV6 K.O. is very low fertility of male TRPV6 mice (Bianco et al. 2007; Weissgerber et al. 2012). Reduced sperm motility seems to be the cause for infertility but TRPV6 transcripts are not present in sperm. This puzzle was solved by the finding that the ductal lumen of the epididymis contains an inadequate calcium concentration that inhibits sperm maturation. Under wild type conditions luminal calcium is lowered by TRPV6 dependent uptake into epithelia of the epididymis. Interestingly, reduced fertility was also found in TRPV6D521A knockin mice which express homozygously a calcium impermeable channel. Thus, calcium uptake is critical for this observed phenotype (Weissgerber et al. 2011). Furthermore, bone metabolism was also unaffected (van der Eerden et al. 2011).

TRPV5/6-like currents have not been measured in any of the native tissues expressing TRPV5/6 as one can easily detect if TRPV5/6 is overexpressed in cell lines or oocytes. On the other hand, inactivating TRPV5/6 by gene deletion clearly results in a decrease of calcium uptake in TRPV5/6 expressing tissues (usual measured as uptake of ^{45}Ca). The reason for this discrepancy is unclear.

27.6.2 Channel Properties and Related Diseases

Very quickly after identification of the TRPV6 gene it became apparent that in certain malignancies transcript levels are increased. TRPV6 was found to be overexpressed in prostate cancer thereby the expression pattern is correlated with the Gleason grading and the tumor staging (Fixemer et al. 2003; Peng et al. 2001; Wissenbach et al. 2001, 2004; Zhuang et al. 2002). It has to be noted that in contrast to the murine prostate which expresses high level of TRPV6 transcripts shown by q-PCR (Nijenhuis et al. 2003; Weissgerber et al. 2012), in healthy prostatic tissue of humans TRPV6 is undetectable as demonstrated by northern analysis and In-Situ-Hybridization (Fixemer et al. 2003; Wissenbach et al. 2001, 2004). In humans TRPV6 was found in ~90 % of patients with advanced prostate cancer whereas TRPV6 expression in benign prostatic tissue (BPH) and prostatic neoplasia (Peppin et al. 2011) was negative. From the expression pattern and the patient data, it was concluded that TRPV6 may be useful for diagnosis and treatment of prostate cancer. Furthermore, TRPV6 transcripts are upregulated in endometrial, breast, thyroid, colon, and ovarian derived cancer (Wissenbach and Niemeyer 2007; Zhuang et al. 2002). Much research has been done to illuminate the role of TRPV6 in the prostate cancer derived cell line LNCaP by the group of Natalia Prevarskaya (Lehen'kyi et al. 2007, 2011, 2012; Wissenbach and Niemeyer 2007; Zhuang et al. 2002). The authors conclude that TRPV6 upregulation results in a higher proliferation rate, increasing cell survival and apoptosis resistance of these cells. In addition pro-proliferative effects of vitamin D3 are directly mediated by enhanced expression of TRPV6 in LNCaP cells.

Recently, it was demonstrated that both TRPV6 expression in a breast cancer cell line and the channel activity could be reduced with the steroid tamoxifen (Bolanz et al. 2008, 2009). But Peters and co-workers demonstrated that increased TRPV6 expression is a feature of estrogen receptor (ER)-negative breast cancer and these patients have a poor outcome (Peters et al. 2012). Furthermore, the authors conclude that enhanced TRPV6 expression in breast cancer may result from gene amplification.

Schwarz and co-workers demonstrated that TRPV6 overexpression increases cell proliferation of HEK-293 cells in a calcium dependent manner (Schwarz et al. 2006). As shown in Fig. 26.1a, injection of a stably TRPV6-expressing HEK-cell line in nude mice resulted in the development of solid tumors whereas nontransfected HEK cells and cells expressing the calcium permeable channel TRPC4 induced only comparable small tumors. Figure 26.1b demonstrates the presence of the human TRPV6 protein in xenografts excised from the mice in Fig. 26.1a (the bands represent glycosylated and nonglycosylated forms of TRPV6) and in human prostate cancer of four patients (Fig. 26.1c) (all data from Wissenbach, Menger and Flockerzi).

A recent paper investigated if TRPV6 is involved in Loewe syndrome and Dent disease (Wu et al. 2012). Loewe syndrome is characterized by hypercalciuria and Dent disease patients show hyperabsorption of Ca^{2+} with increased Ca^{2+} excretion.

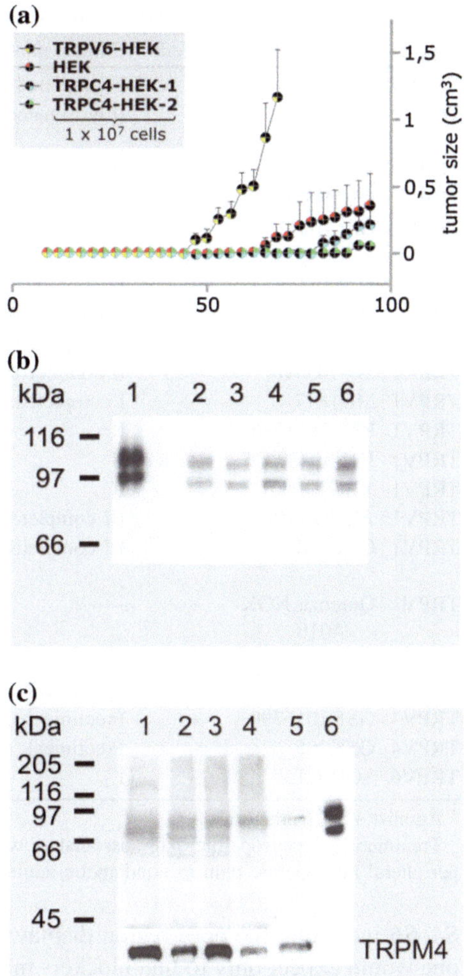

Fig. 26.1 TRPV6-HEK293 xenograft reveal tumorous growth. **a** Stably expressing TRPV6-HEK293, HEK293 and stably TRPC4 expressing cells ($\sim 10^7$ cells) were injected subcutaneously into nude mice and tumor growth was monitored. **b** Five tumor xenografts were subjected to Western blot and analyzed with a TRPV6 antibody. *Lane 1* TRPV6 overexpressed in HEK293 cells, *lanes 2–6* TRPV6 mouse xenografts. **c** Prostate cancer tissue of four patients was analyzed with TRPV6 specific antibody (*lanes 1–4*). *Lane 5* contains murine prostate. As positive control, the western blot was re-probed with an antibody for TRPM4 which is expressed in healthy and diseased prostate

The authors present data indicating that TRPV6 activity is controlled by a phosphatidyl inositol 4,5-bisphosphate [PI(4,5)P(2)] 5-phosphatase (likely mediated by PIP2 level). However, several Dent patients exhibit mutations in the phosphatase resulting in an increase of TRPV6 activity.

The cancer derived data suggest that TRPV6 may be a useful target for the treatment of the most common cancers, namely breast and prostate cancer. Therefore, it is somewhat surprising that screening specific TRPV6 modulators had been a rare event in the pharmaceutical industry. However, it seems that TRPV5 is not overexpressed in malignancies although a leukemia derived cell line expresses TRPV5 and TRPV6 transcripts (Semenova et al. 2009).

It is noteworthy that human TRPV5 and TRPV6 proteins are very similar ($\sim$75 % identical amino acids) and the region of the transmembrane domains

Table 27.1 Modulators of the TRPV-family

Target	Compound	Mode[a]	Clinical phase	Company	Notes
TRPV1	ABT-102	–	I completed	Abott	Unpublished results
TRPV1	AMG-517	–	II terminated	Amgen	Increase of body temperature
TRPV1	AZD-1386	–	II terminated	AstraZeneca	Liver enzyme elevation
TRPV1	DWP-05195	–	I	Daewong	Unpublished results
TRPV1	GRC-6211	–	I terminated	Eli Lilly/ Glenmark	Unpublished results
TRPV1	JTS-653	–	I	Japan Tobacco	On going
TRPV1	MK-2295	–	II completed	Merck/Neurogen	Unpublished results
TRPV1	PHE377	–	I completed	PharmEste	On going
TRPV1	SB-705498	–	II completed	GlaxoSmithKline	Unpublished results
TRPV1	MR1817	–	I completed	Mochida	Unpublished results
TRPV1	PF-03864086	–	II	Pfizer	Unpublished results
TRPV1	JNJ-39439335	–	II	Johnson&Johnson	Unpublished results
TRPV1	XEN-D0501	–	II	Provesica	Unpublished results
TRPV1	ALRGX-4975	+	III completed	Adlea	Unpublished results
TRPV1	Civamide	+	III completed	Winston laboratory	Unpublished results
TRPV1	Qutenza NGX-4010	+	–	NeurogesX	[b]
TRPV3	GRC 15300	–	I completed	Glenmark	Unpublished results
TRPV4	HC-067047	–	Preclinical	Hydrabiosciences	Unpublished results
TRPV4	GSK1016790A	+	Preclinical	GlaxoSmithKline	
TRPV4	GSK205	–	Preclinical	GlaxoSmithKline	
TRPV6	SOR-C13	–	I	Soricimed	On going

[a] Agonist + and antagonist –

[b] Treatment of neuropathic pain associated with postherpetic neuralgia (US), treatment of peripheral neuropathic pain in nondiabetic adults (EU)

S4–S6 including the pore region displays 89 % identical amino acids. From this, one would expect only to find blockers inhibiting both channels to a similar extent. However, this is not the case (see below).

27.6.3 TRPV5/6 Antagonists

The first published blocker of TRPV6 was an isolate of marine sponge, Xestospongin C (Vassilev et al. 2001). Xestospongin C blocks in the lower micromolar range. However, the compound also binds to IP_3 receptors. The unspecific blocker ruthenium red was found to block TRPV5 with an IC50 of $\sim$121 nM but has a 100-fold lower affinity for TRPV6 (Hoenderop et al. 2001). Econazole and miconazole commonly used as antifungal drugs block TRPV5/6 in the lower micromolar range (Nilius et al. 2001a). Econazole is not a specific drug and blocks several ion channels (Franzius et al. 1994). However, ketoconazole and clotrimazole did not block TRPV6 channels in micromolar amounts (Wissenbach unpublished data). A compound TH-1177 was found to inhibit tumor growth of prostate cancer cell lines in severe combined immunodeficient mice (Haverstick et al. 2000). Later it was found that the compound blocks TRPV5/6 and derivatives thereof were synthesized with an increased specificity to TRPV6 (Landowski et al. 2011). Compound #03 inhibits growth of a prostate cancer derived cell line LNCaP, a cell line known to express TRPV6 transcripts (Bodding et al. 2003). The compound has an IC50 value for growth inhibition by blocking endogenous Ca^{2+} entry channels of 0.44 μM and inhibited TRPV6 more than five times better than TRPV5. 2-APB (2-aminoethoxydiphenyl borate) known for long time as an unspecific modulator of TRP channels and IP_3 receptors (Gregory et al. 2001; Iwasaki et al. 2001; Taufiq et al. 2005) blocks TRPV6 but surprisingly not TRPV5 (Kovacs et al. 2012). 2-APB displays IC-50 values of $\sim$70 μM and blocks human, rat and mouse TRPV6.

Very interesting results are published by a Canada based company (Soricimed). Soricidin a mammalian peptide component of shrew venom, was isolated. The peptide has two important properties. First, the peptide has an analgetic effect that is not mediated by opioid receptors and may be used to treat several types of pain. Second, the peptide inhibits growth of some cancer cells. The paralytic domain and the anti-oncology domain could be separated leading to a series of C-terminal peptides. It turned out that the C-terminal part of the peptide inhibits TRPV6 (see http://www.soricimed.com/). One peptide, SOR-C13, entered clinical phase I in 2012 (http://clinicaltrials.gov/ct2/show/NCT01578564).

27.6.4 TRPV5/6 Agonists

No TRPV5/6 agonist has been described so far.

27.7 Conclusion

Although all of the TRPV channels are clearly interesting targets, most companies focused on TRPV1 with frustrating results although in the future one would expect that screening programs are expanded on other TRPV channels. Table 27.1 summarizes TRPV channel modulators. At present, capsaicin which was introduced in pain therapy by Raubenheimer 1928 is still the only substance which is commonly used in the US and EU for the treatment of pain. None of the published TRPV modulators managed to make it into the market. However, TRPV1 is clearly a promising pharmacological target and I would like to follow the suggestions of Peter Holzer (Holzer 2008) to design for instance modality specific TRPV1 blockers which inhibit certain types of stimuli thereby protecting sensitivity to others to reduce unwanted side effects.

References

Ahern GP, Brooks IM, Miyares RL, Wang XB (2005) Extracellular cations sensitize and gate capsaicin receptor TRPV1 modulating pain signaling. J Neurosci 25(21):5109–5116

Ahern GP, Wang X, Miyares RL (2006) Polyamines are potent ligands for the capsaicin receptor TRPV1. J Biol Chem 281(13):8991–8995

Alessandri-Haber N, Yeh JJ, Boyd AE, Parada CA, Chen X, Reichling DB, Levine JD (2003) Hypotonicity induces TRPV4-mediated nociception in rat. Neuron 39(3):497–511

Alessandri-Haber N, Dina OA, Yeh JJ, Parada CA, Reichling DB, Levine JD (2004) Transient receptor potential vanilloid 4 is essential in chemotherapy-induced neuropathic pain in the rat. J Neurosci 24(18):4444–4452

Auer-Grumbach M, Olschewski A, Papic L, Kremer H, McEntagart ME, Uhrig S, Fischer C, Frohlich E, Balint Z, Tang B, Strohmaier H, Lochmuller H, Schlotter-Weigel B, Senderek J, Krebs A, Dick KJ, Petty R, Longman C, Anderson NE, Padberg GW, Schelhaas HJ, van Ravenswaaij-Arts CM, Pieber TR, Crosby AH, Guelly C (2010) Alterations in the ankyrin domain of TRPV4 cause congenital distal SMA, scapuloperoneal SMA and HMSN2C. Nat Genet 42(2):160–164

Axelsson HE, Minde JK, Sonesson A, Toolanen G, Hogestatt ED, Zygmunt PM (2009) Transient receptor potential vanilloid 1, vanilloid 2 and melastatin 8 immunoreactive nerve fibers in human skin from individuals with and without Norrbottnian congenital insensitivity to pain. Neuroscience 162(4):1322–1332

Bang S, Kim KY, Yoo S, Lee SH, Hwang SW (2007) Transient receptor potential V2 expressed in sensory neurons is activated by probenecid. Neurosci Lett 425(2):120–125

Bianco SD, Peng JB, Takanaga H, Suzuki Y, Crescenzi A, Kos CH, Zhuang L, Freeman MR, Gouveia CH, Wu J, Luo H, Mauro T, Brown EM, Hediger MA (2007) Marked disturbance of calcium homeostasis in mice with targeted disruption of the Trpv6 calcium channel gene. J Bone Miner Res 22(2):274–285

Bodding M, Fecher-Trost C, Flockerzi V (2003) Store-operated Ca2+ current and TRPV6 channels in lymph node prostate cancer cells. J Biol Chem 278(51):50872–50879

Boels K, Glassmeier G, Herrmann D, Riedel IB, Hampe W, Kojima I, Schwarz JR, Schaller HC (2001) The neuropeptide head activator induces activation and translocation of the growth-factor-regulated Ca(2+)-permeable channel GRC. J Cell Sci 114(Pt 20):3599–3606

Bolanz KA, Hediger MA, Landowski CP (2008) The role of TRPV6 in breast carcinogenesis. Mol Cancer Ther 7(2):271–279

Bolanz KA, Kovacs GG, Landowski CP, Hediger MA (2009) Tamoxifen inhibits TRPV6 activity via estrogen receptor-independent pathways in TRPV6-expressing MCF-7 breast cancer cells. Mol Cancer Res 7(12):2000–2010

Brauchi S, Orta G, Mascayano C, Salazar M, Raddatz N, Urbina H, Rosenmann E, Gonzalez-Nilo F, Latorre R (2007) Dissection of the components for PIP2 activation and thermosensation in TRP channels. Proc Natl Acad Sci USA 104(24):10246–10251

Buchholz CF (1816) Chemische Untersuchung der trockenen reifen spanischen Pfeffers, vol 37. Weimar

Camacho N, Krakow D, Johnykutty S, Katzman PJ, Pepkowitz S, Vriens J, Nilius B, Boyce BF, Cohn DH (2010) Dominant TRPV4 mutations in nonlethal and lethal metatropic dysplasia. Am J Med Genet A 152A(5):1169–1177

Casas S, Novials A, Reimann F, Gomis R, Gribble FM (2008) Calcium elevation in mouse pancreatic beta cells evoked by extracellular human islet amyloid polypeptide involves activation of the mechanosensitive ion channel TRPV4. Diabetologia 51(12):2252–2262

Caterina MJ, Schumacher MA, Tominaga M, Rosen TA, Levine JD, Julius D (1997) The capsaicin receptor: a heat-activated ion channel in the pain pathway. Nature 389(6653):816–824

Caterina MJ, Rosen TA, Tominaga M, Brake AJ, Julius D (1999) A capsaicin-receptor homologue with a high threshold for noxious heat. Nature 398(6726):436–441

Charrua A, Cruz CD, Narayanan S, Gharat L, Gullapalli S, Cruz F, Avelino A (2009) GRC-6211, a new oral specific TRPV1 antagonist, decreases bladder overactivity and noxious bladder input in cystitis animal models. J Urol 181(1):379–386
Chizh BA, Sang CN (2009) Use of sensory methods for detecting target engagement in clinical trials of new analgesics. Neurotherapeutics 6(4):749–754
Cho TJ, Matsumoto K, Fano V, Dai J, Kim OH, Chae JH, Yoo WJ, Tanaka Y, Matsui Y, Takigami I, Monges S, Zabel B, Shimizu K, Nishimura G, Lausch E, Ikegawa S (2012) TRPV4-pathy manifesting both skeletal dysplasia and peripheral neuropathy: a report of three patients. Am J Med Genet A 158A(4):795–802
Chung MK, Lee H, Caterina MJ (2003) Warm temperatures activate TRPV4 in mouse 308 keratinocytes. J Biol Chem 278(34):32037–32046
Chung MK, Lee H, Mizuno A, Suzuki M, Caterina MJ (2004a) 2-Aminoethoxydiphenyl borate activates and sensitizes the heat-gated ion channel TRPV3. J Neurosci 24(22):5177–5182
Chung MK, Lee H, Mizuno A, Suzuki M, Caterina MJ (2004b) TRPV3 and TRPV4 mediate warmth-evoked currents in primary mouse keratinocytes. J Biol Chem 279(20):21569–21575
Cordeiro S, Seyler S, Stindl J, Milenkovic VM, Strauss O (2010) Heat-sensitive TRPV channels in retinal pigment epithelial cells: regulation of VEGF-A secretion. Invest Ophthalmol Vis Sci 51(11):6001–6008
Cui M, Honore P, Zhong C, Gauvin D, Mikusa J, Hernandez G, Chandran P, Gomtsyan A, Brown B, Bayburt EK, Marsh K, Bianchi B, McDonald H, Niforatos W, Neelands TR, Moreland RB, Decker MW, Lee CH, Sullivan JP, Faltynek CR (2006) TRPV1 receptors in the CNS play a key role in broad-spectrum analgesia of TRPV1 antagonists. J Neurosci 26(37):9385–9393
Delany NS, Hurle M, Facer P, Alnadaf T, Plumpton C, Kinghorn I, See CG, Costigan M, Anand P, Woolf CJ, Crowther D, Sanseau P, Tate SN (2001) Identification and characterization of a novel human vanilloid receptor-like protein, VRL-2. Physiol Genomics 4(3):165–174
Deng HX, Klein CJ, Yan J, Shi Y, Wu Y, Fecto F, Yau HJ, Yang Y, Zhai H, Siddique N, Hedley-Whyte ET, Delong R, Martina M, Dyck PJ, Siddique T (2010) Scapuloperoneal spinal muscular atrophy and CMT2C are allelic disorders caused by alterations in TRPV4. Nat Genet 42(2):165–169
Dimke H, Hoenderop JG, Bindels RJ (2011) Molecular basis of epithelial Ca2+ and Mg2+ transport: insights from the TRP channel family. J Physiol 589(Pt 7):1535–1542
Earley S (2005) Vanilloid and melastatin transient receptor potential channels in vascular smooth muscle. Microcirculation 17(4):237–249
Everaerts W, Zhen X, Ghosh D, Vriens J, Gevaert T, Gilbert JP, Hayward NJ, McNamara CR, Xue F, Moran MM, Strassmaier T, Uykal E, Owsianik G, Vennekens R, De Ridder D, Nilius B, Fanger CM, Voets T (2010) Inhibition of the cation channel TRPV4 improves bladder function in mice and rats with cyclophosphamide-induced cystitis. Proc Natl Acad Sci USA 107(44):19084–19089
Fian R, Grasser E, Treiber F, Schmidt R, Niederl P, Rosker C (2007) The contribution of TRPV4-mediated calcium signaling to calcium homeostasis in endothelial cells. J Recept Signal Transduct Res 27(2–3):113–124
Fixemer T, Wissenbach U, Flockerzi V, Bonkhoff H (2003) Expression of the Ca2+-selective cation channel TRPV6 in human prostate cancer: a novel prognostic marker for tumor progression. Oncogene 22(49):7858–7861
Franzius D, Hoth M, Penner R (1994) Non-specific effects of calcium entry antagonists in mast cells. Pflugers Arch 428(5–6):433–438
Gavva NR (2008) Body-temperature maintenance as the predominant function of the vanilloid receptor TRPV1. Trends Pharmacol Sci 29(11):550–557
Gavva NR, Bannon AW, Hovland DN Jr, Lehto SG, Klionsky L, Surapaneni S, Immke DC, Henley C, Arik L, Bak A, Davis J, Ernst N, Hever G, Kuang R, Shi L, Tamir R, Wang J, Wang W, Zajic G, Zhu D, Norman MH, Louis JC, Magal E, Treanor JJ (2007a) Repeated administration of vanilloid receptor TRPV1 antagonists attenuates hyperthermia elicited by TRPV1 blockade. J Pharmacol Exp Ther 323(1):128–137

Gavva NR, Bannon AW, Surapaneni S, Hovland DN Jr, Lehto SG, Gore A, Juan T, Deng H, Han B, Klionsky L, Kuang R, Le A, Tamir R, Wang J, Youngblood B, Zhu D, Norman MH, Magal E, Treanor JJ, Louis JC (2007b) The vanilloid receptor TRPV1 is tonically activated in vivo and involved in body temperature regulation. J Neurosci 27(13):3366–3374

Gavva NR, Treanor JJ, Garami A, Fang L, Surapaneni S, Akrami A, Alvarez F, Bak A, Darling M, Gore A, Jang GR, Kesslak JP, Ni L, Norman MH, Palluconi G, Rose MJ, Salfi M, Tan E, Romanovsky AA, Banfield C, Davar G (2008) Pharmacological blockade of the vanilloid receptor TRPV1 elicits marked hyperthermia in humans. Pain 136(1–2):202–210

Gevaert T, Vriens J, Segal A, Everaerts W, Roskams T, Talavera K, Owsianik G, Liedtke W, Daelemans D, Dewachter I, Van Leuven F, Voets T, De Ridder D, Nilius B (2007) Deletion of the transient receptor potential cation channel TRPV4 impairs murine bladder voiding. J Clin Invest 117(11):3453–3462

Gomtsyan A, Bayburt EK, Schmidt RG, Surowy CS, Honore P, Marsh KC, Hannick SM, McDonald HA, Wetter JM, Sullivan JP, Jarvis MF, Faltynek CR, Lee CH (2008) Identification of (R)-1-(5-tert-butyl-2,3-dihydro-1H-inden-1-yl)-3-(1H-indazol-4-yl)urea (ABT-102) as a potent TRPV1 antagonist for pain management. J Med Chem 51(3):392–395

Gregory RB, Rychkov G, Barritt GJ (2001) Evidence that 2-aminoethyl diphenylborate is a novel inhibitor of store-operated Ca2+ channels in liver cells, and acts through a mechanism which does not involve inositol trisphosphate receptors. Biochem J 354(Pt 2):285–290

Gu Q, Lin RL, Hu HZ, Zhu MX, Lee LY (2005) 2-aminoethoxydiphenyl borate stimulates pulmonary C neurons via the activation of TRPV channels. Am J Physiol Lung Cell Mol Physiol 288(5):L932–L941

Guler AD, Lee H, Iida T, Shimizu I, Tominaga M, Caterina M (2002) Heat-evoked activation of the ion channel, TRPV4. J Neurosci 22(15):6408–6414

Gullapalli TA, Lingam PR, Kattige V, Gudi GS, Khairatkar-Joshi N (2008) GRC 15133—a novel, selective TRPV3 antagonist with anti-hyperalgesic effects in inflammatory and neuropathic pain. CHI-World Pharmaceutical Congress, http://wwwglenmarkpharma-com/research

Haverstick DM, Heady TN, Macdonald TL, Gray LS (2000) Inhibition of human prostate cancer proliferation in vitro and in a mouse model by a compound synthesized to block Ca2+ entry. Cancer Res 60(4):1002–1008

Hirnet D, Olausson J, Fecher-Trost C, Bodding M, Nastainczyk W, Wissenbach U, Flockerzi V, Freichel M (2003) The TRPV6 gene, cDNA and protein. Cell Calcium 33(5–6):509–518

Hoenderop JG, van der Kemp AW, Hartog A, van de Graaf SF, van Os CH, Willems PH, Bindels RJ (1999) Molecular identification of the apical Ca2+ channel in 1, 25-dihydroxyvitamin D3-responsive epithelia. J Biol Chem 274(13):8375–8378

Hoenderop JG, Vennekens R, Muller D, Prenen J, Droogmans G, Bindels RJ, Nilius B (2001) Function and expression of the epithelial Ca(2+) channel family: comparison of mammalian ECaC1 and 2. J Physiol 537(Pt 3):747–761

Hoenderop JG, Dardenne O, Van Abel M, Van Der Kemp AW, Van Os CH, St -Arnaud R, Bindels RJ (2002) Modulation of renal Ca2+ transport protein genes by dietary Ca2+ and 1,25-dihydroxyvitamin D3 in 25-hydroxyvitamin D3-1alpha-hydroxylase knockout mice. Faseb J 16(11):1398–1406

Hoenderop JG, van Leeuwen JP, van der Eerden BC, Kersten FF, van der Kemp AW, Merillat AM, Waarsing JH, Rossier BC, Vallon V, Hummler E, Bindels RJ (2003) Renal Ca2+ wasting, hyperabsorption, and reduced bone thickness in mice lacking TRPV5. J Clin Invest 112(12):1906–1914

Holzer P (1991) Capsaicin: cellular targets, mechanisms of action, and selectivity for thin sensory neurons. Pharmacol Rev 43(2):143–201

Holzer P (2008) The pharmacological challenge to tame the transient receptor potential vanilloid-1 (TRPV1) nocisensor. Br J Pharmacol 155(8):1145–1162. doi:10.1038/bjp.2008.351

Honore P, Chandran P, Hernandez G, Gauvin DM, Mikusa JP, Zhong C, Joshi SK, Ghilardi JR, Sevcik MA, Fryer RM, Segreti JA, Banfor PN, Marsh K, Neelands T, Bayburt E, Daanen JF, Gomtsyan A, Lee CH, Kort ME, Reilly RM, Surowy CS, Kym PR, Mantyh PW, Sullivan JP,

Jarvis MF, Faltynek CR (2009) Repeated dosing of ABT-102, a potent and selective TRPV1 antagonist, enhances TRPV1-mediated analgesic activity in rodents, but attenuates antagonist-induced hyperthermia. Pain 142(1–2):27–35

Hu HZ, Gu Q, Wang C, Colton CK, Tang J, Kinoshita-Kawada M, Lee LY, Wood JD, Zhu MX (2004) 2-aminoethoxydiphenyl borate is a common activator of TRPV1, TRPV2, and TRPV3. J Biol Chem 279(34):35741–35748

Huang J, Zhang X, McNaughton PA (2006) Inflammatory pain: the cellular basis of heat hyperalgesia. Curr Neuropharmacol 4(3):197–206

Ichikawa H, Fukunaga T, Jin HW, Fujita M, Takano-Yamamoto T, Sugimoto T (2004) VR1-, VRL-1- and P2X3 receptor-immunoreactive innervation of the rat temporomandibular joint. Brain Res 1008 (1):131–136

Iwasaki H, Mori Y, Hara Y, Uchida K, Zhou H, Mikoshiba K (2001) 2-Aminoethoxydiphenyl borate (2-APB) inhibits capacitative calcium entry independently of the function of inositol 1,4,5-trisphosphate receptors. Receptors Channels 7(6):429–439

Iwata Y, Katanosaka Y, Arai Y, Komamura K, Miyatake K, Shigekawa M (2003) A novel mechanism of myocyte degeneration involving the Ca2+-permeable growth factor-regulated channel. J Cell Biol 161(5):957–967

Iwata Y, Katanosaka Y, Arai Y, Shigekawa M, Wakabayashi S (2009) Dominant-negative inhibition of Ca2+ influx via TRPV2 ameliorates muscular dystrophy in animal models. Hum Mol Genet 18(5):824–834

Jones VM, Moore KA, Peterson DM (2011) Capsaicin 8% topical patch (Qutenza)—a review of the evidence. J Pain Palliat Care Pharmacother 25(1):32–41

Jung J, Shin JS, Lee SY, Hwang SW, Koo J, Cho H, Oh U (2004) Phosphorylation of vanilloid receptor 1 by Ca2+/calmodulin-dependent kinase II regulates its vanilloid binding. J Biol Chem 279(8):7048–7054

Kanzaki M, Zhang YQ, Mashima H, Li L, Shibata H, Kojima I (1999) Translocation of a calcium-permeable cation channel induced by insulin-like growth factor-I. Nat Cell Biol 1(3):165–170

Kitagawa Y, Miyai A, Usui K, Hamada Y, Deai K, Wada M, Koga Y, Sakata M, Hayashi M, Tominaga M, Matsushita M (2012a) Pharmacological characterization of (3S)-3-(hydroxymethyl)-4-(5-methylpyridin-2-yl)-N-[6-(2,2,2-trifluoroethoxy)pyrid in-3-yl]-3,4-dihydro-2H-benzo[b][1,4]oxazine-8-carboxamide (JTS-653), a novel transient receptor potential vanilloid 1 antagonist. J Pharmacol Exp Ther 342(2):520–528

Kitagawa Y, Wada M, Kanehisa T, Miyai A, Usui K, Maekawa M, Sakata M, Matsuo A, Hayashi M, Matsushita M (2012b) JTS-653, an orally active and selective TRPV1 antagonist, blocks the afferent nerve firing and attenuates bladder overactivity without affecting the normal voiding function. J Urol

Klausen TK, Pagani A, Minassi A, Ech-Chahad A, Prenen J, Owsianik G, Hoffmann EK, Pedersen SF, Appendino G, Nilius B (2009) Modulation of the transient receptor potential vanilloid channel TRPV4 by 4alpha-phorbol esters: a structure-activity study. J Med Chem 52(9):2933–2939

Kovacs G, Danko T, Bergeron MJ, Balazs B, Suzuki Y, Zsembery A, Hediger MA (2010) Heavy metal cations permeate the TRPV6 epithelial cation channel. Cell Calcium 49(1):43–55

Kovacs G, Montalbetti N, Simonin A, Danko T, Balazs B, Zsembery A, Hediger MA (2012) Inhibition of the human epithelial calcium channel TRPV6 by 2-aminoethoxydiphenyl borate (2-APB). Cell Calcium

Lambert GA, Davis JB, Appleby JM, Chizh BA, Hoskin KL, Zagami AS (2009) The effects of the TRPV1 receptor antagonist SB-705498 on trigeminovascular sensitisation and neurotransmission. Naunyn Schmiedebergs Arch Pharmacol 380(4):311–325

Landoure G, Zdebik AA, Martinez TL, Burnett BG, Stanescu HC, Inada H, Shi Y, Taye AA, Kong L, Munns CH, Choo SS, Phelps CB, Paudel R, Houlden H, Ludlow CL, Caterina MJ, Gaudet R, Kleta R, Fischbeck KH, Sumner CJ (2010) Mutations in TRPV4 cause Charcot-Marie-Tooth disease type 2C. Nat Genet 42(2):170–174

Landowski CP, Bolanz KA, Suzuki Y, Hediger MA (2011) Chemical inhibitors of the calcium entry channel TRPV6. Pharm Res 28(2):322–330
Leffler A, Linte RM, Nau C, Reeh P, Babes A (2007) A high-threshold heat-activated channel in cultured rat dorsal root ganglion neurons resembles TRPV2 and is blocked by gadolinium. Eur J Neurosci 26(1):12–22
Lehen'kyi V, Flourakis M, Skryma R, Prevarskaya N (2007) TRPV6 channel controls prostate cancer cell proliferation via Ca(2+)/NFAT-dependent pathways. Oncogene 26(52):7380–7385
Lehen'kyi V, Raphael M, Oulidi A, Flourakis M, Khalimonchyk S, Kondratskyi A, Gordienko DV, Mauroy B, Bonnal JL, Skryma R, Prevarskaya N (2011) TRPV6 determines the effect of vitamin D3 on prostate cancer cell growth. PLoS ONE 6(2):e16856
Lehen'kyi V, Raphael M, Prevarskaya N (2012) The role of the TRPV6 channel in cancer. J Physiol 590(Pt 6):1369–1376
Lehto SG, Tamir R, Deng H, Klionsky L, Kuang R, Le A, Lee D, Louis JC, Magal E, Manning BH, Rubino J, Surapaneni S, Tamayo N, Wang T, Wang J, Wang J, Wang W, Youngblood B, Zhang M, Zhu D, Norman MH, Gavva NR (2008) Antihyperalgesic effects of (R, E)-N-(2-hydroxy-2,3-dihydro-1H-inden-4-yl)-3-(2-(piperidin-1-yl)-4-(trifluorom ethyl)phenyl)-acrylamide (AMG8562), a novel transient receptor potential vanilloid type 1 modulator that does not cause hyperthermia in rats. J Pharmacol Exp Ther 326(1):218–229
Lewinter RD, Skinner K, Julius D, Basbaum AI (2004) Immunoreactive TRPV-2 (VRL-1), a capsaicin receptor homolog, in the spinal cord of the rat. J Comp Neurol 470(4):400–408
Liedtke W (2005) TRPV4 as osmosensor: a transgenic approach. Pflugers Arch 451(1):176–180
Liedtke W, Friedman JM (2003) Abnormal osmotic regulation in trpv4-/- mice. Proc Natl Acad Sci USA 100(23):13698–13703
Liedtke W, Choe Y, Marti-Renom MA, Bell AM, Denis CS, Sali A, Hudspeth AJ, Friedman JM, Heller S (2000) Vanilloid receptor-related osmotically activated channel (VR-OAC), a candidate vertebrate osmoreceptor. Cell 103(3):525–535
Liedtke W, Tobin DM, Bargmann CI, Friedman JM (2003) Mammalian TRPV4 (VR-OAC) directs behavioral responses to osmotic and mechanical stimuli in *Caenorhabditis elegans*. Proc Natl Acad Sci USA 100(Suppl 2):14531–14536
Lukacs V, Thyagarajan B, Varnai P, Balla A, Balla T, Rohacs T (2007) Dual regulation of TRPV1 by phosphoinositides. J Neurosci 27(26):7070–7080
Macpherson LJ, Hwang SW, Miyamoto T, Dubin AE, Patapoutian A, Story GM (2006) More than cool: promiscuous relationships of menthol and other sensory compounds. Mol Cell Neurosci 32(4):335–343
Masuyama R, Vriens J, Voets T, Karashima Y, Owsianik G, Vennekens R, Lieben L, Torrekens S, Moermans K, Vanden Bosch A, Bouillon R, Nilius B, Carmeliet G (2008) TRPV4-mediated calcium influx regulates terminal differentiation of osteoclasts. Cell Metab 8(3):257–265
McKemy DD, Neuhausser WM, Julius D (2002) Identification of a cold receptor reveals a general role for TRP channels in thermosensation. Nature 416(6876):52–58
Mendoza SA, Fang J, Gutterman DD, Wilcox DA, Bubolz AH, Li R, Suzuki M, Zhang DX (2010) TRPV4-mediated endothelial Ca2+ influx and vasodilation in response to shear stress. Am J Physiol Heart Circ Physiol 298(2):H466–H476
Micko K (1898) Zur Kenntnisnahme des Capsaicins. Zeitschrift für Untersuchung der Nahrungs- und Genussmittel 1:818–829
Mizoguchi F, Mizuno A, Hayata T, Nakashima K, Heller S, Ushida T, Sokabe M, Miyasaka N, Suzuki M, Ezura Y, Noda M (2008) Transient receptor potential vanilloid 4 deficiency suppresses unloading-induced bone loss. J Cell Physiol 216(1):47–53
Mizuno A, Matsumoto N, Imai M, Suzuki M (2003) Impaired osmotic sensation in mice lacking TRPV4. Am J Physiol Cell Physiol 285(1):C96–101
Mohapatra DP, Nau C (2005) Regulation of Ca2+-dependent desensitization in the vanilloid receptor TRPV1 by calcineurin and cAMP-dependent protein kinase. J Biol Chem 280(14):13424–13432

Monet M, Lehen'kyi V, Gackiere F, Firlej V, Vandenberghe M, Roudbaraki M, Gkika D, Pourtier A, Bidaux G, Slomianny C, Delcourt P, Rassendren F, Bergerat JP, Ceraline J, Cabon F, Humez S, Prevarskaya N (2010) Role of cationic channel TRPV2 in promoting prostate cancer migration and progression to androgen resistance. Cancer Res 70(3):1225–1235

Moqrich A, Hwang SW, Earley TJ, Petrus MJ, Murray AN, Spencer KS, Andahazy M, Story GM, Patapoutian A (2005) Impaired thermosensation in mice lacking TRPV3, a heat and camphor sensor in the skin. Science 307(5714):1468–1472

Moran WD, Zhen X, Mandel-Brehm J, Witek J, Yaksh T, McCumber D, Zhao M, Chong JA, Fanger CM (2007) Potent and selective antagonists validate TRPV3 as a target for analgesic therapeutics. Soc Neurosci Abstr 33 (143.5)

Moussaieff A, Rimmerman N, Bregman T, Straiker A, Felder CC, Shoham S, Kashman Y, Huang SM, Lee H, Shohami E, Mackie K, Caterina MJ, Walker JM, Fride E, Mechoulam R (2008) Incensole acetate, an incense component, elicits psychoactivity by activating TRPV3 channels in the brain. Faseb J 22(8):3024–3034

Muraki K, Iwata Y, Katanosaka Y, Ito T, Ohya S, Shigekawa M, Imaizumi Y (2003) TRPV2 is a component of osmotically sensitive cation channels in murine aortic myocytes. Circ Res 93(9):829–838

Nagy JI, Iversen LL, Goedert M, Chapman D, Hunt SP (1983) Dose-dependent effects of capsaicin on primary sensory neurons in the neonatal rat. J Neurosci 3(2):399–406

Nelson EK (1919) The constitution of capsaicin, the pungent principle of capsicum. J Am Chem Soc 41:1115–1121

Nijenhuis T, Hoenderop JG, van der Kemp AW, Bindels RJ (2003) Localization and regulation of the epithelial Ca2+ channel TRPV6 in the kidney. J Am Soc Nephrol 14(11):2731–2740

Nilius B, Voets T (2004) Diversity of TRP channel activation. Novartis Found Symp 258:140–149; discussion 149–159, 263–146

Nilius B, Vennekens R, Prenen J, Hoenderop JG, Bindels RJ, Droogmans G (2000) Whole-cell and single channel monovalent cation currents through the novel rabbit epithelial Ca2+ channel ECaC. J Physiol 527(Pt 2):239–248

Nilius B, Prenen J, Vennekens R, Hoenderop JG, Bindels RJ, Droogmans G (2001a) Pharmacological modulation of monovalent cation currents through the epithelial Ca2+ channel ECaC1. Br J Pharmacol 134(3):453–462

Nilius B, Prenen J, Wissenbach U, Bodding M, Droogmans G (2001b) Differential activation of the volume-sensitive cation channel TRP12 (OTRPC4) and volume-regulated anion currents in HEK-293 cells. Pflugers Arch 443(2):227–233

Nishimura G, Dai J, Lausch E, Unger S, Megarbane A, Kitoh H, Kim OH, Cho TJ, Bedeschi F, Benedicenti F, Mendoza-Londono R, Silengo M, Schmidt-Rimpler M, Spranger J, Zabel B, Ikegawa S, Superti-Furga A (2010) Spondylo-epiphyseal dysplasia, Maroteaux type (pseudo-Morquio syndrome type 2), and parastremmatic dysplasia are caused by TRPV4 mutations. Am J Med Genet A 152A(6):1443–1449

Nolano M, Simone DA, Wendelschafer-Crabb G, Johnson T, Hazen E, Kennedy WR (1999) Topical capsaicin in humans: parallel loss of epidermal nerve fibers and pain sensation. Pain 81(1–2):135–145

Othman AA, Nothaft W, Awni WM, Dutta S (2012) Effects of the TRPV1 antagonist ABT-102 on body temperature in healthy volunteers: pharmacokinetic/pharmacodynamic analysis of three phase 1 trials. Br J Clin Pharmacol

Peier AM, Reeve AJ, Andersson DA, Moqrich A, Earley TJ, Hergarden AC, Story GM, Colley S, Hogenesch JB, McIntyre P, Bevan S, Patapoutian A (2002) A heat-sensitive TRP channel expressed in keratinocytes. Science 296(5575):2046–2049

Peng JB, Chen XZ, Berger UV, Vassilev PM, Tsukaguchi H, Brown EM, Hediger MA (1999) Molecular cloning and characterization of a channel-like transporter mediating intestinal calcium absorption. J Biol Chem 274(32):22739–22746

Peng JB, Chen XZ, Berger UV, Weremowicz S, Morton CC, Vassilev PM, Brown EM, Hediger MA (2000) Human calcium transport protein CaT1. Biochem Biophys Res Commun 278(2):326–332

Peng JB, Zhuang L, Berger UV, Adam RM, Williams BJ, Brown EM, Hediger MA, Freeman MR (2001) CaT1 expression correlates with tumor grade in prostate cancer. Biochem Biophys Res Commun 282(3):729–734

Peppin JF, Majors K, Webster LR, Simpson DM, Tobias JK, Vanhove GF (2011) Tolerability of NGX-4010, a capsaicin 8% patch for peripheral neuropathic pain. J Pain Res 4:385–392

Peters AA, Simpson PT, Bassett JJ, Lee JM, Da Silva L, Reid LE, Song S, Parat MO, Lakhani SR, Kenny PA, Roberts-Thomson SJ, Monteith GR (2012) Calcium channel TRPV6 as a potential therapeutic target in estrogen receptor-negative breast cancer. Mol Cancer Ther 11(10):2158–2168

Premkumar LS, Qi ZH, Van Buren J, Raisinghani M (2004) Enhancement of potency and efficacy of NADA by PKC-mediated phosphorylation of vanilloid receptor. J Neurophysiol 91(3):1442–1449

Prescott ED, Julius D (2003) A modular PIP2 binding site as a determinant of capsaicin receptor sensitivity. Science 300(5623):1284–1288

Qin N, Neeper MP, Liu Y, Hutchinson TL, Lubin ML, Flores CM (2008) TRPV2 is activated by cannabidiol and mediates CGRP release in cultured rat dorsal root ganglion neurons. J Neurosci 28(24):6231–6238

Remadevi R, Szallisi A (2008) Adlea (ALGRX-4975), an injectable capsaicin (TRPV1 receptor agonist) formulation for longlasting pain relief. IDrugs 11(2):120–132

Rock MJ, Prenen J, Funari VA, Funari TL, Merriman B, Nelson SF, Lachman RS, Wilcox WR, Reyno S, Quadrelli R, Vaglio A, Owsianik G, Janssens A, Voets T, Ikegawa S, Nagai T, Rimoin DL, Nilius B, Cohn DH (2008) Gain-of-function mutations in TRPV4 cause autosomal dominant brachyolmia. Nat Genet 40(8):999–1003

Round P, Priestley A, Robinson J (2011) An investigation of the safety and pharmacokinetics of the novel TRPV1 antagonist XEN-D0501 in healthy subjects. Br J Clin Pharmacol 72(6):921–931

Saper JR, Klapper J, Mathew NT, Rapoport A, Phillips SB, Bernstein JE (2002) Intranasal civamide for the treatment of episodic cluster headaches. Arch Neurol 59(6):990–994

Schnitzer TJ, Pelletier JP, Haselwood DM, Ellison WT, Ervin JE, Gordon RD, Lisse JR, Archambault WT, Sampson AR, Fezatte HB, Phillips SB, Bernstein JE (2012) Civamide cream 0.075% in patients with osteoarthritis of the knee: a 12-week randomized controlled clinical trial with a longterm extension. J Rheumatol 39(3):610–620

Schwarz EC, Wissenbach U, Niemeyer BA, Strauss B, Philipp SE, Flockerzi V, Hoth M (2006) TRPV6 potentiates calcium-dependent cell proliferation. Cell Calcium 39(2):163–173

Semenova SB, Vassilieva IO, Fomina AF, Runov AL, Negulyaev YA (2009) Endogenous expression of TRPV5 and TRPV6 calcium channels in human leukemia K562 cells. Am J Physiol Cell Physiol 296(5):C1098–C1104

Sherkheli MA, Gisselmann G, Hatt H (2012) Supercooling agent icilin blocks a warmth-sensing ion channel TRPV3. ScientificWorld J 2012:982725

Sidhaye VK, Guler AD, Schweitzer KS, D'Alessio F, Caterina MJ, King LS (2006) Transient receptor potential vanilloid 4 regulates aquaporin-5 abundance under hypotonic conditions. Proc Natl Acad Sci USA 103(12):4747–4752

Sidhaye VK, Schweitzer KS, Caterina MJ, Shimoda L, King LS (2008) Shear stress regulates aquaporin-5 and airway epithelial barrier function. Proc Natl Acad Sci USA 105(9):3345–3350

Smith GD, Gunthorpe MJ, Kelsell RE, Hayes PD, Reilly P, Facer P, Wright JE, Jerman JC, Walhin JP, Ooi L, Egerton J, Charles KJ, Smart D, Randall AD, Anand P, Davis JB (2002) TRPV3 is a temperature-sensitive vanilloid receptor-like protein. Nature 418(6894):186–190

Smith PL, Maloney KN, Pothen RG, Clardy J, Clapham DE (2006) Bisandrographolide from Andrographis paniculata activates TRPV4 channels. J Biol Chem 281(40):29897–29904

Spath E, Darling SF (1930) Synthesis of capsaicin. Ber Dtsch Chem Ges 63B:737–740

Stein AT, Ufret-Vincenty CA, Hua L, Santana LF, Gordon SE (2006) Phosphoinositide 3-kinase binds to TRPV1 and mediates NGF-stimulated TRPV1 trafficking to the plasma membrane. J Gen Physiol 128(5):509–522. doi:10.1085/jgp.200609576

Steiner AA, Turek VF, Almeida MC, Burmeister JJ, Oliveira DL, Roberts JL, Bannon AW, Norman MH, Louis JC, Treanor JJ, Gavva NR, Romanovsky AA (2007) Nonthermal activation of transient receptor potential vanilloid-1 channels in abdominal viscera tonically inhibits autonomic cold-defense effectors. J Neurosci 27(28):7459–7468

Stotz SC, Vriens J, Martyn D, Clardy J, Clapham DE (2008) Citral sensing by transient [corrected] receptor potential channels in dorsal root ganglion neurons. PLoS ONE 3(5):e2082

Strotmann R, Harteneck C, Nunnenmacher K, Schultz G, Plant TD (2000) OTRPC4, a nonselective cation channel that confers sensitivity to extracellular osmolarity. Nat Cell Biol 2(10):695–702

Surowy CS, Neelands TR, Bianchi BR, McGaraughty S, El Kouhen R, Han P, Chu KL, McDonald HA, Vos M, Niforatos W, Bayburt EK, Gomtsyan A, Lee CH, Honore P, Sullivan JP, Jarvis MF, Faltynek CR (2008) (R)-(5-tert-butyl-2,3-dihydro-1H-inden-1-yl)-3-(1H-indazol-4-yl)-urea (ABT-102) blocks polymodal activation of transient receptor potential vanilloid 1 receptors in vitro and heat-evoked firing of spinal dorsal horn neurons in vivo. J Pharmacol Exp Ther 326(3):879–888

Swanson DM, Dubin AE, Shah C, Nasser N, Chang L, Dax SL, Jetter M, Breitenbucher JG, Liu C, Mazur C, Lord B, Gonzales L, Hoey K, Rizzolio M, Bogenstaetter M, Codd EE, Lee DH, Zhang SP, Chaplan SR, Carruthers NI (2005) Identification and biological evaluation of 4-(3-trifluoromethylpyridin-2-yl)piperazine-1-carboxylic acid (5-trifluoromethylpyridin-2-yl)amide, a high affinity TRPV1 (VR1) vanilloid receptor antagonist. J Med Chem 48(6):1857–1872

Szallasi A, Sheta M (2012) Targeting TRPV1 for pain relief: limits, losers and laurels. Expert Opin Investig Drugs 21(9):1351–1369

Tamayo N, Liao H, Stec MM, Wang X, Chakrabarti P, Retz D, Doherty EM, Surapaneni S, Tamir R, Bannon AW, Gavva NR, Norman MH (2008) Design and synthesis of peripherally restricted transient receptor potential vanilloid 1 (TRPV1) antagonists. J Med Chem 51(9):2744–2757

Tanaka R, Muraki K, Ohya S, Yamamura H, Hatano N, Itoh Y, Imaizumi Y (2008) TRPV4-like non-selective cation currents in cultured aortic myocytes. J Pharmacol Sci 108(2):179–189

Taniguchi J, Tsuruoka S, Mizuno A, Sato J, Fujimura A, Suzuki M (2007) TRPV4 as a flow sensor in flow-dependent K + secretion from the cortical collecting duct. Am J Physiol Renal Physiol 292(2):F667–F673

Taufiq AM, Fujii S, Yamazaki Y, Sasaki H, Kaneko K, Li J, Kato H, Mikoshiba K (2005) Involvement of IP3 receptors in LTP and LTD induction in guinea pig hippocampal CA1 neurons. Learn Mem 12(6):594–600

Thorneloe KS, Cheung M, Bao W, Alsaid H, Lenhard S, Jian MY, Costell M, Maniscalco-Hauk K, Krawiec JA, Olzinski A, Gordon E, Lozinskaya I, Elefante L, Qin P, Matasic DS, James C, Tunstead J, Donovan B, Kallal L, Waszkiewicz A, Vaidya K, Davenport EA, Larkin J, Burgert M, Casillas LN, Marquis RW, Ye G, Eidam HS, Goodman KB, Toomey JR, Roethke TJ, Jucker BM, Schnackenberg CG, Townsley MI, Lepore JJ, Willette RN (2008) An orally active TRPV4 channel blocker prevents and resolves pulmonary edema induced by heart failure. Sci Transl Med 4(159):159ra148

Thresh JC (1846) Isolation of capsaicin. Pharm J Trans 6(3):941–947

Tian W, Fu Y, Garcia-Elias A, Fernandez–Fernandez JM, Vicente R, Kramer PL, Klein RF, Hitzemann R, Orwoll ES, Wilmot B, McWeeney S, Valverde MA, Cohen DM (2009) A loss-of-function nonsynonymous polymorphism in the osmoregulatory TRPV4 gene is associated with human hyponatremia. Proc Natl Acad Sci USA 106(33):14034–14039

Trevisani M, Smart D, Gunthorpe MJ, Tognetto M, Barbieri M, Campi B, Amadesi S, Gray J, Jerman JC, Brough SJ, Owen D, Smith GD, Randall AD, Harrison S, Bianchi A, Davis JB, Geppetti P (2002) Ethanol elicits and potentiates nociceptor responses via the vanilloid receptor-1. Nat Neurosci 5(6):546–551

Ufret-Vincenty CA, Klein RM, Hua L, Angueyra J, Gordon SE (2011) Localization of the PIP2 sensor of TRPV1 ion channels. J Biol Chem 286(11):9688–9698 van der Eerden BC, Weissgerber P, Fratzl-Zelman N, Olausson J, Hoenderop JG, Schreuders-Koedam M, Eijken M, Roschger P, de Vries TJ, Chiba H, Klaushofer K, Flockerzi V, Bindels RJ, Freichel M, van Leeuwen JP (2011) The transient receptor potential channel TRPV6 is dynamically expressed in bone cells but is not crucial for bone mineralization in mice. J Cell Physiol 227(5):1951–1959

Vassilev PM, Peng JB, Johnson J, Hediger MA, Brown EM (2001) Inhibition of CaT1 channel activity by a noncompetitive IP3 antagonist. Biochem Biophys Res Commun 280(1):145–150

Vennekens R, Hoenderop JG, Prenen J, Stuiver M, Willems PH, Droogmans G, Nilius B, Bindels RJ (2000) Permeation and gating properties of the novel epithelial Ca(2+) channel. J Biol Chem 275(6):3963–3969

Verma P, Kumar A, Goswami C (2010) TRPV4-mediated channelopathies. Channels (Austin) 4(4):319–328

Vincent F, Duncton MA (2011) TRPV4 agonists and antagonists. Curr Top Med Chem 11(17):2216–2226

Vincent F, Acevedo A, Nguyen MT, Dourado M, DeFalco J, Gustafson A, Spiro P, Emerling DE, Kelly MG, Duncton MA (2009) Identification and characterization of novel TRPV4 modulators. Biochem Biophys Res Commun 389(3):490–494

Voets T, Prenen J, Vriens J, Watanabe H, Janssens A, Wissenbach U, Bodding M, Droogmans G, Nilius B (2002) Molecular determinants of permeation through the cation channel TRPV4. J Biol Chem 277(37):33704–33710

Voets T, Janssens A, Prenen J, Droogmans G, Nilius B (2003) Mg2+-dependent gating and strong inward rectification of the cation channel TRPV6. J Gen Physiol 121(3):245–260

Vogt-Eisele AK, Weber K, Sherkheli MA, Vielhaber G, Panten J, Gisselmann G, Hatt H (2007) Monoterpenoid agonists of TRPV3. Br J Pharmacol 151(4):530–540

Vriens J, Watanabe H, Janssens A, Droogmans G, Voets T, Nilius B (2004) Cell swelling, heat, and chemical agonists use distinct pathways for the activation of the cation channel TRPV4. Proc Natl Acad Sci USA 101(1):396–401

Vriens J, Owsianik G, Janssens A, Voets T, Nilius B (2007) Determinants of 4 alpha-phorbol sensitivity in transmembrane domains 3 and 4 of the cation channel TRPV4. J Biol Chem 282(17):12796–12803

Wallace M, Pappagallo M (2011) Qutenza(R): a capsaicin 8% patch for the management of postherpetic neuralgia. Expert Rev Neurother 11(1):15–27

Watanabe H, Davis JB, Smart D, Jerman JC, Smith GD, Hayes P, Vriens J, Cairns W, Wissenbach U, Prenen J, Flockerzi V, Droogmans G, Benham CD, Nilius B (2002) Activation of TRPV4 channels (hVRL-2/mTRP12) by phorbol derivatives. J Biol Chem 277(16):13569–13577

Watanabe H, Vriens J, Prenen J, Droogmans G, Voets T, Nilius B (2003) Anandamide and arachidonic acid use epoxyeicosatrienoic acids to activate TRPV4 channels. Nature 424(6947):434–438

Watanabe H, Murakami M, Ohba T, Ono K, Ito H (2009) The pathological role of transient receptor potential channels in heart disease. Circ J 73(3):419–427

Weissgerber P, Kriebs U, Tsvilovskyy V, Olausson J, Kretz O, Stoerger C, Vennekens R, Wissenbach U, Middendorff R, Flockerzi V, Freichel M (2011) Male fertility depends on Ca(2)+ absorption by TRPV6 in epididymal epithelia. Sci Signal 4 (171):ra27

Weissgerber P, Kriebs U, Tsvilovskyy V, Olausson J, Kretz O, Stoerger C, Mannebach S, Wissenbach U, Vennekens R, Middendorff R, Flockerzi V, Freichel M (2012) Excision of Trpv6 gene leads to severe defects in epididymal Ca2+ absorption and male fertility much like single D541A pore mutation. J Biol Chem 287(22):17930–17941

Willette RN, Bao W, Nerurkar S, Yue TL, Doe CP, Stankus G, Turner GH, Ju H, Thomas H, Fishman CE, Sulpizio A, Behm DJ, Hoffman S, Lin Z, Lozinskaya I, Casillas LN, Lin M, Trout RE, Votta BJ, Thorneloe K, Lashinger ES, Figueroa DJ, Marquis R, Xu X (2008) Systemic activation of the transient receptor potential vanilloid subtype 4 channel causes endothelial failure and circulatory collapse: part 2. J Pharmacol Exp Ther 326(2):443–452

Wissenbach U, Niemeyer BA (2007) Trpv6. Handb Exp Pharmacol 179:221–234

Wissenbach U, Bodding M, Freichel M, Flockerzi V (2000) Trp12, a novel Trp related protein from kidney. FEBS Lett 485(2–3):127–134

Wissenbach U, Niemeyer BA, Fixemer T, Schneidewind A, Trost C, Cavalie A, Reus K, Meese E, Bonkhoff H, Flockerzi V (2001) Expression of CaT-like, a novel calcium-selective channel, correlates with the malignancy of prostate cancer. J Biol Chem 276(22):19461–19468

Wissenbach U, Niemeyer B, Himmerkus N, Fixemer T, Bonkhoff H, Flockerzi V (2004) TRPV6 and prostate cancer: cancer growth beyond the prostate correlates with increased TRPV6 Ca2+ channel expression. Biochem Biophys Res Commun 322(4):1359–1363

Wong GY, Gavva NR (2009) Therapeutic potential of vanilloid receptor TRPV1 agonists and antagonists as analgesics: recent advances and setbacks. Brain Res Rev 60(1):267–277

Woudenberg-Vrenken TE, Lameris AL, Weissgerber P, Olausson J, Flockerzi V, Bindels RJ, Freichel M, Hoenderop JG (2012) Functional TRPV6 channels are crucial for transepithelial Ca2+ absorption. Am J Physiol Gastrointest Liver Physiol 303(7):G879–G885

Wu L, Gao X, Brown RC, Heller S, O'Neil RG (2007) Dual role of the TRPV4 channel as a sensor of flow and osmolality in renal epithelial cells. Am J Physiol Renal Physiol 293(5):F1699–F1713

Wu G, Zhang W, Na T, Jing H, Wu H, Peng JB (2012) Suppression of intestinal calcium entry channel TRPV6 by OCRL, a lipid phosphatase associated with Lowe syndrome and Dent disease. Am J Physiol Cell Physiol 302(10):C1479–C1491

Xu H, Ramsey IS, Kotecha SA, Moran MM, Chong JA, Lawson D, Ge P, Lilly J, Silos-Santiago I, Xie Y, DiStefano PS, Curtis R, Clapham DE (2002) TRPV3 is a calcium-permeable temperature-sensitive cation channel. Nature 418(6894):181–186

Xu H, Blair NT, Clapham DE (2005) Camphor activates and strongly desensitizes the transient receptor potential vanilloid subtype 1 channel in a vanilloid-independent mechanism. J Neurosci 25(39):8924–8937

Xu H, Delling M, Jun JC, Clapham DE (2006) Oregano, thyme and clove-derived flavors and skin sensitizers activate specific TRP channels. Nat Neurosci 9(5):628–635

Yang XR, Lin MJ, McIntosh LS, Sham JS (2006) Functional expression of transient receptor potential melastatin- and vanilloid-related channels in pulmonary arterial and aortic smooth muscle. Am J Physiol Lung Cell Mol Physiol 290(6):L1267–L1276

Yin J, Kuebler WM (2008) Mechanotransduction by TRP channels: general concepts and specific role in the vasculature. Cell Biochem Biophys 56(1):1–18

Zanou N, Iwata Y, Schakman O, Lebacq J, Wakabayashi S, Gailly P (2009) Essential role of TRPV2 ion channel in the sensitivity of dystrophic muscle to eccentric contractions. FEBS Lett 583(22):3600–3604

Zhang L, Jones S, Brody K, Costa M, Brookes SJ (2004) Thermosensitive transient receptor potential channels in vagal afferent neurons of the mouse. Am J Physiol Gastrointest Liver Physiol 286(6):G983–G991

Zhang X, Huang J, McNaughton PA (2005) NGF rapidly increases membrane expression of TRPV1 heat-gated ion channels. EMBO J 24(24):4211–4223. doi:10.1038/sj.emboj.7600893

Zhuang L, Peng JB, Tou L, Takanaga H, Adam RM, Hediger MA, Freeman MR (2002) Calcium-selective ion channel, CaT1, is apically localized in gastrointestinal tract epithelia and is aberrantly expressed in human malignancies. Lab Invest 82(12):1755–1764

Zimon M, Baets J, Auer-Grumbach M, Berciano J, Garcia A, Lopez-Laso E, Merlini L, Hilton-Jones D, McEntagart M, Crosby AH, Barisic N, Boltshauser E, Shaw CE, Landoure G, Ludlow CL, Gaudet R, Houlden H, Reilly MM, Fischbeck KH, Sumner CJ, Timmerman V, Jordanova A, Jonghe PD (2010) Dominant mutations in the cation channel gene transient receptor potential vanilloid 4 cause an unusual spectrum of neuropathies. Brain 133(Pt 6):1798–1809

Zygmunt PM, Petersson J, Andersson DA, Chuang H, Sorgard M, Di Marzo V, Julius D, Hogestatt ED (1999) Vanilloid receptors on sensory nerves mediate the vasodilator action of anandamide. Nature 400(6743):452–457

[illegible] calcium [illegible] Ca2+ [illegible]

[illegible]

[illegible]

[illegible]

Part III
Pathologies of Ligand-Gated Calcium Channels

Chapter 28
IP_3 Receptors in Neurodegenerative Disorders: Spinocerebellar Ataxias and Huntington's and Alzheimer's Diseases

Masayoshi Tada, Masatoyo Nishizawa and Osamu Onodera

Abstract Modulation of intracellular calcium concentration is a ubiquitous signaling system involved in numerous biological processes in diverse cell types. Alterations of intracellular calcium homeostasis have been implicated in age-related neurodegenerative diseases, including Alzheimer's disease, Parkinson's disease, amyotrophic lateral sclerosis, Huntington's disease, and spinocerebellar ataxias (SCAs). Inositol 1,4,5-trisphosphate (IP_3) receptors (IP_3Rs), calcium release channels in the ER membrane, play a key role in regulating intracellular calcium concentration. IP_3R type 1 (IP_3R1), a major neuronal type of IR_3R, is expressed ubiquitously and is involved in diverse biological processes. Cerebellar Purkinje cells are mainly affected by alterations in IP_3R1. Heterozygous deletion or missense mutations in *ITPR1,* the *IP3R1* gene, result in autosomal dominantly inherited ataxias, including SCA type 15 or 29. In addition, mutations in carbonic anhydrase-related protein VIII, which suppresses the binding ability of IP_3 to IP_3R1, cause recessively, inherited ataxia. These results indicate that IP_3R1-mediated calcium signaling has an important role in maintaining the function of Purkinje cells. Moreover, cytosolic calcium overload with excessive IP_3R1 activity has been implicated in pathogenesis of other neurodegenerative diseases, including SCA type 2, SCA type 3, Huntington's disease, and Alzheimer's disease, where dysregulation of IP_3R1-mediated calcium signaling may link to the pathogenesis.

Abbreviations (alphabetical)

AD	Alzheimer's disease
$A\beta$	Amyloid β

M. Tada · M. Nishizawa
Department of Neurology, Brain Research Institute, Niigata University, Niigata, Japan
e-mail: tadamasa@bri.niigata-u.ac.jp

O. Onodera (✉)
Department of Molecular Neuroscience, Center for Bioresources,
Brain Research Institute, Niigata University, Niigata, Japan
e-mail: onodera@bri.niigata-u.ac.jp

N. Weiss and A. Koschak (eds.), *Pathologies of Calcium Channels,*
DOI: 10.1007/978-3-642-40282-1_28,

AMPA	α-amino-3-hydroxy-5-methyl-4-isoxazolepropionic acid
ATXN2	Ataxin-2
ATXN3	Ataxin-3
CAMRQ3	Cerebellar ataxia and mental retardation with or without quadrupedal locomotion 3
CARP	Carbonic anhydrase-related protein VIII
ER	Endoplasmic reticulum
HAP1	Htt-associated protein1
HD	Huntington's disease
Htt	Huntingtin
mHtt	Mutant huntingtin
IP_3	Inositol 1,4,5-trisphosphate
IP_3R	IP_3 receptor
IP_3R1	Inositol 1,4,5-trisphosphate receptor type 1
LTD	Long-term depression
mGluR	Metabotropic glutamate receptors
MSN	Medium spiny neuron
nAChR	Nicotinic acetylcholine receptor
NCX	Sodium-calcium exchanger
NMDA	*N*-methyl-D-aspartic acid
PMCA	Plasma membrane calcium ATPase
PS	Presenilin
RyR	Ryanodine receptor
SCA	Spinocerebellar ataxia
SERCA	Sarco-/endoplasmic reticulum calcium ATPase
SUMF1	Sulfatase modifying factor 1
VGCC	Voltage-gated calcium channel

28.1 Introduction

Modulation of cytoplasmic free calcium (Ca^{2+}) concentration is a universal intracellular signaling system involved in numerous biological processes, including learning and memory, membrane transport, cell excitability, synaptic transmission, axonal transport, cell division, apoptosis, and cell development, in diverse cell types (Foskett et al. 2007; Bezprozvanny 2010; Finch et al. 2012; Stutzmann and Mattson 2011; Goto and Mikoshiba 2011). Inositol 1,4,5-trisphosphate (IP_3) receptors (IP_3Rs) form a group of Ca^{2+} release channels localized in the endoplasmic reticulum (ER) membrane (Foskett et al. 2007; Bezprozvanny 2005). They function to release Ca^{2+} from ER, the major Ca^{2+} storage organelle, into the cytoplasm in response to IP_3, an intracellular second messenger, which is generated through hydrolysis of phosphatidyl-inositol 4,5-bisphosphate, a component of the plasma membrane, by phospholipases $C\beta$ and $C\gamma$. This hydrolysis is triggered by the ligand binding to G protein-coupled or tyrosine phosphorylation-coupled receptors followed by the sequential activation of the phospholipases (Fig. 28.1a).

IP_3Rs are ubiquitously expressed in all cell types. Three IP_3R isoforms, IP_3R type 1 (IP_3R1), IP_3R type 2 (IP_3R2), and IP_3R type 3 (IP_3R3), are expressed in mammals including humans (Furuichi et al. 1994; Mikoshiba et al. 1993; Foskett et al. 2007; Taylor et al. 1999; Taylor et al. 2004; Stutzmann and Mattson 2011; Goto and Mikoshiba 2011). The three IP_3R isoforms are 60–70 % identical in sequence (Furuichi et al. 1994; Michikawa et al. 1996). Most tissues express more than one and often all three IP_3R isoforms at different ratios (Taylor et al. 1999; Foskett et al. 2007), and the expression level of each IP_3R isoform can be regulated according to cellular states.

IP_3R1 is the major neuronal form of IP_3R family in the central nervous system (Yamada et al. 1994) and is abundant in the cerebellum, particularly in cerebellar Purkinje cells. It is also expressed in other brain areas, including the cerebral cortex, hippocampus, basal ganglia, and thalamus, as well as in peripheral tissues (Furuichi et al. 1994; Foskett et al. 2007; Nakanishi et al. 1991; Sharp et al. 1999). IP_3R1 is 2,758 residues in length and forms a homotetramer. The primary structure of IP_3R1 consists of three domains, including an IP_3-binding domain near the N terminus, a coupling/regulatory domain in the middle of the molecule, and a transmembrane-spanning domain near the C terminus (Fig. 28.2). In addition, there is at least two consensus protein kinase A phosphorylation sites and at least one consensus ATP-binding site (Nucifora et al. 1995; Foskett et al. 2007). See also Chap. 11 for the details of IP_3R.

Because of the ubiquitous expression of IP_3Rs and their roles in diverse biological processes, it is likely that IP_3R can be implicated in a number of disease conditions. In this chapter, we will discuss the neurological disorders, spinocerebellar ataxia type 15 (SCA15) and 29, caused by alterations in the *IP3R* gene. In addition, we will highlight other neurological disorders, including some SCAs, Huntington's disease (HD), and Alzheimer's disease (AD), where alterations in IP_3R-mediated Ca^{2+} signaling may link to their pathogenesis.

28.2 Dominantly Inherited SCAs Caused by Alteration in *IP3R*

SCA15 (MIM 606658) is an autosomal dominant neurodegenerative disorder characterized by very slowly progressive, pure cerebellar ataxia (Storey et al. 2001; Gardner et al. 2005). A family of Australian origin with SCA15 was the first report on this condition, and the locus was mapped to 3pter-p24.2 (Knight et al. 2003). Then, in two Japanese families with benign SCA, Hara et al. narrowed the

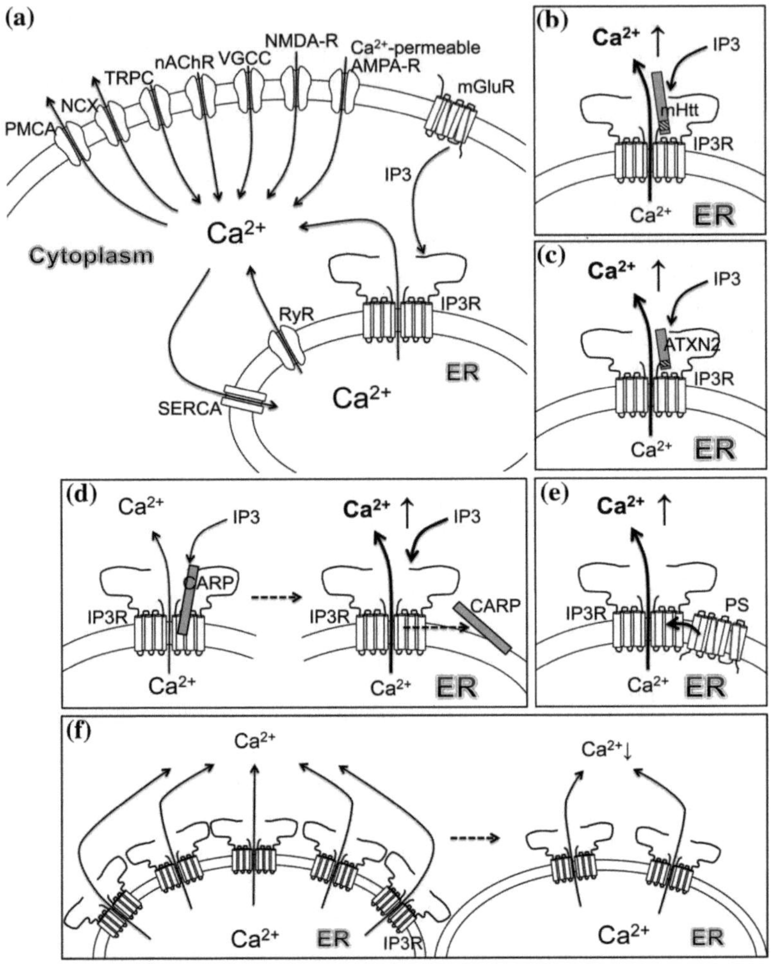

Fig. 28.1 A model of deranged inositol 1,4,5-trisphosphate receptor (IP_3R)-mediated Ca^{2+} signaling in Huntington's disease, spinocerebellar ataxia type 2 (SCA2), type 3 (SCA3), type 15 (SCA15), and type 29 (SCA29), cerebellar ataxia and mental retardation with or without quadrupedal locomotion 3 (CAMRQ3), and Alzheimer's disease. **a** Sources of Ca^{2+} influx are Ca^{2+}-permeable α-amino-3-hydroxy-5-methyl-4-isoxazolepropionic acid (AMPA) and *N*-methyl-D-aspartate (NMDA) glutamate receptors, voltage-gated Ca^{2+} channels (VGCCs), nicotinic acetylcholine receptors (nAChR), and transient receptor potential type C (TRPC) channels. Ca^{2+} release from internal stores is mediated by inositol triphosphate receptors (IP_3R) and ryanodine receptors (RyR). Inositol triphosphate (IP_3) can be generated by metabotropic glutamate receptors (mGluR). Ca^{2+} efflux is mediated by the sodium–calcium exchanger (NCX), the plasma membrane calcium ATPase (PMCA), and the sarco-/endoplasmic reticulum calcium ATPase (SERCA). **b** In HD, mutant huntingtin (mHtt) binds to the C-terminal region of IP_3R and enhances its affinity to IP_3. **c** In AD, presenilins (PSs) can directly increase the activity of IP_3R. ER = endoplasmic reticulum. **d** In healthy individuals, the carbonic anhydrase-related protein VIII (CARP) binds to the modulatory domain of IP_3R and suppresses its affinity to IP_3. In SCA29 or CAMRQ3, CARP cannot bind to IP_3R, resulting in increased affinity of IP_3R to IP_3. **e** In SCA2 or SCA3, mutant ataxin-2 (ATXN2) or ataxin-3 (ATXN3) also binds to the C-terminal region of IP_3R and enhances its affinity to IP_3. **f** In SCA15, the reduced IP_3R levels results in dysregulation of IP_3R-mediated Ca^{2+} signaling

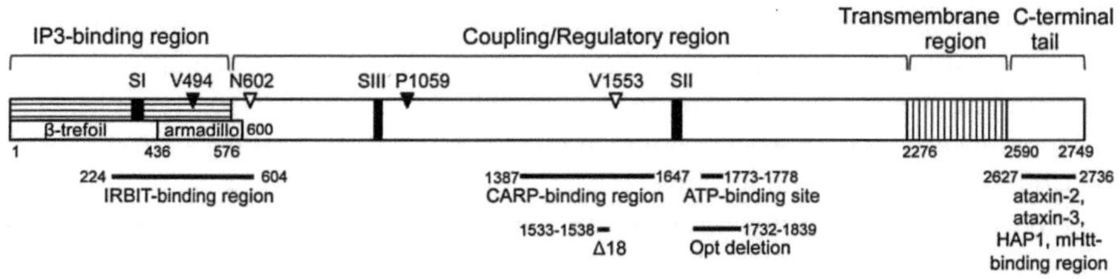

Fig. 28.2 Domain structure of inositol 1,4,5-trisphosphate receptor type 1 (IP$_3$R1). IP$_3$R1 consists of three major domains, including the amino-terminal IP$_3$-binding region, coupling/regulatory region, and transmembrane region. The structural features shown are as follows: alternative splicing sites SI (318–332), SII (1692–1731), and SIII (917/918). β-trefoil domain, armadillo repeat domain, IRBIT-binding region, carbonic anhydrase-related protein VIII (CARP)-binding region, Opt deletion, Δ18 deletion, ATP-binding site, and ataxin-1-, ataxin-2-, HAP1- and mHtt-binding regions

region to 3p26.1-p25.3 (Hara et al. 2004). In a large four-generation Japanese family, initial studies showed a linkage to chromosome 8, and the condition was formally designated SCA16. However, additional studies revealed a linkage to 3pter-p26.2 (Miura et al. 2006). In 2007, heterozygous large deletions (200–400 kb) in genes encoding IP$_3$R1 and sulfatase modifying factor 1 (SUMF1) were identified in affected members of the Australian and two other British families (van de Leemput et al. 2007). In addition, a heterozygous large deletion only in *ITPR1*, the *IP3R1* gene, as well as a point mutation in *ITPR1* was identified in the Japanese families. These have been reported to link to the same locus, indicating that *ITPR1* is the causative gene for SCA in humans (Hara et al. 2008; Iwaki et al. 2008).

28.2.1 Clinical Features of SCA15

SCA15 is clinically characterized by autosomal dominant inheritance, very slow progression, and pure cerebellar ataxia. Age at onset varies between 7 and 66 years (usually between 30 and 50) (Storey et al. 2001; van de Leemput et al. 2007; Hara et al. 2008; Iwaki et al. 2008). The disease usually begins with gait ataxia. Tremor may begin simultaneously with or even occasionally precede gait ataxia. Deterioration in handwriting, motion-induced instability, and myoclonus were also the initial symptoms in some individuals. Cerebellar signs and symptoms including truncal and limb ataxia, ataxic speech, and gaze-evoked nystagmus are core features in combination with head tremor (titubation), upper limb postural tremor, action tremor, and impaired oculocephalic reflex. Hyperreflexia, but neither Babinski reflex nor spasticity, may be noted as a pyramidal sign. Fatal complications such as severe bulbar palsy do not develop. Cognitive function seems to be intact. There have been no reports describing epilepsy in individuals affected with SCA15.

Brain magnetic resonance imaging (MRI) reveals marked atrophy of the cerebellar vermis with mild atrophy of the cerebellar hemispheres (Hara et al. 2004; Knight et al. 2003; van de Leemput et al. 2007; Novak et al. 2010; Synofzik et al. 2011). Nerve conduction studies are typically normal, but mild slowing of conduction velocities of sural sensory and median motor nerves were shown in affected members of a Japanese family with SCA15 (Hara et al. 2008). Disease progression is notably slow. Most patients with SCA15 can ambulate independently or with a cane 10–40 years after onset (Storey et al. 2001; van de Leemput et al. 2007; Hara et al. 2008; Iwaki et al. 2008). Neuropathological findings are not available in SCA15.

28.2.2 SCA15 Diagnosis

SCA15 is defined by the presence of a pathogenic mutation in *ITPR1*. SCA15 diagnosis should be considered in individuals who exhibit the clinical features of SCA15 and in whom the diagnosis of SCA1, 2, 3, 5, 6, 8, 12, and 14 have been excluded by genetic testing. Most patients with SCA15 are diagnosed by gene dosage analysis for *ITPR1*. Because most *ITPR1* mutations are exonic deletions, genetic testing should begin with gene dosage analysis followed by sequence analysis if a deletion is not identified.

28.2.3 SCA15 Prevalence

In the Australian population, pathogenic *ITPR1* deletions were found in approximately 2.7 % of families with autosomal dominant SCA who were negative for common SCA repeat expansions in coding exons (Ganesamoorthy et al. 2009). In the Caucasian population, an *ITPR1* deletion was found in 1.8 % of 333 families (Marelli et al. 2011). On the other hand, *ITPR1* deletions were found in 8.9 % of 56 central European families negative for common SCA repeat expansions (Synofzik et al. 2011). The precise prevalence of SCA15 is however still obscure, because most previous studies used quantitative PCR for genetic testing of SCA15 and this method cannot detect small deletions, insertions, or nonsense mutations in *ITPR1*. Hara et al. analyzed *ITPR1* deletions using custom high-definition comparative genomic hybridization microarrays covering the entirety of *ITPR1* at an average interval of 200 bp for the probes in 54 Japanese families with undetermined autosomal dominant SCA and did not find *ITPR1* deletions, indicating that SCA15 is a quite rare ataxia in the Japanese population (Hara et al. 2008).

28.2.4 Molecular Genetics of SCA15

28.2.4.1 *ITPR1* Deletions

ITPR1 consists of 58 exons. Heterozygous deletions encompassing exons 1–10, 1–40, and 1–44 of *ITPR1* were identified in three unrelated Australian and British families with SCA15 (van de Leemput et al. 2007), and a heterozygous deletion of entire exons in the gene was found in another Japanese family (Hara et al. 2008). Most patients with SCA15 also have deletions in the adjacent *SUMF1* (van de Leemput et al. 2007; Hara et al. 2008; Ganesamoorthy et al. 2009; Novak et al. 2010; Di Gregorio et al. 2010; Castrioto et al. 2011). Although individuals with homozygous *SUMF1* deletions show mental retardation, seizure, and leukodystrophy, individuals with heterozygous *SUMF1* deletions are healthy (Cosma et al. 2003). These findings suggest that partial *SUMF1* deletion does not contribute to SCA15 pathogenesis. In addition, a heterozygous deletion of exons 1–48 in *ITPR1*, but not in *SUMF1*, was identified in a Japanese family (Iwaki et al. 2008), indicating that the pathogenic mechanism underlying SCA15 is *ITPR1* haploinsufficiency. Although it is expected that micro deletions, insertions, or nonsense mutations cause SCA15, these mutations have not yet been fully identified and evaluated.

28.2.4.2 Missense Mutations in *ITPR1*

Two heterozygous missense mutations, P1059L (c.8581C > T) and V494I (c.1480G > A), have been also identified in Japanese and Australian families with SCA15, respectively (Hara et al. 2008; Ganesamoorthy et al. 2009). Proline at position 1059 in the amino acid sequence is highly conserved in IP_3R1 among species, although it is not the same in human ITPR2 and ITPR3. This residue is located in the coupling domain (Fig. 28.2), whose function remains poorly understood. Valine at position 494 is located in the IP_3-binding domain consisting of β-trefoil and armadillo repeat domains (Ganesamoorthy et al. 2009; Foskett et al. 2007). The proper coordination of both domains is necessary for the binding of IP_3 to IP_3R1. Valine at position 494 is not particularly conserved among species. Although it seems likely that the missense mutations may affect IP_3R1 function, the Ca^{2+} release properties of IP_3R1 with P1059I mutation is largely unaffected (Yamazaki et al. 2011). To show that these missense mutations contribute to disease pathogenesis, it should be clarified how these mutations affect the functional properties or the kinetics of biogenesis and turnover of IP_3R1.

A recent study demonstrated that two other heterozygous missense mutations in *ITPR1*, V1553 M (c.4657G > A) and N602D (c.1804A > G), caused another neurological disease in families with autosomal dominantly inherited congenital nonprogressive ataxia, designated as SCA29 (OMIM 117360) (Huang et al. 2012). Mild and very slow progressive ataxia observed in SCA29 is similar to that in

SCA15. However, these families exhibited several characteristic clinical features, including delayed motor milestones (suggesting the existence of congenital ataxia) and mild intellectual impairment (Dudding et al. 2004, Huang et al. 2012). In addition, the severity of ataxia and intellectual impairment was variable in each affected member even in the same family. Valine at position 1553 is located in the carbonic anhydrase-related protein VIII (CARP)-binding region, and asparagine at position 602 is in the IRBIT-binding region, respectively. Both IRBIT and CARP compete with IP_3 for binding to IP_3R1 and suppress IP_3R1 activity (Hirota et al. 2003; Ando et al. 2006). Interestingly, homozygous mutations in *CA8*, which encodes CARP, cause an autosomal recessive congenital ataxia associated with mild intellectual impairment (Turkmen et al. 2009). Therefore, the two missense mutations in the families might increase the sensitivity of IP_3R1 in response to IP_3 and therefore be a cause of the disease in these families.

28.2.5 Roles of IP_3R1 in SCA15 Pathogenesis

Western blot analysis of IP_3R1 protein levels in immortalized lymphoblasts from affected individuals carrying *ITPR1* deletions revealed remarkable reduction in IP_3R1 protein levels (van de Leemput et al. 2007; Novak et al. 2010). In addition, RT-PCR analysis showed that the mRNA expression levels of *ITPR1* in fibroblasts obtained from an affected individual with SCA15 were half of the levels measured in normal controls (Hara et al. 2008). These findings suggest that SCA15 is caused by *ITPR1* haploinsufficiency and that cerebellar Purkinje cells are particularly vulnerable to the dosage of *ITPR1* (Fig. 28.1b).

How does IP_3R1 haploinsufficiency cause cerebellar ataxia in patients with SCA15? IP_3R1, the major neuronal IP_3R, is expressed ubiquitously in various regions of the central nervous system including CA1, basal ganglia, and the thalamus and particularly in the cerebellar Purkinje cells (Nakanishi et al. 1991; Sharp et al. 1999). Intracellular Ca^{2+} homeostasis is important for maintaining the function of neurons, particularly Purkinje cells (Hartmann and Konnerth 2005; Mikoshiba 2007). As described later, mice homozygous for null *ITPR1* develop ataxia and epilepsy (Matsumoto and Nagata 1999; Matsumoto et al. 1996), whereas mice heterozygous for null *ITPR1* develop only mild motor discoordination (Ogura et al. 2001). Thus, *ITPR1* haploinsufficiency may result in dysfunction restricted to the cerebellar Purkinje cells, whereas complete loss of IP_3R1 results in more severe dysfunction of not only Purkinje cells but also cortical neurons.

Indeed, none of the individuals with SCA15 with heterozygous *ITPR1* deletions had epilepsy or abnormal electroencephalogram, and the clinical phenotype was restricted to pure cerebellar ataxia even in the elderly (Gardner et al. 2005; Hara et al. 2004; Knight et al. 2003; van de Leemput et al. 2007). These findings indicate that Purkinje cells are particularly vulnerable to abnormalities in IP_3R1. The reduced IP_3R1 levels may cause dysregulation of intracellular Ca^{2+} homeostasis, leading to persistent long-standing dysfunction of Purkinje cells and eventually degeneration of the selective neuronal populations. The neuropathological findings of affected individuals with deletion or missense mutations in *ITPR1* will confirm this speculation.

28.3 IP_3R Mutant/Deficient Mice

28.3.1 ITPR1 Knockout Mice

Homozygous IP_3R1 knockout mice, in which cytosolic IP_3-induced Ca^{2+} release is almost completely deficient, are rarely born alive, indicating that IP_3R1 has some function during embryonic development. Even if they survive, the mice exhibit severe ataxia and tonic or tonic–clonic seizure and die by 3–4 weeks after birth (Matsumoto et al. 1996; Matsumoto and Nagata 1999). Cultured Purkinje cells from the mice shows abnormal dendritic development and enlarged parallel fiber terminals with many vesicles (Hisatsune et al. 2006). IP_3R1 in granule cells, not in Purkinje cells, is crucial for the outgrowth of the Purkinje cell dendrites. Brain-derived neurotrophic factor (BDNF) production in cerebellar granule cells induced by IP_3R1-mediated signaling, modifies the parallel fiber-Purkinje cell synaptic efficacy, resulting in the formation of Purkinje cell dendrites (Hisatsune et al. 2006).

In neurophysiological analysis, long-term depression (LTD) is completely diminished in cerebellar Purkinje cells (Inoue et al. 1998). However, the effect of IP_3R1 on synaptic plasticity in the hippocampus is a little complicated. A classical form of LTD induced by sustained low-frequency stimulation is not affected at the CA3-CA1 synapses (Fujii S et al. 2000), whereas it is diminished at mossy fiber-CA3 synapses (Itoh et al. 2001). In addition, although paring stimulations of pre- and postsynaptic sites in a post - > pre order induces homo- and heterosynaptic LTD at the normal hippocampal CA3-CA1 synapses, the homosynaptic LTD is converted to long-term potentiation (LTP) and heterosynaptic LTD is disappeared (Nishiyama et al. 2000; Nagase et al. 2003). LTP induced by the short tetanus (100 Hz, 10 pulses) is enhanced in CA1 synapses. Moreover, IP_3R1 is indispensable to the induction of depotentiation and suppression of LTP (Fujii et al. 2000). These results indicated that IP_3R1-mediated Ca^{2+} signaling plays an important role for the regulation of synaptic plasticity in different ways in each situation.

28.3.2 Opisthotonus (Opt) Mouse

The *Opt* mouse has a spontaneously generated allele of deletion of exons 43 and 44 in *ITPR1* that results in an in-frame deletion of residues 1732–1839 (107 amino acids) in the regulatory domain (Street et al. 1997). Homozygous *Opt* mice are small at birth, lack their normal mobility, exhibit seizures 2 weeks after birth, and die by 4 weeks of age. The phenotype of *Opt* mice largely overlaps that of homozygous *ITPR1* knockout mice. The recombinant *Opt* IP_3R1 mice showed reduction in ATP sensitivity compared with wild-type IP_3R1 mice, consistent with the fact that *Opt* deletion involves the ATP-binding site, yet the recombinant *Opt* IP_3R1 remains functional (Tu et al. 2002). A strong Ca^{2+} release from intracellular stores was elicited in the cerebellar Purkinje cells of homozygous *Opt* mice treated with the mGluR agonist quisqualate (Street et al. 1997). Nevertheless, *Opt* IP_3R1 expression levels in the brain tissues of heterozygous *Opt* mice were reduced compared with those of the wild-type mice, and *Opt* IP_3R1 was almost undetectable in the homozygous *Opt* mice (Street et al. 1997; Foskett et al. 2010). Although mechanisms underlying the reduced IP_3R1 levels remain to be elucidated, it is presumed that cellular protein quality control mechanisms may recognize *Opt* IP_3R1 as aberrant and degrade it promptly. Further studies are needed to clarify the precise mechanisms underlying neurological deficits in *Opt* mice.

28.3.3 The Δ18 Mouse

The Δ18 mouse has a spontaneously generated deletion of 18 nucleotides in exon 36 of *ITPR1* that results in an in-frame deletion of six amino acid residues (residues 1533–1538; Glu-Ser-Cys-Ile-Arg-Val) in the regulatory domain (van de Leemput et al. 2007). The homozygous Δ18 mice show severe neurological symptoms, small weight at birth, abnormal mobility, and die by 4 weeks of age. Their phenotype is similar to those of *ITPR1* knockout and *Opt* mice (van de Leemput J et al. 2007; Street et al. 1997; Matsumoto et al. 1996). The functional significance of the six deleted residues, which are not particularly conserved among isoforms and species, remains to be examined. As observed in *Opt* mice, immunostaining of the cerebellar Purkinje cells and western blotting of the whole brain lysates revealed that IP_3R1 expression levels were markedly reduced in the homozygous Δ18 mice (van de Leemput et al. 2007). It is interesting that the recombinant expression of the mutant IP_3R1 proteins produces functional ion channels, including *Opt*, *Δ18*, and P1059L, but appears to cause disease because of reduced IP_3R1 levels, perhaps due to rapid degradation by cellular quality control mechanisms. Thus, it will be interesting, in future studies, to investigate the effects of these mutations not only on ion channel properties but also on the kinetics of channel biogenesis and turnover.

28.4 CARP and Ataxias

Despite the abundant expression of IP_3R1 in Purkinje cells, IP_3R1-mediated Ca^{2+} release in response to IP_3 in these cells is lower than that in other tissues. CARP may, in part, account for this mechanism. CARP binds to the modulatory domain of IP_3R1 (residues 1387–1647) and suppresses the binding ability of IP_3 to IP_3R1 (Hirota et al. 2003). In addition, CARP is expressed exclusively in the Purkinje cells. These results suggest that CARP regulates IP_3R1-mediated Ca^{2+} signaling particularly in the Purkinje cells. Therefore, it would be speculated that the loss of function of CARP results in an enhanced sensitivity of IP_3 binding to IP_3R1, consequently leading to dysregulation of IP_3R1-mediated Ca^{2+} signaling in the Purkinje cells (Fig. 28.1c).

Interestingly, CARP was identified as the antigen of auto-antibody observed in a patient with paraneoplastic cerebellar degeneration (Bataller et al. 2004). Moreover, Turkmen et al. identified a homozygous missense mutation (S100P, c.298T > G) in *CA8*, which encodes CARP, in affected patients with recessively inherited ataxia (Turkmen et al. 2009). They exhibited mild mental retardation and congenital ataxia with quadrupedal gait. Another homozygous missense mutation, G162R (c.484G > A), was identified in three related Arabian families with ataxia and mild cognitive impairment without quadrupedal gait (Kaya et al. 2011). A whole brain MRI showed varying degrees of cerebellar atrophy. Fluorodeoxy-glucose positron emission tomography revealed hypometabolic cerebellar hemispheres, temporal lobes, and mesial cortex. These families are designated as cerebellar ataxia and mental retardation with or without quadrupedal locomotion 3 (CAMRQ3, MIM 613227). The reduced levels of S100P CARP in cell culture experiments suggest that the loss of function of CARP caused ataxia. Indeed, in CAMRQ3, waddles (*wdl*) mice, harboring a spontaneously occurring 19-base pair deletion in *CA8*, exhibited ataxia and appendicular dystonia without pathological abnormalities in the central nervous system (Jiao et al. 2005).

28.5 Deranged IP_3R-Mediated Ca^{2+} Signaling in Ataxias Caused by Expanded Polyglutamine (polyQ) Stretches

SCA type 2 (SCA2) and type 3 (SCA3), polyQ diseases, are autosomal dominantly inherited ataxias caused by the expansion of CAG repeats that encode abnormally expanded polyQ in the ataxin-2 (ATXN2) and ataxin-3 (ATXN3) proteins, respectively (Zoghbi and Orr 2000; Williams and Paulson 2008; La Spada and Taylor 2010; Costa Mdo and Paulson 2012). The diseases are progressive in nature and generally feature degeneration of the cerebellum, brainstem, and spinocerebellar tracts. Mutant polyQ proteins including mutant ATXN2 and ATXN3 are prone to undergo a conformational change that favors β sheet-rich structures and to aggregate in cells, leading to the formation of neuronal inclusion bodies, a prominent pathological hallmark of polyQ diseases (Muchowski and Wacker 2005; Williams and Paulson 2008; Nagai et al. 2007; Paulson et al. 1997). PolyQ expansions usually act in a dominant toxic manner associated with altered interactions with other proteins, resulting in altered cellular processes such as perturbed proteostasis, transcriptional dysregulation, oxidative stress, impaired neurotransmission, insufficient trophic support, and aberrant cellular excitability (Williams and Paulson 2008; Costa Mdo and Paulson 2012; Takahashi et al. 2010). Increasing evidence suggests that deranged neuronal Ca^{2+} signaling plays a role in the pathogenesis of polyQ diseases (Chen et al. 2008; Liu et al. 2009). Cerebellar Purkinje cells seem to be particularly vulnerable to fluxes in cytosolic Ca^{2+} levels. Several neuronal genes abundantly expressed in Purkinje cells that are involved in Ca^{2+} signaling or homeostasis are downregulated in the cerebellum of SCA1 mutant mice before the occurrence of motor deficits or pathology (Serra et al. 2004; Lin et al. 2000). Unlike SCA15, other SCAs are affected by exaggerated Ca^{2+} flux but not by suppressed cytosolic Ca^{2+} signaling. Among these SCAs, to date, only ATXN2 and ATXN3 have been reported to directly affect IP_3R1 function.

28.5.1 SCA2

SCA2 is clinically characterized by progressive cerebellar ataxia of gait, limbs, and speech associated with slow saccades, early hyporeflexia, severe tremor of postural or action type, peripheral neuropathy, cognitive disorders, and other multisystemic features (Lastres-Becker et al. 2008; Magana et al. 2012). Cerebellar Purkinje cells are predominantly affected in SCA2. The disease-causing protein ATXN2 is expressed ubiquitously. Increasing evidence suggests that ATXN2 is involved in multiple cellular processes including RNA post-transcriptional and translational regulation, stress-granule formation, endocytosis, cytoskeletal reorganization, and Ca^{2+}-mediated signaling (Albrecht et al. 2004; van de Loo et al. 2009; Neuwald and Koonin 1998; Satterfield and Pallanck 2006; Lastres-Becker et al. 2008; Ralser et al. 2005a; Ralser et al. 2005b; Shibata et al. 2000; Kozlov et al. 2010; Ciosk et al. 2004; Satterfield et al. 2002; Liu et al. 2009), although the precise physiological function of ATXN2 is unknown (Pulst et al. 2005). The mechanisms underlying Purkinje cell degeneration in SCA2 are also poorly understood.

The presence of ATXN2 in ER suggests its participation in intracellular Ca^{2+} signaling pathways. Supporting this hypothesis, pull-down and co-immunoprecipitation assays revealed that mutant, but not wild-type, ATXN2 (58Q) specifically binds to the cytosolic C-terminal region (residues 2627–2749) of IP_3R1 (Liu et al. 2009). Association of mutant ATXN2 (58Q) with the receptor increases the sensitivity of IP_3R1 to activation by IP_3 in bilayer reconstitution experiments (Fig. 28.1d). In Ca^{2+} imaging experiments, a significant increase in Ca^{2+} release from ER through IP_3R1 was observed in primary Purkinje cells cultured from SCA2 transgenic mice (58Q), which express human *ATXN2* with 58 CAG repeats under the control of the Purkinje cell-specific promoter (Huynh et al. 2000). Ryanodine or dantrolene, inhibitors of ryanodine receptors (RyR), alleviated the adverse effects of mutant ATXN2 such as excessive Ca^{2+} release and glutamate-induced cell death in 58Q Purkinje cell cultures (Liu et al. 2009). In addition, long-term feeding of SCA2 mice (58Q) with dantrolene ameliorated age-dependent motor discoordination and loss of Purkinje cells. More recently, long-term suppression of IP_3R1-mediated Ca^{2+} signaling by viral expression of the inositol 1,4,5-phosphatase enzyme in the Purkinje cells of SCA2 transgenic mice (58Q) rescued age-dependent dysfunction in the firing pattern of SCA2 Purkinje cells and motor deficits and cell death in SCA2 mice (Kasumu et al. 2012). These findings support the idea that excitotoxic Ca^{2+} signaling through IP_3R1 plays a key role in SCA2 pathogenesis.

28.5.2 SCA3

SCA type 3 (SCA3), also known as Machado–Joseph disease, is the most common inherited SCA and one of the nine known polyQ diseases (Costa Mdo and Paulson 2012; Tsuji et al. 2008; Paulson 2012). SCA3 is clinically characterized by progressive cerebellar ataxia and variable findings including a dystonic-rigid syndrome, a Parkinsonian syndrome, or a combined syndrome of dystonia and peripheral neuropathy. The most affected brain regions are the dentate and pontine nuclei, internal portion of globus pallidus, substantia nigra, subthalamic nucleus, and spinocerebellar tracts (Stevanin et al. 2000; Yamada et al. 2008; Yamada et al. 2000). The cerebellar cortex is relatively spared in SCA3 compared with other SCAs. The disease-causing protein ATXN3 is ubiquitously expressed and abundant in cerebellar Purkinje cells. ATXN3 is a 43-kDa cytosolic protein containing the amino-terminal Josephin domain and three ubiquitin-interacting motifs and functions as a deubiquitinating enzyme (Costa Mdo and Paulson 2012). Similar to other SCAs, the precise mechanisms of SCA3 remain poorly understood.

In SCA3, deranged Ca^{2+} signaling has also been implicated in pathogenesis (Bezprozvanny 2011). Inhibition of Ca^{2+}-dependent protease calpain suppressed aggregation of mutant ATXN3 in transfected cells (Haacke et al. 2007). In a SCA3 fly model, knockdown of expression of *PICK1*, which is a regulator of traffic of ion channels involved in Ca^{2+} homeostasis (Chung et al. 2000; Hanley 2006; McGurk and Bonini 2012), suppressed external eye degeneration, insoluble aggregations, and inclusions. Mutant, but not wild-type, ATXN3 specifically binds to the cytosolic C-terminal region of IP_3R1 (Chen et al. 2008), as cases in mutant huntingtin (mHtt) and mutant ATXN2 (Fig. 28.1d and e). Association of mutant ATXN3 with the receptor increases the sensitivity of IP_3R1 to activation by IP_3 in bilayer reconstitution and Ca^{2+} imaging experiments. In addition, long-term feeding of SCA3-YAC-84Q transgenic mice with dantrolene ameliorated gae-dependent motor deficits and prevented neuronal cell loss in the pontine nuclei and substantia nigra regions. These findings indicate that deranged IP_3R1-mediated Ca^{2+} signaling may play an important role in SCA3 pathogenesis.

28.6 Deranged IP_3R-Mediated Ca^{2+} Signaling in HD

HD is a dominantly inherited neurodegenerative disorder caused by polyQ expansions in Htt, which primarily results in the selective degeneration of the striatal medium spiny neurons (MSNs) (Bonelli and Beal 2012). The disease is clinically characterized by movement disorders, cognitive decline, and psychiatric symptoms. A pathological hallmark of HD is cytoplasmic and nuclear aggregates containing htt and other proteins. Although the physiological function of wild-type htt is unknown, its amino acid sequence indicates that it possesses HEAT repeats, protein interaction domains, suggesting that it may function as a scaffold protein (Bonelli and Beal 2012). Several lines of evidence indicate that a toxic gain of function of mHtt accounts for HD pathogenesis, although the molecular mechanisms that underlie this pathogenesis and selective neurodegeneration remain unknown.

Deranged Ca^{2+} signaling has also been implicated in HD pathogenesis (Bezprozvanny 2011; Bezprozvanny 2009). mHtt binds specifically to the C-terminal region of IP_3R1 (Tang et al. 2003; Tang et al. 2005). A comprehensive high-throughput screening confirmed binding of mHtt to IP_3R1 (Kaltenbach et al. 2007). Mutant, but not wild-type, Htt sensitizes IP_3R1 activation by IP_3 in planar lipid bilayer experiments and facilitates IP_3R1-mediated intracellular Ca^{2+} release in rat striatal MSNs (Tang et al. 2003) (Fig. 28.1e). The effect of mHtt on IP_3R1 is facilitated when mHtt is associated with Htt-associated protein 1 (HAP1), which has also been shown to interact with IP_3R1 (Tang et al. 2004), suggesting that HAP1 plays an important role in functional interactions between Htt and IP_3R1. Specific inhibitors of IP_3R1, 2-aminoethoxydiphenyl borate and enoxaparin, provided protection in the same model (Tang et al. 2005). Genetic knockdown and chemical inhibition of IP_3R1 also reduced mHtt aggregation in cultured cells (Bauer et al. 2011). Expression of the GFP-fused C-terminal fragment of IP_3R1 in MSNs from HD transgenic mice stabilized exaggerated Ca^{2+} signaling and protected HD MSNs from glutamate excitotoxicity (Tang et al. 2009). Infection of adeno-associated viruses expressing the recombinant IP_3R1 C-terminal fragment in the striatum ameliorated motor deficits and loss of MSNs in a HD mouse model (Tang et al. 2009). In addition, long-term feeding of HD mice with dantrolene, a relevant Ca^{2+} signaling stabilizer, alleviated motor deficits, formation of nuclear inclusion bodies, and loss of MSNs (Chen et al. 2011). Thus, deranged IP_3R1-mediated Ca^{2+} signaling also plays an important role in HD pathogenesis.

ER stress has been implicated in the pathogenesis of numerous neurodegenerative diseases including HD. It was demonstrated that ER stress induced IP_3R1 dysfunction through an impaired interaction of IP_3R1 with an ER chaperone GRP78, which positively regulates IP_3R1 tetrameric assembly in an energy dependent manner (Higo et al. 2010). Stabilizing Ca^{2+} signaling by targeting IP_3R1 appears as an attractive therapeutic strategy for HD.

28.7 Deranged IP_3R-Mediated Ca^{2+} Signaling in AD

AD is the most common form of age-related dementia, clinically characterized by a decline in memory, particularly in short-term and working memory, apathy, depression, impaired judgment, and changes in behavior (Forman et al. 2004; Brookmeyer et al. 2007). The key pathological hallmarks of AD are accumulation of extracellular amyloid β (Aβ) plaques, intracellular neurofibrillary tangles, and neuronal loss accompanied by extensive neurodegeneration of the median temporal lobe, parietal lobe, selective regions of the frontal cortex, and cingulate gyrus (Wenk 2003; Giannakopoulos et al. 2009; Forman et al. 2004). Mutations in presenilins (PS1 and PS2) and amyloid precursor protein (APP) cause most early-onset, autosomal dominant familial cases of AD (Tanzi and Bertram 2005).

Numerous lines of evidence indicate that altered Ca^{2+} signaling also plays an important role in AD pathogenesis. Aβ oligomers can form Ca^{2+}-permeable channels in neuron plasma membranes (Arispe et al. 1993; Lee et al. 2002; Kuchibhotla et al. 2008). Aβ oligomers also perturb neuronal Ca^{2+} homeostasis through modulation of the activities of *N*-methyl-D-aspartic acid receptors (De Felice et al. 2007; Shankar et al. 2007), AMPA receptors (Hsieh et al. 2006), and P/Q-type voltage-gated Ca^{2+} channels (Nimmrich et al. 2008). Another key connection between Ca^{2+} signaling and AD pathogenesis is based on studies demonstrating that mutations in *PSs* found in familial AD cause dysregulation of Ca^{2+} signaling (Ito et al. 1994; Leissring et al. 1999; Stutzmann et al. 2004; Stutzmann et al. 2006; Stutzmann 2005; Yoo et al. 2000; LaFerla 2002). Despite some differences in the proposed mechanisms, most studies have shown that various PS mutations result in exaggerated Ca^{2+} release from ER through IP_3R1 or RyR (Leissring et al. 1999; Cai et al. 2006; Cheung et al. 2008; Stutzmann et al. 2006; Chan et al. 2000; Rybalchenko et al. 2008; Chakroborty et al. 2009; Smith et al. 2005; Berridge 2010; Supnet and Bezprozvanny 2011) (Fig. 28.1f). Exaggerated Ca^{2+} signaling in AD may negatively affect reactive oxygen species generation, mitochondrial function, gene transcription, and Aβ production. Aged neurons are particularly vulnerable to cytosolic Ca^{2+} overload because of their lower capacity of buffering Ca^{2+} (reviewed in Supnet and Bezprozvanny 2011; Berridge 2010; Hermes et al. 2010).

28.8 Future Perspectives

As stated above, increasing evidence indicates that deranged IP_3R1-mediated Ca^{2+} signaling has been implicated in neurological diseases including AD, HD, and SCAs. Despite many advances in understanding disease mechanisms, no preventive treatment exists for these fatal neurological disorders. In SCA15, point mutations as well as large deletion mutations in *ITPR1* cause diseases because of reduced IP_3R1 levels, perhaps due to rapid degradation by cellular quality control mechanisms. On the other hand, increasing evidence supports the idea that exaggerated Ca^{2+} influx through IP_3R1 plays an important role in pathogenesis of other neurological diseases, such as AD, HD, and some SCAs. Neurons abundantly expressing IP_3R1 are vulnerable to alterations of intracellular Ca^{2+} homeostasis, particularly exaggerated Ca^{2+} signaling. Understanding the molecular mechanisms underlying neurodegeneration caused by reduced IP_3R1 levels or exaggerated IP_3R1-mediated Ca^{2+} signaling will provide new insights into disease pathogenesis and eventually the development of new therapeutic approaches. Modulation of Ca^{2+} signaling by targeting IP_3R1 appears as an attractive therapeutic strategy for these neurological disorders.

Acknowledgments This study was supported in part by a Grant-in-Aid for Scientific Research on Priority Areas (C) from the Ministry of Education, Culture, Sports, Science and Technology, Japan, and a Grant-in-Aid for the Research Committee for Ataxic Diseases from the Ministry of Health, Labor and Welfare, Japan.

References

Albrecht M, Golatta M, Wullner U, Lengauer T (2004) Structural and functional analysis of ataxin-2 and ataxin-3. Eur J Biochem 271:3155–3170

Ando H, Mizutani A, Kiefer H, Tsuzurugi D, Michikawa T, Mikoshiba K (2006) IRBIT suppresses IP3 receptor activity by competing with IP3 for the common binding site on the IP3 receptor. Mol Cell 22:795–806

Arispe N, Rojas E, Pollard HB (1993) Alzheimer disease amyloid beta protein forms calcium channels in bilayer membranes: blockade by tromethamine and aluminum. Proc Natl Acad Sci U S A 90:567–571

Bataller L, Sabater L, Saiz A, Serra C, Claramonte B, Graus F (2004) Carbonic anhydrase-related protein VIII: autoantigen in paraneoplastic cerebellar degeneration. Ann Neurol 56:575–579

Bauer PO, Hudec R, Ozaki S, Okuno M, Ebisui E, Mikoshiba K, Nukina N (2011) Genetic ablation and chemical inhibition of IP3R1 reduce mutant huntingtin aggregation. Biochem Biophys Res Commun 416:13–17

Berridge MJ (2010) Calcium hypothesis of Alzheimer's disease. Pflugers Arch 459:441–449

Bezprozvanny I (2005) The inositol 1,4,5-trisphosphate receptors. Cell Calcium 38:261–272

Bezprozvanny I (2009) Calcium signaling and neurodegenerative diseases. Trends Mol Med 15:89–100

Bezprozvanny I (2011) Role of inositol 1,4,5-trisphosphate receptors in pathogenesis of Huntington's disease and spinocerebellar ataxias. Neurochem Res 36:1186–1197

Bezprozvanny IB (2010) Calcium signaling and neurodegeneration. Acta Naturae 2:72–82

Bonelli RM, Beal MF (2012) Huntington's disease. Handb Clin Neurol 106:507–526

Brookmeyer R, Johnson E, Ziegler-Graham K, Arrighi HM (2007) Forecasting the global burden of Alzheimer's disease. Alzheimers Dement 3:186–191

Cai C, Lin P, Cheung KH et al (2006) The presenilin-2 loop peptide perturbs intracellular Ca^{2+} homeostasis and accelerates apoptosis. J Biol Chem 281:16649–16655

Castrioto A, Prontera P, Di Gregorio E et al (2011) A novel spinocerebellar ataxia type 15 family with involuntary movements and cognitive decline. Eur J Neurol 18:1263–1265

Chakroborty S, Goussakov I, Miller MB, Stutzmann GE (2009) Deviant ryanodine receptor-mediated calcium release resets synaptic homeostasis in presymptomatic 3xTg-AD mice. J Neurosci 29:9458–9470

Chan SL, Mayne M, Holden CP, Geiger JD, Mattson MP (2000) Presenilin-1 mutations increase levels of ryanodine receptors and calcium release in PC12 cells and cortical neurons. J Biol Chem 275:18195–18200

Chen X, Tang TS, Tu H, Nelson O, Pook M, Hammer R, Nukina N, Bezprozvanny I (2008) Deranged calcium signaling and neurodegeneration in spinocerebellar ataxia type 3. J Neurosci 28:12713–12724

Chen X, Wu J, Lvovskaya S, Herndon E, Supnet C, Bezprozvanny I (2011) Dantrolene is neuroprotective in Huntington's disease transgenic mouse model. Mol Neurodegener 6:81

Cheung KH, Shineman D, Muller M et al (2008) Mechanism of Ca^{2+} disruption in Alzheimer's disease by presenilin regulation of InsP3 receptor channel gating. Neuron 58:871–883

Chung HJ, Xia J, Scannevin RH, Zhang X, Huganir RL (2000) Phosphorylation of the AMPA receptor subunit GluR2 differentially regulates its interaction with PDZ domain-containing proteins. J Neurosci 20:7258–7267

Ciosk R, DePalma M, Priess JR (2004) ATX-2, the C. elegans ortholog of ataxin 2, functions in translational regulation in the germline. Development 131:4831–4841

Cosma MP, Pepe S, Annunziata I, Newbold RF, Grompe M, Parenti G, Ballabio A (2003) The multiple sulfatase deficiency gene encodes an essential and limiting factor for the activity of sulfatases. Cell 113:445–456

Costa Mdo C, Paulson HL (2012) Toward understanding Machado-Joseph disease. Prog Neurobiol 97:239–257

De Felice FG, Velasco PT, Lambert MP, Viola K, Fernandez SJ, Ferreira ST, Klein WL (2007) Abeta oligomers induce neuronal oxidative stress through an N-methyl-D-aspartate receptor-dependent mechanism that is blocked by the Alzheimer drug memantine. J Biol Chem 282:11590–11601

Di Gregorio E, Orsi L, Godani M et al (2010) Two Italian families with IP3R1 gene deletion presenting a broader phenotype of SCA15. Cerebellum 9:115–123

Dudding TE, Friend K, Schofield PW, Lee S, Wilkinson IA, Richards RI (2004) Autosomal dominant congenital non-progressive ataxia overlaps with the SCA15 locus. Neurology 63:2288–2292

Finch EA, Tanaka K, Augustine GJ (2012) Calcium as a trigger for cerebellar long-term synaptic depression. Cerebellum 11:706–717

Forman MS, Trojanowski JQ, Lee VM (2004) Neurodegenerative diseases: a decade of discoveries paves the way for therapeutic breakthroughs. Nat Med 10:1055–1063

Foskett JK (2010) Inositol trisphosphate receptor Ca^{2+} release channels in neurological diseases. Pflugers Arch 460:481–494

Foskett JK, White C, Cheung KH, Mak DO (2007) Inositol trisphosphate receptor Ca^{2+} release channels. Physiol Rev 87:593–658

Fujii S, Matsumoto M, Igarashi K, Kato H, Mikoshiba K (2000) Synaptic plasticity in hippocampal CA1 neurons of mice lacking type 1 inositol-1,4,5-trisphosphate receptors. Learn Mem 7:312–320

Furuichi T, Kohda K, Miyawaki A, Mikoshiba K (1994) Intracellular channels. Curr Opin Neurobiol 4:294–303

Ganesamoorthy D, Bruno DL, Schoumans J et al (2009) Development of a multiplex ligation-dependent probe amplification assay for diagnosis and estimation of the frequency of spinocerebellar ataxia type 15. Clin Chem 55:1415–1418

Gardner RJ, Knight MA, Hara K, Tsuji S, Forrest SM, Storey E (2005) Spinocerebellar ataxia type 15. Cerebellum 4:47–50

Giannakopoulos P, Kovari E, Gold G, von Gunten A, Hof PR, Bouras C (2009) Pathological substrates of cognitive decline in Alzheimer's disease. Front Neurol Neurosci 24:20–29

Goto J, Mikoshiba K (2011) Inositol 1,4,5-trisphosphate receptor-mediated calcium release in Purkinje cells: from molecular mechanism to behavior. Cerebellum 10:820–833

Haacke A, Hartl FU, Breuer P (2007) Calpain inhibition is sufficient to suppress aggregation of polyglutamine-expanded ataxin-3. J Biol Chem 282:18851–18856

Hanley JG (2006) Molecular mechanisms for regulation of AMPAR trafficking by PICK1. Biochem Soc Trans 34:931–935

Hara K, Fukushima T, Suzuki T et al (2004) Japanese SCA families with an unusual phenotype linked to a locus overlapping with SCA15 locus. Neurology 62:648–651

Hara K, Shiga A, Nozaki H et al (2008) Total deletion and a missense mutation of IP3R1 in Japanese SCA15 families. Neurology 71:547–551

Hartmann J, Konnerth A (2005) Determinants of postsynaptic Ca^{2+} signaling in Purkinje neurons. Cell Calcium 37:459–466

Hermes M, Eichhoff G, Garaschuk O (2010) Intracellular calcium signalling in Alzheimer's disease. J Cell Mol Med 14:30–41

Higo T, Hamada K, Hisatsune C, Nukina N, Hashikawa T, Hattori M, Nakamura T, Mikoshiba K (2010) Mechanism of ER stress-induced brain damage by IP(3) receptor. Neuron 68:865–878

Hirota J, Ando H, Hamada K, Mikoshiba K (2003) Carbonic anhydrase-related protein is a novel binding protein for inositol 1,4,5-trisphosphate receptor type 1. Biochem J 372:435–441

Hisatsune C, Kuroda Y, Akagi T, Torashima T, Hirai H, Hashikawa T, Inoue T, Mikoshiba K (2006) Inositol 1,4,5-trisphosphate receptor type 1 in granule cells, not in Purkinje cells, regulates the dendritic morphology of Purkinje cells through brain-derived neurotrophic factor production. J Neurosci 26:10916–10924

Hsieh H, Boehm J, Sato C, Iwatsubo T, Tomita T, Sisodia S, Malinow R (2006) AMPAR removal underlies Abeta-induced synaptic depression and dendritic spine loss. Neuron 52:831–843

Huang L, Warman J, Carter MT, Friend KL, Dudding TE, Schwartzentruber J, Zou R, Schofield PW, Douglas S, Bulman DE, Boycott KM (2012) Missense mutations in IP3R1 cause autosomal dominant congenital nonprogressive spinocerebellar ataxia. Orphanet J Rare Dis 7:67

Huynh DP, Figueroa K, Hoang N, Pulst SM (2000) Nuclear localization or inclusion body formation of ataxin-2 are not necessary for SCA2 pathogenesis in mouse or human. Nat Genet 26:44–50

Inoue T, Kato K, Kohda K, Mikoshiba K (1998) Type 1 inositol 1,4,5-trisphosphate receptor is required for induction of long-term depression in cerebellar Purkinje neurons. J Neurosci 18:5366–5373

Ito E, Oka K, Etcheberrigaray R, Nelson TJ, McPhie DL, Tofel-Grehl B, Gibson GE, Alkon DL (1994) Internal Ca^{2+} mobilization is altered in fibroblasts from patients with Alzheimer disease. Proc Natl Acad Sci U S A 91:534–538

Itoh S, Ito K, Fujii S, Kaneko K, Kato K, Mikoshiba K, Kato H (2001) Neuronal plasticity in hippocampal mossy fiber-CA3 synapses of mice lacking the inositol-1,4,5-trisphosphate type 1 receptor. Brain Res 901:237–246

Iwaki A, Kawano Y, Miura S et al (2008) Heterozygous deletion of IP3R1, but not SUMF1, in spinocerebellar ataxia type 16. J Med Genet 45:32–35

Jiao Y, Yan J, Zhao Y, Donahue LR, Beamer WG, Li X, Roe BA, Ledoux MS, Gu W (2005) Carbonic anhydrase-related protein VIII deficiency is associated with a distinctive lifelong gait disorder in waddles mice. Genetics 171:1239–1246

Kaltenbach LS, Romero E, Becklin RR et al (2007) Huntingtin interacting proteins are genetic modifiers of neurodegeneration. PLoS Genet 3:e82

Kasumu AW, Liang X, Egorova P, Vorontsova D, Bezprozvanny I (2012) Chronic suppression of inositol 1,4,5-triphosphate receptor-mediated calcium signaling in cerebellar purkinje cells alleviates pathological phenotype in spinocerebellar ataxia 2 mice. J Neurosci 32:12786–12796

Kaya N, Aldhalaan H, Al-Younes B et al (2011) Phenotypical spectrum of cerebellar ataxia associated with a novel mutation in the CA8 gene, encoding carbonic anhydrase (CA) VIII. Am J Med Genet B Neuropsychiatr Genet 156B:826–834

Knight MA, Kennerson ML, Anney RJ, Matsuura T, Nicholson GA, Salimi-Tari P, Gardner RJ, Storey E, Forrest SM (2003) Spinocerebellar ataxia type 15 (sca15) maps to 3p24.2-3pter: exclusion of the IP3R1 gene, the human orthologue of an ataxic mouse mutant. Neurobiol Dis 13:147–157

Kozlov G, Safaee N, Rosenauer A, Gehring K (2010) Structural basis of binding of P-body-associated proteins GW182 and ataxin-2 by the Mlle domain of poly(A)-binding protein. J Biol Chem 285:13599–13606

Kuchibhotla KV, Goldman ST, Lattarulo CR, Wu HY, Hyman BT, Bacskai BJ (2008) Abeta plaques lead to aberrant regulation of calcium homeostasis in vivo resulting in structural and functional disruption of neuronal networks. Neuron 59:214–225

La Spada AR, Taylor JP (2010) Repeat expansion disease: progress and puzzles in disease pathogenesis. Nat Rev Genet 11:247–258

LaFerla FM (2002) Calcium dyshomeostasis and intracellular signalling in Alzheimer's disease. Nat Rev Neurosci 3:862–872

Lastres-Becker I, Rub U, Auburger G (2008) Spinocerebellar ataxia 2 (SCA2). Cerebellum 7:115–124

Lee G, Pollard HB, Arispe N (2002) Annexin 5 and apolipoprotein E2 protect against Alzheimer's amyloid-beta-peptide cytotoxicity by competitive inhibition at a common phosphatidylserine interaction site. Peptides 23:1249–1263

Leissring MA, Paul BA, Parker I, Cotman CW, LaFerla FM (1999) Alzheimer's presenilin-1 mutation potentiates inositol 1,4,5-trisphosphate-mediated calcium signaling in Xenopus oocytes. J Neurochem 72:1061–1068

Lin X, Antalffy B, Kang D, Orr HT, Zoghbi HY (2000) Polyglutamine expansion down-regulates specific neuronal genes before pathologic changes in SCA1. Nat Neurosci 3:157–163

Liu J, Tang TS, Tu H, Nelson O, Herndon E, Huynh DP, Pulst SM, Bezprozvanny I (2009) Deranged calcium signaling and neurodegeneration in spinocerebellar ataxia type 2. J Neurosci 29:9148–9162

Magana JJ, Velazquez-Perez L, Cisneros B (2012) Spinocerebellar ataxia type 2: clinical presentation, molecular mechanisms, and therapeutic perspectives. Mol Neurobiol 47(1):90–104

Marelli C, van de Leemput J, Johnson JO et al (2011) SCA15 due to large IP3R1 deletions in a cohort of 333 white families with dominant ataxia. Arch Neurol 68:637–643

Matsumoto M, Nagata E (1999) Type 1 inositol 1,4,5-trisphosphate receptor knock-out mice: their phenotypes and their meaning in neuroscience and clinical practice. J Mol Med (Berl) 77:406–411

Matsumoto M, Nakagawa T, Inoue T et al (1996) Ataxia and epileptic seizures in mice lacking type 1 inositol 1,4,5-trisphosphate receptor. Nature 379:168–171

McGurk L, Bonini NM (2012) Protein interacting with C kinase (PICK1) is a suppressor of spinocerebellar ataxia 3-associated neurodegeneration in Drosophila. Hum Mol Genet 21:76–84

Michikawa T, Miyawaki A, Furuichi T, Mikoshiba K (1996) Inositol 1,4,5-trisphosphate receptors and calcium signaling. Crit Rev Neurobiol 10:39–55

Mikoshiba K (2007) IP3 receptor/Ca^{2+} channel: from discovery to new signaling concepts. J Neurochem 102:1426–1446

Mikoshiba K, Furuichi T, Miyawaki A et al (1993) Structure and function of inositol 1,4,5-trisphosphate receptor. Ann N Y Acad Sci 707:178–197

Miura S, Shibata H, Furuya H et al (2006) The contactin 4 gene locus at 3p26 is a candidate gene of SCA16. Neurology 67:1236–1241

Muchowski PJ, Wacker JL (2005) Modulation of neurodegeneration by molecular chaperones. Nat Rev Neurosci 6:11–22

Nagai Y, Inui T, Popiel HA, Fujikake N, Hasegawa K, Urade Y, Goto Y, Naiki H, Toda T (2007) A toxic monomeric conformer of the polyglutamine protein. Nat Struct Mol Biol 14:332–340

Nagase T, Ito KI, Kato K, Kaneko K, Kohda K, Matsumoto M, Hoshino A, Inoue T, Fujii S, Kato H, Mikoshiba K (2003) Long-term potentiation and long-term depression in hippocampal CA1 neurons of mice lacking the IP(3) type 1 receptor. Neuroscience 117:821–830

Nakanishi S, Maeda N, Mikoshiba K (1991) Immunohistochemical localization of an inositol 1,4,5-trisphosphate receptor, P400, in neural tissue: studies in developing and adult mouse brain. J Neurosci 11:2075–2086

Neuwald AF, Koonin EV (1998) Ataxin-2, global regulators of bacterial gene expression, and spliceosomal snRNP proteins share a conserved domain. J Mol Med (Berl) 76:3–5

Nimmrich V, Grimm C, Draguhn A et al (2008) Amyloid beta oligomers (A beta(1-42) globulomer) suppress spontaneous synaptic activity by inhibition of P/Q-type calcium currents. J Neurosci 28:788–797

Nishiyama M, Hong K, Mikoshiba K, Poo MM, Kato K (2000) Calcium stores regulate the polarity and input specificity of synaptic modification. Nature 408:584–588

Novak MJ, Sweeney MG, Li A et al (2010) An IP3R1 gene deletion causes spinocerebellar ataxia 15/16: a genetic, clinical and radiological description. Mov Disord 25:2176–2182

Nucifora FCJ, Li SH, Danoff S, Ullrich A, Ross CA (1995) Molecular cloning of a cDNA for the human inositol 1,4,5-trisphosphate receptor type 1, and the identification of a third alternatively spliced variant. Brain Res Mol Brain Res 32:291–296

Ogura H, Matsumoto M, Mikoshiba K (2001) Motor discoordination in mutant mice heterozygous for the type 1 inositol 1,4,5-trisphosphate receptor. Behav Brain Res 122:215–219

Paulson H (2012) Machado-Joseph disease/spinocerebellar ataxia type 3. Handb Clin Neurol 103:437–449

Paulson HL, Perez MK, Trottier Y et al (1997) Intranuclear inclusions of expanded polyglutamine protein in spinocerebellar ataxia type 3. Neuron 19:333–344

Pulst SM, Santos N, Wang D, Yang H, Huynh D, Velazquez L, Figueroa KP (2005) Spinocerebellar ataxia type 2: polyQ repeat variation in the CACNA1A calcium channel modifies age of onset. Brain 128:2297–2303

Ralser M, Albrecht M, Nonhoff U, Lengauer T, Lehrach H, Krobitsch S (2005a) An integrative approach to gain insights into the cellular function of human ataxin-2. J Mol Biol 346:203–214

Ralser M, Nonhoff U, Albrecht M, Lengauer T, Wanker EE, Lehrach H, Krobitsch S (2005b) Ataxin-2 and huntingtin interact with endophilin-A complexes to function in plastin-associated pathways. Hum Mol Genet 14:2893–2909

Rybalchenko V, Hwang SY, Rybalchenko N, Koulen P (2008) The cytosolic N-terminus of presenilin-1 potentiates mouse ryanodine receptor single channel activity. Int J Biochem Cell Biol 40:84–97

Satterfield TF, Jackson SM, Pallanck LJ (2002) A Drosophila homolog of the polyglutamine disease gene SCA2 is a dosage-sensitive regulator of actin filament formation. Genetics 162:1687–1702

Satterfield TF, Pallanck LJ (2006) Ataxin-2 and its Drosophila homolog, ATX2, physically assemble with polyribosomes. Hum Mol Genet 15:2523–2532

Serra HG, Byam CE, Lande JD, Tousey SK, Zoghbi HY, Orr HT (2004) Gene profiling links SCA1 pathophysiology to glutamate signaling in Purkinje cells of transgenic mice. Hum Mol Genet 13:2535–2543

Shankar GM, Bloodgood BL, Townsend M, Walsh DM, Selkoe DJ, Sabatini BL (2007) Natural oligomers of the Alzheimer amyloid-beta protein induce reversible synapse loss by modulating an NMDA-type glutamate receptor-dependent signaling pathway. J Neurosci 27:2866–2875

Sharp AH, Nucifora FCJ, Blondel O, Sheppard CA, Zhang C, Snyder SH, Russell JT, Ryugo DK, Ross CA (1999) Differential cellular expression of isoforms of inositol 1,4,5-triphosphate receptors in neurons and glia in brain. J Comp Neurol 406:207–220

Shibata H, Huynh DP, Pulst SM (2000) A novel protein with RNA-binding motifs interacts with ataxin-2. Hum Mol Genet 9:1303–1313

Smith IF, Hitt B, Green KN, Oddo S, LaFerla FM (2005) Enhanced caffeine-induced Ca^{2+} release in the 3xTg-AD mouse model of Alzheimer's disease. J Neurochem 94:1711–1718

Stevanin G, Durr A, Brice A (2000) Clinical and molecular advances in autosomal dominant cerebellar ataxias: from genotype to phenotype and physiopathology. Eur J Hum Genet 8:4–18

Storey E, Gardner RJ, Knight MA, Kennerson ML, Tuck RR, Forrest SM, Nicholson GA (2001) A new autosomal dominant pure cerebellar ataxia. Neurology 57:1913–1915

Street VA, Bosma MM, Demas VP, Regan MR, Lin DD, Robinson LC, Agnew WS, Tempel BL (1997) The type 1 inositol 1,4,5-trisphosphate receptor gene is altered in the opisthotonos mouse. J Neurosci 17:635–645

Stutzmann GE (2005) Calcium dysregulation, IP3 signaling, and Alzheimer's disease. Neuroscientist 11:110–115

Stutzmann GE, Caccamo A, LaFerla FM, Parker I (2004) Dysregulated IP3 signaling in cortical neurons of knock-in mice expressing an Alzheimer's-linked mutation in presenilin1 results in exaggerated Ca^{2+} signals and altered membrane excitability. J Neurosci 24:508–513

Stutzmann GE, Mattson MP (2011) Endoplasmic reticulum $Ca(^{2+})$ handling in excitable cells in health and disease. Pharmacol Rev 63:700–727

Stutzmann GE, Smith I, Caccamo A, Oddo S, Laferla FM, Parker I (2006) Enhanced ryanodine receptor recruitment contributes to Ca^{2+} disruptions in young, adult, and aged Alzheimer's disease mice. J Neurosci 26:5180–5189

Supnet C, Bezprozvanny I (2011) Presenilins as endoplasmic reticulum calcium leak channels and Alzheimer's disease pathogenesis. Sci China Life Sci 54:744–751

Synofzik M, Beetz C, Bauer C et al (2011) Spinocerebellar ataxia type 15: diagnostic assessment, frequency, and phenotypic features. J Med Genet 48:407–412

Takahashi T, Katada S, Onodera O (2010) Polyglutamine diseases: where does toxicity come from? what is toxicity? where are we going? J Mol Cell Biol 2:180–191

Tang TS, Guo C, Wang H, Chen X, Bezprozvanny I (2009) Neuroprotective effects of inositol 1,4,5-trisphosphate receptor C-terminal fragment in a Huntington's disease mouse model. J Neurosci 29:1257–1266

Tang TS, Slow E, Lupu V, Stavrovskaya IG, Sugimori M, Llinas R, Kristal BS, Hayden MR, Bezprozvanny I (2005) Disturbed Ca^{2+} signaling and apoptosis of medium spiny neurons in Huntington's disease. Proc Natl Acad Sci U S A 102:2602–2607

Tang TS, Tu H, Chan EY, Maximov A, Wang Z, Wellington CL, Hayden MR, Bezprozvanny I (2003) Huntingtin and huntingtin-associated protein 1 influence neuronal calcium signaling mediated by inositol-(1,4,5) triphosphate receptor type 1. Neuron 39:227–239

Tang TS, Tu H, Orban PC, Chan EY, Hayden MR, Bezprozvanny I (2004) HAP1 facilitates effects of mutant huntingtin on inositol 1,4,5-trisphosphate-induced Ca release in primary culture of striatal medium spiny neurons. Eur J Neurosci 20:1779–1787

Tanzi RE, Bertram L (2005) Twenty years of the Alzheimer's disease amyloid hypothesis: a genetic perspective. Cell 120:545–555

Taylor CW, da Fonseca PC, Morris EP (2004) IP(3) receptors: the search for structure. Trends Biochem Sci 29:210–219

Taylor CW, Genazzani AA, Morris SA (1999) Expression of inositol trisphosphate receptors. Cell Calcium 26:237–251

Tsuji S, Onodera O, Goto J, Nishizawa M (2008) Sporadic ataxias in Japan–a population-based epidemiological study. Cerebellum 7:189–197

Tu H, Miyakawa T, Wang Z, Glouchankova L, Iino M, Bezprozvanny I (2002) Functional characterization of the type 1 inositol 1,4,5-trisphosphate receptor coupling domain SII($\pm$) splice variants and the Opisthotonos mutant form. Biophys J 82:1995–2004

Turkmen S, Guo G, Garshasbi M et al (2009) CA8 mutations cause a novel syndrome characterized by ataxia and mild mental retardation with predisposition to quadrupedal gait. PLoS Genet 5:e1000487

van de Leemput J, Chandran J, Knight MA et al (2007) Deletion at IP3R1 underlies ataxia in mice and spinocerebellar ataxia 15 in humans. PLoS Genet 3:e108

van de Loo S, Eich F, Nonis D, Auburger G, Nowock J (2009) Ataxin-2 associates with rough endoplasmic reticulum. Exp Neurol 215:110–118

Wenk GL (2003) Neuropathologic changes in Alzheimer's disease. J Clin Psychiatry 64(Suppl 9):7–10

Williams AJ, Paulson HL (2008) Polyglutamine neurodegeneration: protein misfolding revisited. Trends Neurosci 31:521–528

Yamada M, Sato T, Tsuji S, Takahashi H (2008) CAG repeat disorder models and human neuropathology: similarities and differences. Acta Neuropathol 115:71–86

Yamada M, Tsuji S, Takahashi H (2000) Pathology of CAG repeat diseases. Neuropathology 20:319–325

Yamada N, Makino Y, Clark RA et al (1994) Human inositol 1,4,5-trisphosphate type-1 receptor, InsP3R1: structure, function, regulation of expression and chromosomal localization. Biochem J 302:781–790

Yamazaki H, Nozaki H, Onodera O, Michikawa T, Nishizawa M, Mikoshiba K (2011) Functional characterization of the P1059L mutation in the inositol 1,4,5-trisphosphate receptor type 1 identified in a Japanese SCA15 family. Biochem Biophys Res Commun 410:754–758

Yoo AS, Cheng I, Chung S et al (2000) Presenilin-mediated modulation of capacitative calcium entry. Neuron 27:561–572

Zoghbi HY, Orr HT (2000) Glutamine repeats and neurodegeneration. Annu Rev Neurosci 23:217–247

Chapter 29
RyR2 in Cardiac Disorders

Ineke Nederend, Christian van der Werf and Arthur A. M. Wilde

Abstract Catecholaminergic polymorphic ventricular tachycardia (CPVT) is an inherited arrhythmia syndrome characterized by the occurrence of adrenergically induced polymorphic ventricular arrhythmias. Mutations in the cardiac ryanodine receptor (*RYR2*) underlie the majority of CPVT cases and show an autosomal dominant inheritance pattern. Mutations in *RYR2* and other genes involved in CPVT cause spontaneous diastolic calcium release from the sarcoplasmatic reticulum (SR), which eventually lead to triggered arrhythmias. CPVT is usually diagnosed by use of exercise testing. β-blockers are the mainstay of drug therapy in CPVT, whereas flecainide and left cardiac sympathetic denervation can be added or performed in patients with significant ventricular arrhythmias or arrhythmic events on β-blocker therapy.

29.1 Introduction

In cardiac muscle, excitation–contraction coupling is initiated by a small influx of external calcium through the voltage-dependent L-type calcium channels into the cytosol during an action potential (Fabiato 1983). This calcium triggers the opening of the cardiac ryanodine receptor (RyR2), which regulates a large calcium release from intracellular calcium stores to the cytoplasm, ultimately leading to contraction. Mutations in the gene encoding RyR2 (OMIM #180902) have been associated with an autosomal dominant form of catecholaminergic polymorphic ventricular tachycardia (CPVT) (Priori et al. 2001), an inherited arrhythmia syndrome characterized by adrenergically induced ventricular arrhythmias including ventricular fibrillation (VF) (Leenhardt et al. 1995). Ventricular arrhythmias or

I. Nederend · C. van der Werf · A. A. M. Wilde (✉)
Department of Clinical and Experimental Cardiology, Academic Medical Centre, Amsterdam, The Netherlands
e-mail: a.a.wilde@amc.uva.nl

N. Weiss and A. Koschak (eds.), *Pathologies of Calcium Channels*,
DOI: 10.1007/978-3-642-40282-1_29,

symptoms typically occur during physical or emotional stress in patients with CPVT. The exact prevalence of CPVT is unknown but is estimated to be 1 in 10,000.

The first case report on CPVT dates from 1960, describing three sisters out of five siblings who suffered from frequent attacks of multifocal ventricular extra-systoles without any structural cardiac abnormalities (Berg KJ 1960). This case was followed by another case report in 1975 (Reid et al. 1975), and two series published by Coumel and coworkers from Paris (Coumel et al. 1978; Leenhardt et al. 1995). The link between mutations in *RYR2* and CPVT was discovered in 2001 (Laitinen et al. 2003; Priori et al. 2001). In the past decade, much has been learned about *RYR2* mutations and CPVT across the full spectrum from bench to bedside. This knowledge is also relevant, because proarrhythmic mechanisms similar to CPVT are believed to play a role in atrial fibrillation (AF) (Voigt et al. 2012) and ventricular arrhythmias in heart failure (George 2008).

29.2 Catecholaminergic Polymorphic Ventricular Tachycardia

29.2.1 The Cardiac RYR2

There are three isoforms of RyRs: RyR1 is mainly expressed in skeletal muscle, *RYR2* in cardiac muscle, and *RYR3* in the brain. RyR2 is located in the membrane of the (SR) of,cardiomyocytes and is a homotetramer (Fig. 29.1). Each monomer consists of a large cytosolic domain and a smaller transmembrane domain. The channel pore is encompassed by four polypeptides, forming a tetramer. RyR2 is, among others, associated with FKBP12.6, protein kinase A, calcium/calmodulin-dependent kinase II (CaMKII), and calmodulin (CALM) at the N-terminal cyto-plasmic domain (Priori and Chen 2011), and calsequestrin (CASQ2), junctin (ASPH), and triadin (TRDN) at the C-terminus (Faggioni et al. 2012).

29.2.1.1 *RYR2* Mutations

The gene encoding the 4967-amino acid RyR2 channel is located on the long arm of chromosome 1 (1q42-43) and contains 105 exons, making it one of the largest genes in the human genome. Mutations in *RYR2* are identified in approximately 60 % of patients with a definite clinical diagnosis of CPVT (Medeiros-Domingo et al. 2009). To date, over 130 unique disease-causing mutations have been identified in *RYR2* (Medeiros-Domingo et al. 2009). These include almost exclusively missense mutations and approximately 20 % of *RYR2* mutations are de novo (Medeiros-Domingo et al. 2009). Mutations in *RYR2* have a propensity to be clustered in three known hotspots: the N-terminal domain (codons 44–466), the

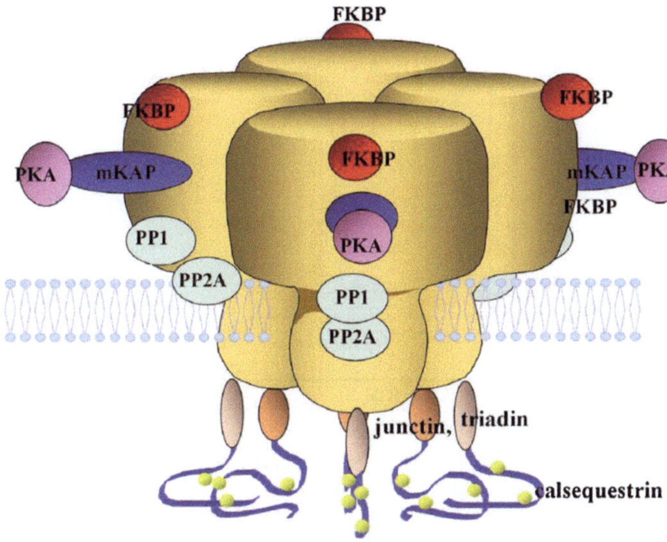

Fig. 29.1 Schematic illustration of the *RYR2* macromolecular complex. Calmodulin (CALM), FKBP12.6 (calstabin 2), protein kinase A (PKA), phosphatase 1 (PP1), and phosphatase 2A (PP2A) bind to the cytoplasmic region of *RYR2*; junctin and triadin anchor calsequestrin to *RYR2* and bind to the intraluminal portion of *RYR2* (Reproduced with permission from Yano et al. 2005. Pharmacol Ther 107(3):377–391)

central domain (codons 2246–2534), and the C-terminal channel forming domain (codons 3778–4959) (Medeiros-Domingo et al. 2009; Priori and Chen 2011).

Mutations in *RYR2* have also been identified with more complex phenotypes than classic CPVT. In two separate families with a large genomic deletion in *RYR2*, involving exon 3, sinoatrial node and atrioventricular node dysfunction, atrial fibrillation, atrial standstill, and left ventricular dysfunction and dilatation were identified in addition to the classic CPVT phenotype (Bhuiyan et al. 2007). In addition, fibrofatty myocardial replacement in the right ventricle and intracellular calcium deposits have been identified in patients carrying a *RYR2* mutation (Tiso et al. 2001).

29.2.2 Pathological Background

29.2.2.1 Excitation–Contraction Coupling

Calcium-induced calcium release (CICR) plays an important role in cardiac excitation–contraction coupling. CICR is initiated by a small influx of external calcium through the voltage-dependent L-type calcium channels into the cytosol during the plateau phase of the action potential (Fabiato 1983). The entered

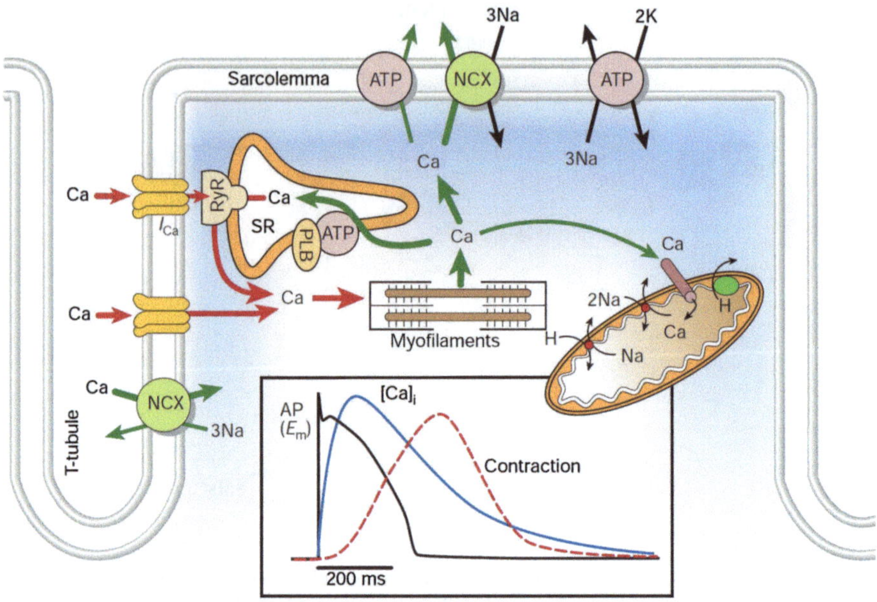

Fig. 29.2 Calcium transport in ventricular myocytes. Inset shows the time course of an action potential, calcium transient, and contraction measured in a rabbit ventricular myocyte. NCX, sodium/calcium exchanger; ATP, ATPase; PLB, phospholamban; SR, sarcoplasmic reticulum. Reproduced with permission from Bers 2002. Nature. 415, (6868):198–205

calcium triggers subsequent calcium release from the SR, the main intracellular calcium storage of the cardiac myocyte, through activation of the cytosolic calcium sensor of RyR2. Calcium then binds to troponin C and eventually results in myocardial contraction. During relaxation, the bulk of the released calcium is transferred back into the SR by sarcoplasmatic reticulum calcium adenosine triphosphate (SERCA). The remainder is removed from the cytosol by the sodium/calcium exchanger (NCX) (Fig. 29.2).

RyR2 activity is regulated by several signaling pathways (Priori and Chen 2011). Under adrenergic stimulation, β-adrenergic receptors stimulate adenylyl cyclase to produce cAMP, which in turn activates protein kinase A (PKA) as well as other mediators. PKA phosphorylates RyR2 and other central proteins related to excitation–contraction coupling, such as phospholamban and the L-type calcium channels. The overall result is CICR gain of function in response to adrenergic activation.

29.2.2.2 Pathophysiology of CPVT Related to Mutations in RYR2

RYR2 mutations result in inappropriate calcium leakage from the SR, which leads to elevated calcium concentrations in the cytoplasm (Fatima et al. 2011; Suetomi et al. 2011). This activates the electrogenic NCX. As NCX exchanges one calcium ion outward for three sodium ions inward, there is a net inward electrical current (named transient inward current; I_{Na}), generating delayed after potentials (DADs). If this depolarization wave amplitude reaches the threshold potential, full depolarization and triggered arrhythmias may occur. Adrenergic stimulation increases spontaneous calcium leak (Cerrone et al. 2007).

Several mechanisms by which *RYR2* mutations lead to aberrant diastolic calcium release have been hypothesized: a reduced binding affinity of the channel-stabilizing protein calstabin 2 (FKBP12.6) (Wehrens et al. 2004), destabilization of the closed state of the channel by mediation of defective interdomain interaction (Suetomi et al. 2011), and store overload induced calcium release, in which calcium accumulates in the SR (Jiang et al. 2004). Several mouse studies suggest that the His-Purkinje system is the critical contributor to, i.e., the origin of ventricular arrhythmias in CPVT (Cerrone et al. 2007; Herron et al. 2010; Kang et al. 2010).

29.2.2.3 Other CPVT Types

CASQ2 is the main calcium buffering protein in the SR. Mutations in *CASQ2* (OMIM #114251) probably cause CPVT by a loss of Ca buffering (Viatchenko-Karpinski et al. 2004). In addition, loss of CASQ2 may lead to a reduced direct inhibitory effect on RyR2. Finally, CASQ2 loss may lead to a reduction of the CASQ2 binding proteins TRDN and ASPH and remodeling of SR structure. Mutations in *TRDN* (OMIM #603283) could cause CPVT by an impaired FKBP12.6-*RYR2* interaction or a reduction of CASQ2 (Roux-Buisson et al. 2012). Finally, mutations in *CALM* may cause CPVT through a dominant-negative effect on the RyR2 channel complex, leading to inappropriate calcium leakage from the SR (Nyegaard et al. 2012).

29.2.3 Clinical Characteristics of CPVT

29.2.3.1 Clinical Presentation

The classic CPVT patient is a child experiencing emotion or exercise-induced syncope, aborted cardiac arrest or sudden cardiac death (SCD) with a family history of similar events in young relatives. In some cases, children are initially diagnosed with epilepsy, because CPVT-related syncope may be accompanied by convulsive movements and urinary or fecal incontinence, and CPVT may be diagnosed during follow-up when symptoms persist despite antiepileptic drug therapy. However, CPVT patients with a more benign course and a debut of symptoms during adulthood are increasingly being identified (Sy et al. 2011). Conversely, *RYR2* mutations have been identified in victims of SIDS, suggesting a wide range of phenotype severity among patients carrying a *RYR2* mutation (Tester et al. 2004).

29.2.3.2 Electrophysiological Characteristics

Resting 12-lead electrocardiograms of CPVT patients are normal, including a normal corrected QT-interval. However, sinus bradycardia (Leenhardt et al. 1995; Postma et al. 2005) and prominent U-waves (Leenhardt et al. 1995; Viitasalo et al. 2008) may be present. In a series containing 116 relatives carrying a mutation in *RYR2*, which were identified by predictive genetic testing of the mutation identified in the proband, sinus bradycardia was observed in 19 % (van der Werf et al. 2012a). In addition, other supraventricular dysrhythmias were present in 16 %, and mainly included intermittent ectopic atrial rhythm identified by Holter monitoring. In one study including eight CPVT patients in whom electrophysiological study was performed, evidence of sinus node dysfunction was demonstrated in four (Sumitomo et al. 2007). Indeed, a recent study found that increased diastolic calcium release through mutant *RYR2* led to a decrease in sino-atrial node automaticity by a reduced L-type calcium channel current and SR calcium depletion during diastole (Neco et al. 2012).

During exercise testing, a characteristic increase in severity of polymorphic ventricular arrhythmias is often observed, starting with isolated ventricular premature beats (VPB), followed by bigeminal VPBs, couplets, and eventually runs of multiple VPBs (Fig. 29.3). In the minority of patients, bidirectional ventricular tachycardia (VT) can be observed: a hallmark of CPVT defined as VT with a beat-to-beat alternating QRS axis (Sy et al. 2011). When exercise testing is ended, ventricular arrhythmias rapidly recede in most patients, and the reverse heart rate–dependent sequence can sometimes be observed during recovery. VPBs usually occur at a heart rate of 110–130 beats per min. and, in the absence of therapeutic interventions, the ventricular arrhythmia threshold heart rate is remarkably reproducible in an individual patient. VPBs with a left bundle branch block morphology and inferior axis and a right bundle block morphology and superior axis are predominant (Sumitomo et al. 2003; Sy et al. 2011). The occurrence of exercise-induced supraventricular tachyarrhyhmias in CPVT patients has been reported, but is not commonly observed (see above) (van der Werf et al. 2012a).

29.2.3.3 Genotype-Phenotype Correlations

Among patients with a mutation in *RYR2*, approximately 50 % display phenotypic features of CPVT at the first cardiological examination (van der Werf et al. 2012a). This number increases to over 60 % when mutation-carriers are repeatedly being examined during follow-up (van der Werf et al. 2012a).

Hitherto, *RYR2* mutation-carriers with a CPVT phenotype might have an increased risk of arrhythmic events as compared with mutation-carriers without a CPVT phenotype (so-called silent mutation-carriers), but this has not convincingly been shown (Hayashi et al. 2012; van der Werf et al. 2012a). An association between the presence of a CPVT phenotype among *RYR2* mutation-carriers and risk of arrhythmic events is, however, expected in future studies with larger

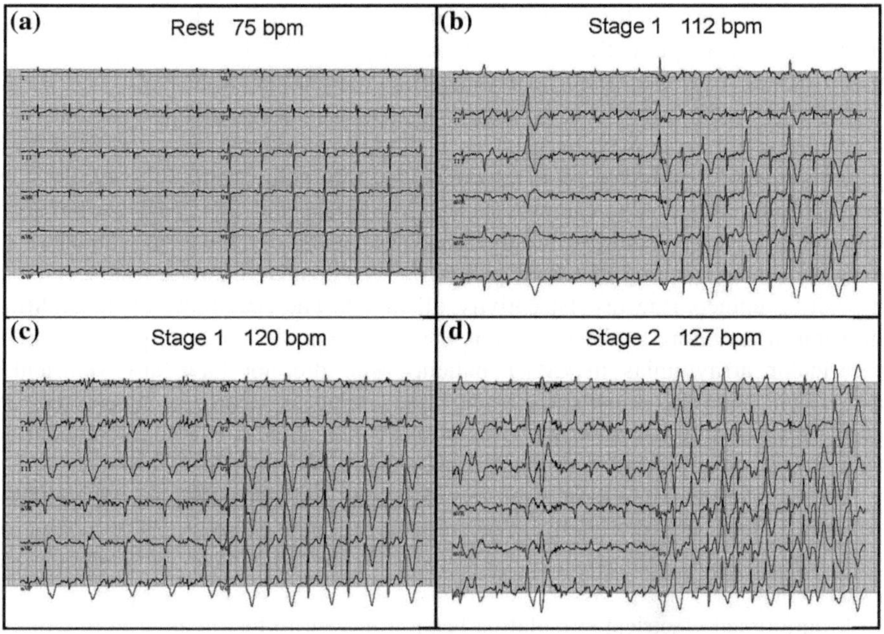

Fig. 29.3 Ventricular arrhythmias in catecholaminergic polymorphic ventricular tachycardia. Polymorphic ventricular arrhythmias during an treadmill exercise test of *RyR2* mutation-associated catecholaminergic polymorphic ventricular tachycardia patient. **a** Normal resting ECG (**b–d**) ECGs during exercise, showing an increasing polymorphic ventricular arrhythmia burden, starting with isolated and bigeminal ventricular premature beats, and ending with bidirectional and polymorphic couplets and non-sustained ventricular tachycardia (Reproduced with permission from van der Werf and Wilde 2013. Heart 99(7):497–504)

numbers of patients. Because arrhythmic events may also occur in silent mutation-carrier, the therapeutic approach to these individuals should nowadays probably be similar to patients with a CPVT phenotype (van der Werf et al. 2012b).

29.2.4 Diagnosis of CPVT

29.2.4.1 Clinical Diagnosis and Differential Diagnosis

The clinical diagnosis of CPVT is made in case of documented exercise- or catecholamine-induced bidirectional of polymorphic VT in the absence of resting ECG abnormalities, structural heart disease, and coronary artery disease, particularly in patients under the age of 40 years. Older patients or patients with polymorphic ventricular arrhythmias, but not VT, who meet all criteria have a possible diagnosis of CPVT.

The gold standard for diagnosis is incremental exercise testing. In a study including 67 asymptomatic relatives of seven probands, the sensitivity and specificity of the exercise testing for predicting carriership of a *RYR2* or *CASQ2* mutation was 50 and 97 %, respectively (Hayashi et al. 2012). Alternatively, epinephrine infusion may be used as a diagnostic tool. A study in 36 CPVT patients and 45 unaffected relatives concluded that epinephrine infusion has low sensitivity as compared to exercise testing (Marjamaa et al. 2012). Maximum heart rate achieved upon epinephrine challenge was markedly lower as compared to exercise testing. Among 25 CPVT patients with a positive exercise test, seven had a positive epinephrine test (sensitivity of 28 %). The specificity of epinephrine infusion in the entire study population was 98 %. On the contrary, provocation of ventricular arrhythmias in CPVT patients who did not have any ventricular arrhythmia on Holter monitoring or exercise testing, have also been reported (Sy et al. 2011). In resuscitated patients who are not able to exercise, epinephrine testing is the test of choice to reach a diagnosis.

Holter or implantable loop recorder monitoring can be useful in young children or other patients who are unable to perform an adequate exercise test or in patients with adrenergically triggered unexplained syncope who have an unremarkable exercise test.

Programmed electrical stimulation has not proven to be effective in provoking arrhythmia in CPVT (Leenhardt et al. 1995). However, recently prominent postpacing changes of the QT-interval in mutation-carriers from one family with the M4109R *RYR2* mutation were reported (Nof et al. 2011).

The differential diagnosis of CPVT includes long QT syndrome (LQTS), in particular type 1, in patients with ventricular arrhythmias of symptoms under conditions of increased sympathetic activity. In case of an inconclusive resting ECG, incremental exercise testing may help in discriminating between these two channelopathies. Ventricular arrhythmia beyond single VPBs is far more common in CPVT compared to LQTS (Horner and Ackerman 2008). Another alternate diagnosis is Andersen-Tawil syndrome, which may very much mimic the CPVT phenotype, including the presence of bidirectional VT (Tristani-Firouzi and Etheridge 2010). Careful inspection of the resting ECG, focused on mild prolongation of the QT-interval or the presence of prominent U-waves, may help in distinguishing between both entities.

29.2.4.2 Genetic Testing

Comprehensive CPVT genetic testing is recommended in index patients in whom a cardiologist has established a clinical index of suspicion for CPVT based on examination of the patient's clinical history, family history, and expressed electrocardiographic phenotype during provocative stress testing with cycle, treadmill, or catecholamine infusion (Ackerman et al. 2011). In addition, mutations in *RYR2* may be regarded as a cause of adrenergically mediated idiopathic VF, which may justify genetic testing in such instances (Ackerman et al. 2011). Genetic testing in

CPVT is particularly important to identify asymptomatic relatives following identification of a CPVT-causative mutation in the index patient. This allows initiating prophylactic treatment in mutation-carriers, as SCD may be the first phenotypic manifestation. In addition, genetic testing may be useful to confirm the diagnosis in patients with a possible clinical diagnosis of CPVT.

In patients with a strong CPVT phenotype in whom *RYR2* genetic testing is negative, genetic testing of the other CPVT-associated genes may be considered: *CALM* (autosomal-dominant inheritance pattern) (Nyegaard et al. 2012), *CASQ2* (autosomal recessive) (Lahat et al. 2001), *TRDN* (autosomal recessive) (Roux-Buisson et al. 2012), and *KCNJ2* (underlying Andersen-Tawil syndrome, autosomal dominant). At present, the yield of mutational analysis of these genes in CPVT is unknown, but seems to be low.

29.2.5 Management of CPVT

Because very little is known on risk stratification in patients with CPVT, the current consensus is to have a similar therapeutic approach to every patient with a CPVT phenotype or to every patient with a pathogenic CPVT-associated mutation (van der Werf et al. 2012b). All patients are advised to avoid competitive sports and strenuous exercise.

As CPVT is induced under conditions of increased sympathetic activity, β-blockers are the cornerstone of therapy. Nadolol is probably the most effective and should therefore be the first choice (Hayashi et al. 2009). In countries where nadolol is not available, propranolol is presumably the preferred choice, although studies comparing different ß-blockers are not available to date. Unfortunately, β-blocker therapy does not completely prevent arrhythmic events. The pooled 8-year overall, near-fatal, and fatal event rates of 354 CPVT patients on β-blocker therapy were 35.9, 14.3, and 6.4 %, respectively (van der Werf et al. 2012b). Importantly, a significant proportion of events is probably due to non-compliance rather than true therapy failure.

The class 1c antiarrhythmic drug flecainide has proven to be effective in reducing ventricular arrhythmias in CVPT mouse models as well as in human CPVT patients, possibly through direct inhibition of RyR2-mediated calcium release as well as to the well-known I_{Na} blocking effect (van der Werf et al. 2011; Watanabe et al. 2009). The precise antiarrhythmic effect of flecainide in this setting is, however, disputed (Liu et al. 2011). In patients who are symptomatic despite β-blocker therapy, flecainide can be effective and is recommended. In addition, flecainide may be used in ventricular arrhythmia storms in CPVT patients (Hong et al. 2012).

Left cardiac sympathetic denervation appears to be an effective therapy for CPVT patients who are not adequately controlled by pharmacological treatment (Collura et al. 2009; Wilde et al. 2008). The anti-arrhythmic effect may not be achieved immediately after the procedure, but it might take up to a couple of months before maximal response is reached (Gopinathannair et al. 2010).

Current guidelines recommend implantable cardioverter-defibrillator (ICD) in patients with aborted cardiac arrest or patients with resistant ventricular arrhythmia despite β-blocker therapy (Zipes et al. 2006). Given the adrenergic nature of arrhythmias in CPVT, catecholamine release as a result of pain or fear following an appropriate or inappropriate discharge of the ICD may be followed by ventricular arrhythmias and subsequent shocks, which may result in an arrhythmic storm. Several cases of this potentially proarrhythmic effect of ICDs in CPVT patients have been reported (Makanjee et al. 2009; Mohamed et al. 2006; Palanca et al. 2006, 2008), turning ICDs into an unattractive option in patients with CPVT. Even if ICD implantation is considered indicated in a CPVT patient, optimal drug therapy and/or LCSD is crucial to reduce the number of appropriate and inappropriate discharges.

29.3 Atrial Fibrillation and Heart Failure

Susceptibility to spontaneous diastolic calcium release from the SR through *RYR2* appears higher in AF and might trigger of maintain AF (Dobrev et al. 2011). This is most probably caused by altered RyR2 function (Vest et al. 2005). In addition, an increase in NCX expression and function has been observed in AF patients (Neef et al. 2010). DADs, caused by a mechanism similar to the mechanism underlying CPVT, may lead to triggered activity that contributes to AF maintenance.

VT in heart failure appears to be due to triggered activity primarily from DADs that arise from altered cellular calcium handling (Pogwizd and Bers 2004). HF is a chronic hyperadrenergic state, and it has been suggested that ß-adrenergic activation of PKA destabilized RyR2, contributing to SR calcium leak and consequent systolic dysfunction and arrhythmogenesis (Marx et al. 2000). In addition, CaMKII, which is upregulated and more active in heart failure, CaMKII-dependent *RYR2* phosphorylation may be a critical mediator of arrhythmias in heart failure (van Oort et al. 2010). New drugs specifically targeting diastolic calcium SR leak are being developed to reduce atrial and ventricular arrhythmogenesis in these settings.

References

Ackerman MJ, Priori SG, Willems S, Berul C, Brugada R, Calkins H, Camm AJ, Ellinor PT, Gollob M, Hamilton R, Hershberger RE, Judge DP, Le MH, McKenna WJ, Schulze-Bahr E, Semsarian C, Towbin JA, Watkins H, Wilde A, Wolpert C, Zipes DP (2011) HRS/EHRA expert consensus statement on the state of genetic testing for the channelopathies and cardiomyopathies: this document was developed as a partnership between the Heart Rhythm Society (HRS) and the European Heart Rhythm Association (EHRA). Europace 13(8):1077–1109

Berg KJ (1960) Multifocal PVCs with Adams-Stokes syndrome in siblings. Am Heart J 60(6):965–970

Bhuiyan ZA, van den Berg MP, van Tintelen JP, Bink-Boelkens MT, Wiesfel C, Alders M, Postma AV, van I, Mannens MM, Wilde AA (2007) Expanding spectrum of human *RYR2*-related disease: new electrocardiographic, structural, and genetic features. *Circulation* 116(14):1569–1576

Cerrone M, Noujaim SF, Tolkacheva EG, Talkachou A, O'Connell R, Berenfeld O, Anumonwo J, Pandit SV, Vikstrom K, Napolitano C, Priori SG, Jalife J (2007) Arrhythmogenic mechanisms in a mouse model of catecholaminergic polymorphic ventricular tachycardia. Circ Res 101(10):1039–1048

Collura CA, Johnson JN, Moir C, Ackerman MJ (2009) Left cardiac sympathetic denervation for the treatment of long QT syndrome and catecholaminergic polymorphic ventricular tachycardia using video-assisted thoracic surgery. Heart Rhythm 6(6):752–759

Coumel P, Fidelle J, Lucet V, Attuel P, Bouvrain Y (1978) Catecholamine-induced severe ventricular arrhythmias with Adam-Stokes in children: report of four cases. Br Heart J 40:28–37

Dobrev D, Voigt N, Wehrens XH (2011) The ryanodine receptor channel as a molecular motif in atrial fibrillation: pathophysiological and therapeutic implications. Cardiovasc Res 89(4):734–743

Fabiato A (1983) Calcium-induced release of calcium from the cardiac sarcoplasmic reticulum. Am J Physiol 245(1):C1–C14

Faggioni M, Kryshtal DO, Knollmann BC (2012) Calsequestrin mutations and catecholaminergic polymorphic ventricular tachycardia. Pediatr, Cardiol

Fatima A, Xu G, Shao K, Papadopoulos S, Lehmann M, Arnaiz-Cot JJ, Rosa AO, Nguemo F, Matzkies M, Dittmann S, Stone SL, Linke M, Zechner U, Beyer V, Hennies HC, Rosenkranz S, Klauke B, Parwani AS, Haverkamp W, Pfitzer G, Farr M, Cleemann L, Morad M, Milting H, Hescheler J, Saric T (2011) In vitro modeling of ryanodine receptor 2 dysfunction using human induced pluripotent stem cells. Cell Physiol Biochem 28(4):579–592

George CH (2008) Sarcoplasmic reticulum Ca2+ leak in heart failure: mere observation or functional relevance? Cardiovasc Res 77(2):302–314

Gopinathannair R, Olshansky B, Iannettoni M, Mazur A (2010) Delayed maximal response to left cardiac sympathectomy for catecholaminergic polymorphic ventricular tachycardia. Europace 12(7):1035–1039

Hayashi M, Denjoy I, Extramiana F, Maltret A, Buisson NR, Lupoglazoff JM, Klug D, Hayashi M, Takatsuki S, Villain E, Kamblock J, Messali A, Guicheney P, Lunardi J, Leenhardt A (2009) Incidence and risk factors of arrhythmic events in catecholaminergic polymorphic ventricular tachycardia. Circulation 119(18):2426–2434

Hayashi M, Denjoy I, Hayashi M, Extramiana F, Maltret A, Roux-Buisson N, Lupoglazoff J M, Klug D, Maury P, Messali A, Guicheney P, Leenhardt A (2012) The role of stress test for predicting genetic mutations and future cardiac events in asymptomatic relatives of catecholaminergic polymorphic ventricular tachycardia probands. Europace

Herron TJ, Milstein ML, Anumonwo J, Priori SG, Jalife J (2010) Purkinje cell calcium dysregulation is the cellular mechanism that underlies catecholaminergic polymorphic ventricular tachycardia. Heart Rhythm 7(8):1122–1128

Hong RA, Rivera KK, Jittirat A, Choi JJ (2012) Flecainide suppresses defibrillator-induced storming in catecholaminergic polymorphic ventricular tachycardia. Pacing Clin, Electrophysiol

Horner JM, Ackerman MJ (2008) Ventricular ectopy during treadmill exercise stress testing in the evaluation of long QT syndrome. Heart Rhythm 5(12):1690–1694

Jiang D, Xiao B, Yang D, Wang R, Choi P, Zhang L, Cheng H, Chen SR (2004) *RyR2* mutations linked to ventricular tachycardia and sudden death reduce the threshold for store-overload-induced Ca2+ release (SOICR). Proc Natl Acad Sci USA 101(35):13062–13067

Kang G, Giovannone SF, Liu N, Liu FY, Zhang J, Priori SG, Fishman GI (2010) Purkinje cells from *RyR2* mutant mice are highly arrhythmogenic but responsive to targeted therapy. Circ Res 107(4):512–519

Lahat H, Pras E, Olender T, Avidan N, Ben-Asher E, Man O, Levy-Nissenbaum E, Khoury A, Lorber A, Goldman B, Lancet D, Eldar M (2001) A missense mutation in a highly conserved region of CASQ2 is associated with autosomal recessive catecholamine-induced polymorphic ventricular tachycardia in Bedouin families from Israel. Am J Hum Genet 69(6):1378–1384

Laitinen PJ, Swan H, Kontula K (2003) Molecular genetics of exercise-induced polymorphic ventricular tachycardia: identification of three novel cardiac ryanodine receptor mutations and two common calsequestrin 2 amino-acid polymorphisms. Eur J Hum Genet 11(11):888–891

Leenhardt A, Lucet V, Denjoy I, Grau F, Ngoc DD, Coumel P (1995) Catecholaminergic polymorphic ventricular tachycardia in children. A 7-year follow-up of 21 patients. Circulation 91(5):1512–1519

Liu N, Denegri M, Ruan Y, Avelino-Cruz JE, Perissi A, Negri S, Napolitano C, Coetzee WA, Boyden PA, Priori SG (2011) Short communication: flecainide exerts an antiarrhythmic effect in a mouse model of catecholaminergic polymorphic ventricular tachycardia by increasing the threshold for triggered activity. Circ Res 109(3):291–295

Makanjee B, Gollob MH, Klein GJ, Krahn AD (2009) Ten-year follow-up of cardiac sympathectomy in a young woman with catecholaminergic polymorphic ventricular tachycardia and an implantable cardioverter defibrillator. J Cardiovasc Electrophysiol 20(10):1167–1169

Marjamaa A, Hiippala A, Arrhenius B, Lahtinen AM, Kontula K, Toivonen L, Happonen JM, Swan H (2012) Intravenous epinephrine infusion test in diagnosis of catecholaminergic polymorphic ventricular tachycardia. J Cardiovasc Electrophysiol 23(2):194–199

Marx SO, Reiken S, Hisamatsu Y, Jayaraman T, Burkhoff D, Rosemblit N, Marks AR (2000) PKA phosphorylation dissociates FKBP12.6 from the calcium release channel (ryanodine receptor): defective regulation in failing hearts. Cell 101(4):365–376

Medeiros-Domingo A, Bhuiyan ZA, Tester DJ, Hofman N, Bikker H, van Tintelen JP, Mannens MM, Wilde AA, Ackerman MJ (2009) The *RYR2*-encoded ryanodine receptor/calcium release channel in patients diagnosed previously with either catecholaminergic polymorphic ventricular tachycardia or genotype negative, exercise-induced long QT syndrome: a comprehensive open reading frame mutational analysis. J Am Coll Cardiol 54(22):2065–2074

Mohamed U, Gollob MH, Gow RM, Krahn AD (2006) Sudden cardiac death despite an implantable cardioverter-defibrillator in a young female with catecholaminergic ventricular tachycardia. Heart Rhythm 3(12):1486–1489

Neco P, Torrente AG, Mesirca P, Zorio E, Liu N, Priori SG, Napolitano C, Richard S, Benitah JP, Mangoni ME, Gomez AM (2012) Paradoxical effect of increased diastolic Ca(2+) release and decreased sinoatrial node activity in a mouse model of catecholaminergic polymorphic ventricular tachycardia. Circulation 126(4):392–401

Neef S, Dybkova N, Sossalla S, Ort KR, Fluschnik N, Neumann K, Seipelt R, Schondube FA, Hasenfuss G, Maier LS (2010) CaMKII-dependent diastolic SR Ca2+ leak and elevated diastolic Ca2+ levels in right atrial myocardium of patients with atrial fibrillation. Circ Res 106(6):1134–1144

Nof E, Belhassen B, Arad M, Bhuiyan ZA, Antzelevitch C, Rosso R, Fogelman R, Luria D, El-Ani D, Mannens MM, Viskin S, Eldar M, Wilde AA, Glikson M (2011) Postpacing abnormal repolarization in catecholaminergic polymorphic ventricular tachycardia associated with a mutation in the cardiac ryanodine receptor gene. Heart Rhythm 8(10):1546–1552

Nyegaard M, Overgaard MT, Sondergaard MT, Vranas M, Behr ER, Hildebrandt LL, Lund J, Hedley PL, Camm AJ, Wettrell G, Fosdal I, Christiansen M, Borglum AD (2012) Mutations in calmodulin cause ventricular tachycardia and sudden cardiac death. Am J Hum Genet 91(4):703–712

Palanca V, Quesada A, Trigo A, Jimenez J (2006) Arrhythmic storm induced by AICD discharge in a patient with catecholaminergic polymorphic ventricular tachycardia. Rev Esp Cardiol 59(10):1079–1080

Pizzale S, Gollob MH, Gow R, Birnie DH (2008) Sudden death in a young man with catecholaminergic polymorphic ventricular tachycardia and paroxysmal atrial fibrillation. J Cardiovasc Electrophysiol 19(12):1319–1321

Pogwizd SM, Bers DM (2004) Cellular basis of triggered arrhythmias in heart failure. Trends Cardiovasc Med 14(2):61–66

Postma AV, Denjoy I, Kamblock J, Alders M, Lupoglazoff JM, Vaksmann G, Dubosq-Bidot L, Sebillon P, Mannens MM, Guicheney P, Wilde AA (2005) Catecholaminergic polymorphic ventricular tachycardia: *RYR2* mutations, bradycardia, and follow up of the patients. J Med Genet 42(11):863–870

Priori SG, Chen SR (2011) Inherited dysfunction of sarcoplasmic reticulum Ca2+ handling and arrhythmogenesis. Circ Res 108(7):871–883

Priori SG, Napolitano C, Tiso N, Memmi M, Vignati G, Bloise R, Sorrentino V, Danieli GA (2001) Mutations in the cardiac ryanodine receptor gene (*hRyR2*) underlie catecholaminergic polymorphic ventricular tachycardia. Circulation 103(2):196–200

Reid DS, Tynan M, Braidwood L, Fitzgerald GR (1975) Bidirectional tachycardia in a child. A study using His bundle electrography. Br. Heart J. 37(3):339–344

Roux-Buisson N, Cacheux M, Fourest-Lieuvin A, Fauconnier J, Brocard J, Denjoy I, Durand P, Guicheney P, Kyndt F, Leenhardt A, Le MH, Lucet V, Mabo P, Probst V, Monnier N, Ray PF, Santoni E, Tremeaux P, Lacampagne A, Faure J, Lunardi J, Marty I (2012) Absence of triadin, a protein of the calcium release complex, is responsible for cardiac arrhythmia with sudden death in human. Hum Mol Genet 21(12):2759–2767

Suetomi T, Yano M, Uchinoumi H, Fukuda M, Hino A, Ono M, Xu X, Tateishi H, Okuda S, Doi M, Kobayashi S, Ikeda Y, Yamamoto T, Ikemoto N, Matsuzaki M (2011) Mutation-linked defective interdomain interactions within ryanodine receptor cause aberrant Ca(2) release leading to catecholaminergic polymorphic ventricular tachycardia. Circulation 124(6):682–694

Sumitomo N, Harada K, Nagashima M, Yasuda T, Nakamura Y, Aragaki Y, Saito A, Kurosaki K, Jouo K, Koujiro M, Konishi S, Matsuoka S, Oono T, Hayakawa S, Miura M, Ushinohama H, Shibata T, Niimura I (2003) Catecholaminergic polymorphic ventricular tachycardia: electrocardiographic characteristics and optimal therapeutic strategies to prevent sudden death. Heart 89(1):66–70

Sumitomo N, Sakurada H, Taniguchi K, Matsumura M, Abe O, Miyashita M, Kanamaru H, Karasawa K, Ayusawa M, Fukamizu S, Nagaoka I, Horie M, Harada K, Hiraoka M (2007) Association of atrial arrhythmia and sinus node dysfunction in patients with catecholaminergic polymorphic ventricular tachycardia. Circ J 71(10):1606–1609

Sy RW, Gollob MH, Klein GJ, Yee R, Skanes AC, Gula LJ, Leong-Sit P, Gow RM, Green MS, Birnie DH, Krahn AD (2011) Arrhythmia characterization and long-term outcomes in catecholaminergic polymorphic ventricular tachycardia. Heart Rhythm 8(6):864–871

Tester DJ, Spoon DB, Valdivia HH, Makielski JC, Ackerman MJ (2004) Targeted mutational analysis of the *RyR2*-encoded cardiac ryanodine receptor in sudden unexplained death: a molecular autopsy of 49 medical examiner/coroner's cases. Mayo Clin Proc 79(11):1380–1384

Tiso N, Stephan DA, Nava A, Bagattin A, Devaney JM, Stanchi F, Larderet G, Brahmbhatt B, Brown K, Bauce B, Muriago M, Basso C, Thiene G, Danieli GA, Rampazzo A (2001) Identification of mutations in the cardiac ryanodine receptor gene in families affected with arrhythmogenic right ventricular cardiomyopathy type 2 (ARVD2). Hum Mol Genet 10(3):189–194

Tristani-Firouzi M, Etheridge SP (2010) Kir 2.1 channelopathies: the Andersen-Tawil syndrome. Pflugers Arch 460(2):289–294

van der Werf C, Kannankeril PJ, Sacher F, Krahn AD, Viskin S, Leenhardt A, Shimizu W, Sumitomo N, Fish FA, Bhuiyan ZA, Willems AR, van der Veen MJ, Watanabe H, Laborderie J, Haissaguerre M, Knollmann BC, Wilde AA (2011) Flecainide therapy reduces exercise-induced ventricular arrhythmias in patients with catecholaminergic polymorphic ventricular tachycardia. J Am Coll Cardiol 57(22):2244–2254 van der Werf C, Nederend I, Hofman N, van GN, Ebink C, Frohn-Mulder IM, Alings AM, Bosker HA, Bracke FA, van den Heuvel F, Waalewijn RA, Bikker H, van Tintelen JP, Bhuiyan ZA, van den Berg MP, Wilde AA (2012a) Familial evaluation in catecholaminergic polymorphic ventricular tachycardia: disease penetrance and expression in cardiac ryanodine receptor mutation-carrying relatives. Circ Arrhythm Electrophysiol 5(4):748–756

van der Werf C, Zwinderman AH, Wilde AA (2012b) Therapeutic approach for patients with catecholaminergic polymorphic ventricular tachycardia: state of the art and future developments. Europace 14(2):175–183

van Oort RJ, McCauley MD, Dixit SS, Pereira L, Yang Y, Respress JL, Wang Q, De Almeida AC, Skapura DG, Anderson ME, Bers DM, Wehrens XH (2010) Ryanodine receptor phosphorylation by calcium/calmodulin-dependent protein kinase II promotes life-threatening ventricular arrhythmias in mice with heart failure. Circulation 122(25):2669–2679

Vest JA, Wehrens XH, Reiken SR, Lehnart SE, Dobrev D, Chandra P, Danilo P, Ravens U, Rosen MR, Marks AR (2005) Defective cardiac ryanodine receptor regulation during atrial fibrillation. Circulation 111(16):2025–2032

Viatchenko-Karpinski S, Terentyev D, Gyorke I, Terentyeva R, Volpe P, Priori SG, Napolitano C, Nori A, Williams SC, Gyorke S (2004) Abnormal calcium signaling and sudden cardiac death associated with mutation of calsequestrin. Circ Res 94(4):471–477

Viitasalo M, Oikarinen L, Vaananen H, Kontula K, Toivonen L, Swan H (2008) U-waves and T-wave peak to T-wave end intervals in patients with catecholaminergic polymorphic ventricular tachycardia, effects of beta-blockers. Heart Rhythm 5(10):1382–1388

Voigt N, Li N, Wang Q, Wang W, Trafford AW, Abu-Taha I, Sun Q, Wieland T, Ravens U, Nattel S, Wehrens XH, Dobrev D (2012) Enhanced sarcoplasmic reticulum Ca2+ leak and increased Na+–Ca2+ exchanger function underlie delayed afterdepolarizations in patients with chronic atrial fibrillation. Circulation 125(17):2059–2070

Watanabe H, Chopra N, Laver D, Hwang HS, Davies SS, Roach DE, Duff HJ, Roden DM, Wilde AA, Knollmann BC (2009) Flecainide prevents catecholaminergic polymorphic ventricular tachycardia in mice and humans. Nat Med 15(4):380–383

Wehrens XH, Lehnart SE, Reiken SR, Deng SX, Vest JA, Cervantes D, Coromilas J, Landry DW, Marks AR (2004) Protection from cardiac arrhythmia through ryanodine receptor-stabilizing protein calstabin2. Science 304(5668):292–296

Wilde AA, Bhuiyan ZA, Crotti L, Facchini M, De Ferrari GM, Paul T, Ferrandi C, Koolbergen DR, Odero A, Schwartz PJ (2008) Left cardiac sympathetic denervation for catecholaminergic polymorphic ventricular tachycardia. N Engl J Med 358(19):2024–2029

Zipes DP, Camm AJ, Borggrefe M, Buxton AE, Chaitman B, Fromer M, Gregoratos G, Klein G, Moss AJ, Myerburg RJ, Priori SG, Quinones MA, Roden DM, Silka MJ, Tracy C, Smith SC, Jr Jacobs AK, Adams CD, Antman EM, Anderson JL, Hunt SA, Halperin JL, Nishimura R, Ornato JP, Page RL, Riegel B, Blanc JJ, Budaj A, Dean V, Deckers JW, Despres C, Dickstein K, Lekakis J, McGregor K, Metra M, Morais J, Osterspey A, Tamargo JL, Zamorano JL, American College of Cardiology/American Heart Association Task Force, European Society of Cardiology Committee for Practice Guidelines, European Heart Rhythm Association, Heart Rhythm Society (2006) ACC/AHA/ESC 2006 Guidelines for Management of Patients With Ventricular Arrhythmias and the prevention of sudden cardiac death: a report of the American College of Cardiology/American Heart Association Task Force and the European Society of Cardiology Committee for Practice Guidelines (writing committee to develop Guidelines for Management of Patients With Ventricular Arrhythmias and the Prevention of Sudden Cardiac Death): developed in collaboration with the European Heart Rhythm Association and the Heart Rhythm Society. Circulation 114(10):e385-e484

Chapter 30
P2X Receptors and Pain

François Rassendren and Lauriane Ulmann

Abstract P2X receptors are ATP-gated channels and form the third major family of ligand-gated channels. Unlike other neurotransmitter-gated channels, P2X receptors have a widespread distribution outside the brain and are involved in many physiological functions. A main function associated to P2X receptors is chronic pain. This chapter focuses on the mechanisms underlying P2X receptors involvement in pain, with a particular emphasis on the contribution of P2X4 and P2X7 receptors expressed by immune cells in the establishment of network hyperexcitability and chronic pain.

30.1 Introduction

The first suggestion that ATP could be a mediator of nociception arised from the work of Pamela Holton in the 1950s, in which she demonstrated that antidromic sensory nerve stimulation triggers the release of extracellular ATP (Holton and Holton 1953, 1954). It took almost 20 years to firmly establish that extracellular ATP could evoke pain sensation in human, and coincidently to define the concept of purinergic transmission (Bleehen and Keele 1977; Burnstock 1977). Later on, two groups demonstrated that both sensory neurons and spinal neurons could be depolarized by direct applications of ATP, likely through the opening of a ligand-gated ion channel (Jahr and Jessell 1983; Krishtal et al. 1983). This later hypothesis was supported by the demonstration that ATP acts as a fast excitatory

F. Rassendren (✉) · L. Ulmann
CNRS, UMR 5203, Institut de Génomique Fonctionnelle, LabEx ICST, Université Montpellier, 1, 2, UMR-S661, Montpellier, France
e-mail: francois.rassendren@igf.cnrs.fr

F. Rassendren · L. Ulmann
Inserm, U661, Montpellier, France

N. Weiss and A. Koschak (eds.), *Pathologies of Calcium Channels*,
DOI: 10.1007/978-3-642-40282-1_30,

neurotransmitter in central and enteric nervous systems (Edwards et al. 1992; Evans et al. 1992). However, it is only with the molecular identification of the different purinergic receptors that the role of ATP in pain processing was unambiguously established (Burnstock and Wood 1996).

Twenty years after the cloning era of purinergic receptors, numerous studies have contributed to precise the mechanisms by which ATP is involved in pain processing. From the initial, and rather simple observation that ATP was able to depolarize sensory neurons, it now appears that purinergic signaling is involved in both nociception and chronic pain states. In addition, the role of purinergic receptors in the establishment of chronic pain state has crossed the boundaries of neurons and recent findings have demonstrated that ATP also acts through receptors expressed in immune and glial cells, highlighting its function as a danger signal.

The purinergic receptors family is divided into two main branches P2Y receptors that are G-protein coupled receptors and P2X receptors that are ATP-gated channels. This chapter will focus on the mechanisms underlying P2X receptors contribution to pain processing, with a particular emphasis on their regulation of immune cell functions in inflammatory and neuropathic pain.

30.2 P2X Receptor Family

The existence of P2X receptors was initially proposed by Burnstock based on the specific pharmacological profile of a subtype of P2 receptors (Burnstock 1976). A few years later, ATP-gated channels were unambiguously identified first in cultured sensory neurons (Jahr and Jessell 1983; Krishtal et al. 1983) and then in smooth muscle (Benham et al. 1987). Yet, the diversity of P2X receptor family was only appreciated later with the molecular identification of seven cDNAs encoding subunits of P2X receptors (Buell et al. 1996). Through expression cloning in *Xenopus* oocytes, two groups concomitantly cloned from bladder and NGF-differentiated PC12 cells, two cDNAs encoding proteins with significant homologies, called P2X1 and P2X2, respectively (Valera et al. 1994; Brake et al. 1994). Homology cloning subsequently leaded to the identification of five additional P2X cDNAs, P2X3-P2X7. The seven P2X proteins share around 45 % homology and the same membrane topology with two transmembrane spanning domains that connect a large extracellular domain constrained by five disulfide bridges, both extremities being located intracellularly (Buell et al. 1996). P2X receptors are made by the association of three subunits, either identical (homomeric receptors) or different (heteromeric receptors) (Murrell-Lagnado and Qureshi 2008). So far, only heteromeric receptors integrating two different subunits have been described and the real subunit stoichiometry of these trimeric receptors remains to be elucidated, although in one case it has been proposed that the two potential stoichiometry's of a given heteromeric receptor are present at the plasma membrane (Compan et al. 2012).

All P2X subunits, but P2X6, form functional homomeric ATP-gated channels with specific biophysical and pharmacological properties when recombinantly expressed in mammalian cells (North 2002). As stated above, several heteromeric receptors with specific properties compared to homomeric receptor have been characterized. Yet, among all potential heteromeric P2X receptors, only a few of them (P2X1/5, P2X2/3, and P2X2/5) have been unambiguously validated in native tissues either functionally or biochemically (Compan et al. 2012).

P2X receptors are cationic channels with a substantial calcium permeability, which is specific to each receptor type. The amount of calcium flowing through individual P2X receptor has been characterized in details by measuring fractional calcium currents (Pf %) (Egan and Khakh 2004). It appears that compared to other ligand-gated channels, P2X receptors are among the most calcium permeant. In particular, P2X1 and P2X4 display Pf % comparable to that of the NMDA receptors (NR1/NR2A), which is widely considered as the most calcium-permeant ligand-gated channel. Importantly, this high calcium permeability of P2X1 and P2X4 was also found in human receptors, demonstrating a conserved feature of these channels.

The structural determinants of the calcium permeability of P2X1 and P2X4 compared to the other P2X receptor is mainly due to a ring of negative charged amino acids at the extracellular mouth of the pore forming region; these acids residues are located near the extracellular part of both transmembrane domains. This negatively charged ring might serve as a binding site for calcium ions as described for other ligand-gated channels. In addition to this, residues located in the narrow region of the pore also contribute to calcium permeation. Indeed, in the P2X2 receptor, this region is composed of polar residues, whereas in P2X1 and P2X4 only one such residue is present. It is possible that the polar nature of this region might acts as selectivity filter, reducing the accessibility of calcium ions to the lower portion of the permeability pathway (Samways and Egan 2007). The recent structure of P2X4 receptor in its open state seems to favor this hypothesis (Hattori and Gouaux 2012).

High calcium permeability of ion channels is often associated with cytotoxicity, which is prevented by diverse regulatory mechanisms, such as the voltage-dependence block by magnesium of the NMDA receptors. Since resting extra-cellular ATP levels surrounding cells might be sufficient to open P2X receptors at resting membrane potential, a tight control of their activity is thus necessary to avoid calcium-mediated cytotoxicity. In the case of the P2X1 receptor, this is mainly achieved by a rapid and long lasting desensitization of the channel, which is induced by nanomolar concentrations of ATP. However, this does not apply for P2X4 receptors that do not show rapid desensitization. Rather, P2X4 exposition to the cell surface appears to be tightly regulated through endocytosis (Royle et al. 2005). As a consequence, most of P2X4 receptors are located in intracellular compartments and requires an yet unidentified signal that triggers their transport to the plasma membrane. Although the mechanisms regulating P2X4 trafficking are still poorly understood, a current model suggests that these proteins are rapidly removed from the plasma membrane through a clathrin-dependent mechanism into endosomes. These will enter a recycling pathway and ultimately merge with the secretory lysosomal pathway (Qureshi et al. 2007). One recently proposed hypothesis suggests that P2X4 on the surface of lysosomal secretory vesicle may contribute to calcium entry necessary to exocytotic vesicle post-fusion phase (Miklavc et al. 2011). Whether this function of P2X4-evoked calcium entry is restricted to certain cell types or a general mechanism remains to be elucidated.

30.3 Physiological Functions of P2X Receptors

One of the unique characteristic of P2X receptors is without contest their tissue distribution. Unlike other ligand-gated channels, whose expression is mainly restricted to the nervous system, P2X receptor are express in a large variety of tissue and cell type outside the nervous system, including smooth muscle, different epithelia, immune cells and others non-excitable cells (Khakh and North 2006). According to this pattern of expression, P2X receptors are directly involved in diverse physiological functions such as smooth muscle contraction, secretion, vascular tone regulation, and renal function.

Importantly, P2X receptors appear to play an important role in the immune system (Feske et al. 2012). Three receptors, P2X1, P2X4, and P2X7 are expressed in most immune cell types.

Due to its unique functions, the role of P2X7 receptors in immune cell has been extensively characterized (Chen and Brosnan 2006). One of the most striking function of P2X7 is its ability to open a large pore permeant to large cations. In myeloid cells, this property is directly link to micro vesicle shedding as well as to the activation of the inflammasome and the subsequent maturation and secretion of IL1ß (Di Virgilio 2007). In addition to this central function, activation of P2X7 has also been linked to the activation of diverse intracellular pathways important for the production of proinflammatory mediators such as prostaglandin, reactive oxygen species. In lymphocyte, P2X7 seem to regulate different facets of adaptive immune response (Di Virgilio 2012).

The function of P2X4 in immune cells is slowly emerging. In tissue-resident macrophage, P2X4 is directly responsible, through calcium influx and activation of the p38 MAPK pathway, for the activation of cytosolic phospholipase A2 and the subsequent synthesis and release of PGE2. As a consequence, P2X4 knock-out mice is unable to produce PGE2 in inflammatory conditions (Ulmann et al. 2010). Although the functions of P2X4 in other immune cells have not been investigated in details, several reports indicate that P2X4-evoked calcium entry may have important role in the regulation of diverse immune functions. Interestingly, P2X4 and P2X7 seem to interact physically in myeloid cells. Although the existence of P2X4/7 heteromeric receptor was initially proposed (Guo et al. 2007), its seems that this interaction is likely due to the presence within the same membrane delimited compartment (Nicke 2008). One interesting hypothesis is that both P2X4 and P2X7 form a specific signaling complex in which the two proteins collaborate to activate specific pathways.

The role of P2X1 in immune cells is still poorly, yet its expression is induced in lymphocyte entering apoptosis (Valera et al. 1994) and because of its high calcium permeability, P2X1 may have a crucial role in triggering activation of immune cells.

30.4 P2X Receptors in the Nervous System

In the nervous system, P2X receptors were first identified as excitatory receptors triggering membrane depolarization, action potentials, and fast synaptic transmission (Robertson et al. 2001). Yet, despite the expression of all P2X subunit in the brain, there are still little data on the role of individual subunit or receptors in brain functions. In neurons, it is now well admitted that P2X receptors may act both at pre- and post-synaptic levels where they directly influence synaptic transmission. However, these direct synaptic effects appear to be the exception rather than the generality. Indeed, electron microscopy revealed that the P2X2 and P2X4 receptors are not localized in the synaptic cleft, but rather at the periphery of the spine, suggesting that these receptors do not directly contribute to synaptic transmission but rather modulate somehow postsynaptic excitability. A current consensus in that neuronal P2X receptors have neuromodulatory functions in the nervous system, finely tuning excitability (Khakh and North 2012).

In addition to their expression in neurons, P2X receptors are also expressed in glial cells. Cortical astrocytes express the heteromeric P2X1/5 receptor which may play a role in cortical network activity (Palygin et al. 2010). Several studies have also proposed that distinct populations of astrocyte, including somatosensory cortex, hippocampus, and bergman glia express P2X7 (Butt 2011). These data are still debated, since not always reproducible and often lacking specificity. Beside potential explanation such as species differences, it is important to note that other P2X receptors, such as the P2X2/5 receptor, have functional properties that encompass most that of the P2X7 receptor. Still, the role of P2X receptors in astrocytes is still poorly understood and will require the development of specific genetic tools to reveal their functions.

30.5 Neuronal P2X Receptors and Pain Processing

The function of extracellular ATP as a mediator of pain was established well before the demonstration of the presence of ATP-gated channels in sensory neurons. Of interest, several studies performed in human volunteers support the peripheral nociceptive properties of ATP (Bleehen and Keele 1977). However, the whole field really took some spin with the molecular identification of the P2X3 subunit and its almost specific expression in a subset of sensory neurons (Chen et al. 1995; Lewis et al. 1995). Subsequently, it was found that mRNAs encoding the seven P2X receptor subunits are expressed in dorsal root ganglion neurons (Collo et al. 1996). Even though only a subset of sensory neurons are involved in pain processing, the potential expression of all P2X receptors in DRG neurons further support a central role of this family of receptor in the transfer of nociceptive information. Although the mechanisms by which extracellular ATP is released is still a matter of debate, an old consensus is that ATP levels correlates with specific physiological or pathological conditions, supporting that ATP was not mainly involved in nociception but rather in pathological pain.

Since the identification of the P2X3 receptor almost 20 years ago, a considerable amount of work has been achieved, and from the simple observation that peripheral ATP application generates itching sensation, there is now amounting evidences that ATP through P2X receptors is a major mediator of pathological pain. So far, to the exception of P2X1 and P2X6 all other P2X subunits have been shown to process pain. However, a set of unanticipated results recently established that P2X receptors expressed by immune cells play a central role in pathological pain, namely neuropathic and inflammatory pains. In the following section, the respective role of the different P2X receptor in pain processing is presented.

30.5.1 P2X3 Receptors

As soon as it was identified, P2X3 generated a tremendous interest in the field of pain physiology (North 2004). It is expressed almost exclusively in a subpopulation of small and intermediate seize sensory neurons that correspond to nociceptive neurons generating C and A∂ fibers. Further, immunohistological studies revealed that P2X3 subunits were expressed in nonpeptidergic, IB4 positive neurons that terminate in inner lamina II of the dorsal horn of the spinal cord (Vulchanova et al. 1998). Interestingly, P2X3 subunits are targeted to both peripheral and central termini of these nociceptors and their expression is upregulated following peripheral nerve injury.

In addition to this pattern of expression, P2X3 subunits presented specific biophysical and pharmacological properties that greatly facilitated their functional characterization in vivo. Homomeric P2X3 receptors are activated by α,β-methyleneadenosine 5′-triphosphate (α,βmeATP), a non-hydrolysable analog of ATP. In addition, P2X3-evoked currents show a very rapid desensitization. Among the seven P2X subunits, these two properties are only shared by the P2X1 receptor which, however, is poorly expressed in sensory neurons. Second, P2X3 forms heteromeric receptor by association with P2X2 subunit that is sensitive to α,βmeATP but generates non-desensitizing currents (Lewis et al. 1995). This specific signature of the heteromeric P2X2/3 current closely resembles that observed in sensory neurons. Because α,βmeATP is much more stable than ATP when administered in vivo, it represented a very useful tool to address the involvement of P2X3 to nociception. In addition, TNP-ATP, a high affinity antagonist specific of P2X1 and P2X3 containing receptors, was soon discovered allowing for a relatively specific pharmacological characterization of P2X3 receptors function in vivo (Honore et al. 2002). Several studies demonstrated that P2X3 containing receptors are likely to participate to ATP-evoked nociception and to contribute to both inflammatory and neuropathic pain phenotypes. These results were confirmed in P2X3-deficient mice, with the development of specific P2X3 inhibitors and by siRNA knock down of the subunit (North 2003).

An important feature of P2X3 subunit is its expression at both peripheral and central termini of C and A∂ fibers. The peripheral expression confers to P2X3 a crucial role in sensing extracellular ATP in both physiological and pathological conditions. P2X3 is highly express in sensory nerve fibers innervating the bladder and contributes to elimination reflexes by sensing ATP released by epithelial cells lining the bladder upon distension (Cockayne et al. 2000). A similar role is attributed to P2X3 in many visceral organs, where the receptor functions as a sensor of ATP released following mechanical deformation of tissues. Peripheral ATP sensing by P2X3 is also relevant in inflammatory conditions, in which ATP levels increase secondary to tissue injury or through release by recruited immune cells. In addition, in inflammatory conditions, several mechanisms contribute to enhance P2X3 response. First, different studies have reported an increased expression of the protein in DRG neurons in peripheral inflammatory conditions. Second, following inflammation P2X3 receptors are sensitized by different intracellular pathways, likely through phosphorylation of the protein.

The expression of P2X3 at central termini of primary afferent was somehow unexpected. Yet, many studies have convincingly demonstrated that P2X3 subunits are localized in afferent fibers innervating the inner lamina II of the dorsal horn. This presynaptic localization of P2X3 is thought to be involved in the release of glutamate and thus to participate to central sensitization of these synapses associated with chronic or long lasting pain (Gu and MacDermott 1997).

Several highly specific P2X3 or P2X2/3 antagonists have been developed by pharmaceutical companies (North and Jarvis 2013). These molecules efficiently reduced behavioral pain phenotypes associated with a variety of preclinical pain models. Some of these molecules have advanced to phase II clinical trial and it will be of interest to see whether P2X3 is a sustainable therapeutic target.

As stated above, P2X2 subunits form heteromeric receptor by association with P2X3 but also homomeric receptors. Whether, homomeric or heteromeric P2X2 receptors are involved in pain processing remains unclear. In rodent, studies based on specific P2X2/3 antagonist or genetically engineered mice strongly support that heteromeric P2X2/3 receptors participate to the elaboration of nociceptive information. However, this seems to be different in primate in which the co-localization of P2X2 and P2X3 subunits is clearly different than in rodent (Serrano et al. 2012). In addition, electrophysiological recording clearly demonstrated that P2X2 is not expressed in human DRG neurons. It appears than in primate and human, P2X3 is the main neuronal P2X subunit responsible for sensing ATP-evoked alarm signals and conveying nociceptive neuronal excitability to the spinal cord.

30.5.2 P2X Receptors, Immune Cells, and Pain

One of the main advances in the field of purinergic signaling in pain comes from the demonstration that P2X receptors expressed by immune cells plays a central role in the establishment and the maintenance of chronic pain states. Cells from both innate and adaptive immunity express different P2X receptors among which P2X1, P2X4, and P2X7 have been functionally characterized. While the role of P2X1 in these cells remains poorly characterized, P2X4 and P2X7 receptors expressed by peripheral immune cells emerge as central to the development of the innate inflammatory response, and as a consequence to inflammatory pain. In addition to this peripheral site of action, evidence are accumulating that P2X receptors expressed by microglia, the brain-resident macrophage, are involved in the regulation of network excitability in the dorsal horn of the spinal cord, underlying persistent pain states.

30.5.3 P2X7 Receptors

Before its molecular identification, P2X7 was named P2Z receptor because of its atypical properties compared to other P2 receptors. P2Z was characterized by a specific pharmacological profile, a low sensitivity to ATP, a relatively good apparent affinity for 2′(3′)-O-(4-Benzoylbenzoyl)adenosine 5′-triphosphate (BzATP). In addition, P2Z receptors showed unusual pore properties with the ability to open a pore permeant to large cations or small molecules, eventually leading to plasma membrane permeabilization. Finally, P2Z was highly expressed in immune cells (Murgia et al. 1993). All these properties are recapitulated in P2X7 receptors that show a prominent expression in myeloid cells (Surprenant et al. 1996). The peculiar properties of the P2X7 receptor combined to its expression in immune cells prompted a great interest among the immunologist community and lead to the discovery that P2X7 activation is necessary for IL1ß processing and release (Ferrari et al. 1997).

Soon after its molecular elucidation, it was proposed that P2X7 could be involved in inflammatory pain; although this study was based on the use of oxidized ATP, a purinergic antagonist with low specificity. The first comprehensive demonstration for the direct involvement of P2X7 receptors to pain was provided by the phenotypic characterization of P2X7-deficient mice. Development of both inflammatory and neuropathic pain behaviors (mechanical and thermal hypersensitivity) as well as BzATP-evoked release of IL1ß from peritoneal macrophages are impaired in P2X7-deficient mice (Solle et al. 2001). These alterations of pain phenotypes appear to be directly related to the reduced processing of IL1ß since similar phenotypes are observed in mice lacking IL1αß. Yet, profiling of paw cytokines after local injection of Freund Complete Adjuvant (FCA) revealed that P2X7 gene deletion also affects the expression of IL10 and IL6, suggesting that gene inactivation can induced subtle compensatory mechanisms (Chessell et al. 2005). The potential involvement of P2X7 receptors in numerous pathologies fostered the interest of pharmaceutical companies to develop specific antagonists, which have rapidly been released. In rodent, these molecules with nanomolar affinities have potent analgesic effects on both inflammatory and neuropathic pain behaviors (North and Jarvis 2013). Despite the numerous studies demonstrating the involvement of P2X7 receptors in pathological pain, the mechanism by which these receptors are able to drive excitatory input remain poorly characterized. P2X7 expression in neurons remains debated and there is little experimental data suggesting that sensory or dorsal horn neurons express these receptors (Sim et al. 2004). Rather, most results support the involvement of P2X7 expressed by macrophages and microglia in inflammatory and neuropathic pain, respectively. It is also widely accepted that P2X7 exerts its effects by triggering IL1ß maturation and release at the site of inflammation. Yet, the experimental results directly linking endogenous P2X7 receptor activation to IL1ß release in pain models remain sparse. There are also few data explaining how IL1ß drives neuronal excitability or synaptic remodeling accompanying persistent pain. In addition to its crucial role in IL1ß release, P2X7 might also contribute to pain through the evoked release of other extracellular signaling molecules such as chemokines, prostaglandins, cathepsins, or through the activation of specific intracellular pathways leading to transcriptional regulation of gene expression.

There is no doubt that in rodents P2X7 receptors contribute to hyperalgesia in different experimental models of inflammatory and neuropathic pain. A central question is whether this is also true in human. So far, clinical trials testing the efficacy of P2X7 antagonists in human pathologies have been unsuccessful. However, these clinical trials were designed to assess the efficacy of these molecules on the signs and symptoms of active rheumatoid arthritis using ACR20 criteria which do not obligatorily include pain assessment. Yet, *P2X7* receptor gene is highly polymorphic in the human population, with numerous gain- or loss-of-function polymorphisms. Current clinical trials assessing this variability are currently ongoing. Nevertheless, two polymorphisms known to enhanced or reduced P2X7 receptors function have been correlated to higher and lower pain sensations in female population following mastectomy (Sorge et al. 2012). Further studies in populations carrying polymorphism resulting in a complete loss-of-function should provide more information on the real involvement of P2X7 receptors to chronic pain in humans.

30.5.4 P2X4 Receptors

Among the different P2X subunits, P2X4 shows the largest tissue repartition; P2X4 receptors are expressed in numerous tissues and cell types, including immune cells, endothelia and epithelia, neurons among others. Yet, our understanding of the physiological roles of P2X4 receptors remains sparse. There are several reasons for this. First, in contrast to P2X7 receptors, specific P2X4R antagonists that can be used in vivo are still missing; second, despite an abundant tissue distribution of the receptors, mice in which *p2x4* gene has been genetically deleted show no obvious phenotypes; Third, membrane expression of P2X4R is tightly regulated. Indeed, in immune cells (macrophage and microglia), P2X4R are mostly localized in intracellular organelles such as endosomes and lysosomes (Qureshi et al. 2007); they reach the plasma membrane upon a yet unidentified triggering signal (potentially intracellular calcium rise), and once activated by extracellular ATP, are rapidly internalized and recycled (Royle et al. 2005). This model of trafficking might explain why endogenous the failure to record P2X4-evoked currents in ex vivo tissue preparation such as acute brain slices. Indeed, during tissue processing large amounts of extracellular ATP are likely released which contribute to P2X4 endocytosis. This, combined to the known dialysis effect of whole cell recording on intracellular compartment, is likely that such recording approach prevents forward trafficking of the receptor to the plasma membrane (Fountain and North 2006). Nevertheless, P2X4-dependent modulation of synaptic activity has been reported in hippocampus using extracellular field recording as well as P2X4-evoked current in different types of primary cell in culture (Sim et al. 2006). Similar results have been obtained through calcium imaging. In these experiments, the specific P2X4 signature is provided by the use of ivermectin a positive allosteric modulator of P2X4- and of P2X4-deficient mice (Ulmann et al. 2010).

Most of the knowledge obtained on the physiological functions of P2X4 receptors emanate from studies based on protein knockdown or P2X4-deficient mice. These studies have provided some insight on the mode of action of P2X4 receptors in the regulation of different physiological functions. Three major phenotypes associated with P2X4 deletion have been described, decreased synaptic plasticity (Sim et al. 2006), altered vascular tone regulation (Yamamoto et al.2006), and lack of inflammatory and neuropathic pain (Tsuda et al. 2009; Ulmann et al. 2010). These phenotypes are related to non-homeostatic conditions supporting that P2X4 receptors are activated during episodes of strong or pathological activities during which extracellular ATP is likely released. This suggests that P2X4 receptors activation is brought into play in specific physiological situation and may explain the lack of obvious phenotype of P2X4-deficient mice despite a widespread tissue expression.

As already mentioned, P2X4 receptors have a remarkably high calcium permeability equivalent to that of the NMDA receptors (Egan and Khakh 2004). This property combined to the relatively slow inactivation of ATP-evoked current at P2X4 receptors implies that P2X4 receptors are important contributors of calcium influx, particularly in non-excitable cells. This confers to P2X4 receptors the opportunity to activate multiple intracellular calcium-dependent signaling pathways with short-term or long-lasting effects such as secretion and transcription, respectively. Several demonstrations of these properties have been provided. ATP-evoked intracellular calcium signals are significantly impaired in macrophages P2X4-deficient mice compared to wild type, which is directly linked to an impaired activation of the calcium-dependent phospholipase A2 and the subsequent synthesis and release of proinflammatory PGE2 (Ulmann et al. 2010). P2X4-evoked calcium entry is also involved in the release of BDNF from microglia (Trang et al. 2009) and certainly, although not demonstrated, for the release of NO from vascular endothelial cells. In addition, recent data show that P2X4 receptors control the transcriptional remodeling in the hippocampus following the induction of a status epilepticus. Because other purinergic receptors, such as metabotropic P2Y receptors also trigger intracellular calcium rise, it is still difficult to estimate the exact contribution of P2X4-evoked calcium influx in cell physiology. Such studies would require the use of specific genetically encoded tools allowing the direct quantification of calcium influx through P2X4 channels (Richler et al. 2008).

One unexpected physiological implication of P2X4 receptors is their involvement in two types of pathological pain: neuropathic and inflammatory pains. Indeed, P2X4-deficient mice do not develop hyperalgesia following sciatic nerve injury (a model of neuropathic pain) or following induction of peripheral inflammation. Interestingly, studies have demonstrated that in both models immune cells expressing P2X4 receptors are primarily responsible for generating hyperalgesia, however, through distinct mechanisms.

30.5.5 P2X4 Receptors and Neuropathic Pain

Neuropathic pain is a pathological type of chronic pain that results from alterations of the peripheral or central nervous system. Most often neuropathic pain is caused by trauma, infection, pathological degeneration of peripheral nerve, or chemotherapy. It is a debilitating condition that poorly responds to traditional analgesic treatments. Hallmarks of neuropathic pains are hyperalgesia, allodynia (pain evoked by innocuous stimuli), and spontaneous pain, which are thought to result from a reduction of the inhibitory control of excitatory nociceptive network in the spinal cord.

The initial suggestion for a role of P2X4R in neuropathic pain was established by the demonstration that allodynia developing after sciatic nerve ligation was inhibited by intrathecal injection of TNP-ATP but not by PPADS, two P2X antagonists that discriminate P2X4 from other P2X receptors in rat (Tsuda et al. 2003). This study provided two key results: first, that P2X4 receptor expression is induced *de novo* in the dorsal horn ofthe sinal cord following peripheral nerve injury and, second that this up regulation of P2X4 is specifically found in activated microglial cell. Remarkably, intraspinal injection of ATP-primed cultured microglia induces allodynia in naive animals that is prevented by pharmacological inhibition of P2X4 receptors. In addition to the discovery that P2X4 receptors are responsible for the induction of allodynia, this study provided the first demonstration that microglia was able to generate hyperexcitability of spinal network. This original work was followed by two key studies that provided a mechanistic explanation on the involvement of microglial cells to neuropathic pain. In a first study, the involvement of KCC2 in neuropathic pain was proposed. KCC2 is a chloride/potassium co-transporter that regulates intracellular chloride homeostasis in mature neurons. By maintaining chloride driving force in neurons, KCC2 is thus critical to assure the inhibitory function of chloride permeant ligand-gated channel such as GABAA or glycine receptors. In hippocampus, KCC2 downregulation is mediated by BDNF producing an impairment of neuronal chloride extrusion, a process linked to seizure activity and local hyperactivity (Rivera et al. 2002). Flowing peripheral nerve injury, a model of neuropathic pain, KCC2 is downregulated in the dorsal horn of thespinal cord ispi lateral to the lesion. As demonstrated in the hippocampus, this downregulation produces a shift in anion gradient in lamina I excitatory neurons that results in their des-inhibition and local hyperexcitability (Coull et al. 2003). Importantly, this loss of KCC2 turns GABAA receptors into excitatory channels. Such a disinhibition of lamina I neurons is likely underlying the development of neuropathic pain.

The second study finally linked microglia to spinal disinhibition to neuropathic pain, by demonstrating that the downregulation of KCC2 associated with spinal disinhibition depends on BDNF signaling. Furthermore, the origin of BDNF was identified as being produced by local activated microglia in response to purinergic stimulation (Coull et al. 2005). Ultimately the link between microglial P2X4, the release of BDNF, and neuropathic pain was provided by the characterization of P2X4-deficient mice that do not develop behavioral hypersensitivity following nerve ligation and in which microglia is unable to release BDNF upon ATP stimulation (Ulmann et al. 2008). It was further demonstrated that P2X4 receptors are directly involved in BDNF secretion from microglia through a calcium and p38 MAPK-dependent pathway (Trang et al. 2009). This series of studies established the existence of a microglia-to-neuron pathway involving P2X4-BDNF-KCC2 that results in hyperexcitability of dorsal horn network and underlies neuropathic pain (Trang et al. 2011). Unexpectedly, this microglial P2X4-BDNF pathway was found to be responsible for morphine-induced hyperalgesia, a common and paradoxical side effect of analgesic morphine treatment. Daily injection of morphine induces a progressive thermal and mechanical hypersensitivity that is independent of the classical tolerance effect. This morphine-induced hypersensitivity presents all the hallmark of microglia P2X4-BDNF-KCC2 pathways. In fact direct stimulation of microglia upregulates P2X4 expression through μ-opioid receptors activation, while for unexplained reasons, microglial BDNF release is triggered through a μ opioid-independent pathway (Ferrini et al. 2013).

30.5.6 P2X4 Receptors in Inflammatory Pain

Pain phenotyping of P2X4-deficient mice revealed a strong reduction of hyperalgesia induced by different paradigm of peripheral inflammatory pain, yet normal nociception is unaffected (Tsuda et al. 2009). Interestingly, this absence of hyperalgesia is observed in acute inflammatory models (i.e., carrageenan paw injection) in which hypersensitivity is monitored within a few minutes and for a few hours following the inflammatory challenge, but also in long lasting or chronic inflammatory models (i.e. Complete Freund Adjuvant injection) in which hyperalgesia lasts several days after the peripheral insult (Ulmann et al. 2010). Since upregulation of P2X4 receptor in spinal microglia is absent following peripheral inflammation caused by CFA, this suggested the involvement of a different mechanism than microglia-evoked spinal disinhibition. One hallmark of peripheral inflammation is the sequential recruitment of inflammatory cells at the inflammatory site, among which tissue-resident macrophages are the first responders. P2X4 receptors like P2X7 are known to be expressed in peritoneal macrophages. However, peritoneal macrophages are activated and data obtained from microglia support that P2X4 is not be expressed in non-activated tissue-resident macrophages. In addition, the mechanisms by which P2X4 could contribute to peripheral inflammatory pain were unclear. These issues were resolved through the use of the $CX3CR1^{+/eGFP}$ mice in which GFP is expressed in all myeloid cells, including tissue-resident macrophages and microglia (Ulmann et al. 2010).

By immunohistochemistry, GFP positive cells can clearly be identified in paw tissue. These cells also express the macrophage-specific marker F4/80 and P2X4 receptors, indicating that unlike microglia, tissue-resident macrophage express P2X4 receptors in resting conditions. These receptors are functional since calcium video-microscopy revealed that repeated ATP stimulations evoke calcium increase that can be potentiated by the positive allosteric P2X4 modulator ivermectin. In macrophage from P2X4-deficient mice, similar repetitive ATP stimulations induce a strong rundown of the signal, indicating that P2X4 receptors evoked substantial calcium influx in these cells.

A next question is do P2X4 receptors from macrophages contribute to inflammatory pain and how. P2X4 are highly permeant to calcium. One possibility was that P2X4-evoked calcium signals can promote the synthesis and or release of proinflammatory molecules. In this scheme, one potential candidate was the prostaglandin PGE2. Indeed, PGE2 is synthesized through the initial production of arachidonic acid (AA) that itself is released by the calcium-dependent enzyme phospholipase A2. ATP stimulation of macrophage triggers the release of AA that is impaired in P2X4-deficient macrophages. Furthermore, PGE2 production triggered by ATP, is dependent on extracellular calcium influx, p38 MAPK, and cyclooxygenase enzyme, and is impaired in P2X4-deficient macrophage. This demonstrates that in macrophage maintained in vitro P2X4 stimulation activates the calcium-dependent PLA2-COX pathway and the subsequent release of PGE2. Yet, this does not provide any conclusive answer to the real in vivo inflammatory pain condition. A definite answer was provided by directly measuring PGE2 in paw exudate following local inflammatory challenge. PGE2 levels rapidly increase in exudate following carrageenan or CFA peripheral injection. This increase is completely blunted in P2X4-deficient mice, yet basal PGE2 levels remain unaffected, indicating that P2X4 receptors only contribute to inflammatory PGE2 release, but have no effect on homeostatic PGE2 production. Finally, a direct demonstration that P2X4 receptors expressed by macrophages contribute to PGE2 evoked hyperalgesia was provided by transfer experiments. Indeed, paw administration of ATP-stimulated macrophages induces mechanical hypersensitivity that is potentiated by ivermectin but inhibited by indomethacin a COX inhibitor. However, transfer of P2X4-deficient macrophage to wild-type mice does not confers hypersensitivity, yet the converse experiment does, demonstrating that PGE2 production by macrophage is impaired in P2X4-deficient mice, but that sensitivity to PGE2 remains unaltered. This later observation is confirmed by direct injection of PGE2 in paw that triggers similar hyperalgesic levels in both wild-type and P2X4-deficient mice.

In macrophage, P2X4 receptors primarily stimulate the activity of PLA2 and the production of AA, which is the precursor of several proinflammatory lipids including leukotrienes (LT) that are produced through lipoxygenase pathway (Cunha et al. 2003). Particularly, LTB4 promotes inflammation through its chemoattracting activity for neutrophils to the inflammatory site. However, P2X4 receptors does not seems to be involved in LT production since neutrophils recruitment to inflammatory site is not affected in P2X4-deficient mice, although no attempt have been made to measure its production in inflamed tissues. This suggests the existence of a intracellular signaling pathway that specifically links P2X4 receptors to PLA2 and COX metabolism. So far, there are very few clues about the determinant of such a specific coupling, particularly given that P2X4 receptors rapidly cycle from the plasma membrane too intracellular compartments. One possibility would be that P2X4 dynamically associate with a signaling complex within the plasma membrane. This hypothesis is supported by the interaction that exists between P2X4 and P2X7 in macrophages (Nicke 2008). In this scheme, P2X7 could represent and anchoring protein of the complex and P2X4 the ATP-sensor. Such a dynamic complex could result in a synergy between the two receptors that would contribute to activate specific intracellular pathway underlying inflammation and pain.

So far no specific P2X4 antagonist has been reported, and most of the data demonstrating the involvement of these receptors in pathological pain rely on P2X4-deficient mice or knockdown of the protein. Reports have suggested that serotonin reuptake inhibitor specific antidepressants such as paroxetine or fluoxetine inhibit rat and human P2X4-evoked calcium entry in recombinant system (Nagata et al. 2009). These molecules also show anti-allodynic effects in neuropathic rats. In addition, amitriptyline, a tricyclic antidepressant, was shown in a mouse microglial cell line to impair P2X4 receptors trafficking to the cell surface (Toulme et al. 2010). These results suggest that the efficacy of both class of antidepressant, which are commonly used in clinics to treat neuropathic pain, could be due to a direct effect on P2X4 receptors activity rather than through their classical mode of action. However, the effect of amitriptyline on human P2X4 could not be reproduced in recombinant system (Sim and North 2010). These discrepancies could be due to species differences or to the use of a recombinant system in the latter study. In this case, this could suggest that amitriptyline may act on a specific pathway involved in P2X4 receptor trafficking. Further studies will be necessary to assess whether SSRI or tricyclic antidepressant are effective P2X4 antagonists and to understand their mode of action.

One central question regarding P2X4 receptors involvement in chronic pain is whether this is also true in human. There are no direct data that relates P2X4 and pain processing in human. However, a functional polymorphism in human P2X4 encoding Tyr315Cys variation resulting in a receptor with hypo-function, is associated with increased pulse pressure (Stokes et al. 2011). Interestingly, this phenotype is close to what is observed in P2X4-deficient mice that show defective vascular tone regulation (Yamamoto et al. 2006). In addition, this polymorphism together with the Gly150Arg P2X7 functional polymorphism results in a reduced phagocytic capacity of human monocyte and are linked to the development of age-related macular degeneration (Gu et al. 2013). Because, in primate both P2X7 and P2X4 receptor are co-expressed in macrophage and microglia strongly support a role for P2X4 receptors in human microglial functions and pain processing. It will thus be interesting to seek whether P2X4 functional polymorphisms are linked to differential chronic pain perception in human cohorts.

30.6 Conclusion

In 20 years since the molecular identification of the P2X receptors family, these receptors have emerged as important contributors to pain processing. Not only P2X receptors classically contribute to the excitability of primary sensory neurons, but they also shape local network excitability though their expression in microglia and the release of diverse signaling molecules. Progress made in the development of specific antagonists targeting human P2X receptors have led to several clinical trials some of them being encouraging in very specific field of pain. One of the greatest current challenge is to determine whether targeting P2X receptors can efficiently reduce neuropathic pain without comprising the normal immune response. Answering this question represents a major goal in the field of purinergic signaling in pain processing.

References

Benham CD, Bolton TB, Byrne NG, Large WA (1987) Action of externally applied adenosine triphosphate on single smooth muscle cells dispersed from rabbit ear artery. J Physiol 387:473–488

Bleehen T, Keele CA (1977) Observations on the algogenic actions of adenosine compounds on the human blister base preparation. Pain 3(4):367–377

Brake AJ, Wagenbach MJ, Julius D (1994) New structural motif for ligand-gated ion channels defined by an ionotropic ATP receptor. Nature 371(6497):519–523. doi:10.1038/371519a0

Buell G, Collo G, Rassendren F (1996) P2X receptors: an emerging channel family. Eur J Neurosci 8(10):2221–2228

Burnstock G (1976) Purinergic receptors. J Theor Biol 62(2):491–503
Burnstock G (1977) The purinergic nerve hypothesis. Ciba Found Symp 48:295–314
Burnstock G, Wood JN (1996) Purinergic receptors: their role in nociception and primary afferent neurotransmission. Curr Opin Neurobiol 6(4):526–532
Butt AM (2011) ATP: a ubiquitous gliotransmitter integrating neuron-glial networks. Semin Cell Dev Biol 22(2):205–213. doi:10.1016/j.semcdb.2011.02.023
Chen CC, Akopian AN, Sivilotti L, Colquhoun D, Burnstock G, Wood JN (1995) A P2X purinoceptor expressed by a subset of sensory neurons. Nature 377(6548):428–431. doi:10.1038/377428a0
Chen L, Brosnan CF (2006) Regulation of immune response by P2X7 receptor. Crit Rev Immunol 26(6):499–513
Chessell IP, Hatcher JP, Bountra C, Michel AD, Hughes JP, Green P, Egerton J, Murfin M, Richardson J, Peck WL, Grahames CB, Casula MA, Yiangou Y, Birch R, Anand P, Buell GN (2005) Disruption of the P2X7 purinoceptor gene abolishes chronic inflammatory and neuropathic pain. Pain 114(3):386–396. doi:10.1016/j.pain.2005.01.002
Cockayne DA, Hamilton SG, Zhu QM, Dunn PM, Zhong Y, Novakovic S, Malmberg AB, Cain G, Berson A, Kassotakis L, Hedley L, Lachnit WG, Burnstock G, McMahon SB, Ford AP (2000) Urinary bladder hyporeflexia and reduced pain-related behaviour in P2X3-deficient mice. Nature 407(6807):1011–1015. doi:10.1038/35039519
Collo G, North RA, Kawashima E, Merlo-Pich E, Neidhart S, Surprenant A, Buell G (1996) Cloning OF P2X5 and P2X6 receptors and the distribution and properties of an extended family of ATP-gated ion channels. J Neurosci Official J Soc Neurosci 16(8):2495–2507
Compan V, Ulmann L, Stelmashenko O, Chemin J, Chaumont S, Rassendren F (2012) P2X2 and P2X5 subunits define a new heteromeric receptor with P2X7-like properties. J Neurosci Official J Soc Neurosci 32(12):4284–4296. doi:10.1523/JNEUROSCI.6332-11.2012
Coull JA, Beggs S, Boudreau D, Boivin D, Tsuda M, Inoue K, Gravel C, Salter MW, De Koninck Y (2005) BDNF from microglia causes the shift in neuronal anion gradient underlying neuropathic pain. Nature 438(7070):1017–1021. doi:10.1038/nature04223
Coull JA, Boudreau D, Bachand K, Prescott SA, Nault F, Sik A, De Koninck P, De Koninck Y (2003) Trans-synaptic shift in anion gradient in spinal lamina I neurons as a mechanism of neuropathic pain. Nature 424(6951):938–942. doi:10.1038/nature01868
Cunha JM, Sachs D, Canetti CA, Poole S, Ferreira SH, Cunha FQ (2003) The critical role of leukotriene B4 in antigen-induced mechanical hyperalgesia in immunised rats. Brit J Pharmacol 139(6):1135–1145. doi:10.1038/sj.bjp.0705346
Di Virgilio F (2007) Liaisons dangereuses: P2X(7) and the inflammasome. Trends Pharmacol Sci 28(9):465–472. doi:10.1016/j.tips.2007.07.002
Di Virgilio F (2012) Purines, purinergic receptors, and cancer. Cancer Res 72(21):5441–5447. doi:10.1158/0008-5472.CAN-12-1600
Edwards FA, Gibb AJ, Colquhoun D (1992) ATP receptor-mediated synaptic currents in the central nervous system. Nature 359(6391):144–147. doi:10.1038/359144a0
Egan TM, Khakh BS (2004) Contribution of calcium ions to P2X channel responses. J Neurosci Official J Soc Neurosci 24(13):3413–3420. doi:10.1523/JNEUROSCI.5429-03.2004
Evans RJ, Derkach V, Surprenant A (1992) ATP mediates fast synaptic transmission in mammalian neurons. Nature 357(6378):503–505. doi:10.1038/357503a0
Ferrari D, Chiozzi P, Falzoni S, Hanau S, Di Virgilio F (1997) Purinergic modulation of interleukin-1 beta release from microglial cells stimulated with bacterial endotoxin. J Exp Med 185(3):579–582
Ferrini F, Trang T, Mattioli TA, Laffray S, Del'Guidice T, Lorenzo LE, Castonguay A, Doyon N, Zhang W, Godin AG, Mohr D, Beggs S, Vandal K, Beaulieu JM, Cahill CM, Salter MW, De Koninck Y (2013) Morphine hyperalgesia gated through microglia-mediated disruption of neuronal Cl(-) homeostasis. Nat Neurosci 16(2):183–192. doi:10.1038/nn.3295
Feske S, Skolnik EY, Prakriya M (2012) Ion channels and transporters in lymphocyte function and immunity. Nat Rev Immunol 12(7):532–547. doi:10.1038/nri3233

Fountain SJ, North RA (2006) A C-terminal lysine that controls human P2X4 receptor desensitization. J Biol Chem 281(22):15044–15049. doi:10.1074/jbc.M600442200

Gu BJ, Baird PN, Vessey KA, Skarratt KK, Fletcher EL, Fuller SJ, Richardson AJ, Guymer RH, Wiley JS (2013) A rare functional haplotype of the P2RX4 and P2RX7 genes leads to loss of innate phagocytosis and confers increased risk of age-related macular degeneration. FASEB J: Official Publication of the Federation of American Societies for Experimental Biology 27(4):1479–1487. doi:10.1096/fj.12-215368

Gu JG, MacDermott AB (1997) Activation of ATP P2X receptors elicits glutamate release from sensory neuron synapses. Nature 389(6652):749–753. doi:10.1038/39639

Guo C, Masin M, Qureshi OS, Murrell-Lagnado RD (2007) Evidence for functional P2X4/P2X7 heteromeric receptors. Mole Pharmacol 72(6):1447–1456. doi:10.1124/mol.107.035980

Hattori M, Gouaux E (2012) Molecular mechanism of ATP binding and ion channel activation in P2X receptors. Nature 485(7397):207–212. doi:10.1038/nature11010

Holton FA, Holton P (1953) The possibility that ATP is a transmitter at sensory nerve endings. J Physiol 119(4):50P–51P

Holton FA, Holton P (1954) The capillary dilator substances in dry powders of spinal roots; a possible role of adenosine triphosphate in chemical transmission from nerve endings. J Physiol 126(1):124–140

Honore P, Mikusa J, Bianchi B, McDonald H, Cartmell J, Faltynek C, Jarvis MF (2002) TNP-ATP, a potent P2X3 receptor antagonist, blocks acetic acid-induced abdominal constriction in mice: comparison with reference analgesics. Pain 96(1–2):99–105

Jahr CE, Jessell TM (1983) ATP excites a subpopulation of rat dorsal horn neurones. Nature 304(5928):730–733

Khakh BS, North RA (2006) P2X receptors as cell-surface ATP sensors in health and disease. Nature 442(7102):527–532. doi:10.1038/nature04886

Khakh BS, North RA (2012) Neuromodulation by extracellular ATP and P2X receptors in the CNS. Neuron 76(1):51–69. doi:10.1016/j.neuron.2012.09.024

Krishtal OA, Marchenko SM, Pidoplichko VI (1983) Receptor for ATP in the membrane of mammalian sensory neurones. Neurosci Lett 35(1):41–45

Lewis C, Neidhart S, Holy C, North RA, Buell G, Surprenant A (1995) Coexpression of P2X2 and P2X3 receptor subunits can account for ATP-gated currents in sensory neurons. Nature 377(6548):432–435. doi:10.1038/377432a0

Miklavc P, Mair N, Wittekindt OH, Haller T, Dietl P, Felder E, Timmler M, Frick M (2011) Fusion-activated Ca2 + entry via vesicular P2X4 receptors promotes fusion pore opening and exocytotic content release in pneumocytes. Proc Natl Acad Sci USA 108(35):14503–14508. doi:10.1073/pnas.1101039108

Murgia M, Hanau S, Pizzo P, Rippa M, Di Virgilio F (1993) Oxidized ATP. An irreversible inhibitor of the macrophage purinergic P2Z receptor. J Biol Chem 268(11):8199–8203

Murrell-Lagnado RD, Qureshi OS (2008) Assembly and trafficking of P2X purinergic receptors (Review). Mole Membr Biol 25(4):321–331. doi:10.1080/09687680802050385

Nagata K, Imai T, Yamashita T, Tsuda M, Tozaki-Saitoh H, Inoue K (2009) Antidepressants inhibit P2X4 receptor function: a possible involvement in neuropathic pain relief. Mole Pain 5:20. doi:10.1186/1744-8069-5-20

Nicke A (2008) Homotrimeric complexes are the dominant assembly state of native P2X7 subunits. Biochem Biophys Res Commun 377(3):803–808. doi:10.1016/j.bbrc.2008.10.042

North RA (2002) Molecular physiology of P2X receptors. Physiol Rev 82(4):1013–1067. doi:10.1152/physrev.00015.2002

North RA (2003) The P2X3 subunit: a molecular target in pain therapeutics. Curr Opin Investig Drugs 4(7):833–840

North RA (2004) P2X3 receptors and peripheral pain mechanisms. J Physiol 554(Pt 2):301–308. doi:10.1113/jphysiol.2003.048587

North RA, Jarvis MF (2013) P2X receptors as drug targets. Mole Pharmacol 83(4):759–769. doi:10.1124/mol.112.083758

Palygin O, Lalo U, Verkhratsky A, Pankratov Y (2010) Ionotropic NMDA and P2X1/5 receptors mediate synaptically induced Ca2+ signalling in cortical astrocytes. Cell Calcium 48(4):225–231. doi:10.1016/j.ceca.2010.09.004

Qureshi OS, Paramasivam A, Yu JC, Murrell-Lagnado RD (2007) Regulation of P2X4 receptors by lysosomal targeting, glycan protection and exocytosis. J Cell Sci 120(Pt 21):3838–3849. doi:10.1242/jcs.010348

Richler E, Chaumont S, Shigetomi E, Sagasti A, Khakh BS (2008) Tracking transmitter-gated P2X cation channel activation in vitro and in vivo. Nat Methods 5(1):87–93. doi:10.1038/nmeth1144

Rivera C, Li H, Thomas-Crusells J, Lahtinen H, Viitanen T, Nanobashvili A, Kokaia Z, Airaksinen MS, Voipio J, Kaila K, Saarma M (2002) BDNF-induced TrkB activation down-regulates the K + -Cl- cotransporter KCC2 and impairs neuronal Cl- extrusion. J Cell Biol 159(5):747–752. doi:10.1083/jcb.200209011

Robertson SJ, Ennion SJ, Evans RJ, Edwards FA (2001) Synaptic P2X receptors. Curr Opin Neurobiol 11(3):378–386

Royle SJ, Qureshi OS, Bobanovic LK, Evans PR, Owen DJ, Murrell-Lagnado RD (2005) Non-canonical YXXGPhi endocytic motifs: recognition by AP2 and preferential utilization in P2X4 receptors. J Cell Sci 118(Pt 14):3073–3080. doi:10.1242/jcs.02451

Samways DS, Egan TM (2007) Acidic amino acids impart enhanced Ca2+ permeability and flux in two members of the ATP-gated P2X receptor family. J Gen Physiol 129(3):245–256. doi:10.1085/jgp.200609677

Serrano A, Mo G, Grant R, Pare M, O'Donnell D, Yu XH, Tomaszewski MJ, Perkins MN, Seguela P, Cao CQ (2012) Differential expression and pharmacology of native P2X receptors in rat and primate sensory neurons. J Neurosci: Official J Soc Neurosci 32(34):11890–11896. doi:10.1523/JNEUROSCI.0698-12.2012

Sim JA, Chaumont S, Jo J, Ulmann L, Young MT, Cho K, Buell G, North RA, Rassendren F (2006) Altered hippocampal synaptic potentiation in P2X4 knock-out mice. J Neurosci: Official J Soc Neurosci 26(35):9006–9009. doi:10.1523/JNEUROSCI.2370-06.2006

Sim JA, North RA (2010) Amitriptyline does not block the action of ATP at human P2X4 receptor. Brit J Pharmacol 160(1):88–92. doi:10.1111/j.1476-5381.2010.00683.x

Sim JA, Young MT, Sung HY, North RA, Surprenant A (2004) Reanalysis of P2X7 receptor expression in rodent brain. J Neurosci: Official J Soc Neurosci 24(28):6307–6314. doi:10.1523/JNEUROSCI.1469-04.2004

Solle M, Labasi J, Perregaux DG, Stam E, Petrushova N, Koller BH, Griffiths RJ, Gabel CA (2001) Altered cytokine production in mice lacking P2X(7) receptors. J Biol Chem 276(1):125–132. doi:10.1074/jbc.M006781200

Sorge RE, Trang T, Dorfman R, Smith SB, Beggs S, Ritchie J, Austin JS, Zaykin DV, Vander Meulen H, Costigan M, Herbert TA, Yarkoni-Abitbul M, Tichauer D, Livneh J, Gershon E, Zheng M, Tan K, John SL, Slade GD, Jordan J, Woolf CJ, Peltz G, Maixner W, Diatchenko L, Seltzer Z, Salter MW, Mogil JS (2012) Genetically determined P2X7 receptor pore formation regulates variability in chronic pain sensitivity. Nat Med 18(4):595–599. doi:10.1038/nm.2710

Stokes L, Scurrah K, Ellis JA, Cromer BA, Skarratt KK, Gu BJ, Harrap SB, Wiley JS (2011) A loss-of-function polymorphism in the human P2X4 receptor is associated with increased pulse pressure. Hypertension 58(6):1086–1092. doi:10.1161/HYPERTENSIONAHA.111.176180

Surprenant A, Rassendren F, Kawashima E, North RA, Buell G (1996) The cytolytic P2Z receptor for extracellular ATP identified as a P2X receptor (P2X7). Science 272(5262):735–738

Toulme E, Garcia A, Samways D, Egan TM, Carson MJ, Khakh BS (2010) P2X4 receptors in activated C8-B4 cells of cerebellar microglial origin. J Gen Physiol 135(4):333–353. doi:10.1085/jgp.200910336

Trang T, Beggs S, Salter MW (2011) Brain-derived neurotrophic factor from microglia: a molecular substrate for neuropathic pain. Neuron Glia Biol 7(1):99–108. doi:10.1017/S1740925X12000087

Trang T, Beggs S, Wan X, Salter MW (2009) P2X4-receptor-mediated synthesis and release of brain-derived neurotrophic factor in microglia is dependent on calcium and p38-mitogen-activated protein kinase activation. J Neurosci: Official J Soc Neurosci 29(11):3518–3528. doi:10.1523/JNEUROSCI.5714-08.2009

Tsuda M, Kuboyama K, Inoue T, Nagata K, Tozaki-Saitoh H, Inoue K (2009) Behavioral phenotypes of mice lacking purinergic P2X4 receptors in acute and chronic pain assays. Mole Pain 5:28. doi:10.1186/1744-8069-5-28

Tsuda M, Shigemoto-Mogami Y, Koizumi S, Mizokoshi A, Kohsaka S, Salter MW, Inoue K (2003) P2X4 receptors induced in spinal microglia gate tactile allodynia after nerve injury. Nature 424(6950):778–783. doi:10.1038/nature01786

Ulmann L, Hatcher JP, Hughes JP, Chaumont S, Green PJ, Conquet F, Buell GN, Reeve AJ, Chessell IP, Rassendren F (2008) Up-regulation of P2X4 receptors in spinal microglia after peripheral nerve injury mediates BDNF release and neuropathic pain. J Neurosci: Official J Soc Neurosci 28(44):11263–11268. doi:10.1523/JNEUROSCI.2308-08.2008

Ulmann L, Hirbec H, Rassendren F (2010) P2X4 receptors mediate PGE2 release by tissue-resident macrophages and initiate inflammatory pain. The EMBO J 29(14):2290–2300. doi:10.1038/emboj.2010.126

Valera S, Hussy N, Evans RJ, Adami N, North RA, Surprenant A, Buell G (1994) A new class of ligand-gated ion channel defined by P2x receptor for extracellular ATP. Nature 371(6497):516–519. doi:10.1038/371516a0

Vulchanova L, Riedl MS, Shuster SJ, Stone LS, Hargreaves KM, Buell G, Surprenant A, North RA, Elde R (1998) P2X3 is expressed by DRG neurons that terminate in inner lamina II. Eur J Neurosci 10(11):3470–3478

Yamamoto K, Sokabe T, Matsumoto T, Yoshimura K, Shibata M, Ohura N, Fukuda T, Sato T, Sekine K, Kato S, Isshiki M, Fujita T, Kobayashi M, Kawamura K, Masuda H, Kamiya A, Ando J (2006) Impaired flow-dependent control of vascular tone and remodeling in P2X4-deficient mice. Nat Med 12(1):133–137. doi:10.1038/nm1338

Chapter 31
P2X7 in Bipolar and Depressive Disorders

Melissa L. Barron, Eryn L. Werry, Iain S. McGregor and Michael Kassiou

Abstract Psychiatric illnesses affect a substantial number of people worldwide, with major depressive disorder (MDD) and bipolar disorder (BD) amongst the most common disorders. Current medications for MDD target monoamine systems, principally serotonin, norepinephrine and dopamine, to augment synaptic levels of some or all of these neurotransmitters. However, these treatments exhibit comparatively low efficacy relative to placebo, a slow onset of action, and a multiplicity of adverse side effects resulting in poor adherence. Similarly, the pharmacological treatment of BD involves drugs such as lithium, anticonvulsants and antipsychotics that have a range of troubling side effects. Novel therapeutic interventions are badly needed for MDD and BD to improve efficacy, minimise side effects and to maximise patient adherence. Recent reports suggest that mood disorders may sometimes involve a neuroinflammatory state, with particular involvement of cytokines such as interleukin (IL)-1β. Generation, maturation and release of IL-1β can occur via several pathways. One particular pathway is via the activation of the P2X7 receptor, a purinergic ligand-gated ion channel. This chapter focuses on recent research carried out on the P2X7 ion channel receptor, including intracellular signalling events following activation of this receptor, genetic linkage studies of P2X7 and results from studies involving P2X7 knockout mice. Our aim is to provide further insight into the possible role of P2X7 in MDD and BD and the use of P2X7 acting drugs as novel therapies for BD and MDD.

M. L. Barron · E. L. Werry · M. Kassiou
Brain and Mind Research Institute and Bosch Institute, University of Sydney, Sydney NSW 2050, Australia

M. L. Barron · E. L. Werry · M. Kassiou
Faculty of Health Sciences, University of Sydney, Sydney NSW 2141, Australia

I. S. McGregor
School of Psychology, University of Sydney, Sydney NSW 2006, Australia

M. Kassiou (✉)
School of Chemistry, University of Sydney, Sydney NSW 2006, Australia
e-mail: michael.kassiou@sydney.edu.au

N. Weiss and A. Koschak (eds.), *Pathologies of Calcium Channels*,
DOI: 10.1007/978-3-642-40282-1_31,

31.1 Bipolar and Depressive Disorders

Psychiatric disorders represent a major component of the overall global burden of disease, with major depressive disorder (MDD), otherwise known as unipolar depression, and bipolar disorder (BD) among the most common psychiatric conditions. Core symptoms of MDD include anhedonia (lack of sensitivity to pleasure), depressed or negative mood, loss of appetite, sleep disturbances, poor concentration and little interest in social activities. For a MDD diagnosis, these symptoms must persist beyond a 2-week period without being secondary to any current medical condition (DSM-IV-TR 2000). The overall burden of depression is on the rise, with an estimated 350 million people affected worldwide, and more than one million suicides annually (WHO 2011). It is predicted that MDD will become the leading cause of disability by the year 2030. Current first-line treatments for depression are ineffective in up to two-thirds of patients with MDD (Trivedi et al. 2006), with 15–33 % failing to respond to multiple interventions (Little 2009; Berlim et al. 2008). The chances of remission decrease considerably when patients fail to respond to two or more treatment strategies (Shelton et al. 2010).

Those who suffer from BD experience episodes of manic (bipolar I) or hypomanic (bipolar II) behaviour. Bipolar behaviour is characterised by elevated mood often accompanied by inflated self-esteem, decreased need for sleep, flight of ideas and impaired judgement, and is present for over 1 week with a diagnosis of mania or over 4 days with hypomania (DSM-IV-TR 2000). Patients with bipolar II disorder experience dominant periods of depression, with occasional hypomania, while patients with bipolar I disorder experience fully fledged mania and may or may not experience periods of depression. Bipolar disorder is ranked among the 10 most disabling medical conditions worldwide (WHO 2003), and approximately 15–20 % of people with BD end their own lives (Gonda et al. 2012). It is estimated that current treatments for BD are ineffective in about 50 % of patients (Fountoulakis et al. 2012a, b). Hence, there is an obvious need to identify new drug targets for more effective treatment of bipolar and major depressive disorders.

31.2 Pathways Linked to Psychiatric Disorders

A number of pathophysiological mechanisms have been proposed in MDD and BD. These two disorders generate a wide spectrum of symptoms; from tiredness, feelings of guilt and lack of interest to insomnia, loss of appetite, flight of ideas and suicidal ideation. Therefore, various neurobiological systems may be involved and are possibly affected by a series of different processes within each individual.

31.2.1 Depression

There are many apparent triggers for mood disorders, especially in the case of MDD, and this is reflected in a strong co-morbid association with other illnesses, such as epilepsy (Epps and Weinshenker 2012), as well as cardiovascular disease and type 2 diabetes (Katon et al. 2005). Moreover, there is evidence that some medications, such as levodopa and corticosteroids, may cause or exacerbate depression, complicating the treatment options for these patients (Little 2009) and highlighting the heterogeneity of the disease.

Prolonged stress or susceptibility to stress is a significant contributor to the development of mood disorders. In fact, it has been suggested that a primary link between stress and depression is an aberration in neuronal plasticity, whereby functional and structural alterations occur as a result of repetitive environmental insults (Calabrese et al. 2009). This remodelling of the brain's neural networks is considered one of the major pathological features in patients with psychiatric disorders (Knable et al. 2002). Chronic stress, anxiety and depression all similarly affect a number of brain areas, particularly the hippocampus and prefrontal cortex, with circuitry changes in neuronal and glial populations being the most consistent. A reduction in overall volume of these brain areas is also particularly noticeable, associated with altered glutamatergic, monoaminergic and γ-aminobutyric acid (GABA)ergic signalling pathways within the limbic system (Knable et al. 2004).

One of the earliest theories of MDD pathophysiology was the 'monoamine theory' of depression, which suggests neurochemical imbalances involving serotonin, norepinephrine and/or dopamine. These theories were constructed on the premise that current first-line drug treatments for MDD involved increasing synaptic levels of these transmitters. However, with many patients insensitive to treatments aimed at restoring these neurochemical imbalances, and the slow onset of antidepressant actions, the monoamine theory appears insufficient.

The hypothalamic–pituitary–adrenal (HPA) axis is often found to be dysfunctional in depressed patients and this has led to the corticosteroid hypothesis of depression (Holsboer 2000). The HPA complex regulates the adaptive hormonal response to stress, involving the interplay between corticotrophin-releasing hormone (CRH), adrenocorticotropic hormone (ACTH) and cortisol. Abnormal and excessive activation of the HPA axis in response to stress, measured by the increased production and secretion of these hormones, is regularly seen in depressed patients and is considered a good biomarker for the disorder (Holsboer 2000; De Kloet et al. 2005; Paez-Pereda and Panhuysen 2008). These irregularities provide evidence in support of a link between stress and mood. Some of the pathological consequences of chronic stress include neuronal cell death or neuronal damage involving reduced dendritic spines and branching within the hippocampus, an overall reduction in hippocampal size, as well as high glucocorticoid and cortisol levels (Joels et al. 2004).

Neuroplastic changes primarily in the hippocampus, amygdala and the prefrontal cortex, and reduced production of neurotrophins (proteins required for neuronal survival) also occur following chronic stress (Bennett 2011). These abnormalities have been consistently observed in patients with depression (McEwen 2005) and are the pathological mechanisms underpinning the neurotrophin hypothesis of depression. Some studies indicate that stress-induced elevated levels of glucocorticoid cause neuron damage and cell death in depressed patients, reducing hippocampal size as a consequence (Joels et al. 2004; Geuze et al. 2005; Gianaros et al. 2007; Paez-Pereda and Panhuysen 2008). However, these findings are not always consistent and not selective for depression, being observed in many other disease states (Nestler et al. 2002).

Currently used antidepressants target the monoamine system and this approach has barely changed in several decades. The earliest classes of medications were known as the monoamine oxidase inhibitors (MAOIs) and the tricyclic antidepressants (TCAs), and both elevate serotonin and/or norepinephrine signalling. The TCAs and MAOIs are still in use, but are generally only administered following failure of first-line treatment, due to an unfavourable side effect burden. Second-generation medications, like selective serotonin reuptake inhibitors (SSRIs) and serotonin and norepinephrine reuptake inhibitors (SNRIs), are more widely prescribed today due to their relatively low toxicity and high tolerability (Rush et al. 2006). However, they are no more effective than the TCAs or MAOIs in treating MDD. As noted earlier, patient responsiveness and drug effectiveness is markedly low; in the large STAR-D study only one-third of patients achieved remission with citalopram, a commonly used SSRI (Trivedi et al. 2006). In addition, a delay of 2–3 weeks for therapeutic relief and uncomfortable side effects results in poor compliance (Nakajima et al. 2010; Furukawa et al. 2002).

31.2.2 Bipolar Disorder

Bipolar disorder is a disorder of emotional regulation, the causes of which are unclear. Emotional regulation is associated with brain regions such as the anterior cingulate cortex, the prefrontal cortex, insula and limbic areas such as the amygdala (Townsend and Altshuler 2012; Phillips et al. 2003). Imaging studies in BD show abnormalities in these regions, including hypoactivation of the venterolateral prefrontal cortex in both mania and bipolar depression, hyperactivation of the amygdala in mania and diminished frontal-limbic connectivity (Townsend and Altshuler 2012). Voxel-based morphometry studies also suggest a decrease in grey matter in the anterior cingulate cortex and fronto-insular cortex in BD (Berk et al. 2011; Bora et al. 2010).

While the pathological mechanisms that contribute to these changes in emotional brain activity and volume are not fully resolved and are likely to be multifactorial, a predominant theory attributes these changes to a disturbed balance between inhibitory and excitatory neurotransmission (Lan et al. 2009). There is evidence for decreased GABAergic interneurons in BD with reduced markers for GABAergic neurons reported in the prefrontal cortex and anterior cingulate cortex

(Sibille et al. 2011; Beasley et al. 2002; Benes and Berretta 2001; Impagnatiello et al. 1998). A polymorphism in $GABA_A$ receptor genes is associated with a subtype of BD (Green et al. 2010; Breuer et al. 2011). GABAergic interneurons play a key role in cortico-limbic function (Lan et al. 2009), so a loss of these neurons may be detrimental to the functioning of brain areas that regulate emotion. GABAergic interneurons also ensure glutamatergic neurotransmission is not excessive, and accordingly, increased levels of glutamate are found in the anterior cingulate cortex and prefrontal cortex in BD (Lan et al. 2009; Maddock and Buonocore 2012; Yuksel and Ongur 2010). Excess glutamate can result in over-activation of one of its receptors, N-methyl D-aspartate (NMDA), and a subsequent excitotoxic increase in intracellular Ca^{2+} (Savitz et al. 2012). There are reports of increased intracellular Ca^{2+} and markers of apoptosis in the frontal cortex in BD (Berk et al. 2011; Rao et al. 2010), suggesting a mechanism for the decreased volume in key emotional regulation brain areas in BD.

The current gold-standard treatment for BD is lithium. While its mechanisms of action are not well understood, part of its effects are likely due to its ability to decrease intracellular Na^+ and Ca^{2+} levels (Catalan and Paz Quezada 2012) and increase GABA levels (Lan et al. 2009), stabilising neuronal electrical activity and the overall balance of excitatory versus inhibitory neurotransmission. Lithium may have further impact as a result of its neurotrophic effects and effects on intracellular signalling pathways (Gould et al. 2004).

The two other classes of drugs currently used in the treatment of BD are the anticonvulsants valproate and carbamazepine, and the second-generation antipsychotics such as quetiapine, asenapine and olanzapine (Fountoulakis et al. 2012a). Amongst other effects, valproate facilitates the activity of GABA and blocks Na^+ channels and carbamazepine blocks Na^+ channels and Ca^{2+} influx (Gould et al. 2004). The mechanism through which second-generation antipsychotics produce mood-stabilising effects in BD is uncertain (Gould et al. 2004).

The three main bipolar drug classes show efficacy in the treatment of acute mania, with clinical trials suggesting that around 50 % of patients respond to each of these drugs, compared to a 25 % response to placebos (Fountoulakis et al. 2012a). There is more limited efficacy in bipolar depression, with clinical trial data inconclusive at best for lithium and second-generation antipsychotics, as well as for common antidepressants such as the SSRI fluoxetine (Fountoulakis et al. 2012b). Anticonvulsants such as lamotrigine show some efficacy against bipolar depression, but this data is derived from a mixed population of MDD and bipolar depression so the selective effect on bipolar depression is unclear (Fountoulakis et al. 2012b). Some efficacy is also seen by combining the antipsychotic olanzapine with fluoxetine (Fountoulakis et al. 2012b).

Although current BD treatments show effectiveness in some patients, there are still many non-responders (around 50 %; (Fountoulakis et al. 2012a)). Of the patients that do respond, medication adherence can be low due to side effects (Gould et al. 2004; Nemeroff 2003). Over 80 % of people on common BD treatments experience adverse side effects, which can include kidney damage and hypothyroidism for lithium, teratogenic effects and weight gain for valproate, and weight gain, hyperlipidaemia and diabetes with the second-generation antipsychotics (Fountoulakis et al. 2012a). Furthermore, antidepressant drugs and lithium withdrawal may induce a switch from depression to mania (Belmaker 2004; Fountoulakis et al. 2012a). Lithium also has a narrow therapeutic window and relapse can be common (Gould et al. 2004; Fountoulakis et al. 2012a). Given these drawbacks, there is a pressing need for BD treatments that are safer and more efficacious.

Further advancement in MDD and BD drug development might be achieved by developing drugs that exert their action via different mechanisms to those currently on the market. Interestingly, elevated release of pro-inflammatory cytokines occurs with MDD, BD and with stress, and intervening to prevent this inflammatory response may represent a novel approach for alleviating these disorders. This cytokine hypothesis of mood disorders will be discussed below in more detail.

31.3 Cytokines in Mood Disorders

It is well recognised that inflammatory mediators play an important role in the body's immune response, but mounting evidence indicates that inflammation may also have a significant role in MDD and BD (Miller et al. 2009). The release of pro-inflammatory mediators, such as interleukin (IL)-1, IL-6 and tumour necrosis factor-α (TNF-α), usually occurs as a consequence of injury or infection. There are consistent findings that these particular cytokines are markedly raised in the plasma and cerebral spinal fluid (CSF) in a subpopulation of patients with MDD (Levine et al. 1999; Maes et al. 1995; Hestad et al. 2003). An increase in prostaglandin E_2 (PGE_2) levels has also been identified in the plasma, as well as in lymphocytes and saliva of depressive patients (Lieb et al. 1983; Ohishi et al. 1988), with antidepressants having some inhibitory effect on the synthesis of PGE_2 (Yaron et al. 1999) and several cytokines (Kim et al. 2007).

IL-1β levels are increased in the post-mortem BD frontal cortex (Rao et al. 2010), while serum and plasma levels of TNF- α are increased in both the manic and depressive phases of BD (Ortiz-Dominguez et al. 2007; Su et al. 2011; O'Brien et al. 2006; Kauer-Sant'Anna et al. 2009). Furthermore, increased monocyte mRNA levels of a range of pro-inflammatory mediators including IL-1β and TNF-α is considered one of the best biomarkers of BD (Padmos et al. 2008). From these observations, it is apparent that cytokines are altered in mood disorders, but the extent to which they participate remains to be elucidated.

Patients undergoing cytokine therapy for treatment of various cancers or chronic hepatitis C display symptoms of depression (Schiepers et al. 2005; Bonaccorso et al. 2001) or mania (Strite et al. 1997; Greenberg et al. 2000; Onyike et al. 2004; Constant et al. 2005). Furthermore, activation of the immune system is observed in many patients with depression, and those with immune dysfunction are more likely to suffer from depression (Sluzewska et al. 1996; Lane et al. 1991; Kronfol 2002). The cytokine hypothesis of depression was further substantiated when researchers were able to induce depression-like symptoms in mice through administration of pro-inflammatory molecules, such as lipopolysaccharides (LPS) and IL-1 (predominantly IL-1β) (Dantzer et al. 2008; Bluthé et al. 1994). The behavioural changes seen with these treatments are known as "sickness behaviour" and have many similarities to depression, chiefly the neurobehavioural signs such as fatigue, decreased interest in, and exploration of, the environment, decreased libido, blunted motivation and impaired cognition (Dantzer 2001; Schiepers et al. 2005). Animal studies that model depression using cytokine-induced sickness behaviour require careful extrapolation to human depressive states, given that sickness behaviour and depression are distinct, albeit related, phenomena. Reports both affirm and contest the use of these animal paradigms (McArthur and Borsini 2006), with some accounts of novel ideas for improvements that may help bridge the gap between animal and human studies of BD and MDD (Cryan and Slattery 2007). While some antidepressants have a therapeutic effect on cytokine-induced sickness behaviour in mice, not all have this effect, suggesting only a partial and incomplete map onto the pathophysiology of mood disorders (Dunn et al. 2005).

31.4 The Importance of IL-1β in Depressive Disorders

Recent reports indicate that IL-1β may be one of the major cytokines responsible for depressive symptoms. IL-1β is a pro-inflammatory mediator and is known to decrease adult hippocampal neurogenesis, a well-documented characteristic of depression (Santarelli et al. 2003; Warner-Schmidt and Duman 2006). Understanding the mechanisms behind this effect remains to be elucidated. It has been suggested by Zunszain et al. (2012) that IL-1β acts by affecting the kynurenine pathway, which involves the metabolism of tryptophan, resulting in neurotoxicity (Zunszain et al. 2012). This pathway follows the conversion of tryptophan into kynurenine, catalysed by the enzyme indoleamine-2,3-dioxygenase (IDO) in response to inflammation, and is notably activated during cytokine-induced depression (Capuron et al. 2002; Wichers and Maes 2004). This has been reliably observed in animal models of depression. In previous studies with mice, chronic exposure to IL-1β results in decreased numbers of neurons (Goshen et al. 2008). In support of this, administration of an IL-1 antagonist in rats blocks the decrease in levels of neurogenesis seen in animal models of depression (Koo and Duman 2008). Additionally, when LPS (an activator of microglia and IL-1β release) is administered, a reduction in neurogenesis is observed, further implicating the role of IL-1β in the pathophysiology of depression.

Increased levels of IL-1β may also contribute to the excitotoxic damage seen in BD. IL-1β can increase extracellular glutamate levels by stimulating the cystine/glutamate anti-porter (Jackman et al. 2010) and decreasing levels of GLT-1, the main astroglial glutamate transporter (Prow and Irani 2008). IL-1β-activation of IDO, as mentioned above, may also contribute to excitotoxicity in BD as increased levels of kynurenine have been reported in the BD anterior cingulate cortex (Miller et al. 2006). Kynurenine can be metabolised to quinolinic acid, which activates the NMDA receptor (Savitz et al. 2012). Given this, it is not surprising that polymorphisms in the IL-1β gene are associated with grey matter deficits in BD (Papiol et al. 2008).

IL-1β is one of many cytokines that is synthesised as a precursor molecule (pro-IL-1β) before being processed into a mature form, gaining access to the extracellular environment only through secretory mechanisms. The IL-1β-converting enzyme, also known as caspase-1, is an intracellular cysteine protease that cleaves pro-IL-1β (Fantuzzi and Dinarello 1999; Rothwell and Luheshi 2000), transforming it into the mature, biologically active protein IL-1β. Upon inhibition of caspase-1, IL-1β generation, maturation and secretion are reduced (Ghayur et al. 1997). Release of IL-1β is dependent on a number of signalling molecules, particularly adenosine triphosphate (ATP), which acts as the body's energy source as well as a specific neurotransmitter of the nervous system (Burnstock 1972). During inflammation or cellular damage, ATP is locally released where it has been shown to induce mature IL-1β release from LPS-primed cells, predominantly microglia, monocytes and macrophages (Di Virgilio et al. 2001). Additionally, IL-1β in return potently elicits ATP release from ex vivo rat hippocampal sections (Sperlágh et al. 2004), creating the possibility of a self-sustaining source of IL-1β in mood disorders. Emerging data from recent reports link ATP to the release of IL-1β through one promising target, the P2X7 receptor (P2X7R).

31.5 The P2X7 Receptor: Pharmacology and Physiology

The P2X7R (previously identified as P2Z) is a trimeric ligand-gated ion channel activated by ATP, which is abundantly expressed on cells of the immune system. In brief, the P2X7R belongs to the family of purinergic receptors, classified into two groups; P1 and P2, endogenously activated by adenosine and ATP, respectively. The P2 plasma membrane receptors can be further categorised into P2X, which are ion channel receptors and P2Y, which are G-protein coupled receptors. Currently, seven P2X receptor subunits have been characterised, P2X1–7. The P2X7R, the most studied of all P2 receptors, has a 595 amino acid sequence, with an average of 40 % homology with the other P2X members (Kaczmarek-Hájek et al. 2012). Interestingly, it also features a 240 amino acid carboxyl-terminal tail that is significantly longer than all other P2X receptor structures (Gunosewoyo et al. 2007) (Fig. 31.1). Another important difference is that the P2X7R requires a relatively high concentration of the endogenous ligand ATP for activation (Surprenant et al. 1996; Rassendren et al. 1997; Chessell et al. 1998). A synthetic analogue of ATP, 2′,3′-O-(4-benzoyl)benzoyl-ATP (BzATP), is a useful, higher potency agonist to P2X7R that allows easier manipulation for in vitro systems (Guile et al. 2009), although it is not selective (Kaczmarek-Hájek et al. 2012; Gunosewoyo and Kassiou 2010).

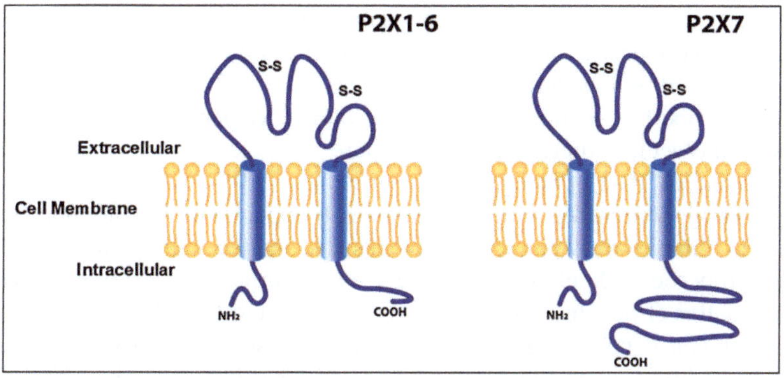

Fig. 31.1 Structural differences between P2X7 and the other P2X receptors, with emphasis on the longer C-terminus domain for P2X7

Fig. 31.2 The channel properties of the P2X7 receptor; short stimulation with ATP for the ion channel and prolonged ATP application for pore formation

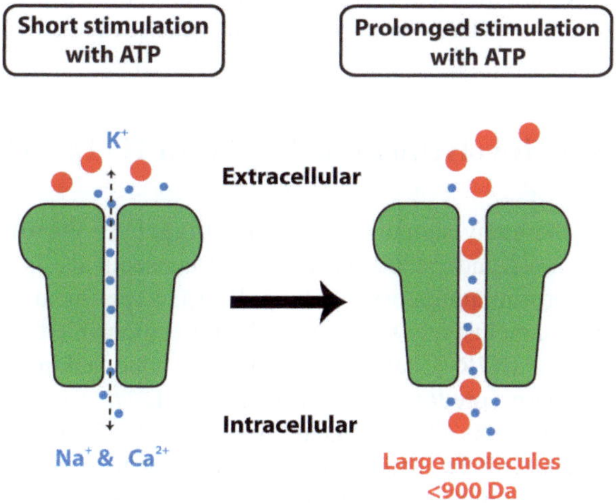

In addition to being a typical cation-selective ion channel that allows the influx of Ca^{2+} and Na^{+} and efflux of K^{+}, the P2X7R has the rare ability to form pores that are permeable to molecules of up to 900 Da when exposed to repeated or prolonged application of its nucleotide agonist (Gallagher 2004; Gunosewoyo et al. 2009). This makes the P2X7R a bifunctional receptor, which acts through non-selective pores allowing transmembrane fluxes of large molecules to occur (Falzoni et al. 2000) (Fig. 31.2). Recent evidence has shown that P2X4 also has this ability (Bernier et al. 2012). How this pore formation transpires is quite controversial and remains under investigation. There have been at least two conflicting hypotheses postulated to explain pore formation: (1) the P2X7 pore is formed as a result of cationic channel dilation, increasing its permeability from small to large molecules (Virginio et al. 1999a, b) and (2) the P2X7 pore represents a distinct channel, possibly activated by second messengers, such as Ca^{2+} and MAP kinases, and not directly by extracellular nucleotides (Faria et al. 2005; Liang and Schwiebert 2005).

Recent work has identified a protein, known as pannexin-1, as the critical accessory molecule required for P2X7R pore formation to occur and the consequent release of IL-1β (Pelegrin and Surprenant 2006). A member of a hemichannel gap-junction-like family, pannexin-1 appears to be the pore-forming protein, as inhibition of pannexin-1 blocks dye uptake without altering the ion channel properties (Ma et al. 2009; Pelegrin and Surprenant 2009). However, this is still a matter of debate as another study has reported that effective inhibition of pannexin-1 can occur using agonists and antagonists of P2X7R (Qiu and Dahl 2009). In addition, pannexin-1 knockout mice have been shown to still display P2X7 pore formation (Qu et al. 2011). Dissociation of ATP from a non-muscle myosin from the P2X7R complex has also been suggested as a possible protein-anchoring mechanism for pore formation (Gu et al. 2009). Regardless of how the pores are induced, this rare ability allows the P2X7R to initiate multiple signalling events and thus varied functional consequences, including IL-1β release.

31.6 Involvement of P2X7R in IL-1β Release

Of the many signalling pathways triggered following P2X7R activation, perhaps the most understood physiological consequence is the maturation and release of the pro-inflammatory cytokine, IL-1β. Experiments performed in vitro and in vivo have conclusively shown the effector responsible for ATP-dependent IL-1β release is the P2X7R (Chakfe et al. 2002). In mice lacking the receptor, the inflammatory response, predominantly driven by IL-1β release, is attenuated and cytokine production is altered (Labasi et al. 2002; Solle et al. 2001; Honore et al. 2009). The mechanism by which the P2X7R leads to caspase-1 activation and thus IL-1β release is not well understood. It has been postulated that this process involves K^+ efflux, as inhibition of K^+ transport prevents caspase-1 activation and subsequent IL-1β release, while enhanced K^+ efflux is associated with an increase in IL-1β release following P2X7R activation (Sanz and Virgilio 2000). While K^+ efflux occurs, it has also been documented that this efflux is concomitant upon the release of Ca^{2+} from endoplasmic reticulum stores (Brough et al. 2003). According to these studies, it appears that increases in intracellular Ca^{2+}, and consequently intracellular K^+ depletion, results in the activation of the NALP3 inflammasome protein complex, which in turn converts pro-caspase-1 to its active caspase-1 form, culminating with mature IL-1β release (Di Virgilio 2007; Bhattacharya et al. 2011; Wiley et al. 2011). Furthermore, in a study using NALP3-deficient mice, serum IL-1β levels were significantly impaired when compared to wildtype, demonstrating the importance of the NALP3 inflammasome in ATP-driven activation of caspase-1, and subsequently, IL-1β release (Sutterwala et al. 2006).

Since IL-1β lacks a signal peptide for self-secretion, it is dependent on its extracellular surroundings consisting of non-classical pathways. Some authors have proposed mechanisms to account for ATP-induced IL-1β release, suggesting ATP-induced cell lysis and cell death play a role, while other researchers postulate cell lysis is not required (Bhattacharya et al. 2011) and that ATP possibly triggers exocytosis of secretory lysosomes enclosing both IL-1β and caspase-1 (Qu and Dubyak 2009). In human monocytes, lysosome exocytosis and IL-1β release can be blocked by inhibitors targeting Ca^{2+}-dependent and Ca^{2+}-independent phospholipase A_2 and C, implicating these enzymes in the inflammation pathways required for IL-1β secretion (Andrei et al. 2004). However, when extracellular Ca^{2+} is removed from murine macrophages, the ATP-induced lysosome exocytosis is completely abolished whereas the IL-1β release still occurs, suggesting that lysosome secretion can arise independent of the non-classical P2X7R-regulated IL-1β release pathway (Qu et al. 2007).

31.7 The P2X7 Receptor and Depressive Disorders

A link between P2X7R and disease is not a novel concept. The receptor is ubiquitous in nearly all tissues and organs of the body, with its expression principally on immune cells, such as lymphocytes, monocytes and macrophages. It has been linked to disorders such as hypertension (Ji et al. 2012), diabetes (Sugiyama et al. 2004), osteoporosis (Ohlendorff et al. 2007; Jørgensen et al. 2012) and tuberculosis (Franco-Martínez et al. 2006; Singla et al. 2012). In the brain, it is found on glial cells including astrocytes and microglia, as well as neurons (Sperlágh et al. 2006). With the receptor having such a large representation on brain cells, it is no surprise that it has been implicated in neurological disorders. Recent studies have linked P2X7R to Alzheimer's disease (Rampe et al. 2004), Parkinson's disease (Marcellino et al. 2010), neuropathic pain (Chessell et al. 2005), multiple sclerosis (Yiangou et al. 2006), as well as depressive disorders (Lucae et al. 2006).

31.7.1 Molecular Pathways of P2X7 Involvement

The P2X7R makes a promising target for research into MDD and BD aetiology. As previously described, both disorders have an association with inflammation, with increased production of pro-inflammatory cytokines including IL-1β, which are frequently observed in patients (Maes et al. 1993; Owen et al. 2001; Thomas et al. 2005). Since the P2X7R has a prevalent expression on microglia, the brain's key immune cells, and one of its well-established functional roles is its participation in IL-1β release, the P2X7R may play a role in MDD and BD through release of IL-1β (Fig. 31.3). In addition, it has been well established that apoptosis is an effect of P2X7R pore activation (Humphreys et al. 2000). With evidence suggesting cell

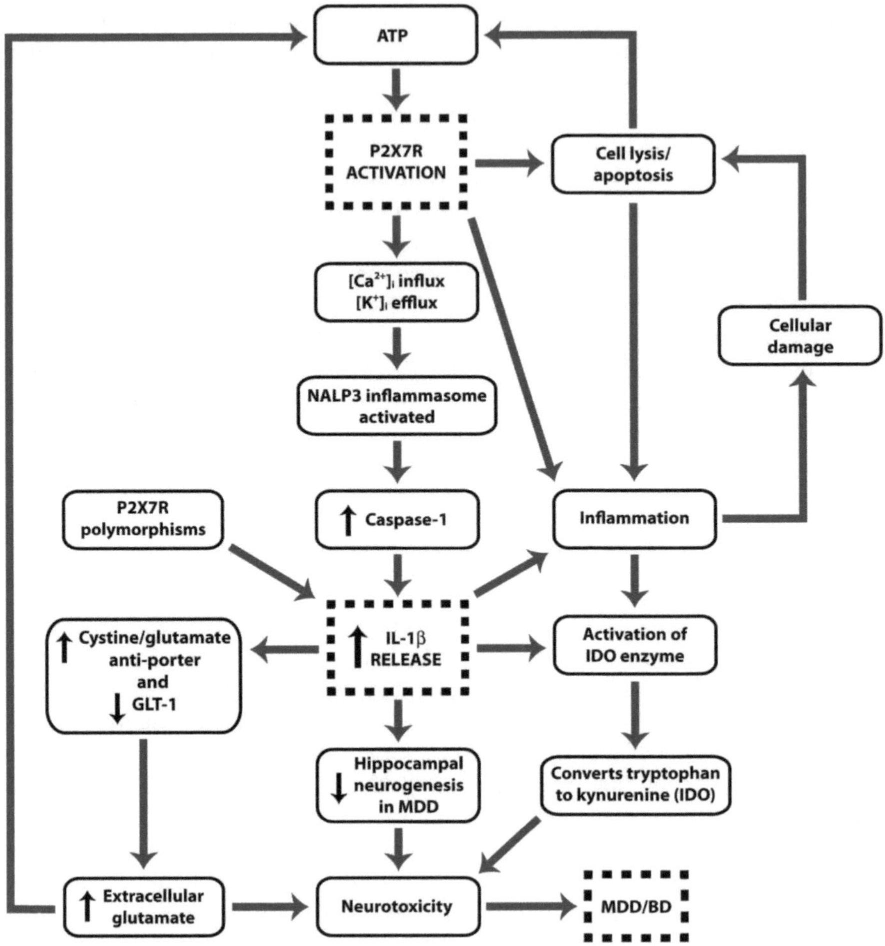

Fig. 31.3 A schematic diagram of the proposed pathomechanisms in bipolar and depressive disorders involving P2X7R-activation and IL-1β release

loss in certain brain areas in depressive disorders, there may also be a link for P2X7R involvement through other signalling pathways independent of IL-1β release (Le Feuvre et al. 2002; Brough et al. 2002). Alternatively, it has been hypothesised that an increase in central glutamatergic synaptic transmission within the neuronal networks of the hippocampus can evoke ATP release from astrocytes and microglia, providing a regenerative-loop for P2X7R activation (Bennett 2007), especially since IL-1β release increases extracellular glutamate (Jackman et al. 2010). Furthermore, recent investigations demonstrate a strong correlation between the gene encoding the P2X7R and affective disorders (Nagy et al. 2008; Soronen et al. 2011; Lucae et al. 2006; Barden et al. 2006), as well as results with

P2X7R knockout mice that are consistent with a role in depressive disorders (Basso et al. 2009; Boucher et al. 2011).

31.7.2 Genetic Studies

There is convincing data demonstrating that mood disorders, particularly bipolar and major depression, are triggered by the interaction of inherited factors together with environmental factors. The heritability of MDD is considered to be around 40–60 % (Sullivan et al. 2000), while the heritability for BD is estimated to be higher—at close to 90 % (Craddock and Sklar 2009). Evidence from familial, twin and adoption studies have assisted in determining the susceptibility genes linked with these affective diseases (Craddock and McGuffin 1993; Craddock and Owen 1994; Zubenko et al. 2003). The first linkage study was reported by Dawson et al. (1995), identifying the chromosome 12q23-24.1 as a suggested 'candidate' gene associated with BD (Dawson et al. 1995). Additional bipolar-related linkage studies soon followed. Ewald et al. (1998) investigated 16 microsatellite markers across chromosome 12q22-24 in two Danish families, demonstrating a strong correlation to BD, although the exact location of the affective locus was uncertain (Ewald et al. 1998). Further evidence in support of a linkage of chromosome 12q23-24 to BD has been reported in populations from the United Kingdom and Iceland (Curtis et al. 2003), Faroe Island (Degn et al. 2001) and Quebec (Morissette et al. 1999; Shink et al. 2005).

A correlation with chromosome 12q23-24 is also established in MDD by means of familial linkage studies. A genome-wide investigation was conducted on 1,890 Utah individuals with a strong family history of depression (Abkevich et al. 2003). In this study, the presence of one or more genes on the q arm of chromosome 12 appeared to be involved in depressive illnesses. Interestingly, the authors reported for the first time the existence of a sex-specific predisposition gene to MDD at chromosome 12q22-23.2. Another genome-wide linkage analysis was conducted on a sample consisting of 497 siblings, providing persuasive evidence that chromosome 12q carries a genetic connection to recurrent depression (McGuffin et al. 2005).

The P2X7R gene (P2RX7) is 13 exons in length and located on the chromosome region 12q24.31, a position identified as a susceptibility candidate locus for mood disorders by linkage and association studies, as described above. P2X7R is highly polymorphic with 32 non-synonymous single nucleotide polymorphisms (SNPs) carrying amino acid changes currently documented (Fuller et al. 2009). Several of these SNPs have been described as having allele variations affecting receptor function. Currently, eight loss-of-function polymorphisms in the P2X7R gene have been well characterised, with three of these resulting in total loss of receptor function. Another four non-synonymous variants have been characterised as gain-of-function mutations (Sluyter and Stokes 2011). A recent case–control study, conducted by McQuillin et al. (2009), shows convincing evidence that at least one of these functional polymorphisms in the P2X7R is distinctly connected to both bipolar and unipolar-affective disorders (McQuillin et al. 2009). This SNP, rs2230912, is located in exon 13 of the P2RX7 gene, resulting in a non-conservative amino acid change from a glutamine to arginine at position 460 (Gln460Arg), situated in the long intracellular C-terminal domain of the P2X7R (Denlinger et al. 2001). Functionally, this SNP displays a partial loss of pore formation during ATP-induced ethidium bromide uptake in P2X7R-mutated, transfected HEK293 cells (Fuller et al. 2009) and is likely to affect P2RX7 dimerization and protein–protein interactions (Lucae et al. 2006). However, another study has associated Gln460Arg with a gain-of-function haplotype as it was found to enhance pore activity (Denlinger et al. 2006). Nevertheless, the Gln460Arg mutation has thus far shown to hold the strongest association to affective disorders (Barden et al. 2006; Lucae et al. 2006).

The functional importance of P2X7R and its allele variants in mood disorders is not well established, which is perhaps partly due to the heterogeneic nature of depressive illnesses. One study examined whether patients with diabetes mellitus were genetically predisposed to develop MDD as both frequently occur concurrently (Nagy et al. 2008). In a Hungarian diabetic population, candidate polymorphisms of serotonin, tryptophan and P2X7R genes were analysed in patients not taking antidepressants, with the authors reporting a significant correlation between the P2X7R polymorphism Gln460Arg (rs2230912) and depression. Two earlier studies, involving French Canadian (Barden et al. 2006) and German Caucasian populations (Lucae et al. 2006), as well as another more recent clinical cohort study carried out in Finland (Soronen et al. 2011), also reported a strong linkage between the Gln460Arg haplotype and MDD. A separate research group using an equivalent Hungarian cohort to Nagy et al. (2008) also reported an association between Gln460Arg and MDD, but only when based on a symptom severity scale rather than a control comparison (Hejjas et al. 2009). In the same investigation, the authors also reported a similar association between the P2X7R haplotype and anxiety, strengthening the concept that anxiety and depression occur co-morbidly (Clarke and Currie 2009), as well as supporting the work done by Nagy et al. (2008). Interestingly, it was noted by Hejjas et al. (2009) that patients with the Gln460Arg genotype showed higher level depression scores; however, control subjects carrying the same P2X7R polymorphism had no affect on their level of depression (Hejjas et al. 2009), indicating that additional factors may be involved. Furthermore, the lack of significant difference between patient and control groups in the Hungarian population is supported by another research group in the UK, documenting no correlation (Green et al. 2009). However, a dimensional approach based on symptom severity was not conducted in this study, and the association reported by Hejjas et al. (2009) was only determined using this severity form of analysis. In addition, a more recent case–control study has suggested a link between Gln460Arg and BD, which is also based on a depression score that determines severity of symptoms. However, when the authors compared genotype frequencies between patients and controls, there was no correlation (Halmai et al. 2013).

Although a number of studies show an association between the Gln460Arg SNP and affective disorders, several more recent genetic studies have not supported this link (Green et al. 2009; Grigoroiu-Serbanescu et al. 2009; Lavebratt et al. 2010; Viikki et al. 2011). In fact, one research group failed to note any detectable changes in ATP-activated ion channel and pore formation ability of the P2X7R in cells with this mutation (Roger et al. 2010). Interestingly, in this study, Gln460Arg was the only mutation not to have a functional effect on its own. It may be possible that the correlation between P2X7R and mood disorders could be due to interactions between multiple polymorphisms; for example, the Ala348Thr (rs1718119) mutation has been shown to be critical for a gain-of-function effect when associated with Gln460Arg, with smaller contributions from SNPs His155Tyr and His270Arg (Stokes et al. 2010). This could explain the lack of significance for Gln460Arg as a susceptibility gene for MDD and BD, as recorded in more recent studies (Green et al. 2009; Grigoroiu-Serbanescu et al. 2009; Lavebratt et al. 2010; Viikki et al. 2011). Even so, evidence still points to the P2X7R as playing a pivotal role in psychiatric illnesses. In particular, the presence of Ala348Thr not only exhibits a gain-of-function as mentioned above, but also enhances the secretion of IL-1β (Stokes et al. 2010), one of the proposed hallmark biomarkers of depression (Müller et al. 2011). Moreover, the Ala348Thr mutation on its own has demonstrated an association with anxiety disorders (Erhardt et al. 2007), further implicating the P2X7R in mood disorders. Three more recent studies have shown a correlation between P2RX7 gene variants and specific symptoms or indicators relating to mood disorders, in particular, rapid cycling (a form of bipolar) that affects sleep (Backlund et al. 2012), manic behaviour associated with BD (Backlund et al. 2011) and neuroticism (Mantere et al. 2012). While there are a number of other mutations that have an influence on P2X7R functionality that require further examination, some convincing evidence has been documented in knockout mouse models linking P2X7R and mood disorders.

31.7.3 Studies with Knockout Mice

Animal models of depression are based on the assumption that a specific behavioural phenotype resembles a human depressive-like symptom that will be sensitive to clinically effective antidepressant drugs (Dunn et al. 2005). These models can provide insight into the mechanisms underlying affective disorders, particularly the role of pro-inflammatory cytokines. As previously stated, administration of cytokines, such as IL-1β, into animals can induce a 'sickness behaviour', a somewhat similar behavioural pattern to that seen in anxiety and depression. A recent study using IL-1 receptor null mice provided evidence that the IL-1β signalling pathway induces anxiety-related behaviours (Koo and Duman 2009). Since the P2X7R plays a crucial role in IL-1β release, examination of depressive-like symptoms in P2X7R knockout mice has been reported by three research groups (Basso et al. 2009; Boucher et al. 2011; Csolle et al. 2012).

The forced swim test (FST) and the tail suspension test (TST) are the two most common paradigms used to induce depression-like behaviour in animals. In brief, the FST entails placing mice into individual cylinders containing water, where the mice are observed swimming, becoming immobile over a 6 min period. Immobility time is measured in the final 4 min of the test and an antidepressant response is characterised by a decrease in time spent immobile (Dalvi and Lucki 1999). The TST is similar to FST, in that it measures immobility, but in this test the mice are suspended by the tail to a horizontal bar using adhesive tape. Following an initial period of agitation, the animals soon become immobile, an effect that is reversed upon antidepressant therapy (Dalvi and Lucki 1999). The first investigation on P2X7R knockout mice using these two behavioural models reported an antidepressant-like phenotype, with significantly less time spent immobile compared to wildtype controls (Basso et al. 2009). The null mice also showed a greater response to the antidepressant drug imipramine at a sub-threshold dose of 15 mg/kg in the FST, indicating a relationship between lack of P2X7R and improved responsiveness to antidepressant treatments. Interestingly, there were no significant differences between the knockout and wildtype mice for all anxiety-related behaviours examined (Basso et al. 2009).

In contrast, a more recent study with P2X7R knockout mice showed no difference between null and control mice following FST on the first day, but did demonstrate significantly greater immobility after 3 consecutive days of FST in control mice, providing further evidence that deleting the P2X7R engenders resilience against depression-like behaviours when subjected to chronic stress (Boucher et al. 2011). This repeated exposure to stress triggering depressive symptoms has also been reported in humans (Kendler et al. 1999). A very recent study carried out by Csolle et al. (2012) documented a similar difference in immobility between wildtype mice and P2X7R knockout mice, following repeated FST and TSP tests (Csolle et al. 2012). Their investigation also examined possible pathways involved in the antidepressant-like phenotype profile, with IL-1β protein significantly reduced in the knockout mice within the amygdala. Furthermore, these P2X7R knockout mice had an attenuated response to the amphetamine-induced hyperlocomotion test model used to model BD-like symptoms (Csolle et al. 2012).

In conclusion, even though animal models for depression have limitations and caution must be taken when extrapolating these results to human depression, these studies and future work will help provide a better understanding of the aetiology and mechanisms behind affective disorders, allowing current treatments to be improved and developed.

31.8 P2X7R Drug Development: Improving Current Treatments

There has been little improvement in antidepressant therapy over the last several decades, and with a considerable amount of patients not responding to current treatments or relapsing after remission, new therapeutic targets are urgently needed. Emerging evidence suggests that compounds targeting the P2X7R have potential utility in providing relief to MDD and BD sufferers, by inhibiting cytokine release, especially the release of IL-1β. At present, there are a few research groups developing P2X7R-specific antagonists, with one in particular focusing on novel compounds that are able to cross the blood–brain barrier (Gunosewoyo et al. 2008). Others have either developed prospective drugs or tested known P2X7R antagonists for therapeutic use across a wide spectrum of illnesses, such as neuropathic pain (Honore et al. 2006; Donnelly-Roberts and Jarvis 2007), arthritis (Dell'Antonio et al. 2002) and Huntington's disease (Díaz-Hernández et al. 2009). There are also some reports of P2X7R antagonists being used to target inflammation (Broom et al. 2008; Dell'Antonio et al. 2002), and more importantly, IL-1β release (Honore et al. 2009). The scene is thus set for promising new research involving the development of P2X7R antagonists as a novel treatment of major depressive and bipolar disorders.

Acknowledgments The authors would like to recognise and thank Dr. Marcus J. Robinson, at The University of Sydney, for his assistance in producing the figures included in this chapter.

References

Abkevich V, Camp NJ, Hensel CH, Neff CD, Russell DL, Hughes DC, Plenk AM, Lowry MR, Richards RL, Carter C (2003) Predisposition locus for major depression at chromosome 12q22–12q23. 2. Am J Hum Genet 73(6):1271–1281

Andrei C, Margiocco P, Poggi A, Lotti LV, Torrisi M, Rubartelli A, Dinarello CA (2004) Phospholipases C and A2 Control Lysosome-Mediated IL-1β secretion: implications for inflammatory processes. In: Proceedings of the National Academy of Sciences of the United States of America, pp 9745–9750

Backlund L, Lavebratt C, Frisén L, Nikamo P, Hukic SD, Träskman-Bendz L, Landén M, Edman G, Vawter M, Osby U (2012) P2RX7: expression responds to sleep deprivation and associates with rapid cycling in bipolar disorder Type 1. PloS one 7(8):e43057

Backlund L, Nikamo P, Hukic DS, Ek IR, Träskman-Bendz L, Landén M, Edman G, Schalling M, Frisén L, Ösby U (2011) Cognitive manic symptoms associated with the P2RX7 gene in bipolar disorder. Bipolar Disord 13(5–6):500–508

Barden N, Harvey M, Gagné B, Shink E, Tremblay M, Raymond C, Labbé M, Villeneuve A, Rochette D, Bordeleau L (2006) Analysis of single nucleotide polymorphisms in genes in the chromosome 12Q24. 31 region points to P2RX7 as a susceptibility gene to bipolar affective disorder. Am J Med Genet Part B: Neuropsychiatric Genet 141(4):374–382

Basso A, Bratcher N, Harris R, Jarvis M, Decker M, Rueter L (2009) Behavioral profile of P2X7 receptor knockout mice in animal models of depression and anxiety: relevance for neuropsychiatric disorders. Behav Brain Res 198(1):83

Beasley CL, Zhang ZJ, Patten I, Reynolds GP (2002) Selective deficits in prefrontal cortical GABAergic neurons in schizophrenia defined by the presence of calcium-binding proteins. Biol Psychiatry 52(7):708–715

Belmaker RH (2004) Bipolar disorder. New Engl J Med 351(5):476–486

Benes FM, Berretta S (2001) GABAergic interneurons: implications for understanding schizophrenia and bipolar disorder neuropsychopharmacology. Official Publ Am Coll Neuropsychopharmacol 25(1):1–27

Bennett MR (2007) Synaptic P2X7 receptor regenerative-loop hypothesis for depression. Aust N Z J Psychiatry 41(7):563–571

Bennett MR (2011) The prefrontal–limbic network in depression: a core pathology of synapse regression. Prog Neurobiol 93(4):457–467

Berk M, Kapczinski F, Andreazza AC, Dean OM, Giorlando F, Maes M, Yucel M, Gama CS, Dodd S, Dean B, Magalhaes PV, Amminger P, McGorry P, Malhi GS (2011) Pathways underlying neuroprogression in bipolar disorder: focus on inflammation, oxidative stress and neurotrophic factors. Neurosci Biobehav Rev 35(3):804–817

Berlim MT, Fleck MP, Turecki G (2008) Current trends in the assessment and somatic treatment of resistant/refractory major depression: an overview. Ann Med 40(2):149–159

Bernier LP, Ase AR, Boué-Grabot É, Séguéla P (2012) P2X4 receptor channels form large noncytolytic pores in resting and activated microglia. Glia

Bhattacharya A, Neff R, Wickenden A (2011) The physiology, pharmacology and future of P2X7 as an analgesic drug target: hype or promise? Curr Pharm Biotechnol 12(10):1698

Bluthé RM, Walter V, Parnet P, Layé S, Lestage J, Verrier D, Poole S, Stenning BE, Kelley KW, Dantzer R (1994) Lipopolysaccharide induces sickness behaviour in rats by a vagal mediated mechanism. C R Acad Sci 317(6):499

Bonaccorso S, Puzella A, Marino V, Pasquini M, Biondi M, Artini M, Almerighi C, Levrero M, Egyed B, Bosmans E (2001) Immunotherapy with interferon-alpha in patients affected by chronic hepatitis C induces an intercorrelated stimulation of the cytokine network and an increase in depressive and anxiety symptoms. Psychiatry Res 105(1–2):45

Bora E, Fornito A, Yucel M, Pantelis C (2010) Voxelwise meta-analysis of gray matter abnormalities in bipolar disorder. Biol Psychiatry 67(11):1097–1105

Boucher A, Arnold J, Hunt G, Spiro A, Spencer J, Brown C, McGregor I, Bennett M, Kassiou M (2011) Resilience and reduced c-Fos expression in P2X7 receptor knockout mice exposed to repeated forced swim test. Neuroscience 189:170–177

Breuer R, Hamshere ML, Strohmaier J, Mattheisen M, Degenhardt F, Meier S, Paul T, O'Donovan MC, Muhleisen TW, Schulze TG, Nothen MM, Cichon S, Craddock N, Rietschel M (2011) Independent evidence for the selective influence of GABA(A) receptors on one component of the bipolar disorder phenotype. Mol Psychiatry 16(6):587–589

Broom DC, Matson DJ, Bradshaw E, Buck ME, Meade R, Coombs S, Matchett M, Ford KK, Yu W, Yuan J (2008) Characterization of *N*-(adamantan-1-ylmethyl)-5-[(3R-aminopyrrolidin-1-yl) methyl]-2-chloro-benzamide, a P2X7 antagonist in animal models of pain and inflammation. J Pharmacol Exp Ther 327(3):620–633

Brough D, Le Feuvre RA, Iwakura Y, Rothwell NJ (2002) Purinergic (P2X7) receptor activation of microglia induces cell death via an interleukin-1-independent mechanism. Mol Cell Neurosci 19(2):272–280

Brough D, Le Feuvre RA, Wheeler RD, Solovyova N, Hilfiker S, Rothwell NJ, Verkhratsky A (2003) Ca^{2+} stores and Ca^{2+}entry differentially contribute to the release of IL-1β and IL-1α from murine macrophages. J Immunol 170(6):3029–3036

Burnstock G (1972) Purinergic nerves. Pharmacol Rev 24(3):509

Calabrese F, Molteni R, Racagni G, Riva MA (2009) Neuronal plasticity: a link between stress and mood disorders. Psychoneuroendocrinology 34(Suppl 1):S208–S216

Capuron L, Ravaud A, Neveu P, Miller A, Maes M, Dantzer R (2002) Association between decreased serum tryptophan concentrations and depressive symptoms in cancer patients undergoing cytokine therapy. Mol Psychiatry 7(5):468–473

Catalan JF, Paz Quezada M (2012) Lithium: molecular interactions, clinical actions. Transl Neurosci 3(1):61–66

Chakfe Y, Seguin R, Antel JP, Morissette C, Malo D, Henderson D, Séguéla P (2002) ADP and AMP induce interleukin-1β release from microglial cells through activation of ATP-primed P2X7 receptor channels. J Neurosci 22(8):3061–3069

Chessell I, Hatcher J, Bountra C, Michel A, Hughes J, Green P, Egerton J, Murfin M, Richardson J, Peck W (2005) Disruption of the P2X7 purinoceptor gene abolishes chronic inflammatory and neuropathic pain. Pain 114(3):386

Chessell I, Michel A, Humphrey P (1998) Effects of antagonists at the human recombinant P2X7 receptor. Br J Pharmacol 124(6):1314–1320

Clarke DM, Currie KC (2009) Depression, anxiety and their relationship with chronic diseases: a review of the epidemiology, risk and treatment evidence. Med J Aust 190(7):54

Constant A, Castera L, Dantzer R, Couzigou P, de Ledinghen V, Demotes-Mainard J, Henry C (2005) Mood alterations during interferon-alfa therapy in patients with chronic hepatitis C: evidence for an overlap between manic/hypomanic and depressive symptoms. J Clin Psychiatry 66(8):1050–1057

Craddock N, McGuffin P (1993) Approaches to the genetics of affective disorders. Ann Med 25(4):317–322

Craddock N, Owen M (1994) Chromosomal aberrations and bipolar affective disorder. Br J Psychiatry 164(4):507–512

Craddock N, Sklar P (2009) Genetics of bipolar disorder: successful start to a long journey. Trends Genet 25(2):99–105

Cryan JF, Slattery DA (2007) Animal models of mood disorders: recent developments. Curr Opin Psychiatry 20(1):1

Csolle C, Andó RD, Kittel A, Goloncser F, Baranyi M (2012) The absence of P2X7 receptors (P2rx7) on non-haematopoietic cells leads to selective alteration in mood-related behaviour with dysregulated gene expression and stress reactivity in mice. Int J Neuropsychopharmacol 1(1):1–21

Curtis D, Kalsi G, Brynjolfsson J, McInnis M, O'Neill J, Smyth C, Moloney E, Murphy P, McQuillin A, Petursson H (2003) Genome scan of pedigrees multiply affected with bipolar disorder provides further support for the presence of a susceptibility locus on chromosome 12q23-q24, and suggests the presence of additional loci on 1p and 1q. Psychiatr Genet 13(2):77–84

Dalvi A, Lucki I (1999) Murine models of depression. Psychopharmacology 147(1):14–16

Dantzer R (2001) Cytokine-induced sickness behavior: where do we stand? Brain Behav Immun 15(1):7–24

Dantzer R, O'Connor JC, Freund GG, Johnson RW, Kelley KW (2008) From inflammation to sickness and depression: when the immune system subjugates the brain. Nat Rev Neurosci 9(1):46–56

Dawson E, Parfitt E, Roberts Q, Daniels J, Lim L, Sham P, Nöthen M, Propping P, Lanczik M, Maier W (1995) Linkage studies of bipolar disorder in the region of the Darier's disease gene on chromosome 12q23-24.1. Am J Med Genet 60(2):94–102

De Kloet ER, Joëls M, Holsboer F (2005) Stress and the brain: from adaptation to disease. Nat Rev Neurosci 6(6):463–475

Degn B, Lundorf M, Wang A, Vang M, Mors O, Kruse T, Ewald H (2001) Further evidence for a bipolar risk gene on chromosome 12q24 suggested by investigation of haplotype sharing and allelic association in patients from the Faroe Islands. Mol Psychiatry 6(4):450

Dell'Antonio G, Quattrini A, Dal Cin E, Fulgenzi A, Ferrero M (2002) Antinociceptive effect of a new P2Z/P2X7 antagonist, oxidized ATP, in arthritic rats. Neurosci Lett 327(2):87–90

Denlinger LC, Coursin DB, Schell K, Angelini G, Green DN, Guadarrama AG, Halsey J, Prabhu U, Hogan KJ, Bertics PJ (2006) Human P2X7 pore function predicts allele linkage disequilibrium. Clin Chem 52(6):995–1004

Denlinger LC, Fisette PL, Sommer JA, Watters JJ, Prabhu U, Dubyak GR, Proctor RA, Bertics PJ (2001) Cutting edge: the nucleotide receptor P2X7 contains multiple protein-and lipid-

interaction motifs including a potential binding site for bacterial lipopolysaccharide. J Immunol 167(4):1871–1876
Di Virgilio F (2007) Liaisons dangereuses: P2X7 and the inflammasome. Trends Pharmacol Sci 28(9):465
Di Virgilio F, Chiozzi P, Ferrari D, Falzoni S, Sanz JM, Morelli A, Torboli M, Bolognesi G, Baricordi OR (2001) Nucleotide receptors: an emerging family of regulatory molecules in blood cells. Blood 97(3):587–600
Díaz-Hernández M, Díez-Zaera M, Sánchez-Nogueiro J, Gómez-Villafuertes R, Canals JM, Alberch J, Miras-Portugal MT, Lucas JJ (2009) Altered P2X7-receptor level and function in mouse models of Huntington's disease and therapeutic efficacy of antagonist administration. FASEB J 23(6):1893–1906
Donnelly-Roberts D, Jarvis M (2007) Discovery of P2X7 receptor-selective antagonists offers new insights into P2X7 receptor function and indicates a role in chronic pain states. Br J Pharmacol 151(5):571–579
DSM-IV-TR (2000) Mood disorders. In: Diagnostic and statistical manual of mental disorders, test revision,4th edn. American Psychiatric Association, Virginia
Dunn AJ, Swiergiel AH, Beaurepaire R (2005) Cytokines as mediators of depression: what can we learn from animal studies? Neurosci Biobehav Rev 29(4):891–909
Epps SA, Weinshenker D (2012) Rhythm and blues: Animal models of epilepsy and depression comorbidity. Biochem Pharmacol 85(2):135–146
Erhardt A, Lucae S, Unschuld PG, Ising M, Kern N, Salyakina D, Lieb R, Uhr M, Binder EB, Keck ME (2007) Association of polymorphisms in P2RX7 and CaMKKb with anxiety disorders. J Affect Disord 101(1–3):159–168
Ewald H, Degn B, Mors O, Kruse T (1998) Significant linkage between bipolar affective disorder and chromosome 12q24. Psychiatric Genet 8:131–140
Falzoni S, Chiozzi P, Ferrari D, Buell G, Di Virgilio F (2000) P2X7 receptor and polykarion formation. Mol Biol Cell 11(9):3169–3176
Fantuzzi G, Dinarello CA (1999) Interleukin-18 and interleukin-1β: two cytokine substrates for ICE (caspase-1). J Clin Immunol 19(1):1–11
Faria RX, DeFarias FP, Alves LA (2005) Are second messengers crucial for opening the pore associated with P2X7 receptor? Am J Physiol-Cell Physiol 288(2):C260–C271
Fountoulakis KN, Kasper S, Andreassen O, Blier P, Okasha A, Severus E, Versiani M, Tandon R, Moller HJ, Vieta E (2012a) Efficacy of pharmacotherapy in bipolar disorder: a report by the WPA section on pharmacopsychiatry. Eur Arch Psychiatry Clin Neurosci 262(Suppl 1):1–48
Fountoulakis KN, Kelsoe JR, Akiskal H (2012b) Receptor targets for antidepressant therapy in bipolar disorder: an overview. J Affect Disord 138(3):222–238
Franco-Martínez S, Niño-Moreno P, Bernal-Silva S, Baranda L, Rocha-Meza M, Portales-Cervantes L, Layseca-Espinosa E, González-Amaro R, Portales-Pérez D (2006) Expression and function of the purinergic receptor P2X7 in patients with pulmonary tuberculosis. Clin Exp Immunol 146(2):253–261
Fuller SJ, Stokes L, Skarratt KK, Gu BJ, Wiley JS (2009) Genetics of the P2X7 receptor and human disease. Purinergic Signalling 5(2):257–262
Furukawa TA, McGuire H, Barbui C (2002) Meta-analysis of effects and side effects of low dosage tricyclic antidepressants in depression: systematic review. BMJ 325(7371):991
Gallagher J (2004) ATP P2 receptors and regulation of bone effector cells. J Musculoskelet Neuronal Interact 4(2):125
Geuze E, Vermetten E, Bremner J (2005) MR-based in vivo hippocampal volumetrics: 2 Findings in neuropsychiatric disorders. Mol Psychiatry 10(2):160–184
Ghayur T, Banerjee S, Hugunin M, Butler D, Herzog L, Carter A, Quintal L, Sekut L, Talanian R, Paskind M (1997) Caspase-1 processes IFN-γ-inducing factor and regulates LPS-induced IFN-γ production
Gianaros PJ, Jennings JR, Sheu LK, Greer PJ, Kuller LH, Matthews KA (2007) Prospective reports of chronic life stress predict decreased grey matter volume in the hippocampus. NeuroImage 35(2):795–803

Gonda X, Pompili M, Serafini G, Montebovi F, Campi S, Dome P, Duleba T, Girardi P, Rihmer Z (2012) Suicidal behavior in bipolar disorder: epidemiology, characteristics and major risk factors. J Affect Disord 143(1–3):16–26

Goshen I, Kreisel T, Ben-Menachem-Zidon O, Licht T, Weidenfeld J, Ben-Hur T, Yirmiya R (2008) Brain interleukin-1 mediates chronic stress-induced depression in mice via adreno-cortical activation and hippocampal neurogenesis suppression. Mol Psychiatry 13(7):717–728

Gould TD, Quiroz JA, Singh J, Zarate CA, Manji HK (2004) Emerging experimental therapeutics for bipolar disorder: insights from the molecular and cellular actions of current mood stabilizers. Mol Psychiatry 9(8):734–755

Green EK, Grozeva D, Moskvina V, Hamshere ML, Jones IR, Jones L, Forty L, Caesar S, Gordon-Smith K, Fraser C, Russell E, St Clair D, Young AH, Ferrier N, Farmer A, McGuffin P, Holmans PA, Owen MJ, O'Donovan MC, Craddock N (2010) Variation at the GABAA receptor gene, Rho 1 (GABRR1) associated with susceptibility to bipolar schizoaffective disorder. Am J Med Genet Part B, Neuropsychiatric Genet: Official Publ Int Soc Psychiatric Genet 153B(7):1347–1349

Green EK, Grozeva D, Raybould R, Elvidge G, Macgregor S, Craig I, Farmer A, McGuffin P, Forty L, Jones L (2009) P2RX7: a bipolar and unipolar disorder candidate susceptibility gene? Am J Med Genet Part B: Neuropsychiatric Genet 150(8):1063–1069

Greenberg DB, Jonasch E, Gadd MA, Ryan BF, Everett JR, Sober AJ, Mihm MA, Tanabe KK, Ott M, Haluska FG (2000) Adjuvant therapy of melanoma with interferon-alpha-2b is associated with mania and bipolar syndromes. Cancer 89(2):356–362

Grigoroiu-Serbanescu M, Herms S, Mühleisen TW, Georgi A, Diaconu CC, Strohmaier J, Czerski P, Hauser J, Leszczynska-Rodziewicz A, Jamra RA (2009) Variation in P2RX7 candidate gene (rs2230912) is not associated with bipolar I disorder and unipolar major depression in four European samples. Am J Med Genet Part B: Neuropsychiatric Genet 150(7):1017–1021

Gu BJ, Rathsam C, Stokes L, McGeachie AB, Wiley JS (2009) Extracellular ATP dissociates nonmuscle myosin from P2X7 complex: this dissociation regulates P2X7 pore formation. Am J Physiol-Cell Physiol 297(2):C430–C439

Guile SD, Alcaraz L, Birkinshaw TN, Bowers KC, Ebden MR, Furber M, Stocks MJ (2009) Antagonists of the P2X7 receptor. From lead identification to drug development. ChemInform 40(36):

Gunosewoyo H, Coster MJ, Bennett MR, Kassiou M (2009) Purinergic P2X7 receptor antagonists: chemistry and fundamentals of biological screening. Bioorg Med Chem 17(14):4861–4865

Gunosewoyo H, Coster MJ, Kassiou M (2007) Molecular probes for P2X7 receptor studies. Curr Med Chem 14(14):1505–1523

Gunosewoyo H, Guo J, Bennett M, Coster M, Kassiou M (2008) Cubyl amides: novel P2X7 receptor antagonists. Bioorg Med Chem Lett 18(13):3720

Gunosewoyo H, Kassiou M (2010) P2X purinergic receptor ligands: recently patented compounds. Expert Opin Ther Pat 20(5):625–646

Halmai Z, Dome P, Vereczkei A, Abdul-Rahman O, Szekely A, Gonda X, Faludi G, Sasvari-Szekely M, Nemoda Z (2013) Associations between depression severity and purinergic receptor P2RX7 gene polymorphisms. Journal of affective disorders

Hejjas K, Szekely A, Domotor E, Halmai Z, Balogh G, Schilling B, Sarosi A, Faludi G, Sasvari-Szekely M, Nemoda Z (2009) Association between depression and the Gln460Arg polymorphism of P2RX7 gene: a dimensional approach. Am J Med Gen Part B: Neuropsychiatric Genet 150(2):295–299

Hestad KA, Tønseth S, Støen CD, Ueland T, Aukrust P (2003) Raised plasma levels of tumor necrosis factor [alpha] in patients with depression: normalization during electroconvulsive therapy. J ECT 19(4):183–188

Holsboer F (2000) The corticosteroid receptor hypothesis of depression Neuropsychopharma-cology. Official Publ Am Coll Neuropsychopharmacol 23(5):477–501

Honore P, Donnelly-Roberts D, Namovic M, Hsieh G, Zhu C, Mikusa J, Hernandez G, Zhong C, Gauvin D, Chandran P (2006) A-740003 [N-(1-{[(cyanoimino)(5-quinolinylamino) methyl] amino}-2, 2-dimethylpropyl)-2-(3, 4-dimethoxyphenyl) acetamide], a novel and selective P2X7 receptor antagonist, dose-dependently reduces neuropathic pain in the rat. J Pharmacol Exp Ther 319(3):1376

Honore P, Donnelly-Roberts D, Namovic M, Zhong C, Wade C, Chandran P, Zhu C, Carroll W, Perez-Medrano A, Iwakura Y (2009) The antihyperalgesic activity of a selective P2X7 receptor antagonist, A-839977, is lost in IL-1alphabeta knockout mice. Behav Brain Res 204(1):77–81

Humphreys BD, Rice J, Kertesy SB, Dubyak GR (2000) Stress-activated protein kinase/JNK activation and apoptotic induction by the macrophage P2X7 nucleotide receptor. J Biol Chem 275(35):26792–26798

Impagnatiello F, Guidotti AR, Pesold C, Dwivedi Y, Caruncho H, Pisu MG, Uzunov DP, Smalheiser NR, Davis JM, Pandey GN, Pappas GD, Tueting P, Sharma RP, Costa E (1998) A decrease of reelin expression as a putative vulnerability factor in schizophrenia. Proc Natl Acad Sci USA 95(26):15718–15723

Jackman NA, Uliasz TF, Hewett JA, Hewett SJ (2010) Regulation of system x(c)(-)activity and expression in astrocytes by interleukin-1beta: implications for hypoxic neuronal injury. Glia 58(15):1806–1815. doi:10.1002/glia.21050

Ji X, Naito Y, Weng H, Endo K, Ma X, Iwai N (2012) P2X7 deficiency attenuates hypertension and renal injury in deoxycorticosterone acetate (DOCA)-salt hypertension. Am J Physiol-Renal Physiol 48(3):150–154

Joels M, Karst H, Alfarez D, Heine VM, Qin Y, Riel E, Verkuyl M, Lucassen PJ, Krugers HJ (2004) Effects of chronic stress on structure and cell function in rat hippocampus and hypothalamus. Stress: Int J Biol Stress 7(4):221–231

Jørgensen NR, Husted LB, Skarratt KK, Stokes L, Tofteng CL, Kvist T, Jensen JEB, Eiken P, Brixen K, Fuller S (2012) Single-nucleotide polymorphisms in the P2X7 receptor gene are associated with post-menopausal bone loss and vertebral fractures. Eur J Hum Genet 20(6):675–681

Kaczmarek-Hájek K, Lörinczi É, Hausmann R, Nicke A (2012) Molecular and functional properties of P2X receptors—recent progress and persisting challenges. Purinergic Signalling 8(3):375–417

Katon WJ, Rutter C, Simon G, Lin EHB, Ludman E, Ciechanowski P, Kinder L, Young B, Von Korff M (2005) The association of comorbid depression with mortality in patients with type 2 diabetes. Diabetes Care 28(11):2668–2672

Kauer-Sant'Anna M, Kapczinski F, Andreazza AC, Bond DJ, Lam RW, Young LT, Yatham LN (2009) Brain-derived neurotrophic factor and inflammatory markers in patients with early- vs. late-stage bipolar disorder. Int J Neuropsychopharmacol 12(4):447–458

Kendler K, Karkowski L, Prescott C (1999) Causal relationship between stressful life events and the onset of major depression. Am J Psychiatry 156(6):837

Kim YK, Na KS, Shin KH, Jung HY, Choi SH, Kim JB (2007) Cytokine imbalance in the pathophysiology of major depressive disorder. Prog Neuropsychopharmacol Biol Psychiatry 31(5):1044–1053

Knable M, Barci B, Bartko J, Webster M, Torrey E (2002) Molecular abnormalities in the major psychiatric illnesses: classification and regression tree (CRT) analysis of post-mortem prefrontal markers. Mol Psychiatry 7(4):392

Knable MB, Barci BM, Webster MJ, Meador-Woodruff J, Torrey EF (2004) Molecular abnormalities of the hippocampus in severe psychiatric illness: postmortem findings from the Stanley neuropathology consortium. Mol Psychiatry 9(6):609–620

Koo JW, Duman RS (2008) IL-1β is an essential mediator of the antineurogenic and anhedonic effects of stress. Proc Natl Acad Sci 105(2):751–756

Koo JW, Duman RS (2009) Interleukin-1 receptor null mutant mice show decreased anxiety-like behavior and enhanced fear memory. Neurosci Lett 456(1):39–43

Kronfol Z (2002) Immune dysregulation in major depression: a critical review of existing evidence. Int J Neuropsychopharmacol 5(04):333–343

Labasi JM, Petrushova N, Donovan C, McCurdy S, Lira P, Payette MM, Brissette W, Wicks JR, Audoly L, Gabel CA (2002) Absence of the P2X7 receptor alters leukocyte function and attenuates an inflammatory response. J Immunol 168(12):6436–6445

Lan MJ, McLoughlin GA, Griffin JL, Tsang TM, Huang JT, Yuan P, Manji H, Holmes E, Bahn S (2009) Metabonomic analysis identifies molecular changes associated with the pathophysiology and drug treatment of bipolar disorder. Mol Psychiatry 14(3):269–279

Lane TJ, Manu P, Matthews DA (1991) Depression and somatization in the chronic fatigue syndrome. Am J Med 91(4):335–344

Lavebratt C, Aberg E, Sjöholm L, Forsell Y (2010) Variations in FKBP5 and BDNF genes are suggestively associated with depression in a Swedish population-based cohort. J Affect Disord 125(1–3):249

Le Feuvre RA, Brough D, Iwakura Y, Takeda K, Rothwell NJ (2002) Priming of macrophages with lipopolysaccharide potentiates P2X7-mediated cell death via a caspase-1-dependent mechanism, independently of cytokine production. J Biol Chem 277(5):3210–3218

Levine J, Barak Y, Chengappa K, Rapoport A, Rebey M, Barak V (1999) Cerebrospinal cytokine levels in patients with acute depression. Neuropsychobiology 40(4):171–176

Liang L, Schwiebert EM (2005) Large pore formation uniquely associated with P2X7 purinergic receptor channels. Focus on "Are second messengers crucial for opening the pore associated with P2X7 receptor?". Am J Physiol-Cell Physiol 288(2):C240–C242

Lieb J, Karmali R, Horrobin D (1983) Elevated levels of prostaglandin E2 and thromboxane B2 in depression. Prostaglandins, Leukot Med 10(4):361–367

Little A (2009) Treatment-resistant depression. Am Fam Physician 80(2):167–172

Lucae S, Salyakina D, Barden N, Harvey M, Gagné B, Labbé M, Binder EB, Uhr M, Paez-Pereda M, Sillaber I (2006) P2RX7, a gene coding for a purinergic ligand-gated ion channel, is associated with major depressive disorder. Hum Mol Genet 15(16):2438–2445

Ma W, Hui H, Pelegrin P, Surprenant A (2009) Pharmacological characterization of pannexin-1 currents expressed in mammalian cells. J Pharmacol Exp Ther 328(2):409–418

Maddock RJ, Buonocore MH (2012) MR spectroscopic studies of the brain in psychiatric disorders. Cur Top Behav Neurosci. doi:10.1007/7854_2011_197

Maes M, Bosmans E, Meltzer HY, Scharpé S, Suy E (1993) Interleukin-1 beta: a putative mediator of HPA axis hyperactivity in major depression? Am J Psychiatry 150(8):1189

Maes M, Meltzer HY, Bosmans E, Bergmans R, Vandoolaeghe E, Ranjan R, Desnyder R (1995) Increased plasma concentrations of interleukin-6, soluble interleukin-6, soluble interleukin-2 and transferrin receptor in major depression. J Affect Disord 34(4):301–309

Mantere O, Soronen P, Uher R, Ketokivi M, Jylhä P, Melartin T, Paunio T, Isometsä E (2012) Neuroticism mediates the effect of P2RX7 on outcomes of mood disorders. Depression Anxiety 29(9):816–823

Marcellino D, Suárez-Boomgaard D, Sánchez-Reina MD, Aguirre JA, Yoshitake T, Yoshitake S, Hagman B, Kehr J, Agnati LF, Fuxe K (2010) On the role of P2X7 receptors in dopamine nerve cell degeneration in a rat model of Parkinson's disease: studies with the P2X7 receptor antagonist A-438079. J Neural Transm 117:681–687

McArthur R, Borsini F (2006) Animal models of depression in drug discovery: a historical perspective. Pharmacol Biochem Behav 84(3):436–452

McEwen BS (2005) Glucocorticoids, depression, and mood disorders: structural remodeling in the brain. Metabolism 54(5):20–23

McGuffin P, Knight J, Breen G, Brewster S, Boyd PR, Craddock N, Gill M, Korszun A, Maier W, Middleton L (2005) Whole genome linkage scans of recurrent depressive disorder from the depression network study. Hum Mol Genet 14(22):3337–3345

McQuillin A, Bass N, Choudhury K, Puri V, Kosmin M, Lawrence J, Curtis D, Gurling H (2009) Case–control studies show that a non-conservative amino-acid change from a glutamine to arginine in the P2RX7 purinergic receptor protein is associated with both bipolar-and unipolar-affective disorders. Mol Psychiatry 14(6):614–620

Miller AH, Maletic V, Raison CL (2009) Inflammation and its discontents: the role of cytokines in the pathophysiology of major depression. Biol Psychiatry 65(9):732–741

Miller CL, Llenos IC, Dulay JR, Weis S (2006) Upregulation of the initiating step of the kynurenine pathway in postmortem anterior cingulate cortex from individuals with schizophrenia and bipolar disorder. Brain Res 1073–1074:25–37

Morissette J, Villeneuve A, Bordeleau L, Rochette D, Laberge C, Gagne B, Laprise C, Bouchard G, Plante M, Gobeil L (1999) Genome-wide search for linkage of bipolar affective disorders in a very large pedigree derived from a homogeneous population in Quebec points to a locus of major effect on chromosome 12q23-q24. Am J Med Genet 88(5):567–587

Müller N, Myint AM, Schwarz MJ (2011) Inflammatory biomarkers and depression. Neurotox Res 19(2):308–318

Nagy G, Ronai Z, Somogyi A, Sasvari-Szekely M, Rahman OA, Mate A, Varga T, Nemoda Z (2008) P2RX7 Gln460Arg polymorphism is associated with depression among diabetic patients. Prog Neuropsychopharmacol Biol Psychiatry 32(8):1884–1888

Nakajima S, Suzuki T, Watanabe K, Kashima H, Uchida H (2010) Accelerating response to antidepressant treatment in depression: a review and clinical suggestions. Progress in neuro-psychopharmacology & biological psychiatry

Nemeroff CB (2003) Safety of available agents used to treat bipolar disorder: focus on weight gain. J Clin Psychiatry 64(5):532–539

Nestler EJ, Barrot M, DiLeone RJ, Eisch AJ, Gold SJ, Monteggia LM (2002) Neurobiology of depression. Neuron 34(1):13–25

O'Brien SM, Scully P, Scott LV, Dinan TG (2006) Cytokine profiles in bipolar affective disorder: focus on acutely ill patients. J Affect Disord 90(2–3):263–267

Ohishi K, Ueno R, Nishino S, Sakai T, Hayaishi O (1988) Increased level of salivary prostaglandins in patients with major depression. Biol Psychiatry 23(4):326–334

Ohlendorff SD, Tofteng CL, Jensen JEB, Petersen S, Civitelli R, Fenger M, Abrahamsen B, Hermann AP, Eiken P, Jørgensen NR (2007) Single nucleotide polymorphisms in the P2X7 gene are associated to fracture risk and to effect of estrogen treatment. Pharmacogenet Genomics 17(7):555–567

Onyike CU, Bonner JO, Lyketsos CG, Treisman GJ (2004) Mania during treatment of chronic hepatitis C with pegylated interferon and ribavirin. Am J Psychiatry 161(3):429–435

Ortiz-Dominguez A, Hernandez ME, Berlanga C, Gutierrez-Mora D, Moreno J, Heinze G, Pavon L (2007) Immune variations in bipolar disorder: phasic differences. Bipolar Disord 9(6):596–602

Owen B, Eccleston D, Ferrier I, Young H (2001) Raised levels of plasma interleukin-1β in major and postviral depression. Acta Psychiatr Scand 103(3):226–228

Padmos RC, Hillegers MH, Knijff EM, Vonk R, Bouvy A, Staal FJ, de Ridder D, Kupka RW, Nolen WA, Drexhage HA (2008) A discriminating messenger RNA signature for bipolar disorder formed by an aberrant expression of inflammatory genes in monocytes. Arch Gen Psychiatry 65(4):395–407

Paez-Pereda M, Panhuysen M (2008) Strategies to identify biomarkers for depression. In: Turck C (ed) Biomarkers for psychiatric disorders. Springer, pp 299–314

Papiol S, Molina V, Desco M, Rosa A, Reig S, Sanz J, Palomo T, Fananas L (2008) Gray matter deficits in bipolar disorder are associated with genetic variability at interleukin-1 beta gene (2q13). Genes, Brain Behav 7(7):796–801

Pelegrin P, Surprenant A (2006) Pannexin-1 mediates large pore formation and interleukin-1β release by the ATP-gated P2X7 receptor. EMBO J 25(21):5071–5082

Pelegrin P, Surprenant A (2009) The P2X 7 receptor–pannexin connection to dye uptake and IL-1β release. Purinergic Signalling 5(2):129–137

Phillips ML, Drevets WC, Rauch SL, Lane R (2003) Neurobiology of emotion perception I: The neural basis of normal emotion perception. Biol Psychiatry 54(5):504–514

Prow NA, Irani DN (2008) The inflammatory cytokine, interleukin-1 beta, mediates loss of astroglial glutamate transport and drives excitotoxic motor neuron injury in the spinal cord during acute viral encephalomyelitis. J Neurochem 105(4):1276–1286

Qiu F, Dahl G (2009) A permeant regulating its permeation pore: inhibition of pannexin 1 channels by ATP. Am J Physiol-Cell Physiol 296(2):C250–C255

Qu Y, Dubyak GR (2009) P2X7 receptors regulate multiple types of membrane trafficking responses and non-classical secretion pathways. Purinergic Signalling 5(2):163–173

Qu Y, Franchi L, Nunez G, Dubyak GR (2007) Nonclassical IL-1β secretion stimulated by P2X7 receptors is dependent on inflammasome activation and correlated with exosome release in murine macrophages. J Immunol 179(3):1913–1925

Qu Y, Misaghi S, Newton K, Gilmour LL, Louie S, Cupp JE, Dubyak GR, Hackos D, Dixit VM (2011) Pannexin-1 is required for ATP release during apoptosis but not for inflammasome activation. J Immunol 186(11):6553–6561

Rampe D, Wang L, Ringheim GE (2004) P2X7 receptor modulation of β-amyloid-and LPS-induced cytokine secretion from human macrophages and microglia. J Neuroimmunol 147(1):56–61

Rao JS, Harry GJ, Rapoport SI, Kim HW (2010) Increased excitotoxicity and neuroinflammatory markers in postmortem frontal cortex from bipolar disorder patients. Mol Psychiatry 15(4):384–392

Rassendren F, Buell GN, Virginio C, Collo G, North RA, Surprenant A (1997) The permeabilizing ATP receptor, P2X7. J Biol Chem 272(9):5482–5486

Roger S, Mei Z, Baldwin J, Dong L, Bradley H, Baldwin S, Surprenant A, Jiang L (2010) Single nucleotide polymorphisms that were identified in affective mood disorders affect ATP-activated P2X7 receptor functions. J Psychiatr Res 44(6):347

Rothwell NJ, Luheshi GN (2000) Interleukin 1 in the brain: biology, pathology and therapeutic target. Trends Neurosci 23(12):618–625

Rush AJ, Trivedi MH, Wisniewski SR, Stewart JW, Nierenberg AA, Thase ME, Ritz L, Biggs MM, Warden D, Luther JF (2006) Bupropion-SR, sertraline, or venlafaxine-XR after failure of SSRIs for depression. N Engl J Med 354(12):1231–1242

Santarelli L, Saxe M, Gross C, Surget A, Battaglia F, Dulawa S, Weisstaub N, Lee J, Duman R, Arancio O (2003) Requirement of hippocampal neurogenesis for the behavioral effects of antidepressants. Science Signalling 301(5634):805

Sanz JM, Virgilio FD (2000) Kinetics and mechanism of ATP-dependent IL-1ß release from microglial cells. J Immunol 164(9):4893–4898

Savitz J, Preskorn S, Teague TK, Drevets D, Yates W, Drevets W (2012) Minocycline and aspirin in the treatment of bipolar depression: a protocol for a proof-of-concept, randomised, double-blind, placebo-controlled, 2x2 clinical trial. BMJ Open 2(1):e000643

Schiepers OJG, Wichers MC, Maes M (2005) Cytokines and major depression. Prog Neuropsychopharmacol Biol Psychiatry 29(2):201–217

Shelton RC, Osuntokun O, Heinloth AN, Corya SA (2010) Therapeutic options for treatment-resistant depression. CNS drugs 24(2):131–161

Shink E, Morissette J, Sherrington R, Barden N (2005) A genome-wide scan points to a susceptibility locus for bipolar disorder on chromosome 12. Molecular psychiatry 10(6):545–552

Sibille E, Morris HM, Kota RS, Lewis DA (2011) GABA-related transcripts in the dorsolateral prefrontal cortex in mood disorders. Int J Neuropsychopharmacol 14(6):721–734

Singla N, Gupta D, Joshi A, Batra N, Singh J (2012) Genetic polymorphisms in the P2X7 gene and its association with susceptibility to tuberculosis. Int J Tuberc Lung Dis 16(2):224–229

Sluyter R, Stokes L (2011) Significance of P2X7 receptor variants to human health and disease. Recent Pat DNA Gene Sequences 5(1):41–54

Sluzewska A, Rybakowski J, Bosmans E, Sobieska M, Berghmans R, Maes M, Wiktorowicz K (1996) Indicators of immune activation in major depression. Psychiatry Res 64(3):161–167

Solle M, Labasi J, Perregaux DG, Stam E, Petrushova N, Koller BH, Griffiths RJ, Gabel CA (2001) Altered cytokine production in Mice Lacking P2X7Receptors. J Biol Chem 276(1):125–132

Soronen P, Mantere O, Melartin T, Suominen K, Vuorilehto M, Rytsälä H, Arvilommi P, Holma I, Holma M, Jylhä P (2011) P2RX7 gene is associated consistently with mood disorders and

predicts clinical outcome in three clinical cohorts. Am J Med Genet Part B: Neuropsychiatric Genet 156(4):435–447
Sperlágh B, Baranyi M, Haskó G, Vizi ES (2004) Potent effect of interleukin-1β to evoke ATP and adenosine release from rat hippocampal slices. J Neuroimmunol 151(1):33–39
Sperlágh B, Vizi E, Wirkner K, Illes P (2006) P2X7 receptors in the nervous system. Prog Neurobiol 78(6):327
Stokes L, Fuller SJ, Sluyter R, Skarratt KK, Gu BJ, Wiley JS (2010) Two haplotypes of the P2X7 receptor containing the Ala-348 to Thr polymorphism exhibit a gain-of-function effect and enhanced interleukin-1β secretion. FASEB J 24(8):2916–2927
Strite D, Valentine AD, Meyers CA (1997) Manic episodes in two patients treated with interferon alpha. J Neuropsychiatry Clin Neurosci 9(2):273–276
Su SC, Sun MT, Wen MJ, Lin CJ, Chen YC, Hung YJ (2011) Brain-derived neurotrophic factor, adiponectin, and proinflammatory markers in various subtypes of depression in young men. Int J Psychiatry Med 42(3):211–226
Sugiyama T, Kobayashi M, Kawamura H, Li Q, Puro DG (2004) Enhancement of P2X7-induced pore formation and apoptosis: an early effect of diabetes on the retinal microvasculature. Invest Ophthalmol Vis Sci 45(3):1026–1032
Sullivan PF, Neale MC, Kendler KS (2000) Genetic epidemiology of major depression: review and meta-analysis. Am J Psychiatry 157(10):1552–1562
Surprenant A, Rassendren F, Kawashima E, North R, Buell G (1996) The cytolytic P2Z receptor for extracellular ATP identified as a P2X receptor (P2X7). Science (New York, NY) 272(5262):735
Sutterwala FS, Ogura Y, Szczepanik M, Lara-Tejero M, Lichtenberger GS, Grant EP, Bertin J, Coyle AJ, Galán JE, Askenase PW (2006) Critical role for NALP3/CIAS1/Cryopyrin in innate and adaptive immunity through its regulation of caspase-1. Immunity 24(3):317–327
Thomas AJ, Davis S, Morris C, Jackson E, Harrison R, O'Brien JT (2005) Increase in Interleukin-1ß in Late-Life Depression. Am J Psychiatry 162(1):175–177
Townsend J, Altshuler LL (2012) Emotion processing and regulation in bipolar disorder: a review. Bipolar Disord 14(4):326–339
Trivedi MH, Rush AJ, Wisniewski SR, Nierenberg AA, Warden D, Ritz L, Norquist G, Howland RH, Lebowitz B, McGrath PJ, Shores-Wilson K, Biggs MM, Balasubramani GK, Fava M (2006) Evaluation of outcomes with citalopram for depression using measurement-based care in STAR*D: implications for clinical practice. Am J Psychiatry 163(1):28–40
Viikki M, Kampman O, Anttila S, Illi A, Setälä-Soikkeli E, Huuhka M, Mononen N, Lehtimäki T, Leinonen E (2011) P2RX7 polymorphisms Gln460Arg and His155Tyr are not associated with major depressive disorder or remission after SSRI or ECT. Neurosci Lett 493(3):127
Virginio C, MacKenzie A, North R, Surprenant A (1999a) Kinetics of cell lysis, dye uptake and permeability changes in cells expressing the rat P2X7 receptor. J Physiol 519(2):335–346
Virginio C, MacKenzie A, Rassendren F, North R, Surprenant A (1999b) Pore dilation of neuronal P2X receptor channels. Nat Neurosci 2(4):315–321
Warner-Schmidt JL, Duman RS (2006) Hippocampal neurogenesis: opposing effects of stress and antidepressant treatment. Hippocampus 16(3):239–249
WHO (2003) The world health report 2003—shaping the future. World Health Organistion, Geneva
WHO (2011) Global burden of mental disorders and the need for a comprehensive, coordinated response from health and social sectors at the country level. World Health Organization, Geneva
Wichers MC, Maes M (2004) The role of indoleamine 2, 3-dioxygenase (IDO) in the pathophysiology of interferon-α-induced depression. J Psychiatry Neurosci 29(1):11
Wiley J, Sluyter R, Gu B, Stokes L, Fuller S (2011) The human P2X7 receptor and its role in innate immunity. Tissue Antigens 78(5):321–332
Yaron I, Shirazi I, Judovich R, Levartovsky D, Caspi D, Yaron M (1999) Fluoxetine and amitriptyline inhibit nitric oxide, prostaglandin E2, and hyaluronic acid production in human synovial cells and synovial tissue cultures. Arthritis Rheum 42(12):2561–2568

Yiangou Y, Facer P, Durrenberger P, Chessell IP, Naylor A, Bountra C, Banati RR, Anand P (2006) COX-2, CB2 and P2X7-immunoreactivities are increased in activated microglial cells/macrophages of multiple sclerosis and amyotrophic lateral sclerosis spinal cord. BMC Neurol 6(1):12

Yuksel C, Ongur D (2010) Magnetic resonance spectroscopy studies of glutamate-related abnormalities in mood disorders. Biol Psychiatry 68(9):785–794

Zubenko GS, Maher B, Hughes III HB, Zubenko WN, Stiffler JS, Kaplan BB, Marazita ML (2003) Genome-wide linkage survey for genetic loci that influence the development of depressive disorders in families with recurrent, early-onset, major depression. Am J Med Genet Part B: Neuropsychiatric Genetics 123(1):1–18

Zunszain PA, Anacker C, Cattaneo A, Choudhury S, Musaelyan K, Myint AM, Thuret S, Price J, Pariante CM (2012) Interleukin-1β: A new regulator of the kynurenine pathway affecting human hippocampal neurogenesis. Neuropsychopharmacol: Official Publ Am Coll Neuropsychopharmacol 37(4):939–949

Chapter 32
Targeting NMDA Receptors in Epilepsy

Mehdi Ghasemi and Ahmad Reza Dehpour

Abstract *N*-methyl-D-aspartate receptors (NMDARs) are one of the most widely-studied receptors in the central nervous system (CNS). These receptors mediate Ca^{2+} influx into the neurons and play pivotal roles in the pathophysiology of many neurological diseases including epilepsy. Animal models of seizure and status epilepticus and clinical studies have indicated that NMDAR functioning can be altered during epilepsy and thereafter in some specific types of seizure. Different classes of NMDAR antagonists also have antiepileptic effects in both clinical and preclinical studies, although some contradictory effects in both preclinical and clinical studies have been reported. In this book chapter, we review the evidence for the involvement of NMDARs in the pathophysiology of epilepsy and provide an overview of NMDAR antagonists being investigated in clinical trials and animal models of seizure.

32.1 Introduction

Epilepsy is an important neurologic disorder that affects approximately 2–4 million people in the United States or 1 in 50 children and 1 in 100 adults (Schachter 2009). Over the past two decades, a new generation of antiepileptic drugs (AEDs) has emerged for the pharmacological management of seizures (Schachter 2007). However, the prospect of freedom from seizures and adverse effects remains

M. Ghasemi (✉)
Department of Neurology, Johns Hopkins University School of Medicine, Baltimore, MD 21205, USA
e-mail: m82_ghasemi@yahoo.com

A. R. Dehpour
Department of Pharmacology, School of Medicine, Tehran University of Medical Sciences, Tehran, Iran
e-mail: dehpour@yahoo.com

N. Weiss and A. Koschak (eds.), *Pathologies of Calcium Channels*,
DOI: 10.1007/978-3-642-40282-1_32,

elusive for a considerable number of patients with epilepsy despite concerted attempts by their physicians to utilize available pharmacotherapies to their full advantage (Ghasemi and Schachter 2011). Approximately 25–30 % of patients continue to suffer from seizures despite being treated state of the art (Dichter and Brodie 1996). It is well established that alterations in the central inhibitory (e.g. γ-aminobutyric acid or GABA) and excitatory (e.g. glutamate) neurotransmission play a crucial role in the etiology of epilepsy. Overactivation of glutamatergic transmission/glutamate receptors may be of significant relevance for the manifestation of this disease (Urbanska et al. 1998). Targeting the glutamatergic neurotransmission is a highly sophisticated matter due to the ambiguous functions of glutamate in several physiological and pathophysiological conditions. Among glutamate receptors, *N*-methyl-D-aspartate receptors (NMDARs) have been the focus of much basic and clinical research during the past three decades, producing an overwhelming body of evidence that inhibition of NMDARs can prevent and, in some cases, reverses the pathology in various models of neurological diseases such as epilepsy. The repertoire of NMDAR-targeted drugs in neurology is expected to grow in the near future. In this book chapter we review the rationale for the development of such drugs.

32.2 The NMDAR Structure and Signaling

Glutamate is the major excitatory neurotransmitter in the central nervous system (CNS) and acts on ionotropic and metabotropic glutamate receptors located at the presynaptic terminal and in the postsynaptic membrane at synapses in the brain and spinal cord. NMDARs are tetrameric structures of seven subunits which including at least one copy of an obligatory subunit, GluN1, and varying expression of a family of GluN2 (GluN2A-D) or GluN3 (GluN3A-B) subunits, with multiple binding sites including ones for glutamate, polyamine, Mg^{2+}, and glycine (Fig. 32.1). Both the GluN1 and GluN2 subunits are involved in the formation of the NMDAR ion channel. The GluN3 subunit does not form functional receptors alone, but can co-assemble with GluN1/GluN2 complexes (Das et al. 1998; Perez-Otano et al. 2001). The glutamate-binding site is located on the GluN2 subunits, and the glycine-binding site is on the GluN1 subunits. Unlike GluN2, but similar to GluN1, GluN3 binds glycine and D-serine rather than glutamate; the binding of each occurs with much higher affinities, as is the case for GluN1 (Nishi et al. 2001; Pachernegg et al. 2012; Yao and Mayer 2006). The glycine (and/or D-serine) co-agonist site must be occupied before glutamate activates the ion channel. The NMDAR channels can conduct Na^{+} and Ca^{2+}. The Ca^{2+} influx through NMDARs is the critical factor that mediates many of its roles in health and disease (Cull-Candy and Leszkiewicz 2004). Under basal conditions the channel is blocked by magnesium ion (Mg^{2+}) within the channel pore. The Mg^{2+} blockade is relieved by cellular depolarization, with implications for synaptic plasticity, especially long-term potentiation (LTP). Continuous strong stimulation

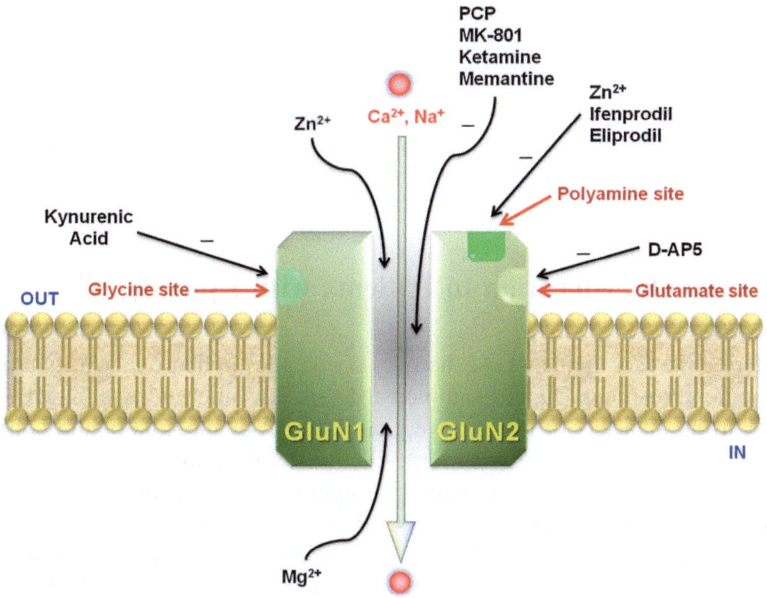

Fig. 32.1 NMDAR model showing binding sites for agonists and antagonists. Glycine and D-serine are agonists for GluN1 subunits, whereas glutamate and NMDA acts on GluN2 subunit agonistic site. Kynurenic acid is an endogenous ligand for GluN1. Zn^{2+} is an endogenous ligand for GluN2A and GluN2B modulatory domain. Low concentrations (nanomolar) of Zn^{2+} have affinity to GluN2A, whereas higher concentration (micromolar) has affinity to GluN2B subunits. Zn^{2+} at higher concentrations can also acts as an NMDAR pore blocker. Ifenprodil and its derivatives bind at the modulatory domain in an GluN2B selective manner. D-AP5 is an antagonist for agonistic site of GluN2 subunit. Mg^{2+}, MK-801, ketamine, PCP, and memantine act as noncompetitive antagonists whose binding sites are within the ion channel pore region

optimally activates NMDARs and plays an important role in LTP. With neurotoxic insults, disruption of energy metabolism diminishes the driving force for the Na^+ pump that maintains the resting membrane potential of cells so that neurons become depolarized, relieving the Mg^{2+} block of NMDARs. Therefore, NMDAR-mediated responses contribute to the later components in paroxysmal depolarizing shifts and provide for much of the Ca^{2+} entry associated with seizure discharges (Ghasemi and Schachter 2011). Ca^{2+} ions play a central role in the control of neuronal excitability (DeLorenzo 1986; Heinemann and Hamon 1986). Several lines of investigation have also shown that abnormalities in Ca^{2+}-regulated processes are related to the hyperexcitability of neuronal populations and seizure activity in different brain regions such as hippocampus and neocortex (Avoli et al. 1990; Kriegstein et al. 1987; Stefani et al. 1997). Excess Ca^{2+} entry then could lead to neuronal excitotoxicity and even cell death. The brain damage is more pronounced when the seizures are prolonged such as status epilepticus (Fujikawa 2005). Forty to sixty minutes of continuous electrographic seizure discharges is required to cause neuronal death in selectively vulnerable brain regions in rats

(Fujikawa 1996; Nevander et al. 1985). Similarly, in a focal seizure model in rats, hippocampal CA3 and hilar neuronal death was produced by 45 min of continuous seizures (Henshall et al. 2000). This process is mediated by activation of a number of enzymes by excess Ca^{2+}, including phospholipases, endonucleases, and proteases such as calpain. These enzymes lead to damage cell structures such as components of the cytoskeleton, membrane, and DNA (Fujikawa 2005).

32.3 NMDAR Alteration in Patients with Epilepsy

Considering the fact that NMDAR activity contributes in neuronal excitation in the CNS, some researchers have examined the possible alterations of NMDARs in epilepsy using a variety of methods (e.g. assessment of subunits binding affinities, immunoblotting, and gene expression). Glutamatergic impulses from the entorhinal cortex constitute the major excitatory input to the hippocampus and a shift in glutamate-mediated excitability may contribute to the pathogenesis of epileptic discharges (Carter et al. 2010). Moreover, NMDARs could be responsible for the seizure-induced excitotoxic cell death of certain hippocampal neuronal populations because NMDAR antagonists provide protection against such damage (Meldrum 1993). Using non-radioactive in situ hybridization studies have demonstrated that in the hippocampal specimens of patients with chronic temporal lobe epilepsy there is a loss of GluN1-positive cells that is related to the overall neuronal loss in the respective specimen and to Ammon's horn sclerosis (Bayer et al. 1995). This loss of GluN1 expression may be partly a reflection of pyramidal cell loss (Bayer et al. 1995). Other studies showed that GluN2 subunit mRNA levels are elevated in the hippocampus of patients with hippocampal sclerosis (HS) (Mathern et al. 1997). In the dentate gyrus, there is an increase in GluN2 immunoreactivity which is associated with abnormal mossy fiber sprouting in this region (Mathern et al. 1996). Mossy fiber sprouting in the human hippocampus is in association with physiologically active NMDARs. Accordingly, it was shown that inhibition of glutamate binding site of GluN2 subunit decreased granule cell hyperexcitability in cases with mossy fiber sprouted hippocampi, whereas granule cells in non-sprouted hippocampi were not affected by such treatment (Franck et al. 1995; Isokawa and Levesque 1991; Masukawa et al. 1991). Chronic temporal lobe seizures are associated with differential changes in hippocampal GluN1 and GluN2A-D hybridization densities that vary by subfield and clinical-pathological category (Mathern et al. 1996). In patients with temporal lobe epilepsy, these findings support the hypothesis that in dentate granule cells NMDARs are increased, and excitatory postsynaptic potentials should be strongly NMDA mediated compared with non-seizure autopsies (Mathern et al. 1999; Neder et al. 2002). Using human focal cortical dysplasia specimens, it was shown that GluN2B and GluN2C subunit mRNA was increased, and GluN2A subunit mRNA was decreased in dysplastic compared with pyramidal and heterotopic neurons (Crino et al. 2001). Other studies (Andre et al. 2004; Liu et al. 2007) reported an upregulation of GluN2B subunit composition and an altered Mg^{2+} sensitivity in pediatric cortical dysplasia or adult temporal lobe tissues.

Some studies have also shown a role for NMDARs in the epileptogenesis in cortical tubers in the tuberous sclerosis complex (White et al. 2001). The level of GluN2B and 2D subunit mRNAs and functional GluN2B-containing receptors are increased in tubers (White et al. 2001). Elevated expression of GluN2B and 2C subunit mRNAs was noted in the dysplastic neurons, whereas only the GluN2D mRNA was upregulated in giant cells, indicating that dysplastic neurons and giant cells make differential contributions to epileptogenesis in the tuberous sclerosis complex (White et al. 2001). It seems that the type of epilepsy, its anatomical region, and the underlying diseases that could be responsible for seizure development (such as tuberous sclerosis) could play a role in the differential alterations in subunits of NMDARs (especially GluN1 and GluN2 subunits) in different regions of CNS. Although alterations in GluN1 and GluN2 subunits have been widely investigated in previous studies, the possible alterations in GluN3 subunits have not been yet reported and this issue warrants further investigations.

32.4 NMDAR Alteration in Animal Studies of Epilepsy

32.4.1 Animal Models of Epilepsy

A wide variety of animal models of epilepsy and status epilepticus have been used by many researchers worldwide to find out the mechanisms underlying epileptic disorders. These include electrical stimulation models, chemoconvulsant-induced models (e.g. kainic acid; KA, pilocarpine, picrotoxin, or bicuculline), physical models (e.g. hyperthermia, or photic or auditory stimulation), genetic models (e.g. mutant, transgenic, or knockout), and spontaneous seizure models (e.g. post-kindling or post-chemoconvulsant). The maximal electroshock seizure (MES) test has predictive value for agents with activity against generalized seizures of tonic–clonic (grand mal) type. The subcutaneous (s.c. pentylenetetrazole (PTZ or metrazol) test is also be predictive of anticonvulsant activity of many drugs against non-convulsive (absence or myoclonic) seizures (Krall et al. 1978). Chronic models of epilepsy can be divided into models of acquired (symptomatic) epilepsy and models of genetic (idiopathic) epilepsy (Fig. 32.2). Models with electrical induction of epilepsy or epilepsy-like conditions include the kindling model and models in which recurrent spontaneous seizures develop after a self sustained status epilepticus (SSSE), which is elicited by sustained electrical stimulation of the hippocampus, the lateral or basolateral nucleus of the amygdala, or other limbic brain regions. The kindling, pilocarpine, and kainite (KA) models are well-known animal models of limbic epileptogenesis. SSSE can also be induced by chemical convulsants, such as pilocarpine and KA.

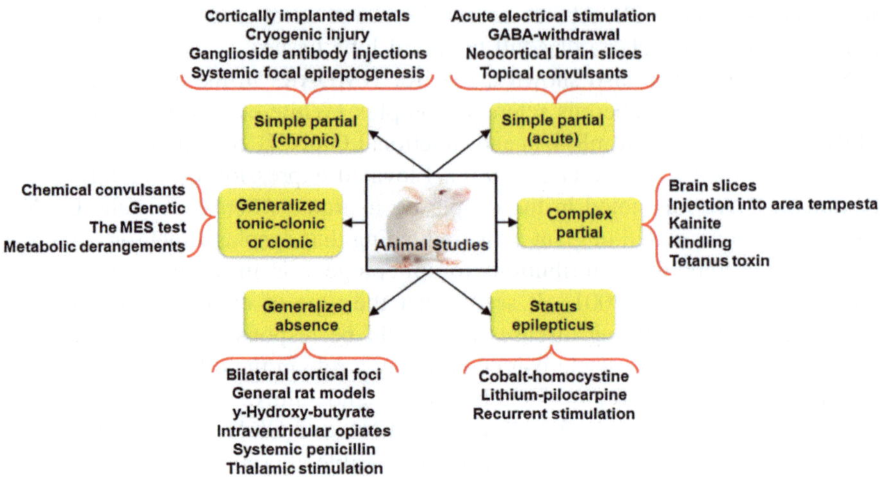

Fig. 32.2 Preclinical animal studies in epilepsy research

Genetic animal models of epilepsy can be divided up into animals with spontaneous mutations and animals with induced mutations. Models with spontaneous mutations can be further subdivided into two groups:

A. *Mutant animals with "reflex epilepsy"*. In this group seizures are elicited by specific sensory stimulation. Audiogenic seizure-susceptible rats and DBA/2 mice, photosensitive baboons, or the E1 mouse (in which seizures are induced by vestibular stimulation) are some examples for this group category.
B. *Animals with "spontaneous recurrent" seizures.* For example, Genetic Absence Epilepsy Rat from Strasbourg (GAERS) is a group of rats with spontaneous spike-wave discharges. Mutant animals, such as transgenic or knockout mice, are rarely used for drug studies.

32.4.2 NMDARs Alteration in Animal Experimental Studies

Using several animal models of seizure or epilepsy many researchers worldwide have investigated the possible alterations in NMDARs in epilepsy. There are several differences in their results which might be because of the differences in the animal models, the brain regions, and the NMDAR subunits that were examined (Table 32.1). Pratt et al. (1993) reported that in partially kindled rats (10 stimulations), while the GluN1 subunit mRNA remained unaltered after a period of 2 h, the GluN2A and GluN2B subunit mRNAs were bilaterally reduced in dentate gyrus granule cells. In fully kindled animals (40 stimulations), there was a progressive reduction in GluN1 subunit mRNA levels in the dentate gyrus, being maximal after 4 h. At the same time point, GluN2A and GluN2B transcript levels were transiently increased (Pratt et al. 1993). When studied 28 days after the last

Table 32.1 Alteration in mRNA expression of NMDAR subunits in experimental seizure studies

NMDAR subunit	Animal model of seizure	Brain region	Result	Reference
GluN1	Rapid partial hippocampal kindling in rats (10 stimulations)	Granule cells of the dentate gyrus	↔	Pratt et al. (1993)
	Rapid fully hippocampal kindling in rats (40 stimulations)		↓ progressive	
	Hippocampal kindling in rats	CA3 pyramidal cells of hippocampus	↔	Kraus et al. (1994)
	Chronic ethanol exposure in rats	Cortex or hippocampus	↔	Morrow et al. (1994)
	Picrotoxin-induced epileptiform discharge	Rat hippocampal slice cultures (after 2 days)	↔	(Gerfin-Moser et al. 1995)
	Amygdaloid kindling in rats	Supraopitc nucleus (after one month)	↔	Al-Ghoul et al. (1997)
		Ipsilateral frontal & temporal cortices at 4 weeks after the last generalized seizure	↑	(Kikuchi et al. 2000)
		Bilateral piriform cortices at 4 weeks after the last generalized seizure	↓	
	Kainate-induced seizure (model of limbic seizure) in rats	Hippocampal dentate gyrus (after 24 and 72 h)	↔	Lason et al. (1997)
		Hippocampal CA1 and CA3 pyramidal cells (after 24 and 72 h)	↓	
	Pilocarpine model of limbic seizure in rats	Hippocampal dentate gyrus, CA1 (after 24 and 72 h) and CA3 (only after 72 h) pyramidal cells	↓	
	Audiogenic seizure in albino Swiss mice acoustically primed at 10 days	Central nucleus of the inferior colliculus	↑ Transient	(Marianowski et al. 1995)

(continued)

Table 32.1 (continued)

NMDAR subunit	Animal model of seizure	Brain region	Result	Reference
GluN2A	Rapid partial hippocampal kindling in rats (10 stimulations)	Granule cells of the hippocampal dentate gyrus	↓	Pratt et al. (1993)
	Hippocampal kindling in rats	CA3 pyramidal cells of hippocampus	↔	(Kraus et al. 1994)
	Picrotoxin-induced epileptiform discharge	Rat hippocampal slice cultures (after 2 days)	↓	Gerfin-Moser et al. (1995)
	Self sustained limbic status epilepticus rats	Molecular layer of hippocampus	↑	Mathern et al. (1998a)
GluN2B	Rapid partial hippocampal kindling in rats (10 stimulations)	Granule cells of the hippocampal dentate gyrus	↓	Pratt et al. (1993)
	Rapid fully hippocampal kindling in rats (40 stimulations)		↑ Transient	
	Hippocampal kindling in rats	CA3 pyramidal cells of hippocampus	↔	Kraus et al. (1994)
	Picrotoxin-induced epileptiform discharge	Rat hippocampal slice cultures (after 2 days)	↓	(Gerfin-Moser et al. 1995)
	Amygdaloid kindling in rats	Supraopitc nucleus and adjacent piriform cortex, lateral nucleus of the olfactory tract, anterior cortical amygdaloid nucleus, and cortex-amygdala transition zone (after one month)	↑	Al-Ghoul et al. (1997)
	Self sustained limbic status epilepticus rats	Molecular layer of hippocampus	↑	Mathern et al. (1998a)
	Pilocarpine model of spontaneous limbic epilepsy in rats	Hippocampal fascia dentate, Subiculum	↑	Mathern et al. (1998b)

(continued)

Table 32.1 (continued)

NMDAR subunit	Animal model of seizure	Brain region	Result	Reference
GluN2C	Hippocampal kindling in rats	CA3 pyramidal cells of hippocampus	↔	Kraus et al. (1994)
	Amygdaloid kindling in rats	Supraopitc nucleus (after one month)	↔	Al-Ghoul et al. (1997)
	Audiogenic seizure in albino Swiss mice acoustically primed at 10 days	Central nucleus of the inferior colliculus	↑ transient	Marianowski et al. (1995)
GluN2D	Hippocampal kindling in rats	CA3 pyramidal cells of hippocampus	↔	Kraus et al. (1994)
	Amygdaloid kindling in rats	Supraopitc nucleus (after one month)	↓	Al-Ghoul et al. (1997)

evoked seizure, (Kraus et al. 1994) observed that kindling induced a 2.8-fold elevation in the number of binding sites for the competitive NMDAR antagonist 3-[(±)-2-(carboxypiperazine-4-yl)][1,2-3H-]propyl-1-phosphonic acid (^{3}H-CPP). In situ hybridization also showed that in amygdaloid kindled rats, there was an increase in the GluN2B and decrease in GluN2D subunit mRNA levels in the neuroendocrine cells of the supraoptic nucleus 1 month after kindling, whereas GluN1 and GluN2C mRNA levels did not change (Al-Ghoul et al. 1997). Seizure can rapidly and significantly increase GluN1 mRNA and protein expression in rat cerebral cortex (Jensen et al. 1997). Northern blots showed a sustained increase in cortical mRNA for GluN1 in the amygdaloid Kindled rats (Kikuchi et al. 2000). Using in situ hybridization histochemistry Gerfin-Moser et al. (Gerfin-Moser et al. 1995) demonstrated that picrotoxin (500 μM, for 2 days) decreased GluN2A and GluN2B mRNA levels in the rat hippocampal slice cultures, without any change in the GluN1 mRNA. Mathern et al. (1998a), using both kindled animals and self sustained limbic status epilepticus (SSLSE) rats, found that the molecular layer of hippocampus in SSLSE rats with spontaneous seizures demonstrated higher levels of GluN2A and GluN2B immunoreactivity in comparison with kindled animals and controls. They suggested that hippocampal GluN2A and GluN2B mRNAs and proteins are differentially increased in association with spontaneous, but not kindled, seizures (Mathern et al. 1998a). Based on the other observation, hippocampal GluN1 and GluN2B mRNA levels changed as rats progressed from the latent to chronic seizure phase in pilocarpine model of spontaneous limbic epilepsy and GluN1 subunit alterations correlated with mossy fiber sprouting (Mathern et al. 1998b). In the KA-induced model of limbic seizure in rats, KA-induced seizures decreased GluN1 mRNA level in CA1 and CA3 pyramidal cells, but not dentate gyrus, at 24 and 72 h after drug injection (Lason et al. 1997). Pilocarpine-induced seizures significantly reduced both CA1 pyramidal cells and dentate gyrus at 24 and 72 h, and CA3 at only 72 h after drug injection. The binding of [^{3}H]MK-801 was not changed at 3 h after pilocarpine injection, whereas it was decreased in stratum lucidum at 3 and 24 h after drug injection (Lason et al. 1997). These data indicate some differences in hippocampal NMDAR regulation in pilocarpine and KA models of limbic seizures.

DBA/2 mice are prone to audiogenic seizures and in these animals GluN1 antisense administration causes a short-term anticonvulsant protection; with seizure response gradually returning to control levels 12–24 h following the termination of antisense administration. Autoradiography has also shown that GluN1 subunit levels significantly decrease in the retrosplenial cortex and the overall cortex after GluN1 antisense administration (Chapman et al. 1996). In albino Swiss mice acoustically primed at 10 days, some subsequently show audiogenic seizures at 20 days. These sound sensitive mice show a marked transient expression of GluN2C in the central nucleus of the inferior colliculus (not seen in the non-sensitive mice) and a doubling of the expression of GluN1 (Marianowski et al. 1995). Therefore, the audiogenic susceptibility may be associated with the transient expression of the GluN2C subunit during a brief neonatal period during which synaptic reorganization happens (Marianowski et al. 1995). In the cortex of rats with absence type seizures, the response to NMDA appears to be elevated (Pumain et al. 1992). Study on ethanol withdrawal seizures revealed that chronic ethanol administration results in increased levels of [^{3}H]MK-801 recognition sites on NMDARs (Morrow et al. 1994). Because MK-801 labels all NMDARs so the change presumably reflects a change in the number of receptors not only in the recognition sites. However, GluN1 subunit mRNAs did not change following chronic ethanol exposure in rat cortex or hippocampus (Morrow et al. 1994).

32.5 NMDAR Antagonists for Control of Seizure

32.5.1 Felbamate

Felbamate is submitted to the antiepileptic drug development programme within the epilepsy branch of the National Institute of Neurological Disorders and Stroke in 1982 (Graves 1993). Several mechanisms might be involved in the anticonvulsant effects of felbamate:

(a) Blockage of voltage-sensitive Na^+ or Ca^{2+} channels (Stefani et al. 1996; Taglialatela et al. 1996).
(b) Increase in the GABA-induced chloride currents (Rho et al. 1994).
(c) The excitatory glutamatergic NMDAR can also be involved in the antiepileptic effects of felbamate and the drug acts as an NMDAR inhibitor in the CNS. However, the exact site of action is still unclear. Felbamate may bind to the strychnine-insensitive glycine site (McCabe et al. 1993; Subramaniam et al. 1995; White et al. 1995) or to a site within the NMDAR channel pore and act as an open-channel blocker (Rho et al. 1994). There is also evidence that felbamate is a non-competitive, allosteric inhibitor with some modest selectivity for GluN2B-containing receptors that might also associate with the channel pore (Harty and Rogawski 2000; Kleckner et al. 1999).

Although the exact mechanisms underlying the anticonvulsant effects of felbamate has not been demonstrated, some investigators have suggested that felbamate can decrease seizure susceptibility mainly through inhibition of Ca^{2+} currents through either NMDAR or voltage-gated Ca^{2+} channels (for review see (Stefani et al. 1997)). Several clinical trials of felbamate as a second generation antiepileptic drug was done before its approval by the U.S. Food and Drug Administration (FDA) in 1993 (Faught et al. 1993; Leppik et al. 1991; Sachdeo et al. 1992; Theodore et al. 1991). Felbamate has also been studied in Lennox-Gastaut syndrome in children (Dodson 1993; Kalviainen et al. 1993). Felbamate-treated patients with Lennox-Gastaut syndrome demonstrate a robust decrease in the frequency of atonic seizures (drop attacks) and total seizure frequency compared to placebo groups (Dodson 1993). Currently, felbamate is not used as a first-line AED and generally is used for patients with intractable partial seizures, infantile spasms, or Lennox-Gastaut syndromes who are resistant to primary AEDs (French et al. 1999; Hosain et al. 1997; Hurst and Rolan 1995; Zupanc et al. 2010). However, it has two rare but serious adverse effects, aplastic anemia and hepatotoxicity.

32.5.2 Magnesium Sulfate

Clinical observations in patients (Griffiths 1947; Hanna et al. 1960; Randall et al. 1959) and experimental investigations in animals (Kruse et al. 1932) have demonstrated that magnesium (Mg^{2+}) depletion causes a marked irritability of the nervous system, eventually leading to epileptic seizures or increasing susceptibility to seizure-inducing stimuli (Buck et al. 1976; Greenberg and Tufts 1934). Mg^{2+} level in serum and CSF were significantly lower in patients with grand mal epilepsy than in controls (Babel et al. 1973; Govil et al. 1981; Sinert et al. 2007). Serum and CSF Mg^{2+} levels fell with increasing duration and frequency of seizures. Within the brain, Mg^{2+} preferentially accumulates within the cortex and hippocampus (Hallak 1998). Magnesium sulfate enters the cerebrospinal fluid and brain after systemic administration (Hallak 1998). The significant rise in brain Mg^{2+} concentration is associated with an elevation of the seizure threshold and a marked resistance of the animal to electrically as well as NMDA-stimulated hippocampal seizures (Hallak 1998). Mg^{2+} blocks Ca^{2+} within NMDAR channel. When the Mg^{2+} blockade is relieved by cellular depolarization, the channel is unblocked and Ca^{2+} and Na^{+} enter the postsynaptic neuron as potassium exits (Stephenson 2006).

As summarized in Table 32.2, systemic administration of Mg^{2+} exerts an anticonvulsant effect against experimentally induced epileptic foci in cats and dogs (Borges and Gucer 1978), status epilepticus (Mikati et al. 2006), audiogenic and PTZ-induced clonic seizure (Bac et al. 1993; Ghasemi et al. 2010a; Safar et al. 2010), and NMDA-induced hippocampal seizures (Cotton et al. 1993) in rodents. In humans, Mg^{2+} is an effective treatment for the seizures of neonatal tetany (Turner et al. 1977) and eclampsia (Euser and Cipolla 2009) and possibly for those associated with ethanol withdrawal and acute intermittent porphyria (Sadeh et al. 1991; Taylor 1981). In spite of its effectiveness in controlling the hyperexcitability and seizures associated with toxemia of pregnancy, there is still some controversy over magnesium's merits as an anticonvulsant (Kaplan et al. 1990; Tso and Barish 1992). Mg^{2+} may be helpful in potentiating the therapeutic effects of adrenocorticotropic hormone on infantile spasms (or West syndrome) (Zou et al. 2010). Intravenous Mg^{2+} therapy could improve refractory epilepsy with recurrent status epilepticus and episodes of epilepsia partialis continua in two patients with juvenile-onset Alpers' syndrome (Visser et al. 2011).

32.5.3 Remacemide

Remacemide hydrochloride is a racemate. The two enantiomers of remacemide and those of APL 12495AA have been tested for their anticonvulsive properties and the potency of the drugs were different depending on the route of administration and type of seizure test (Garske et al. 1991; Palmer et al. 1991). Both remacemide hydrochloride and APL 12495AA displace [^{3}H]MK-801 binding from the synaptic membrane fractions of the rat cerebral cortex and hippocampus (Ray et al. 1992). Another binding studies using [^{3}H]desglycinylremacemide demonstrated a binding site within the NMDAR ion channel, which is supposed to be associated with the benzomorphan attachment site of the NMDAR (Ahmed et al. 1999). Both remacemide and APL 12495AA inhibit NMDA-induced responses in a variety of preparations such as cultured rat hippocampal neurons (Subramaniam et al. 1993) and cortical wedges prepared from audiogenic seizure-prone DBA/2 mice (Hu and Davies 1995). Besides acting on NMDARs, both remacemide and its metabolite are potent Na^+ fast-channel blockers (Norris and King 1997; Santangeli et al. 2002). It is likely that synergism between these two actions occur. Additionally, because of the relatively low affinity of remacemide and its metabolite to the NMDAR, the anticonvulsant effects of these agents may be due to their effects on Na^+ channels. A possible partial blockade of potassium channels has also been suggested (Norris and King 1997).

Remacemide reduces pyramidal cell damage in the CA3 and CA1 subregions of the hippocampus in a model of rat status epilepticus (Halonen et al. 1999). As summarized in Table 32.2, remacemide and APL 12495AA have anticonvulsant property in a number of animal models. However, remacemide had no significant protection against bicorneal kindled seizures and those induced by PTZ, bicuculline, picrotoxin, and strychnine (Garske et al. 1991; Palmer et al. 1992; Stagnitto et al. 1990) or by acute heat stress (Palmer et al. 1998). Many clinical studies on refractory epilepsy patients have also investigated remacemide as an adjuvant. In a double blind placebo-controlled cross-over study with remacemide as adjunctive therapy in 28 patients with refractory seizures, one-third of the patients showed a reduction in seizure frequency by 50 % or more and four patients were free of seizures during remacemide medication (Bialer et al. 1999). Other study (Owen et al. 1992) also found a significant improvement in the appearance of the EEG recordings of epileptic patients after remacemide treatment. Other placebo-controlled trial on adult patients with refractory epilepsy showed that remacemide as an add-on therapy significantly increased the percentage of responders (defined as, at least, 50 % reduction in seizure frequency) compared with placebo (Chadwick et al. 2002). Adjunctive remacemide treatment was associated with a higher, dose-related responder rate than that observed in placebo-administered patients (Jones et al. 2002). Remacemide had therapeutic activity as monotherapy in patients with refractory epilepsy following pre-surgical assessment (Devinsky et al. 2002). However, there is some evidence that remacemide is not beneficial in newly diagnosed epilepsy compared to other AEDs such as carbamazepine (Brodie

Table 32.2 Anticonvulsant effects of some NMDAR antagonists in animal behavioral experiments

Drug	Effect	Animal model of seizure	Reference
Magnesium sulfate	Non-selective NMDAR channel blocker	Audiogenic and PTZ-induced clonic seizure in rodents	Bac et al. (1993), Ghasemi et al. (2010a), Safar et al. (2010)
		Experimentally induced epileptic foci in cats and dogs	Borges and Gucer (1978)
		NMDA-induced rat hippocampal seizures	Cotton et al. (1993)
Remacemide and APL 12495AA	Non-selective NMDAR channel blocker	MES test in rodents	Garske et al. (1991), Palmer et al. (1992) Stagnitto et al. (1990)
		NMDA, kainic acid, electrical kindling, and 4-aminopyridine-induced seizures	Cramer et al. (1994) Garske et al. (1991)
		Spike-and-wave discharges in the WAG/Rij rats and in the GAERS rats	Nehlig and Boehrer (2003), van Luijtelaar and Coenen (1995)
Ketamine	Non-selective NMDAR channel blocker	Electrically precipitated tonic hind-limb extension	Deutsch et al. (1997)
		Generalized tonic–clonic seizures induced by metrazol in rats	Velisek et al. (1989)
		Kindled amygdaloid seizures in rats	Bowyer et al. (1983b)
		Kindling epileptogenesis and seizure expression in developing rats	Trommer and Pasternak (1990)
		Latency of onset to seizures induced by ether flurothyl	Wurpel et al. (1992)
		Limbic status epilepticus induced by 90 min continuous electrical stimulation of the hippocampus	Bertram and Lothman (1990)
		Lithium-pilocarpine seizure model in rats	Cook (2000)
		MES test in mice	Khanna and Bhalla (1999),Manocha et al. (2001)
		Morphine-induced hind-limb myoclonic seizures	Kolesnikov et al. (1997)
		Partial status epilepticus in rat	Navarro et al. (2007)
		Pilocarpine-induced status epilepticus in rats	Freitas et al. (2006)
		Prolonged status epilepticus in rats and dog	Borris et al. (2000), Serrano et al. (2006)
		Seizures induced by NMDA, guanidinosuccinate, p-toluidino-3-propylamino-2-propanol, mercaptopropionate, *N*-(3,5-dimethoxy-4-propoxyphenylethyl)-aziridine, lidocaine, picrotoxin, bicuculline, strychnine or PTZ in rodents	Chen et al. (1966), D'Hooge et al. (1993), Ghasemi et al. (2010b, 2010c), Guler et al. (2005), Herink (1997), Irifune et al. (2000), Kubova and Mares (1994), Parsons et al. (1995), Sofia et al. (1994), Stafstrom and Sasaki-Adams (2003), Taberner (1976), Tricklebank et al. (1989), Velisek et al. (1993); Veliskova et al. (1990), Wardley-Smith et al. (1988)
		Seizures kindled by repetitive electrical stimulation of the rat motor cortex	Bowyer et al. (1983a)
		Sound-induced convulsions in epilepsy prone rats	Bourn et al. (1983), De Sarro and De Sarro (1993)
		Sound-induced seizures in DBA/2 mice	Chapman and Meldrum (1989)
		Young chick model of epilepsy	Reder et al. (1980)

(continued)

Table 32.2 (continued)

Drug	Effect	Animal model of seizure	Reference
Amantadine	Non-selective NMDAR channel blocker	MES in mice	Parsons et al. (1995)
		NMDA-induced lethality in mice	Geter-Douglass and Witkin (1999), Parsons et al. (1995)
		PTZ-induced tonic convulsions in mice	Kleinrok et al. (1981), Parsons et al. (1995)
Memantine	Non-selective NMDAR channel blocker	Audiogenic convulsions in Krushinskii-Molodkina strain rats	Vataev et al. (2009), (2010)
		Cocaine-induced convulsions in mice	Brackett et al. (2000)
		Electrically precipitated tonic hind-limb extension in mice	Deutsch et al. (1997)
		Ethanol withdrawal-associated audiogenic seizures in rats	Bienkowski et al. (2001)
		MES in mice	Chojnacka-Wojcik et al. (1983), Czuczwar et al. (1996), Kleinrok et al. (1995), Parsons et al. (1995),Urbanska et al. (1992)
		Soman-induced seizures in rats	Deshpande et al. (1995), McLean et al. (1992), Shih et al. (1999)
		Tonic (but not clonic) seizures in mice induced by PTZ, bicuculline, picrotoxin, 3-mercaptopropionic acid and NMDA in mice	Bisaga et al. (1993), Geter-Douglass and Witkin (1999), Lukomskaia et al. (2003), (2004), Mares and Mikulecka (2009) Meldrum et al. (1986) Parsons et al. (1995)

et al. 2002), questioning efficacy of remacemide as monotherapy. Remacemide has some side effects such as dizziness, somnolence, and gastrointestinal symptoms, when the drug is given as monotherapy. Adjunctive therapy with the conventional AEDs carbamazepine and/or phenytoin might result in diplopia and fatigue (Bialer et al. 1999).

32.5.4 Ketamine

Ketamine has a low affinity for the NMDAR at the phencyclidine site within the ionotropic channel (Harrison and Simmonds 1985; Mealing et al. 1999), with slow non-selective open-channel blocking/unblocking kinetics, and a specific type of channel closure, called "trapping block" (Machado-Vieira et al. 2009). As early as 1965, it was found that ketamine suppresses convulsions induced by electrical stimuli as well as those induced by intravenous infusion of CNS stimulants such as PTZ or caffeine (McCarthy et al. 1965). Since then many studies have observed that ketamine has anticonvulsant effects in many animal models of seizure (Table 32.2). Infusions of ketamine into the substantia nigra pars reticulata of adult rats also increases the latency of onset to seizures induced by ether flurothyl (Wurpel et al. 1992). Ketamine can act synergistically with AEDs such as valproate and carbamazepine to decrease the MES-induced seizures in mice (Borowicz et al. 2004). Other studies also demonstrated synergistic anticonvulsant action of diazepam and ketamine in status epilepticus induced by either lithium-pilocarpine (Martin and Kapur 2008) or kainic acid (Vermoesen et al. 2010). Ketamine in association with atropine sulfate could be highly effective in the delayed treatment of severe soman intoxication (Dorandeu et al. 2007).

Ketamine consistently suppressed focal seizures on neocortical and hippocampal epileptogenic loci in cats but was ineffective in modifying interictal epileptogenic activity both at the primary and mirror focus (Celesia and Chen 1974). It (50–100 mg/kg, i.p.) also reduces the intensity of picrotoxin convulsions and eliminates metrazol-induced seizures (Myslobodsky et al. 1981). Both application of ketamine blocked penicillin-induced, synchronized after discharges in immature rat CA3 hippocampal neurons, and depressed intracellular depolarizations produced by iontophoretic application of NMDA (Brady and Swann 1986). Ketamine decreases epileptiform activity in both CA1 and CA3 regions of the rat hippocampus (Lee and Hablitz 1990; Mikolasova et al. 1994). Moreover, ketamine reduces cell loss in the hippocampus of post-status epilepticus rats (Cunha et al. 2009), blocks epileptiform activity induced by alkalosis in rat neocortical slices (Aram and Lodge 1987), decreases 4-aminopyridine- or picrotoxin-or "magnesium free"-induced epileptogenic activity in hippocampal slices of rats (Kohr and Heinemann 1989; Sagratella et al. 1987), attenuates epileptogenic responses in a synaptic and a non-synaptic model of epileptogenesis in the CA1 region of the hippocampal slice (Ashton et al. 1988), and has anticonvulsant effects on different patterns of epileptiform activity in rat temporal cortex slices (Zhang et al. 1994).

Degeneration of thalamic neurons caused by persistent seizure activity in the corticothalamic tract (putative glutamergic transmitter pathway) is prevented by ketamine or MK-801, despite the failure of these agents to eliminate persistent electrographic seizure activity recorded from cortex and thalamus (Clifford et al. 1989). Ketamine as well as other NMDAR antagonists (MK-801 and phencyclidine) could protect against kainic acid-induced seizure-related rat brain damage (Clifford et al. 1990). Another study reported a 23-year-old man whose refractory status epilepticus was refractory to standard AEDs and barbiturates, but was successfully terminated only with intravenous ketamine (Hsieh et al. 2010).

Regarding the effects of ketamine in human, Thompson in 1972 (Thompson 1972) reported that during a study of anesthesia for outpatient surgery, a generalized grand mal seizures in a 26-year-old woman (who had no history of neurologic diseases) appeared 2 min after an induction dose of 130 mg ketamine (i.v.) given for elective abortion. Another study reported the occurrence of extensor-type seizures in 4 asthmatic patients receiving theophylline within minutes following induction of anesthesia with ketamine (Hirshman et al. 1982). In patients who had depth electrodes implanted in the limbic and temporal regions, ketamine (2–4 mg/kg, i.v.) caused the pronounced activity in these brain regions which progressed to electrical seizure phenomena in some of patients (Ferrer-Allado et al. 1973). However, another study (Corssen et al. 1974) found no evidence that ketamine (1 mg./lb. body weight, i.v.) precipitates generalized convulsions, even in patients with both a history of epilepsy and an abnormal EEG, but, on the contrary, they observed that ketamine suppresses seizure or seizure-like EEG discharges in epileptic patients during clinical seizures. Other study showed that ketamine neither precipitates nor aggravates seizures and is less effective than natural sleep as an activator of epileptic discharges in epileptic patients (Celesia et al. 1975). The anticonvulsant activity of ketamine was found in one study in 1968 (Corssen et al. 1968). Davis and Tolstoshev in 1976 also reported the benefit of ketamine in severe febrile convulsions (Davis and Tolstoshev 1976). Other study (Sheth and Gidal 1998) described a 13-year-old girl with seizures refractory to standard status epilepticus protocols, alternative therapeutic strategies, and 4 weeks of pentobarbital coma who achieved control of both clinical and electrographic seizures after ketamine treatment (2 mg/kg bolus; continuous infusion to a maximum rate of 7.5 mg/kg/h for 14 days). Oral ketamine, administered to five children with severe epilepsy (Lennox-Gastaut Syndrome, myoclonic-astatic epilepsy, progressive myoclonic epilepsy, and Pseudo-Lennox Syndrome) during an episode of non-convulsive status epilepticus (NCSE) resulted in the resolution of NCSE in all cases within 24–48 h (Mewasingh et al. 2003). A retrospective series of 6 fatal cases of refractory status epilepticus treated with ketamine suggested some modest efficacy, although there was no mention of the dose regimen used or duration of therapy (Nathan et al. 2002). A series of seven patients treated with ketamine for refractory status epilepticus reported electrographic seizure control in four patients on 0.3–5.8 mg/kg/h infusion (Bleck et al. 2002). In another study (Ubogu et al. 2003) it was observed that ketamine infusion following low-dose propofol sedation in 44-year-old man with treated neurosyphilis presented with subclinical refractory status epilepticus gradually controlled electrographic seizures over 72 h. More recent study reported development of refractory status epilepticus in a 22-year-old woman with mitochondriopathy and pre-existing epilepsy which was treated with supplemental administration of continuous ketamine infusion to midazolam (Pruss and Holtkamp 2008). Therefore, ketamine seems to be useful in controlling malignant status epilepticus in patients with disorders particularly predisposing to death from treatment-resistant status epilepticus.

32.5.5 Amantadine

Amantadine was originally licensed to treat influenza A infections (Davies et al. 1964), and later parkinsonism (Schwab and England 1969). Some evidence indicates that amantadine can inhibit NMDARs (Kornhuber et al. 1991; Lupp et al. 1992). When amantadine is bound in the channel of NMDARs, it increases the rate of channel closure. As a result, the predominant inhibitory mechanism of amantadine is not blockade of current flow through open channels but rather increasing occupancy of channel closed states (Blanpied et al. 2005). Early evidence in the 1970s showed that amantadine is able to increase seizure susceptibility in electroshock test and PTZ-induced seizures in mice (Kleinrok et al. 1978; Lazarova and Roussinov 1979; Lazarova and Roussinov 1978). Amantadine (25 and 100 mg/kg) decreased the anticonvulsant effect of diphenylhydantoin on electroshock test (Kleinrok et al. 1980) and inhibited the anticonvulsant effects of serotonin (5-HT) on the PTZ-induced convulsions in mice at 100 mg/kg (Lazarova and Roussinov 1979). These effects of amantadine were explained by its possible effects on the central dopaminergic system. However, next study (Kleinrok et al. 1981) demonstrated that amantadine (up to 100 mg/kg) did not affect PTZ-induced convulsions in mice, but combined treatment of the GABAergic agonist baclofen (5 mg/kg) and amantadine (100 mg/kg) significantly decreased the number of animals with tonic seizures induced by PTZ. Subsequent studies found that amantadine had anticonvulsant property against MES and PTZ-induced tonic convulsions and NMDA-induced lethality in mice (Geter-Douglass and Witkin 1999; Parsons et al. 1995). Oral amantadine administration (50 and 100 mg/kg) decreased the human motor cortex excitability in normal subjects (Reis et al. 2006). The NMDAR antagonism seems to be the most relevant effect on cortical excitability (Reis et al. 2006). Other study (Rohrbacher et al. 1994) revealed that amantadine (100 μM) decreased the synaptic excitation intrastriatally evoked in rat neostriatal slices. High doses of amantadine at 200 mg/kg could induce convulsion in mice (Vamvakides 1990).

Amantadine treatment (12–16 week) in ten children with medically refractory seizures could control myoclonic or atypical absence seizures without alteration (or even worsened) in tonic–clonic and atonic seizures. Tonic seizures were controlled in one patient, but worsened in another (Shields et al. 1985). Other study in ten adolescents and adults with refractory generalized tonic–clonic, myoclonic, or absence seizures showed that amantadine combined with AEDs regimens in weekly increments to 400 mg/day decreased myoclonic or absence seizures in four patients, whereas it worsened tonic–clonic seizures in three patients (Drake et al. 1991). It was also reported that add-on amantadine administration to four children with refractory absence epilepsy led to complete resolution of absence episodes within 1 week (Shahar and Brand 1992). All patients remained free of symptoms for 27–36 months. An attempt to discontinue the use of this medication in three children caused a prompt relapse (Shahar and Brand 1992). Other case report (Matsushige et al. 2007) similarly described a girl refractory childhood absence epilepsy who improved by add-on amantadine therapy. These reports suggest that amantadine may be beneficial as an augmentation for controlling refractory absence seizures especially in children. However, high doses of amantadine even induce seizure in patients (Claudet and Marechal 2009) and in animals (Vamvakides 1990).

32.5.6 Memantine

Memantine is a low affinity, uncompetitive, non-selective, open-channel NMDAR antagonist which enters the receptor channel and blocks the NMDAR-associated ion channel similar to Mg^{2+}, by binding to or near the Mg^{2+} binding site. Indeed, memantine can inhibit the prolonged influx of Ca^{2+} ions, which forms the basis of neuronal excitotoxicity. Preliminary animal studies in the early 1980s suggested that memantine can act as an anticonvulsant drug alone or combined with other AEDs (Chojnacka-Wojcik et al. 1983; McLean 1987). Subsequent preclinical studies revealed that memantine had anticonvulsant effects in MES-induced seizures (Chojnacka-Wojcik et al. 1983; Czuczwar et al. 1996; Kleinrok et al. 1995; Parsons et al. 1995; Urbanska et al. 1992), tonic (but not clonic) seizures in mice induced by PTZ, bicuculline, picrotoxin, 3-mercaptopropionic acid and NMDA (Bisaga et al. 1993; Geter-Douglass and Witkin 1999; Lukomskaia et al. 2003; Lukomskaya et al. 2004; Mares and Mikulecka 2009; Meldrum et al. 1986; Parsons et al. 1995), cocaine-induced convulsions in mice (Brackett et al. 2000), ethanol withdrawal-associated audiogenic seizures in rats (Bienkowski et al. 2001), soman-induced seizures in rats (Deshpande et al. 1995; McLean et al. 1992; Shih et al. 1999), electrically precipitated tonic hind-limb extension in mice (Deutsch et al. 1997), and audiogenic convulsions in Krushinskii-Molodkina strain rats at 5–10 mg/kg (Vataev et al. 2009; Vataev et al. 2010). Memantine can potentiate anticonvulsant effects of AEDs such as valproate in MES test (Urbanska et al. 1992). Memantine also reduces the spontaneous absence-like paroxysms in the cortical EEG of rats (Frey and Voits 1991), the synaptic excitation intrastriatally evoked in rat neostriatal slices (Rohrbacher et al. 1994), and NMDA- or Mg^{2+} free-induced epileptiform activity in area CA1 of guinea pig hippocampal slices (Apland and Cann 1995). Memantine protects against diisopropylphosphorofluoridate (an organophosphate compound)-induced seizures and dendritic degeneration of pyramidal neurons in the CA1 hippocampal area in rats (Zaja-Milatovic et al. 2009). Acute memantine (20 mg/kg) significantly retarded the progression of the dystonic attack in a mutant hamster model of paroxysmal dystonia (Richter et al. 1991). Although memantine was ineffective against amygdala kindling in rats when given alone (5–10 mg/kg) (Loscher and Honack 1990, 1994), co-administration of memantine with the AMPA receptor antagonist NBQX (2,3-dihydroxy-6-nitro-7-sulfamoylbenzo[f]quinoxaline) had an over-additive anticonvulsant effect in the kindling model of epilepsy (Loscher and Honack 1994). This observation may suggest that blockade of both NMDA and AMPA receptors could be a better therapeutic approach for controlling seizures. Accordingly, Gmiro and Serdyuk (Gmiro and Serdyuk 2008) have recently reported that although memantine decreased 4-fold the incidence of PTZ-induced generalized tonic–clonic seizures in rats, IEM-1913 (combined blockade of NMDA and AMPA receptors in the brain) decreased the incidence of PTZ-induced clonic and tonic–clonic seizures in rats by 4–8 times. However, some studies have reported that memantine may cause seizure in patients. For instance, a case report (Peltz et al. 2005) described that a 72-year-old Caucasian woman, taking memantine for Alzheimer's disease, was admitted to the hospital with new-onset, generalized tonic–clonic seizure. After memantine was discontinued, the disturbance resolved and increased Δ waves on the patient's EEG were improved (Peltz et al. 2005). The clinical data regarding the anticonvulsant effects of memantine is scarce and more trials are needed in this regard.

32.6 Conclusion

A huge number of investigation have indicated that NMDAR-mediated Ca^{2+} signaling plays a pivotal role in seizure phenomena, although more studies are clearly needed to demonstrate that pathology of NMDARs underlies epilepsy. Considering animal seizure models, NMDAR antagonists may be effective anticonvulsants. The early and most obvious approach to the development of NMDAR-targeting drugs for the treatment of neurological diseases such as epilepsy was to directly target the NMDAR itself. Three major classes of antagonists can be distinguished based on their site of action: competitive antagonists that act at the glutamate or glycine-binding site, non-competitive NMDAR allosteric inhibitors that act at other extracellular sites, and NMDAR channel blockers that bind to sites within the NMDAR channel pore. Many compounds modulate NMDAR activity by binding to these various extracellular sites and their use in basic neuroscience research has contributed substantially to our understanding of NMDAR function. However, few NMDAR antagonists have been used in clinical trials for evaluation of their efficacy on controlling seizures in patients with epilepsy. Some of NMDAR antagonists such as ketamine and magnesium sulfate can be used clinically for controlling status epilepticus and eclampsia-related seizures, respectively, whereas many other agents have not been evaluated in the clinical trials yet. Moreover, some studies have indicated that combination of AEDs with NMDAR antagonists (such as memantine) may be a promising strategy to improve the anticonvulsive effects, both for the conventional AEDs and the NMDAR antagonist.

References

Ahmed MS, Mather A, Enna SJ (1999) Binding of [3H]desglycinyl remacemide to rat brain membranes: association with the benzomorphan attachment site of the *N*-methyl-D-aspartic acid receptor channel. Brain Res 827:46–50

Al-Ghoul WM, Meeker RB, Greenwood RS (1997) Amygdala kindling alters *N*-methyl-D-aspartate receptor subunit messenger RNA expression in the rat supraoptic nucleus. Neuroscience 77:985–992

Andre VM, Flores-Hernandez J, Cepeda C, Starling AJ, Nguyen S, Lobo MK, Vinters HV, Levine MS, Mathern GW (2004) NMDA receptor alterations in neurons from pediatric cortical dysplasia tissue. Cereb Cortex 14:634–646

Apland JP, Cann FJ (1995) Anticonvulsant effects of memantine and MK-801 in guinea pig hippocampal slices. Brain Res Bull 37:311–316

Aram JA, Lodge D (1987) Epileptiform activity induced by alkalosis in rat neocortical slices: block by antagonists of *N*-methyl-D-aspartate. Neurosci Lett 83:345–350

Ashton D, Willems R, De Prins E, Wauquier A (1988) Selective inhibition of synaptic versus non-synaptic epileptogenesis by NMDA antagonists in the in vitro hippocampus. Epilepsy Res 2:219–222

Avoli M, Drapeau C, Perreault P (1990) Epileptiform activity induced by low chloride medium in the CAI subfield of the hippocampal slice. J Neurophysiol 6:1747–1757

Babel CS, Bedi HK, Devpura JC, Pandey SK (1973) Serum and cerebrospinal fluid magnesium and calcium contents in idiopathic grandmal epilepsy. J Assoc Physicians India 21:481–487

Bac P, Herrenknecht C, Binet P, Durlach J (1993) Audiogenic seizures in magnesium-deficient mice: effects of magnesium pyrrolidone-2-carboxylate, magnesium acetyltaurinate, magnesium chloride and vitamin B-6. Magnes Res 6:11–19

Bayer TA, Wiestler OD, Wolf HK (1995) Hippocampal loss of *N*-methyl-D-aspartate receptor subunit 1 mRNA in chronic temporal lobe epilepsy. Acta Neuropathol 89:446–450

Bertram EH, Lothman EW (1990) NMDA receptor antagonists and limbic status epilepticus: a comparison with standard anticonvulsants. Epilepsy Res 5:177–184

Bialer M, Johannessen SI, Kupferberg HJ, Levy RH, Loiseau P, Perucca E (1999) Progress report on new antiepileptic drugs: a summary of the fourth Eilat conference (EILAT IV). Epilepsy Res 34:1–41

Bienkowski P, Krzascik P, Koros E, Kostowski W, Scinska A, Danysz W (2001) Effects of a novel uncompetitive NMDA receptor antagonist, MRZ 2/579 on ethanol self-administration and ethanol withdrawal seizures in the rat. Eur J Pharmacol 413:81–89

Bisaga A, Krzascik P, Jankowska E, Palejko W, Kostowski W, Danysz W (1993) Effect of glutamate receptor antagonists on *N*-methyl-D-aspartate- and (S)-alpha-amino-3-hydroxy-5-methyl-4-isoxazolepropionic acid-induced convulsant effects in mice and rats. Eur J Pharmacol 242:213–220

Blanpied TA, Clarke RJ, Johnson JW (2005) Amantadine inhibits NMDA receptors by accelerating channel closure during channel block. J Neurosci 25:3312–3322

Bleck TP, Quigg M, Nathan BR, Smith TL, Kapur J (2002) Electroencephalographic effects of ketamine treatment for refractory status epilepticus. Epilepsia 43:282

Borges LF, Gucer G (1978) Effect of magnesium on epileptic foci. Epilepsia 19:81–91

Borowicz KK, Luszczki J, Czuczwar SJ (2004) Interactions between non-barbiturate injectable anesthetics and conventional antiepileptic drugs in the maximal electroshock test in mice–an isobolographic analysis. Eur Neuropsychopharmacol 14:163–172
Borris DJ, Bertram EH, Kapur J (2000) Ketamine controls prolonged status epilepticus. Epilepsy Res 42:117–122
Bourn WM, Yang DJ, Davisson JN (1983) Effect of ketamine enantiomers on sound-induced convulsions in epilepsy prone rats. Pharmacol Res Commun 15:815–824
Bowyer JF, Albertson TE, Winters WD (1983a) Cortical kindled seizures: modification by excitant and depressant drugs. Epilepsia 24:356–367
Bowyer JF, Albertson TE, Winters WD, Baselt RC (1983b) Ketamine-induced changes in kindled amygdaloid seizures. Neuropharmacology 22:887–894
Brackett RL, Pouw B, Blyden JF, Nour M, Matsumoto RR (2000) Prevention of cocaine-induced convulsions and lethality in mice: effectiveness of targeting different sites on the NMDA receptor complex. Neuropharmacology 39:407–418
Brady RJ, Swann JW (1986) Ketamine selectively suppresses synchronized afterdischarges in immature hippocampus. Neurosci Lett 69:143–149
Brodie MJ, Wroe SJ, Dean AD, Holdich TA, Whitehead J, Stevens JW (2002) Efficacy and safety of remacemide versus carbamazepine in newly diagnosed epilepsy: comparison by sequential analysis. Epilepsy Behav 3:140–146
Buck DR, Mahoney AW, Hendricks DG (1976) Effect of magnesium deficiency on nonspecific excitability level (NEL) and audiogenic seizure susceptibility. Pharmacol Biochem Behav 5:529–534
Carter DS, Deshpande LS, Rafiq A, Sombati S and Delorenzo RJ (2010) Characterization of spontaneous recurrent epileptiform discharges in hippocampal-entorhinal cortical slices prepared from chronic epileptic animals. Seizure
Celesia GG, Chen RC (1974) Effects of ketamine on EEG activity in cats and monkeys. Electroencephalogr Clin Neurophysiol 37:354–353
Celesia GG, Chen RC, Bamforth BJ (1975) Effects of ketamine in epilepsy. Neurology 25:169–172
Chadwick DW, Betts TA, Boddie HG, Crawford PM, Lindstrom P, Newman PK, Soryal I, Wroe S, Holdich TA (2002) Remacemide hydrochloride as an add-on therapy in epilepsy: a randomized, placebo-controlled trial of three dose levels (300, 600 and 1200 mg/day) in a Q.I.D. regimen. Seizure 11:114–123
Chapman AG, Meldrum BS (1989) Non-competitive *N*-methyl-D-aspartate antagonists protect against sound-induced seizures in DBA/2 mice. Eur J Pharmacol 166:201–211
Chapman AG, Woodburn VL, Woodruff GN, Meldrum BS (1996) Anticonvulsant effect of reduced NMDA receptor expression in audiogenic DBA/2 mice. Epilepsy Res 26:25–35
Chen G, Ensor CR, Bohner B (1966) The neuropharmacology of 2-(omicron-chlorophenyl)-2-methylaminocyclohexanoe hydrochloride. J Pharmacol Exp Ther 152:332–339
Chojnacka-Wojcik E, Tatarczynska E, Maj J (1983) The influence of memantine on the anticonvulsant effects of the antiepileptic drugs. Pol J Pharmacol Pharm 35:511–515
Claudet I, Marechal C (2009) Status epilepticus in a pediatric patient with amantadine overdose. Pediatr Neurol 40:120–122
Clifford DB, Olney JW, Benz AM, Fuller TA, Zorumski CF (1990) Ketamine, phencyclidine, and MK-801 protect against kainic acid-induced seizure-related brain damage. Epilepsia 31:382–390
Clifford DB, Zorumski CF, Olney JW (1989) Ketamine and MK-801 prevent degeneration of thalamic neurons induced by focal cortical seizures. Exp Neurol 105:272–279
Cook LL (2000) Ventricular enlargement and the lithium/pilocarpine seizure model: possible explanation for agonistic behaviour. Int J Neurosci 100:117–123
Corssen G, Little SC, Tavakoli M (1974) Ketamine and epilepsy. Anesth Analg 53:319–335
Corssen G, Miyasaka M, Domino EF (1968) Changing concepts in pain control during surgery: dissociative anesthesia with CI-581. A progress report. Anesth Analg 47:746–759

Cotton DB, Hallak M, Janusz C, Irtenkauf SM, Berman RF (1993) Central anticonvulsant effects of magnesium sulfate on *N*-methyl-D-aspartate-induced seizures. Am J Obstet Gynecol 168:974–978

Cramer CL, Stagnitto ML, Knowles MA, and Palmer GC (1994) Kainic acid and 4-aminopyridine seizure models in mice: evaluation of efficacy of anti-epileptic agents and calcium antagonists. Life Sci 54:PL271–PL275

Crino PB, Duhaime AC, Baltuch G, White R (2001) Differential expression of glutamate and GABA-A receptor subunit mRNA in cortical dysplasia. Neurology 56:906–913

Cull-Candy SG, Leszkiewicz DN (2004) Role of distinct NMDA receptor subtypes at central synapses. Sci STKE 16:1–9

Cunha AO, Mortari MR, Liberato JL, dos Santos WF (2009) Neuroprotective effects of diazepam, carbamazepine, phenytoin and ketamine after pilocarpine-induced status epilepticus. Basic Clin Pharmacol Toxicol 104:470–477

Czuczwar SJ, Turski WA, Kleinrok Z (1996) Interactions of excitatory amino acid antagonists with conventional antiepileptic drugs. Metab Brain Dis 11:143–152

D'Hooge R, Pei YQ, De Deyn PP (1993) *N*-methyl-D-aspartate receptors contribute to guanidinosuccinate-induced convulsions in mice. Neurosci Lett 157:123–126

Das S, Sasaki YF, Rothe T, Premkumar LS, Takasu M, Crandall JE, Dikkes P, Conner DA, Rayudu PV, Cheung W et al (1998) Increased NMDA current and spine density in mice lacking the NMDA receptor subunit NR3A. Nature 393:377–381

Davies WL, Grunert RR, Haff RF, McGahen JW, Neumayer EM, Paulshock M, Watts JC, Wood TR, Hermann EC, Hoffmann CE (1964) Antiviral activity of 1-adamantanamine (amantadine). Science 144:862–863

Davis RW, Tolstoshev GC (1976) Ketamine–use in severe febrile convulsions. Med J Aust 2:465–466

De Sarro GB, De Sarro A (1993) Anticonvulsant properties of non-competitive antagonists of the *N*-methyl-D-aspartate receptor in genetically epilepsy-prone rats: comparison with CPPene. Neuropharmacology 32:51–58

DeLorenzo RJ (1986) A molecular approach to the calcium signal in brain: relationship to synaptic modulation and seizure discharge. Adv Neurol 44:435–464

Deshpande SS, Smith CD, Filbert MG (1995) Assessment of primary neuronal culture as a model for soman-induced neurotoxicity and effectiveness of memantine as a neuroprotective drug. Arch Toxicol 69:384–390

Deutsch SI, Mastropaolo J, Riggs RL, Rosse RB (1997) The antiseizure efficacies of MK-801, phencyclidine, ketamine, and memantine are altered selectively by stress. Pharmacol Biochem Behav 58:709–712

Devinsky O, Vazquez B, Faught E, Leppik IE, Pellock JM, Schachter S, Alderfer V, Holdich TA (2002) A double-blind, placebo-controlled study of remacemide hydrochloride in patients with refractory epilepsy following pre-surgical assessment. Seizure 11:371–376

Dichter MA, Brodie MJ (1996) New antiepileptic drugs. N Engl J Med 334:1583–1590

Dodson WE (1993) Felbamate in the treatment of Lennox-Gastaut syndrome: results of a 12-month open-label study following a randomized clinical trial. Epilepsia 34(Suppl 7):S18–S24

Dorandeu F, Baille V, Mikler J, Testylier G, Lallement G, Sawyer T, Carpentier P (2007) Protective effects of S + ketamine and atropine against lethality and brain damage during soman-induced status epilepticus in guinea-pigs. Toxicology 234:185–193

Drake ME Jr, Pakalnis A, Denio LS, Phillips B (1991) Amantadine hydrochloride for refractory generalized epilepsy in adults. Acta Neurol Belg 91:159–164

Euser AG, Cipolla MJ (2009) Magnesium sulfate for the treatment of eclampsia: a brief review. Stroke 40:1169–1175

Faught E, Sachdeo RC, Remler MP, Chayasirisobhon S, Iragui-Madoz VJ, Ramsay RE, Sutula TP, Kanner A, Harner RN, Kuzniecky R et al (1993) Felbamate monotherapy for partial-onset seizures: an active-control trial. Neurology 43:688–692

Ferrer-Allado T, Brechner VL, Dymond A, Cozen H, Crandall P (1973) Ketamine-induced electroconvulsive phenomena in the human limbic and thalamic regions. Anesthesiology 38:333–344

Franck JE, Pokorny J, Kunkel DD, Schwartzkroin PA (1995) Physiologic and morphologic characteristics of granule cell circuitry in human epileptic hippocampus. Epilepsia 36:543–558

Freitas RM, Sousa FC, Viana GS, Fonteles MM (2006) Effect of gabaergic, glutamatergic, antipsychotic and antidepressant drugs on pilocarpine-induced seizures and status epilepticus. Neurosci Lett 408:79–83

French J, Smith M, Faught E, Brown L (1999) Practice advisory: the use of felbamate in the treatment of patients with intractable epilepsy: report of the quality standards subcommittee of the American academy of neurology and the American epilepsy society. Neurology 52:1540–1545

Frey HH, Voits M (1991) Effect of psychotropic agents on a model of absence epilepsy in rats. Neuropharmacology 30:651–656

Fujikawa DG (1996) The temporal evolution of neuronal damage from pilocarpine-induced status epilepticus. Brain Res 725:11–22

Fujikawa DG (2005) Prolonged seizures and cellular injury: understanding the connection. Epilepsy Behav 7:S3–S11

Garske GE, Palmer GC, Napier JJ, Griffith RC, Freedman LR, Harris EW, Ray R, McCreedy SA, Blosser JC, Woodhead JH et al (1991) Preclinical profile of the anticonvulsant remacemide and its enantiomers in the rat. Epilepsy Res 9:161–174

Gerfin-Moser A, Grogg F, Rietschin L, Thompson SM, Streit P (1995) Alterations in glutamate but not GABAA receptor subunit expression as a consequence of epileptiform activity in vitro. Neuroscience 67:849–865

Geter-Douglass B, Witkin JM (1999) Behavioral effects and anticonvulsant efficacies of low-affinity, uncompetitive NMDA antagonists in mice. Psychopharmacology 146:280–289

Ghasemi A, Saberi M, Ghasemi M, Shafaroodi H, Moezi L, Bahremanhd A, Montaser-Kouhsari L, Ziai P, Dehpour AR (2010a) Administration of lithium and magnesium chloride inhibited tolerance to the anticonvulsant effect of morphine on pentylenetetrazole-induced seizures in mice. Epilepsy Behav. doi:10.1016/j.yebeh.2010.09.004

Ghasemi M, Schachter SC (2011) The NMDA receptor complex as a therapeutic target in epilepsy: a review. Epilepsy Behav 22:617–640

Ghasemi M, Shafaroodi H, Nazarbeiki S, Meskar H, Ghasemi A, Bahremand A, Ziai P, Dehpour AR (2010b) Inhibition of NMDA receptor/NO signaling blocked tolerance to the anticonvulsant effect of morphine on pentylenetetrazole-induced seizures in mice. Epilepsy Res 91:39–48

Ghasemi M, Shafaroodi H, Nazarbeiki S, Meskar H, Heydarpour P, Ghasemi A, Talab SS, Ziai P, Bahremand A, Dehpour AR (2010c) Voltage-dependent calcium channel and NMDA receptor antagonists augment anticonvulsant effects of lithium chloride on pentylenetetrazole-induced clonic seizures in mice. Epilepsy Behav 18:171–178

Gmiro VE, Serdyuk SE (2008) Combined blockade of AMPA and NMDA receptors in the brain of rats prevents pentylenetetrazole-induced clonic and tonic-clonic seizures without ataxia. Bull Exp Biol Med 145:728–730

Govil MK, Mangal BD, Alam SM, Mahendru RK, Srivastava DK, Mudgal JC (1981) Serum and cerebrospinal fluid calcium and magnesium levels in cases of idiopathic grand mal epilepsy and induced convulsions. J Assoc Physicians India 29:695–699

Graves NM (1993) Felbamate. Ann Pharmacother 27:1073–1081

Greenberg DM, Tufts EV (1934) Effect of a diet low in magnesium on the rat. Proc Soc Exp Biol Med 5:674–675

Griffiths WJ Jr (1947) Audiogenic fits produced by magnesium deficiency in tame domestic Norway rats and in wild Norway and Alexandrine rats. Am J Physiol 149:135–141

Guler G, Erdogan F, Golgeli A, Akin A, Boyaci A (2005) Ketamine reduces lidocaine-induced seizures in mice. Int J Neurosci 115:1239–1244

Hallak M (1998) Effect of parenteral magnesium sulfate administration on excitatory amino acid receptors in the rat brain. Magnes Res 11:117–131

Halonen T, Nissinen J, Pitkanen A (1999) Neuroprotective effect of remacemide hydrochloride in a perforant pathway stimulation model of status epilepticus in the rat. Epilepsy Res 34:251–269

Hanna S, Harrison M, Macintyre I, Fraser R (1960) The syndrome of magnesium deficiency in man. Lancet 2:172–176

Harrison NL, Simmonds MA (1985) Quantitative studies on some antagonists of *N*-methyl D-aspartate in slices of rat cerebral cortex. Br J Pharmacol 84:381–391

Harty TP, Rogawski MA (2000) Felbamate block of recombinant *N*-methyl-D-aspartate receptors: selectivity for the NR2B subunit. Epilepsy Res 39:47–55

Heinemann U, Hamon B (1986) Calcium and epileptogenesis. Exp Brain Res 65:1–10

Henshall DC, Chen J, Simon RP (2000) Involvement of caspase-3-like protease in the mechanism of cell death following focally evoked limbic seizures. J Neurochem 74:1215–1223

Herink J (1997) Effect of alprazolam and ketamine on seizures induced by two different convulsants. Acta Medica (Hradec Kralove) 40:9–11

Hirshman CA, Krieger W, Littlejohn G, Lee R, Julien R (1982) Ketamine-aminophylline-induced decrease in seizure threshold. Anesthesiology 56:464–467

Hosain S, Nagarajan L, Carson D, Solomon G, Mast J, Labar D (1997) Felbamate for refractory infantile spasms. J Child Neurol 12:466–468

Hsieh CY, Sung PS, Tsai JJ, Huang CW (2010) Terminating prolonged refractory status epilepticus using ketamine. Clin Neuropharmacol 33:165–167

Hu RQ, Davies JA (1995) The effect of the desglycinyl metabolite of remacemide on cortical wedges prepared from DBA/2 mice. Eur J Pharmacol 287:251–256

Hurst DL, Rolan TD (1995) The use of felbamate to treat infantile spasms. J Child Neurol 10:134–136

Irifune M, Sato T, Kamata Y, Nishikawa T, Dohi T, Kawahara M (2000) Evidence for GABA(A) receptor agonistic properties of ketamine: convulsive and anesthetic behavioral models in mice. Anesth Analg 91:230–236

Isokawa M, Levesque MF (1991) Increased NMDA responses and dendritic degeneration in human epileptic hippocampal neurons in slices. Neurosci Lett 132:212–216

Jensen PJ, Millan N, Mack KJ (1997) Cortical NMDAR-1 gene expression is rapidly upregulated after seizure. Brain Res Mol Brain Res 44:157–162

Jones MW, Blume WT, Guberman A, Lee MA, Pillay N, Weaver DF, Veloso F, Holdich TA (2002) Remacemide hydrochloride as an add-on therapy in epilepsy: a randomized, placebo-controlled trial of three dose levels (300, 600 and 800 mg/day) in a B.I.D. regimen. Seizure 11:104–113

Kalviainen R, Keranen T, Riekkinen PJ Sr (1993) Place of newer antiepileptic drugs in the treatment of epilepsy. Drugs 46:1009–1024

Kaplan PW, Lesser RP, Fisher RS, Repke JT, Hanley DF (1990) A continuing controversy: magnesium sulfate in the treatment of eclamptic seizures. Arch Neurol 47:1031–1032

Khanna N, Bhalla S (1999) Role of ketamine in convulsions. Indian J Med Sci 53:475–480

Kikuchi S, Iwasa H, Sato T (2000) Lasting changes in NMDAR1 mRNA level in various regions of cerebral cortex in epileptogenesis of amygdaloid-kindled rat. Psychiatry Clin Neurosci 54:573–577

Kleckner NW, Glazewski JC, Chen CC, Moscrip TD (1999) Subtype-selective antagonism of *N*-methyl-D-aspartate receptors by felbamate: insights into the mechanism of action. J Pharmacol Exp Ther 289:886–894

Kleinrok Z, Czuczwar S, Wojcik A, Przegalinski E (1978) Brain dopamine and seizure susceptibility in mice. Pol J Pharmacol Pharm 30:513–519

Kleinrok Z, Czuczwar SJ, Kozicka M (1980) Effect of dopaminergic and GABA-ergic drugs given alone or in combination on the anticonvulsant action of phenobarbital and diphenylhydantoin in the electroshock test in mice. Epilepsia 21:519–529

Kleinrok Z, Czuczwar SJ, Kozicka M, Zarkowski A (1981) Effect of combined GABA-ergic and dopaminergic stimulation on the action of some antiepileptic drugs in pentetrazol-induced convulsions. Pol J Pharmacol Pharm 33:13–23

Kleinrok Z, Turski WA, Czuczwar SJ (1995) Excitatory amino acid antagonists and the anticonvulsive activity of conventional antiepileptic drugs. Pol J Pharmacol 47:247–252

Kohr G, Heinemann U (1989) Effects of NMDA antagonists on picrotoxin-, low Mg^{2+}- and low Ca^{2+}-induced epileptogenesis and on evoked changes in extracellular Na^{+} and Ca^{2+} concentrations in rat hippocampal slices. Epilepsy Res 4:187–200

Kolesnikov Y, Jain S, Wilson R, Pasternak GW (1997) Blockade of morphine-induced hindlimb myoclonic seizures in mice by ketamine. Pharmacol Biochem Behav 56:423–425

Kornhuber J, Bormann J, Hubers M, Rusche K, Riederer P (1991) Effects of the 1-amino-adamantanes at the MK-801-binding site of the NMDA-receptor-gated ion channel: a human postmortem brain study. Eur J Pharmacol 206:297–300

Krall RL, Penry JK, White BG, Kupferberg HJ, Swinyard EA (1978) Antiepileptic drug development: II Anticonvulsant drug screening. Epilepsia 19:409–428

Kraus JE, Yeh GC, Bonhaus DW, Nadler JV, McNamara JO (1994) Kindling induces the long-lasting expression of a novel population of NMDA receptors in hippocampal region CA3. J Neurosci 14:4196–4205

Kriegstein AR, Suppes T, Prince DA (1987) Cellular and synaptic physiology and epileptogenesis of developing rat neocortical neurons in vitro. Dev Brain Res 34:161–171

Kruse HD, Orent ER, McCollum EV (1932) Studies on magnesium deficiency in animals I. Symptomatology resulting from magnesium deprivation. J Biol Chem 96:519–539

Kubova H, Mares P (1994) Effects of MK-801 (dizocilpine) and ketamine on strychnine-induced convulsions in rats: comparison with benzodiazepines and standard anticonvulsants. Physiol Res 43:313–320

Lason W, Turchan J, Przewlocki R, Machelska H, Labuz D, Przewlocka B (1997) Effects of pilocarpine and kainate-induced seizures on *N*-methyl-D-aspartate receptor gene expression in the rat hippocampus. Neuroscience 78:997–1004

Lazarova M, Roussinov K (1979) On some relationships between dopaminergic and serotoninergic mechanisms in pentylenetetrazol convulsions in albino mice. Acta Physiol Pharmacol Bulg 5:67–74

Lazarova MB, Roussinov KS (1978) On certain effects of dopaminergic agents in pentylenetetrazol convulsions. Acta Physiol Pharmacol Bulg 4:50–55

Lee WL, Hablitz JJ (1990) Effect of APV and ketamine on epileptiform activity in the CA1 and CA3 regions of the hippocampus. Epilepsy Res 6:87–94

Leppik IE, Dreifuss FE, Pledger GW, Graves NM, Santilli N, Drury I, Tsay JY, Jacobs MP, Bertram E, Cereghino JJ et al (1991) Felbamate for partial seizures: results of a controlled clinical trial. Neurology 41:1785–1789

Liu FY, Wang XF, Li MW, Li JM, Xi ZQ, Luan GM, Zhang JG, Wang YP, Sun JJ, Li YL (2007) Upregulated expression of postsynaptic density-93 and *N*-methyl-D-aspartate receptors subunits 2B mRNA in temporal lobe tissue of epilepsy. Biochem Biophys Res Commun 358:825–830

Loscher W, Honack D (1990) High doses of memantine (1-amino-3,5-dimethyladamantane) induce seizures in kindled but not in non-kindled rats. Naunyn Schmiedebergs Arch Pharmacol 341:476–481

Loscher W, Honack D (1994) Over-additive anticonvulsant effect of memantine and NBQX in kindled rats. Eur J Pharmacol 259:R3–R5

Lukomskaia N, Rukoiatkina NI, Gorbunova LV, Gmiro VE, Magazanik LG (2003) Role of NMDA and AMPA glutamate receptors in the mechanism of korazol-induced convulsions in mice. Ross Fiziol Zh Im I M Sechenova 89:292–301

Lukomskaya NY, Rukoyatkina NI, Gorbunova LV, Gmiro VE, Magazanik LG (2004) Studies of the roles of NMDA and AMPA glutamate receptors in the mechanism of corasole convulsions in mice. Neurosci Behav Physiol 34:783–789

Lupp A, Lucking CH, Koch R, Jackisch R, Feuerstein TJ (1992) Inhibitory effects of the antiparkinsonian drugs memantine and amantadine on *N*-methyl-D-aspartate-evoked acetylcholine release in the rabbit caudate nucleus in vitro. J Pharmacol Exp Ther 263:717–724

Machado-Vieira R, Salvadore G, Diazgranados N, Zarate CA Jr (2009) Ketamine and the next generation of antidepressants with a rapid onset of action. Pharmacol Ther 123:143–150

Manocha A, Sharma KK, Mediratta PK (2001) Possible mechanism of anticonvulsant effect of ketamine in mice. Indian J Exp Biol 39:1002–1008

Mares P, Mikulecka A (2009) Different effects of two *N*-methyl-D-aspartate receptor antagonists on seizures, spontaneous behavior, and motor performance in immature rats. Epilepsy Behav 14:32–39

Marianowski R, Pollard H, Moreau J, Despres G, Ben Ari Y, Tran Ba Huy P, Romand R (1995) *N*-Methyl-D-aspartate receptor subunits NR1 and NR2C are overexpressed in the inferior colliculus of audiogenic mice. Neurosci Lett 189:190–194

Martin BS, Kapur J (2008) A combination of ketamine and diazepam synergistically controls refractory status epilepticus induced by cholinergic stimulation. Epilepsia 49:248–255

Masukawa LM, Higashima M, Hart GJ, Spencer DD, O'Connor MJ (1991) NMDA receptor activation during epileptiform responses in the dentate gyrus of epileptic patients. Brain Res 562:176–180

Mathern GW, Leite JP, Babb TL, Pretorius JK, Kuhlman PA, Mendoza D, Fried I, Sakamoto AC, Assirati JA, Adelson PD, Peacock WJ (1996) Aberrant hippocampal mossy fiber sprouting correlates with greater NMDAR2 receptor staining. NeuroReport 7:1029–1035

Mathern GW, Pretorius JK, Kornblum HI, Mendoza D, Lozada A, Leite JP, Chimelli LM, Fried I, Sakamoto AC, Assirati JA et al (1997) Human hippocampal AMPA and NMDA mRNA levels in temporal lobe epilepsy patients. Brain 120(Pt 11):1937–1959

Mathern GW, Pretorius JK, Leite JP, Kornblum HI, Mendoza D, Lozada A, Bertram EH 3rd (1998a) Hippocampal AMPA and NMDA mRNA levels and subunit immunoreactivity in human temporal lobe epilepsy patients and a rodent model of chronic mesial limbic epilepsy. Epilepsy Res 32:154–171

Mathern GW, Pretorius JK, Mendoza D, Leite JP, Chimelli L, Born DE, Fried I, Assirati JA, Ojemann GA, Adelson PD et al (1999) Hippocampal *N*-methyl-D-aspartate receptor subunit mRNA levels in temporal lobe epilepsy patients. Ann Neurol 46:343–358

Mathern GW, Pretorius JK, Mendoza D, Lozada A, Kornblum HI (1998b) Hippocampal AMPA and NMDA mRNA levels correlate with aberrant fascia dentata mossy fiber sprouting in the pilocarpine model of spontaneous limbic epilepsy. J Neurosci Res 54:734–753

Matsushige T, Matsufuji H, Ichiyama T, Furukawa S (2007) A girl with refractory childhood absence epilepsy improved by add-on amantadine therapy. No To Hattatsu 39:309–310

McCabe RT, Wasterlain CG, Kucharczyk N, Sofia RD, Vogel JR (1993) Evidence for anticonvulsant and neuroprotectant action of felbamate mediated by strychnine-insensitive glycine receptors. J Pharmacol Exp Ther 264:1248–1252

McCarthy DA, Cheng G, Kaump DH, Ensor C (1965) General anesthetic and other pharmacological properties of 2-(O-chlorophenyl)-2-methylamino-cyclohexanone HCl (CI-581). J New Drugs 5:21–33

McLean MJ (1987) In vitro electrophysiological evidence predicting anticonvulsant efficacy of memantine and flunarizine. Pol J Pharmacol Pharm 39:513–525

McLean MJ, Gupta RC, Dettbarn WD, Wamil AW (1992) Prophylactic and therapeutic efficacy of memantine against seizures produced by soman in the rat. Toxicol Appl Pharmacol 112:95–103

Mealing GA, Lanthorn TH, Murray CL, Small DL, Morley P (1999) Differences in degree of trapping of low-affinity uncompetitive *N*-methyl-D-aspartic acid receptor antagonists with similar kinetics of block. J Pharmacol Exp Ther 288:204–210

Meldrum BS (1993) Excitotoxicity and selective neuronal loss in epilepsy. Brain Pathol 3:405–412

Meldrum BS, Turski L, Schwarz M, Czuczwar SJ, Sontag KH (1986) Anticonvulsant action of 1,3-dimethyl-5-aminoadamantane. Pharmacological studies in rodents and baboon Papio papio. Naunyn Schmiedebergs Arch Pharmacol 332:93–97
Mewasingh LD, Sekhara T, Aeby A, Christiaens FJ, Dan B (2003) Oral ketamine in paediatric non-convulsive status epilepticus. Seizure 12:483–489
Mikati MA, Injibar H, Kurdi RM, El Hokayem J, Abou Rialy S, Lteif L, Abdul Jawad M, Francis E, Geha G, Farhat F (2006) Effects of magnesium sulfate in kainic acid-induced status epilepticus. J Med Liban 54:200–204
Mikolasova R, Velisek L, Vorlicek J, Mares P (1994) Developmental changes of ketamine action against epileptic afterdischarges induced by hippocampal stimulation in rats. Brain Res Dev Brain Res 81:105–112
Morrow AL, Devaud LL, Bucci D, Smith FD (1994) GABAA and NMDA receptor subunit mRNA expression in ethanol dependent rats. Alcohol Alcohol Suppl 2:89–95
Myslobodsky MS, Golovchinsky V, Mintz M (1981) Ketamine: convulsant or anti-convulsant? Pharmacol Biochem Behav 14:27–33
Nathan BR, Smith TL, Bleck TP (2002) The use of ketamine in the treatment of refractory status epilepticus. Neurology Suppl 3:A197
Navarro V, Le Van Quyen M, Martinerie J, Rudrauf D, Baulac M, Menini C (2007) Loss of phase synchrony in an animal model of partial status epilepticus. Neuroscience 148:304–313
Neder L, Valente V, Carlotti CG Jr, Leite JP, Assirati JA, Paco-Larson ML, Moreira JE (2002) Glutamate NMDA receptor subunit R1 and GAD mRNA expression in human temporal lobe epilepsy. Cell Mol Neurobiol 22:689–698
Nehlig A, Boehrer A (2003) Effects of remacemide in two models of genetically determined generalized epilepsy, the GAERS and the audiogenic Wistar AS. Epilepsy Res 52:253–261
Nevander G, Ingvar M, Auer R et al (1985) Status epilepticus in welloxygenated rats causes neuronal necrosis. Ann Neurol 18:281–290
Nishi M, Hinds H, Lu HP, Kawata M, Hayashi Y (2001) Motoneuron-specific expression of NR3B, a novel NMDA-type glutamate receptor subunit that works in a dominantnegative manner. J Neurosci 21:RC185
Norris SK, King AE (1997) Electrophysiological effects of the anticonvulsant remacemide hydrochloride and its metabolite ARL 12495AA on rat CA1 hippocampal neurons in vitro. Neuropharmacology 36:951–959
Owen L, Cresswell P, Gifford C, McDade G, Mawer G (1992) Influence of remacemide on EEG in chronic epilepsy. Epilepsy 1:7
Pachernegg S, Strutz-Seebohm N, Hollmann M (2012) GluN3 subunit-containing NMDA receptors: not just one-trick ponies. Trends Neurosci 35:240–249
Palmer GC, Borrelli AR, Hudzik TJ, Sparber S (1998) Acute heat stress model of seizures in weanling rats: influence of prototypic anti-seizure compounds. Epilepsy Res 30:203–217
Palmer GC, Murray RJ, Wilson TC, Eisman MS, Ray RK, Griffith RC, Napier JJ, Fedorchuk M, Stagnitto ML, Garske GE (1992) Biological profile of the metabolites and potential metabolites of the anticonvulsant remacemide. Epilepsy Res 12:9–20
Palmer GC, Stagnitto ML, Ordy JM, Griffith RC, Napiera JJ, Gentile RJ, Woodhead JH, White HS, Swinyard EA (1991) Preclinical profile of stereoisomers of the anticonvulsant remacemide in mice. Epilepsy Res 8:36–48
Parsons CG, Quack G, Bresink I, Baran L, Przegalinski E, Kostowski W, Krzascik P, Hartmann S, Danysz W (1995) Comparison of the potency, kinetics and voltage-dependency of a series of uncompetitive NMDA receptor antagonists in vitro with anticonvulsive and motor impairment activity in vivo. Neuropharmacology 34:1239–1258
Peltz G, Pacific DM, Noviasky JA, Shatla A, Mehalic T (2005) Seizures associated with memantine use. Am J Health Syst Pharm 62:420–421
Perez-Otano I, Schulties CT, Contractor A, Lipton SA, Trimmer JS, Sucher NJ, Heinemann SF (2001) Assembly with NR1 subunit is required for surface expression of NR3A-containing NMDA receptors. J Neurosci 21:175–218

Pratt GD, Kokaia M, Bengzon J, Kokaia Z, Fritschy JM, Mohler H, Lindvall O (1993) Differential regulation of *N*-methyl-D-aspartate receptor subunit messenger RNAs in kindling-induced epileptogenesis. Neuroscience 57:307–318

Pruss H, Holtkamp M (2008) Ketamine successfully terminates malignant status epilepticus. Epilepsy Res 82:219–222

Pumain R, Louvel J, Gastard M, Kurcewicz I, Vergnes M (1992) Responses to *N*-methyl-D-aspartate are enhanced in rats with petit mal-like seizures. J Neural Transm Suppl 35:97–108

Randall RE Jr, Rossmeisl EC, Bleifer KH (1959) Magnesium depletion in man. Ann Intern Med 50:257–287

Ray R, Julien R, Gordon J, Blosser J (1992) FPL 12495, a metabolite of remacemide, is a noncompetitive NMDA antagonist. Soc Neurosci Abstr 17:391

Reder BS, Trapp LD, Troutman KC (1980) Ketamine suppression of chemically induced convulsions in the two-day-old white leghorn cockerel. Anesth Analg 59:406–409

Reis J, John D, Heimeroth A, Mueller HH, Oertel WH, Arndt T, Rosenow F (2006) Modulation of human motor cortex excitability by single doses of amantadine. Neuropsychopharmacology 31:2758–2766

Rho JM, Donevan SD, Rogawski MA (1994) Mechanism of action of the anticonvulsant felbamate: opposing effects on *N*-methyl-D-aspartate and gamma-aminobutyric acidA receptors. Ann Neurol 35:229–234

Richter A, Fredow G, Loscher W (1991) Antidystonic effects of the NMDA receptor antagonists memantine, MK-801 and CGP 37849 in a mutant hamster model of paroxysmal dystonia. Neurosci Lett 133:57–60

Rohrbacher J, Bijak M, Misgeld U (1994) Suppression by memantine and amantadine of synaptic excitation intrastriatally evoked in rat neostriatal slices. Neurosci Lett 182:95–98

Sachdeo R, Kramer LD, Rosenberg A, Sachdeo S (1992) Felbamate monotherapy: controlled trial in patients with partial onset seizures. Ann Neurol 32:386–392

Sadeh M, Blatt I, Martonovits G, Karni A, Goldhammer Y (1991) Treatment of porphyric convulsions with magnesium sulfate. Epilepsia 32:712–715

Safar MM, Abdallah DM, Arafa NM, Abdel-Aziz MT (2010) Magnesium supplementation enhances the anticonvulsant potential of valproate in pentylenetetrazol-treated rats. Brain Res 1334:58–64

Sagratella S, Frank C, de Carolis AS (1987) Effects of ketamine and (+)cyclazocine on 4-aminopyridine and "magnesium free" epileptogenic activity in hippocampal slices of rats. Neuropharmacology 26:1181–1184

Santangeli S, Sills GJ, Thompson GG, Brodie MJ (2002) Na^+ channel effects of remacemide and desglycinyl-remacemide in rat cortical synaptosomes. Eur J Pharmacol 438:63–68

Schachter SC (2007) Currently available antiepileptic drugs. Neurotherapeutics 4:4–11

Schachter SC (2009) Seizure disorders. Med Clin North Am 93:343–351

Schwab RS, England AC Jr (1969) Amantadine HCL (Symmetrel) and its relation to Levo-Dopa in the treatment of Parkinson's disease. Trans Am Neurol Assoc 94:85–90

Serrano S, Hughes D, Chandler K (2006) Use of ketamine for the management of refractory status epilepticus in a dog. J Vet Intern Med 20:194–197

Shahar EM, Brand N (1992) Effect of add-on amantadine therapy for refractory absence epilepsy. J Pediatr 121:819–821

Sheth RD, Gidal BE (1998) Refractory status epilepticus: response to ketamine. Neurology 51:1765–1766

Shields WD, Lake JL, Chugani HT (1985) Amantadine in the treatment of refractory epilepsy in childhood: an open trial in 10 patients. Neurology 35:579–581

Shih T, McDonough JH Jr, Koplovitz I (1999) Anticonvulsants for soman-induced seizure activity. J Biomed Sci 6:86–96

Sinert R, Zehtabchi S, Desai S, Peacock P, Altura BT, Altura BM (2007) Serum ionized magnesium and calcium levels in adult patients with seizures. Scand J Clin Lab Invest 67:317–326

Sofia RD, Gordon R, Gels M, Diamantis W (1994) Comparative effects of felbamate and other compounds on *N*-methyl-D-aspartic acid-induced convulsions and lethality in mice. Pharmacol Res 29:139–144

Stafstrom CE, Sasaki-Adams DM (2003) NMDA-induced seizures in developing rats cause long-term learning impairment and increased seizure susceptibility. Epilepsy Res 53:129–137

Stagnitto ML, Palmer GC, Ordy JM, Griffith RC, Napier JJ, Becker CN, Gentile RJ, Garske GE, Frankenheim JM, Woodhead JH et al (1990) Preclinical profile of remacemide: a novel anticonvulsant effective against maximal electroshock seizures in mice. Epilepsy Res 7:11–28

Stefani A, Calabresi P, Pisani A, Mercuri NB, Siniscalchi A, Bernardi G (1996) Felbamate inhibits dihydropyridine-sensitive calcium channels in central neurons. J Pharmacol Exp Ther 277:121–127

Stefani A, Spadoni F, Bernardi G (1997) Voltage-Activated Calcium Channels: Targets of Antiepileptic Drug Therapy? Epilepsia 38:959–965

Stephenson FA (2006) Structure and trafficking of NMDA and GABAA receptors. Biochem Soc Trans 34:877–881

Subramaniam S, Donevan SD, Rogawski MA (1993) 1.2-Diphenyl-2-propylamine. A major metabolite of the anti-convulsant remacemide, produces a stereoselective block of NMDA receptor currents. Soc Neurosci Abstr 19:717

Subramaniam S, Rho JM, Penix L, Donevan SD, Fielding RP, Rogawski MA (1995) Felbamate block of the *N*-methyl-D-aspartate receptor. J Pharmacol Exp Ther 273:878–886

Taberner PV (1976) The anticonvulsant activity of ketamine agains siezures induced by pentylenetetrazol and mercaptopropionic acid. Eur J Pharmacol 39:305–311

Taglialatela M, Ongini E, Brown AM, Di Renzo G, Annunziato L (1996) Felbamate inhibits cloned voltage-dependent Na^+ channels from human and rat brain. Eur J Pharmacol 316:373–377

Taylor RL (1981) Magnesium sulfate for AIP seizures. Neurology 31:1371–1372

Theodore WH, Raubertas RF, Porter RJ, Nice F, Devinsky O, Reeves P, Bromfield E, Ito B, Balish M (1991) Felbamate: a clinical trial for complex partial seizures. Epilepsia 32:392–397

Thompson GE (1972) Ketamine-induced convulsions. Anesthesiology 37:662–663

Tricklebank MD, Singh L, Oles RJ, Preston C, Iversen SD (1989) The behavioural effects of MK-801: a comparison with antagonists acting non-competitively and competitively at the NMDA receptor. Eur J Pharmacol 167:127–135

Trommer BL, Pasternak JF (1990) NMDA receptor antagonists inhibit kindling epileptogenesis and seizure expression in developing rats. Brain Res Dev Brain Res 53:248–252

Tso EL, Barish RA (1992) Magnesium: clinical considerations. J Emerg Med 10:735–745

Turner TL, Cockburn F, Forfar JO (1977) Magnesium therapy in neonatal tetany. Lancet 1:283–284

Ubogu EE, Sagar SM, Lerner AJ, Maddux BN, Suarez JI, Werz MA (2003) Ketamine for refractory status epilepticus: a case of possible ketamine-induced neurotoxicity. Epilepsy Behav 4:70–75

Urbanska E, Dziki M, Czuczwar SJ, Kleinrok Z, Turski WA (1992) Antiparkinsonian drugs memantine and trihexyphenidyl potentiate the anticonvulsant activity of valproate against maximal electroshock-induced seizures. Neuropharmacology 31:1021–1026

Urbanska EM, Czuczwar SJ, Kleinrok Z, Turski WA (1998) Excitatory amino acids in epilepsy. Restor Neurol Neurosci 13:25–39

Vamvakides A (1990) Are there glutamatergic, anti-GABAergic or antiglycinergic components in the action mechanism of adamantamines? Study of convulsions induced by amantadine or 1-adamantylcyclopentanamine in mice. Ann Pharm Fr 48:312–320

van Luijtelaar EL, Coenen AM (1995) Effects of remacemide and its metabolite FPL 12495 on spike-wave discharges, electroencephalogram and behaviour in rats with absence epilepsy. Neuropharmacology 34:419–425

Vataev SI, Zhabko EP, Lukomskaia N, Oganesian GA, Magazanik LG (2009) Effects of memantine on convulsive reactions and sleep-waking cycle in Krushinskii-Molodkina strain

rats with the inherited predisposition to audiogenic convulsions. Ross Fiziol Zh Im I M Sechenova 95:802–812
Vataev SI, Zhabko EP, Lukomskaya NY, Oganesyan GA, Magazanik LG (2010) Effects of memantine on convulsive reactions and the organization of sleep in krushinskii-molodkina rats with an inherited predisposition to audiogenic convulsions. Neurosci Behav Physiol 40:913–919
Velisek L, Mikolasova R, Blankova-Vankova S, Mares P (1989) Effects of ketamine on metrazol-induced seizures during ontogenesis in rats. Pharmacol Biochem Behav 32:405–410
Velisek L, Vondrickova R, Mares P (1993) Models of simple partial and absence seizures in freely moving rats: action of ketamine. Pharmacol Biochem Behav 45:889–896
Veliskova J, Velisek L, Mares P, Rokyta R (1990) Ketamine suppresses both bicuculline- and picrotoxin-induced generalized tonic-clonic seizures during ontogenesis. Pharmacol Biochem Behav 37:667–674
Vermoesen K, Smolders I, Massie A, Michotte Y, Clinckers R (2010) The control of kainic acid-induced status epilepticus. Epilepsy Res 90:164–166
Visser NA, Braun KP, Leijten FS, van Nieuwenhuizen O, Wokke JH, van den Bergh WM (2011) Magnesium treatment for patients with refractory status epilepticus due to POLG1-mutations. J Neurol 258:218–222
Wardley-Smith B, Little HJ, Halsey MJ (1988) Lack of correlation between the anaesthetic and anti-convulsant potencies of althesin, ketamine and methohexitone. Br J Anaesth 60:140–145
White HS, Harmsworth WL, Sofia RD, Wolf HH (1995) Felbamate modulates the strychnine-insensitive glycine receptor. Epilepsy Res 20:41–48
White R, Hua Y, Scheithauer B, Lynch DR, Henske EP, Crino PB (2001) Selective alterations in glutamate and GABA receptor subunit mRNA expression in dysplastic neurons and giant cells of cortical tubers. Ann Neurol 49:67–78
Wurpel JN, Sperber EF, Moshe SL (1992) Age-dependent differences in the anticonvulsant effects of 2-amino-7-phosphono-heptanoic acid or ketamine infusions into the substantia nigra of rats. Epilepsia 33:439–443
Yao YN, Mayer ML (2006) Characterization of a soluble ligand binding domain of the NMDA receptor regulatory subunit NR3A. J Neurosci 26:4559–4566
Zaja-Milatovic S, Gupta RC, Aschner M, Milatovic D (2009) Protection of DFP-induced oxidative damage and neurodegeneration by antioxidants and NMDA receptor antagonist. Toxicol Appl Pharmacol 240:124–131
Zhang CL, Gloveli T, Heinemann U (1994) Effects of NMDA- and AMPA-receptor antagonists on different forms of epileptiform activity in rat temporal cortex slices. Epilepsia 35(Suppl 5):S68–S73
Zou LP, Wang X, Dong CH, Chen CH, Zhao W, Zhao RY (2010) Three-week combination treatment with ACTH + magnesium sulfate versus ACTH monotherapy for infantile spasms: a 24-week, randomized, open-label, follow-up study in China. Clin Ther 32:692–700
Zupanc ML, Roell Werner R, Schwabe MS, O'Connor SE, Marcuccilli CJ, Hecox KE, Chico MS, Eggener KA (2010) Efficacy of felbamate in the treatment of intractable pediatric epilepsy. Pediatr Neurol 42:396–403

Chapter 33
Nicotinic AChR in Congenital Myasthenic Syndromes

Francesca Grassi and Sergio Fucile

Abstract About 20 % of the identified cases of congenital myasthenic syndromes are due to defects causing overstimulation of endplate ACh receptors, with consequent excessive Ca^{2+} entry, endplate structural damage, and impairment of the neuromuscular transmission. Overstimulation arises from extended dwelling of ACh in the synaptic cleft because of absence of acetylcholine esterase or from prolonged activation of "slow-channel" mutant ACh receptors. The high Ca^{2+} permeability of human endplate ACh receptor, recently described, likely predisposes to excitotoxic damage. The good knowledge of ACh receptor function has allowed molecular understanding of the defects introduced in channel kinetics and Ca^{2+} permeability by slow-channel mutations and the design of efficient therapeutic strategies. These forms of congenital myasthenic syndrome are treated by limiting ACh-induced cation entry, thus preventing endplate degeneration. Several molecules of wide clinical use, such as fluoxetine and quinidine, but also verapamil and salbutamol modulate ACh receptor kinetics and ion selectivity and may be used to lower ACh-evoked responses in these patients. In this contribution we summarize the main findings in the field.

33.1 Congenital Myasthenic Syndromes

Congenital myasthenic syndromes (CMS; phenotype MIM #608931) are rare inherited disorders in which the safety margin at the neuromuscular junction is compromised by specific defects in genes encoding proteins involved in neuromuscular transmission (Fig. 33.1a). To date, about 500 kinships bearing CMS have

F. Grassi (✉) · S. Fucile
Department of Physiology and Pharmacology, Sapienza University, piazzale Aldo Moro 5 00185 Rome, Italy
e-mail: francesca.grassi@uniroma1.it

S. Fucile
e-mail: sergio.fucile@uniroma1.it

N. Weiss and A. Koschak (eds.), *Pathologies of Calcium Channels*,
DOI: 10.1007/978-3-642-40282-1_33,

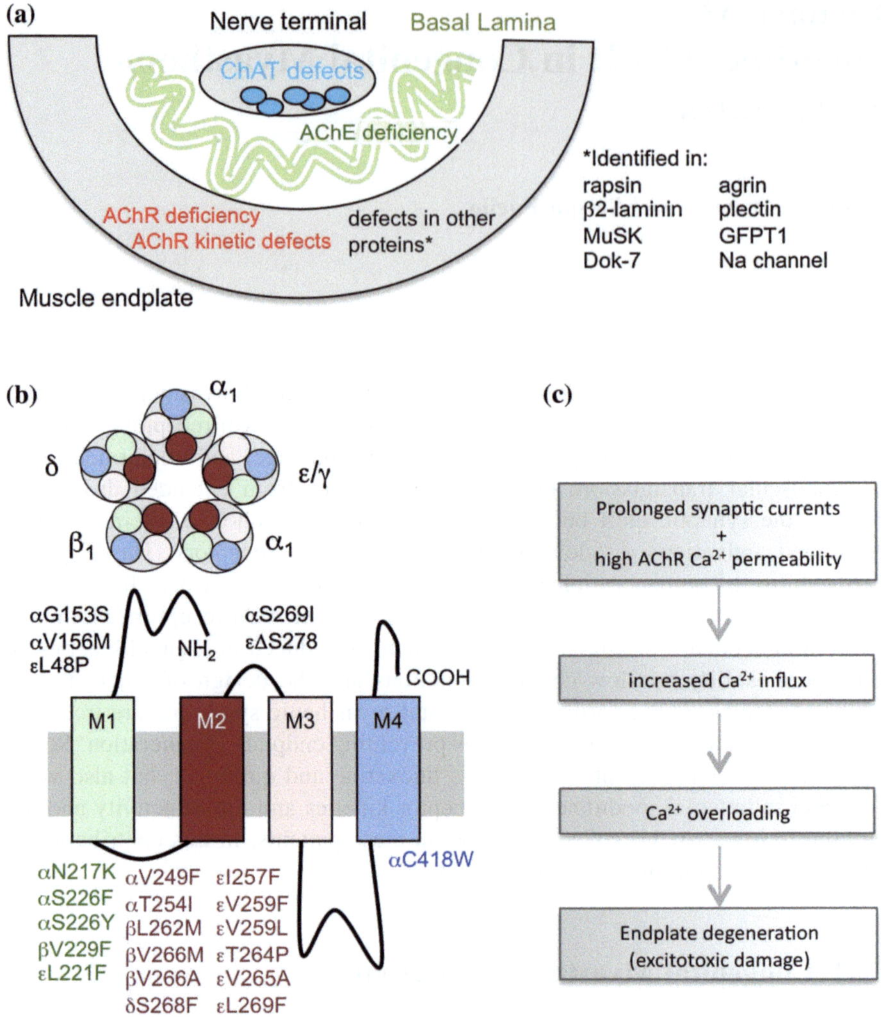

Fig. 33.1 **a** Presynaptic, synaptic and postsynaptic proteins causing congenital myasthenic syndromes identified by end 2012. ChAT: choline acetyltransferase; MuSK: muscle specific protein kinase; Dok-7: downstream of tyrosine kinase 7; GFPT1: glutamine–fructose-6-phosphate transaminase 1, **b** Schematic structure of AChR pentamer and of each subunit. Slow-channel mutations are listed by the region in which they fall, according to the same color code. Mutation εΔS278 was initially reported as εL78P, but later corrected (Croxen et al. 2002; Shelley and Colquhoun 2005), **c** Cascade leading to endplate myopathy and myasthenic symptoms in AChE deficiency and slow-channel CMS forms

been identified worldwide (Beeson et al. 2005), with mutations in at least 14 genes (Engel 2012b), and over 1,000 cases indexed in the main referral centers in Europe and USA (Chaouch et al. 2012). All CMS forms recognized to date are caused by loss-of-function mutations, with autosomal-recessive inheritance, except for slow-channel syndromes, caused by gain-of-function mutations in acetylcholine (ACh) receptor (AChR) protein (Beeson et al. 2005; Engel 2012b). In general, patients experience fatigable weakness involving ocular, bulbar, and limb muscles. Decremental amplitude of compound muscle action potential (CMAP) upon repetitive nerve stimulation at 2- to 3-Hz is also a common finding. Single fiber electromyography documents abnormal jitter in most patients. Symptoms are usually—but not always—present since infancy or early childhood and similarly affected relatives may exist, in line with the genetic origin of the disease (Chaouch et al. 2012; Engel 2012a).

Pathogenic mutations occur in proteins of the presynaptic region (mostly choline acetyltransferase enzyme, ChAT), the synaptic cleft (causing deficiency of acetylcholine esterase, AChE), or in postsynaptic proteins, usually AChR subunits or proteins involved in endplate assembly (Fig. 33.1a). About 50 % of known CMS cases are caused by mutations in AChR genes (Chaouch et al. 2012), causing reduced expression and/or kinetic defects.

Muscle nicotinic AChRs are pentameric proteins made up by $2(\alpha_1)\beta_1\varepsilon\delta$ subunits at mature endplate (ε-AChR), by $2(\alpha_1)\beta_1\gamma\delta$ subunits (γ-AChR) in noninnervated muscle fibers. Each subunit, encoded by a distinct gene, is formed by 4 transmembrane α-helixes, with M2 segment lining the channel pore (Fig. 33.1b).

Reduced expression is caused by frameshift or missense mutations that alter subunit translation or pentameric assembly, resulting in reduced AChR membrane incorporation. Loss-of-function mutations occurring in the γ subunit are responsible for some cases of Multiple Pterygium Syndrome, a severe prenatal condition causing deformity, disability or even stillbirth (Vogt et al. 2012). AChR kinetic defects can be broadly divided into two classes: those shortening channel activity (fast-channel syndromes) and those leading to prolonged activity (slow-channel forms).

Loss-of-function mutations become clinically apparent when both alleles of a single AChR subunit carry pathogenic mutations, preventing formation of fully functional AChRs. In several cases, patients are homozygous, possibly because of consanguineous marriages (Burke et al. 2004). Low-expression mutations occur predominantly, although not exclusively, in the ε subunit, as compensatory expression of the γ subunit attenuates the phenotype. Apparently, low-expression mutations in both alleles of α, β, or δ subunits are lethal in early embryo development, except for few, very severe cases (Burke et al. 2004; Engel et al. 2010).

Unsurprisingly, reduction of AChR expression or activity results in deficient neuromuscular transmission and myasthenic symptoms, which have no specific clinical signature (Engel 2012a), and are adequately treated by drugs that block the activity of AChE or increase ACh release, thereby returning endplate depolarization to supra-threshold levels.

How congenital AChE deficiency, accounting for about 15 % of all CMS cases (Engel 2012b) or slow-channel AChR mutations, causing around 5 % of known CMS cases (Chauch et al. 2012), result in myasthenia may be counterintuitive, since both conditions produce endplate overstimulation. Indeed, in these CMS forms pathology is secondary to excitotoxic endplate damage: sustained cation influx through endplate AChR-channels causes Ca^{2+} overloading, which triggers Ca^{2+}-

induced degenerative events of the endplate fine structure (Fig. 33.1c), degeneration of the junctional folds and AChR loss. These CMS forms also share a key diagnostic signature: in most patients a single nerve stimulus evokes repetitive CMAP. AChE inhibitors worsen this abnormal response in case of slow-channel mutations, while they are fully ineffective in case of AChE deficiency (Engel 2012a).

In this contribution, we focus on these "excitotoxic" CMS forms, the most pertinent to a book on Ca^{2+}-related pathologies. For a broader view of CMS, we refer the readers to the many excellent reviews dealing with the subject (Engel et al. 2003, 2010; Palace and Beeson 2008; Barišić et al. 2011; Lorenzoni et al. 2012).

33.2 Excitotoxic CMS Forms

Clinical evidence of CMS forms due to AChE deficiency or slow-channel AChR mutations was presented at the end of 1970s (Engel et al. 1977, 1979). At the same time it was shown that endplate AChR-channels are permeable to Ca^{2+} ions (Bregestovski et al. 1979) and that experimental blockade of AChE in mice causes cationic overloading of the endplate and degenerative changes of its fine structure in a Ca^{2+}-dependent manner (Salpeter et al. 1979; Leonard and Salpeter 1979). Ca^{2+} accumulation at endplates of patients with slow-channel syndrome was soon recognized (Engel et al. 1982).

In the initial works, electrophysiologic recordings were performed on patients' muscles, documenting the slowed decay of endplate currents (Engel et al. 1977, 1979), and later the kinetic abnormalities of ACh-gated channels in the case of slow-channel CMS (Ohno et al. 1995; Sine et al. 1995). In recent years, most advances in the field have been obtained by heterologous expression of cDNAs with mutations identified in patients and by the development of transgenic or knockout animal models of the pathology (Gomez et al. 2002a; Zhu et al. 2011; Chevessier et al. 2012), although there are important differences in the Ca^{2+} permeability of mouse and human endplate AChR-channels (see Sect. 33.2.2.3).

33.2.1 AChE Deficiency

This form of CMS derives from abnormal activation of normally functioning, wild-type AChRs, due to lack of AChE at the neuromuscular junction, because of mutations in the COLQ gene, encoding the collagen tail domain that anchors AChE within the basal lamina of the synaptic cleft (Ohno et al. 1998). Enzyme absence prevents ACh clearance from synaptic space, resulting in repeated activations of AChR-channels—normally activated only once (Katz and Miledi 1973)—sustained endplate depolarization and repetitive CMAP in response to single nerve stimulations. Reduced or absent tendon reflex together with slow pupillary response to light is often present in these patients (Mihaylova et al.

2008). Comprehensive descriptions of this pathology have been published (Mihaylova et al. 2008; Palace and Beeson 2008; Engel 2012b; Wargon et al. 2012). Here, it is important to note that alterations in endplate ultrastructure have been invariably detected, in patients (Engel et al. 1977; Hutchinson et al. 1993; Wargon et al. 2012) as well as in COLQ knockout mice (Feng et al. 1999). Presynaptic abnormalities have also been observed (Engel et al. 1977; Hutchinson et al. 1993), so that, ultimately, the safety margin of neuromuscular transmission is compromised by reduction of both ACh quantal release and postsynaptic response (Engel 2012b). Desensitization of AChR from overexposure to ACh may also contribute to pathology (Engel 2012b). It must be noted that pre and postsynaptic defects may also act in a compensatory manner: reduced ACh release results in reduced endplate Ca^{2+} overloading, while prolonged ACh action allows maintenance of neuromuscular transmission at damaged junctions. These homeostatic adjustments would also explain why disease severity is highly variable during the life of single individuals and across patients, even when carrying the same mutation (see for example reports by Mihaylova et al. 2008; Wargon et al. 2012).

33.2.2 Slow-Channel AChR Mutations

The name of this form of CMS refers to the slow time course of endplate currents that outlast the presence of free ACh in the synapse, due to abnormally long channel openings (Fig. 33.2a). In transfected cells, this phenomenon is evident in the form of deactivation currents, which are still measurable tens of milliseconds after ACh removal (Fig. 33.2b). In vivo, stimulation at physiologic rates causes superimposition of prolonged endplate potentials, resulting in a progressive depolarization block of the postsynaptic membrane. The sustained endplate

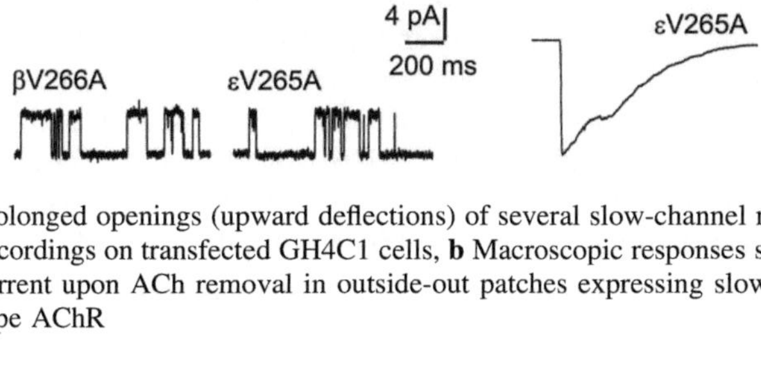

Fig. 33.2 **a** Prolonged openings (upward deflections) of several slow-channel mutant AChRs in cell-attached recordings on transfected GH4C1 cells, **b** Macroscopic responses show a prolonged deactivation current upon ACh removal in outside-out patches expressing slow-channel mutant but not wild-type AChR

currents cause cationic overloading of the postsynaptic region and endplate myopathy, clearly visible in muscle biopsies (Engel 2012b).

In the last two decades, many mutations leading to slow-channel syndrome have been identified (Fig. 33.1b) and their functional consequences described in details. Most are found in the pore-lining M2 segment of the ε subunit (Fig. 33.1b). Preferential localization in the ε subunit might be due to the fact that it is expressed late in development (after the 16th gestational week in humans, Hesselmas et al. 1993), and therefore mutations do not interfere with early embryonic development. Moreover, expression of γ-AChR, often observed in slow-channel CMS patients (Ohno et al. 1995), can have a compensatory role.

Slow-channel mutations have little effect on AChR expression (Engel 2012b). A slight increase (less than 10 %) of channel conductance has been reported for some slow-channel mutations (Croxen et al. 1997; Milone et al. 1997), but the contribution of this alteration to the pathogenic cascade—if any—has not been examined.

33.2.2.1 Causes of Kinetic Defects

As expected for gain-of-function mutants, ACh efficacy is increased for all slow-channel mutations tested (Hatton et al. 2003; Navedo et al. 2006; Di Castro et al. 2007; Moriconi et al. 2010), as revealed by a leftward shift in the ACh concentration—current response curve. Equivalent shifts were found by ACh competition assays of α-bungarotoxin binding (Sine et al. 1995; Engel et al. 1996; Milone et al. 1997; Fidzianska et al. 2005; Shen et al. 2006). In various cases, the molecular bases for this effect have been mechanistically derived from single channel studies.

Several mutations affect ACh binding, reducing the rate of agonist dissociation from AChR, thus leading to multiple channel openings before ACh dissociates from AChR. This behavior was initially demonstrated for mutations at N-terminal domain of the α subunit such as αG153S (Sine et al. 1995) and αN217K (Wang et al. 1997), but soon observed also for mutations in other domains, for instance the αV249F mutation in M2 (Milone et al. 1997), or even in M1 domain of other subunits, for instance εL221F (Hatton et al. 2003) and βV229F (Navedo et al. 2006) or εΔS278 in the extracellular M2–M3 linker (Shelley and Colquhoun 2005). Channel gating is also enhanced in several instances, increasing the opening rate, as has been documented for mutations in several domains of the alpha subunit (see Fig. 33.1 for the localization): αV249F (Milone et al. 1997), αS269I (Grosman et al. 2000), αC418W (Shen et al. 2006), and possibly also αV156M and αT254I (Croxen et al. 1997). Alterations in channel gating have been reported also for βV229F (Navedo et al. 2006). Spontaneous channel openings, indicative of destabilization of the AChR-closed conformation, have been documented for εT264P, βV266M, and αV249F mutations (Ohno et al. 1995; Engel et al. 1996; Milone et al. 1997), and contribute to pathology by causing continuous cation leakage into the endplate (Engel et al. 2003). However, other mutations, namely αV156M, αT254I, and αS269I (Croxen et al. 1997) do not favor spontaneous openings, indicating that this is not an obligate consequence of slow-channel mutations.

Slow-channel kinetics together with delayed channel openings has been reported for δS268F mutation, at least when engineered in mouse AChR (Gomez et al. 2002b). No mechanistic insight has been provided for some mutations, including the recently reported εV259L (Outteryck et al. 2009) or βL262M (Gomez et al. 1996), which involves a very conserved leucine residue with a strong impact on the function and Ca^{2+} permeability of neuronal AChRs (Fucile 2004; Fucile et al. 2000). However, data summarized above clearly show that the precise mechanism of action of each mutation correlates poorly with its position.

33.2.2.2 AChR Desensitization

In the continuous presence of ACh, AChR undergoes desensitization, a reversible decrease in response (Katz and Thesleff 1957; for a recent review see Giniatullin et al. 2005). In the case of endplate AChR, desensitization is unlikely to play important roles under physiological conditions, but it contributes to shape synaptic response in AChE deficiency CMS, when ACh remains in the synaptic cleft for several milliseconds. In the case of slow-channel CMS, the situation is more complicated. Katz and Thesleff (1957) showed that ACh affinity is higher for the desensitized than for active conformations. Thus, the high affinity observed in α-bungarotoxin binding studies has been interpreted as evidence of an increased rate of desensitization for slow-channel mutant AChRs (see for instance Sine et al. 1995; Engel et al. 1996). However, when looking at this issue from a functional point of view, different conclusions can be reached. Current decay because of AChR desensitization can be faster, slower, or equal to that of wild-type AChR responses, depending on the mutated subunit and on the specific mutation (Di Castro et al. 2007; Moriconi et al. 2010; Piccari et al. 2011). In general, no univocal relation exists between current decay rate and enhancement in α-bungarotoxin binding. For instance, binding is enhanced 10 times for both εL269F- and εV259F-AChRs (Engel et al. 1996; Fidzanska et al. 2005), but current decay rate is accelerated only for the latter (Di Castro et al. 2007; Piccari et al. 2011). As a whole, the altered kinetics of mutant AChRs is such that entrance into a desensitized state may occur at a physiologically relevant rate even in the absence of ACh (Elenes et al. 2006), influencing synaptic responses. It must be noted that although slow-channel AChR desensitization is a factor that contributes to failure of neuromuscular transmission, it also mitigates cationic overload, thus protecting endplates.

33.2.2.3 Ca^{2+} Permeability of Muscle AChR

Two independent experimental methods are commonly used to measure the Ca^{2+} permeability of ligand-gated channels, including AChR-channels. The first (from an historical point of view) approach measures the shift in the reversal potential of transmitter-evoked currents upon changes of the extracellular Ca^{2+} concentration, to obtain the permeability ratio P_{Ca}/P_{Cs} or P_{Ca}/P_{Na} using the Goldmann-Hodgkin-

Katz constant field assumptions (Lassignal and Martin 1977; Lewis 1979). The other method measures the percentage of the total transmitter-evoked current carried by Ca^{2+} ions, the so-called "fractional Ca^{2+} current" (usually indicated as P_f), by simultaneously recording fluorescence signals, and transmembrane currents (Zhou and Neher 1993; Neher 1995). Being independent of theoretical assumptions and considering that the hypothesis of ion independence—central to Goldman-Hodgkin-Katz model—does not hold for some receptors, including muscle AChRs (Lewis 1979), P_f values are usually considered as more reliable measures of Ca^{2+} permeability (Burnashev et al. 1995).

Both methods univocally show that for wild-type AChRs, Ca^{2+} permeability depends on subunit composition (Fucile 2004). Muscle AChRs are no exception: γ-AChRs have lower Ca^{2+} permeability than ε-AChRs (Table 33.1). The difference has been described for mouse and rat ($P_{Ca}/P_{Cs} \sim 0.2$, $P_f \sim 2$ % vs. $P_{Ca}/P_{Cs} \sim 0.45$, $P_f \sim 4$–7 %) in both native and heterologous systems (Villarroel and Sakmann 1996; Cens et al. 1997; Ragozzino et al. 1998) and confirmed for human AChRs (Fucile et al. 2006b). Moreover, the wild-type human ε-AChR is much more permeable to Ca^{2+} than its mouse counterpart, as shown by both higher P_f and P_{Ca}/P_{Cs} (Fucile et al. 2006b; Di Castro et al. 2007). This difference depends entirely on the ε subunit, as the chimeric receptor containing human α, β, and δ subunits but mouse ε subunit has the same P_f of mouse ε-AChR (Fucile et al. 2006b). Given this difference, we here refer only to studies of Ca^{2+} permeability performed on human mutant slow-channel AChRs.

In agreement with the pivotal role of the ε subunit in controlling divalent cation permeability, several slow-channel mutations occurring in the ε subunit (but not in α or β subunits; δ mutants have not been tested yet) affect Ca^{2+} permeability of human mutant AChR-channels (Table 33.1). Out of five slow-channel mutations in the human ε subunit tested to date, only one (εV265A) has no effect, while three (εV259F, εT264P, εL269F) enhance and one reduces (εI257F) Ca^{2+} permeability of AChR (Fucile et al. 2006b; Di Castro et al. 2007; Piccari et al. 2011). The central role of ε subunit is further emphasized by the observation that P_f is not affected by mutation αV249F (Fucile et al. 2006a, b), although it involves the residue corresponding to εV259F, which has a large impact on Ca^{2+} permeability. All these mutations reside in the M2 transmembrane domain, known to be relevant for ion selectivity of muscle AChRs (Konno et al. 1991) and to influence Ca^{2+} permeability of neuronal AChRs (Bertrand et al. 1993).

That human ε-AChR has the highest P_f value reported for AChRs containing non-α subunits (Fucile 2004) likely predisposes to endplate cationic overloading and consequent damage when AChR activation is prolonged. Indeed, Ca^{2+} permeability of human ε-AChR is so high that slow-channel mutations remain pathogenic even when P_f is almost halved, as in the case of εI257F-AChR. This is not surprising if we consider that P_f for εI257F-AChR is similar to that of mouse ε-AChR, and endplate myopathy occurs in mouse models of slow-channel CMS (Gomez et al. 2002a; Zhu et al. 2011).

Table 33.1 Ca^{2+} permeability of human wild-type and slow-channel mutant muscle AChRs

	Mutation	P_f (%)	P_{Ca}/P_{Cs}	References
γ-AChR				
Human myotubes	Wild-type	2.7		Fucile et al. (2006b)
GH4C1	Wild-type	2.9		Fucile et al. (2006b)
GH4C1	αV249F	3.3		Fucile et al. (2006b)
GH4C1	αG153S	2.9		Fucile et al. (2006b)
ε-AChR				
GH4C1	wild-type	7.2–7.8	0.73	Fucile et al. (2006b, Di Castro et al. (2007), Piccari et al. (2011)
GH4C1	αV249F	7.6		Fucile et al. (2006b)
GH4C1	αG153S	7.1		Fucile et al. (2006b)
GH4C1	αC418 W	7.5		Piccari et al. (2011)
GH4C1	βV266A	7.6		Piccari et al. (2011)
GH4C1	βV266 M	6.2		Piccari et al. (2011)
GH4C1	εI257F	4.6[a]		Piccari et al. (2011)
GH4C1	εV259F	15.4[a]	1.2	Di Castro et al. (2007)
GH4C1	εT264P	11.8[a]	1.35	Di Castro et al. (2007)
GH4C1	εV265A	7.5		Piccari et al. (2011)
GH4C1	εL269F	10.2[a]		Piccari et al. (2011)

P_{Ca}/P_{Cs} Bi-ionic permeability ratio, calculated from the shift in the reversal potential of ACh-evoked currents upon $[Ca^{2+}]_{out}$ changes from 0 to 5 mM, P_f percentage of total ACh-evoked current carried by Ca^{2+} ions, obtained by simultaneously recording fluorescence signals and transmembrane currents

[a] Significantly different from wild type

33.3 Pathways to Excitotoxic Damage

As pointed out above, it is currently accepted that absence of AChE or slow-channel mutations in AChR lead to enhanced cation influx through endplate AChR-channels and Ca^{2+} accumulation. The high Ca^{2+} permeability of human ε-AChR channel contributes to disrupt Ca^{2+} homeostasis and trigger Ca^{2+}-induced toxic events (Fig. 33.1c). Given the large size of muscle fibers and the complex dynamic of Ca^{2+} transients in sarcomer/triad compartment, damage remains restricted to the perisynaptic region. In slow-channel CMS patients and mouse models, localized and possibly reversible activation of caspases and apoptotic pathways has been reported (Vohra et al. 2004, 2006). Endplate damage observed in AChE deficiency patients suggests that a similar cascade may be activated in these cases, too, but no details are available.

In a transgenic mouse carrying a slow-channel AChR mutation, Ca^{2+} influx through endplate AChR is amplified by inositol trisphospate (IP_3) mediated Ca^{2+} release from intracellular stores (Zayas et al. 2007; Zhu et al. 2011). Silencing IP_3 receptor by siRNA reverts the endplate damage (Zhu et al. 2011), although the mechanism linking AChR activation to IP_3 mobilization remains unknown. Some differences between mouse and human pathology suggest that caution (and possibly direct experimental assessment) is necessary before concluding that the same cascade applies to human disease. First, the human ε-AChR has higher Ca^{2+} permeability than mouse receptor (Fucile et al. 2006b), so that downstream amplification of Ca^{2+} influx might be less relevant, or otherwise it might be more potent. In slow-channel animals spontaneous ACh release and the quantal content of evoked responses are reduced, indicative of presynaptic remodeling; this defect is also reversed upon IP_3 receptor blockade (Zhu et al. 2011). Conversely, quantal content was normal in slow-channel patients (see for instance Engel et al. 1996), coherent with the fact that this is a postsynaptic pathology. Endplate damage is observed only in transgenic mouse lines with the greatest AChR functional alterations (Gomez et al. 2002a), while human pathology shows less correlation with AChR functional changes, and large interindividual differences exist even in patients carrying the same molecular defect (Palace and Beeson 2008). Possibly, compensatory mechanisms are different in human and mouse muscles, but Ca^{2+} signaling might also be differentially fine-tuned in the two species. Thus, it remains to be established whether participation of IP_3 receptor to the pathogenic cascade of ACh-induced endplate damage is as important in human muscles as it is in the mouse, and whether IP_3-mediated Ca^{2+} release contributes to endplate damage also in AChE deficiency.

33.4 Therapy: Clinics and Basic Science

Strictly speaking, "cure" for excitotoxic CMS requires a gene therapy able to silence the expression of the slow channel-mutated subunit or to restore AChE deficiency. This goal has been attained to some extent in vitro (Abdelgany et al. 2003; Shen et al. 2006) and in mice (Ito et al. 2012), but remains very distant in time for human patients. Pharmacological treatment therefore represents the only currently feasible approach. At variance with all other forms of myasthenia (congenital or acquired), excitotoxic CMS cannot be treated using AChE blockers, because the primary pathogenic event is excessive endplate stimulation (so, it is irrelevant that the target is missing in AChE deficiency). Therefore, therapeutic interventions must be aimed at limiting ACh-induced cation entry, to prevent endplate degeneration. Controlling Ca^{2+} permeability of slow-channel mutant AChRs might also contribute to decrease the excitotoxic damage, although even a strong reduction of P_f does not prevent disease (see the case of εI257F mutation).

Building on basic evidence that quinine and fluoxetine are open channel blockers of AChR (Sieb et al. 1996; Garcia-Colunga et al. 1997), treatment of slow-channel CMS patients has been introduced, with satisfactory results (Harper and Engel 1998; Harper et al. 2003). Therapy of AChE deficiency is more empirical: several patients benefit of treatment with ephedrine (whose use is restricted in many countries) or salbutamol, but the mechanism of action of these drugs is presently unknown. In fact, open channel block has been demonstrated for wild-type or mutant AChRs only at concentrations far beyond clinically usable range (Bouzat 1996; Milone and Engel 1996; Piccari et al. 2011).

Since all drugs have limitations and relatively major side effects, enlarging the panel of potentially usable compounds is important for patients. Considering that in excitotoxic CMS the main pathogenic event is ACh-induced endplate damage, we have recently introduced an in vitro assay of ACh-induced death of HEK cells transfected with human muscle AChR. Cell death can be inhibited by AChR blockers such as curare and removal of extracellular Ca^{2+} (Piccari et al. 2011), suggesting that this experimental model can be used to provide a rationale for off-label medicament use (Piccari et al. 2011).

Slow-channel mutant AChRs are more sensitive than wild-type AChR to open channel block by quinidine (Fukudome et al. 1998) or fluoxetine (Harper et al. 2003). However, it has recently been reported that patients with AChE deficiency CMS can be treated with fluoxetine (Wargon et al. 2012) and we have recently shown (Deflorio et al. 2013) that fluoxetine prevents ACh-induced death in cells expressing ε-AChR.

Verapamil, an established blocker of voltage-gated Ca^{2+} channels, has a well-documented direct action on AChRs (Bregestovski et al. 1980; Wachtel 1987; Edeson 1988; Sharifullina et al. 2002; Moriconi et al. 2010). At clinically attainable concentrations verapamil induces open channel block only for the εV265A mutant AChR, but not for AChRs bearing the same mutation at corresponding positions in the β subunit (βV266A, also occurring in patients) or in the α and δ subunit (Moriconi et al. 2010). However, verapamil reduces channel opening frequency for wild-type and all slow-channel mutant AChRs tested (Moriconi et al. 2010). This effect develops over several milliseconds of ACh application, so verapamil has no effect on brief ACh-evoked currents in patches expressing wild-type ε-AChR (at normal endplates ACh action lasts just 1–2 ms) but effectively curtails deactivation currents in those expressing slow-channel mutant AChRs (Moriconi et al. 2010). Furthermore, verapamil reduces the P_f at least for εL269F-AChR, though not for wild-type ε-AChR (Piccari et al. 2011), an effect that likely contributes to protection against ACh-induced cell death.

Salbutamol, a β_2-adrenergic receptor agonist also reduces P_f of slow-channel mutant εL269F-AChR (but not wild-type AChR), and its action is partially prevented by blockade of β-adrenergic receptors (Piccari et al. 2011), supporting the hypothesis that activation of β_2-adrenergic receptors plays a role in the process. Stimulation of β_2-adrenergic receptors results in the activation of cAMP-dependent protein kinase A, a pathway that controls the modulation of Ca^{2+} permeability of glutamate receptors (Skeberdis et al. 2006; Sobczyk and Svoboda 2007) and several functions of muscle AChRs (Huganir and Greengard 1990; Caratsch et al. 1992; Swope et al. 1999). In particular, salbutamol and other agents that raise cAMP levels, reduce ACh-evoked $^{22}Na^{+}$ influx in TE671 rhabdomyosarcoma cells that express γ-AChR (Li et al. 1996). Reduction of Ca^{2+} influx surely plays a role in protection of HEK cells expressing slow-channel AChRs from ACh-induced toxicity, but the basis of beneficial action of salbutamol in patients with AChE deficiency CMS (i.e., with wild-type AChRs at their endplate) remains to be elucidated.

Finally, riluzole has been shown to block γ- and ε-AChRs with a time course too slow to interfere with normal neuromuscular transmission (Palma et al. 2011; Deflorio et al. 2012). However, the protective effect observed in transfected cells raises the possibility that it might turn out to be have a therapeutic action in patients with excitotoxic CMS forms.

33.5 Conclusions and Future Directions

Research on congenital myasthenic syndromes represents a paradigmatic example of collaborative efforts between basic science and the clinics. For instance, studies on CMS forms related to kinetic defects in AChRs highlighted several physiological aspects of the receptor-channel complex (Engel et al. 2010), yielding important clues to the role of specific protein regions to overall function. On the other side, understanding the underlying molecular defect has allowed the design of effective therapy, as in the case of fluoxetine therapy for slow-channel syndrome.

Studying excitotoxic CMS forms has brought to light the pathogenic role of Ca^{2+} influx through endplate AChR-channels in causing excitotoxic damage upon AChR over-activation. This idea was somewhat unsupported by experimental findings, as it has long been accepted that muscle AChRs have low Ca^{2+} permeability, indicated by P_{Ca}/P_{Na} values as low as 0.2 (Le Novere et al. 2002). These values were obtained for γ-AChR, but values for ε-AChR are invariably <1, while they are usually larger for neuronal AChRs. A deeper analysis, partly prompted by findings on excitotoxic CMS, has demonstrated that, when Ca^{2+} permeability is measured as P_f, ranking can be different. The endplate ε-AChR is second only to the highly Ca^{2+} permeable $\alpha 7$ and $\alpha 9\alpha 10$ neuronal AChRs: 11.4 % for recombinant human $\alpha 7$-AChRin GH4C1 cells (Fucile et al. 2003) or 22 % for $\alpha 9\alpha 10$-AChR (Fucile et al. 2006a) versus 7.5 % for human ε-AChR. Given that excitotoxic damage is usually associated to over-activation of receptors with large Ca^{2+}-permeability (such as glutamate NMDA receptors), these experimental results appear to be more consistent with the existence of excitotoxic CMS forms. The molecular determinants of the Ca^{2+} permeability of muscle AChRs remain largely unknown. However, work on slow-channel mutants has remarked the specificity of ε subunit, in particular of its M2 domain, in controlling Ca^{2+} permeability, in analogy with observations on neuronal AChRs. Thus, these studies also emphasized the importance of the methodological approach used to measure AChR Ca^{2+} permeability.

Hopefully, synergy between different approaches to the issue of AChR involvement in various CMS forms will yield even more important results in the future.

Acknowledgments Work on slow-channel AChR mutants in our lab was funded by AFM-Association Française contre les Myopathies.

References

Abdelgany A, Wood M, Beeson D (2003) Allele-specific silencing of a pathogenic mutant acetylcholine receptor subunit by RNA interference. Hum Mol Genet 12:2637–2644

Barišić N, Chaouch A, Müller JS, Lochmüller H (2011) Genetic heterogeneity and pathophysiological mechanisms in congenital myasthenic syndromes. Eur J Paediatr Neurol 15:189–196

Beeson D, Hantaï D, Lochmüller H, Engel AG (2005) 126th International Workshop: congenital myasthenic syndromes. Neuromuscul Disord 15:498–512

Bertrand D, Galzi JL, Devillers-Thiéry A, Bertrand S, Changeux JP (1993) Mutations at two distinct sites within the channel domain M2 alter calcium permeability of neuronal alpha 7 nicotinic receptor. Proc Natl Acad Sci U S A 90:6971–6975

Bouzat C (1996) Ephedrine blocks wild-type and long-lived mutant acetylcholine receptor channels. NeuroReport 8:317–321

Bregestovski PD, Miledi R, Parker I (1979) Calcium conductance of acetylcholine-induced endplate channels. Nature 279:638–639

Bregestovski PD, Miledi R, Parker I (1980) Blocking of frog endplate channels by the organic calcium antagonist D600. Proc R Soc Lond B Biol Sci 211:15–24

Burke G, Cossins J, Maxwell S, Robb S, Nicolle M, Vincent A, Newsom-Davis J, Palace J, Beeson D (2004) Distinct phenotypes of congenital acetylcholine receptor deficiency. Neuromuscul Disord 14:356–364

Burnashev N, Zhou Z, Neher E, Sakmann B (1995) Fractional calcium currents through recombinant GluR channels of the NMDA, AMPA and kainate receptor subtypes. J Physiol 485:403–418

Caratsch CG, Grassi F, Eusebi F (1992) Functional regulation of nicotinic acetylcholine receptor channels in muscle. Ion Channels 3:177–206

Cens T, Nargeot J, Charnet P (1997) Ca^{2+} permeability of muscle nicotinic acetylcholine receptor is increased by expression of the ε subunit. Receptors Channels 5:29–40

Chaouch A, Beeson D, Hantaï D, Lochmüller H (2012) 186th ENMC international workshop: congenital myasthenic syndromes. Neuromuscul Disord 22:566–576

Chevessier F, Peter C, Mersdorf U, Girard E, Krejci E, McArdle JJ, Witzemann V (2012) A new mouse model for the slow-channel congenital myasthenic syndrome induced by the AChR eL221F mutation. Neurobiol Dis 45:851–861

Croxen R, Hatton C, Shelley C, Brydson M, Chauplannaz G, Oosterhuis H, Vincent A, Newsom-Davis J, Colquhoun D, Beeson D (2002) Recessive inheritance and variable penetrance of slow-channel congenital myasthenic syndromes. Neurology 59:162–168 (Partial retraction in: Neurology 72:294 (2009))

Croxen R, Newland C, Beeson D, Oosterhuis H, Chauplannaz G, Vincent A, Newsom-Davis J (1997) Mutations in different functional domains of the human muscle acetylcholine receptor alpha subunit in patients with the slow-channel congenital myasthenic syndrome. Hum Mol Genet 6:767–774

Deflorio C, Palma E, Conti L, Roseti C, Manteca A, Giacomelli E, Catalano M, Limatola C, Inghilleri M, Grassi F (2012) Riluzole blocks human muscle acetylcholine receptors. J Physiol 590:2519–2528

Deflorio C, Catalano M, Fucile S, Limatola C, Grassi F (2013) Fluoxetine prevents acetylcholine-induced excitotoxicity blocking human endplate acetylcholine receptor. Muscle Nerve, doi: 10.1002/mus.23870

Di Castro A, Martinello K, Grassi F, Eusebi F, Engel AG (2007) Pathogenic point mutations in a transmembrane domain of the epsilon subunit increase the Ca^{2+} permeability of the human endplate ACh receptor. J Physiol 579:671–677

Edeson RO (1988) Verapamil, neuromuscular transmission and the nicotinic receptor. Eur J Pharmacol 151:301–306

Elenes S, Ni Y, Cymes GD, Grosman C (2006) Desensitization contributes to the synaptic response of gain-of-function mutants of the muscle nicotinic receptor. J Gen Physiol 128:615–627

Engel AG (2012a) Congenital myasthenic syndromes in 2012. Curr Neurol Neurosci Rep 12:92–101

Engel AG (2012b) Current status of the congenital myasthenic syndromes. Neuromuscul Disord 22:99–111

Engel AG, Lambert EH, Gomez MR (1977) A new myasthenic syndrome with end-plate acetylcholinesterase deficiency, small nerve terminals, and reduced acetylcholine release. Ann Neurol 1:315–330

Engel AG, Lambert EH, Mulder DM, Torres CF, Sahashi K, Bertorini TE, Whitaker JN (1979) Investigations of 3 cases of a newly recognized familial, congenital myasthenic syndrome. Trans Am Neurol Assoc 104:8–12

Engel AG, Lambert EH, Mulder DM, Torres CF, Sahashi K, Bertorini TE, Whitaker JN (1982) A newly recognized congenital myasthenic syndrome attributed to a prolonged open time of the acetylcholine-induced ion channel. Ann Neurol 11:553–569

Engel AG, Ohno K, Milone M, Wang HL, Nakano S, Bouzat C, Pruitt JN 2nd, Hutchinson DO, Brengman JM, Bren N, Sieb JP, Sine SM (1996) New mutations in acetylcholine receptor subunit genes reveal heterogeneity in the slow-channel congenital myasthenic syndrome. Hum Mol Genet 5:1217–1227

Engel AG, Ohno K, Sine S (2003) Sleuthing molecular targets for neurological diseases at the neuromuscular junction. Nat Rev Neurosci 4:339–352

Engel AG, Shen XM, Selcen D, Sine SM (2010) What have we learned from the congenital myasthenic syndromes. J Mol Neurosci 40:143–153

Feng G, Krejci E, Molgo J, Cunningham JM, Massoulié J, Sanes JR (1999) Genetic analysis of collagen Q: roles in acetylcholinesterase and butyrylcholinesterase assembly and in synaptic structure and function. J Cell Biol 144:1349–1360

Fidzianska A, Ryniewicz B, Shen XM, Engel AG (2005) IBM-type inclusions in a patient with slow-channel syndrome caused by a mutation in the AChR epsilon subunit. Neuromuscul Disord 15:753–759

Fucile S (2004) Ca^{2+} permeability of nicotinic acetylcholine receptors. Cell Calcium 35:1–8

Fucile S, Palma E, Mileo AM, Miledi R, Eusebi F (2000) Human neuronal threonine-for-leucine-248 $\alpha 7$ mutant nicotinic acetylcholine receptors are highly Ca^{2+} permeable. Proc Natl Acad Sci USA 97:3643–3648

Fucile S, Renzi M, Lax P, Eusebi F (2003) Fractional Ca(2+) current through human neuronal alpha7 nicotinic acetylcholine receptors. Cell Calcium 34:205–209

Fucile S, Sucapane A, Eusebi F (2006a) Ca^{2+} permeability through rat cloned a9-containing nicotinic acetylcholine receptors. Cell Calcium 39:349–355

Fucile S, Sucapane A, Grassi F, Eusebi F, Engel AG (2006b) The human adult subtype ACh receptor has high Ca^{2+} permeability and predisposes to endplate Ca^{2+} overloading. J Physiol 573:35–43

Fukudome T, Ohno K, Brengman JM, Engel AG (1998) Quinidine normalizes the open duration of slow-channel mutants of the acetylcholine receptor. NeuroReport 9:1907–1911

García-Colunga J, Awad JN, Miledi R (1997) Blockage of muscle and neuronal nicotinic acetylcholine receptors by fluoxetine (Prozac). Proc Natl Acad Sci U S A 94:2041–2044

Giniatullin R, Nistri A, Yakel J (2005) Desensitization of nicotinic ACh receptors: shaping cholinergic signaling. Trends Neurosci 28:371–378

Gomez CM, Maselli R, Gammack J, Lasalde J, Tamamizu S, Cornblath DR, Lehar M, McNamee M, Kuncl RW (1996) A beta-subunit mutation in the acetylcholine receptor channel gate causes severe slow-channel syndrome. Ann Neurol 39:712–723

Gomez CM, Maselli RA, Groshong J, Zayas R, Wollmann RL, Cens T, Charnet P (2002a) Active calcium accumulation underlies severe weakness in a panel of mice with slow-channel syndrome. J Neurosci 22:6447–6457

Gomez CM, Maselli RA, Vohra BP, Navedo M, Stiles JR, Charnet P, Schott K, Rojas L, Keesey J, Verity A, Wollmann RW, Lasalde-Dominicci J (2002b) Novel delta subunit mutation in slow-channel syndrome causes severe weakness by novel mechanisms. Ann Neurol 51:102–112

Grosman C, Salamone FN, Sine SM, Auerbach A (2000) The extracellular linker of muscle acetylcholine receptor channels is a gating control element. J Gen Physiol 116:327–340

Harper CM, Engel AG (1998) Quinidine sulfate therapy for the slow-channel congenital myasthenic syndrome. Ann Neurol 43:480–484

Harper CM, Fukodome T, Engel AG (2003) Treatment of slow-channel congenital myasthenic syndrome with fluoxetine. Neurology 60:1710–1713

Hatton CJ, Shelley C, Brydson M, Beeson D, Colquhoun D (2003) Properties of the human muscle nicotinic receptor, and of the slow-channel myasthenic syndrome mutant epsilon L221F, inferred from maximum likelihood fits. J Physiol 547:729–760

Hesselmans LF, Jennekens FG, Van den Oord CJ, Veldman H, Vincent A (1993) Development of innervation of skeletal muscle fibers in man: relation to acetylcholine receptors. Anat Rec 236:553–562

Huganir RL, Greengard P (1990) Regulation of neurotransmitter receptor desensitization by protein phosphorylation. Neuron 5:555–567

Hutchinson DO, Walls TJ, Nakano S, Camp S, Taylor P, Harper CM, Groover RV, Peterson HA, Jamieson DG, Engel AG (1993) Congenital endplate acetylcholinesterase deficiency. Brain 116:633–653

Ito M, Suzuki Y, Okada T, Fukudome T, Yoshimura T, Masuda A, Takeda S, Krejci E, Ohno K (2012) Protein-anchoring strategy for delivering acetylcholinesterase to the neuromuscular junction. Mol Ther 20:1384–1392

Katz B, Miledi R (1973) The binding of acetylcholine to receptors and its removal from the synaptic cleft. J Physiol 231:549–574

Katz B, Thesleff S (1957) A study of the desensitization produced by acetylcholine at the motor end-plate. J Physiol 138:63–80

Konno T, Busch C, Von Kitzing E, Imoto K, Wang F, Nakai J, Mishina M, Numa S, Sakmann B (1991) Rings of anionic amino acids as structural determinants of ion selectivity in the acetylcholine receptor channel. Proc Biol Sci 244:69–79

Lassignal NL, Martin AR (1977) Effect of acetylcholine on post junctional membrane permeability in eel electroplaque. J Gen Physiol 70:23–36

Leonard JP, Salpeter MM (1979) Agonist-induced myopathy at the neuromuscular junction is mediated by calcium. J Cell Biol 82:811–819

Le Novère N, Corringer PJ, Changeux JP (2002) The diversity of subunit composition in nAChRs: evolutionary origins, physiologic and pharmacologic consequences. J Neurobiol 53:447–456

Lewis CA (1979) Ion-concentration dependence of the reversal potential and the single channel conductance of ion channels at the frog neuromuscular junction. J Physiol 286:417–445

Li Z, Forester N, Vincent A (1996) Modulation of acetylcholine receptor function in TE671 (rhabdomyosarcoma) cells by non-AChR ligands: possible relevance to seronegative myasthenia gravis. J Neuroimmunol 64:179–183

Lorenzoni PJ, Scola RH, Kay CS, Werneck LC (2012) Congenital myasthenic syndrome: a brief review. Pediatr Neurol 46:141–148

Mihaylova V, Müller JS, Vilchez JJ, Salih MA, Kabiraj MM, D'Amico A, Bertini E, Wölfle J, Schreiner F, Kurlemann G, Rasic VM, Siskova D, Colomer J, Herczegfalvi A, Fabriciova K, Weschke B, Scola R, Hoellen F, Schara U, Abicht A, Lochmüller H (2008) Clinical and molecular genetic findings in COLQ-mutant congenital myasthenic syndromes. Brain 131:747–759

Milone M, Engel AG (1996) Block of the endplate acetylcholine receptor channel by the sympathomimetic agents ephedrine, pseudoephedrine, and albuterol. Brain Res 740:346–352

Milone M, Wang HL, Ohno K, Fukudome T, Pruitt JN, Bren N, Sine SM, Engel AG (1997) Slow-channel myasthenic syndrome caused by enhanced activation, desensitization, and agonist binding affinity attributable to mutation in the M2 domain of the acetylcholine receptor alpha subunit. J Neurosci 17:5651–565

Moriconi C, Di Castro MA, Fucile S, Eusebi F, Grassi F (2010) Mechanism of verapamil action on wild-type and slow-channel mutant human muscle acetylcholine receptor. J Neurochem 114:1231–1240

Navedo MF, Lasalde-Dominicci JA, Báez-Pagán CA, Díaz-Pérez L, Rojas LV, Maselli RA, Staub J, Schott K, Zayas R, Gomez CM (2006) Novel beta subunit mutation causes a slow-channel syndrome by enhancing activation and decreasing the rate of agonist dissociation. Mol Cell Neurosci 32:82–90

Neher E (1995) The use of fura-2 for estimating Ca buffers and Ca fluxes. Neuropharmacology 34:1423–1442

Ohno K, Brengman J, Tsujino A, Engel AG (1998) Human endplate acetylcholinesterase deficiency caused by mutations in the collagen-like tail subunit (ColQ) of the asymmetric enzyme. Proc Natl Acad Sci U S A 95:9654–9659

Ohno K, Hutchinson DO, Milone M, Brengman JM, Bouzat C, Sine SM, Engel AG (1995) Congenital myasthenic syndrome caused by prolonged acetylcholine receptor channel openings due to a mutation in the M2 domain of the epsilon subunit. Proc Natl Acad Sci U S A 92:758–762

Outteryck O, Richard P, Lacour A, Fournier E, Zéphir H, Gaudon K, Eymard B, Hantaï D, Vermersch P, Stojkovic T (2009) Novel epsilon subunit mutation of the muscle acetylcholine receptor causing a slow-channel congenital myasthenic syndrome. J Neurol Neurosurg Psychiatry 8:450–451

Palace J, Beeson D (2008) The congenital myasthenic syndromes. J Neuroimmunol 201–202:2–5

Palma E, Inghilleri M, Conti L, Deflorio A, Frasca V, Manteca A, Pichiorri F, Roseti C, Torchia G, Limatola C, Grassi F, Miledi R (2011) Physiological characterization of human muscle acetylcholine receptors from ALS patients. Proc Natl Acad Sci USA 108:20184–20188

Piccari V, Deflorio C, Bigi R, Grassi F, Fucile S (2011) Modulation of the Ca^{2+} permeability of human endplate acetylcholine receptor-channel. Cell Calcium 49:272–278

Ragozzino D, Barabino B, Fucile S, Eusebi F (1998) Ca^{2+} permeability of mouse and chick nicotinic acetylcholine receptors expressed in transiently transfected human cells. J Physiol 507:749–757

Salpeter MM, Kasprzak H, Feng H, Fertuck H (1979) End-plates after esterase inactivation in vivo: correlation between esterase concentration, functional response and fine structure. J Neurocytol 8:95–115

Sharifullina ER, Afzalov RA, Talantova MV, Vyskochil F, Giniatullin RA (2002) Pre- and postsynaptic effects of the calcium channel blocker verapamil at neuromuscular junctions. Neurosci Behav Physiol 32:309–315

Shelley C, Colquhoun D (2005) A human congenital myasthenia-causing mutation (epsilon L78P) of the muscle nicotinic acetylcholine receptor with unusual single channel properties. J Physiol 564:377–396. Erratum in: J Physiol 586:5277 (2008)

Shen XM, Deymeer F, Sine SM, Engel AG (2006) Slow-channel mutation in acetylcholine receptor alphaM4 domain and its efficient knockdown. Ann Neurol 60:128–136

Sieb JP, Milone M, Engel AG (1996) Effects of the quinoline derivatives quinine, quinidine, and chloroquine on neuromuscular transmission. Brain Res 712:179–189

Sine SM, Ohno K, Bouzat C, Auerbach A, Milone M, Pruitt JN, Engel AG (1995) Mutation of the acetylcholine receptor alpha subunit causes a slow-channel myasthenic syndrome by enhancing agonist binding affinity. Neuron 15:229–239

Skeberdis VA, Chevaleyre V, Lau CG, Goldberg JH, Pettit DL, Suadicani SO, Lin Y, Bennett MV, Yuste R, Castillo PE, Zukin RS (2006) Protein kinase A regulates calcium permeability of NMDA receptors. Nat Neurosci 9:501–510

Sobczyk A, Svoboda K (2007) Activity-dependent plasticity of the NMDA-receptor fractional Ca^{2+} current. Neuron 53:17–24

Swope SL, Moss SJ, Raymond LA, Huganir RL (1999) Regulation of ligand-gated ion channels by protein phosphorylation. Adv Second Messenger Phosphoprotein Res 33:49–78

Villarroel A, Sakmann B (1996) Calcium permeability increase of endplate channels in rat muscle during postnatal development. J Physiol 496:331–338

Vogt J, Morgan NV, Rehal P, Faivre L, Brueton LA, Becker K, Fryns JP, Holder S, Islam L, Kivuva E, Lynch SA, Touraine R, Wilson LC, MacDonald F, Maher ER (2012) CHRNG genotype-phenotype correlations in the multiple pterygium syndromes. J Med Genet 49:21–26

Vohra BP, Groshong JS, Maselli RA, Verity MA, Wollmann RL, Gomez CM (2004) Focal caspase activation underlies the endplate myopathy in slow-channel syndrome. Ann Neurol 55:347–352

Vohra BPS, Groshong JS, Zayas R, Wollmann RL, Gomez CM (2006) Activation of apoptotic pathways at muscle fiber synapses is circumscribed and reversible in a slow-channel syndrome model. Neurobiol Dis 23:462–470

Wachtel RE (1987) Effects of diltiazem and verapamil on responses to acetylcholine. Br J Pharmacol 92:561–566

Wang HL, Auerbach A, Bren N, Ohno K, Engel AG, Sine SM (1997) Mutation in the M1 domain of the acetylcholine receptor alpha subunit decreases the rate of agonist dissociation. J Gen Physiol 109:757–766

Wargon I, Richard P, Kuntzer T, Sternberg D, Nafissi S, Gaudon K, Lebail A, Bauche S, Hantaï D, Fournier E, Eymard B, Stojkovic T (2012) Long-term follow-up of patients with congenital myasthenic syndrome caused by COLQ mutations. Neuromuscul Disord 22:318–324

Zayas R, Groshong JS, Gomez CM (2007) Inositol-1,4,5-triphosphate receptors mediate activity-induced synaptic Ca^{2+} signals in muscle fibers and Ca^{2+} overload in slow-channel syndrome. Cell Calcium 41:343–352

Zhou Z, Neher E (1993) Calcium permeability of nicotinic acetylcholine receptor channels in bovine adrenal chromaffin cells. Pflugers Arch 425:511–517

Zhu H, Bhattacharyya BJ, Lin H, Gomez CM (2011) Skeletal muscle IP3R1 receptors amplify physiological and pathological synaptic calcium signals. J Neurosci 31:15269–15283

Chapter 34
CatSper in Male Infertility

Takuya Nishigaki, Ana Laura González-Cota and Gerardo José Orta Salazar

Abstract In 2001, two groups independently reported different components of a novel Ca^{2+} channel named CatSper, which is expressed only in the testis and localized in the sperm flagellum. Now, we know that CatSper is a sperm-specific Ca^{2+} channel composed of four distinct pore-forming subunits accompanied with, at least, three auxiliary subunits. Although there is no heterologous expression system to study this CatSper channel, the elimination of any single subunit ever tried in transgenic mice results in male infertility, which indicates that each individual subunit is essential for the correct channel assembly. Whole-cell patch clamp recordings directly taken from spermatozoa revealed that CatSper is a moderately voltage-dependent Ca^{2+} channel and is activated by intracellular alkalinization and several extracellular ligands, i.e., progesterone and prostaglandin E in human spermatozoa. The spermatozoa of CatSper null mice exhibit a defect in hyperactivated flagellar motility, a vigorous flagellar movement required for fertilization under physiological conditions. In agreement with this, there are some families suffering from male infertility correlated with mutations in CatSper-related genes.

34.1 Introduction

The motility propelled by the flagellum is a special feature of spermatozoa and any defects in this function can lead to male infertility (Darszon et al. 2011). The flagellar beat is generated by the axoneme, which is also found in the cilium (or cilia) of epithelial cells in the whole body (Lindemann and Lesich 2010). Therefore, a defective mutation of a certain component in the axoneme can cause a

T. Nishigaki (✉) · A. L. González-Cota · G. J. Orta Salazar
Departamento de Genética del Desarrollo y Fisiología Molecular, Instituto de Biotecnología, Universidad Nacional Autónoma de México (UNAM), Avenida Universidad #2001, Colonia Chamilpa 62210 Cuernavaca, Morelos, México
e-mail: takuya@ibt.unam.mx

N. Weiss and A. Koschak (eds.), *Pathologies of Calcium Channels*,
DOI: 10.1007/978-3-642-40282-1_34, © Springer-Verlag Berlin Heidelberg 2014

malfunction in sperm flagellar beating leading to male infertility accompanied with diminution of ciliary motion such as primary ciliary dyskinesia (Munro et al. 1994). On the other hand, there are several external factors that modulate the axoneme function such as ATP, pH, Ca^{2+}, and cAMP (protein phosphorylation depending on cAMP). Spermatozoa possess multiple sperm-specific proteins that control these factors: glycolytic enzymes (Miki et al. 2004), sodium-proton exchanger (Wang et al. 2003), Ca^{2+} channel (Ren et al. 2001), Ca^{2+}-ATPase (Okunade et al. 2004), adenylyl cyclase (Buck et al. 1999), and protein kinase (Nolan et al. 2004), etc. Therefore, a mutation in those proteins can cause male infertility as a result of the abnormal regulation of sperm motility without any other defects in the body. In this chapter, we review the structure and the function of CatSper, a sperm-specific Ca^{2+} channel essential to male fertility, and some reports of male infertility related to mutations that affect this channel.

34.2 Structure of CatSper

As seen in Fig. 34.1, CatSper is composed of four pore-forming α subunits (Ren et al. 2001; Quill et al. 2001; Arias et al. 2003; Lobley et al. 2003), which have six transmembrane segments, and at least three auxiliary subunits, β (Liu et al. 2007), γ (Wang et al. 2009), and δ (Chung et al. 2011). In the fourth segment (S4) of $\alpha1$ and $\alpha2$ subunit, four or five positively charged residues are found every three amino acids, which is a typical feature of the S4 segment of voltage-gated channels. In contrast, $\alpha3$ and $\alpha4$ have only two (Fig. 34.1c), suggesting a correlation with the moderate voltage dependence of this channel described later. One of the striking features of CatSper is that the pore of the channel is composed of four separated polypeptides instead of a single polypeptide as known in other voltage-gated Ca^{2+} channels. Another interesting feature is that $\alpha1$ subunit has many histidine residues in its N-terminal cytoplasmic domain (Fig. 34.1b), which is proposed to function as a pH sensor (Kirichok et al. 2006). All α subunits have a pore-forming loop with the typical Ca^{2+} selective channel motif ([T/S] $\times$ [D/E] $\times$ W) (Fig. 34.1d). Interestingly, CatSper uses only aspartic acids in this motif (DDDD) instead of a mixture of aspartic acids and glutamic acids (EEDD) found in T-type Ca^{2+} channel (Ca_v3) or only glutamic acids (EEEE) found in other voltage-gated Ca^{2+} channels (Ca_v1 and Ca_v2) in vertebrates (Senatore et al. 2013), which suggests a distinct cation selectivity of CatSper compared with typical voltage-gated Ca^{2+} channels.

At present, three membrane proteins (β, γ, and δ) have been reported as auxiliary subunits of CatSper. The β subunit has two transmembrane segments and γ and δ have a single transmembrane segment (Liu et al. 2007; Wang et al. 2009; Chung et al. 2011) (Fig. 34.1a). All three auxiliary subunits have large extracellular domains, which may serve as receptors for extracellular ligands, in contrast to small cytoplasmic tails.

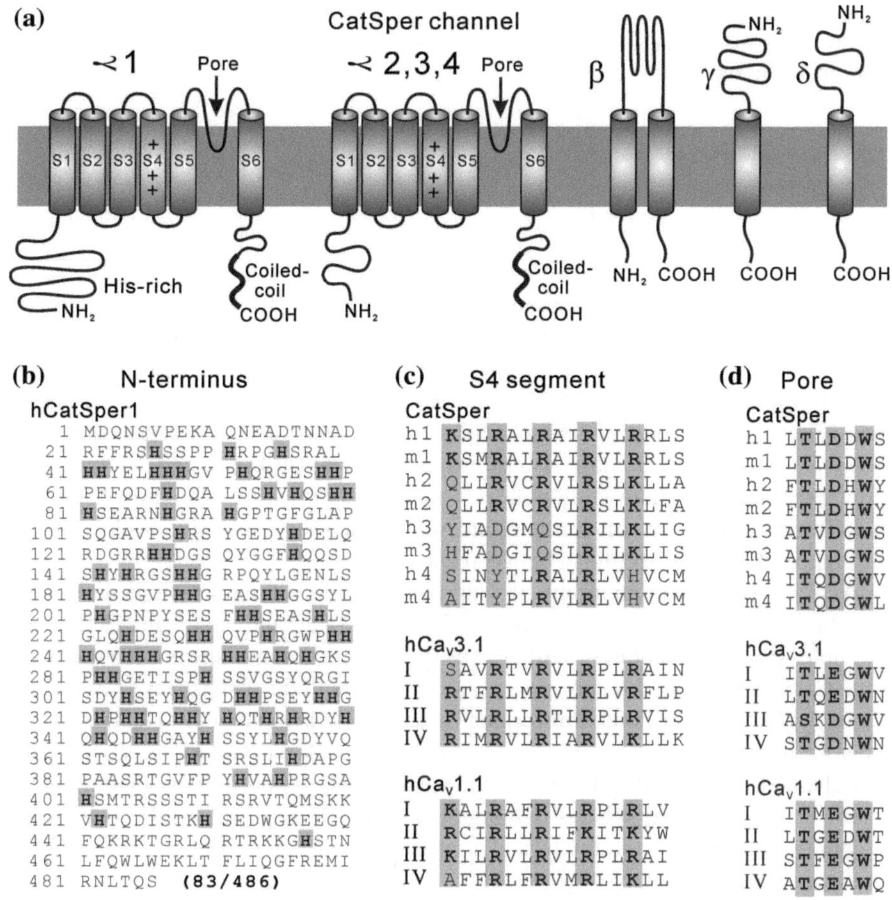

Fig. 34.1 Structure of CatSper. **a** Topology of pore-forming α subunits (1 – 4) and auxiliary subunits (β, γ, and δ). **b** Amino acid sequence of cytoplasmic N-terminal region of mouse CatSper is shown with histidine residue as *bold* and *gray* shadow. **c** Amino acid sequences of S4 segments of human CatSper (h1 – h4), mouse CatSper (m1 – m4), human $Ca_v3.1$ ($hCa_v3.1$, I – IV) (Perez-Reyes et al. 1998) and human $Ca_v1.1$ ($hCa_v1.1$, I – IV) (Tanabe et al. 1987) are aligned with *gray* shadows in every three residues. Arginine and Lysine are expressed as *bold*. **d** Amino acid sequences of the pore region are aligned as in (**c**). Three important residues for Ca^{2+} selective channels (T/S, E/D and W) are shown as *bold* and *gray* shadow

34.3 Aspects of Evolution

Amino acid sequences of all subunits of CatSper show a high level of diversity among different species (Table 34.1), which is a common feature of proteins specifically found in gametes (Swanson and Vacquier 2002). In general, transmembrane segments conserve their amino acid sequences, but relatively large cytoplasmic domains found in CatSper 1, 2, and 3 ($\alpha 1$, $\alpha 2$, and $\alpha 3$) and even large extracellular

Table 34.1 Amino acid identity of CatSper subunits between mouse and human

CatSper subunit	% of identity
α1	55
α2	65
α3	66
α4	72
β	56
γ	55
δ	51

domains found in all auxiliary subunits (β, γ, and δ) show increased diversity in their amino acid sequences. As a consequence, CatSper4 (α4) subunit, which have relatively small cytoplasmic domains, conserve their amino acid sequences better than any other subunits (Jin et al. 2005) (Table 34.1). The diversity found in α1 subunit is mainly due to a variety of amino acid sequences found in the histidine-rich N-terminal domain. This domain also shows many indels (insertion and deletion) even among species of primates although its biological significance is unknown (Podlaha and Zhang 2003). The diversity found in the auxiliary subunits could reflect a difference of ligand specificity between species.

Another very peculiar point of CatSper in terms of evolution is its mosaic distribution in metazoa as shown in Fig. 34.2 (Cai and Clapham 2008). For example, among vertebrates, neither aves (birds) and amphibians nor teleosts (fish) have CatSper (all α and β subunits) although cartilaginous fishes (ray) possess this channel. On the other hand, arthropoda (insects) and nematoda (*C. elegans*) do not have CatSper, but echinoderms (sea urchin) and cnidaria (sea anemone) conserve this channel. It is worth noting that the δ subunit is found only in mammals and reptiles but not in other species (Chung et al. 2011), indicating that this subunit has a unique function in mammals and reptiles and is not an essential subunit for CatSper channel in many other species.

In the case of the fruit fly, *Drosophila melanogaster*, TRPP2 channel (PKD2 channel) is localized on the distal tip of the sperm flagellum and a targeted mutation of this channel results in male infertility without apparent alterations in sperm morphology (Watnick et al. 2003). A detailed analysis of sperm motility in this mutant revealed that those spermatozoa swim backwards in the female reproductive tract (Kottgen et al. 2011), which clearly indicates a crucial role of TRPP2 channel for sperm motility regulation in this species. It would be interesting to identify what type of Ca^{2+} channels are involved in sperm motility regulation in species that lack CatSper in their genome.

34.4 Biophysical Properties

Elimination of any single subunit of CatSper that has been attempted (all α and δ subunits), in transgenic mice, results in male infertility accompanied with the lack of CatSper current in the spermatozoa (Qi et al. 2007; Chung et al. 2011; Jin et al.

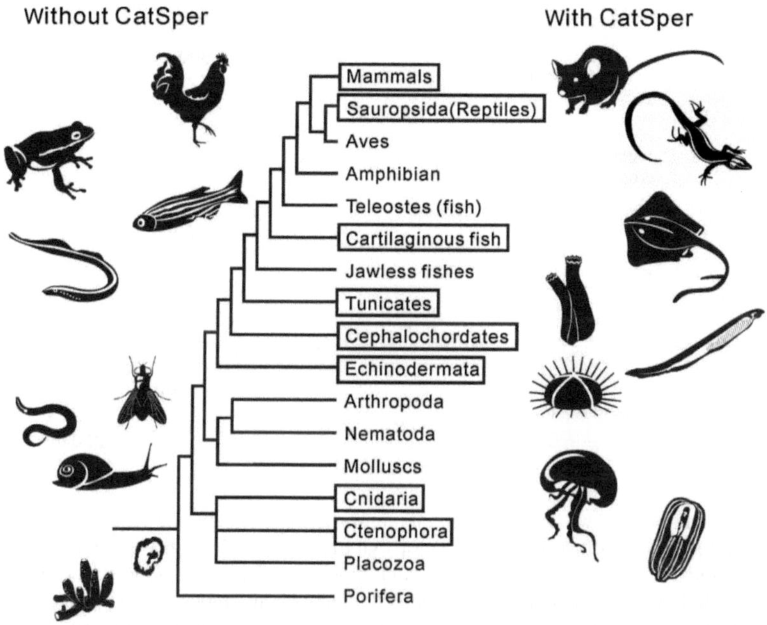

Fig. 34.2 Distribution of CatSper $\alpha(1-4)$ and β subunits in metazoan. This figure indicates the distribution of CatSper orthologs in whole metazoan species, which was revealed by the bioinformatic analysis of genome DNA sequences (Cai and Clapham 2008). The species which conserve CatSper genes are in box. In order to recognize each species easily, animal images are illustrated separating species possessing CatSper on the *right side* and those lacking this channel on the *left*. Some animal images do not correspond to the exact species whose genome sequences are determined, but they represent popular species from the same taxonomic phylum or class

2007; Kirichok et al. 2006), indicating that each subunit is indispensable for the proper channel assembly and function. In CatSper1 (α1 subunit) null mice, all other subunits including β, γ, and δ are missing in the mature spermatozoa (Liu et al. 2007; Wang et al. 2009; Chung et al. 2011), suggesting that only the correctly assembled CatSper channel can be transferred to the appropriate place, the plasma membrane of the principal piece in the flagellum. When they fail to assemble a functional channel, they are likely to be degraded during spermatogenesis. On the other hand, all attempts ever made to express a functional CatSper channel in heterologous systems have been in vain, which indicates that there might still be an unidentified auxiliary subunit, including germ cell-specific chaperons. Otherwise, a certain particular environment created by a particular lipid compositions or special scaffolding proteins might be necessary for a proper CatSper assembly and function.

While there is no heterologous expression system for CatSper yet, the biophysical properties of this channel had remained unknown until the development of the whole-cell patch clamp recording directly from mouse spermatozoa was reported (Kirichok et al. 2006). The trick of this technique is to use the

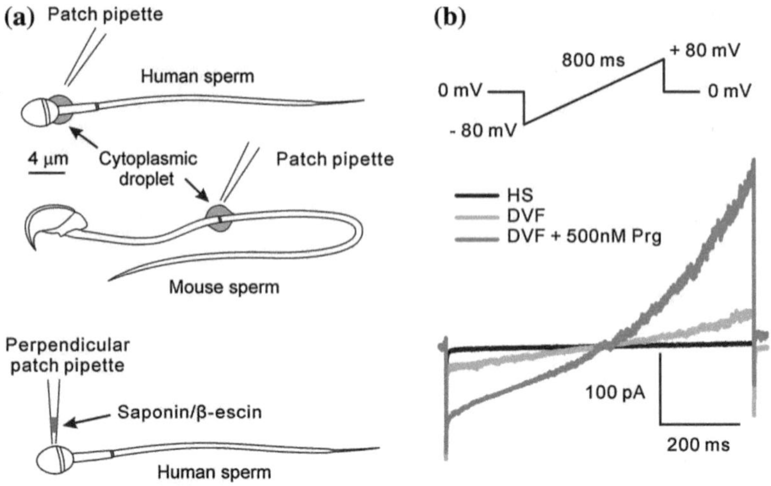

Fig. 34.3 Techniques of whole-cell patch clamp to spermatozoa and the typical CatSper current ($I_{CatSper}$). **a** *Upper* part illustrates the sites of cytoplasmic droplet (*gray shadow*) where patch pipettes should be attached to, in human and mouse spermatozoa. *Lower* part illustrates a perforated whole-cell patch clamp after on-cell patch clamp mode (Orta et al. 2012). **b** A representative $I_{CatSper}$ of human spermatozoon in response to a voltage-ramp, indicated above, under different conditions: standard solution for human sperm (HS, *black*), divalent cation-free solution (DVF, *gray*), and DVF in presence of 500 nM of progesterone (DVF + 500 nM Prg, *dark gray*)

cytoplasmic droplet, a residual cytoplasm occasionally found in mature spermatozoa (Fig. 34.3a), as a target for the patch clamp pipette. Even so, it is still difficult to do due to the small size and the motility of the cell (Lishko et al. 2013). Alternatively, it is possible to record a whole-cell current from mature spermatozoa using a perforated patch clamp on the sperm head (Orta et al. 2012). Although this method may be even more difficult to perform, it has the advantage of retaining the intracellular components of the cell. In addition, this method can be applied independently of the cytoplasmic droplet, which is not always present in mature spermatozoa.

Because of its small Ca^{2+} conductance and Ca^{2+}-dependent inactivation, CatSper current has been studied using monovalent cation current (Na^+ or Cs^+) in divalent cation-free (DVF) media (Fig. 34.3b). In this condition, the voltage dependence of the channel is not so prominent (Kirichok et al. 2006), which is at least partially explained by the surface-potential hypothesis; a shift of voltage-dependent curve toward the negative side in the absence of divalent cations (Hille 1991). Monovalent cation currents through CatSper ($I_{CatSper}$) can be inhibited by addition of low concentration of Ca^{2+} to the medium with IC_{50} of 65 nM in mouse (Kirichok et al. 2006) and 1.2 μM in human (Smith et al. 2013), which is a typical feature of a Ca^{2+} selective channel. Voltage dependence of CatSper was determined in the presence of Ba^{2+}, which does not induce the rapid inactivation of the channel as Ca^{2+}. Using the tail current of Ba^{2+}, it was determined that the CatSper channel is a moderate voltage-dependent channel with a slope factor k ~30 for mouse (Kirichok et al. 2006) and ~20 for human (Lishko et al. 2011). It is a striking feature, in both murine and human CatSper, that the voltage-dependent curve markedly shifts to a more negative values by intracellular alkalinization, which is considered a physiological signal to activate this channel (Kirichok et al. 2006; Lishko et al. 2010).

Recently, it was reported that progesterone and prostaglandin E activate CatSper in human sperm but not in murine sperm (Fig. 34.3b) (Lishko et al. 2011; Strunker et al. 2011; Smith et al. 2013). Both hormones are physiological ligands for human sperm and have been known to immediately increase the intracellular concentration of Ca^{2+} ($[Ca^{2+}]i$) (Thomas and Meizel 1989; Shimizu et al. 1998; Schaefer et al. 1998). The action of these ligands on spermatozoa had been a great mystery for a long time in the field of reproduction. On the other hand, several artificial compounds, such as odorants (bourgeonal and undecanal) and cyclic nucleotide analogs (8-Br-cGMP and 8-Br-cAMP), also activate human CatSper, indicating that human CatSper functions as a polymodal chemosensor (Brenker et al. 2012). In the case of murine sperm, it is demonstrated that albumin (BSA) and zona pellucida increase the $[Ca^{2+}]i$ in a CatSper-dependent manner (Xia and Ren 2009a, b). The difference of ligand specificity between these species may be attributed to the diversity of the amino acid sequences of auxiliary subunits of CatSper (β, γ, and δ), which have large extracellular domains and are supposed to function as receptors for extracellular ligands.

At present, there are no specific blockers for CatSper. HC-056456 was reported as the first CatSper blocker (IC_{50}: ~3 μM) although its specificity is unknown (Carlson et al. 2009). Mibefradil and NNC 55-0396, both T-type voltage-gated Ca^{2+} channel blockers, in the 10–20 μM range inhibit efficiently the CatSper current (Strunker et al. 2011; Lishko et al. 2011) and Na^{+} influx induced by removal of external Ca^{2+} in human spermatozoa (Torres-Flores et al. 2011). However, Mibefradil and NNC 55-0396 elevate the intracellular pH of spermatozoa by an unknown mechanism (Brenker et al. 2012). MDL12330A, known as an adenylyl cyclase inhibitor, was recently found to block CatSper (Brenker et al. 2012). Therefore, a more specific blocker for CatSper without any secondary effects, which potentially functions as a male contraceptive, is eagerly desired in the field of reproduction.

34.5 Physiological Function

Spermatozoa recovered from the oviduct show a vigorous flagellar movement called hyperactivated motility, which is indispensable for the fertilization and characterized by asymmetric, large amplitude, and low frequent flagellar beating (Suarez 2008) as illustrated in Fig. 34.4a. Several experimental evidences suggest that hyperactivated motility has four major roles in the process of the fertilization

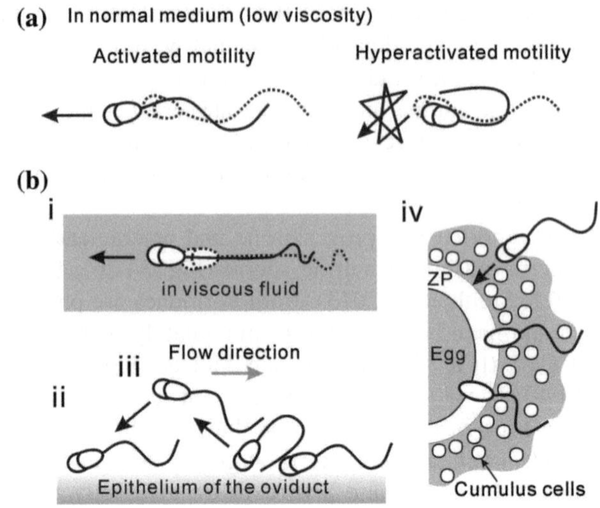

Fig. 34.4 Roles of hyperactivated motility in fertilization. **a** In normal experimental medium, immediately after spermatozoa are ejaculated they manifest a symmetric flagellar bend with high beat frequency called activated motility. In contrast, the hyperactivated motility is characterized by an asymmetric flagellar bend with low beat frequency. **b** Hyperactivated motility is required for four processes in mammalian fertilization: **i** Generate a propulsive force in viscous (or viscoelastic) environments, **ii** escape from the initial part of the isthmus of the oviduct, named sperm reservoir, **iii** swim against fluid flow, called rheotaxis, and **iv** penetrate through extracellular matrix of cumulus cells and the egg, named zona pellucida (ZP)

(Fig. 34.4b); (i) keep progressive motility in viscous environment (Suarez et al. 1991), (ii) detach from the initial part of the oviduct, lower isthmus, also called sperm reservoir (Suarez 1987), (iii) swim against fluid flow direction (rheotaxis) (Miki and Clapham 2013), and (iv) penetrate through extracellular matrix of the oocyte (Stauss et al. 1995). Murine spermatozoa lacking CatSper lose all these capacities together with fertility (Quill et al. 2003; Ren et al. 2001; Ho et al. 2009; Miki and Clapham 2013; Carlson et al. 2003), which confirms the physiological significance of hyperactivated sperm motility in this species. Therefore, the most obvious physiological role of CatSper is to induce and maintain the hyperactivated motility of the spermatozoa (Carlson et al. 2003). However, it is not completely understood what the physiological trigger(s) of sperm hyperactivation may be. There are several possibilities: (1) an increase in bicarbonate (HCO_3^-) concentration (Maas et al. 1977), which activates soluble adenylyl cyclase (Buck et al. 1999), (2) an increase in the fluid pH (Maas et al. 1977), (3) a specific ligand such as progesterone and prostaglandin E in human spermatozoa (Lishko et al. 2011; Strunker et al. 2011), (4) a decrease in Zn^{2+} concentration, which releases inhibition of voltage-gated proton channel (Lishko et al. 2010), and (5) a decrease in temperature (Bahat et al. 2005), which activates TRPM8 channel (De Blas et al. 2009). It is most likely that hyperactivated sperm motility is induced by several factors simultaneously and/or intermittently, which are generated concomitantly upon ovulation. Some of these factors are common among different species, but some are species-specific as the case of progesterone and prostaglandin E, which activate human CatSper but not murine. Further studies are required to understand the mechanism of induction of hyperactivation under physiological conditions.

Although the physiological roles of CatSper in mouse sperm are well established, those in human sperm are still not fully understood. In humans, not only Ca^{2+} influx through CatSper but also its release from intracellular Ca^{2+} stores is indispensable to maintain hyperactivation (Harper et al. 2004; Kirkman-Brown et al. 2004; Alasmari et al. 2013b). So far, it is known that CatSper is critical for sperm hyperactivation induced by progesterone (Lishko et al. 2011; Strunker et al. 2011; Smith et al. 2013; Alasmari et al. 2013a; Senatore et al. 2013; Servin-Vences et al. 2012) and this channel also seems essential for sperm to penetrate viscous media (Alasmari et al. 2013a). The positive correlation found in the success rate of in vitro fertilization (IVF) and progesterone-induced intercellular Ca^{2+} increase (Alasmari et al. 2013a) would be a helpful information to understand the role of CatSper and to establish a diagnostic test to evaluate sperm capacity for fertilization.

On the other hand, progesterone and prostaglandin E were initially identified as physiological ligands to induce the acrosome reaction (AR) in human sperm (Osman et al. 1989; Schaefer et al. 1998; Shimizu et al. 1998). In murine model, it was reported that the egg coat (zona pellucida) induces an increase in intracellular Ca^{2+} in wild type but not in CatSper1 null mice spermatozoa (Xia and Ren 2009b) although the AR still can be induced by zona pellucida in the same transgenic mice. Even though there are some contradictory results, CatSper could play a role in the AR, which may provide another reason for the positive correlation between CatSper activity and the success rate of IVF (Alasmari et al. 2013a).

34.6 Channelopathy

In mouse, all α subunits and δ subunit were demonstrated to be essential for CatSper channel assembly and the absence of this channel results in male infertility due to a defect in sperm motility regulation, namely, lack of hyperactivated sperm motility. Therefore, a similar function of human orthologs might be expected in human sperm. As anticipated, mutations of two loci that encode CATSPER1 and CATSPER2 have been reported from some families associated with male infertility (Avidan et al. 2003; Avenarius et al. 2009; Zhang et al. 2007). Coincidentally, all of the patients were found to have a homozygous mutation, an autosomal recessive mutation, through a consanguineous marriage. This is probably owing to the fact that the pore-forming α subunit of CatSper is composed of hetero-tetramers but not homo-tetramers like a voltage-gated K^+ channel. Thus, a dominant negative mutation of α subunit of CatSper should be difficult to encounter.

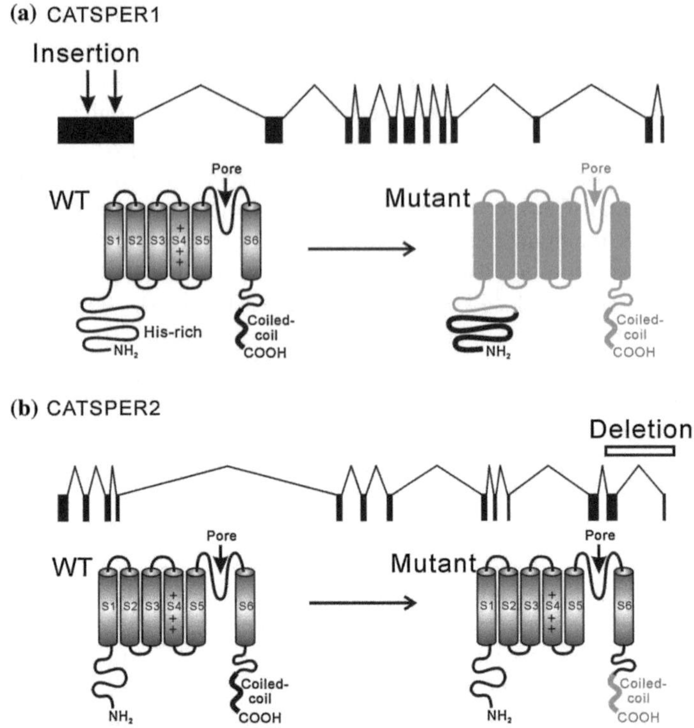

Fig. 34.5 Sites of CATSPER mutations correlated to human infertility and their predicted proteins. **a** The genome structure of CATSPER1 is illustrated as 12 exons (*black bars*) and introns (*polygonal lines*). The *arrows* indicate the two separated insertion mutations found in male infertile patients. Both mutants have some extra bases in the first exon, which encodes the His-rich N-terminal cytoplasmic domain, and are supposed to produce truncated proteins by a frame shift of mRNA as illustrated (*light gray*). **b** Genome structure of CATSPER2 (13 exons). The *white* box over the last two exons represents the deleted region of this gene found in infertile patients (French family). This region encodes a coiled-coil motif of the cytoplasmic C-terminal domain (*light gray*)

In 2009, two consanguineous Iranian families were identified as carriers of a similar, but distinct, insertion mutation in exon 1 of CATSPER1 gene (Fig. 34.5a) localized at chromosome 11q13.1 (c.539-540insT or c.948-949insATGGC) (Avenarius et al. 2009), which encodes the histidine-rich N-terminal cytoplasmic domain. Both mutations cause frame shifts and generate premature stop codons. As a consequence, they are predicted to produce proteins lacking all the trans-membrane segments and the pore-forming loop, as illustrated in Fig. 34.5a, almost equivalent to a deletion of the whole $\alpha 1$ subunit. As is expected, the male homozygous carriers of these mutations are infertile although there is no detailed information about the sperm function of these patients.

Another locus of mutation of male infertility related to CatSper, as the most studied case, is found on the chromosome 15p15, which is also characterized with non-syndromic deafness. This syndrome (male infertility and deafness) was first identified in a French family carrying a homozygous deletion of ~70 kb in the chromosome 15p15 which affects three contiguous genes: the first 24 exons of IP6 K, a kinase expressed ubiquitously; the entire coding sequence of SRTC, a protein mainly expressed on inner ear stereocilia which is related to the deafness; and the last two exons of CATSPER2 (Avidan et al. 2003) (Fig. 34.5b). This mutation removes the last 225 bases of the mRNA which encode the cytoplasmic C-terminal region of CatSper2 including a coiled-coil region (Fig. 34.5b) predicted to be necessary for protein–protein interactions to form the channel pore complex (Quill et al. 2001; Lobley et al. 2003). As mentioned above, all pore-forming subunits ($\alpha 1 - 4$) of CatSper are needed to assemble the functional murine CatSper channel (Qi et al. 2007). Therefore, CatSper2 and, in consequence, the entire CatSper channel complex had been expected to be absent in the CatSper2-deficient patient. Recently, this prediction was finally confirmed by detailed analysis of spermatozoa of one of these patients (II-2 in Avidan et al. 2003) using immunostaining and electrophysiological techniques (Smith et al. 2013). Namely, CatSperβ was unable to be detected by anti-CatSperβ antibody and CatSper current was absent from the spermatozoa of this patient. Furthermore, progesterone did not amplify any ionic current although the activities of other channels, K^+ channel (Slo3) and voltage-gated H^+ channel (Hv), were retained (Smith et al. 2013). Therefore, this study supports the idea that the principal Ca^{2+} channel in human sperm is CatSper and it is activated by progesterone. This locus is likely to be a hot spot to provoke a deletion mutation because a duplicated copy of these genes is located adjacently as pseudogenes (Avidan et al. 2003). In agreement with this, it was reported that there are three more Iranian families with similar delete mutations in this locus (Zhang et al. 2007) although the entire CATSPER2 gene was deleted in all three cases of Iranian families (Table 34.2). Combined, these reports provide strong evidences that link CatSper deficiencies to male infertility.

However, the phenotype of spermatozoa from the patients possessing the deletion mutation of the chromosome 15p15 is not exactly coincident with the phenotype of spermatozoa of transgenic mice lacking CatSper-related genes. In murine models, CatSper mutations did not cause any defects in spermatogenesis (number of matured spermatozoa) or sperm shape (Ren et al. 2001; Qi et al. 2007; Chung et al. 2011). Moreover, those mouse spermatozoa manifest normal-activated flagellar motility characterized by symmetric, low amplitude, and high frequent flagellar beating although they do not undergo hyperactivation (Carlson et al. 2003). In contrast, those patients suffer from asthenoteratospermia; disorders in sperm morphology and motility. Taking the phenotype of the transgenic mice into account, defects in sperm morphology and basal motility found in the patients with CATSPER2 mutation (Avidan et al. 2003; Zhang et al. 2007) could be attributed to a defect of another gene, including IP6 K, rather than different functions of CatSper between mouse and human. To date, there are only a few reported clinical cases that link CatSper to male infertility; two having defects on CATSPER1 (Avenarius et al. 2009) and four on CATSPER2 (Avidan et al. 2003; Zhang et al. 2007). Since all reported patients come from consanguineous families,

Table 34.2 Clinical cases of infertility correlated to CatSper mutations

Mutated gene	Affected families	Number of patient	Mutation (Locus)	Associated defects	Reference
CatSper1 (α1)	Iranian L-1025	2	c.539-540insT (11q13.1)	Asthenoteratospermia/ low sperm count	Avenarius et al. (2009)
CatSper1 (α1)	Iranian L-968	1	c.948-949insATGGC (11q13.1)	Asthenoteratospermia/ low sperm count	Avenarius et al. (2009)
CatSper2 (α2)	French	3	Del(15) ~70 kb Deletion of the last two exons (12 and 13) (15q15.1-15q15.3)	Deafness/ asthenoteratospermia/ no CatSper current	Avidan et al. (2003); Smith et al. (2013)
CatSper2 (α2)	Iranian D_SM	4	Del(15) ~100 kb Deletion of the entire gene (15q15.1-15q15.3)	Deafness/ asthenoteratospermia	Zhang et al. (2007)
CatSper2 (α2)	Iranian L-705	1	Del(15) ~100 kb Deletion of the entire gene (15q15.1-15q15.3)	Deafness/ asthenoteratospermia	Zhang et al. (2007)
CatSper2 (α2)	Iranian L-1014	2	Del(15) ~90 kb Deletion of the entire gene (15q15.1-15q15.3)	Deafness/ Asthenoteratospermia	Zhang et al. (2007)

they could also carry defects on other genes. Considering some distinct phenotypes between the transgenic mice and human patients, it is difficult to conclude the deficiencies on CatSper as the direct cause of male infertility in these patients. Further information about patients of male infertility and CatSper-related gene mutations is required to answer this question.

On the other hand, although genetic analysis has not been carried out, a systematic analysis of sperm Ca^{2+} responses and hyperactivation from healthy donors and sub-fertile patients were recently reported (Alasmari et al. 2013a). In this work, sub-fertile patients were classified into two groups; patients who could fertilize by in vitro fertilization (IVF) (Steptoe and Edwards 1978) and those who required intra-cytoplasmic sperm injection (ICSI) (Palermo et al. 1992). ICSI is an advanced technique of assisted reproduction and usually applied for male patients who failed in IVF. While spermatozoa from almost all donors and IVF patients manifest progesterone-induced intracellular Ca^{2+} increase, spermatozoa of 27 % ICSI patients failed to respond to progesterone. Considering that CatSper is activated by progesterone in human spermatozoa (Lishko et al. 2011; Strunker et al. 2011), roughly 27 % of ICSI patients have some defects in CatSper. Genetic analysis of those patients would contribute to the understanding of the correlation between mutations in CatSper-related genes and male infertility.

34.7 Conclusion

Although it is possible to record CatSper channel current by whole-cell patch clamping directly from spermatozoa, it is still a difficult technique. On the other hand, in vitro spermatogenesis is not so efficient and is a time-consuming process (~40 days) (Sato et al. 2011). Currently, mutagenesis of a CatSper-related gene requires an approach using transgenic animals (murine models), which is a very expensive project. Therefore, to establish a heterologous expression system for CatSper channel is of primary interest and the most important obstacle to overcome in order to promote the study of CatSper channel research.

CatSper is an essential channel for male fertility, but there are only a few reports about male infertility correlated to alteration of CatSper genes. There are several reasons to explain this situation. One of them would be owing to the fact that genetic defects in male fertility had not been inherited to the next generation, which could have prevented genetic research about male infertility in general. This situation has been altered by application of intra-cytoplasmic sperm injection (ICSI), one of advanced assisted reproduction techniques (ARTs). Since ICSI is getting to be a popular ART these days (Wong and Ledger 2013), CatSper mutations will accumulate in the future (Devroey and Van Steirteghem 2004). Nevertheless, this defect is supposed to affect only spermatozoa not other organs in contrast to the case of patients of ciliopathies (Munro et al. 1994). Considering the advance of technology for whole-genome sequencing, the lists of gene alterations of CatSper channel are supposed to be amplified soon in the future, which may also contribute to the understanding of the structure and function relationship of CatSper channel.

Acknowledgments We thank Dr. Yuriy Kirichok for his explanation of the shift of voltage dependence of CatSper in divalent cation-free medium. We appreciate the help of Guadalupe Itzel Galán Enríquez for providing us animal illustrations in Fig. 34.2. We also thank Robyn Duckworth and Dr. Alberto Darszon for critical reading of our manuscript. This work is supported by grant of CONACyT (CB2012-177138) and PAPIIT (IN203513).

References

Alasmari W, Barratt CL, Publicover SJ, Whalley KM, Foster E, Kay V, Martins da Silva S, Oxenham SK (2013a) The clinical significance of calcium-signalling pathways mediating human sperm hyperactivation. Hum Reprod 28(4):866 – 876

Alasmari W, Costello S, Correia J, Oxenham SK, Morris J, Fernandes L, Ramalho-Santos J, Kirkman-Brown J, Michelangeli F, Publicover S, Barratt CL (2013b) Ca2 + signals generated by CatSper and Ca2 + stores regulate different behaviors in human sperm. J Biol Chem 288(9):6248–6258

Arias JM, Murbartian J, Perez-Reyes E (2003) Cloning of a novel one-repeat calcium channel-like gene. Biochem Biophys Res Commun 303(1):31–36

Avenarius MR, Hildebrand MS, Zhang Y, Meyer NC, Smith LL, Kahrizi K, Najmabadi H, Smith RJ (2009) Human male infertility caused by mutations in the CATSPER1 channel protein. Am J Hum Genet 84(4):505–510

Avidan N, Tamary H, Dgany O, Cattan D, Pariente A, Thulliez M, Borot N, Moati L, Barthelme A, Shalmon L, Krasnov T, Ben-Asher E, Olender T, Khen M, Yaniv I, Zaizov R, Shalev H, Delaunay J, Fellous M, Lancet D, Beckmann JS (2003) CATSPER2, a human autosomal nonsyndromic male infertility gene. Eur J Hum Genet 11(7):497–502

Bahat A, Eisenbach M, Tur-Kaspa I (2005) Periovulatory increase in temperature difference within the rabbit oviduct. Hum Reprod (Oxford, England) 20 (8):2118 – 2121

Brenker C, Goodwin N, Weyand I, Kashikar ND, Naruse M, Krahling M, Muller A, Kaupp UB, Strunker T (2012) The CatSper channel: a polymodal chemosensor in human sperm. EMBO J 31(7):1654–1665

Buck J, Sinclair ML, Schapal L, Cann MJ, Levin LR (1999) Cytosolic adenylyl cyclase defines a unique signaling molecule in mammals. Proc Natl Acad Sci USA 96(1):79–84

Cai X, Clapham DE (2008) Evolutionary genomics reveals lineage-specific gene loss and rapid evolution of a sperm-specific ion channel complex: CatSpers and CatSperbeta. PLoS ONE 3(10):e3569

Carlson AE, Burnett LA, del Camino D, Quill TA, Hille B, Chong JA, Moran MM, Babcock DF (2009) Pharmacological targeting of native CatSper channels reveals a required role in maintenance of sperm hyperactivation. PLoS ONE 4(8):e6844

Carlson AE, Westenbroek RE, Quill T, Ren D, Clapham DE, Hille B, Garbers DL, Babcock DF (2003) CatSper1 required for evoked $Ca2^{+}$ entry and control of flagellar function in sperm. Proc Natl Acad Sci USA 100(25):14864–14868

Chung JJ, Navarro B, Krapivinsky G, Krapivinsky L, Clapham DE (2011) A novel gene required for male fertility and functional CATSPER channel formation in spermatozoa. Nat Communi 2:153

Darszon A, Nishigaki T, Beltran C, Trevino CL (2011) Calcium channels in the development, maturation, and function of spermatozoa. Physiol Rev 91(4):1305–1355

De Blas GA, Darszon A, Ocampo AY, Serrano CJ, Castellano LE, Hernandez-Gonzalez EO, Chirinos M, Larrea F, Beltran C, Trevino CL (2009) TRPM8, a versatile channel in human sperm. PLoS ONE 4(6):e6095

Devroey P, Van Steirteghem A (2004) A review of ten years experience of ICSI. Hum Reprod Update 10(1):19–28

Harper CV, Barratt CL, Publicover SJ (2004) Stimulation of human spermatozoa with progesterone gradients to simulate approach to the oocyte. Induction of $[Ca(2^{+})](i)$ oscillations and cyclical transitions in flagellar beating. J Biol Chem 279(44):46315–46325

Hille B (1991) Chapter 17 Modifiers of gating. In: Ionic channel of excitable membranes, 2nd edn. Sinauer Associates Inc., Massachusetts

Ho K, Wolff CA, Suarez SS (2009) CatSper-null mutant spermatozoa are unable to ascend beyond the oviductal reservoir. Reprod Fertil Dev 21(2):345–350

Jin J, Jin N, Zheng H, Ro S, Tafolla D, Sanders KM, Yan W (2007) CatSper3 and CatSper4 are essential for sperm hyperactivated motility and male fertility in the mouse. Biol Reprod 77(1):37–44

Jin JL, O'Doherty AM, Wang S, Zheng H, Sanders KM, Yan W (2005) Catsper3 and catsper4 encode two cation channel-like proteins exclusively expressed in the testis. Biol Reprod 73(6):1235–1242

Kirichok Y, Navarro B, Clapham DE (2006) Whole-cell patch-clamp measurements of spermatozoa reveal an alkaline-activated $Ca2^{+}$ channel. Nature 439(7077):737–740

Kirkman-Brown JC, Barratt CL, Publicover SJ (2004) Slow calcium oscillations in human spermatozoa. Biochem J 378(Pt 3):827–832

Kottgen M, Hofherr A, Li W, Chu K, Cook S, Montell C, Watnick T (2011) Drosophila sperm swim backwards in the female reproductive tract and are activated via TRPP2 ion channels. PLoS ONE 6(5):e20031

Lindemann CB, Lesich KA (2010) Flagellar and ciliary beating: the proven and the possible. J Cell Sci 123(Pt 4):519–528

Lishko P, Clapham DE, Navarro B, Kirichok Y (2013) Sperm patch-clamp. Methods Enzymol 525:59–83

Lishko PV, Botchkina IL, Fedorenko A, Kirichok Y (2010) Acid extrusion from human spermatozoa is mediated by flagellar voltage-gated proton channel. Cell 140(3):327–337
Lishko PV, Botchkina IL, Kirichok Y (2011) Progesterone activates the principal $Ca2^+$ channel of human sperm. Nature 471(7338):387–391
Liu J, Xia J, Cho KH, Clapham DE, Ren D (2007) CatSperbeta, a novel transmembrane protein in the CatSper channel complex. J Biol Chem 282(26):18945–18952
Lobley A, Pierron V, Reynolds L, Allen L, Michalovich D (2003) Identification of human and mouse CatSper3 and CatSper4 genes: Characterisation of a common interaction domain and evidence for expression in testis. Reprod Biol Endocrinol 1(1):53
Maas DH, Storey BT, Mastroianni L Jr (1977) Hydrogen ion and carbon dioxide content of the oviductal fluid of the rhesus monkey (Macaca mulatta). Fertil Steril 28(9):981–985
Miki K, Clapham DE (2013) Rheotaxis guides Mammalian sperm. Curr Biol 23(6):443–452
Miki K, Qu W, Goulding EH, Willis WD, Bunch DO, Strader LF, Perreault SD, Eddy EM, O'Brien DA (2004) Glyceraldehyde 3-phosphate dehydrogenase-S, a sperm-specific glycolytic enzyme, is required for sperm motility and male fertility. Proc Natl Acad Sci USA 101(47):16501–16506
Munro NC, Currie DC, Lindsay KS, Ryder TA, Rutman A, Dewar A, Greenstone MA, Hendry WF, Cole PJ (1994) Fertility in men with primary ciliary dyskinesia presenting with respiratory infection. Thorax 49(7):684–687
Nolan MA, Babcock DF, Wennemuth G, Brown W, Burton KA, McKnight GS (2004) Sperm-specific protein kinase A catalytic subunit Calpha2 orchestrates cAMP signaling for male fertility. Proc Natl Acad Sci USA 101(37):13483–13488
Okunade GW, Miller ML, Pyne GJ, Sutliff RL, O'Connor KT, Neumann JC, Andringa A, Miller DA, Prasad V, Doetschman T, Paul RJ, Shull GE (2004) Targeted ablation of plasma membrane $Ca2^+$-ATPase (PMCA) 1 and 4 indicates a major housekeeping function for PMCA1 and a critical role in hyperactivated sperm motility and male fertility for PMCA4. J Biol Chem 279(32):33742–33750
Orta G, Ferreira G, Jose O, Trevino CL, Beltran C, Darszon A (2012) Human spermatozoa possess a calcium-dependent chloride channel that may participate in the acrosomal reaction. J Physiol 590(Pt 11):2659–2675
Osman RA, Andria ML, Jones AD, Meizel S (1989) Steroid induced exocytosis: the human sperm acrosome reaction. Biochem Biophys Res Commun 160(2):828–833
Palermo G, Joris H, Devroey P, Van Steirteghem AC (1992) Pregnancies after intracytoplasmic injection of single spermatozoon into an oocyte. Lancet 340(8810):17–18
Perez-Reyes E, Cribbs LL, Daud A, Lacerda AE, Barclay J, Williamson MP, Fox M, Rees M, Lee JH (1998) Molecular characterization of a neuronal low-voltage-activated T-type calcium channel. Nature 391(6670):896–900
Podlaha O, Zhang J (2003) Positive selection on protein-length in the evolution of a primate sperm ion channel. Proc Natl Acad Sci USA 100(21):12241–12246
Qi H, Moran MM, Navarro B, Chong JA, Krapivinsky G, Krapivinsky L, Kirichok Y, Ramsey IS, Quill TA, Clapham DE (2007) All four CatSper ion channel proteins are required for male fertility and sperm cell hyperactivated motility. Proc Natl Acad Sci USA 104(4):1219–1223
Quill TA, Ren D, Clapham DE, Garbers DL (2001) A voltage-gated ion channel expressed specifically in spermatozoa. Proc Natl Acad Sci USA 98(22):12527–12531
Quill TA, Sugden SA, Rossi KL, Doolittle LK, Hammer RE, Garbers DL (2003) Hyperactivated sperm motility driven by CatSper2 is required for fertilization. Proc Natl Acad Sci USA 100(25):14869–14874
Ren D, Navarro B, Perez G, Jackson AC, Hsu S, Shi Q, Tilly JL, Clapham DE (2001) A sperm ion channel required for sperm motility and male fertility. Nature 413(6856):603–609
Sato T, Katagiri K, Gohbara A, Inoue K, Ogonuki N, Ogura A, Kubota Y, Ogawa T (2011) In vitro production of functional sperm in cultured neonatal mouse testes. Nature 471(7339):504–507

Schaefer M, Hofmann T, Schultz G, Gudermann T (1998) A new prostaglandin E receptor mediates calcium influx and acrosome reaction in human spermatozoa. Proc Natl Acad Sci USA 95(6):3008–3013

Senatore A, Monteil A, van Minnen J, Smit AB, Spafford JD (2013) NALCN ion channels have alternative selectivity filters resembling calcium channels or sodium channels. PLoS ONE 8(1):e55088

Servin-Vences MR, Tatsu Y, Ando H, Guerrero A, Yumoto N, Darszon A, Nishigaki T (2012) A caged progesterone analog alters intracellular $Ca2^+$ and flagellar bending in human sperm. Reproduction 144(1):101 – 109

Shimizu Y, Yorimitsu A, Maruyama Y, Kubota T, Aso T, Bronson RA (1998) Prostaglandins induce calcium influx in human spermatozoa. Mol Hum Reprod 4(6):555–561

Smith JF, Syritsyna O, Fellous M, Serres C, Mannowetz N, Kirichok Y, Lishko PV (2013) Disruption of the principal, progesterone-activated sperm $Ca2^+$ channel in a CatSper2-deficient infertile patient. Proc Natl Acad Sci USA 110(17):6823

Stauss CR, Votta TJ, Suarez SS (1995) Sperm motility hyperactivation facilitates penetration of the hamster zona pellucida. Biol Reprod 53(6):1280–1285

Steptoe PC, Edwards RG (1978) Birth after the reimplantation of a human embryo. Lancet 2(8085):366

Strunker T, Goodwin N, Brenker C, Kashikar ND, Weyand I, Seifert R, Kaupp UB (2011) The CatSper channel mediates progesterone-induced $Ca2^+$ influx in human sperm. Nature 471(7338):382–386

Suarez SS (1987) Sperm transport and motility in the mouse oviduct: observations in situ. Biol Reprod 36(1):203–210

Suarez SS (2008) Control of hyperactivation in sperm. Hum Reprod update 14(6):647–657

Suarez SS, Katz DF, Owen DH, Andrew JB, Powell RL (1991) Evidence for the function of hyperactivated motility in sperm. Biol Reprod 44(2):375–381

Swanson WJ, Vacquier VD (2002) The rapid evolution of reproductive proteins. Nat Rev 3(2):137–144

Tanabe T, Takeshima H, Mikami A, Flockerzi V, Takahashi H, Kangawa K, Kojima M, Matsuo H, Hirose T, Numa S (1987) Primary structure of the receptor for calcium channel blockers from skeletal muscle. Nature 328(6128):313–318

Thomas P, Meizel S (1989) Phosphatidylinositol 4,5-bisphosphate hydrolysis in human sperm stimulated with follicular fluid or progesterone is dependent upon $Ca2^+$ influx. Biochem J 264(2):539–546

Torres-Flores V, Picazo-Juarez G, Hernandez-Rueda Y, Darszon A, Gonzalez-Martinez MT (2011) Sodium influx induced by external calcium chelation decreases human sperm motility. Hum Reprod 26 (10):2626 – 2635

Wang D, King SM, Quill TA, Doolittle LK, Garbers DL (2003) A new sperm-specific Na +/ H + exchanger required for sperm motility and fertility. Nat Cell Biol 5(12):1117–1122

Wang H, Liu J, Cho KH, Ren D (2009) A novel, single, transmembrane protein CATSPERG is associated with CATSPER1 channel protein. Biol Reprod 81(3):539–544

Watnick TJ, Jin Y, Matunis E, Kernan MJ, Montell C (2003) A flagellar polycystin-2 homolog required for male fertility in Drosophila. Curr Biol 13(24):2179–2184

Wong MY, Ledger WL (2013) Is ICSI Risky? Obstetrics and gynecology international 2013:473289

Xia J, Ren D (2009a) The BSA-induced $Ca2^+$ influx during sperm capacitation is CATSPER channel-dependent. Reprod Biol Endocrinol 7:119

Xia J, Ren D (2009b) Egg coat proteins activate calcium entry into mouse sperm via CATSPER channels. Biol Reprod 80(6):1092–1098

Zhang Y, Malekpour M, Al-Madani N, Kahrizi K, Zanganeh M, Lohr NJ, Mohseni M, Mojahedi F, Daneshi A, Najmabadi H, Smith RJ (2007) Sensorineural deafness and male infertility: a contiguous gene deletion syndrome. J Med Genet 44(4):233–240

Index

N. Weiss and A. Koschak (eds.), *Pathologies of Calcium Channels*,
DOI: 10.1007/978-3-642-40282-1, © Springer-Verlag Berlin Heidelberg 2014

MIX
Papier aus verantwortungsvollen Quellen
Paper from responsible sources
FSC® C105338

If you have any concerns about our products,
you can contact us on
ProductSafety@springernature.com

In case Publisher is established outside the EU,
the EU authorized representative is:
Springer Nature Customer Service Center GmbH
Europaplatz 3, 69115 Heidelberg, Germany

Printed by Libri Plureos GmbH
in Hamburg, Germany